D0083538

The Geology of Earthquakes

The Geology of Earthquakes

Robert S. Yeats
Kerry Sieh
Clarence R. Allen

New York *Oxford*
OXFORD UNIVERSITY PRESS
1997

Oxford University Press

Oxford New York
Athens Auckland Bogota Bangkok Bombay Buenos Aires
Calcutta Cape Town Dar es Salaam Delhi
Florence Hong Kong Istanbul Karachi
Kuala Lumpur Madras Madrid Melbourne
Mexico City Nairobi Paris Singapore
Taipei Tokyo Toronto

and associated companies in

Berlin Ibadan

Published by Oxford University Press, Inc.,
198 Madison Avenue, New York, New York 10016

Library of Congress Cataloging in Publications Data

Yeats, Robert S.
Geology of earthquakes / Robert S. Yeats, Kerry Sieh, Clarence R. Allen
p. cm. Includes bibliographical references (p. 500–557) and index.
ISBN 0-19-507827-6 (cloth : alk. paper)
1. Earthquakes. I. Sieh, Kerry E. II. Allen, Clarence R.
(Clarence Roderic), 1925- . III. Title.
QE534.2.Y43 1996 551.2'2—dc20 95-4903 CIP

Cover Photo: Right-lateral offset of an asphalt road near the epicenter of the 1992 Landers earthquake, the largest to strike California in 40 years. Although obscured on the flanks of the road within hours of the earthquake, by traffic skirting the disruption, the fault zone is apparent across the road as a principal rupture, a minor rupture, and intervening right-lateral warp. Total offset was about 1.5 meters and occurred within the first few seconds of the M 7.3 earthquake, before most of the faults that produced the earthquake had begun to slip.

Photo by Kerry Sieh, 10 am, June 28th, 1992, five hours after the earthquake. View toward the southwest along Aberdeen Drive, near Landers, San Bernardino County, southern California.

1 3 5 7 9 8 6 4 2
Printed in the United States of America
on acid-free paper

Preface

The history of the practice of earthquake geology extends more than a century into the past, but it is still only a fledgling field as a modern discipline. In fact, most of what we know has been learned only in the past quarter century. And so it is not surprising that this textbook is the first attempt to survey systematically the purposes, methods, and scientific principles of the geological investigation of earthquakes.

Our impetus for writing this book was, in part, the frustration we felt in teaching, year after year, earthquake geology classes without the benefit of a textbook. This book began, then, as so many do, as sets of lecture notes and class handouts for graduate courses in earthquake geology. In 1986, we, together with Professors Richard Sibson and David Nash, offered a short course in earthquake geology at the annual meeting of the Geological Society of America in Phoenix, Arizona. The response to that course, as well as to the requests of students in our university courses, led to our combining forces to write this book. We hope that this will help our colleagues at other universities who also labor to teach this material without the benefit of a textbook. A second motivation for our effort was the need we perceived in industry, in particular the geotechnical community, but also others involved in assessing the hazards posed by earthquakes, for a book that summarizes current knowledge and methods and provides references to more detailed information.

Charles Richter strongly influenced the philosophy of this book. His own book, *Elementary Seismology*, was written to be understood by a large readership, including geologists, engineers, and planners. We wanted this book to be as useful and as understandable to a broad community as was his. Our challenge was to make the book useful to a non-geologist, who might have no previous coursework in geology, without boring the geologist with introductory detail. Our method in accomplishing this was to divide the book into a set of introductory chapters providing the necessary background in geology (Chapters 1, 2, 3, 6, and 7), seismology *sensu stricto* (Chapter 4), and geodesy (Chapter 5) prior to discussing the various environments in which earthquakes occur. We also included a Glossary to prevent scientific jargon from getting in the way of learning, particularly for a non-earth scientist. The book also includes an extensive bibliography and a compilation of historical earthquakes with surface faulting as resources to those readers who wish to pursue particular topics in greater depth.

Our own experience with earthquake faults outside North America and the work of our colleagues in other countries convinced us that the earthquake geology of North America is not comprehensive enough to provide an adequate coverage of the geological environment of earthquakes. By selecting examples from Japan, China, New Zealand, the Mediterranean region, and elsewhere, we include earthquake environments that differ from those studied in North America. In addition, overseas studies add insights and analogs that are relevant to conclusions about, say, the earthquake hazard facing southern California. We have drawn extensively upon the work of scientists outside North America, who bring their own techniques and working philosophies to enrich our own.

We have been assisted in these international exchanges by International Geological Correlation Program Project 206, Worldwide Characterization of Major Active Faults, chaired by Robert Bucknam, Ding Guoyu, and Zhang Yuming, and Inter-Union Commission of the Lithosphere Task Group II-2, World Map of Active Faults, chaired by Vladimir Trifonov and Michael Machette, and Task Group II-3, Great Earthquakes of the Late Holocene, chaired by R.S.Y. and Yoshihiro Kinugasa, with K.E.S. as a task-group member.

Most of the chapters of the book were written by R.S.Y. K.E.S. wrote Chapter 8 on strike-slip faults, and C.R.A. wrote Chapter 13 on seismic hazard analysis. Each of us reviewed the chapters written by the others, and, in addition, C.R.A. served as an overall coordinator to ensure that the contributions by each author meshed into a coherent book.

In writing this book, we became painfully aware of our own limited experience in the broad field of earthquake geology and related fields. We were fortunate to find specialists in those fields farthest from our own experience to review and add their own insights, especially to our introductory chapters. In addition, the chapters that cover subjects within our own areas of expertise benefited greatly from extensive and detailed reviews by our colleagues. Our thanks and gratitude go to Kelvin Berryman, Sarah Beanland, Mike Bevis, Roger Bilham, Ron Bruhn, Bob Bucknam, Bill Bull, Tony Crone, James Jackson, David Keefer, Ed Keller, Geoff King, Vern Kulm, Shaul Levi, Mike Machette, Rob McCaffrey, Peter Molnar, George Moore, Dan Muhs, John Nábělek, Charles Naeser, Steve Obermeier, Gilles Peltzer, Nick Pisias, Leon Reiter, Woody Savage, Rick Sibson, Ross Stein, Art Sylvester, John Van Couvering, and Nobuyuki Yonekura. The earthquake table represents the collective efforts of the international community; we thank Nick Ambraseys, Rolando Armijo, Aykut Barka, Manuel Berberian, Kelvin Berryman, Tony Crone, Deng Qidong, Agust Gudmundsson, Yoshi Kinugasa, Mike Machette, Bertrand Meyer, Andrei Nikonov, Daniela Pantosti, Luca Valensise, Xu Xiwei, and Haruo Yamazaki. R.S.Y. received help and perspective in writing the personal vignettes from Diane Bright, Iaacov Karcz, Patrick LeFort, Tokihiko Matsuda, Takashi Nakata, Amos Nur, Paul Tapponnier, Hiroyuki Tsutsumi, and Bob Wallace. We have learned a lot from these workers.

We thank those authors who not only granted permission to use their published illustrations but also, where possible, provided originals and, in some cases, unpublished photographs.

R.S.Y. thanks Yoshi Kinugasa of the Seismotectonic Research Section of the Geological Survey of Japan and Paul Tapponnier of the Laboratoire de Tectonique, Méchanique de la Lithosphère, Institut de Physique du Globe de Paris for providing space and facilities for writing and for valuable discussions about the earthquake geology of faraway places. KES thanks Larry Edwards of the Department of Geology and Geophysics, University of Minnesota and also Paul Tapponnier, for a similar opportunity. Support for RSY came from the Japan Program of the National Science Foundation, the Institut de Physique du Globe de Paris, the Centre National de la Recherche Scientifique, and Oregon State University.

Much of the American data and concepts presented in this book are the products of research supported by the Earthquake Hazards Reduction Program, administered by the U.S. Geological Survey and the U.S. National Science Foundation. Although this is an applied research program, it has had a profound effect on the development of our field and upon understanding the fundamental nature of earthquakes.

Cheryl Hummon, Margaret Mumford, and Carrie Sieh did the computer illustrations, and Sue Pullen and Angela Yeats helped with the references.

Finally, we appreciate the patience and understanding of our editor, Joyce Berry, for organizing our writing into this book and for being sympathetic to our view that studying two unanticipated large earthquakes in California (1992 Landers and 1994 Northridge) had to take precedence over manuscript deadlines.

R.S.Y. *Corvallis, Ore.*
K.E.S. *Pasadena, Calif.*
C.R.A. *Pasadena, Calif.*
September 1995

Contents

The Geology of Earthquakes

Introduction

In its relation to man, an earthquake is a cause. In its relation to the earth, it is chiefly an incidental effect of an incidental effect.

G. K. Gilbert, 1912, preface to U.S. Geological Survey Professional Paper 69

Most books about earthquakes are written by geophysicists rather than geologists. In part, this is because geophysicists study the earthquake waves directly, as recorded on a seismograph, whereas geologists study the surface expression of earthquake faults. Prior to the middle of the nineteenth century, earthquakes were not perceived to be related to faults, and no one could predict what geological expression, if any, an earthquake should have.

Charles Lyell (1797–1875), in his *Principles of Geology,* was the first to recognize that earthquakes abruptly modify the ground surface, based on reports of the 1819 Rann of Cutch, India, earthquake and the 1855 Wairarapa, New Zealand, earthquake. An earthquake in the northwestern Peloponnesos Peninsula of Greece produced surface rupture described in 1875 by an astronomer, J. F. J. Schmidt of Germany. G. K. Gilbert and I. C. Russell visited a fault scarp produced by the 1872 Owens Valley, California, earthquake. Gilbert recognized similar surface features on the Wasatch fault zone near Salt Lake City, Utah, and he concluded that these features were also produced by earthquakes. The correlation between earthquakes and surface faulting was strengthened in 1888, in New Zealand, when Alexander McKay visited the site of the Marlborough earthquake of that year and recognized strike-slip offset of features by the Hope fault during the earthquake. This was followed by the discovery by Bunjiro Koto of surface faulting related to the Mino-Owari earthquake of 1891 in Japan, and the observation by C. L. Griesbach of surface rupture on the Chaman fault accompanying the Baluchistan earthquake of 1892. The San Francisco earthquake of 1906 led to the careful mapping of displacements on the San Andreas fault, and earthquake geology seemed to become firmly established as a subdiscipline of structural geology.

However, for more than a half century following the description of the San Francisco earthquake, the geophysical study and geological study of earthquakes tended to follow separate paths, and most geologists showed no interest in the geological expression of earthquakes. The principal reason for this was the invention of the seismograph, which permitted the direct study of earthquake waves. Accordingly, the geophysical study of earthquakes, principally using seismograms, came to be known as *seismology,* although this term could be broadened to include the study of earthquakes based on their geological or geodetic expression. The Seismological Society of America reaches out to geophysicists, geologists, and engineers for its membership, taking a broad definition of seismology. Richter (1958), in his text, *Elementary Seismology,* included descriptions of the geological expression of earthquakes. We, too, take this broader view and consider that we have written a textbook on seismology, focusing on the geological aspects of earthquakes.

The geologist, unfortunately, stands at the surface, many kilometers above the point of earthquake rupture, and geological inferences about the earthquake must be filtered through intervening layers of rock and sediment. The geophysicist, however, is limited by the length of time seismographs have been in operation, less than a century, and knowledge of the pre-twentieth century earthquake history of a fault or region depends on historical records and on geology, particularly in regions where the history of record-keeping is short. So each needs the other.

This book is written from the perspective of the geologist, who sees an earthquake as a quantum in building a mountain or pulling apart a valley. An earthquake is part of a long-term geological process, and it is also an event in the geologic history of a fault or region. A large earthquake commonly (but not always) leaves clues about itself in the rocks and in the landforms for the geologist to find and interpret. The geologist believes in *uniformitarianism:* the natural processes that operate today operated in the same way during earth's long past. A complex fold or a great fault displacement can be formed in very small increments over millions of years. James Hutton (1726–1797) first advanced this view, but it was Charles Lyell who documented Hutton's ideas through careful field observations.

But an earthquake and its accompanying rupture of the ground surface is essentially a catastrophic event, occurring so rapidly that we almost never observe a ground rupture being formed. A rare exception was reported by Pelton et al. (1984) and by Wallace (1984). On the morning of October 28, 1983, Don Hendricksen and John Turner were on a dirt road in Arentson Gulch, Idaho, in a 4-wheel drive vehicle, looking for elk, when Don, after feeling light-headed and dizzy, saw the road fall away in front of the vehicle, as if a sinkhole had formed. This was followed by the formation of a surface rupture about 20 m in front of the vehicle, with the only sound being the crumbling of earth in front of them. This was followed by violent shaking and a deafening rumbling noise, the entire episode lasting 10 or 15 seconds. At the same time, Mrs. Lawana Knox was seated not too far away on a slope north of Thousand Springs Valley, Idaho, looking for her husband, a hunter, when a 1- to 1.5-meter-high fault scarp formed in front of her at about 300 meters distance, reaching its full height in about one second. The scarp seemed to tear from the northwest to the southeast along the flank of the mountain "just as though one took a paint brush and painted a line along the hill." The scarp took only a few seconds to extend several miles along the range front, but it did not form until the peak of strong shaking had begun to subside, at least 10 seconds afterwards, according to Wallace (1984). Don Hendricksen, John Turner, and Lawana Knox had witnessed the rupture of the ground surface by the Borah Peak earthquake (M_S7.3).

A large-displacement fault is the cumulative result of hundreds of sudden slip events accompanying earthquakes, with each event separated by long periods of dormancy. In California, there are cases of fault motion unaccompanied by large earthquakes (fault creep), and fault displacement during an earthquake may continue for a time after the earthquake (afterslip). But most of the displacement is catastrophic, a series of sudden jerks separated by long periods of quiescence which may last for thousands of years.

An earthquake does its work in a few seconds or tens of seconds, but the crustal movement that builds up strain and causes the earthquake is so slow that earthquakes at the same spot may be separated by tens of thousands of years. Even this great length of time is brief and understandable to the geologist, an earth historian, for whom changes in scenery may take millions of years. But of what use is this to society, where a lifetime is less than a hundred years? How should the Oregon state legislature, for example, respond when it is told that yes, Oregon has had great earthquakes off its coastline in the past and will undoubtedly have them again in the future, but no, we can't tell you whether the next earthquake will strike tomorrow or several hundred years from now?

This book places earthquakes in a geological perspective. Where are earthquakes most likely to occur, and what controls their distribution? What is the geologic setting of an earthquake source zone, and how is the earthquake expressed at the surface? What other geological phenomena can be attributed to earthquake shaking, and how can these phenomena be used to identify earthquakes in the geologic record? How can geological information about earthquakes and fault displacement be used to evaluate hazards to society?

Earthquake geology is commonly regarded as synonymous with *neotectonics,* defined in the *Glossary of Geology* as "the study of the post-Miocene structures and structural history of the Earth's crust." Another definition is "the study of recent deformation of the crust, generally Neogene (post-Oligocene)." These definitions are not very satisfactory, because many structures that formed millions of years ago in the Miocene or Pliocene are clearly inactive today. Another suggested definition is "tectonic processes now active, taken over the geologic time span during which they have been acting in the presently observed sense, and the resulting structures" (J. G. Dennis, written commun., 1982).

Wallace (1986) dropped the term *neotectonics* altogether and introduced a new term, *active tectonics,* for "tectonic movements that are expected to occur within a future time span of concern to society." Society is not interested in a fault that will not move in the future, even if the fault last moved only recently. Wallace's definition, then, is one that is more useful to the general public.

Another problem is defining an *active fault,* because such a term has legal significance. Various U.S. government agencies have defined active faults differently: (1) it moved in the last 10,000 years, (2) it moved in the last 35,000 years, (3) it moved in the last 150,000 years, and (4) it moved twice in the last 500,000 years. These diverse definitions, based on the agency's perception of risk, lead to confusion.

Because active tectonics involves dynamic processes operating today, it can contribute a data set that involves a considerably thicker slice of the earth than the study of tectonic processes which are no longer active. A surface fault may be mapped to depths of 10-20 km based on the planar distribution of earthquakes on the fault. Seismicity contributes much data to active tectonics, although the instrumental record involves such a small amount of time (less than a century) that the absence of instrumental seismicity cannot be assumed to mean an absence of earthquakes over the time span of interest in assessing earthquake hazard. *Seismotectonics* is not defined in any recent glossary, but it could be considered as that subfield of active tectonics concentrating on the seismicity, both instrumental and historical, and dealing also with geological and other geophysical data sets. Other geophysical data sets are useful: refraction and reflection seismology, gravity, magnetics, and heat flow. The field of geodesy allows the direct measurement of deformation by releveling and by retrilateration, and the strain field can be used to model crustal structure and to map buried earthquake faults. Strainmeters and tiltmeters measure deformation at a more local scale.

Other fields of geology, in addition to structural geology, contribute to the study of earthquakes. One is volcanology. Many nineteenth century geologists, including Alexander von Humboldt, believed that earthquakes were caused by volcanoes. Active volcanism, like active faulting, is a measure of geodynamic processes, and so most neotectonic maps include the distribution of active volcanoes and of volcanic rocks of Quaternary age. Some of the same instruments used to study active tectonic deformation are also used to predict future volcanic activity. Deformation of volcanic carapaces generally takes place at strain rates several orders of magnitude higher than rates of tectonic deformation.

The field of geomorphology also contributes to an understanding of earthquakes. Most of the large-scale landforms of the earth—basins and mountain ranges—are controlled by internal processes producing subsidence and uplift, respectively. Active faults and folds produce a wide variety of landforms, and their study contributes information about the tectonic processes themselves. Closely allied to geomorphology is Quaternary stratigraphy, in which tectonic activity is reflected in the record of Quaternary deposits. The geochronology of the Quaternary, the last 1,800,000 years, is a specialized field involving dating techniques of little or no use in the older geologic record: radiocarbon dating, open-system uranium-series dating, amino-acid stereochemistry, thermoluminescence, and others. Added to this are tephrochronology, secular changes in the Earth's magnetic field, degree of development of soils, degree of weathering of sedimentary clasts, and the degree of dissection of surfaces and scarps.

This book first reviews the disciplines that are the foundation of earthquake geology: plate tectonics, structural geology, seismic waves, geodesy, Quaternary stratigraphy and dating, and geomorphology. This is followed by a survey of expressions of earthquakes at the earth's surface as faults and folds, including a chapter on subduction-zone megathrusts. Secondary effects are then discussed; these effects are the result of the violent shaking and tsunami generation accompanying earthquakes. We then consider how scientific information is used to evaluate earthquake hazards. The book closes with a catalog of earthquakes with evidence for faulting at the ground surface.

SUGGESTIONS FOR FURTHER READING

Wallace, R. E. 1986, ed. Studies in geophysics—Active tectonics: Washington, D.C.: National Academy Press, 176 p.

PART ONE

Background

1

Earthquake Geography and Plate Tectonics

STRUCTURE AND COMPOSITION OF THE EARTH'S CRUST

The surface of the Earth is at two predominant levels, *continents,* which are high-standing, with the mean land surface 840 m above sea level, and *ocean basins,* which are at a mean depth of 3700 m below sea level (Fig. 1-1). The different elevations reflect internal composition. Continents are composed of *silicic* or *granitic rocks,* lighter crustal materials rich in the minerals quartz and feldspar (orthoclase and plagioclase), whereas ocean-basin crust is denser, composed of *mafic* or *basaltic rocks* rich in the denser minerals pyroxene and olivine, together with plagioclase feldspar. The boundaries between these two classes of crust occupy relatively little surface area of the Earth and are relatively steep, forming the submerged slopes around the continents. Both continental crust and oceanic crust are separated from the underlying, dense *mantle* by the *Mohorovicic discontinuity,* or *Moho,* a relatively sharp compositional boundary between lighter crust richer in feldspar and denser mantle composed of olivine and pyroxene, with feldspar virtually absent.

The granitic crust stands high with respect to basaltic ocean crust because its lower density makes it buoyant, thereby supporting the high topography. The continents are underlain by low-density roots; continents rise above the basaltic ocean floor for the same reason that an iceberg rises above the surface of the sea. Mountains rise above the continental plateaus because they are underlain by even deeper low-density roots (Fig. 1-1). The lower density of the root results in a relative mass that is low relative to the more dense rock around it. This relatively low mass more or less compensates for the mass of the high-standing mountains such that the topography is a reflection of hydrostatic equilibrium. These concepts comprise the theory of *isostasy,* and continents and most mountain ranges are said to be *isostatically compensated.*

But how does this compensation occur in crust and mantle that is solid rock? Under high temperatures, crystalline rock is so weak that it creeps very slowly.

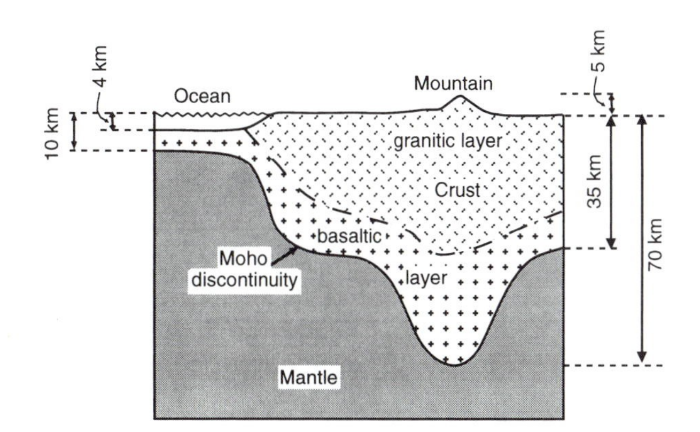

Figure 1-1. Cross section of oceanic crust (left) and continental crust (center and right). Continent is composed of silicic (granitic) rock which is lighter and thus more buoyant than oceanic crust, which is underlain by mafic (basaltic) rock. The continental crust is believed to be underlain by mafic crust; both continental and oceanic crust overlie the mantle, composed mainly of pyroxene and olivine. The top of the mantle is the Mohorovicic (Moho) discontinuity, first detected by seismic waves. The continent stands high with respect to the ocean basin, and for it to be in isostatic balance, it is underlain by a deep root of lighter crust. Mountain ranges stand above the continent and are underlain by still deeper roots. The continental margin depicted here is a passive margin. Vertical scale is greatly exaggerated.

It behaves, over time, like a viscous fluid. This topic is discussed in greater detail in Chapter 3.

EARTHQUAKE GEOGRAPHY

Many of the continental slopes, including most of the slopes surrounding continents facing the Atlantic, Indian, and Arctic oceans, have very few earthquakes, and thus they, together with the continents and ocean basins adjacent to them, are inherently stable. We refer to such slopes as *passive continental margins*. Other continental slopes, including most of those bordering continents facing the Pacific Ocean, have many earthquakes. These continental margins are flanked by high mountains and deep trenches, departing from the mean elevations of continents and ocean basins, and they also contain active volcanoes. We refer to these slopes as *active continental margins*, active because of the presence of large earthquakes, active volcanoes, and extreme variations in topographic relief. We discuss active margins in detail in Chapter 11.

Figure 1-2 shows a world map of earthquake epicenters for the period 1963-1987 based on the Worldwide Standard Seismographic Network (WWSSN) of about 120 seismographic stations in 60 countries. The difference between active and passive margins is most clearly seen in North and South America. The west coasts of both continents have many earthquakes, whereas the east coasts have relatively few earthquakes. Note that many bands of earthquakes occur far from continents, particularly in the western Pacific Ocean. Nearly all of these zones of earthquakes are characterized by active volcanoes on islands and by deep-sea trenches, indicating that an active margin does not require a continent on one side. These zones are called *island arcs*, reflecting the broad curvature on the map many of them show.

Viewing the distribution of earthquakes on Figure 1-2 without regard to the location of continents, it can be seen that most earthquakes occur in two broad belts, one skirting the margins of the Pacific Ocean (Pacific Ring of Fire) and the other extending from the Pacific Ocean westward through Indonesia,

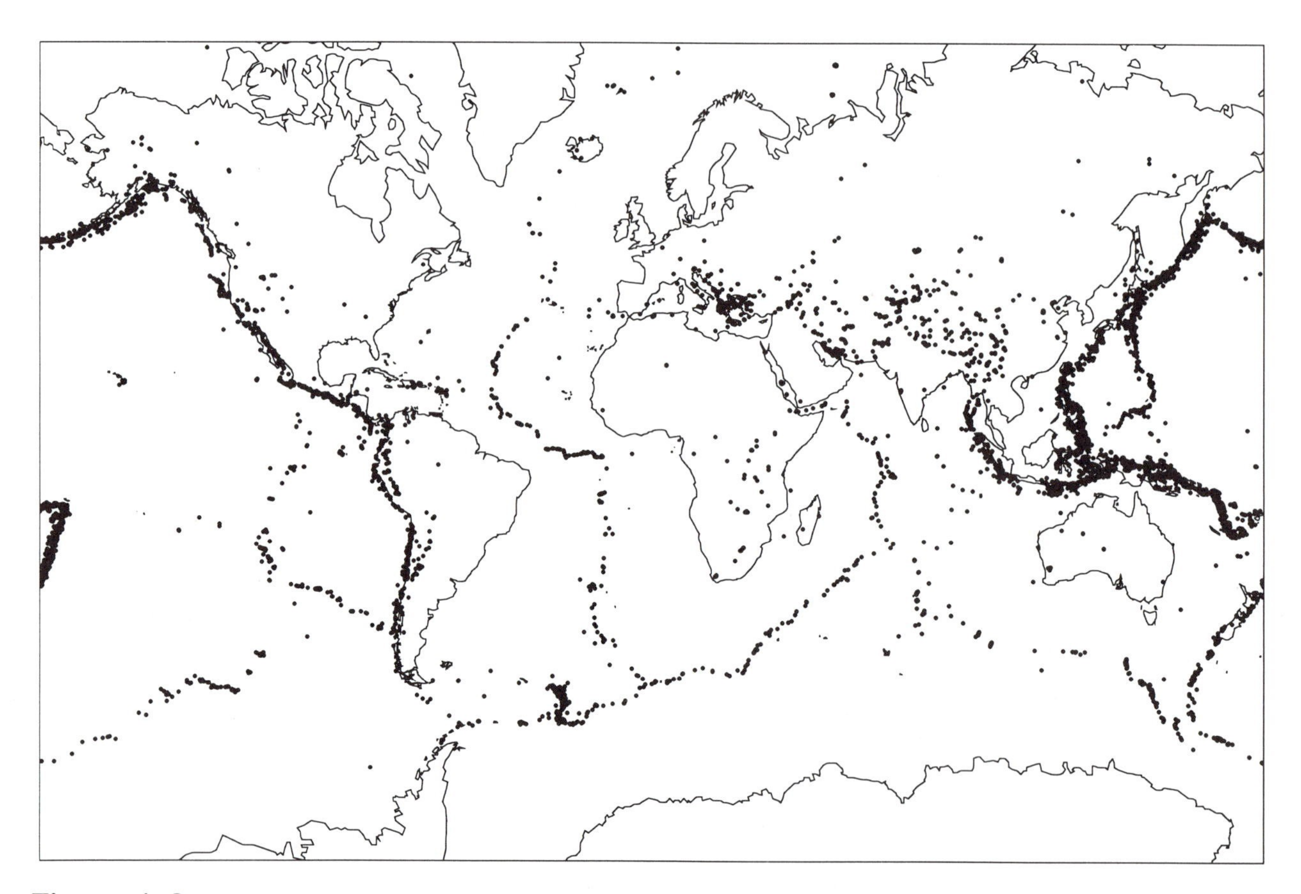

Figure 1-2. Epicenters for earthquakes shallower than 50 km and with magnitude equal to or greater than 5.5, in the period 1963–1987, from the catalog of the National Earthquake Information Center. From Gordon and Stein (1992).

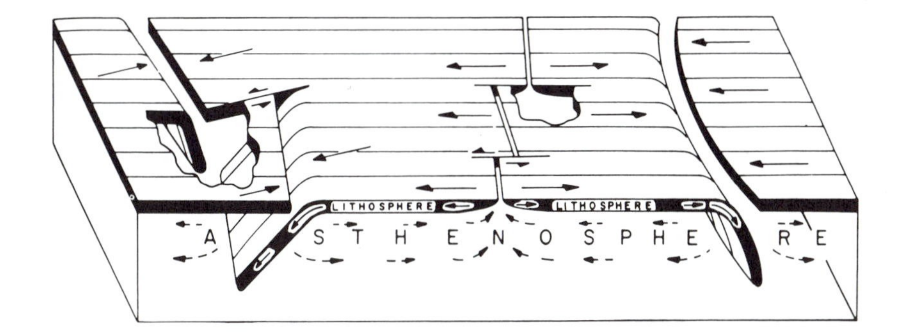

Figure 1-3. Block diagram showing the lithosphere and asthenosphere and the motion of plates with respect to one another. Arrows on lithosphere show relative motion of adjacent plates. New lithosphere is formed at mid-ocean ridges and thickens as it cools and moves away. At points of collision of plates, one plate, usually oceanic, descends (is subducted) beneath the other such that cold, earthquake-producing lithosphere penetrates several hundred kilometers into the otherwise aseismic asthenosphere. These collision zones are the sites of active volcanoes and trenches. Both mid-ocean ridges and trenches may be offset along zones where plates move parallel to each other along transform faults. The area depicted is the South Pacific Ocean between Australia and northern South America. Modified slightly from Isacks et al. (1968).

China, Central Asia, the Near East, and the Mediterranean Sea. Other concentrations of earthquakes in the Caribbean Sea between North and South America and in the South Atlantic Ocean between South America and Antarctica are also island arcs.

A narrow band of earthquakes extends down the Atlantic Ocean midway between the Americas and Europe-Africa and continues into the Indian Ocean where it bifurcates. One band extends north into the Red Sea, and the other band extends south of Australia into the South Pacific Ocean. This band divides again, with one arm extending north to Mexico and other arms extending east to the South American coast. These narrow bands mark the *mid-ocean spreading ridges*. The Mid-Atlantic Ridge and the ridge between Australia and Antarctica maintain a position midway between the active margins of continents. The East Pacific Rise off the west coast of South and Central America does not, because the adjacent continental margins are active, not passive. The East Pacific Rise comes ashore on the Pacific coast of North America. A diffuse band of earthquakes in east Africa contains active volcanoes and long, narrow lakes; this is the East African Rift System, which has many characteristics of mid-ocean spreading ridges except that it lies in continental crust.

The high-quality locations of the WWSSN show that most earthquakes occur predominantly at deep-sea trenches, mid-ocean spreading ridges, and active mountain belts on continents. However, a few, such as a very diffuse pattern of earthquakes in Australia and the eastern United States, do not.

PLATE TECTONICS

The tendency of earthquakes to occur in relatively narrow belts, first discovered by the WWSSN, was one of the key lines of evidence leading in the 1960s to the theory of *plate tectonics*. The Earth's surface consists of a series of tectonic plates, with each plate consisting of the crust and the more rigid part of the upper mantle, termed the *lithosphere*. The lithosphere contains all the world's earthquakes, and it is underlain by a weaker zone called the *asthenosphere*, with the lithosphere-asthenosphere boundary controlled by temperature. The asthenosphere is solid, but it yields by hot creep, allowing solid-state flow that compensates for the motion of the lithospheric plates. The lithosphere may be more than 100 km thick beneath older continental and oceanic regions, but it is only a few kilometers thick beneath mid-ocean ridges, where it first forms. Lithospheric plates are in motion with respect to one another, and the disruption produced at the boundaries between plates results in earthquakes.

Figure 1-3 shows a block diagram of part of the South Pacific Ocean region illustrating some of the main characteristics of plate tectonics. This diagram is from a paper by Isacks et al. (1968; reproduced in Cox, 1973), which showed how the distribution of earthquakes could be used to demonstrate plate motion as an essential element in global tectonics.

Mid-ocean spreading ridges such as the East Pacific Rise illustrated in the figure are formed by the upwelling of hot magma that cools and solidifies as lithospheric plates move away from the ridge crest, and newly-formed oceanic crust is carried down the flanks of the ridge (Fig. 1-3). The history of ocean-floor spreading is recorded in the magnetization of basaltic crust into magnetic stripes that are symmetric about the ridge axis. These magnetic stripes reflect the alternation between reversed and normal polarity of the earth's magnetic field through time. Because these alternations between reversed and normal polarity have been dated (cf. Chapter 6), the pattern of magnetic stripes records the rate at which sea-floor spreading is taking place, like a magnetic tape recorder. Earthquakes on spreading ridges are limited to the ridge crest, where new crust is being formed. The crust is so hot that only a thin, near-surface zone is brittle enough to produce earthquakes. These earthquakes tend to be relatively small and occur at shallow depths (cf. Chapter 9).

Deep-sea trenches form where lithosphere converges, forcing one slab beneath the other. This convergent boundary, called a *subduction zone,* is de-

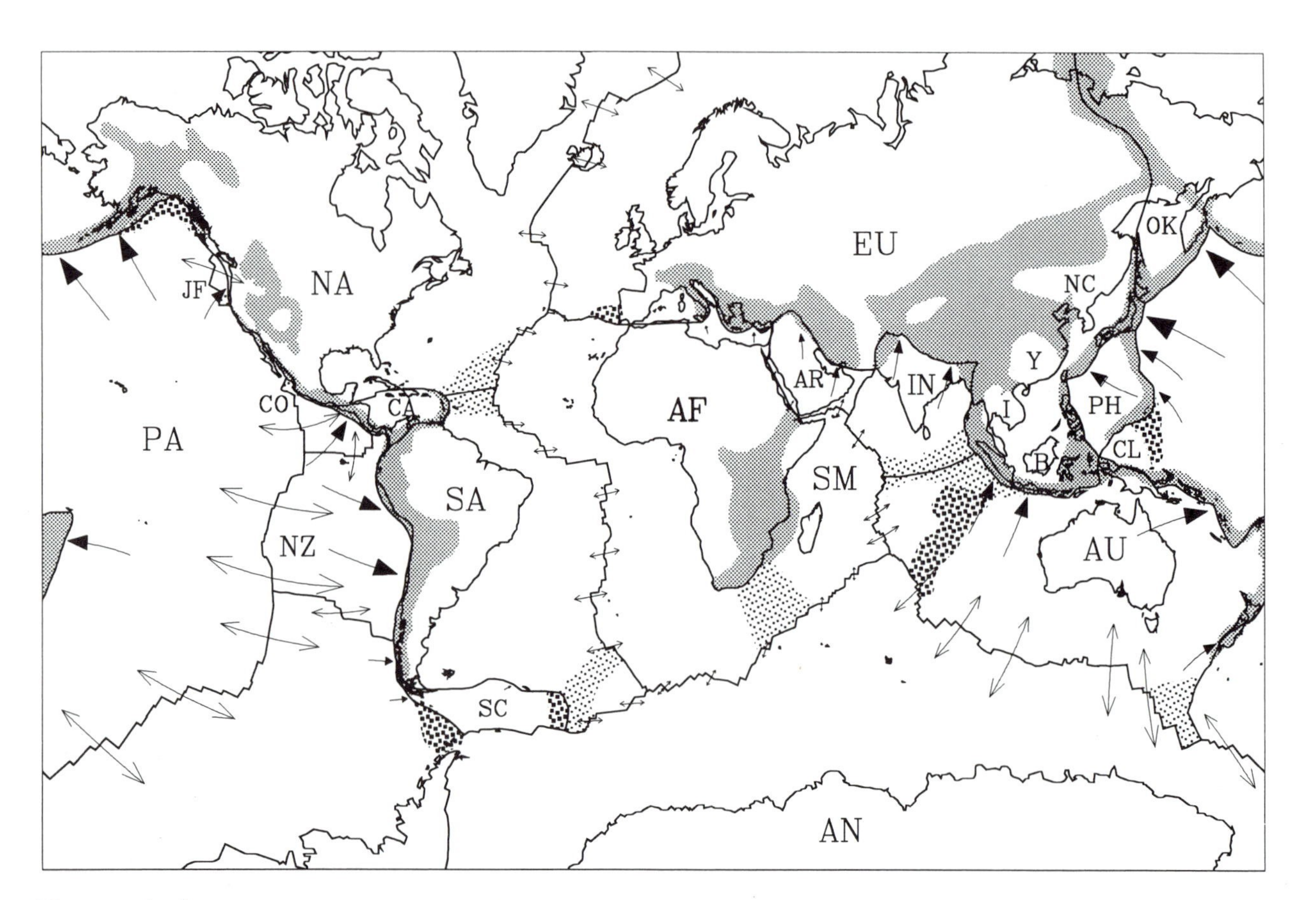

Figure 1–4. Map showing division of the lithosphere into large plates moving with respect to one another: AF, Africa; AN, Antarctica; AR, Arabia; AU, Australia; B, Borneo; CA, Caribbean; CL, Caroline; CO, Cocos; EU, Eurasia; JF, Juan de Fuca; I, Indochina; IN, India; NA, North America; NC, North China; NZ, Nazca; OK, Okhotsk; PA, Pacific; PH, Philippine Sea; SA, South America; SM, Somalia; SC, Scotia Sea; Y, Yangtze. Plate velocities are shown by arrows, with the length of the arrows equal to the predicted displacement for the next 25 million years. Plate convergence rate shown by arrows with solid arrowhead on underthrust plate pointing toward overthrust plate. Shaded patterns identify deforming regions within diffuse plate-boundary zones. Fine stipple pattern identifies mainly subaerial regions where deformation is inferred from earthquakes and evidence of active faulting. Medium stipple pattern identifies mainly submarine regions where deformation is based on lack of closure of plate-tectonic models and commonly on earthquakes. Coarse stipple pattern marks mainly submarine regions in which evidence of deformation is based mainly on earthquakes. After Gordon (1994).

fined by a zone of earthquakes (*Wadati-Benioff zone*) that marks where lithosphere extends into the mantle to depths up to 700 kilometers. The shape of the subduction zone in Figure 1-3 is determined on the basis of seismicity within the dipping zone of earthquakes beneath active volcanic areas, as described in the 1930s by K. Wadati of Japan and subsequently by H. Benioff of the United States, both prior to the discovery of plate tectonics. Subduction zones dip at angles less than 90°, resulting in a broad map pattern for earthquakes around the Pacific Ocean (Figure 1-2). The prominence of the earthquake zone around the Pacific has another explanation: a large percentage of the seismic energy released worldwide is generated at subduction zones. Subduction zones are discussed in greater detail in Chapter 11.

Earthquakes also occur along zones where the slabs move parallel to one another; these zones are called *transform faults*. Two types of transform faults are shown in Figure 1-3. The mid-ocean spreading ridge appears to be offset along transform faults such that lithospheric plates move past one another between the offsets. For relatively long transform faults, the lithosphere far from either offset spreading ridge has cooled enough to be relatively thick and rigid and is thereby able to generate large earthquakes. East-west-trending zones of earthquakes along the Mid-Atlantic Ridge mark transform faults offsetting the ridge; these offsets are shown in Figure 1-4. The second type of transform fault shown in Figure 1-3 separates trenches, one subducting in one direction and one subducting in the opposite direction. Figure 1-3 shows diagrammatically the Fiji transform between the Tonga trench, subducting westward, and the Vanuatu (New Hebrides) trench, subducting eastward. Transform faults are discussed further in Chapter 8.

The global distribution of earthquakes defines a mosaic of crustal plates up to 100 km thick that are moving with respect to one another (Fig. 1-4). Note that most of the plate boundaries in Figure 1-4 are imaged by earthquake epicenters in Figure 1-2. These plate boundaries do not correspond to continental slopes. Indeed, the continents are high-standing passengers riding on plates where much of the dynamic activity involves oceanic crust at spreading ridges and subduction zones. Note also that the block diagram in Figure 1-3 does not separate continents from ocean basins; continents are not essential elements in the basic concepts of plate tectonics.

Motions between plates over the last few million years have been modeled by Minster and Jordan (1978) and other workers, most recently by DeMets et al. (1990; Fig. 1-4), based on correlation of magnetic anomaly patterns on the sea floor with the geomagnetic time scale, on orientations of major transform faults, and on first-motion data (discussed in Chapter 4) from earthquakes on plate boundaries. We now know the relative motion of all plates with respect to one another averaged over a few million years. Spreading rates vary from 12 to 160 mm/yr, and convergence rates at trenches vary from 20 to more than 110 mm/yr (DeMets et al., 1990). These rates have been confirmed by measurements averaged over a few years using space geodetic techniques (Gordon and Stein, 1992; cf. discussion in Chapter 5).

Because plates move on the surface of a sphere, relative motions are considered as rotations about a pole which may lie outside the plates whose relative motion is being described (Fig. 1-5). Relative motion on a plate boundary varies from zero at the pole of rotation to a maximum at the equator of rotation, 90° from the poles. This is analogous to considering the motion of any point on the earth due to the earth's rotation. Motion is zero at the spin axis of the earth, and it is at a maximum at the equator. Transform faults defining boundaries of pure strike slip (where the plates neither converge nor diverge) would occupy small circles of rotation about the pole, analogous to lines of latitude with respect to the earth's spin axis. "Instantaneous" models of plate motion assume con-

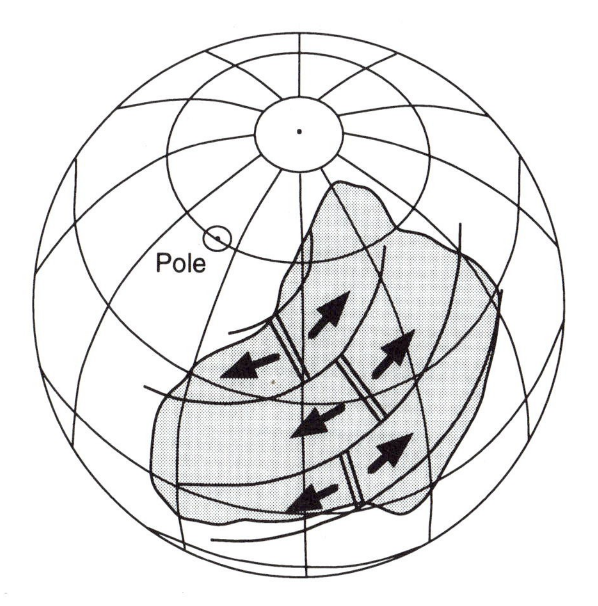

Figure 1-5. When plates (dot pattern) on a globe rift and move apart, points that were originally together move apart along curved lines. The plates rotate apart around a pole. An ocean spreading ridge (double lines) is offset along transform faults, which have the same orientation with respect to the rotation pole of the plates as lines of latitude have with respect to the spin axis of the globe.

tinuous, homogeneous motion. This motion is accommodated at plate boundaries, accompanied by creep where the boundary is weak and by stick slip, expressed by earthquakes, where the boundary is strong.

The assumption that all plate motion is taken up at plate boundaries implies that plates are rigid and undergo no internal deformation during plate motion. The diffuse distribution of earthquakes across plate boundaries within continents, such as those in central Asia and the western United States and within some ocean basins such as the Indian Ocean between India and Australia (Figs. 1-2 and 1-4) shows that the assumption of perfectly rigid plates is not correct. Plate motion can be distributed across a zone hundreds or even thousands of kilometers wide, as shown in Figure 1-4. These diffuse boundary zones contain most of the earth's great earthquakes; they may occupy as much as 15% of the earth's surface (Gordon and Stein, 1992).

Some diffuse plate boundaries contain relatively rigid microplates whose relative motion with respect to the major plates must be calculated to understand slip rates across major faults within the region. For example, displacement rates across the San Andreas fault of California based on geology are lower than the rate predicted by relative motion between the Pacific and North American plates, producing a slip-rate discrepancy. However, when displacement across the Great Basin between the Sierra Nevada of California and the Wasatch Range of Utah is taken into account, the discrepancy is largely removed by considering California from the San Andreas fault to the east edge of the Sierra Nevada as a rigid microplate (Gordon and Stein, 1992). This is discussed further and illustrated in Chapter 5.

The island of Hawaii has frequent earthquakes, even though it is entirely within the Pacific plate, far from a plate boundary. Hawaii is the emergent portion of several huge basaltic shield volcanoes that rise from the ocean floor. Some of these earthquakes are related to the injection of magma into the volcanic edifice, whereas others may be due to gravitational collapse of the flanks of the volcano (cf. Chapter 9). The volcanoes are related to a thermal plume or "hotspot" within the mantle, which warms and elevates the oceanic crust in addition to producing the volcano. The thermal plume is more stationary with respect to the lower mantle than the Pacific plate, which is moving west-northwest with respect to the center of the earth. The other islands of the Hawaiian chain west-northwest of Hawaii have progressively older volcanoes and represent times when that part of the Pacific plate passed over the stationary mantle plume. In this way, the volcanic islands record former positions of the plate with respect to the plume, much as a piece of paper passed over the flame of a stationary candle would show a linear pattern of scorch marks. Similar age patterns of hotspot tracks relate the Reunion and Kerguelen hotspots in the southern Indian Ocean to basaltic lava fields in the Indian peninsula (Duncan, 1991).

Yellowstone Park in the western United States provides an example of a mantle hotspot rising beneath a continent. Yellowstone Park is characterized by young volcanic rocks, by hot groundwater manifested as geysers, reflecting anomalously high heat flow through the crust, and by elevated topography. If the Yellowstone hotspot is stationary with respect to the center of the earth, it should have left a track of older volcanic rocks west of it as North America moved westward away from the Mid-Atlantic Ridge. Volcanic rocks of the Snake River Plains of southern Idaho and the Columbia Plateau of Washington and Oregon may be ancient expressions of the Yellowstone hotspot.

SUMMARY

In summary, plate tectonics provides a dynamic framework for the buildup of strain in the lithosphere that is in part released by earthquakes, principally at plate boundaries. The origin of earthquakes is ultimately the jostling between moving plates which produces the strains within the lithosphere that must be relieved by earthquakes. The distribution of earthquakes and the sense of displacement documented by earthquakes comprise one of the data sets essential to plate tectonic theory. More important to our subject, displacement rates on an earthquake-producing structure may be compared with displacement rates on the plate boundary of which that structure is a part. Over a long period of time, the sum of displacement rates across a diffuse plate boundary should equal the displacement rate between the plates themselves.

Suggestions for Further Reading

Cox, A., ed. 1973. Plate Tectonics and Geomagnetic Reversals. San Francisco: W. H. Freeman, 702 p. This volume is a collection of the papers leading to the new paradigm of plate tectonics with comments by the editor. Included are papers by Isacks et al. (1968) and Morgan (1968) discussed in this chapter.

DeMets, C., Gordon, R. G., Stein, S., and Argus, D. F. 1990. Current plate motions. Geophys. J. Int. 101:425–78.

Duncan, R. A. 1991. Oceanic drilling and the volcanic record of hotspots. GSA Today, 1:213–16, 219.

Fowler, C. M. R. 1990. The solid earth, an introduction to global geophysics. Cambridge: Cambridge University Press, 472 p.

Glen, W. 1982. The road to Jaramillo: Critical years of the revolution in earth science. Stanford, California, Stanford University Press, 459 p. History of the development of the theory of plate tectonics.

Gordon, R. G. 1994. Present plate motions and plate boundaries. *In* Handbook of Physical Constants, ed. T. Ahrens. American Geophysical Union Monograph, in press.

Gordon, R. G., and Stein, S. 1992. Global tectonics and space geodesy. Science 256:333–42.

Kearey, P., and Vine, F. J. 1990. Global Tectonics. Oxford: Blackwell Scientific Publications, 302 p.

Uyeda, S. 1978. The New View of the Earth. San Francisco: W. H. Freeman, 217 p. (In Japanese, Atarashi Chikyukan, published in 1972 by Iwanami Shoten, Tokyo.) Description of plate tectonics and the primary structural features of the earth, easily understood by the non-geologist and non-geophysicist.

Zechariah (ca. 520 B.C.E.)

Israel

Palestine lies astride a plate boundary. Across the Jordan River, the Arabian plate is driving northward against Eurasia, forcing up the Zagros Mountains of Iran and other high mountains to the west in Turkey and the Caucasus. The plate boundary on the west is a left-lateral strike-slip fault extending northward from the Red Sea through the Gulf of Aqaba into the plain of the Jordan River and into Lebanon. This region has been inhabited by people keeping written records for thousands of years. Perhaps the oldest city in the world is Jericho, and its destruction by an earthquake around 1560 B.C.E. may have been caused by movement on this plate boundary. The memory of this earthquake was probably still on the minds of the citizens of Jericho when the city was taken by Joshua more than 300 years later, around 1230 B.C.E. The two events may have been confused when oral traditions were put in writing in the seventh century B.C.E., to the effect that at Jericho, "the walls came tumbling down."

If an earthquake produced surface rupture on the Jericho fault, the changes in the land surface should have attracted the attention of people living on the Jordan plain. Springs and sand blows should have erupted, a line of newly disturbed ground would have formed, and goat paths and wagon tracks would have been offset by strike slip.

Surface rupture accompanying an earthquake in 759 B.C.E. may have been observed and recorded and could have been known to the prophet Zechariah more than two centuries later, when he issued what may be the first earthquake forecast:

"And his feet shall stand in that day upon the mount of Olives, which is before Jerusalem on the east, and the mount of Olives shall cleave in the midst thereof toward the east and toward the west; and there shall be a very great valley; and half of the mountain shall remove toward the north, and half of it toward the south.

And ye shall flee to the valley of the mountains; for the valley of the mountains shall reach unto Azal: yea, ye shall flee, like as ye fled from before the earthquake in the days of Uzziah king of Judah." (Zech. 14:4–5)

The forecast earthquake did not arrive until 31 B.C.E., when Josephus recorded a great earthquake that killed 10,000 men and destroyed many buildings in Judea, with damage to the Second Temple in Jerusalem.

Most biblical scholars now believe that only the first eight chapters of the Book of Zechariah can be attributed to that author in the period 520–518 B.C.E. The last six chapters, including the passage cited above, may have been written by more than one author as early as the late eighth century B.C.E. (soon after the earthquake of 759 B.C.) and as late as the third century B.C.E.

A controversy has arisen over whether the cleaving in the midst of the Mount of Olives reflected fault rupture on the plate boundary or whether it was a large landslide (Wachs and Levitte, 1984), possibly accompanying the earthquake of 759 B.C.E. There is, in fact, no strike-slip fault on the Mount of Olives (Bentor, 1989)! The phrase "And ye shall flee to the valley of the mountains; for the valley of the mountains shall reach into Azal" could be translated, "And the valley of my mountain shall be stopped up / For the valley of the mountains shall touch the side of it," which could be interpreted as a description of a large landslide. There is, indeed, a large landslide on the west slope of the Mount of Olives, but it is not known if it dates back to the eighth century B.C.E.

The earthquake of 759 B.C.E., which Ben-Menahem (1991) considers to be of $M_L = 7.3$, was also mentioned in the *Book of Amos (1:1)*: "The words of Amos, who was among the herdmen of Tekoa, which he saw concerning Israel in the days of Uzziah king of Judah, and in the days of Jeroboam the son of Joash king of Israel, two years before the earthquake."

Suggestions for Further Reading

Ben-Menahem, A. 1991. Four thousand years of seismicity along the Dead Sea rift. Jour. Geophys. Res., 96:20,195–20–216.

Ben-Menahem, A., Nur, A., and Vered, M. 1976. Tectonics, seismicity and structure of the Afro-Eurasian junction—the breaking of an incoherent plate. Physics of the Earth and Planetary Interiors, 12:1–50.

Bentor, Y. K. 1989. Geological events in the bible. Terra Nova, 1:326–38.

Nur, A. 1990. Earthquakes in the Holyland. Documentary film: EOS, Trans. Am. Geophys. Union, 71:1222.

Wachs, D., and Levitte, D. 1984. Earthquakes in Jerusalem and the Mount of Olives landslide. Israel—Land and Nature 9(3):118–21.

2

Rock Deformation and Structure

Structural geology is the study of deformation, including rigid-body translation, of the earth's crust at all scales from mountain ranges to individual crystals in rock. The study of mountain ranges, where structural geology began as a field of study nearly two centuries ago, is included in *tectonics*. It is necessary to have some background in structural geology to understand earthquakes, because earthquakes are themselves a manifestation of rock deformation. This background includes an understanding of the physics of solid-state deformation: the different ways in which rocks respond to stress. We also need some knowledge of structures that are produced when rocks are deformed, particularly faults and folds.

It is relatively easy to describe the folds and faults in a road cut or on a cliff face. It is harder to relate what we see on the earth's surface to the physics of deformation. There are two reasons for this difficulty. First, the structures seen in the field may have taken hundreds of thousands of years to form, and it is impossible to duplicate such a long time in the laboratory. Second, the rocks observed at the surface may have formed at depths of many kilometers, and we need to consider their deformation under pressures and temperatures existing at those depths. Earthquakes nucleate at depth, and we have the same problem in visualizing how earthquakes occur because we tend to think in terms of rocks at the surface. We have to make assumptions, and these assumptions lead to uncertainties.

Earthquake geology makes an important contribution to structural geology, because it is real-time structural geology, the structural geology of the present day. If a folded rock layer took hundreds of thousands of years to form, then we may see a very small part of the folding process taking place during an earthquake. Furthermore, faults and folds that are generated far beneath the surface are expressed at the surface as tectonic landforms. By careful analysis of these landforms (discussed in Chapter 7) and their rates of formation, we can learn about the rates of folding and faulting at depth.

Structural geology is presented here in two chapters. Chapter 2 discusses stress, strain, and the different ways in which rocks deform, and, in addition, describes the rock structures that are important to an understanding of earthquakes. Chapter 3 deals with the geology of that deeply buried part of the brittle crust where most earthquakes are generated.

STRESS AND STRAIN

Earth materials respond to *stress,* or force per unit area. *Strain* is the change of length, volume, or shape of a body with respect to its original length, volume, or shape; it is dimensionless. For elastic bodies, stress is proportional to strain, with a proportionality constant called the *modulus of elasticity,* expressed in units of stress. All elastic strain is recoverable if the stress is removed. We consider two types of strain, *change in volume* and *change in shape*.

For a change in volume, a body is subjected to stresses that are uniform in all directions, such as the stresses imposed on a submarine below the surface of the sea. The load of seawater acting downward on the submarine is exactly counteracted by pressures acting upward and from the sides. This is *hydrostatic pressure*. This pressure will cause the submarine to decrease in volume slightly, but this pressure will not change the shape of or distort the submarine. When the submarine returns to the surface, its original volume is restored, meaning that the volume strain on the submarine is elastic and recoverable. A change in volume is elastic for a liquid as well as a solid; the volume of the liquid increases as the confining pressure is removed.

Similarly, a rock buried in the crust is subjected to the confining pressure of the load of a column of rock on top of it, which is accompanied by an equal stress acting upward on its base. However, the confining crust is a rigid solid, not a liquid, and so the stresses acting from the sides may not be equal to those imposed from above and below. For this reason, the rock may undergo some change in shape as well as volume. But there is still a decrease in volume which is recoverable when the stress is removed, as long as the rock deforms elastically.

For change in shape to occur, the imposed stresses acting on the body must be unequal. Change of shape of a liquid is not recoverable when the stresses are removed, hence a liquid has no elastic response to change in shape. But if a solid such as a rubber eraser is squeezed, it is deformed elastically because the imposed stresses are unequal. When these stresses are removed, the eraser returns to its original shape. In the discussion that follows, stresses acting on a solid body are unequal, and the response of the body to these stresses is elastic.

For convenience, a stress acting on a surface can be resolved into a normal component (σ) acting perpendicular to the surface, and a shear component (τ) acting parallel to the surface. A stress acting normal to a surface across which shear stress is zero is a *principal stress*. There are three mutually perpendicular principal stresses, σ_1, σ_2, and σ_3 (Fig. 2-1). In a stressed specimen, the maximum stress = σ_1, the intermediate stress = σ_2, and the minimum stress = σ_3. In the crust, σ_1, at least, is compressive. Commonly all three principal stresses are compressive, although near the surface, σ_3 may be tensile.

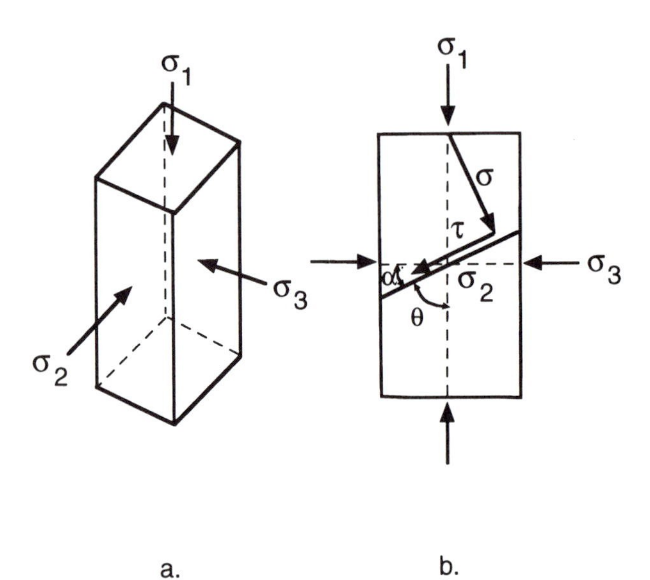

Figure 2–1. (a) Principal stresses acting on a body, $\sigma_1 > \sigma_2 > \sigma_3$. Shear stress = 0 on the surface on which each of these stresses acts. (b) For a surface making an angle α with the plane normal to the principal stress σ_1, stresses are resolved into a normal stress σ and a shear stress τ.

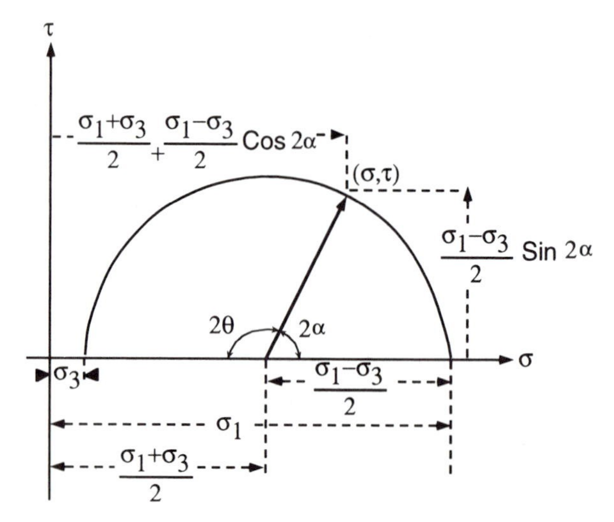

Figure 2–2. Mohr stress circle for positive (counterclockwise) shear stress. Below the abscissa, the other half of the circle represents clockwise shear stress. This circle is the locus of values of σ and τ acting on a surface that forms an angle α with the plane normal to σ_1 (Fig. 2–1b). As the surface rotates from $\alpha = 0°$ to $\alpha = 90°$ (Fig. 2–1b), the radius vector rotates such that the angle with the abscissa changes 180°; thus α in Figure 2–1b is plotted as 2α in Figure 2–2.

THE MOHR FAILURE ENVELOPE

Consider the shear and normal stresses on any plane perpendicular to the σ_1 σ_3 plane and making an angle α with the σ_3 axis (Fig. 2-1b). Let normal stress σ and shear stress τ be the abscissa and ordinate, respectively, of a graph (Fig. 2-2). Each mutually perpendicular principal stress axis acts in a direction perpendicular to a plane across which shear stress $\tau = 0$. Thus σ_1 and σ_3 must be plotted on the abscissa, where $\tau = 0$. They represent points on a circle in which the abscissa includes a diameter, and the center is located at $\frac{\sigma_1+\sigma_3}{2}$ This is called a Mohr stress circle. If from this midpoint between σ_3 and σ_1, a radius vector of length $\frac{\sigma_1-\sigma_3}{2}$ is erected at a counterclockwise angle 2α with the σ axis, its coordinates would be σ and τ, the normal and shear stresses, respectively, acting on the plane at an angle α to the minimum principal stress σ_3.

Thus: $$\tau = \frac{\sigma_1 - \sigma_3}{2} \sin 2\alpha$$

$$\sigma = \frac{\sigma_1 + \sigma_3}{2} + \frac{\sigma_1 - \sigma_3}{2} \cos 2\alpha$$

The Mohr stress circle gives at a glance the values of the normal and shear stresses across the plane of reference for any angle α (Fig. 2-1b) and any combination of values of σ_1 and σ_3, regardless of whether failure occurs or not. Maximum shear stress, τ_{max}, occurs where angle $\alpha = 45°$, angle $2\alpha = 90°$, and $\sin 2\alpha = 1$. $\tau_{max} = \frac{\sigma_1 - \sigma_3}{2}$, the radius of the circle. τ_{max} occurs on a plane including σ_2 and bisecting the angle between σ_1 and σ_3, where $\alpha = 45°$. $\frac{\sigma_1 + \sigma_3}{2}$ is the mean stress, or hydrostatic component of stress, that which produces change in volume. The radius of the circle, $\frac{\sigma_1 - \sigma_3}{2}$ is the maximum possible value of shear stress, that which produces change in shape.

Mohr stress circles may also be drawn for the $\sigma_1\sigma_2$ plane, and for the $\sigma_2\sigma_3$ plane. Figure 2-2 shows only half of the Mohr diagram, that for counterclockwise shear stress, here considered as positive. That part of the Mohr circle below the abscissa would represent clockwise shear stress.

SHEAR-FRACTURE CRITERIA

Rocks may be subjected to differential stresses in the laboratory to determine the conditions under which they fail. A common procedure is to place a cylindrical rock sample in a deformable jacket of copper or rubber and subject it to stress parallel to the axis of the cylinder and a confining pressure applied uniformly around the sides of the cylinder by a fluid that is isolated from the rock by the deformable jacket (Fig. 2-3). The rock sample should have no preexisting planes of weakness so that a failure plane would represent a newly formed fracture. The confining pressure represents the pressure on a rock imposed by the load of overlying rock, which is a function of the density of the overlying rock. At a given confining pressure, the axial stress is increased by a moving piston or ram (Fig. 2-3) until the rock fractures. The experiment is repeated with another cylinder of the same rock under a higher confining pressure. If the axial pressure exceeds the confining pressure, the axial pressure = σ_1 and the confining pressure = σ_3. If the axial pressure is less than the confining pressure, the axial pressure = σ_3. Figure 2-4 shows diagrammatically the deformation style of cylinders of rock that have been subjected to compression and extension tests under brittle to plastic conditions.

A particular Mohr circle may be plotted for each pair of values of σ_3 and σ_1 for which failure occurs.

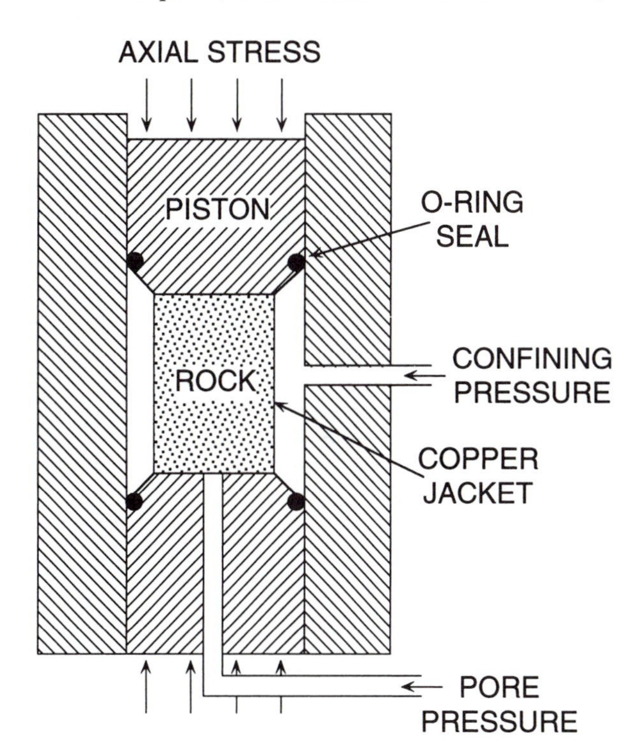

Figure 2–3. Simplified diagram of press used to deform rocks in laboratory. Rock sample, in an impermeable jacket, is subjected to axial stress and confining pressure. For compression tests, axial stress > confining pressure; for extension tests, confining pressure > axial stress. Fluid pressure within the rock is controlled independently from confining pressure.

The successive tests at failure for increasing values of σ_3 and σ_1 will be a family of circles whose centers lie at successively greater distances out on the σ axis (Fig. 2-5). For a given value of σ_3 (confining pressure on a jacketed specimen), axial stress σ_1 begins at σ_3 and is gradually increased until failure occurs. As axial stress increases, the rock cylinder deforms homogeneously by becoming shorter and wider, but the strain is elastic. If the axial stress is removed prior to failure, the cylinder would return to its original length and width; the elastic strain would be recoverable. But normally the axial stress is increased until the rock cylinder fractures, and the Mohr circle is constructed for the values of σ_1 and σ_3 at failure.

For a single test, there is an infinite number of values of σ and τ, represented by points on the Mohr circle, for which failure could have occurred. But in a series of tests made under increasing values of σ_3, failure is shown to occur at values of (σ, τ) at the point of tangency between the Mohr circles and a straight line whose slope is μ, which is called the *coefficient of internal friction,* and which intersects the ordinate at τ_o, the *cohesion,* or inherent shear strength when

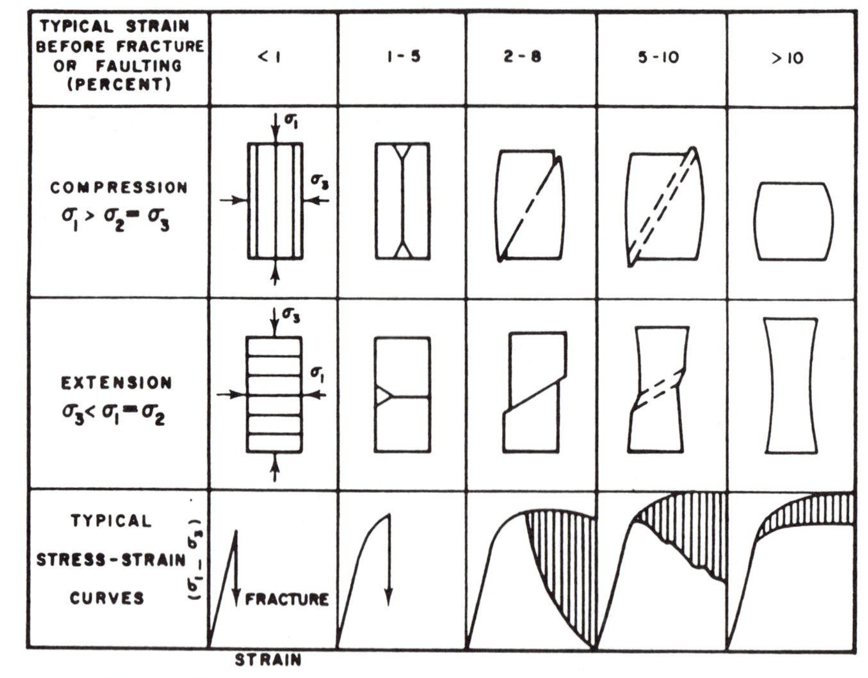

Figure 2–4. Diagrammatic representation of rock cylinders deformed in a press under brittle (left) to plastic (right) conditions, with expected stress-strain curves. In compression tests, axial stress > confining pressure; in extension tests, confining pressure > axial stress. Modified from Griggs and Handin (1960).

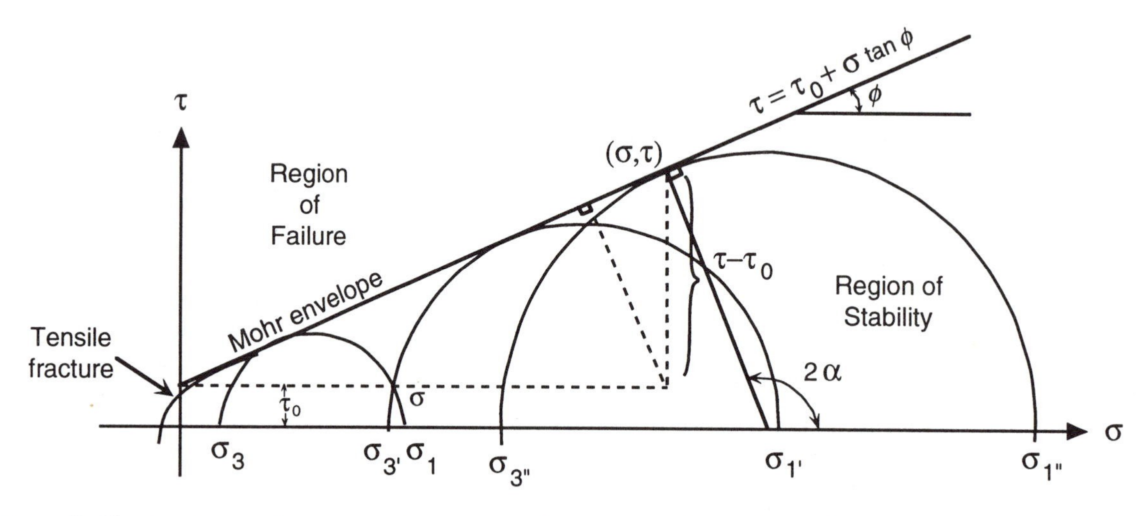

Figure 2–5. Mohr envelope of failure for a rock specimen that fails in three successive tests at principal stresses $\sigma_1\sigma_3$, $\sigma_1'\sigma_3'$ and $\sigma_1''\sigma_3''$. The rock specimen contains no preexisting planes of weakness. The failure envelope is tangent to the three Mohr circles and has the equation $\tau = \tau_0 + \sigma \tan \phi$, where τ_0 = cohesion (shear stress where normal stress = 0) and ϕ is the angle between the Mohr envelope and the abscissa. Tan $\phi = \mu$, the coefficient of internal friction. This is the *Coulomb fracture criterion.* The failure envelope in the tensile field ($\sigma < 0$) does not follow this fracture criterion due to a nonlinear decrease in rock strength in tension.

normal stress = 0. $\mu = \tan \phi$, where ϕ is the *angle of internal friction,* so the equation of this line is commonly written $\tau = \tau_o + \sigma \tan \phi$. This straight-line relationship is the *Coulomb fracture criterion,* which was first applied to the strength of cohesionless soils. As applied to rocks, both μ and τ_o are no more than empirical relationships. The slope of the Mohr envelope illustrates the property of most rocks to increase in strength as confining pressure is increased.

The Mohr envelope may be concave toward the σ axis under high and very low values of σ_3. Under values of $\sigma_3 < 0$, failure is by tensile fracture perpendicular to σ_3 so that the Coulomb fracture criterion does not apply (cf. discussion in following text on Griffith cracks). In the tensile field, the failure envelope is similar to a parabolic curve.

GRIFFITH CRACKS

The Coulomb fracture criterion does not describe the processes of failure, it only predicts the shear stress necessary to cause failure under different confining pressures. In particular, it does not predict the low tensile strength of rock compared to theoretical values based on the strength of atomic bonds of rock constituents. Microscopic examination of rocks and minerals reveals the presence of many cracks and flaws that cause materials to fail at stresses less than their theoretical strengths. These flaws, called *Griffith cracks* after A. A. Griffith, who studied this problem in the 1920s, may become enlarged and may propagate by tensile fracture under the influence of differential stress. These cracks propagate unstably due to high stress concentrations around their edges or crack tips. As tensile fractures increase in cross-section area, the tensile force is concentrated on a decreasing cross-section area of rock that has not failed, thereby increasing tensile stress and making macroscopic tensile failure more likely. Under compression, the cracks may be closed, and shear along the closed cracks will be accompanied by friction (see the following text) and will result in tensile stress at the ends of these shear cracks.

The displacement of cracks is described by Scholz (1990) in three modes (Fig. 2-6). In Mode I, the tensile mode, crack displacements are normal to the crack wall. An example of a Mode I crack is a sandstone dike, in which unconsolidated, liquefied sand is injected into a tensional crack during an earthquake. Mode II and Mode III are both shear modes, with displacements in the plane of the crack. In Mode II, displacements are normal to the crack edge (edge dislocation), and in Mode III, displacements are parallel to the edge (screw dislocation). In fracture mechanics, it is assumed that the crack has no cohesion, however, in Modes II and III, friction may be a significant factor.

FRICTION

Friction is the resistance to motion of a body sliding past another body along a surface of contact. Surfaces have topographic irregularities so that when they are in contact, they touch at only a few points called *asperities*. The surface area of asperity contact is small with respect to the total area of the sliding surface, resulting in a high compressive stress, high enough that *adhesion* may occur due to interpenetration and locking and chemical bonding of parts of the surface. As projections on one surface plow through another, and locked surfaces ride up on one another, resistance to sliding is produced. This process generates heat.

Consider a block of rock resting on a surface acted upon by a normal force F and a parallel force V (Fig. 2-7), transmitted through a spring, the end of which moves at constant velocity u. Normal stress $\sigma_n = \frac{F}{A}$ and shear stress $\tau = \frac{V}{A}$. As the end P continues to move, parallel force V increases until it exceeds a value V_A necessary to overcome *static friction*. The

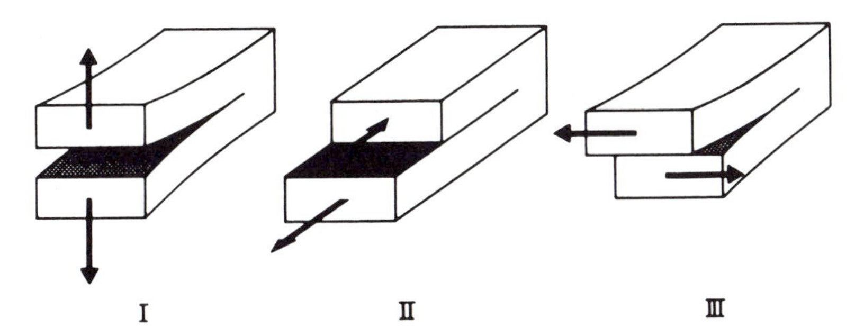

Figure 2–6. The three modes of crack propagation. I, tensile mode. II, shear mode: edge dislocation (displacement normal to line defect). III, shear mode, screw dislocation (displacement parallel to line defect). After Scholz (1990).

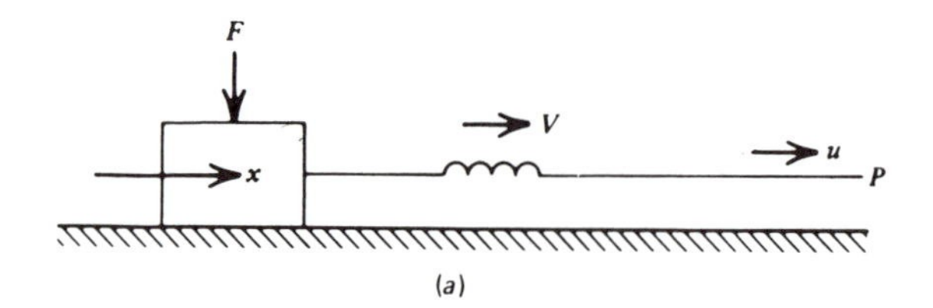

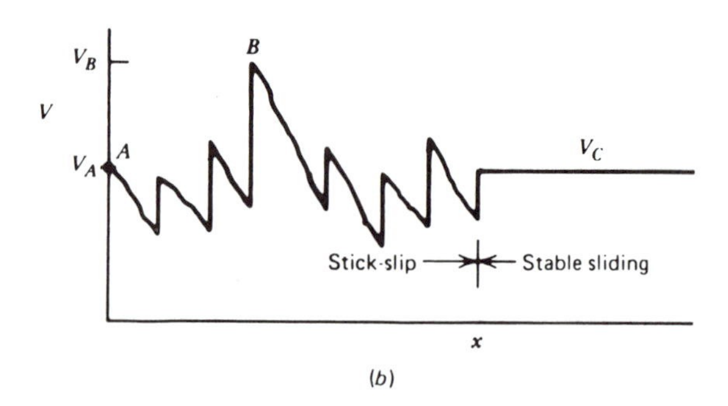

Figure 2–7. (a) A typical friction experiment. The block of rock is pressed against the surface by the force F. The displacement of point P at a uniform rate transmits a force V to the block by the spring. (b) The force V varies with the displacement x of the block. Sliding initially occurs at point A and the corresponding force V_A defines the coefficient of static friction, μ_s. The maximum horizontal force V_B at point B defines the maximum value of the friction coefficient. The horizontal force V_C after stable sliding has been established defines the coefficient of dynamic friction, μ_d. From Byerlee (1977).

coefficient of static friction $\mu_s = \frac{V_A}{F} = \frac{\tau_A}{\sigma_n}$, if τ_A is the stress at the limiting force V_A. μ_s depends upon the type of rocks being pressed together and the nature of the surfaces in contact, but it is independent of the normal force F. Experiments show that μ_s increases with the time the surfaces are in stationary contact.

After the block starts to move, it is resisted by *dynamic friction*. The coefficient of dynamic friction μ_d tends to decrease as sliding velocity increases (Rabinowicz, 1965). During movement of the block, the parallel force V may decrease until friction between block and surface stops the block. This is called *stick-slip* behavior and is illustrated in Figure 2-7. If velocity u and the coefficient of friction μ_s remained constant, the time between successive movements of the block would be the same. Because surface irregularities will not be the same each time, μ_s varies, and so the time between movements varies.

This theory of stick-slip behavior may be illustrated empirically by considering the behavior of different segments of the San Andreas fault, California. The Pacific plate moves with respect to North America with a constant velocity u. The *average recurrence interval* for nine successive displacements at Pallett Creek in the Transverse Ranges section of the San Andreas fault over the last 1300 years is 132 years (Sieh et al., 1989). Individual recurrence intervals may depart from this value in part because of variations in μ_s. Farther northwest, parts of the San Andreas fault southeast of San Francisco move by fault creep; μ_s here must be very low, and displacement approaches stable sliding (Fig. 2-7). This is, admittedly, a simplistic analogy, because of other variables in the San Andreas fault zone such as fluid pressure and motion on other faults.

The frictional strengths of many kinds of rocks under varying normal stresses were tabulated by Byerlee (1978). Under low normal stresses, the frictional strength is related to normal stress σ_n by $\tau = 0.85\sigma_n$ with a large amount of scatter due to the effect of roughness of the surface. At higher normal stresses, surface roughness is relatively unimportant, and the relationship is $\tau = 0.5 + 0.6\sigma_n$ (Fig. 2-8). This relationship is, in a general way, independent of rock type, ductility, sliding velocity, and surface roughness at temperatures up to 400°C, and is commonly known as *Byerlee's law*. Because it has such general application, it may be used to estimate the frictional strength of faults. However, its validity at depths greater than 5 km is controversial, and certain clays do not appear to follow the relationship.

EFFECTS OF PORE FLUIDS

The presence of pore water, even where porosity is very low, produces a pore pressure p_f which acts outward, whereas confining pressure σ_n acts inward. This can be simulated in the laboratory by introducing fluid pressure into the rock cylinder which acts outward against the deformable jacket (Fig. 2-3). In Coulomb-Mohr theory, this moves the center of the Mohr circle toward the origin but does not change the radius, or shear-stress component (Fig. 2-9). This can change conditions along a fault from stable to unstable such that displacement occurs. For a fluid-saturated fault, $\tau = \tau_o + \mu_s(\sigma_n - p_w)$, in which τ_o = cohesion, which should be so small that the term can be ignored. Thus, in comparison with dry rocks, the frictional resistance of the rock can be overcome by a lower threshold value of τ.

If the pore water is connected to the surface, the water pressure is hydrostatic, about 45% of the value of overburden pressure. Hubbert and Rubey (1959) introduced a dimensionless number λ, the ratio of fluid pressure to the vertical load due to rock overburden:

$$\lambda = \frac{p_f}{\sigma_v} = \frac{p_f}{\rho g z}$$

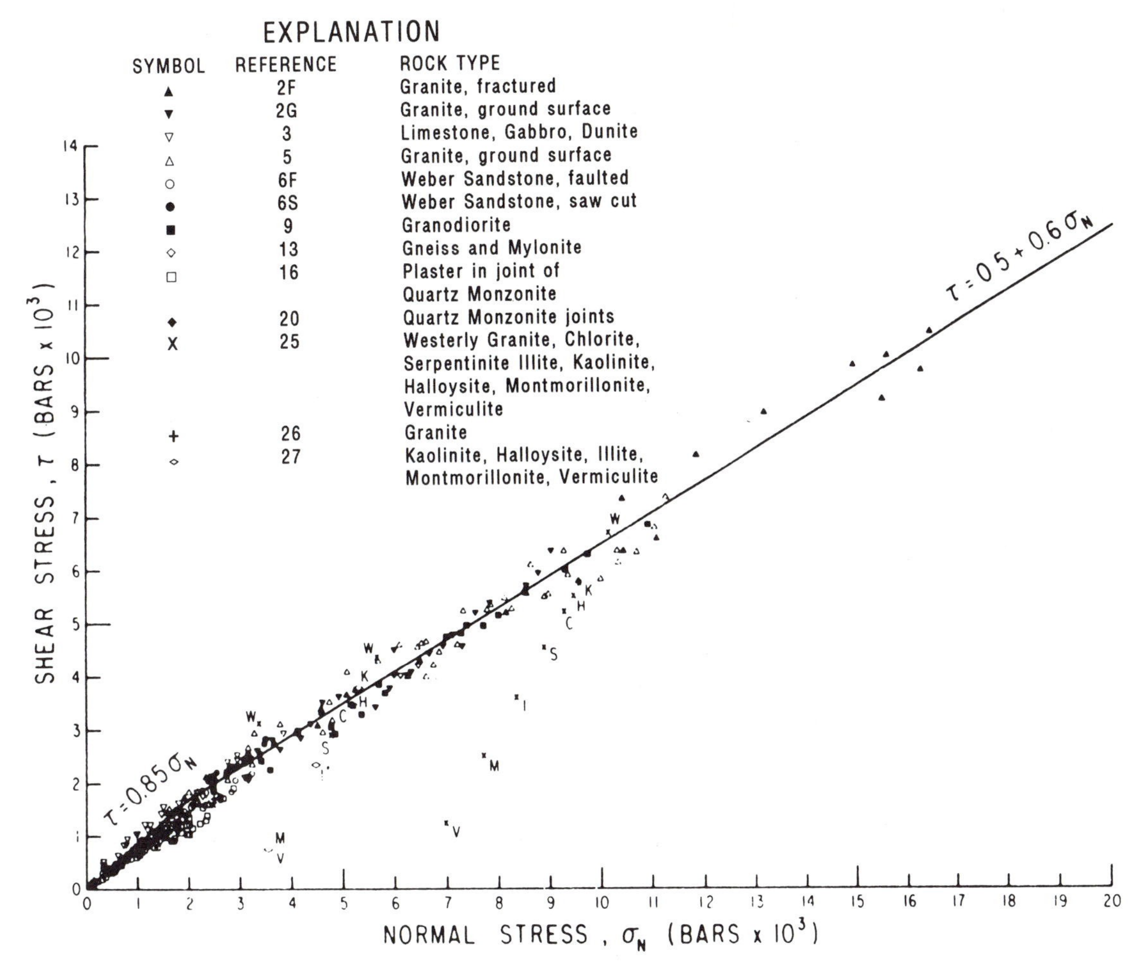

Figure 2–8. Maximum shear stress to initiate sliding as a function of normal stress for a variety of rock types. The linear fit defines a maximum coefficient of static friction f_s equal to 0.85. From Byerlee (1978).

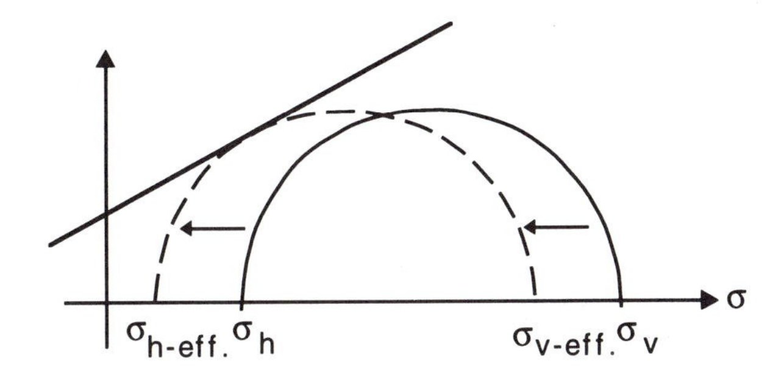

σ_v = vertical stress (dry rock)

$\sigma_{v\text{-eff.}}$ = vertical stress including pore pressure

σ_h = horizontal stress (dry rock)

$\sigma_{h\text{-eff.}}$ = horizontal stress including pore pressure

Figure 2–9. Effect of pore pressure in changing a stable stress field for dry rock to a stress field in which failure occurs. The effect is to change the position of the center of the circle but not the length of the radius. Stresses taking into consideration pore pressures are called *effective stresses*. ($\sigma_{v\text{-eff}}, \sigma_{h\text{-eff}}$).

where σ_v = vertical load or overburden pressure, ρ = density of overburden, z = depth, and g = gravitational acceleration. They showed that pore water not connected with the surface may be subjected to confining pressures such that the pore pressures exceed hydrostatic ($\lambda > 0.45$). In some instances, fluid pressures may approach overburden pressure ($\lambda \rightarrow 1$), almost "floating" the rock and reducing to almost zero the shearing stress necessary to produce displacement. On the other hand, an increase in shear strain may cause growth of microscopic cracks, or it may cause asperities to ride up over one another along sliding surfaces, increasing the pore space. This local volume increase causes the pore pressure to decrease, which moves the Mohr circle of Figure 2-9 away from the origin, a process called *strain dilatancy*.

Pore fluids, principally water, react with and weaken rock, a process that is time-dependent, based on reaction rate. Water may cause silicates to become plastic, a process called *hydrolytic weakening*. Chemical reactions are enhanced by the presence of high stress fields at crack tips. The reaction replaces Si-O bonds with weaker hydrogen bonds. This induces time-dependent *stress-corrosion cracking*. Fluid-rock reactions, because they weaken rock surfaces of sliding, also have an effect on friction.

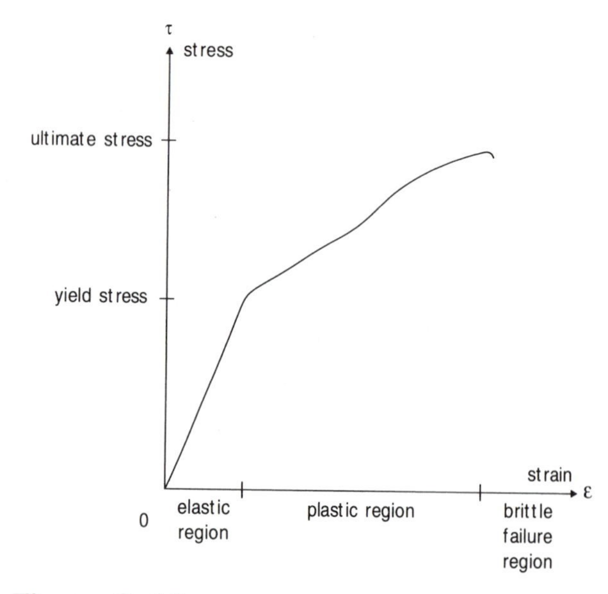

Figure 2–10. Stress-strain curve showing *elastic behavior* in which stress is proportional to strain, and strain is recoverable, *plastic behavior* in which behavior is by ductile flow, and *brittle fracture*.

RHEOLOGICAL MODELS

Figure 2-10 is a stress-strain curve showing three types of rheological behavior: the *elastic region*, which behaves according to Hooke's law (stress is proportional to strain); the *plastic region*, which behaves according to ductile flow; and failure by *brittle fracture*. Many stress-strain curves, particularly for rocks under low confining pressure, show only the elastic and brittle failure regions (for example, the left-hand stress-strain curve in Fig. 2-4). *Elastic limit* is the maximum stress under which the material exhibits elastic strain. Beyond this value, the material undergoes permanent deformation by ductile flow or by brittle fracture. The highest value on the stress-strain curve is the *ultimate stress* or *ultimate strength*.

Simple, idealized models may be used to explain rheological behavior (Fig. 2-11). Elastic behavior is modeled by a spring (Fig. 2-11a). When the spring is pulled and stretched, elastic strain energy is stored so that the spring recovers its original shape when the deforming stress is removed. Plastic flow is modeled by a block on a flat surface which is subjected to a lateral force (St. Venant model; Fig. 2-11b). The block does not move until frictional resistance to the underlying surface is overcome, its *threshold strength* or *yield stress*. Movement is not recoverable. The model depicts ideal plastic flow, whereas the actual behavior of earth materials is commonly described as quasi-plastic.

It is closer to nature to attach a spring to the block (Prandtl model; Fig. 2-11c). When the block is pulled, elastic-strain energy is stored in the spring. When the yield stress is exceeded, the block moves, and this strain is not recoverable. When movement stops, some elastic strain is still stored in the spring. In the earlier versions of the elastic-rebound theory of earthquakes, it was believed that the earthquake released all the elastic-strain energy. Many people now believe that some strain energy is still stored in the rock after the earthquake, like the Prandtl model.

The Newtonian model of viscous flow is a cylinder filled with a viscous fluid and containing a loosely fitting piston which is pulled through the fluid at a rate proportional to the applied stress (Fig. 2-11d). Even the smallest amount of stress will produce a displacement on the piston. A Newtonian body differs from a plastic body in that it has no threshold strength which must be exceeded before displacement can occur. This is the essential difference between viscous flow and ideal plastic flow.

The visco-elastic, or Maxwell, model consists of an elastic spring mounted in series with a loosely fitted piston (Fig. 2-11e). Permanent strain begins to accumulate as soon as a stress is applied, a viscous property of the model. Maxwell bodies of high viscosity behave like an elastic body for loads of very short du-

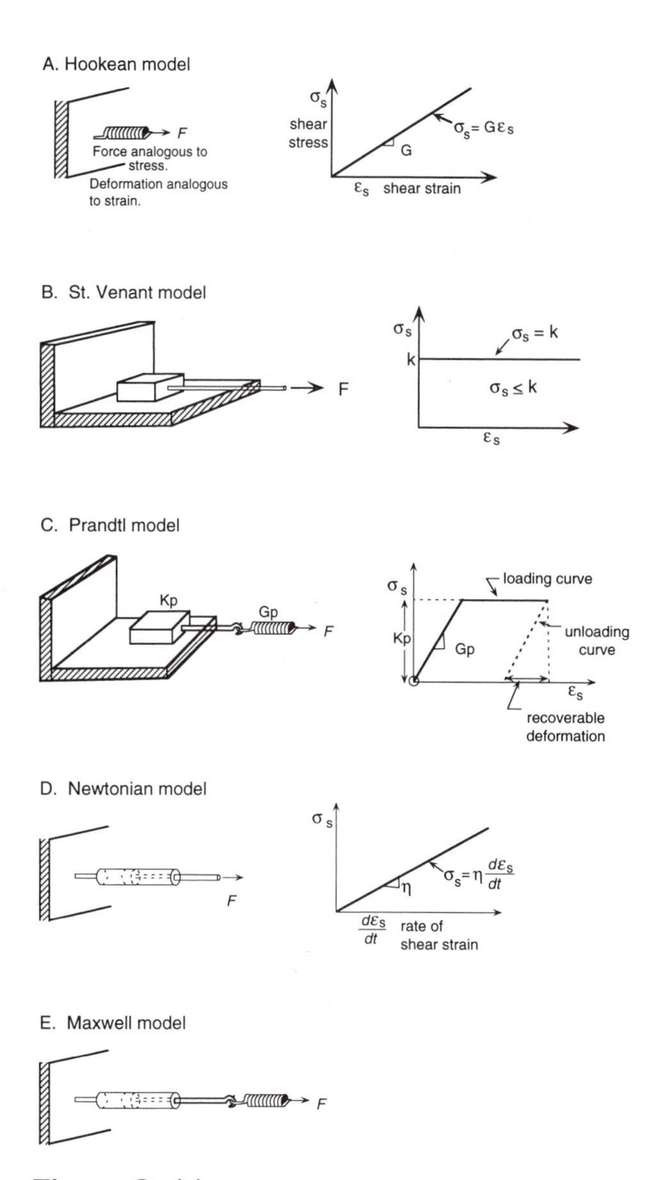

Figure 2–11. Rheological models. After Johnson (1970).

ration, but like a viscous body for long-duration loads. For constant strain, the initial response is elastic, converted over time to permanent viscous deformation as the associated stress decays with time. Silly Putty can bounce like a ball, but it will flow slowly over time. The earth's asthenosphere transmits compressional seismic waves elastically, but flows slowly by creep as predicted by plate tectonics.

FAULTS: RHEOLOGY

Faulting produces localized offset parallel to a plane surface of high shear stress (not necessarily the plane of maximum shear stress, as discussed in the following text). Faulting may or may not involve total loss of cohesion, actual separation, release of stored elastic-strain energy, or loss of resistance to differential stress. In contrast, a *joint* produces offset perpendicular to the fracture surface. A joint always involves loss of cohesion and release of stored elastic-strain energy.

Deformation may be *brittle,* produced by *cataclasis* (microscopic granulation, displacement within and between grains, in which friction is important), or *plastic,* in which friction is not important. Localized plastic deformation is controlled by imperfections or defects in crystals which make them orders of magnitude weaker than they would be otherwise. Imperfections include (1) *substitutional impurities,* a different atom in the crystal lattice than called for in the ideal crystal structure; (2) *interstitials,* atoms occupying sites not normally occupied; and (3) *vacancies,* unfilled sites in the crystal lattice. Most plastic deformation is associated with motion of defects through the crystal in response to stress. The crystal may glide by slip along an internal crystallographic plane (*translation gliding*) or along a *twin plane* (*twin gliding*). Deformation may also take place by *pressure-solution flow,* in which material is dissolved by greater compressive stress and reprecipitated in regions of lower stress, and *grain-boundary diffusion,* including recrystallization. Modes of plastic deformation are dependent on temperature and shear stress. Temperature influences the velocity of deformation and the mode of deformation, whether by grain boundary diffusion or by solid-state, high-temperature creep. On the other hand, brittle deformation tends to be independent of temperature but dependent on confining pressure.

FAULTS: GEOMETRY

A fault with dip $<90°$ is bounded by a *hangingwall* overlying the fault and a *footwall* beneath it (Fig. 2-12). The *fault trace* is its intersection with the earth's surface. *Fault slip* is a vector measuring relative displacement of formerly adjacent points on opposite sides of the fault, measured in the fault surface. It may be broken into strike-slip and dip-slip components, or horizontal and vertical components (*heave* and *throw,* respectively). Because it is a linear feature, fault slip can be measured by *trend* (the strike of a vertical plane that includes the line) and *plunge* (the angle between the line and the horizontal in that vertical plane). It can also be measured by the *rake,* (the angle a line in the fault plane makes with a horizontal line *in that plane*). Rake equals *pitch*. Grooves on the fault surface called *slickensides* show the rake of most recent slip.

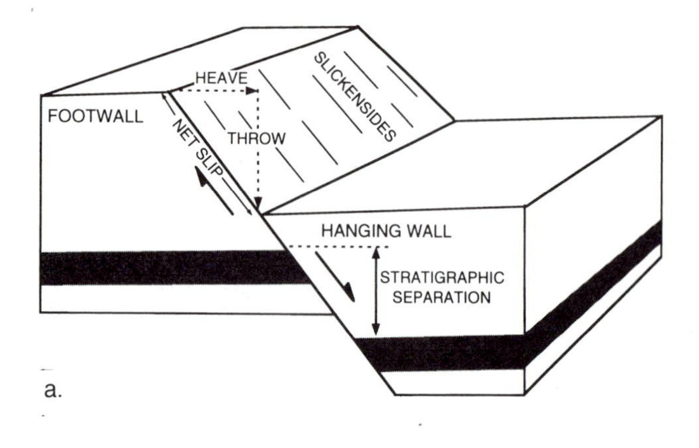

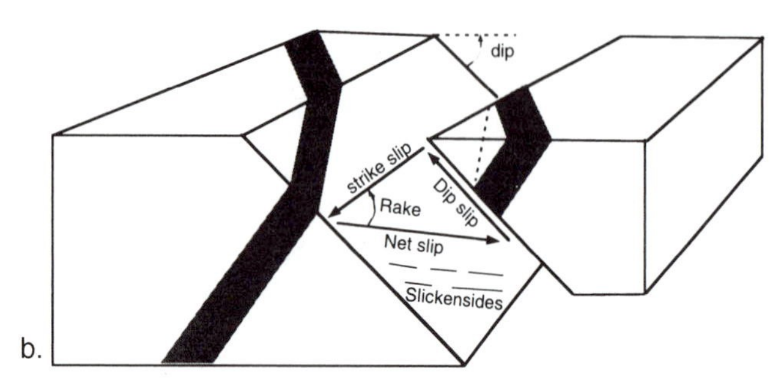

Figure 2–12. Fault nomenclature for (a) normal fault, (b) oblique-slip normal fault. Shaded band is a sedimentary bed offset by fault.

In contrast to slip, *separation* is the distance between two parts of a disrupted plane measured in any direction. *Stratigraphic separation* is measured perpendicular to bedding.

Anderson (1942) assumed that faults are formed in previously-unfaulted rock by Coulomb fracture theory, and he then worked out the geometric relations between principal stresses and the orientation and slip of conjugate faults. The predicted relations are based upon the assumption that faulting will occur parallel to planes of high shear stress. As shown in Figure 2-2, τ_{max} occurs at an angle of 45° to the orientation of the maximum (σ_1) and minimum (σ_3) principal compressive stresses. Accordingly, there are two conjugate planes of maximum shear stress which intersect along a line parallel to σ_2, the intermediate principal stress. This theory is used, with some reservations, in the calculation of earthquake focal mechanisms using the double-couple model (described in Chapter 4).

As shown in Figure 2-13, a stress field in which the maximum principal compressive stress (σ_1) is vertical will result in *normal faults* (hangingwall moves down with respect to footwall; Fig. 2-13a). *Reverse faults* (hangingwall moves up with respect to footwall) result where the minimum principal compressive stress (σ_3) is vertical (Fig. 2-13b); ***thrust faults*** are those reverse faults that dip $<45^\circ$. ***Strike-slip faults*** (relative displacement is horizontal, in the direction of strike) result where the intermediate principal compressive stress (σ_2) is vertical (Fig. 2-13c).

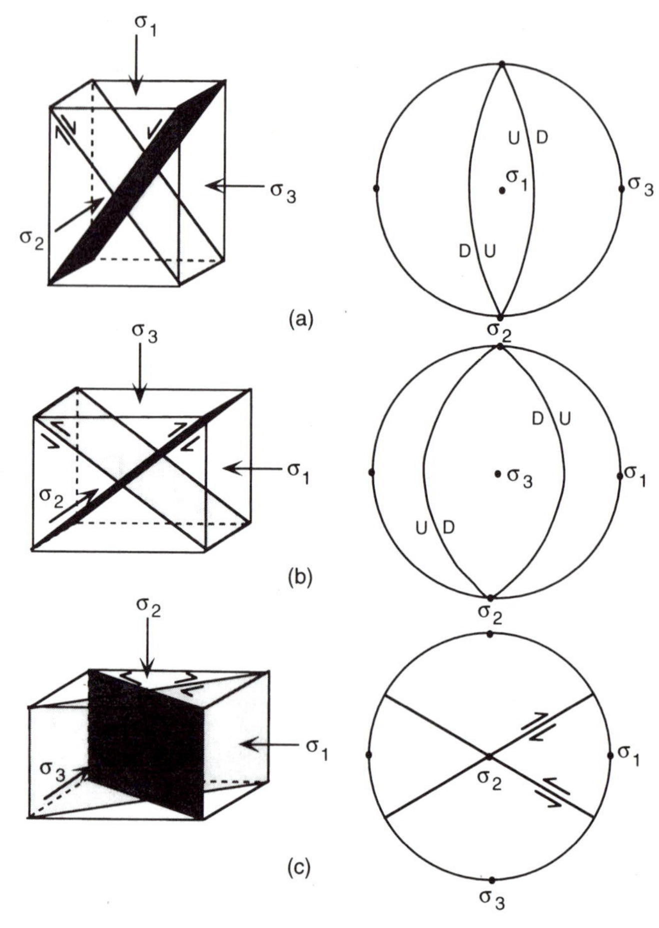

Figure 2–13. Orientation of conjugate fault planes (shaded) with respect to principal stress directions (σ_1, σ_2, σ_3) in isotropic rock for (a) normal faults, (b) reverse faults, (c) strike-slip faults. The fault planes are shown on left as block diagrams, and on right in stereographic projection. To visualize the stereographic projection, imagine that you are looking at the inside of a half sphere, its lower hemisphere concave downward into the page. The circle is the circumference of the sphere intersecting a horizontal plane. The lines are the intersection of planes that pass through the center of the sphere with the lower hemisphere surface. Strike-slip faults dip 90° and show as a straight line; normal and reverse faults dip $<90^\circ$ and show as curved lines.

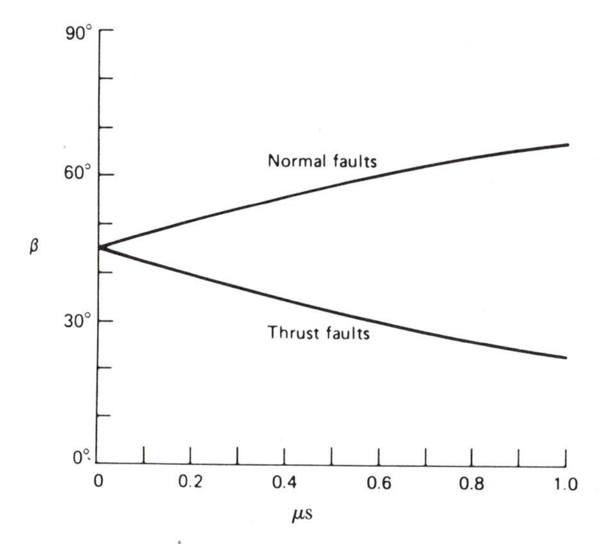

Figure 2–14. Relation of angle of fault dip β to coefficient of friction μ_s for normal and thrust faults. Where $\mu_s = 0$, $\beta = 45°$, the plane of maximum shear stress when either σ_1 or σ_3 is vertical. After Turcotte and Schubert (1982).

If none of the principal stresses is vertical, faults will move by oblique slip. Anderson's theory rests upon the assumption that the body undergoing faulting does not contain structural anisotropies such as bedding, foliation, cleavage, or preexisting fractures that may influence fault orientation.

However, it is common knowledge that conjugate faults do not intersect at 90°, as do the conjugate planes of maximum shear stress. The angle of intersection is about 60°, with the acute bisector σ_1, the maximum principal stress direction. The reason for this is that the fault angle with respect to principal stresses is also influenced by the coefficient of static friction μ_s. Turcotte and Schubert (1982, pp. 354-55) combine three relationships described above, (1) the expression for τ and σ for stresses oriented at an angle θ to the maximum principal stress σ_1, (2) pore pressure p_w, and (3) coefficient of static friction $\mu_s = \frac{\tau}{\sigma_n - p_w}$, ignoring the term τ_o, which would be very small.

Based upon these relationships, the dip angle of thrust and normal faults is given as a function of μ_s (Fig. 2-14). The angle is 45° where $\mu_s = 0$, but at $\mu_s = 0.6$ (cf. Fig. 2-8), the dip angles are about 60° for normal faults and about 30° for thrust faults. This corresponds reasonably well with field observations, as illustrated in Figures 2-15 and 2-16. Figure 2-15 shows a normal fault with a dip of about 50°, and Figure 2-16 shows a reverse fault with a dip of 15°.

It is rare to find a structurally flawless rock mass in the upper crust in which a fault was initiated whose orientation is controlled solely by the orientation of the principal stresses and the coefficient of static friction. Rocks have bedding, cleavage, and joints, and once a fault is formed, a recurrence of displacement is likely along the same fault surface, even if the stress field changes to some degree.

Once a fault has formed, the frictional resistance to reactivation of the fault is less than the shearing stress necessary to form a new fault in the same

Figure 2–15. Normal fault cutting Pleistocene strata at Cape Kidnappers, North Island, New Zealand (note persons for scale). Footwall, on right, has moved upward with respect to hangingwall. White layer on hangingwall in center of photograph is displaced to upper right corner of photograph in footwall, allowing determination of stratigraphic separation. Fault zone is simple overall but complex in detail, including a subsidiary strand in hangingwall near top. Photo by Alan Hull, Institute of Geological and Nuclear Sciences, New Zealand.

Figure 2–16. San Cayetano reverse fault at Silverthread oil field, Ventura basin, California. Highly fractured Miocene strata are thrust over Pliocene beds along a relatively narrow fault zone. Photo by Robert Yeats.

rock. This may be represented by a Mohr diagram with the envelope for frictional failure on a preexisting fault along with the Coulomb failure envelope for initiation of a new fault in flawless rock (Fig. 2-17). Preexisting faults with a range of orientations with respect to maximum and minimum principal stresses will rupture before a new fault forms by Coulomb failure. For this reason, one cannot directly use focal-plane solutions for earthquakes based on the double-couple method to infer principal stresses (cf. Chapter 4). The earthquake may nucleate on a preexisting fault which may vary considerably from the ideal orientation predicted by Anderson based on Coulomb failure.

Minor faults with slickensides may be analyzed for orientation of principal stresses if the following are known: dip and strike of the fault, trend and plunge of the slickensides, and sense of displacement on the fault. It is also necessary to know that the slickensides being analyzed are part of the present-day strain regime if it is planned to compare the data with *in situ* stress data (cf. Chapter 5) or earthquake focal mechanisms (cf. Chapter 4). Each fault is plotted as a great circle on the lower hemisphere of a Wulff stereonet, and the slickensides are plotted as a point on the great circle. An arrow on this circle points in the direction of displacement of the hangingwall. Figure 2-18 shows examples from Quaternary fault systems near Cuzco, Peru, from Mercier et al. (1992). For further details, see Angelier

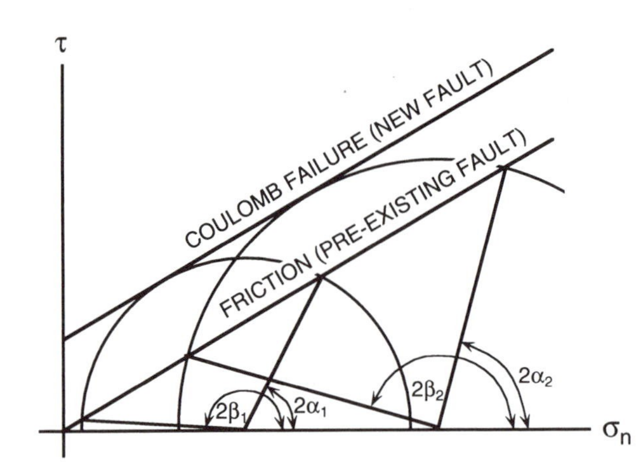

Figure 2–17. Comparison of failure envelopes for formation of a new fault in a structurally flawless material (Coulomb failure) and for slip on a preexisting fault (frictional resistance to fault reactivation according to Byerlee's law). If the rock is flawless, failure occurs at the point of tangency of the Mohr circle to the Coulomb failure envelope (cf. Fig. 2–5). The rock would fail on a preexisting fracture if it was oriented so that α, the angle between the fracture and a plane normal to σ_1 (Fig. 2–1b) is between α_1 and α_2. The rock will fail along that fracture plane at a shear stress lower than that necessary for Coulomb failure. Accordingly, a preexisting fracture may fail even if its orientation is different from the ideal orientation for Coulomb failure.

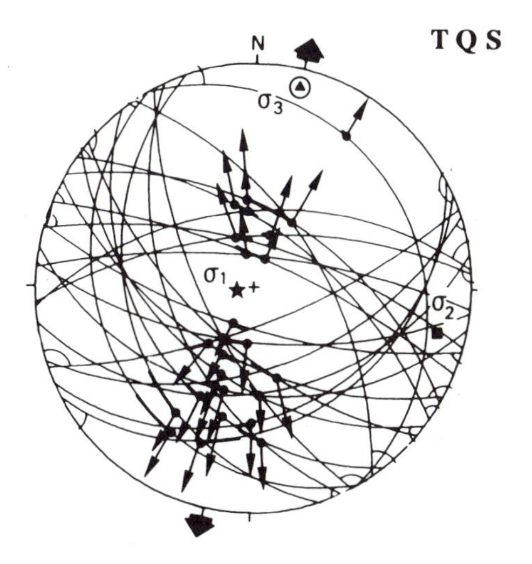

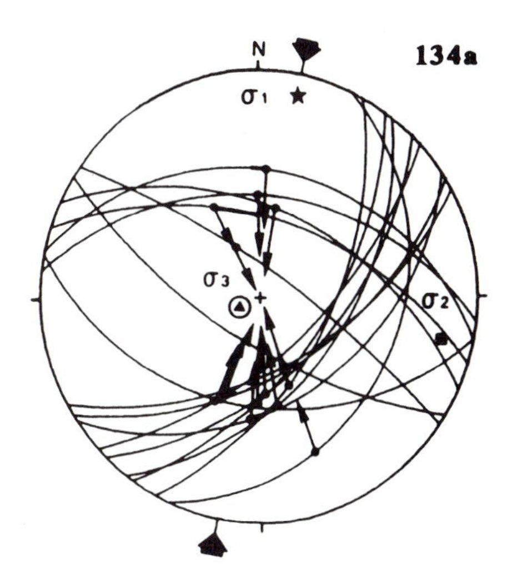

Figure 2–18. Slip-vector data from Quaternary faults in the Cuzco region, Peru, plotted on lower hemisphere of a Wulff stereonet. Arrows attached to fault traces correspond to trend and plunge of measured slip vector (slickensides) on fault surface. Arrows point toward direction of relative displacement of hanging wall. Departure of individual slip vectors from ideal 30° from σ_1 in $\sigma_1\sigma_3$ plane is due to control of fault by rock anisotropy such as bedding or preexisting fracture. TQs, normal-fault example; 134a, reverse-fault example. After Mercier et al. (1992).

(1979), and for a computer-aided method to determine principal stress directions, see Carey (1979).

FOLDS

Layered deposits may be deformed by folding in addition to (or instead of) faulting. Folds are not known to generate large earthquakes, but in recent years, they have been shown to overlie and mask earthquake faults that are *blind,* that is, faults that never reach the surface. The geometry of the fold provides information about the unseen fault as well as information about the behavior of the rock as it was deformed (Suppe and Medwedeff, 1990). A bed folded upwards is an *anticline,* and a bed folded downwards is a *syncline.*

The *hinge* or *hinge line* is that part of a folded surface where the *radius of curvature* (r_c) is minimum. r_c is the radius of a circle whose arc most nearly matches fold curvature. Figure 2-19 illustrates the concept of radius of curvature at the hinge and a limb of a fold. If the limb is planar, r_c of the limb = infinity. The *axis* of a fold is a line within bedding that is parallel to the hinge. The geometry of a *cylindrical fold* can be mimicked by a straight line (fold axis) moving parallel to itself through space. All cross sections through a cylindrical fold normal to the fold axis are identical. We use *axial surface* rather than *axial plane* to describe the surface connecting all the hinges, because commonly this surface is curved.

Symmetrical folds are upright, with a vertical, planar axial surface. *Asymmetrical folds* have one limb steeper than the other, and the axial surface dips less than 90°. The fold has *vergence* toward the steeper limb. *Overturned folds* have one limb overturned (dip > 90°). Vergence is toward the overturned limb.

Folds may be divided into two classes on the basis of their geometry, which is itself an expression of the ductility and ductility contrast of the rocks being folded. These two classes are illustrated in Figure

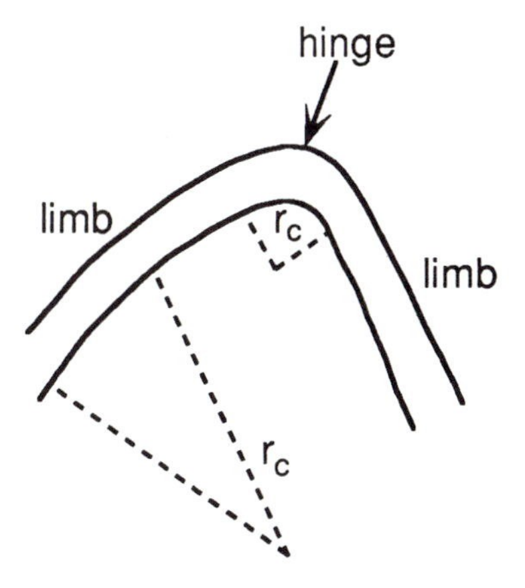

Figure 2–19. Cross section of anticline illustrating limbs, hinge, and radius of curvature r_c.

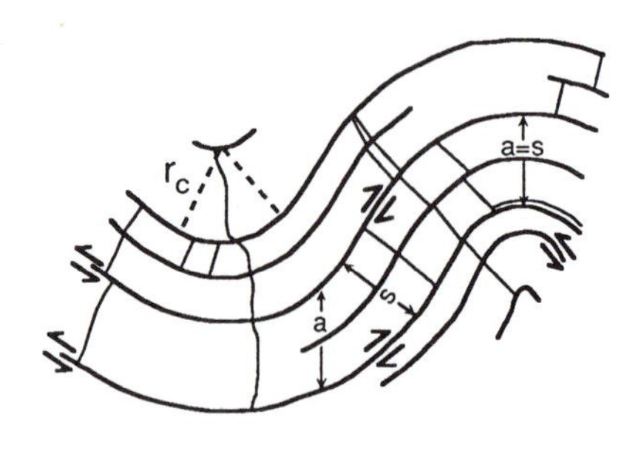

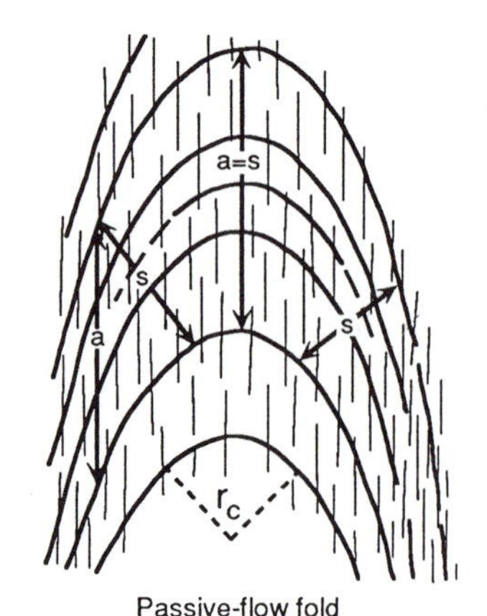

Figure 2–20. Two classes of folds based on ductility. Flexural-slip folds maintain constant bedding thickness *s* between limbs and hinges. Passive-flow folds maintain constant thickness *a* of a line parallel to the axial surface between limbs and hinges.

2-20. *Parallel folding or flexural-slip folding* involves slip between flexed competent layers. Beds behave as flexed beams of low ductility. Stratigraphic thickness *s* is constant from hinge to limbs. Apparent thickness measured parallel to axial surface $a = s$ at the hinge and decreases in limbs. High ductility contrast exists between the flexed layers and ductile layer boundaries where slip occurs. Because the beds behave as flexed beams, convex surfaces of folded layers tend to be in tension, and concave surfaces tend to be in compression; that is, bending moment is important. Rock anisotropy is effective. Flexural-slip folds are commonly *concentric folds;* that is, the folded surfaces of successive beds define circular arcs all of which have the same center of curvature. This requires that the fold must die out downward; there is a surface below which there is no folding. *Similar folding* or *passive-flow folding* involves distributed shear parallel to the axial surface. Stratigraphic thickness *s* is maximum in the hinge, minimum in limbs. Apparent thickness *a* parallel to the axial surface is constant from limb to hinge. The fold is characterized by low ductility contrast and high ductility. Bending moment is unimportant, and anisotropy is ineffective.

Five kinds of folds can be identified that are a direct result of fault displacement. These are *drag folds, fault-bend folds, fault-propagation folds, basement-involved compressive structures,* and *décollement folds* (Fig. 2-21; cf. Suppe, 1985; Dahlstrom, 1990; Narr and Suppe, 1994).

Drag folds exhibit asymmetry appropriate to the sense of displacement on the adjacent fault, and they respond to frictional drag on strata that are cut by the fault. They are probably much smaller-scale features than the other kinds of folds described below. The axial surface of a drag fold tends to be parallel to the fault, but the mechanical properties of the folded beds also affect the orientation of the axial surface. Drag folding may be accompanied by slip along bedding planes of contrasting anisotropy, changing the angle between bedding and the controlling fault. In determining total displacement across a fault at depth, it is important to include the effects of drag folding as well as fault displacement, because both are near-surface, low-seismic expressions of reverse faulting in high-strength rocks at depth in the seismogenic zone. For example, in Figure 2-21a, if fault slip is in the plane of the section, displacement is measured along the fault between the top of the shaded bed in the hangingwall and the top of the shaded bed in the footwall (A-A′). However, one may extend the top of this bed to the fault from where it is planar away from the fault in both the hangingwall and footwall (B and B′). This measurement, (B-B′), is much more significant in determining displacements at seismogenic depths.

This principle is illustrated in Figure 2–22, although the structure is, in a strict sense, not truly a drag fold. The top of the late Pleistocene Saugus Formation is preserved in the footwall block of the Oak Ridge fault, and it can be projected in the air from the Long Canyon syncline, south of the fault. The piercing-point separation of this horizon, DE, varies considerably in the four cross sections shown. Distance AC, taking out the effects of drag, is about the same in the four

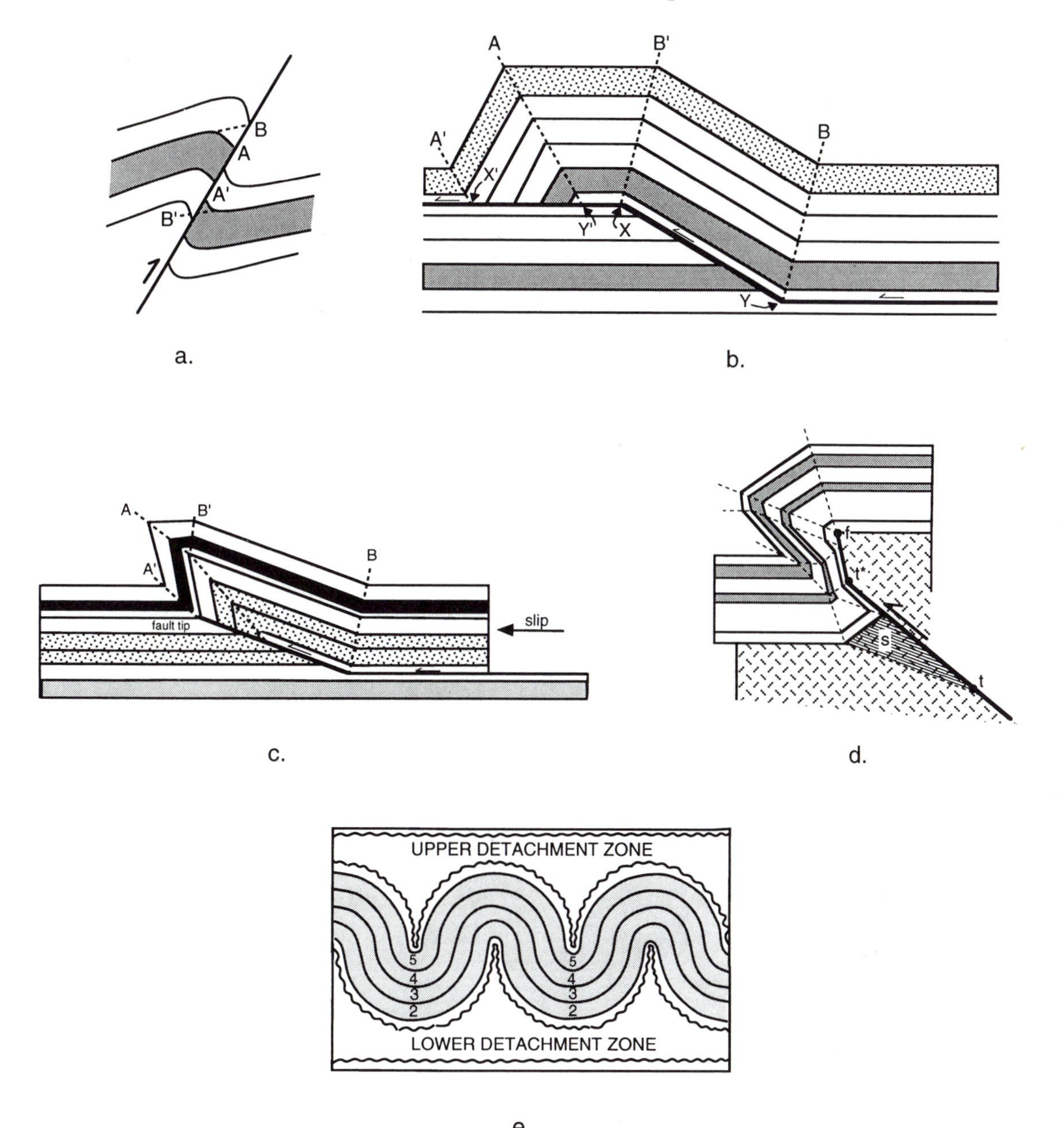

Figure 2–21. Classification of fault-related folds. (a) drag fold; (b) fault-bend fold; (c) fault-propagation fold; (d) basement-involved compressive structure; (e) décollement fold. (b) and (c) from Suppe (1985), (d) from Narr and Suppe (1993), (e) from Dahlstrom (1990).

sections; this is interpreted by Yeats (1988) to be more representative of slip on the fault in the earthquake source region.

Fault-bend folds (Fig. 2-21b) are formed because fault surfaces are not perfectly planar in the direction of slip (Suppe, 1983). Movement across a nonplanar fault surface should produce deformation of one or both of the blocks adjacent to the fault or cause the fault itself to break through one of the blocks to produce a straighter trace in the direction of slip. Fault-bend folds are most common in the hangingwalls of faults, either in tilting of strata toward the surface of a low-angle, concave-upward listric normal fault or in bending of strata as they override a thrust surface characterized by ramps. These folds are produced by bending rather than by

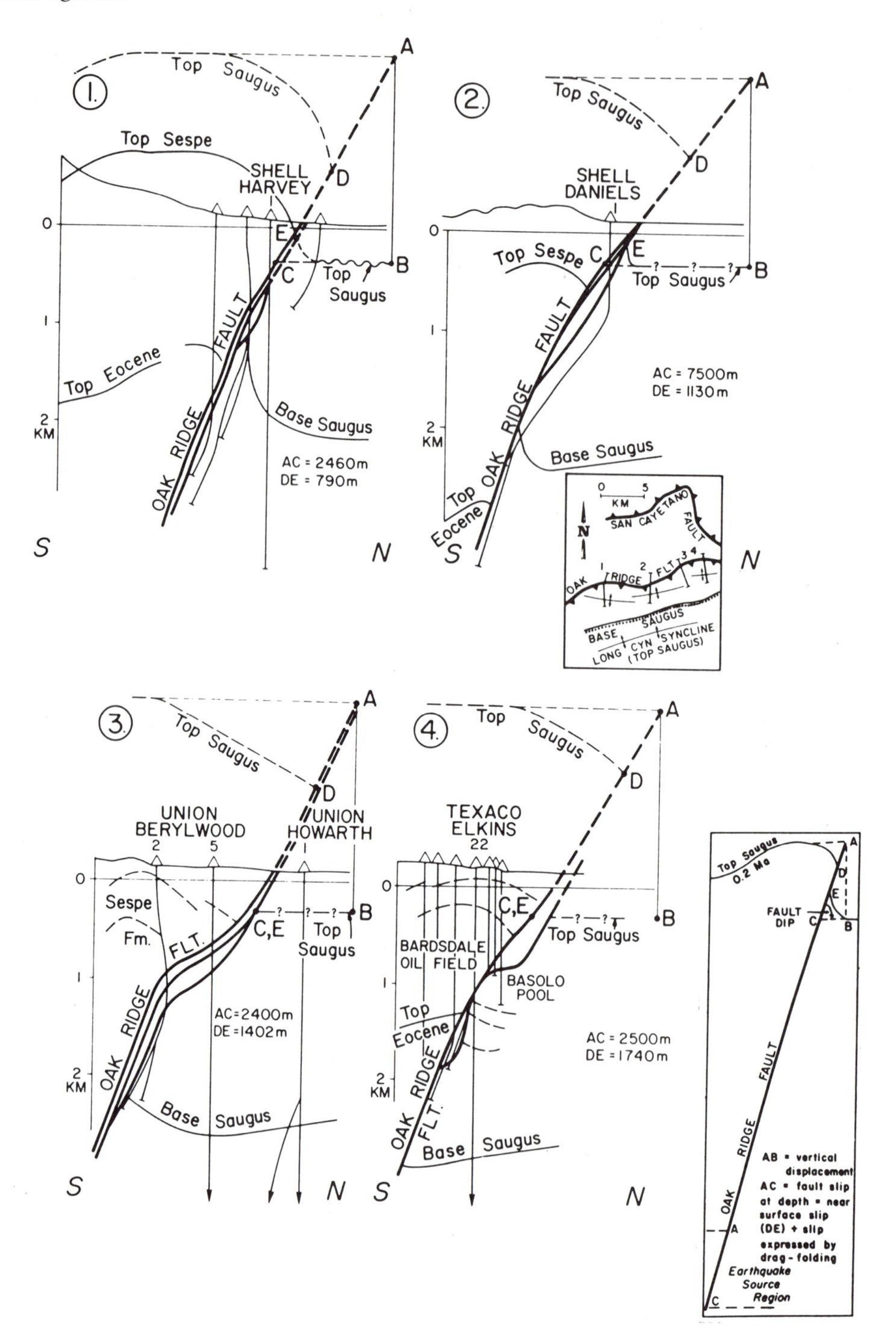

Figure 2–22. Four cross sections across Oak Ridge fault from South Mountain oil field (section 1) to Bardsdale oil field (section 4), Ventura basin, California. DE is piercing-point offset of top of Saugus Formation; AC is calculated offset on fault in earthquake source region considering piercing-point offset and near-surface folding of strata into fault. Radius of curvature of folds indicates that these folds are near-surface phenomena and would not project downward to the earthquake source region. After Yeats (1988).

frictional drag. The geometry of the fold provides information about the buried thrust surface. The length of the backlimb, BB′, which can be observed directly, is a lower bound to the slip on the fault, YY′, even though the fault is not observed directly. In northern Pakistan, the Salt Range thrust brings platform strata southward over Precambrian basement along a décollement horizon in salt (Fig. 2-23). The basement is cut by a high-angle fault, and the thrust sheet must bend upwards to override this block, producing a fault-bend fold that controls the position of the north flank of the Salt Range (Baker et al., 1988).

Fault-propagation folds (Fig. 2-21c) emanate from the tip of a propagating fault (Suppe and Medwedeff, 1990). In some cases, the fault-propagation fold process locks up, and the fault propagates through the existing fold. In these cases, fault propagation is rapid compared to the rate of buildup of slip, whereas in the general case, fault-tip propagation is slow relative to the rate of buildup of slip. During the process of folding, a limb may fail, and

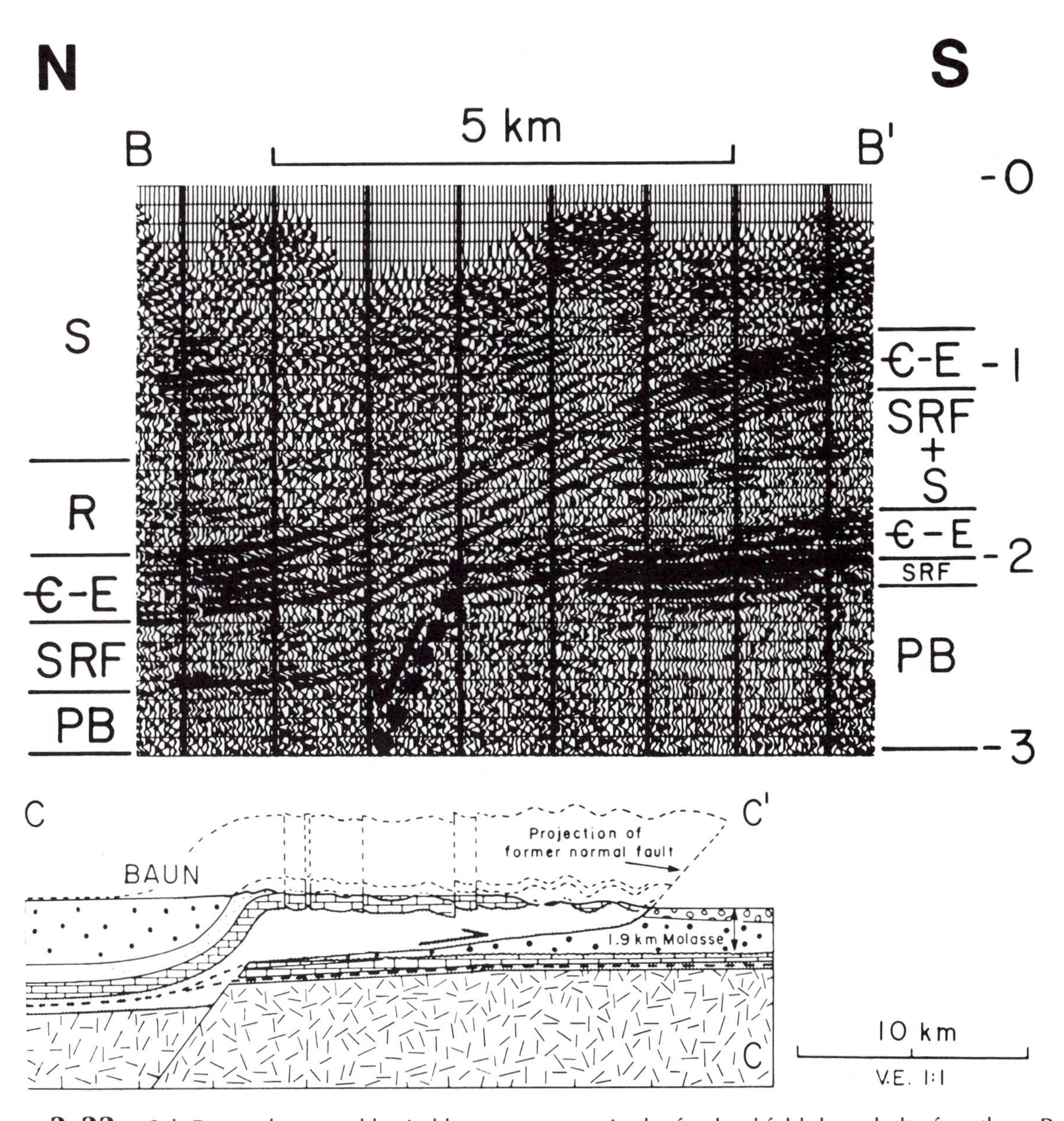

Figure 2–23. Salt Range thrust and buried basement ramp in the foreland fold-thrust belt of northern Pakistan. PB, Precambrian basement; SRF, Eocambrian salt and evaporite; C-E, platform cover sequence; R, S, Miocene to Quaternary continental deposits. Platform sequence is deformed as a fault-bend fold. After Baker et al. (1988).

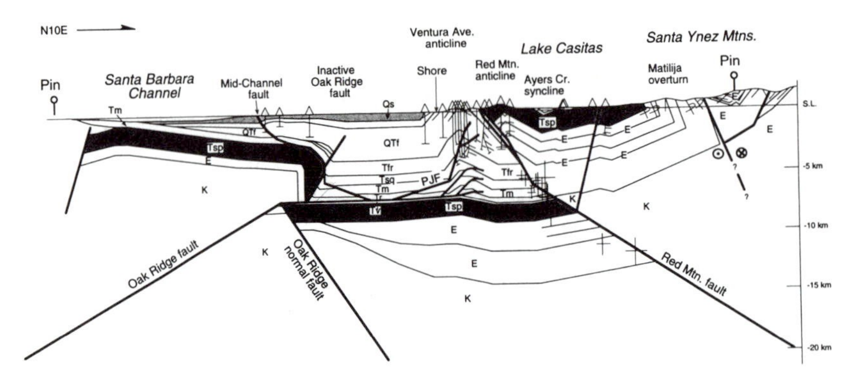

Figure 2–24. Balanced cross section across central Ventura basin, California, showing seismogenic reverse faults of opposite dip, in part controlled by wells (open triangles with lines extending below them). Pins are assumed to move toward one another. Change in dip on Red Mountain fault is necessary to accommodate backlimb in syncline to north. Qs, Pleistocene; QTf, Pliocene-Pleistocene; Tfr, Pliocene; Tsq, Tm, Tr, Miocene shale; Tv, Tsp, Oligocene; E, Eocene; K, Cretaceous; Plusses locate selected earthquakes. After Huftile and Yeats (1995).

the resulting fault will propagate toward the surface. Stratigraphic separation across the fault decreases to zero as the fault tip is approached, and an increasing amount of displacement is taken up by folding. As in the case of fault-bend folds, the fold geometry provides information about the buried blind thrust. In the Ventura basin, the Ventura Avenue anticline is a fault-propagation fold related to the subjacent Padre Juan fault (Grigsby, 1986; Fig. 2-24). The Padre Juan fault ramps upward from Miocene shales across Pliocene and Pleistocene deep-water sandstone and shale as the Red Mountain fault is approached. Separation across the fault diminishes upsection as more displacement is taken up by folding, and near the surface, where the fault flattens, separation is near zero. This example is complicated by the fact that the near-surface trace of the Padre Juan fault has been folded into an orientation of lower shear stress, and a new fault, the Javon Canyon fault, has broken through to the surface and displaced Holocene terrace deposits (Sarna-Wojcicki et al., 1987).

Basement-involved compressive structures are found in the hangingwalls of faults in basement that propagate upward into a layered, flat-lying, and less-cohesive cover sequence (Narr and Suppe, 1994). Orientation of the basement faults is influenced by preexisting zones of weakness in the basement as well as the stress field at the time of faulting. The fault propagates into the cover sequence as a complex monocline (*drape fold*) over the protruding edge of the basement (Fig. 2-21d). The basement fault may be seismogenic, and so structural analysis of the cover sequence and the top of basement may provide important information about the fault at seismogenic depths.

Décollement folds are concentric folds above a décollement zone below which there is no folding (Fig. 2-21e). Such folds do not have a master fault rising into the structure; instead, ductile material moves from synclinal troughs into the cores of anticlines (Dahlstrom, 1990).

Figure 2–25. Concentric fold (syncline) in Bloomsburg Formation of Silurian age at Hancock, Maryland. Photo by Robert Yeats.

Figure 2–26. Ventura Avenue flexural-slip anticline in Pliocene deep-water sandstones near Ventura, California. The anticline began forming about 500,000 years ago and is still active. Photo by Robert Yeats.

Field examples of folds are shown in Figures 2-25, 2-26, and 2-27. The folds shown in Figures 2-25 and 2-27 are no longer active, whereas the Ventura Avenue anticline in Figure 2-26 is active.

BALANCED CROSS SECTIONS

Figure 2-21b, c, and d and Figure 2-24 are examples of *balanced cross sections*. These are sections that can be retro-deformed to an undeformed state along the line of tectonic transport without loss of cross-section area (cf. Woodward et al., 1989). If stratigraphic thicknesses are well known, as in Figures 2-21b, c, and d, then retro-deforming can be done by *line-length balancing*. Points far away from the zone of deformation are assumed to be *pinned*, that is, a vertical pin through the entire section prior to deformation would still be vertical after deformation (cf., Fig. 2-24). Line lengths of bedding planes between the two pins should be the same before and after deformation. Another method is *area-balancing*. In Figure 2-21c, the area of rock above the uppermost rock layer should be equal to the area of the rectangle defined by shortening at the right of the figure.

A cross section that is balanced is *admissible*, that is it could be correct. (If a cross section cannot be balanced, it cannot be correct if displacement is in the plane of the section.) However, there exists a family of admissible cross sections that may yield different values for horizontal shortening and different estimates of the position of a blind fault at depth. Balanced cross sections across fold-thrust belts in California cannot take into account the effects of strike-slip faulting if such faulting involves tectonic transport into or out of the section. (However, John Suppe pointed out to one of us that strike-slip faulting could be considered by a "balanced map," be-

Figure 2–27. Ductile anticline in Attock-Cherat Range, south of Peshawar, Pakistan. Strata are thinly-bedded limestones of Paleocene-early Eocene age. Photo by Robert Yeats.

cause tectonic transport takes place in the plane of the map rather than the plane of a cross section. Balancing in three dimensions is currently an active research area.) Similarly, a balanced cross section cannot consider the effects of tectonic rotation about a vertical axis.

Fault-bend folds and fault-propagation folds carry the assumption that convergence involves only shallow layers of the crust, or, at best, the entire brittle crust. No consideration is given to shortening in the plastically deforming lower crust. However, the Moho at the base of the crust (cf. Chapter 1) is a compositional discontinuity, and if the brittle crust is shortened, the shortening may be taken up by thickening above the Moho, which could be depicted by area-balancing the lower crust.

GROWTH STRATA AND DETERMINING SLIP RATE AND DIP OF BLIND FAULTS

In the Transverse Ranges of California, blind faults have accumulated slip during the accumulation of sediments. Sediments that are deposited during faulting are referred to here as *growth strata*. We present examples in which growth strata are used to determine the slip rate and dip of blind thrusts that are too deeply buried for direct observation at the surface or in wells.

Figure 2-28 shows the geometry of growth strata deposited during fault-bend folding (Shaw and Suppe, 1994). Figure 2-28b shows a fault-bend fold with kink bands AA′ and BB′. These kink bands are bounded by surfaces A and B, which are called *active axial surfaces*. They are pinned to the two bends in the fault ramp, and the surfaces move through the beds as displacement takes place. In contrast, surfaces A′ and B′ are *inactive axial surfaces* that form at the initial fault bends and are translated away passively from the active axial surfaces as displacement takes place, widening the kink band AA′ with time. Strata deposited prior to displacement on the blind thrust show no change of thickness across the fault-bend fold, and the axial surfaces bounding the kink bands are parallel to each other.

In contrast, axial surfaces in growth strata converge upward. At the initiation of faulting, A′ is located at A, so that the bed length AA′ is a minimum approximation of fault slip. The width of the kink band narrows to zero at P because P is both an active and an inactive axial surface. The narrowing kink band AA′P forms a *growth triangle*. If the ages of the oldest and youngest growth strata are known, the length of the kink band AA′ at the base of the growth triangle can be divided by the difference in age between the top and the base of the growth section to determine the slip rate on the blind thrust. The dip of the thrust is the dip of the backlimb of the fault-bend fold. Note that the depth of the fault

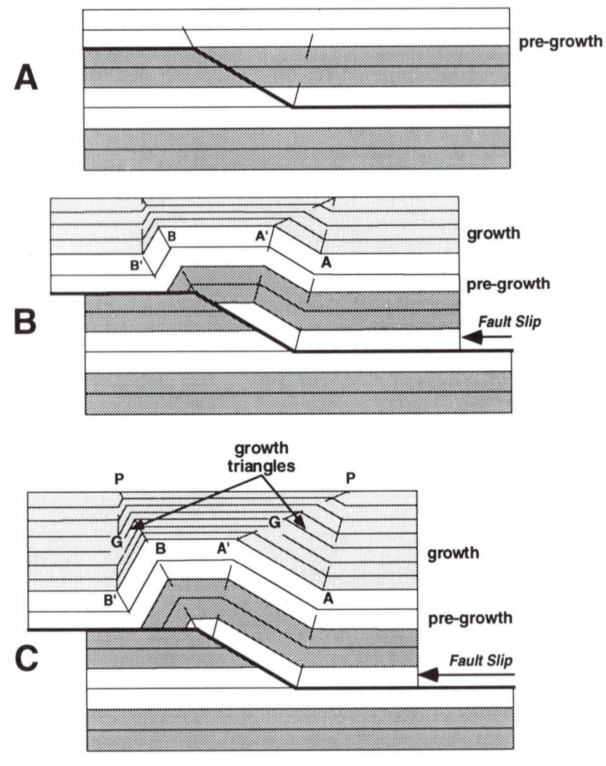

Figure 2–28. Kinematic development of a growth fault-bend fold after Suppe et al. (1992). (a) Pre-growth strata and position of future thrust ramp. (b) Slip on thrust fault folds the hangingwall along active axial surfaces A and B that are pinned to fault bends. Inactive axial surfaces A′ and B′ form at fault bends and are translated away from active axial surfaces by slip. Kink-band widths AA′ and BB′ equal slip on fault, with the difference between AA′ and BB′ representing slip consumed by folding. (c) continued slip widens kink bands, which narrow upward in the growth strata and converge on point P, producing a growth triangle. Limb widths of growth horizons represent fault slip since they were deposited. Dip of backlimb is equal to dip of blind thrust. After Shaw and Suppe (1994).

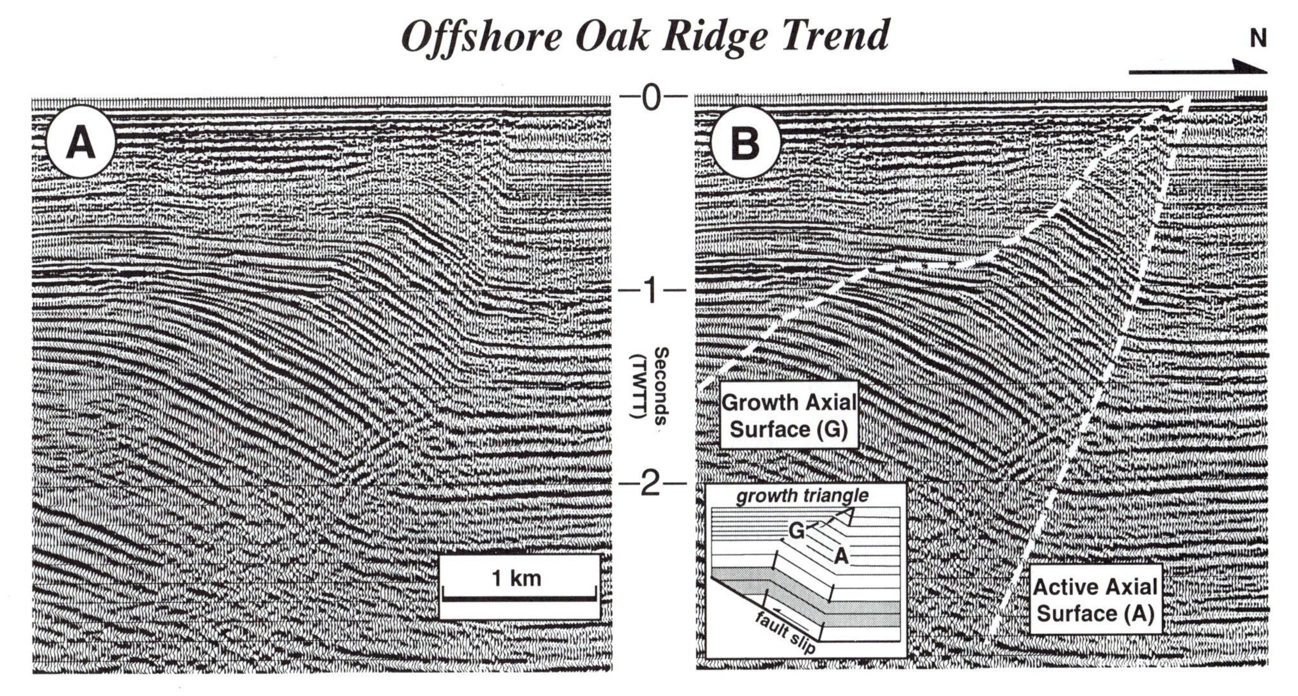

Figure 2–29. A growth triangle in the Santa Barbara Channel, southern California, imaged by a migrated seismic-reflection profile. (a) uninterpreted; (b) interpreted. Note the parallelism of dipping strata within the growth triangle. After Shaw and Suppe (1994).

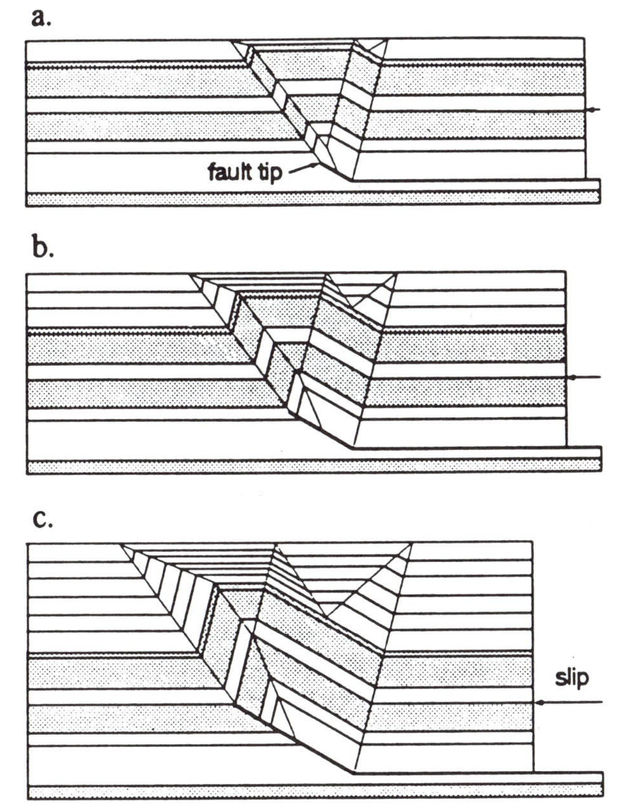

Figure 2–30. Progressive development of a fault-propagation fold with pre-growth strata (alternating shaded and clear pattern) and growth strata (uppermost clear pattern). In contrast to the fault-bend fold example (Fig. 2–28), growth strata in the forelimb are the same thickness as those ahead of the advancing tip of the thrust. Both axial surfaces in the backlimb are active, moving out with respect to rock of the upper plate. The triangle of flat-lying strata over the backlimb records its progressive widening. After Suppe et al. (1992).

cannot be determined directly using this method, because there is no way to measure the thickness of the pre-growth strata above the fault. Distinguishing features of growth fault-bend folding are the presence of growth triangles on both forelimb and backlimb of the fold and the parallelism of inclined growth strata throughout the growth triangle. Figure 2-29 is an example of a growth triangle above a fault-bend fold from the Santa Barbara Channel in southern California.

Growth strata deposited during fault-propagation folding produce more complex geometry, depending on the rate of sedimentation with respect to rate of slip on the blind thrust. Growth triangles developed with fault-propagation folding have much steeper bases in the forelimb than the backlimb, and there may be two growth triangles in the backlimb, reflecting the fact that both axial surfaces bounding the kink band in the backlimb are active (Fig. 2-30). Suppe and Medwedeff (1990) present the method of determining dip and slip rate on blind thrusts underlying fault-propagation folds.

In several reverse-fault earthquakes, including the 1994 Northridge, California, earthquake, the seismicity illuminates a fault with a moderate dip that extends through the seismogenic zone and does not flatten as do faults underlying fault-bend folds and fault-propagation folds. Active reverse faults in basement may propagate into the overlying sedimentary section as basement-involved compressive structures (Narr and Suppe, 1994). Figure 2-31 shows the south flank of the Las Cienegas monoclinal fold on the

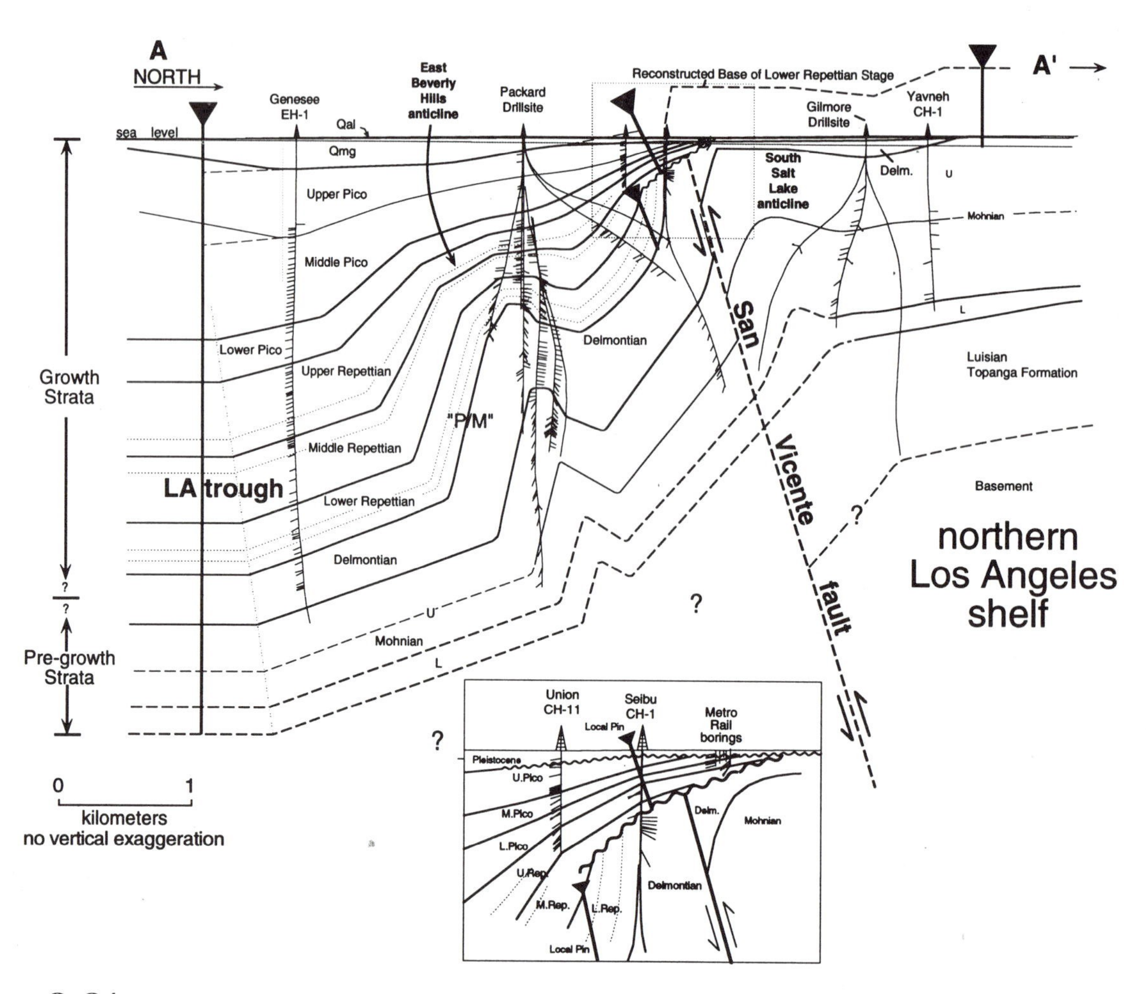

Figure 2–31. Monocline at the northern edge of the Los Angeles trough, southern California. Pre-growth strata show no thinning across this structure. Growth strata thin across the structure, producing rotation of older strata about a horizontal axis. In contrast to growth triangles, dip of growth strata increases with age. After Schneider et al. (1996).

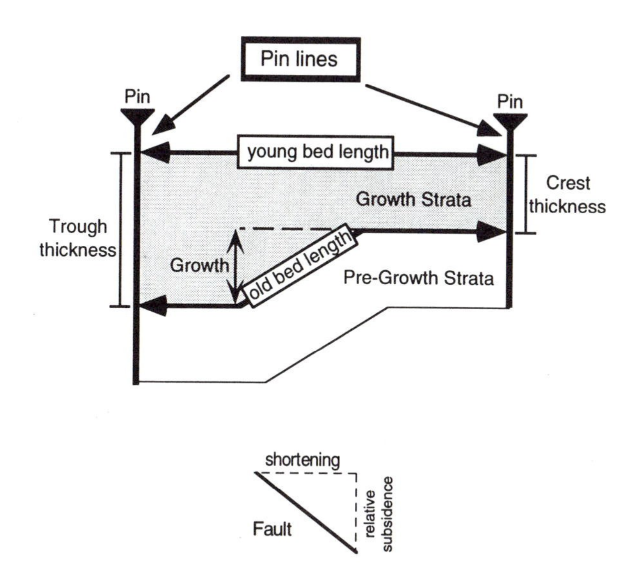

Shortening = old bed length-young bed length.
Relative subsidence = Trough thickness -Crest thickness

$$\text{Fault dip} = \text{Tan}^{-1}\left(\frac{\text{Relative Subsidence}}{\text{Shortening}}\right)$$

Figure 2–32. Method of calculating displacement of blind fault based on growth strata in which dip of base of layer is greater than that of top of layer. Difference in thickness between crest and trough gives vertical component of displacement (relative subsidence), whereas difference in bed length between the base and top of layer gives horizontal component. After Schneider et al. (1996).

northern margin of the Los Angeles basin, California (Schneider et al., 1996). Pre-growth strata (Delmontian and older) show no tendency to decrease in thickness between the footwall and hangingwall, whereas growth strata of Pliocene and Pleistocene age thin as they pass through the monocline. Distinctive features of this type of structure are an absence of a backlimb and a steepening of dips with greater age of the growth strata, reflecting rotation of the strata about a horizontal axis. No growth triangle develops, because strata do not pass through a migrating axial surface.

Because the growth strata decrease in thickness across the structure, the difference in thickness of a growth unit across the structure is used to determine the vertical component of relative displacement on the fault, after taking into account compaction of sediments (Fig. 2-32). The bed length of the base of the bed is greater than the top of the bed, because the base is folded with respect to the top. The difference in bed length between the base and the top is used to determine the horizontal component of relative displacement on the fault. The vertical component divided by the horizontal component is the tangent of the angle of fault dip. This can be displayed graphically by showing the displacement of the lower block with respect to the upper block (Figure 2-33). If the ages of the growth sequence are known, this displacement vector can be age-calibrated and a slip rate on the blind thrust determined. This displacement rate is a minimum, because this method does not perceive bedding slip in the trough. If there is bedding slip in the trough, then the assumption of a pin in the trough would be incorrect.

GEOLOGIC MAP SYMBOLS

Figure 2-34 shows the more important map symbols used in geologic maps to display faults, joints, folds, inclined bedding, and inclined foliation (preferred orientation of mineral grains or rock layers of different composition due to metamorphism of rock under differential stress or to flow of magma).

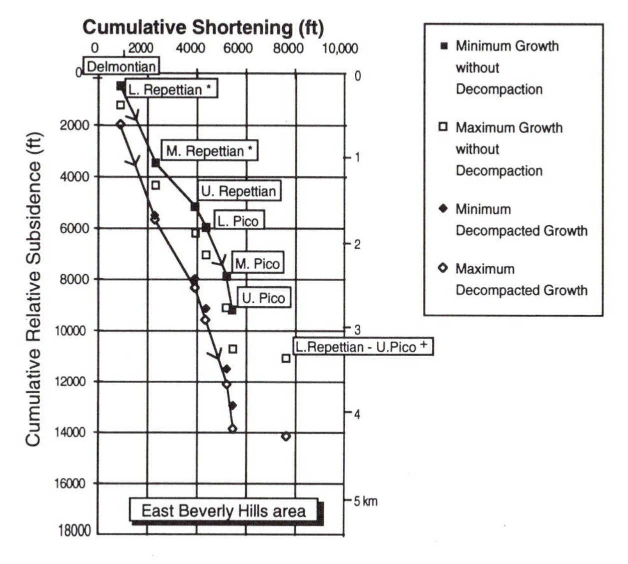

Figure 2–33. Cumulative shortening vs. cumulative relative subsidence for structure shown in Figure 2–31, shown as the relative displacement of a particle in the lower block with respect to the upper block. Displacement shown both with and without considering the effects of compaction. After Schneider et al. (1996).

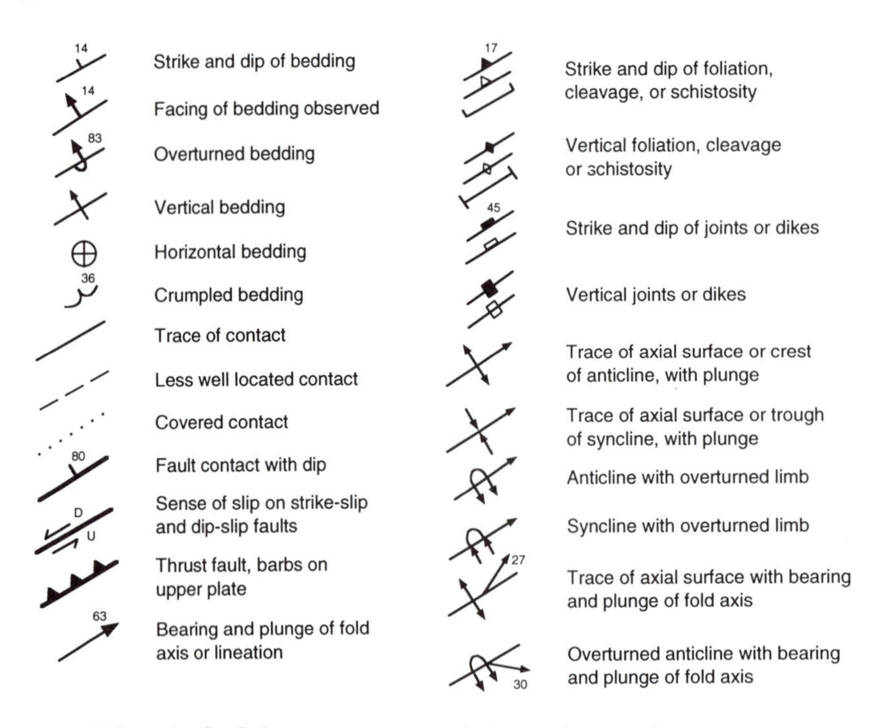

Figure 2–34. Common symbols used on geologic maps.

SUMMARY

This chapter describes the basic principles that control the deformation of rocks and the more important structures that result from rock deformation. Earthquakes occur principally by the sudden displacement on faults, and an understanding of the long-term history of faults is necessary to understand the earthquake environment. For a more thorough treatment of structural geology, the reader is referred to a standard text in undergraduate structural geology such as Suppe (1985) or Twiss and Moores (1992). For a treatment of rock mechanics as applied to earthquakes, see Scholz (1990). The following chapter focuses on the geology and mechanics of the nucleation sites of crustal earthquakes.

Suggestions for Further Reading

Dahlstrom, C. D. A. 1990. Geometric constraints derived from the law of conservation of volume and applied to evolutionary models for detachment folding. American Association of Petroleum Geologists Bulletin 74:336–44.

Jaeger, J. C., and Cook, N. G. W. 1979. Fundamentals of rock mechanics, 3rd ed. London: Methuen, 593 p.

Marshak, S., and Mitra, G. 1988. Basic Methods of Structural Geology. Englewood Cliffs, N. J.: Prentice-Hall, 446 p. Lab manual with problems in structural geology.

Narr, W., and Suppe, J. 1994. Kinematics of basement-involved compressive structures. American Journal of Science 294:802–60.

Scholz, C. H. 1990. The Mechanics of Earthquakes and Faulting. New York: Cambridge University Press, 439 p. Chapters 1-3 describe in detail brittle fracture, friction, and fault mechanics with a full set of references.

Suppe, J. 1983. Geometry and kinematics of fault-bend folding. American Journal of Science 283:684–721.

Suppe, J. 1985. Principles of Structural Geology. Englewood Cliffs, N. J.: Prentice-Hall, pp. 110–68, pp. 179–90.

Suppe, J., and Medwedeff, D. A. 1990. Geometry and kinematics of fault-propagation folding. Eclogae Geologicae Helvetiae 83:409–54.

Turcotte, D. L., and Schubert, G. 1982. Geodynamics. New York, N.Y.: John Wiley, pp. 348–61.

Twiss, R. J., and Moores, E. M. 1992. Structural Geology. New York: W. H. Freeman and Co., 532 p.

Uemura, T., and Mizutani, S. 1984. Geological Structures. J. Wiley, 309 p. Originally published in Japanese in 1979 by Iwanami Shoten, Tokyo. Structural geology text from Japanese perspective, many Japanese examples.

Woodward, N. B., Boyer, S. E., and Suppe, J. 1989. Balanced geological cross-sections: An essential technique in geological research and exploration. Washington, D.C.: American Geophysical Union Short Course.

3

Geology of the Earthquake Source Region

It is generally accepted that crustal earthquakes are caused by sudden displacement on faults, and many earthquakes of M > 6 are accompanied by faulting at the earth's surface. However, faulting at the surface may not be representative of fault conditions at 10–20 km depth where crustal earthquakes nucleate. High pressure and temperature cause rocks to respond far differently in the earthquake source region than they do at the surface. Most earth materials at the surface are too weak to store enough elastic strain energy to produce a damaging earthquake, and the displacement and rupture length at the surface may be considerably less than they are at nucleation depths.

Understanding what happens at the nucleation site of a large earthquake is complicated by the fact that we are as yet unable to sample rocks directly at earthquake-nucleation depths, although deep drilling projects by several countries may resolve the sampling problem in the next decade or two. We are limited now to (1) interpretations based on seismograms of earthquakes and on deep seismic reflection profiles, (2) study of ancient mid-crustal rocks that have been uplifted and eroded so that they may be studied directly, (3) experimental deformation of rocks in the laboratory under mid-crustal pressures and temperatures, and (4) theoretical modeling.

EVIDENCE FROM EARTHQUAKE DISTRIBUTION

What can we infer from the distribution of earthquakes in the earth's crust? In continents, earthquakes occur in the upper crust, with the deepest earthquakes in the oldest crust. In ocean basins, the deepest earthquakes occur in the oldest lithosphere (excluding those in subduction zones). Because the oldest crust is the coldest in continents and ocean basins alike, the depth of earthquakes appears to be limited by temperature.

The sharp cutoff of earthquakes with depth is most apparent along fault zones characterized by high seismicity and monitored by extensive, high-quality seismic networks. Figure 3-1 shows long-term seismicity parallel to strike of the San Jacinto and central San Andreas faults of California. The base of earthquakes along the San Jacinto fault rises to the southeast, reflecting the higher geothermal gradient near the Salton Sea. The base of earthquakes along the San Andreas fault rises and falls, suggesting control by rock type in the fault zone in addition to temperature. The shallowest levels tend to have fewer earthquakes than deeper levels; note the San Jacinto fault between 10 and 40 km south and 80 to 150 km south of point A. (The shallow aseismic layer between 80 and 130 km may be in a seismic gap analogous to the gap filled by the Loma Prieta earthquake, as shown in Figure 3-1b.) The shallowest levels of the San Andreas fault have relatively low seismicity near San Francisco, San Juan Bautista, and Parkfield.

Aftershocks of the Loma Prieta, California, earthquake of October 17, 1989 (M7.1) illustrate seismicity commonly observed for earthquakes on strike-slip faults recorded by a dense seismic network (Fig. 3-2). First, the base and top of the aftershock zone are rather sharply defined, as in Figure 3-1. There are virtually no earthquakes along the shallowest 1.5–2 km of the fault plane (cross sections, Fig. 3-2). Second, the distribution of aftershocks is nonuniform, as noted on cross section A-A′ of Figure 3-2, parallel to the fault. Some parts of the fault have concentrations of aftershocks, whereas other parts of the fault with areas of several km^2 are aseismic. Finally, there are a significant number of aftershocks off the San Andreas fault, as noted on the map and cross section B-B′ of Figure 3-2. Note the planar zone of aftershocks west

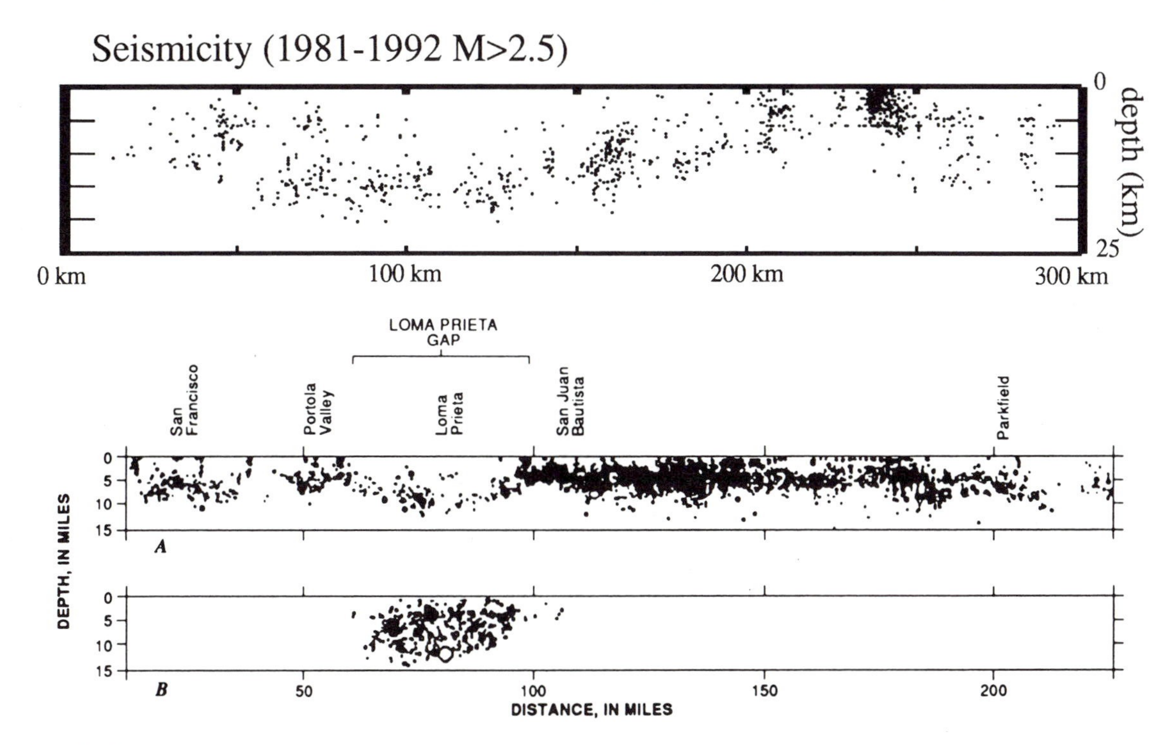

Figure 3–1. Depth distribution of earthquakes parallel to strike of two major strike-slip faults in California. Top diagram shows earthquakes on the San Jacinto fault from 1981 to 1992. Northwest is to left. Bottom diagram shows (a) a 20-year record of earthquakes on central San Andreas fault prior to the October 17, 1989 Loma Prieta earthquake; and (b) aftershocks of the Loma Prieta earthquake. Northwest is to left. San Jacinto seismicity cross section from Peterson and Wesnousky (1994). San Andreas cross sections from U.S. Geological Survey Staff (1990).

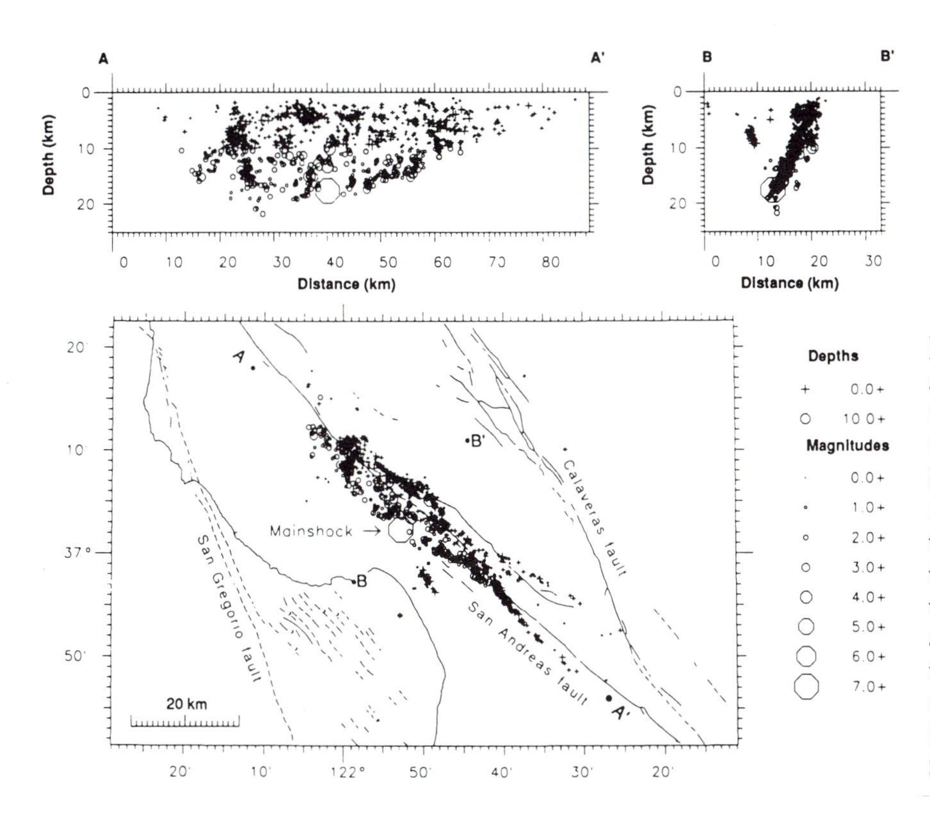

Figure 3–2. Distribution of aftershocks of Loma Prieta, California, earthquake of October 17, 1989. Mainshock is near the base of earthquakes. Note the near-absence of earthquakes in the shallow 1.5–2 km of section, the nonuniform distribution of earthquakes within the aftershock zone, and the off-fault seismicity as observed on map and cross section B-B′. From U.S. Geological Survey Staff (1990).

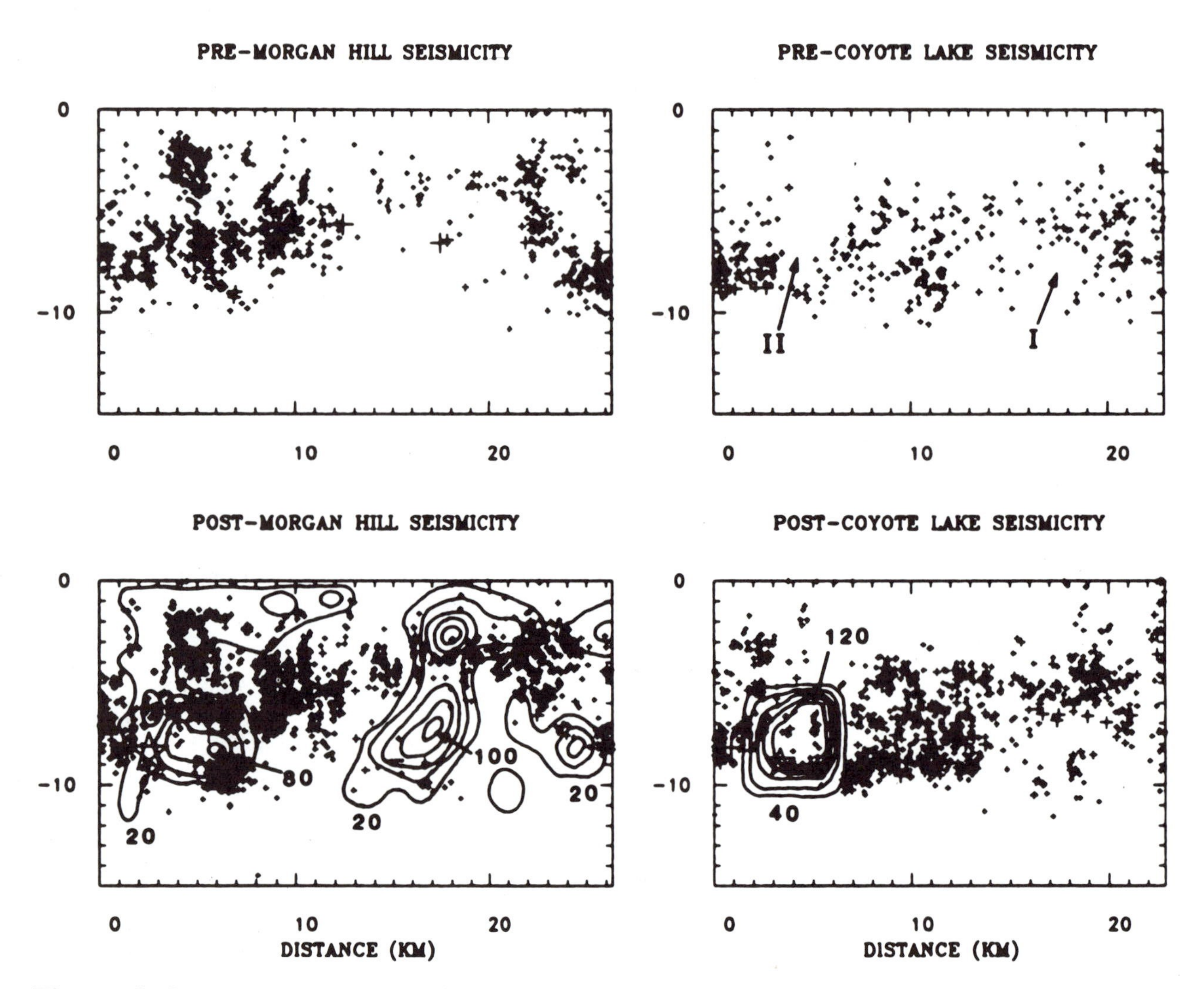

Figure 3–3. Comparisons of pre-earthquake and post-earthquake seismicity with slip distribution on the 1979 Coyote Lake (M5.9) and 1984 Morgan Hill (M6.2) earthquakes on the Calaveras fault. Data are shown in cross sections parallel to the fault. Stars locate mainshock hypocenters. Slip contours are in centimeters. From Oppenheimer et al. (1990).

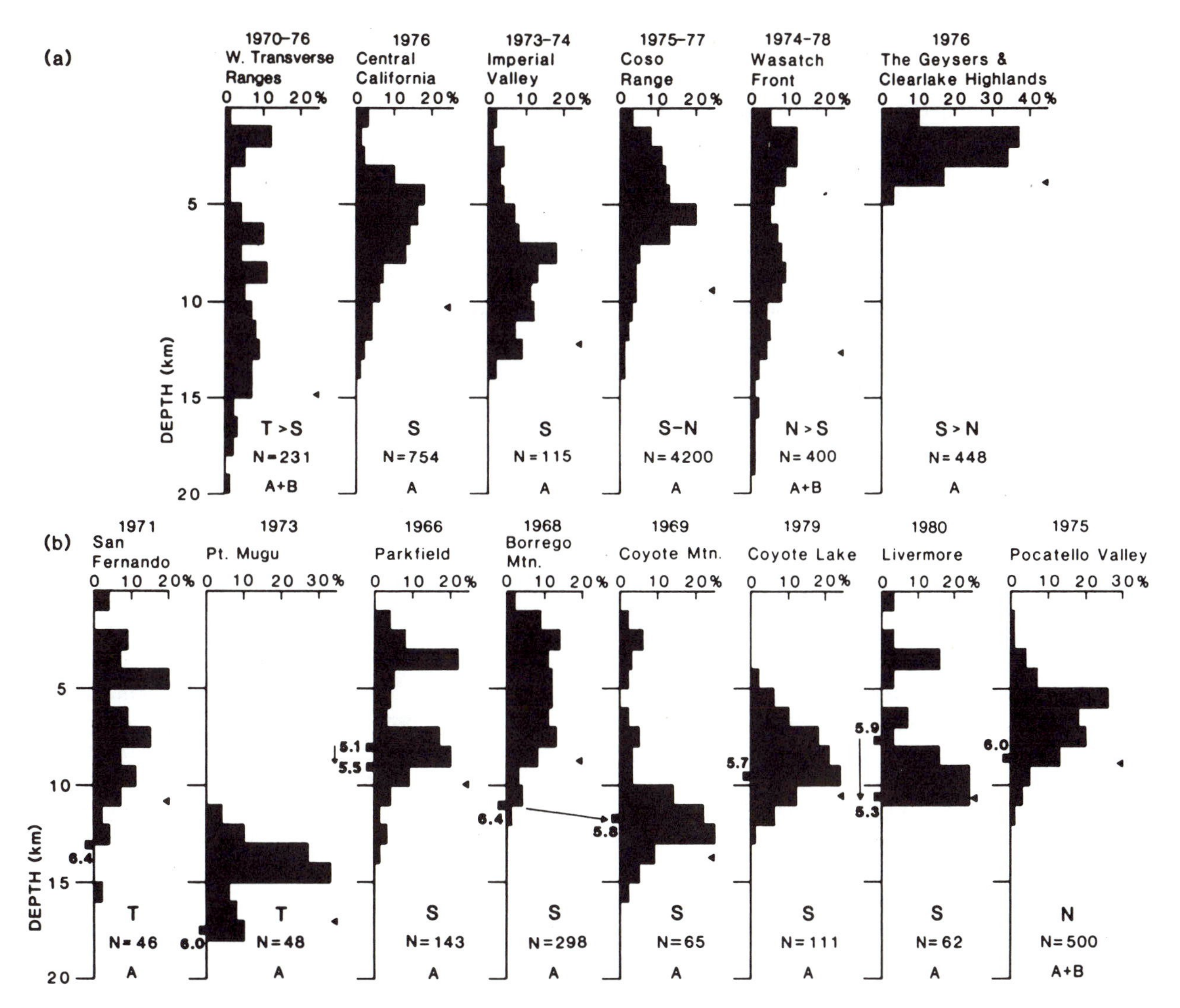

Figure 3–4. Earthquake distribution by depth in the western United States. Solid triangles locate depth above which 90% of earthquake activity occurs. The dominant faulting mode is T, thrust; S, strike-slip; N, normal. N shows the number of data points, and A,B gives data quality. Depth and local magnitude (M_L) of mainshock(s) are shown by solid squares. From Sibson (1984).

of the San Andreas fault and south-southeast of the mainshock, and the greater "thickness" of the aftershock zone in cross section B-B′ at 2–10 km depth in comparison to that at greater depths.

The distribution of slip during two moderate-size earthquakes on the Calaveras strike-slip fault in California is compared with the distribution of aftershocks in Figure 3-3, which shows cross sections parallel to the fault. That part of the fault characterized by slip during the earthquake has relatively few aftershocks. The concentration of aftershocks around the perimeter of the mainshock area may reflect the high stress concentrations there (cf. discussion of Griffith cracks in Chapter 2).

Figure 3-4 shows the depth distribution of small earthquakes in several seismically active areas in the western United States and the mainshock focus and aftershock depth distribution of eight moderate-size earthquakes in the same region. A solid triangle shows the depth above which 90% of the earthquake activity occurs. At the bottom of each diagram is given the dominant faulting mode (T, thrust or reverse; S, strike-slip; N, normal), the number of data points, and the quality of the data. The deepest earthquakes are in the western Transverse Ranges where the geothermal gradient is the lowest in southern California. The shallowest position of the base of earthquakes is in the area including the Geysers geothermal field in northern California. In the eight earthquake sequences presented, the mainshock focus is at or near the base of earthquakes. Earthquakes are absent or less abundant in the shallowest part of each diagram, even for those earthquakes that were accompanied by surface rupture such as San Fernando, Parkfield, and Borrego Mountain.

Thus, in summary, a geological explanation of the distribution of relatively well-located earthquakes must include the following: (1) Earthquakes terminate downward at a relatively sharp boundary within the crust, and the mainshock tends to nucleate near the base of the zone of aftershocks. (2) This boundary is shallower in areas of higher geothermal gradient and in younger oceanic and continental crust, suggesting that it is controlled by temperature. (3) The near-surface portion of a fault zone has no earthquakes or fewer earthquakes than the fault at greater depth, even in places where there is coseismic surface rupture. (4) Aftershocks are not distributed uniformly along the fault but are concentrated in some areas and absent in others. (5) Some aftershocks are located off the fault. (6) Aftershocks tend to concentrate around, but not within, the mainshock zone.

RHEOLOGY OF THE SEISMOGENIC CONTINENTAL CRUST

We cannot observe the seismogenic zone of earthquakes directly, but we can study sections of old continental crust that have been uplifted and exhumed by erosion. Based on the deformation features that are present at various depths within the crust, we can make some inferences about whether deformation was accompanied by earthquakes or not. In addition to studying these sections of deep crust in the field, we can deform rock samples in the laboratory under varying confining pressures, differential stresses, fluid pressures, temperatures, and strain rates and try to relate these results to deformation and crustal seismicity.

Experimental data show that the strength of most rocks is lowered by increasing temperature, and nonelastic deformation is more likely to be plastic at higher temperatures and brittle at lower temperatures.

Stress-depth profiles for various crustal rocks (Fig. 3-5a-d) and mantle rocks (Fig. 3-5e-f) are calculated for strain rates of 10^{-15} s^{-1} characteristic of cratonic areas (stable continental shields) and 10^{-14} s^{-1} for orogenic areas (mountain belts). An average continental geothermal gradient (Mercier, 1980) is used. In the middle crust, rock strength *decreases* sharply with increasing depth and temperature, with the critical temperature at which this decrease in strength occurs a characteristic of the material. This is in contrast to the shallow crust where Byerlee's law holds: rock strength as measured by the resistance to frictional sliding *increases* with depth because it is dependent on confining pressure. For granite and quartzite, rocks common in the upper crust, the critical temperature is 300–500°C, whereas for lower crustal rocks (diabase and pyroxenite), it is 400–600°C. Using the continental geothermal gradient, these temperatures correspond to depths of 10–23 km, well above the continental Moho discontinuity. However, the critical temperature for mantle rocks is 700–900°C, corresponding to depths below the Moho.

Figure 3-6 shows the Coulomb-Mohr failure envelope for diabase (assumed to be representative of lower-crust composition) at various temperatures. This figure takes into account the experimental observation that at shallow depths, *brittle* strength is largely independent of temperature. Furthermore, *plastic* strength is theoretically independent of confining pressure if we ignore for the moment the pressure-dependent weakening effects of fluids, so that the slope of the strength envelope under plastic conditions is zero. There is a relatively sharp changeover at the transition in which increasing pressure tending to strengthen the rock gives way to increasing temperature tending to weaken the rock and to produce ductile flow.

Because confining pressure increases with vertical load, rheological changes can be shown with respect to depth instead of confining pressure. At shallow depths, Byerlee's law for frictional sliding (cf. Chapter 2), based on laboratory results for pressures to 600 MPa (megapascals), shows that maximum shearing stress increases linearly with depth, as shown in the shallow part of Figure 3-7. However, rock frictional resistance is complex enough that Byerlee's law may be linear only to about 5 km depth. Accordingly, the frictional relation in Figure 3-7 is dashed below that depth. Figure 3-7 also shows that when a temperature threshold (critical temperature) is reached, rock strength decreases sharply with depth, in agreement with the data of Figure 3-5. The lower the geothermal gradient, the deeper the level at which this critical temperature is reached.

Because earthquakes form by unstable frictional sliding, the zone of frictional sliding is, to a first approximation, the zone of earthquakes. The upper crust can be viewed rheologically as a series of laminations of material in which each lamination is stronger than the one above it, with the strongest lamination closest to the transition between frictional failure and quasi-plastic flow, commonly known as the brittle-plastic transition. Therefore, this deepest brittle lamination is the load-bearing zone concentrating strain energy at that level, and when failure occurs at this level, the entire upper crust fails. This accounts for the tendency of the mainshock to occur near the base of the zone of aftershocks. The strongest crust near the brittle-plastic transition (or, more precisely,

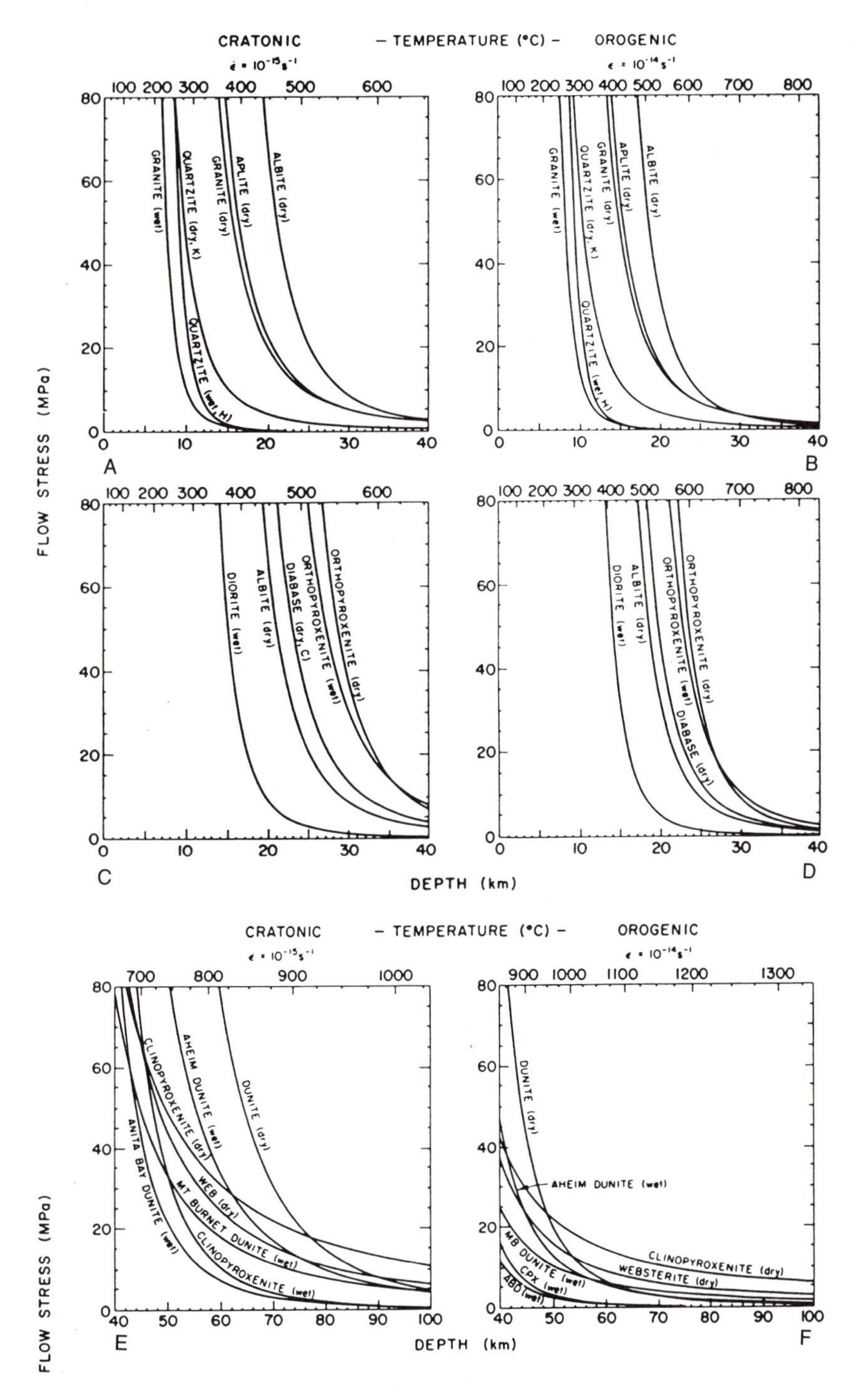

Figure 3–5. Flow stress–depth–temperature profiles for felsic rocks typical of upper continental crust (a, b), mafic rocks typical of lower crust (c, d), and ultramafic rocks typical of upper mantle (e, f), assuming geothermal gradients characteristic of stable continental cratons and orogenic belts. Flow stress is a measure of rock strength. From Carter and Tsenn (1987).

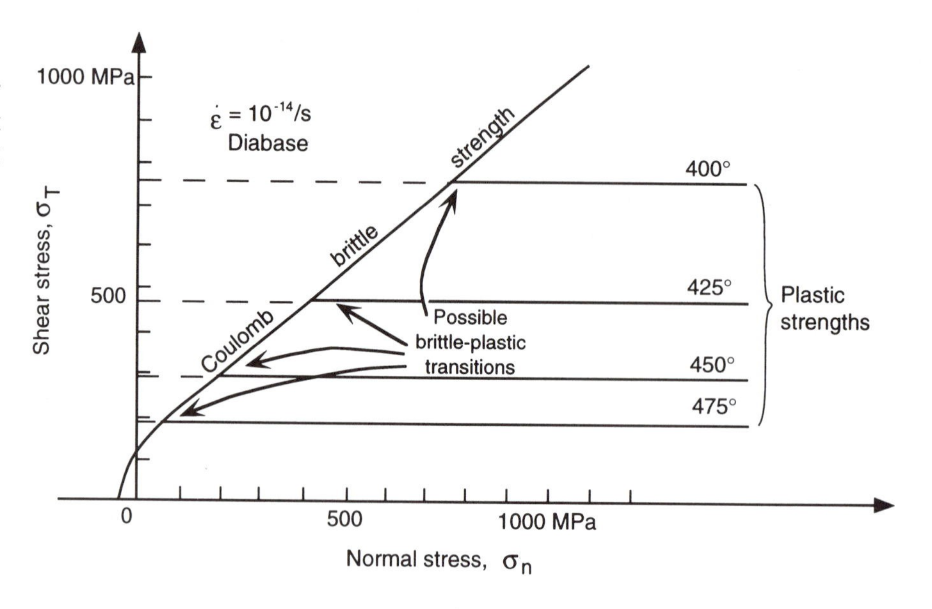

Figure 3–6. Coulomb-Mohr failure envelopes for diabase at various temperatures showing brittle-plastic transition, which occurs at higher stresses with lower temperature. The model assumes that brittle strength is independent of temperature (otherwise, the slope of the brittle-strength curve would change), and plastic strength is independent of pressure (otherwise the slope of the envelope in plastic conditions would not be zero). From Figure 5–14 of Suppe (1985).

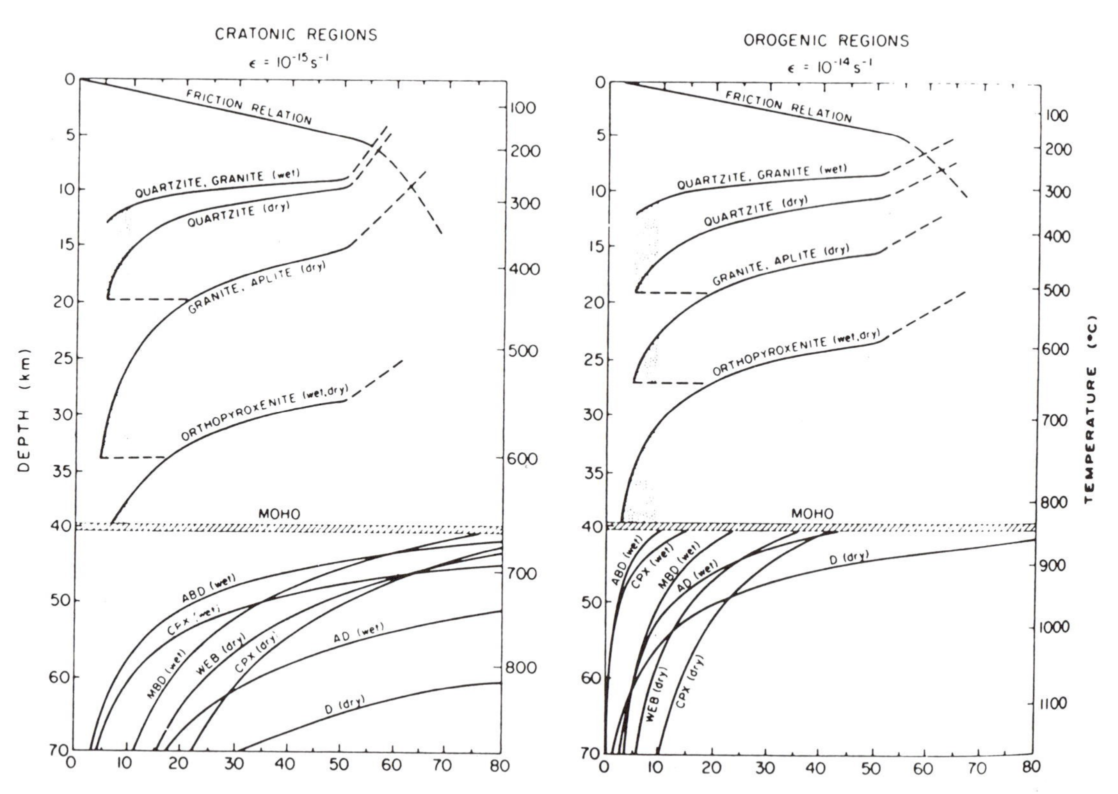

Figure 3–7. Comparison of pressure-dependent friction relation (Byerlee's law) with flow-stress profiles for various crustal and mantle rocks. Dashed lines in crustal region indicate uncertainty in the nature of the breakdown of the friction relation with increasing depth and temperature and the breakdown of high-temperature flow-stress laws with decreasing depth and temperature. Note that crustal rocks have low strength above the Moho whereas mantle rocks have high strength below the Moho. ABD, AD, D are dunite; CPX, clinopyroxenite; WEB, websterite (all ultramafic rocks). After Carter and Tsenn (1987).

the frictional-quasi-plastic transition) ruptures, then the rupture radiates upward toward the surface and laterally along strike, due to high stress concentrations at the rapidly-moving rupture tip. Figure 3-8 shows an earthquake rupture in cross section. The mainshock focus, shown by a star, is near the brittle-plastic transition, and, in this case, is located near one end of the rupture zone. The earthquake ruptures upward toward the surface and laterally along strike. It also may rupture downward into the transitional zone between frictional and plastic behavior; this is discussed below.

Figures 3-5 and 3-7 show that mantle rocks behave in a brittle fashion at temperatures 300 to 400°C higher than crustal rocks such as diabase and quartzite. If the Moho marks the compositional change to olivine-rich rocks, it is possible and indeed likely that mantle rocks below the Moho may yield by brittle fracture whereas under the same P-T conditions, overlying crustal rocks yield by quasi-plastic flow. Therefore, a zone of seismicity may be encountered just below the top of the Moho, as suggested by Figure 3-7.

Chen and Molnar (1983) documented this sub-Moho seismic zone. They determined the distribution of focal depths for intraplate earthquakes and observed that the deepest events are in the oldest crust. Intraplate earthquakes in continents are restricted to the upper crust and the upper mantle; the lower crust is aseismic. Tibet and the Karakorum of central Asia and the High Atlas Mountains of North Africa show shallow crustal seismicity, an aseismic zone in the lower crust presumably dominated by plastic flow, and a deeper seismic zone just below the Moho. The aseismic zone extends to shallower crust under higher geothermal gradients such as those found in younger, hotter crust. Under higher geothermal gradients, the entire mantle may behave aseismically, which may explain in part the absence of mantle earthquakes in California. However, the absence of mantle earthquakes beneath the western Transverse Ranges of California, where geothermal gradients are low, is still unexplained.

The physical boundary between shallow, brittle crust and deeper, plastic crust may allow a concentration of strain at the boundary such that brittle crystalline slabs can detach from the underlying lower crust and be emplaced as huge crystalline thrust plates, as observed in the Alps, in the ranges of Scandinavia, and in the San Gabriel Mountains of southern California (Chen and Molnar, 1983; Yeats, 1983). This implies that high-angle seismogenic faults in the upper crust may, in some cases, merge into a flat décollement at the brittle-plastic transition. Webb and Kanamori (1985) noted that some deeper earthquakes in the Transverse Ranges of southern California are characterized by focal mechanisms (discussed in Chapter 4) that include a flat-thrust nodal plane, which they presented as seismic evidence for a flat décollement.

However, some high-angle faults in the upper crust continue as steep faults in the lower crust, as demonstrated by the presence of high-angle ductile shear zones in exhumed lower crust. In addition, the variability in depth of the seismogenic zone (Fig. 3-1) is difficult to explain if the base of seismogenic crust, a rheological boundary, is marked by a décollement which should reduce such irregularities. This problem is discussed further in Chapter 10.

MODEL OF A CONTINENTAL FAULT ZONE

A model of a fault zone in continental crust may be developed based on experimental data on rock deformation, on textures and structures observed in the field, particularly in deeply-exhumed fault zones, and on the distribution of earthquakes (Sibson, 1977; 1983; Scholz, 1990). Sibson's model for a strike-slip fault is shown as Figure 3-8. The general characteristics of this model are a shallow regime of cataclastic deformation influenced by confining pressure and involving stick-slip frictional sliding accompanied by earthquakes and a deep regime of plastic deformation influenced by temperature and involving deformation that is aseismic, not accompanied by earthquakes. The brittle-plastic transition is gradational and broad, based on the overlap of deformation mechanisms for different rock-forming minerals and for different grain sizes.

In the fault-zone model of Scholz (1990), part of which is shown as Figure 3-9, the brittle-plastic transition is discussed in greater detail. The transition must occur below the onset of quartz plasticity at 300°C (his T_1) and the above plasticity of feldspar at 450°C (his T_2). T_1 corresponds to a change in the fault zone from cataclastic rocks to mylonites and a change in wear mechanism from *abrasive wear,* in which there is a sufficient difference in hardness to allow the harder material to plow through the softer material, breaking off pieces by brittle fracture and creating gouge, to *adhesive wear,* in which asperities are sheared off and transferred from one side of the fault to the other (Scholz, 1990).

T_1, the base of the seismogenic zone, marks the boundary between *velocity-weakening,* unstable frictional-rate behavior on the fault, and *velocity-strengthening,* stable behavior at greater depths. As

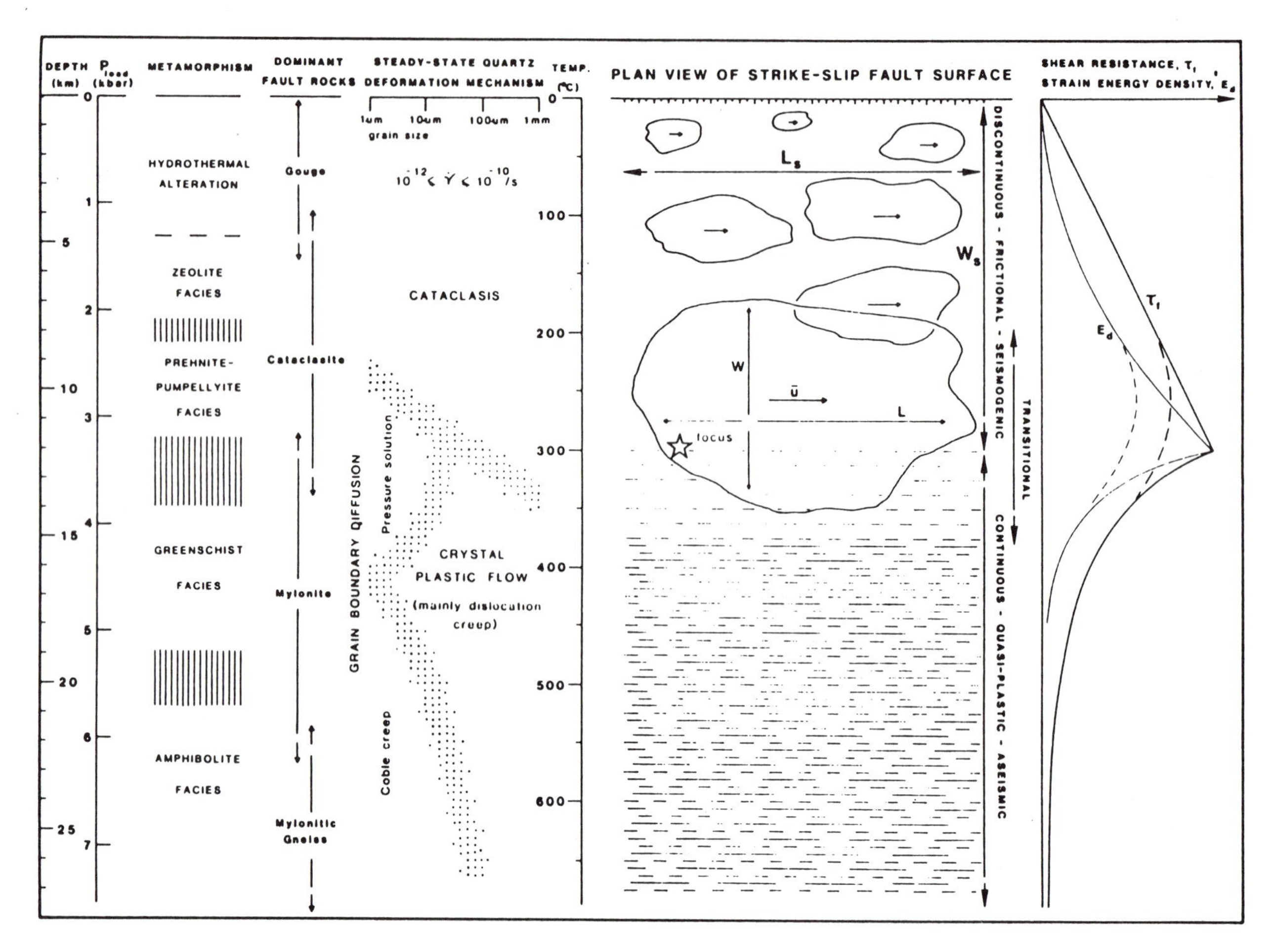

Figure 3–8. Deformation mechanisms accompanying faulting in continental crust. After Sibson (1983). Metamorphic facies represent mineral assemblages stable under increasing temperature and confining pressure.

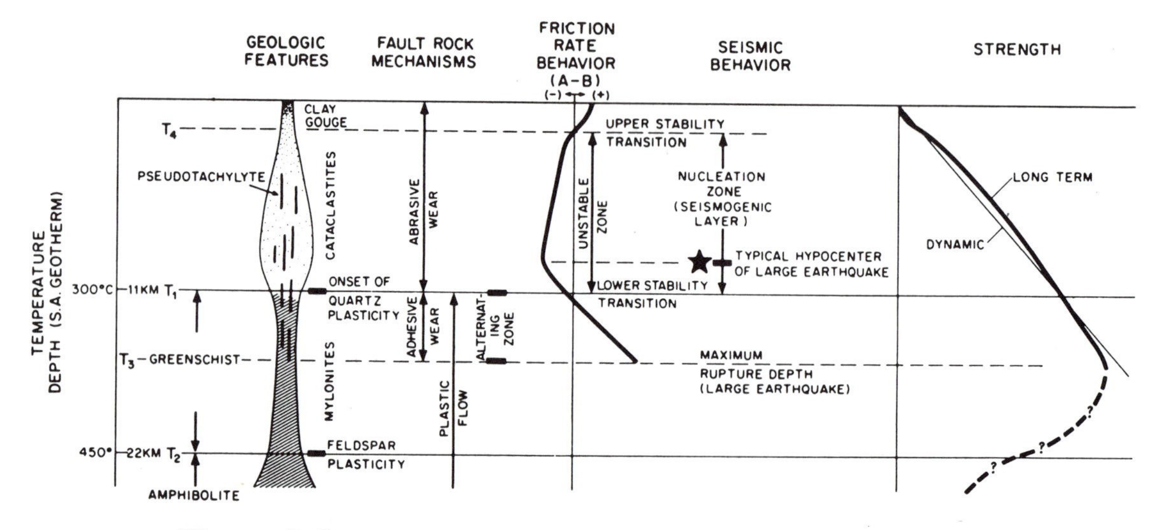

Figure 3–9. Model of a continental fault zone, after Scholz (1990).

discussed by Scholz (1990), velocity-weakening is the condition in which dynamic friction is lower than static friction. An instability may develop because frictional resistance force decreases with displacement at a rate faster than the system can respond to it; this leads to stick-slip behavior. If dynamic frictional stress is higher than static strength, the condition is velocity-strengthening, and there is a tendency to terminate rupture. This is the condition of stable sliding. T_1 is the lower boundary of the region where earthquakes can nucleate, but large earthquakes may propagate downward to a depth T_3, shallower than the depth at which feldspar deforms plastically. The transitional zone behaves plastically at low strain rates but in a brittle fashion at high strain rates accompanying propagation of earthquake rupture. (This is illustrated by Silly Putty and by the Maxwell rheological model.)

Scholz notes that frictional-rate behavior on faults should be velocity-strengthening, or stable, near the surface above a depth T_4, an observation based on the cutoff of seismicity near the surface (Figs. 3-1 and 3-2). The shallowest portion of the fault is most likely to contain relatively unconsolidated fault gouge or to be in contact with sediments with low strength. As is the case with the base of the seismogenic zone, a large earthquake may propagate through T_4 and produce rupture at the ground surface. This surface rupture involves earth materials that are too weak to produce seismicity separate from that at greater depth, accounting for the low seismicity in the shallowest parts of the San Jacinto and San Andreas faults (Fig. 3-1).

The crust is saturated with fluid, and the fluid pressure, which is isotropic, acts in a direction opposite to the confining rock pressure, which is not isotropic. As shown in Figure 2-9, this moves the Mohr circle toward the failure envelope, effectively weakening the rock. The fluid pressure is taken into consideration using the fluid pressure ratio: $\lambda = \frac{pf}{\sigma_v}$. As $\lambda \rightarrow 1$, fault strength decreases.

The presence of fluids also may cause recrystallization to mineral assemblages with lower strength and subcritical fracture growth due to stress corrosion, processes that are time-dependent (cf. Journal of Geophysical Research, 1995). These processes have been described on the basis of study of fluid inclusions and hydrothermal mineral assemblages at deeper levels of normal faults in the Great Basin that have been exhumed by uplift and erosion of the footwall block (Bruhn et al., 1990; Parry et al., 1991). Subcritical growth of cracks affected by fluids takes place below the Griffith fracture criteria (see Chapter 2). Working against crack growth is crack sealing due to the precipitation of dissolved material under hydrothermal conditions, a process influenced by salinity of the fluid, partial pressure of CO_2, and pH. Knowledge of the P-T conditions under which successive mineral assemblages evolve leads to the conclusion that fluid pressures may vary from hydrostatic to near-lithostatic (close to overburden pressure). In addition, precipitation of hydrous minerals in the fault zone changes fault-zone rheology, generally making the fault zone weaker than surrounding rock.

In the brittle crust, faults move by stick slip, characterized by nearly instantaneous displacement followed by a long period during which strain builds up for the next event. It may be inappropriate to treat earthquake rupture using equations of mechanical equilibrium in which all forces add up to zero, because at the time of rupture, forces are unbalanced, producing an acceleration. Inertia and momentum must be taken into consideration.

Nearly all earthquakes nucleate on existing faults, and the strength that must be overcome to produce an earthquake is not that of flawless rock but that of fault-zone rock (Fig. 2-17). In the shallower part of brittle crust, this consists of clay-rich *fault gouge* (Figs. 3-10 and 3-11) with a low coefficient of friction and a low permeability, the latter contributing to the development of geopressures within fault zones as compaction takes place and the fault zone is sealed away from surrounding rock (Blanpied et al., 1992). In fault gouge, visible fragments are $<30\%$ of rock mass. Gouge develops by frictional wear of fault-wall rock, drag of asperities into the opposite side of the fault and shearing of the asperities, leading to development of clay. Some of these clay surfaces develop lineations in the direction of fault slip called *slickensides* (Fig. 3-10). At greater depths, fault-zone rock consists of *fault breccia* (visible fragments $> 30\%$ of rock mass) produced by brittle fragmentation of rocks and intragranular and intergranular cracking of grains, grain rotation, and sliding, leading to the development of more cohesive *cataclastic rocks* (Figs. 3-12 and 3-13).

The depth of clay-rich fault gouge is a matter of controversy, but the relatively low coefficient of friction of some clays such as montmorillonite suggests that sudden slip in clay-rich fault gouge may be relatively aseismic, thereby explaining the low seismicity of the shallowest portions of fault zones as illustrated in Figures 3-1 through 3-4. Cataclastic rocks, fault breccias, and fault gouge low in clay content should have much higher coefficients of friction than clay-rich fault gouge, comparable to most crustal rocks that follow Byerlee's law (Byerlee, 1978; Fig. 2-8). However, the control of seismic vs. aseismic slip may depend not on the magnitude of the coefficient of friction but

Figure 3–10. Clay-rich fault gouge with slickensides formed during the 1992 Landers, California, earthquake. View to northeast. Image is 20 cm across. Photo by Hervé Leloup, Institut de Physique du Globe de Paris.

on whether the fault rock is velocity-strengthening or velocity-weakening.

Much of the displacement in a fault zone is concentrated on *principal slip surfaces,* which may be curved, discontinuous, en échelon, or stepping from one side of a fault zone to the other. Parts of the fault zone are stronger than others, producing asperities. These differences in fault-zone strength are probably the cause of uneven concentration of aftershocks along the fault zone, as illustrated in the distribution of aftershocks of the Loma Prieta earthquake (Fig. 3–2). Asperities may explain why zones of slip during major earthquakes on the Calaveras fault have very few microearthquakes in the interseismic period (Fig. 3–3). These patches are stuck and do not produce microearthquakes, whereas other patches are weaker and move, accompanied by microseismicity.

High fluid pressures may develop due to local increases in temperature as frictional heat dissipates during increments of rapid (seismic?) slip. Evidence of this is in clastic dikes filled with gouge and breccia adjacent to sliding surfaces and networks of veins associated with slickensides. Figure 3–14 shows *pseudotachylite,* rock melted by the high levels of energy dissipation associated with seismic slip and solidified as rock glass (aseismic slip would generate heat too slowly to melt rock). In Figure 3–15, the dark matrix is pseudotachylite; the stellate clusters of crystals form as the melt quenches after being injected into wall rock adjacent to a fault. The P-T conditions corresponding to frictional (seismic) sliding are the same as those accompanying hydrothermal alteration,

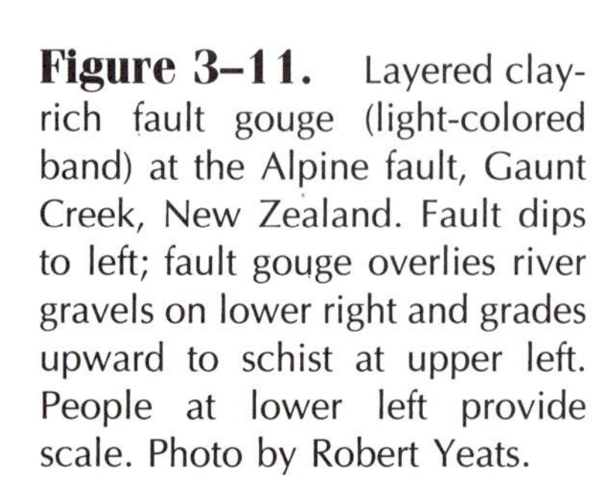

Figure 3–11. Layered clay-rich fault gouge (light-colored band) at the Alpine fault, Gaunt Creek, New Zealand. Fault dips to left; fault gouge overlies river gravels on lower right and grades upward to schist at upper left. People at lower left provide scale. Photo by Robert Yeats.

Figure 3–12. San Andreas fault-zone breccia, Leona Valley, California, consisting of angular fragments of siltstone and sandstone of Pliocene Anaverde Formation. Photo by Robert Yeats.

zeolite-facies metamorphism, and prehnite-pumpellyite-facies metamorphism (Fig. 3–8).

At greater depths and temperatures of greenschist-facies and amphibolite-facies metamorphism, deformation occurs by dislocation creep and glide (plastic flow of crystals), grain-boundary diffusion (Coble creep), and pressure solution, developing mylonite as the principal fault-zone rock. Mylonite is a layered and streaked rock in which quartz grains have been converted to a very fine grain size by dynamic recrystallization accompanying plastic deformation (Figs. 3–16 and 3–17). Deformation processes are strongly influenced by temperature as well as by the presence of water (*hydrolytic weakening*).

Deformation is aseismic, so earthquakes add no information about the nature of displacement within the aseismic zone. However, there must be two regimes (Fig. 3–18): a deeper regime where displacement is steady and continuous, and a shallower regime where displacement is episodic but still aseismic. This has been studied by Tse and Rice (1986), and one of their models is shown as Figure 3–19, a plot of depth distribution of slip at increments of time for three earthquake cycles. The contour interval in time is not uniform, as noted. At some depth below the diagram, slip vs. depth would be uniform and steady, driven by the velocity of the plate boundary. Loading begins with slip below the transition between frictional failure and quasi-plastic flow, extending upward toward the locked zone, although in their model, some slip also occurs near the surface. Nucleation begins at 300 days before the earthquake mainshock in this particular

Figure 3–13. Fault breccia, Dixie Valley, Nevada. Photo by Jonathan Caine, courtesy of Ronald Bruhn.

Figure 3–14. Photomicrograph of mylonitic shear zone containing deformed pseudotachylite, an example of mixed brittle and ductile behavior in the transition zone. Field of view about 15 mm. Photo by Richard Sibson.

Figure 3–15. Photomicrograph of pseudotachylite showing quench texture: stellate clusters of plagioclase feldspar that have nucleated on breccia fragments. Outer Hebrides thrust zone, Scotland. Field of view about 4 mm. Photo by Richard Sibson.

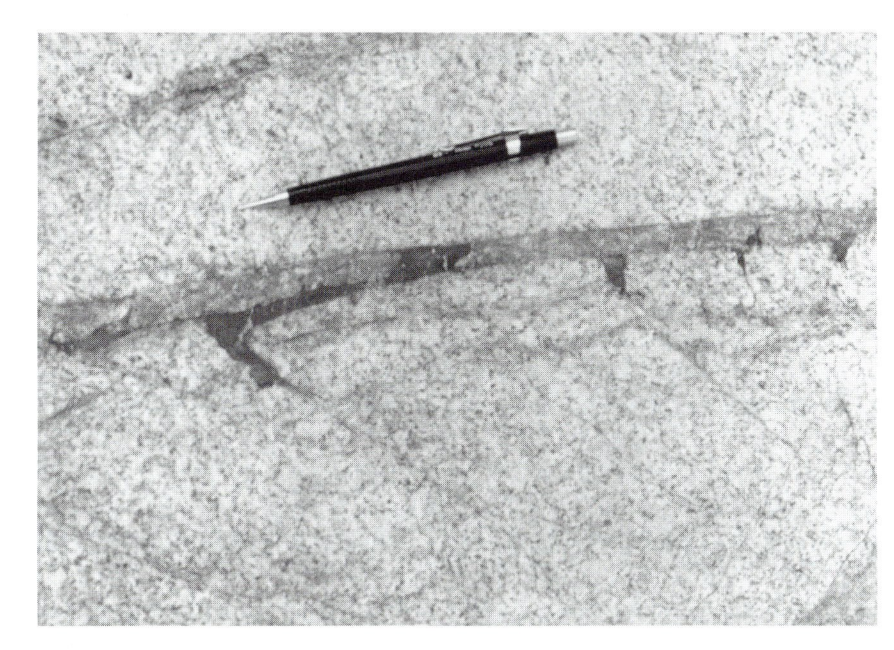

Figure 3–16. Pseudotachylite (dark, cross-cutting injection veins) formed by local frictional melting of rock during seismic slip. Veins cut light-colored granitic rock, locally mylonitized, in Cucamonga Canyon, east San Gabriel Mountains, California. Photo by Richard Sibson.

model (slip between A and B). The coseismic slip B to B′ produces a stress drop in the seismogenic zone that transfers stress to the weak zone below. Slip penetrates into the weak zone to 13–15 km depth, then to 20 km depth, followed by steady deep slip and reloading of the seismogenic zone.

Deep seismic-reflection profiling permits the direct imaging of continental crust to the depths of the Moho and below. In many continental regions, crustal reflectivity increases abruptly in the middle crust and terminates at the Moho, evidence of low-dipping velocity contrasts in the plastic lower crust (Mooney and Meissner, 1992). The sets of closely-spaced reflections have been referred to as *seismic lamellae* or *laminae*. These laminae undoubtedly have more than one explanation, including mylonite zones, metamorphic layering, igneous intrusions, and variation in rock type (Mooney and Meissner, 1992).

Faults in normal oceanic crust would not be expected to have a weak, aseismic zone in the lower

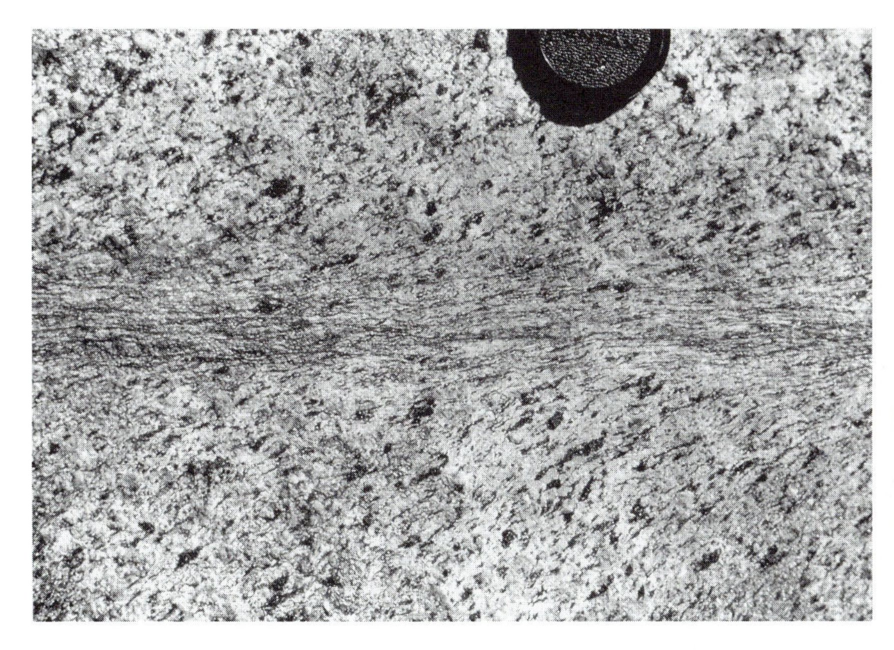

Figure 3–17. Ductile shear zone developed under amphibolite-facies metamorphic conditions in a late Paleozoic (Hercynian) granite, Laghetti area, Ticino, Switzerland. Photo by Richard Sibson.

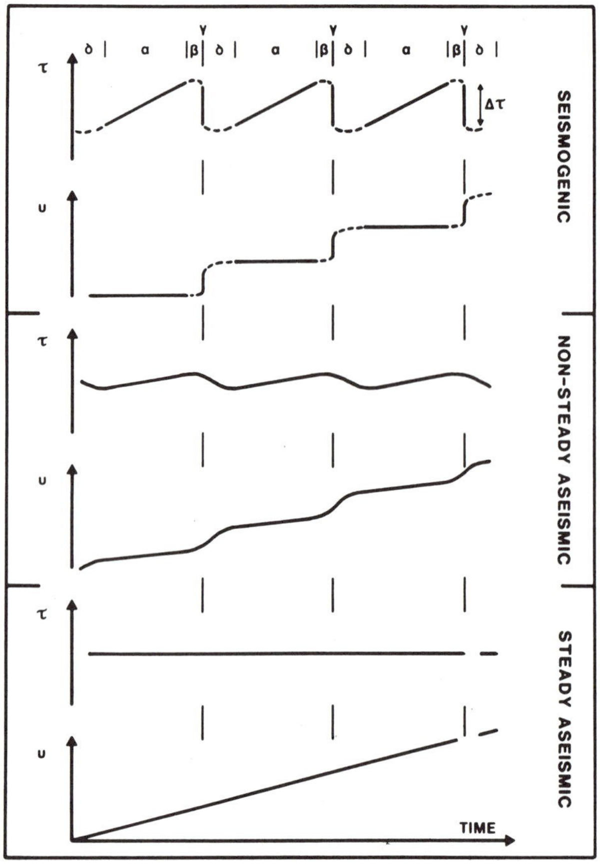

Figure 3–18. Hypothetical stress (τ)–time and displacement (u)–time relations in seismogenic crust, non-steady aseismic crust, and steady aseismic crust, plotted with respect to the earthquake cycle: α = elastic strain accumulation, β = nonelastic pre-seismic deformation. γ = coseismic mainshock slip and rupture propagation, and δ = post-seismic afterslip and aftershock activity decaying asymptotically with time. From Sibson (1986).

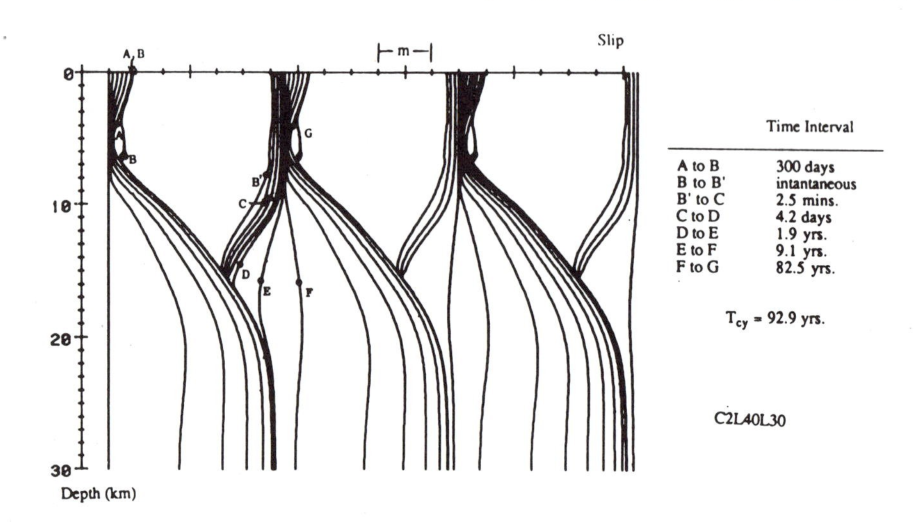

Figure 3–19. Plots of depth distribution of slip at constant time through three earthquake cycles. Note that the interval between successive time lines varies, as shown at right. After Tse and Rice (1986).

crust because in most cases, the Moho is shallower than the brittle-plastic transition for diabase. Accordingly, the brittle seismogenic layer would include the uppermost mantle, and the brittle-plastic transition would be controlled by the temperature at which olivine becomes ductile, about 750–900°C. The depth of seismicity is shallower for younger oceanic crust. For oceanic crust newly generated at a mid-ocean ridge, the seismogenic layer is only a few kilometers thick.

Bruhn et al. (1990) present a model of the earthquake cycle in terms of geological processes operating in the fault zone. An earthquake takes place, producing a stress drop on a segment of the fault and a stress concentration at the tip of the fault activated by the earthquake. Stress continues to relax after the earthquake due to aftershocks and aseismic creep. Stress then accumulates linearly, reaching a point where a dramatic increase in propagation of microcracks takes place by subcritical growth due to the effect of fluids. Healing of cracks is important immediately after the earthquake because of permeability to fluids created by coseismic fracturing. This permeability results in *fault-valve behavior* (Sibson, 1992) and may result in sudden fluctuations in fluid pressure. After the cracks heal, the fault seals itself off from fluid flow, and stress accumulates linearly again.

We must now consider the zone of transition between frictional failure and quasi-plastic flow, the brittle-plastic transition zone. We assumed earlier that this is a relatively sharp transition, but we must now reconsider it as a broader, more diffuse zone of transition (Figs. 3-8 and 3-9). For granitic crust, plastic deformation of quartz occurs in the shallower greenschist metamorphic facies, where biotite mica is kinked, and feldspar undergoes brittle fracture (Fig. 3-20). In the deeper greenschist facies, plagioclase feldspar still shows only minor plasticity, (although alteration to micaceous minerals is occurring under these conditions), whereas quartz, biotite, and orthoclase feldspar are completely recrystallized. Only under amphibolite facies and beyond, with temperatures >550°C, does plastic deformation occur in all crustal minerals.

Even for monomineralic rocks, the brittle-plastic transition is gradual. When the threshold temperature for plastic deformation is exceeded, brittle deformation (crack propagation) occurs together with dislocation creep, which is controlled by the orientation of the crystal lattice. If the crystal lattice is oriented close to a plane of high shear stress, dislocation creep will occur; if it is not, brittle deformation will occur. This transition zone is called the *semi-brittle field*. Peak shear strength may continue to increase with depth in this zone, decreasing only when deformation is fully plastic.

Finally, slip velocity has an effect on frictional behavior. As noted in Figure 3-5, plastic strength is lower with lower strain rates. Thus the pre-seismic buildup of unrelieved slip within the brittle-ductile transition zone is very slow, measured in millimeters per year, and it may be aseismic. Coseismic slip, measured at rates of kilometers per second, may be brittle under the same crustal conditions. Thus an earthquake may propagate downward to depths and temperatures where deformation is otherwise aseismic. The intercalation of mylonite, a rock formed under quasi-plastic flow, with pseudotachylite, a product of frictional faulting, may be evidence of this rate-related behavior (Fig. 3-16).

Figure 3-20. Photomicrograph of cataclastic textures in granite in the Willard thrust zone, Utah, showing brecciated twin lamellae in plagioclase feldspar. Photo by R. Hanson, courtesy of Ronald Bruhn.

SUMMARY

The mainshock zone of most earthquakes may not be sampled directly, but an understanding of its geological characteristics is emerging based upon the distribution of well-located earthquakes, study of exhumed fault zones, laboratory experiments at mainshock depths and pressures, and theoretical modeling. Rock strength increases with confining pressure to a point in the middle crust where temperature-controlled weakening takes over (brittle-plastic transition). The temperature at which weakening occurs is dependent on rock type and is shallowest for granitic crust and deepest for mantle materials. This results in the lower continental crust being aseismic, separating seismic upper crust and seismic uppermost mantle in some regions. Earthquakes tend to nucleate a short distance above the onset of weakening, where crust is strongest, and to propagate laterally and upward toward the surface. Below the brittle-plastic transition, earthquakes terminate abruptly except for large shocks which may propagate into this transition because this transition is brittle at coseismic strain rates. Fault zones tend to be aseismic near the surface where they contain weak fault gouge and where faults cut weak sediments. Concentrations of earthquakes along certain parts of fault zones may indicate asperities, or stronger sections of the fault zone.

Direct evidence of earthquakes in exhumed fault zones is limited to pseudotachylite, in which enough frictional heat is generated to melt rock. This is considered possible only under conditions of seismic slip.

Thus, even though mainshocks occur in rocks too deep to sample, conclusions reached by indirect means are consistent with the distribution of earthquakes based on the seismographic record.

Suggestions for Further Reading

Byerlee, J. D. 1978. Friction of rocks. Pure and Applied Geophysics 116:615–26.

Carter, N. L., and Tsenn, M. C. 1977. Flow properties of continental lithosphere. Tectonophysics, 136:27–63.

Carter, N., Friedman, M., Logan, J., and Stearns, D. 1981. Mechanical behavior of crustal rocks. American Geophysical Union Monograph 24.

Higgins, M. W. 1971. Cataclastic rocks. U.S. Geological Survey Professional Paper 687, 97 p.

Journal of Geophysical Research, 1995. Special section: Mechanical involvement of fluids in faulting: v. 100:12,837–13,132.

Kirby, S. H. 1983. Rheology of the lithosphere. Reviews of Geophysics and Space Physics, 21:1458–87.

Mooney, W. D., and Meissner, R. 1992. Multigenetic origin of crustal reflectivity: A review of seismic reflection profiling of the continental lower crust and Moho. *In* Continental Lower Crust. Fountain, D. M., Arculus, R., and Kay, R. W., eds. Amsterdam: Elsevier, pp. 45–79.

Paterson, M. S. 1978. Experimental Rock Deformation: The brittle field. Berlin: Springer-Verlag, 278 p.

Schmid, S. M. 1982. Microfabric studies as indicators of deformation mechanisms and flow laws operative in mountain building. Pp. 95–110 *In* Mountain Building Processes. Hsü, K. J., ed. London: Academic Press, 263 p.

Scholz, C. H. 1990. The Mechanics of Earthquakes and Faulting. Cambridge: Cambridge University Press, 439 p.

Sibson, R. H. 1986. Earthquakes and rock deformation in crustal fault zones. Annual Review of Earth and Planetary Sciences, 14:149–75.

Suppe, J. 1985. Principles of Structural Geology. Englewood Cliffs, N.J.: Prentice-Hall, Inc., 537 p.

Twiss, R. J., and Moores, E. M. 1992. Structural Geology. New York: W. H. Freeman and Co., 532 p.

Johann F. J. Schmidt (1825-1884)

Germany

The Greek mainland is nearly cut in two by the Gulf of Corinth, which separates the Peloponnesos Peninsula from the rest of Greece to the north. The Gulf is one of the most seismically active regions of the Mediterranean, with damaging earthquakes recorded there for more than 2500 years. The Gulf is a graben 200 km long, bounded by active normal faults on both sides. Extension across the Gulf of Corinth is several millimeters per year, very high relative to other normal-faulted areas of Greece. (Damaging earthquakes in 1981 at the eastern end of the Gulf of Corinth are described in Chapter 9.)

The town of Egion, at the western end of the Gulf of Corinth on the Peloponnesos coast, was heavily damaged by an earthquake of M 7 on December 26, 1861. Shortly afterward, a young German astronomer, Johann Schmidt, arrived on the scene to survey the damage. Schmidt had been appointed three years before as director of the astronomical observatory in Athens.

As an astronomer, Schmidt wrote about variable stars and sunspots, and later in his life, he made a map of the surface of the moon. But he also developed an interest in earthquakes, and, at the age of 21, calculated the seismic wave velocity of an earthquake in the Rhine region in 1846. He is best known for his book, *Studien über Erdbeben,* in which the term "epicenter" was first used. Schmidt had an astronomer's interest in periodicity, and he compiled a catalogue of earthquakes to see if there were astronomical controls, or a relationship to lunar tidal cycles. Clearly, when he came to Egion, he already had a distinguished reputation in geophysics in addition to astronomy.

Schmidt's contribution to earthquake geology is that he was the first trained observer to map a surface fault accompanying an earthquake. At Egion, he observed that "a crack about eight feet [2.4 m] high and six feet [1.8 m] wide appeared in the earth and a strip of plain slipped slowly under the sea." It is not clear that Schmidt recognized this "crack" as a fault. More likely, he thought it was a landslide, and some observers today still interpret this feature as a landslide. Other cracks to the north, close to the sea, were also mapped by Schmidt, and these are clearly slumps.

But there is another story at Egion, a story of a lost city that was mentioned by Homer in the *Iliad* and was the center of a prominent religious cult . This city was Helice, destroyed by an earthquake in 373 B.C.E. The buildings of the city collapsed, and a great wave swept over it, either a tsunami, or the effect of sudden subsidence below sea level. The ruins of Helice were visible for more than 500 years, but now they have vanished without a trace. It may be that the fate of Helice was sealed by the same fault that ruptured at Egion in 1861, and Schmidt's fault has been renamed the Helice fault. The search for the remains of ancient Helice by archeologists still continues.

Unfortunately, disaster still stalks the city of Egion. On June 15, 1995, it was struck once more by an earthquake of M_S 6.5 in which at least 22 persons lost their lives.

Suggestions for Further Reading

Davison, C. 1978. The founders of seismology. New York: Arno Press, p. 240.

Mouyaris, N., Papastamatiou, D., and Vita-Finzi, C. 1992. The Helice fault? Terra Nova 4:124–29.

Tams, E. 1952. Materialen zur Wende vom 19- zum 20-Jahrhundert. Neues Jahrbuch Geol. Paläont. Abh. 95(2):165–292 (pp. 184–87).

4

Seismic Waves

Most earthquakes are caused by release of elastic strain accompanying sudden displacements on faults (Reid, 1910). *Seismographs* record the ground shaking that results from this release of energy and causes great loss of life and property. The resulting *seismograms* provide information about the earthquake process itself and about the earth materials the elastic seismic waves pass through.

The study of seismic waves began with the construction of the first instruments that could record ground motion. The first device, built by Chang Heng in A.D. 132, crudely recorded only the local intensity and the apparent direction from which the first ground shaking came. No subsequent progress was made in the analysis of seismic waves for more than 1800 years, until the construction of seismographs shortly before the beginning of the twentieth century. For this reason, only earthquakes occurring in the past century can be studied by analysis of seismic waves.

There is an earlier, preinstrumental history of earthquakes, however, which spans the time of written historical records. The length of that period varies markedly from place to place. This period is more than two thousand years long in northeastern China and in the eastern Mediterranean, but is only about 200 years in California. In Oregon, Washington, southern Alaska, and New Zealand, it is about 150 years in length, and in the seismically active island nation of Vanuatu, in the southwest Pacific, written records have been kept for less than a century. The preinstrumental, historical record is difficult to interpret, however, because it is a record of shaking and destruction, which is not easily translated into displacement on specific faults (see discussion in Appendix preceding the Table of Historical Earthquakes with Surface Rupture).

It was stated in the Introduction that the geologist is separated from a nearby earthquake source by many kilometers of the earth's crust. This is also true for the seismograph, which records the shaking produced by the release of elastic strain at the source, but only after the signal generated at the source has passed through that same thickness of heterogeneous crust. For seismic waves recorded at distant seismographs (*teleseismic waves*), the earth filter may be thousands of kilometers in thickness, including the crust, mantle, and core. The emerging field of *seismic tomography* uses the variation in travel time of seismic waves from different directions and distances to map in three dimensions the heterogeneous mantle and crust the waves pass through. Seismic waves are the earth's equivalent of an x ray in medicine, able to see through rock and to map the complex geology deep within the earth.

ELASTIC WAVES

Because earthquakes generate elastic waves, we begin with a brief review of wave motion, which is the way in which energy is transmitted from the earthquake source to the earth's surface. A wave is a disturbance that passes through a medium, but the medium as a whole does not progress in the direction of motion of the wave. An analogy is the wave train produced by a stone dropped into a pond. The water is disturbed, and waves radiate out from the point of entry, reaching the far corners of the pond. However, the disturbed particles of water are not permanently displaced in the direction the wave moves.

Elastic waves are classified based on the motion of individual particles with respect to the direction of propagation of the wave. Where the motion of individual particles is in the direction of propagation (Fig. 4-1a), the waves are *longitudinal* or *compressional (P) waves,* and when the motion is at right angles to the direction of propagation (Fig. 4-1b), the waves are *transverse* or *shear (S) waves*. Compressional waves move by alternately compressing and rarefying the elastic medium through which they pass. This re-

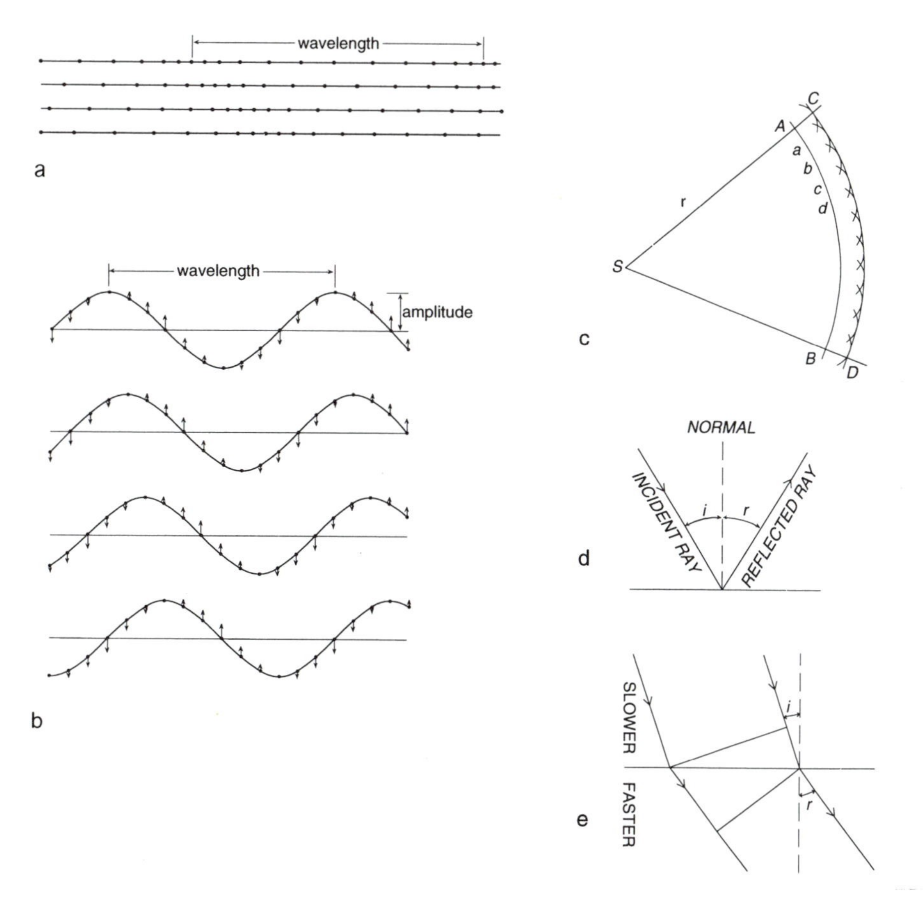

Figure 4–1. Elastic waves. (a) Longitudinal or compressional (P) waves, in which wave propagation occurs by alternately compressing and rarefying particles. Particle motion is in the direction of wave propagation. (b) Transverse or shear (S) waves, in which wave propagation occurs by displacement of particles from side to side, at right angles to propagation direction. For both examples, read from the top down so that the wave moves to the right. Individual particles vibrate but do not move permanently in the direction of wave propagation. (c) In a homogeneous medium, the wave front is the surface of a sphere with a center S and radius r. At any point a, b, c, or d on the surface, a new spherical wavefront is generated (Huygen's principle). Lines AS and BS are rays drawn in the direction of wave propagation. (d) When a wave is reflected from a medium boundary, the angle with the normal to the medium boundary (angle of incidence) is equal to the angle of reflection. (e) When a wave passes into a medium with a higher speed, the angle of refraction is increased according to Snell's law.

quires volume elasticity, which is possessed by gases, liquids, and solids. The elastic modulus appropriate for a longitudinal wave is related to the bulk modulus, or the modulus of incompressibility. Sound is transmitted through the air by compressional waves. An analog to a shear wave can be demonstrated by flipping a cord tied at the far end. Individual parts of the cord move up and down as the waves pass from one end to the other. This type of wave in the earth requires shear elasticity, which is possessed by solids but not liquids or gases. The elastic modulus appropriate for shear is the modulus of rigidity. If the transverse motion occurs in a single plane, the shear wave is said to be *polarized*.

Return to the waves produced by the stone dropped into the pond. The distance between two successive crests of the waves in the pond is the *wavelength* (Fig. 4-1a, b). The maximum displacement of a particle from an equilibrium position as the wave passes through it is the *amplitude* of the wave (Fig. 4-1b). The length of time it takes one wavelength to pass is the *period*. The reciprocal of the period, the number of wavelengths that passes a point per unit time, is the *frequency*. During one period, a particle moves in one direction, slows down, then moves back through its original, undisturbed position, continues back, slows down, then moves forward again through the equilibrium position. This is an expression of the *phase* of the wave. If two stones are dropped into different parts of the pond, their wavefronts will interfere with each other. If the wavefronts reach a particle such that both tend to displace the particle in the same direction, then through the equilibrium position at the same time, the wave interference is said to be constructive and in phase, resulting in a larger amplitude of the combined wave. If, however, one wave front displaces the particle in one direction, and the other wave front displaces it in the opposite direction at the same time, the interference is destructive, the waves are out of phase, and the resultant wave amplitude is less than the amplitudes of its constituent waves.

A wave passes out from a point source in the earth as an expanding sphere, the surface of which contains the energy of the wave at any given time (Fig. 4-1c). At a later time, the same energy passes through a larger spherical surface, with the energy per square centimeter of surface inversely proportional to the area, $4\pi r^2$, of the surface. The amplitude, then, is inversely proportional to the square of the distance from the source, the square of the radius, r, of the sphere. The reduction in amplitude of a wave with time or distance from the source is called *attenuation*. *Huygen's principle* states that every point on a wavefront is the source of a new wave that travels out from this point as a spherical surface also (Fig. 4-1c).

When a wave reaches the boundary of a medium with different elastic constants and a different density, it may continue into the second medium with a different speed and wavelength, it may be absorbed in the second medium, or it may reflect from the boundary with the same speed and wavelength as before. Commonly all three occur. In discussing these phenomena, it is convenient to use the concept of a *ray*, a line drawn in the direction the wave is traveling, such as line AS or BS in Figure 4-1c. In a homogeneous medium, the ray is a straight line.

A ray intersecting a surface between two media, such as the Moho discontinuity between the crust and the mantle, makes an angle with the normal to that surface called the *angle of incidence*. If the ray is reflected from the surface, it makes an *angle of reflection* that is equal to the angle of incidence (Fig. 4-1d). If a ray that is not normal to the boundary surface passes through into the second medium with a higher speed and different wavelength, the direction of the ray is *refracted* such that the angle between the ray and the normal to the surface is increased according to the equation:

$$\frac{\sin(\text{angle of incidence})}{\sin(\text{angle of refraction})} = \frac{\text{speed of wave in upper medium}}{\text{speed of wave in lower medium}}$$

This is *Snell's law* (Fig. 4-1d). If the medium through which the wave propagates produces a gradational increase in speed, the ray travels a curved rather than a straight-line path, as illustrated in Figure 4-5.

In seismic waves that pass through the earth, the longitudinal wave is called the *primary wave* or *P wave*, and the transverse wave is the *secondary wave* or *S wave*. The P wave arrives first, and because it is compressional, it can be transmitted into the air, thus making a sound. The S wave travels at a slower speed and thus arrives later, producing up-and-down and sideways shaking. The P and S waves together are called *body waves*, because they pass through the earth. A third type of wave, called a *surface wave*, is only observed close to the surface of the earth, analogous to the ripples on the pond after the stone was thrown in. The amplitude of surface-wave motion decreases with depth below the surface. Surface waves are further subdivided into *Love waves*, in which the wave motion is horizontal and transverse, and *Rayleigh waves*, in which particles move elliptically in a vertical plane oriented in the direction of wave propagation. Love waves are analogous to S waves in that they are transverse, shear waves. Rayleigh waves, on the other hand, can propagate at the surface of a liquid.

Earthquakes generate an array of P, S, and surface waves due to complexity in the strain-release pattern at the source and complexity in the earth materials through which the waves pass to reach the seismograph. These complexities include conversion of P waves to S waves or S waves to P waves as the wave passes into different media. One problem facing the geophysicist is how to use seismograms to work out the earthquake rupture process, based on various types of waves that pass through and are influenced by a heterogeneous Earth, the geological properties of which are not well understood.

To appreciate this problem, consider a music lover at sea, listening on a short-wave radio to a Beethoven symphony played by the New York Philharmonic Orchestra. The performance is magnificent, but what the listener hears on the short-wave radio is not. There is static, and the deep basses are barely audible at all, so that the listening experience is less than satisfactory. This is caused by (1) distortion of the music produced by the orchestra due to interference from both the transmitter and the radio waves between the transmitter and the ship and (2) distortion produced by the radio receiver on the ship. A third problem is that the listener is hard of hearing, and his ears ring! The radio-wave problem is analogous to the distortion of the seismic signal as it passes through heterogeneous earth materials whose properties are not well known. The problems with the listener's ears and the radio receiver are analogous to the limitations of the seismograph in recording faithfully the earthquake.

If the listener next hears a new symphony performed for the first time on unfamiliar musical instruments, we get a good idea of the geophysicist's problem. The listener knows how the Beethoven symphony ought to sound and is able to make a mental correction for the distortion. If the music and instruments are unfamiliar, then the listener does not know how the music is supposed to sound and is not sure how to compensate for the distortion. This is the difficulty facing the interpreter of a seismogram, who is not entirely sure which is signal and which is noise! The geophysicist attempts to get around this problem by designing source models that duplicate the natural seismic waves as closely as possible, producing a *synthetic seismogram*.

SEISMOGRAPHS

How do we measure the motion of the ground when we and our measuring devices are attached to the ground? We use the principle of inertia. If we set a glass of water on a piece of cardboard which rests on a table, we can quickly yank the cardboard out from under the glass without spilling the water. This is due to the inertia of the glass and its water content. The seismograph has a mass attached very loosely to the earth by means of a pendulum or spring (Fig. 4-2). A pen is attached to the mass such that it is in contact with a moving chart which is firmly attached to the ground. When the ground moves suddenly, the chart moves, but the mass and its pen remain more or less stationary, just as the glass and its water content remain stationary when the cardboard is pulled away.

However, this presents another problem. After the shaking stops, the pendulum and the spring will continue to move freely, making waves on the chart that are related to the characteristic oscillation rate of the pendulum or spring and not to seismic waves. Accordingly, the pendulum and spring must be damped in order to produce a record closer to that of the actual ground motion.

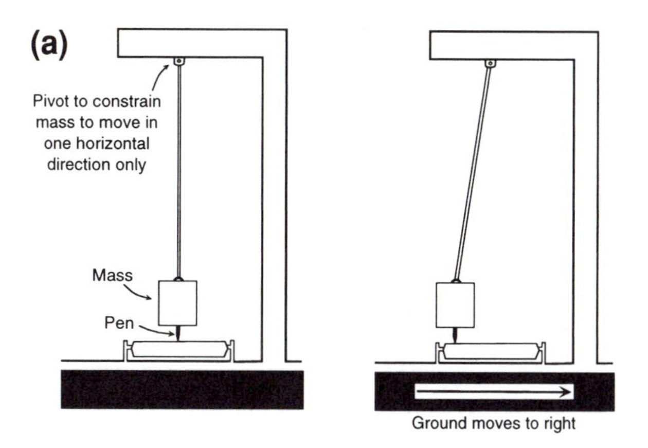

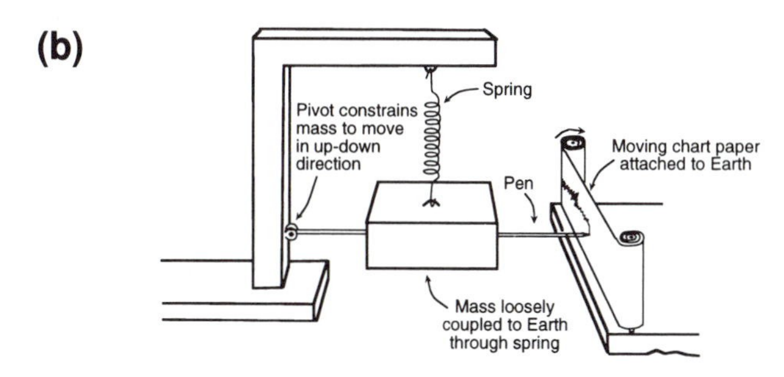

Figure 4–2. Seismographs. (a) A pendulum seismograph to record horizontal motion. When the frame moves to right, the chart moves with it. Because of inertia, the mass does not keep up with the motion of the frame, and the pen traces the difference between the positions of the mass and the frame on a moving chart. (b) A spring-mounted seismograph to record vertical motion. The frame and the chart move with the earthquake, and inertia in the spring allows the pen to trace the difference between motion of the frame and chart and that of the mass.

The natural period of the pendulum controls the periods of seismic waves that can be recorded best. Short-period seismographs (about 1 second) are useful in detecting local earthquakes and the initial body waves of more distant shocks, whereas seismographs with periods of 20 seconds are better at measuring the longer-period surface waves of more distant earthquakes. These longer-period seismographs are not particularly sensitive to the ambient noise of the earth's surface caused by wind, ocean waves, and human activity. The most common noise from these sources has a period of about 6 seconds.

The seismographs described above to illustrate the principles of recording seismic waves are now museum pieces due to electronic improvements in the last few decades. The pen on the old-style seismograph produced resistance against the chart, and occasionally it would run out of ink. The system was improved by having the record made by a light beam on photographic film. Voltage can be generated across a coil attached to the mass of the pendulum which moves through a magnetic field generated by magnets attached to the ground. The system generates an electric current that is recorded on magnetic tape. The signal may be magnified thousands of times before it is recorded. The signal can also be recorded digitally so that the complex seismic signal may be analyzed by computer to break out the various interfering waves of differing wavelength and period. Digitization of the signal has led to construction of a new array of *broad-band seismographs*, so named because they can record and study a much broader spectrum of waves of the earthquake symphony: short-period and long-period, low-amplitude and high-amplitude.

To describe the ground motion completely, three separate seismographs are needed; two pendulum instruments operating at right angles to each other to record horizontal motion, and a spring-mounted mass to record vertical motion. Love waves, which shake the ground from side to side, are recorded only by horizontal components of a seismograph station. Rayleigh waves, which cause a particle to move both vertically and horizontally in an ellipse in the vertical plane, are also recorded on the vertical component, as is the P wave.

Many seismograph networks consist of seismographs recording only the vertical component of motion over a restricted, narrow portion of the frequency spectrum, say 1 to 10 Hertz (cycles per second). These instruments are useful for picking the first arrivals of seismic waves that are important in locating earthquakes. Many of these simple, inexpensive instruments can be installed for the same price as a smaller number of broad-band, three-component instruments. There is a trade-off for the same investment: a more geographically extensive coverage with the vertical-component instruments vs. more complete information about a given earthquake and the earth materials through which the seismic wave passes obtained from the three-component, broad-band instruments.

Modern instruments can detect ground displacements as small as 10^{-8} cm, far smaller than the ambient noise at the earth's surface caused by automobiles, wind, and ocean waves. The seismograph left by the Apollo astronauts on the surface of the moon, where there is no such noise, is sensitive enough to record seismic waves generated by a 1-kg meteorite striking anywhere on the moon (Press and Siever, 1982). The moon, however, has much lower attenuation of seismic waves than does the earth, so even a perfectly quiet Earth could not match that recording feat on the moon.

A *strong-motion accelerograph* is a low-magnification seismograph designed especially to record the strong shaking of the ground that is of particular interest to engineers in their studies of man-made structures such as buildings and dams. Such strong ground motion is too large for ordinary seismographs because it causes them to go off scale. Rather than measuring displacements, accelerographs measure ground acceleration as a percentage of g, the acceleration of gravity. These instruments do not operate continuously, but are triggered by the arrival of the first strong waves. Such instruments, which are relatively cheap and can be installed in large numbers, are particularly valuable for evaluating the response of near-surface materials to earthquake waves and to study the vibration patterns in man-made structures. Some specially instrumented buildings, for example, may contain several strong-motion accelerographs at different plan locations and floor elevations. The records produced by such instruments are analyzed by structural engineers in order to improve the design of similar structures elsewhere.

LOCATING AN EARTHQUAKE

The *epicenter* is that point on the surface of the earth that is directly above the *focus*, that place within the earth where earthquake rupture starts. How is the epicenter located? Prior to seismographs, the epicenter was generally assumed to be the place where damage was the greatest. In many cases, however, the degree of damage is more strongly influenced by near-surface ground conditions and by the progression of the fault rupture than by proximity to the focus. Seismographs provide a better way to locate the epicenter and focus.

Assume as a first approximation that the speed of elastic waves in rock varies only with depth. The travel time between an earthquake focus and a seismograph is a function of distance, and graphs plotting distance against travel time are available for both P and S waves. Because the S wave travels more slowly, the difference in arrival time of the P and S wave at the seismograph increases with distance from the earthquake. This difference in arrival time can be used to determine how far the seismograph is from the earthquake, even though a single-component seismograph is unable to tell the direction from which the wave came. The distance is shown on a map as a circle of known radius centered on the seismograph a.

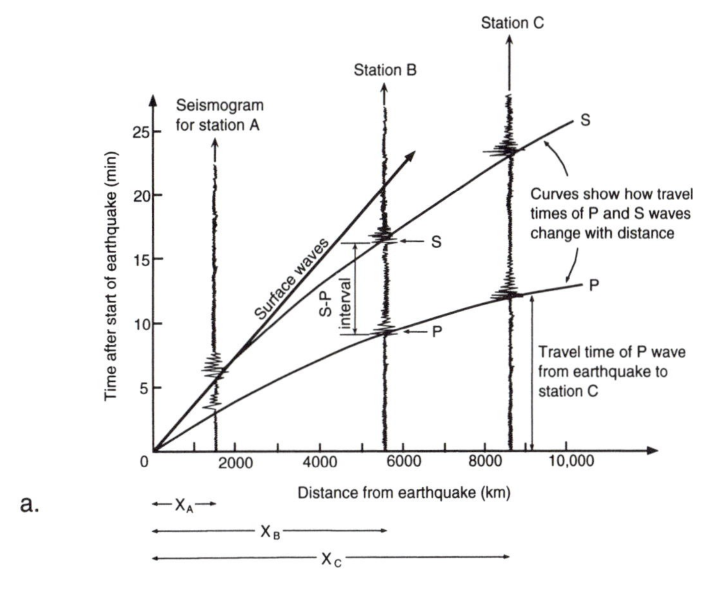

b.

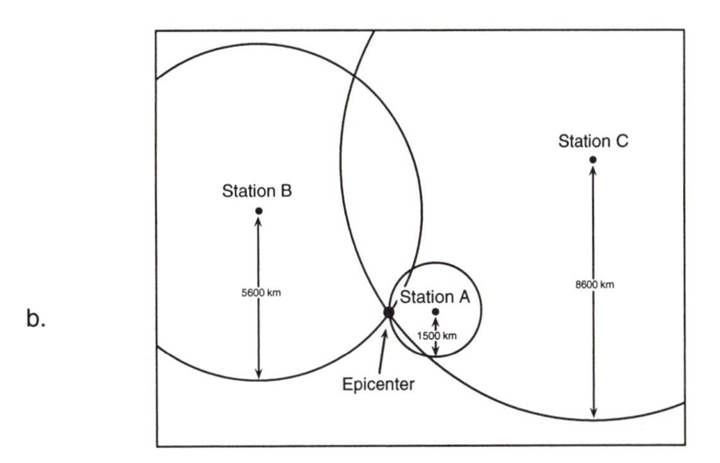

Figure 4–3. (a) Seismograms of the same earthquake as recorded by seismographs located at A, B, and C. The time interval between the passage of the P wave and that of the S wave allows the calculation of the distance between the seismograph and the earthquake, because the speeds of P waves and S waves are known. (b) This is shown on a map as the radius of a circle with the seismograph at the center. The earthquake is located where the three circles intersect. Modified from Press and Siever (1994).

(Fig. 4-3b). It takes three single-component seismographs to locate the earthquake based on where the circles drawn from each seismograph intersect (Fig. 4-3). However, a modern, multiple-component seismograph may yield the same information by calculating the direction from which the signal arrives in addition to the distance.

The speed of seismic waves in the earth varies horizontally as well as vertically. Assumptions about the speed of seismic waves can be constrained by geologic structure and distribution of rock units that transmit seismic waves at different speeds.

SEISMIC WAVES, FAULT ORIENTATION, AND SENSE OF SLIP

Earthquakes occur on faults that are assumed to be oriented in a plane of high shear stress (not neces-

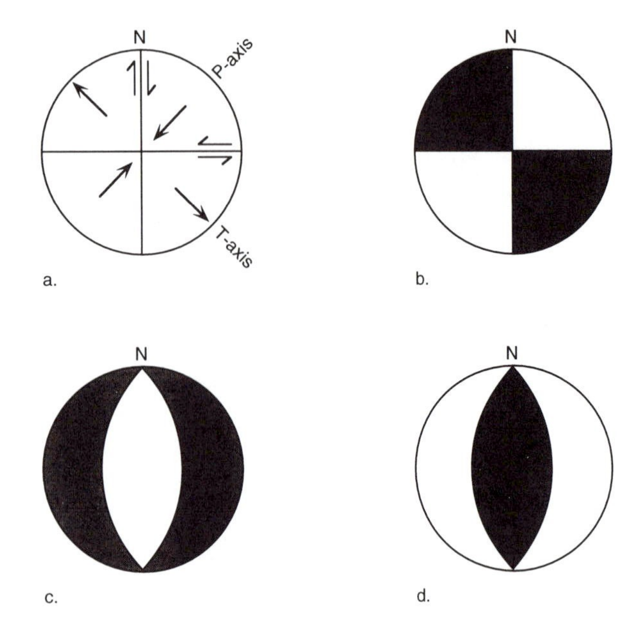

Figure 4–4. Double-couple fault-plane solutions. (a) Map of a north-trending right-lateral strike-slip fault and an east-trending left-lateral strike-slip fault. Motion on either fault causes the northeast and southwest quadrants to move toward an earthquake source (center of circle) and the northwest and southeast quadrants to move away. (b) Lower-hemisphere stereonet projection of focal mechanism determined from earthquake waves received from the earthquake source at (a) at various seismograph stations. Quadrants receiving compressional first arrivals are shaded. (c) Focal mechanisms of earthquake on a north-trending normal fault. (d) Focal mechanism of earthquake on a north-trending reverse fault.

sarily the maximum shear stress) with respect to the maximum and minimum principal compressive stresses (σ_1 and σ_3, respectively). As discussed in Chapter 2, there are two conjugate planes of maximum shear stress for any pair of values for σ_1 and σ_3 where $\sigma_1 > \sigma_3$. Rupture may occur on or close to one of these shear planes if the material is structurally isotropic, that is, if there are no preexisting planes of weakness.

However, the crust of the earth is riddled with preexisting planes of weakness. Slip may occur on a plane of weakness such as a fault even if the stress field has rotated from its position when the fault first formed in flawless rock (cf. Fig. 2-17). Because nearly all earthquakes occur on faults that have moved previously, earthquakes do not provide direct evidence for the orientation of principal stresses, but instead provide evidence for the orientation of axes of strain.

The view that deformation at an earthquake source can be represented by slip on a fault plane leads us to look for evidence for that fault plane in the earthquake signal. We are able to do this because the seismograph can distinguish between a wave where the first motion is a compression and a wave where the first motion is a rarefaction or dilatation. Consider the north-striking right-lateral strike-slip fault in Figure 4-4a. If your station is in the southeast or northwest quadrant with respect to the source, the first motion of a P wave would be a compression, in which the recorded signal would be an upward displacement (Fig. 4-5). However, if you were in the northeast or southwest quadrant, the first motion would be a dilatation, and the signal would be a downward displacement. If your station were on strike with the fault, that is, north or south of the earthquake focus, P waves moving toward the station would be nearly canceled out by P waves moving away. If your station were east or west of the earthquake, perpendicular to the direction of fault movement, P waves would not propagate in those directions, and so the signal would be of low amplitude, difficult to distinguish from that received at stations on strike with the fault.

If we now consider the P waves from an earthquake rupturing an east-striking left-lateral fault (Fig. 4-4a),

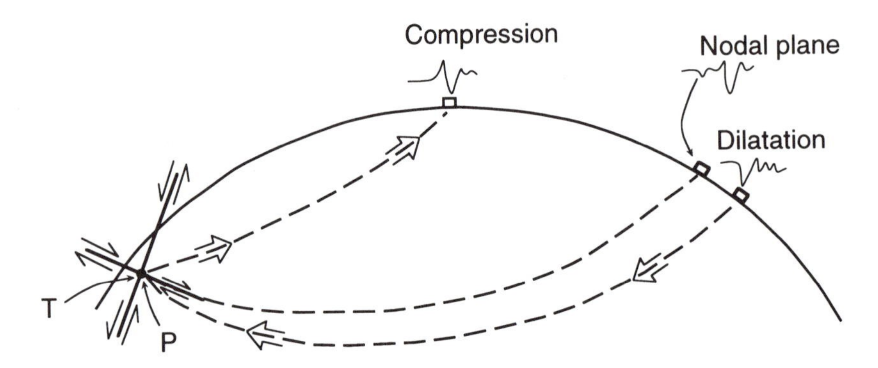

Figure 4–5. First-motion of earthquake, including a compression, dilatation, and nodal-plane solution.

we find that the radiation pattern is exactly the same as that from the north-striking right-lateral fault. First motions of P waves are, therefore, unable to discriminate between these two fault orientations and senses of slip, although they can distinguish these from other senses of slip and other fault orientations. This is the *double-couple assumption;* the possible fault orientations are limited to two planes perpendicular to one another. The planes are bisected by the *P axis* and the *T axis*. These axes are defined kinematically by fault slip. Thus, they are principal strain axes and not principal stress directions σ_1 and σ_3. At the instant of rupture, rock in the quadrants including the P axis as a bisector moves *toward* the earthquake source, which would cause rock to move *away* from the seismograph, producing a *dilatation* (Fig. 4-5). At the same time, rock in the quadrants including the T axis as a bisector moves *away* from the source and *toward* the seismograph, producing a *compression*. As shown on Figure 4-5, the location of the station on a stereonet is based on the path of the wave (curved because it is refracted by rock that transmits seismic waves at differing speeds) to the station from the source. First motions from a large number of seismographs at different azimuths and distances from the focus can locate the two possible fault planes on a stereonet, based on the distribution by quadrants of compressional and dilatational first arrivals. An example of a fault-plane solution is given in Figure 4-6.

Because the distribution of P wave first motions defines two possible nodal planes, either of which could be the source fault, other information is needed. In Figure 4-6, the Tangshan earthquake occurred on a fault striking NE-SW in the North China basin, thus the NE-trending nodal plane with a right-lateral strike-slip displacement is the more likely solution. In other cases, the aftershock pattern is linear and subparallel to one of the nodal planes. In some cases, slip may occur on faults parallel to *both* nodal planes. The 1927 Tango, Japan, earthquake was caused by rupture of two strike-slip faults at right angles, one left-lateral and one right-lateral (Fig. 8-33a).

Notice that the slip direction for each nodal plane can be selected by constructing a plane perpendicular to the two nodal planes, thereby including the P axis and T axis. The trend and plunge of the slip direction will be the intersection of this plane with a nodal plane.

To understand the earthquake source, modern seismology uses not only P wave first motions, but also the amplitudes of direct P and S waves, waves reflected many times from the surface of the earth, and surface waves. With data from seismographs transmitted to a central source by satellite, an earthquake can be located in a few minutes to a few hours, and a fault-plane solution can be determined a few minutes to a few hours after the earthquake occurs.

Although the double-couple assumption accounts for the pattern of strain release in most earthquakes, for some, it does not. Seismic waves would be produced by an implosion, an isotropic radial collapse toward the source region that might accompany a sudden mineral phase transition. If this type of source is present at all, observations suggest that it must be small relative to double-couple radiation patterns. Another type of source mechanism involves uniform inward or outward motion on a plane accompanied by shortening or extension on the plane, a source called a *compensated linear vector dipole (CLVD)*. CLVD sources may be explained by opening of tensile cracks due to high groundwater or magmatic pressure or by contractional cooling. This mechanism is probably limited to depths shallow enough for tensile fractures to form, and to magma chambers underlying regions of active volcanism such as Iceland and the Mammoth Lakes area of eastern California. The most attractive explanation of non-double-couple radiation patterns is slip along curved fault surfaces, multiple subevents on faults not parallel to one another, or shear-strain release in a volume of rock rather than a fault plane (Frohlich, 1994).

Still other earthquakes are caused by massive landslides, such as the M = 5.1 earthquake that occurred at the beginning of the eruption of Mt. St. Helens on May 18, 1980 (Kanamori and Given, 1982). This event involved the collapse and slumping away of the north side of the volcano, partially unroofing a shallow magma chamber. A M = 7.2 earthquake on the Grand Banks off the east coast of Canada in November, 1929 may have been caused by a submarine landslide which produced a deep-sea turbidity current, observed because it snapped submarine cables (Hasegawa and Kanamori, 1987; cf. Chapter 12). In both these cases, the focal mechanisms suggested a single couple rather than a double couple, reflecting the fact that the upper plate (the slide) was not "tied" to the rest of the earth, as is normally the case with the two sides of a fault. (A major controversy in the early days of seismology, between Harold Jeffreys and B. B. Galitzin, revolved around whether the seismic signals associated with a massive landslide in 1911 in the Pamir of central Asia could be explained adequately by the slide itself or whether they required an underlying triggering earthquake.) Finally, rare geologic events such as meteor impacts and cavern collapses occasionally produce seismic waves, sometimes damaging, that are obviously unrelated to faulting.

Different types of sources are best analyzed by de-

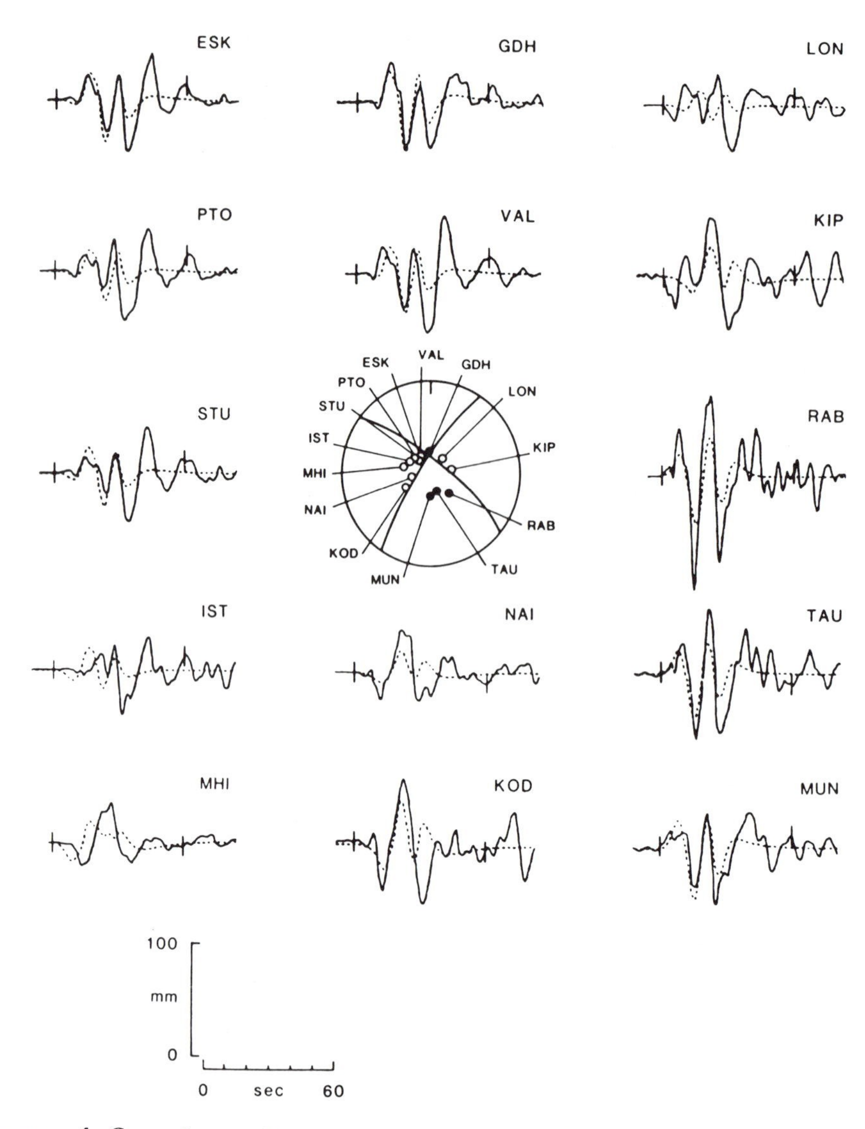

Figure 4–6. Observed P-wave seismograms (solid lines) of the Tangshan earthquake in northeast China and the theoretical seismograms (dotted lines) using a single-point-source model. Also shown is the fault-plane solution plotted on lower-hemisphere, equal-area stereonet projection. The circles on the stereonet are located based on the takeoff direction from the source. Compressional first-motions (upward on seismogram) are solid circles, dilatational (downward) are open circles. Upward first-motions are clearly represented at stations MUN, TAU, and RAB: downward first-motions at stations MHI, LON, and KIP. Tangshan fault strikes NE, so the NE-trending nodal plane is the preferred rupture surface. After Náběelek et al. (1987).

termination of seismic moment (see the following discussion) and the use of *centroid moment tensor (CMT)* solutions (Dziewonski and Woodhouse, 1983), which take into account the relative amplitudes of seismic waves leaving the earthquake source in different directions. The moment tensor is proportional to that part of the strain tensor released seismically.

Three-component seismograph stations, because their components are mutually perpendicular, permit the investigation of *shear-wave splitting,* or *polarization,* analogous to the polarization of light in anisotropic minerals (Fig. 4–7). Li et al. (1994) analyzed three-component records of 74 earthquakes in the northern Los Angeles basin, California, and showed that the first shear waves to arrive are polarized in an approximate N-S direction, a phenomenon they attribute to oriented microcracks in the seismogenic layer (Fig. 4–7). These microcracks should be in the plane of extension normal to σ_3, the least principal compressive stress, and parallel to the greatest horizontal stress. Shear-wave splitting thus is a method to determine *in situ* stress (discussed further in Chapter 5).

MAGNITUDE SCALES

With the development of seismographs, it became clear that there was a need for a quantitative measure of the size, or *magnitude,* of an earthquake based on the seismogram rather than on the amount of damage done. But what part of the earthquake signature should be used in determining its magnitude? Different magnitude scales measure different parts of the radiation pattern of earthquakes, and one magnitude scale is not necessarily right and another wrong. There is no single number that describes everything about an earthquake, just as no single number describes the quality of wine or the beauty of a sunset. The discussion below describes the more important magnitude scales now in use based on those characteristics of the earthquake that each measures.

Following K. Wadati of Japan, who worked on the idea of an earthquake scale in 1931, Charles Richter developed the first magnitude scale in 1935. Richter's magnitude is the logarithm to the base 10 of the maximum seismic-wave amplitude, in thousandths of a millimeter, recorded on a special type of seismograph (Wood-Anderson seismograph) at a distance of 100 km from the earthquake epicenter. Richter took into account the decreased amplitude with increased distance from the earthquake (increased time between P and S waves) such that a Richter magnitude could be calculated if both the amplitude and the distance were known. The Richter magnitude is known as the local magnitude and is referred to as M_L.

Richter designed the scale so that magnitude 0 corresponded approximately to the smallest earthquakes

(a)

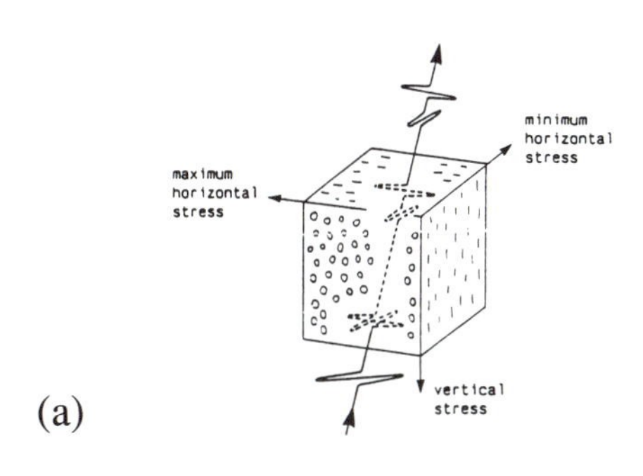

(b)

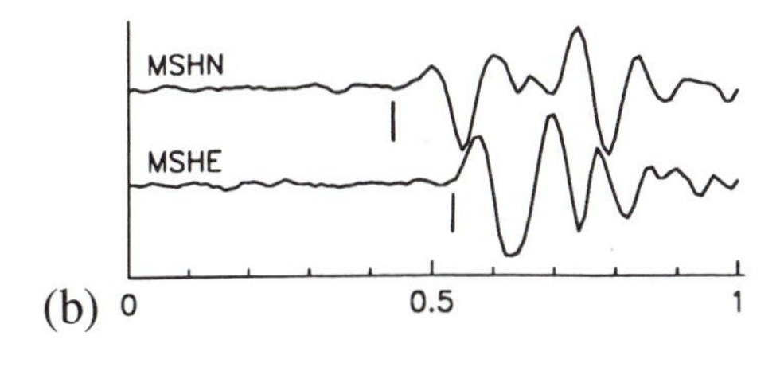

(c)

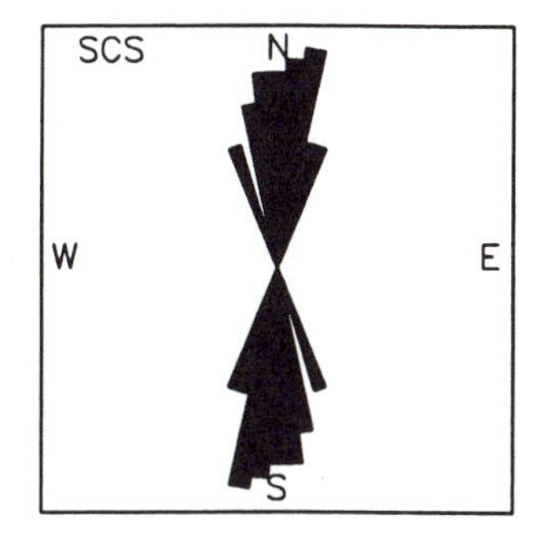

Figure 4–7. Shear-wave splitting. (a) A shear wave traveling on a path within 35° of the vertical commonly splits into two components generally polarized at right angles to each other (analogous to double refraction of polarized light in birefringent minerals). The polarization of the slower component is parallel to the direction of minimum horizontal stress, a phenomenon attributed to microcracks that preferentially form perpendicular to the minimum stress direction. (b) Time-expanded seismograms of horizontal components from station SCS, part of an array in Los Angeles basin, California. Vertical lines show difference in arrival times of shear waves from the same earthquake. (c) Equal-area Rose diagram of distribution of polarization of fastest shear waves from 74 events in the Los Angeles basin array. (a) from Crampin and Lovell (1991), (b) and (c) from Li et al. (1994).

then being recorded, but with more sensitive instruments, negative magnitudes are possible. $M_L = -2$ roughly reflects the energy of a brick hitting the ground when dropped from a tabletop (Bolt, 1993). In principle, there is no upper or lower limit to the magnitude scale. What does an increase of one unit in magnitude signify? Because different vibrational periods dominate at different magnitudes, and because part of the tectonic energy of an earthquake is taken up as heat due to friction and as permanent deformation, the increase in energy for an increase of one unit of magnitude is roughly 30-fold and is different for different magnitude intervals.

Richter recognized that his scale was not directly tied to fundamental physical parameters, and he was careful to say that it represented the "size" of an earthquake at its source rather than a specific parameter such as energy. He wanted quantitatively to separate large, moderate, and small local earthquakes rather than having to depend on "the uncertainties of personal estimates or the accidental circumstances of reported effects." Different stations commonly assign slightly different Richter magnitudes to the same earthquake, which is not surprising considering the complexity of the rock formations through which the seismic waves must pass, as well as the fact that the causative fault may not release the same amount of stored elastic strain energy in all directions.

The Wood-Anderson seismograph, which is the instrument used in determining M_L, has a natural oscillation period of about 0.2 seconds, and waves of longer period are increasingly diminished on the record even if they are present in the ground. Thus a large earthquake in Japan, for example, would hardly be seen on a Wood-Anderson seismograph in Europe because the high-frequency seismic waves would have been attenuated (or damped) by the long travel path, and only long-period waves would remain. Similarly, even a large nearby earthquake would be dominated by long-period seismic waves that would not be appropriately measured by a local Wood-Anderson instrument.

Thus the need arose for a magnitude scale appropriate for use with larger earthquakes and distant earthquakes (*teleseisms*), with their dominance of long-period energy. One such scale, the *surface-wave magnitude* or M_s scale, is obtained by measuring the largest amplitude in a surface-wave train with a period close to 20 seconds. Another is the *body-wave magnitude* or m_b, which is based on the maximum amplitude of teleseismic P waves with a period of about 1 second. Both of these magnitude scales were initially calibrated so that they were thought to be compatible with M_L in the transition range from small to large earthquakes, although they are, in fact, measuring different parameters that are not entirely comparable to one another.

None of the magnitude scales described above is very successful at describing earthquakes of large source dimensions. For example, the 1906 San Francisco earthquake, the 1964 Alaskan earthquake, and the 1960 Chilean earthquake all had a surface-wave magnitude $M_s = 8.3$. However, the San Francisco earthquake resulted from a much smaller energy release than the other two, based on the relative size of the affected areas and the associated tectonic changes. We say that the magnitude scales described above are *saturated* for large earthquakes; they cannot distinguish these earthquakes by size based simply on the amplitude of a particular seismic wave on the seismogram. The body-wave magnitude scale becomes saturated at m_b = about 6, and the surface-wave magnitude scale becomes saturated at M_s = about 7.3.

Another way of visualizing the problem of measuring the size of an earthquake based on waves of a particular period is to compare the source dimensions with the period and speed of earthquake waves. A wave with a 20-second period traveling at 4 km/sec would have a wavelength of 80 km. An earthquake with source dimensions smaller than this (the 1994 Northridge earthquake source dimensions were about 16 km along strike and 14 km down-dip) would appear as a point source, and so 20-second waves provide a good measure of its size. On the other hand, 20-second waves measured by M_s would not be useful in measuring the size of a large earthquake with a rupture length of several hundred kilometers. Seismic waves from different parts of the fault source would arrive at the seismograph at slightly different times and be out of phase, so that M_s would underestimate the size. If the source dimensions are small compared with the wavelength, then the seismic waves from all parts of the fault will not interfere with one another, whereas if the wavelength is short compared with the source dimensions, waves from different parts of the source would interfere destructively.

A better measure of the size of a large earthquake is its *seismic moment*, M_o.

Consider two people on opposite sides of a large table that they want to move. If they stand on opposite sides of the table and push toward each other with the same force, the table will not move. If, however, one person pushes in the same direction, but from one end of the table, and the other pushes from the other end, the table will rotate, producing a *couple*. The reason the table rotates is that the opposing forces, although equal in magnitude, are separated from the axis of rotation of the table by a *moment arm*. The *moment* is one of the forces multiplied by the length of the moment arm, which is the distance between the forces measured perpendicular to the direction of the force. The units of measure would be force times length, with force measured in newtons (1 newton, or 1N, is that force that gives a mass of 1 g an acceleration of $1 m/s^2$) and length of moment arm measured in meters. (Moment can also be measured in dyne-cm.)

A rupture along a fault also involves equal and opposite forces that produce a couple. The *seismic moment* $M_o = \mu uA$, where μ is the shear modulus of elasticity (taken for most moment calculations as 3×10^{10} Nm^{-2} for the crust and 7×10^{10} Nm^{-2} for the mantle), and u is the average slip over the ruptured segment of a fault with area A. M_o measures the energy radiated from the entire fault rather than an assumed point source, and it is independent of the frictional energy that is dissipated during faulting.

The seismic moment is measured from seismograms using very long-period waves for which even a fault with a very large rupture area appears as a point source. Aki (1966) was the first to measure earthquakes in this way in his study of the 1964 Niigata, Japan, earthquake. Because seismic moment is a measure of strain energy released from the entire rupture surface, a magnitude scale based on seismic moment most accurately describes the size of the largest earthquakes. Hiroo Kanamori designed such a scale, called a *moment magnitude,* or M_w, scale which is related to seismic moment as follows:

$$M_w = \frac{2}{3}\log_{10} M_o - 6.0, \text{ where } M_o \text{ is in Nm.}$$

The moment may also be estimated from the geology using the surface-rupture length and the average slip associated with the earthquake measured along the fault. The depth of aftershocks times the length of surface rupture gives the area A that has undergone rupture. (A caveat here is that the aftershock zone may represent a considerable areal expansion of that part of the fault that ruptured during the mainshock.) Determining moment from surface geology works best for earthquakes on strike-slip faults with rupture lengths measured in hundreds of kilometers, where surface slip is likely to be as much as slip at the mainshock. It works less well for earthquakes on strike-slip or normal faults with short rupture length where surface slip is generally less than mainshock slip, and the length of surface rupture may be much less than that at mainshock depths. Geologic estimates are least useful for reverse faults where the seismic rupture may not reach the surface at all (blind fault), or may be exposed at the surface as a warp. However, the geodetic dimensions of the warp at the surface can be used to model the blind fault (cf. Chapters 5 and 10).

Using the moment-magnitude scale, the 1906 San Francisco earthquake has $M_w = 7.7$, and the 1964 Alaskan earthquake has $M_w = 9.2$. The 1960 Chilean earthquake has $M_w = 9.5$ and is the largest earthquake of the twentieth century. A fault rupture extending all the way around the world would have $M_w = 10.6$, and if this rupture somehow produced brittle fracture extending through the center of the earth (impossible due to the nonbrittle mantle and liquid outer core), its M_w would be 12.1 (assuming displacements comparable to those recorded in $M_w = 9$ to 9.5 earthquakes).

Moment magnitude is being increasingly used worldwide for moderate and large earthquakes because (1) it can be quickly calculated with modern instruments and computerized analytical techniques, (2) it is tied directly to physical parameters such as fault area, fault slip, and energy, rather than to amplitudes of particular seismographic records in particular frequency bands, (3) it can be independently estimated by geodetic, field-geologic and by seismographic methods, and (4) it is the only magnitude scale that estimates adequately the size of the source of very large earthquakes.

INTENSITY SCALE

Intensity is a measure of the violence of earthquake shaking at a given site, which is calculated by the amount of damage done to structures, the degree to which the earthquake was felt by individuals, and the presence of secondary effects such as landslides, liquefaction of soils, and ground cracking (Richter, 1958). A given earthquake has many intensities, usually highest near the area of maximum fault displacement (not necessarily the epicenter) and successively lower farther away. A map showing intensities resulting from an individual earthquake relies on reports on the intensity of shaking from a large number of locations. An intensity map is contoured in *isoseismals,* lines of equal intensity. The central area of violent shaking and great damage, even in areas underlain by bedrock, is called the *meizoseismal region*.

Following work in the nineteenth century by Robert Mallet, M. S. de Rossi, and François Forel, an intensity scale was proposed by Giuseppi Mercalli in 1902, and it has since undergone several revisions, resulting in the Modified Mercalli scale (MM) now in use. The scale goes from I to XII, using Roman numerals to avoid confusion with magnitude scales. Other intensity scales have been designed by the Japan Meteorological Agency (JMA scale) and by the Russians and the Chinese. These scales are in part based on traditional local construction practices and may be more appropriate for estimating the intensity of ground shaking in those countries than scales based on construction practices elsewhere in the world.

Figure 4-8 is the intensity map of the Loma Prieta, California, earthquake of October 17, 1989. Each reporting locality is shown by a number that records the intensity at that locality. Isoseismals of this earthquake are elongated NW-SE, parallel to the strike of the fault source, the zone of aftershocks shown in Figure 3-2, and the regional structural grain. Note that there are data points that do not fit the isoseismals, such as intensity IX at Oakland and at the north tip of the San Francisco Peninsula. These areas experienced the collapse of the double-decker Nimitz Expressway at Oakland and the severe damage to the Marina District in San Francisco. The unusually severe damage at these localities was due in part to ground conditions: liquefaction of artificial fill (Marina District) and amplified shaking of soft sediments (Nimitz Expressway). Another factor may have been a focusing of waves reflected from the Moho discontinuity at a critical distance from the mainshock; other areas around San Francisco Bay closer to the source and with similar ground conditions sustained much less severe damage. Intensities were only VI in nearby ar-

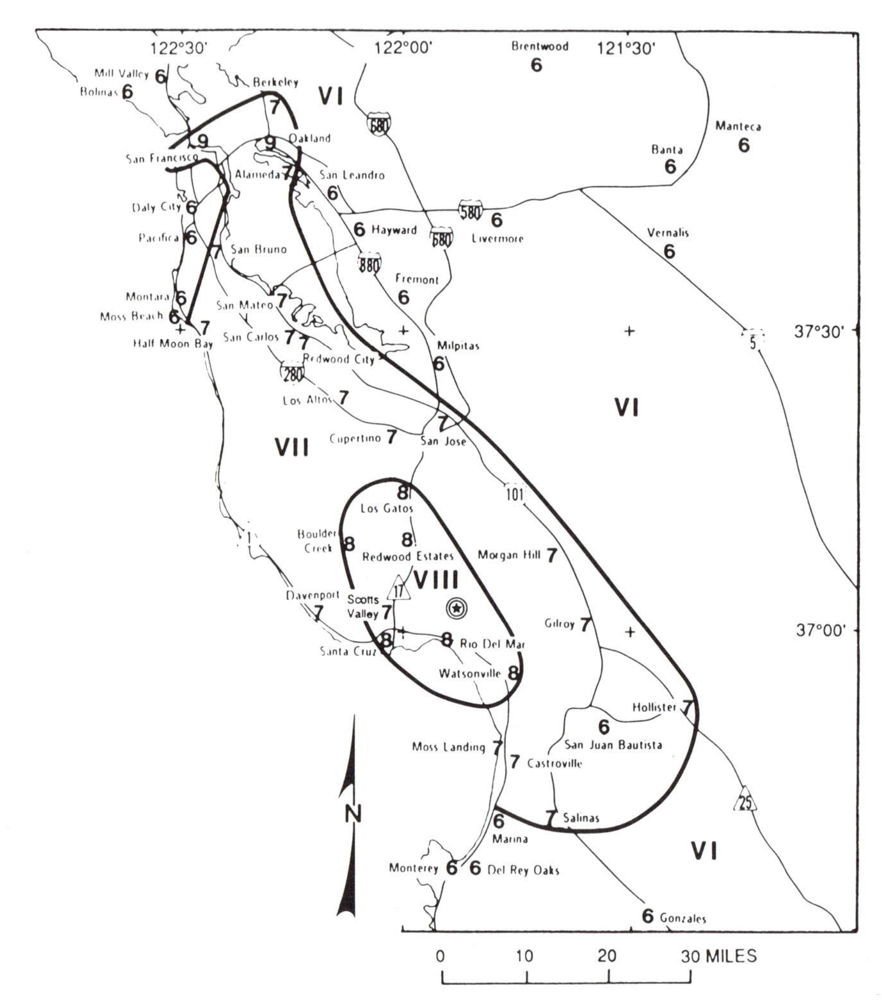

Figure 4–8. Distribution of modified Mercalli intensity for the Loma Prieta, California, earthquake of October 17, 1989. Numbers show observed intensity values; Roman numerals show intensity levels between isoseismals. Epicenter is located by circled star. Isoseismals are elongated NW-SE. Local intensities of 9 at San Francisco and Oakland were due to local conditions where seismic waves were amplified. From Plafker and Galloway (1989).

eas underlain by bedrock. Intensity VI would be considered as more representative of the earthquake for that part of San Francisco Bay area region as a whole.

Isoseismal maps are the only quantitative scale possible for earthquakes recorded before the development of seismographs: accordingly, it is useful to correlate MM intensity maps of an earthquake with its source size and, thus, with its magnitude, taking into account variability in ground conditions and focal depth (for example, see Table 4-1). For preinstrumental earthquakes, one can take the meizoseismal zone as an indicator of the area of fault rupture. With assumptions about amount of slip, a moment magnitude can be estimated. Magnitudes of preinstrumental earthquakes in the Table of Historic Earthquakes with Surface Rupture (Appendix) were estimated in this way.

LIMITATIONS OF SEISMICITY DATA

Maps of instrumental seismicity sometimes take on a life of their own, particularly when earthquake epicenters are shown on the same map as active surface faults or other geological data. Instrumental locations prior to the 1930s were relatively poor due to the primitive nature of the instruments and too few stations. The modern teleseismic age began with the Worldwide Standardized Seismograph Network (WWSSN), established in the early 1960s, to detect underground nuclear tests; 120 stations were operating in 60 countries by 1969. Each station consisted of three short-period seismographs and three long-period seismographs, and the entire network was tied together by very accurate clocks. Whereas the poor locations of epicenters on early global seismicity maps resulted in diffuse seismicity patterns around the Pacific, in the Mediterranean, and across southern Asia, the WWSSN seismicity maps sharply define tectonic plate boundaries, including the mid-ocean spreading centers (cf. Fig. 1-2). These epicenter maps and the accompanying fault-plane solutions were a major contributor to the plate-tectonics revolution of the earth sciences in the 1960s, as summarized by Isacks et al. (1968).

A similar development took place at the local level in areas with regional seismograph networks. Compare, for example, seismicity maps of southern California for 1932 (Fig. 4-9) and for a 12-month period in 1990-91 (Fig. 4-10). Both maps were prepared at the California Institute of Technology. Epicenters on the 1932 map are based on records from only seven seismographic stations. Epicenters on the 1990-91 map are based on records from 220 stations. The larger number of earthquakes mapped in 1990-91 does not reflect an increase in seismic activity but, instead, reflects far greater sensitivity of the newer seismographs and the ability to process many more records using high-speed computers. The 1932 map shows a rather diffuse scatter of events, contrary to the expectation at that time that the network would show alignments of epicenters along the active faults of California. There is a concentration of events on the San Jacinto fault which is probably real, but the lineup of events between the San Jacinto and Elsinore faults is probably an artifact of poor location of epicenters.

In contrast, the 1990-91 map shows many more events, even though this reporting period was not particularly active. Like the 1932 map, the 1990-91 map

Table 4-1

Approximate Relation between Richter Magnitude (M_L) and Maximum Intensity. After Gere and Shah (1984).

Richter Magnitude	Maximum MM Intensity	Typical Effects
2.0 and under	I-II	Not generally felt by people.
3.0	III	Felt indoors by some people; no damage.
4.0	IV-V	Felt by most people; objects disturbed; no structural damage.
5.0	VI-VII	Some structural damage, such as cracks in walls and chimneys.
6.0	VII-VIII	Moderate damage, such as fractures of weak walls and toppled chimneys.
7.0	IX-X	Major damage, such as collapse of weak buildings and cracking of strong buildings.
8.0 and over	XI-XII	Damage total or nearly total.

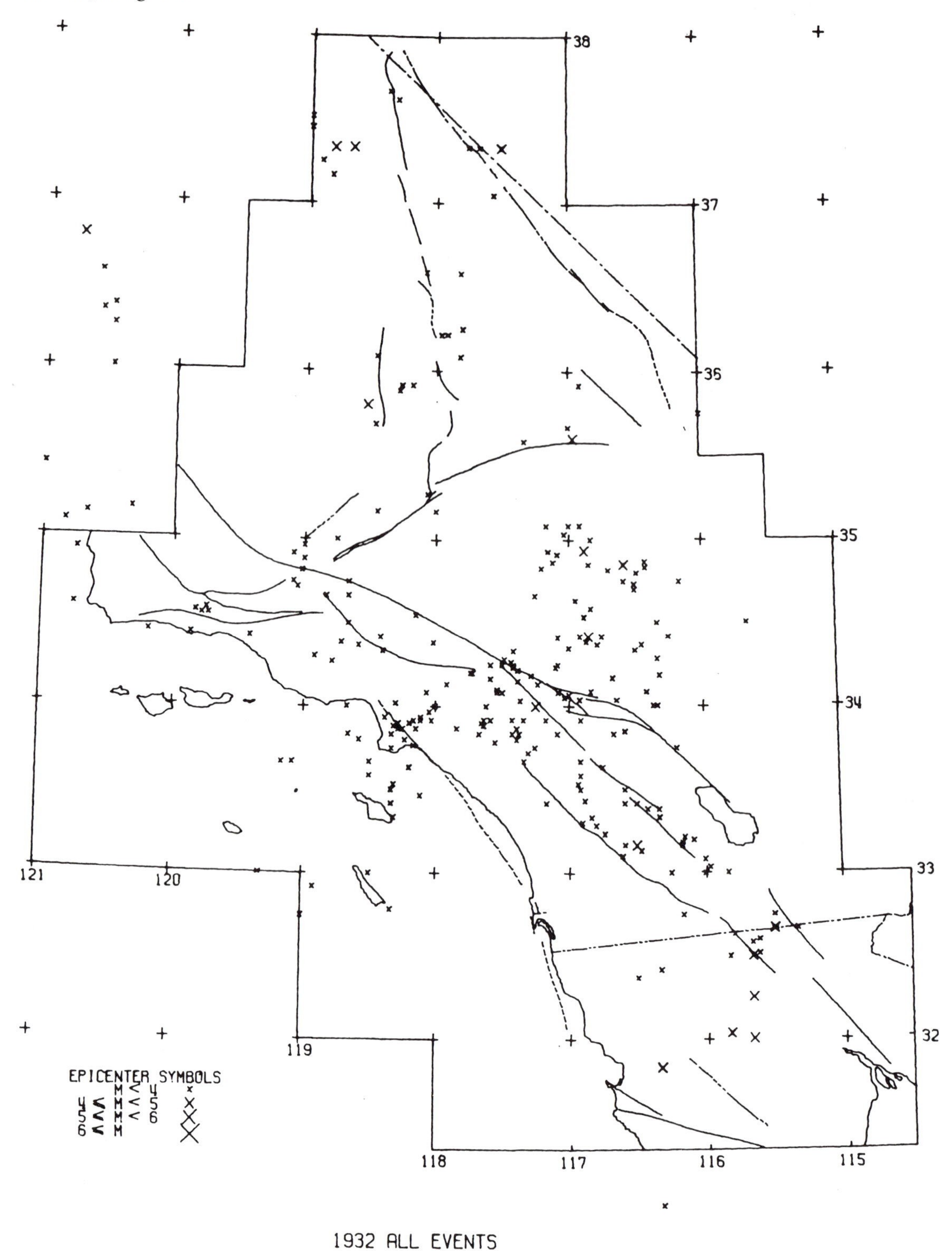

Figure 4–9. Seismicity of southern California during 1932 based on seven seismograph stations. From Hileman et al. (1973).

shows a diffuse distribution of epicenters, but, unlike the earlier map, several geologic structures are clearly marked by linear trends of epicenters. In particular, the San Jacinto fault zone in southern California and the San Andreas fault in central California are well defined by seismicity. A diffuse pattern persists in the offshore area, the Mojave desert, and the Ventura and Los Angeles basins. Unlike the diffuse patterns on the

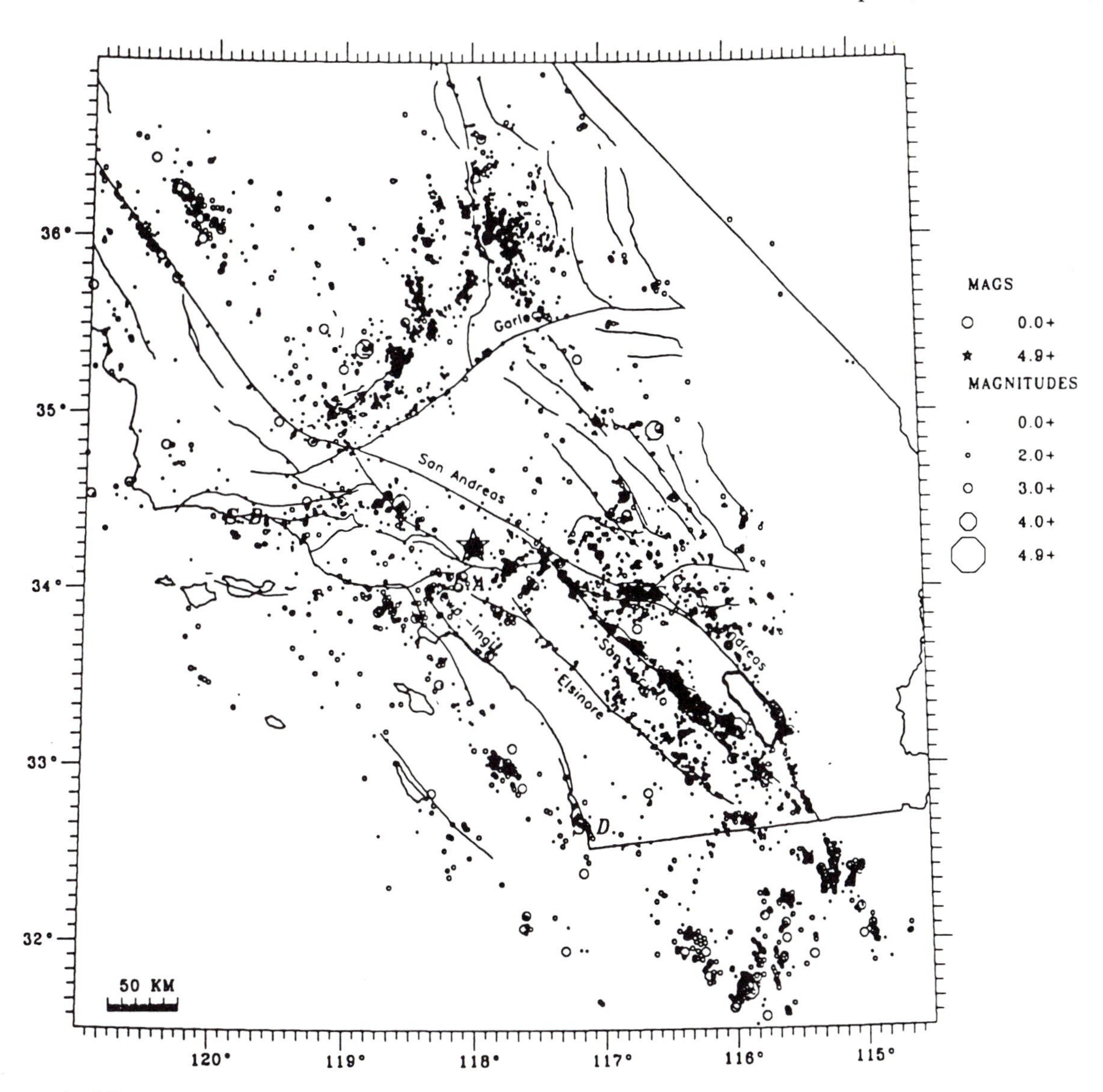

Figure 4–10. Seismicity of southern California from October, 1990 to September, 1991, based on 220 stations. From Clayton and Hauksson (1992).

1932 map, these patterns show a geographic diffuseness of small earthquake epicenters, most of which are not clearly correlated to known faults. In this geographic sense, the "seismograph revolution" has not clarified local tectonics as well as it has plate tectonics at a global scale.

One reason for this is that oceanic crustal structure is relatively simple, whereas continental crustal structure is extremely complex. The structure is particularly complex in the shallow 10–20 km in which the crust behaves in a brittle fashion. The location of earthquakes, particularly the depth determination, is sensitive to the assumptions made about the speed of earthquake waves between the earthquake source and individual seismograph stations. This, in turn, is based upon the geology, particularly the presence of sedimentary basins that tend to slow down earthquake waves. The geophysicist can improve the velocity model by setting off a series of calibration shots after a major earthquake, and the geologist can also help in locating zones within the crust that transmit earthquake waves faster or slower than the average. Based on this information, computer programs can be written to locate earthquakes.

Figures 4–11 and 4–12 illustrate the sensitivity of seismicity maps and cross sections to the crustal model used. Locations in panel (a) are calculated using a computer program that assumes a particular crustal model. The epicenters fall well to the east of the surface trace of the active Calaveras fault. Those of Group I define a plane dipping to the northeast. Aftershock set II, however, defines a vertical fault plane 2 km east of the surface trace of the fault. In addition, three calibration explosions within the network are located, using the same crustal model, about 0.5 km northeast of their known location. Panel (b) displays locations obtained by assuming that the events in Group II define a vertical plane vertically beneath the surface trace of the Calaveras fault (Fig. 4–12). Segment I still dips northeast, but more steeply, and it projects upward to the surface trace of the fault. The calibration shots also moved southwest with the new model, but now they are 0.8 km southwest of their known location!

Figures 4–13 and 4–14 illustrate another technique for displaying seismicity data using panel (b) locations of Figure 4–12. Stereographic views are shown for the seismicity (Fig. 4–13) and for the fault-plane solutions (Fig. 4–14) in three dimensions, assuming that the nodal plane closest to the surface orientation of the Calaveras fault is correct. Use a pocket stereoscope for these two figures.

We now consider the controversial Santa Barbara, California, earthquake of August 13, 1978, controversial because different investigators have come up with different locations of the mainshock and aftershocks, which have led to different conclusions about the tectonics. Part of the controversy arises from the fact that there were too few stations offshore, south of the earthquake, leading to potential errors in hypocenter locations that need to be evaluated in discussions of the seismicity.

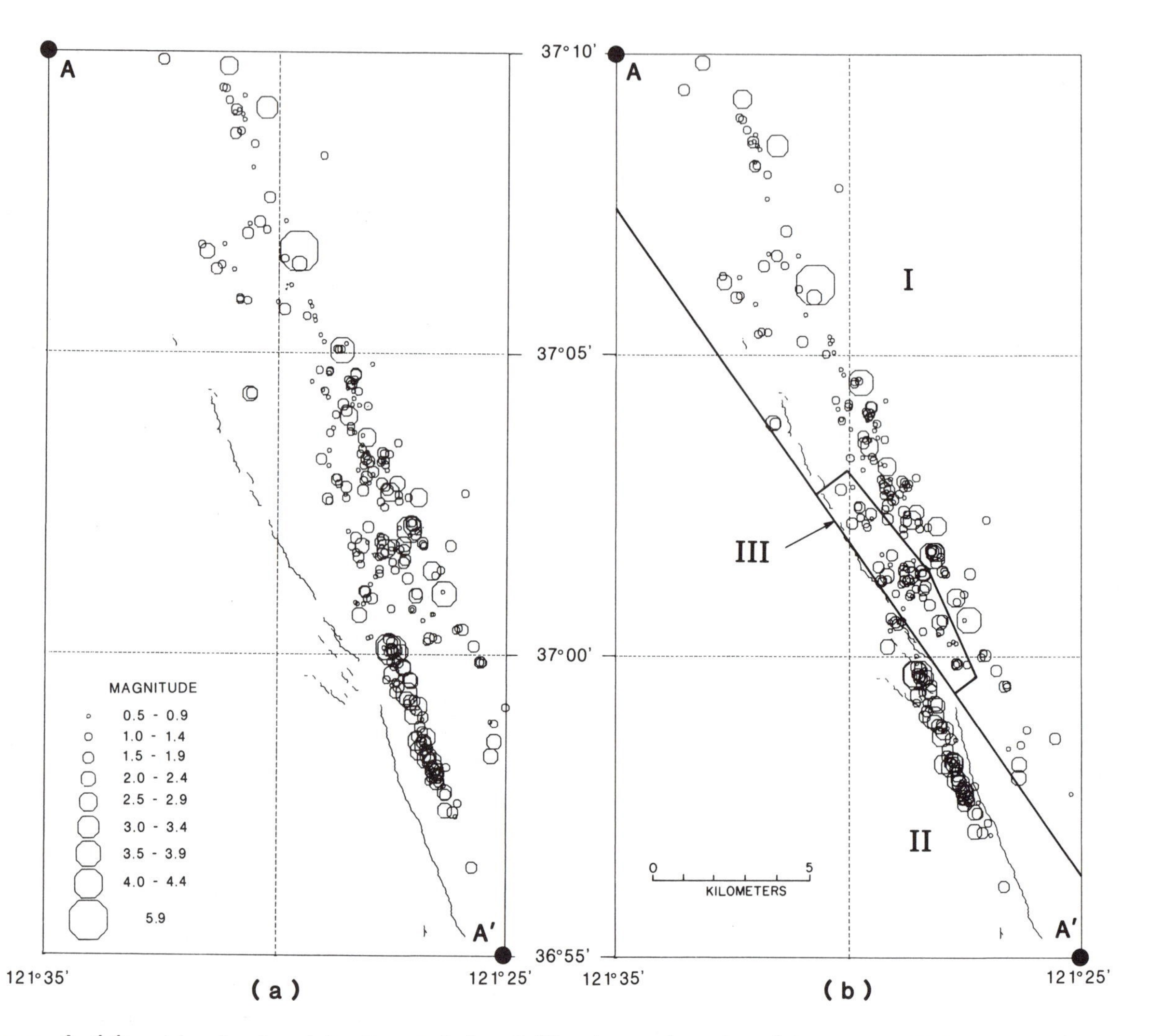

Figure 4–11. Aftershocks of the Coyote Lake, California, earthquake of August 6, 1979. (a) Locations using an unconstrained model. (b) Locations using a model that constrains some hypocenters to the fault plane. Solid and broken lines show the surface trace of the Calaveras fault. I, II, and III are different sets of aftershocks. After Reasenberg and Ellsworth (1982).

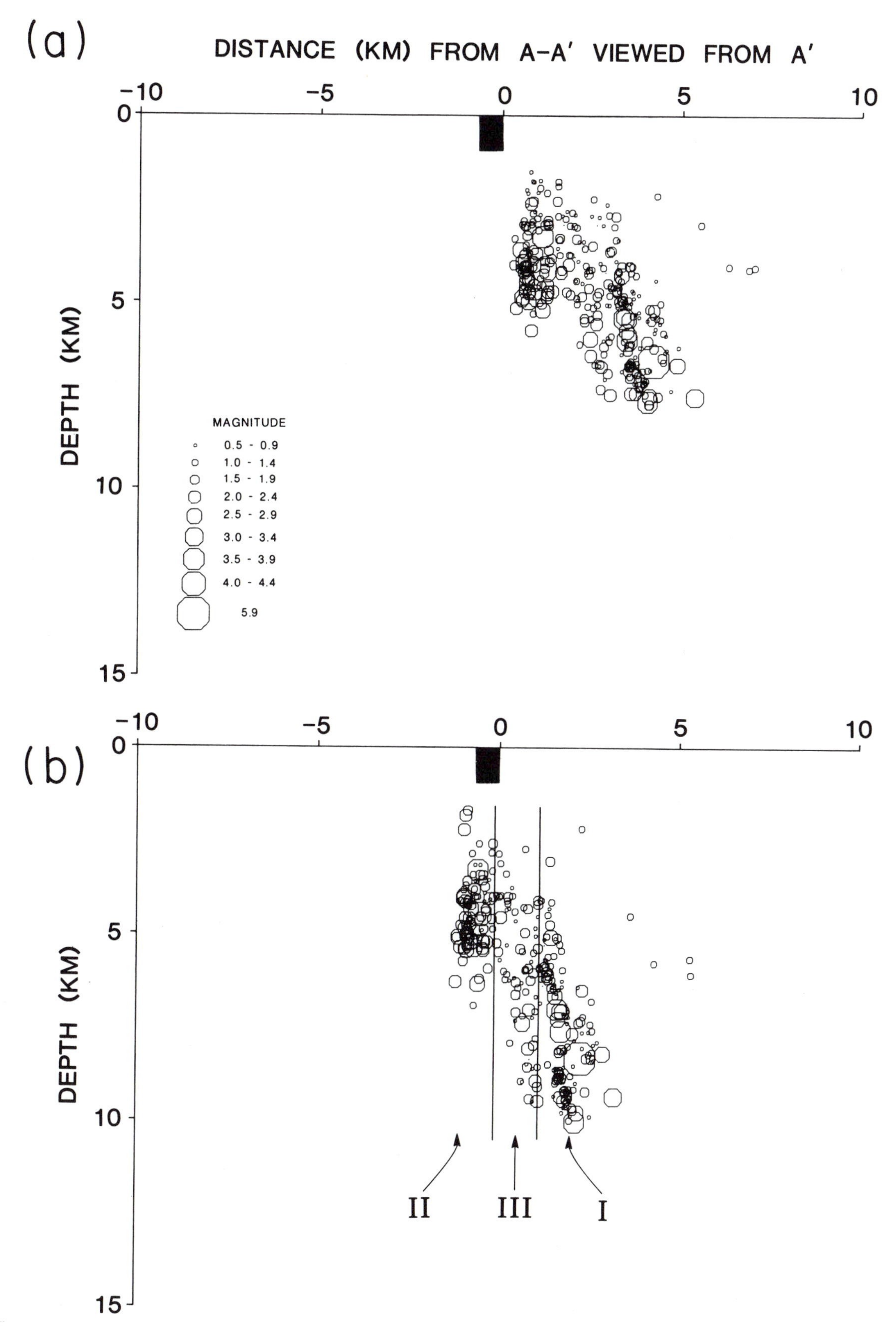

Figure 4–12. Aftershocks of the Coyote Lake earthquake projected onto a vertical plane perpendicular to line A-A′ in Figure 4–11: view from southeast. (a) Locations using unconstrained model. (b) Locations using constrained model. Shaded rectangle contains surface trace of Calaveras fault. After Reasenberg and Ellsworth (1982).

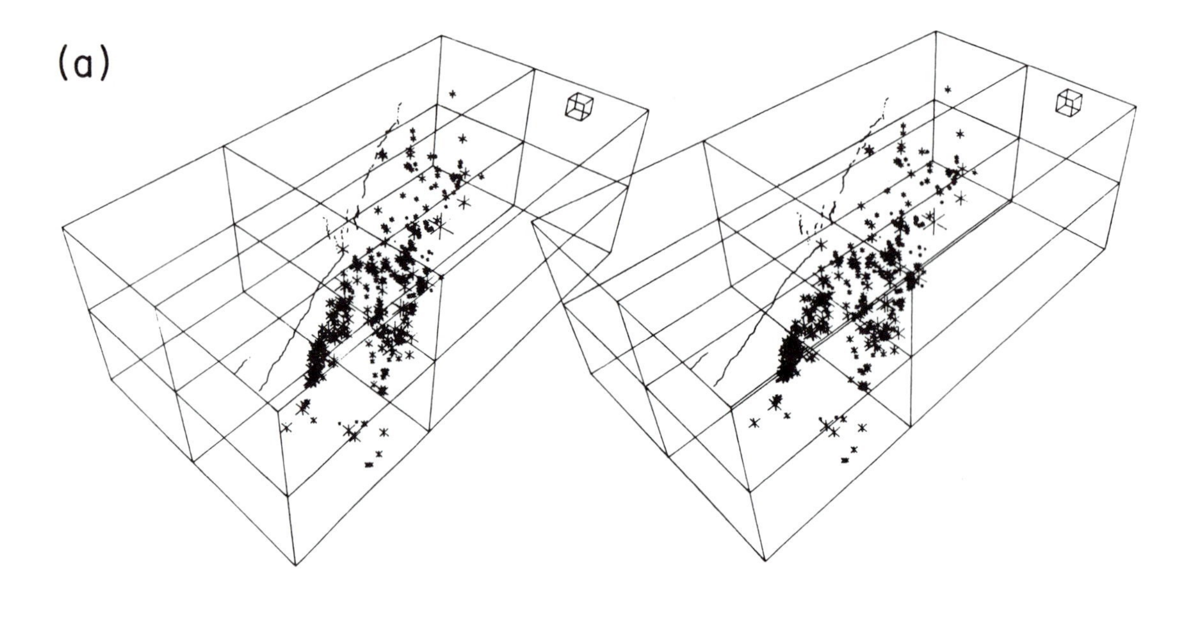

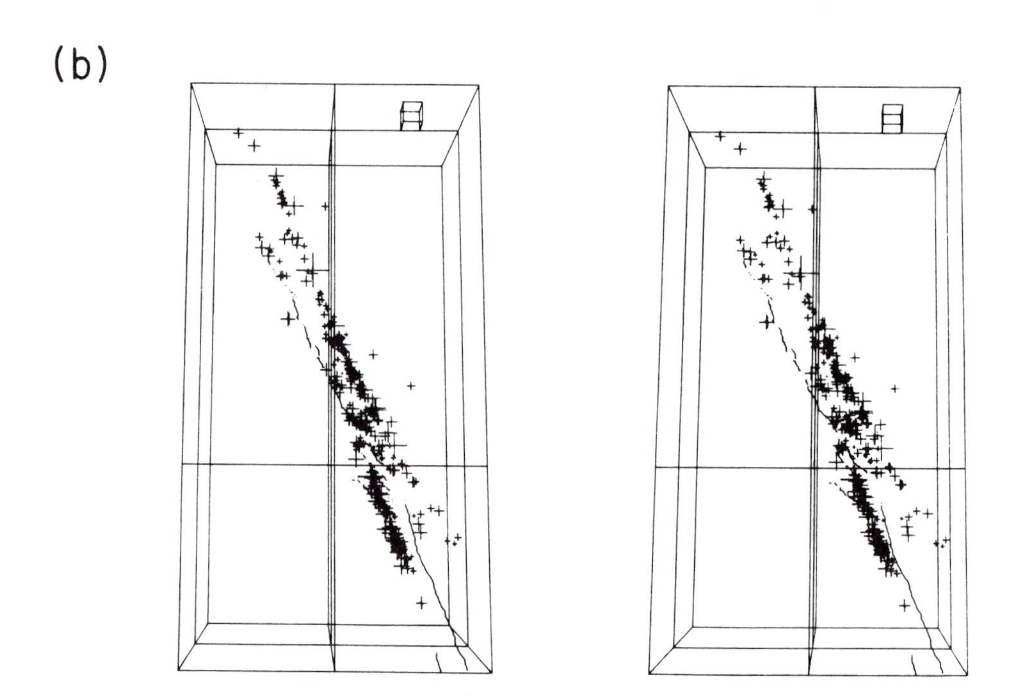

Figure 4–13. Stereoscopic views of aftershocks of the Coyote Lake earthquake using location model (b). Area covered is about the same as for Figure 4–11; depth of the box is 10 km. Reference cube has 1-km sides. (a) View from southeast. (b) View from above. After Reasenberg and Ellsworth (1982).

The locations of the mainshock and aftershocks by Lee et al. (1978) are shown on Figures 4-15 and 4-16. These were calculated using a computer program, HYPO71, written at the U.S. Geological Survey (Lee and Lahr, 1975). This computer program minimizes the residual travel-times between observed and calculated arrivals and assumes a horizontal, multilayered crust. Figure 4-15 shows a somewhat diffuse pattern of aftershocks trending WNW, and lying just offshore. The distribution in cross section, Figure 4-16, sug-

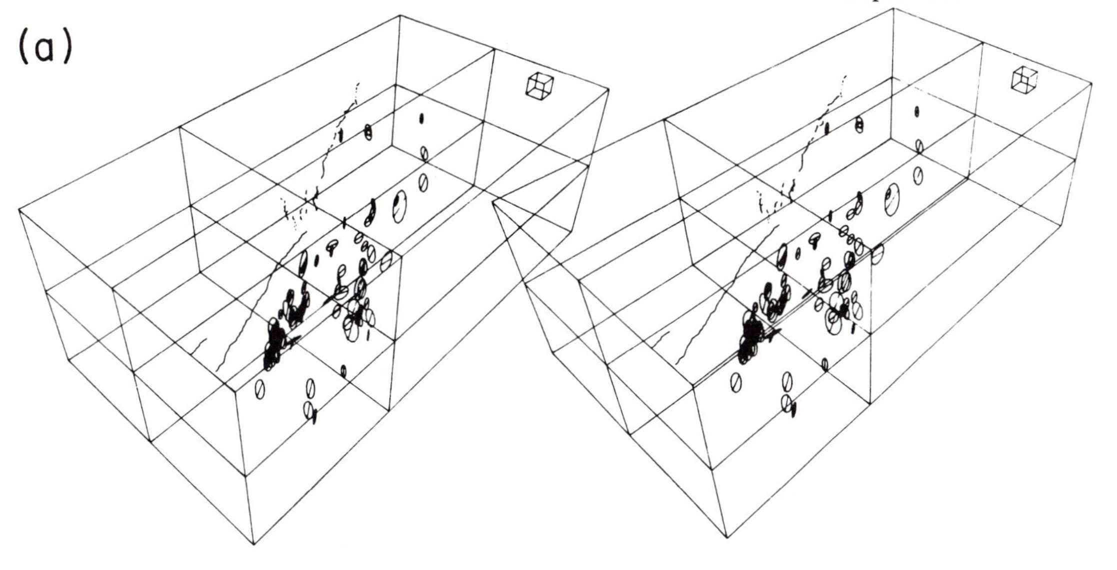

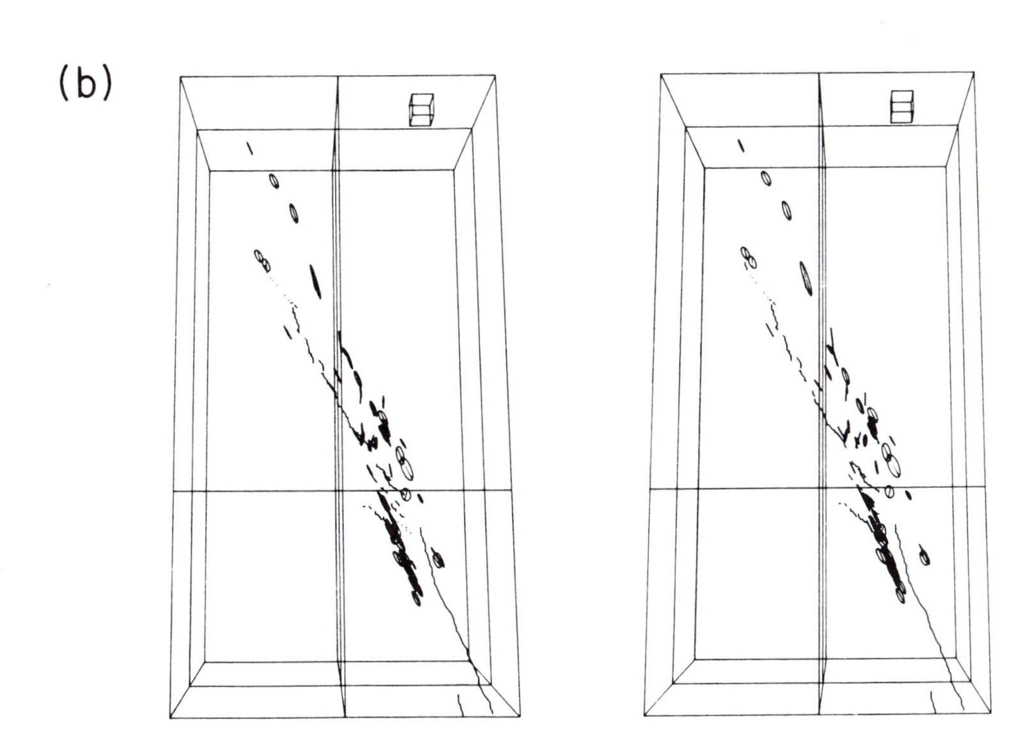

Figure 4–14. Stereoscopic views of fault-plane solutions for the mainshock and 76 aftershocks of the Coyote Lake earthquake using location model (b). Circle symbol is centered on a hypocenter and is within the slip plane assuming that rupture occurred on the Calaveras fault or a fault subparallel to it. After Reasenberg and Ellsworth (1982).

gests rupture on a fault which dips moderately to steeply north. The fault-plane solution of the mainshock is also compatible with a north-dipping reverse fault. The data of Lee et al. (1978) could be used as evidence that the earthquake sequence was caused by rupture on an active fault reaching the surface in the Santa Barbara Channel such as the Pitas Point fault or another fault farther south (Fig. 4–16). But the fault-plane solution of the mainshock also permits activity on one or more south-dipping reverse faults reaching

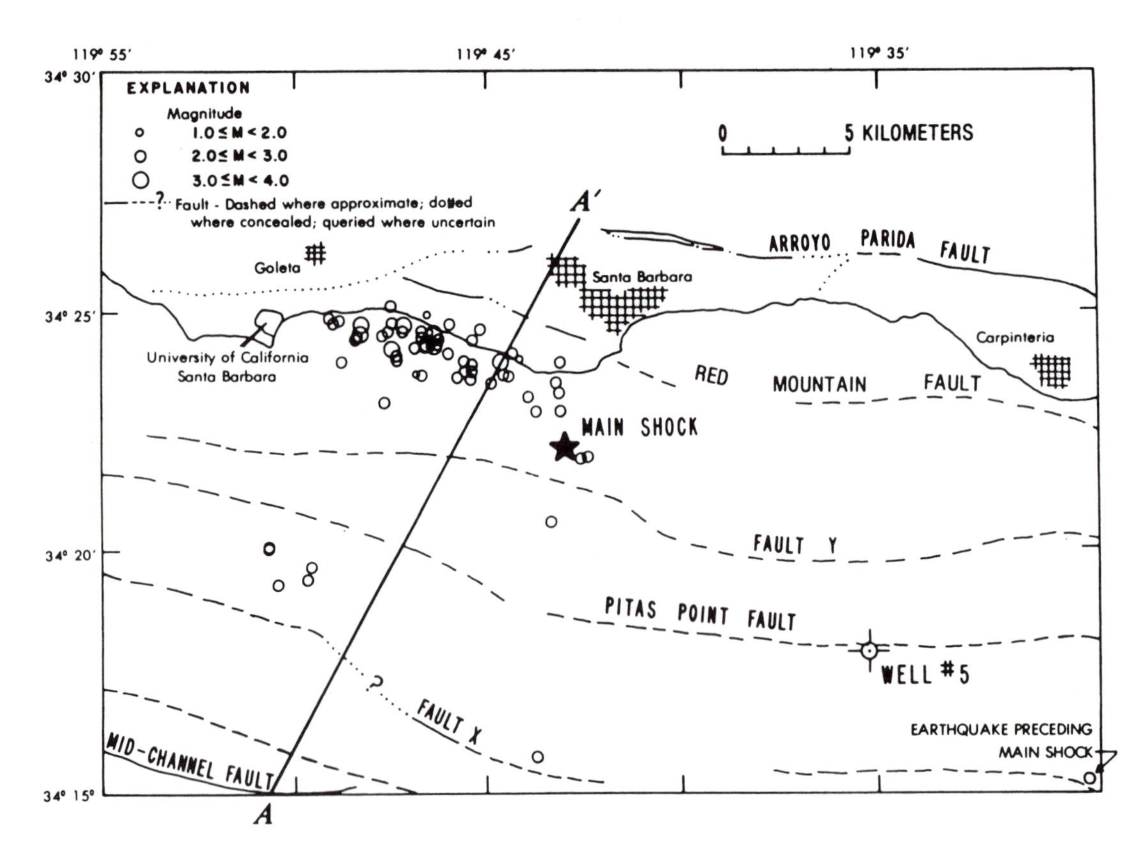

Figure 4–15. Epicenters of Santa Barbara earthquake and its major aftershocks. After Lee et al. (1978).

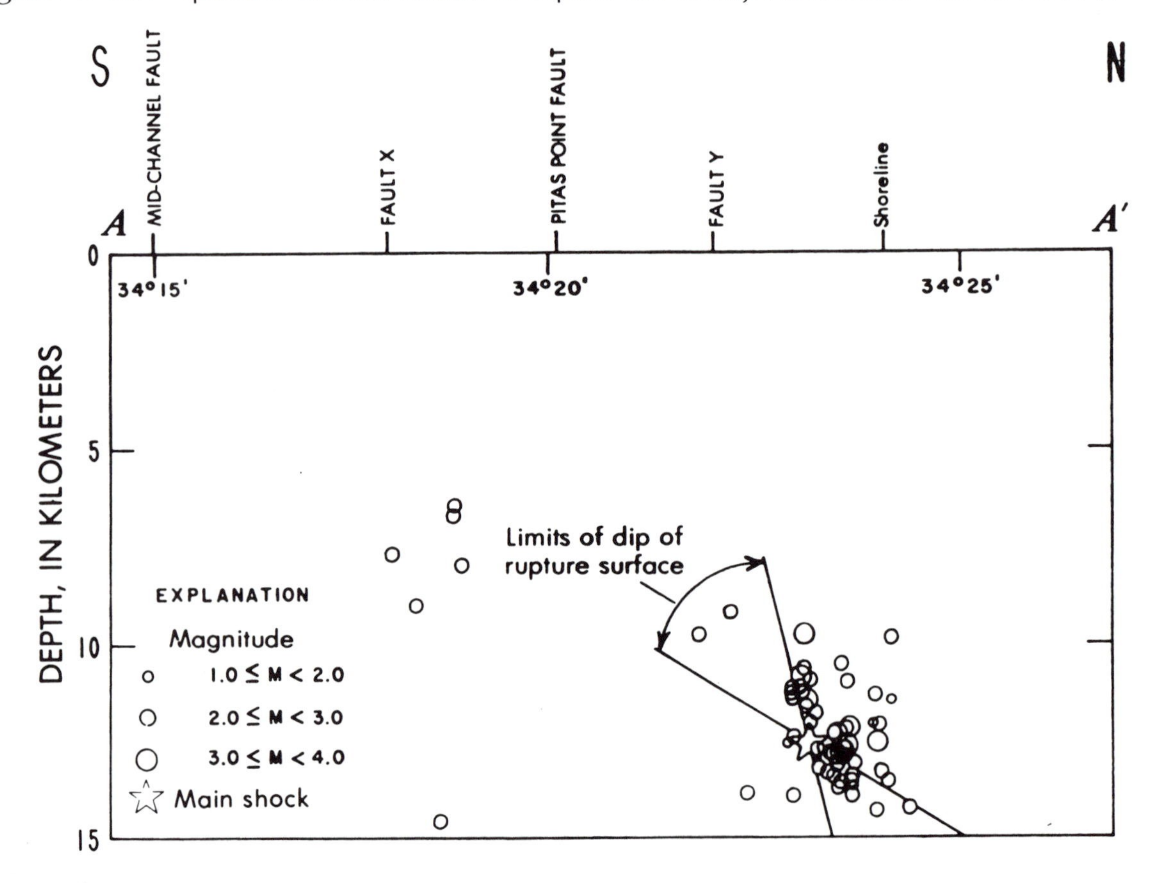

Figure 4–16. Cross section along A-A′ of figure 4–13, showing hypocenter distributions and faults. After Lee et al. (1978).

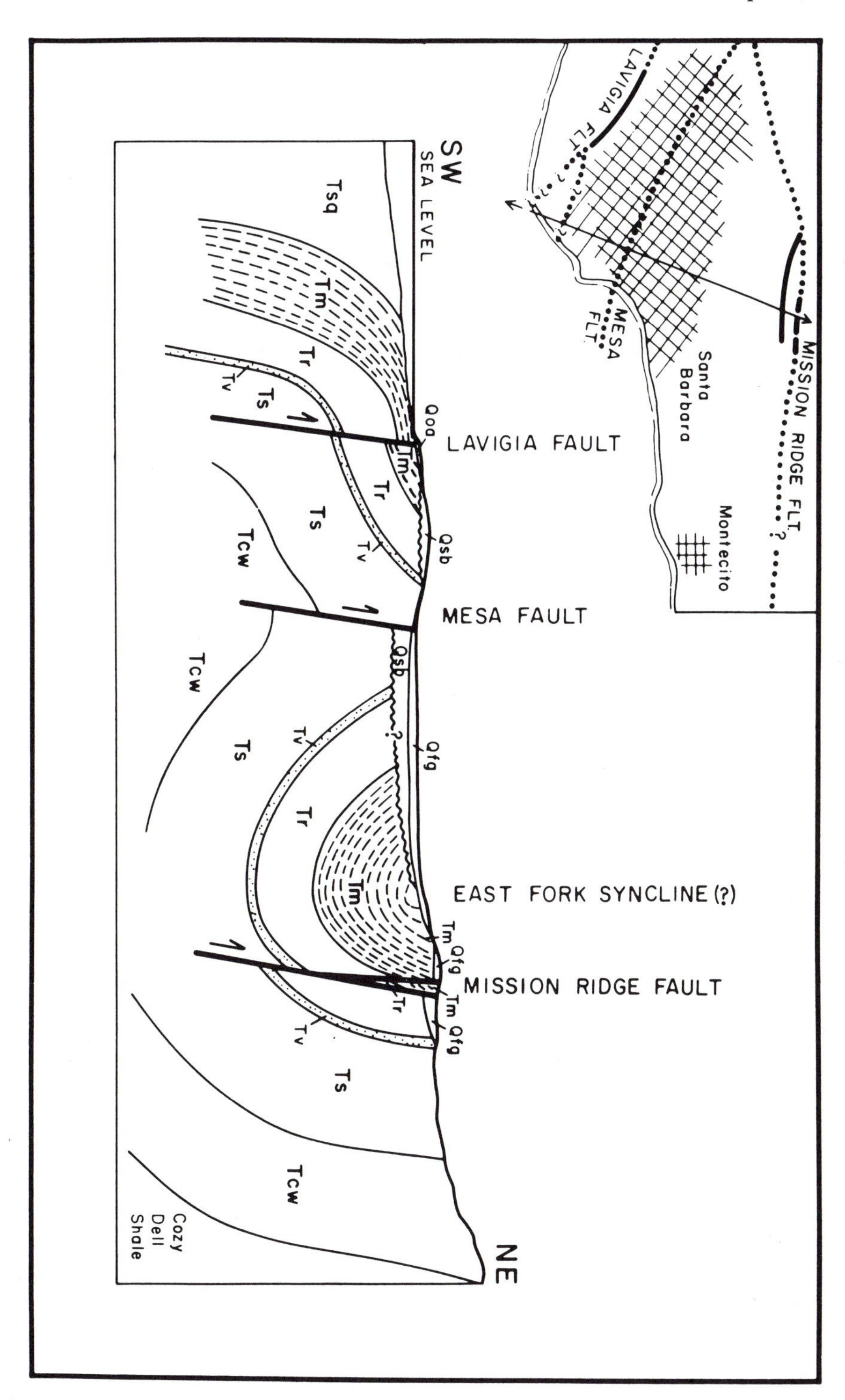

Figure 4–17. Diagrammatic cross section across Mission Ridge and Santa Barbara Plain. From Yeats and Olson (1984).

the surface in downtown Santa Barbara (Fig. 4-17) (Yeats and Olson, 1984).

Corbett and Johnson (1982) used a hybrid crustal model which included data from 10 shots used for a seismic-refraction line in the Santa Barbara Channel and other location parameters designed at the California Institute of Technology. Because the locations of the seismic-refraction shot points are known, the arrival times of Corbett and Johnson (1982) can be compared with arrival times calculated using various

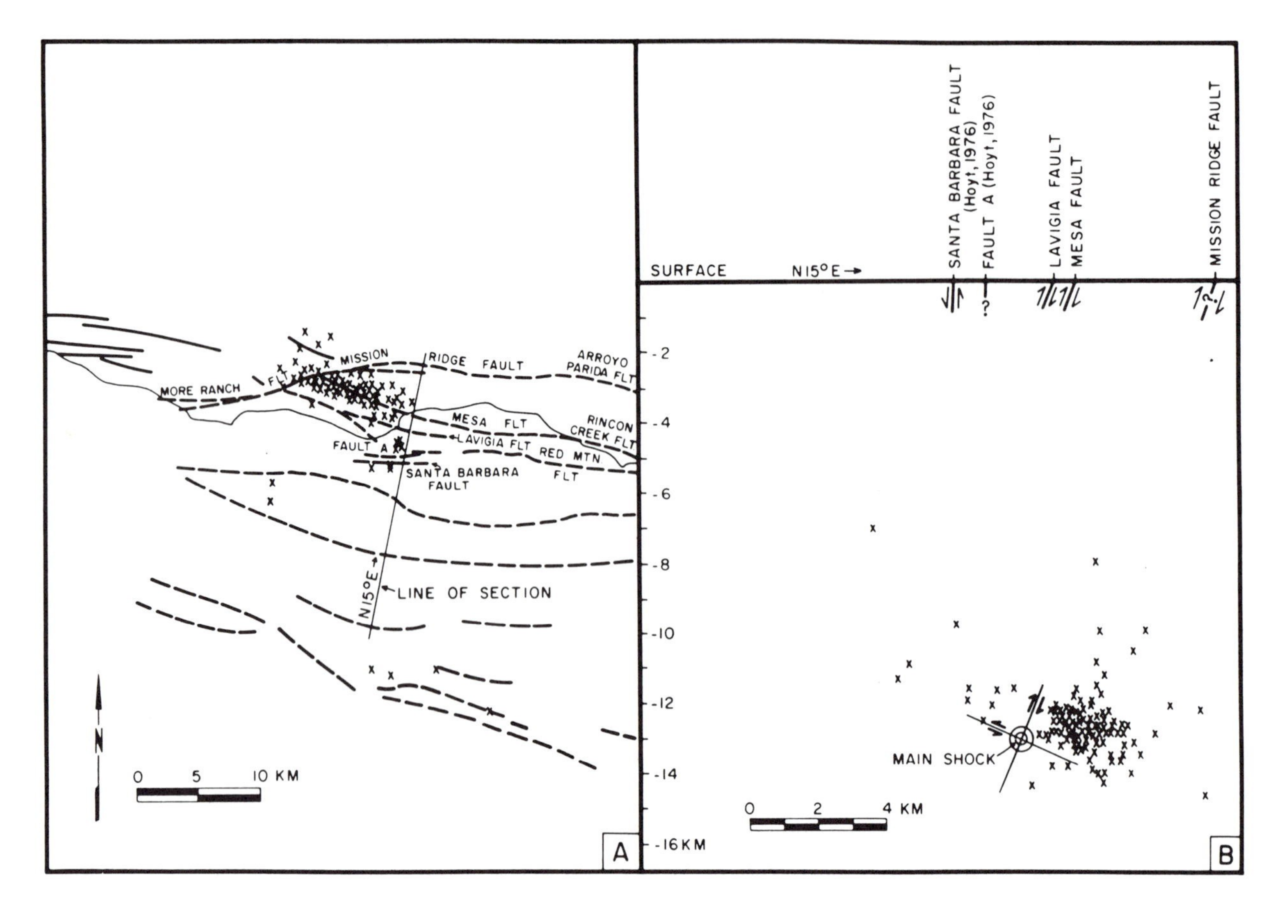

Figure 4–18. (a) Aftershock distribution following 1978 Santa Barbara earthquake. (b) Cross section across Santa Barbara with local faults plotted. Modified from Corbett and Johnson (1982).

crustal models. They found that there is a considerable difference between crustal structure in the Santa Barbara Channel and that at the coast. Their portrayal of seismicity differs from Lee et al. (1978) in that the linear zone of earthquake epicenters is directly beneath the city of Santa Barbara (Fig. 4-18), several kilometers north of the locations of Lee et al. (1978). Their fault-plane solution and the aftershock distribution suggest a relatively low-angle north-dipping thrust, north over south, which supports the suggestion of Yeats (1981) that the Transverse Ranges are underlain by a low-angle décollement at depths similar to those of the Santa Barbara aftershocks. However, the linear pattern of aftershocks is more typical of a steeply-dipping fault rather than a low-dipping fault, and if the *other* nodal plane of the fault-plane solution is used, it would be parallel to the south-dipping Lavigia, Mesa, and Mission Ridge faults (cf. Figures 4-17 and 4-18). The significance of this alternative nodal-plane interpretation is rather profound for the residents of Santa Barbara, although evidence for late Quaternary activity indicates that these faults pose a significant seismic hazard whether or not the 1978 earthquake was located on one of them.

The earthquake was investigated further by William A. Prothero, Barbara Bogaert, and Barry Keller at the University of California, Santa Barbara, using additional data from seven portable stations as well as the permanent array used by earlier workers. Their pattern is more diffuse than Corbett and Johnson's, even though individual hypocenters are better constrained, and the linear zone has again moved offshore. Prothero concludes that their data on aftershock distribution and focal mechanisms are "consistent with the idea of a horizontal décollement, but (they offer) very little real support for it." The diffuse pattern of aftershocks is similar to that produced by a reverse fault that does not reach the surface (*blind thrust*) but is expressed at the surface as a broad fold, still another interpretation discussed further in Chapter 10.

In summary, the seismicity data have not resolved this critical problem in environmental geology of an urban area. The pattern of Lee et al. (1978) supports a moderately north-dipping thrust fault. Corbett and Johnson's (1982) pattern support a low-angle fault which may reach the surface far to the south of the mainland, or may not reach the surface at all.

Prothero's pattern, when evaluated in terms of Yeats and Olson's (1984) geology, is consistent with the possibility that the earthquake was produced by steep south-dipping reverse faults which reach the surface in the middle of Santa Barbara. Or the aftershocks may have been produced by a blind thrust extending upward as a fold.

The lesson to be learned is that maps and cross sections based on instrumental seismicity, like geological cross sections, are interpretive. Earthquake hypocenter locations are strongly dependent on the assumed distribution of rocks in the crust that transmit earthquake waves at different speeds. Uncertainties are created by the necessity of simplifying the geologic structure in order to create a crustal model that can incorporate the data. Earthquakes at the edge of a network, particularly at coastlines, are even less well located, as can be seen by noting the horizontal and vertical error ellipses for each event as listed in the catalog.

SPECIAL TOPICS

Seismic Inversions

In *forward* seismic modeling, one assumes the distribution and timing of slip on a fault plane and computes what the resulting seismograms would look like at various distant localities by producing *synthetic seismograms*. In the *inverse* problem, actual seismograms from a variety of stations are used (inverted) to infer the timing and distribution of the slip on the fault plane that caused the event. Explanation of the actual techniques employed are beyond the scope of this chapter, but the method is a very powerful one, and the results are particularly relevant to the geologic understanding of individual earthquakes and the nature of fault rupture.

As an example, the Landers earthquake ($M_w = 7.3$) was caused by strike-slip faulting on three overlapping en échelon fault segments with a total rupture length of about 70 km and a maximum surface offset of 6 meters (Sieh et al., 1992; cf. Chapter 8). The earthquake was well recorded by numerous seismographs internationally as well as by 16 nearby strong-motion instruments. Additionally, the static deformation field was well documented by the Global Positional System (GPS), synthetic-aperture radar, and trilateration measurements (cf. Figs. 5-25 and 5-26), and the surface displacements along several faults were carefully recorded (cf. Chapter 8). Seismographic, geologic, and geodetic data were inverted by Wald and Heaton (1994) to calculate displacements and rupture times to each of 186 individual $2^1/_2 \times 3$ km areas on the overlapping fault planes. Solving for amount and timing of slip at 186 locations produces solutions that are nonunique. However, each of three independent data sets (teleseismic, strong-motion, geodetic), when inverted by themselves, gave somewhat similar results. Inverting all the data together gives considerable added resolution and greater confidence in the results.

Wald and Heaton point out several important conclusions regarding the nature of the fault rupture. (1) Slip on the fault was extremely heterogeneous, both along strike and down-dip (Fig. 4-19). (2) Fault slip measured at the ground surface is only crudely related to slip at depth. The maximum surface displacement of 6 m, which occurred on the Camp Rock-Emerson fault, diminished rapidly down-dip. Similarly large displacement occurred locally at depth on the Homestead Valley fault but diminished rapidly updip to

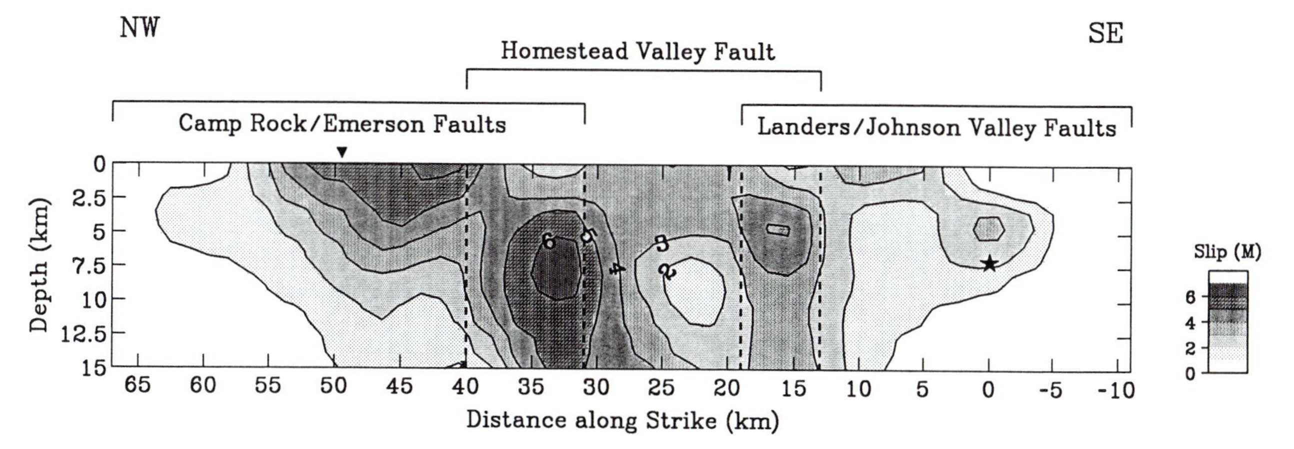

Figure 4–19. Cross section of the strike-slip distribution for the 1992 Landers, California, earthquake determined from modeling all data sets combined. Contour interval is 1 m, and the first contour is 1 m. Star locates hypocenter. From Wald and Heaton (1994).

smaller values at the surface. (3) The rupture front, which propagated northward at an average speed of 2 km/sec, slowed down as it transferred from one fault segment to another. (4) As the rupture propagated northward, only a portion of the fault was slipping at any given time (Fig. 4-20), and although the total rupture duration was 24 seconds, the total rupture duration at any one locality was less than 4 seconds. (5) There was very little afterslip. (6) The hypocenter was in an area of only modest slip and shaking.

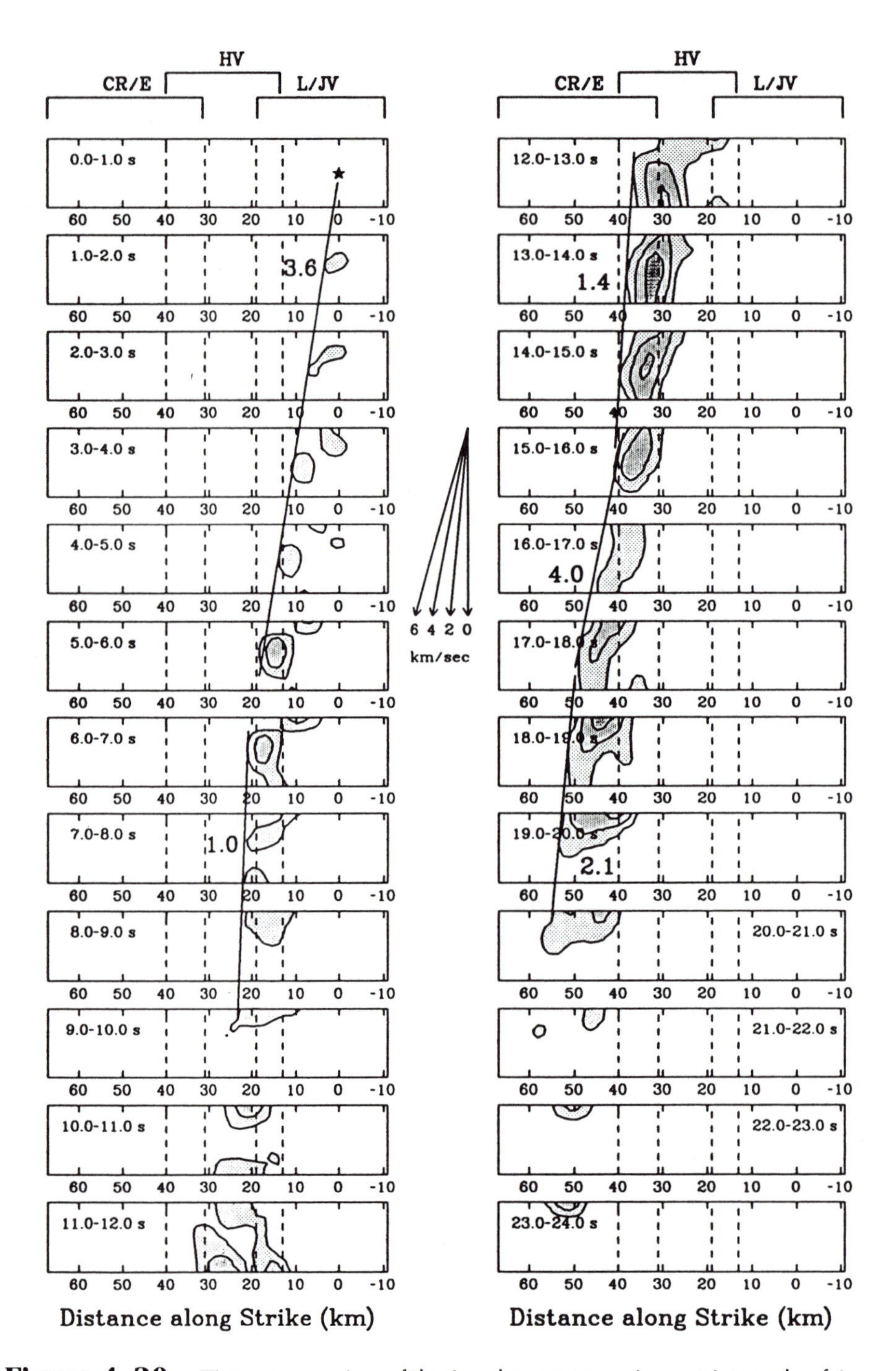

Figure 4–20. Time progression of the Landers rupture given at intervals of 1 second, as labeled. Contour interval is 0.5 m of slip. For reference, a rupture speed rosette is given with arrows. Solid lines indicate selected regions of fairly constant rupture speed, with the speed shown in large numerals. From Wald and Heaton (1994).

At present, these inversions are useful as first-order approximations, that is, meters vs. centimeters of slip, or hundreds vs. tens of kilometers of rupture length. The resolution of parameters in the subsurface using this technique is at least an order of magnitude poorer than coseismic parameters at the surface.

Slow, Quiet, and Silent Earthquakes—A Contradiction in Terms?

When we think of an earthquake source, we think of a crack that propagates through the crust close to the shear-wave speed, which is generally several kilometers per second. Fault rupture is sudden, accompanied by violent shaking of the ground. But what about creep events on the San Andreas, Hayward, and Calaveras faults in California, during which propagation along a fault occurs at rates less than one meter per second, and slip occurs at rates of millimeters per year? These, too, are part of the continuum of earth deformation, as are nonelastic strain events within ductile crust. Earth deformation occurs at rates that differ widely, from fast ruptures that suddenly release stored elastic-strain energy (*ordinary earthquakes*) to *slow earthquakes* (speeds of hundreds of meters per second), *silent earthquakes* (speeds of tens of meters per second), *creep events,* and finally *strain migration episodes* with speeds in centimeters or millimeters per second (Beroza and Jordan, 1990).

Slow earthquakes are the bass violins of the earthquake symphony orchestra. They include episodes of high-speed rupture propagation that produce an ordinary seismogram of high-frequency body waves. However, slow earthquakes take an unusually long time to rupture in comparison to ordinary earthquakes of similar moment magnitude. Oceanic transform faults have produced several slow earthquakes, such as the June 6, 1960 Chilean transform fault earthquake that ruptured for about an hour as a series of small events.

If slow earthquakes are the bass violins of the earthquake symphony, silent earthquakes are the equivalent of music below the audible range. Silent earthquakes are not accompanied by high-speed rupture propagation events, and thus they do not generate high-frequency waves that are recorded teleseismically. Conventional seismographs do not record these events. Alan Linde and Selwyn Sacks of the Department of Terrestrial Magmatism, Carnegie Institute of Washington, however, have used strainmeters (discussed further in Chapter 5) to record them. Strainmeters also document creep events on the San Andreas fault system (10 mm/sec). Low-frequency waves (10 m/sec) were recorded prior to the 1976 Friuli, Italy, earthquake, and silent earthquakes preceded a M7.7 earthquake in the Japan Sea in 1983. This suggests that silent earthquakes may offer promise as precursors to ordinary, stick-slip earthquakes. Laboratory experiments suggest that stick slip is preceded by aseismic slip with propagation speeds of 20 to 200 m/sec. In the near field, silent earthquakes may be recorded geodetically and with strainmeters; it is necessary to use digital, broadband seismographs to record waves of such low frequency.

Free oscillations, in which the earth rings like a bell, were recorded 1500 times over a ten-year period, mostly triggered by very large, ordinary earthquakes. But in some cases, the triggering earthquake did not appear to be large enough to produce free oscillations. Gregory Beroza, now of Stanford University, and Thomas Jordan of the Massachusetts Institute of Technology suggest that the free oscillations were triggered by slow earthquakes. One hundred sixty-four of the 1500 free-oscillation episodes were not accompanied by a recorded earthquake at all. Beroza and Jordan attribute these free-oscillation episodes to *quiet earthquakes,* which are faster than silent earthquakes, producing faint, low-frequency seismic waves. These ultra-low-frequency earthquakes may release enough seismic energy to make the earth ring, despite the fact that they do not produce high-speed, damaging seismic waves.

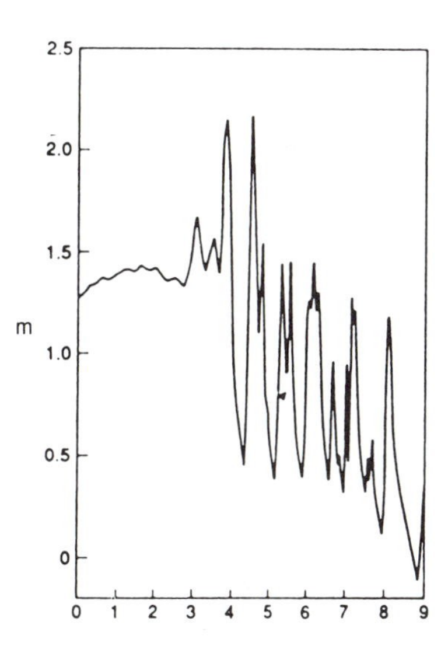

Figure 4–21. Tide-gauge record of the tsunami generated by the 1960 Chilean earthquake and recorded at Miyako, Japan. The earthquake occurred nearly 24 hours earlier. After Satake (1992).

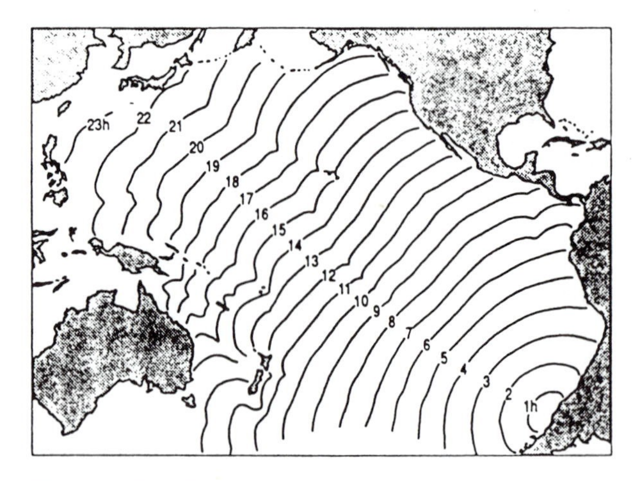

Figure 4–22. Refraction diagram for the tsunami generated by the 1960 Chilean earthquake made by the Japan Meteorological Agency. Numbers refer to time after the earthquake in hours. After Satake (1992).

Tsunamis

A gravity wave may propagate in seawater from a large disturbance of the sea floor such as an earthquake, eruption of a submarine volcano, or a massive landslide. Such a wave is called a *tsunami,* a Japanese word meaning "harbor wave." Most tsunamis are the result of earthquakes within subduction zones, in which sudden uplift or subsidence of the sea floor by faulting causes a vertical displacement of the water above the sea floor. Strike-slip faults are less likely to generate a tsunami because the fault motion is horizontal, not vertical, and there is no sudden change in sea-floor depth. Earthquakes generating tsunamis are predominantly shallow, and a fault in a sedimentary layer will generate a larger tsunami than a fault in rigid basement rock. The "seismographs" that measures tsunami wave forms are tide gauges and pressure gauges on the sea floor (Fig. 4-21). A pressure gauge on the sea floor measures the height of water above the gauge; this has the advantage of relatively continuous recording.

Tsunami intensity is measured by the Imamura-Iida scale **m,** which is a logarithmic scale measuring either the run-up height or the wave amplitude on tide-gauge records. Tsunami magnitude M_t was defined by K. Abe as:

$$M_t = \log H + C + 9.1 \text{ for a trans-Pacific tsunami, and}$$

$$M_t = \log H + \log \Delta + 5.8 \text{ for tsunamis from 100 to 3500 km away.}$$

H is the maximum amplitude recorded on the tide gauge, in meters, C is a distance factor depending on the source and observation points, and Δ is the distance in kilometers. These formulas have been calibrated with the moment magnitude scale M_W so that the 1960 Chilean earthquake with $M_W = 9.5$ produced a tsunami with $M_t 9.4$, and the 1964 Alaska earthquake with $M_W 9.2$ produced a tsunami with $M_t = 9.1$.

The tsunami wave speed in deep water is around 0.2 km/sec in contrast to seismic wave speeds of 5–10 km/sec. This difference in wave speeds permits an effective tsunami warning system for tsunamis generated by distant earthquakes. The 1960 Chilean earthquake generated a destructive tsunami in Japan that did not arrive until nearly 24 hours after the earthquake (Figs. 4-21 and 4-22). The tsunami wave speed is slower in shallow water, producing the refraction of the wave front shown in Figure 4-22. The slowing

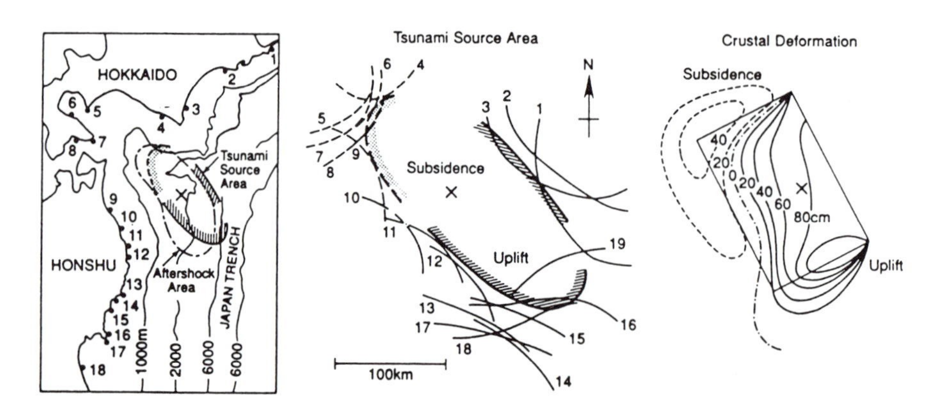

Figure 4–23. Tsunami source area and crustal deformation pattern based on tide-gauge records from 18 stations in northeast Japan. Solid arcs refer to upward first-motion, and dashed arcs refer to downward first-motion. After Satake (1992).

of the wave front also increases the wave height, so that a tsunami that is barely noticeable in the deep sea may generate a large wave as it approaches the coastline.

Tsunami wave forms recorded at a large number of tide-gauge stations can be used to determine the source dimensions of an earthquake. Just as a refraction diagram can be computed to show the wave fronts from a distant tsunami (Fig. 4-22), an inverse refraction diagram may be drawn from the tide gauge back to the source if the travel time of the wave from source to tide gauge is known. Figure 4-23 shows the source dimensions of the 1968 Tokachi-oki earthquake based on 18 tide-gauge stations. Just as the upward or downward first motion of an earthquake wave can be used to determine a double-couple fault-plane solution, the upward or downward first motion of the tide gauge can be used to determine whether the sea floor uplifted or subsided, respectively.

Suggestions for Future Reading

Bolt, B. A. 1993. Earthquakes. New York: W. H. Freeman and Co., 331 p.

Boore, D. M. 1977. The motion of the ground in earthquakes. Scientific American, 237:69-78.

Bolt, B. A. 1982. Inside the Earth: Evidence from earthquakes. San Francisco: W. H. Freeman and Co., 191 p.

Frohlich, C. 1994. Earthquakes with non-double couple mechanisms. Science, 264:804-9.

Gere, J. M., and Shah, H. C. 1984. Terra non firma. New York: W. H. Freeman and Co., 203 p.

Isacks, B., Oliver, J., and Sykes, L. R. 1968. Seismology and the new global tectonics. Jour. Geophys. Res., 73:5855-99.

Press, F. and Siever, R. 1982. Earth, 3rd ed. New York: W. H. Freeman and Co., 613 p. See chapter 17, pp. 393-418.

Richter, C. F. 1958. Elementary Seismology. San Francisco: W. H. Freeman and Co., 768 p.

Satake, K. 1992. Tsunamis. Encyclopedia of Earth System Science, 4:389-92.

Scholz, C. H. 1990. The mechanics of earthquakes and faulting. New York: Cambridge University Press, 439 p.

5

Tectonic Geodesy

Geodesy is one of the three major disciplines, together with geology and the study of seismic waves, that are employed in the study of earthquakes. Unlike the study of seismic waves, which was directed toward earthquakes, geodesy was developed as a way to measure more precisely distances and elevations of the land, and only later was it found that geodetic changes could be used in the study of deformation of the earth, including that by earthquakes. In addition to providing an independent dataset in describing earthquake fault rupture, geodesy is the only one of the three disciplines mentioned above that permits the description of deformation between successive earthquakes.

The first earthquake to be studied geodetically was the May 17, 1892 Tapanuli, Sumatra, earthquake which struck as a triangulation survey was taking place. The surveyor, J. J. A. Müller, found that the angles between survey monuments observed after the earthquake did not correspond to those measured before the earthquake. He concluded that horizontal motion of at least two meters had taken place along a structure we now recognize as a branch of the Great Sumatran fault.

The great earthquakes of June 12, 1897 and April 4, 1905 struck the foothills of the Himalaya, where extensive triangulation and leveling surveys had taken place during the last half of the nineteenth century. Oldham (1899) examined the changes in angles and in relative elevations and concluded that they were due to tectonic movements associated with the 1897 earthquake. Leveling data collected before and after the 1905 earthquake convinced Middlemiss (1910) that these changes were related to that earthquake. Middlemiss' data set was of high enough quality that it could be used 80 years later to model the earthquake source as a blind thrust.

H. F. Reid's study of the 1906 San Francisco earthquake was the first widespread use of geodetic data in earthquake studies. He compared three sets of triangulation surveys across the San Andreas fault, one conducted in 1851–1865, another made in 1874–1892, and a third completed a short time after the earthquake. Comparison of the two surveys conducted prior to the earthquake showed that widely separated points on opposite sides of the fault had moved 3.2 meters prior to 1906, with the west side moving north with respect to the east side. From this observation, Reid then concluded that the crust had accumulated elastic strain in the years prior to the earthquake. By comparing the surveys before and after the 1906 earthquake, Reid was able to relate the release of elastic strain during the earthquake to its buildup prior to the earthquake.

Just as a bent stick breaks when the bending forces exceed its strength, so does the crust rupture when its strength is exceeded. The catastrophic motion on the fault marked the rebound of the crust as the elastic strain was released. In 1910, Reid combined his observations of the San Francisco earthquake with those gathered earlier by Müller after the Tapanuli earthquake to propose his *elastic rebound theory* to explain both earthquakes. Reid even suggested that study of triangulation records could lead to a forecast of the next earthquake.

Reid's ideas about earthquake prediction led to a proposal by Arthur L. Day of the Carnegie Institute of Washington that the U.S. Coast and Geodetic Survey conduct periodic triangulation surveys along the California coast. This program, led by William Bowie, led to the observation by C. A. Whitten that the crust west of the San Andreas fault system was moving (by "slow drift") at a rate of 50 mm/yr with respect to the crust east of the fault. Geodetic observations were made after the 1940 Imperial Valley earthquake and the 1952 Tehachapi earthquake.

In Japan, the need for earthquake research had been recognized by the Japanese Diet after the 1891 Nobi earthquake, and the research program that developed after that earthquake included extensive use of leveling data at coastal sites in southwest Japan and across active faults on Honshu. This work was published in the 1920s and 1930s by N. Yamasaki, A. Imamura, C. Tsuboi, N. Miyabe, and others. Leveling data were used in studies of the 1927 Tango earthquake and the 1944 and 1946 subduction-zone earthquakes off southwest Japan (cf. Chapter 11). In New Zealand, geodetic surveys were conducted as part of the investigations following the 1929 Murchison earthquake.

Conventional techniques of geodesy (triangulation, trilateration, and leveling) came into increased use for tectonic studies in the 1960s, after discovery of creep along some faults in the San Francisco Bay area, and they were the principal methods of tectonic geodesy until the late 1980s, when space-based tectonic geodesy became operational. Space-geodetic techniques include Very Long Baseline Interferometry (VLBI), Satellite Laser Ranging (SLR), and the Global Positioning System (GPS), all of which are described in this chapter. VLBI has established a global reference frame that has confirmed that tectonic rates measured in years and decades are in general agreement with plate-tectonic rates measured in timescales of 10^5–10^6 years. GPS, which measures strain in three dimensions, has revolutionized tectonic geodesy. It is inexpensive, lightweight, and relatively simple to use, and dense GPS networks are being set up in many parts of the world. Long-term monitoring of GPS networks near heavily populated regions at seismic risk may have potential as a method of earthquake forecasting in these regions.

In this chapter, we begin our discussion of tectonic geodesy with a discussion of conventional techniques and conclude with a review of space-based techniques.

CONVENTIONAL TECHNIQUES

Precise geodetic surveys, when repeated, measure crustal strain and are a major source of information in studying the earthquake environment. These consist of *triangulation* (measurement of angles between survey markers, which was the dominant method of the early days of geodesy) and *trilateration* (measurement of line lengths, which is the preferred method of the modern era) to measure horizontal deformation. They also include *leveling* to measure vertical deformation and tilting. Triangulation, trilateration, and leveling networks are occupied repeatedly, and the difference between the older and newer surveys is a measure of the crustal deformation in the time between the two surveys. Because the strains measured during interseismic periods are very small, the nontectonic effects of earth tides, heavy rainfall, air temperature, and humidity must be removed from the data. If a sufficiently long time interval has elapsed, or if a large earthquake has occurred, deformation amounts and deformation rates may be calculated and compared with rates based on geological data.

Reoccupation of terrestrial geodetic survey networks is a time-consuming and costly operation, particularly in remote areas. The use of mobile radio receivers and satellite transmitters as part of GPS has revolutionized geodesy by reducing the temporal and financial cost of surveys.

HORIZONTAL DEFORMATION

Consider four stations, A, B, C, and D in a quadrilateral across a right-lateral strike-slip fault (Fig. 5-1). One station A is held fixed, as is the azimuth of the line between stations A and B. When the stations are re-surveyed, the quadrilateral is deformed by shear to ABC′D′, representing shear strain of $\Delta L/L$ or the tangent of an angle CAC′ where that angle is very small. Shear strain is dimensionless and can be given in parts per million (ppm) or μstrains ($\Delta L/L \times 10^{-6}$). The angle whose tangent is $\Delta L/L$ (or CC′/CA in the case of Figure 5-1) is commonly expressed in microradians (μrad). The deformation of the quadrilateral causes diagonal AD to lengthen to AD′ and diagonal BC to shorten to BC′. Repeated surveys at given time increments allow the strain rate to be calculated in ppm/yr, μstrain/yr, or μrad/yr.

With a larger number of stations within the quadrilateral, details about the nature of the deformation can be worked out. A creeping fault (Fig. 5-1b) slips freely and does not accumulate elastic strain; the networks would show two blocks slipping past each other with no elastic distortion within either block. Alternatively, the network may show accumulation of elastic strain and no movement on the fault (Fig. 5-1c). In this case, we visualize a layer of crust, including the fault, that is accumulating elastic strain underlain by a layer of crust that is slipping freely along the fault. If the elastic strain is limited to an area fairly close to the fault, then this may be used as evidence that the fault is locked and accumulating strain only to a shallow depth, and that it slips freely below this depth. If the elastic strain is distributed more broadly, then the fault is considered to be locked to a greater depth.

Since the early 1970s, trilateration networks established by the U.S. Geological Survey across major strike-slip faults of California have been re-surveyed repeatedly using laser geodimeters and Geodolites. Distance is determined by projecting a laser beam from the instrument to a reflector from which the

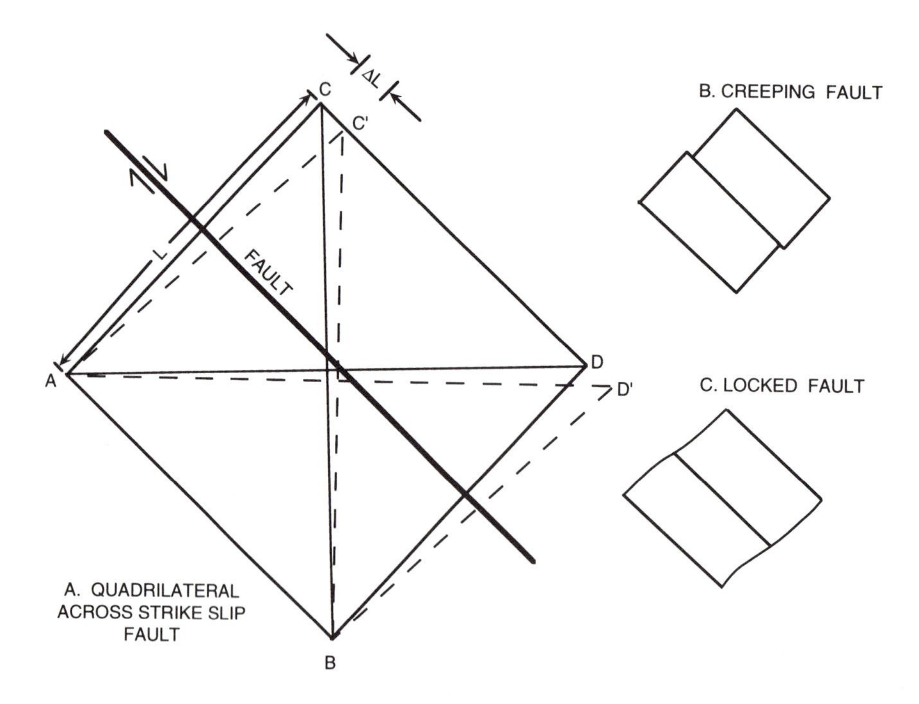

Figure 5–1. (a) Deformation of a quadrilateral across a right-lateral strike-slip fault. ABCD is deformed by shear to ABC′D′; shear is $\Delta L/L$. Diagonal BC is shortened to BC′, whereas diagonal AD is lengthened to AD′. (b) Fault may slip freely from surface downward, or (c) it may be locked to a certain depth, slipping freely below that depth.

beam returns to the instrument. The beam is modulated, and a comparison of the modulated phases of the projected beam and reflected beam determines the length of the path traveled by the beam in terms of modulation lengths. Corrections must be made for refraction or bending of the beam by the atmosphere; thus, the temperature and humidity of the atmosphere along the line of sight and the air pressure at instrument and reflector must be measured. The shear strain on the network is calculated on the basis of changes in line length based on several measurements through time. Because of measurement errors and nontectonic influences, the more measurements and the longer the time interval sampled, the more reliable the shear-strain rate will be, because the long-term tectonic trend will show through the nontectonic noise.

To illustrate the use of a high-quality trilateration network, we now discuss a network in the San Francisco Bay area of California, resurveyed repeatedly between 1970 and 1980 (Prescott et al., 1981). This network crosses the San Andreas, Hayward, and Calaveras right-lateral strike-slip faults (Fig. 5-2). The basic data are the changes in length of individual legs of the network, as illustrated in Figure 5-3. The least-squares fit of a straight line to these measurements gives the rate of change of length of a given line. Line lengths Allison-American and Hamilton-Llagas are analogous to diagonal BC in Figure 5-1a; they shorten with time due to accumulation of right-lateral shear strain. Line length Allison-Hamilton is analogous to diagonal AD; it lengthens with time.

The rotation and the horizontal translation of the entire network are unknown, because the network is not tied to a global reference frame (a problem solved later by VLBI and GPS). To deal with this deficiency, rates of change of line lengths for the entire network are combined to give the displacement rates of individual stations with respect to the center of mass of the network, which is arbitrarily assigned a velocity of zero. Displacement vectors are selected that minimize the station velocity components normal to the faults. This works well if deformation is by simple shear along a plane parallel to the local strike of the San Andreas fault system. (Simple shear is analogous to the "deformation" of a deck of cards by sliding each card the same amount with respect to its neighbor.)

The results are shown in Figure 5-4. In general, the displacements are parallel to fault strike, although there are extensional and rotational components east of the Calaveras fault. Figure 5-5a shows displacement rates resolved into components parallel to and

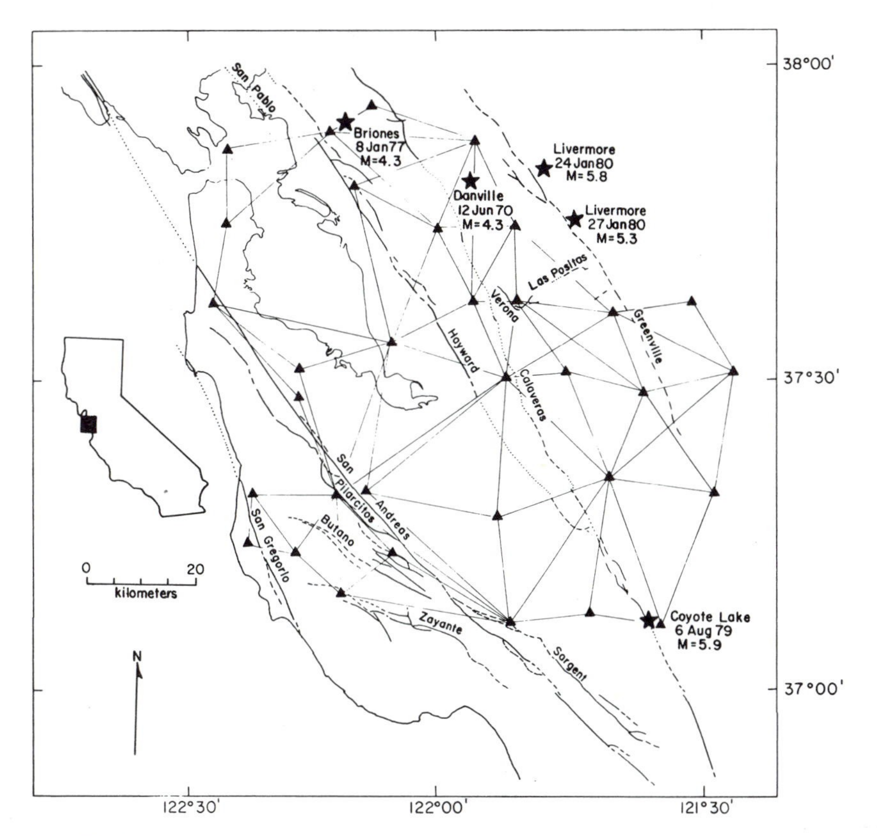

Figure 5–2. Map of San Francisco Bay area locating trilateration stations (triangles), faults, and major earthquakes during the 1970–1980 decade. From Prescott et al. (1981).

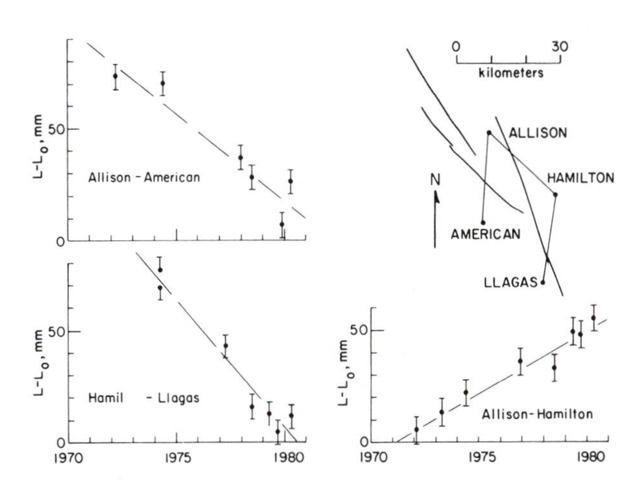

Figure 5–3. Change in line lengths ($L-L_O$) vs. time for three fault-crossing lines southeast of San Francisco Bay. Allison-American crosses the Hayward fault, and the other two lines cross the Calaveras fault. Error bars indicate plus and minus 1 standard deviation. Straight lines are least-squares best fits. Note that Allison-American and Hamilton-Llagas cut diagonally across the fault and are shortened (cf. BC, Fig. 5-1). Allison-Hamilton is lengthened (cf. AD, Fig. 5-1). After Prescott et al. (1981).

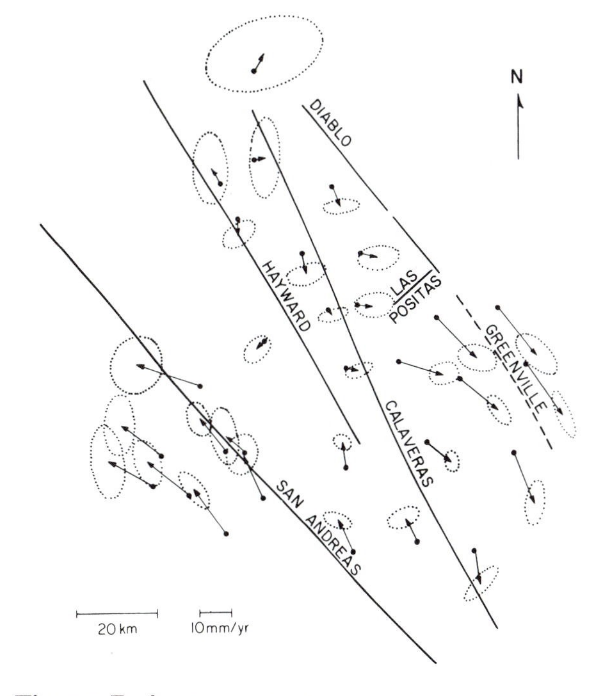

Figure 5–4. Vector displacement diagram for the San Francisco Bay network. Vectors give displacement rates in mm/yr. Error ellipses give 95% confidence region. After Prescott et al. (1981). For location, see Figure 5–2.

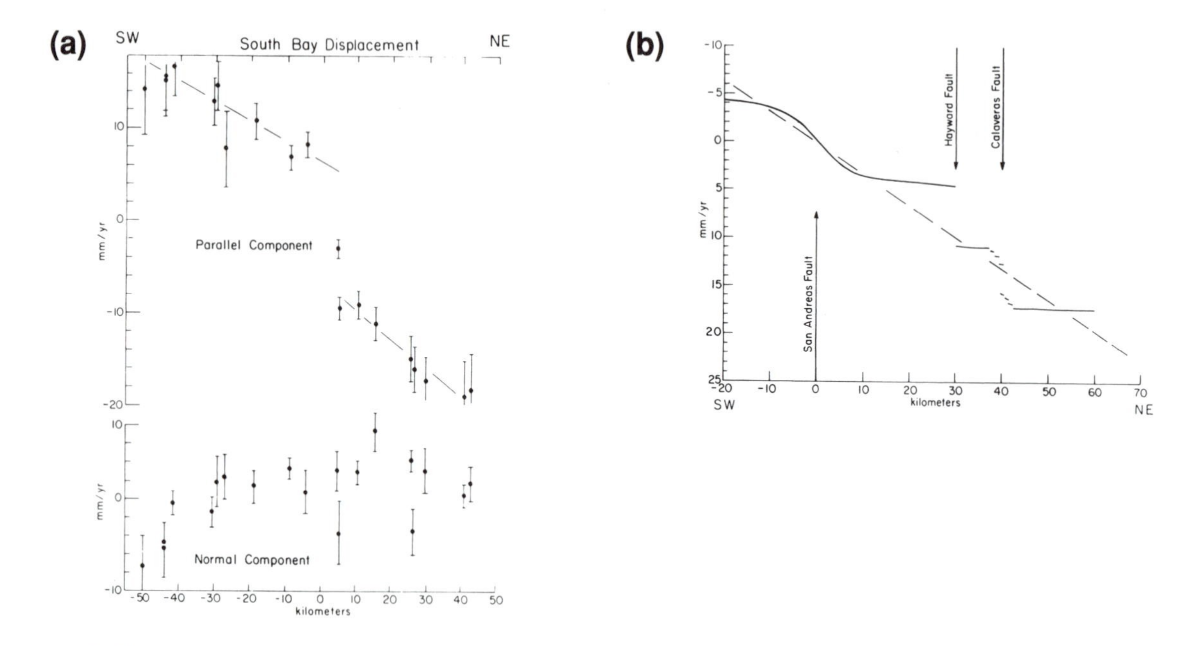

Figure 5–5. (a) Displacement vectors of Figure 5-4 resolved into components parallel to and perpendicular to N35°W in the South Bay area. Top diagram plots displacement in N35°W direction vs. distance from Hayward fault; bottom is displacement in N53°E direction as a function of distance normal to fault strike. (b) Smoothed version of top diagram of Figure 5–5a with zero velocity at San Andreas fault. After Prescott et al. (1981).

perpendicular to fault strike (N35°W) in the South San Francisco Bay area. Figure 5-5b is a smoothed and idealized version of the fault-parallel component of Figure 5-5a. The solid line assumes 12.2 mm/yr slip rate below 6.7 km on the San Andreas fault, rigid-block surface slip rate of 6.3 mm/yr on the Hayward fault, and 6.2 mm/yr slip on the Calaveras fault across a band 5 km wide. The dashed straight line shows the deformation of 0.40 μrad/yr, which characterizes the broad-scale pattern of shear across the entire region.

Horizontal deformation can also be measured using closely-spaced geodetic arrays and strainmeters, a technique known as *near-field tectonic geodesy*. Near-field tectonic geodesy has proven useful in the study of deformation of volcanic carapaces, starting in Iceland in 1938. Deformation accompanying volcanism commonly occurs at strain rates that are much higher than rates of tectonic deformation, and can, therefore, be effectively measured over short periods by near-field techniques.

Near-field tectonic geodesy began along the San Andreas fault beginning in 1960, soon after it was discovered that a segment of the fault southeast of San Francisco was slipping steadily (or nearly so), unaccompanied by large earthquakes, in a process termed *fault creep*. Techniques were designed to monitor this displacement across creeping segments of the San Andreas, Calaveras, and Hayward faults of the San Francisco Bay region. These techniques subsequently were used on the Imperial and Garlock faults of California and on the North Anatolian fault of Turkey.

A simple inexpensive technique for measuring horizontal deformation on a creeping fault is a line of nails in pavement across the fault (Fig. 5-6). Each nail can be located to within one millimeter using precision calipers tied to a straight line of sight with a theodolite (Sylvester, 1986). In Figure 5-6, the line is long enough to establish the width of the creeping fault zone with respect to rocks outside the fault zone.

An extensometer measures changes in distance between two fixed points. Information from two and preferably three extensometers oriented at different azimuths can be used to measure shear strain, as illustrated diagrammatically in Figure 5-1.

Constant-tension wire creepmeters have been mounted across creeping faults in central California. These consist of two piers on opposite sides of the fault connected by a wire fixed at one pier and extending through an underground conduit to the other pier. This wire is held under constant uniform tension. Temperature is recorded along with changes in line length to take into account the coefficient of thermal expansion of the wire. More expensive strainmeters, which use laser interferometry, are able to measure changes in line length much more precisely than creepmeters (0.15 μm in 800 m (0.0005 μstrain).

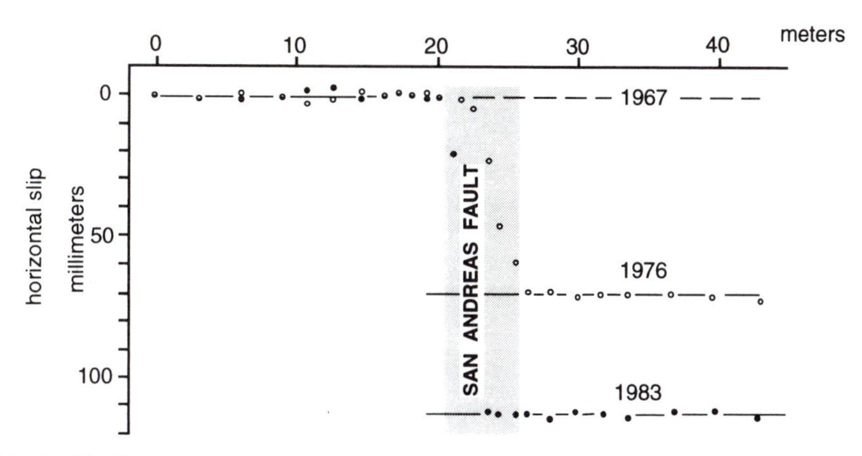

Figure 5–6. Offset of a nail line across the San Andreas fault (shaded) near San Juan Bautista, California. Line was straight in 1967. Vertical scale shows horizontal slip. After Sylvester (1986).

Short line-length measurements may measure only part of the displacement field across a fault zone. An advantage of such measurements made from fixed installations, however, is that measurements can be taken very frequently, even continuously. This allows nontectonic signals to be identified, thereby enhancing the detectability of tectonic strain signals, including silent earthquakes (discussed in Chapter 4). Near-field strain across faults in New Zealand is measured on fault-monitoring arrays of four to eight concrete monuments that are resurveyed periodically using an electronic distance meter (EDM) (Wood and Blick, 1986). An example of a New Zealand fault-monitoring pattern is shown in Figure 5-7. For a more detailed discussion, see Agnew (1986).

A two-color (red and blue) laser geodimeter array has been in operation since mid-1984 by the U.S. Geological Survey across the Parkfield segment of the San Andreas fault (Langbein et al., 1990; Fig. 5-8). Distance measurement is made with the red wavelength, and atmospheric correction is measured with the blue wavelength, so independent temperature and humidity observations are not necessary. For two-color laser geodimeters, typical standard errors of single measurements are 0.5-0.7 mm for 4-6 km of line length. Measurements are made several times a week to 18 stations from a central station at CARR. Precision of length measurements approaches 0.1 ppm. MID-E, BUCK, BARE, and CAN are north of CARR, so line lengths shorten with time across the right-lateral strike-slip fault zone, analogous to diagonal BC in Figure 5-1a POMO-CARR is analogous to diagonal A.D. However, a seasonal oscillation produced largely by winter rainfall is also apparent, especially for line MID-E. Filtering out the nontectonic seasonal oscillation aids in attaining the objective of detecting shorter-period tectonic variations that may be produced by an earthquake precursor, such as nonelastic pre-seismic deformation predicted by the model illustrated in Figure 3-19. Comparison of shorter and longer line-length variations may distinguish between near-surface slip and deeper slip along this part of the fault (Langbein et al., 1990).

VERTICAL DEFORMATION

The most important source of information about vertical deformation comes from repeated leveling surveys between benchmarks, commonly following highways, railroads, aqueducts, and pipelines. Highway leveling surveys made as early as the beginning of the twentieth century have been useful in studying long-term vertical changes, and tectonic studies using leveling data have been carried out in Japan since the

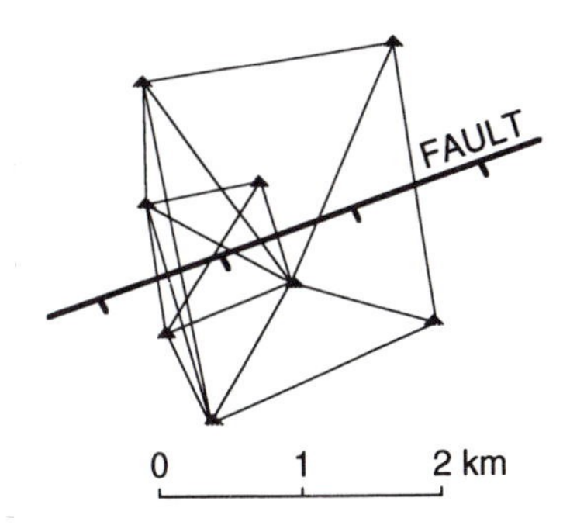

Figure 5–7. Typical New Zealand Geological Survey fault-monitoring pattern for measuring strain across an active fault (line with hachures). After Wood and Blick (1986).

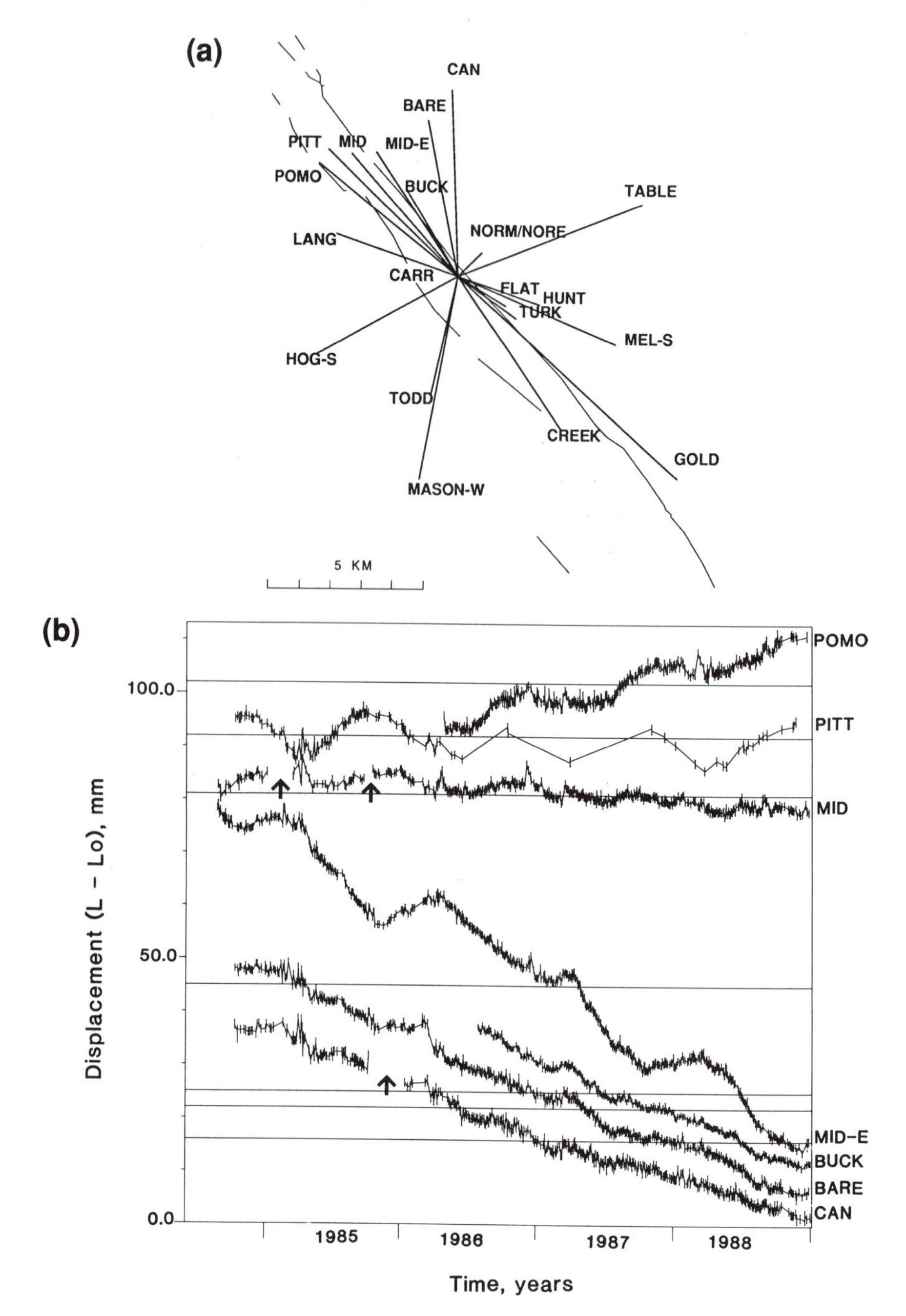

Figure 5–8. (a) Location of two-color geodimeter network at Parkfield, California, site of an experiment by the U.S. Geological Survey to predict an earthquake on the San Andreas fault. Active surface traces shown in solid lines. All network stations connect to a central instrument at CARR. (b) Line length changes between CARR and several stations across the main trace of the San Andreas fault. Error bars (vertical lines) represent one standard deviation level of each distance measurement. Arrow indicates break in record. After Langbein et al. (1990).

late 1920s. Standards of surveying for both procedure and instrumentation have been established by the U.S. Department of Commerce, and leveling surveys in the United States are the responsibility of the U.S. National Geodetic Survey.

Leveling surveys determine height differences between adjacent benchmarks, not horizontal differences. A line of such height measurements is often referred to a base station such as a tide gauge. Leveling is classified as first-order, second-order, and third-order, with first-order being the most accurate. First-order surveys must be double run (height differences between benchmarks measured in both directions), with maximum distance between leveling stations 50 m (meaning that the distance between the leveling rod and the leveling instrument is a maximum of 25 m). Because of these stringent measures and other strict procedures for first-order surveys, first-order leveling is slow and expensive. Leveling surveys with lower accuracy (second-order, third-order) are sometimes undertaken with less stringent observational procedures (longer distances between leveling staffs and asymmetrical viewing distances). Older surveying standards permitted longer sight lengths. When surveys of varying ages (differing surveying epochs) are compared for the same route, the older surveys generally have fewer stations, in part because many of the older stations cannot be relocated precisely. Present standards for leveling surveys were established in 1984.

There are several sources of error. An error in length of the 3-m-long invar rod produces height-dependent errors. (Invar is a steel-nickel alloy, characterized by a low coefficient of thermal expansion.) Calibration errors due to rod distortion are generally estimated as 10^{-5}, but may in cases be as large as 10^{-4}.

Errors caused by refraction of the line of sight between instrument and rods are also dependent on height and slope, because refraction varies with the density of the atmosphere, which is itself dependent on pressure, temperature, and humidity. These errors are strongly affected by height, because the uphill rod is read closer to the ground than the downhill rod, and temperature variations are most severe close to the ground. A survey showing a close correlation between height and height changes may be contaminated by rod and refraction error.

If levelings over different routes are compared, a correction related to differences in gravity between the two routes is needed, because the local horizontal (the equipotential surface) is distorted by gravity variation. These sources of error are discussed more thoroughly by Jackson et al. (1980), Strange (1981), and Castle et al. (1984, pp. 14–24).

Height differences along U.S. Highway 101 on the coast of Oregon and northern California based on leveling surveys in 1931, 1941, and 1987–1988 are compared in Figure 5-9a. No station is tied to a tide gauge, so that all stations could be undergoing uplift or all could be undergoing subsidence. The relative uplift rates vary along the coast during these time periods. They are lowest between 43.5° and 45.5°N latitude and highest farther south. The 1941 to 1930s comparison covers such a short time interval that systematic leveling errors dominate. It is included to show the negative cusp at 42.2°N corresponding to the topography which rises from 50 to 500 m, suggesting rod and/or refraction error (Vincent, 1989; Mitchell et al., 1994).

Several tide gauges along the survey route allow the relative elevation changes to be tied to sea level, after taking into account the effects of worldwide sea-level rise, monthly variations in sea level at tide-gauge stations, and changes in discharge of the Columbia River in the last few decades. Figure 5-9b shows that the tide gauges record the same variation in uplift rates that the leveling data do. Calibration to tide gauges also permits the determination of absolute rates of uplift, which vary from more than 5 mm/yr at 42°–43°N to zero at 45°N.

Figure 5-10 shows elevation differences across the Red Mountain reverse fault in the Ventura basin, California. In this case, benchmark I30, at sea level, was assumed to be invariant, although it is not tied to a tide-gauge record. The survey line north of the Red Mountain fault (north of benchmark G173) is rising, possibly due to the accumulation of strain on the active Red Mountain fault. The drop in benchmark F173, located in the center of the Ventura Avenue oil field, is due to subsidence accompanying fluid withdrawal from the oil field, a common source of error in sedimentary basins. Benchmarks I30, N30, and N173 may have subsided to a lesser degree, perhaps in response to compaction following withdrawal of groundwater from Ventura River gravels. Benchmark I30 is not surveyed to a tide gauge, and so our assumption that this benchmark is invariant may be wrong. If all benchmarks south of the Red Mountain fault have subsided due to fluid withdrawal, then the leveling lines provide no evidence for elastic-strain accumulation on the Red Mountain fault.

Another problem is benchmark instability caused by movement of the benchmark due to slope failure or soil creep.

Releveling of the region surrounding the source of the San Fernando, California, earthquake of February 9, 1971, soon after the earthquake permitted comparisons with observations in 1960–1963 and in 1929 for parts of the San Gabriel Mountains to the north as well as the San Fernando Valley. Figure 5-11 shows contours of vertical displacements surrounding much of the surface-rupture zone relative to a local benchmark north of Hansen Lake, which was arbitrarily held at zero. Leveling between this benchmark and a benchmark tied to a tide gauge at San Pedro, in Los Angeles harbor, indicated that the Hansen Lake reference benchmark probably moved downward about 2 cm with respect to sea level.

The area of surface faulting coincides with the

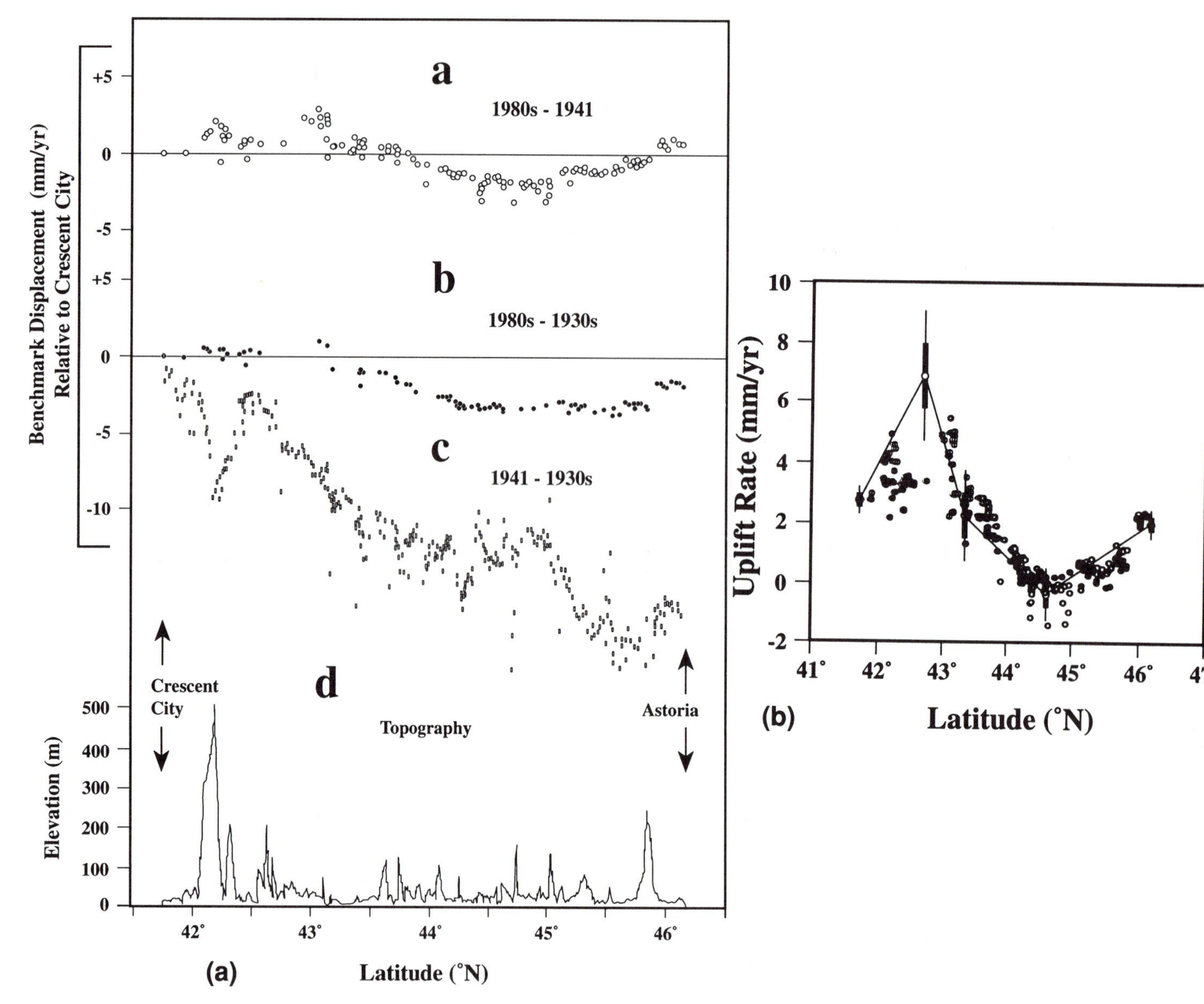

Figure 5–9. (a) Uplift rates relative to Crescent City, California, based on repeated first-order leveling of benchmarks along U.S. Highway 101, parallel to the Cascadia subduction zone. Topography along leveling route is shown at bottom. Leveling shows northward tilt toward 45.2°N and southward tilt toward 45.7°N. Note the effect of topography on 1931–1941 releveling at 42.2°N. After Vincent (1989). (b) 1980s to 1941 and 1980s to 1960s releveling data calibrated by tide gauges, shown by vertical bars; thick bars indicate one standard error based on monthly subsets of data. After Mitchell et al. (1994).

closely-spaced contours from zero to +1 meter. The contours show a narrow zone of tectonic subsidence south of the surface rupture and a much broader zone of uplift north of the surface rupture. The epicenter was north of the area covered by Figure 5-11. The data can be explained by an elastic-rebound model (Fig. 5-12) in which the crust had already accumulated elastic strain by the time of the pre-1971 surveys. Elastic rebound produced the localized uplift and subsidence close to the fault. The uplift may have been localized by steeper faults splaying off the main fault at depth. The elevation differences, when combined with horizontal changes, allow comparison of the observed deformation (Fig. 5-13a, b) with dislocation models (Fig. 5-13c, d) specifying fault dip; fault length parallel to strike, fault width down-dip, depth to the top of the fault, and dip-slip and strike-slip components. Two dislocation models are shown with two

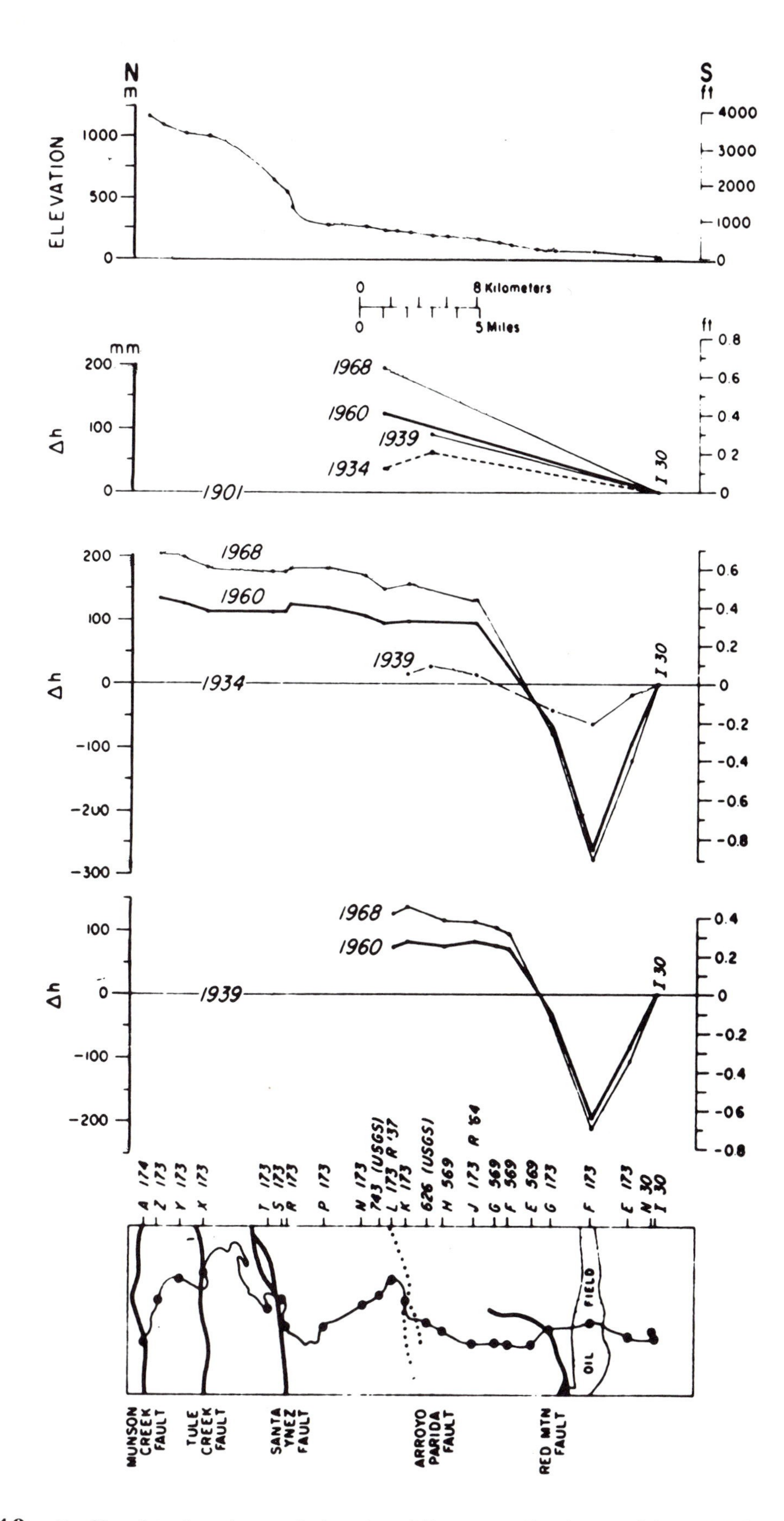

Figure 5–10. Profiles showing observed elevation differences (Δh) along California Highway 33 north from Ventura, measured with respect to benchmark I30, which is assumed to be invariant in elevation. Topography also shown. Note depression of benchmark F173 due to oil production from Ventura Avenue oil field. Steady "uplift", relative to benchmark I30, north of benchmark G173, may represent subsidence of all benchmarks south of the fault due to groundwater withdrawal rather than to accumulation of strain on Red Mountain fault. After Buchanan-Banks et al. (1975).

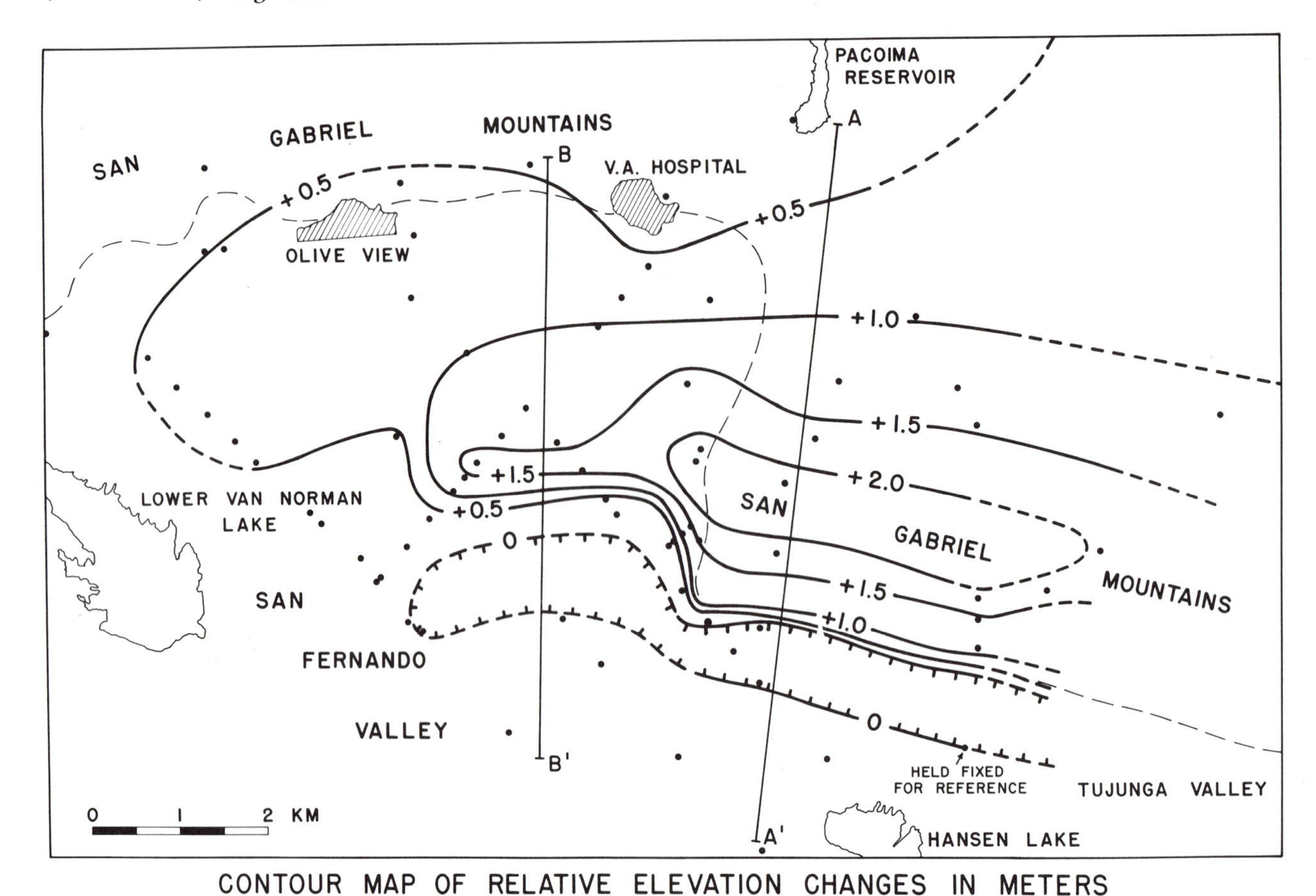

Figure 5–11. Vertical displacement in meters for the central and western area of the 1971 San Fernando, California, earthquake. Elevation differences are relative to a reference benchmark north of Hansen Lake which is arbitrarily held at zero, although this benchmark probably moved downward about 2 cm relative to a tidal benchmark at San Pedro, Los Angeles Harbor. A-A′ and B-B′ show profiles of Figure 5–13. After Savage et al. (1975).

values for fault widths. Thus the geodetic measurement of coseismic deformation permits an estimate of fault parameters independent of those determined from seismicity data.

Can releveling surveys provide information about earthquake faults that do not rupture the surface at all? For the 1983 Coalinga earthquake ($M_w = 6.5$), 1987 Whittier Narrows earthquake ($M_w = 6.0$), and 1994 Northridge earthquake ($M_w = 6.7$), all in California, the answer is yes. All three earthquakes were generated by slip on thrust faults beneath anticlines that grew by millimeters to centimeters during the events (see Chapters 2 and 10). From these geodetic measurements, it is now widely believed that many active folds, such as the Ventura Avenue anticline (Figures 2-24, 2-26), may be the result of hundreds of moderate-size earthquakes on a blind thrust, each raising the anticlinal crest a few millimeters to a few tens of centimeters. The top part of Figure 5-14 shows the coseismic elevation changes accompanying the 1987 Whittier Narrows earthquake in the metropolitan Los Angeles region.

The location of and displacement along a blind thrust is modeled from the geodetic data by considering the earth's crust as an elastic solid containing a planar cut, like a stiff block of rubber that has been partly sliced through by a knife. When the faces of the cut slide past each other, the rubber in the uncut part deforms elastically. The dimensions (area and slip) of the fault are selected so that the surface expression of the predicted deformation coincides with the actual leveling changes observed. Lin and Stein (1989) matched the observations at Whittier Narrows by using a thrust fault dipping 30° ±4°N with a reverse slip of 1.1 ±0.3 m with its upper edge at 12 ±1 km and its lower edge at 17 ±1 km (Fig. 5-14). The fault was further constrained to pass through the mainshock as determined from seismo-

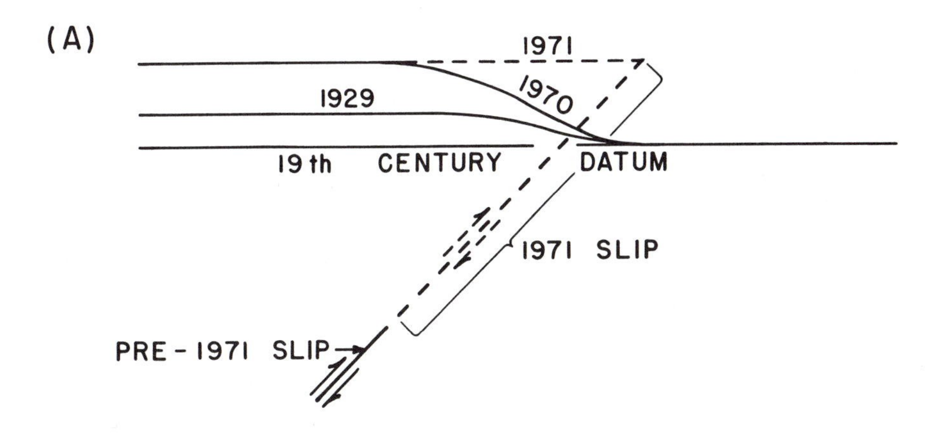

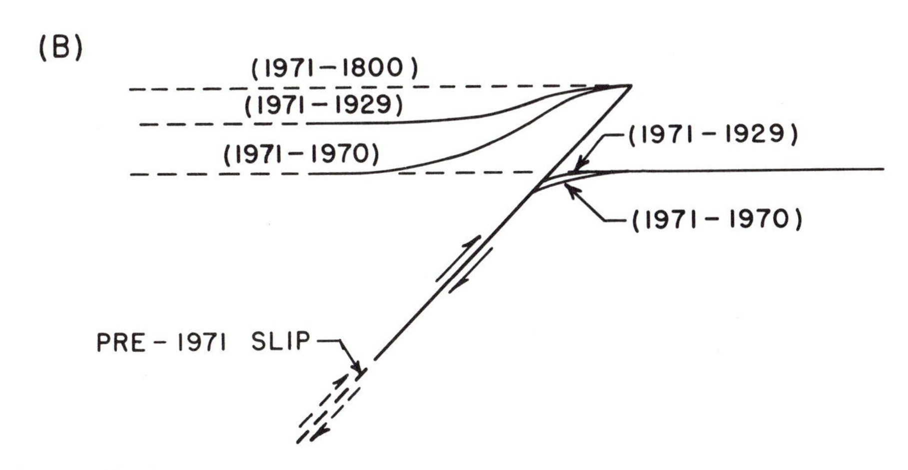

Figure 5–12. Elastic-rebound qualitative model for San Fernando fault and earthquake. (a) A hypothetical nineteenth-century horizontal datum accumulates elastic strain measured in 1929 and 1970, is deformed during earthquake, and is releveled shortly afterward. (b) Comparison of post-earthquake leveling with elevations determined in 1929 and 1970. After Savage et al. (1975).

graphic observations. Because the model results in a value for fault slip and, when combined with other cross sections parallel to strike, results in a value for fault area, it is possible to infer geodetically the seismic moment of the earthquake rupture. Although the model fits the observations reasonably well, note that it does not predict the elevation changes in the southernmost 7 km of the profile. In addition, the data would have fit a south-dipping fault equally well.

Elastic dislocation models for a vertical fault reaching the surface, an inclined reverse fault reaching the surface, and an inclined reverse fault not reaching the surface (a blind fault) are shown in Figure 5-15. Slip of one meter on the vertical fault is expressed as one meter at the ground surface, decreasing asymptotically away from the fault to approach zero. The vertical displacement on a fault dipping 45° is the fault slip multiplied by the cosine of 45°. If the fault is blind, the vertical displacement is distributed broadly above the fault, producing an anticline. Compare the geodetic signal of the inclined fault reaching the surface with that of the 1971 San Fernando earthquake fault (Figs. 5-12 and 5-13) and the geodetic signal of the blind thrust that produced the Whittier Narrows earthquake (Fig. 5-14). Figure 5-15c shows that the predicted elevation changes would be the same for a blind fault dipping to the right as for a fault dipping to the left. See King et al. (1988) and Stein et al. (1988) for additional background on this topic.

Near-field vertical changes are most commonly measured by precise leveling. Fault-crossing arrays have been in operation by the New Zealand Geological Survey and the University of California, Santa Barbara, for many years. An example of such an array shows the vertical separation across the San Andreas fault at San

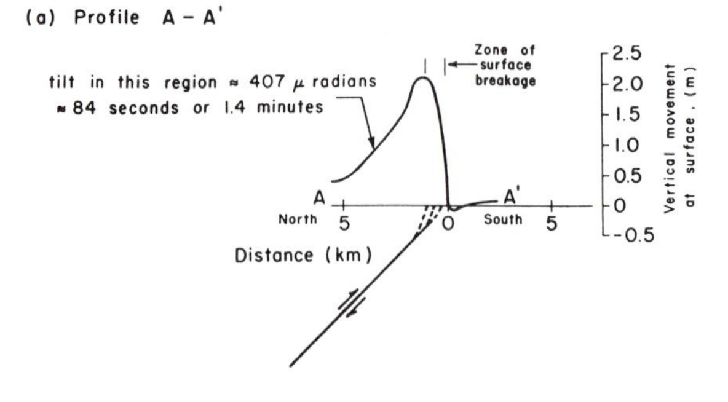

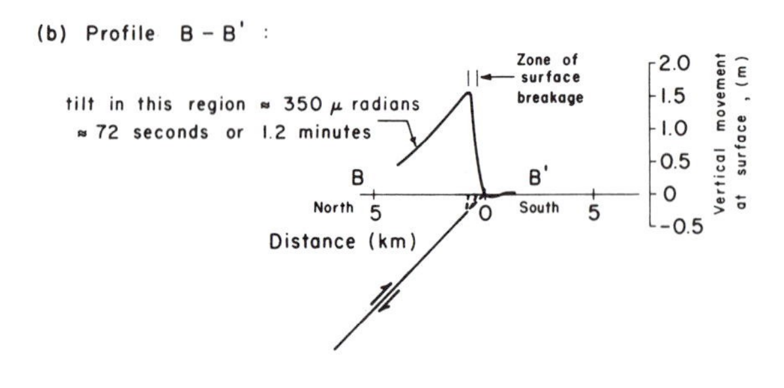

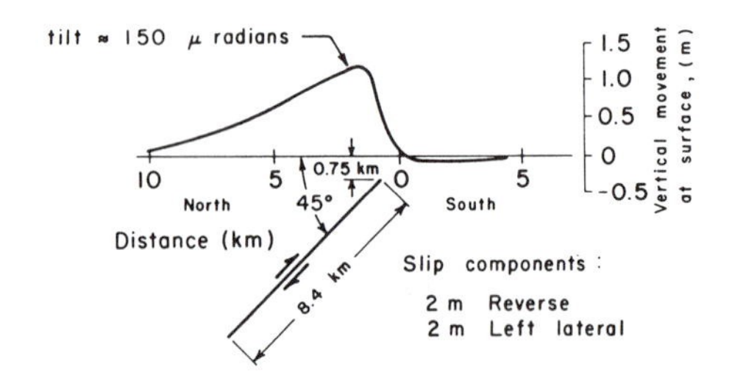

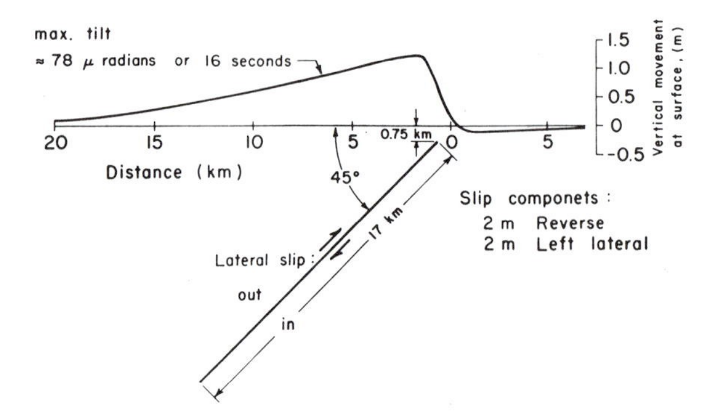

Figure 5–13. Profiles of observed vertical displacements (a,b) and theoretical vertical displacements (c,d) predicted by two dislocation models differing in down-dip width of fault. After Savage et al. (1975).

Juan Bautista for the period 1975–1983 (Fig. 5–16). The topographic profile shows a fault scarp, geomorphic evidence for a vertical component of displacement at the site, which was consistent with the data from periodic releveling. However, lowering of benchmarks northeast of the fault may be due to groundwater withdrawal from the relatively downthrown side (A. Sylvester, pers. commun., 1991), the same problem experienced on a larger scale across the Red Mountain fault (Fig. 5–10).

Tilt in three dimensions can be measured by using an optical spirit-level (dry-tilt), which determines the tilt of a plane defined by at least three benchmarks (Fig. 5–17). The tilt is measured by comparing height differences among the benchmarks in successive surveys, as determined by precision leveling from an instrument at the center of the array. Sites with the least nontectonic noise are on flat ground not perturbed by fluid withdrawal, slope instability, recent construction, or tidal loading. Because of the small apertures, the lack of redundant measurements, and random and systematic errors in leveling, however, the sensitivity of this method is no better than 0.01 mm in 50 mm, and accuracy is 1–10 μradians. Bevis and Isacks (1981) combined arrays with apertures of 1 km with sub-arrays with apertures of about 70 m to measure tilt in the New Hebrides (Vanuatu) island arc over a five-year period. They were able to achieve resolution of about 1 μradian on both the large and small arrays. Because of the nontectonic noise, the dry-tilt technique is best suited for determining large, rapid tilts such as those occurring on the flanks of active volcanoes. For further information on tiltmeters; see Agnew (1986).

SPACE GEODESY

Although conventional land-based techniques have yielded most of the interesting results up to now, a revolution has taken place in the past decade, and space-based geodesy will be used to generate most results in the future. We discuss four techniques: Very Long Baseline Interferometry (VLBI), Satellite Laser Ranging (SLR), Global Positioning System (GPS), and Synthetic Aperture Radar (SAR). VLBI and SLR generate relatively few points, but these points are used to establish a global reference frame for GPS and SAR, which measure more local deformation fields.

Very Long Baseline Interferometry (VLBI)

It seems incredible that the faint radio signals from quasars, objects that are billions of light-years away in

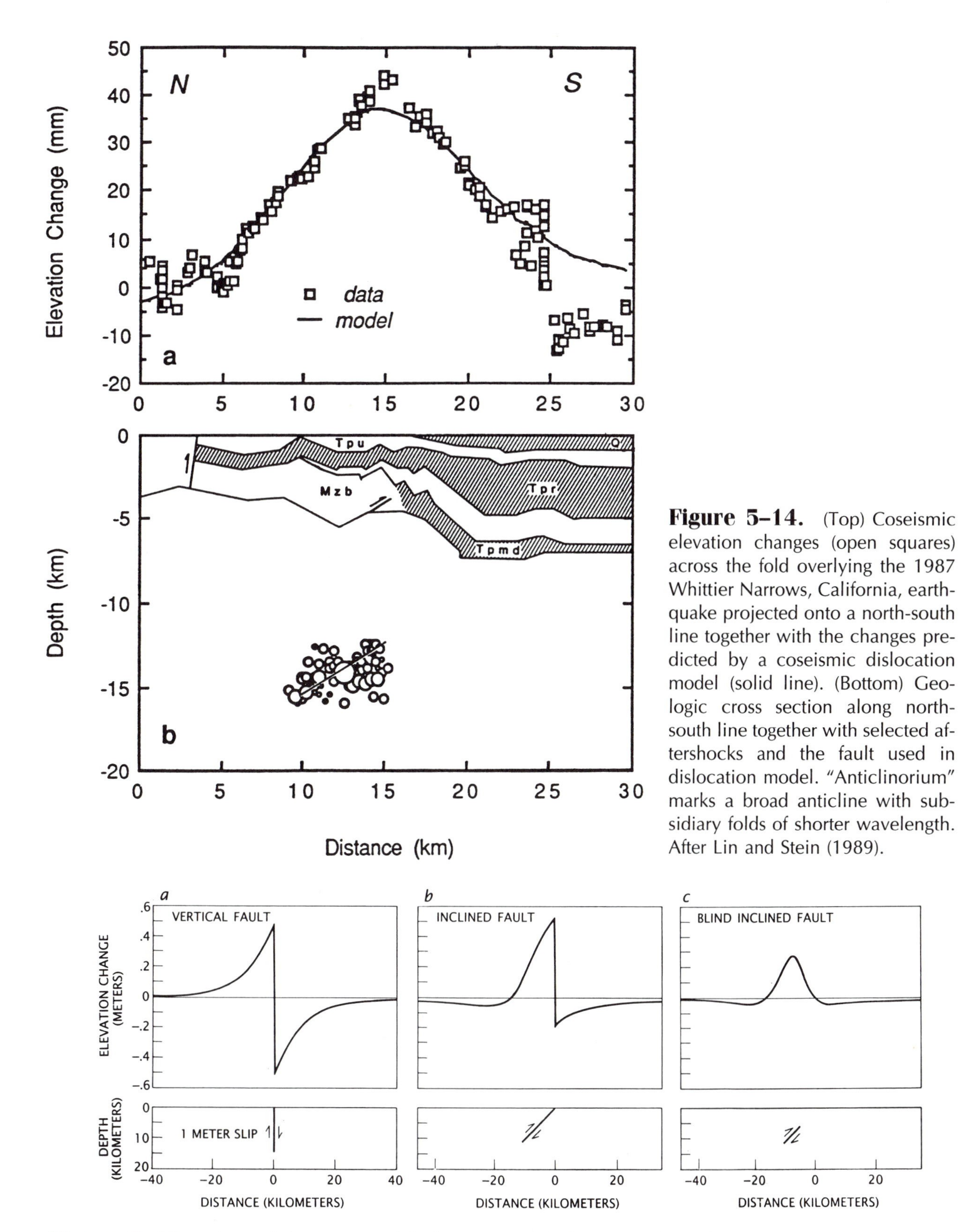

Figure 5–14. (Top) Coseismic elevation changes (open squares) across the fold overlying the 1987 Whittier Narrows, California, earthquake projected onto a north-south line together with the changes predicted by a coseismic dislocation model (solid line). (Bottom) Geologic cross section along north-south line together with selected aftershocks and the fault used in dislocation model. "Anticlinorium" marks a broad anticline with subsidiary folds of shorter wavelength. After Lin and Stein (1989).

Figure 5–15. Elastic dislocation models for a vertical fault cutting the surface, an inclined reverse fault cutting the surface (example: 1971 San Fernando earthquake), and an inclined reverse fault not reaching the surface (example: 1987 Whittier Narrows earthquake). Slip of one meter on the vertical fault is expressed as one meter at the fault, decreasing asymptotically along from the fault to approach zero. Surface uplift on a fault inclined 45° is only 70% of slip, or the slip multiplied by the cosine of 45°. If the fault is blind, the vertical displacement is distributed broadly above the fault, producing an anticline. After Stein and Yeats (1989).

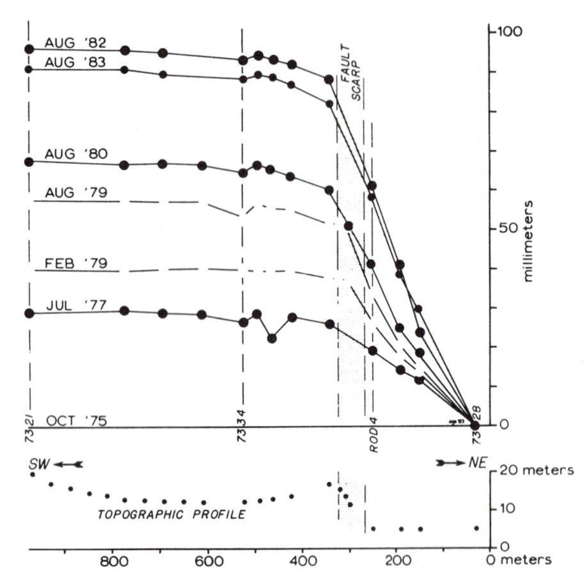

Figure 5–16. Array across the San Andreas fault (shaded) at San Juan Bautista showing change in height of the block southeast of the fault with respect to benchmark 7328 northeast of the fault, which is arbitrarily held fixed. After Sylvester (1986).

outermost space, can be used to measure deformation on the earth. Quasars are so far away that, unlike visible stars, no relative motion (called *proper motion*) has ever been detected for any of them, so that they serve as a celestial reference frame (Carter and Robertson, 1986). Because their radio signals are not blocked by clouds, they can be observed day or night in cloudy or clear weather.

Two or more radio telescopes as much as 10,000 km apart (*baseline distance*) observe the same quasar with a broad-band noise spectrum at frequencies of about 8 and 2 GHz (X-band and S-band, respectively). Because the radio source is so far from earth, the signals arrive as essentially planar wave fronts. Consider two radio telescopes A and B separated by a baseline vector AB (Fig. 5–18). As the earth rotates, the arrival

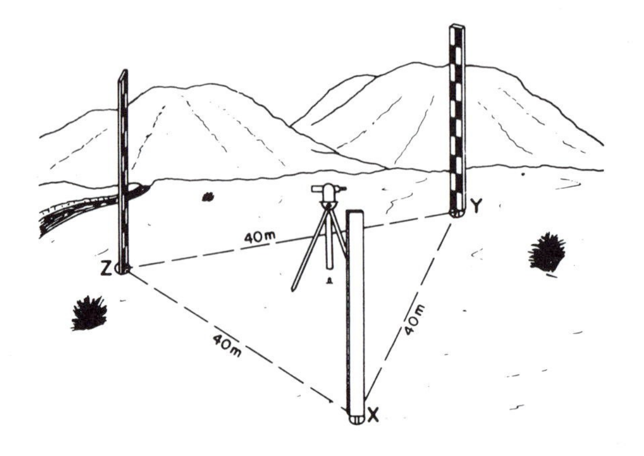

Figure 5–17. Example of dry-tilt array using three benchmarks. After Sylvester (1986).

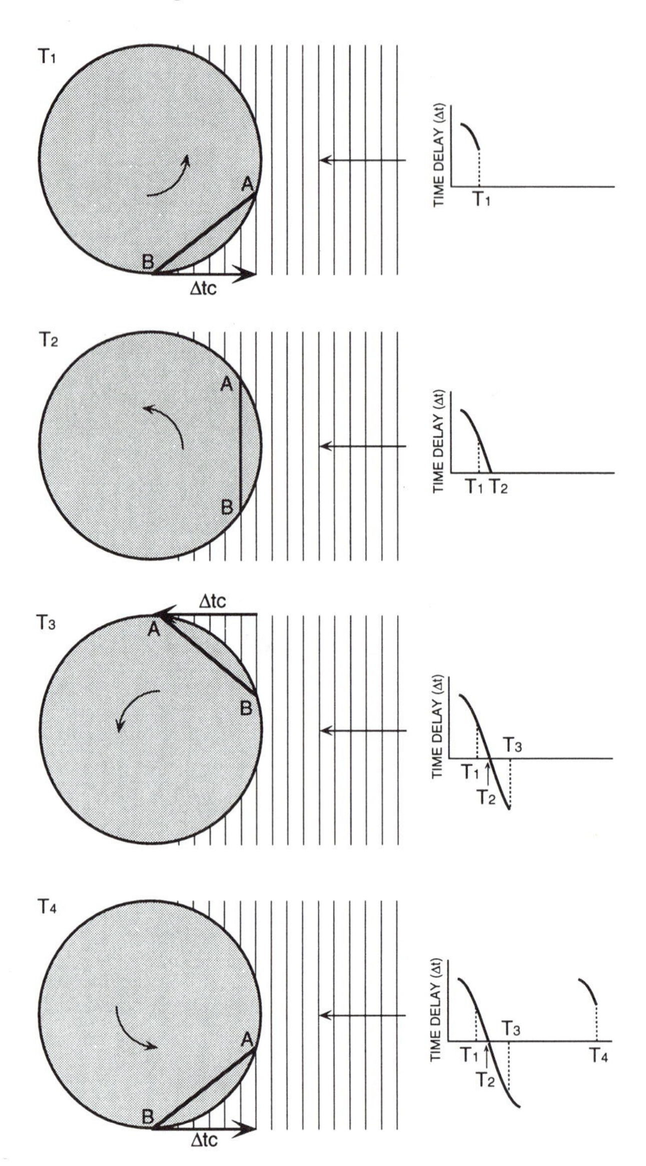

Figure 5–18. Radio telescopes A and B record the arrival of planar wavefronts from a quasar source. At time T_1, A receives signal before B, producing a time delay Δt_c. At time T_2, A and B receive the source signal simultaneously, and at time T_3, B records it before A. At time T_4, the earth has rotated once about its axis, and the time delay is the same as at T_1. The amplitude of the time delay (Δt) is proportional to the baseline distance vector AB. Changes in the time delay after repeated observations record changes in the length of AB due to crustal deformation. After Carter and Robertson (1986).

of a planar wave front occurs at A first, then at A and B simultaneously, then at B first. The time delay between the two radio telescopes is proportional to the baseline distance between them, the chord AB. The arrival time of the source signal is calibrated between radio telescopes by very precise atomic clocks, so that the baseline distance can be determined very accurately. Corrections must be made for earth tides, variations in earth rotation, and ionospheric and atmospheric effects. In a 24-hour observing period, up to 200 observations may be made to more than 24 radio sources (Bilham, 1991). Baselines several thousand kilometers long are surveyed to accuracies of 3 cm horizontally and 7 cm vertically.

By measuring changes in baselines between radio telescopes, VLBI has resolved the discrepancy between the slip rate of 34 mm/yr across the San Andreas fault in central California and the 48 mm/yr velocity of the Pacific plate relative to the North American plate. Motion of a telescope site at Vandenburg (numbered 48 on Figure 5-19) relative to telescope sites on the North American plate east of the Great Basin is the same as that predicted by plate tectonics, 48 mm/yr. The "discrepancy" is caused by extension across the Great Basin and strike slip on other faults east and west of the San Andreas fault (D. F. Argus in Gordon and Stein, 1992).

Figure 5-19. Horizontal velocity vectors of VLBI sites relative to stable sites in North America east of the Great Basin. The velocity at Vandenburg (vector numbered 48) is indistinguishable from the Pacific-North American plate velocity vector predicted by plate tectonics. After Gordon and Stein, (1992), based on work by D. F. Argus.

Satellite Laser Ranging (SLR)

In SLR, laser pulses are reflected from small satellites (LAGEOS, Starlette) orbiting at relatively low altitude, and their two-way transit time from and to the same optical telescope is measured. Simultaneous observations of the same satellite from several ground locations and at different times allow trilateration between the ground locations. Very accurate clocks, high-power laser pulse systems, and extremely sensitive optical detectors are needed for observations that achieve accuracies of 2-3 cm. Several weeks or months of data are needed for a three-dimensional location accurate to about 1 cm. SLR measurements are degraded by cloud cover in temperate latitudes.

Global Positioning System (GPS)

Since the mid-1980s, an increased amount of tectonic geodesy has been done using GPS, based on a group of NAVSTAR satellites orbiting the earth at an altitude of about 20,000 km. In contrast to VLBI, the radio signals are strong and well-structured, so that smaller receivers and less-precise timing may be used, greatly reducing the cost. A position on the earth can be located in three dimensions (horizontally and vertically) to an accuracy of several meters to a few tens of meters by determining the distance of the position from three satellites with known orbits. Each satellite broadcasts on two wavelengths, 19 cm and 24 cm, allowing correction for propagation velocity of the signal in the ionosphere.

For tectonic applications, *relative* position can be determined with an uncertainty three orders of magnitude smaller than position location. Location is three-dimensional (horizontal and vertical), so that GPS is the space-geodetic counterpart of both ground-based trilateration and leveling. Position can be established with respect to the VLBI reference frame, so that rotation can also be measured. Change of distance between source and receiver is measured rather than the distance. This is done by phase measurements of the carrier signal, keeping track of the number of cycles since the signal was acquired. Measurements of vertical position are several times less accurate than those of horizontal position.

Prescott et al. (1989) determined that measurements with GPS and Geodolite agree within 0.2 ppm over baseline distances of 10 to 40 km. Uncertainties for baseline lengths of 200 km are larger, but measurements agree well with other space-geodetic techniques. Although GPS surveying is affected less than terrestrial surveying by bad weather and variable atmospheric conditions, the main sources of error are the precise orbital parameters of the satellite and uncertainties in path delays in the troposphere. Selection of GPS control points is much simpler, particularly in remote areas where terrestrial surveying is

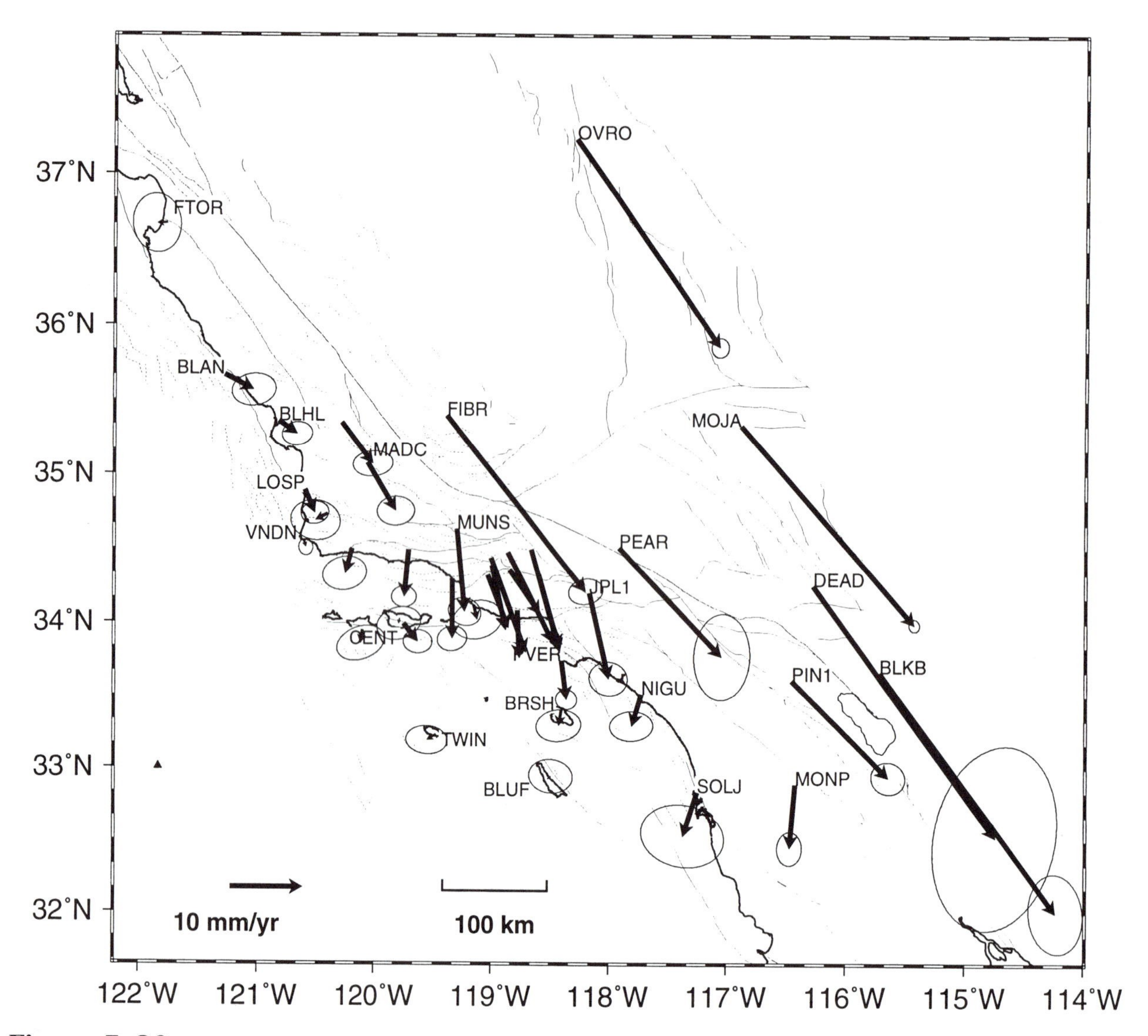

Figure 5–20. Velocities of stations with respect to the Pacific plate based on GPS and VLBI data. Ellipses show region of 95% confidence; not shown in the Ventura region (lat. 34.5°N, long. 118.5–119.5°W). After Feigl et al. (1993).

unreliable or difficult to obtain. Accurate data can be obtained with only a few hours of observation, and new techniques allow accurate solutions with less than a minute of observation.

Coastal California has been the best location to study active deformation, because (1) the initial GPS satellite orbits were designed to provide accuracy for testing there, (2) several VLBI radio telescopes in the region allow GPS measurements to be calibrated with respect to the celestial reference frame, and (3) deformation rates are high. VLBI data have been obtained for the period 1984–1992, and GPS data for the period 1986–1992 (Feigl et al., 1993). Figure 5-20 shows velocities of stations assuming that the Pacific plate is fixed. This shows that the motion of North America with respect to the Pacific plate is only partly accommodated at the San Andreas fault. If the predicted strike-slip displacement along the San Andreas fault system is removed (Fig. 5-21), the residual velocity field shows that additional strike slip is taken up on faults in the southern Coast Ranges west of the San Andreas fault. In addition, the western Transverse Ranges are shortening north-south at a rate of about 5 mm/yr. The residual velocity data can also be displayed as axes of horizontal strain rates (Fig. 5-22), and, because the network is tied in to a global reference frame, rotation rate can also be estimated (Fig. 5-23).

The uncertainty of GPS measurements is dependent on the number of times the network can be occupied. This has led to the establishment of a permanent GPS geodetic array (PGGA) to monitor crustal deformation in near real-time to millimeter accuracy. Position of the initial PGGA stations has been monitored daily since August 1991 with respect to the International Terrestrial Reference Frame, which is based on a global network. An example of PGGA data is shown in Figure 5-24. Figure 5-25 shows the absolute displacements of PGGA stations associated with the 1992 Landers and Big Bear earthquakes with respect to a global reference frame.

Bilham (1987) suggested that GPS networks are the preferred method for long-range earthquake forecasting in heavily populated megacities located in tectonically active areas. GPS also allows the calculation of vertical changes in addition to horizontal changes, although less precisely, thereby supplementing leveling surveys. The NAVSTAR satellites were designed for military navigation, and their use could be downgraded by the military by scrambling systems or by encrypting the satellite transmission.

Synthetic-Aperture Radar (SAR)

SAR acquired by the ERS-1 satellite was used to map the displacement field of the 28 June 1992 Landers, California, earthquake (Massonnet et al., 1993). SAR measures the ground reflectivity and the distance (range) between the radar antenna and the ground. Images from an altitude of 785 km, pointed west at

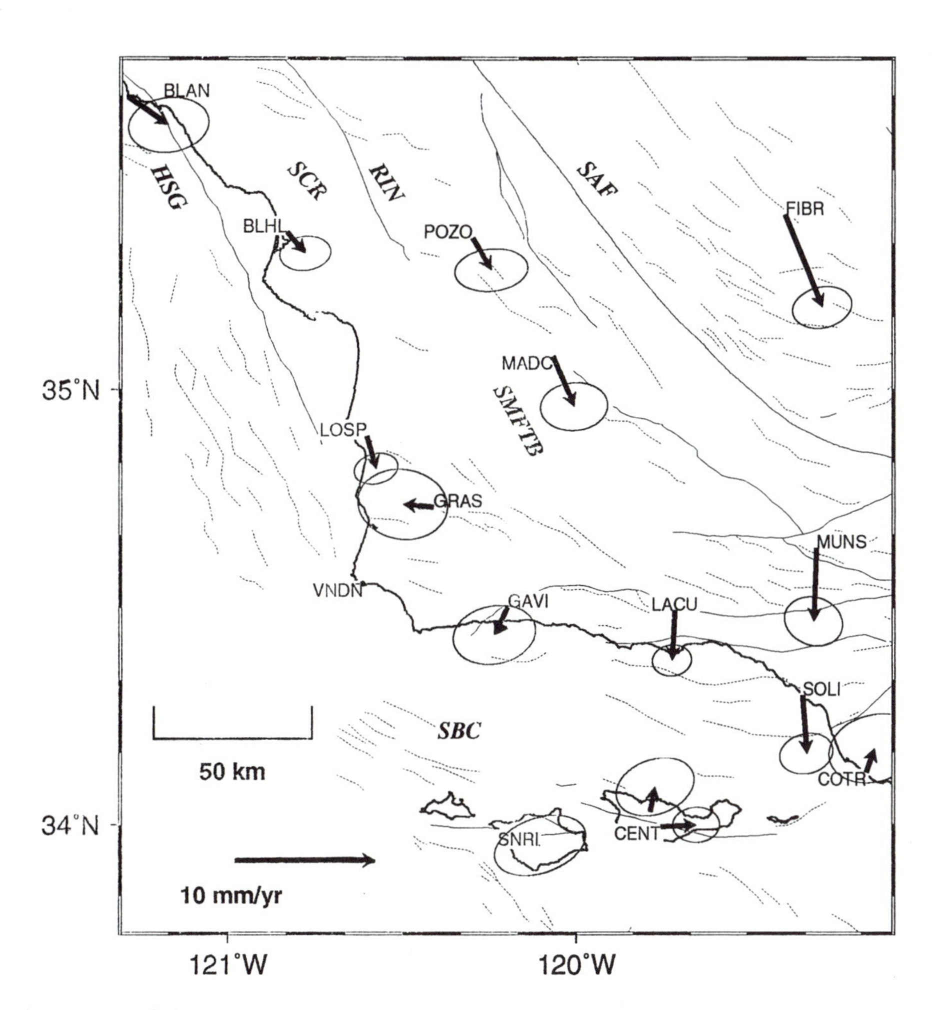

Figure 5–21. Residual velocities of stations relative to VNDN, site of a VLBI radio telescope, after removal of strike-slip displacement on the San Andreas fault system. This shows convergence across the Santa Barbara Channel (SBC) and Ventura basin to the east at about 5 mm/yr, as well as strike slip on faults west of the San Andreas fault (SAF), including the Rinconada (RIN) and Hosgri (HSG) faults. After Feigl et al. (1993).

an angle 23° from the vertical, were acquired before (24 April) and after the earthquake (7 August). The two images were taken from similar orbits and under conditions of similar ground reflectivity so that they could be superimposed, canceling out the topographic differences except for the component of displacement that affected the range. The resulting image (Fig. 5-26) contains interferometric fringes that are a contour map of the changes of range relative to points far enough away that they are assumed to be unaffected by displacements accompanying the earthquake. Each fringe is equivalent to 28 mm relative change in range, half the wavelength of the ERS-1 SAR. The fringes are incoherent close to the surface rupture because complexities in fault geometry produced displacements that are too local and too abrupt to be resolved by SAR. The displacements are consistent with those measured by GPS and ground stations and with an elastic dislocation model.

Summary

Plate tectonics shows that deformation on the earth's surface is best described by relative motions of rigid plates with narrow to very broad boundaries (Chapter 1). Relative motion between plates is based on geological and geophysical data averaged over timescales of a million years or more. Space geodesy shows that motions averaged over long timescales are similar to those during a period of only a few years, a demonstration that plate motion is steady (Gordon and Stein, 1992).

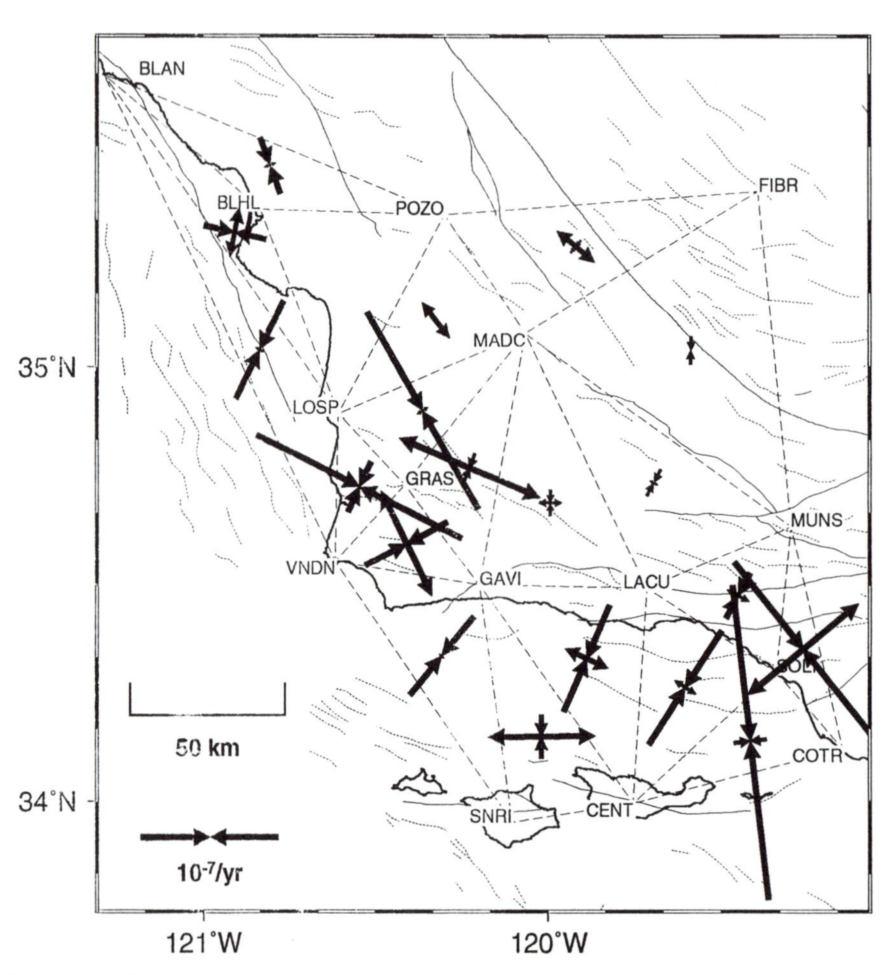

Figure 5–22. Principal axes of horizontal strain rate based on the residual velocity vectors of Figure 5–21. Inward-pointing arrows show compression; outward-pointing arrows show extension. Length of arrow is proportional to strain rate. After Feigl et al. (1993).

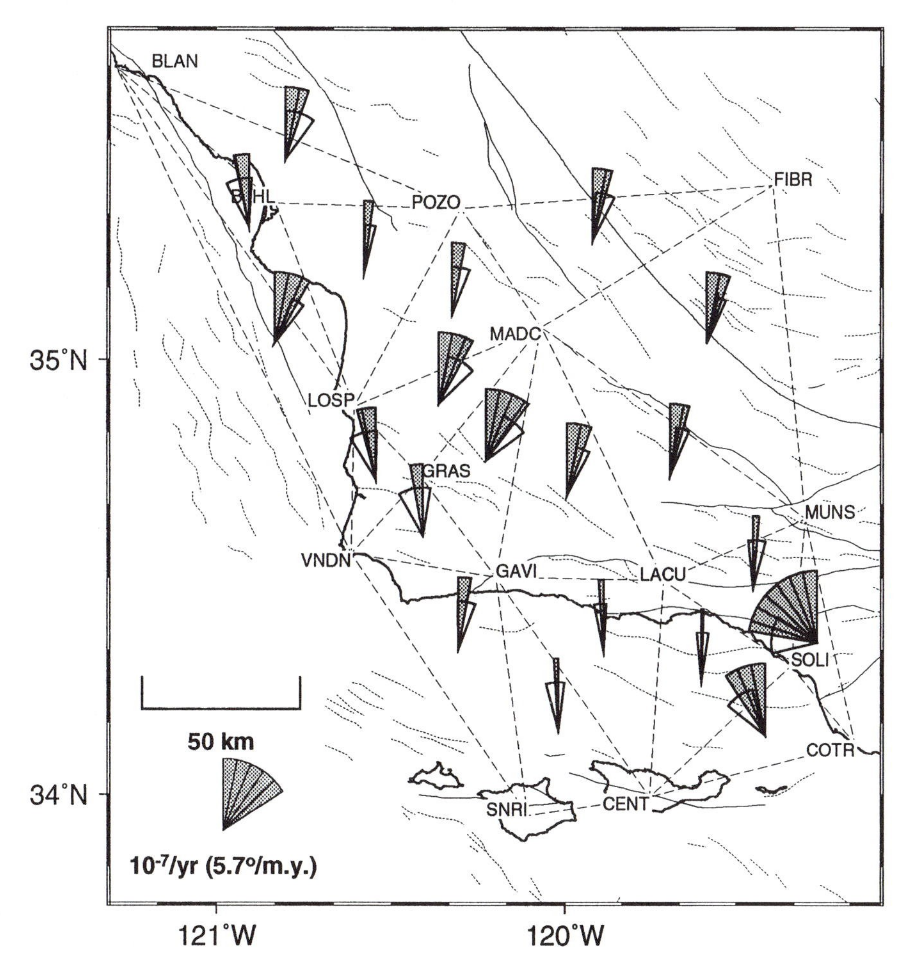

Figure 5–23. Average rotation rates inferred for the residual velocity field of Figure 5–21. Rates are shown as gray fans, in which each fold of the fan denotes 1.1°/m.y. White wedge represents plus or minus one standard deviation. After Feigl et al. (1993).

MEASUREMENTS OF IN SITU STRESS

The previous sections of this chapter have discussed the measurement of crustal strain. An important objective of these measurements is the determination of the orientation and magnitude of the principal stresses σ_1, σ_2, and σ_3 in the crust (in situ stress). However, as in the previous sections, we are actually measuring strain, and this section could be called ultra-near-field tectonic geodesy.

In situ stress may be *tectonic,* caused by forces at plate boundaries, large-scale flexure by the lithosphere, or uplift accompanying isostasic compensation, or it may be *local,* caused by topographic relief, heterogeneity in rock properties, or bending moment in local folds. Tectonic stresses are uniform over distances much greater than plate thickness, whereas local stresses are uniform over distances much less than plate thickness (Hickman, 1991).

Estimates of stress orientations from strain data yield surprisingly consistent results for large regions (Zoback and Zoback, 1980). These results led to a program of the Inter-Union Commission on the Lithosphere to construct a world map showing the distribution of principal stress axes (Fig. 5-27; Zoback, 1992). Information comes from fault-plane solutions for earthquakes (discussed in Chapter 4, along with their limitations), earthquake shear-wave polarization (see Chapter 4), orientation and displacement on conjugate fault systems (see Fig. 2-13), orientation of dikes and elongation of volcanoes, assumed to be in a plane normal to σ_3 (Nakamura, 1977), and trend of fold axes and strike of reverse faults, assumed to be normal to σ_1. This section focuses on direct *in situ* measuring techniques, generally used in boreholes.

In *overcoring*, a hole is drilled in rock that is free from anisotropies such as existing faults, joints, or bedding planes (Fig. 5-28). Gauges are placed in the hole to measure strain in three mutually perpendicular directions. The hole is then overcored such that the annulus of the new hole has a diameter greater than the original hole, presumably relieving the stress in the inner hole previously measured with the strain gauges. The displacements on the strain gauges as a result of overcoring are assumed to represent the recovery of elastic strain after the *in situ* stress is removed, so that the *in situ* stresses can be estimated. The principal drawback to this technique is the length of the hole, only a few meters. The orientation of the near-surface stresses may be very different from orientation of stresses at depth. Another technique is to cut a slot in a rock after first measuring the distance between two points across the slot. Cutting the slot causes a decrease in the distance between the two points. A rubber insert is then placed in the slot, and pressure is increased in the insert, using a hydraulic pump, until the points are the same distance apart they were before the slot was cut. This pressure is called the *slot stress*. These measurements are influenced by rock inhomogeneity, including residual stresses, and topography.

A second technique is *hydrofracturing*, used in the petroleum industry to increase production of oil in wells. A section of the hole is isolated by inflatable rubber packers and pressurized by pumping fluid into the hole until the rock fails by tensile fractures, with the fracture plane assumed to be perpendicular to σ_3. If the drill hole is vertical, and the load stress is a prin-

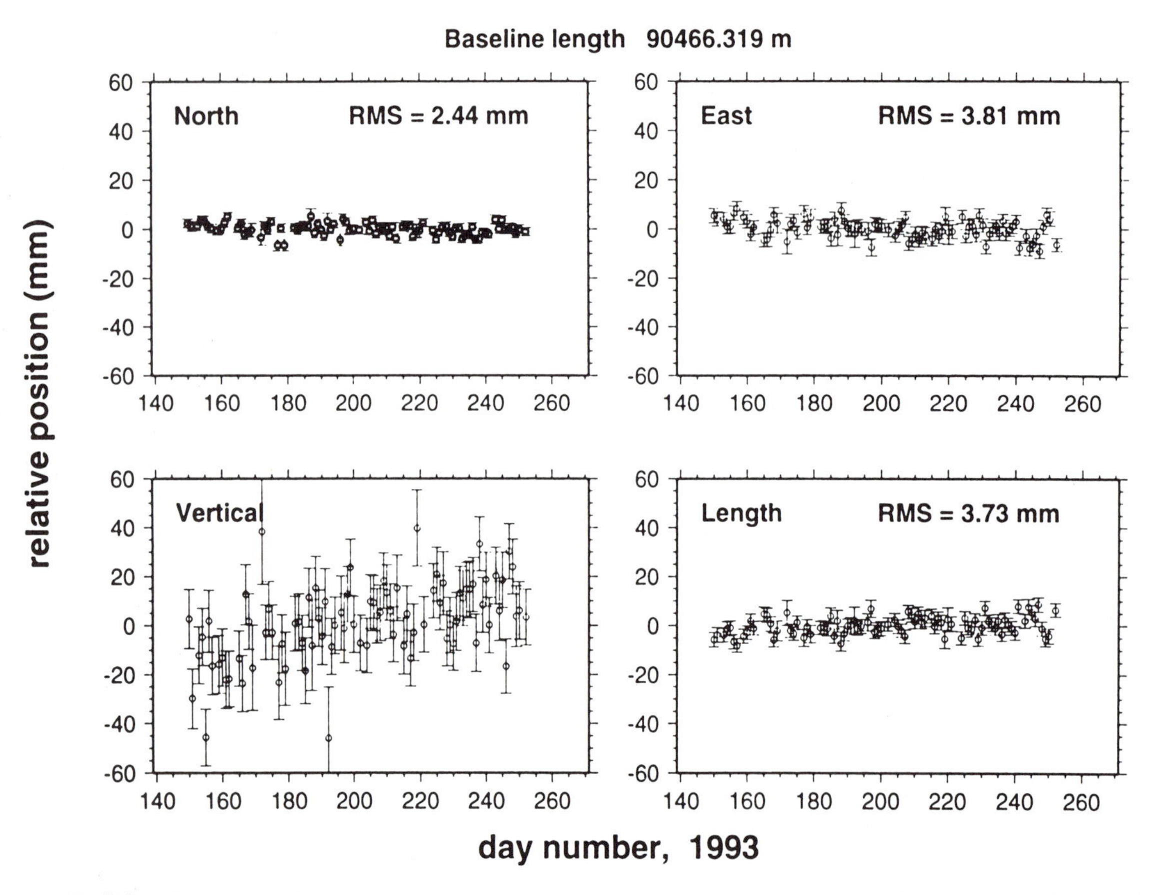

Figure 5–24. Daily baseline determinations between permanently occupied GPS sites at Lake Mathews (MATH) and Palos Verdes (PVEP), California, in terms of north, east, and vertical components, and baseline length. Note the larger error bars for vertical determinations. Each point represents a solution based on 24 hours of data. After Bock (1994).

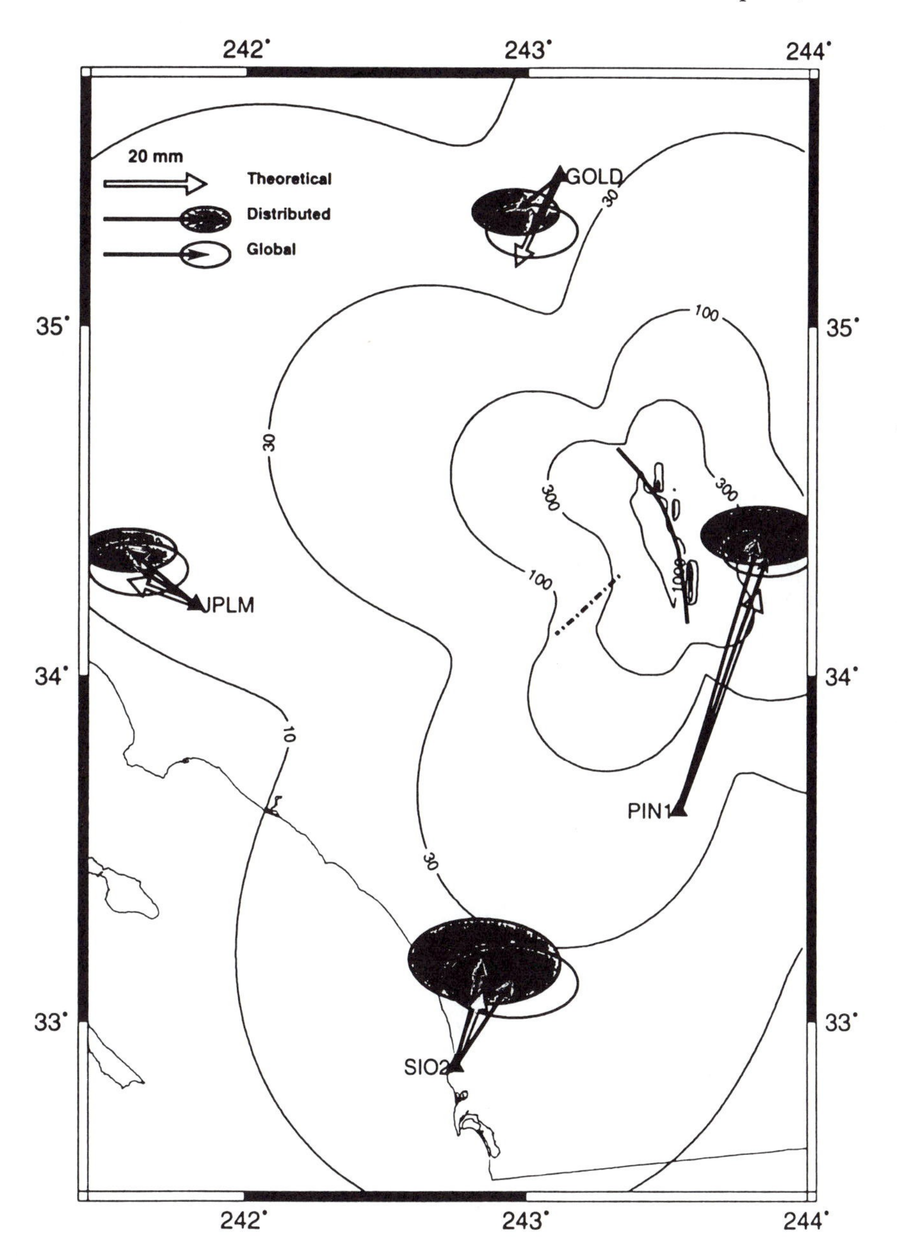

Figure 5–25. Observed (solid arrows) and modeled (blank arrow) displacement at four permanent GPS stations in southern California due to 1992 Landers (solid line) and Big Bear (dash-dot line) earthquakes. Two solutions are shown (shaded and open ellipses). Displacements are with respect to a global reference frame. Contours are in millimeters. After Bock (1994).

cipal stress, the fracture strikes in the direction of σ_1 or σ_2 (Haimson and Fairhurst, 1970). Repeated measurements at different depths in the same borehole have a range of $\pm15°$ (Zoback and Zoback, 1980). Magnitudes of principal stresses can be calculated from the pressures at which the rock fractures and from the shut-in pressure immediately after hydrofracturing, but these measurements are of limited accuracy because of assumptions that must be made about the rock condition prior to hydrofracturing.

A third technique is the use of ***borehole breakouts,*** the elongation of a once-circular vertical borehole due to differential horizontal stress (Plumb and Hickman, 1985; Zoback et al., 1985). Drilling relieves *in situ* stress in the walls of the borehole, resulting in spalling of the walls and elongation of the hole in the direction of least horizontal stress (Fig. 5-29), as confirmed by study of the borehole walls using a downhole televiewer. Borehole elongation has been shown to be relatively independent of lithology and bedding dip. Noncircular borehole cross sections may be oriented using four-arm, dual-caliper logs run in connection with continuous dipmeter logs run in near-vertical boreholes.

A stress map based on borehole breakouts in California is shown as Figure 5-30a. The arrows show the orientation of maximum horizontal stress, a principal stress if the load stress is also a principal stress. The

Figure 5–26. Interferometric fringes obtained by superimposing ERS-1 synthetic-aperture radar images taken before and after the June, 1992, Landers, California, earthquake. One cycle of gray shading represents a range difference of 28 mm between the two images. Irregular white line is surface rupture accompanying the earthquake. Compare with Figure 5–25. After Massonnet et al. (1993).

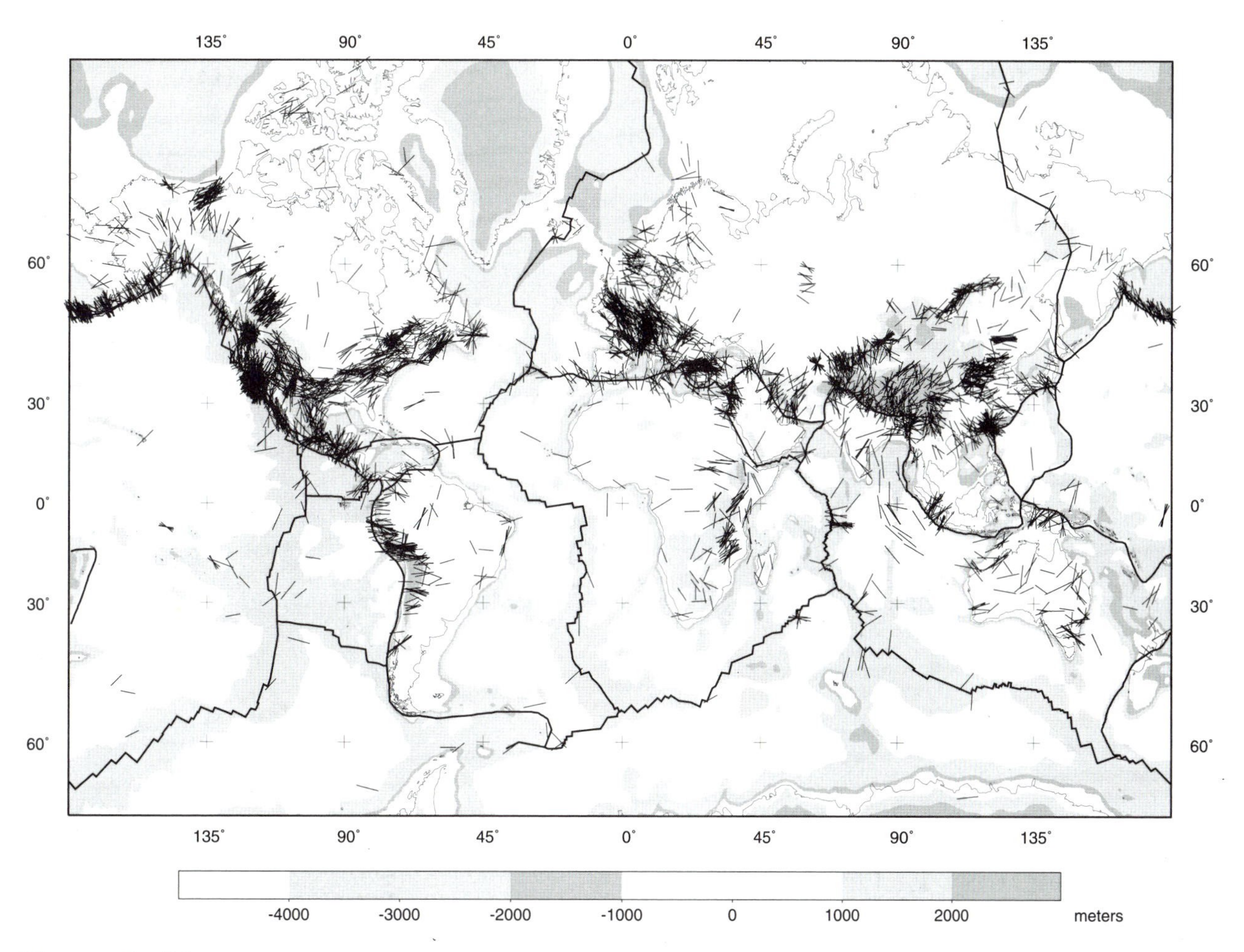

Figure 5–27. Maximum horizontal stress orientations on a base of average topography of the earth. From Zoback (1992).

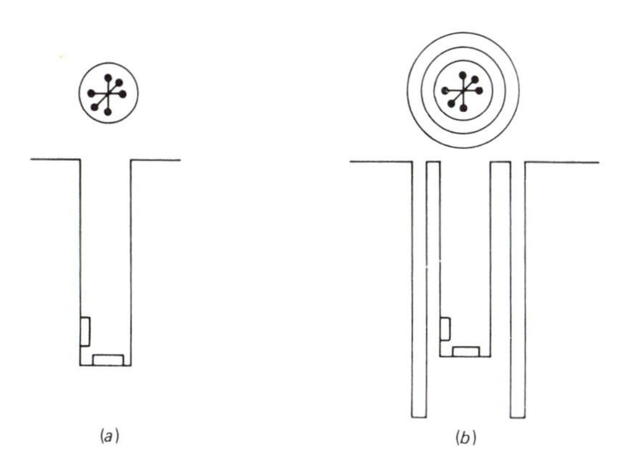

Figure 5–28. Overcoring for measurements of *in situ* stress. (a) A hole is drilled, and strain gauges are placed on sidewall and base. (b) Annular hole is cored such that the core includes the first hole with its strain gauges. The annular hole is assumed to relieve *in situ* stress, which is recorded as strain changes which are then used to estimate stress. After Turcotte and Schubert (1982).

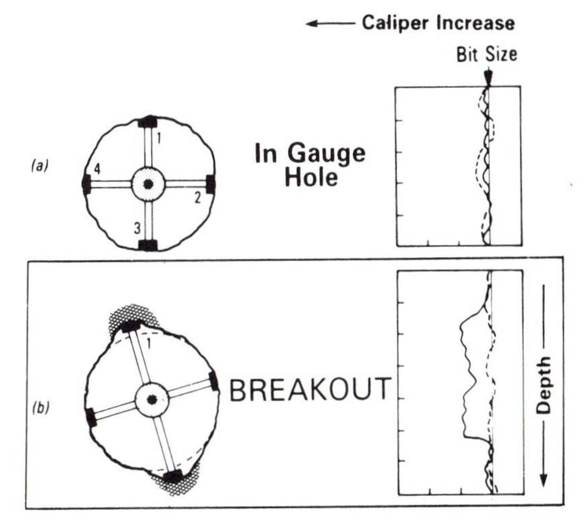

Figure 5–29. Cross section of borehole with four-arm dual-caliper in place, with caliper logs on right side. Solid line, calipers 1 and 3; dashed line, calipers 2 and 4. (a) Hole is in gauge with bit size. (b) Hole is elongated due to stress-induced breakout. After Plumb and Hickman (1985).

MAXIMUM HORIZONTAL COMPRESSION

BOREHOLE ELONGATIONS

MAXIMUM HORIZONTAL COMPRESSION

BOREHOLE ELONGATIONS HYDROFRAC & GEOLOGIC FOCAL MECHANISMS

(Modified from Zoback, Zoback, Mount, Suppe, et al., 1988)

Figure 5–30. (a) Maximum horizontal stress directions, indicated by short bars, inferred from 118 oil and gas wells. Longer bars indicate higher-quality data. (b) Stress orientations inferred from breakouts together with those inferred from earthquake fault-plane (focal-mechanism) solutions, measurements of hydraulic-fracture orientations, and geologic data such as trend of fold axes. After Mount and Suppe (1992).

maximum horizontal stress is σ_1 except in those cases in which σ_1 is the load stress; in such cases, the maximum horizontal stress is σ_2. The maximum horizontal stress based on borehole breakouts is consistent with that based on earthquake fault-plane solutions, hydraulic fracture measurements, and orientation of fold axes (Fig. 5–30b) Borehole breakouts in California are fairly consistent, despite the complex geology, and they show that σ_1 is oriented nearly at right angles to the strike of the San Andreas fault (Mount and Suppe, 1992). This result came as somewhat of a surprise, because it was assumed that the San Andreas fault followed a plane of high shear stress, and therefore σ_1 should be oriented north-south. The orientation is consistent with the view that the San Andreas fault is a fairly weak fault (Zoback et al., 1987).

SUMMARY

The geodetic study of earthquakes has come of age since Reid used the displacement of benchmarks east and west of the San Andreas fault to propose his elastic-rebound theory of earthquakes in 1910. Early work focused on triangulation and leveling, but trilateration became more important in the 1960s. In the last decade, space geodesy, principally GPS supplemented by VLBI, has taken over as the technique of choice.

Geodesy has aided in the study of individual earthquakes since geodetic evidence for coseismic displacement was first noted during the 1892 Sumatran earthquake. In addition, geodesy has played an important role in understanding the earthquake cycle. Geodesy allows strains to be measured far from individual faults, and in addition, space geodesy using VLBI has confirmed that plate motions measured in geologic time scales of millions of years are comparable to those measured in decades.

Measurement baselines vary from 10,000 kilometers (VLBI) to the diameter of a borehole (*in situ* stress). Measurement of error is in many cases scale dependent. As early workers in the field found out, the determination of errors in measurement is a major part of tectonic geodesy.

Suggestions for Further Reading

Agnew, D. C. 1986. Strainmeters and tiltmeters. Reviews of Geophysics, 24:579-624.

Bilham, R. 1991. Earthquakes and sea level: Space and terrestrial metrology on a changing planet. Reviews of Geophysics, 29:1-29.

Bomford, G. 1971. Geodesy. Oxford: Clarendon Press, 731 p. (the classical text).

Carter, W. E., and Robertson, D. S. 1986. Studying the Earth by very-long-baseline interferometry. Scientific American, 255(5):46-54.

Castle, R. O., Elliot, M. R., Church, J. P., and Wood, S. H. 1984. The evolution of the southern California uplift, 1955 through 1976. U.S. Geological Survey Professional Paper 1342, 136 p.

Dixon, T. H. 1991. An introduction to the Global Positioning System and some geological applications. Reviews of Geophysics, 29:249-76.

Gordon, R. G., and Stein, S. 1992. Global tectonics and space geodesy. Science, 256:333-42.

Hager, B. H., King, R. W., and Murray, M. H. 1991. Measurement of crustal deformation using the Global Positioning System. Annual Review of Earth and Planetary Sciences, 19:351-82.

Hickman, S. H. 1991. Stress in the lithosphere and the strength of active faults: Contributions in Tectonophysics. U.S. National Report 1987-1990. American Geophysical Union, pp. 759-75.

Journal of Geophysical Research. 1980. Special Issue on Stress in the Lithosphere, 85:6083-435.

Journal of Geophysical Research. 1992. Special Issue on the World Stress Map Project, 97:11,703-12,013.

King, G. C. P., Stein, R. S., and Rundle, J. B. 1988. The growth of geological structures by repeated earthquakes: 1. Conceptual framework. Jour. Geophys. Res. 93:13,307-318.

Lambeck, K. 1988. Geophysical Geodesy: The slow deformation of the earth. Oxford: Clarendon Press, 718 p.

Prescott, W. H., Davis, J. L., and Svarc, J. L. 1989. Global positioning system measurements for crustal deformation: Precision and accuracy. Science, 244:1337-40.

Stein, R. S., King, G. C. P., and Rundle, J. B. 1988. The growth of geological structures by repeated earthquakes: 2. Field examples of continental dip-slip faults. Jour. Geophys. Res. 93:13,319-31.

Sylvester, A. G. 1986. Near-field tectonic geodesy. *In* Active Tectonics. Washington, D.C., National Academy Press, pp. 164-80.

Thatcher, W. 1986. Geodetic measurement of active-tectonic processes. *In* Active Tectonics. Washington, D.C., National Academy Press, pp. 155-63.

Zoback, M. L. 1992. First- and second-order patterns of stress in the lithosphere: The World Stress Map Project. Jour. Geophys. Res., 97:11,703-28.

Bunjiro Koto (1856–1935)

Japan

Japan teeters on a tectonic precipice. The geological legacy of the Home Islands is not mineral resources but natural disasters: tsunamis, volcanic eruptions, and earthquakes.

In the late nineteenth century, however, calamities of nature were far from the public mind. Japan was caught up in the great ferment of the Meiji Restoration, transforming itself from the isolated feudal society of the Tokugawa shoguns to a modern nation ready to take its place alongside those in Europe and North America. The country began to industrialize, build railroads and modern harbors, reform its educational system, and send its brightest young scholars overseas to learn the new ways of the West.

But Japan still had to reckon with its geological affliction, and it had been a long time since the Great Earthquakes of Ansei in 1854–1855. On the morning of October 28, 1891, disaster struck. A calamitous earthquake destroyed much of the provinces of Mino and Owari and part of Echizen, which lay between the two cultural nuclei of Tokyo and Kyoto and included the major cities of Gifu and Nagoya. The region contains some of the few broad lowlands in the country and is a great rice-growing area, among the most densely populated in Japan. The zone of destruction virtually split the country in two, extending from Ise Bay on the Pacific coast northward almost to the Japan Sea. More than 7000 people lost their lives, and 200,000 buildings were destroyed. Just as Japan was moving toward a position of world leadership, the earthquake came as a reality check. What caused Japan's terrible earthquakes, and what could be done?

A few days after the disaster, Professor Bunjiro Koto, newly appointed to the faculty of the Imperial University of Tokyo, reached the devastated region. Koto himself was a product of the Restoration. Son of a samurai family of Tsuwano-han province in southwestern Japan, he had distinguished himself by his studies, which led to a provincial scholarship to study in Tokyo. In 1879, he graduated from the University of Tokyo and joined the Geological Survey of Japan. In the following year, he became one of those fortunate few selected to receive additional education in Europe. He spent four years studying petrology to become more knowledgeable about the great volcanoes that dominated the landscape of Japan. While he was in Europe, he read the works of Charles Lyell and Edward Suess, and in that way he learned about earthquakes.

After Koto returned to Japan, he was sent by the Geological Survey to work in the volcanic region of Kumamoto on Kyushu. On July 28, 1889, while he was engaged in this project, a violent earthquake struck the area. In the investigation that followed, he found that the damage was concentrated on three "seismic lines," which he recognized as faults. He remembered that Lyell in *Principles of Geology* had described a fault in the lower Wairarapa Valley of New Zealand that formed during the Wellington earthquake of 1855, and that in India, the Rann of Cutch earthquake of 1819 had been accompanied by upheaval of a great tract of land called the Mound of God, or Allah-bund. So he was prepared for what he would find in the tea gardens and rice paddies of Mino and Owari.

Just as he had found on Kyushu, the damaged zone was strongly linear, extending from the country southeast of Nagoya northwest through Gifu almost to Fukui, near the Japan Sea. In the Neo Valley, between the cities of Gifu and Fukui, he found more, an earth-rent, or fault, that cut across rice fields and hillslopes alike. At Midori, the valley floor was split along its length, with one side raised 5.5 to 6 meters vertically and displaced 4 meters horizontally with respect to the other. At most places, however, the fault moved only horizontally; the two sides were at the same level, but the north side had moved to the west, as observed by offset of the earthen dikes (*aze*) bounding individual paddy fields.

Koto described the surface features of the fault, which resembled "the pathway of a gigantic mole or the track of a plough-share" when motion was purely horizontal. In his paper, published in 1893, he illustrated a pressure ridge, including the longitudinal tensional cracks that are commonly found associated with pressure ridges. He pointed out that "the sudden elevations, depressions, or lateral shiftings of large tracts of country which take place at the time of destructive earthquakes—[were] the actual cause of the great earthquake" and not a secondary effect as was commonly believed.

A century later, Japan had achieved and far exceeded the goals of the Meiji Restoration. The nation had become an economic superpower, and many of its products dominated the markets of the world. The earthquake legacy had been forgotten; the last earthquake to cause great loss of life was the Fukui earthquake of 1948, ironically on the northern extension of the 1891 fault. But on January 17, 1995, disaster struck the port city of Kobe, one of the jewels of the Kansai, the great metropolitan area of western Japan. The earthquake was the result of strike slip on the Arima-Takatsuki Tectonic Line, an active fault zone that had been mapped and described by Japanese geological descendants of Koto. An earthquake of M6.1 had struck this fault in 1916, and a great earthquake, probably larger than the 1995 event, had devastated the region in 1596. But that knowledge was not enough: more than 5000 people died, and many tens of billions of dollars in damage resulted.

Suggestions for Further Reading

Fujii, Y. 1967. Seismology in Japan. Tokyo: Kinokuniya Book Co., p. 239 (in Japanese).

Imai, I. 1966. The dawning of the Japanese geology. Tokyo: Lattice Publishing Co., p. 193 (in Japanese).

Koto, B. 1893. On the cause of the great earthquake in central Japan, 1891. Jour. College of Science, Imperial University of Japan, 5:295–353.

6

Quaternary Timescales and Dating Techniques

During the eighteenth century, James Hutton recognized that the earth must be much older than had been imagined previously. It was known at that time that certain sequences of sedimentary rocks contain characteristic assemblages of fossils that are unlike organisms living today. As one proceeded from older formations to younger, overlying formations, the fossil assemblages changed or evolved in a regular way. These same changes were found in different sedimentary rock sequences that were separated by great distances. Using these regular changes, paleontologists assigned formations with a particular assemblage of fossils to geologic periods, and the geologic timescale was born, starting with the Cambrian, which contains the oldest strata with abundant fossils. The underlying rocks were shown to be relatively unfossiliferous, and these were referred to simply as Precambrian. The periods were further grouped into the Paleozoic, Mesozoic, and Cenozoic eras. Later, with the discovery of radioactivity, it was found that the age of each period in millions of years could be determined. For example, the Cambrian began about 570 million years ago, and the earth itself is 4.5 to 4.6 billion years old (Fig. 6-1).

The Cenozoic Era is divided into two periods, the Tertiary and the Quaternary, a holdover from the days when older rocks were called Primary and Secondary. The Quaternary Period was named for the fourth and least consolidated group of deposits, including those that overlie all other formations, are largely unconsolidated, and are related, more or less, to the present landscape. The two epochs of the Quaternary were named the Pleistocene, to include the ice ages, and the Holocene (or Recent), the time since the retreat of the last great continental ice sheets. Subsequent work showed that there are unconsolidated sediments and glacial deposits of pre-Pleistocene age, and some beds of Pleistocene age are consolidated into rocks. Thus, it became necessary to establish ages independent of climatic change and degree of consolidation for the beginning of the Pleistocene.

The Pliocene-Pleistocene boundary has been established at the base of a particular claystone in Vrica, near Crotone in southern Italy (Aguirre and Pasini, 1985). This claystone overlies the top of the Olduvai Magnetic Subchron (see the following text), and the Pleistocene-Holocene boundary has been established at 10,000 years B.P. (before present, with "present" defined as A.D. 1950). However, neither boundary has yet been accepted as "official" by the International Commission on Stratigraphy. Until these boundaries are established internationally, we adopt the timescale of the Decade of North American Geology, Geological Society of America, modified based on a revised age of the top of the Olduvai, which establishes the age of the Pleistocene as 1.8 Ma (Ma = million years) to 10 ka (ka = kiloannum, 1000 years), and the Holocene from 10 ka to the present.

Morrison (1991) pointed out that a better beginning for the Pleistocene would be the time of advance of major continental glaciers in the Northern Hemisphere, about 2.6 Ma, which took place at a geomagnetic reversal (see the following text), but although this suggestion has merit, it has not been accepted by most scientists.

Radiometric dating is based on rates of radioactive decay whereby atoms give off particles from their nuclei and thereby change into atoms of other elements or isotopes of the same element. These rates, independent of other geologic processes and conditions, are known to such great precision that radioactive decay rates serve as geological clocks that are used to calibrate in years the timescale, which is based on fossils. In dating Quaternary deposits, several techniques

Era	Period		Epoch	Duration in millions of years	Millions of years ago
CENOZOIC	Quaternary		Pleistocene	1.8	1.8
	Tertiary	Neogene	Pliocene	3.7	5.3
			Miocene	18.4	23.7
		Paleogene	Oligocene	12.9	36.6
			Eocene	21.2	57.8
			Paleocene	8.6	66.4
MESOZOIC	Cretaceous			78	144
	Jurassic			64	208
	Triassic			37	245
PALEOZOIC	Permian			41	286
	Carboniferous		Pennsylvanian	34	320
			Mississippian	40	360
	Devonian			48	409
	Silurian			30	438
	Ordovician			67	505
	Cambrian			65	570
PRECAMBRIAN					

Figure 6–1. Geologic timescale for Cambrian and younger periods. Precambrian begins with the age of the earth, about 4.6 billion years.

may be used together to arrange the geological formations according to their relative ages; these include fossils, radiometric dating, and other nonradiometric timescales such as those developed from time series of $^{18}O/^{16}O$ ratios in deep-sea sediments and reversals of the earth's magnetic field.

This chapter focuses on the Quaternary Period because events in this period are most relevant to establishing the history of active faults and folds in the present-day tectonic regime. The chapter discusses the various Quaternary timescales that are in use, then considers the more important Quaternary dating techniques, concentrating on those most useful in determining history of deformation.

TIMESCALES USEFUL IN THE QUATERNARY

Oxygen-Isotope Timescale

The Quaternary has been a time of climate change, and for more than a century, attempts have been made to subdivide the Pleistocene on the basis of advances and retreats of glacial ice. Emiliani (1955) suggested that the $^{18}O/^{16}O$ ratio in calcium carbonate in marine fossils reflects both the isotopic composition and the temperature of the seawater in which the organisms grew. Glacial ice preferentially stores the lighter isotope of oxygen. Measurements of the $^{18}O/^{16}O$ ratio are reported with reference to an international standard (Peedee belemnite, or PDB) as $\delta^{18}O$ in parts per thousand (‰). Because significant variation in isotopic composition of the ocean is due to global increases and decreases in the volume of ice (which also controls eustatic changes in sea level), isotopic data from cores of deep-sea sediments of a given age are globally similar (Mix, 1987). The similarity of oxygen-isotope curves for deep-sea cores from low and middle latitudes of the world oceans (Fig. 6–2) and for ice cores from Greenland and Antarctica (Delmas, 1992) indicates that the isotope-ratio changes they measure are global and are synchronous within the time it takes for mixing of oceanic water masses to occur, which is considered to be about 1000 years (Shackleton and Opdyke, 1973).

Using this global record of ice-volume change, Emiliani (1955) and Shackleton and Opdyke (1973) constructed an oxygen-isotope scale of 22 stages in which odd-numbered stages correspond to interglacial (warm) intervals, which are isotopically light, and even-numbered stages correspond to glacial (cool) intervals. The youngest 22 of these stages are shown in Figure 6–2a, where they are plotted against age based on a chronology developed by Imbrie et al. (1984) and modified by Bassinot et al. (1994). Stage 1 is the Holocene, and Stage 2 is the last glacial interval. $\delta^{18}O$ values in Stage 5 are similar to those today, indicating that this stage represents the peak of the last major interglacial interval (the Sangamon), a sea-level high stand that is recorded in Papua New Guinea by coral reefs that have been dated radiometrically by the uranium-series method (see discussion in the following text) as 118 and 134 ka (Stein et al., 1993), thereby providing an age-calibration point for the oxygen-isotope scale. Uranium-thorium ages from the Bahamas suggest that this sea-level high stand began at 132 ka and ended at 120 ka (Chen et al., 1991). A second calibration point is the Brunhes-Matuyama magnetic chron boundary, which has been located on the basis of a change from reversed- to normal-magnetic polarity in several cores near the base of Stage 19 (see following section).

The climatic fluctuations as recorded by the $\delta^{18}O$

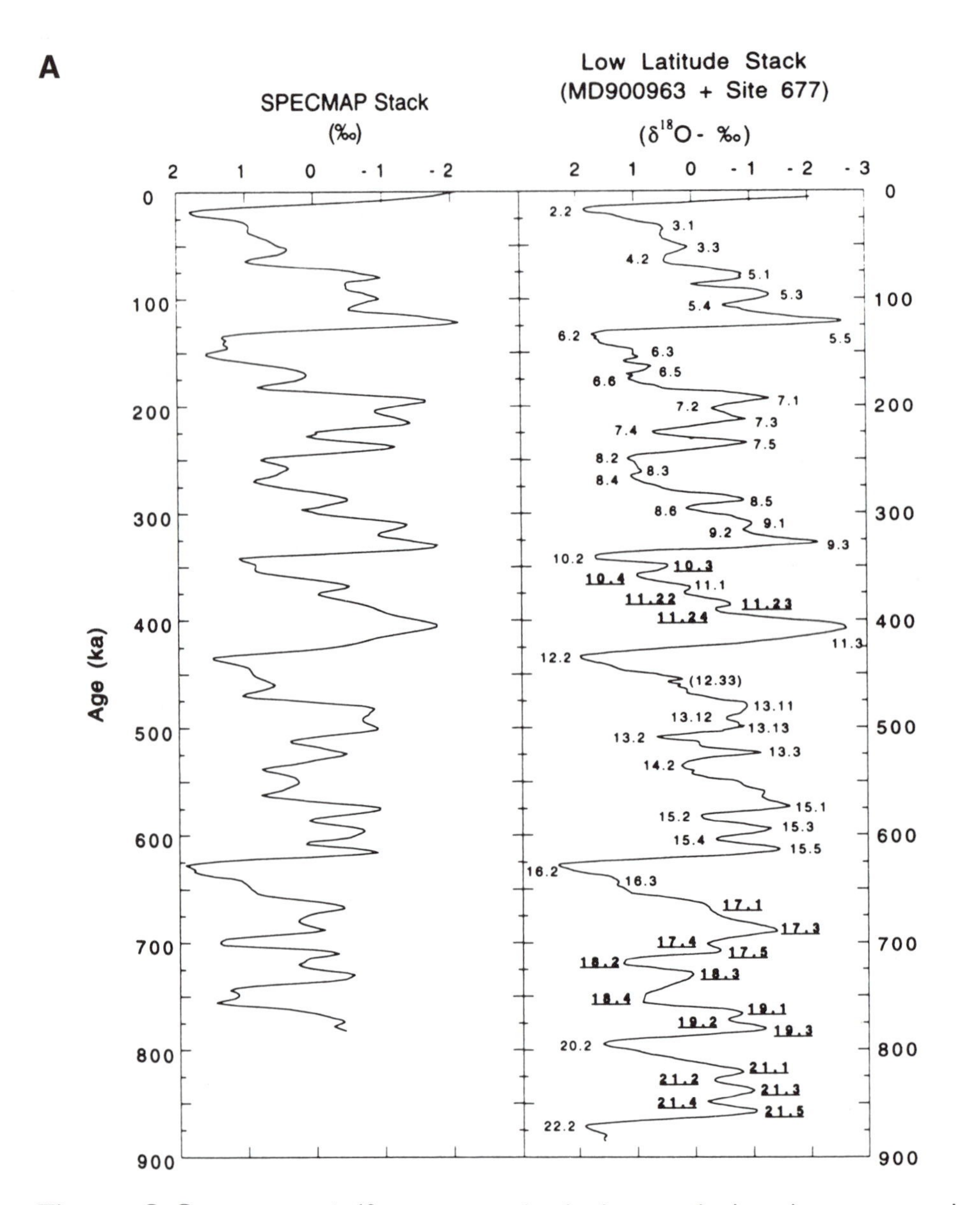

Figure 6–2. (a) Age of $\delta^{18}O$ variations for the last 870 ka based on an age calibration at Stage 5.5 and a comparison with paleomagnetic declinations in Core MD 900963 in the Indian Ocean, with additional data from Ocean Drilling Project Site 677, revised by Bassinot et al. (1994) from Imbrie et al. (1984). Even-numbered stages correspond to glacial maxima, and odd-numbered stages to interglacial intervals. Substages 5.1, 5.3, and 5.5 also referred to as Substages 5a, 5c, and 5e, respectively.

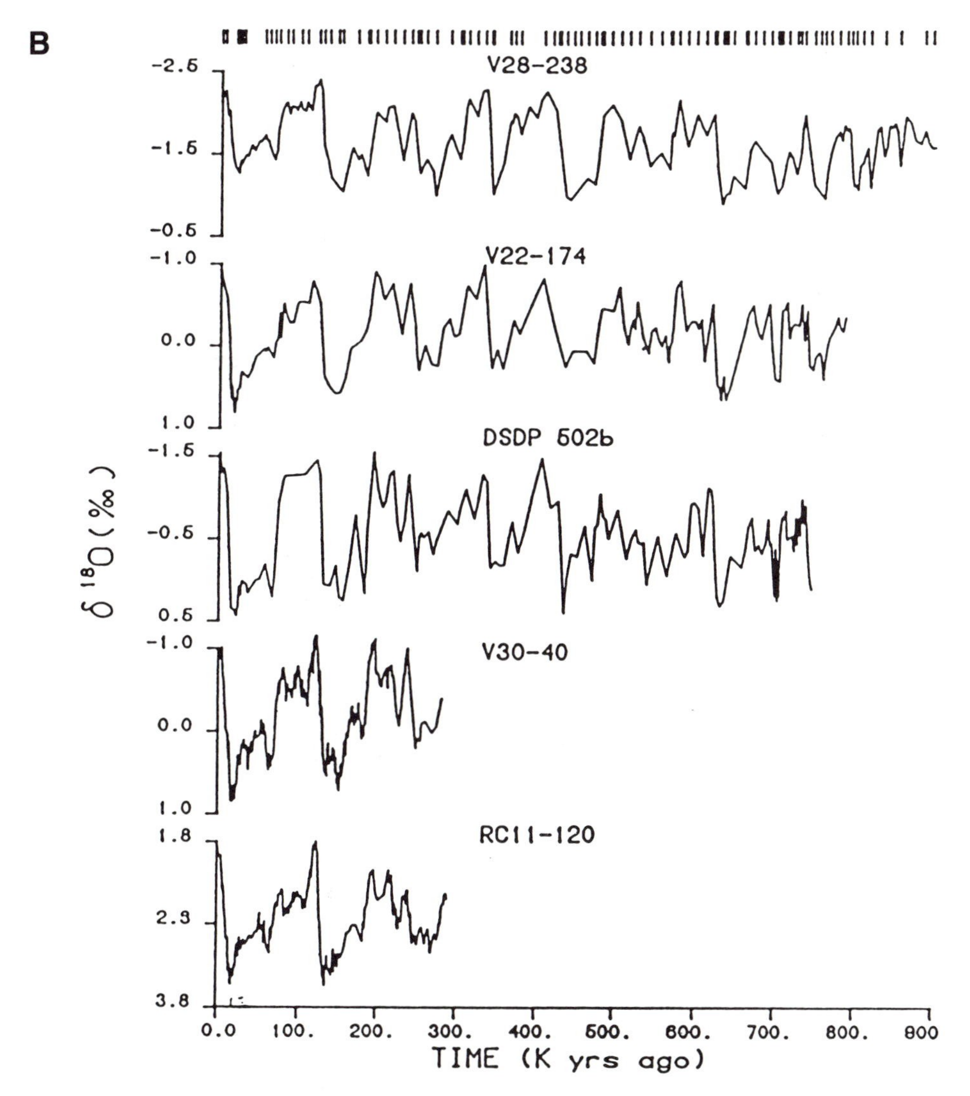

Figure 6–2. (b) Variations in $\delta^{18}O$ from five deep-sea sections normalized for stage correlation. Top of each core is at left. From Imbrie et al. (1984).

record appear to respond to regular orbital variations that affect the distribution of solar radiation on the earth's surface (Hays et al., 1976), verifying a hypothesis proposed by M. Milankovitch in 1941. These orbital variations, with periods of 100,000, 41,000, 23,000, and 19,000 years, have served as a pacemaker or timer to calibrate further by orbital tuning the $\delta^{18}O$ timescale, taking into consideration the length of time it takes the oceans to respond to these variations (Imbrie et al., 1984; Prell et al., 1986; Pisias et al., 1990). However, high-precision uranium-series dates of corals deposited during the last interglacial high stand do not correlate precisely with the predicted time based on the Milankovitch hypothesis (Stein et al., 1993).

The disadvantage of the oxygen-isotope timescale for earthquake studies is, of course, that a continuous, correlatable sequence is needed in the vicinity of structures that are to be dated, and this is rarely found in subaerial exposures. The method has applications in the study of cave travertine, and it has great potential in dating faults on the deep-sea floor that cut Quaternary sequences where continuous deposition can be demonstrated. Oxygen-isotope ratios for molluscs in late Pleistocene marine terraces serve as a check against other dating techniques that give ambiguous measurements (Muhs et al., 1990). The uncertainty in the age for a given core, using the orbitally-tuned $\delta^{18}O$ timescale is 3000–5000 years (Imbrie et al., 1984).

Magnetic-Reversal Timescale

The polarity of the earth's magnetic field has reversed many times in the past. The geomagnetic field is thought to be generated by fluid motion of the conducting Fe-Ni liquid in the earth's outer core. The field is a vector quantity whose direction at each point on the globe is usually described by the *inclination* and *declination*. The *inclination* is the dip of the vector with respect to the horizontal; downward and upward dips are arbitrarily defined as positive and negative inclinations, respectively, in the Northern Hemisphere (Fig. 6-3). The *declination* is the angle of the horizontal projection of the magnetic vector with respect to the geographic north pole. The field usually occurs in one of two stable states: *normal polarity,* defined as being similar to the present field, in which the compass needle points in a northerly direction; and *reversed polarity* in which the compass needle points in the opposite (southerly) direction (Fig. 6-3). When averaged over sufficient time—10,000 to 100,000 years—both normal and reversed states behave as a geocentric axial dipole, analogous to a bar magnet embedded along the earth's axis. About 90% of the magnetic field can be expressed as a dipole; the rest is *secular variation,* which records more complex, smaller-scale portions of the magnetic field. When the magnetic field reverses, a process that takes between 1000 and 10,000 years, the magnitude of the field is thought to diminish substantially relative to its stable configuration, leaving only the secular field. Because polarity reversals originate in the earth's outer core, they are global phenomena, and magnetic reversals can be used to construct a timescale. On the other hand, the secular component of the field is generally not a global phenomenon.

The magnetic-reversal timescale for the past four million years was calibrated by radiometrically-dated lava flows using the K/Ar method (Mankinen and Dalrymple, 1979). However, paleomagnetic boundaries have also been established in upper Cenozoic sedimentary deposits tied to the orbitally-tuned timescale (see discussion in preceding section). This work (Hilgen, 1991; Zijderveld et al., 1991) showed that the age calibration of the paleomagnetic timescale based on K/Ar dates is about 5 to 7% younger than the orbitally-tuned paleomagnetic timescale. According to the revised timescale, illustrated in Figure 6-3, the current normal-polarity epoch, the Brunhes Chron, began at around 780 ka (Spell and McDougall, 1992; Bassinot et al., 1994). The preceding reversed-polarity epoch, the Matuyama Chron, began in the Pliocene at 2.60 Ma. The Matuyama contains three subepochs of normal polarity, the Reunion Subchron (2.15-2.14 Ma), the Olduvai Subchron (1.95-1.79 Ma), and the Jaramillo Subchron (1.01-0.915 Ma).

The detailed structure of the Matuyama Chron and the precise age of magnetic reversals within the chron are subjects of active ongoing research (cf. Spell and McDougall, 1992). The Brunhes-Matuyama boundary is older than a sample of normally-magnetized Bishop Tuff radiometrically dated as 759 ±3 ka (Pringle et al., 1992), a confirmation of the orbitally-tuned age of 780 ka (the Mankinen and Dalrymple age is 730 ka). Furthermore, there is, at Vrica, a short reversed-polarity zone near the top of the Olduvai Subchron with an orbitally-tuned age of 1.84-1.80 Ma (Hilgen, 1991; Zijderveld et al., 1990; Fig. 6-4), thereby complicating the paleomagnetic age of the base of the Pleistocene. There may be additional,

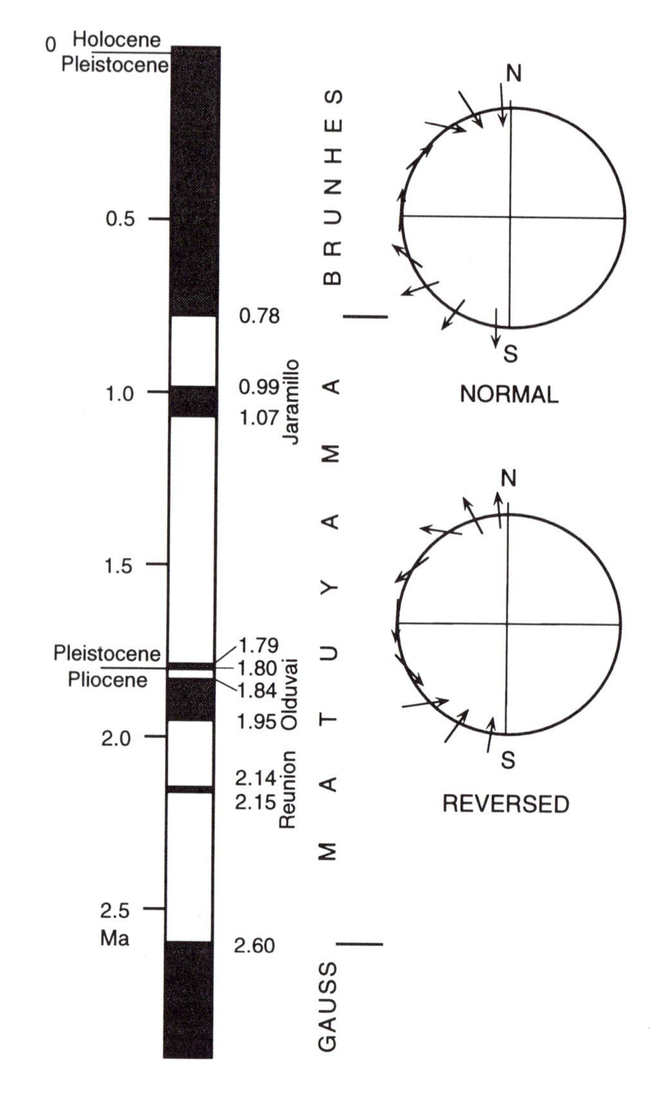

Figure 6-3. The magnetic-polarity timescale for the last 2.6 million years, modified from Mankinen and Dalrymple (1979) on the basis of orbital tuning (Hilgen, 1991). Black: normal polarity; clear: reversed polarity. The insets show the inclination of magnetic vectors during normal polarity and reversed polarity.

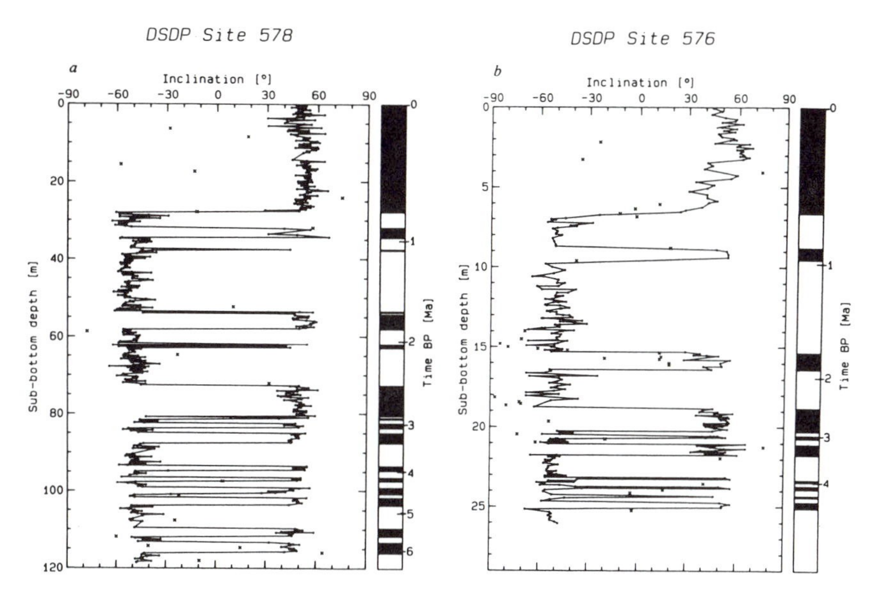

Figure 6–4. Magnetostratigraphy of two coreholes in the northwest Pacific Ocean obtained by the Deep-Sea Drilling Project. Positive inclinations are downward and represent normal polarity; negative (upward) inclinations represent reversed polarity. The Brunhes Chron is characterized entirely by positive inclinations; the Matuyama Chron is characterized by negative inclinations with several positive subchrons. The normal excursion at 38 m depth at DSDP Site 578 may represent a brief excursion below the Jaramillo Subchron. Scatter of up to 30° in data together with crosses may represent secular variation or post-depositional magnetic overprints; the data shown as crosses were excluded from the curve. From Arason and Levi (1990).

yet unrecognized, short polarity subchrons in both the Brunhes and Matuyama Chrons.

The pattern of magnetic reversals (Fig. 6-3) has been confirmed by the symmetrical pattern of magnetic anomalies that flank spreading ridges on the sea floor and also by patterns of reversals in marine sedimentary sequences deposited on oceanic crust of known age (Fig. 6-4). Studies of marine magnetic anomalies have now extended the magnetic polarity timescale to the Late Jurassic, about 165 million years ago, the approximate age of the oldest existing ocean basins that are still preserved and have not yet been consumed at subduction zones.

Sedimentary sequences contain microfossils, and evolutionary changes in fossils or fossil assemblages may be dated using magnetic reversals. Marine sediments contain radiolarians, foraminifera, and nannofossils, while nonmarine sediments are more likely to have pollen and vertebrate remains. Correlation of fossil assemblages with magnetic reversals has been most successful in deep-sea sediments, where deposition, though slow (on the order of mm/ka), is more likely to be continuous.

Sediments and sedimentary rocks retain a much weaker paleomagnetic signal than lava flows, and in many cases, the primary magnetic remanence, which is acquired by the magnetic minerals during their deposition, is overprinted by younger remanences imposed by more recent fields subsequent to deposition. The primary remanance is usually more stable in fine-grained sediments, but even then, magnetic overprinting must be removed. Most commonly, this is done by one of two procedures. In *thermal demagnetization,* samples are heated in a zero magnetic field to a particular temperature and then cooled to room temperature, where the remanence is measured. This procedure is repeated at progressively higher temperatures. Assuming that the primary remanence is more stable at higher temperatures, it will persist after the secondary overprints have been removed by thermal demagnetization. In *alternating-field (af) demagnetization,* the sample is subjected to alternating magnetic fields of successively higher intensity, under the assumption that the primary remanence is the most likely to remain after partial af demagnetization.

A disadvantage of using magnetic stratigraphy in active-fault studies is that the field has been predominantly normal during the Brunhes Chron, which occupied the last 780,000 years, the period of most interest in earthquake studies. However, the declination of magnetic vectors, when measured over a large enough stratigraphic interval to exclude the possibility of secular variation, may differ significantly from the modern declination, thereby providing evidence of tectonic rotation about a vertical axis. For example, a sample traverse at right angles to the San Andreas fault at Pallett Creek showed that late Holocene sediments close to the fault were rotated clockwise, evidence of nonbrittle deformation outside the fault zone itself (Salyards et al., 1992). This required an increase in the estimate of displacement of these sediments beyond the piercing-point offset at the fault.

Secular variations in magnetic field direction and intensity in the normally-polarized Brunhes Chron offer the possibility of correlation between sedimentary sequences in a given region. This has been done most successfully for the Holocene of Japan, using baked clays at archeological sites for the last 2000 years and shallow-marine and lacustrine sediments for the Holocene and late Pleistocene (Hirooka, 1991). Magnetic stratigraphy of sediments in Biwa Lake in central Japan document seven *magnetic excursions* (large departures of direction and intensity from the usual dipole field, but not full reversals) during the past 380 ka (Kawai, 1984). Two excursions have been dated at 80–70 ka and 60–50 ka based on dated tephra layers in western and central Honshu and in Hokkaido. The younger excursion has also been recognized in tephra layers in coastal sand dunes, and it may correlate with the 49-ka excursion in Lake Biwa (Hirooka, 1991). Work is also underway in correlating secular variations between late Pleistocene lake basins in the western United States. The Mono Lake excursion, dated by ^{14}C as 28 ka, has been identified in several basins in western Nevada and eastern California (Liddicoat, 1992). The Laschamps excursion, first identified in the Massif Central of France, is best dated around 36 ka, but other excursions may have occurred at 17 and 28 ka (Thouveny and Creer, 1992). For a more thorough discussion of secular variation, see McElhinny and Senanayake (1982).

Secular variation was used to date a liquefaction deposit of silt in a trench across the Atotsugawa fault in central Honshu. The fault cuts a sequence of sediments dated as 12 to 5 ka. At 3 m depth in the trench, the silt was forced up along the fault zone. The magnetization of the distorted silt layer was that expected from the modern geomagnetic field and different from that in the Holocene strata in the trench, suggesting that the magnetic minerals reoriented themselves in the earth's magnetic field at the time of strong shaking and liquefaction (*shock remanence*). In detail, the magnetization of the silt was in agreement with the orientation of the field at A.D. 1880 $\pm$ 60, suggesting that the liquefaction event may have occurred during a destructive earthquake in A.D. 1858 (Takeuchi and Sakai, 1985; Hirooka, 1991).

Strontium-Isotope Timescale

The ratio of ^{87}Sr to ^{86}Sr in seawater is the same in all oceans, but the ratio has increased during the last 35 million years based on analyses of marine calcium carbonate (Burke et al., 1982; DePaolo and Ingram, 1985). The $^{87}Sr/^{86}Sr$ ratio, which has an analytical precision better than one part in 100,000 (DePaolo, 1986), is the same for calcium carbonate of the same age worldwide, showing that ancient oceans were characterized by uniform $^{87}Sr/^{86}Sr$ ratios just as modern oceans are. Accordingly, it is possible to construct a timescale based on these ratios, and this timescale correlates well with that based on planktonic microfossils (Fig. 6-5). The $^{87}Sr/^{86}Sr$ ratios increased relatively sharply starting around 2.6 Ma, the time of initiation of major ice sheets in the Northern Hemisphere (Hodell et al., 1991) and of increased amplitudes of $\delta^{18}O$ fluctuations. Thus, the rapid increase in $^{87}Sr/^{86}Sr$ ratios during the Pleistocene makes it possible to correlate $^{87}Sr/^{86}Sr$ to the timescale, with resolution of about 0.1 million years.

Biostratigraphic Timescale

The Pleistocene was first defined by Charles Lyell in 1839 on the basis of fossils, and only later did it become associated with the ice ages. For the next century, most biostratigraphic work concentrated on megafossils (especially molluscs), vertebrates, and leaves of trees, and stages for the Cenozoic presumed to be bounded by time lines were established for all three. These schemes presented three problems. First, megafossil populations are relatively small and commonly are not well preserved, and the relative abundance of preserved fossils may differ greatly from the relative abundance of the living organisms. Limited preservation means that the change from one faunal succession to another cannot be monitored continuously. Second, these populations are provincial, as illustrated by the local development of marsupials in Australia or the exclusion of hominids from the New World until the latest Pleistocene. Floras and faunas develop in isolation from one another so that it is difficult to find an evolutionary transition that takes place at the same time worldwide. Third, what appears to be an evolutionary change may actually be an environmental change. Changes in floral successions, whether marked by leaves or by pollen, reflect principally changes in climate. Molluscan communities have migrated poleward and equatorward in response, respectively, to Pleistocene warming and cooling intervals, and these climatic changes modify oceanic circulation patterns. Successions of Pliocene and Pleistocene bottom-dwelling foraminifera in oil-producing strata in basins of southern California reflect a change in depth of water through time; many of the key fossil species used in stratigraphic subdivision have living representatives in deep water off the coast.

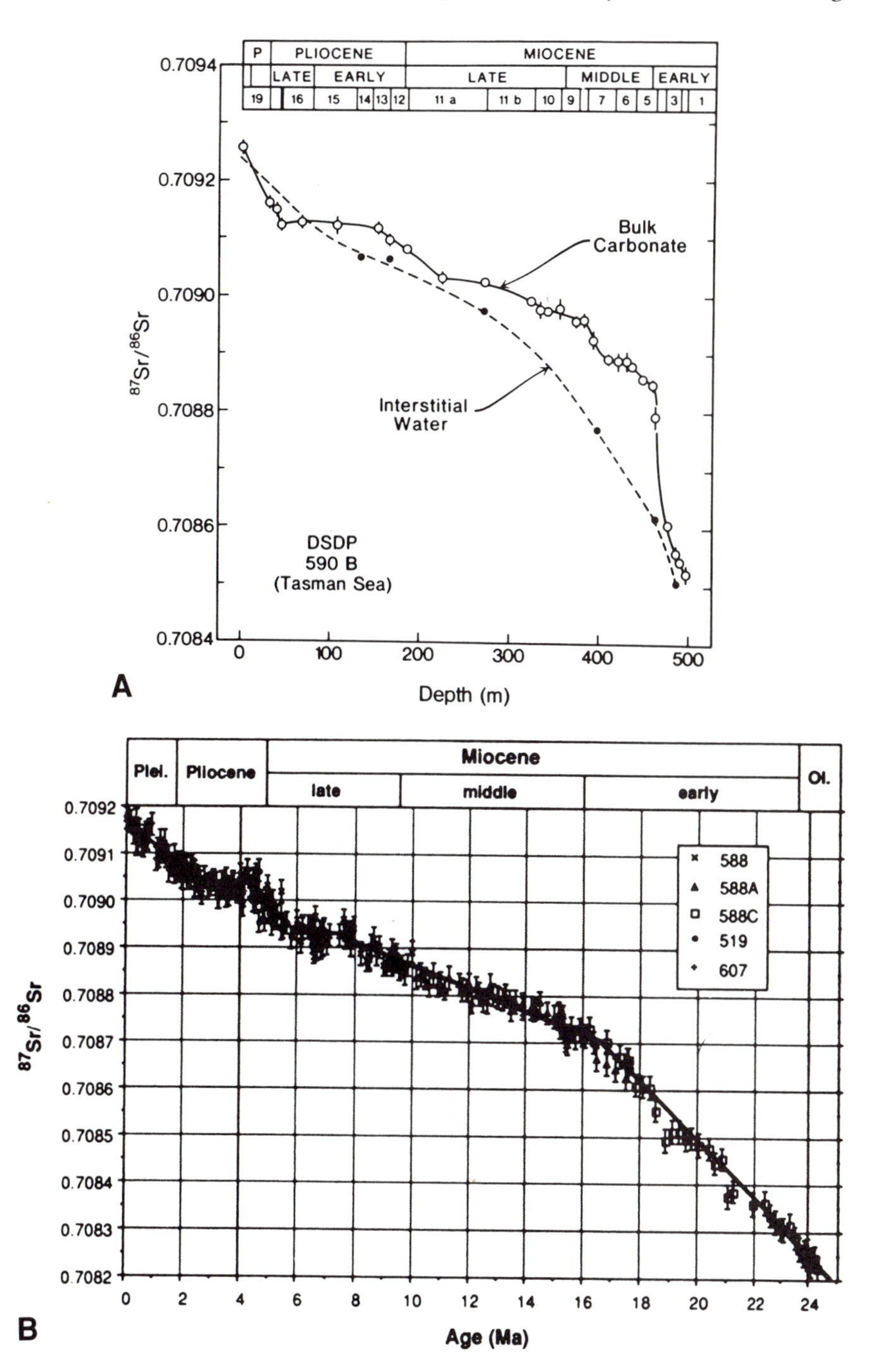

Figure 6–5. Two versions of the $^{87}Sr/^{86}Sr$ ratio changes for the Neogene. (a) $^{87}Sr/^{86}Sr$ ratios versus subbottom core depth for samples of calcium carbonate sediment and interstitial water from a site in Tasman Sea. Numbers on upper scale are calcareous nannofossil biostratigraphic zones. Note the relatively sharp increase in $^{87}Sr/^{86}Sr$ ratio in the late Pliocene and Pleistocene (labeled P). After DePaolo (1986). (b) Age vs. $^{87}Sr/^{86}Sr$ for 261 measurements on calcium carbonate from planktonic foraminifera from 24 Ma to present. Regression lines plotted for six segments, with the most recent segment 2.6 Ma to present. After Hodell et al. (1991).

The organisms that can contribute most toward a timescale based on biostratigraphy are the free-swimming microplankton that live in that part of the sea that receives sunlight. These organisms have calcareous or siliceous shells that accumulate on the sea floor after death. A sample of deep-sea sediment may contain millions of these shells, and evolutionary changes can be monitored in closely-spaced samples from cores. The micropaleontologist bases an age determination not on a few fossils but on large populations, and a biostratigraphic stage boundary may be based on changes in abundance rather than on a first or last appearance of a single fossil. Microfossil assemblages from the open ocean may be calibrated by $\delta^{18}O$ stratigraphy, $^{87}Sr/^{86}Sr$ stratigraphy, and/or magnetostratigraphy, commonly from the same core sample.

Yet there are still difficulties, largely stemming from the control of changes in fossil assemblages by shifts in oceanic circulation patterns. The most successful microfossil zonation schemes are those established in the tropical and subtropical oceans. Different schemes are sometimes necessary for Arctic and Antarctic latitudes. The evolution from one key species to another, documented by a detectable change in shell morphology, is not instantaneous; the change may take more than 10,000 years to spread around the world. Accordingly, many of the zonation schemes may be applied to some parts of the ocean but not to others.

The most successful biostratigraphic zonation schemes have been established for planktonic foraminifera, calcareous and siliceous nannofossils, radiolarians, and diatoms. These schemes are reviewed by Berggren et al. (1980; 1985) and Jenkins et al. (1985). Timescales have been established for each of these groups of microfossils, based principally on the initial occurrence of guide fossils useful in zoning (first-appearance datum or FAD) or the last occurrence (last-appearance datum or LAD). Figure 6-6 shows an example of such a scale for Pleistocene calcareous nannofossils. Most of these changes are evolutionary, but in some regions, zoning consists of alternation of coiling ratios in certain planktonic foraminifera between left-coiling and right-coiling.

An example of how microfossil stage boundaries compare with another timescale appears in Figure 6-7, which compares relative abundance of two coccolith (nannofossil) species with oxygen-isotope stratigraphy. The FAD for *Emiliania huxleyi* is consistently in oxygen-isotope Stage 8, a glacial maximum, for low-latitude and high-latitude cores alike. However, there are differences in abundance of *E. huxleyi* between low-latitude and high-latitude cores. Figure 6-7 also shows the shift in worldwide dominance in coccolith floras from *Gephyrocapsa caribbeanica* in Stage 9 to *E. huxleyi* in Stages 3 and younger. The transition in dominance takes place in Stage 4 in higher-latitude cores and in Stage 5 in lower-latitude cores, indicating that the dominance reversal was time-transgressive, occurring about 12,000 years earlier in the tropics (Thierstein et al., 1977). The age resolution of these biostratigraphic zones appears to be about the same as that for timescales based on oxygen-isotope stratigraphy.

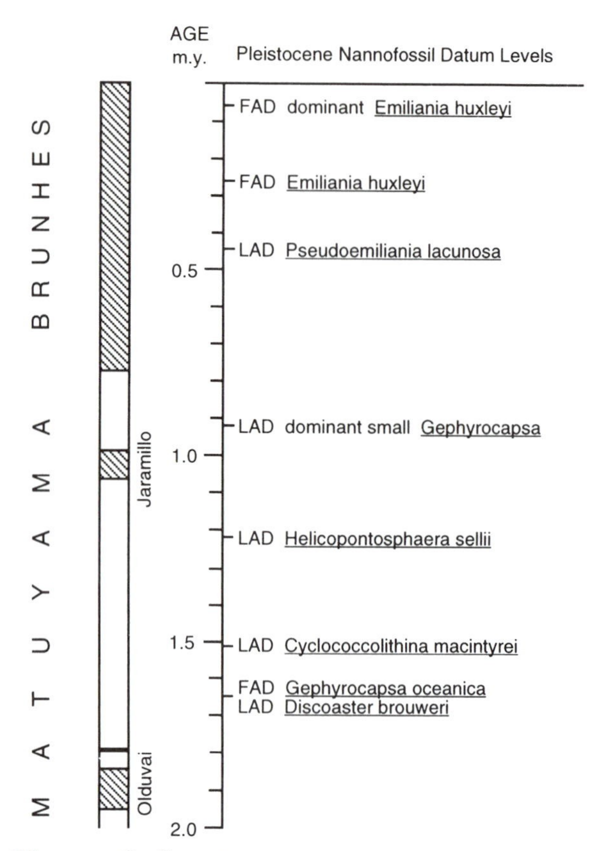

Figure 6–6. Pleistocene calcareous nannofossil datum levels compared with magnetic-reversal timescale. FAD, first-appearance datum; LAD, last-appearance datum. Normal polarity in lined pattern; reversed polarity clear. Modified from Berggren et al. (1980) to reflect new age calibration of magnetic timescale.

DATING TECHNIQUES

Measurements to the Nearest Year

Historical records in ancient Egypt and Mesopotamia go back more than 5000 years, and earthquakes during the reign of a particular king can be calibrated against astronomical events such as eclipses. Records in the eastern Mediterranean region and the Yellow River basin of northern China are available for about 3000 years and for the western Mediterranean region for about 2000 years. Northwestern Europe and other parts of China produced records somewhat later. In the New World, more than 450 years of records are available for the Pacific coast of Latin America. For other active seismogenic areas, records are available for two centuries or less.

Dendrochronology, the study of annual growth rings of trees, has the potential to date events nearly 10,000 years old when rings on living trees are correlated with those on dead trees by use of relative ring thickness in good vs. poor growth years. Even if the connection to living wood is broken, and the calendar year cannot be established, growth rings from different trees can be correlated with one another, establishing that an unusual ring pattern (or the death of the tree, if bark is preserved) occurred in the same growth year. Correlation of rings is more likely in semiarid regions like the American Southwest, where variations in width of rings may be caused by drought (rings are *moisture-stressed*) than in humid regions with less climatic variations, resulting in more uniform ring width (samples are *complacent*). Burn scars from major forest fires also help in ring correlation within the same forest.

Trees may be tilted or may have their roots sheared off by a fault, or shaking accompanying an earthquake may knock off major branches. These will put stress on tree growth, resulting in very narrow or nonconcentric growth rings. Trees in coastal forests may be drowned by the sea as a result of sudden, earthquake-induced subsidence, as documented for the 1964 Alaskan earthquake.

Tree-ring studies are being used in several earthquake-related investigations. Old trees growing in the San Andreas fault zone provided evidence that an earthquake in 1812 was produced by slip on the fault. East of Seattle, Washington, correlation among rings of trees (including bark) drowned by three massive landslides into Lake Washington suggests that the landslides occurred simultaneously and may have accompanied a large earthquake on the Seattle fault 1000–1100 years ago (Jacoby et al., 1992). Seven bark-bearing trees were correlated with one another. Six were from Lake Washington, and one had been

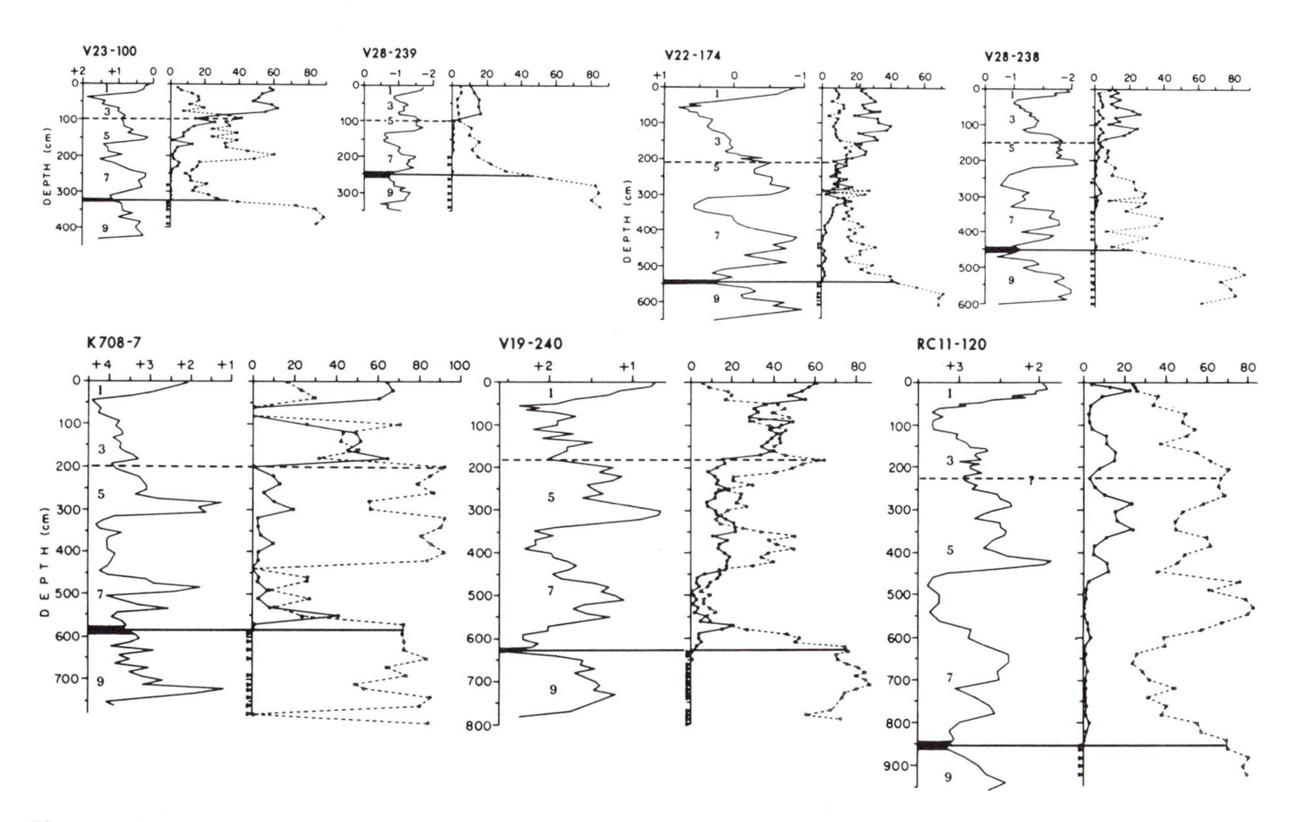

Figure 6–7. Comparison of $\delta^{18}O$ ratios (left) with relative abundances of two coccolith species (right) in seven cores. Cores V23-100, V28-239, V22-174, and V28-238 are from tropical or subtropical regions; others are from high-latitude regions. Solid lines, relative abundance in percent of *Emiliania huxleyi;* broken lines, relative abundance of *Gephyrocapsa caribbeanica.* Numbers are $\delta^{18}O$ stages, broken horizontal line is position of dominance reversal level of *G. caribbeanica–E. huxleyi.* After Thierstein et al. (1977).

trapped in an abruptly subsided tidal marsh that was inundated by a tsunami on the shore of Puget Sound, north of Seattle (Atwater and Moore, 1992). Radiocarbon dating showed that trees in both areas died 1000–1100 years ago. The ring correlation between the two areas showed that the trees from Lake Washington and Puget Sound died in the fall, winter, or early spring of the same year. Both sites are close to the Seattle fault, but otherwise the environmental settings are different enough to preclude a nontectonic origin of the simultaneous death of trees in both areas (Jacoby et al., 1992).

Study of growth rings of trees drowned suddenly about 300 years ago on the coast of Washington, presumably by a subduction-zone earthquake, is hampered because the rings, including bark on tree roots, have not been correlated to calendar years. Another problem is that trees can continue to grow for a few years after drowning (G. Jacoby and D. Yamaguchi, work in progress). Dendrochronology in earthquake studies is limited to those areas where large trees are found, and where, in addition, background information is known about other factors influencing ring width.

Annual light-and-dark layers, or *varves,* in lake deposits have the potential to record earthquakes on the basis of disturbances by shaking of near-surface layers. This shaking may produce *turbidites,* sediments resulting from the turbulent flow of heavy, sediment-laden bottom currents (for an example in Lake Washington providing independent evidence of an earthquake 1000–1100 years ago, see Karlin and Abella, 1992 and Chapter 12). In using this method, one must be concerned about undetected unconformities (breaks in sedimentation), as is the case in some late Pleistocene lake deposits in the Great Basin of the western United States. Independent age checks using radiocarbon, volcanic ash beds, or secular magnetic variations, are essential to search for breaks in the depositional record before using this method for dating.

Tephrochronology is the use of tephra deposits (volcanic ash and tuff) to correlate and date sedimentary and volcanic sequences and faulting events (Sarna-Wojcicki and Davis, 1991). Tephra deposits may be dated radiometrically, using K/Ar or fission-track techniques (see the following text), and they may be correlated (or "fingerprinted") on the basis of major-element and trace-element chemistry, on mineralogy, and on the color and morphology of glass shards. An ash bed marks an instant in geologic time during which a climactic, plinian volcanic ash eruption occurred—a day, or perhaps a series of eruptions over a period of a year or more. Widespread tephra layers offer the possibility of correlation among depositional sequences in different sedimentary environments. For example, the Mazama ash bed from Crater Lake, Oregon, dated as 6700–6850 years B.P., is widespread in the Basin and Range province of the western United States, in the Rocky Mountains of southwestern Canada, and in abyssal-plain sediments on the Juan de Fuca plate west of Oregon. Quaternary tephrochronologic correlations are widely used in the western United States (Sarna-Wojcicki et al., 1991), Japan (Machida, 1991), and New Zealand.

Radiocarbon Dating

One out of every 10^{12} carbon atoms is ^{14}C, formed by bombardment of N-atoms in the upper atmosphere by cosmic rays. ^{14}C combines with oxygen to form CO_2, which diffuses through the lower atmosphere and enters the geologic cycle, mainly via surface waters. ^{14}C becomes incorporated in living organic matter largely by photosynthesis and thence throughout the food chain. A constant isotopic ratio of ^{14}C to ^{12}C is maintained throughout the life of a plant or animal, but at death, no more ^{14}C can enter the system, provided the system is closed. The isotopic ratio begins to decrease because ^{14}C decays and reverts back to ^{14}N, with a half-life of about 5730 years. Radiation is emitted as it decays. The half-life means that half the ^{14}C in a sample is eliminated in 5730 years, half the remaining ^{14}C is eliminated in another 5730 years, and so on. The maximum radiocarbon age measured so far is about 75 ka; the ^{14}C measured was about 10^{-4} that of the level at death of the organism (Stuiver, 1991).

In *conventional ^{14}C dating,* the amount of ^{14}C is estimated by measuring the radiation emitted as it decays naturally to ^{14}N. It may take several days to record enough radiation to provide a precision of ± 80 years. If the carbon sample is small, or if it is older than about 30 ka, then it is fairly difficult to detect the radiation due to decay of ^{14}C from the background radiation produced by cosmic rays. This background radiation may be reduced by increasing the shielding around the equipment used to measure the radiation, for example, by measuring sample radiation in laboratories that are deep underground. The age resolution can also be improved by measuring larger samples, which permits a larger amount of ^{14}C to decay during the counting period. Larger samples have the potential to extend the age range to 60 ka.

In *accelerator mass spectrometry (AMS)* dating, the sample is ionized, and ions are separated by atomic mass or, more precisely, by the ratio of mass to charge. Whereas only ^{14}C decaying to ^{12}C during the sampling time is measured in conventional dating, all ions of mass 14 are counted directly in AMS dating.

In this way, a sample can be dated with a mass 1/1000 that necessary for conventional dating, as low as 0.01 mg carbon. AMS dating has a dating range of about 50 ka, not significantly different from the time range of conventional dating because of contamination by minute amounts of modern carbon, but better laboratory procedures may increase the range in the future. Currently, the cost of accelerator dating is 1.5 to 2.5 times that of routine conventional dating. However, conventional radiocarbon ages with precision of ± 10 years (Sieh et al., 1989) are more expensive than standard AMS dates.

Very inexpensive radiocarbon dating of shell carbonate relies on liquid scintillation counting of absorbed CO_2 (Vita-Finzi, 1991). This dating, called first-order dating by Vita-Finzi, compares well with conventional dates for the past 7000 years. The low cost means that a large number of samples can be dated at one's own laboratory, so that the results can be obtained quickly.

A basic assumption is that the carbon sample is uncontaminated, and that it has been a closed system with respect to isotopic carbon since death of the organism. Collection of samples far beneath the ground surface reduces the possibility of contamination. Contamination due to saturation of charcoal by humic acids can be dealt with by removing the humic components by an alkali pretreatment. Shells, tufa, and soil carbonate are generally unreliable for dating, because they do not form a closed system. Shells exchange C with groundwater, and the resulting date is generally the time that aragonite in the shell recrystallized to calcite. The amount of recrystallization can be determined by x-ray diffraction analysis. Commonly the outer layer of a shell is dissolved with hydrochloric acid on the assumption that the innermost layers are most likely to have been a closed system. Soil carbonates on the underside of stones have been studied by dating individually the outer, middle, and inner parts of the layer (Pierce, 1986). Theoretically, the innermost layer gives the oldest age and is closest to the true age. Contamination can also be introduced during sampling, handling, and storage, including the growth of terrestrial bacteria in water-saturated marine sediment cores during storage.

A sample of 50 ka age contains 10^{-3} of its original ^{14}C, and the addition of this amount of modern ^{14}C contaminant will reduce the age by one half-life, about 5730 years. However, contamination by older, dead carbon such as coal or graphite is less of a problem. Adding 10 mg of dead C contaminant to a 1 g C sample reduces the ^{14}C activity by 1%, making the sample 80 years too old (Stuiver, 1991). Contamination is also less for young samples than for old samples. The contamination problem combined with the low amount of ^{14}C present should lead to great caution in using ^{14}C dates older than about 30 ka.

The ^{14}C present in oceans or lakes may differ from that in the atmosphere, generally having a lower ^{14}C activity than that in the atmosphere. Oceanic upwelling brings older waters to the surface; in the Antarctic Ocean, shells of the nineteenth century have yielded ^{14}C ages of 1000–1300 years (Stuiver, 1991). Lakes also contain water with carbon ratios yielding an age that is too old, particularly where there is a contribution of dissolved bicarbonate from groundwater. Modern shells from springs along the margin of Great Salt Lake, Utah, have yielded dates of 25 ka (D. Curry, personal commun. to M. Machette). A *reservoir correction* must be made for these effects. It may be difficult to estimate what this correction should be, unless it is possible to compare carbon ratios from plants (including diatoms) that are in equilibrium with the atmosphere, and lake carbonates that are in equilibrium with the lake water.

Other errors are introduced by the precision of laboratory measurements, which are evaluated by repeat analyses and analyses of parts of the same sample by different laboratories. There is also a statistical uncertainty in a given radioactivity measurement.

Biasi and Weldon (1994) narrow the statistical confidence intervals in radiocarbon dates at Pallett Creek and Wrightwood on the San Andreas fault by giving weight to other information known about a suite of dated samples. Other information includes stratigraphic order of samples, cross-cutting relationships, sediment accumulation rates, thicknesses of dated peat layers, and historically dated materials such as artifacts. An example of a set of fully ordered dates is shown as Figure 6–8.

An initial assumption was that the ratio of ^{14}C to ^{12}C in the atmosphere remained constant with time. Unfortunately, this ratio has varied with time due to variations in the strength of the magnetic shield around the earth; the magnetic shield deflects some cosmic rays away from the earth's atmosphere. These variations are caused by slow changes in the strength of the earth's internal magnetic field and in shorter-period, more oscillatory changes in the field produced by the solar wind. A calibration curve for terrestrial samples for the past 4500 years (Stuiver and Kra, 1986; Stuiver and Pearson, 1986), based on a comparison of the calendar age and radiocarbon age of a calibrated series of tree rings, shows that the radiocarbon timescale may differ from the calendar scale by as much as 500 years (for part of this curve, see Fig. 6–9). During certain times, ^{14}C levels in the atmosphere decreased at the same rate as ^{14}C decreased

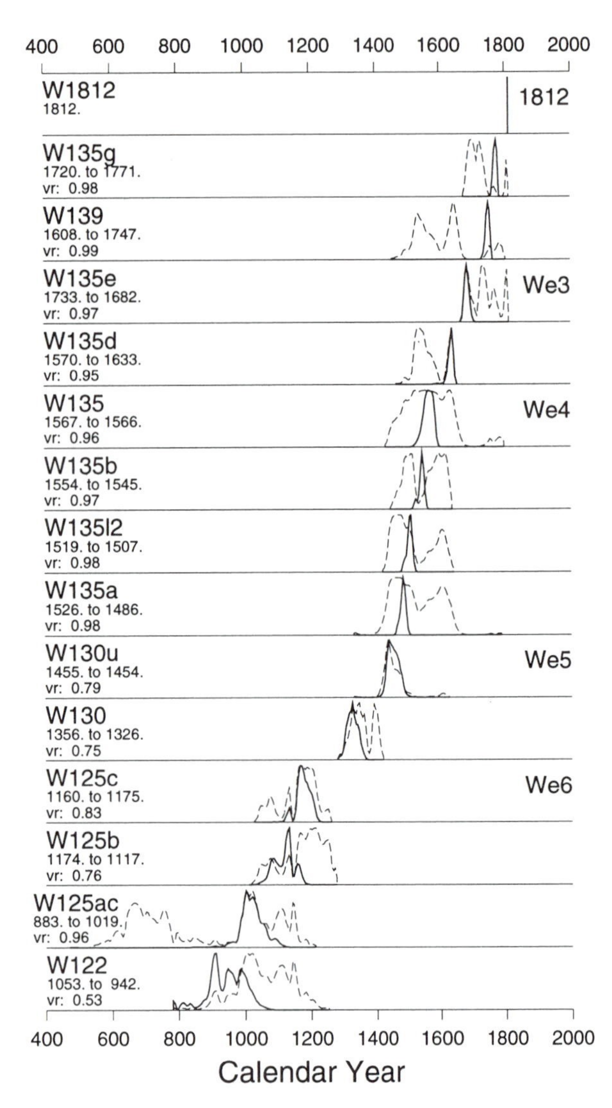

Figure 6–8. Dated horizons at Wrightwood on the San Andreas fault in California, in stratigraphic order. Dashed lines show calibrated age distribution prior to consideration of stratigraphic relations, and solid lines show fully-constrained ages. Means of calibrated and fully-constrained age distributions shown at left under name of stratigraphic layer. 1812 event is historical. We3, We4, etc., refer to individual earthquakes. After Biasi and Weldon (1994).

by radioactive decay, leading to an effect of clustered radiocarbon ages. For example, samples dated independently as 420 B.C.E. to 750 B.C.E. (calendar dates) all have a radiocarbon age of 2430–2470 years B.P. (Stuiver, 1991).

An illustration of this problem is provided by Alan Nelson and Stephen Personius of the U.S. Geological Survey, who obtained ^{14}C ages from carefully selected and pretreated peat samples from South Slough in the southern coastal region of Oregon. They obtained the ages to test their hypothesis that they were sampling the same peat horizon in both cores. Their results are shown in Figure 6–10. The weighted mean age of four ages from the peat in core TC is 2487 ± 35 ^{14}C years, which gives a 260-year spread in calendar years at one standard deviation because the ages fall on a relatively unfavorable segment of the calibration curve. The weighted-mean age from core WC is 2350 ± 36 ^{14}C years, which gives only a 20-year spread in calendar years because the ages fall on a favorable segment of the curve. The spreads do not overlap. ^{14}C dating supports the hypothesis that the two peats are the same only if an error multiplier of 2 is used, which produces a large overlap in the calibrated ages. Thus, despite great care in sample selection and treatment, the eight ^{14}C ages were not able to support or refute the physical correlation of the peat between the two cores.

The tree-ring calibration curve has been extended to 9700 years (Stuiver et al., 1991). Calibration back to 30,000 years B.P. has been achieved by comparing radiocarbon ages of coral from Barbados with high-precision uranium-thorium ages obtained by isotope-

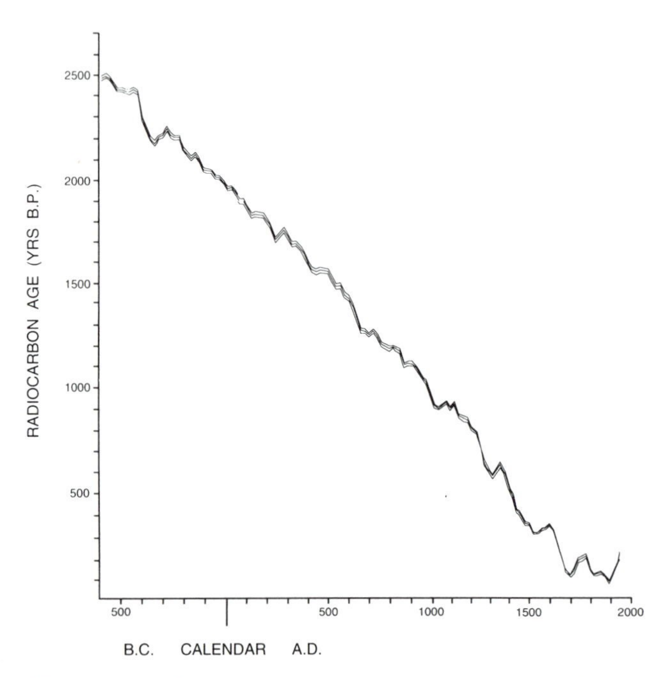

Figure 6–9. Tree-ring calibration of the radiocarbon timescale from 500 B.C. to A.D. 1950. From Stuiver and Pearson (1986).

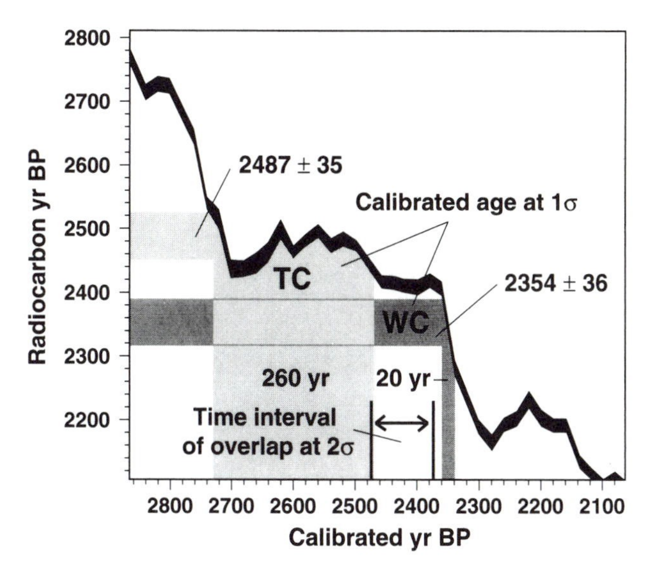

Figure 6–10. Calibration of the mean AMS ages for the 2.3-ka soil in cores WC-12 and TC-01 from radiocarbon years to calibrated years (approximately calendar years before A.D. 1950). Heavy black line shows the calibration curve at one standard deviation from Stuiver and Pearson (1986). Shaded area shows how the mean age from core TC-01 (at one standard deviation) corresponds to a calibrated-year time interval of 260 yr; the mean age from WC-12, which falls on a more favorable portion of the calibration curve, corresponds with a calibrated-year time interval of only 20 yr. Note that at one standard deviation the calibrated age intervals do not overlap, but at two standard deviations they do (time interval of overlap is marked by thick vertical lines above the horizontal axis). Although the AMS ^{14}C laboratory that determined the ages included all analytical errors in the standard deviations of the ages, the standard deviations may not include a small additional error due to possible systematic differences in ages between the AMS laboratory and the ^{14}C laboratories where the calibration curve was developed. Furthermore, the samples are detrital plant fragments that may differ in age by as much as 50–100 yr. These factors would tend to increase the standard deviations on the ages (and hence the overlap of the calibrated ages) by an unknown but probably small number of years. In any case, because the ages overlap we cannot infer that the soils in both cores were submerged at different times. From Nelson (1992).

dilution mass spectrometry (see the following text). This calibration (Bard et al., 1990, Fig. 6-11) shows an age difference as large as 3500 years for corals with radiocarbon ages of about 20 ka. This means that the $^{14}C/^{12}C$ ratio during the time of maximum glaciation was 1.4–1.5 times that of today, indicating a much lower strength of the magnetic field during glacial time (Broecker, 1992). Because of these variations, ages must be specified as either uncorrected radiocarbon years or as calendar years.

These cautions are not discussed to detract from the value of radiocarbon dating but rather to encourage great caution not only in obtaining and preparing the sample but also in interpreting the results.

Uranium-Series Dating

The elements ^{238}U, ^{235}U, and ^{232}Th produce daughter products of varying properties; these are present in trace amounts in most geological materials. If the system with the parent isotopes and daughter nuclides remains undisturbed for around two million years, *secular equilibrium* between parent and daughter products is established. Disturbance of this equilibrium produces fractionation. Uranium-series (U-series) dating measures the degree to which secular equilibrium is established after the disturbance. The material must remain a closed system for the nuclides of concern, and it must be pure and relatively uniform. Three common methods exist: (1) ^{230}Th deficiency or U-Th method (disequilibrium between ^{230}Th and ^{234}U) with an age range of 5 to 350 ka (^{230}Th is so insoluble in seawater that it is nearly absent in living corals), (2) ^{231}Pa deficiency method

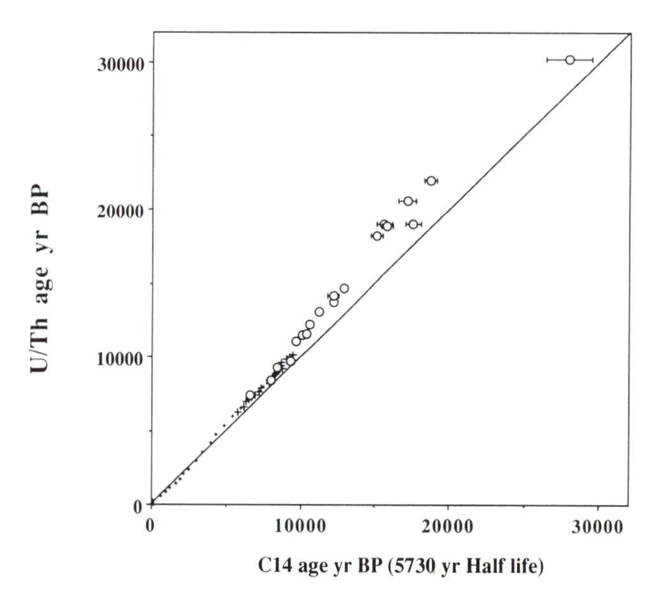

Figure 6–11. Calibration of the radiocarbon timescale for the past 30,000 years. Calibration is based on high-precision U-Th ages obtained by isotope-dilution mass spectrometry (circles), varved sediments (large crosses), and tree rings (small crosses). From Bard et al. (1990).

(disequilibrium between ^{231}Pa and ^{235}U) with an upper age limit of 180 ka, and (3) ^{234}U excess method (excess between ^{234}U with respect to ^{238}U), commonly used together with ^{230}Th dating. Excess ^{234}U is commonly found in natural waters and secondary minerals and can be used for dating if the initial $^{234}U/^{238}U$ activity ratio can be determined. This ratio is known for seawater under open marine conditions (Chen et al., 1986), but it may be different for enclosed bays. This method can be used for dating in the range 100 ka and 1.5 million years (Szabo and Rosholt, 1991).

Inorganic carbonates such as cave calcite, calcite veins, dense spring-deposited calcite (travertine) and porous calcite (calcareous tufa), clean and dense lacustrine carbonate, and soil carbonate (calcrete, carbonate rinds on clasts, and concretions) can be dated using this method. Presence of detrital materials in lacustrine carbonate and soil carbonate requires age corrections due to contamination by ^{230}Th. Coral contains the most reliable organic carbonate for uranium-series dating. Fossil corals in New Guinea have been used to calibrate the $\delta^{18}O$ time scale; on the west coast of North America, corals have been dated as far north as southwestern Oregon. To be dated, coral material must be composed of unrecrystallized aragonite and must not contain contaminants in pores or post-depositional carbonate cement. Development of new measurement techniques using isotope-dilution mass spectrometry has allowed Edwards et al. (1987) to use the ^{230}Th deficiency method to date a coral 180 years old with an uncertainty of ±5 years and a coral 123.1 ka with an uncertainty of ±1100 years.

Molluscs and sea urchins have been considered for uranium-series dating, but uranium can enter the shells after death of the animal, giving erroneous or, at best, minimum dates.

Detritus-free evaporite deposits have provided U-series ages of the time of deposition. Secondary silica has been dated, but the results have not been checked against other techniques. Fossil bones take up uranium after the death of the organism, and dates of bone and tooth enamel are not considered reliable. Young volcanic rocks may be dated if (1) secular equilibrium was maintained during partial melting and eruption, (2) there is no crustal contamination after eruption, and (3) the volcanic rock continues to be a closed system after it solidifies.

Uranium-Trend Dating

The difficulty with uranium-series dating is that most Quaternary deposits are open systems with respect to ^{238}U, ^{234}U, and ^{230}Th, and thus they cannot be dated using this technique. Uranium-trend dating is possible for a much greater variety of deposits. The technique has been applied to samples from Quaternary alluvium, wind deposits, glacial deposits, volcanic ash, and marine terraces.

Uranium occurs in the fixed state, structurally incorporated in minerals, and in the mobile form, mainly in groundwater. Fractionation is the preferential leaching of ^{234}U from the fixed form. The patterns of isotope fractionation are influenced by time, leading to the possibility of open-system dating. This technique is based on empirical modeling required to determine the degree and maintenance of disequilibria in an open system after deposition of the geologic materials (Muhs et al, 1989; Rosholt, 1985; Szabo and Rosholt, 1991).

In surficial deposits, the clock starts with the beginning of groundwater movement through the sediment rather than the beginning of soil formation. Groundwater moves through the sediment, transporting uranium, but it does so at a decreasing rate due to compaction of the sediment and formation of soil. Important considerations include the quantity of water moving through a deposit, its flow rate, the concentration of uranium in solution, and the concentration in solution relative to that in the fixed state. For a uranium-trend age, several samples with slightly different physical and chemical properties are taken from the same site. When $(^{234}U - ^{238}U)/^{238}U$ is plotted against $(^{238}U - ^{230}Th)/^{238}U$, the samples from the same site should follow a straight line. This line is used to determine tangency to the disequilibrium curve (Fig. 6-12a). It is necessary to calibrate this curve with other dating techniques. The slope of the curve is essentially zero for very young and very old ages, and trend slopes of zero are relatively insensitive to age. The most reliable ages are in the range of 60 ka to 600 ka, with an error of 10%. Larger errors characterize the upper and lower age ranges.

Potassium-Argon Dating

Argon makes up close to 1% of the atmosphere. It has three stable isotopes, ^{36}Ar, ^{38}Ar, and ^{40}Ar, with the last making up 99.6% of atmospheric argon. Potassium-argon dating is based on the radioactive decay of ^{40}K to ^{40}Ar. To date a volcanic rock, it is necessary to determine how much ^{40}Ar is atmospheric and how much is a product of the radioactive decay of ^{40}K. This is possible because the atomic ratio of ^{40}Ar to ^{36}Ar in the atmosphere is known to be a constant 295.5. Measurement of ^{36}Ar allows the determination of the amount of ^{40}Ar that is atmospheric, based on this ratio. The remaining ^{40}Ar is radiogenic, and the age is determined by the ratio of ^{40}K to radiogenic ^{40}Ar.

In the $^{40}Ar/^{39}Ar$ dating technique, ^{39}K is converted to ^{39}Ar in a nuclear reactor, and Ar isotope ratios are determined with a mass spectrometer. ^{40}Ar and ^{39}Ar ratios are measured as Ar is released in incremental heating steps, with each heating step producing an apparent age as deeper parts of the sample release argon. These apparent ages should be about the same, producing a "plateau" on a plot of argon ratios against heating steps. This procedure is used predominantly for older samples where there is a greater chance of resetting the K-Ar clock by younger thermal events. It has not been used extensively in Quaternary dating because Quaternary samples rarely contain a high enough percentage of potassium to produce enough radiogenic argon to measure accurately. However, Pringle et al. (1992) used low-temperature heating steps to remove nonradiogenic argon contamination and was able to date plagioclase from New Zealand

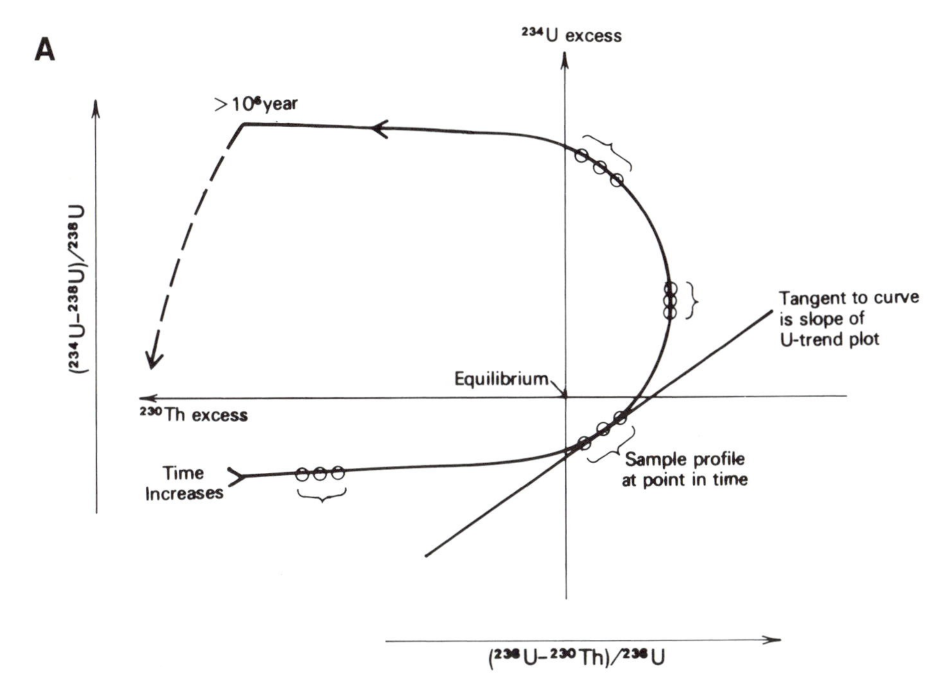

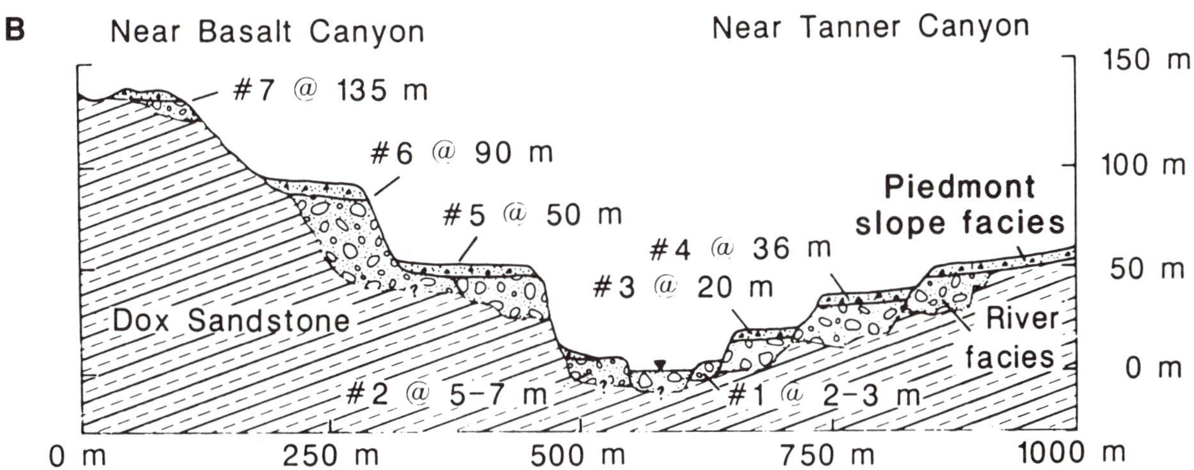

Figure 6–12. (a) Hypothetical development of ^{234}U-^{230}Th disequilibrium for uranium-trend slope for a three-sample deposit. The tangent to the curve is a measure of the age of the deposit, if the curve is age-calibrated by other means. Several samples are necessary to determine the tangent of the slope. From Rosholt (1985) and Muhs et al. (1989). (b) Diagrammatic cross section of Colorado River showing terraces dated by uranium trend.

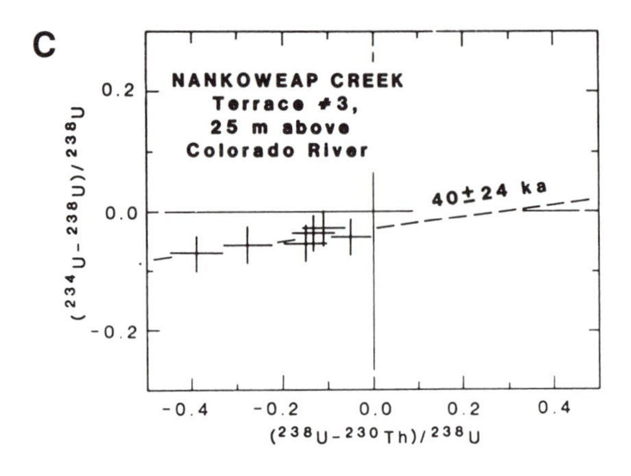

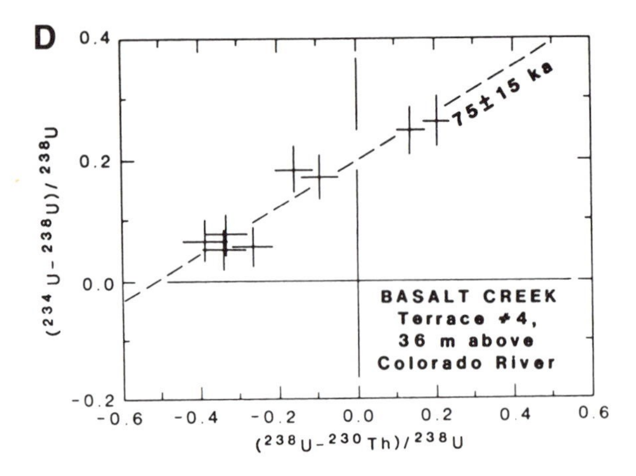

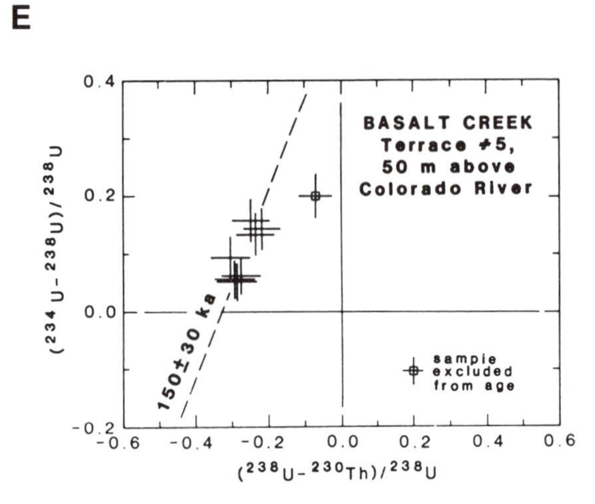

Figure 6–12. (c,d,e) Uranium-trend ages for three terraces shown in (b). Each terrace yielded enough samples to determine a trend line; the steeper slopes represent older ages. (b–e) From Machette and Rosholt (1991).

as young as 330 ka. The success of the technique in dating young rocks is in part the greater precision of modern mass spectrometers.

Dating Quaternary volcanic rocks must confront three problems. (1) Very little ^{40}K decays to ^{40}Ar in a million years, and thus the atmospheric ^{40}Ar correction for Quaternary samples is very large. In addition, atmospheric argon entrapped in magma as it is being erupted may not equilibrate but instead may fractionate, resulting in an apparent age that is too young due to overcompensation for atmospheric ^{40}Ar. (2) A small amount of older material included in the volcanic rock will produce an anomalously old age. (3) The sample may contain excess radiogenic argon, that is, argon with a high ^{40}Ar/^{36}Ar ratio with respect to the atmosphere. Argon readily equilibrates in a magma chamber, and the argon content of a crystal in the magma chamber is controlled by the partial pressure of ^{40}Ar in the magma chamber. Upon eruption, fine-grained crystals in the groundmass (microlites) lose nearly all their excess radiogenic argon, with the loss rate dependent on crystal size and temperature. However, larger crystals (phenocrysts) retain much excess radiogenic argon during cooling, producing ages that are too old, as described above for inclusions of older materials. Accordingly, K-Ar dating on young volcanic rocks is done on microlites in the groundmass rather than phenocrysts, although phenocrysts of sanidine (a variety of potassium feldspar restricted to volcanic rocks), produce reliable ages. It is important to date the microlites and not volcanic glass in the groundmass, which produces unreliable ages if it becomes hydrated . Massive interiors of flows have fewer vesicles (gas bubbles that may contain large amounts of air), are less likely to be contaminated by rocks underlying or overlying the flow, and cool relatively slowly so that excess ^{40}Ar has time to escape.

It was noted previously that atmospheric argon entrained in a magma as it is erupted may fractionate rather than equilibrate, producing apparent ages that are too young. Matsumoto et al. (1989 and in prep.) determined ^{40}Ar/^{36}Ar and ^{38}Ar/^{36}Ar ratios on historical lava flows in Japan and worked out a mass-fractionation correction for atmospheric argon, which they consider to be necessary to date young volcanic rocks. They use a different method (peak height comparison) to estimate the $^{40}Ar/^{36}Ar$ ratio, and they compare it to the $^{38}Ar/^{36}Ar$ ratio, which does not change. Using this method, they are able to date lavas with one to two weight percent K_2O that are as young as 30 ka. Hu et al. (1994) obtained a $^{40}Ar^{39}Ar$ age of 12,560 ± 470 years of the Mono Craters, California, using laser fusion to generate an isochron ($^{36}Ar^{40}Ar$ plotted against $^{39}Ar^{40}Ar$) from 63 sanidine crystals from 5 sites. This permits an independent calibration of ^{14}C ages; future developments may permit $^{40}Ar^{39}Ar$ laser dating of Holocene volcanic materials.

Fission-Track Dating

An atom of ^{238}U most commonly decays by emitting an alpha particle, but more rarely it will decay by spontaneous fission into two highly-charged nuclei that recoil in opposite directions, producing a 10- to 20-μ long linear zone of damage called a *fission track*. After etching with a suitable chemical solution (hydrofluoric acid for volcanic glass, an alkali flux for zircon, nitric acid for apatite), the track is enlarged sufficiently that it can be observed with an optical microscope. The ratio of spontaneous fission of ^{238}U is constant, and therefore the age of a mineral (or glass) can be calculated from the number of spontaneous tracks intersecting a polished surface of the mineral and the amout of uranium that produced those tracks (see Naeser and Naeser, 1988, for an extended review of fission-track dating).

One of the most common applications of fission-track dating of Quaternary samples is dating volcanic ash. Glass or zircon (the two materials most commonly used in dating Quaternary samples) must contain enough uranium so that a statistically meaningful number of tracks can be counted in a reasonable time. Furthermore, for the fission-track age to be geologically significant, the sample must not have been heated sufficiently to cause the tracks to begin to disappear (anneal). This is particularly a problem in volcanic glass in which tracks can anneal at ambient surface temperatures. There are several methods available for correcting glass fission-track ages lowered by partial annealing. The easiest is the modified isothermal plateau annealing method (Westgate, 1989). Another potential problem in dating glass is that fission tracks may be difficult to distinguish from the small vesicles and microlites that are common in some glasses, particularly obsidian.

Zircon is generally considered preferable to glass for fission-track dating, because zircon fission-track ages are not affected by near-surface ambient temperatures. Acidic lavas are more likely to be datable by the fission-track method because there is a greater likelihood of them containing zircon.

In addition to dating volcanic rocks, the annealing of fission tracks, and the resulting effect on fission-track age and track lengths, makes them an excellent tool for determining cooling rates in formerly deeply buried rocks (Fitzgerald and Gleadow, 1990). Under some circumstances, these cooling rates can be correlated to uplift and denudation. Fission tracks in apatite crystals are totally annealed at temperatures between around 105°C for relatively long-term heating of around 10^8 years duration and 150°C for relatively short-term heating of around 10^5 years. Fission tracks in zircon are totally annealed over periods of geological time at temperatures between around 175°C and 250°C. The apparent fission-track age of zircon will be greater than the fission-track age of apatite in the same rock, because zircon begins to retain tracks at higher temperature, hence at greater burial depth, than apatite. If the present geothermal gradient is assumed to have remained relatively constant through time, the depths of burial at the times the zircon and apatite passed through their respective closure temperature may be determined. The depth decreases with time to zero at present, as rocks are progressively brought to the surface. The rate of decrease measures the removal of overburden by erosion, generally accompanying uplift.

In the Pakistan Himalaya, Zeitler (1985), based on fission-track ages, recorded very high denudation rates on the Nanga Parbat massif compared to a region a short distance to the west. The strong gradient in denudation rates was later found to contain the active Raikot fault (Madin et al., 1989). In another study involving a major fault, Naeser et al. (1983) used fission-track ages of apatite to document the long-term uplift of the Wasatch Mountains, Utah. This study showed that the Farmington Canyon complex, northeast of Salt Lake City, has been uplifted at an average rate of 0.4 mm/yr for the past 10 million years. Most of this movement has been along the Wasatch fault (discussed further in Chapter 9).

Luminescence and Electron-Spin-Resonance Dating

Sediments are irradiated naturally by the decay of ^{40}K, ^{238}U, and ^{232}Th. This radiation releases alpha, beta, and gamma particles which are displaced from outer electron shells and trapped in crystal defects (*electron traps*). Radiation is produced by small quantities of ^{40}K, ^{238}U, and ^{232}Th contained in most sediments and, in near-surface sediments, by cosmic rays. Radiation accumulates until electron traps are saturated. The rate of accumulation depends on the amount of radioactive elements in the sediment. Feldspar is much more sensitive to gamma radiation than quartz, but some feldspar samples lose part of their radiation during storage at room temperature, a phenomenon called *anomalous fading*. When heated or exposed to sunlight, the electrons are released from the high-energy traps in mineral grains, a process called *thermoluminescence (TL)*.

TL dating was first applied to pottery, which lost all its previous TL at the time it was fired; the mineral grains in the pottery were *zeroed* by the heating. The measured TL accumulated entirely after firing, and the TL was used to date the time the pottery was fired. One of the ways TL in sediments is lost is by exposure to sunlight, but generally the zeroing process is incomplete; the sediment retains some TL acquired by mineral grains prior to deposition. A nearly ideal sediment for TL dating is wind-deposited silt (loess), which loses much of its TL due to extensive exposure to ultraviolet and visible-spectrum sunlight during transport in the atmosphere. Loess begins to accumulate TL after it is buried by younger sediments. Stream and lake sediments have also provided useful dates if the sediments were exposed to sunlight long enough to lose most of their TL prior to burial. Such sediments generally retain a higher TL because of the attenuation of ultraviolet radiation by water. The retention of an inherited signal results in *partial bleaching*. Sediments which may have only minimal exposure to sunlight, such as mudflows or glacial till, generally yield unreliable ages. Volcanic ash is completely zeroed by heating during eruption, and if it is buried soon after eruption, it can yield reliable TL ages.

The radiation level necessary to reproduce the TL signal is the *equivalent dose (ED)*, measured in grays; 1 gray = 100 rads. To determine the age, the equivalent dose is divided by the dose rate in grays/year; the dose rate measures the environmental radioactivity of the sediment. Dose rate is determined by measuring the concentration of radionuclides in the sediment or by direct measurement of the beta- and gamma-ray dose rates, then calculating the alpha-ray dose rate based on chemical analysis.

There are three methods for determining ED: *regeneration, total-bleach,* and *partial-bleach* (Fig. 6-13). The ED is based on the TL response to a calibrated radioactive source and the reduction of the TL signal due to exposure to laboratory light. The regeneration method compares the natural TL signal to a TL signal regenerated after extended light exposure in the laboratory or in sunlight for more than 7 hours. It is most effective for wind-transported sediments in which the residual TL is relatively low. The TL acquired after increasing ED allows the construction of a growth curve that can be extended back to the residual level. In the total-bleach method, the residual TL after more than 7 hours exposure to laboratory light is compared to a TL growth curve calculated by the

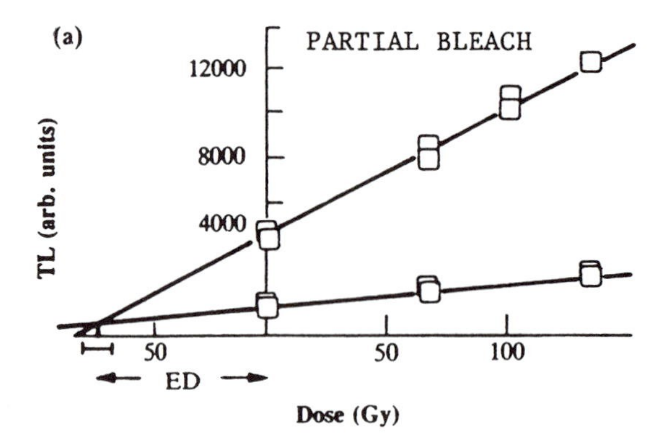

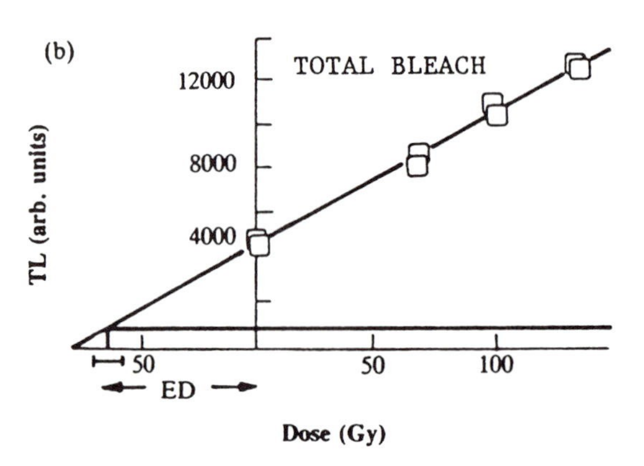

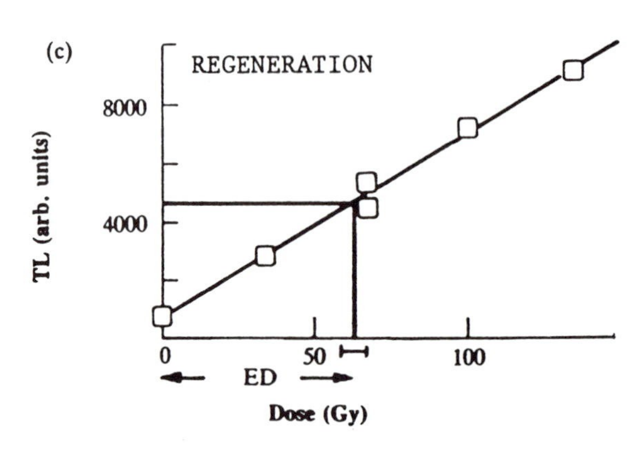

Figure 6–13. Three methods for determining the equivalent dose (ED) in thermoluminescence (TL) measurements. After Wintle and Proszynska (1983) and Forman (1989).

response of the sediment to additive beta or gamma radiation. The ED is the intercept of the additive dose curve with the measured residual TL level. In the partial-bleach method, two sets of samples are irradiated with beta or gamma radiation to produce TL growth curves. Before the TL is measured, one set of samples is exposed to laboratory light. The point of intersection of the two curves is the ED.

One caution in interpreting TL results involves the assumption that the amount of radioactive elements now in the sample is representative of its history since partial zeroing. Groundwater in the sediment may absorb radiation, and the moisture content in the sample must be considered in TL dating. But the present moisture content may not be representative of the past. The climate of the past may have been different than now, resulting in higher or lower moisture content. Another source of error is misinterpreting the depositional environment, which involves assumptions about the time and nature of partial bleaching.

At present, TL dating can be useful for sediments deposited in the past 100 ka with a precision of 10–15%. Sampling of sediments in a trench or borehole must be done without exposure to sunlight. After sampling, the material to be dated must be kept completely in the dark.

Optical-stimulated luminescence (OSL) is similar to TL except that the radiation dose is measured in monochromatic light emitted by an argon laser rather than by heating the sediment. Luminescence measures the release of electrons from light-sensitive traps only, whereas TL measures electrons released from light-sensitive and light-insensitive traps. OSL has the advantage that it requires less time to bleach minerals, but its useful age range is much less than TL. It is best for young deposits.

Electron-spin resonance (ESR) measures the electron-spin concentration trapped by defects in the crystal lattice, which is proportional to the radiation dose (ED) that has been received by the minerals apatite or calcite since formation. The age is calculated from measurements of annual radiation dose and the electron spin concentration that this dose produces. ESR has been applied to cave calcite, shells, coral, tooth enamel, and bone, and it has promise for evaporites and silicates, especially quartz.

ESR has also been used to date fault gouge (Ikeya et al., 1982). Faulting is accompanied by frictional heating and by crushing of mineral grains, both of which lead to complete or partial zeroing of the ESR signal in quartz. The time of last fault movement is determined by measuring the accumulated dose in fine-grained fragments of granulated quartz in gouge. Finer-grained fragments are more completely zeroed than coarser-grained fragments. However, fault gouge from the San Jacinto fault in southern California gave an ESR age of 75 ka, even though the fault has moved within the last century (Buhay et al., 1988). This indicates that a faulting event will only zero the ESR signal at depth in the fault zone, under considerable confining pressure. Fifty to seventy-five meters of overburden may be necessary to zero the quartz. Using ESR to date the most recent faulting event may be practical only in boreholes or deep tunnels. Even then, it may be difficult to determine which strand of a fault zone moved most recently.

Surface-Exposure Dating with Cosmogenic Nuclides

In addition to dating Quaternary deposits that underlie a surface, it is of interest to date the surface itself. If we assume that the cosmic ray flux is constant for timescales greater than 10,000 years, then the concentration of cosmogenic nuclides near the surface is directly related to the time the surface has been exposed. The cosmic-ray flux has not been constant for timescales less than 10,000 years, as noted previously in the section on ^{14}C dating. However, cosmogenic production in rocks integrates these effects over the exposure time, so that these higher-frequency variations are not significant (Lal, 1991; Cerling and Craig, 1993). The production rates of these isotopes are extremely low: <10 to >100 atoms g^{-1} yr^{-1} The development of accelerator mass spectrometry (AMS, discussed previously under ^{14}C dating) makes it possible to measure the low levels of cosmogenic nuclides in surfaces. Furthermore, two isotopes, ^{3}He and ^{21}Ne, have been detected in terrestrial surfaces with conventional mass spectrometers. The four isotopes that are measured routinely and for which the most data are available are ^{3}He, ^{10}Be, ^{26}Al, and ^{36}Cl; research on two more, ^{14}C and ^{21}Ne, is yielding promising results.

Primary cosmic rays, mostly protons, interact with the earth's atmosphere and produce neutrons which bombard the earth's surface, producing cosmogenic isotopes. The rate of isotope production depends on modulation of the galactic cosmic-ray flux by the earth's magnetic field and modulation of the secondary (neutron) flux by the atmosphere. As a result, cosmogenic isotope production is greatest at high altitudes and high latitudes. Altitude and latitude corrections are accurate to within ±10% (Lal, 1991); accuracy is likely to improve with further verification.

Cosmogenic-nuclide dating of surface exposure requires several assumptions: (1) the production rate of the nuclide is well known, (2) the surface was exposed to cosmic rays for the time in question, withough erosion or cover, (3) the sample was not previously exposed to cosmic rays, and (4) the sample has been a closed system during the exposure. Even if the surface has been eroded or covered, a minimum age can be obtained.

Cosmogenic-isotope production is highest at the surface and decreases rapidly with depth. Half the fast cosmic-ray neutrons are absorbed above a depth of 45 cm in rock. An ideal surface is flat rock, and a sample may be collected by hammer and chisel or by a mechanical corer. It is necessary to ensure that the isotope measured is cosmogenic and not an atmospheric contaminant, not produced by radiogenic decay of U or Th, and not inherited from earlier exposure to cosmic rays. Calibration of surface-exposure dating has been done with lava flows (Kurz, 1986a, b), in which the depth of erosion can be estimated based on surface morphology, and which could not have had any previous exposure. Glacially-striated surfaces have also been used, where the age of glaciation has been determined using ^{14}C. Polished glacial striations indicate that the surface is well preserved, but preservation is enhanced by soil cover which could have been removed prior to sampling. An ideal sampling locality would be a steeply inclined striated surface well above any possible soil cover, such as a surface used for calibration between ^{10}Be and ^{26}Al (Nishiizumi et al., 1989).

Antarctica has been an ideal place to study surface-exposure dating (Nishiizumi et al., 1991; Brook et al., 1995, in press). It has a high latitude, and most sampling sites are at high altitude, so that cosmic-ray bombardment is high, and desert conditions ensure very low rates of erosion.

Anthony and Poths (1992) present an example of surface-exposure dating of very young lava flows in the Rio Grande Rift, New Mexico, using ^{3}He. They sampled only the uppermost 3 cm of flows where primary surface flow features were well preserved. They sampled and measured isotopes in the minerals olivine and clinopyroxene. Their calculations took into consideration a magmatic component of ^{3}He as well as atmospheric contamination. Their ages were based on a cosmogenic ^{3}He production rate of 434 atoms g^{-1} yr^{-1} at 39° N and 1445 m altitude. Erosion, which would tend to produce ages that were too low, was estimated as 1–5 cm. Their surface-exposure dates were reproducible within 14% for different samples from the same surface and within 8% for samples which represent synchronous events based on geologic field relations. Ages range from 80 ka to 17 ka and are consistent with degree of soil development, geomorphology, and K/Ar age. Still younger ages were obtained from lava flows in Hawaii (Kurz et al., 1990).

Amino-Acid Racemization

Protein of living organisms contains amino acids that are almost entirely of the L (left-handed) configuration. After death, the skeletons are fossilized, and amino acids reflect diagenetic alteration of the original proteins. During diagenesis, proteins are hydrolyzed into polypeptides and free amino acids, and in the process, as much as 50% of the amino acids are lost (Wehmiller, 1993). The part that remains is usually stable. Racemization is the alteration of amino acids of the L configuration to a mixture of the L and D (right-handed) configuration. In most materials studied, the initial (zero-age) abundance of D-amino acids is zero, so the ratio of the D to L configuration must increase from zero to an equilibrium value, usually 1.0 for most materials. (An equilibrium value would represent an "infinite" age.) D/L ratios do not increase linearly with time, and the nonlinear response is different for different fossil genera. Racemization is also affected by mean annual temperature, the annual temperature range, and the nature of diagenesis, and an understanding of diagenetic temperature history is necessary to interpret amino-acid ratios. Because the reaction kinetics after death and the variations among species are not well known, amino-acid racemization is best used for relative age determination and lateral correlation. Stratigraphic units with molluscs having similar D/L ratios have been called *aminozones*.

Molluscs are common in some marine terraces but are not closed systems for uranium-series dating. For this reason, amino-acid age estimates may be the only way to estimate the age of a terrace, particularly if its age is beyond the range of radiocarbon dating. Individual shells are studied, and age estimates from several shells in the same deposit can test against mixing of older and younger fossil assemblages. Sample size needed is small, 100–400 mg, and sometimes as small as 5 mg.

Suggestions for Further Reading

Oxygen-isotope Stratigraphy

Delmas, R. J. 1992. Environmental information from ice cores. Reviews of Geophysics, 30:1–21.

Emiliani, C. 1955. Pleistocene temperatures. Journal of Geology, 63:538–78.

Imbrie, J., Hays, J. D., Martinson, D. G., McIntyre, A., Mix, A. C., Morley, J. J., Pisias, N. G., Prell,

W. L., and Shackleton, N. J. 1984. The orbital theory of Pleistocene climate: Support from a revised chronology of the marine $\delta^{18}O$ record. *In* Milankovitch and Climate, Part 1. Berger, A. L., et al. eds. D. Reidel, pp. 269-305.

Martinson, D. G., Pisias, N. G., Hays, J. D., Imbrie, J., Moore, T. C., and Shackleton, N. J. 1987. Age dating and the orbital theory of the ice ages: Development of a high-resolution 0 to 300,000-year chronostratigraphy. Quaternary Research, 27:1-29.

Mix, A. C. 1987. The oxygen-isotope record of glaciation. *In* North American and adjacent oceans during the last deglaciation. Ruddiman, W. F., and Wright, H. E., Jr., eds. Boulder, Colorado: Geological Society of America, Decade of North American Geology, K-3:111-35.

Pisias, N. S., Martinson, D. G., Moore, T. C., Shackleton, N. J., Prell, W., Hays, J., and Boden, G. 1984. High resolution stratigraphic correlation of benthic oxygen isotopic records spanning the last 300,000 years. Marine Geology, 56:119-36.

Magnetostratigraphy

Liddicoat, J. C. 1991. Paleomagnetic dating. *In* Quaternary Nonglacial Geology: Conterminous U.S. Morrison, R. B., ed. Geological Society of America, Decade of North American Geology, K-2:60-61.

Mankinen, E. A., and Dalrymple, G. B. 1979. Revised geomagnetic polarity time scale for the interval 0-5 m.y. B.P. Jour. Geophys. Res., 84:615-26.

Strontium-Isotope Stratigraphy

DePaolo, D. J., and Ingram, B. L. 1985. High-resolution stratigraphy with strontium isotopes. Science, 227:938-41.

Hodell, D. A., Mueller, P. A., and Garrido, J. R. 1991. Variations in the strontium isotopic composition of seawater during the Neogene. Geology, 19:24-27.

Biostratigraphy

Berggren, W. A., Hilgen, F. J., Langereis, C. G., Kent, D. V., Obradovich, J. D., Raffi, I., Raymo, M. E., and Shackleton, N. J. 1995. Late Neogene chronology: new perspectives in high-resolution stratigraphy. Geol. Soc. America Bull. 107:1272-87.

Berggren, W. A., Kent, D. V., and Van Couvering, J. A. 1985. The Neogene: Part 2: Neogene geochronology and chronostratigraphy. *In* The Chronology of the Geological Record. Snelling, N.J. Geological Society Memoir 10, Blackwell Scientific Publications, pp. 211-60.

Jenkins, D. G., Bowen, D. Q., Adams, C. G., Shackleton, N. J., and Brassell, S. C. 1985. The Neogene: Part 1. *In* The Chronology of the Geological Record. Snelling, N. J. Geological Society Memoir 10, Blackwell Scientific Publications, pp. 199-210.

Van Couvering, J. A., ed. The Pleistocene Boundary and the Beginning of the Quaternary Definition. Cambridge University Press, in press.

Dendrochronology

Cook, E. R., and Kairiutstis, L. A., eds. Methods of Dendrochronology. London: Kluwer,

Shroder, J. F., Jr. 1980. Dendrogeomorphology—Review and new techniques of tree-ring dating. Progress in Physical Geography, 4:161-88.

Tephrochronology

Machida, H. 1991. Recent progress in tephra studies in Japan. The Quaternary Research, 30:141-49.

Sarna-Wojcicki, A. M., and Davis, J. O. 1991. Quaternary tephrochronology. *In* Quaternary nonglacial geology: Conterminous U.S. Morrison, R. B., ed. Geological Society of America, Decade of North American Geology, K-2:93-116.

Sarna-Wojcicki, A. M., Lajoie, K. R., Meyer, C. E., Adam, D. P., and Rieck, H. J. 1991. Tephrochronologic correlation of upper Neogene sediments along the Pacific margin: Conterminous U.S. Geological Society of America, Decade of North American Geology, K-2:117-40.

Radiocarbon Dating

Stuiver, M. 1991. Radiocarbon dating. *In* Quaternary nonglacial geology: Conterminous U.S. Morrison, R. B., ed. Geological Society of America, Decade of North American Geology, K-2:46-49.

Vita-Finzi, C. 1991. First-order ^{14}C dating Mark II: Quaternary Proceedings No. 1. Cambridge: Quaternary Research Association, pp. 11-17.

Uranium-Series Dating

Chen, J. H., Edwards, R. L., and Wasserburg, G. J. 1986. ^{238}U, ^{234}U and ^{232}Th in seawater. Earth and Planetary Science Letters, 80:241-51.

Edwards, R. L., Chen, J. H., and Wasserburg, G. J. 1987. ^{238}U-^{234}U-^{230}Th-^{232}Th systematics and the precise measurement of time over the past 500,000 years. Earth and Planetary Science Letters, 81:175–92.

Szabo, B. J., and Rosholt, J. N. 1991. Conventional uranium-series and uranium-trend dating. *In* Quaternary nonglacial geology: Conterminous U.S. Morrison, R. B., ed. Geological Society of America, Decade of North American Geology, K-2: 55–60.

Uranium-Trend Dating

Muhs, D. R., Rosholt, J. N., and Bush, C. A. 1989. The uranium-trend dating method: Principles and application for southern California marine terrace deposits. Quaternary International, 1:19–34.

Rosholt, J. N. 1985. Uranium-trend systematics for dating Quaternary sediments. U.S. Geological Survey Open-File Report 85-298, 34 p.

Szabo, B. J., and Rosholt, J. N. 1991. Conventional uranium-series and uranium-trend dating. *In* Quaternary nonglacial geology: Conterminous U.S. Morrison, R. B., ed. Geological Society of America, Decade of North American Geology, K-2: 55–60.

Potassium-Argon Dating

Dalrymple, G. B., and Lanphere, M. A. 1969. Potassium-argon dating. San Francisco: W. H. Freeman and Co., 258 p.

Damon, P. E. 1991. K-Ar dating of Quaternary volcanic rocks. *In* Quaternary nonglacial geology: Conterminous U.S. Morrison, R. B., ed. Geological Society of America, Decade of North American Geology, K-2:49–53.

Matsumoto, A., Uto, K., and Shibata, K. 1989. K-Ar dating by peak comparison method—New technique applicable to rocks younger than 0.5 Ma. Bull. Geological Survey of Japan, 40:565–79.

Fission-Track Dating

Fitzgerald, P. G., and Gleadow, A. J. W. 1990. New approaches in fission-track chronology as a tectonic tool: Examples from the Transantarctic Mountains. Nuclear Tracks and Radiation Measurements, 17:351–57.

Naeser, C. W., and Naeser, N. D. 1988. Fission-track dating of Quaternary events. *In* Dating Quaternary Sediments. Easterbrook, D. J., ed. Geol. Soc. America Special Paper, 227:1–11

Westgate, J. A. 1989. Isothermal plateau fission-track ages of hydrated glass shards from silicic tephra beds: Earth and Planetary Science Letters, 95:226–34.

Luminescence Dating

Berger, G. W. 1988. Dating Quaternary events by luminescence. Geol. Soc. America Special Paper, 227:13–50.

Forman, S. L. 1989. Applications and limitations of thermoluminescence to date Quaternary sediments: Quaternary International, 1:47–59.

Townsend, P. D., Rendell, H. M., Aitken, M. J., Bailiff, I. K., Durrani, S. A., Fain, J., Grun, R., Mangini, A., Me'dahl, V., and Smith, B. W. 1988. Thermoluminescence and electron-spin-resonance dating: Part II—Quaternary Applications. Quaternary Science Reviews, 7:243–536.

Surface-Exposure Dating Using Cosmogenic Nuclides

Bierman, P. R. 1994. Using in situ produced cosmogenic isotopes to estimate rates of landscape evolution: A review from the geomorphic perspective: Jour. Geophys. Res. 99:13,885–96.

Cerling, T. E., and Craig, H. 1994. Geomorphology and in-situ cosmogenic isotopes. Annual Review Earth and Planetary Sciences, 22:273–317.

Kurz, M. D., and Brook, E. J. 1992. Surface exposure dating with cosmogenic nuclides. *In* Dating in Surface Context. Beck, C., ed. University of New Mexico Press, in press.

Amino-Acid Racemization

Muhs, D. R. 1991. Amino acid geochronology of fossil mollusks. *In* Quaternary nonglacial geology: Conterminous U.S. Morrison, R. B., ed. Geological Society of America, Decade of North American Geology, K-2:65–68.

Wehmiller, J. F. 1993. Applications of organic geochemistry for Quaternary research: Aminostratigraphy and aminochronology. Chapter 36 of Organic Geochemistry. Engel, M. H. and Macko, S. A., eds. New York: Plenum Press, pp. 755–83.

General

Aguirre, E., and Pasini, G. 1985. The Pliocene-Pleistocene boundary. Episodes, 8:116–20.

Colman, S. M. and Pierce, K. L. 1991. Summary table of Quaternary dating methods. Plate 2 of Quaternary nonglacial geology: Conterminous U.S. Morrison, R. B. Geological Society of America, Decade of North American Geology, K-2.

7

Tectonic Geomorphology

The surface of the earth is acted on by two opposing forces, one internal, producing uplift and downwarping in response to crustal stresses, and one external, degrading the surface by weathering and erosion. The equilibrium between crustal and surficial processes was recognized as early as 1837 by J. Herschel in a letter to Charles Lyell, and the interplay between internal and external forces was expanded on by Walther Penck in this century. For most of the twentieth century, however, geomorphology and tectonics have followed different tracks, and only recently have the two disciplines begun to converge (see Merritts and Ellis, 1994, for a historical review).

The study of landforms of the earth's surface is called *geomorphology,* which includes the origin and development of landforms, their relations to underlying geologic structure, and the history of geologic changes as recorded in landforms. Bloom (1991) describes geomorphology as "the systematic description, analysis, and understanding of landscapes and the processes that change them."

Tectonic geomorphology is the study of landforms that result from tectonic processes. It can also be defined as the application of geomorphic principles to tectonic problems. The earth's surface reflects the internal dynamics of the crust and mantle as expressed by faulting, folding, uplift, subsidence, and volcanism. But the surface is degraded even as it is constructed by tectonism, and one must understand the amount and rate of degradation of the surface before it is possible to understand completely the tectonic processes that operate beneath the surface in the upper crust. As degradation proceeds, the constructional landform is obscured, and the degree to which it is obscured can be used to determine its relative age with respect to other landforms that have been degraded to a greater or lesser degree, taking into account the relative resistance to erosion of the materials comprising the landform.

In the late nineteenth century, W. M. Davis, influenced by Charles Darwin and his newly developed theory of the evolution of species, developed the view that the cycle of erosion is itself evolutionary. The cycle starts with an initial stage of uplift. This is followed by incision of the uplifted landscape by streams that wear down the landscape to an equilibrium condition of low topographic relief and broad stream valleys, producing a *peneplain*. The landscape is worn down so that it approaches *base level,* that point where the potential energy of falling water reaches zero. Ultimately, this base level is sea level. G. K. Gilbert observed that the longitudinal profile of a stream is concave and approaches a graded condition, with the steepest part of the profile near its headwaters and the gentlest part at the mouth of the stream where it enters the sea and reaches base level. At grade, a stream neither cuts downward nor builds up its bed, and it has barely enough gravitational energy to transport its bedload.

The concepts that erosion cycles are evolutionary, and longitudinal profiles of streams approach grade lead to the conclusion that the erosional stage of development can be used to date, in a relative sense, the time since the cycle was initiated by tectonism. Relative dating is used here because time is only one of the factors that influence erosion rates; others include climate (particularly precipitation) and rock resistance to erosion. The idea of uplift only in the initial stage is simplistic; uplift may repeat or even be continuous throughout the cycle. Furthermore, erosion can take place below sea level. Submarine canyons are carved by bottom-flowing sediments being transported down the continental slope to a different kind of base level: sediments are ultimately ponded on abyssal plains.

The Davis erosion cycle is based on Hutton's principle of uniformitarianism: landscape change takes place very gradually over many thousands of years. Yet just as tectonic landforms can form instantaneously during an earthquake, the processes of degradation can be catastrophic as well, especially in arid regions. For example, the Atacama Desert of northern Chile is one of the driest places on earth, with rainfall occurring only once in several decades. Yet the landscape has been shaped by running water! Virtually all the erosion over a period of a century can take place in a few days or even a few hours during periods of sudden, violent rainfall. Similarly, the erosion of submarine canyons is accomplished by turbulent flows of dense, sediment-laden currents (turbidity currents) that may sweep down the canyons only once every few hundred years.

But the Atacama Desert was not always as dry as it is now; the landscape may have been dissected extensively during times of higher rainfall. If the episode of high rainfall can be dated independently, consistent strike-slip offsets of minor stream gullies formed during this rainy episode may be used to determine the slip rate of the fault. Liu (1993) was able to determine the slip rate of the Karakorum fault in western Tibet based on stream offsets after he had established the age of the last period of high rainfall based on a study of sediment cores and old shorelines of two nearby Tibetan lakes. Winter et al. (1993) used the same reasoning to determine the slip rate of the Pallatanga strike-slip fault in the Andes of Ecuador.

Degradation is commonly not as simple as the Davis erosion cycle would suggest. A local perturbation such as a new fault scarp formed across a stream, a stream capture, or a major landslide may set off a sequence of events in which thresholds are crossed, and the drainage system undergoes major change. Understanding the complex responses of streams is necessary to determine whether perturbations within the drainage system are caused by tectonics, climate, or something else (cf. Bloom, 1991; Bull, 1991; for example, see Merritts et al., 1994).

Recent applications of stages of erosion as a technique to determine the relative age of tectonic landforms have been at a smaller scale than the erosion cycle of Davis, which applies to a broad region. The tectonic event being dated may be the uplift of a faulted mountain range, or it may be the formation of an individual fault scarp or anticline. Mayer (1986) pointed out that the survival time of a mountain front or major escarpment is on the order of 10^6 years, whereas that of an individual fault scarp is 10^4 to 10^5 years. But the concept is similar to that envisioned by Davis. A fault moves during an earthquake, producing a steep scarp that interrupts the longitudinal profiles of streams crossing it. Over time, the scarp is cut by gullies, the sharp crest of the scarp is rounded by gravitational processes, and sediment accumulates at the base of the scarp. Eventually, barring a new surface-faulting event, the scarp is completely degraded, and erosional equilibrium is restored.

It is useful, then, to begin a discussion of tectonic geomorphology with an analysis of a small-scale example, the degradation of fault scarps.

SCARP DEGRADATION

The degradation of late Quaternary fault scarps and wave-cut shorelines of late Pleistocene pluvial lakes in the Basin and Range Province of the western United States was described by Wallace (1977). As shown in Figure 7-1, surface rupture along a fault produces a fault scarp with a *free face* and a sharp *crest.* The free face is a close approximation of the fault plane itself (Fig. 7-1a), but interlocking root systems in the soil may cause the face to overhang, because the soil is more cohesive than the unconsolidated sediments beneath it. Because surficial materials are usually unconsolidated, debris immediately begins to spall off the free face and accumulate at the base of the scarp as a *debris slope* at the angle of repose of the material, 34° to 37° (Fig. 7-2). Below the debris slope, a wedge of alluvium, the *wash slope,* overlaps the debris slope and the lower original surface that was offset by faulting (Fig. 7-1b). In cross section, as viewed from a trench, the debris slope and wash slope comprise a *colluvial wedge* (Schwartz and Coppersmith, 1984) that provides stratigraphic evidence for a surface-faulting event. Over time, the crest becomes rounded, the free face is overtaken by the debris slope (Fig. 7-1c; Fig. 7-3), and both are overtaken by the wash slope (Fig. 7-1d, e), all accompanied by a decrease in the maximum slope angle.

Wallace's work was done on normal-fault scarps formed in relatively unconsolidated gravelly basin fill deposits of the Basin and Range Province, but the mode of degradation applies also to cliffs in similar materials formed by undercutting by wave erosion and to escarpments formed by streams undercutting their banks to a slope angle greater than the angle of repose of the material being eroded. Because the ages of many of the pluvial-lake shorelines in the Basin and Range Province are known, the degree of degradation of wave-cut cliffs in unconsolidated alluvium can be used as a calibration point for the age of degradation of fault scarps. Similarly, the age of a stream-cut terrace scarp, or *terrace riser,* may be constrained by determining the age of the channel bedload of the stream at the time the terrace tread immediately below the riser stopped impinging against and undercutting the terrace riser, a maximum age for when the terrace tread was abandoned.

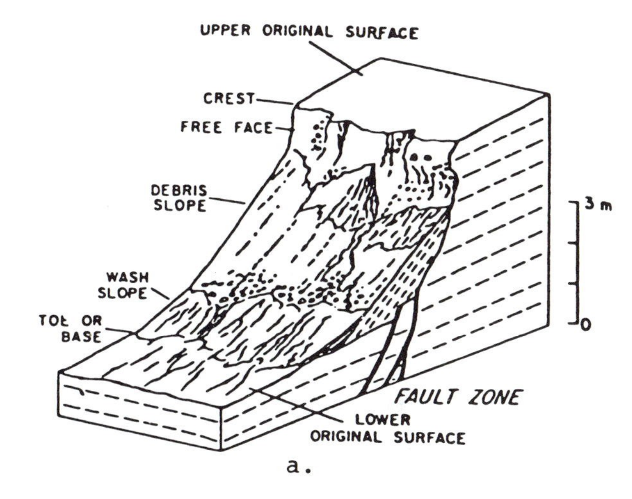

Figure 7–1. Top: Fault scarp in unconsolidated alluvium showing crest, free face, debris slope, and wash slope. Bottom: Stages in degradation of a single-event fault scarp (E + A to E + D). C represents the original scarp crest, and C′ represents the crest of the rejuvenated scarp. Modified from Wallace (1977, fig. 2) and McCalpin (1982, fig. 46).

Quantitative data are derived from topographic profiles measured perpendicular to the trace of the scarp. Scarp morphology parameters such as scarp height (H), maximum scarp-slope angle (θ), original surface slope (α), and vertical component of surface offset (D) were defined by Bucknam and Anderson (1979) and are illustrated in Figure 7-4. Machette (1982) expanded Bucknam and Anderson's work to include scarps formed by multiple faulting events. Machette recognized that a multiple-event fault scarp will have the maximum scarp-slope angle of the most recent event and a scarp height H_m which is the cumulative result of more than one event (Fig. 7-4b). Scarps formed by multiple fault displacements are recognized in the field by pronounced bevels on scarp profiles (Fig. 7-4b) and by scarp heights that are clearly too large for a single event. Figure 7-1, e + a to e + d shows the evolution of a rejuvenated fault scarp. Note that by the time the rejuvenated scarp reaches stage e + d, the evidence for bevelling by multiple fault episodes is obscured. Figure 7-2 shows the bevel on a fault scarp that was rejuvenated by the 1983 Borah Peak, Idaho, earthquake.

Bucknam and Anderson (1979) recognized that the relation between scarp height H and maximum scarp-slope angle θ can be approximated by a logarithmic curve. Figure 7-5 shows slope angles and heights of fault scarps on an alluvial-fan apron on the east flank of the Drum Mountains of western Utah. These scarps are below and younger than the Bonneville-stage shoreline of pluvial Lake Bonneville that is known to be 11,800 years old. Slope angles and scarp heights of the Bonneville shoreline in the same area are also shown. Figure 7-6 plots scarp-slope angle against the logarithm of scarp height for sets of scarps that are 10^3, 10^4, and 10^5 years old. The Drum Mountains scarps appear to have an age of 10^4 years. For a given scarp height of 3 meters, the scarp-slope angle is 10° for a scarp of 10^5 years age, 18° for a scarp of 10^4 years, and 25.5° for a scarp of 10^3 years. On the graph of θ versus log H, older scarps plot below younger scarps. Machette (1982) has cautioned against the over-quantification of this relationship due to differences in materials comprising the scarps, the possibility of unrecognized multiple faulting in the scarps, and differences in climate. Cooler and, presumably, wetter climate in the Pleistocene would result in a different rate of degradation for the scarps dated as older than 10^4 years than for scarps of the Holocene.

A more analytical approach is provided by the diffusion-equation model for scarp degradation (Nash, 1980; Hanks et al., 1984; Colman and Watson, 1983; Andrews and Bucknam, 1987). The diffusion equation is also used in analysis of conductive heat flow, vis-

Figure 7–2. 1983 Borah Peak, Idaho, fault scarp 6 days after the earthquake, showing prominent free face. Person with survey rod is standing at the top of a wash slope of a pre-1983 fault scarp. Photo by Robert Wallace, U.S. Geological Survey.

Figure 7–3. Prehistoric Holocene fault scarp on west side of Stillwater Range, Nevada. No free face remains, scarp is mainly debris slope and wash slope, the latter where people are standing. Note rounded crest. Photo by Robert Yeats.

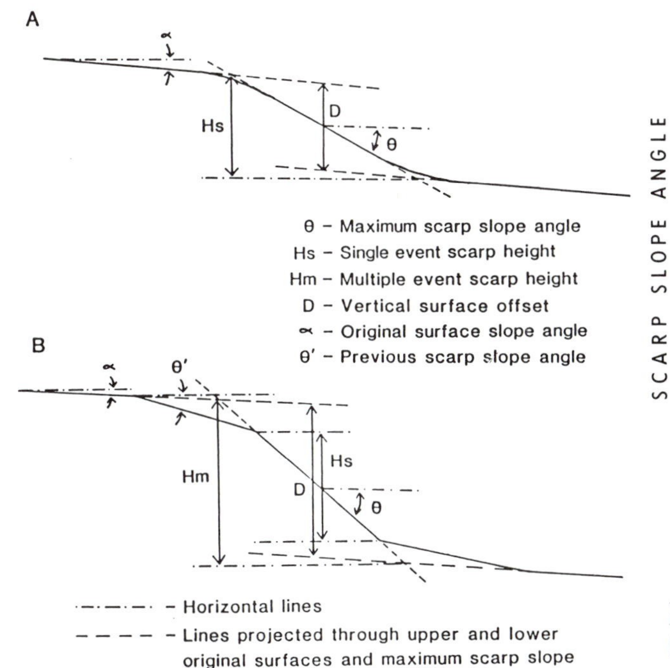

Figure 7–4. Diagrammatic profiles of (a) a single-event fault scarp, modified from Bucknam and Anderson (1979, fig. 1) and (b) a multiple-event fault scarp, modified from Machette (1982, fig. 3) showing scarp morphologic parameters.

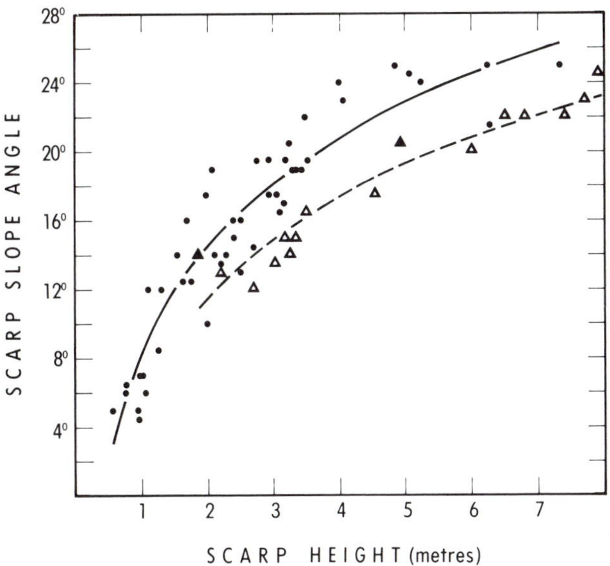

Figure 7–5. Scarp-slope angle versus scarp height for fault scarps (dots) and Bonneville-stage shoreline scarps of pluvial Lake Bonneville (open triangles) near the Drum Mountains, Utah. Solid triangles are Bonneville data overlying fault-scarp data. Solid and dashed lines are regression lines of fault-scarp and Bonneville shoreline scarps, respectively. From Bucknam and Anderson (1979).

cous flow through porous media, and chemical dispersion, as well as other forms of landscape modification.

The diffusion-equation model requires two assumptions. First, mass transport due to wash-controlled processes (soil creep, raindrop impact, and slope wash as opposed to spalling or slumping) proceeds downhill at a rate proportional to the local topographic gradient (Hanks et al., 1984). The second assumption is that there is no net removal of material from or addition of material to the profile at the local scale, and there is no significant change of density of the surficial material.

The change in altitude of a point on a topographic profile is equal to the difference between the amount of material transported to the point and the amount of material transported away from it, assuming that the degradational system is essentially closed (Colman and Watson, 1983). The rate of change in elevation at a point on a slope profile is proportional to the curvature of the profile at that point. Where the profile is concave, as at the base of a scarp, there is positive curvature, and that part of the profile will increase in altitude with time. Where the profile is convex, as at the crest of a scarp, the curvature is negative, and that

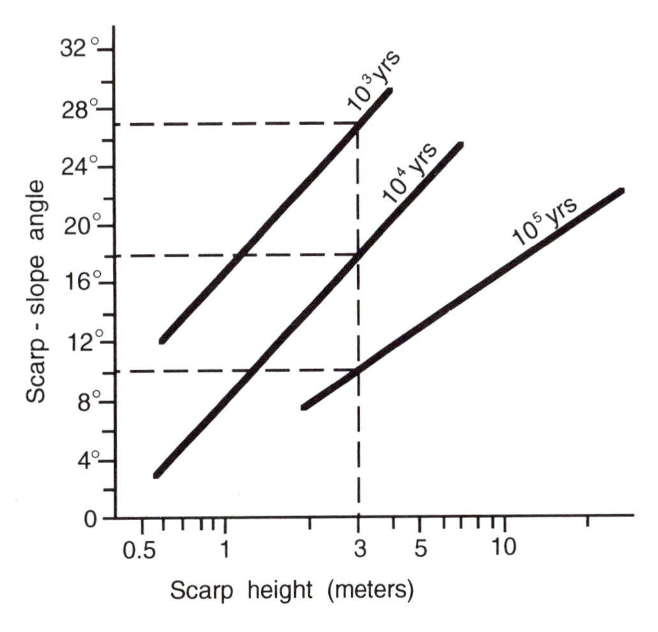

Figure 7–6. Regression lines to fault-scarp measurements in which scarp-slope angle is plotted against log scarp height for three sets of scarps of different ages. Drum Mountains fault scarps are assigned an age of 10^4 years. From Bucknam and Anderson (1979), as redrawn by Keller (1986).

part of the profile will decrease in altitude with time (Fig. 7-7). With time, the crest and the base of the scarp increase their radius of curvature, and the maximum scarp-slope angle θ decreases.

For the diffusion equation to be applicable, the profile must be perpendicular to the scarp, the scarp must be composed of low-cohesion material, the scarp must have been formed by a single event and then degraded under constant conditions, and there must have been no change in base level during degradation.

However, because the diffusion equation only applies to scarp degradation by wash-controlled processes, it cannot apply during the time that the free face is being eroded (Colman and Watson, 1983). Wallace (1977) observed that the amount of time necessary to remove the free face from fault scarps in Nevada is tens to hundreds of years, a short time compared to the elapsed time of thousands of years since scarp formation. Accordingly, Colman and Watson (1983) applied the diffusion equation only to the time after the removal of the free face.

Scarp degradation is a useful indicator of relative scarp age in the age range from a few hundred years (removal of the free face) to about 20,000 years (Machette, 1989).

The Basin and Range model works well for scarps in loosely consolidated materials in arid or semiarid regions, including western China, the Altiplano of the Andes, and the Sinai Peninsula of Egypt. However, it does not work well in the Aegean region of Greece and western Turkey where normal faulting is as prominent as it is in the Basin and Range, but the bedrock is largely limestone, and the climate is wetter (Stewart, 1993). Many active-fault scarps are in bedrock, and the fault zone is commonly marked by breccia sheets that are cemented by calcite and may thus be more resistant to erosion than unfaulted rock. The erosionally-resistant compact breccia sheet may be underlain by relatively uncemented breccia that erodes more readily than does the fault-zone breccia. Much of the degradation of the fault zone takes place by solution of the carbonate cement (Stewart, 1993).

MOUNTAIN-FRONT SINUOSITY

Uplift of a mountain range along a range-front fault produces a mountain front that is relatively straight, because it has not had time to be dissected and embayed by streams. As the range front is eroded, major drainages embay the mountain front and cause it to retreat. Bull and McFadden (1977) were able to compare the recency of tectonic activity north and south of the Garlock fault, California, by comparing the straight-line (or broadly curved) length of the mountain front, L_m, and the length of the embayed boundary between the mountain and the pediment or alluvial fan at its base, L_{mf}. The comparison of these two lengths was defined as mountain-front sinuosity S_{mf}:

$$S_{mf} = \frac{L_{mf}}{L_m}$$

A mountain front that is undergoing rapid, active uplift along a range-front fault would have a sinuosity (S_{mf}) close to 1, because drainages would be downcutting everywhere upstream from the fault. With no further uplift, streams would approach equilibrium and begin to aggrade close to the range front. With time, the aggrading part of each stream would migrate headward and would embay the range front, in part by lateral erosion by streams after they achieve their equilibrium profile. This would cause S_{mf} to increase in value. Mountain-front sinuosity is also affected by

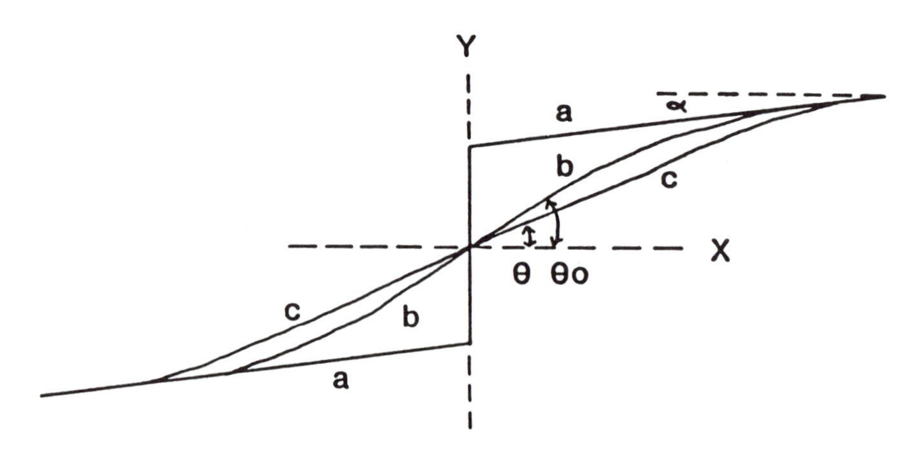

Figure 7–7. Modification of scarps according to diffusion-equation model. From Colman and Watson (1983, fig. 1).

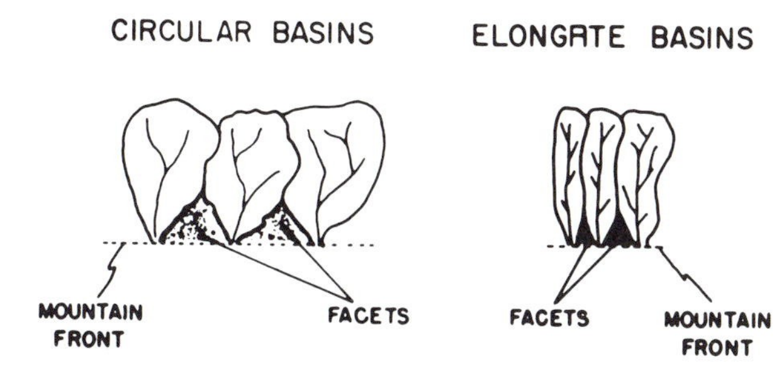

Figure 7–8. Relation between basin shape and morphology of the mountain front. The presence of facets indicates that the overall mountain front has not eroded back appreciably from the range-front fault. After Mayer (1986).

spacing of drainages (Fig. 7-8); the closer the spacing, the higher the sinuosity can be. It is also affected by the width of the range. In general, the spur ridges or drainage divides will not erode back as rapidly as the stream erodes its valley (Mayer, 1986). However, with decreasing stream length, the rate of stream valley erosion is not much less than the rate of spur ridge erosion. Thus, narrow ranges are unlikely to develop sinuosity as rapidly as wide ranges (Mayer, 1986).

Nonetheless, mountain-front sinuosity is useful in a reconnaissance of a large region because the measurements may be made on aerial photographs or satellite imagery. In Figure 7-9, a Landsat image of part of Central Otago in the South Island, New

Figure 7–9. Landsat image 2805-21163-7 of part of Central Otago, a semiarid region of the South Island, New Zealand. The range front at lower left has a low mountain-front sinuosity due to recent uplift along the Dunstan fault. The range front in partial shadow (upper right) is considerably embayed by alluvial fans and has a high mountain-front sinuosity. The narrow dark areas in the bottom center of the image are anticlinal ridges, with closely spaced drainages; these ridges also have a low mountain-front sinuosity. From Yeats (1987).

Zealand, the Dunstan Range (dark, dissected area at lower left) is much wider than the narrow, dark-textured ranges to the east of it, and it could, over time, develop a higher sinuosity. On the other hand, the drainage spacing in the narrow ridges is closer than that in the Dunstan Range, thereby counteracting in part the effect of range width.

The range front between the Dunstan Range and the Manuherikia Basin (light area at lower left, Fig. 7-9) is controlled by the Dunstan range-front fault with field evidence for Holocene dip slip. For this range front, $S_{mf} = 1$. The front of the Hawkdun Range (upper right, Fig. 7-9, partly in shade) is controlled by the Hawkdun fault, on which no late Quaternary displacement has been documented on the ground. This mountain front has been extensively embayed by alluvial fans such that $S_{mf} > 2$. The dark ridges at the bottom center of Figure 7-9 have a low S_{mf} although only one of the ridge flanks is fault-bounded. These anticlinal ridges are the deformed surface of a peneplain developed in bedrock (which shows as dark areas on the image) during Cretaceous and early Tertiary time and uplifted in late Tertiary and Quaternary time.

RATIO OF VALLEY-FLOOR WIDTH TO VALLEY HEIGHT

Bull and McFadden (1977) argued that the width of the valley floor near the mountain front is a measure of the effectiveness of stream downcutting. Streams downcut as a result of a relative lowering of base level, which can itself be produced by uplift. As the stream gradient adjusts to the new, lower base level, downcutting becomes less important, side-slope retreat becomes more important (Mayer, 1986), and the valley becomes wider (Fig. 7-10).

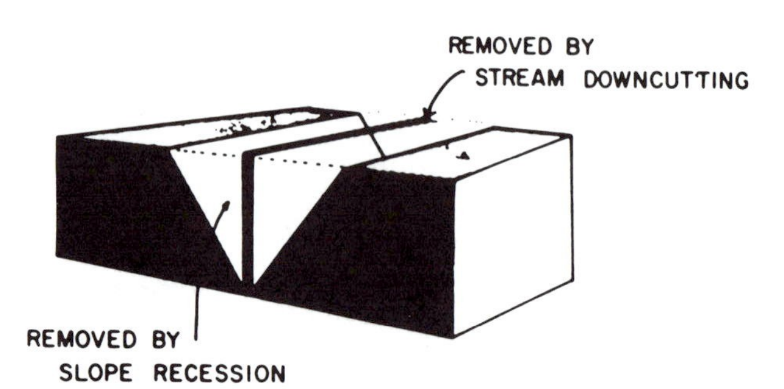

Figure 7-10. Diagrammatic representation of stream downcutting and slope retreat (or recession), thereby determining the shape of the valley. After Mayer (1986).

The ratio of the width of valley floor V_{fw} to valley height V_f is:

$$V_f = \frac{2V_{fw}}{(E_{1d} - E_{sc}) + (E_{rd} - E_{sc})}$$

where E_{ld} and E_{rd} are the altitudes of the left divide and right divide, respectively, and E_{sc} is the altitude of the valley floor along the same transect perpendicular to the drainage. The measurements are taken a short distance upstream from the mountain front. Canyons that are V-shaped in cross section have low values of V_f and reflect relatively recent uplift. Valleys with broad floors reflecting side-slope retreat and lateral shifting of stream channels within their floodplains have high values of V_f: they reflect relative stability of the mountain front (Keller, 1986).

ALLUVIAL FANS

Alluvial fans are deposited due to an abrupt decrease in sediment-carrying capacity because of loss of surface flow of water or widening of the channel. In plan view, an alluvial fan consists of a low-sloping cone with its apex (*fan head*) at the mouth of a canyon embaying a mountain front. Topographic contours describe arcs of circles concentric on the fan head; the fan itself is deposited as the stream shifts from side to side across the fan cone. Fan profiles commonly can be subdivided into segments, which overall are gently concave upward. Segment boundaries are marked by breaks in slope and commonly by contrasts in degree of soil development and development of desert varnish.

Bull (1977) worked out a relation between fan form and rates of tectonic processes, either uplift of the mountain source of fan sediments or tilting of the fan itself. Where the mountain front is uplifted at a high rate relative to the rate of downcutting by streams in the mountain range and the rate of fan deposition, then deposition occurs at the fan head, close to the range front. The youngest fan segment is close to the apex of the fan at the range front. If the rate of uplift is less than the rate of downcutting in the mountain range, then the fan head is itself downcut and incised, and deposition is shifted down the fan, away from the mountain front (Fig. 7-11). The relative age of fans can be determined by comparing the degree of soil development (see the following text).

STREAM-GRADIENT INDEX

The stream gradient (SL) index of Hack (1973) is related in a general way to available stream power, which is the work expended by (or energy loss of)

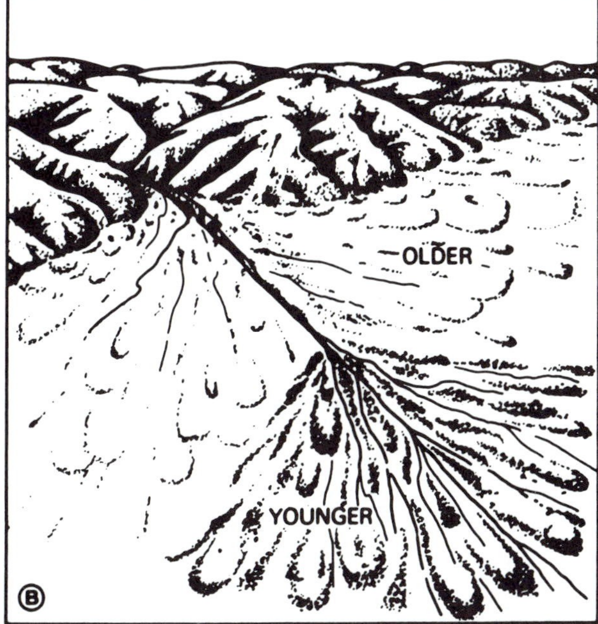

Figure 7–11. Morphology of alluvial fans. (a) Deposition adjacent to mountain front. (b) Deposition down-fan as a consequence of fanhead entrenchment. From Bull (1977).

the stream. Stream power is proportional to the product of stream discharge and water-surface slope; it reflects the ability of the stream to transport its load as well as the channel shape that resists flow.

$$SL = \frac{\Delta H(L)}{\Delta L}$$

where $\Delta H/\Delta L$ is the channel gradient of the reach of the stream being studied, and L is the horizontal projection of the length of the channel from the drainage divide of the longest stream in the drainage basin to the center of the reach being studied (Fig. 7-12).

The stream-gradient index is sensitive to rock resistance to erosion and to uplift rate, increasing with increased uplift rate. Anomalously high SL indices in rocks of relatively low resistance to erosion may be an indicator of recent vertical uplift. Figure 7-13 shows stream-gradient indices for the San Gabriel Mountains in the Transverse Ranges of southern California. Anomalously high indices are found along the southern and eastern margins of the San Gabriel Mountains, both areas of active faulting. High indices along the Sierra Madre reverse-fault zone correspond to the area of the 1971 San Fernando earthquake and to an area of Holocene displacement on the Cucamonga fault, which is the eastern extension of the Sierra Madre fault northeast of Pomona. On the other hand, the relatively inactive San Gabriel fault is characterized by relatively low stream-gradient indices. But the San Andreas fault, which moved in a great earthquake in 1857, also has low indices, except for the area near its intersection with the active San Jacinto fault. Strike-slip faults commonly are marked by narrow valleys with easily-eroded crushed rock in the valley producing low stream-gradient indices (Keller, 1986). Strike-slip faults do not consistently uplift one side of the fault with respect to the other, resulting in low stream-gradient indices. Low indices also result when faults cross easily-eroded sedimentary-rock terrain.

ALLUVIAL RIVERS AND TECTONIC DEFORMATION

Consider an uneroded landscape such as that generated by a catastrophic volcanic eruption or by rapid

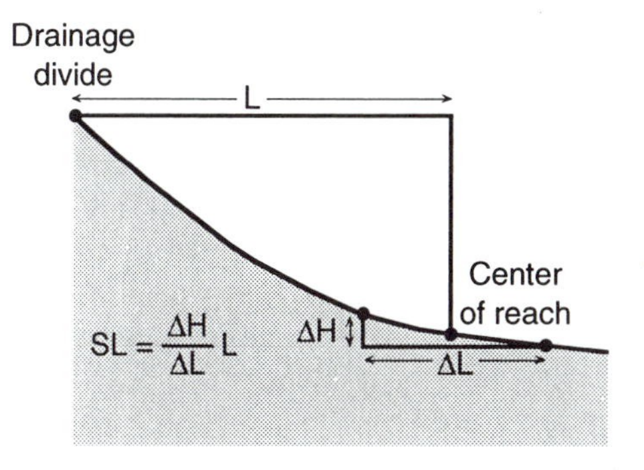

Figure 7–12. Longitudinal profile of a stream showing how the stream-gradient index (SL) is determined. After Hack (1973).

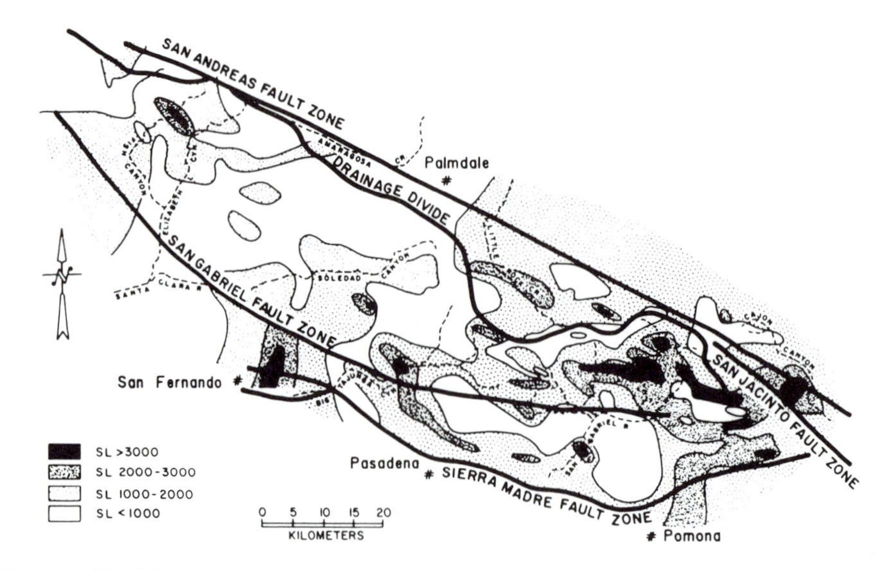

Figure 7–13. Stream-gradient indices (SL) for the San Gabriel Mountains, California. After Keller (1986).

removal of glacial ice from a region. Rainfall leads to the development of streams consequent on this landscape. Flow from these streams may be locally ponded in depressions, producing lakes, and they may cascade over cliffs as waterfalls. The longitudinal profiles of these streams would be highly irregular. But, over time, the streams downcut here and aggrade there and ultimately develop longitudinal profiles of equilibrium, a condition known as *grade*.

The *graded stream* has a slope and channel that provides, "with available discharge, just the velocity required for the transportation of the load supplied from the drainage basin. The graded stream is a system in equilibrium; its diagnostic characteristic is that any change in any of the controlling factors will cause a displacement of the equilibrium in a direction that will tend to absorb the effect of the change" (Leopold and Maddock, 1953, modified from Mackin, 1948). If there is a change in water discharge, in sediment load, or in base level, the river responds to this change by altering other characteristics: channel width, channel depth, roughness of bed, grain size of the sediment load, velocity of flow, and tendency to meander or to form braided channels. The graded condition may be perturbed by a change in climate, which alters the amount of precipitation and the nature of vegetative cover in the drainage basin. It may also be perturbed by deformation, and these tectonic perturbations on the river may be used to identify and describe the deforming structure, even in regions of low relief (Schumm, 1986).

The Ventura River of southern California maintains the lower part of its course across two active structures, the Red Mountain fault and the Ventura Avenue anticline (Fig. 7-14). Because the river course predates uplift accompanying formation of the anticline and fault in its path, it is called an *antecedent stream*. The canyon through which an antecedent stream flows, or which was formerly occupied by a stream, is called a *cluse*, named after a locality in the French Jura Mountains.

Bedrock is exposed in the river bed at the hinge of the Ventura Avenue anticline, and the depth of fill at the river mouth is about 20 m (Rockwell et al., 1988). Remnants of stream terraces of different ages are preserved on both sides of the river, and these are warped over the crest of the anticline, as first pointed out by Putnam (1942). Rockwell et al. (1988) assumed that the longitudinal profile of the river at the time of formation of terraces A through H was the same as it is today. Using this assumption and the altitude and tilt of dated terraces, they were able to determine the rate of growth of the Ventura Avenue anticline.

In another study, Nakata (1989) mapped uplifted and back-tilted terraces of rivers crossing the range front of the Himalaya and showed that these terraces were deformed by movement on the Himalayan Front fault (Fig. 7-15).

Channel patterns of alluvial rivers are so sensitive to change that they may show evidence of tectonic deformation even in areas where deformation rates are relatively low. Channels may be straight, meandering, or braided; meanders may have high sinuosity and equal channel width, or they may be less sinuous and wider at bends than in straight reaches (Schumm, 1986). Figure 7-16 shows a classification

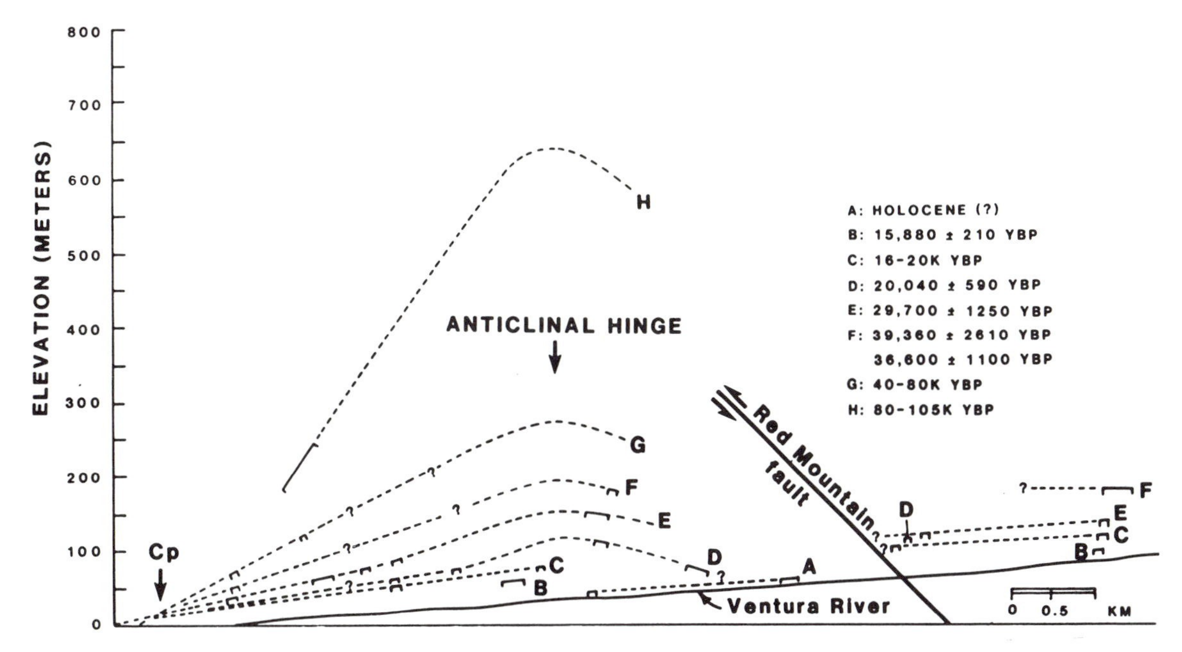

Figure 7–14. Longitudinal profiles of modern Ventura River (solid line) and of river terraces of various ages (dashed lines; brackets show control). Age control is from radiocarbon dates and correlation to a marine terrace that has an age estimate based on amino-acid racemization. Positions of active Red Mountain fault and Ventura Avenue anticline are shown. Vertical exaggeration about 10:1. After Rockwell et al. (1988).

of channels based on channel patterns and type of sediment load. Flume experiments show that a change of valley-floor slope, which may be produced by deformation, will produce a change of channel pattern, and this pattern change may be used to determine the nature of deformation. As an example, we consider tectonic deformation near New Madrid, Missouri, site of three great earthquakes in 1811–1812 (Russ, 1982).

New Madrid is located on the Mississippi River within an upwarped area called the Lake County uplift (Fig. 7-17). The surface of the uplift is as much as 10 m above the remainder of the Mississippi River floodplain, and it contains additional structures including Tiptonville Dome and the Reelfoot fault scarp. The Lake County uplift affects the modern Mississippi meander belt of high sinuosity and three older braided-stream terraces. Profiles show that abandoned river channels and natural levees have been warped, in some cases to the point that they now locally slope upstream. The modern floodplain is also warped, and

Figure 7–15. Back-tilted terrace (Rangamati surface) of the Neora River in the Darjeeling, India, foothills of the Himalaya. Backtilting is related to movement on the Matiali reverse fault to left of terrace. After Nakata (1989).

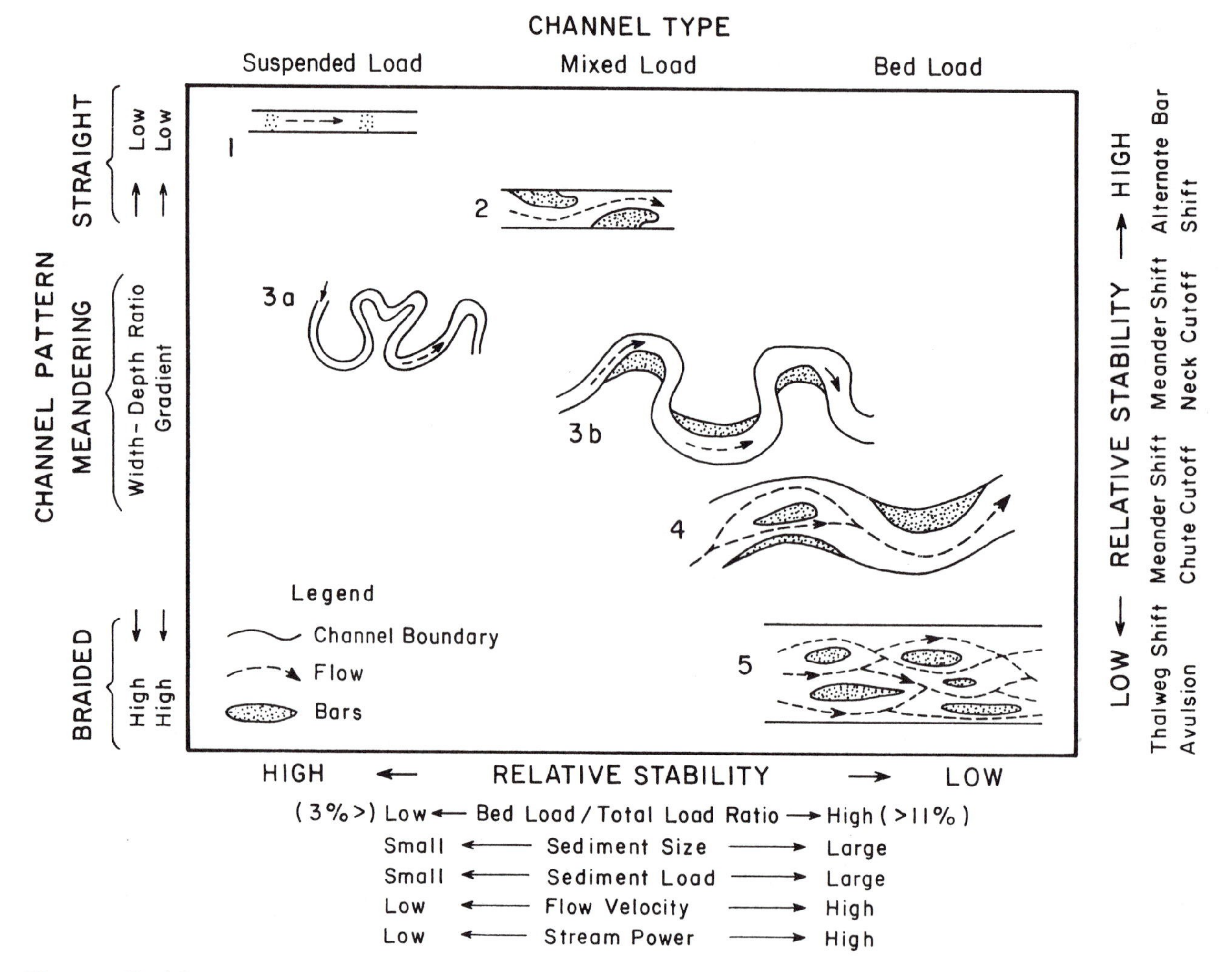

Figure 7–16. Classification of alluvial river channels based on pattern and sediment load and on relative stability. After Schumm (1986).

near New Madrid, its profile is convex, an indicator of uplift.

Upstream from the uplift, between Cairo, Illinois and Hickman, Kentucky, the Mississippi River has a relatively straight course. However, downstream from Hickman, the river has a highly sinuous meander pattern. This may be related to the Lake County uplift. The uplift causes a reduction in channel gradient upstream, and the river may have developed a straight course to maximize its gradient and restore its profile of equilibrium. In contrast, the uplift has increased the channel gradient downstream from the uplift axis. The high meander sinuosity in this region has the effect of reducing the channel gradient to that of its equilibrium profile. Another effect of the uplift may be a shift of the channel westward away from the eastern edge of the meander belt between Hickman, Kentucky and a point south of Blytheville, Arkansas.

MARINE SHORELINES

Near-shore erosion of a coastal hillslope, particularly by breaking waves, produces a *wave-cut marine abrasion platform* and a *sea cliff*. As wave action erodes back the sea cliff, sand and gravel are carried repeatedly landward and seaward by the surf, thereby abrading the platform. The abrasion platform slopes seaward about 1°. Weathering of the platform and cliff face is facilitated by continual wetting and drying accompanying tidal rise and fall of sea level, in addition to activity of shore-zone plants and animals that attach to, scrape, bore, and dissolve rock. If sea level falls relative to the marine abrasion platform, the cliff is removed from the effects of wave erosion, and it undergoes degradation. A thin layer of sand and gravel, in some cases containing marine shells, is the remnant of the surf-borne sediments that abraded the

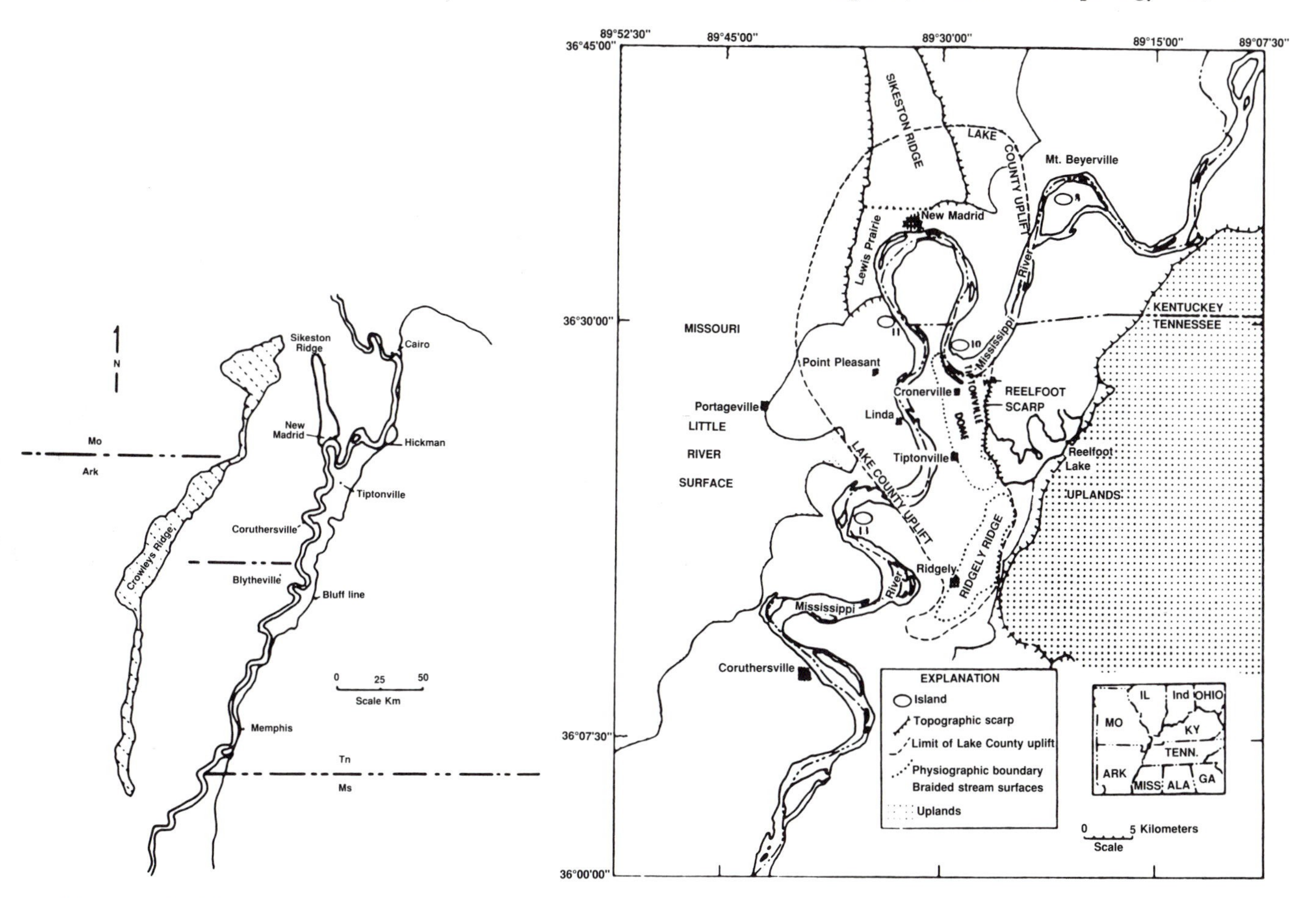

Figure 7–17. Upper left: Central Mississippi Valley. Upper right: New Madrid region and Lake County uplift.

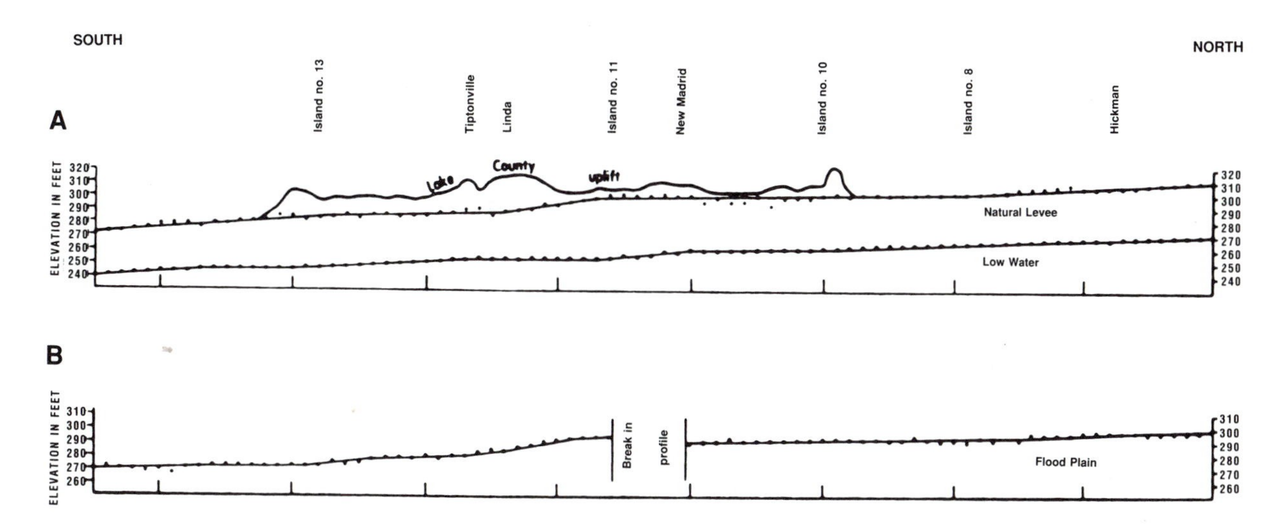

Figure 7–17. Bottom: Longitudinal profiles between miles 845 to 930, Mississippi River. (a) Natural-levee profile and low-water profile. (b) Floodplain profile. From Russ (1982) and Schumm (1986).

marine platform and may be preserved atop the platform. After emergence, a debris slope forms, burying the marine platform, its thin cover of sand and gravel, and the base of the sea cliff. Wind-borne sediments may also contribute to burial of the platform. When the free face of a cliff cut in unconsolidated sediments is removed, the cliff degrades as previously described for fault scarps.

Sea cliffs and marine platforms also form during the erosion of carbonate rocks (limestone and dolomite), but here the processes are more complex. Platforms in the intertidal zone are in part constructed by carbonate-secreting organisms, principally algae and corals. Other organisms dissolve the carbonate rock on both platform and cliff. Uplifted shorelines in tropical areas commonly consist of high-standing coral reefs that grade seaward from abrasion platforms to constructional reef flats.

The altitude with respect to sea level of raised or submerged former shorelines is an important measurement of tectonic deformation of coastlines. The boundary between the fresh-cut sea cliff and the marine abrasion platform, called the *shoreline angle* by some geomorphologists, is, to a first approximation, an indicator of sea level at the time the abrasion platform was cut. However, this angle is generally below sea level, controlled by the depth of effective wave action. Sea level is marked by a bench in the intertidal zone and, in carbonate rock, by a notch which may undercut the cliff (Fig. 7-18). But, before using the change of position of these shoreline features as indicators of deformation, the effect of *eustatic sea-level change* must be considered. The Quaternary Period is characterized by sea-level fluctuations of up to 150 m because of the advance and retreat of glacial ice. During glacial advances, precipitation remained on land as snow and ice and did not return to the sea; over time, this resulted in a lowering of sea level. During interglacial stages, glaciers melted, more water returned to the sea, and sea level was high. Another effect on relative sea-level changes is the isostatic response of the crust to the presence of continental ice sheets, large masses of water, or thick loads of coastal-plain sediments. The isostatic response to ice sheet removal is a particularly important factor in the Quaternary.

The problems of working with Pleistocene shorelines and Holocene shorelines are different enough to warrant a separate treatment for each.

Sea-level changes for the past 150 ka (1 ka, or kiloannum = 1000 years) have been worked out on the basis of uplifted coral reefs along tropical coasts that have been dated radiometrically. These sea-level changes have been corroborated in a general way and refined using changes in $^{18}O/^{16}O$ ratios in shells of microscopic organisms (foraminifera) in deep-sea cores (see discussion in Chapter 6).

A sea cliff and marine platform are cut at a time when the tectonic uplift rate and the rate of eustatic sea-level rise are about the same, so that sea level appears to be stationary. This point is reached just before a eustatic sea-level maximum and just after a sea-level minimum (Fig. 7-19). Because nearly all the

Figure 7–18. Notches marking elevated Holocene shorelines along a limestone coast, Island of Rhodes, Greece. Photo by P. A. Pirazzoli.

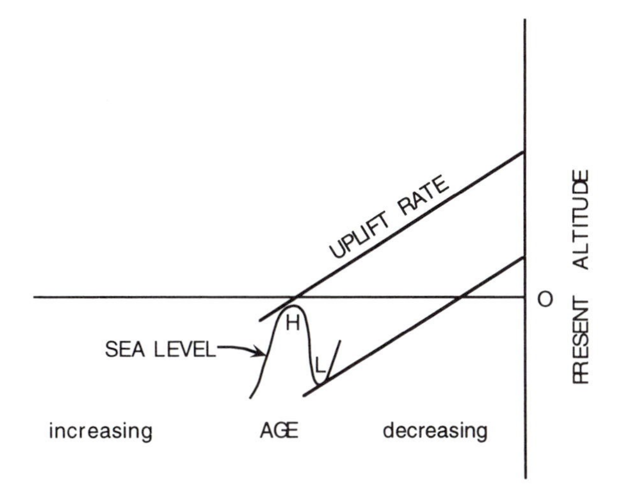

Figure 7–19. Relation between sea-level highstand (H) and lowstand (L) to tectonic uplift rate. A shoreline is preserved when sea-level rise is equal to uplift rate, just before the highstand and after the lowstand. In most cases, only the highstand shoreline is preserved above present sea level.

observations are above present sea level, most observed shorelines occur near sea-level maxima. One of the best-documented records of late Pleistocene sea-level change is found on the Huon Peninsula, Papua New Guinea (Fig. 7–20), where a large fault block has undergone steady uplift for at least the past 140 ka, thereby preserving uplifted fringing reefs and barrier reefs (Chappell, 1974) . The upper surface of each reef is relatively level and consists predominantly of fossil algae and corals; the coralline deposits have been dated radiometrically using the uranium-series method. The last interglacial coral reef, terrace VIIb, has an age of approximately 124 ka and is assumed to have formed when sea level was 6 m higher than it is today. Terraces VI through III have been dated by the uranium-series method, and Terrace II has been dated by radiocarbon (Chappell, 1974; Stein et al., 1993; Ota, 1994). If a constant uplift rate is assumed from the last interglacial highstand to today, the altitudes of the younger, lower terraces with respect to terrace VIIb can be used to determine and date eustatic sea-level highstands between the last interglacial and today.

The sea-level curve based on reefs on the Huon Peninsula and around the island of Barbados in the Caribbean is shown in Figure 7–21. This curve is similar in a general way to the position of sea-level highstands based on oxygen-isotope ratios (Figs. 6–2a, 7–22) which is itself based on an assumption of uniform sedimentation rates in the deep-sea cores. By combining the sea-level curve based on coral reefs with the curve based on oxygen isotopes, the ages of Pleistocene terraces older and younger than the last interglacial highstand may be predicted. In Figure 7–22, the uplift rate (slope R) is too low to preserve terraces younger than oxygen-isotope stage 5 because they would project to a point below present sea level. If uplift rates are higher than 0.3–1 mm/yr (Lajoie, 1986), these younger terraces may be preserved above present-day sea level. Interpolated ages of these terraces are approximately 105, 85, 60, 50, and 40 ka (cf. Lajoie, 1986).

As noted above, sea level would appear to be stationary just after a sea-level minimum, but strandlines formed at sea-level minima would be mainly offshore, and as sea level rose again, the abrasion platform might be covered with younger sediments or younger coral-reef deposits (Chappell, 1974).

Older terraces are not as easily dated by the uranium-series method, and their ages are estimated by interpolation between the 124-ka sea-level highstand and the 780-ka age of the last reversal of magnetic polarity (see discussion in Chapter 6).

The problems with uplifted or subsided Holocene shorelines are different. There is general agreement that eustatic sea level rose rapidly in the early part of the Holocene as glaciers melted, then slowed or stopped rising in the middle of the Holocene. At present, sea level appears to be rising globally at 1–3 mm/yr. But local conditions commonly obscure the eustatic signal (Fig. 7–23). In Greenland and northeastern Canada, the Holocene is dominated by isostatic rebound from the removal of continental ice,

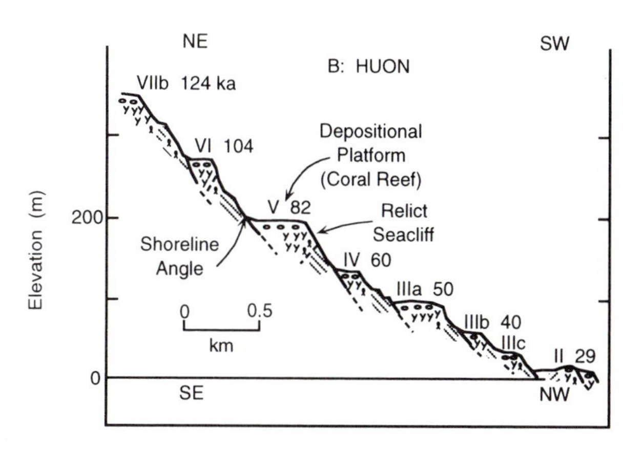

Figure 7–20. Profile of uplifted coral-reef terraces and sea cliffs on Huon Peninsula, Papua New Guinea. From Chappell (1974). Corals are dated by U-series method except for terrace II, which was dated by radiocarbon. Terrace VIIb is the last interglacial terrace, widely preserved around the world.

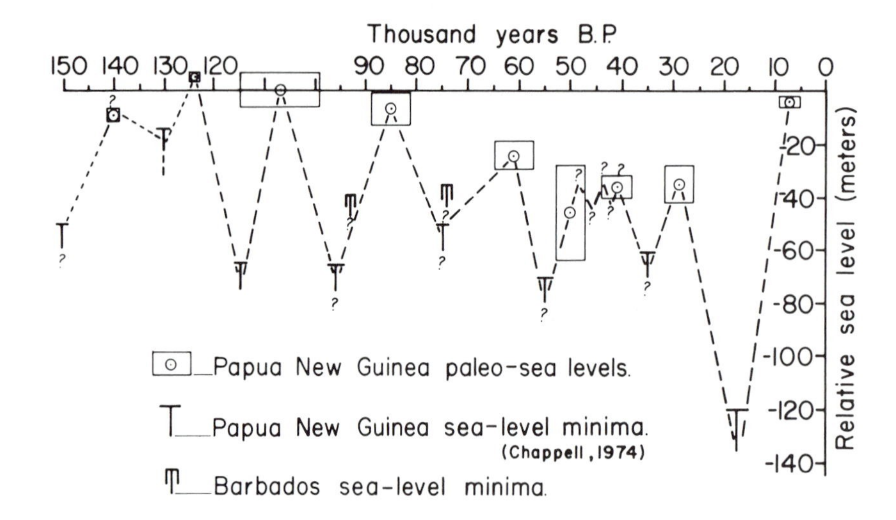

Figure 7–21. Sea-level curve based on the coral-reef terraces on Huon Peninsula and on Barbados. From Bloom and Yonekura (1985).

resulting in the fall of apparent sea level during this time. Farther south, along the Atlantic passive margins of Europe and the United States, the past few thousand years are characterized by an apparent rise in sea level due to coastal subsidence. In some regions, the eustatic rise in sea level causes a change in configuration of coastal bays, thereby changing the tidal heights, which changes the coastal response to sea level. Finally, the tectonic signals being measured are not long-term vertical changes, but jerky uplift or subsidence accompanying individual earthquakes, commonly at intervals of a thousand years or more.

The conditions favoring preservation of Holocene shorelines is the same as for Pleistocene shorelines. The net effect of eustatic sea-level change, isostatic uplift or subsidence, and other factors should produce an apparent sea level that is approximately stationary over a period of hundreds or thousands of years, so that a sea cliff and abrasion platform may be cut. An earthquake may result in sudden tectonic uplift,

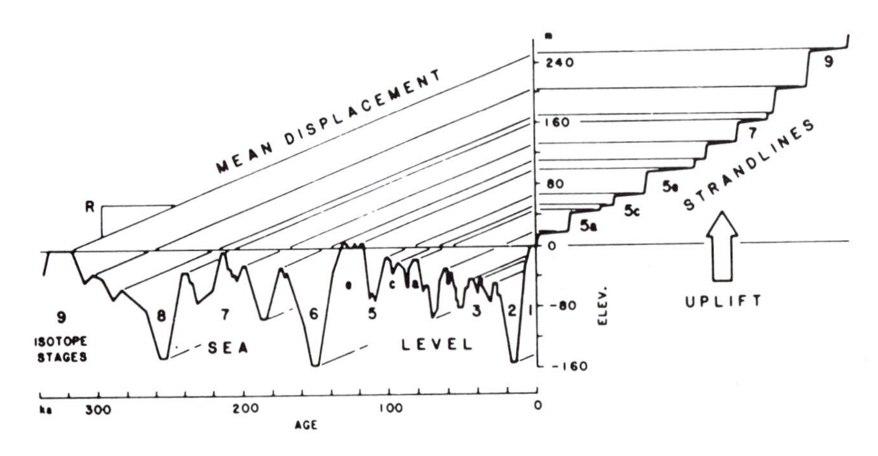

Figure 7–22. Estimating ages of raised Pleistocene marine terraces using sea-level fluctuations based on oxygen-isotope ratios. Numbers identify oxygen-isotope stages. Stages 5a, 5c, and 5e correspond to stages 5.1, 5.3, and 5.5 in Figure 6–2a. Slope R represents uplift rate of about 0.8 mm/yr, assumed to be uniform throughout the period (mean displacement lines are parallel). If the height of the 124-ka highstand is known, the ages of lower, younger terraces may be estimated by interpolation. From Lajoie (1986).

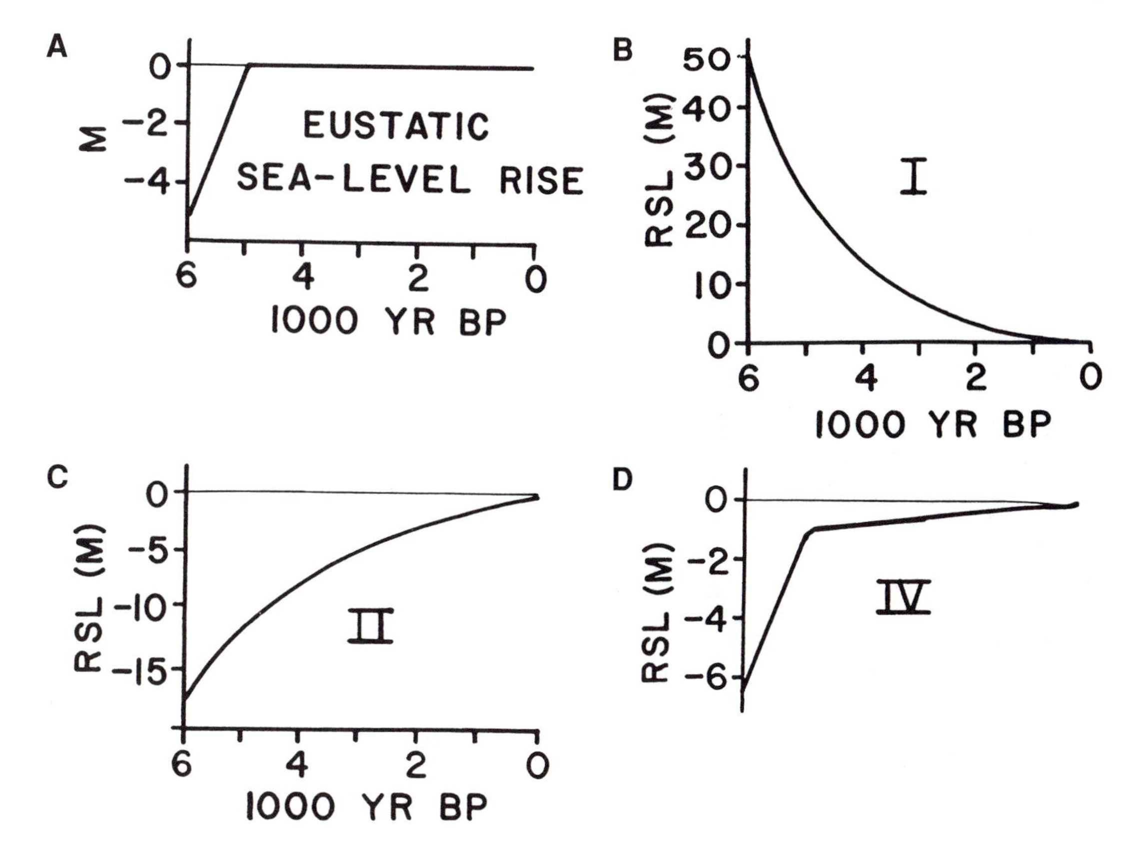

Figure 7–23. Sea-level curves resulting from retreat of Northern Hemisphere ice sheets resulting in eustatic sea-level rise depicted in (a). (b) Northeast Canada and Greenland, dominated by isostatic rebound from the removal of continental ice. (c) Atlantic coast of the United States, influenced by coastal subsidence. (d) Tropical coastline. From Clark and Lingle (1979).

thereby raising the abrasion platform above the effect of wave action (Figs. 7-24, 7-25, and 7-26), or it may produce sudden subsidence, drowning coastal forests and marshes (Fig. 7-27; cf. Chapter 11). Estuaries may contain fossil diatoms or foraminifera that are highly sensitive to water depth, and their position in an estuarine sequence may provide relatively precise information about the magnitude of vertical change accompanying an earthquake.

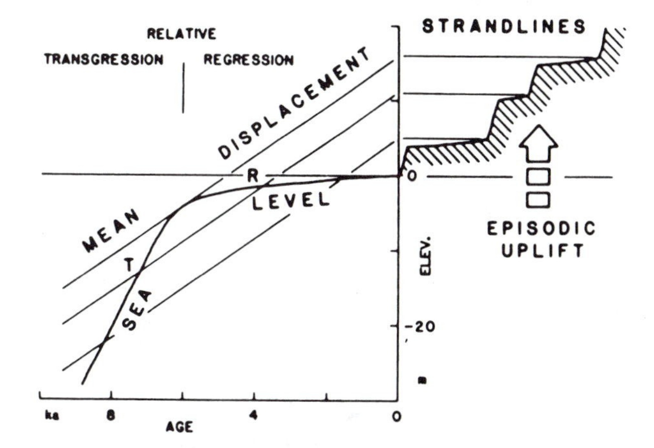

Figure 7–24. Origin of emergent Holocene marine abrasion platforms. Highest platform correlates with the point of tangency between rapidly rising sea-level and approximately stable sea-level between 7 and 5 ka. Lower platforms represent shorelines that were cut during the stable sea-level phase, then uplifted suddenly accompanying a large earthquake. From Lajoie (1986).

WEATHERING

The alteration of rocks and sediments near the earth's surface is called *weathering*. Weathering takes place by *mechanical processes*, in which rock is disaggregated along fractures, aided by differential thermal expansion, pressure release, and wedging by plant roots, burrowing organisms, and low-temperature crystal growth, and by *chemical processes*, through which mineral grains decompose by oxidation and by reaction with carbon dioxide, water, and organic com-

Figure 7–25. Uplifted shore platforms at Mahia Peninsula, North Island, New Zealand. Prominent high platform was cut during the last interglacial highstand. Three low platforms are Holocene, uplifted by great earthquakes. Photo by D. L. Homer, Institute of Geological and Nuclear Sciences, New Zealand.

pounds. A comparison of the degree of weathering of rocks underlying different surfaces may be used to determine their relative ages and, where calibrated by radiometric dating, to estimate their true ages. It is important to note that the degree of weathering dates a geomorphic surface and not the deposits that underlie that surface, an observation that also applies to surface-exposure dating by cosmogenic nuclides (Chapter 6). The degree of weathering is strongly influenced by climate and by the nature of the material being weathered, and these factors must be taken into consideration when making age estimates based on weathering. We consider two weathering processes that have been used in estimating relative ages: (1) formation of weathering rinds on stones, and (2) soil formation.

Weathering Rinds

The Marlborough region of the South Island of New Zealand is underlain by a relatively uniform suite of rocks including graywacke (hard sandstone containing fragments of quartz, feldspar, dark minerals, and rock in a clay matrix) and argillite (hard, somewhat metamorphosed shale). Streams draining this region have cut flights of river terraces including a widespread late-glacial aggradation surface of latest Pleistocene age. Cobbles and boulders on terrace surfaces

Figure 7–26. Numa IV marine terrace platform, on which houses of village of Mera are built, uplifted by the 1703 Genroku earthquake, Boso Peninsula, Japan. Photo by Yoko Ota.

Figure 7–27. Tree roots from spruce forest submerged abruptly 1700 years ago in Niawiakum estuary, Willapa Bay, Washington State, exposed at very low tide. Prominent layer midway up cut bank on left marks peat layer submerged about 300 years ago by a great earthquake. Photo by Robert Yeats.

undergo chemical weathering and develop a light-colored rind that increases in thickness logarithmically with time (Chinn, 1981). With time, quartz veins in the graywacke clasts increase in surface relief, indicating that disintegration and spalling of the surfaces of the clasts accompany inward migration of the rind front. Because of these factors tending to reduce rind thickness, the increase in rind thickness with time is not logarithmic for ages greater than about 20 ka (Whitehouse et al., 1986).

Measurements of rind thickness are made from the surfaces of large clasts that are judged to have been at the ground surface since they were deposited. Knuepfer (1988) measured at least 50 samples from each locality and plotted these on a histogram, as illustrated in Figure 7-28 for five terraces of the Saxton River; terrace 1 is highest and oldest. From the histogram, a weighted running average is constructed to account for differences in postdepositional histories of individual clasts and uncertainty in measuring

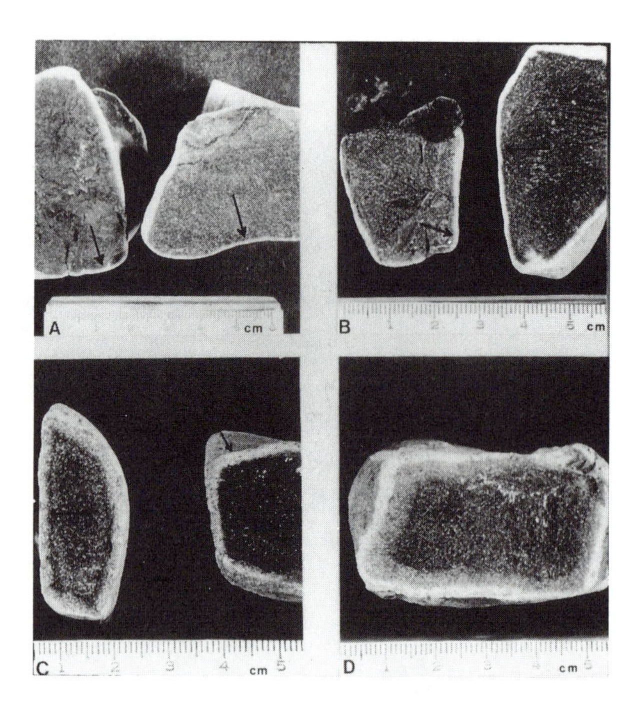

Figure 7-29. Examples of weathering rinds on basaltic rocks from glacial till near McCall, Idaho. (a) Youngest tills (Pinedale). (b) older till (Intermediate). (c,d) Still older tills (c is Bull Lake). Arrows mark places where true rind thickness is measured. From Colman and Pierce (1981, Fig. 2).

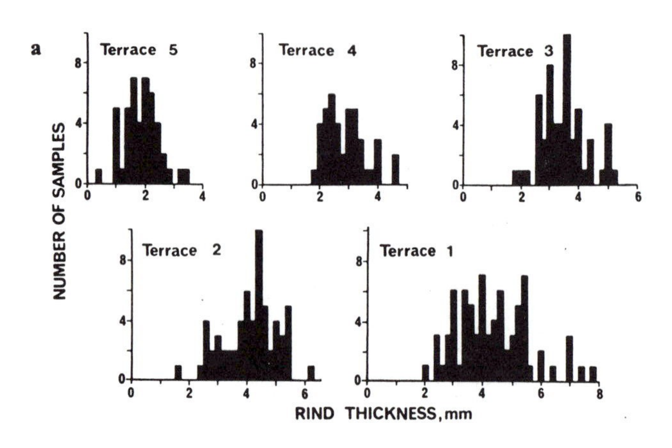

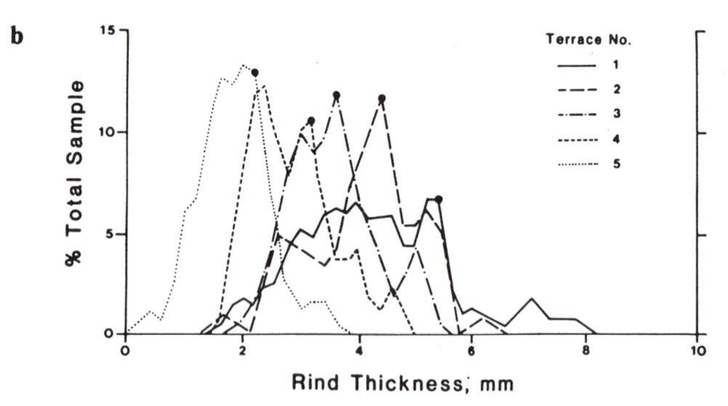

Figure 7-28. Thickness of weathering rinds of graywacke sandstone clasts from the surfaces of five terraces of the Saxton River, Marlborough, South Island, New Zealand. Terrace 1 is oldest; terrace 5 is youngest. (a) Histograms of rind measurements. (b) Measurements smoothed using weighted running mean. From Knuepfer (1988).

individual rind thickness to an accuracy of 0.2 mm. The average rind thickness increases with age, and there is a greater dispersion of rind thickness with age. These river terraces are offset by strike-slip faulting, and the estimated ages of offset terrace risers of different ages have been used to estimate slip rates of faulting.

Colman and Pierce (1981) studied more than 7000 weathering rinds on basaltic and andesitic stones in Quaternary glacial deposits in 17 different areas of the western United States. Sampling sites were limited to terraces or flat moraine crests in areas with only moderate differences in climate. They found that weathering-rind thickness increases with age, as illustrated in Figure 7-29, although variations in rock type and differences in climate are also important. The rate of increase of rind thickness decreases with time. Ages of glacial deposits, based on rind thickness, group into time intervals of 12-22 ka, 35-50 ka, 60-70 ka, 135-145 ka, and possibly older time periods. These time intervals correspond approximately to times of high ice volume worldwide based on the ^{18}O timescale.

Obsidian is nonhydrated volcanic glass with the composition of rhyolite, relatively rich in silica and potassium. Weathering causes a fresh surface of obsidian to absorb water from the atmosphere or soil, and this water diffuses into the obsidian at a rate that is dependent on temperature, somewhat influenced by the chemical composition of the glass, but not dependent on relative humidity of the atmosphere. The diffusion front is abrupt and is recognized in thin sections of obsidian by an abrupt change in index of refraction of the glass. Hydration also takes place along hairline fractures in glass.

Obsidian was used for trade among primitive people, and hydration rinds have been used to estimate the relative age of obsidian artifacts as old as 200–250 ka (Friedman and Smith, 1960). Hydration rinds of obsidian have also been used together with K-Ar dating in determining the age of Pleistocene glacial moraines near Yellowstone Park (Pierce et al., 1976). Obsidian pebbles were collected from the crests of moraines or from bedded gravel deposits in road cuts, quarries, or cliffs. To "date" the moraine deposits, it was necessary to determine that the surface of the obsidian was created by glacial abrasion and has been hydrating ever since, in contrast to surfaces formed by original cooling of the flow, by earthquake shaking, or by frost cracking. Rind thicknesses from a given locality show considerable variation and thus are plotted as a histogram (Fig. 7-30). Variations in rind thickness may be caused by differences in time of glacial abrasion, differences in soil temperature, difference in obsidian composition, or inaccuracy of measurement. Some thick hydration rinds were evidently produced from an earlier glaciation and were excluded from the histograms of Figure 7-30. Based on hydration rinds from surfaces developed by glacial abrasion (Pinedale and Bull Lake moraines) and from surfaces crystallized from flows dated by K-Ar, it appears that the thickness of obsidian hydration rinds increases as the square root of time.

On the southern range front of the San Gabriel Mountains (located on Fig. 7-13), alluvial fans of middle Pleistocene to late Holocene age have been progressively offset by dip slip on the Sierra Madre reverse fault. Clasts of a given lithology show much more pronounced weathering in higher, older fans than they do in younger fans. A quantitative measurement of clast weathering developed by Crook et al. (1987) measures the compressional-wave velocity of individual clasts as measured by a portable instrument. The instrument measures the travel time of a wave generated by a hammer blow on the clast surface and traveling a predetermined distance through the clast. Only boulders larger than 15 cm in diameter were measured. As shown by Table 7-1, average clast velocity decreases with increasing age, reflecting the increased weathering of clasts of the same lithology with age.

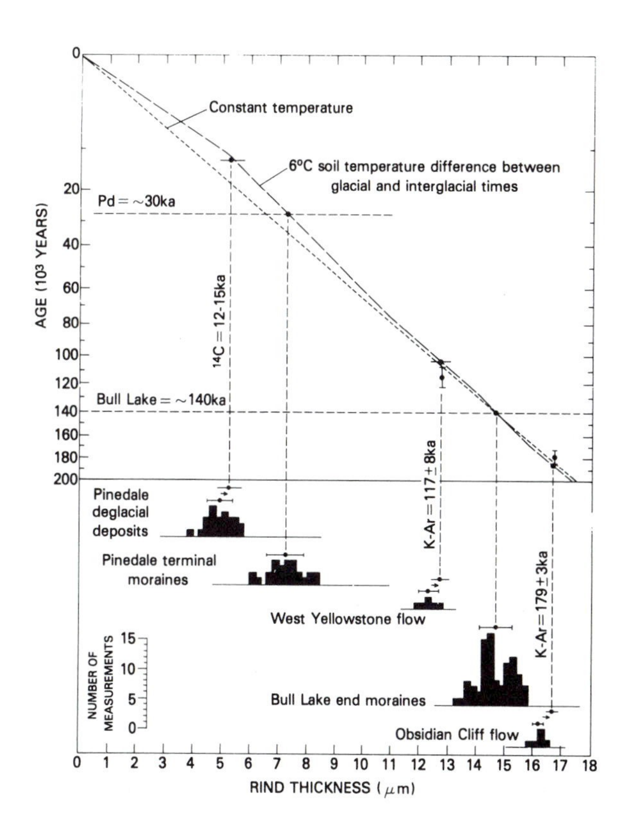

Figure 7–30. Thickness of obsidian-hydration rinds of the Pinedale and Bull Lake Glaciations near West Yellowstone, Montana. Dashed lines show increase in rind thickness with age based on deposits dated by K-Ar or ^{14}C methods. Above histograms, short lines with dots are means and standard deviations of rind-thickness measurements, some of which are offset (small arrows) to account for small differences in temperature between localities. From Pierce (1986).

Soil Development

There are two ways in which soils are used in earthquake-related studies. In an engineering sense, soil is unconsolidated surface material. Its mechanical properties are important in calculating its load-bearing response to engineered structures and its amplification of seismic waves. To a soil scientist, soil is "a natural body consisting of layers or horizons of mineral and/or organic constituents of variable thicknesses, which dif-

Table 7–1

Compressional-Wave Velocity in Clasts of Lowe Granodiorite from Alluvial Deposits of Middle Pleistocene to Late Holocene Age Near the Mouth of Arroyo Seco, Southern Margin of the San Gabriel Mountains. From Crook et al. (1987).

Alluvial-Deposit	Average Clast Velocity, km/s	Standard Deviation of Mean	Number of sequence Clasts Measured
A (oldest)	1.20	0.06	52
B	1.41	.07	28
C	1.58	.06	36
D	1.77	.06	48
E	1.60	.09	17
F (youngest)	2.01	.05	38

fer from the parent material in their morphological, physical, chemical, and mineralogical properties and their biological characteristics" (Joffe, 1949; Birkeland, 1984). Land surfaces develop soils as the underlying material weathers, and the degree of soil development may be used to determine the relative age of geomorphic surfaces in a given area. A genetically-related suite of soils with similar climate, topography, vegetative cover, and parent material is called a *soil chronosequence*. The relative degree of development of soils in a chronosequence may be used to establish the relative ages of surfaces on which the soils developed. Some soil properties develop rapidly at first, then slow down asymptotically later in the weathering process (Fig. 7-31). Thus the changes in soil properties in the last 10^5 years may be as great or greater than changes in the preceding 10^6 years. Calcic soils, which are in part affected by wind-borne additions of Ca^{++}, may take longer for rates to slow down.

Soil profiles become layered with time. The layers that develop as part of soil formation are summarized in Table 7-2. Soil descriptions are described in more detail in Birkeland (1984), who gives references to other work. The more important properties in soil development are summarized in the following text.

Color includes the dark brown to black color of A horizons, yellow-brown to red colors of B horizons, and gray to white colors of calcareous horizons. The three soil-color properties of *hue, value,* and *chroma* are used in the Munsell soil color chart (see also the rock-color chart published by the Geological Society of America). As the soil ages, hue and chroma change from those of the parent material. If iron-oxide pigments are available, soil hues become redder and chromas become brighter; this process is called *rubification*. For example, a chronosequence could show a change in hue from 2.5Y to 10YR to 5YR and a change in chroma from 0 to 6. Where the parent material is red, soil color is not useful as a relative dating technique.

Soil texture is the combined grain size of mineral fractions and grades from end members of clay to silt to sand and various mixtures of these. The term *clay* includes all particles less than 0.002 mm in diameter;

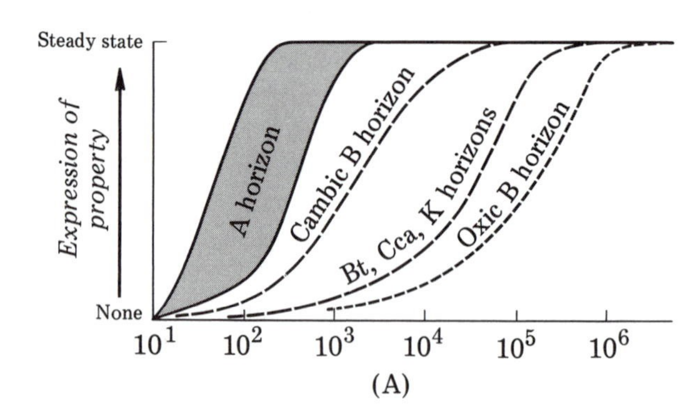

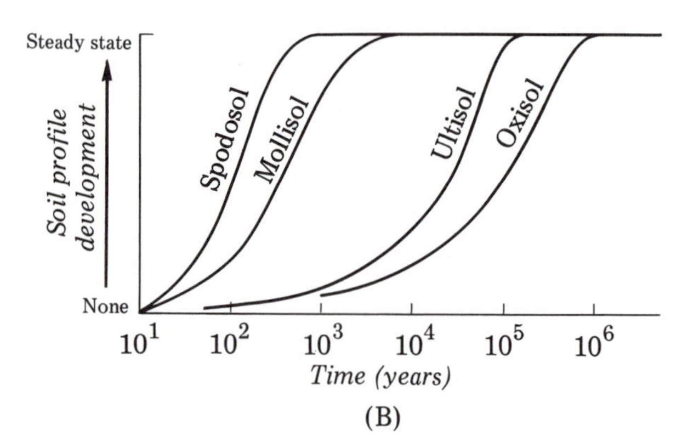

Figure 7–31. Variations in length of time for different soil horizons to reach a point where properties change slowly. Note that the A horizon changes most rapidly in the first few hundred years, whereas the accumulation of clay or carbonate in the B horizon takes thousands of years. After Birkeland (1984).

Table 7–2

Horizon Nomenclature. Simplified from Birkeland (1984).

O horizon	Surface accumulation of mainly organic matter containing little or no mineral matter from the underlying parent material
A horizon	Humified organic matter mixed with mineral matter with mineral matter dominant
E horizon	Less organic matter and/or fewer compounds of iron and aluminum than overlying O or A horizon and/or less clay than the underlying B horizon, giving a relatively light color of parent material
B horizon	Shows little or no evidence of structure or sedimentary layering of parent material, characterized by illuvial accumulations of various kinds of secondary materials, subdivided on basis of type of material accumulated.
	Bh and Bs horizons: amorphous organic matter complexes of iron and aluminum
	Bk horizons: alkaline carbonates, mainly calcium carbonate
	Bn horizon: accumulation of exchangeable sodium
	Bq horizon: secondary silica
	Bt: clay formed in situ or by illuviation
K horizon	Horizon so impregnated with carbonate that the carbonate dominates the morphology
C horizon	Subsurface horizon lacking properties of A or B horizon, but still showing signs of weathering
Cox horizon:	oxidized C horizon overlying apparently unweathered parent member
Cu horizon:	relatively unweathered C horizon

it also refers to the soil texture containing ≥40% clay (Birkeland, 1984) and also to the clay minerals with layered lattices.

Soil structure describes the aggregation of soil particles by bonding, including the accumulation of clay in the B horizon. Structures are described as granular, blocky, prismatic, columnar, and platy (Fig. 7–32, simplified from Table 1-4, Birkeland, 1984). Other types of structures in tropical soils result from the cementation of iron hydroxides. Vesicular structure is found in the upper layer of fine-grained desert soils and in the A horizons of some silty arctic soils.

Other soil characteristics that may be indicative of age include bulk density of an oven-dry sample, soil-moisture retention, cation exchange capacity, and pH.

As a soil develops, the sedimentary layering or other structure of the parent material is obscured, soil horizons develop, and the soil is said to have a profile. The A horizon develops relatively rapidly as vegetation is established on a new surface, such as the recently abandoned channel of a river. Over time, the addition of organic matter is offset by losses due to decomposition, and the A horizon reaches a steady state of development. Because of its organic content, the A horizon of a soil may be dated by radiocarbon. The soil formation process described previously indicates that the material being dated may be of different ages, and the age determination is referred to as *apparent mean residence time* (*AMRT*).

A slower process is the downward translocation of materials, principally compounds of iron and aluminum sesquioxides. Horizons from which these materials are derived are *eluvial;* horizons in which they accumulate are *illuvial*. In Table 7–2, the E horizon is eluvial, and the B horizon is illuvial. There is also a translocation of clay particles so that they are concentrated in the B horizon. Clays in the B horizon come from weathering of material higher in the profile, are transported in suspension by water, are deposited by wind, or are formed in place by weathering. The clays accumulate, they aggregate, and they form *soil peds* with blocky, prismatic, and columnar structure. Soils in dry regions are commonly characterized by carbonate-rich Bk or K horizons that are characteristic of calcic soils or *calcrete* (Machette, 1985). CO_2 and water combine at depth under conditions of high pH or evapotranspiration of soil moisture or other processes to precipitate calcium carbonate.

The development of soil properties occurs with age, but the rate of development is influenced by precipitation, temperature, vegetative cover, parent material, and airborne soil components such as silt, clay, and calcium carbonate. Accordingly, relative dating compares a chronosequence: soils of a given region where the soil-forming processes are similar, and time is the only important variable. Several soil-development indices have been developed that try to quan-

TYPE	SKETCH* AND DESCRIPTION	PROBABLY ORIGIN	USUAL ASSOCIATED SOIL HORIZON
Granular *Crumb*	 Spheroidally shaped aggregates with faces that do not accommodate adjoining ped faces; the two differ only in that crumb is porous	Colloids, mainly organic, bind the particles together; clay and Fe and Al hydroxides may be responsible for some binding, and flocculating capacity of some ions, such as Ca^{2+}, may be helpful; periodic dehydration helps form more stable aggregates	Mollic or Umbric A
Angular blocky *Subangular blocky*	 Approximately equidimensional blocks with planar faces that are accommodated to adjoining ped faces; face intersections are sharp with angular blocky, rounded with subangular blocky	Many faces may be intersecting shear planes developed during swelling and shrinkage that accompany changes in soil moisture	Argillic B
Prismatic *Columnar*	  Particles are arranged about a vertical line, and ped is bounded by planar, vertical faces that accommodate adjoining faces; prismatic has a flat top, and columnar a rounded top	Faces develop as a result of tensional forces during times of dehydration; rounded column tops may be due to some combination of erosion by percolating water and greater amounts of upward swelling of column centers upon wetting	Argillic B (prismatic) Natric B (columnar)
Platy	Particles are arranged about a horizontal plane	May be related to particle-size orientation inherited from parent material or induced by freeze-thaw processes	E, or those with fragipan
		May be related to layering in cementing material, induced during its precipitation (carbonate, silica, Fe hydroxides)	Km, Csim, Spodic B

Figure 7–32. Description of common soil structures. After Birkeland (1984).

tify the degree of soil formation in arriving at relative ages. The Harden index is discussed briefly as an example. It was developed by Harden (1982) for a chronosequence in the San Joaquin Valley and Sierra Nevada foothills of California and has subsequently been applied (with different rates of soil formation and different degrees of success) to other chronosequences in the United States. Harden quantified soil texture, plasticity, stickiness, rubification, melanization, development of clay films, development of soil structure, dry and moist consistency, and soil pH and showed that these changed in a regular way with time. However, the Harden index is not universally accepted among soil scientists; attempts to use it in New Zealand, for example, have had mixed results.

Soils in the arid and semiarid southwestern United States develop by the addition of secondary calcium carbonate. Calcium carbonate is added to the soil by rainfall and by airborne dust, silt, and sand. Ca^{++}deposited on the surface is leached and redeposited, generally in the A and B horizons. Machette (1985) recognized six stages of calcic soil development, beginning with thin, discontinuous coatings on the undersides of pebbles to accumulations of carbonate in the matrix between clasts to development of laminae and pisolites (nodules). In the most advanced stage, the soil is rocklike, with a tabular structure somewhat resembling a limestone. The diagnostic property is the cS index, the weight of calcium carbonate in a 1 cm^2 vertical column through the soil, calculated from thickness, calcium-carbonate concentration, and bulk density of calcic horizons (Machette, 1985). Changes in the amount and seasonal distribution of rainfall, the concentration of calcium carbonate in rainfall, and the calcium-carbonate content of airborne materials will cause variations in the rate of development of calcic soils. Machette was able to recognize four episodes of movement on the County Dump fault near Albuquerque, New Mexico on the basis of displacement of calcic soils with different degrees of development.

Soil evolution must consider the geological context in which the soil is found. An elevated river terrace may have material added to it such as loess or volcanic ash. The planar surface of the terrace would be preserved, but the soil would date from the time of wind-borne cover deposits, not the time of abandonment of the river channel. In some desert regions, weathered surface material may be removed by wind erosion, a process known as *deflation*.

Several case histories of soil development are shown in Figure 7–33. Each has soils of different ages as based on percent clay in the B horizon. Units I, II, and III are deposits of decreasing age. In Case 1, soil on the hilltop is traced into a buried soil overlain by Unit II, indicating that deposition of Unit II postdates the soil, even though it is at the surface at the hilltop. Using the same line of reasoning, soil is developed on Unit II prior to the deposition of Unit III. In contrast, the strongest soil in Case 2 developed after Unit II. Soil *a* at the hilltop is a combination of the soil on top of Unit II and the weaker soil at its base. In Case 3, three river terraces of different ages develop different soils with soil *a* on Unit 1 the oldest and strongest. Soil *a* includes the weathering that developed Soil *b* and *c* plus an earlier stage of weathering prior to deposition of Unit II. Case 4 shows a similar terrace sequence, but Unit I is overlain by ash beds, Unit II was deposited after ash 1 but before ash 2, and Unit III was deposited after both ash beds. The soil is likely to have the same development on the two higher terraces, because it began to develop after the deposition of ash 2. Paleosols may underlie ash 2 and ash 1, however.

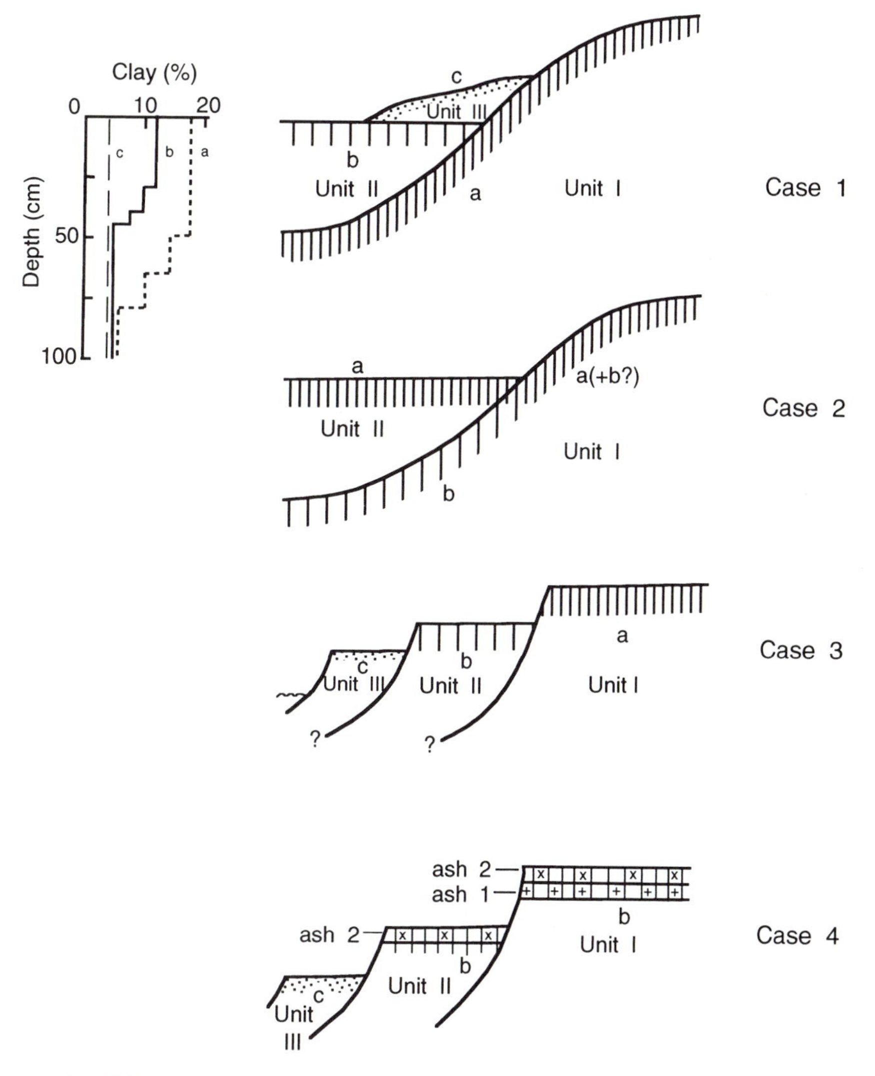

Figure 7–33. Four case histories showing relations between soils and depositional units I (old) to III (young). Soils *a, b,* and *c,* are progressively less strongly developed, as shown in the clay content in profiles. Cases 1–3 are from Birkeland (1984).

SUMMARY

The convergence of tectonics and geomorphology in recent years means that the landscape may be used as a tectonic signal, after the effects of degradation by erosion and weathering have been taken into account. A fault scarp forms, a mountain range uplifts, and surficial processes begin to obscure the feature as it forms. The erosion of fault scarps in the Great Basin of the United States is an example of how the degradation of a tectonic feature can be used to establish its relative age. Other features, the dissection of mountain fronts, the widening of valley floors, the depositional behavior of alluvial fans, and the subtle response of alluvial rivers, may be used to measure crustal deformation, and in some instances to determine the rates at which deformation is taking place. Marine shorelines have been dated directly using the uranium-series and radiocarbon methods, and these dates can be compared with sea-level curves constructed using oxygen isotopes.

Weathering includes the development of weathering rinds on stones and of soils on surfaces. As in erosion degradation studies, development is relative and dependent on many factors in addition to time.

Clearly, the geomorphologist and Quaternary geologist have major roles to play in the study of earthquakes. The response of landforms to earthquake-induced deformation, whether it be formation of a fault scarp or the broad uplift accompanying a blind thrust, is critical to understanding the earthquake environment.

Suggestions for Further Reading

Birkeland, P. W. 1984. Soils and Geomorphology. New York: Oxford University Press, 372 p.

Bloom, A. L. 1991. Geomorphology: A Systematic Analysis of Late Cenozoic Landforms, 2nd ed. Englewood Cliffs, N.J.: Prentice-Hall, 532 p.

Bloom, A. L. and Yonekura, N. 1990. Graphic analysis of dislocated Quaternary shorelines. *In* Sea Level Change. Washington, D.C.: National Academy Press, pp. 104–15.

Bull, W. B. 1977. The alluvial fan environment. Progress in Physical Geography, 1:222–70.

Bull, W. B. 1991. Geomorphic Responses to Climatic Change. New York: Oxford University Press, 326 p.

Colman, S. M., and Pierce, K. L. 1981. Weathering rinds on andesitic and basaltic stones as a Quaternary age indicator, western United States. U.S. Geological Survey Prof. Paper, 1210, 56 p.

Jour. Geophys. Res. 1994. Special Section: Tectonics and Topography, 99: Part 1: 12,135–315; Part 2: 13,871–14,052; Part 3: in press.

Keller, E. A. 1986. Investigation of active tectonics: Use of surficial earth processes. *In* Active Tectonics. Washington D.C.: National Academy Press, pp. 136–47.

Keller, E. A., and Pinter, N. 1996. Active Tectonics—Earthquakes, Uplift, and Landscape: Upper Saddle River, New Jersey, Prentice-Hall, 338 p.

Knuepfer, P. L. K. 1988. Estimating ages of late Quaternary stream terraces from analyisis of weathering rinds and soils. Geol. Soc. America Bulletin, 100:1224–36.

Lajoie, K. R. 1986. Coastal tectonics. *In* Active Tectonics. Washington D.C.: National Academy Press, pp. 95–124.

Machette, M. N. 1985. Calcic soils of the southwestern United States. Geol. Soc. America Special Paper, 203:1–21.

Mayer, L. 1986. Tectonic geomorphology of escarpments and mountain fronts. *In* Active Tectonics. Washington D.C.: National Academy Press, pp. 125–35.

Schumm, S. A. 1986. Alluvial river response to active tectonics. *In* Active Tectonics. Washington D.C.: National Academy Press, pp. 80–94.

Stewart, I. S. 1993. Sensitivity of fault-generated scarps as indicators of active tectonism: Some constraints from the Aegean region. *In* Landscape Sensitivity. Thomas D. S. G., and Allison, R. J., eds. John Wiley & Sons, 129–47.

Wallace, R. E. 1977. Profiles and ages of young fault scarps, north-central Nevada. Geol. Soc. America Bulletin, 88:1267–81.

PART TWO

Earthquake Geology

Most earthquakes occur along active plate margins. The bewildering variety of geometries and kinematics of active plate margins makes it a challenge to keep a discussion of the geology of these earthquakes simple and orderly.

The most basic differences between active plate margins arise primarily from the variability of two principal characteristics: (1) the composition of lithospheres involved and (2) the orientation of lithospheric plate boundaries with respect to the directions of relative plate motion. Lithospheric compositions range from relatively dense oceanic to relatively light continental lithosphere. The relative motions of lithospheric plates are in some places orthogonal to plate boundaries; elsewhere they are parallel, and not uncommonly, they are oblique. Plate interactions, therefore, range from contractional to divergent to translational, and combinations of these are the rule, rather than the exception. Each concoction of these kinematic and compositional variables produces a kinematic design that is characterized by its own assortment of active geological structures and processes.

Hence, we have organized our discussion of Earthquake Geology into four principal chapters, each related to a particular tectonic environment. Basically, these follow the three-fold division of fault types (strike-slip, normal, and reverse). Thus, the first three chapters of this section treat separately those earthquake environments dominated by strike-slip, by normal, and by reverse structures. In addition, because of their special characteristics and surpassing importance, we devote the fourth chapter in this section exclusively to subduction zones.

This four-fold division of earthquake environments is somewhat contrived—after all, the relative-slip vectors of nearly half of the earth's major plate boundaries deviate by more than 22° from pure strike-slip or dip-slip (Woodcock, 1986). The awkwardness that might be engendered by such a division of our discussion is largely mitigated, however, by the fact that this division also seems to be Nature's own strategy: Even in environments of oblique convergence and divergence, segregation or *partitioning* of the components of oblique motion into distinct strike-slip and dip-slip structures is commonplace, especially where the component of boundary-parallel motion is not small. In oceanic lithosphere, for example, truly oblique spreading (that is, where a component of boundary-parallel slip occurs within a spreading center) does not appear to occur where the velocity vector departs by more than 10° to 15° from being boundary-normal (Chase, 1978). Along most obliquely diverging plate boundaries, the creation of strike-slip faults orthogonal to spreading-ridge axes is common. Such partitioning appears to minimize resistance to plate separation (Lachenbruch and Thompson, 1972). The pervasiveness of this natural division of labor implies that it is easier to separate the boundary-normal and boundary-parallel components of slip into purely strike-slip and purely dip-slip faults than to produce oblique spreading. Similarly, mechanical analyses of subduction zones suggest that nonpartitioned oblique slip should be limited to plate boundaries where the angle between the vector of relative plate motion and the plate boundary is high (Woodcock, 1986; Fitch, 1972; Walcott, 1978). This appears to be borne out by the existence of great, boundary-parallel strike-slip faults in the hangingwall block of many obliquely converging plates.

Because of this natural tendency toward partitioning of slip, our four-fold division of this section engenders only minor organizational problems. We trust you, the reader, will bear these graciously.

In the fifth and last chapter of this section we introduce aspects of earthquake geology that are common to all four tectonic environments. These phenomena are the secondary effects of earthquakes: tsunamis, liquefaction, seismically generated landslides and other types of seismically induced ground failure.

8

Strike-Slip Faults

Strike-slip faults occur in a variety of crustal environments, throughout the entire spectrum of geological scales—from grain boundaries to plate boundaries, from less than a thousand microns to more than a thousand kilometers.

The early history of modern seismology, itself, provides an apt illustration of the variety of strike-slip environments on Earth. The recognition that faulting produces earthquakes arose from geologic and geodetic observations in four different strike-slip environments, following four large earthquakes a century ago—the Marlborough, Nobi, Tapanuli, and San Francisco earthquakes of 1888, 1891, 1892, and 1906 (McKay, 1890; Koto, 1893; Reid, 1910; 1913). Each of these four large earthquakes, one in New Zealand, one in Japan, one in Sumatra, and one in California, was generated by a large strike-slip fault.

The 1888 earthquake occurred in New Zealand, along the Hope fault, part of a transform fault system that connects two subduction zones. The 1891 earthquake occurred within the Japanese island arc, above a large subducting slab. The fault zone that produced the earthquake is one of dozens of conjugate sinistral- and dextral-slip fault zones, which are accommodating trench-perpendicular shortening and trench-parallel stretching of the arc lithosphere. The earthquake of 1892 was generated by a large right-lateral fault system that runs through the Sumatran volcanic arc, parallel to the nearby Sumatran subduction zone. This fault accommodates a large component of strike-slip motion across the obliquely convergent Sumatran plate boundary. The great earthquake of 1906 was produced by the San Andreas fault, a transform fault that runs between an oceanic spreading ridge and a triple junction at which two transforms join a trench.

In this chapter, we first introduce these and the other principal tectonic environments on Earth in which strike-slip faults predominate. We then discuss the structural and geomorphic manifestations of these features at local and at regional scales. Third, we present the case histories of two well-documented strike-slip earthquakes, to illustrate how geological, geodetic, and geophysical studies can contribute to a comprehensive understanding of an earthquake. And, finally, we discuss paleoseismologic data that help us understand the long-term seismic behavior of strike-slip faults.

Terminology

One of our more mundane tasks is to adopt, from a century's accumulation of terminology, an efficient set of terms for our discussion. Sylvester (1988) provides an interesting discussion of the origins of many adjectives commonly used today in reference to strike-slip faults—specifically, *strike-slip, wrench, transcurrent,* and *transform.* The term "strike-slip" was coined early in the twentieth century to describe faults with slip vectors parallel to their strike. It is a simple term with obvious meaning and no genetic implications. "Transcurrent" and "wrench" have meanings and birthdates similar to "strike-slip," but were applied originally only to nearly vertical strike-slip faults that cut not only supracrustal sedimentary, but also igneous and metamorphic, rocks. Because of its more-general meaning and clearer connotation, we use "strike-slip" throughout this text and avoid the terms "wrench" and "transcurrent." We also employ the term, "*lateral,*" as a synonym for "strike-slip."

The term "transform fault" was first used by Wilson (1965), in his elucidation of the essential role that a certain "new class" of strike-slip faults played in global tectonics. Wilson put strike-slip faults on the plate-tectonic map—as transform faults. And he demonstrated how the paradigm of plate tectonics could resolve long-standing quandaries concerning the terminations of strike-slip faults. In view of the great value of Wilson's hypothesis about transform faults, it seems wholly appropriate to retain this term for strike-slip faults that conform to his definition.

There exist, however, some ambiguities in Wilson's definition. Transform faults were originally defined as plate-bounding, strike-slip faults that abruptly transform at their termini into either convergent or divergent plate boundaries. In his original formulation, however, neither "plate" nor "abrupt" was strictly defined. Hence, many ambiguities arise in the application of the term "transform fault." In this text, we are not strict constructionists. For example, we consider the San Andreas fault to be a transform fault, even though it does not transform at its northern end into a spreading center, as Wilson envisioned it, but rather into a trench and another strike-slip fault. The *spirit* of Wilson's original definition would seem to allow transformation at a triple junction.

We also recognize that the decision to call a fault a "transform" is, in part, a matter of scale. Many transforms appear as individual faults only on maps of global scale. Viewed regionally, however, not only are they not individual faults, but their terminations are far from abrupt. And so, in this text, we refer to the principal fault of a transform system as a transform fault, but do not refer to lesser faults within these systems as transforms.

Furthermore, many strike-slip faults terminate abruptly into extensional or contractional structures, a characteristic ascribed by Wilson to transforms, but they are not the boundaries of major or even well-defined lesser plates. We choose not to call these structures transform faults, fully realizing that there is not full agreement as to what does and what does not constitute a plate. For example, the great strike-slip fault of Sumatra (#30 in Fig. 8-1) might be considered a transform fault by some, because it intersects the

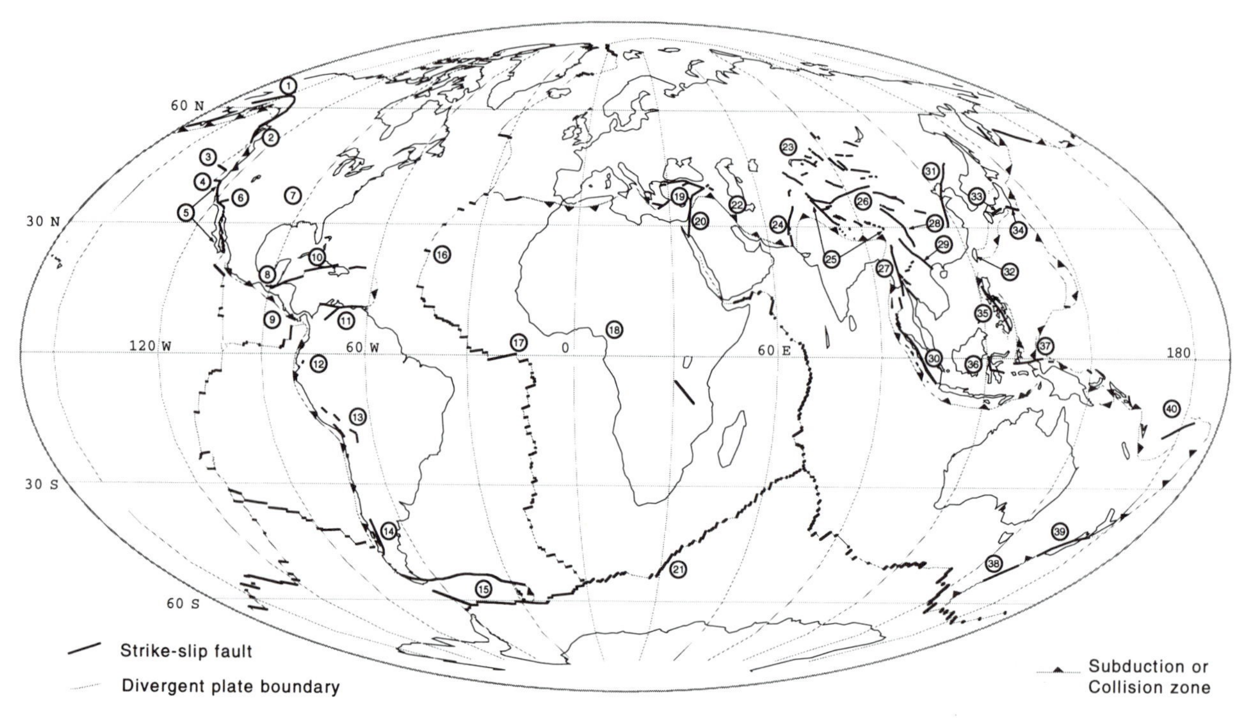

Figure 8–1. Worldwide distribution of major active strike-slip faults. Those mentioned in the text and selected others are numbered: **Ridge-ridge transform faults:** Blanco, 3; Kane, 16; Romanche, 17; Andrew Bain, 21. **Trench-trench transform faults:** Chaman, 24; Macquarie Ridge, 38; Alpine, 39; Fiji, 40; Northern boundary of Caribbean plate (Swan, Oriente, Motagua and Chixoy-Polochic faults, 8; and Septentrional fault, 10) and southern boundary of Caribbean plate (El Pilar and Boconó faults), 11; Northern and southern boundary of the Scotia plate, 15. **Ridge-trench transform faults:** Mendocino, 4; Fairweather and Queen Charlotte Islands, 2; Dead Sea, 20; Sagaing, 27. **Other transform faults:** San Andreas-Gulf of California, 5. **Trench-parallel faults:** Denali, 1; Pallatanga, 12; Atacama and El Tigre, 13; Liquine-Ofqui, 14; Great Sumatran, 30; Median Tectonic Line, 33; Philippine, 35; Sorong, 37. **Indent-linked faults:** North and East Anatolian and Pliny Trough, 19; Dasht-e-Bayaz, Rudbar-Tarom, and others, 22; Fuyun, Talas-Fergana, Kopet-Dagh, and others, 23; Altyn Tagh, Haiyuan and Kunlun, 26; Karakorum-Jiali, 25; Xianshuihe and Xiaojiang, 28; Honghe, 29; Tan-lu, 31. **Other:** conjugate faults of central Honshu, Japan, 33; Gorda Deformation Zone, between 3 and 4; Basin Ranges, 6; New Madrid and Meers, 7; Managua, 9.

Sunda trench at its southern terminus and the spreading centers of the Andaman Sea at its northern end. However, kinematic studies of the wedge of lithosphere west of this strike-slip fault suggest that this forearc sliver "plate" is not an intact plate, because it is becoming elongated parallel to the fault at a rate of several centimeters per year. Furthermore, the fault may not transform at its intersection with the Sunda trench; the rate of slip on the fault appears to diminish southward as it approaches the Sunda trench.

Rather than quibble further about definitions, we prefer to avoid the nomenclatural morass associated with strike-slip faults and to focus, instead, on more substantial and scientifically interesting issues.

STRIKE-SLIP ENVIRONMENTS

The Earth's major strike-slip faults and fault systems are depicted in Figure 8-1. Although most of these structures serve as transform faults at oceanic spreading centers, many fulfill other important tectonic functions. Several important strike-slip fault systems, for example, run between zones of plate convergence. Some of the Earth's largest strike-slip faults accommodate horizontal extrusion in zones of continental collision. Many other large structures are agents for boundary-parallel slip between obliquely convergent oceanic and continental plates. Myriads of lesser strike-slip faults, too small to appear on Figure 8-1, act in roles that support regional fragmentation, contraction or extension.

Basic Settings

Figure 8-2 illustrates, in map view, the basic settings of strike-slip faults. Boxes A through C represent common geometries of transform faults—those that connect two spreading ridges, those that connect two contractional (fold-thrust) belts, and those that connect a spreading ridge with a contractional belt. More complicated arrangements of transforms, ridges, and trenches are also possible and common: Box D depicts just one of several geometries that can arise when a strike-slip fault transforms at a triple junction. This particular arrangement is an idealization of the relationship of the East Pacific Rise, the San Andreas and Mendocino transform faults (4 and 5 in Fig. 8-1), and the Cascadian subduction zone.

Box E represents obliquely contractional plate margins in which a strike-slip fault accommodates the boundary-parallel component of motion. Oblique extensional environments also contain subparallel dip-slip and strike-slip faults, although the structures are not of plate-boundary dimension and normal faults occur in place of reverse faults. The cartoon in Box F illustrates a region where strike-slip faults enable the horizontal extrusion of lithospheric blocks away from a region of continental lithospheric collision, such as is occurring between India and Eurasia and between Arabia and Eurasia. Box G depicts a region in the hangingwall block of a subduction zone in which conjugate left- and right-lateral faults accomplish boundary-parallel extension and boundary-normal contraction.

Box H represents geometries common to normal and reverse faults that are segmented by strike-slip faults in continental lithosphere.

Transform Faults

Many examples of the three basic varieties of transform fault appear in Figure 8-1. Most of these crop out on the ocean floor. The left- and right-lateral transform faults that connect segments of the Mid-Atlantic Ridge, for example, are in this category.

Trench-trench transforms are far less common than the ridge-ridge type. Trench-trench transform systems are exemplified by the Fiji transform, the northern and southern margins of the Scotia plate, between Antarctica and South America, and by the northern and southern boundaries of the Caribbean plate (numbers 40, 15, 8, 10, and 11 on Fig. 8-1). The Alpine fault system of New Zealand is an example of a trench-trench transform in which the strike of the transform fault is far from perpendicular to the strike of adjoining subduction zones.

One example of a transform that runs between an extensional and a contractional plate boundary is the sinistral Dead Sea transform, an element of the Arabian/African plate boundary that transforms southward into the extending oceanic lithosphere of the Red Sea and northward into the contracting continental lithosphere of the Bitlis zone. Similarly, the right-lateral Queen Charlotte Islands-Fairweather fault system is a transform that separates the northeastern margin of the Pacific plate from the North American plate and runs between the Alaskan trench and the Explorer ridge.

Ridge-Ridge Transform Faults

The most common transforms are those that connect spreading ridges (Fig. 8-2a). Only four of these (numbers 3, 16, 17, and 21) are labeled on Figure 8-1. *Ridge-ridge transform systems* have attracted much attention, because of their great importance in establishing the kinematics and evolution of plate tectonic systems. Sykes (1967) first demonstrated the validity of Wilson's hypothesis of transform faults in his study of the focal mechanisms of more than a dozen mod-

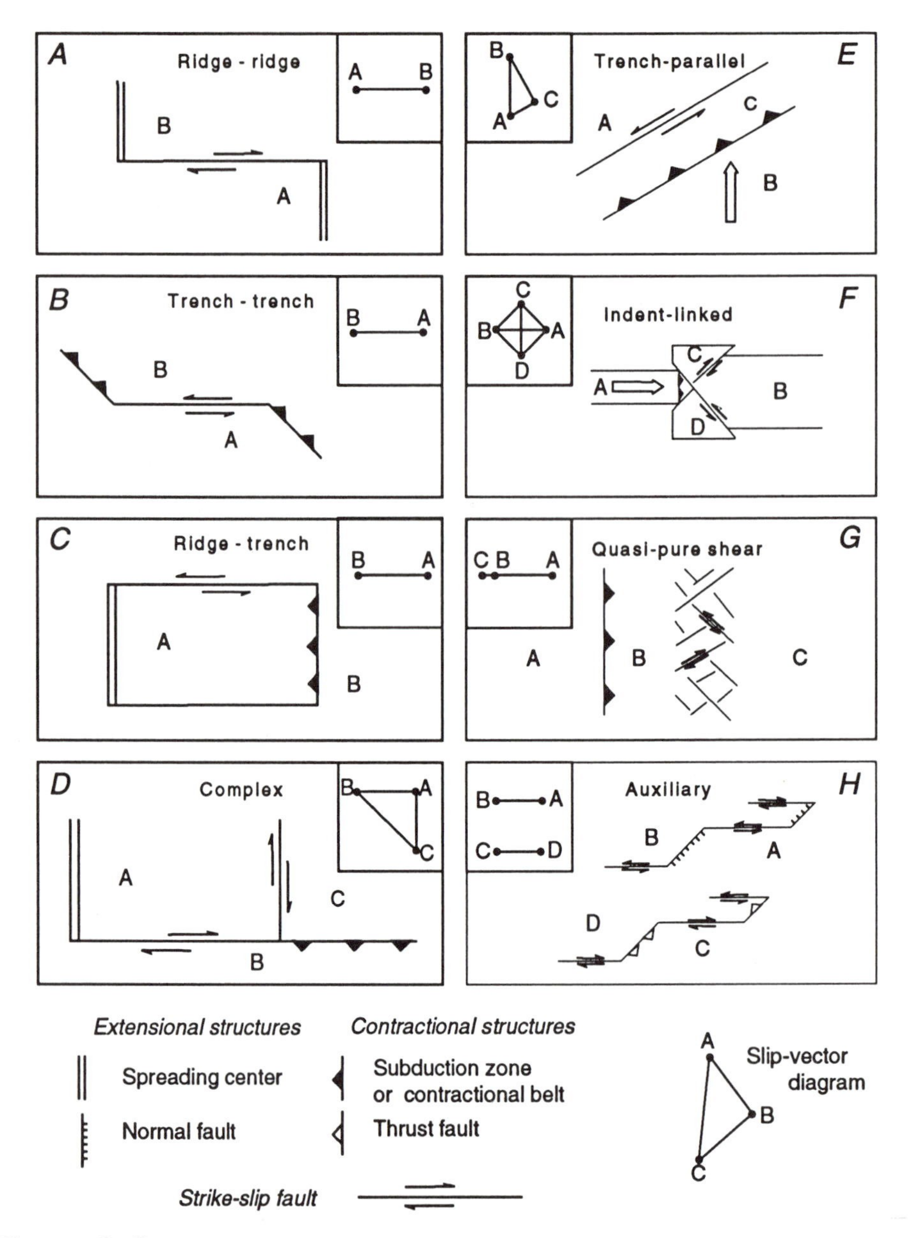

Figure 8–2. Common tectonic settings in which strike-slip faults occur. In settings A through D, strike-slip faults are plate-bounding, transform faults. In settings E through H, they perform other functions. Slip-vector diagrams in the upper, inner corner of each of the eight boxes illustrate the relative motions between plates or blocks separated by the faults.

erate earthquakes, principally along the central Mid-Atlantic Ridge. Subsequent studies extended the validity of the hypothesis to numerous other ridge-ridge transforms (see for example, Bergman and Solomon, 1988). Strike-slip mechanisms predominate along these transforms, although reverse- and normal-fault focal mechanisms are not unknown (for example, Wald and Wallace, 1986; Johnson and Jones, 1978).

Most ridge-ridge transforms have trends that are orthogonal to their adjacent spreading ridges or very nearly so. Woodcock (1986) suggests that, because this nearly complete partitioning of extension and strike-slip is ubiquitous, this geometry probably offers minimum resistance to plate divergence. Transform faults that strike at acute angles to spreading ridges require longer spreading ridges, a situation common only to transforms less than 30 km in length.

Earthquake hypocenters are generally too imprecise to constrain meaningfully the structure of a ridge-ridge transform. Thus, the geometry and geology of ridge-ridge transforms are known primarily from interpretation of bathymetric maps and dredged samples. Let it suffice here to note that typically these transforms traverse young oceanic mantle, atypical in that it is covered only by thin or patchy oceanic crust. The paucity of crustal material enables the chemical interaction of ultramafic mantle rocks and seawater, which produces an abundance of serpentine within the fault zone.

Earthquakes along ridge-ridge transform fault systems pose little threat to humankind because they are typically small and occur in locations remote from human habitation. Despite the great length of many of these faults, only rarely do ridge-ridge transforms produce earthquakes greater than M7 ($M_o > 5 \times 10^{19}$ N-m) (Burr and Solomon, 1978; Bergman and Solomon, 1988). The only instrumentally recorded earthquake on a ridge-ridge transform fault with a magnitude much larger than 7 is a M8.0 strike-slip event, which occurred in 1942 on the low-rate, 440-km-long Andrew Bain transform in the southwest Indian Ocean (Okal and Stein, 1987).

The small size of most earthquakes on ridge-ridge transforms may in part be a consequence of both the short length of most ridge-ridge transforms and the thin character of the brittle lithosphere they traverse. Although some of these transforms are hundreds of kilometers long, most ridge-ridge transform faults run no more than a hundred kilometers between spreading ridge segments. Their earthquakes are characteristically shallow, an indication that the brittle-ductile transition is at shallow depths. Centroid depths of the largest ridge-ridge transforms in the North Atlantic are commonly between 7 and 10 km, and maximum depths of seismic faulting are judged to be no more than twice this (Bergman and Solomon, 1988).

Short lengths and shallow brittle zones are only partially responsible for the small size of earthquakes on ridge-ridge transforms. Dozens of ridge-ridge transform systems have lengths greater than 100 km. Unlike strike-slip faults in continental lithosphere, however, few of these generate earthquakes of $M > 7$, and lengths greater than about 70 km appear to be necessary even to generate earthquakes greater than or equal to M6 (Burr and Solomon, 1978). This suggests that some additional property governs the maximum size of earthquakes. The abundance of serpentine in ridge-ridge transform fault zones suggests one explanation: Ridge-ridge transforms may have lower shear strengths than faults that juxtapose continental lithosphere or unserpentinized oceanic materials.

For faults less than about 400 km long, the maximum size of earthquakes along ridge-ridge transforms is proportional to the fault's total length. Despite this correlation of transform length and maximum earthquake magnitude, none of the earthquakes Burr and Solomon studied appears to have involved rupture of an entire ridge-ridge transform. Furthermore, for lengths greater than 400-km, length and maximum magnitude do not seem to be correlated. The 950-km-long Romanche system, in the equatorial Atlantic Ocean, is, for example, one of the longest ridge-ridge transforms (number 17 in Figure 1). And yet, since 1920, the 20 largest earthquakes this structure has generated range between M6.1 and 7.1. Geometrical segmentation or the shallowness of the brittle/ductile transition may restrict the maximum length of individual coseismic ruptures along this and other very long transforms.

The maximum size of earthquakes produced by a ridge-ridge transform fault also correlates with the inverse of the slip rate of the fault. Very fast-slipping faults (16–18 cm/yr) produce earthquakes up to $M_S = 6.5$ ($M_o = 8 \times 10^{18}$ N-m), whereas slowly slipping faults (2–4 cm/yr) produce earthquakes up to $M_S = 7+$ ($M_o = 4 \times 10^{20}$ N-m). This inverse proportionality reflects the fact that fast-slipping transforms tend to juxtapose younger and warmer (and therefore thinner), brittle lithosphere.

Trench-Trench Boundary Transforms

Transforms that run between trenches are far less common than ridge-ridge transforms. Orthogonal trench-transform-trench geometries analogous to ridge-transform-ridge geometries are, in fact, unknown. Woodcock (1986) has speculated that such convergent geometries are inhibited from forming, because they would require a constant shredding of lithosphere at the trench-transform intersection where the lithosphere is beginning to subduct.

Although they do not occur as closely spaced arrays separating trench segments, trench-trench transforms do occur as the lateral boundaries of large plates. For this reason, Woodcock (1986) places them in his "boundary transform" category. We will call them *trench-trench boundary transforms*.

Trench-trench boundary transforms are typically, though not universally, long. Although some occur in oceanic lithosphere, they also commonly involve continental lithosphere. In both settings, trench-trench transforms do not display the simple geometry suggested by Wilson (1965; Fig. 1-3). Instead, their geometries tend to be rather complex and idiosyncratic. For this reason, several examples follow.

The Macquarie Ridge and Alpine fault systems, along the mutual boundary of the Pacific and Aus-

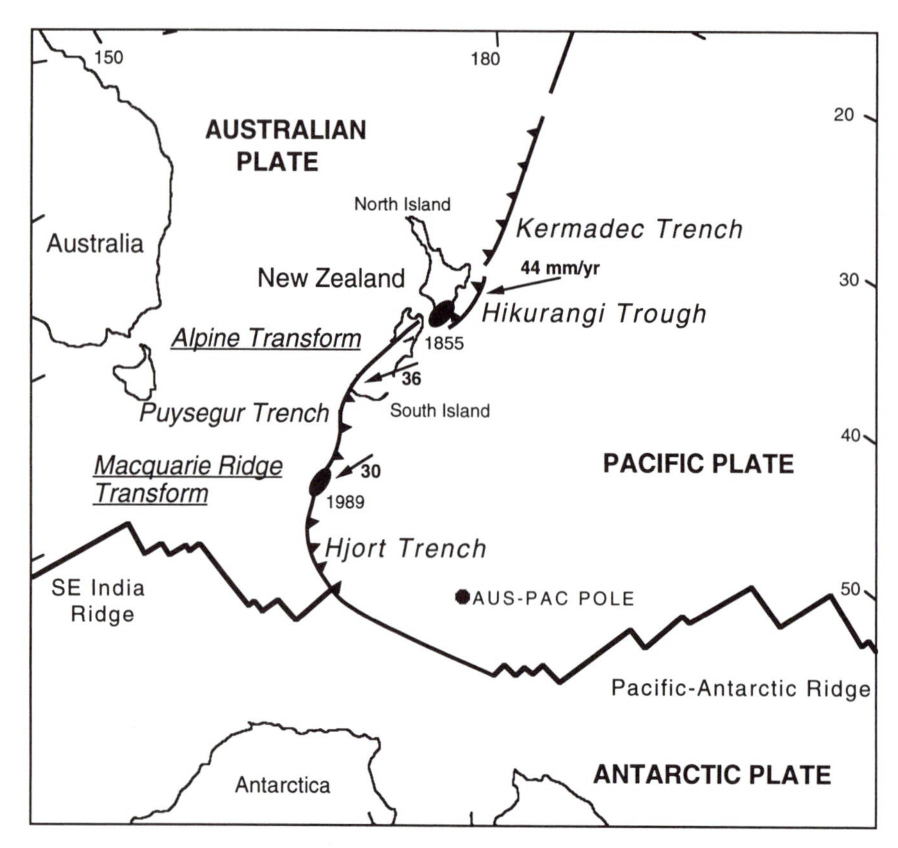

Figure 8–3. The Macquarie Ridge and Alpine transforms are examples of trench-trench boundary transform faults. These fault systems join three subduction zones along the obliquely convergent boundary of the Australian and Pacific plates in and south of New Zealand. Dates and ellipses represent the regions of large earthquakes discussed in the text. Adapted from Berryman et al. (1992).

tralian plates, are both trench-trench boundary transforms (Fig. 8-3). Each is about 800 km long, and each system is currently experiencing a predominance of dextral strike-slip augmented by a significant component of convergence. The Macquarie Ridge system (which is a bathymetric ridge, but not a ridge in the plate-tectonic sense) transforms into trenches with east-dipping subduction zones at both its northern and southern ends. The Alpine system transforms into a trench with a west-dipping subduction zone at its northeast end and a trench with an east-dipping subduction zone at its southwestern end.

Both of these transforms intersect neighboring trenches at an acute angle, a markedly different geometry than the orthogonal geometry of most ridge-ridge transforms and spreading centers. The predominance of strike-slip over reverse slip along the two transforms and the preponderance of dip-slip over strike-slip along the neighboring subduction zones are related to slight variations in the orientation of the plate boundary relative to the direction of relative plate motion.

The Pacific/Australian pole of rotation is currently only 1200 to 2000 km southeast of this portion of the plate boundary (Figure 8-3). The relative plate velocity across the Alpine transform decreases from about 40 mm/yr in the north to about 35 mm/yr in the south, because the southern portions of the transform are closer to the pole of rotation than are the northern reaches (DeMets et al., 1990). At the latitudes of the Macquarie Ridge transform, about 1000 km farther south and still closer to the pole, relative plate velocities range from about 32 to 28 mm/yr. In neither case is the vector of relative plate-motion parallel to the transform. Thus, at the present time, a significant component of convergence exists.

The Macquarie Ridge trench-trench transform acquired celebrity status in 1989, when it produced the great Macquarie Ridge earthquake. This event attracted a great deal of attention because of its large size and its strike-slip character. At M_W 8.1, the earthquake was one of the largest of the decade and the largest strike-slip earthquake yet recorded by modern instruments. From an analysis of long-period body waves and surface waves, Satake and Kanamori (1990) estimated an average dextral offset of about 9 meters and a rupture length of about 120 km on a steeply-dipping fault plane. A smaller (M_S 7.4 to 7.7) strike-slip earthquake, produced by a similarly oriented fault, occurred in 1981 at the north end of Macquarie Ridge (Ruff et al., 1989). Several smaller earthquakes, with reverse-faulting mechanisms, have also occurred along the transform, but bathymetric maps and earthquake locations are too poor to enable discernment of the physical relationship of the convergent and strike-slip structures.

Because of its exposure on land, the Alpine trench-trench transform system is far better known than its cousin to the south. An informative review of the earthquake geology of the Alpine fault, from which most of the following material has been taken, has been made by Berryman et al. (1992). On South Island of New Zealand, the right-lateral component of plate motion is known to be accommodated primarily by the Alpine fault system, whereas the convergent component expresses itself by shortening on structures east of the Alpine fault. Figure 8–4 illustrates numerous geological separations along the fault. Total right-lateral offsets of 480 km have occurred in the past 25 million years. Plate reconstructions suggest that about 70 km of shortening perpendicular to the transform has occurred as well, in the past 7 million years. The spectacular Southern Alps, which bor-

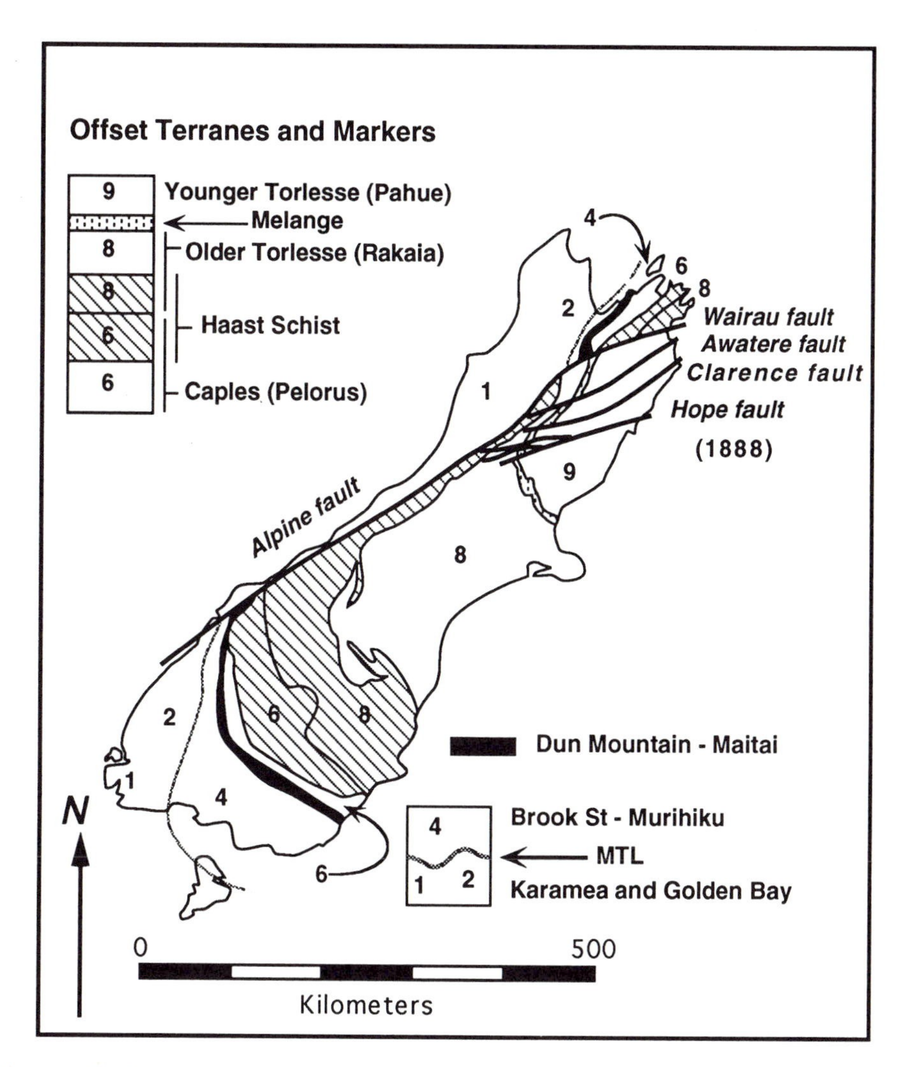

Figure 8–4. Several offset geologic terranes and markers indicate that the Alpine transform fault system has experienced 480 km of dextral offset in the past 25 million years. Except for earthquakes in 1848, 1855 and 1888, however, the transform system has been quiet at the M > 7 level during the period of historical record. Adapted from Berryman et al. (1992).

der the Alpine fault on the east and are the transform's namesake, owe their existence to the convergent component of motion. Fission-track, K-Ar, and radiocarbon ages indicate that rates of uplift adjacent to the Alpine fault are as high as 10 mm/yr. And studies of the Haast Schist indicate that the rocks of these mountains have risen locally as much as 22 km in the past 7 million years.

The Alpine fault, *sensu stricto*, has not experienced surficial fault rupture nor has it generated major earthquakes since at least 1840, when European settlement of New Zealand began. Geological evidence, however, suggests that it has produced large earthquakes, each associated with many meters of dextral slip, in the past few hundred years. A 30-km reach of the Hope fault, the southeasternmost of the band of subparallel dextral-slip faults that connect the Alpine fault with the Hikurangi trench, produced a large earthquake in 1888 and coseismic dextral offsets as great as 2.6 m (Cowan, 1991). The great earthquake of 1855 produced large strike-slip offsets along the Wairarapa fault, another of these faults, on North Island and also up to 2.7 m of uplift along the southern coast of North Island (Fig. 8-3). Darby and Beanland (1992) attempted to model the vertical and horizontal deformation associated with this earthquake. They found the best-fitting source to be an almost vertical Wairarapa fault that becomes shallowly northwest-dipping as it merges down-dip with the northwest-dipping subduction zone. This event probably represents, therefore, the simultaneous rupture of elements of both the strike-slip and subduction systems in the vicinity of their transformation. (For further discussion of this event, see Chapter 11 and Figure 11-46.)

Part of the northern edge of the Philippine Sea plate, near Japan, is an unusual but important example of a complicated trench-trench transform fault (number 34 on Fig. 8-1). Unlike the Wilson model of a trench-trench transform fault, this 300-km-long structure (1) intersects the contiguous Japan and Izu-Bonin trenches on the east, (2) has a complex and obscure connection with the northern end of the subduction zone of the Nankai Trough, on the west, (3) dips shallowly (about 30° to the northeast), and (4) experiences dextral-reverse rather than purely strike-slip motion (Figs. 11-8 and 11-13). This odd trench-trench transform is particularly important because of its proximity to major urban centers of central Honshu, Japan. The great (M_S 8.1) Kanto earthquake of 1923, which resulted in 130,000 deaths, was produced by several meters of right-reverse slip on a segment of this fault more than 100 km long, south of Tokyo (Kanamori, 1971; Matsu'ura et al., 1980; Wald and Somerville, 1994).

The Chaman transform zone of Pakistan and Afghanistan is a trench-trench boundary transform that accommodates sinistral motion between the northward-moving Indian plate and the Afghan block, a part of the Eurasian plate (Lawrence et al., 1992) (number 24 on Fig. 8-1). It transforms at its southern terminus into the 200-km-wide zone of active reverse faults of the ocean-continent collision belt of Pakistan and Iran's Makran coast (Figs. 8-5; 11-45). On its northern end, the Chaman transform zone disappears into a complex zone of faults at the western end of the continent-continent collision belt of the Himalaya. The active Chaman transform system is at least 1100 km long and comprises three major left-lateral components: the Chaman, Ghazaband, and Ornach-Nal faults. Total offset across the Chaman fault may be about 460 km. Several major, destructive earthquakes have occurred along the Chaman system (Quittmeyer and Jacob, 1979). An earthquake in 1505 involved at least 60 km of rupture along the northern portion of the fault, near Kabul (Heuckroth and Karim, 1971). In 1892, another large earthquake was generated by left-lateral slip along a segment of the fault at least 60 km long, near Chaman (Griesbach, 1893; Lawrence et al., 1992). The elongate zone of high intensities suggests a 150-km rupture along the Chaman fault, and at one locality, an offset of about 1 meter was reported. Highest intensities of the M_S 7.5 earthquake of 1935 are along the northern 150 km of the Ghazaband fault, near Quetta, a city that was destroyed during the earthquake. Fissuring along the fault was observed after this earthquake. Quittmeyer and Jacob (1979) documented other large earthquakes in 1931 and 1909 centered east of the structures mapped by Lawrence et al. (1992).

The number of large historical events along the Chaman fault zone is somewhat small for an 1100-km-long fault system that has a slip rate of about 5 cm/yr. If we assume that each of the events in 1892, 1931, and 1935 had ruptures 150 km long and 15 km wide (down-dip) with an average left-lateral slip of 5 m, then the moment rate for the past century has been about 10^{26} dyne-cm/yr. The long-term moment rate for an 1100-km-long fault, with a 15-km down-dip width and a slip rate of 5 cm/yr would be 2.5 times this value. The Chaman fault system may, therefore, have been in a period of relative dormancy during the past several centuries.

Ridge-Trench Boundary Transforms

Transform faults that transform into contractional zones at one end and zones of extension at the other end are akin to those boundary transforms that run between trenches: they occur only as the lateral mar-

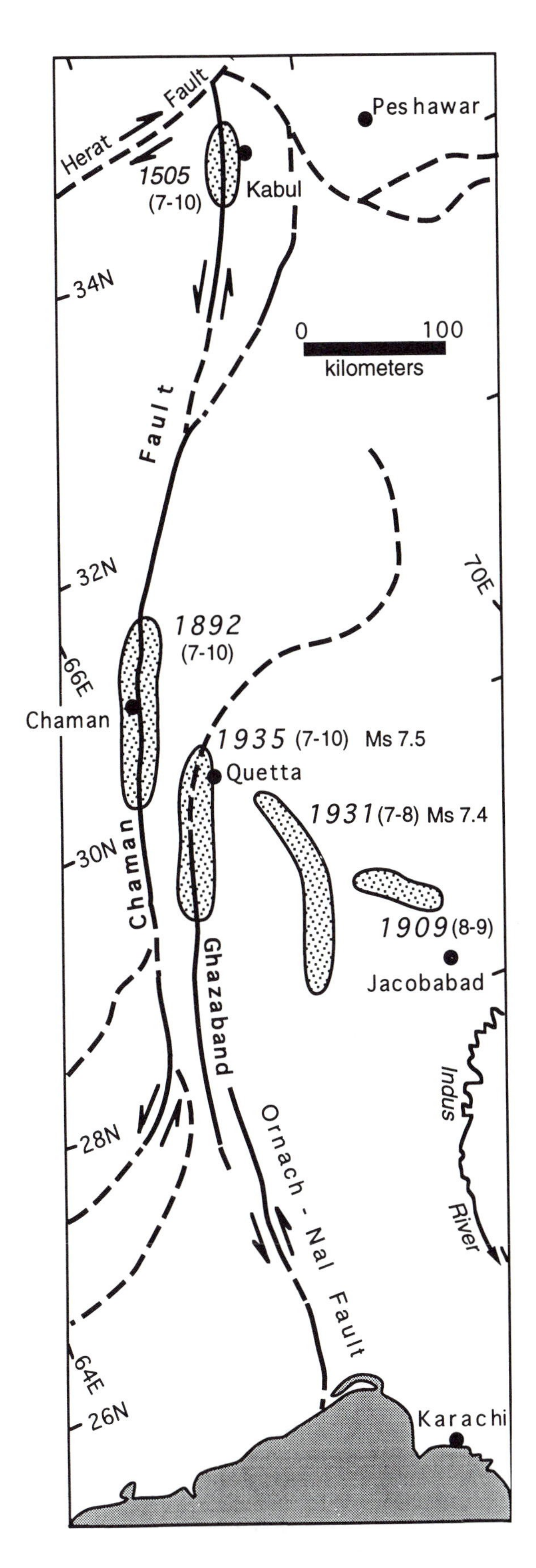

gins of entire plates. We refer to these structures as *ridge-trench boundary transforms*.

Two such transforms are the Dead Sea and Sagaing fault systems. The Dead Sea transform separates the African and Arabian plates and connects the spreading centers of the Red Sea with the Bitlis zone, the oblique collisional zone between the Anatolian and Arabian plates (number 20 on Fig. 8-1; see also Fig. 10-7). The pole of rotation for the African and Arabian plates is in North Africa, only 10° to 20° southwest of the transform fault (Fig. 1-4), so the total sinistral offset across the Dead Sea system, 107 km (Quennell, 1958; 1959) must be less than the total divergence of the Arabian and African plates farther east, across the Red Sea. Girdler (1991) shows that restoration of this amount of slip on the transform does restore offsets across the Red Sea, except for a few tens of kilometers of extension that appear to predate creation of the transform.

Motion between the African and Arabian plates along the Dead Sea transform system occurred in two spurts: 62 km of sinistral slip occurred in the late Oligocene and early Miocene, and 45 km has accumulated during the past few million years. Global kinematic considerations constrain the average rate of slip during the past few million years to about 5 mm/yr (DeMets et al., 1990).

Fragments of the seismic history of the Dead Sea transform are known from four millennia of recorded history and from archeological evidence (Ben Menahem, 1991). A major earthquake destroyed the ancient city of Jericho, on the northern edge of the Dead Sea, in the sixteenth century, B.C.E. This earthquake may have influenced the Old Testament writer who described the collapse of the walls of Jericho with the sounding of Joshua's trumpets about 1230 B.C.E. (Joshua 6:20; Ben Menahem, 1991). The discussion of

Figure 8–5. The Chaman transform of Pakistan and Afghanistan is an active trench-trench transform system composed of three major fault zones. Solid segments have documented geomorphic evidence of activity. Activity of dashed faults is poorly documented. Patterned ellipses indicate regions of strong shaking during large historical earthquakes. Numbers in parentheses indicate the range of Modified Mercalli intensities felt within the ellipses. Dates and magnitudes of these earthquakes are also shown. Surficial fault rupture accompanied the major earthquakes of A.D. 1505, 1892, and 1935. Based upon Lawrence et al. (1992) and Quittmeyer and Jacob (1979).

Biblical earthquakes in the biographical sketch about Zechariah explores this further.

Careful analysis of historical accounts (Ambraseys and Barazangi, 1989; and Ambraseys and Melville, 1988) suggest that the northern 350 km of this system, in Lebanon and Syria, ruptured in a series of eight major destructive earthquakes during the past millennium. These earthquakes occurred in three temporal clusters A.D. 1157–1202, 1404–1407, and 1759–1796. The M7.4 earthquake of 1759 was associated with at least 100 km of rupture, probably along the Yammouneh fault, which traverses the length of Lebanon's Beka'a Valley (Ambraseys and Barazangi, 1989). The earthquake of 1202 appears to have had a very similar source.

The Sagaing fault system is a transform fault system that accommodates right-lateral motion between the Indian plate and the Southeast Asian plate in Burma (number 27 in Fig. 8–1, and Fig. 8–6). The principal fault zone is geomorphically apparent along a distance of at least 1000 km (Le Dain et al., 1984). Its northern terminus has complex connections with the Himalayan collision zone and the Southeast Asian plate, and its southern end is at the spreading centers of the Andaman Sea. The rate of slip across the fault must be somewhat less than the rate of convergence of India and Eurasia, which is about 6 cm/yr.

Le Dain et al. (1984) summarize geologic and geometric evidence for about 460 km of dextral slip along the fault. This amount is far short of the 2000 km of translation of India past Southeast Asia inferred from plate-tectonic reconstructions.

Both crustal blocks adjacent to the Sagaing fault are also experiencing conspicuous active deformation, as evidenced by youthful structures throughout much of Burma (Fig. 8–6). Despite the near parallelism of the northerly strike of the Sagaing fault and the direction of relative motion between the Australian and Southeast Asian plates (N 11 ± 7°E; Tregoning et al., 1994), the difference apparently is enough to cause substantial fault-normal contraction in coastal Myanmar. West of the Sagaing fault system, folds of Pliocene and younger rocks in the coastal Indoburman ranges are evidence of contractional deformation perpendicular to the Sagaing transform. An eastward-dipping zone of seismicity suggests the existence of a slab of oceanic lithosphere beneath this active fold belt. Focal mechanisms of recent earthquakes in this region are not consistent, however, with eastward subduction of the Indian plate beneath Myanmar. Nonetheless, major uplift and subsidence of coastal regions coincident with large historical earthquakes in 1762 and 1858 (Richter, 1958) suggests ongoing subduction. Chhibber (1934) discusses spectacular, young marine platforms now residing meters above their modern equivalents, which he attributes to uplift during these large events. Given the near-parallelism of the Sagaing fault and the relative plate-motion vector, the rate of convergence across this zone west of the fault must be far less than the rate of strike-slip motion across the Sagaing transform.

East of the Sagaing fault, the Southeast-Asian plate contains numerous active left-lateral faults, which trend at a high angle to the transform. In this region, several historical earthquakes in the M7 to M8 range have occurred, but none has been shown to be related to rupture of specific faults. The contractional activity west of the Sagaing fault and left-lateral structures east of the fault may be related to westward expulsion of material from north and east of the Indian/Asian collision.

The Sagaing ridge-trench transform fault has also been very active historically, although data that would constrain the source parameters of large earthquakes are sparse (Le Dain et al., 1984, and Chhibber, 1934). Isoseismals of the great earthquake of 1912 suggest that it was produced by rupture of the Kyaukkyan fault, a probable right-lateral fault about 100 km east of and parallel to the central part of the Sagaing fault. The seismic history of the Sagaing fault itself is quite obscure, although at least six very large events have originated on or near it in the past two centuries.

The Queen Charlotte Islands and Fairweather faults of the northeastern Pacific Ocean constitute a third major example of a ridge-trench boundary transform (number 2 on Fig. 8–1). This system crops out principally on the floor of the ocean and under glacial ice, so little is known directly about its geological history apart from what can be inferred from plate reconstructions. A great earthquake in 1949 was produced by right-lateral slip along a 250- to 500-km-long portion of the Queen Charlotte Islands fault (Sykes, 1971), and a large earthquake was generated by right-lateral slip on the Fairweather fault in 1958.

Collisional Environments

Many strike-slip faults occur in tectonic environments other than in the transform settings exemplified above. These nontransform settings are idealized in Figure 8–2 E through H, and we discuss them on the pages that follow.

Horizontal Extrusion in Collisional Environments

Zones of collision between converging blocks of continental lithosphere commonly are dominated by large reverse faults, which accommodate thickening and el-

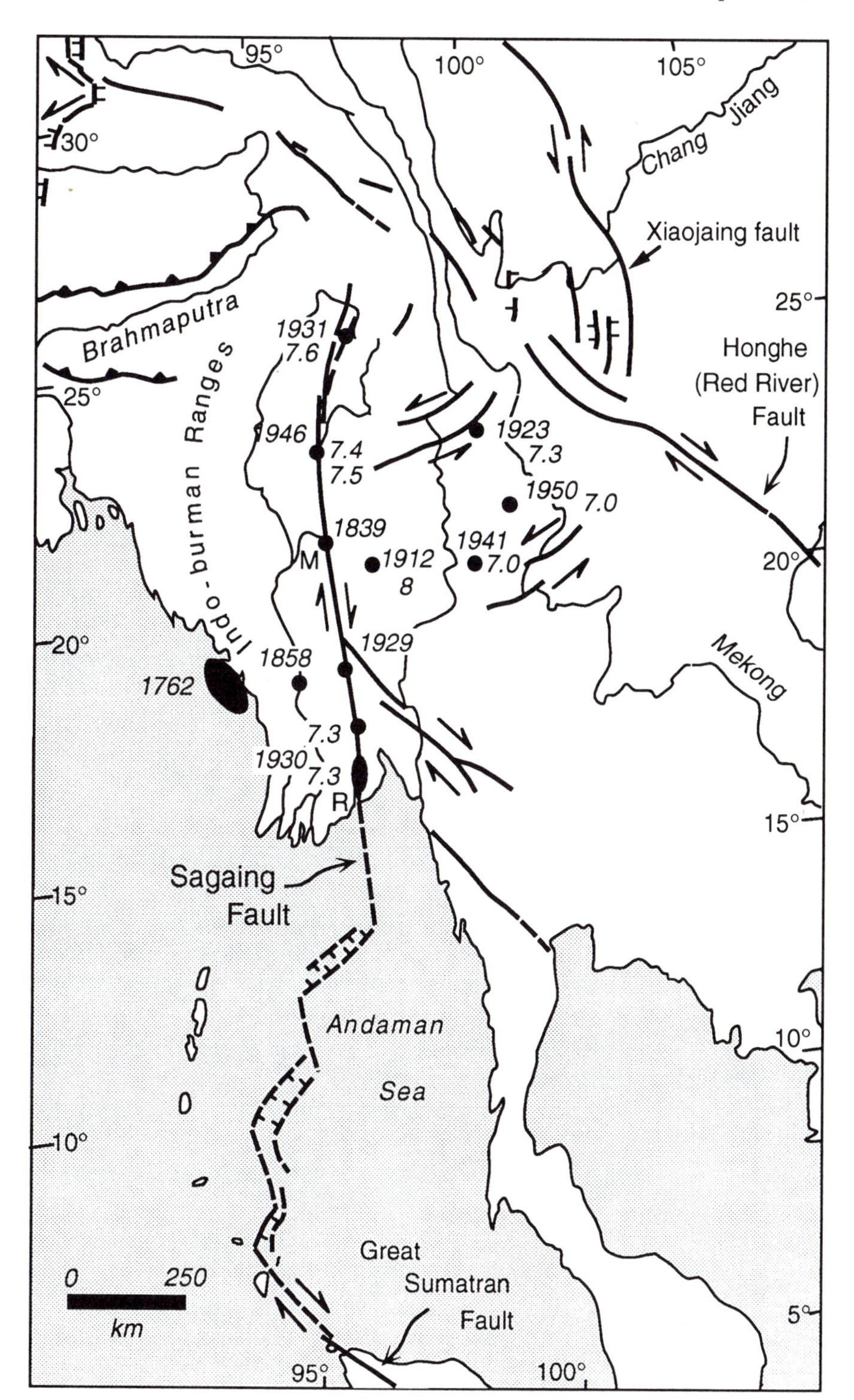

Figure 8–6. The Sagaing fault system is a trench-ridge transform fault and the principal element of the boundary between the Indian/Australian plate and the Southeast Asian plate in Burma. Dots, and ellipses indicate approximate source regions of some large earthquakes in Burma during the past two centuries. Ellipse along coast indicates region of large vertical deformation during earthquake of 1762. R = Rangoon; M = Mandalay. Adapted from Le Dain et al. (1984) and Chhibber (1934).

evation of crustal and lithospheric blocks. Large horizontal extrusions of the lithosphere away from the collision zones are also common within these convergent settings. This horizontal advection of material is accomplished along strike-slip faults. Woodcock (1986) refers to these structures as "indent-linked" strike-slip faults. Figure 8–2 F represents an idealized case.

Asia Minor

Asia Minor (Anatolia) is one region in which horizontal expulsion is taking place. Most of Turkey is extruding westward, away from the Arabian/Eurasian collision and toward the small remnant of oceanic crust underlying the eastern Mediterranean Sea (Fig. 10–26). The arcuate right-lateral North Anatolian fault system constitutes the northern boundary of this Anatolian block, "caught between a rock and a hard place." Despite its great length, this 1500-km-long fault has accommodated only about 40 km of westward expulsion of Anatolia, all in the past 15 million years (Barka, 1992).

The North Anatolian fault has been the source of numerous, destructive historical earthquakes. A remarkable series of seven M_S 7+ earthquakes occurred between 1939 and 1967, as rupture of the fault progressed primarily from east to west (Fig. 8–7). Investigations of historical documents reveal a similar series of events between A.D. 967 and 1035, and a large earthquake involving rupture of much of this part of the fault in 1668 (Ambraseys, 1970; Ambraseys and Finkel, 1988).

The southern margin of the extruding Anatolian block consists of a more complex and more poorly understood group of sinistral strike-slip and reverse faults. Among these are the 580-km-long East Anatolian fault, which has experienced about 25 km of sinistral slip in the past few million years (Saroglu et al., 1992) and the poorly understood faults of the Pliny and Strabo troughs (Fig. 9–8).

The Northeast Anatolian and Ahar blocks, east of the Anatolian block, appear to be crude mirror-images of the Anatolian block, in that they are extruding eastward, between the left-lateral Northeast Anatolian fault of easternmost Turkey and the right-lateral Main Recent fault of Iran (Fig. 10–26).

Asia

The most prominent region of collisional strike-slip faulting in the earth's crust is in central and eastern Asia. Here, numerous tectonic blocks, some hundreds of kilometers wide, are moving eastward and southeastward in order to accommodate the ongoing northward penetration of the Indian subcontinent into Asia (Molnar and Tapponnier, 1975; 1978; Avouac and Tapponnier, 1993). Figures 8–1 and 8–8 illustrate the arrangement of these indent-linked strike-slip faults and related structures. The largest of these faults are left-lateral. These include the Altyn Tagh and Haiyuan faults, a 2000-km-long fault system that arches west to east across most of western China, the Kunlun, Xianshuihe (Freshwater River) and the Xiaojiang (Small River) faults. These left-lateral faults form the northern and eastern boundaries of blocks that are being translated east and south, relative to the rest of Eurasia. Discontinuous right-lateral faults of the en échelon Karakorum-Jiali fault system and the Honghe (Red River) fault zone define the southern and western boundaries of these blocks (Armijo et al., 1986, 1989). Significant tectonic subprovinces distinguished by active reverse faults and by normal faults (Fig. 9–9) are related to the terminations of these left- and right-lateral faults.

The eastward extrusion of China has been extraordinarily seismic: In just the past three centuries, nearly

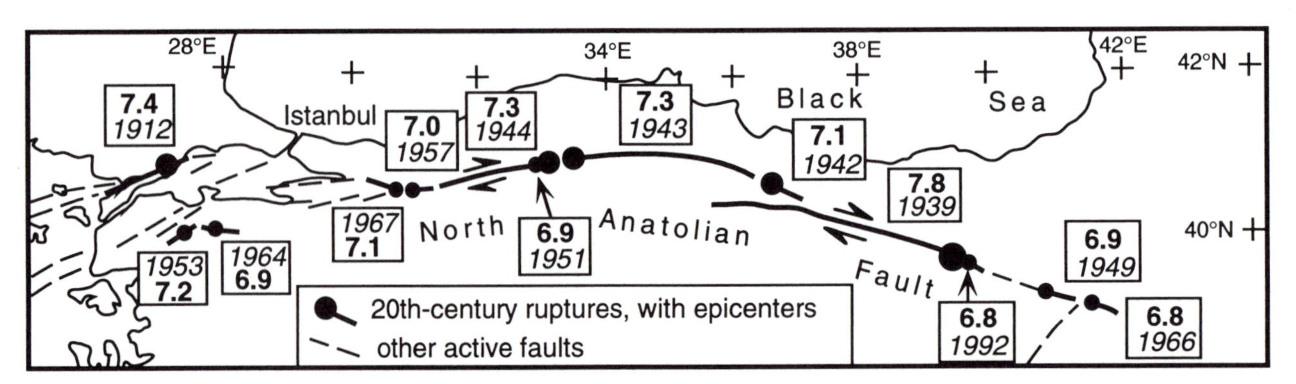

Figure 8–7. The North Anatolian fault is a right-lateral, indent-linked strike-slip fault, which forms the northern boundary of the extruding Anatolian plate. During the twentieth century, ruptures along the North Anatolian fault have been dominated by the east-to-west progression of M_S 7+ earthquakes between 1939 and 1967. Adapted from Barka (1992) and Ambraseys (1988).

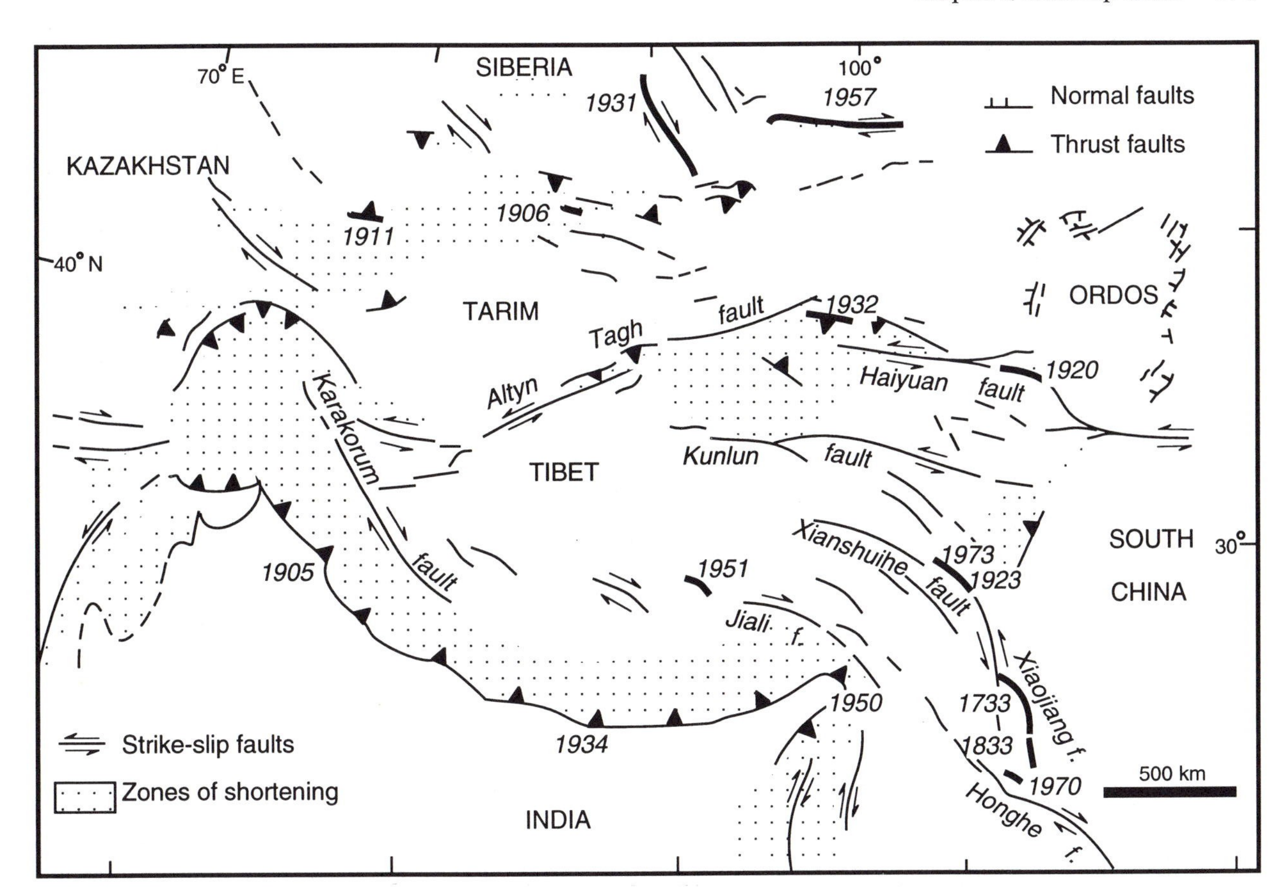

Figure 8–8. The collision of continental Asia and India, which began about 35 million years ago, has resulted in the eastward extrusion of Southeast Asia and China along numerous indent-linked strike-slip faults. The northern and eastern boundaries of the escaping continental fragments are primarily sinistral strike-slip faults, whereas the southern and western boundaries are defined by right-lateral faults. Note the sources and dates of numerous major historical earthquakes.

a dozen earthquakes with magnitudes in the high sevens and low eights have occurred on these strike-slip faults and related secondary reverse and normal faults (Fig. 8–8).

Figure 8–9 illustrates Avouac and Tapponnier's (1993) best-fitting model of the instantaneous motions of the major Eurasian blocks bounded by these strike-slip faults. Slip rates of the faults are derived from measured offset features of estimated Holocene and late Pleistocene age. As much as 50% of the 5 cm/yr of north-south convergence of India and Eurasia appears to be accommodated by this horizontal extrusion of fragments of Eurasia.

Quasi-Pure Shear in Collisional Environments

Strike-slip faults play a prominent role in some island-arc settings. Central Honshu, Japan, provides an impressive example. There a plexus of small, conjugate, left- and right-lateral faults are accommodating deformation of the hangingwall block, well above the subducting Pacific and Philippine Sea plates (Figs. 8–2 G and 8–10). The kinematics of the lateral-slip faults here contrasts markedly with that of any region described previously. The conjugate arrangement of these faults implies that the lithosphere of this portion of the arc is, as a whole, contracting from northwest to southeast and extending from southwest to northeast.

This highly faulted piece of lithosphere consists of continental crust overlain by deposits and edifices of the volcanic arc associated with the subduction of Pacific plate beneath the Eurasian plate (Kinugasa et al., 1992). In Miocene time, this continental crust was rafting eastward, away from China, during formation of the back-arc basin now occupied by the Sea of Japan (Otsuki, 1990). This motion has now reversed, and Japan is now converging toward Asia.

The largest of Japan's modern, non-subduction zone earthquakes, the Nobi earthquake of 1891 (M_W 7.5; Wells and Coppersmith, 1994), was caused by rup-

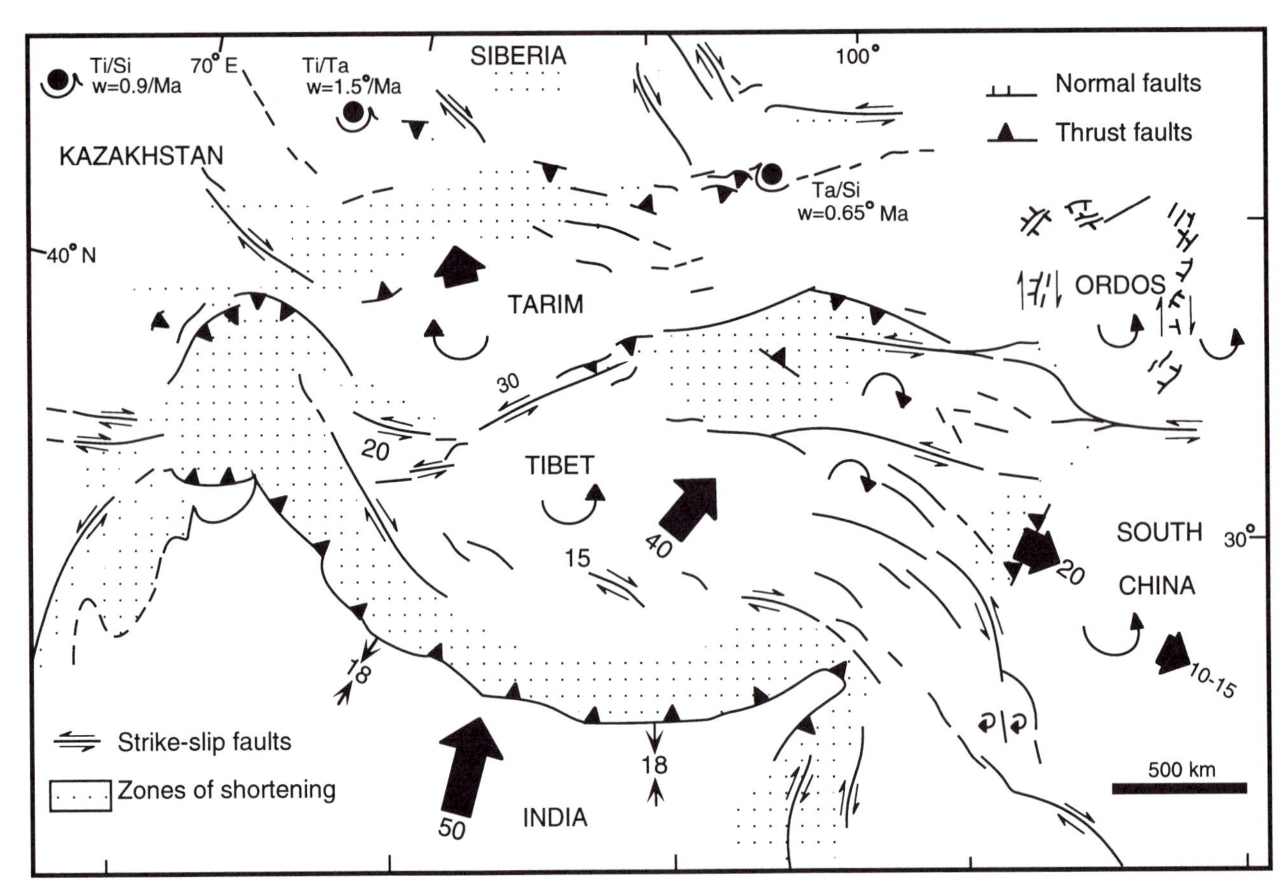

Figure 8–9. A kinematic model of active deformation in central Asia, derived from inversion of various tectonic data. Euler poles, with rates and senses of rotation, indicate motions of Tibet (Ti) and Tarim (Ta) relative to Siberia (Si) and to each other. Broad arrows indicate major block motions in Siberian reference frame. Redrafted from Avouac and Tapponnier (1993).

ture of several faults in one of the larger left-lateral fault zones in this region (Matsuda, 1974). Within just the past century, numerous other faults here have also ruptured. These include earthquakes of M_W 7.1 in 1927 and M_W 7.3 in 1930 (Wells and Coppersmith, 1994), as well as the recent M_W 6.9 Kobe earthquake of 1995, which produced about $200 billion dollars of property damage.

The Izu Peninsula, southwest of Tokyo, is a particularly intensely and recently active portion of this region of conjugate active faults. As is discussed more thoroughly in Chapter 11, this region is the site of collision between the Izu-Bonin volcanic arc, on the Philippine Sea plate, and the Japanese mainland. The largest structure in the region, the right-lateral Kita-Izu fault, failed seismically in 1930 and, perhaps, in A.D. 841 (Fig. 8-36b). Smaller earthquakes have been generated by left-lateral conjugates to the Kita-Izu fault in recent years (1974, 1976, and 1978). Nakamura (1969) and Somerville (1978) pointed out that locally the trends of volcanic fissures and parasitic cones bisect the trends of these left- and right-lateral faults. This is consistent with an orientation of the maximum principal stress (σ_1 between the two sets of faults and a horizontal orientation of the minimum principal stress (σ_3) perpendicular to this.

The mesh of strike-slip faults on central Honshu is astonishingly dense and has been very active in the Quaternary. Nevertheless, trench-perpendicular shortening across this zone amounts to only 3 mm/yr, a mere 5% of the total convergence rate of the Eurasian and Philippine Sea plates (Wesnousky et al., 1984).

Trench-Parallel Strike-Slip Faults

Another important role of strike-slip fault systems is accommodation of the boundary-parallel component of slip along obliquely convergent plate margins (Fitch, 1972). These faults traverse the hangingwall block above subducting slabs and strike roughly parallel to nearby trenches (Figure 8-2e). Woodcock (1986) has named this class of faults "trench-linked" strike-slip faults. We call them *trench-parallel strike-slip faults.*

One prominent example is the active Median Tectonic Line of Japan (number 33 on Fig. 8-1, and Figs. 8-10 and 11-11). This 300-km-long active fault accommodates much of the right-lateral component of motion between the Philippine Sea and Eurasian plates. The fault occupies the central portion of a pronounced, ancient suture, the Median Tectonic Line, the most significant geological break in southwestern Japan. Tsutsumi et al. (1991) and Tsutsumi and Okada (1996) conclude that it must generate infrequent, very large earthquakes. The long-term rate of slip on the fault is 5–10 mm/yr, and the latest large earthquake generated by one segment of the fault occurred in A.D. 1596 and by other segments about 1300 and about 3000 years ago. They estimate that earthquakes separated by such long recurrence intervals and produced by a fault with such a high slip rate have dextral offsets of many meters and moment magnitudes (M_W) as high as 7.5.

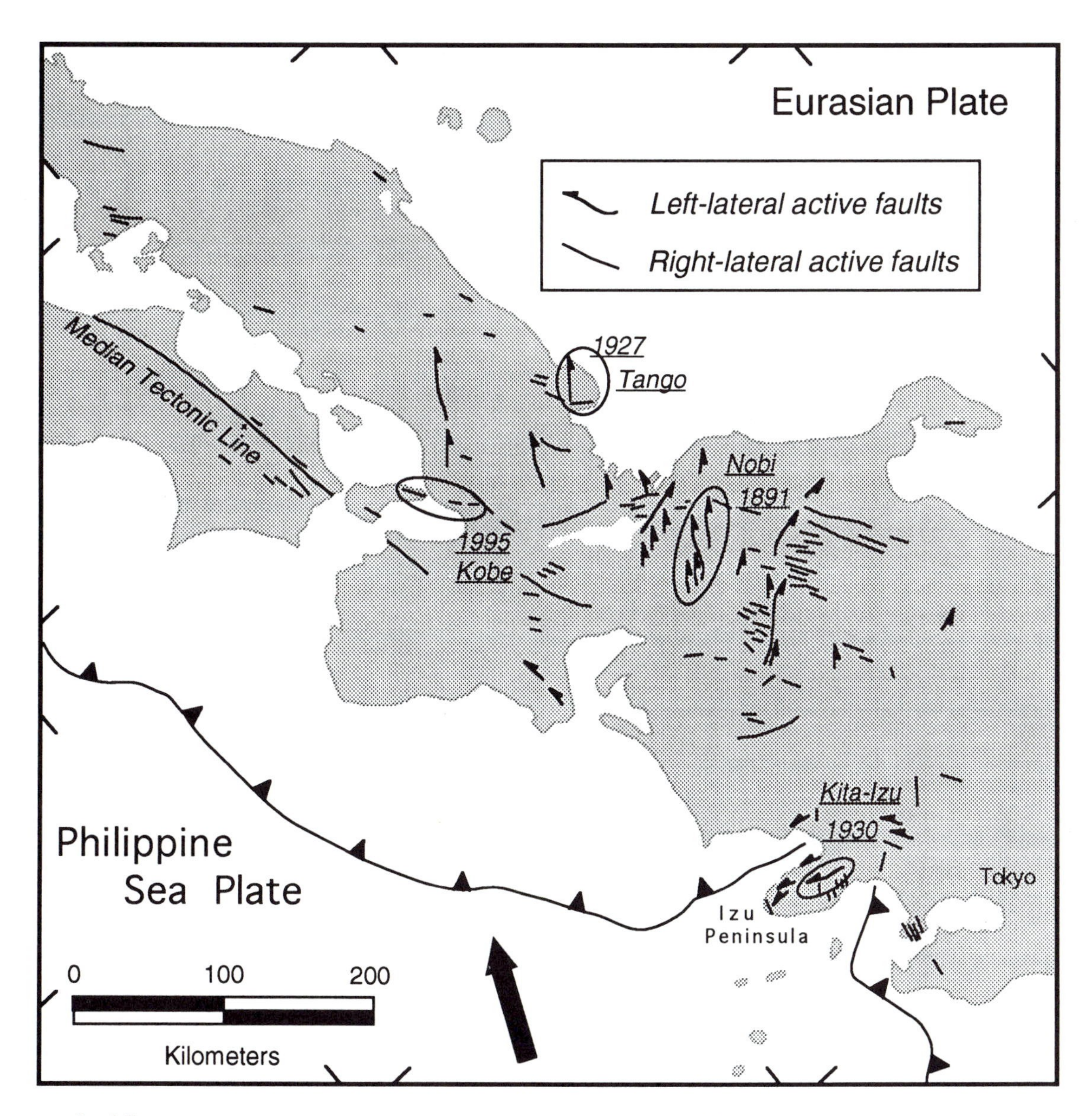

Figure 8–10. Motion on a myriad of dextral- and sinistral-slip faults in central Honshu, Japan, mimics the geometry of faults associated with pure shear. The Median Tectonic Line is a large, active right-lateral trench-parallel fault that accommodates an oblique component of plate convergence. Several regions struck by large twentieth-century earthquakes associated with surficial rupture of these faults are shown as ellipses. Older historic events with surface rupture are also numerous, but are not shown here. Adapted from Faults (1980), supplemented with data from other sources.

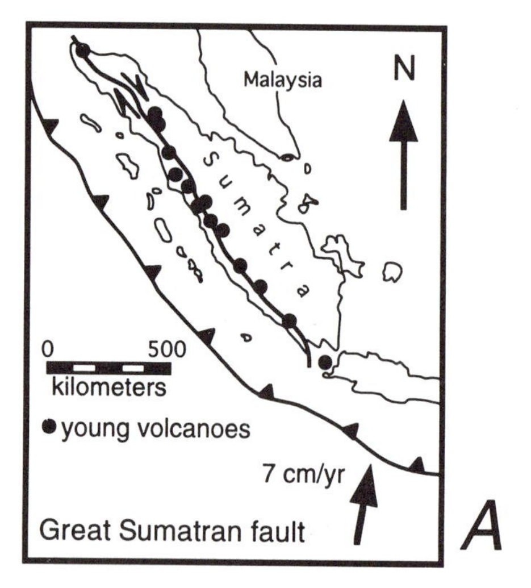

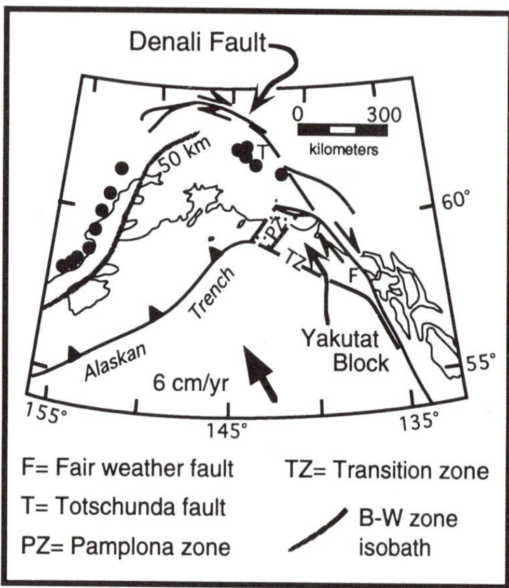

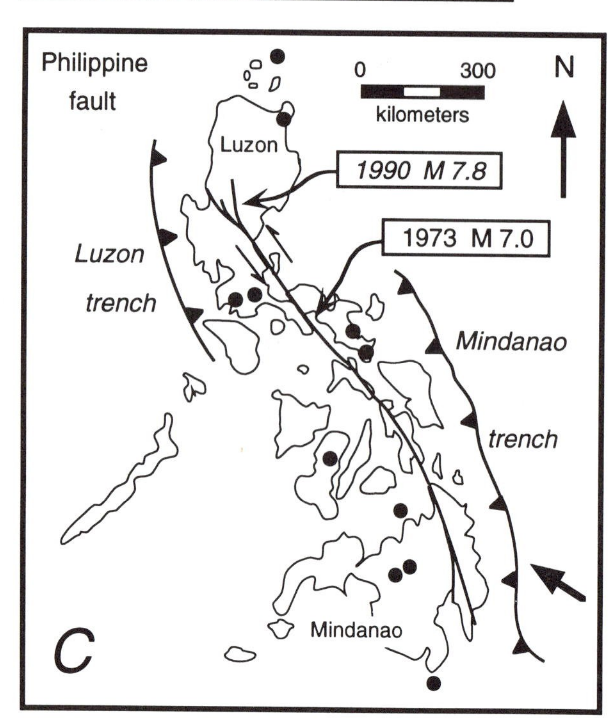

Other examples of trench-parallel strike-slip faults in environments of oblique convergence are the Denali fault (Alaska), the Philippine fault system, and the Great Sumatran fault (numbers 1, 35, and 30 on Fig. 8-1). Little is known about the Philippine fault, other than its first-order geometry and plate-tectonic context, roughly parallel to and sandwiched between the opposing subduction zones of the Mindanao and Luzon trenches, which lie off the east and west coasts of the Philippine archipelago (Fig. 8-11c). The sense of slip of the Philippine fault system was shown to be left-lateral by Allen (1962), on the basis of geomorphologic analysis. This sense of motion is consistent with the left-lateral obliquity of relative plate motion between the Philippine Sea plate and the Southeast Asian plate. The rate of slip of the Philippine fault is estimated to be between 1 and 5 mm/yr (Hirano et al., 1986). A left-lateral sense of slip was confirmed by observations of offsets following major earthquakes, in 1973 and 1990 (Allen, 1975; Nakata et al., 1990). The M_S 7.8 earthquake of 1990 was produced by several meters of left-lateral rupture on a 125-km-long portion of the Digdig fault, an important strand of the Philippine fault system, on the island of Luzon.

The Denali fault is the largest of several right-lateral strike-slip faults in southern Alaska, north of the Alaskan subduction zone (Figs. 8-11b and 11-27). "Denali" is, in fact, the Tanaina word for Mt. McKinley and means "the great one" (G. Plafker, pers. comm., 1995). This 500-km-long, broadly arcuate active fault traverses the lofty, glaciated Alaska range. Total offset across the fault may be as great as 400 km, but this is disputed (Redfield and Fitzgerald,

Figure 8–11. Where plate convergence is oblique, partitioning of slip into dip-slip and strike-slip components is common. (a) Dextral slip along the trench-parallel Great Sumatran fault accommodates most, if not all, of the dextral component of motion between the Indian/Australian plates and the Southeast Asian plate. Slip rates increase northwestward from zero to at least 3 cm/yr, because of the northward-increasing obliquity of convergence. None of the several historical M7+ events are shown, because of the small scale of the map. (b) The Denali fault accommodates about 1 cm/yr of dextral slip, north of the oblique-slip Alaskan trench. It has produced no large events in the historical period. (c) The left-lateral Philippine fault system traverses the length of the Philippine archipelago, in the hangingwall block of nearby subduction zones. Several meters of slip occurred along portions of the fault zone in 1973 and 1990.

1993). The average rate of slip along the fault during the past 12,000 years or so is about 10 mm/yr (Sieh, 1981), but it has produced no large earthquakes during the short period of historical record. Geomorphic evidence suggests offsets of 7 to 15 m may have occurred in association with the most recent earthquakes of the Denali fault. The Denali and its active, right-lateral neighbor to the east, the Totschunda fault (Richter and Matson, 1971), may accommodate a dextral component of motion between the Pacific and North American plates. Alternatively, the Denali and Totschunda faults may be "indent-linked" structures akin to the strike-slip faults of Anatolia and China. Plafker (1984) proposes that a small piece of continental lithosphere, the Yakutat block, is currently colliding with Alaska at the eastern end of the Alaskan subduction zone. Perhaps this collision is driving southern Alaska westward.

The right-lateral Great Sumatran fault, which bisects the island of Sumatra along its entire 1600-km length and continues northward into the Andaman Sea, is another large strike-slip fault parallel to an obliquely convergent plate boundary (Fig. 8–11a). This structure traverses the active volcanic arc in the hangingwall block above a subducting slab. It is curious that, although the fault traverses the volcanic arc, only a few of the volcanic conduits have actually arisen within the fault plane, itself.

Based upon interviews with eyewitnesses, Untung et al. (1985) documented right-lateral slip of up to 2 to 3 meters along at least 60 km of the fault south of the equator during the earthquake of 1943. Muller's (1895) study of geodetic monuments suggested dextral slip of more than 2 meters along a section of the fault just north of the equator in 1892.

Earthquakes along the Great Sumatran fault in the past century have been numerous, but none appear to have been larger than the mid-sevens. This upper limit may be related to high heat flow along the volcanic arc, which could place the brittle-ductile transition higher in the crust than along strike-slip faults that produce larger earthquakes. Or, the high degree of geometrical segmentation of the fault may restrict rupture lengths to a hundred kilometers or less. Alternatively, the historical record, which is only about a century long, may simply be too short to have recorded earthquakes of larger magnitude along the fault.

The kinematic environment of the Sumatran subduction zone requires northward-increasing dextral slip on the Great Sumatran fault (McCaffrey, 1991). The magnitude of the vector of relative plate motion between the Southeast Asian and Australian plates (about 7 cm/yr) and its orientation (about N11°E) does not vary significantly along the entire length of the Sunda trench, because the pole of rotation is far from the plate boundary. The broadly arcuate shape of the trench requires, therefore, that oblique convergence must be occurring along much of the plate boundary. Curiously, however, focal mechanisms of small to moderate earthquakes within the subduction zone are nearly perpendicular to the axis of the trench along its entire arcuate length. The reason for this is unclear, but its consequence is that strike-slip structures are necessary to accommodate strike-slip components of plate motion along most of the plate margin. McCaffrey (1991) suggests that only in central Java, where the trench strikes nearly east-west and strike-slip structures appear to be absent, is there no strike-slip component. This assumption requires that strike-slip increase from zero at about the Sunda Straits, between Java and Sumatra, to about 6 cm/yr in the Andaman Sea, 1900 km to the northwest.

Northwestward-increasing slip does appear to occur along the Great Sumatran fault (Sieh et al., 1991). About 2° north of the equator, major streams incised into a large 73,000-year-old ashflow sheet are dextrally offset 2 km. Thus, the slip rate there is about 28 mm/yr. Three hundred kilometers farther south, a smaller ashflow sheet, about 65,000 years old, is incised by rivers offset only 700 meters. Thus, the slip rate of the fault here is only about 11 mm/yr. The rate of northward increase in slip rate between these two sites is consistent with McCaffrey's model, but the rates predicted by the model are about 10 mm/yr higher than those observed. This discrepancy may be due to the existence of a large submarine strike-slip fault in the forearc basin between the trench and the Great Sumatran fault (Diament et al., 1992).

Dewey and Lamb (1992) have recently uncovered another interesting case of strike-slip faults in an oblique collisional environment. Along the western margin of South America, they find numerous small strike-slip faults in the lithosphere of the block overriding the subduction zone (Fig. 10–8). The sense of slip on these faults varies according to the sense of obliquity of plate convergence, which varies along the coast because of the highly variable strike of the plate margin.

Although many obliquely-convergent boundaries do partition strike-slip and dip-slip components between their subduction zones and large strike-slip faults, not all oblique convergent boundaries behave this way. Examples of nonpartitioned oblique subduction include the Sagami Trough of Japan (Figs. 11–8, 11–9, and 11–13), the Pliny Trough of southeastern Greece (Figs. 8–1 and 9–8), and the westernmost Aleutian Trench (Figs. 8–1 and 11–25).

Intraplate Strike-slip Faults

Fragmentation of Oceanic Crust

Major strike-slip faults also exist in other environments within plates. One important, and recently seismic, example is the set of strike-slip faults within the interior of the southern Juan de Fuca plate, also called the Gorda plate, off the coast of Oregon and northern California (Fig. 11-32). This small fragment of oceanic lithosphere is suffering extreme internal deformation that is accommodated in part by numerous dextral and sinistral strike-slip faults (Fig. 8-12). Wilson (1986; 1989) has shown that the pattern of magnetic anomalies within this "Gorda Deformation Zone" is best explained by diffuse zones of dextral shear immediately east of the Gorda Ridge and sinistral-slip faults parallel to the magnetic lineations farther southeast. This hypothesis is consistent with the occurrence, in 1980, of a M7.2 earthquake off the coast of northern California. The earthquake was generated by left-lateral rupture along a 120 km-long, northeast striking left-lateral fault within the Gorda Deformation Zone.

Fragmentation of Continental Crust

In continental crust, strike-slip faults commonly occur as connecting elements between active reverse or normal faults (Fig. 8-2h). These faults commonly have appreciable components of dip slip where their length is not greater than the thickness of the crust. The extensional Basin and Range Province of the western United States provides many examples. The direction of extension across this large region is about N64°W, but typical trends of the normal faults that dominate the region are closer to north-south (see Fig. 9-11). As a result, northwest-striking right-lateral faults connect many of the normal faults. The Hunter Mountain fault, which runs between the normal faults of Panamint and Saline Valleys, is one example (Fig. 8-13). Farther east, active right-lateral faults predominate over a dip-slip connector in Death Valley. Notice that the angles between the normal faults and the strike-slip faults are closer to 45° than they are to the 90° typical of plate-boundary transform faults. This is consistent with an approximate north-south orientation of the maximum principal stress, σ_1.

Another example of the interaction of continental strike-slip and normal faults comes from southern Tibet. There, Armijo et al. (1989) have documented fresh right-lateral offsets of several meters on the 100-km-long Beng Co fault, which they attribute to slip during a M8+ earthquake in 1951 (Fig. 8-14). The southeast-striking Beng Co fault terminates into north-south-trending graben at both ends. The normal fault that demarcates the western flank of the Gulu graben on the southeastern end of the Beng Co fault displays fresh scarps a meter or so high, which may have formed during a large aftershock in 1952 (Armijo et al., 1986, 1989). The Gulu graben is also the eastern terminus for the active northeast-striking, left-lateral Damxung fault.

Minor strike-slip faults also occur as elements of reverse fault zones. Two short strike-slip faults, for ex-

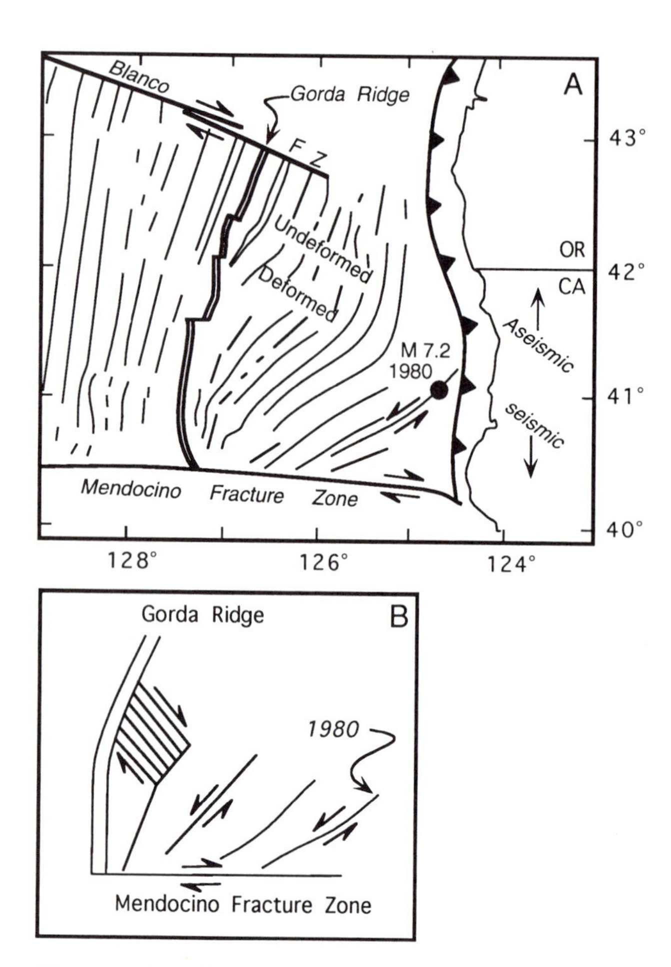

Figure 8–12. Deformation of the southern Juan de Fuca plate, between the Gorda Ridge and Cascadian subduction zone, involves many large, intraplate strike-slip faults. (a) Deformed and undeformed regions of the Juan de Fuca plate are deduced from seismicity and from patterns of magnetic anomalies. The large earthquake of 1980 was produced by a left-lateral strike-slip fault within the deformed part of the plate. (b) Cartoon relating known and hypothetical strike-slip faults within the deforming plate to adjacent plate-boundary elements. Redrawn from Wilson (1989). See Figure 11–32 for broader regional context.

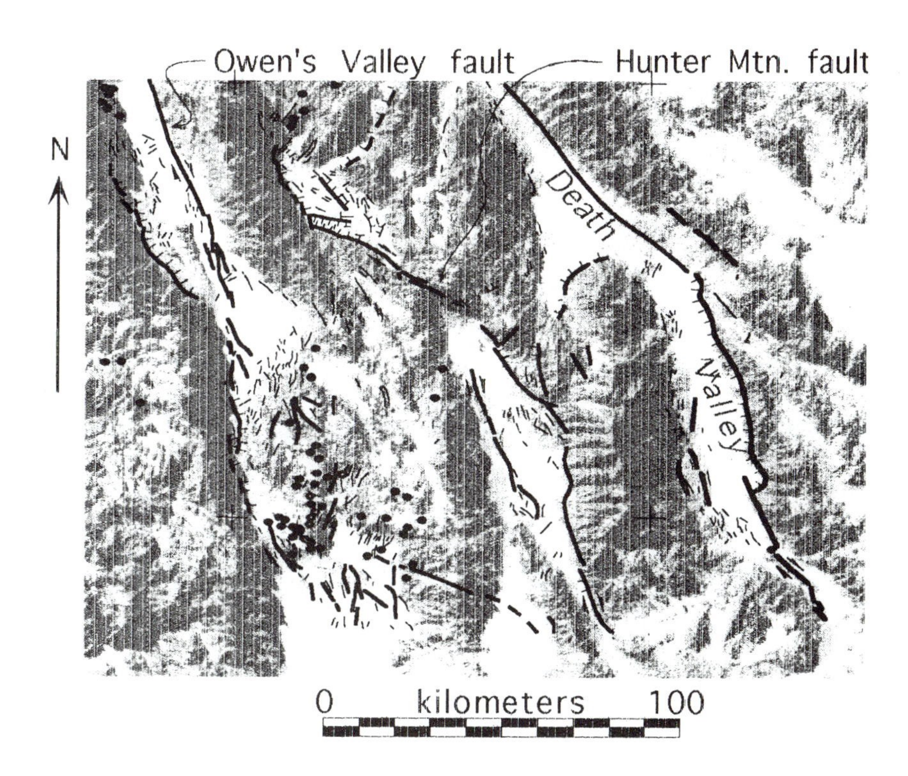

Figure 8–13. Map of active (Quaternary) faults superimposed upon a DEM of the Owens Valley-Death Valley region. In this region, strike-slip faults act both as connectors between normal-fault segments and as elements in partitioned transtensional systems. Unlike transform faults along plate boundaries, these strike-slip connector faults commonly form oblique angles with adjoining dip-slip faults. The Hunter Mountain and Death Valley-Furnace Creek faults are examples of dextral-slip faults that connect normal faults. The right-lateral Owen's Valley fault is an example of slip-partitioning in a transtensional setting. Adapted from Jennings (1992).

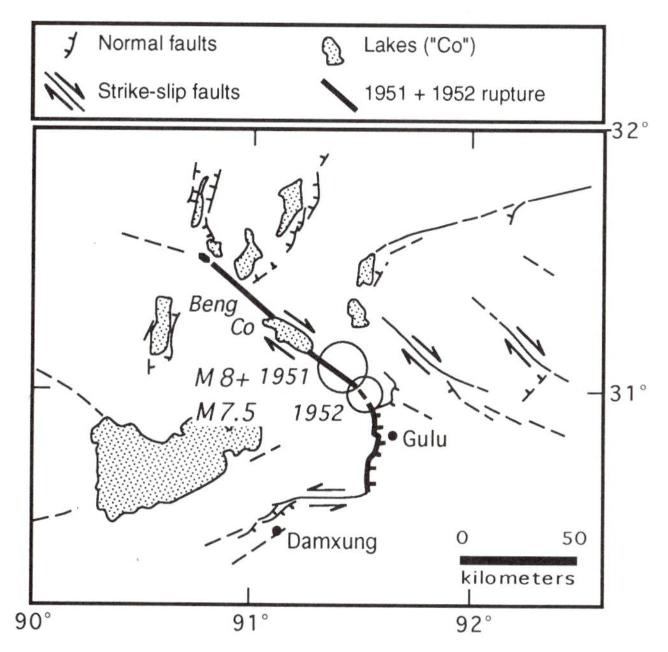

Figure 8–14. The left- and right-lateral Damxung and Beng Co faults in southern Tibet are another example of intracontinental strike-slip faults. They connect to the normal faults of the Gulu graben. Mainshock epicenter of the M8+ Beng Co earthquake of 1951 is shown as large circle. Epicenter of large aftershock is smaller circle. Bold line along the right-lateral fault represents surficial rupture during the 1951 earthquake; normal fault near Gulu may have ruptured in the aftershock of 1952. Adapted from Armijo et al. (1989).

ample, were part of the 1952 M7.7 rupture of the oblique-slip White Wolf fault (Fig. 8–15). The larger of these cut both the hangingwall and footwall blocks of the thrust fault. The orientations of both the reverse-oblique faults and the strike-slip faults are consistent with northwest-southeast shortening and with increasing underthrusting toward the west.

Slip-Partitioning in Oblique Extensional Environments

Partitioning of slip between dip-slip and strike-slip structures is not restricted to regions of oblique convergence. It is also common in regions of crustal extension, although these systems do not achieve the dimensions of their analogs, the trench-parallel strike-slip faults and subduction zones. Good examples exist in both California and China. California's M_W 7.5–7.7 earthquake of 1872 was produced by the Owen's Valley fault, a predominantly right-lateral structure that runs for about 60 km along the axis of the westernmost graben of the Basin and Range extensional province (Fig. 8–13) (Beanland and Clark, 1994; Hobbs, 1910). Only a few kilometers west of the trace of the 1872 rupture is the trace of an active normal fault, at the base of the 3000-m-high escarpment of the Sierra Nevada.

Two other examples of slip partitioning in divergent-oblique settings are in China. Along a portion of the Red River fault of southeastern China, two parallel, active strands of the fault exist only a couple of kilometers apart (Fig. 8–16). The northeastern strand displays right-laterally offset streams, whereas the southwestern strand exhibits dip-slip dislocation of ridge-crests. Figure 8–17 illustrates an example of divergent-oblique slip-partitioning along the Damxung fault zone, in southern Tibet. The block diagram shows that the normal- and strike-slip faults converge into an oblique-slip fault at shallow depth.

COSEISMIC STRUCTURES

The discussion of the preceding section focused on the roles that major strike-slip faults play in each of several tectonic environments around the globe. Many of these structures are hundreds to thousands of kilometers long, significantly greater than the thickness of the lithosphere in which they are embedded. We turn now to an investigation of strike-slip faults at smaller dimensions, that is, at local and regional scales. For it is at these scales that geologists, geodesists, and geophysicists commonly find themselves at work, attempting to understand earthquakes and the structures that produce them.

Structure of the Fault Trace

En Échelon Structure of the Fault Rupture

The appearance of a fresh strike-slip rupture at the ground surface is invariably more complex than an

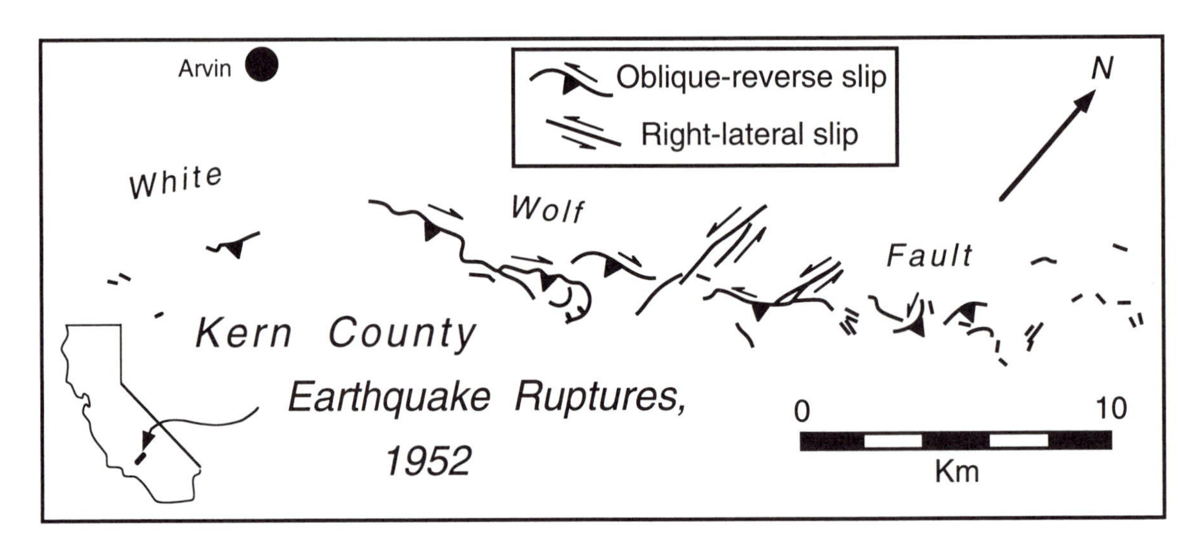

Figure 8–15. Strike-slip faults also exist as auxiliary structures to reverse faults in continental settings. Although faulting along the White Wolf fault in 1952 was predominantly convergent-oblique (right-reverse) in character, minor strike-slip faults were associated with the thrust fault. Adapted from Buwalda and St. Amand (1955).

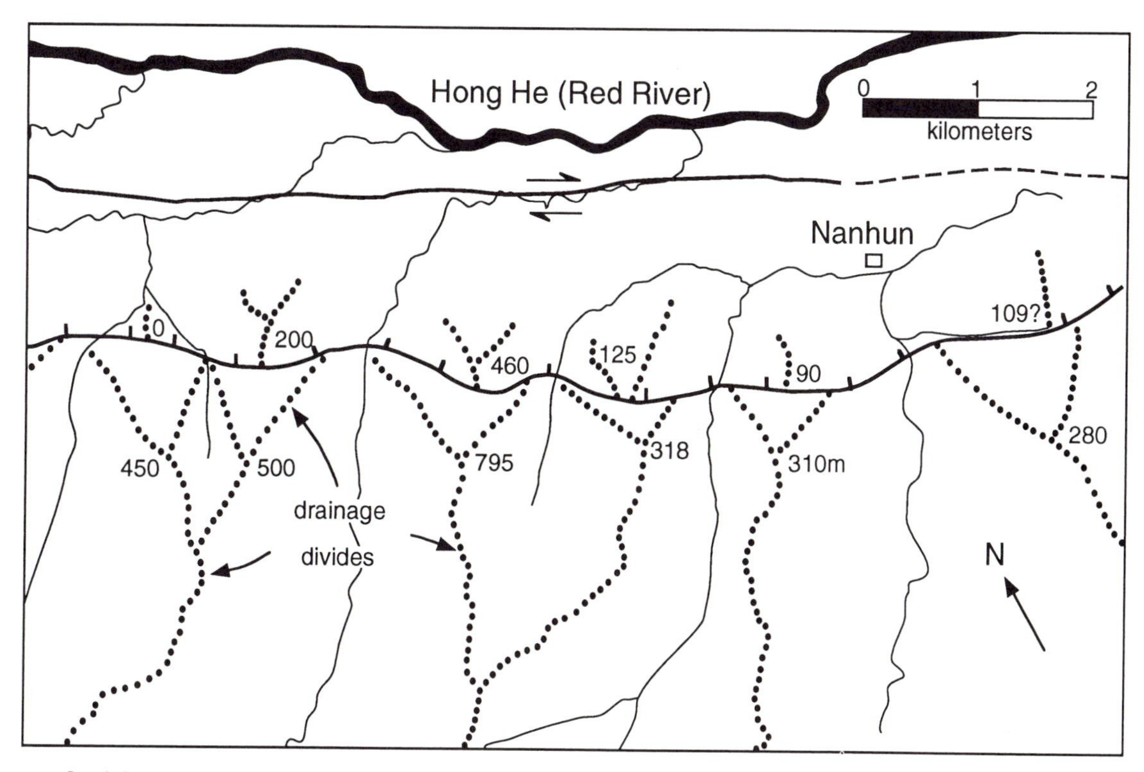

Figure 8–16. Continental fault zones experiencing divergent-oblique slip may also exhibit slip-partitioning into dip-slip and strike-slip faults. Along this portion of the active Honghe (Red River) fault zone in southern China slip is partitioned between the mid-valley right-lateral fault and the range-front normal fault, which are separated by only about a kilometer. The stream valleys (solid lines) and the drainage divides (dotted) show that partitioning is total.

idealized rectiplanar, vertical fault. In indurated sedimentary rock or in crystalline rocks, the surficial traces of a strike-slip rupture almost invariably follow preexisting faults. The geometry of the rupture trace is controlled principally by the geometry of the preexisting faults. This behavior must reflect the great contrast in shear strength between unfractured and fractured rock (cf. Fig. 2–17 and discussion in Chapter 2).

In loose, granular materials, the surficial trace of a fault invariably appears as a zone of fractures with a pronounced en échelon character. Figure 8–18 exhibits three typical examples. These patterns are analogous to those produced in laboratory experiments, where a clay slab is made to fail in shear by horizontal movement of two subjacent boards (Tschalenko, 1970). In both natural and laboratory settings, right-lateral faults generate fractures that are left-stepping—that is, in walking along the fault zone, one steps left from the end of one fracture to get to the next. Along left-lateral fault ruptures, the fractures are predominantly right-stepping. In each case, the stepping is of the sense that allows a small component of tensile opening of the fractures. Three-dimensional excavations of faults show that this stepping geometry and its concomitant complexities may coalesce into a contiguous fault only a few meters down-dip (for example, see Sieh (1984)).

Inspired by the en échelon patterns associated with the 80-km-long rupture that produced the Dasht-e Bayaz earthquake in Iran in 1968, Tschalenko (1970) conducted laboratory experiments to clarify both the mechanics and the evolution of these fault zones. In his repetition of an experiment by Riedel (1929), Tschalenko found that the fracture zone developed in the manner shown in Figure 8–19. The first structures, termed *Riedel* or *R shears,* formed at a counterclockwise angle of about 12° and then about 16° to the shear in the underlying board (panel a). (In cases where water content is low, shears oriented at a high clockwise angle to the underlying fault and with *right-lateral* motion also formed within the shear zone. These are referred to as *conjugate Riedel* or *R′ shears* and appear in Figure 8–20.) As left-lateral shearing continued (panel b), the left-lateral Riedel structures increased in length and new en échelon Riedel shears developed at lower angles to the underlying fault. Still more left-lateral slip on the fault underlying the clay

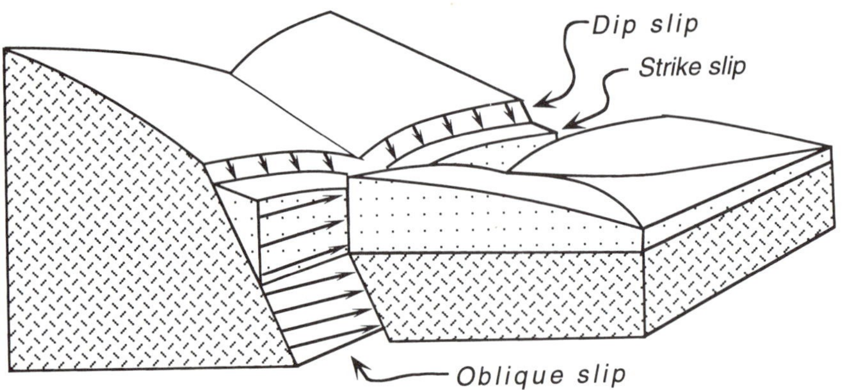

Figure 8–17. Divergent-oblique slip-partitioning between strands of the Damxung fault zone in southern Tibet. (a) Photograph of the two subparallel active fault strands. Multiple normal fault scarps cut hummocky glacial deposits along at the mountain front, whereas the linear fault trace on the piedmont displays a channel that is offset left-laterally. Photograph courtesy of Rolando Armijo. (b) Block diagram illustrates in three dimensions the relationships of the faults in (a). Oblique-slip fault at depth may separate into two structures at bedrock/sediment contact. Modified from Armijo et al. (1989).

produced Riedel shears with even smaller angular deviation and also resulted in the formation of left-lateral *P shears* that formed small *clockwise* angles with the underlying shear (panel c). Panel d illustrates the deformation pattern at yet a later time, when slip on the underlying fault was large enough to force integration of Riedel and P shears into continuous *principal displacement shears.* By the time the fault zone had evolved to the state represented by the panel e, nearly all slip was occurring on these shears.

Because he also conducted shear-box experiments, Tschalenko was able to relate these stages in the development of the en échelon pattern to the evolution of the shear resistance or strength of the clay. Based upon his shear-box experiments, he was able to show that the first shears (Fig. 8-19, panel a) formed just before the peak shear resistance (as in the lowest boxes in Fig 2-4) was reached. Riedel shears continued to form as peak shear resistance was reached and as shear resistance began to drop (panels b and c). Initial formation of the contiguous structures in panel d coincided with a period just prior to leveling off of the shear resistance to a residual level.

The orientations of the Riedel shears formed in the early stages of evolution of a strike-slip fault rupture are adequately predicted by the Coulomb criterion (Fig. 2-5). The general orientation of the later, contiguous structures, however, is clearly constrained by the orientation of the underlying fault, and irregularities of these structures are influenced by the locations of the Riedel and P fractures previously formed. Sieh (1984) has mapped the evolution of such structures through 10 earthquake cycles at one 15- by 55-meter location along the San Andreas fault.

Based upon the "coseismic" structural evolution observed in the laboratory experiments, patterns seen in the field can be related to the stress vs. strain curve. Figure 8-20 shows examples from the Dasht-e Bayaz fault zone of peak, post-peak, and residual structure. In this particular case, peak structure (Riedels and conjugate Riedels) had formed by the time that about 150 cm of sinistral slip had occurred. Post-peak structure (P shears and low-angle Riedels) had formed by the time that about 250 cm of slip had occurred. And residual structures (principal displacement shears) formed after about 300 cm of slip had occurred.

Numerous field studies confirm the usefulness of Tschalenko's basic classification. In California, discrete en échelon Riedel shears have formed in granular materials along the trace of the San Andreas, Superstition Hills, Imperial, and other faults, when total dextral displacements have been only a few millimeters or centimeters (see, for example, Brown and Vedder, 1967; Fuis et al., 1982; Sieh, 1982; and Allen et al., 1972). Slip of even a few centimeters commonly results in visible domino-like rotations of blocks and

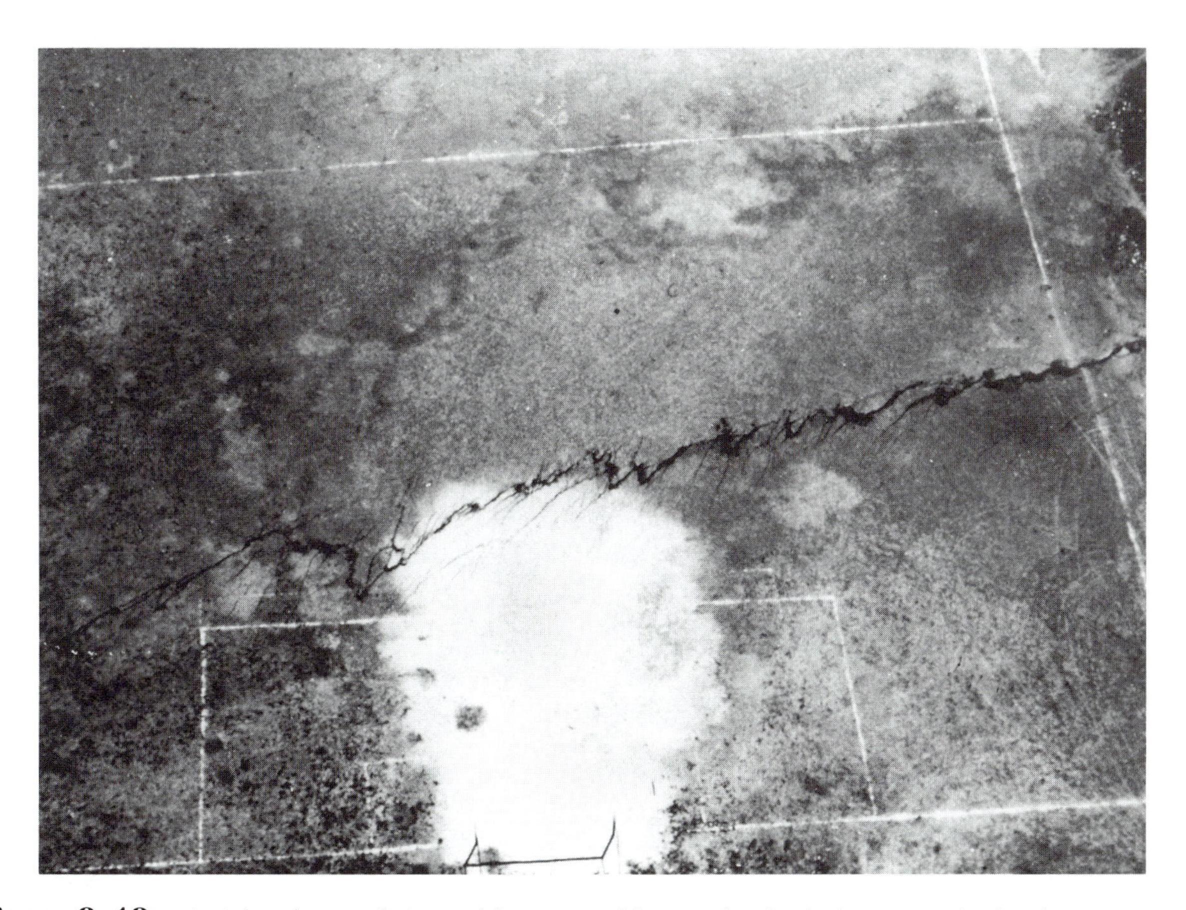

Figure 8–18. Aerial and ground views of the traces of three strike-slip fault ruptures display the pronounced en échelon character of strike-slip fault ruptures. (a) Rupture of the Motagua fault through a soccer field in Guatemala, in 1976. Note chalk line offset left-laterally about 1 m on right. Photo by G. Plafker (February 1976).

Figure 8–18. (b) Oblique aerial view of the Emerson fault, California. Thin motorcycle tracks in foreground (arrows) are offset right-laterally about 3 m. See Figure 8–32 for location. Photo by K. Sieh (April 1994).

Figure 8–18. (c) Left-stepping en échelon fractures and intervening welts along the trace of the Imperial fault, associated with right-lateral offset of about 25 cm in 1979. Photo by K. Sieh (October 1979).

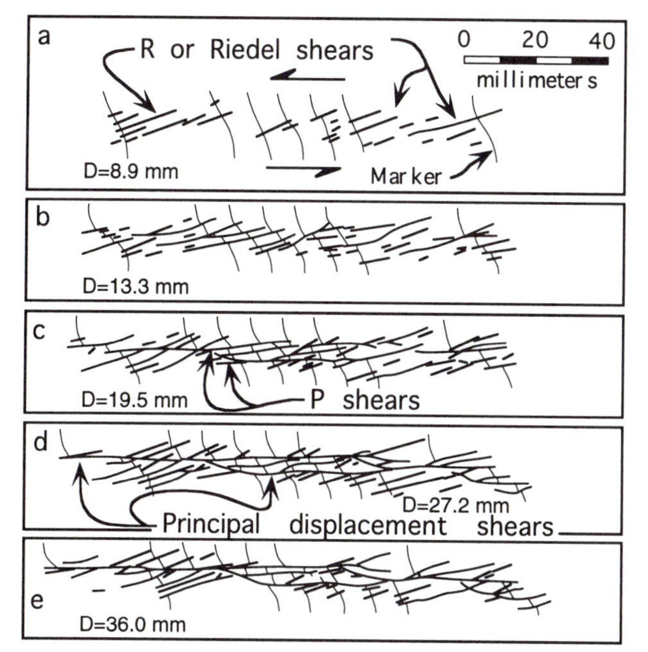

Figure 8–19. These five panels illustrate the sequential development of a fault zone in a Riedel claycake experiment. The patterns and their evolution are very similar to the patterns and evolution of strike-slip fault zones during an earthquake. In each frame, D = total relative movement of underlying boards. Modified from drawings of Tschalenko (1970).

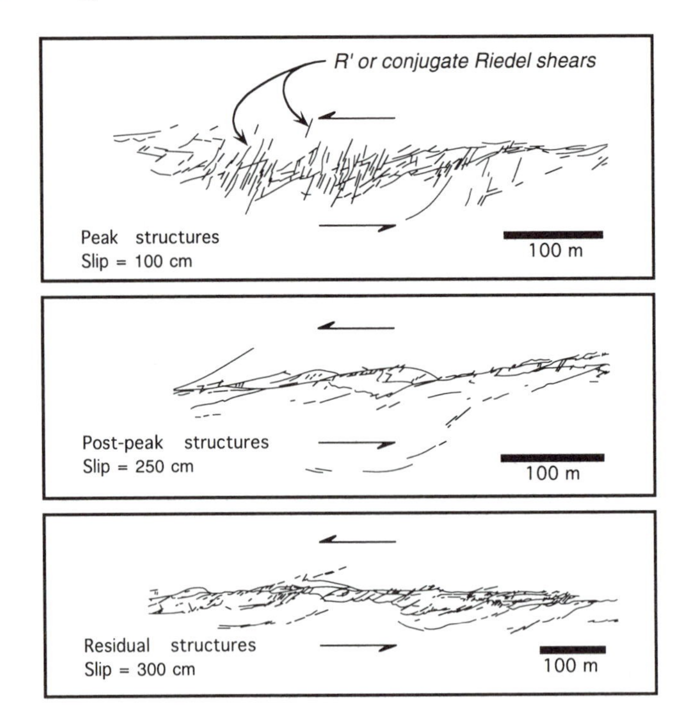

Figure 8–20. Patterns of rupture along strike-slip faults are very similar to those created in the laboratory. These patterns from three localities along the Dasht-e-Bayaz fault rupture correspond structurally to the patterns associated with laboratory peak, post-peak and residual phases. Redrawn from Tschalenko (1970).

topographic welts and fissures within the fault zone (see, for example, Brown and Vedder, 1967). Slip of as little as 10 cm commonly leads to incipient linking of the Riedel fractures by P fractures (for example, Sharp et al., 1982). As one might well expect, P fractures that traverse these mounded areas commonly have appreciable components of reverse slip. Larger amounts of lateral slip promote greater prominence of the mounds and P-fractures between the Riedels, due to the increasing severity of mass-imbalance in these stepover areas. Figure 8–18c shows this mounding quite clearly. When lateral slip greater than about a meter occurs, a jumbled zone of Riedel shears, P fractures, principal displacement shears, and auxiliary fissures and folds is common (for example, Fig. 8–18b). Zones across which strike-slip offsets amount to several meters may be many meters wide (for example, Fig. 8–18b). Viewed as a single entity, such strike-slip ruptures, with all their faults, fractures, troughs and mounds, present the appearance of a large *moletrack*. This was, in fact, the imaginative name given to the fault-rupture zone of the 1891 Nobi earthquake by Japanese villagers (Koto, 1893). Large moletracks are commonly still recognizable decades after their formation (see, for example, Fig. 10 in Armijo et al., 1989).

Measurement Techniques

The measurement of horizontal and vertical offset across strike-slip fault zones is complicated by the development of en échelon structures and moletracks. Thus, the next, short section discusses documentary methods.

Offset along a fault is defined as the magnitude and orientation of the vector separating two *piercing points* formerly contiguous across the fault. In the natural world, the dislocated ends of linear features provide the most common piercing points. These linear "reference lines" may be of human origin (fences, walls, roadways, canals, or vehicle tracks, as in Fig. 8–18a and b, for example) or natural (stream channels, tree roots, burrows, and sharp geological contacts).

In cases where offset is a few centimeters or less, measurement of the magnitude, trend, and plunge of the slip vector across a Riedel fracture is usually straightforward (Fig. 8–21a, b, and c). However, when offsets are large, disruptions within a fault zone commonly obscure or destroy the piercing points (e.g., Figs. 8–18b and 8–21d). In such cases, the horizontal component of the slip vector must be estimated by measuring the separation of the reference line parallel to the strike of the fault. Uncertainties in such cases can be more than 10%. Vertical components of slip

Figure 8–21. Examples of piercing points and their reference lines for strike-slip offsets ranging from centimeters to meters. (a) Offset mud cracks. Note the matching patterns of the edges of the offset mud cracks. The horizontal component of the slip vector`is 3.5 cm N76°W. The scale is marked in centimeters. Imperial fault, 1979. (b) Pencils mark offset thalweg of small gully. Vertical offset is 13 cm; horizontal offset is 17.5 cm. Imperial fault, October 1979.

Figure 8–21. (c) Rows in cultivated onion field offset 15 cm. Imperial fault, October 1979. (d) Dirt road offset 2.5 meters. Power line towers in background displayed an additional meter of right-lateral warping over an aperture of more than 1000 meters. Emerson fault, 28 June 1992.

are best determined from surveyed profiles of the elevation of the reference feature. Table 8-1 is a spreadsheet developed at Caltech for systematically recording data at individual sites along strike-slip faults. It provides for recording the slip vector either by magnitude and plunge, or by horizontal and vertical components.

Nonbrittle, Anelastic Warp

Comparisons of lateral offsets measured across only the obvious fault zone with offsets measured over tens to hundreds of meters are often nearly identical. Such remarkable similarity is an indication that nearly all offset has occurred across the narrow, mappable fault. Not uncommonly, however, a comparison of measurements made across broad and narrow apertures reveals large disparities. The line of transmission towers visible in the background of Figure 8-21d, for example, was offset 2.4 m where it crossed the narrow 1992 rupture of the Emerson fault in the Mojave Desert of California. The dextral offset across an aperture of a couple of hundred meters centered on this narrow fault zone was 3.5 m. This additional meter of offset accrued in a wide zone not traversed by visible fractures.

Numerous other examples of permanent, anelastic, nonbrittle deformation were documented along the San Andreas fault following the 1906 San Francisco earthquake (Fig. 8-22). If such warping were to recur during hundreds and thousands of earthquakes, the cumulative, off-fault coseismic deformations would be large enough to have produced wide zones of pronounced shear fabric flanking the principal zones of cataclasis. This could well be, in fact, one of the causes of such broad zones of cataclasis in bedrock along the San Andreas fault and other large strike-slip faults.

Regional Structure

At local and regional scales, strike-slip faults appear geometrically simpler than dip-slip structures. In contrast to reverse and normal faults, strike-slip faults

Table 8-1 Form from page of field notebook designed for collection of data along a fault rupture.

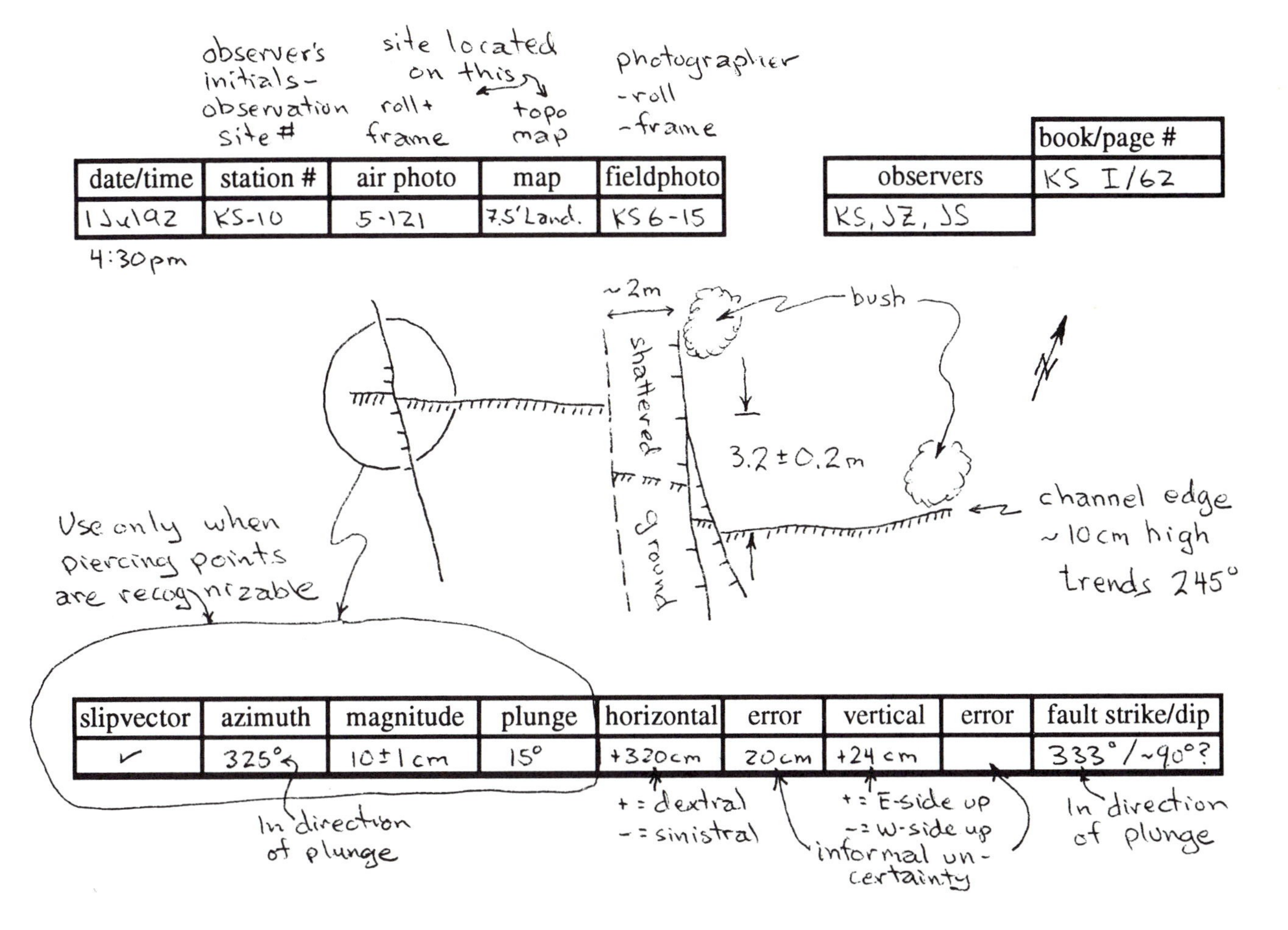

date/time	station #	air photo	map	fieldphoto
1Jul92 4:30pm	KS-10	5-121	7.5' Land.	KS 6-15

observers	book/page #
KS, JZ, JS	KS I/62

slipvector	azimuth	magnitude	plunge	horizontal	error	vertical	error	fault strike/dip
✓	325°	10±1 cm	15°	+320cm	20cm	+24 cm		333°/~90°?

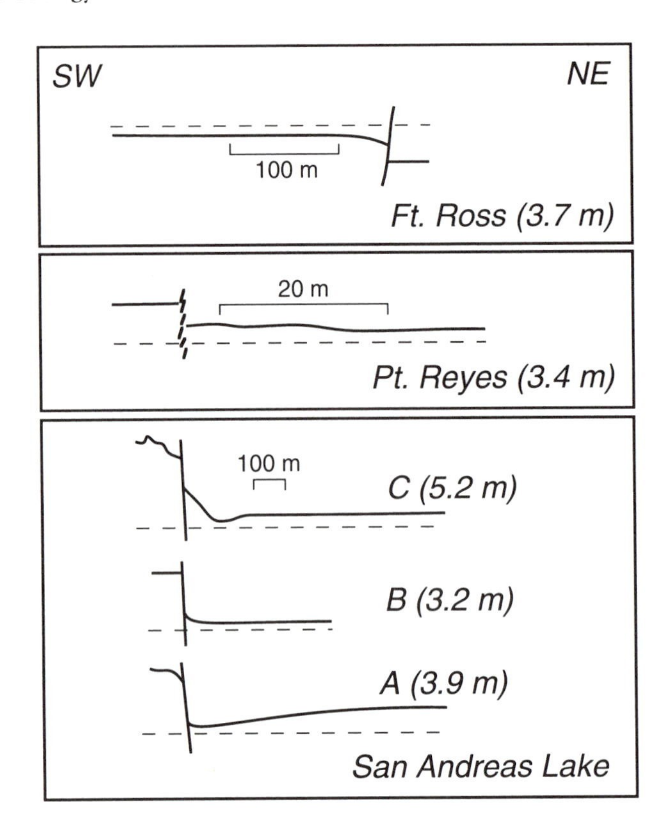

Figure 8–22. An offset measured across a fault rupture zone must often be considered to a minimum value for the offset across the entire fault zone, because substantial coseismic deformation may occur across a broader aperture. These maps of fences offset across the San Andreas fault zone in 1906 show that in some localities a significant percentage of deformation occurred across zones tens to hundreds of meters wider than the mappable fault rupture. Redrafted from data compiled by Thatcher and Lisowski (1987).

commonly have steep dips and have surficial traces that are relatively straight or gently curved. Compare, for example, the traces of the strike-slip San Andreas or Altyn Tagh faults in Figures 8-23, 8-42, or 8-43 with traces of normal faults in Figures 9-12, 9-15, 9-27, and 9-40 and reverse-fault traces in Figures 10-18, 10-31, 10-40, and 10-45.

This difference in surficial expression results from the different orientations of their slip vector relative to the ground surface. Mullions, slickensides, and other outcrop-scale features tell us, of course, that fault planes tend toward smoothness in the direction of slip but commonly exhibit roughness perpendicular to slip (Fig. 3-10, 9-46, 9-48). Because the ground surface traversed by a strike-slip fault is approximately parallel to the fault's slip vector, major corrugations of the fault plane are unlikely to be visible in map view. An unfortunate corollary of this general rule, of course, is that important down-dip irregularities of strike-slip faults, such as horizontal detachments, may have no obvious expression at the surface of the earth!

Despite the relative linearity and smoothness of strike-slip faults in map view, however, any close examination of a strike-slip fault zone reveals deviations of its trace from a solitary straight line or smooth arc. These irregularities create problems of mass balance that result in vertical components of deformation and spawn geologically mappable secondary structures. The larger of these deviations can be seismologically significant. They may influence the initiation, acceleration, deceleration, and termination of earthquake ruptures—in other words, the dynamical characteristics of earthquakes.

These irregularities have causes that are divisible into two categories: Some represent original fault geometries and have resulted from the properties of

Figure 8–23. The surficial traces of strike-slip faults are commonly less irregular than those of normal and reverse faults. The extremely linear trace of the right-lateral San Andreas fault, in the Carrizo Plain of California, is a good example. View is toward the northwest, and closest channel is offset 16 m. Photo by C. R. Allen (1967).

the crust in which they are embedded or from the nature of deformation in the underlying crust or mantle. Others have evolved through the interaction of adjacent structures. Although the evolution of strike-slip structures is a fascinating topic, it is, unfortunately, beyond the scope of this chapter. In the paragraphs below, we restrict ourselves to topics more intimately related to seismic rupture along strike-slip faults.

Almost without exception, when geologists investigate a fault rupture associated with a large earthquake, they find that the fresh, new dislocations have occurred along a preexisting, geologically youthful fault plane. Furthermore, it is almost invariably true that the nature of deformation experienced during the earthquake mimics that reflected in preexisting landforms, structures, and strata. As Charles Darwin was first to note (in 1835), these coincidences demonstrate that earthquakes are the building blocks of tectonic structures.

The *degree* to which these building blocks are alike in length and slip magnitude is currently the subject of lively debate. Regardless of the outcome of that controversy, however, the generally additive nature of individual seismic dislocations is indisputable. If, in fact, this were not the case, large structures would not exist as the geologically mappable features we have described earlier in the chapter.

So, we continue now by focusing our attention on the structural and morphologic characteristics of these building blocks of Earth's plate-boundary structures—the phenomena associated with individual strike-slip earthquakes—with an examination of the shapes and interactions of coseismic ruptures.

Stepovers

It is not uncommon to find geometric complexities along faults that are not related to the phenomenon of Riedel fracturing. These have been called by a variety of names—*stepovers, jogs, overlaps,* and *splays,* to mention a few. Such irregularities along strike-slip faults are, in fact, the rule rather than the exception. Vedder and Wallace's (1970) map of one of the geometrically most simple reaches of the San Andreas fault in the Carrizo Plain of central California, for example, shows that individual, uninterrupted segments of the San Andreas fault nowhere exceed lengths greater than about 15 km. The rock-mechanical reasons for these discontinuities are not well-studied.

Discontinuities and irregularities produce problems of mass balance and stresses that result in the creation of a variety of secondary structures. We consider stepovers first. Figure 8-24 illustrates the two basic types of stepover—dilatational and contractional. For a right-lateral fault, a right stepover results in dilatational secondary structures, and a left stepover produces contractional secondary structures. Segall and Pollard (1980) predicted the orientation, nature and placement of these secondary features by an analysis of the stresses produced by stepping faults with these geometries.

Figure 8-25 illustrates two examples of dilatational stepovers along faults that have ruptured during modern California earthquakes. Figure 8-25a is an oblique aerial photograph of a right step along the right-lateral Emerson fault. Right-lateral slip of three meters during the 1992 Landers earthquake produced a

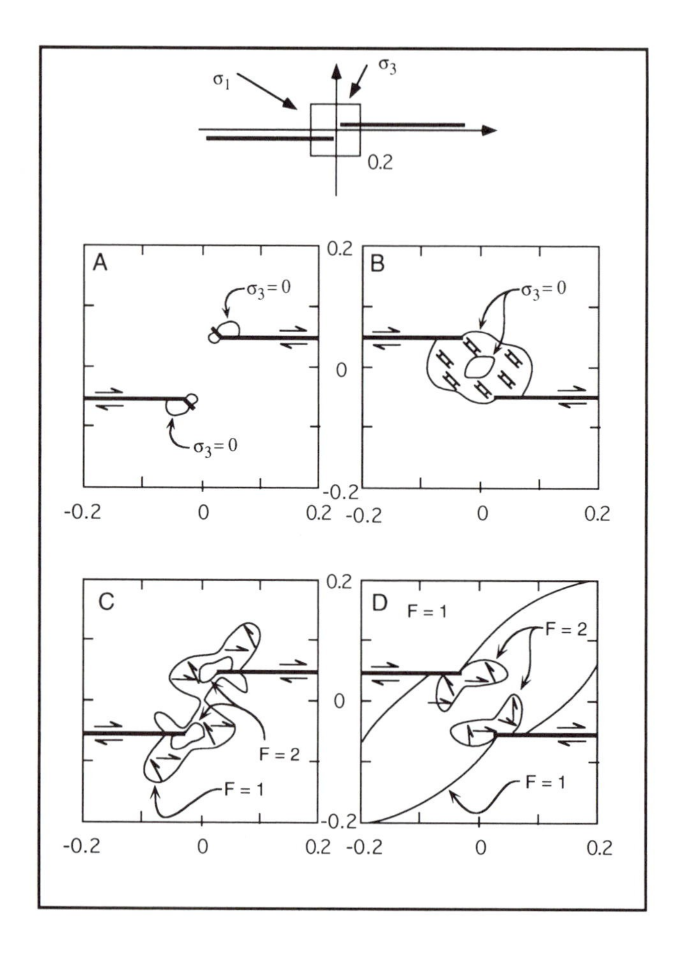

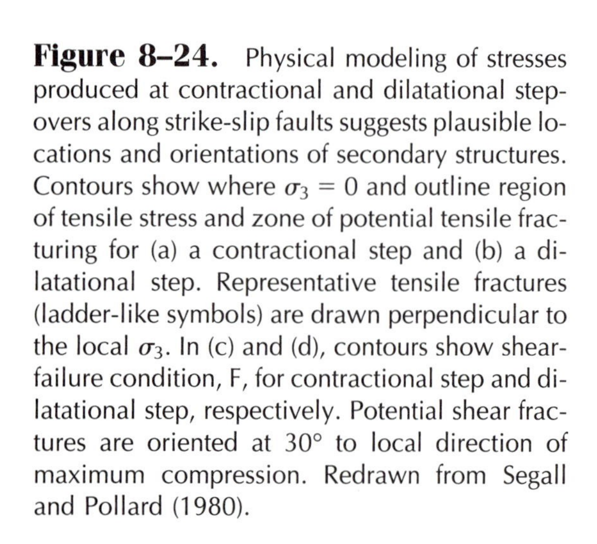

Figure 8–24. Physical modeling of stresses produced at contractional and dilatational stepovers along strike-slip faults suggests plausible locations and orientations of secondary structures. Contours show where $\sigma_3 = 0$ and outline region of tensile stress and zone of potential tensile fracturing for (a) a contractional step and (b) a dilatational step. Representative tensile fractures (ladder-like symbols) are drawn perpendicular to the local σ_3. In (c) and (d), contours show shear-failure condition, F, for contractional step and dilatational step, respectively. Potential shear fractures are oriented at 30° to local direction of maximum compression. Redrawn from Segall and Pollard (1980).

Figure 8–25. Examples of contractional stepovers. (a) Oblique aerial photograph of a graben along the 1992 rupture of the Emerson fault, California. East-facing fault scarps are illuminated by the morning sun; west-facing scarps are shaded. Right-lateral offsets across entire graben were about three meters. The graben floor dropped more than a meter. See Figures 8–32 and 8–30 for location. Photo by K. Sieh (April 1994).

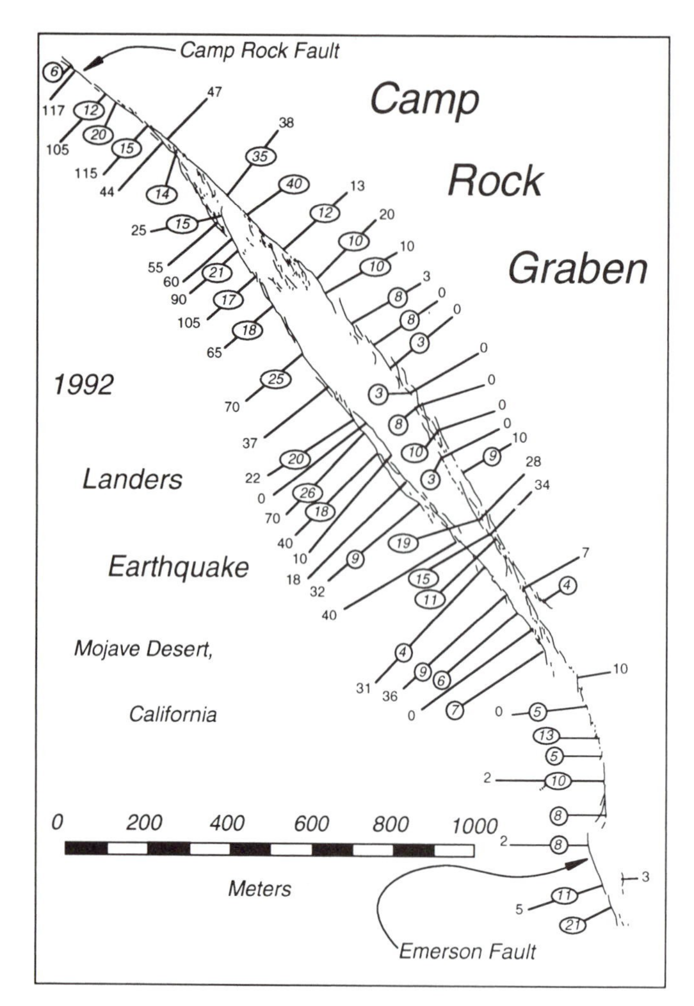

Figure 8–25. (b) Map of a small lenticular graben at a right step between the Camp Rock and Emerson faults, formed during the 1992 Landers earthquake. Dextral offsets appear in centimeters and in plain numerals; vertical offsets are enclosed within circles and ellipses on upthrown side. See Figure 8–30 for location.

graben more than a meter deep with numerous faults cutting the floor of the graben. Figure 8-25b illustrates the geometry and offsets on faults of a graben between the Emerson and Camp Rock faults of the 1992 Landers earthquake.

The Ocotillo Badlands, in southern California, contain a particularly well-documented example of the geomorphic and structural manifestation of coseismic contraction between two strands of a dextral-slip fault. This 2-km-wide hill is an anticlinorium bounded by two en échelon strands of the Coyote Creek fault (Brown et al., 1991; Fig. 8-26). Right-lateral slip of about 30 cm occurred on the bounding faults during the M_W 6.5 Borrego Mountain earthquake of 1968 (Clark, 1972b). Where these faults are adjacent to the Badlands, significant vertical components of slip occurred, consistent with incremental elevation of the Badlands. The folds of the Ocotillo Badlands represent about 800 meters of shortening parallel to the bounding en échelon faults (Brown et al., 1991), so about 2700 nominal 1968 earthquakes would be required to produce the anticlinorium. Paleomagnetic and stratigraphic observations indicate that the structure has been forming for slightly less than the past one million years. This suggests an average interval between nominal 1968 earthquakes of about 370 years. However, dextral slip of about 800 m in about a million years yields an average slip rate of only about one mm/yr, a small fraction of the known rate of the Coyote Creek fault. Because of this discrepancy, Brown et al. conclude that, although the jog has been forming over the past million years, one of the bounding right-lateral faults must continue on past the jog, in order to have accommodated the remainder of slip along the Coyote Creek fault.

Other stepovers are large enough to have influenced the terminations of historical fault ruptures. Sibson (1985) has proposed that suctions induced at fluid-saturated dilatational stepovers may inhibit the propagation of seismic ruptures through them. His analysis was inspired principally by the northward termination of the 1979 and 1940 ruptures of the Imperial fault, in southern California, at a 6-km-wide dilatational step (Fig. 8-27). Harris and Day (1993) present calculations that suggest dilatational stepovers greater than about 5 km in width can, indeed, be expected to impede rupture propagation.

Other large dilatational stepovers have also played important roles in the generation of particular earthquakes. Although it exists only in the upper few kilometers (Eberhart-Phillips and Michael, 1993), a 1-km-wide stepover in the San Andreas fault, in the Cholame Valley of central California, appears to have been the southern terminus of the 38-km-long coseismic rupture of 1966 (Fig. 8-28; Brown and Vedder, 1967). In the hours, days, and weeks following that earthquake, aseismic slip occurred on the strand of the San Andreas that strikes southeastward from this stepover (Wallace and Roth, 1967). Offsets on the main faults were so small that the secondary structures that must connect these two parallel strands of the San Andreas fault across this dilatational stepover did not break the surface in 1966. Sieh (1978a) speculated that the two M6 foreshocks of the great 1857 earthquake were generated by the fault segment northwest of this 1-km surficial stepover, and that these events triggered the great M_W 7.8 earthquake, which was produced an hour later by failure of segments southeast of the stepover.

Another example of a large stepover influencing the terminations of large fault ruptures comes from the North American mid-continent. Macroseismic effects of the three great earthquakes in 1811 and 1812 suggest that the last of these three terrific earthquakes was produced by faults associated with anticlinal warping within a contractional jog between two dextral-slip faults. Uplift of the Tiptonville Dome, athwart the Mississippi River, produced spectacular and deadly effects. From eyewitness accounts, Penick (1976) pieced together what happened. Figure 7-17 illustrates the geography of the region and the locations of places mentioned in Penick's text.

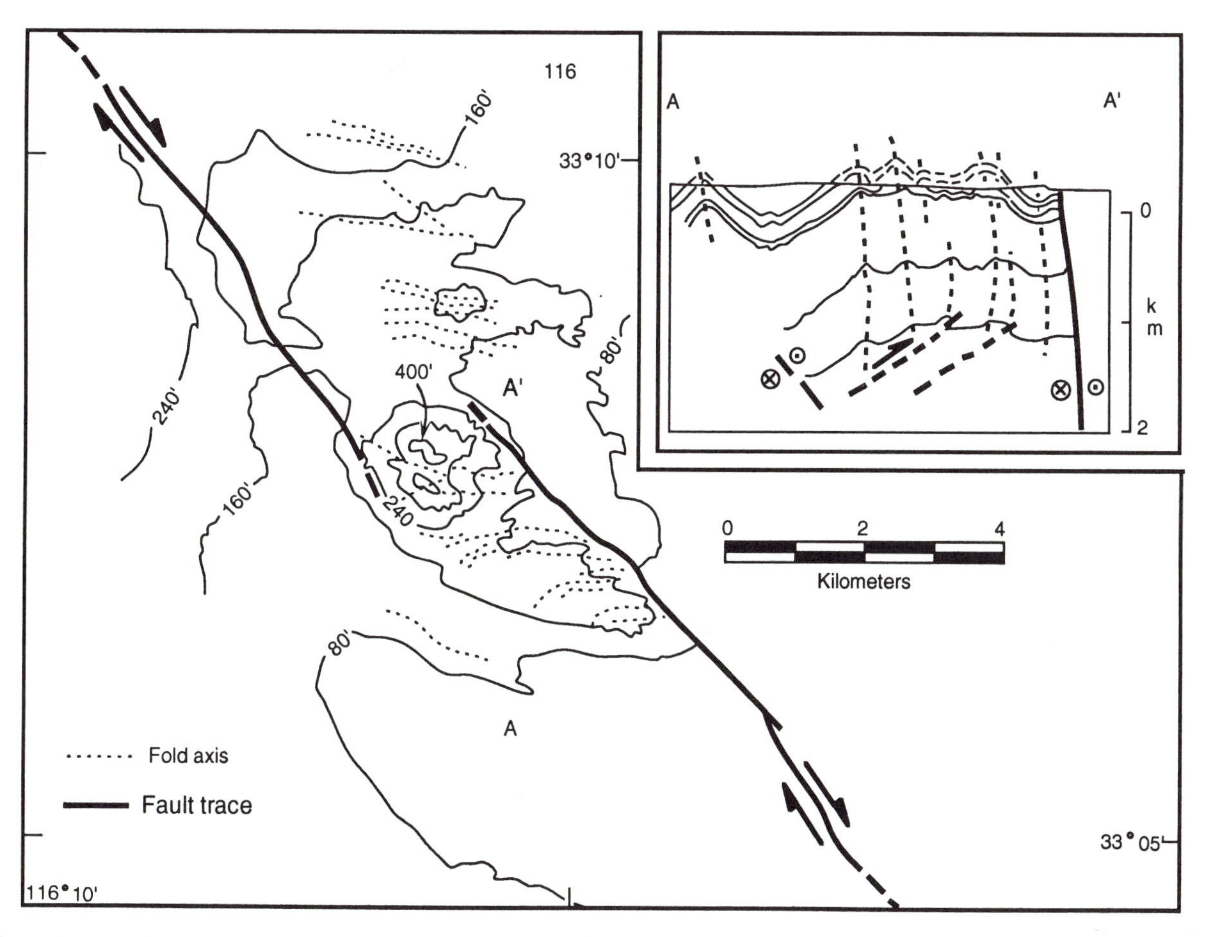

Figure 8–26. Map and cross section of the Ocotillo Badlands, an example of a contractional jog along a strike-slip fault. Contours in feet. Redrafted from Segall and Pollard (1980) and Brown et al. (1991).

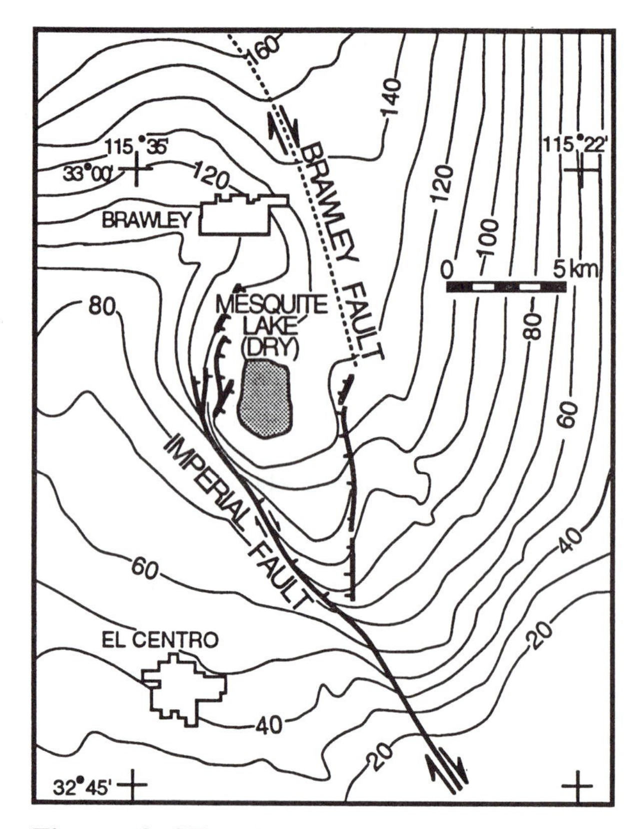

Figure 8–27. The Mesquite Basin in the Imperial Valley of southern California is the topographic expression of a graben between the Imperial and Brawley faults. This structure has impeded through-going right-lateral rupture during two large earthquakes, in 1940 and 1979. Redrafted from figure by Segall and Pollard (1980), based upon Johnson and Hadley (1976). Contours in feet below sea level. See Figure 8–35 for location.

. . . Even more compelling was the eyewitness account of Mathias Speed. His two boats strung together were tied up on the west bank opposite Island Number Nine, a distance of twenty miles above New Madrid, when the first shock came. Forced to get underway by the imminent collapse of the bank, he had the good fortune to survive the hours of darkness. At daylight he saw Island Number Ten before him and realized that he had come a distance of only four miles since three o'clock, "from which circumstance, and from that of an immense quantity of water rushing into the river from the woods—it is evident that the earth at this place, or below, had been raised so high as to stop the progress of the river, and caused it to overflow its banks." The force of this recession was so great at New Madrid that "whole groves of young cotton-wood trees" were swept away. . . .

When the river pushed itself over the barrier created by the elevation of its bed it left behind two sets of falls, the first a half mile above New Madrid and the second about eight miles downriver from the town. As Speed and his crew passed through the channel formed by Island Number Ten "we were affrightened with the appearance of a dreadful rapid of falls in the river just below us: we were so far in the suck that it was impossible now to land—all hopes of surviving was (*sic*) now lost and certain destruction appeared to await us! We . . . passed the rapids without injury, keeping our bow foremost, both boats being still lashed together."

. . . Both falls were estimated to be the size of the falls of the Ohio, a twenty-three foot descent over a distance of two miles. The roar of

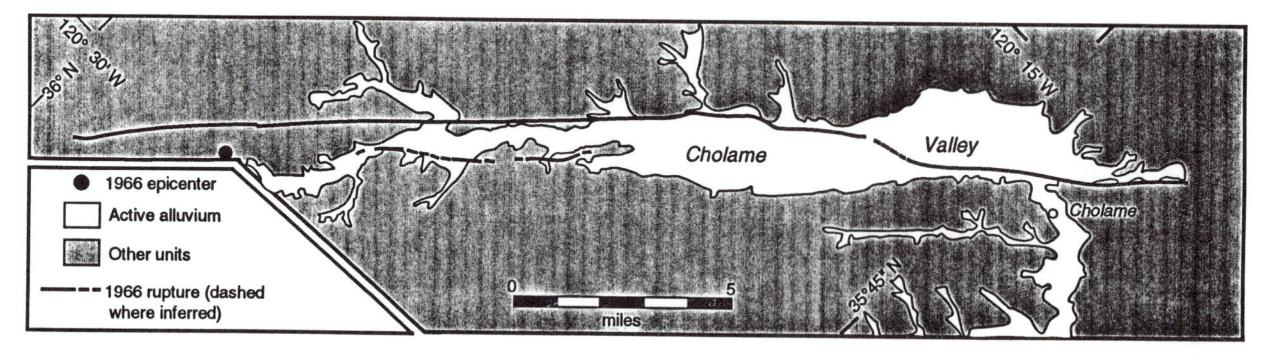

Figure 8–28. A dilatational stepover in the San Andreas fault in the Cholame Valley was terminus of the coseismic rupture of 1966. The initiation of the great 1857 earthquake may also have been here. Redrafted from Brown and Vedder (1967).

the lower falls, although eight miles away, could be heard distinctly at New Madrid. . . . however, the soft sand and mud of the river's bottom was leveled in a few days, and the falls were eliminated.

The two frightening rapids were located 1 km upstream and 13 km downstream from the small town of New Madrid (Fig. 7-17). The overflows and alleged reverse flow of the Mississippi were observed at and upstream from the town. Geological mapping of the elevations of natural levees along the banks of the Mississippi reveals that the river traverses a broad, low structural dome in this vicinity. The occurrence of the rapids on the dome and the location of the flooding and ponding along and just upstream from the upstream flank of the dome indicate that the dome rose during the earthquake.

The dome appears to be the result of right-lateral slip on two adjacent, en échelon right-lateral faults. The location of these faults was deduced from a study of microearthquake locations and focal mechanisms in the region. Figure 8-29 shows that small earthquakes are concentrated in an odd-shaped region between Memphis and New Madrid. One characteristic of the zone is that it contains two linear regions of activity, one protruding to the northeast and the other, toward the southwest. Composite focal mechanisms within these two protrusions are dominantly right-lateral. Thus it appears that the uplift of 1812 occurred in a contractional left-stepover between two right-lateral faults.

Other examples of the influence of stepovers in inhibiting rupture include several from Turkey. Rupture terminations of three of the four large earthquakes along the North Anatolian fault between 1939 and 1944 occurred at stepovers (Barka and Kadinsky-Cade, 1988), all but one of which are too small to see at the scale of Figure 8-7.

Even where faults on both sides of large discontinuities rupture during a large earthquake, the stepover may momentarily impede rupture. The 1992 M_W 7.3 Landers earthquake was produced by rupture of five major dextral-slip faults, separated by several right-jogs and slip gaps . Wald and Heaton's (1994) inversion of strong-ground-motion and teleseismic seismograms of the earthquake and geodetic data suggests that the rupture front was retarded for several seconds at the two largest stepovers, labeled 2 and 3 in Figure 8-30. Figure 4-20, their representation of the evolution of the rupture during the half-minute duration of the source, shows this slowing of the rupture front at each of the stepovers. The ability of seismographic records to resolve such details is controversial, in large part because of the difficulty in modeling the higher frequencies in the records (Cohee and Beroza, 1994). Hough's (1994) modeling suggests a 30-second delay in the rupture front at the slip gap labeled 1 in Figure 8-30.

Transpressional and Transtensional Bends

Irregularities in single faults may also induce secondary dilatation, contraction, and rotation. In any case where the vector representing motion between two blocks is not within the plane of the principal strike-slip fault that separates them, extensional or contractional secondary structures can be expected.

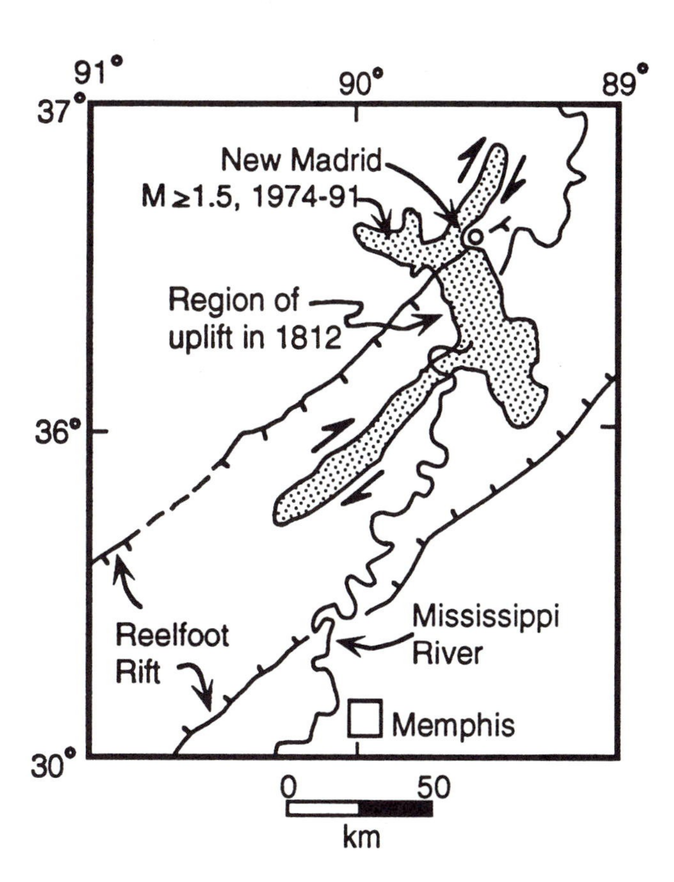

Figure 8–29. Micro-earthquake activity in the region of the great earthquakes of 1811–1812 indicates that the deformation of the bed of the Mississippi River on 9 February 1812, occurred at a contractional stepover between two strike-slip faults. The uplift resulted in spectacular ponding of the Mississippi River upstream of the crest of the uplift and the formation of rapids as the river regraded its bed. Pattern indicates region of high seismicity. Redrawn from figure by Williams et al. (1995).

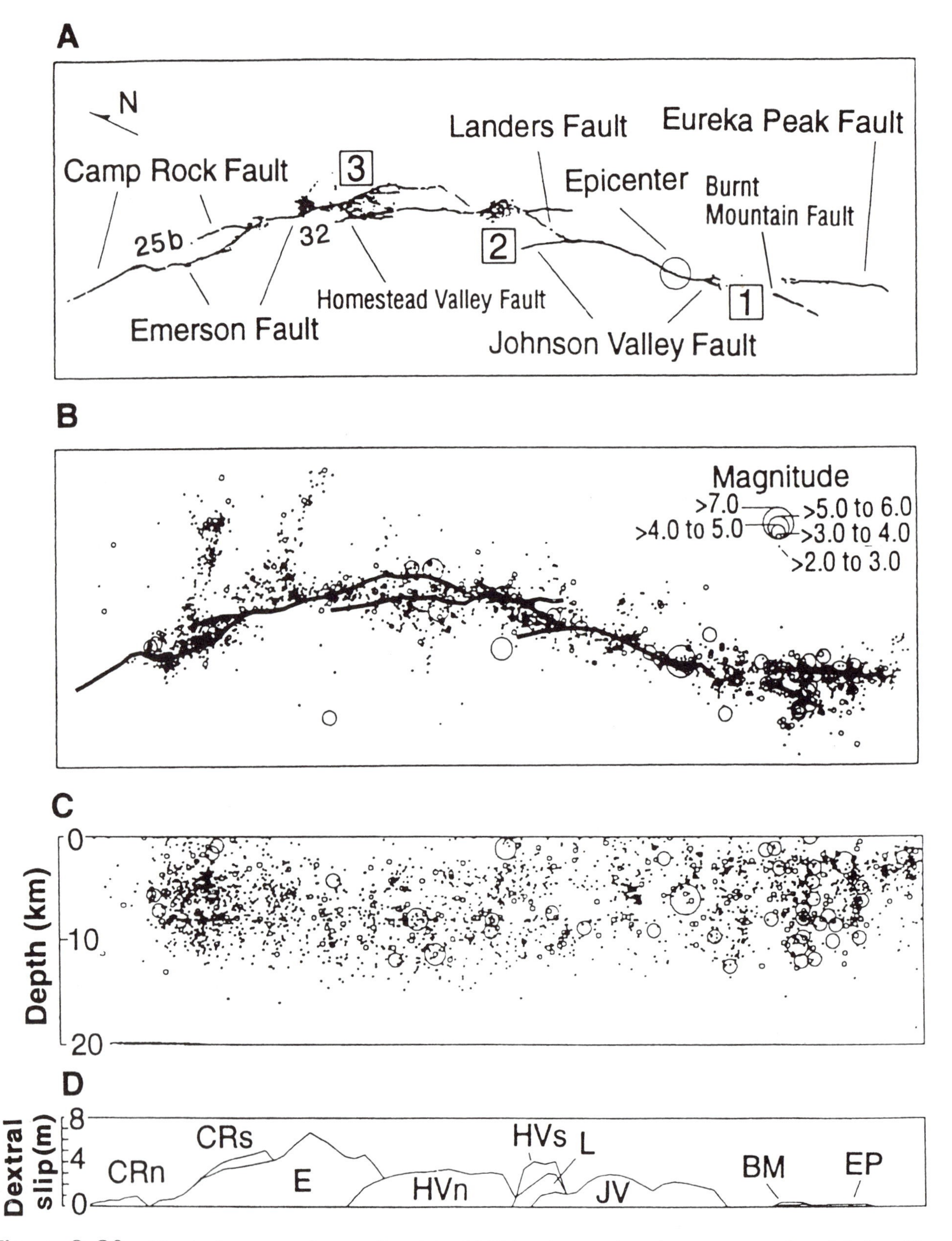

Figure 8–30. The fault ruptures that produced the 1992 Landers earthquake are arranged en échelon. The three largest stepovers and slip gaps that separate the principal ruptures may have impeded the propagation of rupture for several seconds. Locations of Figures 8–32 and 8–25b are indicated. Adapted from Sieh et al. (1993).

Tectonic environments in which these secondary features accompany strike-slip motion have been called *transtensional* or *transpressional*. Though these conflations of the words "transcurrent," "extensional," and "compressional" may demonstrate poor etymological taste, they appear pervasively in the modern literature of tectonics. We, therefore, employ them here.

Figure 8–31 is a map of the faults that ruptured through a dilatational bend or jog in the right-lateral Imperial fault, southern California, during the M_W 6.4 earthquake of 1979. The fault strands northwest and southeast of the jog (labeled A and B) are misaligned by about 200 m, measured perpendicular to their strike. This misalignment resulted in the creation of a broader half-graben between strands A and B. Strike-slip offsets appear above the fault traces, and dip-slip offsets appear below.

Figure 8–32 displays an example of a contractional bend and the secondary features that accompanied it. Right-lateral slip of about 5 meters along this left jog in the Emerson fault in 1992 resulted in an incremental uplift of Stanford Hill (Sieh et al., in preparation; R. Arrowsmith, unpubl. map). On its northeast side, the hill was elevated about 30 cm along several small thrust faults. Along its southwest side, the hill rose up to 1.5 m along the steeply northeast-dipping Emerson fault.

The auxiliary thrust faults of Stanford Hill occur within a kilometer of the Emerson fault. Therefore, they must represent only superficial shortening of the crust adjacent to the jog in the fault. The plexus of right- and left-lateral faults that splays out to the north and northeast from Stanford Hill may represent the response to the jog at deeper levels in the shallow crust. Perhaps at a few kilometers depth, the load stress is great enough to inhibit shortening by reverse faults (which require a vertical orientation of σ_3). That is, σ_3 may be horizontal, not vertical. Thus, instead of upward movement adjacent to the jog, northward shortening and eastward extension occur by way of conjugate left- and right-lateral faults.

Parallel, Contemporaneous Fault Ruptures

Thus far, we have seen Riedel shears, which form an en échelon pattern along the trace of a strike-slip fault zone. And we have seen that, with increasing offset, these coalesce to form moletracks. We have also described fault stepovers and bends, which originate where fault zones consist of noncollinear faults or fault segments, and we have looked at the contrac-

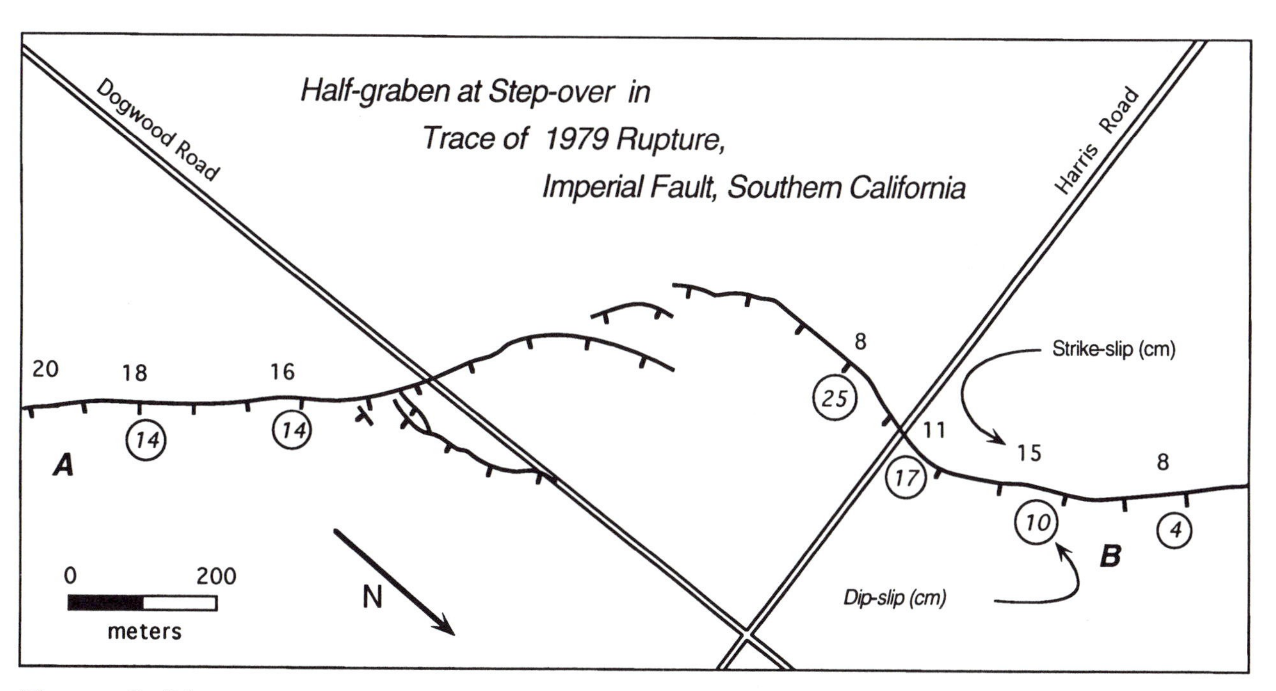

Figure 8–31. Map of an extensional jog along the right-lateral Imperial fault, southern California. This small half-graben along the western edge of the Mesquite Basin, reactivated during the M_W 6.4 earthquake in 1979, shows secondary extensional features associated with the fault jog. Hachures are on the down-dropped block. Values above the fault are horizontal slip (in cm) resolved onto general trend of fault. Circled values below the fault indicate the amount of vertical slip. From 1:9000-scale aerial photograph and field observations.

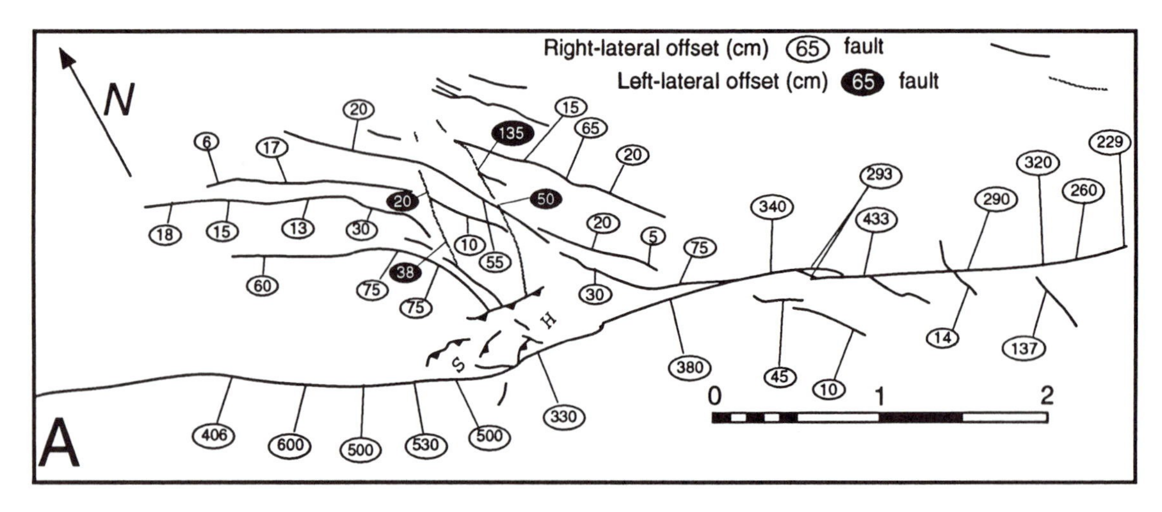

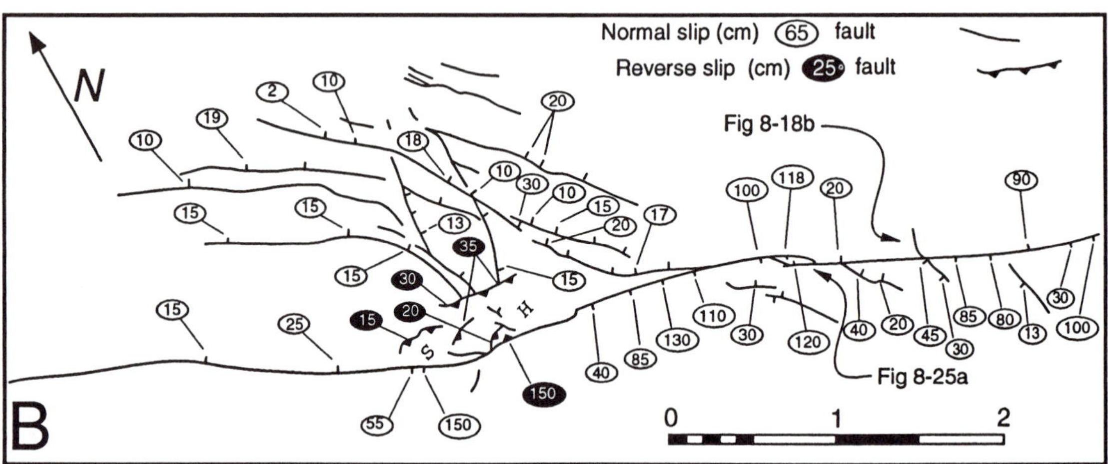

Figure 8–32. Stanford Hill (SH), an anticline bounded by outward-verging faults that were active during the 1992 Landers earthquake, is an example of a landform and superficial structure formed adjacent to a contractional jog in a strike-slip fault. The plexus of auxiliary left- and right-lateral faults that extends northward from Stanford Hill may have resulted from volumetric problems induced by the jog in the Emerson fault at greater depths. Representative sinistral, dextral, and reverse offsets are shown. Unlabelled hachures indicate measurements of dip slip less than 10 cm. See Figure 8–30 for location.

tional and dilatational secondary structures that form because of stepovers and bends in faults. Now we consider parallel and orthogonal coseismic fault ruptures.

Occasionally an earthquake is produced by nearly simultaneous rupture of side-by-side parallel faults. The faults that produced the small, but disastrous Nicaraguan earthquake of 1972 are an infamous example. These four small, parallel faults, equally spaced within a 4-km-wide zone, traverse Managua, the capital city (Fig. 8-33; Brown et al., 1973). The five roughly parallel faults of the M_W 6.2 Elmore Ranch earthquake in southern California are another example. These form an 8-km-wide, 10-km-long zone of left-lateral fractures (Figs. 8-34 and 8-35). Scant attention has been paid to the physical reasons for the simultaneous failure of faults in broad zones such as these.

Intersecting and Orthogonal Fault Ruptures

Within the moletrack of a fresh fault rupture, it is not at all uncommon to find evidence that intersecting faults and fissures were slipping at the same time, during the earthquake. Contemporaneous activity of intersecting faults is also common on grander scales.

During the 1927 Tango earthquake, both the Gomura and the Yamada faults ruptured. Richter (1958) summarizes the voluminous Japanese literature on this M7.3 earthquake. The pair of faults that broke during the earthquake form an angle of nearly 90°

(Fig. 8-36a). Strictly speaking, the two ruptures did not intersect—the 18 km long on-land portion of the Gomura fault rupture ended about 5 km from where it would have intersected the 10-km-long Yamada fault. Left-lateral offset on the Gomura fault was as great as 2.8 m and was accompanied by vertical slip, down on the east. Right-lateral offset on the Yamada fault was about 0.8 m and was accompanied by about 0.7 m of dip slip, down on the south. The dip-slip data imply tilting of the Tango block, east and north of the faults. The opposite senses of strike-slip on the faults indicates northeastward extrusion of the Tango peninsula relative to the Japanese mainland, which is consistent with a NW-SE orientation of the maximum principal stress, σ_1, and a NE-SW orientation of the minimum principal stress, σ_3.

The predominantly left-lateral fault zone of the 1930 M7.3 earthquake on the Izu Peninsula, Japan, also contained a right-lateral rupture oriented at a high angle to the overall trend of the left-lateral zone (Fig. 8-36b).

The pair of fault zones that produced the Superstition Hills and Elmore Ranch earthquakes of November 23 and 24, 1987, constitute another example of intersecting active faults (Fig. 8-35). Five subparallel left-lateral faults within the 8-km-wide Elmore Ranch fault zone experienced cumulative offsets of about 20

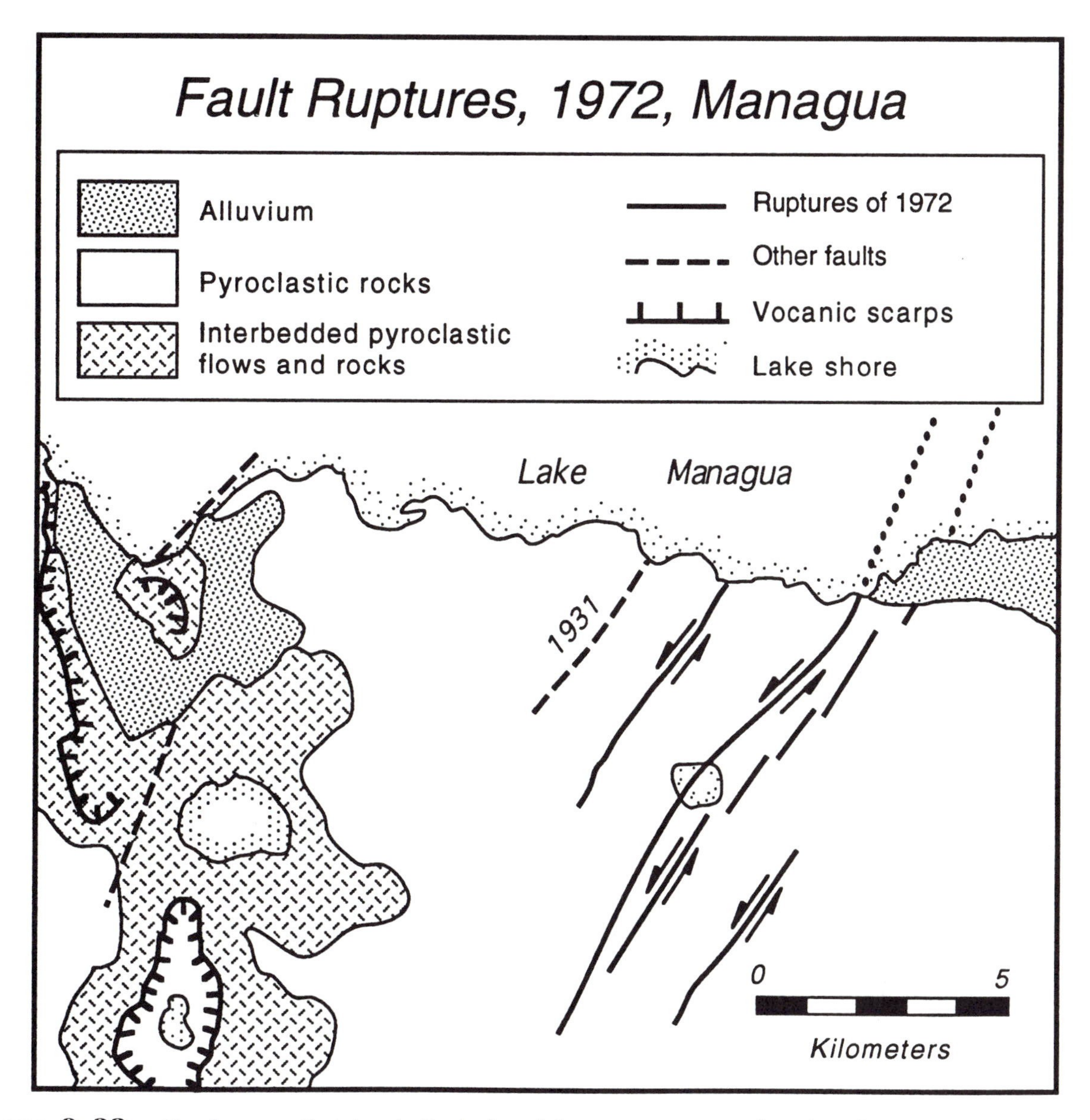

Figure 8–33. The four small sinistral-slip faults of the Nicaraguan earthquake of 1972 are a local example of the parallel arrangement of some strike-slip fault ruptures. Modified from Brown et al. (1973).

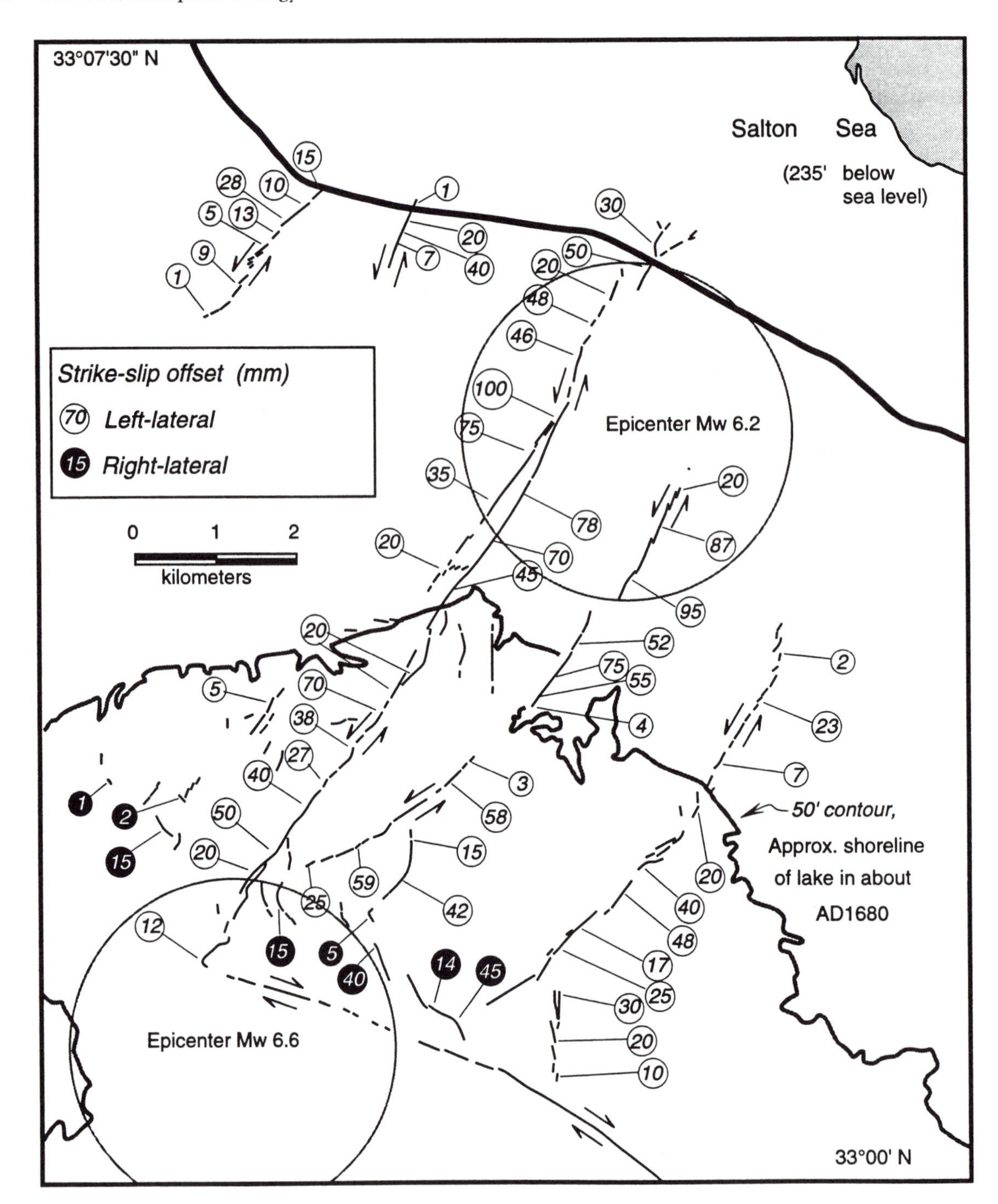

Figure 8–34. The several faults of the 1987 Elmore Ranch earthquake (M_W 6.2) provide another example of coseismic rupture of subparallel faults. Strike-slip offsets on this system are left-lateral (open circles). For the sake of clarity, vertical offsets are not shown. Near its southwestern terminus, left-lateral faults of the Elmore Ranch fault zone intermingle with right-lateral splays of the Superstition Hills fault, which produced a larger (M_W 6.6) earthquake 12 hours later. Right-lateral offsets are shown only for the splays, but were up to a few hundred millimeters on the main fault. For location, see Figure 8–35. Simplified from map by Hudnut et al. (1989).

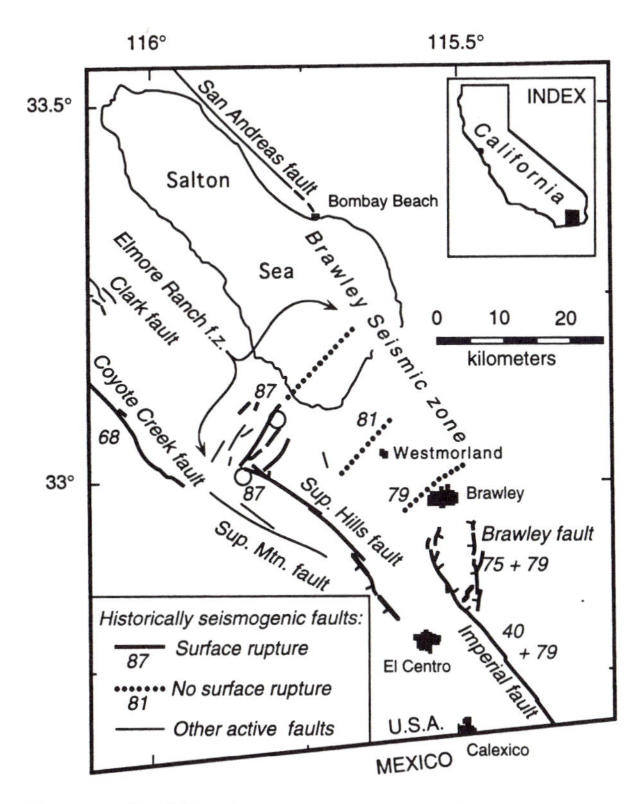

Figure 8–35. The ruptures of the Elmore Ranch faults and the ruptures of the Superstition Hills fault, which formed during M_W 6.2 and 6.6 earthquakes 12 hours apart, constitute an example of conjugate strike-slip fault ruptures. Adapted from Sharp et al. (1989) and data of Hudnut et al. (1989).

cm during a M_W 6.2 earthquake in the late afternoon of November 24 (Hudnut et al., 1989). The next morning, the Superstition Hills fault slipped right-laterally as much as about 50 cm, during a M_W 6.6 event (Williams and Magistrale, 1989). Figure 8–34 shows that the Elmore Ranch fault ruptures ended at the northwestern end of the future Superstition Hills rupture and its epicenter.

Intersecting and parallel faults present interesting questions concerning the long-term, tectonic evolution of structures; unfortunately, discussion of these matters is beyond the scope of this book.

THE LANDSCAPES OF STRIKE-SLIP FAULTS

Now that we have discussed the manifestations of individual strike-slip earthquake ruptures in considerable detail, it is time to consider the additive effects of these individual fault ruptures. For it is not the individual earthquake rupture that geologists usually see, but rather, the geomorphic, stratigraphic, and structural features that have accumulated through many earthquake cycles. Here, the plots become more complex, because we now look not just at the individual building blocks of tectonics, but at various aggregations of the blocks and at the building, itself.

The local manifestations of a strike-slip fault at the

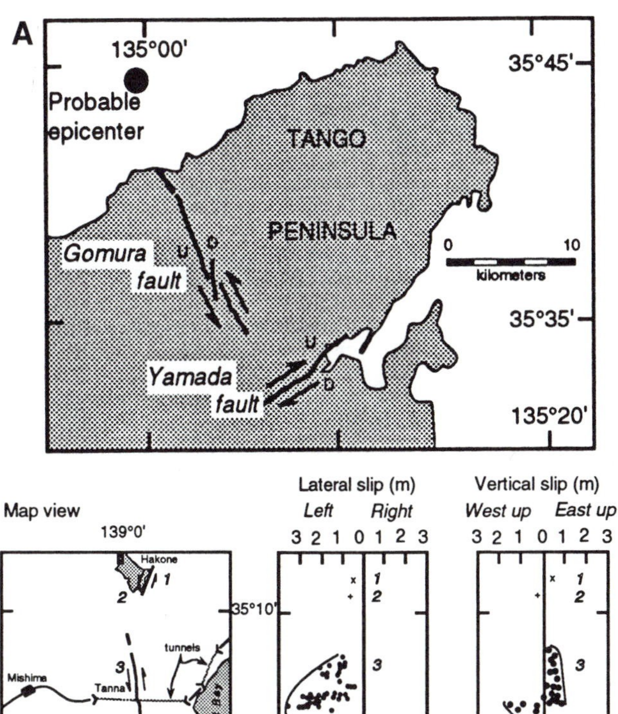

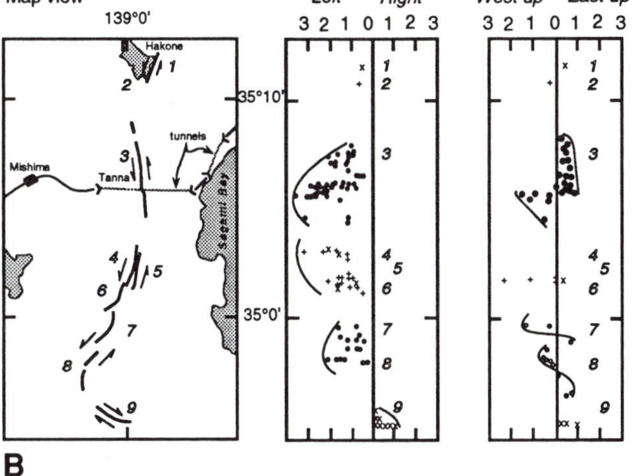

Figure 8–36. Other examples of conjugate strike-slip ruptures. (a) The Gomura and Yamada faults are an example of nearly simultaneous coseismic rupture of two major faults with mutually orthogonal trends. Both broke during the 1927 Tango earthquake, on the northwest coast of Honshu, Japan. Redrawn from figure of Richter (1958), which was based upon Yamasaki and Tada (1928). (b) The complex rupture zone of the 1930 Kita-Izu earthquake consisted predominantly of left-lateral faults (#1 through #8), but a right-lateral fault at the southern end of the rupture zone (#9) also failed. Redrawn from Yoshikawa et al. (1981). For locations, refer to Figure 8–10.

surface of the earth are determined not only by the kinematic peculiarities of the fault. The predominant depositional and erosional processes acting on the surfaces traversed by the fault also profoundly influence its surficial expression.

The nature of these processes also greatly affects the types of information that can be extracted from the fault zone, for the particular landforms and deposits along an active fault are the most accessible libraries from which fragments of the fault's nature and history can be re-

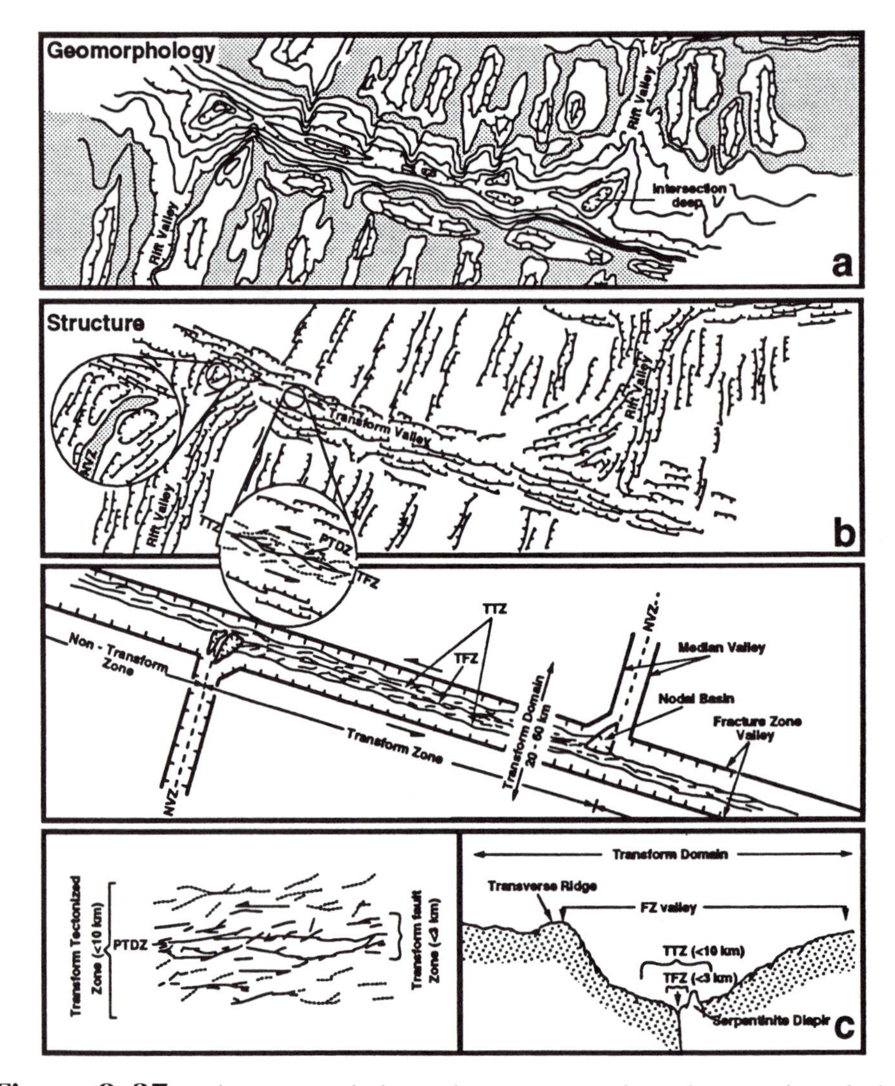

Figure 8–37. The geomorphology of an oceanic ridge-ridge transform fault is closely related to its structure and origin. (a) Geomorphology: The rift valleys of the spreading ridges, in this idealized example, are separated by a transform fault about 100 km long. Ridge-transform intersections are characterized by 3- to 6-km deep closed depressions (intersection deeps). The transform valley strikes at a high angle to the ridge axes, resides at depths of one to several kilometers below adjacent lithosphere and truncates ridges and valleys parallel to the ridge axis. (b) Structure: The rift valley floor is occupied by the linear region of most-recent volcanism, the neovolcanic zone (left insert). The transform valley contains the transform fault zone, a narrow system of braided faults that accommodates strike-slip offsets (right insert). Large valley walls are ornamented with numerous small-throw normal faults. (c) Tectonic elements: The domain of the transform structure is tens of kilometers wide and can be subdivided into several regions: the fracture-zone valley, the transform tectonized zone, the transform fault zone, and the PTDZ. Most of the strike-slip movements occur within a zone only a few kilometers wide (the transform fault zone). From Fox and Gallo (1986).

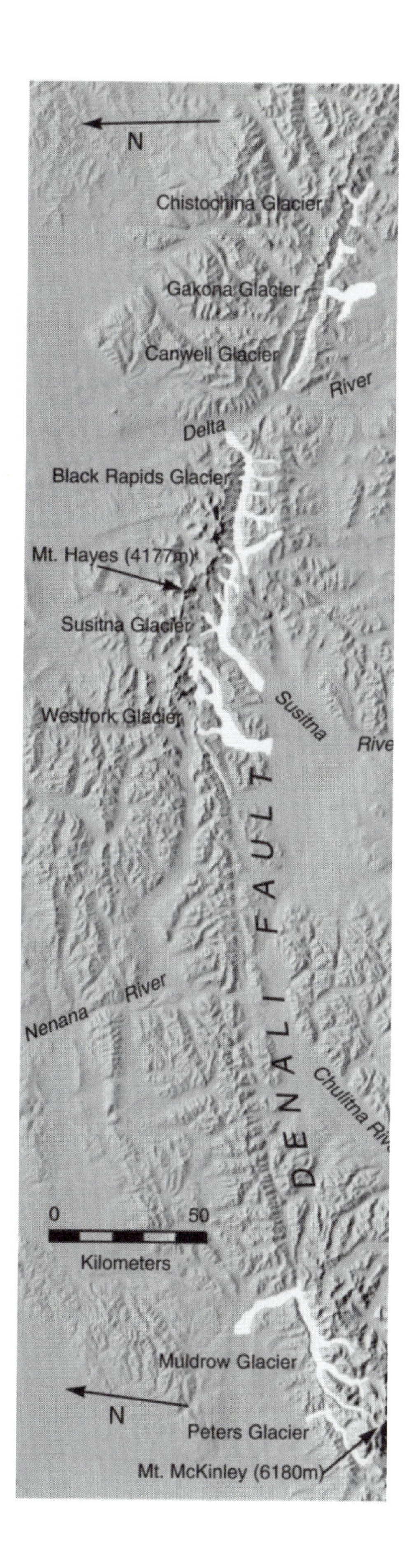

covered. The degree of completeness of these geomorphic and stratigraphic libraries varies tremendously. We begin by contrasting the expressions of strike-slip faults in regions where depositional processes are not dominant with regions where they are.

Submarine Landscapes Dominated by Erosion or Tectonism

Where erosional and tectonic processes are dominant over deposition and have prevailed for a long time, erosional or tectonic landforms are likely to be large and spectacular. Submarine transform faults, for example, commonly are characterized by troughs in the ocean floor, tens of kilometers wide and several kilometers deep (Fig. 8-37). These enormous valleys and related secondary geomorphic features can be directly related to the processes of crustal formation at the intersection of the transform fault and its adjacent spreading ridges (for example, see Pockalny et al., 1988; Tucholke and Schouten, 1988; Gallo et al., 1986). These grand valleys are readily visible as bathymetric features, because rates of sediment accumulation are so abysmally slow in these abyssal settings that tens of millions of years are required to substantially obscure them.

Subaerial Landscapes Dominated by Erosion or Tectonism

In subaerial settings, enormous geomorphic landforms are also visible along lateral faults in environments long-dominated by erosion. Glacial erosion, for example, has dominated the portions of the Alaska Range traversed by the right-lateral Denali fault throughout most of the Quaternary Period. Consequently, most of the fault traverses enormous glacial valleys that display very large right-lateral offsets (Fig. 8-38). Although individual large earthquakes must result from dextral offsets of only several meters, and these large events probably occur no more than a millennium or two apart, little evidence of individual events remains (Sieh, 1981). The longevity and pre-

Figure 8–38. Uplift and glacial erosion have long dominated surficial processes in the portion of the Alaska Range traversed by the Denali fault. Hence, the morphology of the fault zone is dominated by large glacial valleys, which display right-lateral deflections along the fault. Created from U.S. Geological Survey 1:250,000-scale Digital Elevation Model (DEM).

dominance of glacial erosion along the Denali fault have resulted in a paucity of smaller geomorphic or stratigraphic features for study of the detailed recent history of the fault. However, the large size of many southward-flowing valley glaciers has allowed them to maintain the integrity of their channels despite these small, incremental right-lateral offsets, whose cumulative effect has been to add many kilometers to valley length.

The long-lived predominance of fluvial erosion may also lead to creation of large, geomorphically preserved offsets. Gaudemer et al. (1989) have found 10- to 90-km jogs in major river valleys along several of the large strike-slip faults of China. They also propose that in northern India and Pakistan, the Karakorum fault offsets the Indus River 120 km, and that in eastern Turkey, jogs in the Euphrates River of 50 and 10 km are due, respectively, to right-lateral offset by the North Anatolian fault and left-lateral offset along the East Anatolian fault. Several of these geomorphic offsets are illustrated in Figure 8-39. The Xiaojiang fault of southern China displays a spectacular 55-km left-lateral offset of the deeply entrenched Yangzi (Chang Jiang) River (Fig. 8-6).

Large jogs in river valleys alone, however, are seldom adequate proof of offset, because of the *polygenetic* and *diachronous* nature of most landscapes and because of the potential for stream piracy and other nontectonic phenomena. Independent lines of evidence are usually necessary to confirm that these geomorphic jogs are, in fact, offsets. The 50-km right-lateral jog in the Euphrates along the North Anatolian

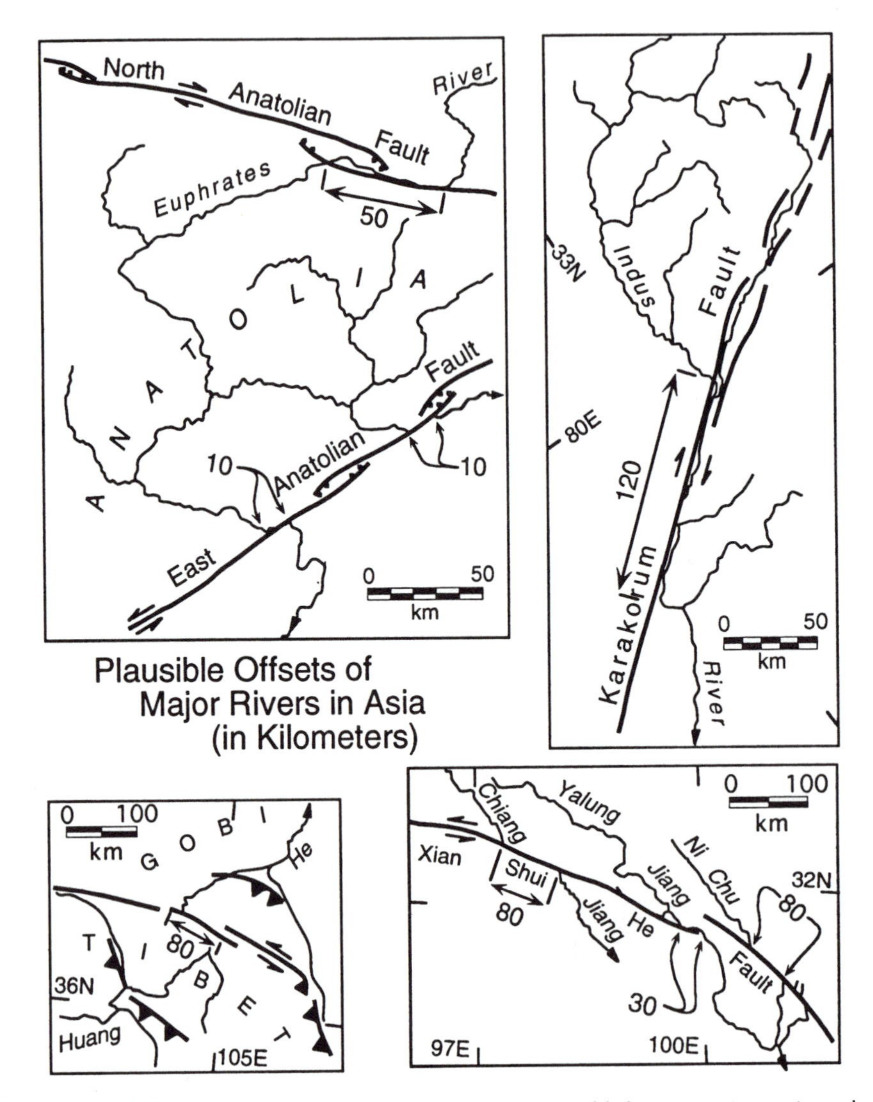

Figure 8–39. Large rivers may be offset of tens of kilometers in regions long-dominated by erosion. For locations see Figure 8–1. Redrafted from drawings of Gaudemer et al. (1989).

fault is similar to the total offset of late Miocene deposits along the fault (Barka, 1992), so this geomorphic example is supported by other geological data. The Honghe (Red River) fault of southern China provides a case where large geomorphic jogs are confirmed to be true offsets by secondary geomorphic criteria. The fault is known from geologic mapping and radiometric dating to have been an active strike-slip fault long before its current period of activity. Lacassin et al. (1993) suggest hundreds of kilometers of left-lateral offset more than 21 million years ago (Tapponnier et al., 1990b). During this previous period of activity, the course of the Honghe assumed its location along the fault. Judging from the sinuous nature of the modern river canyon, the river once meandered upon a broad depositional plain. Uplift of the Yunnan plateau during the past million or more years led to deep entrenchment of the meandering river (Allen et al., 1984). Where the river canyon crosses the trace of the Honghe fault, the canyon appears to be right-laterally offset about 5 km. A careful inspection of the morphology of the canyon and its major tributaries and drainage divides supports the contention that the fault has slipped about 5 km since incision of the drainage basin began.

The San Andreas fault, a couple of hundred kilometers northwest of San Francisco, presents an interesting case of large strike-slip offset of an ancient landscape dominated by fluvial erosion. In synoptic views available on satellite imagery and small-scale maps, more than 50 km of the fault zone is clearly demarcated by two major river valleys, the Garcia and the Gualala (Fig. 8-40e). Earlier in the twentieth century, broad and long stream valleys such as these were considered to be evidence that the San Andreas fault zone was a structural rift, along which a narrow graben had formed by dip-slip motions (Lawson, 1906). These valleys along the fault were, in fact, called *rift valleys*. The development of these landforms, however, is now known to have involved almost no vertical displacement.

Over about the past 2+ million years, right-lateral motion has offset the drainage of the Gualala River about 55 km (Prentice, 1989). In order to lengthen its long-profile throughout the past 2 million years, the Gualala River must have been trapped within a deep channel throughout that period. Panel A of Figure 8-40 depicts the region during the Pliocene Epoch, based upon reconstruction of a shallow marine embayment across the fault, modern outcrops of which appear in panel A as gray patches, for reference. Uplift of the region above sea level initiated formation of an incised, subaerial drainage network, the ancestral Gualala River, conformable to the topography of the shallow embayment (panel B). About 30 km of dextral offset of the Gualala River ensued, to create the channel geometry shown in panel C. *Capture* or *piracy* of part of the headwaters of the Gualala by a deeply incised channel on the block seaward of the fault occurred when that channel and the Gualala were juxtaposed (panel D). Subsequent offset of about 25 km produced the current geometry (panel E), in which the abandoned, downstream portion of the Gualala River has been incorporated into the drainage of the Garcia River.

Some of the most easily interpreted and quantifiable strike-slip landscapes occur in volcanic terrains, where the creation of extensive new surfaces has occurred during large silicic eruptions. One locality along the Great Sumatran fault is a particularly good example. Just south of the equator is the 10- by 20-km caldera of Maninjou volcano (Fig. 8-41). Skirting the volcano is a widespread apron of rhyolitic pyroclastic flow tuff, up to about 200 m in thickness. This thick deposit buried the Great Sumatran fault for about 25 km of its length, east of the volcano. Radiocarbon analyses on wood from plants killed by the emplacement of this 200-m-thick flow suggest that the flow buried the preexisting tropical landscape about 50,000 years ago (Sieh et al., 1991). However, Ar-Ar dates on glass in the flow deposit yield ages of about 68,000 years, which suggest the radiocarbon dates are slightly contaminated by younger carbon (cf. Chapter 6). Preserved, concordant, flat-topped mesas high above the modern river canyons that cut the Maninjou Tuff indicate that the upper surface of the deposit had little relief immediately after emplacement. The Sianok River and its major tributaries, which have deeply dissected the flow deposit, are consistently offset about 700 meters. Since there is little doubt, especially in this tropical environment, that incision of the *pyroclastic flow deposit* began very soon after emplacement, the 700-m offset has accrued in about the past 68,000 years. From these measurements, then, the slip rate of the fault—about 11 mm/yr—can be determined.

Quantification of a fault's kinematic properties is simpler in regions such as this, where large tracts of land have been simultaneously resurfaced by major catastrophic events and then eroded, so that many erosional landforms along the fault display similar amounts of offset.

Landscapes Dominated by Deposition

At the opposite end of the spectrum from long-lived erosional regimes are rapidly aggrading surfaces tra-

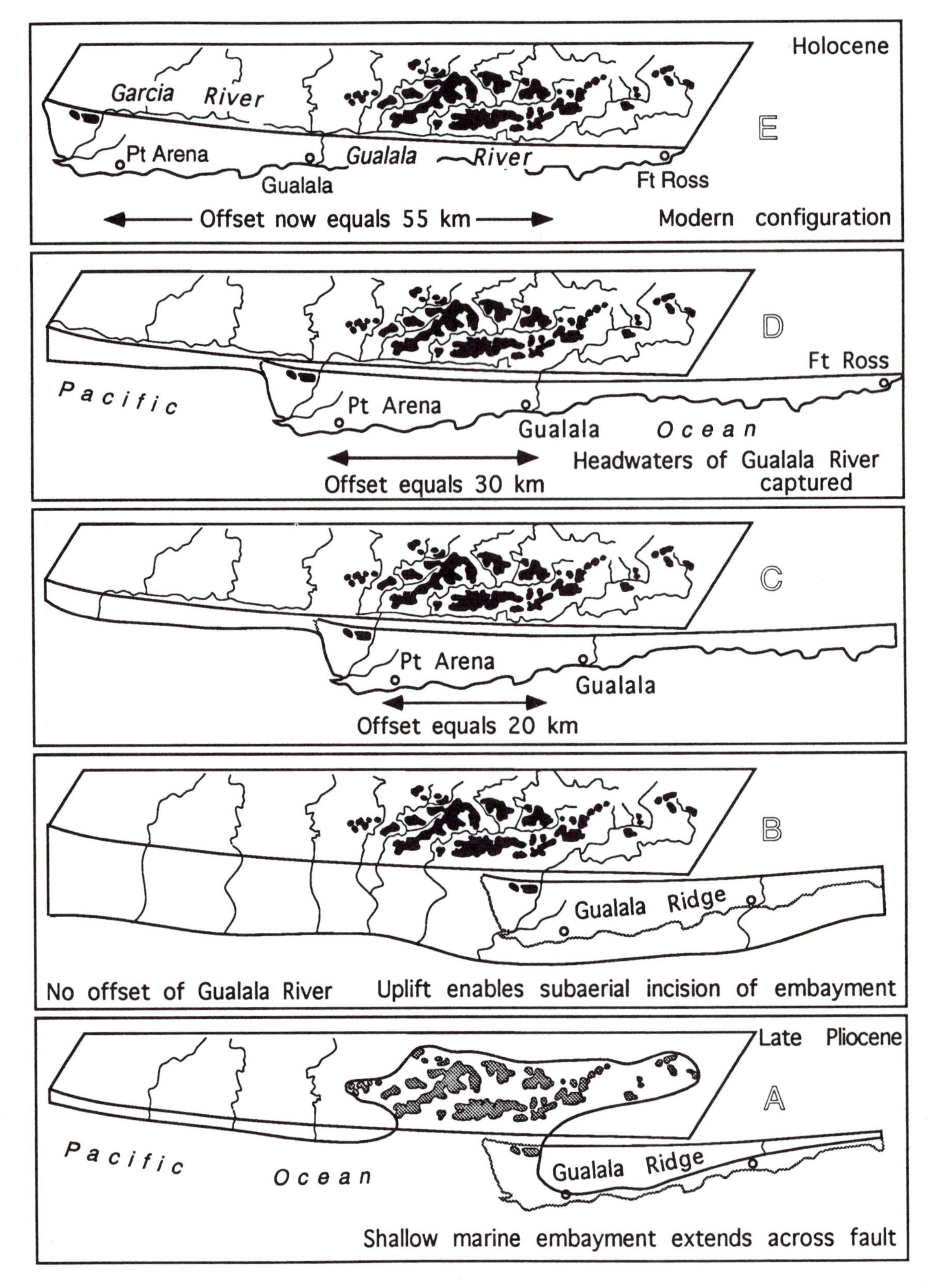

Figure 8–40. Reconstruction of the Gualala river drainage along the coast of northern California, based upon geomorphologic and stratigraphic data, shows the development, from panel A to panel E, of 55 km of offset along the San Andreas fault since the Pliocene Epoch. Black and gray patches represent present outcrops of Pliocene shallow marine sediments. Based upon Prentice (1989).

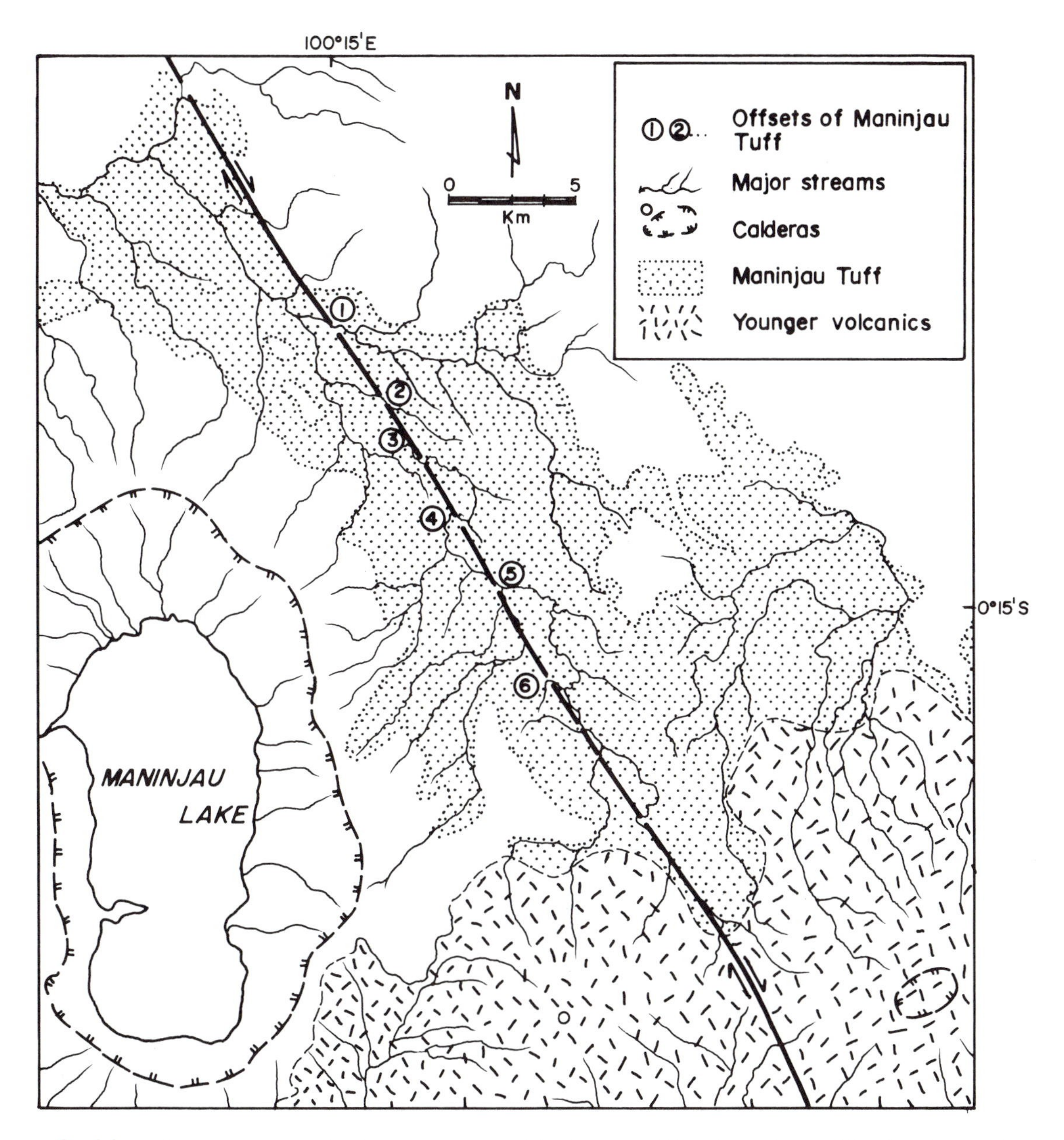

Figure 8–41. The approximately 68,000-year-old Maninjou Tuff has been incised by the Sianok River and its tributaries, which have been offset subsequently about 700 m by the Great Sumatran fault. The offset channels are labeled 1 through 6. Adapted from Kastowo and Leo (1973) and aerial photography.

versed by strike-slip faults. The Imperial fault of northern Baja California and southern California is only one of many good examples. This fault traverses the rapidly aggrading delta of the Colorado River, one of the major rivers of the western United States (Fig. 8-35). Despite its rapid rate of dextral slip, about 3 or 4 cm/yr, the 60-km-long trace of the fault is so obscure that it had not even been recognized prior to investigation of major fault rupture following a large earthquake in 1940. Even now, the trace of the fault is obscure, except along its northern few kilometers, where its dip-slip motion has exceeded its rate of burial by deltaic sediment during the past thousand years or so (Fig. 8-27). The continual burial of the fault trace means that the record of offset along the Imperial fault, though hidden in subsurface stratigraphy, is probably quite complete. Very near-surface deltaic sediments, in fact, have been accessible by excavation, and have extended the seismic history of the fault back about 1000 years (Thomas and Rockwell, in press). Other important examples of lateral faults overwhelmed by rapid deposition include the Tangshan and Tanlu faults of eastern China. Although parts of these faults are geomorphically obscure, they have produced devastating earthquakes in the past few hundred years.

Polygenetic and Diachronous Landscapes

In the middleground geomorphically are faults that traverse surfaces that are polygenetic and diachronous—that is, surfaces that have formed by a variety of erosional and depositional processes over a wide range of time. Most faults traverse such surfaces. Study of faults in these localities has spawned not only an imposing glossary of geomorphic terms and much confusion, but also some very useful kinematic data. These include fault geometries, rates of fault slip, and the dates and characteristics of earthquakes through several earthquake cycles.

Erosional and depositional agents commonly produce surfaces of multiple ages and origins. Faulted nonmarine surfaces may be fluvial, glacial, lacustrine, littoral, aeolian, colluvial or volcanic, and ages may range from less than a year to more than a million years, even along just a short stretch of an active fault. Adjacent tectonic landforms, therefore, may have been constructed over a wide range of ages. Figure 8-42 is an example from western China. Here, the Altyn Tagh fault traverses a landscape composed of alluvial deposits along the flank of eroded mountain front. Left-lateral offsets along this segment of the fault range from meters to hundreds of meters, because of the wide range in age of the offset features.

The older a feature is, the more opportunity agents of erosion or deposition will have had to obscure it, so the larger it must be in order to be preserved as a

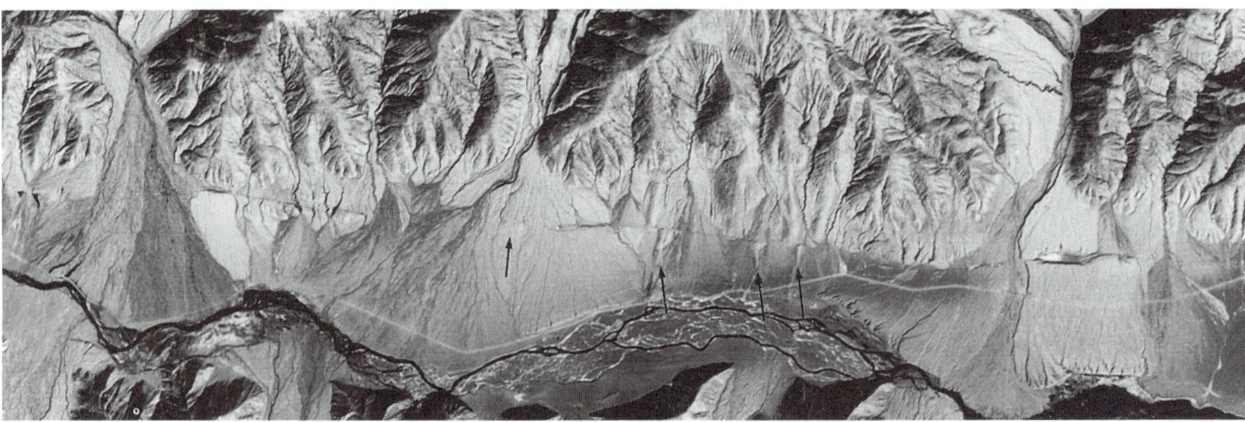

Figure 8–42. Left-lateral offsets along this section of the western Altyn Tagh fault in northwestern Tibet range from meters to hundreds of meters because of the wide range in ages of the offset features. Each panel is about 13 km across. Offset fan edges in top left and bottom center left (arrows) are offset 210 ± 20 and 240 ± 20 m, respectively. Offsets of three transverse ridges (arrows in bottom center right) are 120, 255 and 120 m. SPOT scene KJ 205–277, Sept. 26, 1986. Courtesy of G. Peltzer.

landform. Thus on a faulted landscape, where many ages are represented by various landforms, the larger landforms will usually be older than the smaller ones.

In the next few paragraphs we discuss one example of a polygenetic, diachronous surface cut by a strike-slip fault. In this example, the surfaces range in age from a few years to about 13,000 years. Hence, none of the offsets is very large. We use this example because we do not wish to discuss geomorphic, structural, stratigraphic, and geochronologic aspects of strike-slip faults independently of one another. Rather, we take an integrated approach to the study of landscapes and strata affected by these faults. The ability to interpret polygenetic and diachronous surfaces is critical to deciphering the kinematics and history of active faults. Offsets and fault geometries first suspected from analysis of landforms are difficult to confirm or quantify without the use of geochronology and stratigraphy. In cases of very large and very old offsets, say tens to hundreds of kilometers and millions of years, stratigraphic units may be mapped by conventional geological methods. For smaller and younger offsets, exposures of important strata and surfaces often must be made by artificial excavation.

Figure 8–43 is a topographic map of a portion of the San Andreas fault in central California, where geological mapping, excavations, and radiometric dating are available to elucidate its history (Sieh and Wallace, 1986). Traces of the fault and mappable geologic units have been excluded from this topographic map in order to avoid obscuring the landforms. The location of the fault trace is indicated by small black triangles on the left and right margins of the maps. Features discussed in the text are referenced to a 100-m grid marked from left to right on the lower edge of the map.

The smallest, most ephemeral tectonic features on the topographic map are dextral offsets of approximately 10 m, between the 690- and 1050-m marks. Larger dextral offsets ranging from about 16 meters (at the 2250-m mark) to 130 meters (between the 300- and 500-m marks) are also readily apparent. This range in offsets reflects the difference in ages of the offset features. The young gullies with the least offset were cut across the fault between about A.D. 1480 and 1857 and were offset during the great earthquake of 1857. This interpretation of geomorphic data is confirmed by radiometric dating of stratigraphic units related to the offset landforms (Sieh, 1978b; Grant and Sieh, 1994) and by pre- and post-1857 measurement of distances between survey monuments (Grant and Donnellan, 1994). The larger offsets represent the cumulative offsets of many more earthquakes. For example, the 130-m offset of Wallace Creek, the large stream channel between the 300- and 500-m marks, represents strain relief during the past 3700 years. Sieh and Jahns (1984) proved this by mapping and radiometrically dating several late Pleistocene and Holocene deposits associated with these surfaces. This involved both mapping of geomorphic surfaces and mapping of artificial vertical exposures of the sediments and soils associated with those surfaces. This work resulted in the determination of a late Holocene slip rate for the San Andreas fault of 34 ± 3 mm/yr.

A *beheaded channel* of Wallace Creek, at the 100-m mark, has been offset about 375 meters in about 10,000 radiocarbon years. Other excellent examples of beheaded channels are at the 2000-, 2100- and 2800-m marks. Each of these has been displaced several hundred meters from its source. The history of the former two channels is discussed in more detail by Wallace (1968, pp.17–18) and Sims (1994). A smaller, but very distinct, beheaded channel is visible at the 960-m mark. Very probably it was cut by streams that now are aggrading at the mouth of the canyon at the 1040-m mark. Much smaller beheaded gullies are present between the 570- and 770-m marks. These are barely visible in Figure 8–43, and some are difficult to see and interpret in the field because the channel segments near the fault are choked with debris that has tumbled or washed off of the fault scarp.

Shutter ridges of various sizes are also well represented along this segment of the fault. This term was originally employed by Buwalda (1936, p. 307) to describe topographic highs that had been moved athwart drainage courses by strike-slip motion. Because of this relationship to drainage courses, he saw an analogy between these features and window shutters, which block light and wind from entering through a window. The largest shutter ridge at this site resides between the 1100- and 1900-m marks. Interpreted casually, this low ridge might be thought to represent a broad alluvial apron, offset from sources southeast of the 2100-m mark. The origin of this shutter ridge may, however, have involved more than just horizontal translations: it may also be due to a vertical component of motion that has incrementally raised the feature while it has moved laterally northwest during the past several thousand years. Smaller shutter ridges, invisible at the scale of the topographic map, composed of displaced alluvial and colluvial aprons, deflect offset gullies between the 600- and 900-m marks.

Sediment is commonly *ponded* behind shutter ridges. A large region of active deposition currently exists upslope of the shutter ridge between the 1100- and 1900-m marks. Streams flowing into this region

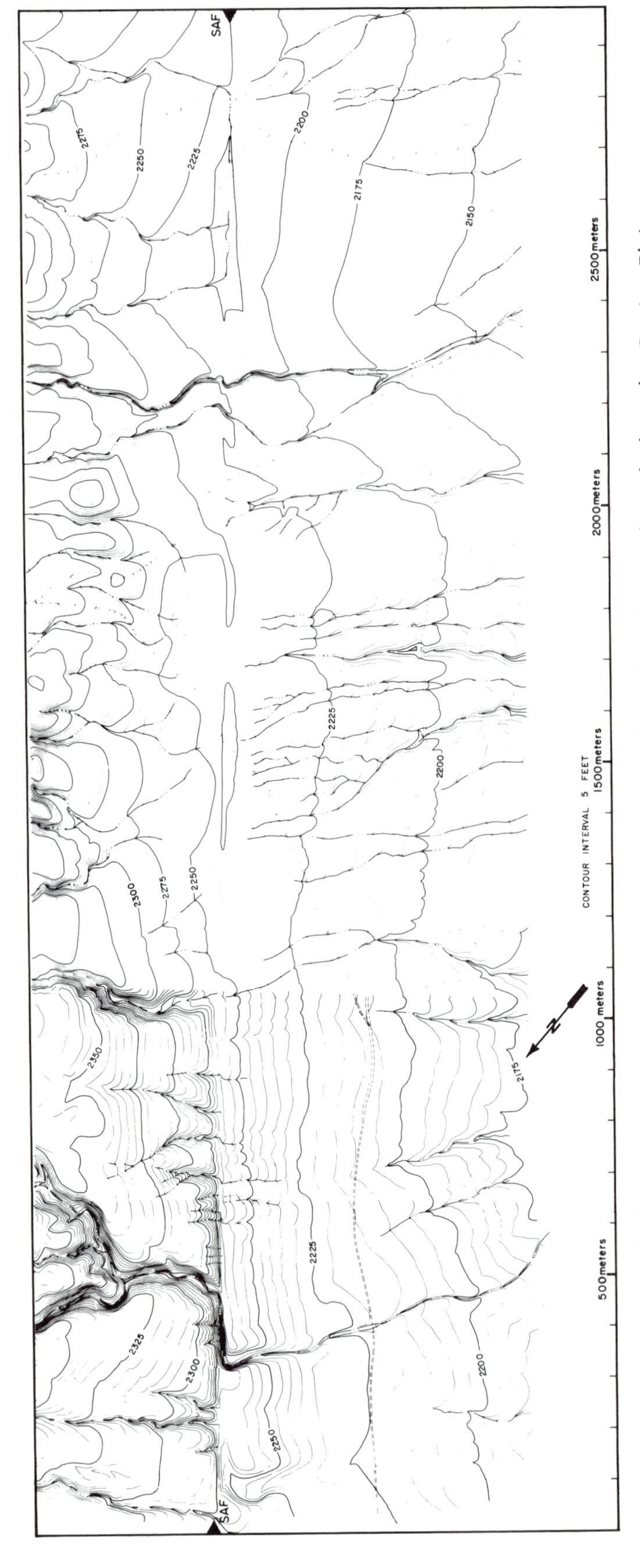

Figure 8–43. This topographic map of a 3-km-long piece of the San Andreas fault in the Carrizo Plain, a desert environment in central California, displays many of the landforms common along strike-slip faults. From Sieh and Wallace (1986).

of topographic closure lose their carrying capacity upon reaching this area of low gradient and drop their bed load. Three small alluvial fans are clearly visible on the northeastern margin of the depression as convex-southward loops of the 2250-foot contour. Water, at times of high discharge, actually ponds in this area of topographic closure behind the shutter ridge. This enables deposition of silty and clayey suspended load as well. Eventually, alluvial fans building southwestward and the quiet-water silts and clays filling the basin may be able to fill the depression, and streams will cross the shutter ridge, enabling incision of the ridge and ponded sediment and the reestablishment of fault-crossing drainages there. This future drainage might well cross the fault at the low point in the shutter ridge at the 1700-mark, reoccupying the ancient beheaded drainage there. Smaller examples of ponded sediment occur immediately upstream from the fault in most of the small drainages between the 600- and 900-m marks. The small right-lateral jogs between these marks have disturbed the equilibrium of these streams (cf. Chapter 7) and have resulted in lesser competence of these ephemeral streams immediately upstream from the fault. Perhaps this is because several meters of length, but virtually no height, are added to the longitudinal profile at the time of dislocation. Thus, deposition may occur, because the dislocations add a section of shallower gradient at the fault. Alternatively, deposition may result from substantial losses in stream- or debris-flow velocity at the fault, due to the abrupt change in direction of flow at the fault.

Between the 2350- and 2700-m marks, two fault traces are arranged en échelon. Dextral slip has resulted in an increase in volume between the overlapping parts of these two faults, and a *sag* or *closed depression* has formed as the surface has dropped between the two fault planes. The filling of the depression is not keeping up with the rate of tectonic subsidence; hence, the existence of the landform. Spectacular evidence of filling was evident immediately following a severe storm in mid-February, 1978. During that storm, the three major channels that terminate in the sag delivered enough water to form an ephemeral *sag pond,* about 3.5 m deep. Substantial erosion of the northwestern and southeastern channel beds also occurred at that time, and alluvial fans with deltaic fronts formed on the margins of the pond. Along the long margins of the pond, many large, elongate pits, some as much as 3 m deep and long, formed as the ponded water catastrophically drained into open fractures along both fault planes. Fresh pond-facing fault scarplets, up to about 15 cm in height, formed along both long margins of the lake as near-surface debris was carried to greater depths by these waters and surficial blocks slumped in to fill the voids. Remnants of the delta fronts, collapse fissures, and fault scarps were still visible a decade later. Such features complicate the interpretation of stratigraphic sequences along strike-slip faults, because they may be confused with evidence for tectonic events. Clark (1972a) discusses collapse pits along active faults in more detail.

Fault scarps of several ages and degrees of activity are present within the area of Figure 8–43. The continuous high scarp that extends from the zero to the 2000-m mark northeast of the fault trace is 10 m high and is still growing, between the 490- and 650-m marks. The trace of the fault lies very near its base, and its lower slopes are much steeper than its upper slopes (about 35° vs. about 5°). Sieh and Jahns (1984) demonstrated that this scarp has been growing since about 13,000 years ago, when the alluvial fan that it truncates on the southwest ceased accumulating sediment. The scarp has risen 3 m in 3700 years, at an average rate of 0.8 mm/yr, about 3% of the strike-slip rate here. Between the 650- and 1900-m marks, the fault scarp has either ceased growth, or it has transformed into a broad monocline. This is indicated by several lines of evidence. First, the fault trace at the 1900-m mark is farther away from the base of the scarp than at the 650-m mark. Second, the scarp is buried by greater volumes of debris toward the southeast (note the large alluvial fans between the 1300- and 1900-m marks). Third, the steepness of the lower portion of the scarp diminishes toward the southeast. Thus, this scarp appears to be growing in height northwest of and decreasing in height southeast of the 650-m mark. Thus, the block upstream from the fault appears now to be bulging upward in the northwest and subsiding in the southeast.

The scarps immediately adjacent to the fault trace alternate along strike from northeast-facing to southwest-facing. This is a common characteristic of strike-slip faults and has been termed *scissoring*. In some localities, scissoring is due to a change in the sense of vertical slip along a fault's strike. Elsewhere, it is clearly a result of purely strike-slip offset of a nonplanar ground surface. In the field, for example, scissoring of the fault scarp can be readily observed across the alluvial fan between the 960- and 1100-m marks. Unfortunately, this is not apparent on Figure 8–43. On the northwestern half of the fan the fault scarp faces uphill, whereas on the southeastern half it faces downslope. This is best explained as strike-slip offset of the convex surface of the alluvial fan, as is illustrated in Figure 8–44. Figure 8–45, a photograph of a fault scarp that resulted from left-lateral strike-slip

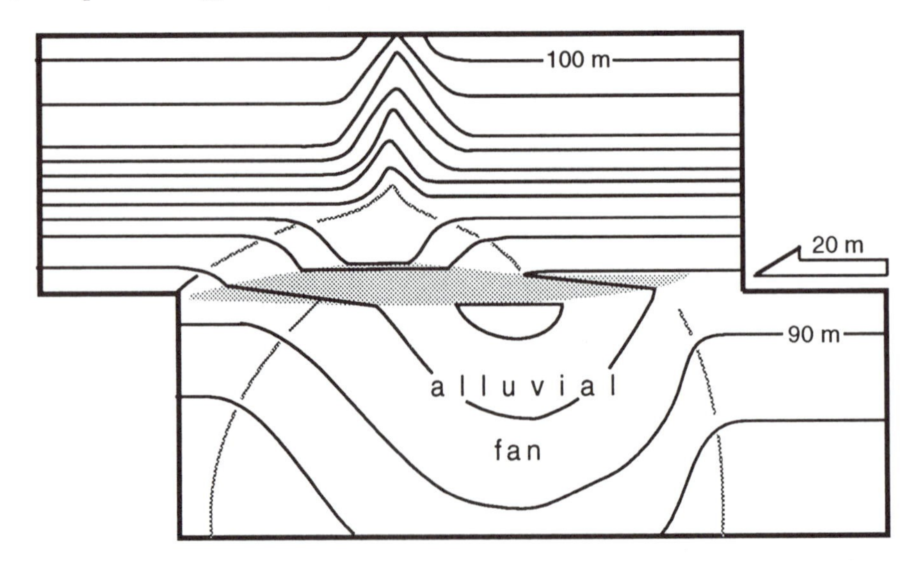

Figure 8–44. Pure strike-slip offset can result in the creation of fault scarps if the ground surface is not flat, as in the case of this hypothetical alluvial fan. Fault scarp is the shaded region.

Figure 8–45. A scarp along the Xianshuihe fault in Sichuan Province, China, produced by a couple of meters of nearly pure sinistral offset of a slope during an earthquake in 1973. Arrows indicate points that were adjacent prior to the earthquake. Photo by C. R. Allen, 1986.

along the Xianshuihe fault in China in 1973, provides another example. The unbroken surface can be restored by subtraction of the meter or two of left-lateral slip that occurred during the earthquake. Scarps of this type are not limited to such small sizes. The 50-km separation of the edifices of the San Gabriel and San Bernardino Mountains in southern California may well represent a 50-km offset of these landforms during the past 2 million years.

REGIONAL STRUCTURE, EVOLUTION, AND HISTORY

If one looks only at the active faulting associated with earthquakes of the past few hundred or even the past few thousand years, it is difficult to appreciate the fact that seismic structures evolve. Centuries and millennia are so much shorter than the life span of most faults that they would represent only a few seconds in a feature-length tectonic film. And yet, an appreciation of the long-term evolution of active faults is desirable, if we wish to understand well the earthquakes that now occur along them.

In the preceding section of this chapter, we completed our discussion of the nature of individual strike-slip events and began to consider their cumulative structural and geomorphic effects. Unfortunately, in this text, limitations of space preclude further discussion of the regional structure and evolution of strike-slip structures, even though this has long been a major focus of geological research. For further information concerning the structure and evolution of strike-slip faults we suggest several references.

Perhaps the clearest understanding of the evolution of strike-slip faults comes from studies of oceanic ridge-ridge transforms (e.g., Pockalny et al., 1988, and Gallo et al., 1986) and oceanic fracture zones (Tucholke and Schouten, 1988). The 20-million-year evolution of the indent-linked strike-slip faults of the Indian-Asian collision are the subjects of many papers by Tapponnier and his colleagues (for example, Avouac and Tapponnier, 1993; Tapponnier et al., 1990a and 1990b; Peltzer and Tapponnier, 1988). Heubeck and Mann (1991) and Mann and Burke (1984) consider the kinematics and evolution of the trench-trench transforms of the Caribbean. The evolution of the North America-Pacific plate boundary, which includes the San Andreas fault, a major ridge-triple junction transform fault, has been studied by many. Atwater (1989), Stock and Molnar (1988), Jahns (1972), and Matti and Morton (1993) are just a few. Shaw et al. (in press) and Anderson (1990) utilize very different approaches in attempting to understand the evolution of the contractional bend in the San Andreas fault in the vicinity of the Loma Prieta earthquake of 1989. Nur et al. (1986), Dokka and Travis (1990), and Dokka (1992), and Garfunkel (1974) consider the evolution of the parallel and conjugate active strike-slip faults of California's Mojave desert.

EARTHQUAKE CASE HISTORIES

Earthquakes are the building blocks of tectonics. The anticlines, synclines, horsts, grabens, and faults that geologists map and interpret are the cumulative results of numerous incremental episodes of folding and faulting. Many, if not most, of these episodes of deformation are seismic. Thus, in understanding the development of these structures, the description of the deformation associated with a few well-documented earthquakes is instructive. Furthermore, to counterbalance the cursory nature of our treatment of topics that is necessary in an introductory text such as this, we now wish to discuss in some detail the geological, geodetic, and geophysical aspects of two specific earthquakes on strike-slip faults.

We have chosen two recent California events—the M_W 6.4 Imperial Valley earthquake of 1979 and the M_W 7.3 Landers earthquake of 1992. The Imperial Valley earthquake is a particularly well-documented and well-behaved example of a moderate earthquake. The Landers event is also unusually well-documented, but represents a much larger and more complex event.

The Imperial Valley Earthquake of 1979

The Imperial fault is a rather simple, 60-km-long, ridge-ridge transform fault within the San Andreas-Gulf of California transform system (Figs. 8-46 and 8-35). Rupture of the entire transform fault produced a M_S 7.1 earthquake in 1940 (Buwalda and Richter, 1941; Ulrich, 1941). A smaller, M_L 6.6 earthquake was generated by the northern half of the fault in 1979.

The Imperial fault traverses the nearly flat delta of the Colorado River, where sedimentation rates are so rapid that the fault trace has almost no geomorphic expression. Only near its northern terminus is the vertical component of slip large enough to exceed sedimentation rates and, thus, produce a scarp more than a couple of earthquake cycles old (Fig. 8-27).

The plane of the Imperial fault exists wholly within the late Cenozoic alluvial and lacustrine clastic sedimentary and metasedimentary pile of the Salton Trough, which is 10 to 12 km thick along the plane of the fault (Fuis et al., 1982).

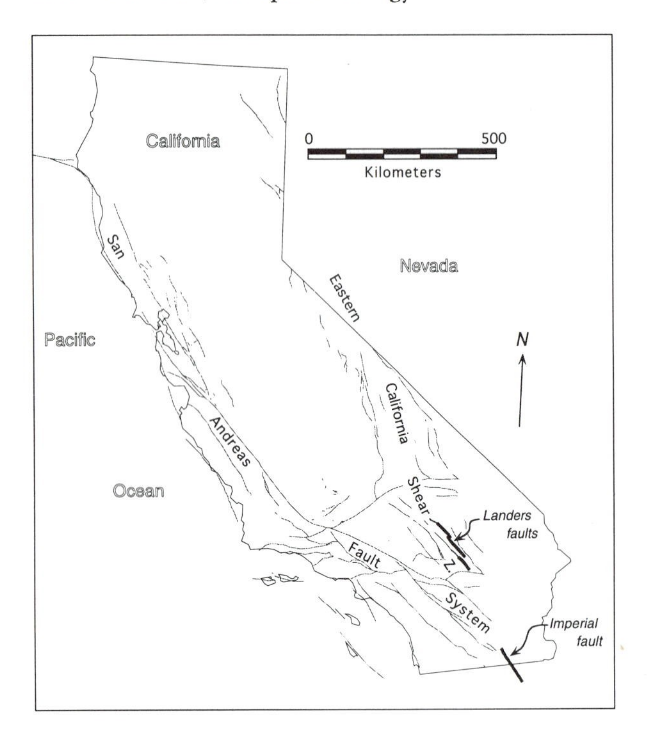

Figure 8–46. The Imperial fault and the faults of the Landers earthquake are components of the San Andreas-Gulf of California transform fault system, the principal element of the Pacific-North America plate boundary in Baja and U.S. California.

Geologic, geodetic, and geophysical investigations place tight constraints on the geometry of the fault and on the kinematics and dynamics of the fault rupture of 1979. Within an hour of the earthquake, the hypocenter of the event had been determined to be on or near the Imperial fault at a depth of about 10 km and at a point a few kilometers south of the International border (Fig. 8-47). Aftershocks within a few hours of the earthquake, however, were concentrated north of the border in a volume of crust surrounding the northern tip of the fault (Johnson and Hutton, 1982). The aftershock region expanded through subsequent days, weeks, and months, but remained concentrated within several kilometers of the northern end of the Imperial fault. This concentration reflects failure of a myriad of small faults in the stepover region between the Imperial fault and the Brawley and Superstition Hills faults, the principal faults of the plate boundary to the north and northwest. The largest aftershock, for example, about 8 hours after the mainshock, was produced by left-lateral subsurficial rupture of a northeast-trending 10-km-long fault immediately north of the Imperial fault. Aftershock locations reveal that the dip of the Imperial fault is nearly vertical from the hypocenter, south of the Border, to a point about 20 km north of the Border. Farther north, however, where it demarcates the western margin of the Mesquite graben, aftershocks show that the fault dips about 70° northeast.

Surficial ruptures associated with the 1979 earthquake were exclusively north of the International Border, 10 to 45 km northwest of the hypocenter. Most of these dislocations were on the Imperial fault (for example, Figures 8-18c and 8-21a,b,c), although minor rupture occurred on two nearby faults as well. Dextral offsets across the Imperial fault, measured within hours of the earthquake, were as large as about 50 cm (Fig. 8-48). Values were greatest near the southern end of the fault and diminished rapidly southward but gradually northward. Dip-slip offsets were greatest along the northern 15 km of the fault, where the fault forms the southwestern margin of the Mesquite graben (Figs. 8-21b, 8-27 and 8-31). Dip slip also occurred along the Brawley fault, a couple of kilometers to the east, on the northeastern margin of the graben. Thus the slip distribution was as one might have expected for a transform fault—predom-

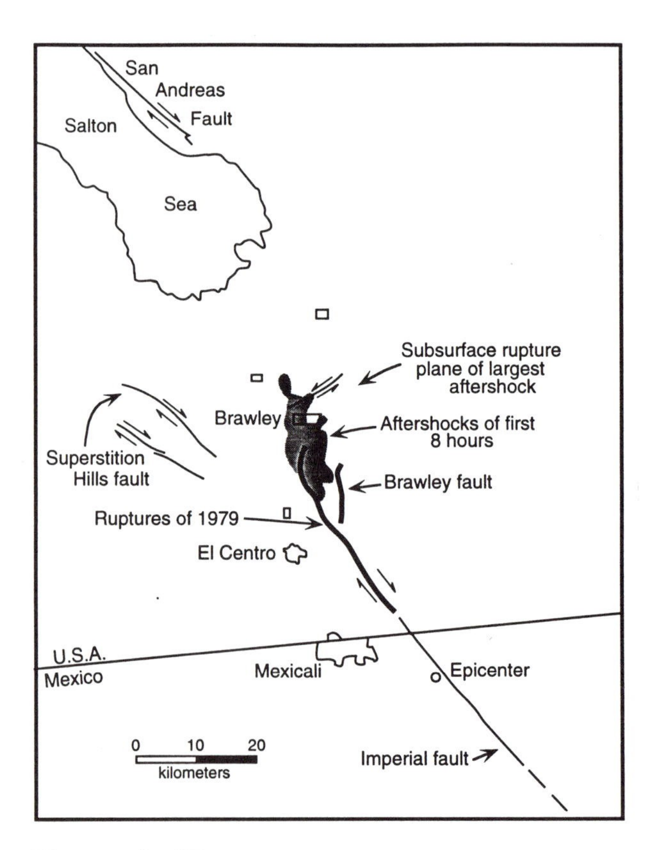

Figure 8–47. Map of the epicenter, aftershocks, and principal fault ruptures of the 1979 Imperial Valley earthquake. Adapted from Johnson and Hutton (1982).

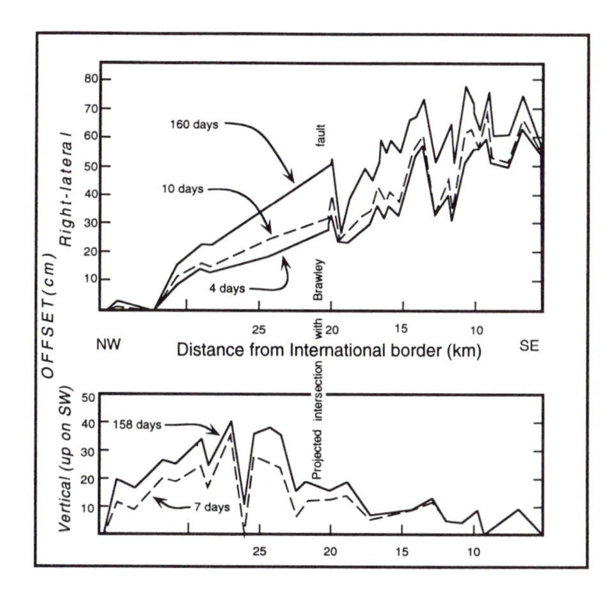

Figure 8–48. Distribution of dextral and dip-slip offsets along the trace of the Imperial fault associated with the 1979 Imperial Valley earthquake. Note the growth of the offsets due to aftercreep in the weeks and months following the earthquake. Also note the greater dip-slip offsets northwest of the projected intersection with the Brawley fault, where the Imperial fault forms the southwestern flank of the Mesquite graben (Figs. 8–27 and 8–35). Redrawn from Sharp et al. (1982). To enable comparisons with geodetic and seismographic inversions, horizontal scale is same as horizontal scale in Figure 8–49.

inantly strike-slip motion along much of its length, but with a large component of dip-slip along the fault where it demarcates the flank of the graben stepover.

Data from dense arrays of near-field geodetic measurements and strong-ground-motion instruments enable an instructive comparison of the surficial pattern of slip with slip on the fault at depth. Crook's (1984) inversion of the geodetic data and Hartzell and Heaton's (1983), Hartzell and Helmberger's (1982) and Archuleta's (1984) inversions of the strong-ground-motion data mimic the surficial observations, inasmuch as they require a concentration of dextral slip well north of the Border, tens of km north of the hypocenter (Fig. 8-49). The magnitude of the maximum slip at depth is 2 to 3 meters, several times greater than slip values across the surficial trace of the fault.

This major discrepancy between determinations of surficial and subsurficial slip poses an interesting neotectonic question: Do the crustal blocks adjacent to the fault move, over many millennia and many earthquake cycles, at a slower rate near the surface than they move at depth? If they do, then rotations about vertical axes and, perhaps, detachment of the surface and subsurface sediments must occur. Alternatively, rigid translation of the adjacent crustal blocks along the fault could occur in the long term, if the surficial slip deficit of 1979 could be eliminated by slip during future earthquakes or aseismic slip.

Slippage along the Imperial fault in the months and years following the 1979 earthquake reduced considerably the magnitude of the discrepancy between the coseismic surface and subsurface offsets but did not eliminate it entirely. The maximum magnitude of this surficial aseismic motion, a phenomenon that commonly is referred to as *creep,* nearly equaled the coseismic surface offset at some localities (Fig. 8-48). Geodetic measurements, here and elsewhere in California, indicate that creep is commonly a superficial phenomenon and does not extend more than a few hundred meters or a few kilometers beneath the surface (lowest panel of Fig. 8-49, and Sharp et al., 1988, and Goulty and Gilman, 1978). Extrapolation of the exponentially decaying rate of aftercreep suggests that the magnitude of coseismic and post-seismic surficial slippage would not reach the maximum subsurficial values of 2.5 to 3.5 m even in several decades. And so the mysterious discrepancy between surficial measurements and subsurficial estimates of slip remains unresolved.

Measurements of offset along the Imperial fault associated with the 1979 earthquake and aftercreep are particularly important, because they enable one of the very few comparisons of measured slip for two major earthquakes generated by the same fault. Such comparison provides a critical piece of information in discussions of the repeatability of earthquake sources.

The M_W 6.9 El Centro earthquake of 1940 was produced by faulting along the entire known 60-km length of the Imperial fault. The M_W 6.4 earthquake of 1979 was associated with slip along only the northern half of the surface trace of the fault. Figure 8-50 shows that surface offsets in 1940 attained values as high as several meters near the International Border. South of the Border, values of about 3 m were typical. Along the northern half of the fault, 1940 offsets were comparable to those of 1979—that is, only a few tens of centimeters. The similarity of the two slip distributions along the northern half of the fault suggests that the physical parameters that control the distribution of slip may not vary appreciably over the timescale of many earthquake cycles (Sieh, in press, 1995). It may also be significant that the location of the steep northward decrease in slip of 1940 coincides with the location of the steep southward de-

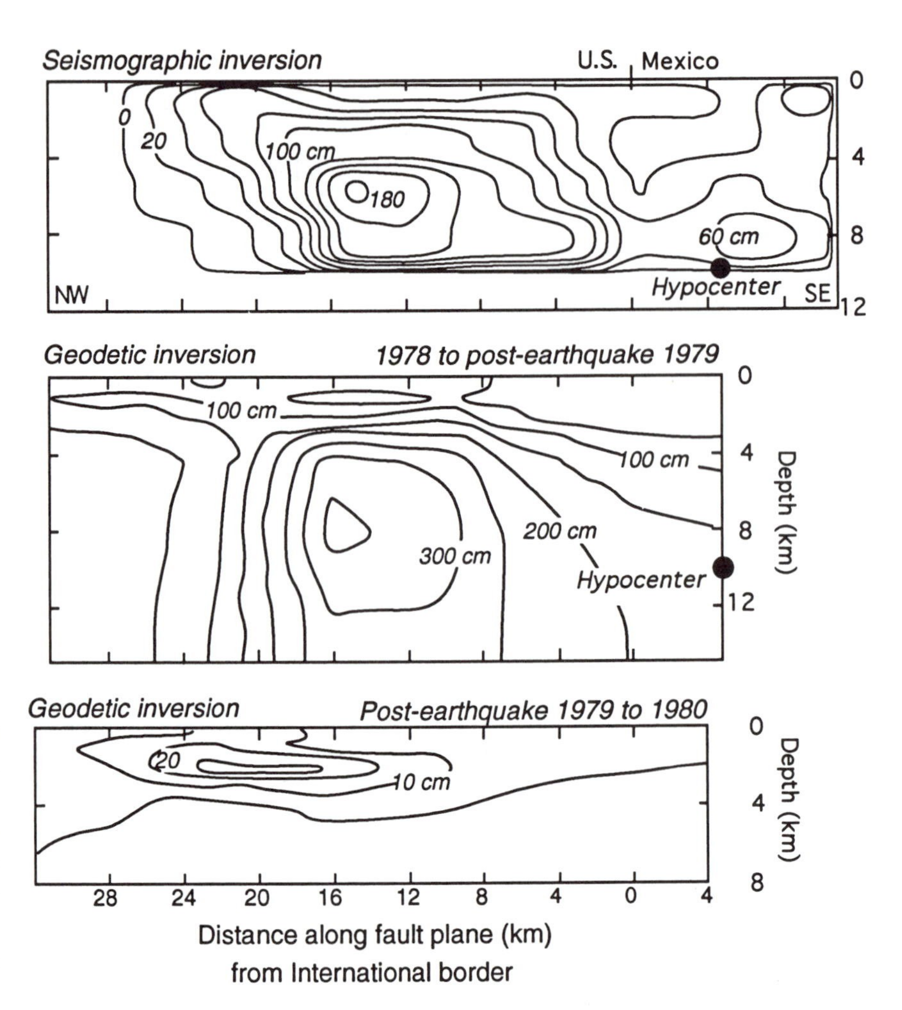

Figure 8–49. Distribution of coseismic dextral slip along the plane of the Imperial fault, determined by Hartzell and Heaton (1983) from seismographic records and by Crook (1984) from geodetic data. Compare with values of coseismic slip at the surface, shown in Figure 8–48.

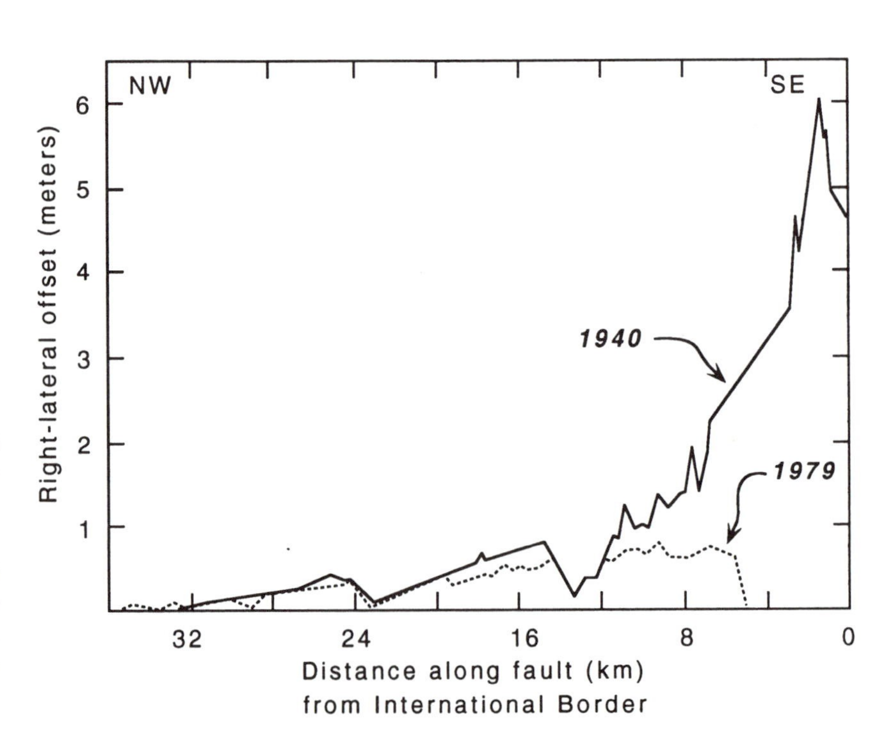

Figure 8–50. Comparison of dextral slip along the Imperial fault in 1940 and 1979 leads to interesting questions about fault segmentation and the repetition of earthquakes on an individual fault plane. Note the similarity of values along the northern two-thirds of the 1979 rupture.

crease of 1979. This coincidence may indicate the persistence of a segment boundary through several earthquake cycles. Because of these similarities in the 1940 and 1979 surficial patterns, it will certainly be interesting to document the next major rupture of the Imperial fault.

The Landers Earthquake of 1992

The M_W 7.3 Landers earthquake differs in several fundamental ways from the 1979 Imperial Valley earthquake. Not only is it a much larger earthquake, but it is also more complex. We discuss it here, because it is currently the best-documented strike-slip earthquake of its size. Seismographic, geologic, and geodetic investigations all provide important information about the sources, context, and processes of this earthquake.

The Landers earthquake was produced by faults within the Eastern California Shear Zone (ECSZ), an 80-km-wide zone of predominantly right-lateral faults that splays northward from the San Andreas fault toward the western Basin and Range Province (Figs. 8-46). The geodetically determined rate of dextral shear across the entire zone is about a centimeter per year (Sauber et al., 1986; 1994).

Two months of intense seismic activity on faults south of the mainshock ruptures, close to the San Andreas fault, preceded the Landers earthquake (Sieh et al., 1993; Hauksson et al., 1993). Seismic harbingers such as this are atypical for large earthquakes in California. The largest of these precursory earthquakes was the Joshua Tree earthquake of April 28, a M_W 6.1 event with aftershocks that delineate a curved mainshock source (Fig. 8-51). This 25-km-long, vertical fault did not rupture the surface, but does project upward to the surficial trace of a mapped minor fault (Rymer, 1992). This large preshock was, itself, preceded by a foreshock—a M_L 4.6 event that occurred about two hours earlier.

Aftershock activity following the Joshua Tree earthquake was unusual, in that it did not die off as quickly as most California aftershock sequences. Activity expanded northward in the two months leading up to the Landers earthquake. By mid-June, small earthquakes were occurring very close to the eventual Landers hypocenter (Fig. 8-51). In the hours prior to the mainshock, several small earthquakes occurred no more than a kilometer from the hypocenter of the upcoming main event.

The Landers earthquake was produced by nearly unilateral, northward rupture of five major right-lateral faults and fault segments during a period of about 25 sec. From south to north, these are the Johnson Valley, Landers, Homestead Valley, Emerson, and Camp Rock faults. Figure 8-30 depicts the geometry of these faults and also the generalized pattern of lateral offset along them. Such complex faulting has been uncommon in California's historical record, but has occurred during large earthquakes elsewhere (for example, the Japanese earthquakes of 1891 and 1930, Figs. 8-10 and 8-36b, and the Haiyuan, China, earthquake of 1920, Fig. 8-8). Harris and Day (1993) provide an analysis of the physical conditions that might allow coseismic rupture across such major steps and discontinuities in active faults.

Figure 8-52 illustrates, in more detail than Figure 8-30, lateral slip values that were measured along the fault zone. The great fluctuations along strike (Fig. 8-52a) imply stresses far greater than the strength of crustal rocks and must reflect variability in off-fault warping (as is depicted for the 1906 San Francisco earthquake in Fig. 8-22). The strong asymmetry in slope evident along the Camp Rock fault in Figure 8-52b is typical for many of the faults of the Landers earthquake. This suggests that failure of the Camp Rock fault was driven by static stresses imposed by failure of the Emerson fault, immediately to the south. Zachariasen and Sieh (1995) observe asymmetry of lateral slip along minor faults in the stepover between the Emerson and Homestead Valley faults to infer the coseismic rupture history depicted in Figure 8-53.

Inversions of seismographic, geodetic, and geologic data indicate that the earthquake consisted of three major concentrations of moment release, the first on the Johnson Valley and Landers faults, the second on the Homestead Valley fault, and the third on the Emerson fault (Wald and Heaton, 1994; Cohee and Beroza, 1994). At the surface, slip on the five major faults was commonly greater than 2 m, and values as high as 6 m were measured locally (Figs. 8-18b, 8-21d, 8-30, and 8-32). Two increments of slip, totaling about 20 cm, also occurred on the Eureka Peak fault, south of the mainshock ruptures, probably about 35 sec. and about 3 min. after the beginning of the mainshock (Fig. 8-51; Hough, 1994).

One inversion of geodetic, strong-ground motion, and teleseismic data, constrained loosely by the surficial rupture, yields a model of the rupture process displayed in Figure 4-20. This model is particularly intriguing from a geological perspective, because the two principal stepovers in the fault system can be related to retardation of the northward-propagating rupture front. During the first 6 sec. of the earthquake, the rupture propagated northward along the Johnson Valley fault at a rate of about 3.6 km/sec. During the next 4 sec., the rupture made very little northward progress, but the magnitude of slip on the faults of the southern stepover (the Landers, northernmost

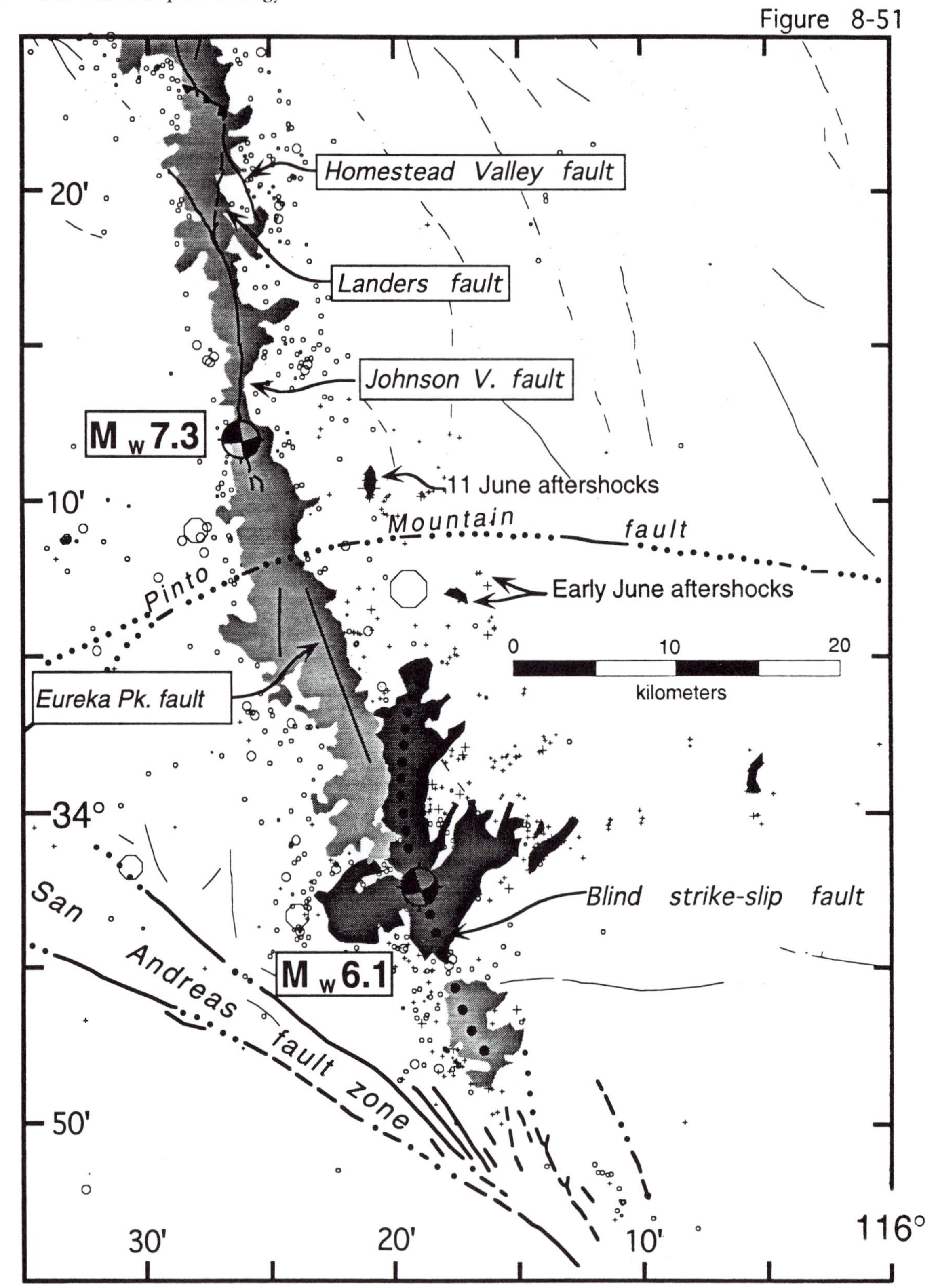

Figure 8–51. Relationship of the Joshua Tree earthquake and subsequent Landers earthquake to each other and to geologic structures. The Landers earthquake (M_W 7.3) was preceded by two months of seismic activity, including the Joshua Tree earthquake (M_W 6.1) as well as the foreshocks and aftershocks of the Joshua Tree earthquake. The Joshua Tree aftershocks were concentrated in the darkly shaded area, which is roughly cruciform in shape and centered on the mainshock's epicenter. Aftershocks that occurred more broadly, between

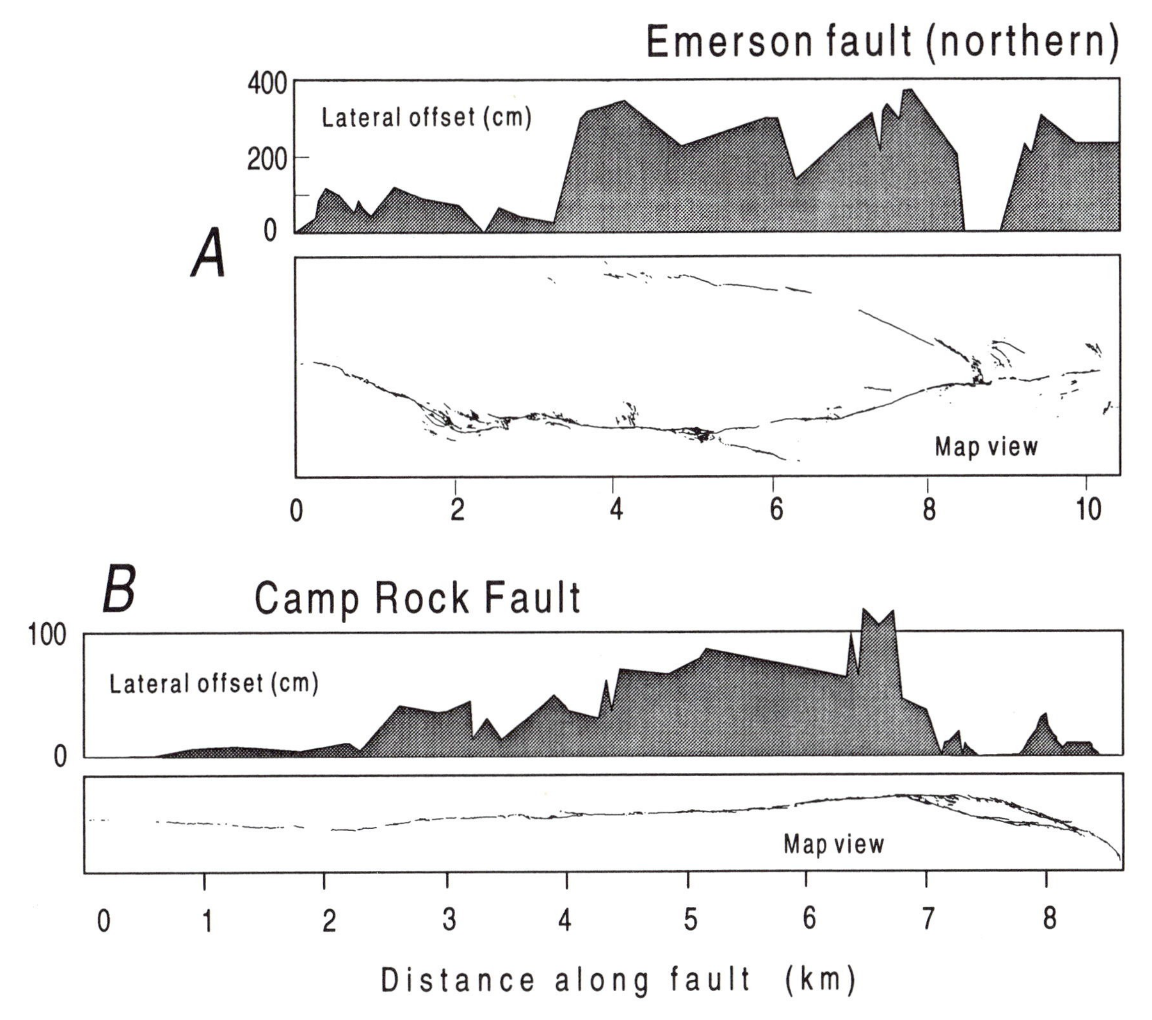

Figure 8–52. Details of lateral slip measured along two sections of the Landers fault zone. See Figure 8–30 for locations. (a) Along the northern portion of the Emerson fault, slip was commonly about 300 cm. This value diminished northward to less than a meter over only a few hundred meters. (b) Slip magnitude rose abruptly to a maximum of about 100 cm from the southern end of the Camp Rock fault but died off gradually toward its northern tip.

Johnson Valley, and southern Homestead Valley ruptures) increased to several meters. Between 10 and 11 sec., a meter or so of slip was released in the shallow crust at the northern end of the southern stepover. This may represent failure of a 3-km-long thrust fault (north edge of Fig. 8–51) that was induced by dextral slip on the Landers fault seconds earlier and may represent the end of the first pulse of the earthquake (Spotila and Sieh, 1995). Also at this time, the first rupture of faults north of the southern stepover began, with deep rupture on the Homestead Valley fault, just north of the southern stepover. In

---

the times of the Joshua Tree and Landers earthquakes, are represented by small crosses. The source of the Joshua Tree earthquake, which did not rupture the earth's surface, is represented by the arcuate line of large dots. The Landers earthquake was produced by dextral slip on the Johnson Valley, Landers and Homestead Valley faults, and other faults north of the edge of the figure. Aftershock activity was intense in the irregular band represented by the lighter shading and was particularly intense between the epicenters of the 7.3 and 6.1 earthquakes. More-broadly dispersed aftershocks of the Landers earthquake appear as open circles. Rupture on the Eureka Peak fault occurred 30 sec. after initiation of the Landers earthquake. Epicenters shown here are for events of $M > 1.8$ that were recorded by the Southern California Seismic Array, provided through the courtesy of E. Hauksson.

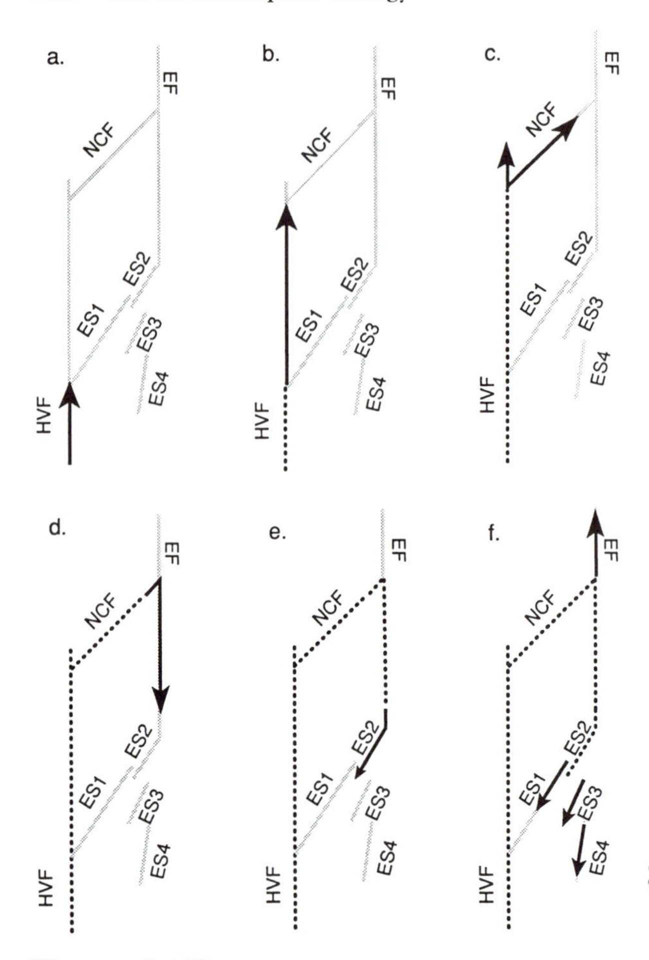

Figure 8–53. Asymmetry of lateral-slip magnitude on minor faults within the 2-km-wide stepover between the Homestead Valley and Emerson faults suggests this sequence of coseismic faulting through the stepover. Minor faults with highest slip values near their southern ends were probably induced to fail by slip on the Homestead Valley fault. Minor faults with highest slip values near their northern ends were probably induced to fail by the Emerson fault. This stepover is labeled "3" in Figure 8–30. From Zachariasen and Sieh (1995).

the following several seconds, rupture propagated upward and, principally, northward. Northward propagation was retarded as the rupture front reached the northern stepover. As before, at the southern stepover, slip continued to grow on faults in the vicinity of this stepover, between the Homestead Valley and Emerson faults. The rupture then continued to the end of the Emerson fault.

This controversial model suggests that the two major stepover regions provided momentary impediments to rupture during the Landers earthquake. Slip on the faults south of each stepover apparently had to grow to several meters before shear stresses exceeded the strengths of other faults within and north of the stepovers.

Careful mapping of the faults of both stepover regions reveals structural reasons for these delays. In the southern stepover, for example, the faults have not yet integrated into a through-going structure. Figure 8–54 illustrates a likely three-dimensional geometry for these noncontiguous faults. The en échelon nature of the Landers fault, and its lack of tectonic landforms, indicates that it has only recently formed between the Johnson Valley and Homestead Valley faults. The thrust fault is not connected to the Landers fault, but has formed in response to stresses induced by dextral slip along the Landers fault. And the en échelon faults west of the thrust fault are evidence of the southern propagation of the northern Homestead Valley fault into the stepover.

Aftershocks of the Landers earthquake define an arcuate 5- to 15-km-wide, 85-km-long vertical zone beneath the traces of the fault (Figs. 8-30 and 8-51).

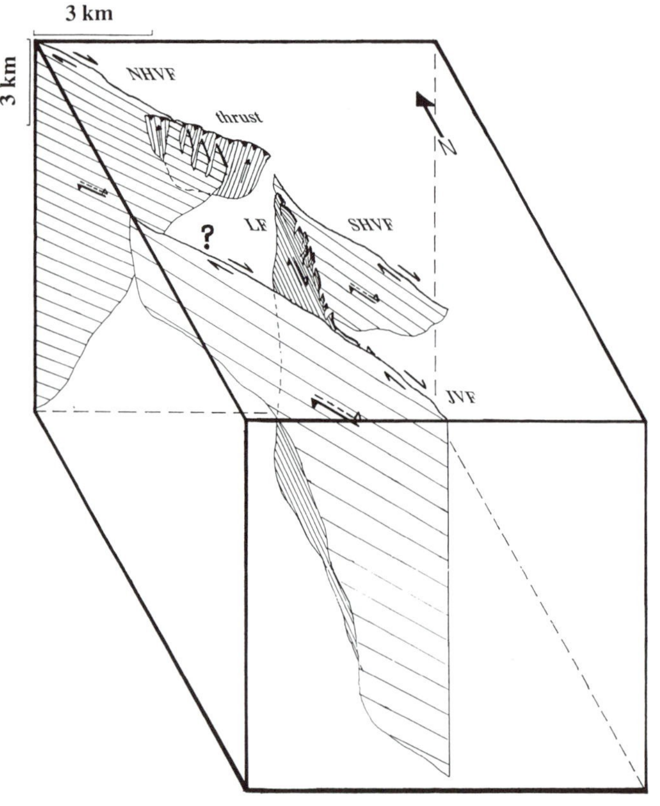

Figure 8–54. Block diagram of the faults of the southern stepover of the Landers earthquake illustrates the lack of through-going faults and evolving nature of the stepover. This stepover is labeled "2" in Figure 8–30. From Spotila and Sieh (1995).

From this pattern alone, the individuality and structural relationships of the six major faults of the earthquake cannot be deduced clearly. The aftershock pattern does reveal other important structural characteristics of the fault zone, however. Several aftershock lineaments reveal fault zones that did not produce surficial rupture. An abundance of aftershocks in the unfaulted rocks between the Eureka Peak and Johnson Valley faults suggests that a myriad of small faults was forced to fail in this region where no mappable connecting fault yet exists. In addition, the aftershock pattern constrains the depth of seismic faulting to about 15 km.

Finally, the band of aftershocks does not extend beneath the southern Homestead Valley, northernmost Johnson Valley, or Camp Rock ruptures, despite the fact that these faults sustained more than a meter of dextral surficial slip. Inversion of geodetic measurements indicates that the large slips seen along the Camp Rock fault must not extend more than a couple of kilometers into the crust. Perhaps the same is true for the other two fault segments that lack aftershocks. Even so, it is difficult to understand the lack of aftershocks along these fault segments, given that each sustained offsets of a meter or more.

During the Landers earthquake, slip occurred on numerous secondary faults near the principal faults. Rupture through a contractional jog along the Emerson fault, for example, resulted in failure of nearby secondary thrust faults and a zone of conjugate, intersecting left- and right-lateral faults (Fig. 8-32). Several thrust and normal faults, each with geomorphic evidence of prior activity also formed within a few hundred meters of the principal faults (see, for example, Zachariasen and Sieh, 1995). This earthquake made it clear that mappable secondary structures along strike-slip faults can fail in concert with the principal structures.

PALEOSEISMOLOGY

Much of this chapter has been devoted to discussion of the tectonic and structural context of strike-slip earthquakes. We discussed strike-slip earthquakes first in their global, plate-tectonic contexts, then with regard to their more regional structural and geomorphic aspects, and then focused on a couple of well-investigated historical strike-slip earthquakes.

We turn now to a brief discussion of *paleoseismology,* a term that denotes the investigation of individual earthquakes well after their occurrence. Many questions of interest to seismologists cannot be addressed without information about earthquakes that occurred prior to the advent of instrumental recordings and modern investigations. These include questions about faults, folds and secondary effects associated with prehistorical earthquakes (and poorly documented historical events). Other questions of importance concern the length of an earthquake cycle, a cycle's regularity or irregularity, and the similarity of source characteristics from one event to the next.

Geomorphic Evidence of Past Earthquakes

In most places, the historical and instrumental records of large earthquakes are so short and incomplete that reliance on paleoseismic data is necessary for understanding the behavior of faults. In California, for example, the occurrence, felt effects, and approximate rupture length are all that is known from historical records about the great earthquake of 1857 (Agnew and Sieh, 1978). Lateral slip along the San Andreas fault associated with the earthquake was first measured more than a century after the earthquake, using small offset landforms, predominantly stream channels and gullies. Several of these small offsets are faintly visible at the scale of the topographic map depicted in Figure 8-43, between the 690- and 1050-m marks. Careful measurement of these small features yield dextral offset values of about 9.5 m. Sieh (1978b) and Wallace (1968) concluded that these offsets represented coseismic slip during the great earthquake of 1857, because (1) they are the smallest geomorphic offsets at that locality, (2) creep has not occurred there in the twentieth century (Brown and Wallace, 1968), and (3) the historical record suggests that new channels are cut across the fault much more frequently than large earthquakes recur.

Geodetic measurements support this conclusion. Recent remeasurement by GPS geodesy of a nearby fault-spanning section line, first surveyed by chain in 1855, suggests 11 $\pm$ 2 m of slip across the fault in 1857 (Grant and Donnellan, 1994). Farther south, small gullies and alluvial fans display offsets of about 6.5 to 7 m (Fig. 8-55). These offsets have also been ascribed to the 1857 event. The difference between the geodetic and geologic measurements is attributable to off-fault warping within a few meters to hundreds of meters of the fault. Figure 8-55b, for example, displays a bend in the downstream segment of an offset stream that may be evidence for several meters of warping.

Offsets about twice as large as these are also common in this area. The gully in the closest foreground of Figure 8-23 is offset about 16 m along the San Andreas fault. The offset probably accumulated during two earthquakes—about 8 m in 1857 and about 8 m in a previous earthquake about A.D. 1480 (K. Sieh and C. Prentice, unpubl. data).

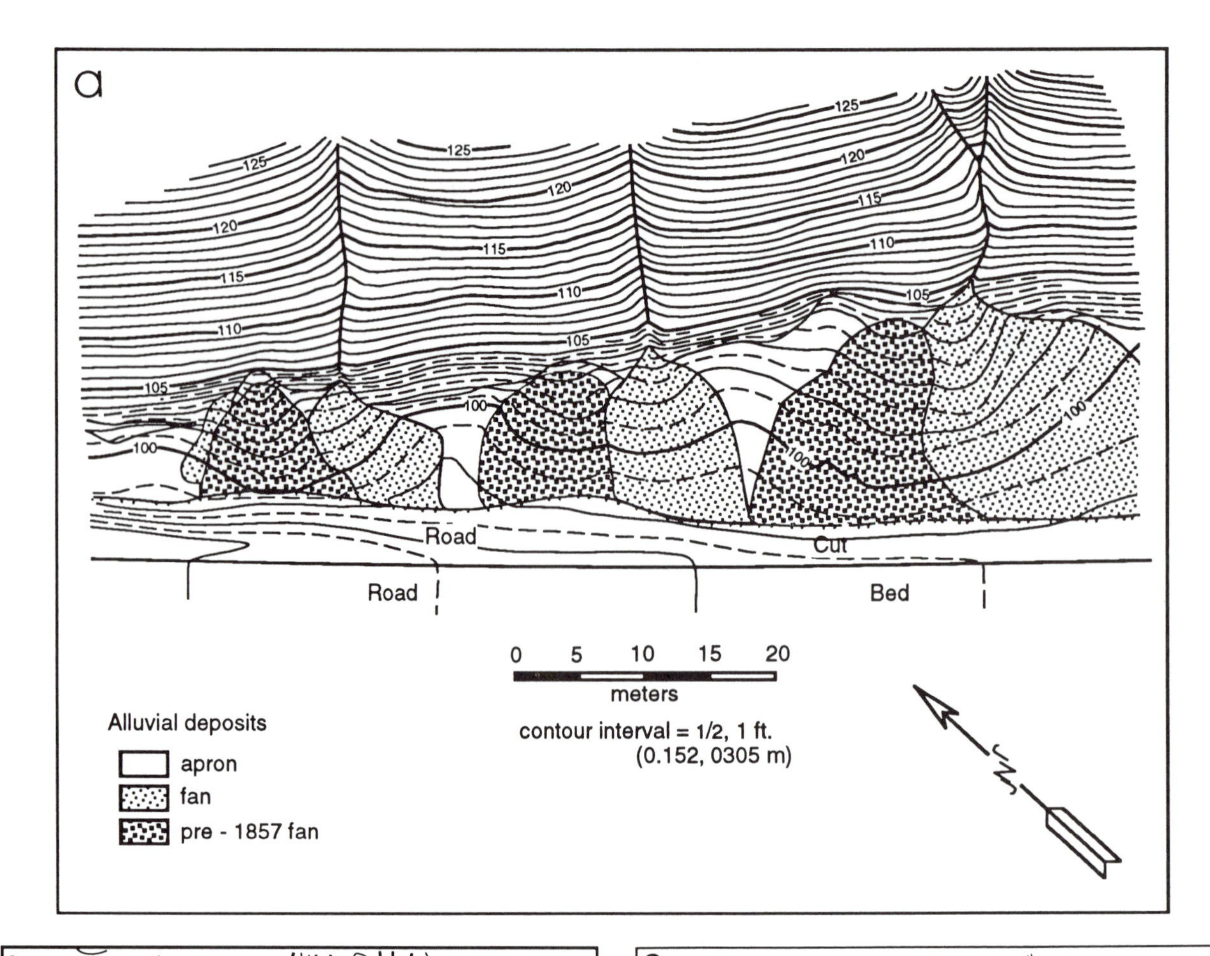

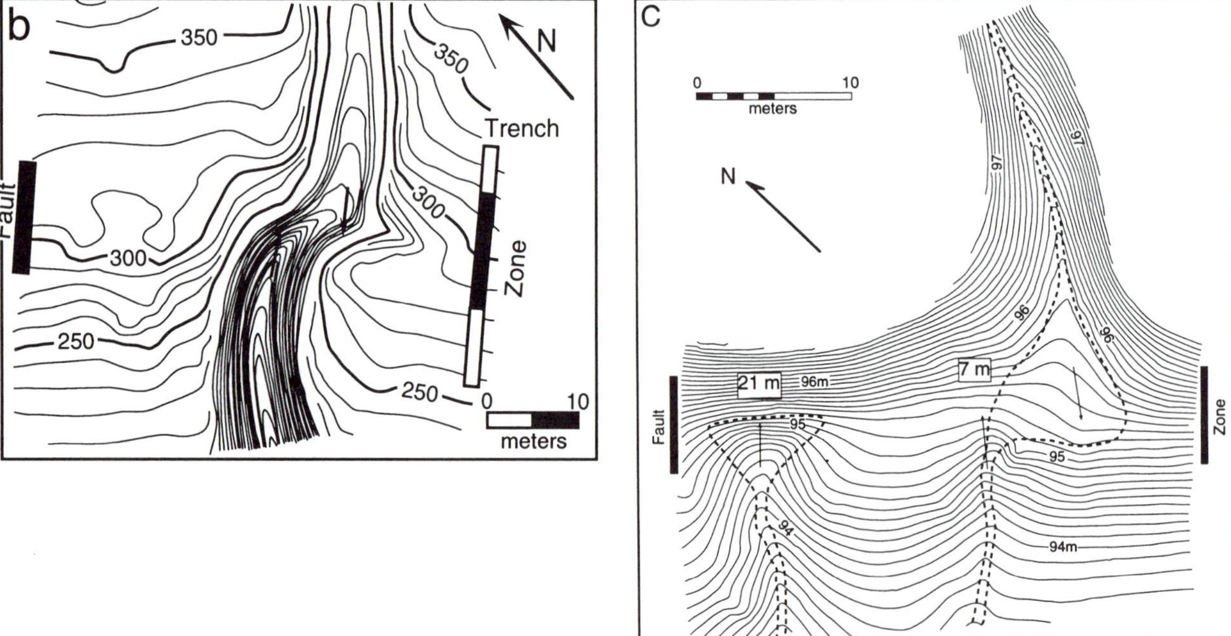

Figure 8–55. Topographic maps of small features, whose offset along the San Andreas fault is ascribed to the earthquake of 1857. (a) Three small alluvial fans, offset from their source gullies about 6.5 m. Modified from Sieh (1978b). (b) Channel incised into young alluvial fan prior to being offset about 7 m. Thin arrows whose tips are separated by 7 m indicate projections of channel segments into the fault zone. Note possible right-lateral warping of downstream segment of channel. Modified from Grant and Sieh (1994). (c) Shallow channels offset about 7 m and 21 m from upstream segment of channel. Thin arrows indicate projections of channels into the fault zone. Numbers in boxes indicate measured offsets. Modified from Grant and Sieh (1993).

The use of small landforms to infer coseismic slip in past events is becoming quite common. Lindvall et al. (1989) recognized many small offset landforms (predominantly rivulets) along the Superstition Hills fault. Many of these features are offset about twice as much as offsets that accrued during and immediately following the M_W 6.6 Superstition Hills earthquake of 1987. Figure 8-56a illustrates one example from their data set. Figure 8-56b displays all of their data, superimposed upon a graph of 1987 slip along the fault.

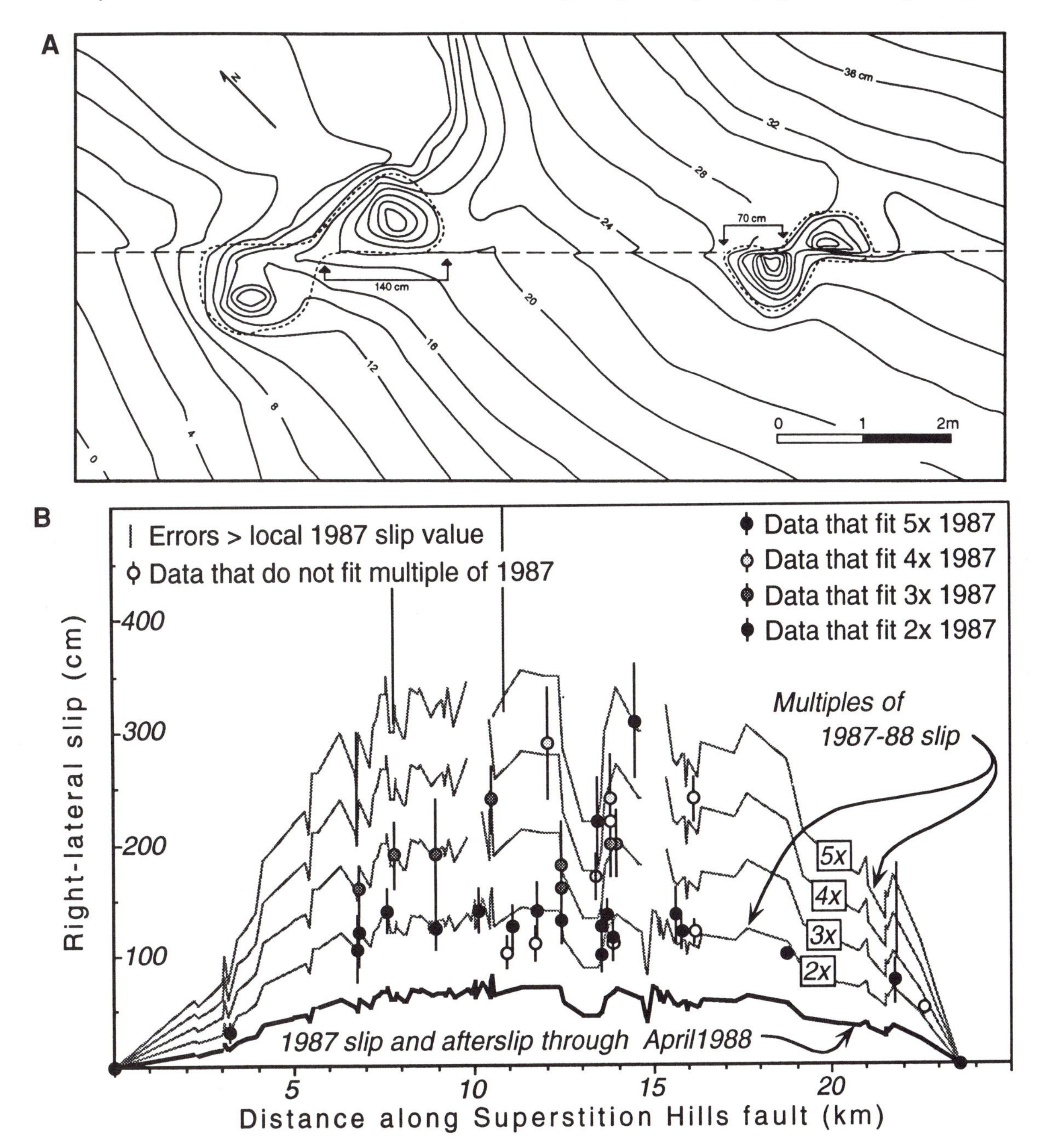

Figure 8–56. This map and plot illustrate one example that suggests faults tend to slip roughly similar ("characteristic") amounts, from earthquake to earthquake. (a) Topographic map of small sand dunes, which straddle the Superstition Hills fault, southern California. Right-lateral offset of the dunes was 70 ± 5 cm during a M_W 6.6 earthquake in 1987. Total offset of the older dune is 140 ± 20, about twice this value, which suggests that slip at this site was similar during at least the latest two events. (b) Small dextral offsets along the Superstition Hills fault at many locations strongly support the hypothesis of characteristic earthquakes along this fault. Redrafted from Lindvall et al. (1989).

The fainter lines represent multiples of the local 1987 values. The degree to which the paleoseismic offsets fall on these multiples is a measure of the degree to which events on this fault have a characteristic slip function. Lindvall and his colleagues concluded that the previous slip event on the fault was of about the same size as the 1987 event. Their data also suggest that even older events had offsets similar to those of 1987. Because substantial aftercreep followed the 1987 earthquake, they had to account for aseismic slip in their discussion of the paleoseismic significance of their measurements.

Other examples of the use of geomorphic evidence to characterize past earthquakes are numerous. Berryman et al. (1992), for example, interpreted offset channels along the historically dormant Alpine fault of New Zealand as evidence for two discrete right-lateral events of about 6 m each. These small offsets also demonstrate that displacement on the fault is predominantly strike-slip. Offset neoglacial moraines and stream channels along the Denali fault in Alaska document individual prehistoric dextral-slip events of 7 to 15 m each (Sieh, 1981). Offsets associated with the series of moderate to large earthquakes on the North Anatolian fault in the past half century (Fig. 8-7) have been recovered years to decades after the events by Barka (in press, 1995). Deng and Zhang (1984) constructed a diagram of right-lateral slip along the Fuyun fault associated with a M8.0 earthquake in China in 1931 from 171 measurements of offset landforms made a half century after the earthquake (Fig. 8-57). Several offsets were greater than 10 m, and one was close to 15 m.

Although geomorphic evidence of earthquakes is, in general, more readily accessible than stratigraphic evidence, interpretation of an offset landform, *per se*, may suffer from significant uncertainties. Without stratigraphic or chronologic constraints, it is difficult to determine the number of discrete events associated with a particular offset landform. Similarly, without supplementary stratigraphic data, afterslip or interseismic creep may be indistinguishable from coseismic slip. Furthermore, accurate measurement of the magnitude of an offset, decades, centuries, or millennia after its formation, may be hindered, if not altogether thwarted, by erosional or depositional modifications of the offset landform.

Stratigraphic Evidence of Past Earthquakes

An offset entombed in a sedimentary sequence may be more reliably measured and more convincingly dated than a geomorphic offset. In geologic environments where rates of sedimentation are faster than rates of vertical deformation and where distinct layering is preserved, discrete *event horizons* may be well-preserved.

Simple Examples

Figure 8-58 illustrates a few idealized examples of paleoseismic data from a strike-slip fault. The map in the center of the figure shows the location of each of the six hypothetical cross sections. All six cross sections display evidence for fault slip and/or seismic shaking that occurred when the top of the thick black layer was the ground surface. In cross sections a and c, movement on simple, vertical strike-slip faults is apparent. Variability of vertical separations of the faulted beds is a clue that a component of strike-slip motion occurred at all three localities. A component of vertical separation has resulted in the creation of a scarp along the fault in panel a. The triangular wedge of debris that underlies the lower half of the scarp rests upon the event horizon, that is, the ground surface at the time of the event. Subsequent burial preserved the event horizon and the fault scarp. Slip on the fault in panel c also resulted in the formation of a scarp, but faulting did not penetrate upward through the surficial layer. The event horizon is clear, however, because (1) the deformed layers do not thicken or thin adjacent to the fault and (2) the overlying layers are undeformed and form a buttress unconformity with

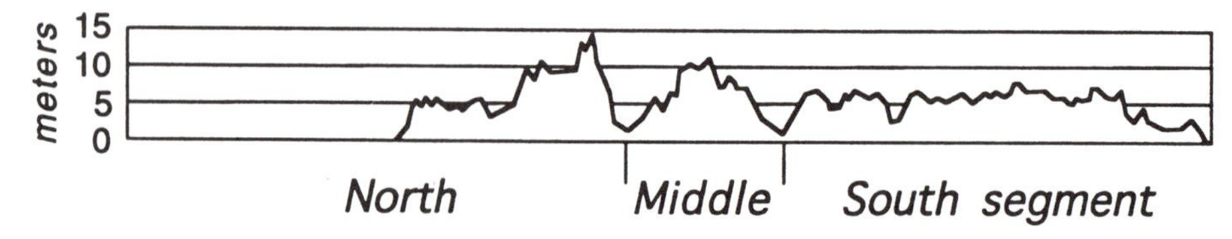

Figure 8–57. Right-lateral offset along the three segments of the 180-km-long rupture of the Fuyun fault in China. The fault ruptured during a M8.0 earthquake in 1931, but offsets were well-enough preserved geomorphically to be measured a half century later. The maximum offset of 14 m is the greatest documented strike-slip offset associated with a single earthquake. From Deng and Zhang (1984).

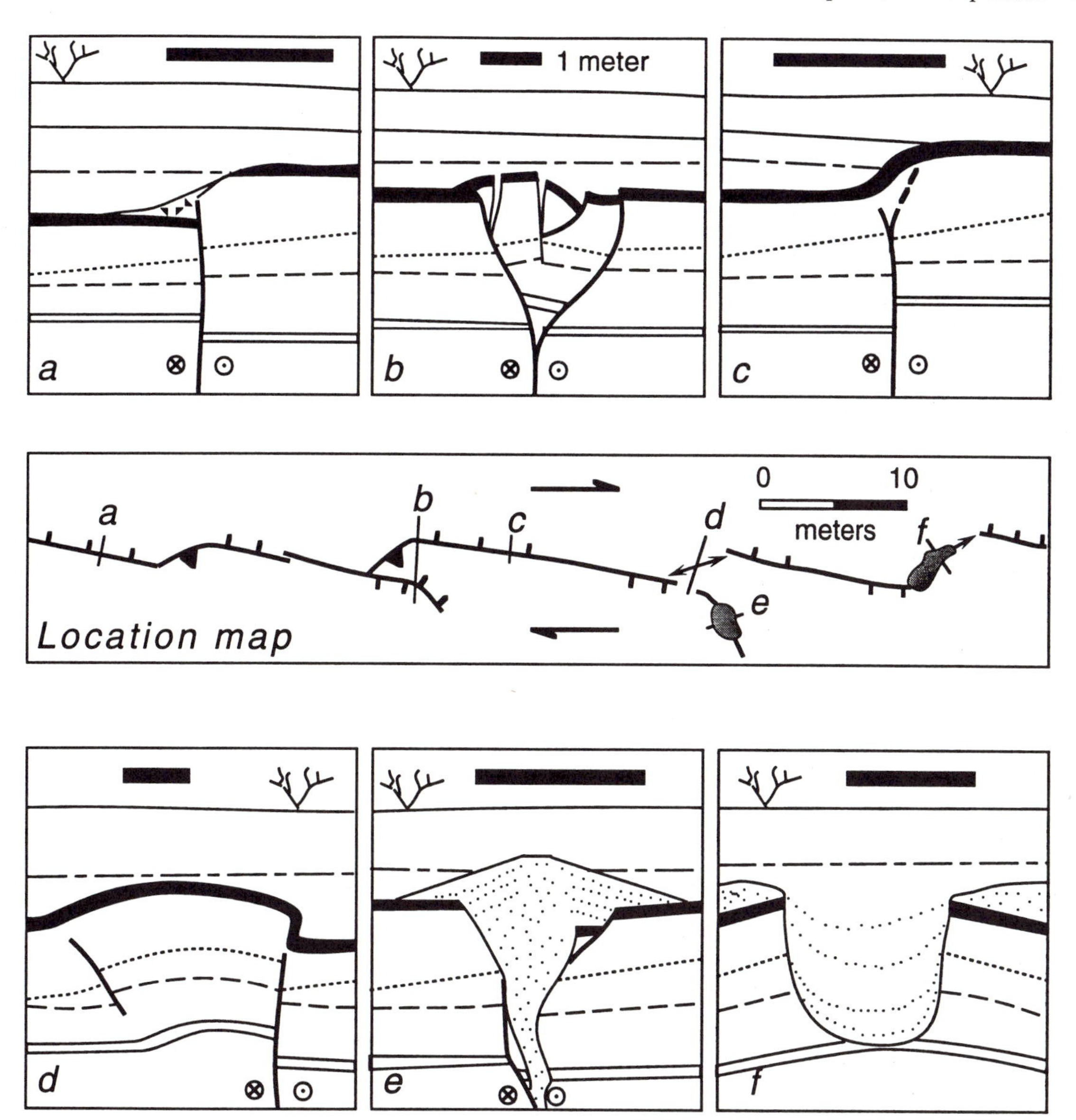

Figure 8–58. These idealized cross sections across strike-slip faults show various kinds of evidence for paleoseismic events. Top of solid black bed is the event horizon. Shaded horizontal bars are 1 meter long. Lines with arrows on location map indicate crests of anticlinal folds. Mismatches of strata across some of the faults is an indication of strike-slip motion.

the scarp. Panel b shows the relationship of two en échelon faults in the subsurface. The cross section shows secondary deformation associated with the stepover between the two faults. The event horizon is clearly defined by the presence of a small moletrack and by cracks, fissures, and tilted blocks that are buried by superjacent sediment.

Folding and tilting is common within zones of strike-slip faulting. Panel d, for example, shows an anticline that formed in the stepover between two en échelon faults when the top of the black layer was the ground surface. Although no faulting, fissuring, or cracking appears at its surface, the top of the black bed is clearly the event horizon. Beds below that level display the same degree of folding, whereas the beds above the event horizon are undeformed. The thickness of the younger bed also varies as a function of the shape and amplitude of the subjacent fold. A structure-contour map drawn for a post-earthquake bed such as this would reveal the geometry and amplitude of vertical deformation along the fault (see, for example, Sieh, 1984).

In panels e and f, liquefaction has accompanied the faulting. Therefore, within the limits of stratigraphic resolution, seismic shaking occurred at the same time as the faulting event. In panel e, liquefied sand has vented onto the event horizon, after the faulting, by traveling upward along the fault. (see Chapter 12 for a discussion of liquefaction). Erosion of the fault zone during the evacuation of the sand is apparent in the cross section. In this particular case, the fault has a component of normal, dip slip. In panel f, a crater has formed along the crest of an anticline within the contractional stepover between two en échelon shears. Unlike the sandblow in panel e, this crater was not fed from below via a pipe or fissure. Sudden slip on the faults produced high pore pressures in the saturated layers. These pressures led to the evacuation of material from the core of the fold.

Figure 8–59 displays two actual examples of evidence for disruption during strike-slip events. Panel a depicts two strands of the San Andreas fault breaking marsh, stream, and aeolian deposits at Pallett Creek, 55 km northeast of downtown Los Angeles. The black beds are silty marsh peats. The other sediments are fluvial sands and gravels, aeolian silts and very fine sands. The top of the peat just below unit 71 was the ground surface at the time of the most obvious event, event V. The most obvious evidence for event V is the buried scarp at the top of the fault on the left. A vertical component of slip resulted in formation of a scarp about 30 cm high. The mismatches of bed thicknesses and facies across the fault are evidence of substantial strike-slip motion. The fault scarp is buried by unfaulted aeolian and fluvial beds. Radiocarbon dates on the top of the thick peat underlying unit 71 and on the very thin peat (unit 72) overlying the event horizon constrain the date of the earthquake to within a few decades of A.D. 1480.

More-subtle, but nevertheless compelling, evidence for event V exists on the other fault in panel a as well. (Vertical separation across this fault from two later earthquakes has been restored in this figure, so there are slight mismatches in unit 71 and younger layers across the fault.) The two thick peats and intervening

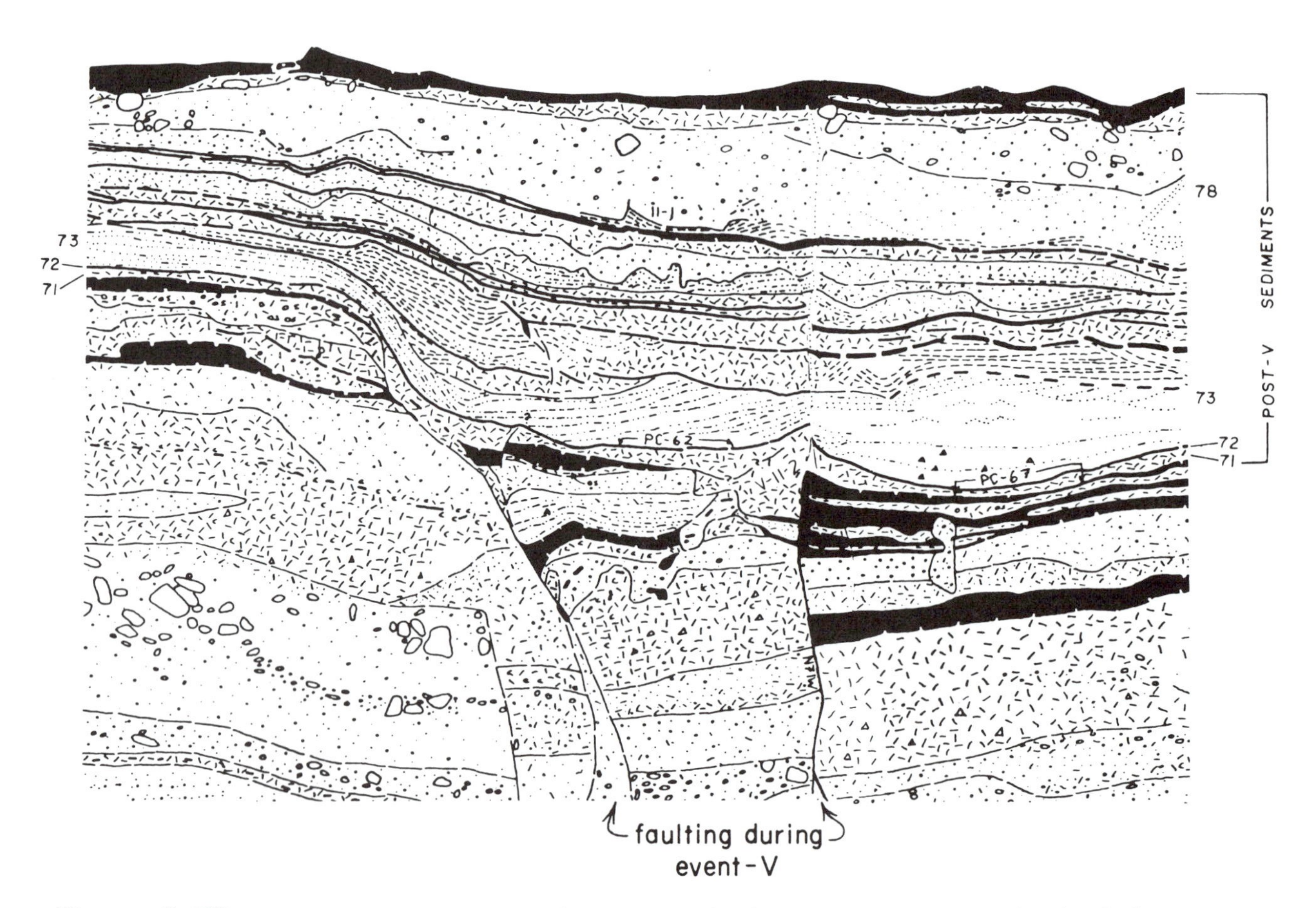

Figure 8–59. Examples of cross-sectional exposures of paleoseismic events in strike-slip fault zones. (a) Fault breaking marsh peats, aeolian silts, and fluvial sands and gravels is associated with oblique dextral slip on the San Andreas fault in about A.D. 1480. One inch equals one-half meter. From Sieh (1978a).

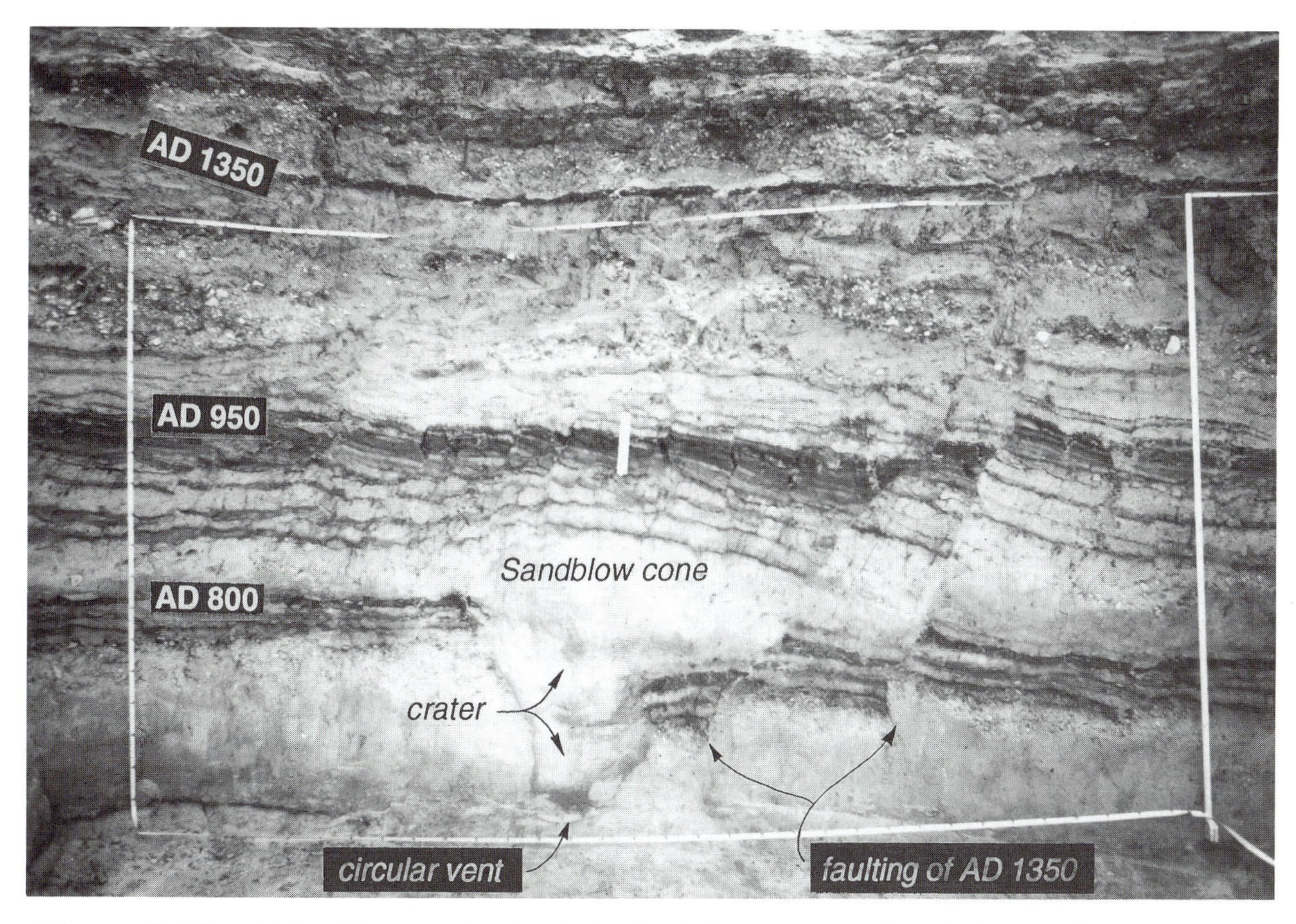

Figure 8–59. (b) Sandblow associated with dip-slip faulting in San Andreas fault zone, about A.D. 800.

sediment below unit 71 are severely disrupted to the left of the fault. This is evidence for tearing of the near-surface section by strike-slip motion during event V. The disturbed beds and irregular event horizon above them are buried by unit 71, an aeolian unit, and peaty unit 72.

In panel b of Figure 8-59 is a sandblow that formed along a dip-slip fault within the San Andreas fault zone about A.D. 800. The event horizon is the top of the thick peat labeled "A.D. 800." Two faults that post-date the sandblow are easily recognized, because they displace units that post-date the sandblows. The cut also exposes a crater that was excavated into the marsh by the evacuating sand. The crater formed along and obliterated a fault with about 20 cm of dip slip, right side down. Visible in panel b are the vent that fed the sandblow and the crater through which sand exited onto the surface of the ground. The sand that was evacuated through the sandblow vent and crater rests upon the A.D. 800 peat.

Multiple Events

Often, more than one event horizon is visible in an exposure. The presence of more than one event in a stratigraphic succession results in a more complex record, yet such records are commonly decipherable.

Figure 8-60 is a map of the wall of a trench cut across the San Andreas fault at Pallett Creek. The cross section displays evidence for eight earthquakes that occurred between about A.D. 750 and 1857. Each piece of evidence is labeled with a small rectangle. The first letter within the rectangle is the letter designation for the earthquake (that is, earthquake Z, X, V, T, R, I, F, and D). The youngest earthquake horizon (A.D. 1857) is the top of unit 88. The earthquake produced flame-like diapirs of unit-88 peat into unit 88. The previous event horizon (A.D. 1812) is the top of unit 81. It has a sandblow resting upon it near the right edge of the cross section. Event V (about A.D. 1480) occurred when unit 68 was at the ground surface. In this cut, a fissure about two meters from the right side of the cross section is the best evidence for event V. Event T (about A.D. 1350) occurred just after unit 61 had been deposited. Slumps, fissures and deformations of unit 61, 2 to 4 m from the left side of the cross section are the clearest evidence for this event. Event I (about A.D. 1000) produced a large sandblow sheet, triangular in cross section, atop unit 45 in the right half of the cross section. The source of the sand was unit 43, a thick sand less than a meter below unit 45. Two kettledrum-shaped craters are the clearest evidence for event F, which occurred about A.D. 800. These are filled with unit-39 sand and silt about 2 to 4 m from the left side of the cross section. The craters appear to be ventless and to have been filled mostly by sediment settling out of water left sitting in the craters. The increasing thickness of unit 38 away from the craters indicates that these two craters formed at the crest of an anticline. The event-F craters obliterated some, but not all, evidence for faulting and liquefaction during event D, only a few decades prior to F. A few flame-like laminations of sand within unit 34 and faults in unit 33 are still visible beneath the craters. This paleoseismic record is discussed in more detail by Sieh (1978a).

Figure 8-61 illustrates another example of multiple events visible in one exposure. In this 4.5-m-deep exposure of the San Andreas fault zone at the Bidart site, central California, evidence for at least six, and perhaps seven, large events is apparent. Most of the event horizons are drawn and labeled with radiocarbon dates in index panel a. The site is on the distal end of a large alluvial fan. Sediments are predominantly massive pebbly gravels, massive clayey, sandy debris-flow beds, normally graded (quiet-water) silty fine sands and laminated to cross-bedded silty fine sands. Weak paleosols, shown as vertical hachures and indicated by bioturbation of the upper few centimeters of a bed, represent dozens of depositional hiatuses, each a decade or so in duration. Depositional hiatuses of a century to a millennia are represented by extensively bioturbated massive pebbly, sandy, silty units—shown in panels b, c, and d as unpatterned areas. Radiocarbon dates on detrital charcoal show that the section spans the period from about 3300 years B.P. to the present.

The ground surface at the time of the most recent disturbance is obvious, about 40 cm beneath the surface of the excavation. It is labeled "1857" in panel a. Evidence for the event is depicted in panel b below the circled "1." The uppermost event horizon consists of four major fault traces across a 16-m-wide moletrack that consists of an anticlinal warp within a graben. This moletrack-style deformation is associated with at least 7 m of strike-slip offset during the great earthquake of 1857 (the offset stream depicted in Fig. 8-55b is nearby).

The cross sections show evidence that previous events within this broad zone also produced tilted blocks, fissure fills, and colluvial wedges. The event

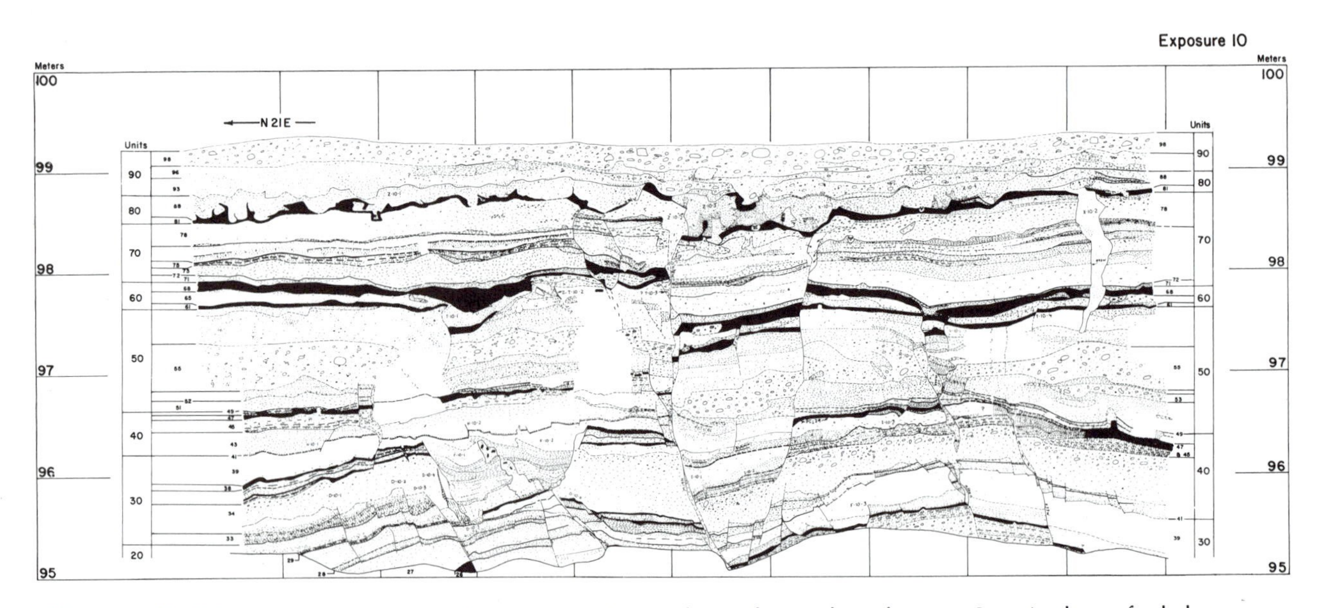

Figure 8–60. Map of trench wall shows evidence for eight earthquakes on San Andreas fault between about A.D. 750 and 1857. Grid spacing is one meter. For large-scale reproduction, see Sieh (1978a).

Figure 8–61. This map of a trench wall shows evidence for several large earthquakes associated with the San Andreas fault in the Carrizo Plain, central California, during the past 3000 years. The uppermost panel, a, shows the locations of panels b, c, and d. Panel d is actually from the opposite wall of the trench and has been inverted. From C. Prentice and K. Sieh (unpubl. data).

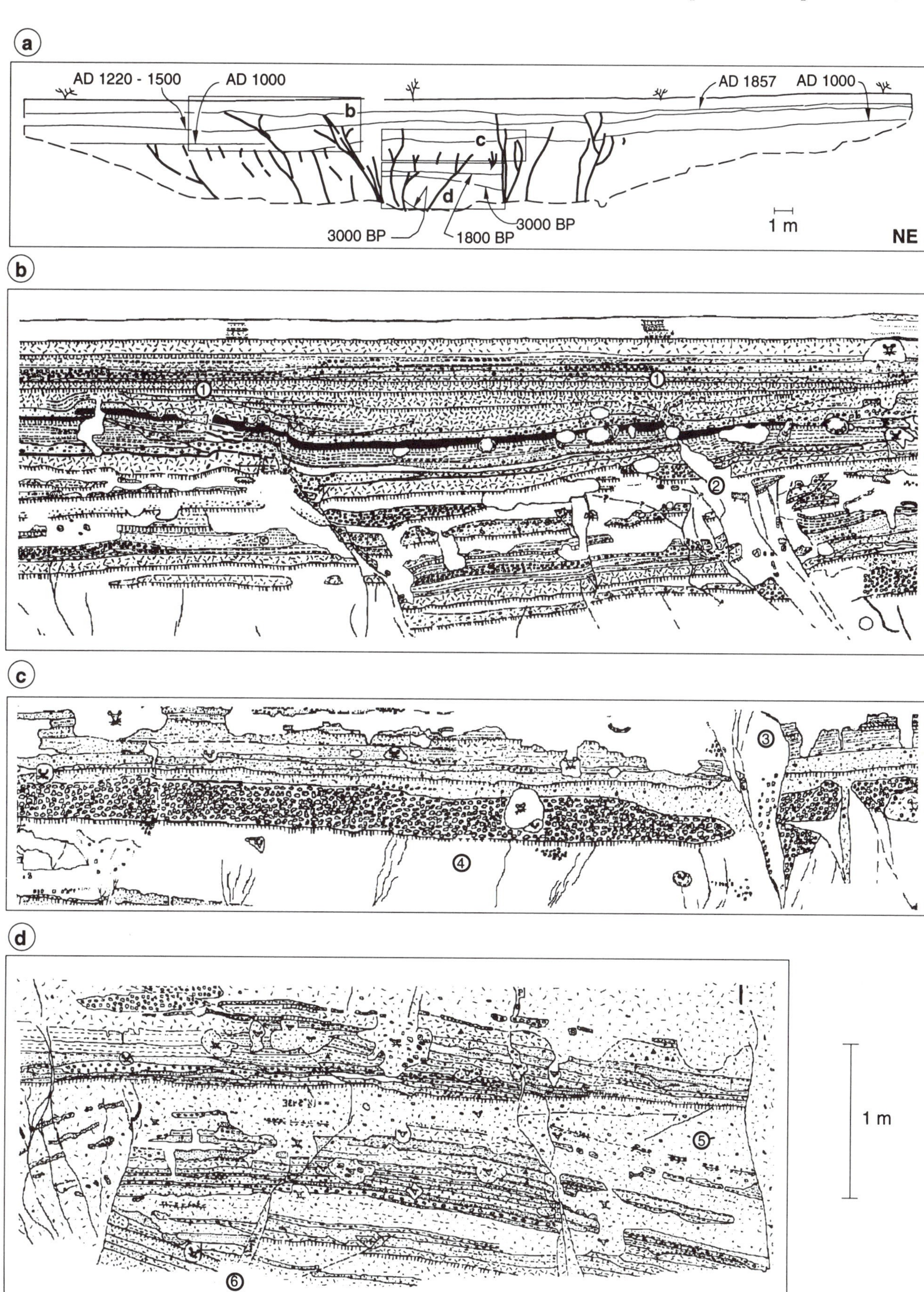
a
AD 1220 - 1500
AD 1000
AD 1857
AD 1000
b
c
d
3000 BP
1800 BP
3000 BP
1 m
NE
b
c
d
1 m

horizon for the youngest of these is labeled "1220–1500" in panel a. Evidence of deformation appears near the right-hand margin of panel b, near the circled "2". There one sees an upward fanning of traces within a highly bioturbated horizon. These faults and the mound formed above these faults were buried by silty and fine sandy beds. Dates from just above the disturbance indicate that it occurred during or before the fifteenth century.

Panel c of Figure 8-61 displays evidence of fissuring and tilting associated with an event about A.D. 1000. Near the "3" in the panel is a complex group of faults and fissures. These clearly antedate the deposition of well-bedded overlying sands and post-date the deposition of a thick, well-sorted gravel.

Also within panel c are faults within a bioturbated unit (labeled "4") that terminate upward at the base of a thick gravel unit. These represent a still-earlier earthquake, which occurred prior to deposition of the thick gravel, well before A.D. 1000.

Panel d displays clear evidence for at least two still-earlier events. The massive, but heterogeneous unit at "5" is a wedge-shaped deposit that formed on the downthrown side of a fault scarp. This *colluvial wedge* formed after formation of the fault scarp, from the spalling of debris off of the upthrown side of the fault. The upper surface of the wedge sloped away from the fault scarp and the source of the colluvium. The scarp has to have been at least as high as the wedge is thick, that is, at least 30 cm. Gradations within the wedge suggest the scarp may have resulted from two events. This evidence is equivocal, but the interpretation of multiple events is consistent with the great length of the period between about 3000 and 1800 years B.P., during which the wedge formed (dormant periods between large earthquakes on this part of the San Andreas fault are commonly less than a few hundred years). The upward truncation of numerous faults at the base of the unit that buries the wedge supports the occurrence of at least one large event during this period. Tilting of the wedge back toward the fault scarp from which it was derived occurred after deposition of the overlying wedge-shaped packet of well-bedded sands and silts. These beds thin to the right, implying that they were deposited against the sloping top of the colluvial wedge. The fact that they now tilt toward the right is an indication that they have been tilted subsequent to deposition. That tilting must have occurred during or before event 4.

The oldest event clearly recorded in this excavation appears at the bottom of the trench, near the numeral "6" in panel d. Here, older sediments were tilted prior to deposition of later sediments. This tilting was probably associated with lateral slip on a nearby fault. The event occurred about 3000 years ago.

Difficulties

Disruption of strata by biological agents—roots, rodents—or low sedimentation rates make sediments imperfect "paleoseismographs." Extensive burrowing by rodents in the area shown in the area represented in Figure 8-61, for example, has homogenized strata deposited between A.D. 1000 and 1500, between 1800 and 1000 years B.P., and between 1800 and 3000 years B.P. Such extensive burrowing occurred only during periods when sedimentation rates were very low. The highly bioturbated units represent long hiatuses in deposition at the site, when deposition was localized elsewhere on the fan.

During these hiatuses, event horizons were obliterated or superimposed. The second-youngest event horizon ("2" in panel b), actually represents four events that occurred between about A.D. 1220 and 1500. Bioturbation and slow accumulation of alluvial strata give the appearance of only one event. Nearby trenches, located on a part of the fan where deposition was more nearly continuous during these three centuries, clearly display evidence for four major events during this period (Grant and Sieh, 1994).

Superimposition of the effects of sequential events may also lead to complex stratigraphic and structural relationships. A sketch of the relationships in an excavation just a few meters along strike from Figure 8-59a, for example, shows complex evidence of three separate events along this fault. The lower-right cross section (f) in Figure 8-62 shows these relationships. The six block diagrams display a plausible sequence of events to explain the relationships seen in that cross section. During an event about A.D. 1100, strike-slip motion was accompanied by formation of a fissure (panel a). The fissure was buried by sediment during the subsequent two and a half centuries (panel b). A second event occurred about A.D. 1350 (panel c). During that event, a fissure also formed, and the right half of the older fissure was moved laterally out of the plane of the exposure. During the next century, the fissures and faults were once again buried by sediment (panel d). The latest event on the fault occurred about A.D. 1480 (panel e). During that event, the second fissure was split in two by the fault, and the right half was moved out of the plane of the exposure. No subsequent ruptures have occurred on this strand of the fault, and accumulating sediment has remained undisturbed (panel f)—except by paleoseismologists!

Excavations in Three Dimensions

The construction of the block diagrams of Figure 8-62 were fabricated to explain just one vertical exposure,

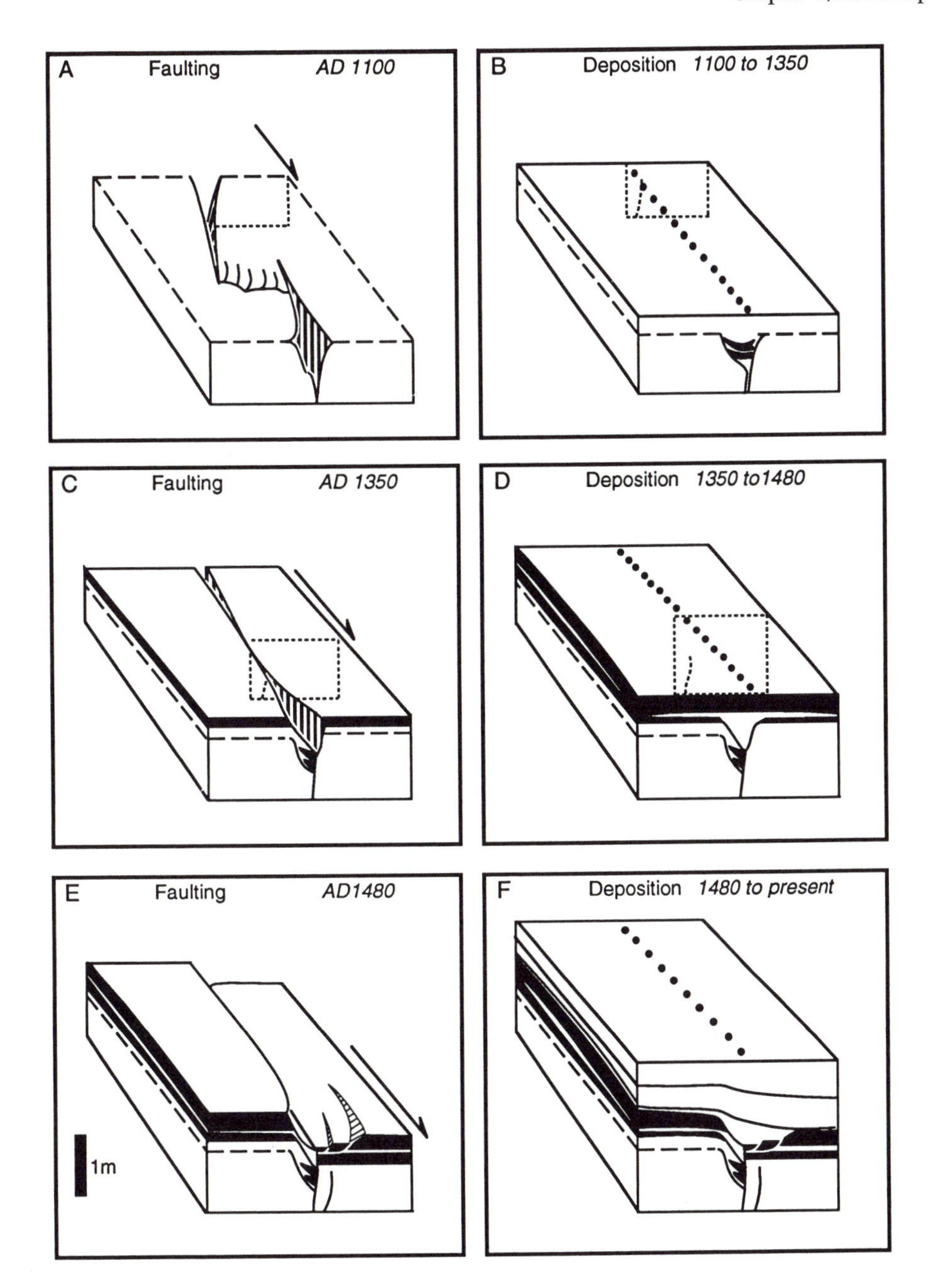

Figure 8–62. These six block diagrams display the hypothetical development of complex relationships between structures and stratigraphy in a paleoseismic excavation. The three diagrams on the left represent three coseismic fault ruptures about A.D. 1100, 1350, and 1480. The diagrams on the right represent intervening periods of fluvial and marsh deposition. The dotted boxes in panels a through d represent the location of the right side of panels e and f as it moved right-laterally to the front of the block. Constructed from data in Sieh (1984) and Sieh et al. (1989).

represented by the front face of panel f. One can well imagine, however, that a series of exposures spaced closely throughout the volume of a faulted block could reveal important information in three dimensions. Buried channels and facies variations might, for example, be mapped in order to determine the magnitude of lateral and vertical offsets associated with one or more events. Fault and fold geometries could also be mapped to understand fault history more completely.

Figures 8-63 and 8-64 illustrate a simple example. Figure 8-63 displays an idealized rendering, based upon real but obscure photographs of a small excavation in progress. The beds are dipping, coarse-sandy, foreset beds resting on nearly flat, silty, bottomset beds of ancient Lake Cahuilla, in the Coachella Valley of southern California. Both units were deposited a couple of meters below the surface of the lake about A.D. 1450. The foreset beds represent the propagation of a bar across the floor of the lake. The traces of two foreset bedding planes on the top of the bottomset bed are sharp linear features. In the figure, several nails mark these traces where the foreset beds have been removed by brush and trowel. The two nails farthest from the viewer have just been positioned, where the two contacts are still visible, at the intersection of the vertical and horizontal excavated surfaces.

Figure 8-64 illustrates the actual, completed excavation. The map shows the en échelon nature of the fault zone, with its Riedel shears and P fractures. Lines A-A′ through D-D′ are the traces of four foresets, mapped on the top of the bottomset bed. The dots on these four lines represent the nails inserted as excavation progressed from left to right toward and across the fault. Dextral offset of the four reference lines ranges from 60–75 cm and vertical offset is about 50 cm.

Important Problems in Paleoseismology

Paleoseismic studies of strike-slip faults have provided fundamental information for understanding earthquakes and active faults. In California, for example, probabilistic hazard maps are based in large part on paleoseismic data (see Chapter 13). Paleoseismic data are also a prominent source of data on the nature of large-earthquake sequences, since the historic and instrumental records are so much shorter than typical repeat times in most regions.

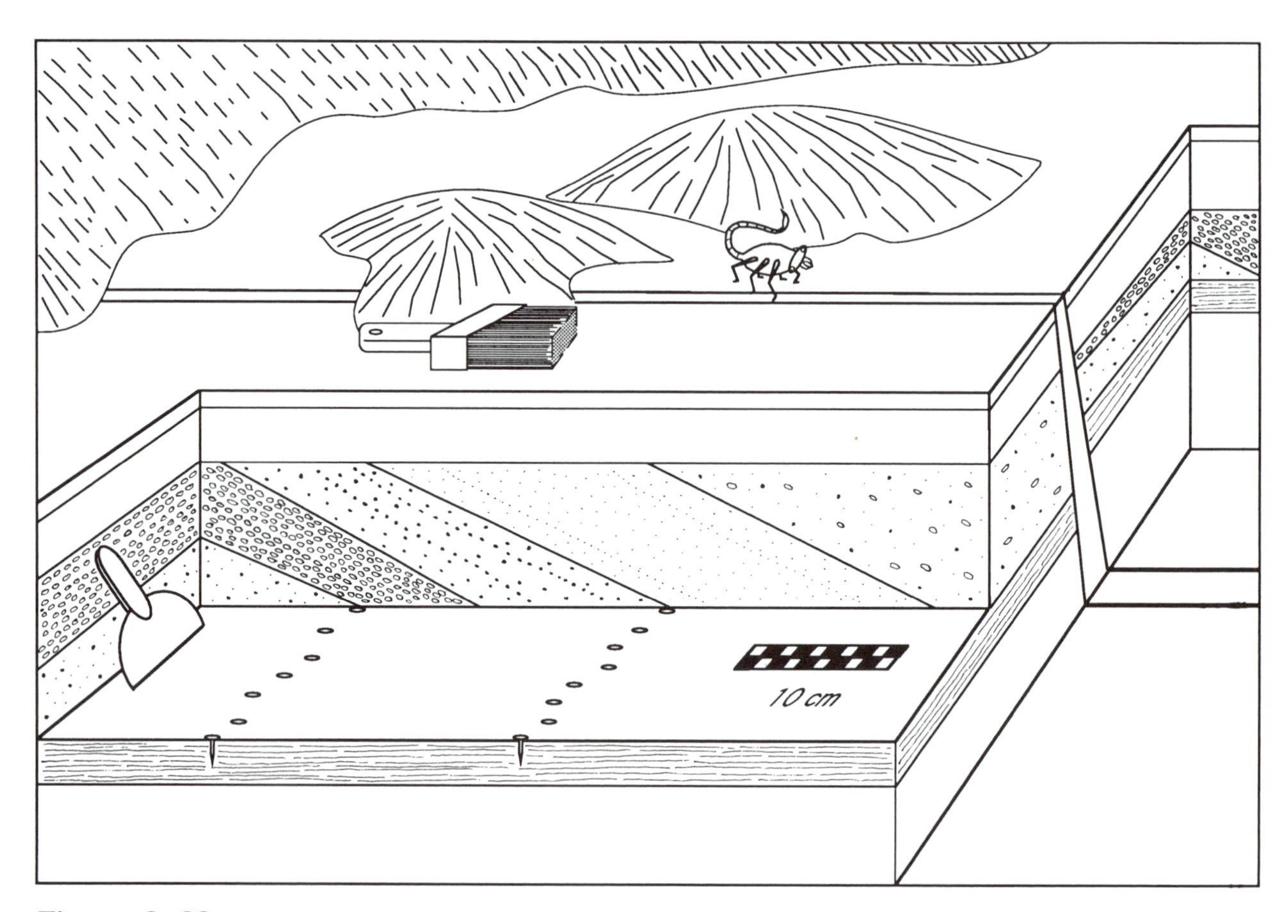

Figure 8–63. This rendering illustrates the sequential excavation of a piercing point in three dimensions. Contact between coarse sandy foreset beds slant upward from right to left in the vertical cuts. These contacts intersect the top of a silty bottomset bed, the top of which is visible, due to removal of superjacent sand. Nails mark the trace of the contact on the top of bottomset bed. Scale is 22 cm long.

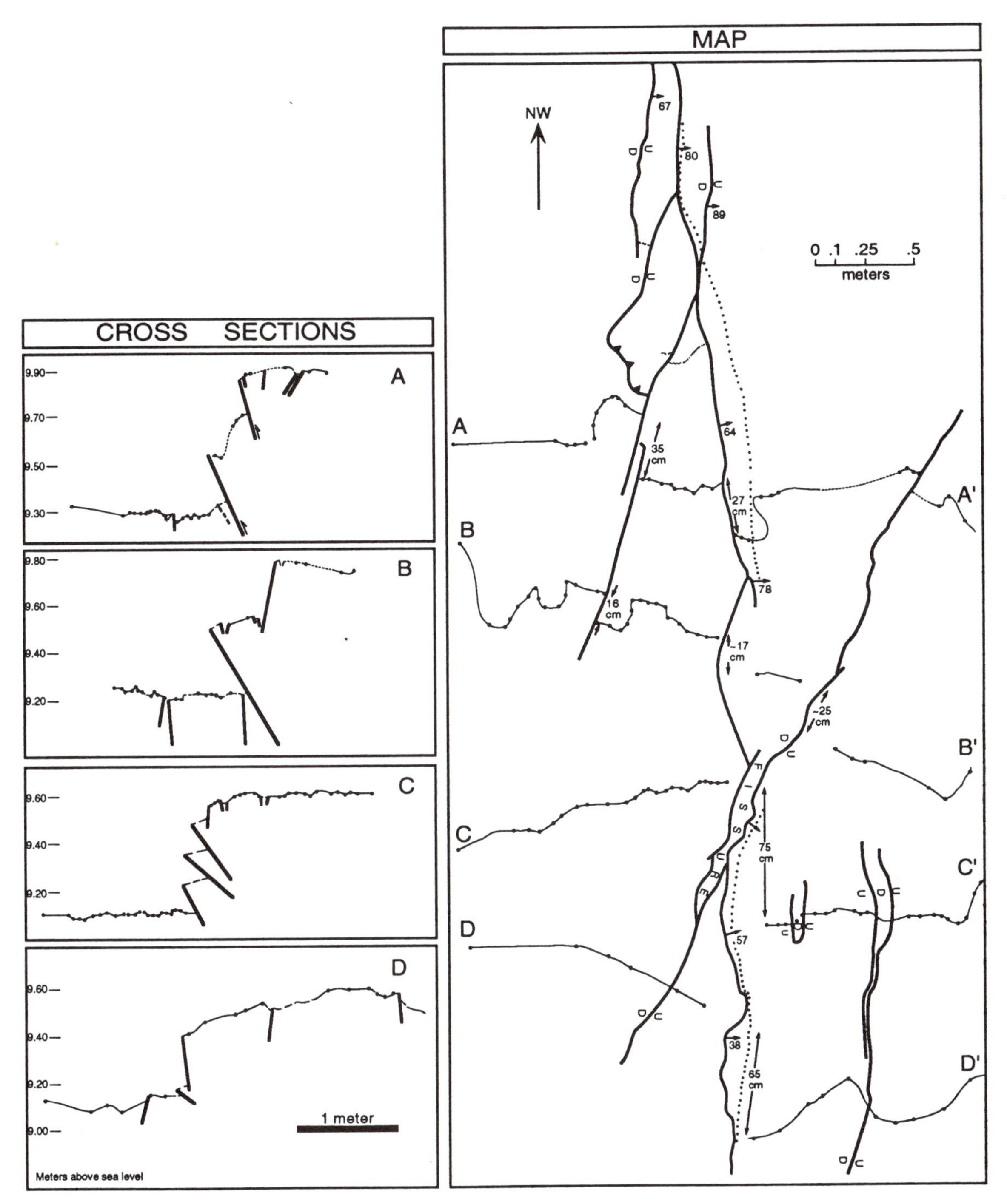

Figure 8–64. Map of the offset traces of four distinctive coarse sandy foreset beds (A through D) on the top of the silty bottomset bed within a lacustrine bar cut by a trace of the San Andreas fault. Dots along lines indicate location of nails marking traces, as in Figure 8–63.

In many cases, the dates of one or more prehistoric slip events have been determined (see, for example, Hudnut and Sieh, 1989, and Rockwell and Sieh, 1994). In other cases, the amount of offset in previous events has also been determined (for example, Salyards et al., 1992).

Along the San Andreas fault, several sites have yielded information about the dates of past earthquakes, and some have been excavated in three dimensions to reveal the amount of slip per event. Figure 8-65 displays a summary of current knowledge about the dates of past large earthquakes along the southern half of the fault. The vertical bars indicate the 95% error ranges of earthquake dates at each of the sites. The horizontal lines are speculative correlations between the sites. The long, uppermost horizontal line between sites indicates the length of the M_W 7.8 to 7.9 rupture of 1857.

The most complete and lengthy record shown on the figure is from the Pallett Creek site. There, the dates of ten large earthquakes are known, generally to within a few decades (Sieh et al., 1989; Biasi and Weldon, 1994). Three-dimensional excavations reveal that all rupture events but the two around A.D. 1000 are associated with 1 to 2 m of dextral slip along the fault. Hence, at least most of these events must represent long fault ruptures. The two lesser events at the site may also represent large earthquakes and long ruptures, if Pallett Creek is located fortuitously near the end of the ruptures, a hypothesis which cannot be excluded by the available data (Sieh, 1984).

Although the latest three events each exhibit only about 2 m of visible dextral offset across fault traces at Pallett Creek, paleomagnetic measurements show that the total offset includes substantial nonbrittle warping within a 50-m-wide zone (Salyards et al., 1992). Including this warp, each event is associated with about 6 m of dextral slip.

One of the most notable characteristics of the large earthquakes depicted in the figure is that their occurrence is highly irregular. At Pallett Creek, intervals of dormancy between earthquakes range from 44 years to about 330 years. Likewise, at the Bidart site in the Carrizo Plain, intervals range from less than one to about four centuries. Furthermore, large events sometimes cluster within particularly active centuries. Such clustering contradicts either the commonly assumed hypothesis of uniform strain accumulation or the concept of complete strain relief. Furthermore, some events at each site occur during times of dormancy at other sites. A south-to-north

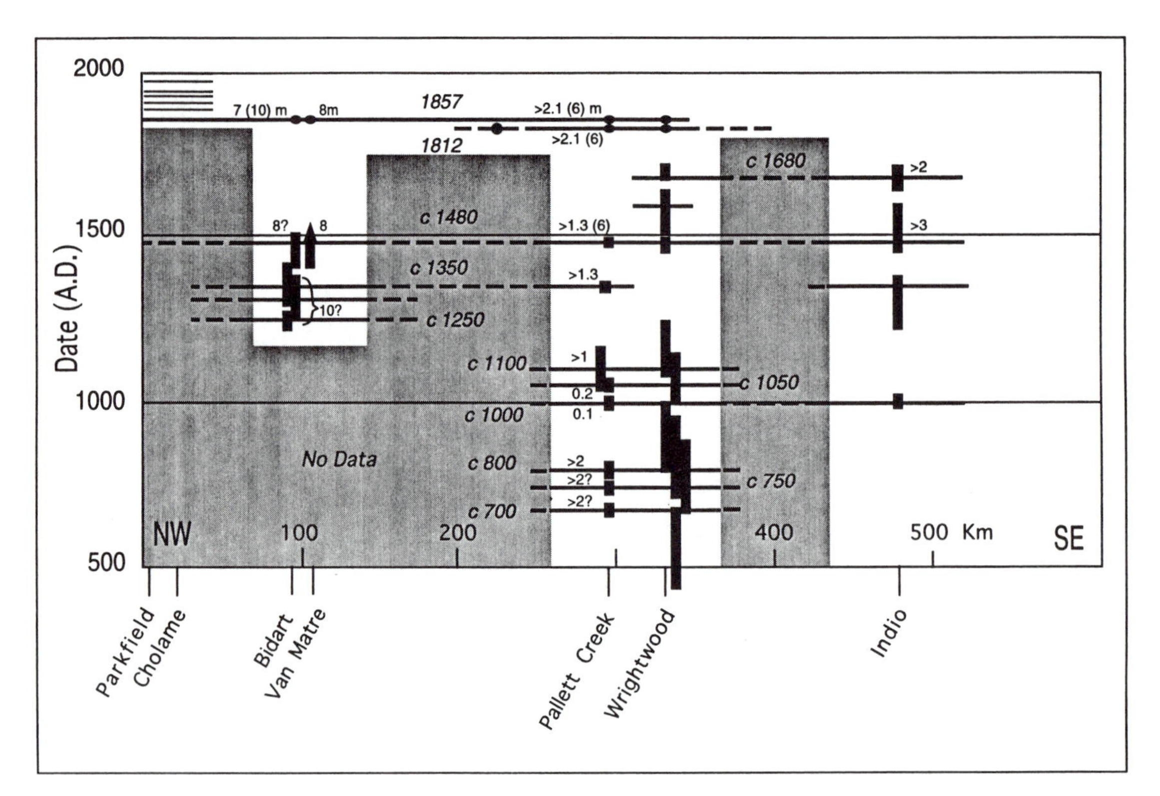

Figure 8–65. This history of large ruptures along the San Andreas fault is based upon data from several paleoseismic sites. Thick horizontal lines represent rupture lengths, based upon proposed correlations between sites. Dextral offsets are indicated (in meters) where available. Offsets in parentheses represent broad-aperture values, whereas others represent offsets measured in 3D excavations only within the fault zone. Values queried where more speculative. Though woefully incomplete, the currently available record demonstrates the clustered nature of earthquake occurrence along the fault and the inappropriateness of the characteristic- or uniform-earthquake model for the San Andreas fault. Modified from Grant and Sieh (1994), with additions from Sieh (1984), Salyards et al. (1992) and other sources.

migration of large earthquakes appears to have occurred between about 1700 and 1857, but other patterns are not so clear. The most likely source of the next major event along this southern half of the San Andreas fault may well be its southernmost 200 km, which has not produced a large event since for about 300 years.

SUMMARY

Strike-slip faults occur in several plate-tectonic settings. In addition to their widely recognized role as oceanic ridge-ridge transform faults, they also exist as plate-boundary transforms running between oceanic subduction zones, continental collision zones, and various triple junctions. Many of the world's most dangerous strike-slip faults, including the Alpine, the San Andreas, and the Motagua, are transforms. At obliquely-convergent plate margins, as in the Philippines, trench-parallel strike-slip faults often accommodate the strike-slip component of relative plate-motion. In regions of continental collision, as in Turkey and central Asia, indent-linked strike-slip faults enable horizontal extrusion of crust away from the zone of collision.

To understand earthquakes, geologists must study active faults at scales more local and regional than plate dimensions. At these scales, meters to tens of kilometers, strike-slip faults possess geometries, behaviors, and histories critical to understanding the earthquakes they produce. The rupture zones of strike-slip faults characteristically display en échelon patterns at scales of less than one to more than hundreds of meters. The maturity and complexity of these patterns (and the difficulty of measuring coseismic slip vectors) are closely related to the amount of coseismic slip. Discontinuities and other geometrical irregularities in strike-slip faults produce a great variety of secondary structures, the dimensions of which vary from less than a meter to hundreds of kilometers. These include a variety of graben at dilatational discontinuities, and horsts at contractional ones. Discontinuities greater than a kilometer or so in dimension commonly are associated with initiations and terminations of earthquake ruptures, but faulting often proceeds through such discontinuities during large earthquakes. Examples include the complex ruptures of the 1891 Nobi earthquake in Japan and the 1992 Landers earthquake in southern California. A strike-slip earthquake may involve rupture on several side-by-side parallel faults, as was the case in Managua in 1972. Rarely, a strike-slip event involves nearly simultaneous slip on conjugate left- and right-lateral faults.

The geomorphic expression of strike-slip faults on the earth's surface depends greatly on the relative rates of tectonism and erosion and deposition. The largest tectonic landforms are preserved where deposition occurs at rates far lower than rates of vertical deformation. Spectacular offsets of major rivers in Asia, Asia Minor, and California occur in uplifted regions where rivers have been eroding their beds for more than a million years. Rates of strike-slip offset can be determined where landforms are clearly offset and can be dated. In most regions, the landscape of strike-slip faulting is polygenetic and diachronous, and tectonic landforms of various sizes and ages reflect a complex interplay between tectonism, erosion, and sedimentation.

The regional geometries of strike-slip faults are the result of both their initial configuration and later deformation. Oceanic transform faults, though seldom posing any seismic risk to humankind, present the clearest examples of the relationship of their shapes and morphologies to the lithosphere in which they are embedded and to major changes in plate motions. Other faults, including many in California, owe seismically important aspects of their current geometries to mutual interactions as well as to changing relative plate motions.

The M_W 6.4 Imperial Valley earthquake of 1979 is a particularly well-understood moderate strike-slip earthquake. Right-lateral and oblique offsets were consistent with the structural setting of the earthquake. Coseismic offsets ranged up to several tens of centimeters at the surface, but inversions of geodetic and seismographic data indicate that dextral offsets as large as 2 or 3 meters occurred a depths of several kilometers. Aftercreep locally raised surficial offsets substantially, but still far less than occurred at depth. A previous earthquake generated by failure of the entire Imperial fault, 39 years earlier, was much larger (M_W 6.9), but produced surficial offsets along the northern half of the fault very similar to those of 1979. The similarities and differences of these two earthquakes suggest that faults may consist of discrete segments, each of which has stable boundaries and a characteristic mode and magnitude of slip through several earthquake cycles.

The M_W 7.3 Landers earthquake exemplifies large, complex strike-slip events. The earthquake was produced by meters of slip on several discrete right-lateral faults within an 80-km-long zone. A robust sequence of precursory earthquakes began two months before and continued to within a few seconds of the mainshock. Inversions of seismographic and geodetic data indicate rupture was primarily unilateral, from south to north. One inversion suggests retardations of the rupture front as it passed through zones of geologically mappable structural complexity. Geologic data show the faults are of various ages and maturity and are in the process of integrating into a throughgoing system.

Geological records of past earthquakes extend the record of fault activity into the centuries and millennia prior to instrumental records and modern field studies. Such paleoseismic evidence is very useful for construction of realistic probabilistic maps of seismic hazard and for understanding the mechanical behavior of faults. Geomorphic evidence for past strike-slip earthquakes consists of offset small landforms, most commonly stream channels and gullies. Stratigraphic evidence is more difficult to access, but more likely to yield precise dates for events and evidence for the discreteness of these events. Three-dimensional excavations have yielded precise information about the amount and character of offset.

Suggestions for Further Reading

Avouac, J. and Tapponnier, P. 1993. Kinematic model of active deformation in central Asia: Geophysical Research Letters, 20(10):895-98.

Lindvall, S., Rockwell, T., and Hudnut, K. 1989. Evidence for prehistoric earthquakes on the Superstition Hills fault from offset geomorphic features: Bulletin of the Seismological Society of America, 79(2):342-61.

Pockalny, R., Detrick, R., and Fox, P. 1988. Morphology and tectonics of the Kane transform from Sea Beam bathymetry data: Journal of Geophysical Research, 93:3179-93.

Sieh, K., 1978. Pre-historic large earthquakes produced by slip on the San Andreas fault at Pallett Creek, California: Journal of Geophysical Research, 83:3907-39.

Sieh, K., 1984. Lateral offsets and revised dates of large earthquakes at Pallett Creek, California: Journal of Geophysical Research, 89:7641-70.

Sieh, K., et al. 1993. Near-field investigations of the Landers earthquake sequence, April to July 1992: Science, 260:171-6.

Sieh, K., and Jahns, R. 1984. Holocene activity of the San Andreas fault at Wallace Creek, California: Geological Society of America Bulletin, 95:883-96.

Spotila, J., and Sieh, K. 1995. Geologic investigations of the "slip gap" in the surficial ruptures of the 1992 Landers earthquake, southern California: Journal of Geophysical Research, 100:543-59.

Sylvester, A., 1988. Strike-slip faults: Geological Society of America Bulletin, 100:1666-1703.

Tapponnier, P., Peltzer, G., Le Dain, A., Armijo, R., and Cobbold, P. 1982. Propagating extrusion tectonics in Asia—New insights from simple experiments with plasticine: Geology, 10:611-16.

Tschalenko, J. 1970. Similarities between shear zones of different magnitudes: Geological Society of America Bulletin, 81:1625-40.

Wilson, J. 1965. A new class of faults and their bearing on continental drift: Nature, 207:343-47.

Grove Karl Gilbert (1843–1918)

United States

Salt Lake City, Utah spans one of the most dramatic geological transitions in North America. To the west stretches the blue water of the Great Salt Lake, succeeded by the dazzling white salt flats that mark the former bed of its larger ancestor, Pleistocene Lake Bonneville. Isolated ranges rise abruptly like islands from the flat desert floor, the beginning of the Basin and Range province, which extends westward across Nevada to the foot of the California Sierra. Above the city to the east are the forested slopes of the Wasatch Range, a mountain wall indented by snow-fed mountain streams that coalesce on the valley floor to create the fertile farmland called Zion by Brigham Young and the Mormon pioneers when they arrived in 1847. Beyond the Wasatch Range to the east lies the Colorado Plateau, a fragment of the ancient North American continent uplifted and deformed in the Cenozoic Era.

G. K. Gilbert first saw the Basin and Range region in 1871 as a 28-year-old field assistant to Lt. George Wheeler, chief of the U.S. Army's Geographical Survey West of the 100th Meridian. Gilbert was a major player in the exploration of the American West. Expeditions such as Wheeler's received the same sort of public acclaim and media adulation as does space exploration today. Geologists, explorers, and surveyors were on center stage. This was the heroic age of American geology: John Wesley Powell's exploration of the Grand Canyon of the Colorado River, Clarence Dutton's mapping of the High Plateaus of Utah and discovery of the concept of isostasy, and Clarence King's travels in the High Sierra of California. From these expeditions, authorized by the U.S. Government, would be born the U.S. Geological Survey in 1879.

A major controversy at the time was the origin of the Basin and Range province, and Gilbert was in the thick of it. It was generally believed that mountains were formed by horizontal contraction, a result of the shrinking of the crust as the earth's interior cooled from a molten state. Accordingly, the Basin and Range province should be a fold belt, similar to the Appalachians, which had already been investigated by geologists. The ranges should be anticlines, and the basins synclines. But in 1875, after spending several field seasons crossing the Great Basin with the Wheeler Survey, Gilbert concluded that the main crustal forces were producing horizontal extension, not contraction. He viewed the ranges as blocks uplifted along faults, not anticlines.

Critical to this interpretation were small fault scarps cutting Quaternary deposits at the front of ranges in the Basin and Range province. And nowhere was this better displayed than at the base of the Wasatch Range near Salt Lake City. But Gilbert saw these fault scarps not only as the answer to a scientific problem but also as evidence for destructive earthquakes, a cause for alarm.

In an article published in the *Salt Lake City Tribune* in September 20, 1883, Gilbert wrote: "When an earthquake occurs, a part of the foot-slope goes up with the mountain, and another part goes down (relatively) with the valley. It is thus divided, and a little cliff marks the line of division. A man ascending the foot-slope encounters here an abrupt hill, and finds the original grade resumed beyond. This little cliff is, in geological parlance, a 'fault scarp.' "

Gilbert found "little cliffs" near Ogden, at Little Cottonwood Canyon, at American Fork Canyon, and elsewhere along the front of the Wasatch Range front bordering Brigham Young's Zion during field work in 1872, 1876, and 1879.

Why did Gilbert make the discovery when others did not? His research was focused on the Pleistocene ancestor of the Great Salt Lake which he named Lake Bonneville in 1875. His interest was the Quaternary history, especially the high shoreline features of the lake, including wave-cut benches and cliffs. Gilbert had to be careful that the wave-cut cliffs not be confused with fault scarps. In addition, Gilbert and a colleague, I. C. Russell, had visited fresh new fault scarps formed by the Ownes Valley, California, earthquake of 1872, and the lessons of these fault scarps were not lost on Gilbert.

Of the Owens Valley scarps, he wrote in 1883: "The height of the scarp varies from five to twenty feet, and its length is forty miles. Various tracts of land were sunk a number of feet below their previous positions, and one tract, several thousand acres in extent, was not only lowered, but carried bodily about fifteen feet northward."

Gilbert also recognized the fault scarps because they were part of the landscape. From the beginning of his career, he had studied the dynamics of landscapes. The forces of running water and ice that eroded landscapes and deposited sediments had to follow the laws of mechanics, and he considered such problems much as a physicist or engineer might have done.

But Gilbert went beyond abstract science and worried about the impact of these fault scarps on Salt Lake City, and this was the reason for his 1883 letter to the *Tribune*. In this letter, he wrote: "The old maxim, 'Lightning never strikes the same spot twice,' is unsound in theory and false in fact; but something similar might truly be said about earthquakes. The spot which is the focus of an earthquake (of the type here discussed) is thereby exempted for a long time. And conversely, any locality on the fault line of a large mountain range, which has been exempt from earthquake for a long time, is by so much nearer to the date of recurrence—and just here is the application of what I have written. Continuous as are the fault-scarps at the base of the Wasatch, there is one place where they are conspicuously absent, and that place is close to [Salt Lake City]. From the Warm Springs to Emigration Canon fault-scarps have not been found, and the rational explanation of their absence is that a very long time has elapsed since their last renewal. In this period the earth strain has been slowly increasing, and some day it will overcome the friction, lift the mountains a few feet, and re-enact on a more fearful scale the catastrophe of Owens Valley."

Gilbert had issued a long-range earthquake forecast, and by singling out that portion of the Wasatch fault where fresh fault scarps were subdued or absent, he anticipated fault segmentation and the study of paleoseismology.

Would the Mormon settlers heed Gilbert's warning? He was not optimistic: "By the time experience has taught us this, Salt Lake City will have been shaken down—to use a homely figure, the horse will have escaped, and the barn door, all too late, will have been closed behind him."

"What are the citizens going to do about it? Probably nothing."

Gilbert's views on the relationship of fault scarps to earthquakes received little notice until much later, in 1906, when the San Francisco earthquake was clearly blamed on the San Andreas fault, in part by Gilbert himself. But his views on the fault origin of the Basin and Range province immediately stirred vigorous dissent, and Gilbert continued to gather data on the Wasatch fault to defend his views, even after the publication of his monograph on Lake Bonneville in 1890. His final statement came in 1928, ten years after his death with publication of a paper on the structure of the Basin and Range, by which time his views were generally accepted.

Gilbert's forecast of an earthquake on the Wasatch fault near Salt Lake City remains unfulfilled today.

Suggestions for Further Reading

Gilbert, G. K. 1875. Report on the geology of portions of Nevada, Utah, California and Arizona examined in the years 1871 and 1872: Report on U.S. Geographical and Geological Surveys West of the 100th Meridian, 3: Geology, pt. 1, 17–187.

Gilbert, G. K. 1883. A theory of the earthquakes of the Great Basin, with a practical application: *Salt Lake City Tribune*, Sept. 30, 1883, reprinted in *American Journal of Science*, 3rd ser., 27:49–53.

Gilbert, G. K. 1890. Lake Bonneville: U.S. Geological Survey Monograph 1, 340 p.

Gilbert, G. K. 1907. The investigation of the California earthquake, in Jordan, D. S., ed., The California earthquake of 1906: San Francisco, A. M. Robertson, pp. 215–56.

Gilbert, G. K. 1928. Studies of basin-range structures: U.S. Geological Survey Professional Paper 153, pp. 1–92.

Machette, M. N., ed. In the footsteps of G. K. Gilbert—Lake Bonneville and neotectonics of the eastern Basin and Range province: Utah Geological and Mineral Survey Misc. Publ. 88-1, 120 p.

Pyne, S. J. 1980. Grove Karl Gilbert, a great engine of research: Austin, University of Texas Press, 306 p.

Wallace, R. E. 1980. G. K. Gilbert's studies of faults, scarps, and earthquakes, in Yochelson, E. L., ed., The Scientific Ideas of G. K. Gilbert: Geol. Soc. America Special Paper 183, pp. 35–44.

Alexander McKay (1841–1917)

New Zealand

It was September, 1888, and Alexander McKay was once again in Marlborough, across Cook Strait from the New Zealand provincial capitol of Wellington. He had been there first as a young man stricken by gold fever, after enduring the long sea passage from Scotland on the *Helenslee*. Landing at Bluff in 1863, he had made his way north to the Whakamarino fields, where, unfortunately, a fortune in gold eluded him. After several years of gold prospecting in New Zealand and Australia, he met Dr. Julius von Haast, Provincial Geologist for Canterbury, in 1868. Von Haast saw in McKay not a Scottish farm boy seeking his fortune but a born naturalist with a feel for the land and a yearning to explore and to understand it.

And so in 1870, Dr. von Haast sent young McKay to explore for coal, to map the mountains of upcountry Canterbury, and to collect vertebrate fossils from the Waipara River and the Moa Bone Cave at Sumner. Working for von Haast, McKay found that a fortune in gold didn't seem to matter so much. Yet he wondered why gold occurred in some places but not others. In 1872, he came to the notice of James Hector of the Colonial Museum and Geological Survey in Wellington, who later employed him to collect fossils for the museum. A prodigious fossil collector, the young man spent long days in the field, packing his fossils in whisky crates from a local pub and sending them off to Wellington.

McKay learned his geology from the hills, not from books, but that was not enough, and he knew it. Hector gave him a copy of Charles Lyell's *Principles of Geology*, already in its twelfth edition, and it became his Bible. He was never without the *Principles*. By day, the book was in his pack; by night, he read its passages aloud in camp, sometimes putting his camp-mates to sleep as they listened to the same words again and again.

In time, McKay learned, and gained confidence in himself as a field geologist. He was able to apply Lyell's vision to his own observations in the previously unstudied hills and mountains of New Zealand.

But the hills had a great surprise for McKay. On September 1, 1888, a massive earthquake rolled through the sheep stations of upcountry Canterbury and Marlborough, and Hector dispatched McKay from Wellington to investigate. He reached the scene of earthquake damage on September 29.

It soon became evident to McKay that the greatest destruction was found in the valley of the Hope River, and that brought him to Glynn Wye Station.

The buildings at the Glynn Wye Station had been erected on a terrace more than 30 meters above the Hope River near its intersection with Kakapo Brook. McKay saw that the stable and storehouse were not badly damaged, although the living quarters and the woolshed were largely wrecked. But what particularly interested him were the wire fences bounding the paddocks west of the station. The fences that ran northward across the terrace toward the Hope River were broken and had shifted locally more than 8 feet (2.6 meters) out of line, the northern side displaced to the east with respect to the southern side. The displacement occurred across a zone of freshly disturbed and broken ground, which he recognized as having formed during the earthquake.

McKay further observed that the new earthquake ruptures followed old "earthquake rents," and he was able to follow the line of disturbance east to Hanmer Springs. He saw that other rents and fissures not on the line of the earthquake rent were not "true rendings, but more the result of excessive vibrations and shaking of the surface." In this way, he distinguished tectonic strike-slip dislocations from landslides and other seismically induced ground failures.

In his exploration of the nearby Clarence and Awatere valleys in 1884–1890, he mapped additional faults, and after the 1888 earthquake, he saw that these faults were also capable of generating earthquakes. Since his work, none of these other active strike-slip faults has produced a surface-rupturing earthquake like the one on the Hope fault in 1888.

In 1890, McKay returned to Wellington and published his observations. He was one of the first persons to document pure strike slip on a fault during an earthquake. A few years earlier, G. K. Gilbert and I. C. Russell had visited the fault scarp of the 1872 earthquake at Owens Valley, California, but Gilbert's description was part of a discussion of a normal fault scarp near Salt Lake City, Utah, which Gilbert claimed to be related to an ancient earthquake.

Suggestions for Further Reading

Burton, M. P. 1965. The New Zealand Geological Survey 1865–1965, p 20–23.

The Cyclopedia of New Zealand, 1897, 1:174–175.

McLintock, A. N. ed. 1966. An Encyclopedia of New Zealand, 2:359–360.

McKay, A. 1890. Reports of Geological Explorations during 1888–1889: New Zealand Geological Survey, 20:1–16.

Obituary: Alexander McKay, 1918, Transactions and Proceedings of the New Zeland Institute, 50:2 p.

9

Normal Faults

TECTONIC SETTING

Normal faults develop in crust undergoing extension in which the maximum principal compressive stress, σ_1, is vertical. Most normal-faulted regions are characterized by high heat flow, a relatively low-velocity upper mantle, and, in some cases, volcanism.

The major regions where active normal faulting is known to occur are shown in Figure 9-1. The most common setting on earth where extension takes place is at sea-floor spreading centers, where new oceanic crust is created. Other extensional plate-tectonic settings are hot spots, back-arc basins (behind island arcs), extensional regions behind continent-continent collision zones, and intracontinental rift systems (Fig. 9-2). Normal faults may also occur near a subduction zone where the downgoing plate, generally oceanic, is flexed by horizontal compression and by the load of the upper plate (Fig. 9-2; see further discussion in Chapter 11).

Spreading Centers

Slow-spreading ridges such as the Mid-Atlantic Ridge are commonly characterized by a central rift valley, or graben. A central rift valley is absent on fast-spreading ridges such as the East Pacific Rise. Most of the displacement on normal faults flanking the central rift valley occurs soon after formation by sea-floor spreading. Later, as the fault blocks are carried away from the spreading center, the faults become inactive, and the fault blocks are covered by sediment. Oceanic

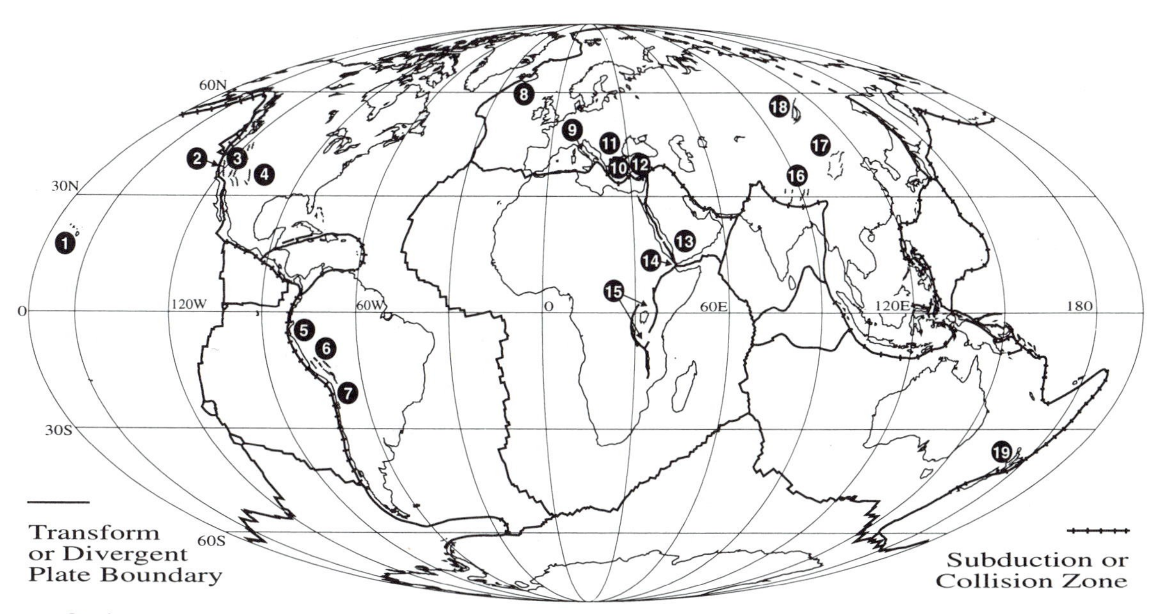

Figure 9–1. Locations of major active onshore normal-fault systems: 1. Hawaii; 2. Sierra foothills; 3. Basin and Range; 4. Rio Grande Rift; 5. Gulf of Guayaquil; 6. Altiplano; 7. Mejillones Peninsula; 8. Iceland; 9. Apennines; 10. Greece; 11. Bulgaria; 12. Western Anatolia; 13. North Yemen; 14. Afar Triangle; 15. East African rift valleys; 16. Southern Tibet; 17. Ordos; 18. Baikal rift system; 19. Taupo Volcanic Zone.

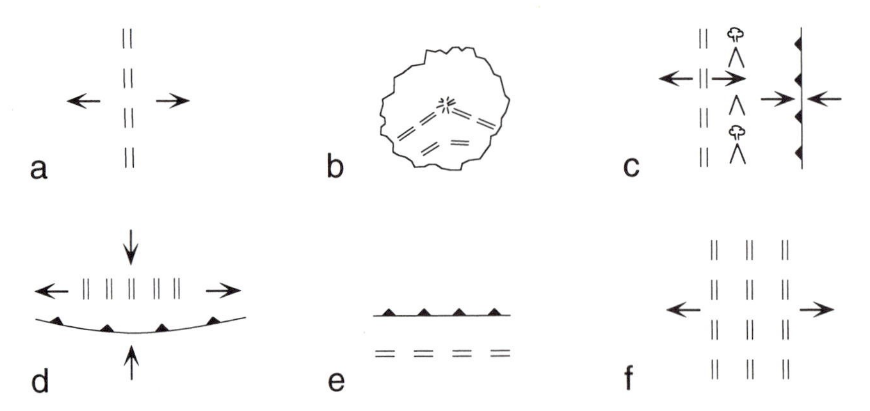

Figure 9–2. Plate-tectonic setting of normal faults. (a) Spreading center; (b) hotspot; (c) back-arc basin, (d) extension behind collision zone; (e) flexing of oceanic plate in front of subduction zone; (f) intracontinental rift. Normal faults shown as thin paired lines.

crust at spreading centers has only recently solidified from magma, therefore it is unusually hot. Accordingly, the brittle-plastic transition is only a few kilometers below the surface, and the seismogenic zone is unable to store much elastic strain energy. Hence, earthquakes formed at spreading centers are relatively small, generally not exceeding magnitude 6.

Submarine spreading centers have been studied in detail using side-scan sonar and submersibles. Figure 9-3 shows a side scan image east of the Juan de Fuca ridge west of Oregon, U.S.A. (Appelgate, 1990). The spreading center consists of a graben in basaltic lava (Normark et al., 1987; Kappel and Normark, 1987). The graben is bounded by sets of closely spaced normal faults, shown in the image. Outside the graben, faults are less abundant and are commonly covered by volcanic mounds. Despite the presence of young faults, very few earthquakes are recorded from the Juan de Fuca ridge, indicating that deformation is overwhelmingly aseismic. Faulting is probably related to rise-crest volcanic activity.

Where spreading centers are composed of normal oceanic crust, they are always more than 2 km below sea level. However, Iceland, which is astride the Mid-Atlantic Ridge, is formed by a hot spot and is subaerial. This permits examination of extensional features called *fissure swarms* in considerable detail (Fig. 9-4). The Icelandic term *gja* (gaping fracture) refers both to extensional fractures and faults. The common view is that crustal extension above a shallow magma reservoir results in extensional fractures filled with congealed magma (*dikes*) at depth and fissures and normal faults closer to the surface. In those areas of Iceland undergoing active extension, there are a large number of closely spaced normal faults occupying fissure swarms, with no single master fault dominant. Near the surface, extension is accommodated by extensional fractures (fissures) and gaping normal faults with near-vertical dips. At greater depths, the normal

Figure 9–3. Side scan sonar image from Sea MARC 1 near the crest of the Juan de Fuca Ridge west of Oregon, U.S.A., showing swarms of west-facing fault scarps (light), some covered by younger lava (gray, obscuring scarps). Image measures backscatter from sound source to the west. From Appelgate (1990).

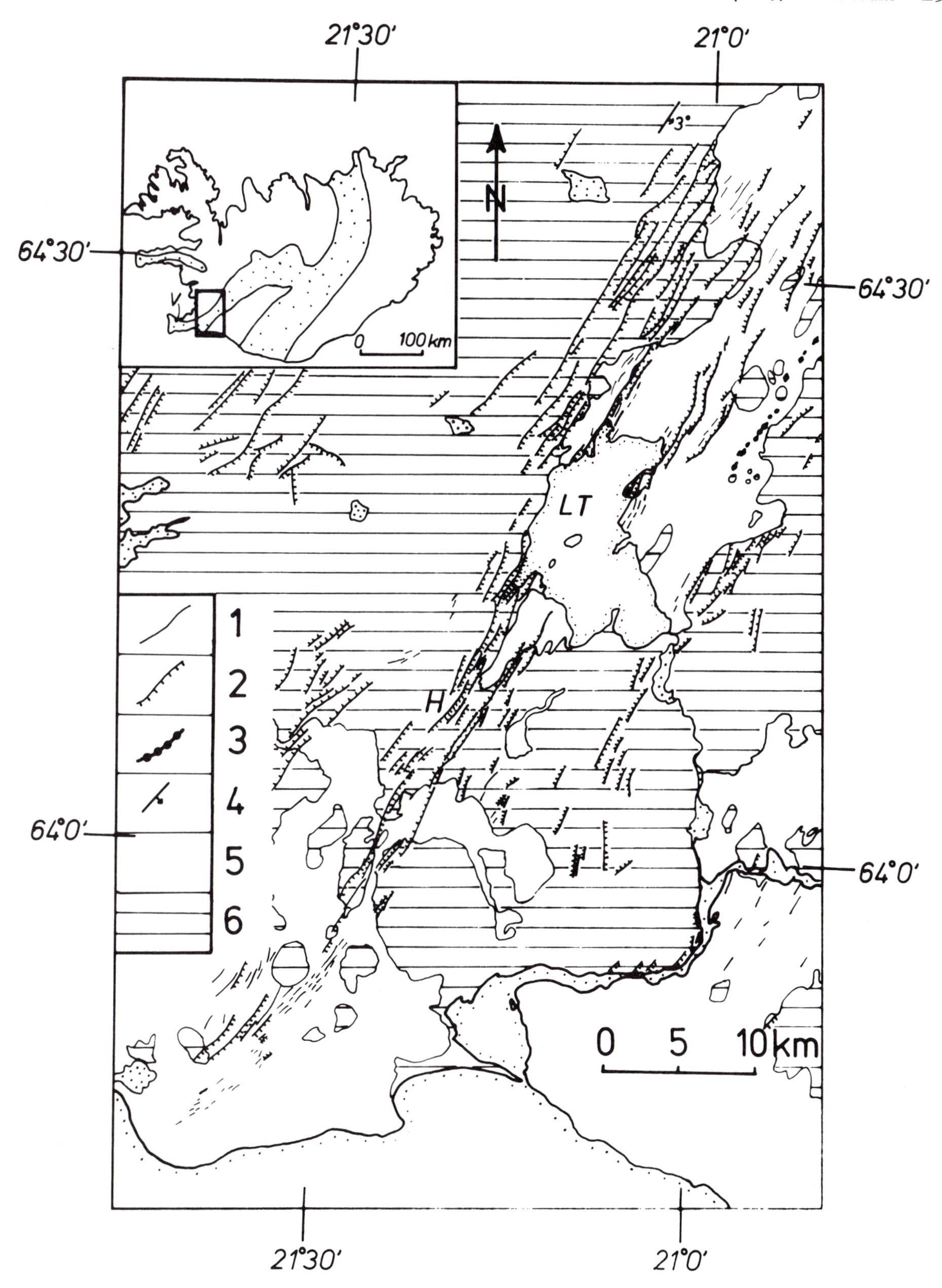

Figure 9–4. (a) Pleistocene Hengill fissure swarm, Iceland, which includes the Holocene Thingvellir fissure swarm at the northern end of Lake Thingvallavatn (LT). H, Hengill central volcano;; V (in inset), Vogar fissure swarm. 1, tectonic fissure; 2, normal fault; 3, volcanic fissure; 4, strike and dip; 5, Holocene lavas; 6, Pleistocene rocks and alluvium. From Gudmundsson (1987).

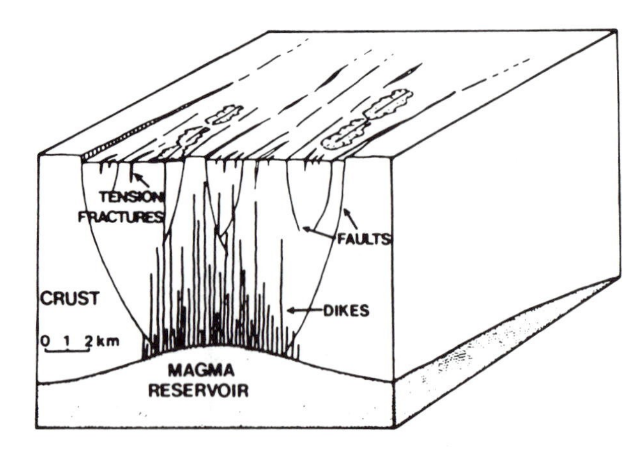

Figure 9–4. (b) Diagram of a typical fissure swarm in the rift zone of Iceland. From Forslund and Gudmundsson (1991).

faults have an average dip of 75° and an average throw of 10 m (Gudmundsson, 1987; Forslund and Gudmundsson, 1991).

The spreading center is offset along a transform fault zone called the South Iceland Seismic Zone (not shown in Fig. 9-4), characterized by short, north-trending faults with right-lateral strike-slip offsets. In contrast to the spreading centers in Iceland, this transform fault zone has produced earthquakes as large as magnitude 7 in historical times (Einarsson et al., 1981; Bjarnason et al., 1993).

The other area where a spreading center comes onshore is in Djibouti, northeast Africa, where oceanic spreading centers in the Red Sea and the Gulf of Aden and the continental East African Rift System meet in the Afar triple junction (Fig. 9-5; Barberi and Varet, 1977; Tapponnier et al., 1990; Acton et al., 1991; Stein et al, 1991). As in Iceland, normal faults and fissures occur in swarms, none of which has large cumulative displacement. In 1978, the region was struck by an earthquake swarm, with the largest magnitude m_b = 5.3. Several faults underwent normal-fault displacements of up to 0.5 m over lengths up to 12 km (Le Dain et al, 1979). The largest events were followed by a fissure eruption of basaltic lava.

Faulting on oceanic spreading centers, including the two onshore examples in Iceland and Djibouti, is largely aseismic, or, at most, is accompanied by earthquakes with magnitudes less than 6. Rupture is distributed among swarms of faults and fissures, and faulting may be accompanied by eruption of lava. Only where spreading centers are separated by a transform fault, as is the case in Iceland, are larger earthquakes to be expected, but even in these cases, rupture is distributed among many closely spaced faults of short length and small displacement.

Hot Spots Not Associated with Spreading Centers

The island of Hawaii, a giant basaltic volcanic edifice, is the surface manifestation of a mantle plume or hot spot (cf. Chapter 1). The edifice consists of five individual Quaternary shield volcanoes, of which the southern two, Mauna Loa and Kilauea, have been ac-

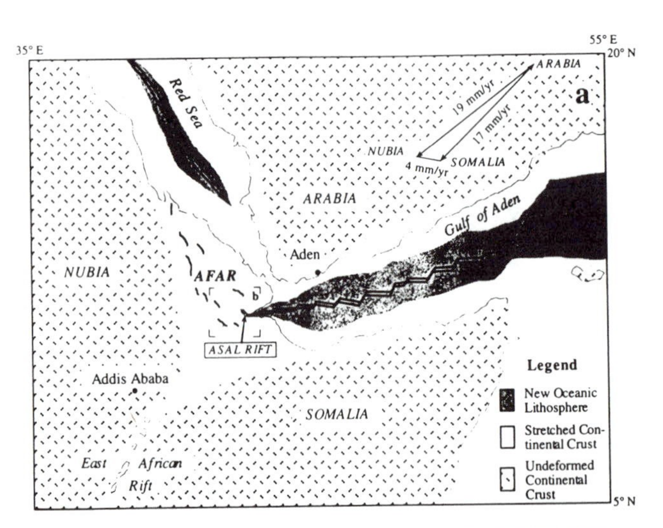

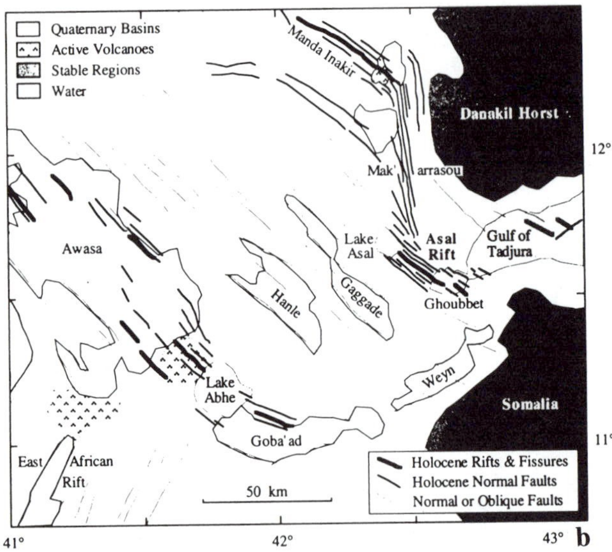

Figure 9–5. (a) Afar triple junction, marking intersection of Red Sea and Gulf of Aden spreading centers (shaded pattern) and East African rift system, after Courtillot et al. (1987). (b) Holocene fissures, normal faults, and active volcanoes in Djibouti. After Tapponnier et al. (1990). Most rifts and fissures in the Asal rift underwent displacement during an earthquake swarm with largest magnitude 5.3 in November, 1978, accompanied by a new basalt flow. After Stein et al. (1991).

tive in historic time. Rift zones that extend southwest and east from the summits of these two volcanoes are the source of flank fissure eruptions (Fig. 9-6). In addition, the Hilina fault system on the south side of Kilauea comprises cliffs (*pali*) which are normal faults with displacements ranging up to more than 100 meters. This fault system also contains gja cutting lava flows no more than a few hundred years old.

Hawaii has produced at least 20 earthquakes of $M \geq 6$ since historical records began to be kept in 1833 (Wyss and Koyanagi, 1992). An earthquake in April, 1868, had a magnitude of 7.9, and the Kalapana earthquake of 1975 had a magnitude of 7.2. Although there was displacement on the Hilina fault system in the Kalapana earthquake, the aftershock distribution and focal-plane solutions suggest that the Hilina fault system was a secondary structure, not the source of the mainshock. The 1975 earthquake was characterized by low-angle nodal planes with slip vectors oriented toward the WSW and SE, away from the summit of Mauna Loa (Wyss and Koyanagi, 1992; Fig. 9-6). The 1975 earthquake may have ruptured a zone of weakness in the

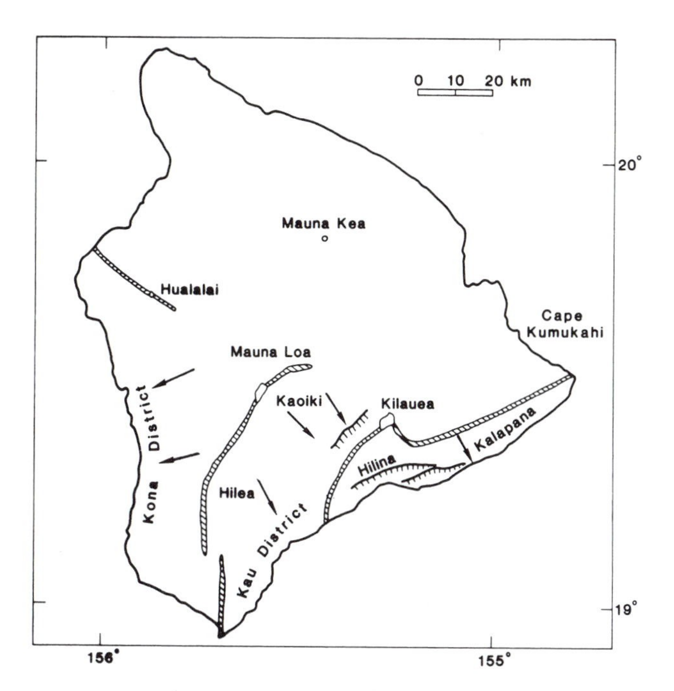

Figure 9–6. Normal faults and fissures on the flanks of Hualalai, Mauna Loa, and Kilauea volcanoes on the island of Hawaii. Double lines indicate volcanic rift zones, which serve as sources for fissure eruptions; single lines indicate south-facing scarps (pali). Arrows show azimuths of slip vectors on earthquakes at about 10 km depth, assuming the near-horizontal nodal plane in fault-plane solutions is the plane of slip. From Wyss and Koyanagi (1992).

oceanic sediments on which the volcano was built (Wyss and Koyanagi, 1992; Gillard et al., 1992). On the other hand, faulting accompanying earthquakes of magnitude 5.5 in 1974 and 6.5 in 1983 on the flank of Mauna Loa volcano was right-lateral, although it took place on several small faults with relatively little topographic expression (Jackson et al., 1992).

Back-Arc Basins

If the upper plate of a subduction zone moves away from, rather than toward, the trench and lower plate (see discussion in Chapter 11), extension may occur in the upper plate behind the island arc, producing new oceanic crust. The extended regions are called *back-arc basins* or *marginal basins*. They are most common in the island arcs of the western Pacific where the subducting oceanic crust is relatively old and cold, but they also occur behind the Caribbean and Scotia arcs in the Atlantic. Most modern back-arc basins are found on the ocean floor, but three are partially on land: the Taupo Volcanic Zone of New Zealand, part of the Aegean-western Anatolian region of Greece and Turkey, and the northeastern end of the Okinawa trough on the island of Kyushu, Japan. The volcanic arcs of central Mexico, Central America, and southern Italy and the area immediately east of the Cascade Range of Oregon also contain young normal faults possibly related to volcanism, but these are not in an obvious back-arc setting.

The Tonga-Kermadec trench is the site of west-dipping subduction of oceanic crust of the Pacific plate beneath island arcs of the Tonga and Kermadec Islands in the southwest Pacific. The upper plate is moving away from this trench, producing extension (discussed further in Chapter 11). West of the island-arc volcanoes, the extension between the island arc and the South Fiji basin resulted in formation of the Lau-Havre trough with new oceanic crust. The Kermadec trench gives way southward to the Hikurangi trench, and the Lau-Havre back-arc basin comes ashore as the Taupo Volcanic Zone of the North Island of New Zealand (Fig. 9-7). The zone is characterized by volcanic rocks, with major eruptions at Lake Taupo 1800 years ago, and at the Tarawera Volcanic Complex in A.D. 1886. Active arc volcanoes are found along the eastern edge of the Taupo Volcanic Zone. The area contains many normal faults with evidence of late Quaternary activity.

Tectonic activity consists of both earthquake swarms and larger earthquakes with a mainshock followed by aftershocks. In 1983, an earthquake swarm was accompanied by as much as 30 mm extension across steeply dipping normal faults north of Lake Taupo (Grindley and Hull, 1986). The Kaiapo fault,

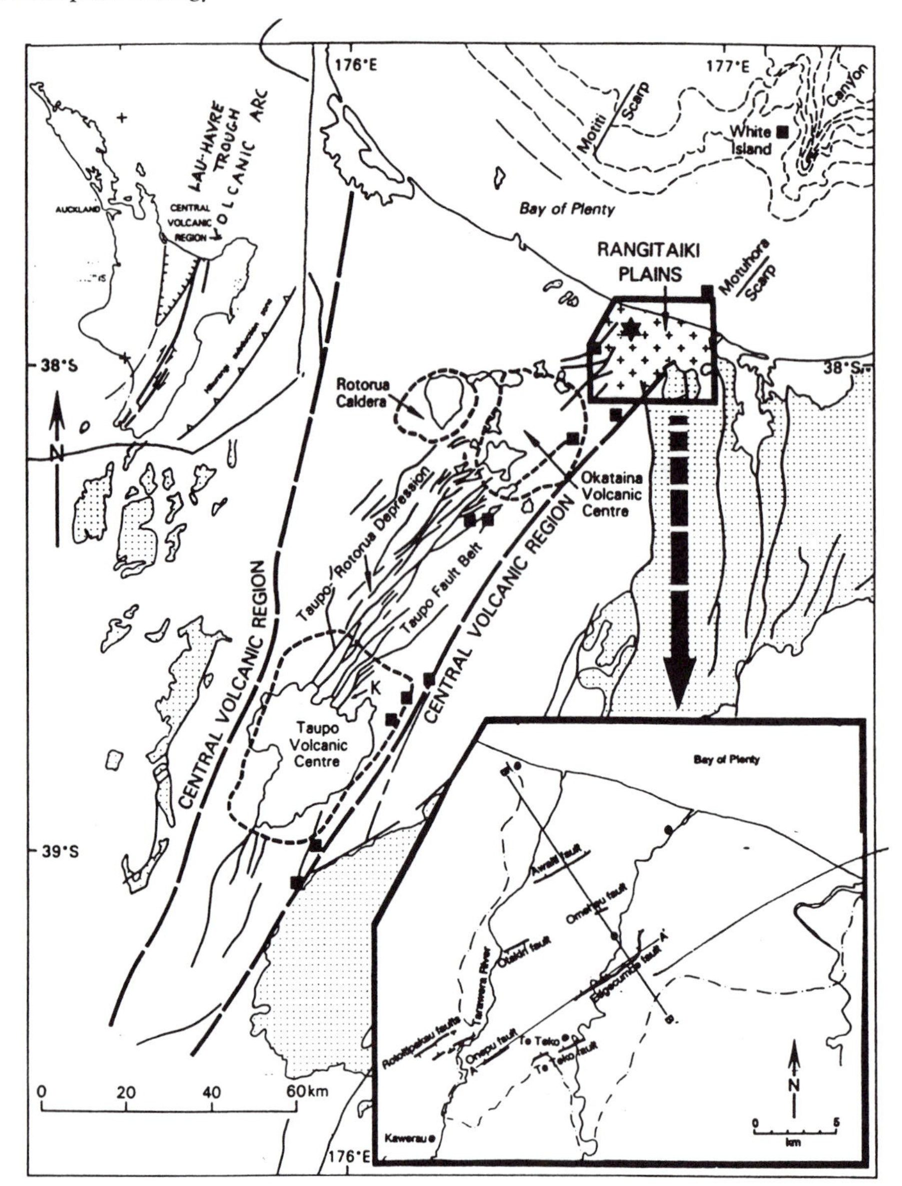

Figure 9–7. Distribution of faults in the Central Volcanic Region (Taupo Volcanic Zone), New Zealand, location of Taupo fault belt, and location of Rangitaiki Plains, site of the 1987 Edgecumbe earthquake (mainshock located by star). The letter K identifies the Kaiapo fault which ruptured in 1922 and 1983. Filled squares mark andesite-dacite arc volcanoes; outcropping Mesozoic graywacke bedrock is stippled. Lower inset shows fault breaks formed during Edgecumbe earthquake; B-B′ locates cross section in Figure 9–32. Inset, upper left, shows relation of Central Volcanic Region to Lau-Havre trough, a back-arc basin. Modified from Berryman and Beanland (1991).

which ruptured in 1983, is located on Figure 9–7. In 1987, the Edgecumbe earthquake (M_S6.6) produced 1.7 m of slip on the Edgecumbe normal fault and lesser displacement on 10 secondary ruptures (Beanland et al., 1990). The Edgecumbe earthquake demonstrated that the crust beneath the Taupo Volcanic Zone is thick enough and brittle enough to generate a damaging normal-fault earthquake, despite the abundance of young volcanism, the presence of closely spaced normal faults with small displacements rather than widely spaced, large-displacement range-front faults, and high heat flow. In this respect, the Taupo Volcanic Zone differs from another volcanic area, the Afar region of Djibouti. The Edgecumbe earthquake is discussed in the following Case Histories section.

In the eastern Mediterranean region, oceanic crust of the African plate is being subducted northward at a trench south of the Aegean Sea between Greece and Turkey (Fig. 9–8). Island-arc volcanoes, including Santorini, which erupted around 1500 B.C.E., occur in the southern Aegean Sea in a zone trending NW-SE. Heat flow is high throughout the region. Normal faults striking approximately east-west, orthogonal to the direction of subduction, occur throughout most of the Greek peninsula north of the Gulf of Corinth, the central and northern Aegean Sea, and western Anatolia. Earthquakes in this region are characterized by normal-fault focal-mechanism solutions with north-south extension. Earthquakes with surface faulting include the 1861 Egion earthquake of magnitude 7 in the northwestern Peloponnesos Peninsula near ancient Helice (H in Fig. 9–8), the 1891 Atalanti earthquake north of Athens, the 1981 Gulf of Corinth earthquakes (discussed under Case Histories) in Greece, and the 1969 Alasehir (M6.5) and 1970 Gediz (M7.2) earthquakes of western Anatolia, Turkey (Eyidogan and Jackson, 1985).

The northern Aegean Sea is dominated by right-lateral strike-slip faults, the western extension of the North Anatolian fault zone of Turkey, but active normal faults occur still farther north, resulting in the 1978 Thessaloniki earthquake of magnitude 6.4 (Soufleris et al., 1982) and earthquakes in Bulgaria in 1904 and 1928 with magnitudes as large as 7.1, all accompanied by surface rupture on west-trending normal faults.

In contrast, normal faults south of the Gulf of Corinth and in the vicinity of Crete, regions between the arc volcanoes and the trench, strike north-south, and earthquake focal mechanisms indicate east-west extension (Armijo et al., 1992). This area was previously characterized by north-south extension, like the rest of Greece; the shift to east-west extension took

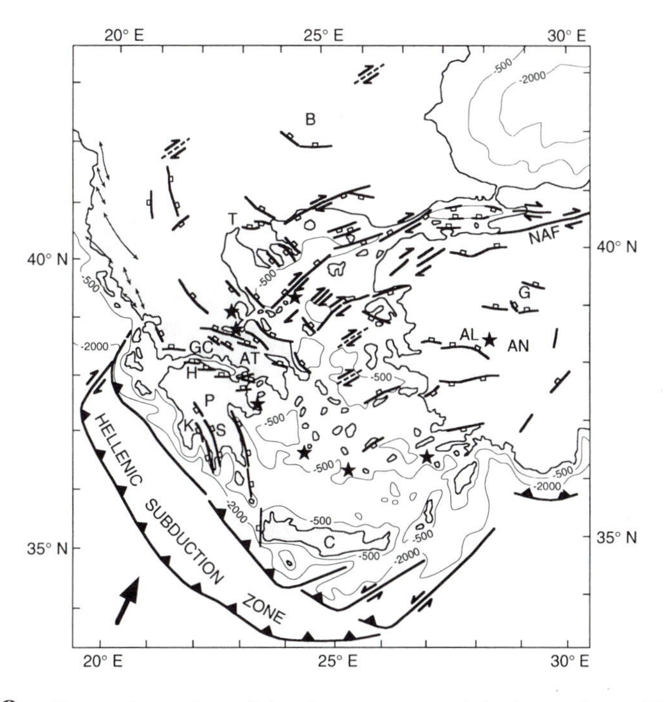

Figure 9–8. Tectonic setting of the Aegean normal-fault province. Normal faults shown with open blocks on downthrown side; many in Aegean have large strike-slip components (heavy black arrows), Strike-slip faults that are dashed are based solely on fault-plane solutions of earthquakes. Quaternary volcanoes shown as stars. Lines with arrows are folds. Only faults with major topographic expression or associated with earthquakes are shown. Five hundred-meter and 2000-meter bathymetric contours shown. Large black arrow is convergence direction between Africa and the Aegean region. C, Crete; AL, Alasehir; AN, Anatolia; AT, Atalanti; B, Bulgaria; G, Gediz; GC, Gulf of Corinth; H, Helice fault; K, Kalamata; NAF, North Anatolian fault; P, Peloponnesos Peninsula; S, Sparta; T, Thessaloniki. Modified from Taymaz et al. (1991).

place in the late Pliocene. The system of north-south-trending faults produced the 1986 Kalamata earthquake on a fault at the western front of the Taygetos Mountains (Lyon-Caen et al., 1988) and an earthquake on the eastern front of this same range that may have destroyed ancient Sparta in 464 B.C.E. (Armijo et al., 1991). The change in extension direction may be due to the beginning of collision with the northern margin of Africa (Armijo et al., 1992).

Extension in the Aegean and west Anatolian regions may be due to the migration of the upper plate away from the trench, or the subsidence of the lower plate at the trench (for further discussion, see Chapter 11). The crust is relatively thin, which may explain why the graben are largely under water, including the Gulf of Corinth.

In the Aegean Sea and western Turkey, the normal faults do not occur in swarms with little topographic expression but as range-front faults, with cross-strike spacing of tens of kilometers. Therefore, despite the presence of active volcanoes and high heat flow, the crust appears to be strong enough to support basin-and-range topography and to sustain earthquakes up to magnitude 7 or even slightly larger.

Active normal faults in the southern Apennines of Italy are unusual in that they occur in mountainous terrain rather than at range fronts. For the most part, normal faults of the Apennines lie east of the active volcanoes of the Tyrrhenian coast of Italy and west of a zone of compression in the eastern, Adriatic Sea side of the Apennines. The southern Apennines were part of a fold-thrust belt until late Tertiary time, and extension has not taken place long enough to develop basin-and-range topography. The Irpinia earthquake of 1980 and the Avezzano earthquake of 1915, both of magnitude 6.9, produced surface rupture (Pantosti and Valensise, 1990; Ward and Valensise, 1989). Extension developed earlier in the northern Apennines, and normal-fault basins are well developed there, but normal-fault earthquakes are less common, and none have produced surface faulting.

The island of Kyushu, Japan, is at the northeastern end of the Ryukyu arc and is the only part of Japan characterized by normal faults. These normal faults may represent the eastern end of the Okinawa trough, a back-arc basin behind the Ryukyu arc. However, the faults occur as swarms within the active volcanic arc rather than behind it. The faults vary in trend from west, where they follow basement trends, to SW, where they are about parallel to the arc. However, many of the NE-trending faults have a large component of strike slip, and these faults may represent a westward continuation of the Median Tectonic Line, the largest strike-slip fault in Japan (cf. Chapters 8 and 11).

Extension of High Plateaus behind Collision Zones

Tibet and the South American Altiplano are so high with respect to their surroundings that active normal faulting in these high plateaus may be related to gravitational collapse toward plateau margins. In addition to being high, these plateaus are characterized by high heat flow and a relatively thin brittle crust. Despite the high heat flow, the plateaus have been struck by large earthquakes with surface rupture.

Continental crust of the Indian plate is being subducted underneath Eurasia along great décollement thrusts at the southern edge of the Himalaya. North of the Himalaya, the Tibetan plateau, at regional altitudes of 4–5 km, is underlain by thick, relatively warm continental crust. The southern part of the plateau is cut by N-S trending normal faults that are the surface expression of regional stretching in an ESE direction, approximately parallel to the Himalaya (Fig. 9-9; Armijo et al., 1986). Intermontane valleys 10–15 km across are grabens or half-grabens, bounded by range-front faults on one or both sides of the valley. These valleys are as much as 200 km apart across strike, in contrast to the Aegean region, where both horsts and grabens are the same width. High heat flow is evident by the abundance of hot springs and the presence of travertine-filled extension fractures. The region of Tibet characterized by abundant normal faults is bounded on the north by a discontinuous zone of strike-slip faults, which suggests that the block north of these faults is not extending. The stretching may be related to "escape-block" tectonics in which Eurasian crust moves eastward toward oceanic crust east of China as Eurasia moves out of the way of the advancing Indian plate (Fig. 10-22).

Hancock and Bevan (1987) suggest that the normal faults are forming in a region of foreland extension, in which brittle fractures strike at a high angle to orogenic margins. Other examples include the Lower Rhine graben north of the Alpine orogenic belt, the Syrian-Turkish foreland south of the Bitlis thrust zone, and the Patagonian platform northeast of the Andean fold and thrust belt in southern South America. Alternatively, the observation that slip vectors on the Himalayan plate-boundary thrust are perpendicular to the arcuate range front (Fig. 10-3) implies that the upper plate is stretching as it acquires curvature during southward thrusting, and this stretching produces extension (Molnar, 1990).

The Altiplano of the Peruvian and Bolivian Andes of South America also contains evidence of active normal faulting (Fig. 9-10; Mercier et al., 1992). The subducting oceanic Nazca plate dips 30° beneath the

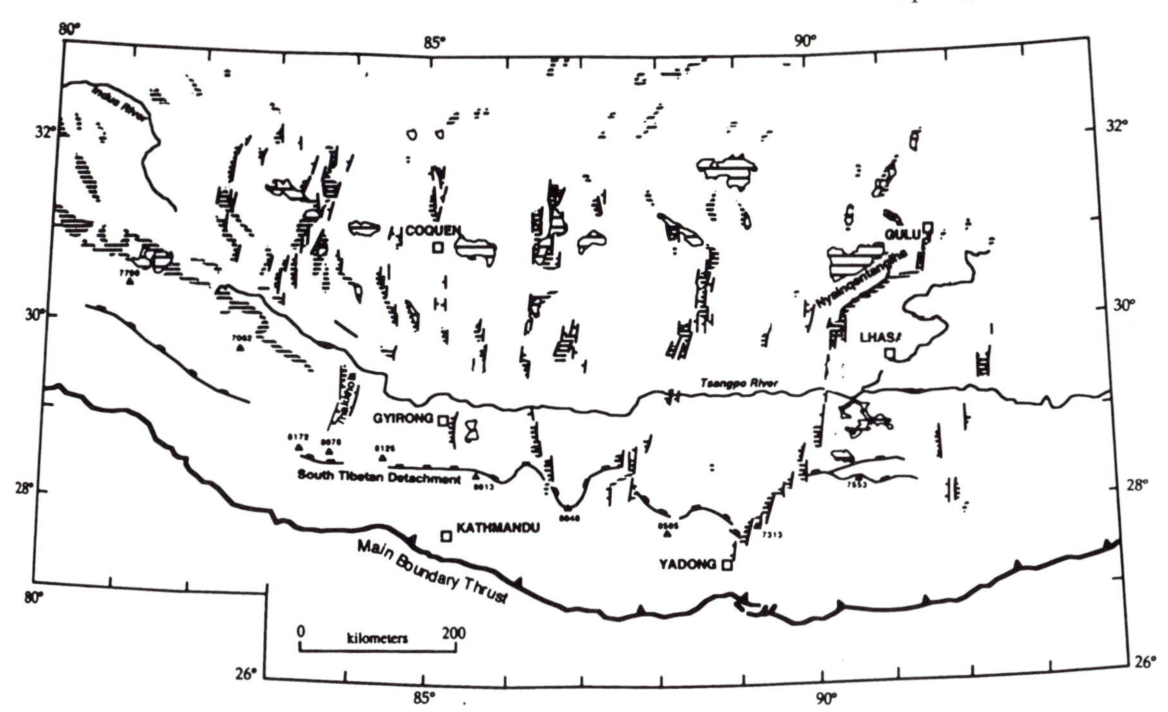

Figure 9–9. Normal faults of southern Tibet, modified from Armijo et al. (1986) by Harrison et al. (1995). Extension is east-west, at approximately right angles to direction of underthrusting of the Indian plate beneath Himalaya. Closely-spaced horizontal lines show graben fill.

South American plate. The convergence direction between these plates is N80°E. The aseismic Nazca ridge is also being subducted, and it separates the oceanic slab into two domains. In the southern domain (southern Peru and northern Chile), the oceanic plate continues down into the asthenosphere, and the Altiplano is characterized by active volcanoes. In the northern domain, north of 14°S (central Peru), the oceanic plate dips 30° as far east as the coastline, east of which it flattens to nearly horizontal beneath the Andes. This part of the Altiplano contains no active volcanoes.

The Altiplano of both domains contains normal faults with evidence for N-S extension, but normal faults are much more extensive in southern Peru, where the subducting plate does not flatten, and the plate boundary is, therefore, less strongly coupled (Fig. 9–10). Faults in the southern Peruvian Altiplano are 5–20 km long with maximum displacements of only a few hundred meters (Sébrier et al., 1985). Above the flat subduction zone in central Peru, the normal faults are found mainly in the Western Cordillera (Sébrier et al., 1988), where they include a normal-fault system nearly 200 km long separating the Cordillera Blanca from the 15-km-wide Callejon de Huaylas basin to the southwest (Schwartz, 1988). Maximum displacement on this normal-fault system is several kilometers. North of the Cordillera Blanca, the Quiches normal fault ruptured during the November 10, 1946 Ancash earthquake (M = 7.25; cf. Heim, 1949; Sébrier et al., 1988). Even though these faults strike NW, parallel to the Peru-Chile trench, the direction of extension is N-S, as in southern Peru.

Normal faulting terminates eastward, and the topographically-lower Subandean Cordillera is characterized by compressional tectonics (Chapter 10). The extensive normal faulting may be a response to an increase in the load stress with increasing altitude of the Altiplano, so that in the highest part of the plateau, the load stress has become the maximum principal stress, σ_1. However, normal faulting is also found in the coastal plain of northern, central, and southern Peru (Mercier et al., 1992) and in the Mejillones Peninsula near Antofagasta, northern Chile. The Mejillones Peninsula normal faults appear to be part of a large, west-facing normal-fault system called the Coastal Scarp (Armijo and Thiele, 1990). The presence of normal faults in the coastal plain suggests that the normal faulting of the Altiplano may have causes other than, or in addition to, high topography.

Intracontinental Rift Zones

Continental breakup is preceded by the development of rift zones in continental crust characterized by

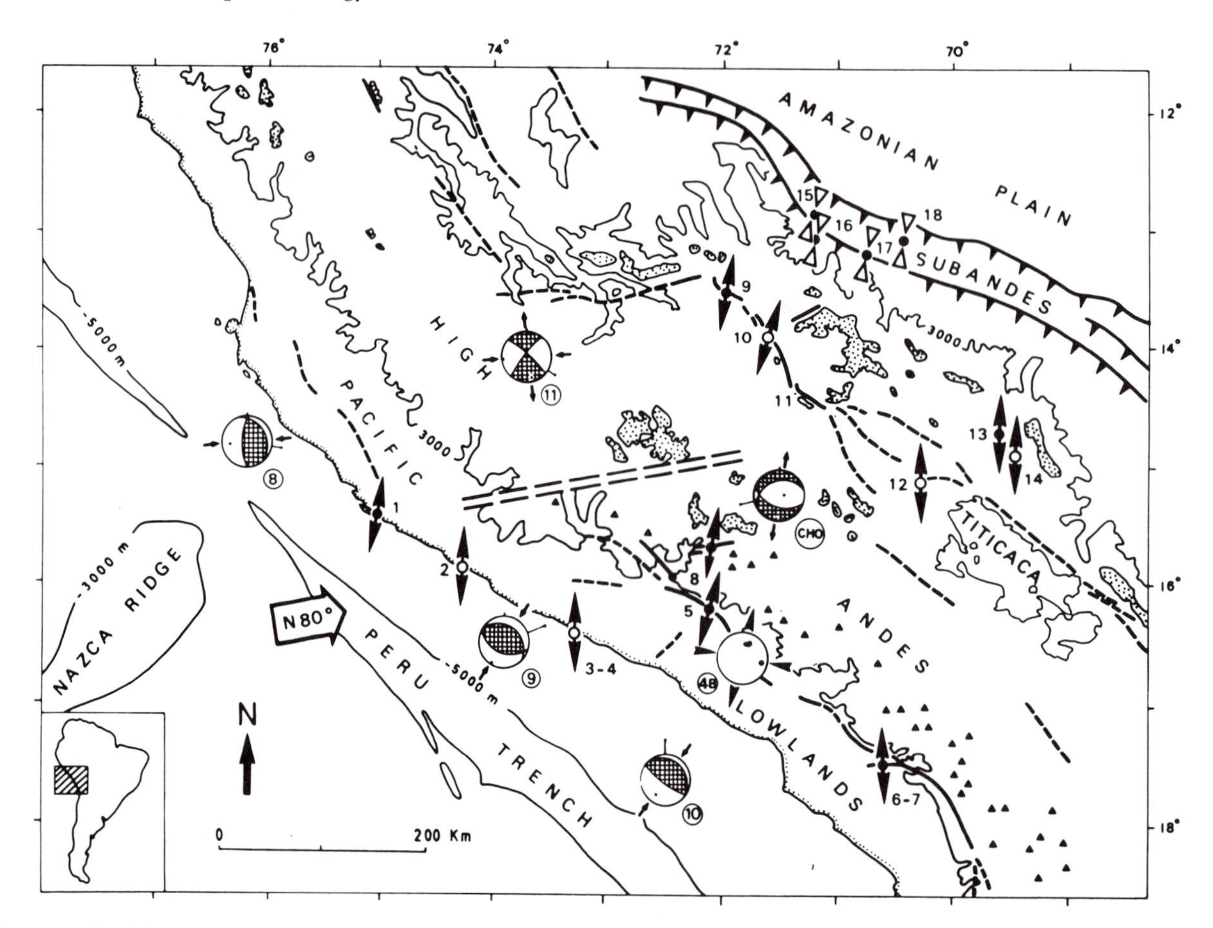

Figure 9–10. Active faulting (heavy lines) in the Altiplano (High Andes) of southern Peru. Extension direction is approximately north-south to N10°E, orthogonal to the Nazca-South American plate convergence vector, but oblique to the generally NW-SE strike of faults. Reverse faults characterize the Subandean province northeast of the Altiplano. Beach balls show fault-plane solutions of selected earthquakes; shaded quadrants are compressional. Solid triangles show Holocene volcanoes. Double dashed line marks boundary between southern domain, with the subducting oceanic plate dipping 30° into the mantle, and northern domain, where the subducting plate flattens to nearly horizontal beneath South American lithosphere. After Sébrier et al. (1985).

grabens bounded by normal faults. Rifting is accompanied by high heat flow from warm, buoyant crust so that the grabens are developed in the centers of elongate, uplifted regions. The grabens contain sedimentary rocks and also commonly contain volcanic rocks that date the onset of rifting.

The East African rift valleys extend from Ethiopia south through Kenya and Tanzania to Malawi. The valleys are commonly occupied by deep lakes, some of which have no outlets and thus are saline. The rift valleys are associated with volcanoes, some of which are active. The rift valleys comprise zones of high seismicity, with earthquakes as large as magnitude 7.4 (1910 Rukwa earthquake in Tanzania). Several earthquakes were accompanied by surface rupture, with the best-known example the 1928 Subukia earthquake (M6.9) in Kenya (Ambraseys, 1991). Brittle crust in the vicinity of the East African rifts is unusually thick, and crustal earthquakes at depths of 25–29 km are known (Jackson and White, 1989). At the northern end, in Ethiopia and Djibouti, the rift valley bifurcates at a triple junction called the Afar Triangle (discussed in the previous text under Spreading Centers).

The Rhine graben of northwest Europe is a rift valley, and the Lower Rhine Embayment, associated with Holocene volcanism in the Rhenish Shield, is the most active part of this structure (Müller et al., 1992). Other rift valleys include the Baikal rift of Siberia and the Rio Grande rift of New Mexico, U.S.A. (Fig. 9–11).

Rift valleys are incipient spreading centers, and there is a complete gradation betwen rift valleys underlain entirely by continental crust (East Africa, Rhine graben, Rio Grande rift) and those underlain by oceanic crust, where continental breakup is com-

Figure 9–11. Topography of the Basin and Range Province, western United States, part of a shaded-relief image prepared by Thelin and Pike (1991) by digitizing elevation values at intervals of 805 m and illuminating the resulting digital elevation model from the west-northwest, 25° above the horizon. North is approximately toward the top of page. Vertical exaggeration 2×. Black area at upper right is Great Salt Lake; flat area farther north is Snake River Plain (cf. Fig. 9–12). Basin and Range Province extends from Wasatch fault, east of Great Salt Lake, to eastern front of Sierra Nevada (lower left corner). Albers Equal-Area Conic Projection.

plete. These latter rift valleys are incipient ocean basins, and they include the Red Sea, Gulf of Aden, and Gulf of California.

Zones of extension that are broader and more diffuse than those discussed above characterize the Basin and Range Province of western United States and northwestern Mexico (Figs. 9-11 and 9-12) and the graben systems of northeastern China. Like the rift zones, these broad regions have high heat flow, local volcanism, and they produce large earthquakes.

The Basin and Range Province extends from California east to the Colorado Plateau in central Utah and from Sonora, Mexico north to southeastern Oregon and western Montana (Fig. 9-11). The province is marked by high heat flow, especially in northwest Nevada, and, locally, by volcanic activity, but its principal characteristic is a series of tilted fault blocks resulting in a horst-and-graben structure in which ranges are bordered by normal faults on one or both sides. These fault blocks are 10-30 km wide (Fig. 9-12), and active faults typically occur at or near range fronts. Extension is oblique, because the Sierra Nevada and areas farther west are moving northwestward with respect to stable North America east of the Basin and

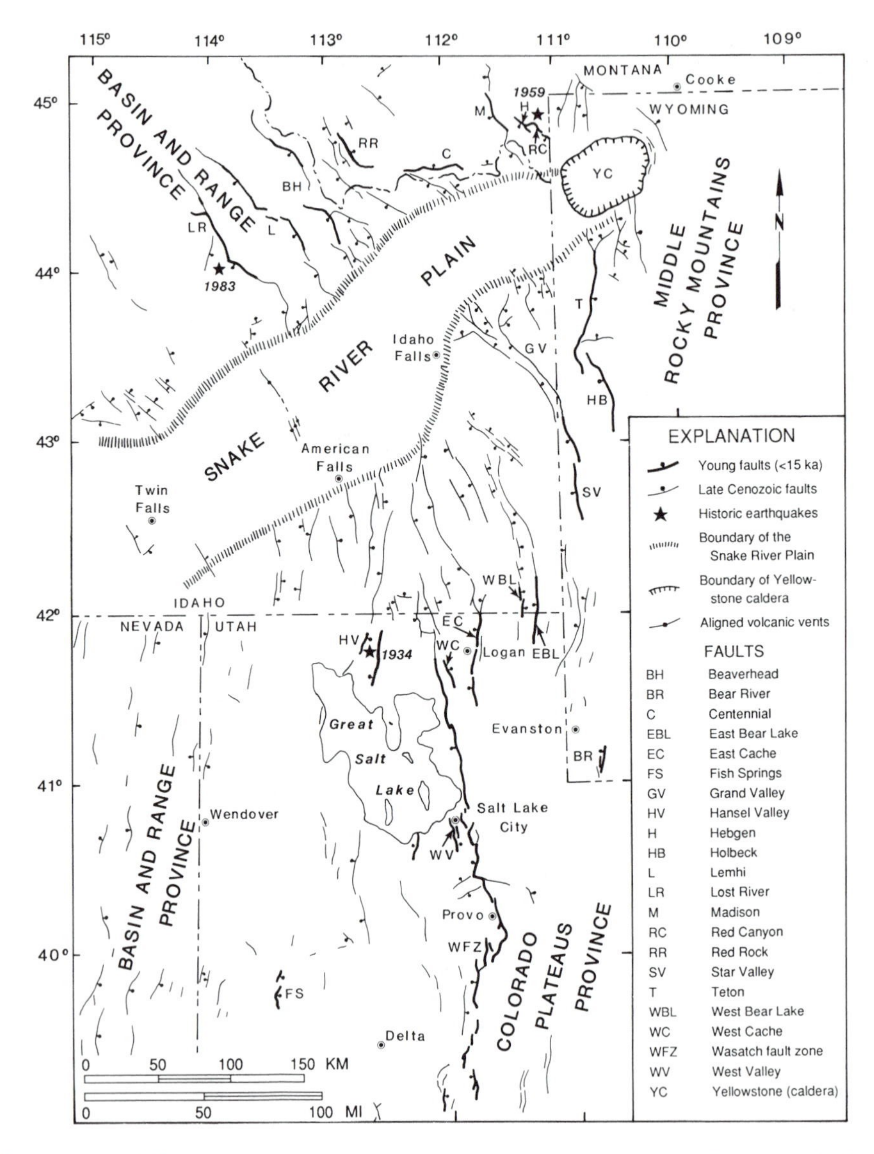

Figure 9-12. Selected major faults in the northeastern Basin and Range Province. The distribution of late Cenozoic faults illustrates the cross-strike spacing of Basin and Range faults and the change in trend across the Snake River plain, the presumed track of the Yellowstone hot spot (YC). Faults active in last 15,000 years are shown in bold lines. From Machette et al. (1991).

Range. Accordingly, there are northwest-trending zones of right-lateral strike slip within the western Basin and Range, and two of the larger historical earthquakes in the region, the 1872 Owens Valley and 1932 Cedar Mountain earthquakes, were predominantly right-lateral strike-slip events.

Although late Cenozoic normal faults are distributed relatively uniformly across the Basin and Range Province, historical and instrumental seismicity is concentrated along a zone of seismicity extending from southern Nevada across central Utah to southwestern Montana (Intermountain Seismic Zone), and a zone extending northwest across eastern California and northeast across Nevada (Central Nevada Seismic Zone). These two zones occupy the western and eastern edges of the Basin and Range province in Figure 9-11. Historical earthquakes with surface rupture are concentrated in the Central Nevada Seismic Zone. It is not yet established if these two active zones have had equally focused activity throughout the Quaternary, or if activity was focused in other zones in the Basin and Range earlier in the Quaternary, as suggested by Pleistocene fault scarps in regions of present-day low seismicity.

The western boundary of the Basin and Range Province is generally considered to be the Sierra Nevada in California. On the other hand, the Sierra Nevada is, in a sense, the largest tilted fault block in the Basin and Range, because normal faults are common in the western foothills of the Sierra and in oil fields producing from strata of Miocene to Pleistocene age on the east side of the San Joaquin Valley adjacent to the Sierra Nevada. In August, 1975, a normal-fault earthquake reactivated a Mesozoic fault zone in the western foothills of the Sierra Nevada near Oroville Dam. The mainshock, with magnitude 5.7, occurred at 5.5 to 8 km depth, and the aftershocks delineated a circular patch on a normal fault dipping about 60°W (Fig. 9-13). Surface rupture in a zone 3.5 km long occurred at that part of the seismicity-defined patch that intersected the surface. Seismic moment measured on the basis of surface rupture alone would be in error; a better measurement of fault area is the area of a circle with diameter 13 km (Fig. 9-13).

Northeastern China is also characterized by normal faults in a broad region, and, like the Basin and Range, there are strike-slip and oblique-slip faults as well (Xu and Ma, 1992). Faults striking northeast are characterized by oblique slip with right-lateral and normal components, whereas faults striking west to northwest are characterized by oblique slip with left-lateral and normal components (Fig. 9-14). Like the Aegean region, northeastern China has been compared to a back-arc basin, in this case behind the Ryukyu arc, but the region appears to be too large and too far from the Ryukyu trench to be a back-arc basin. The region includes two stable plateaus characterized by low seismicity, the Ordos and the Taihang Shan, the latter directly west of the North China Plain in Fig. 9-14. Both platforms are surrounded by grabens characterized by high seismicity. To the east, a northeast-trending, oblique-slip graben system extends to the Gulf of Bohai and into Manchuria. Motion on one of these faults producing the 1976 Tangshan earthquake was largely strike slip (cf. Fig. 4-6).

The Ordos Plateau is a remarkable feature (Fig. 9-14). Its interior comprises much of the great loess uplands of north China, and the Yellow River flows along its northwest, north, and east edges. The margins of the Ordos Plateau are in part reactivated ancient geological structures, indicating that the faults follow old zones of weakness. Seismicity is restricted to its faulted edges so that the interior is as stable seismically as northern Europe or eastern North America. It is analogous to the Colorado Plateau (Fig. 9-11), flanked by zones of strong deformation on both sides with activity during much of the Cenozoic. However, only the western margin of the Colorado Plateau, the Wasatch fault zone (Figs. 9-12, 9-41), appears to be highly active today.

CASE HISTORIES

We now consider in more detail three normal-fault earthquakes, selected in part because of relatively complete data sets and in part because they illustrate different tectonic environments. The first example, the Borah Peak, Idaho, earthquake, is characteristic of an intracontinental rift environment with continental crust, elevated basin-and-range topography, and moderate heat flow, although lacking active volcanism. The second example, the Gulf of Corinth, Greece, earthquakes occurred in a back-arc environment with basin-and-range topography in which the tectonic basin (Gulf of Corinth) is below sea level, suggesting attenuated crust. The final example, the Edgecumbe, New Zealand, earthquake is also in a back-arc environment, but heat flow is much higher, there is contemporaneous volcanism, and basin-and-range topography has not developed.

Borah Peak, Idaho, Earthquake of 28 October 1983 (M_s = 7.3)

The Borah Peak earthquake resulted from rupture of part of the Lost River fault along the front of the Lost

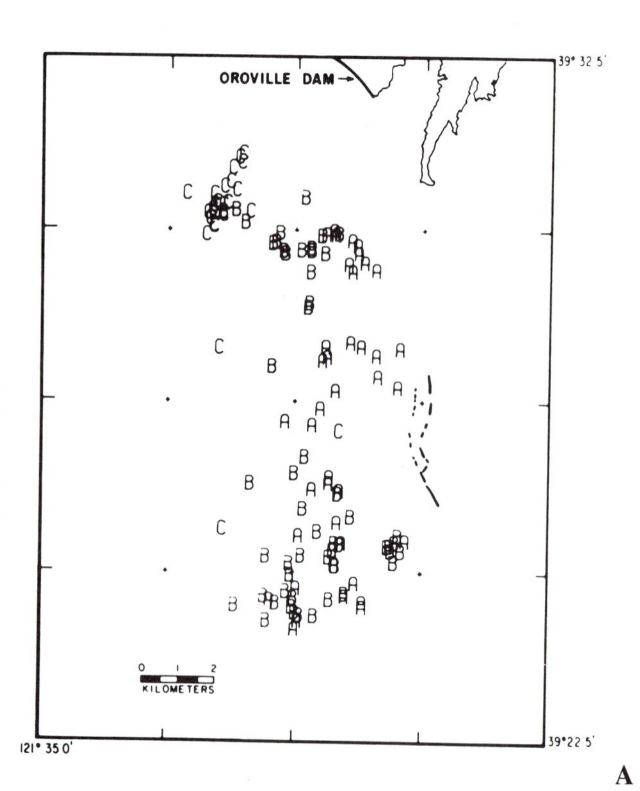

Figure 9–13. Oroville, California, earthquake of August, 1975. (a) Epicentral plot of best-located aftershocks. Aftershocks: A, 0–4 km depth; B, 4–8 km; C, >8 km depth. The aftershocks form an elliptical patch which barely intersects the surface at the zone of surface rupture. After Lahr et al. (1976). (b) Projection of aftershocks onto an east-west cross section. Note the maximum depth of earthquakes at 11 km and the small number of events in the shallow 2 km. A comparison of (a) and (b) shows that the ellipse of (a) is a circular patch on the dipping fault plane. After Clark et al. (1976).

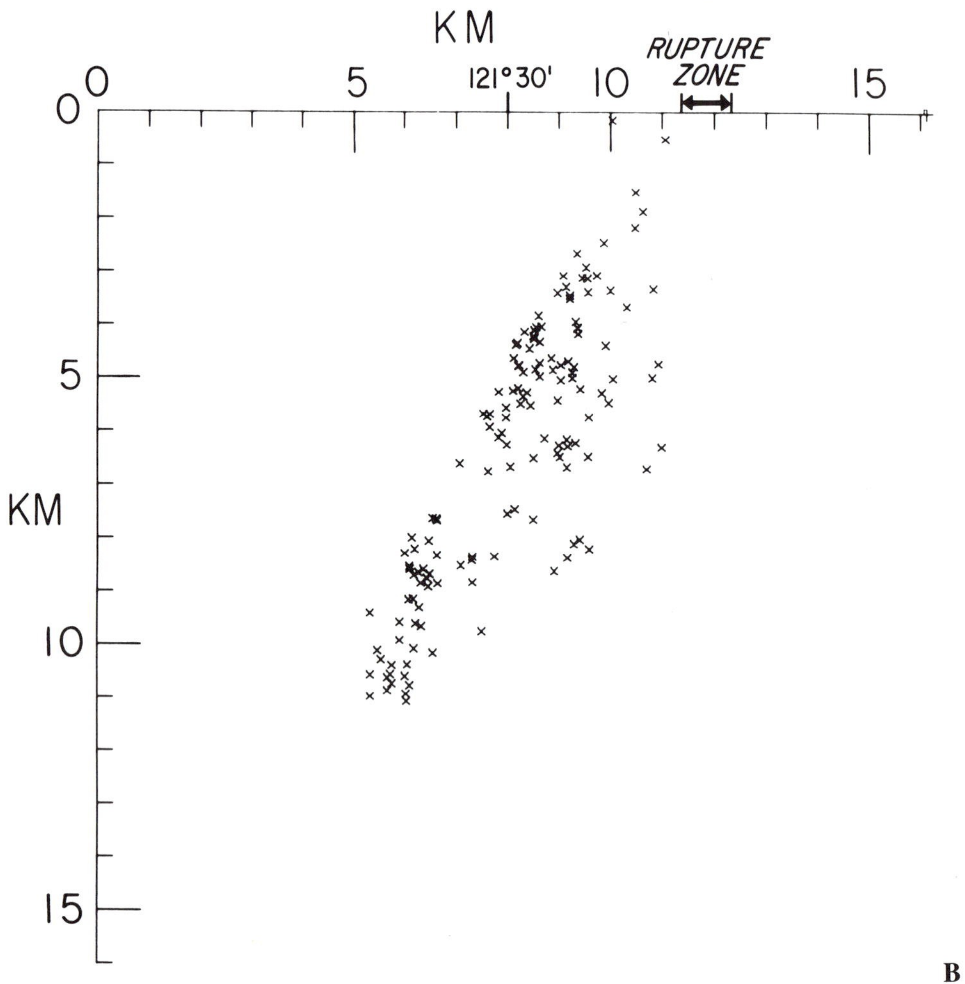

River Range west of Borah Peak, the highest mountain in Idaho (Fig. 9-15). The Lost River Range is within the Basin and Range Province, north of the Snake River Plain (Fig. 9-12). Cross-strike spacing between adjacent active range-front faults is about 30 km, and range-front faults accommodate a regional tilt of individual ranges to the northeast (Fig. 9-16).

Surface rupture occurred on two segments of the Lost River range-front fault. Total length of the surface fault traces is between 33 and 40 km. The variation in the length cited is due to uncertainty about whether breaks south of Devil Canyon in Warm Spring Valley are tectonic or not (Fig. 9-15). The southern segment, 21 km long, constitutes the main rupture, with maximum vertical displacement of 2.7 m (Fig. 9-17) and maximum left-lateral displacement of 0.7 m. The northern rupture, at least 8 km long with maximum displacement of one meter, is separated from the southern segment by a gap 4.8 km long in the Willow Creek Hills, which separates the Thousand Springs Valley from Warm Spring Valley. A zone of surface rupture extends from just south of the Willow Creek Hills WNW across the hills to the southwest side of Warm Spring Valley; maximum displacement there is 1.6 m. The northern and western

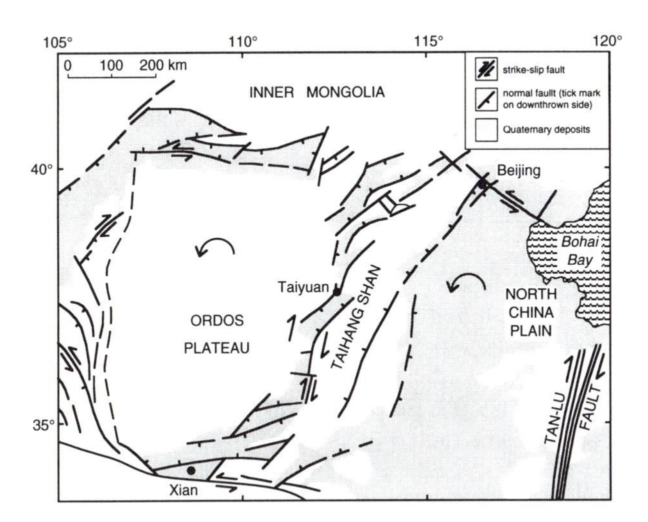

Figure 9-14. Tectonic map of northeast China showing active faults (solid and dashed lines). Normal faults have tick mark on downthrown side, strike-slip faults marked with arrows; some faults have both normal and strike-slip displacement. Late Cenozoic basin fill in dotted pattern. Ordos Plateau, a stable platform surrounded by active faults, is rotating counterclockwise. North China Plain and southwest Manchuria also have active faults with normal and strike-slip displacement. From Xu et al. (1992).

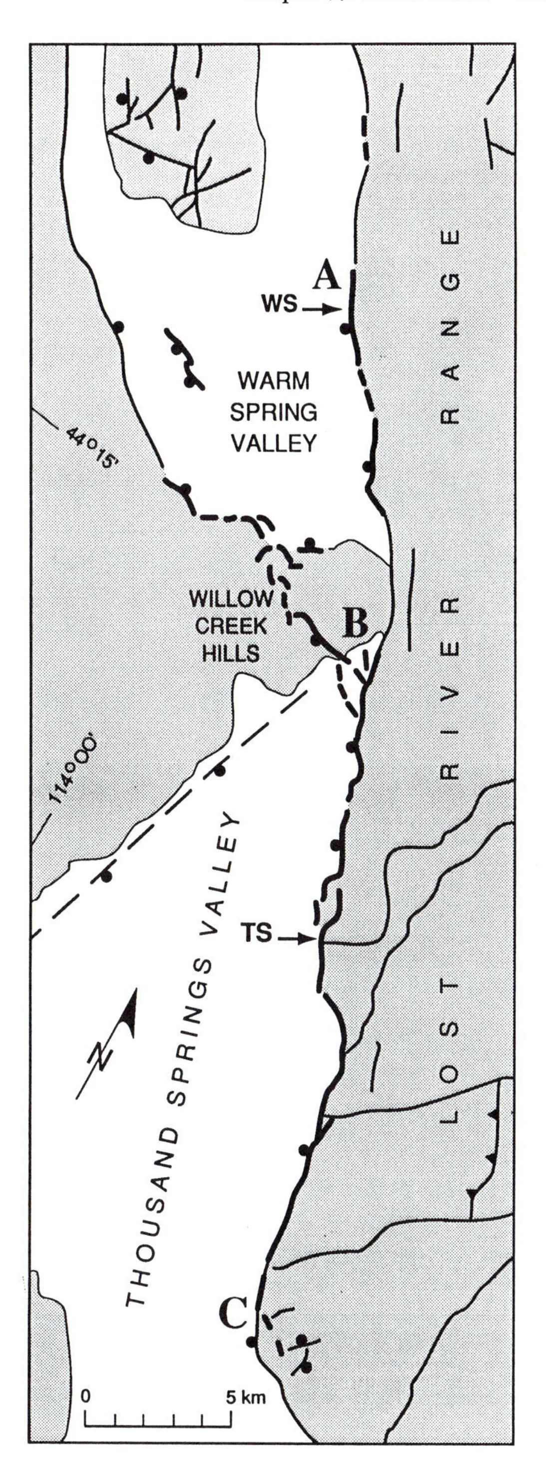

Figure 9-15. Map of fault scarps and ground rupture accompanying the 1983 Borah Peak earthquake (heavy lines, bar and ball on downthrown side). Shaded areas are mountains; alluviated valleys are unpatterned. After Crone et al. (1987) and Crone and Haller (1991).

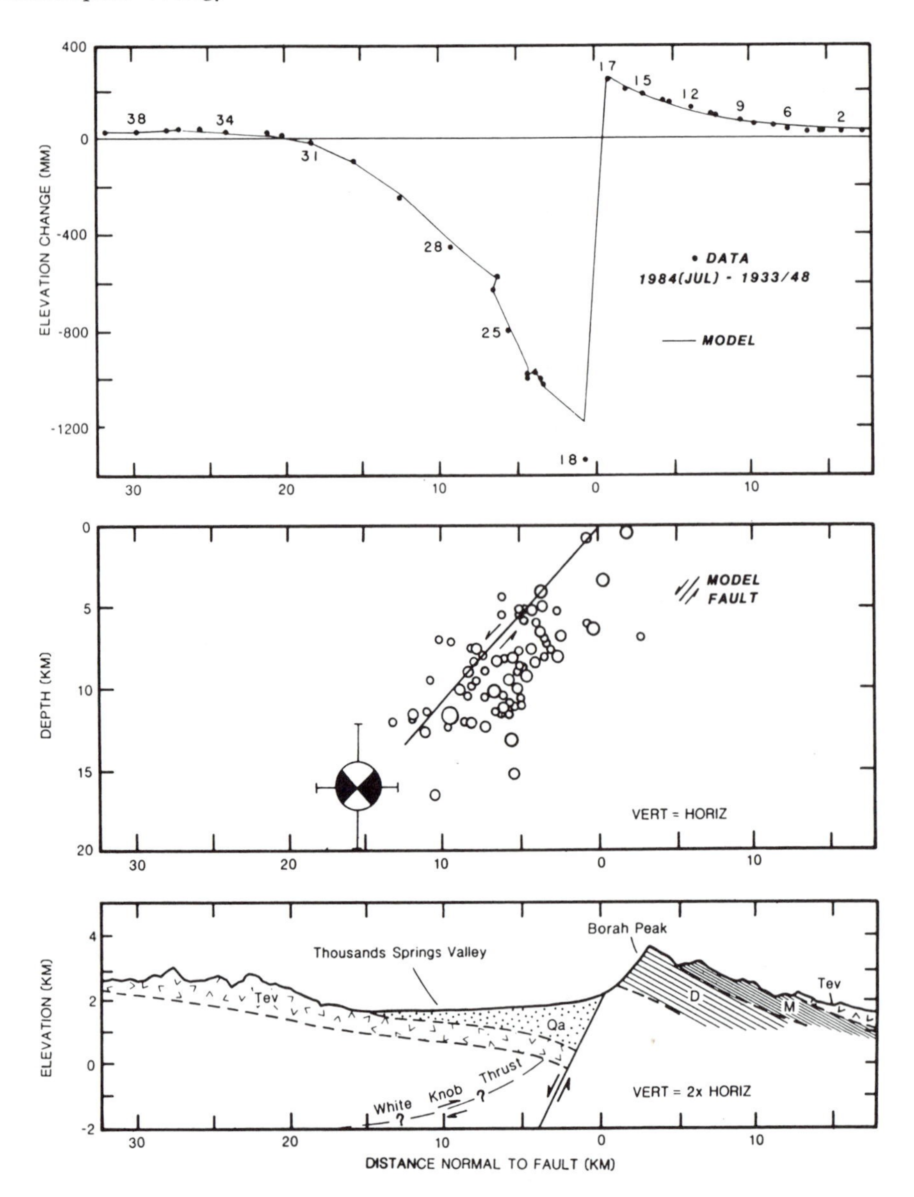

Figure 9–16. Cross sections of Borah Peak earthquake zone showing (top) observed coseismic elevation changes of benchmarks (dots) and the predicted changes based on a model of a planar fault, and mainshock and aftershocks near leveling line together with the location of the fault used in the geodetic model (middle), and a geological cross section including a low-angle thrust which was not reactivated during the earthquake. The distance along the cross section is with respect to the surface trace of the fault (bottom). After Stein and Barrientos (1985).

segments have less-continuous scarps and smaller displacements than the southern segment.

A releveling line parallel to the Lost River fault (Fig. 9-18) showed that nearly all the geodetic changes are associated with the Thousand Springs Valley segment, south of the bifurcation of surface ruptures. The 21-km length of this segment is in agreement with a fault length of 19 km based on source parameters of the mainshock. As a transverse basement block, the Willow Creek Hills served as a boundary between the Thousand Springs segment, which appears to be directly related to the 1983 rupture surface, and the Warm Spring segment, which underwent only slight geodetic change and may represent a sympathetic, secondary rupture, not part of the main rupture surface (see further discussion in the following text).

Along the Thousand Springs segment, the main rupture tended to follow preexisting Holocene scarps at

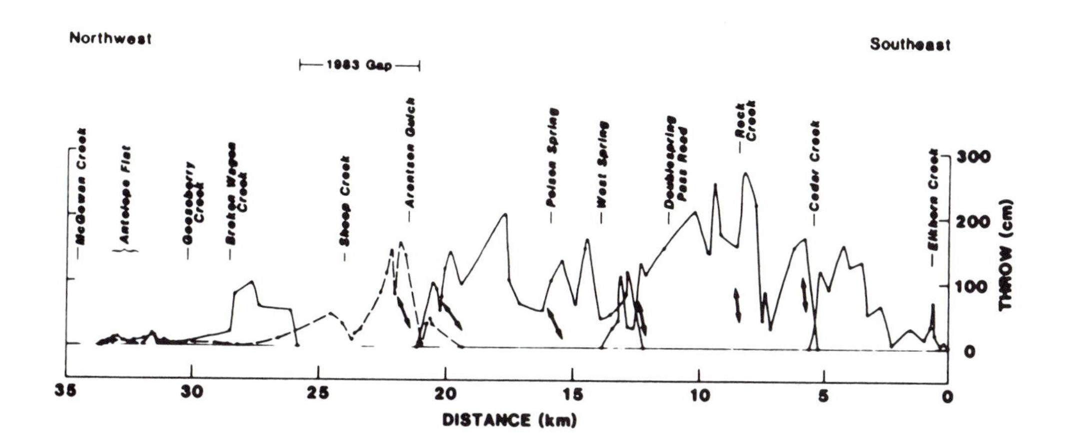

Figure 9–17. Amount of net throw on surface faults accompanying the 1983 Borah Peak earthquake. Solid line: throw on Lost River fault. Dashed line: throw on western section in and near Willow Creek Hills. Arrows show direction of net slip based on surface observations. After Crone et al. (1987).

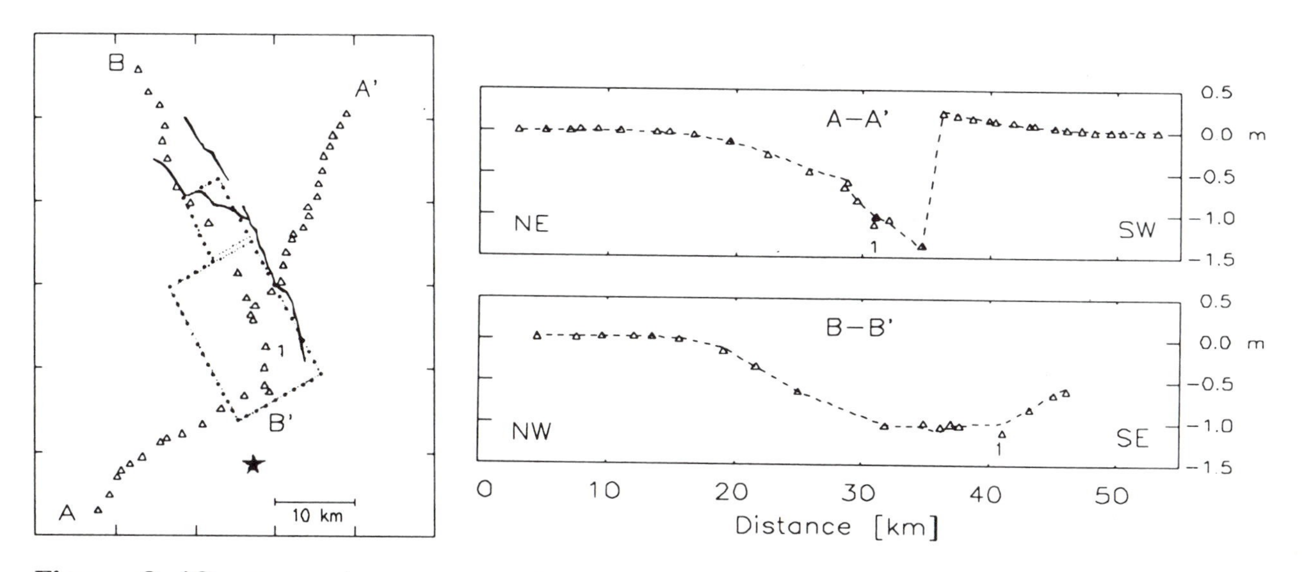

Figure 9–18. Map and cross sections showing leveling lines across (A-A′) and parallel to (B-B′) surface trace of Lost River fault. Box on the map shows the model fault with strike N28°W, dip 49°, and slip 2.2 m. After Fig. 13 of Barrientos et al. (1987).

or close to the alluvium-bedrock contact. This shows that the 1983 earthquake was a quantum in the long-term process of normal faulting and northeastward tilting in this part of the Basin and Range Province. The net throw on the Lost River fault during the Cenozoic is at least 2.5 km, based on the topographic relief between Borah Peak and the valley floor plus the thickness of valley fill. The amount may be more than 5 km based on separation of a Tertiary volcanic formation (Tev in lower diagram of Fig. 9-16). The separation at the Thousand Springs segment is at that part of the Lost River fault where Cenozoic separation is greatest; total separation is less for basins to the southeast and northwest.

In some places, the rupture is a single break, but elsewhere it is accompanied by smaller, short ruptures dipping toward and away from the range (Fig. 9-19). Locally, the main rupture ends, and another rupture is found stepped either to the left or right. Near Doublespring Pass Road, normal faults dip both toward and away from the range. Thrust faults appear locally 30-90 m downslope from the main rupture

Figure 9–19. View south along graben formed by west-dipping and east-dipping normal faults north of Doublespring Pass Road (in middle distance). Main fault scarp is on the left. Photo by R. E. Wallace six days after the 1983 Borah Peak earthquake.

(Fig. 9-20). However, these appear to be shallow features not reflecting movement on the main fault at depth. The trace of the main fault upslope from the thrusts follows an arcuate pattern, concave downslope (Fig. 9-20), analogous to the trace of the headwall scarp of a landslide.

No foreshocks >M2 were recorded during the two months prior to the earthquake, and the region within 25 km of the Borah Peak epicenter had no activity of $m_b \geq 3.5$ for at least two decades before the earthquake. The mainshock occurred at a depth of about 16 km at the base of its eventual aftershock zone (Richins et al., 1987). The first 24 hours of aftershocks delineate a parallelogram with two sides parallel to the surface rupture and two sides striking N-NE. The mainshock is in the southern corner of the parallelogram (Fig. 9-21). A plot of well-located aftershocks recorded in the first 21 days (Fig. 9-22) defines a plane with a dip of 45°, intersecting the zone of surface rupture on the Lost River fault, at the front of the Lost River Range. Most aftershocks originated at depths between 4 and 12 km, although some originated as deep

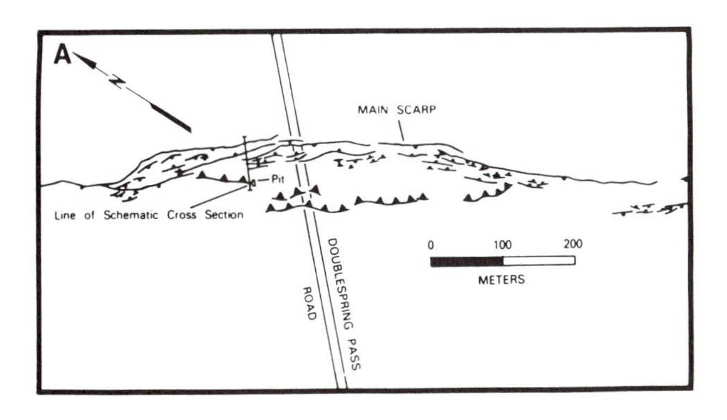

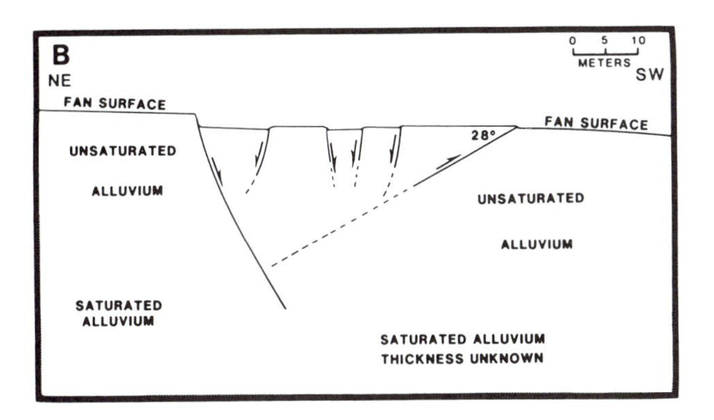

Figure 9-20. Detailed map (a) and cross section (b) of fault scarps near Doublespring Pass Road. Normal faults have ball on downthrown side; thrust faults have barbs on hangingwall. From Crone et al. (1987).

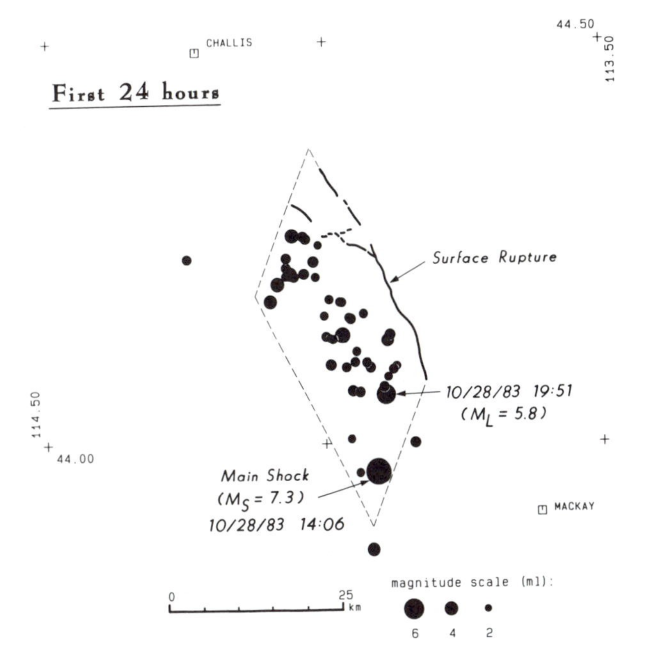

Figure 9-21. Epicenters of the first 24 hours of the Borah Peak, Idaho, earthquake sequence. The dashed lines show a parallelogram within which the mainshock, at its south end, and all but 3 of 47 aftershocks are located. After Richins et al. (1987).

as the mainshock. The zone of concentrated aftershocks strikes parallel to the Lost River fault trace. The small number of aftershocks close to the mainshock indicates that rupture near the hypocenter did not produce stresses that induced subsequent aftershocks in its vicinity.

Fault-plane solutions of the mainshock and aftershocks (Fig. 9-23) show predominantly normal faulting with a component of left-lateral strike slip, so that the axes of extension trend predominantly NNE, at an angle < 90° to the strike of the surface rupture and the zone of concentrated aftershocks (Fig. 9-24). Focal mechanisms in the northwestern sector of Figure 9-23 are more consistently normal than those of other tectonic environments despite the fact that the pattern of surface rupture and the distribution of aftershocks (cross section A-A′, Fig. 9-22) are more complex there. Variation in focal mechanisms (Fig. 9-23) and the diffuse distribution of aftershocks (Fig. 9-22) suggest that part of the deformation occurs off the main fault, and the hangingwall and footwall are internally deformed. The NNE trend of the tension axes (Fig. 9-24) is parallel to the NNE-striking sides of the parallelogram of Figure 9-21, consistent with NNE slip. The location of the mainshock in the southern corner of the aftershock parallelogram suggests that the rupture propagated NW.

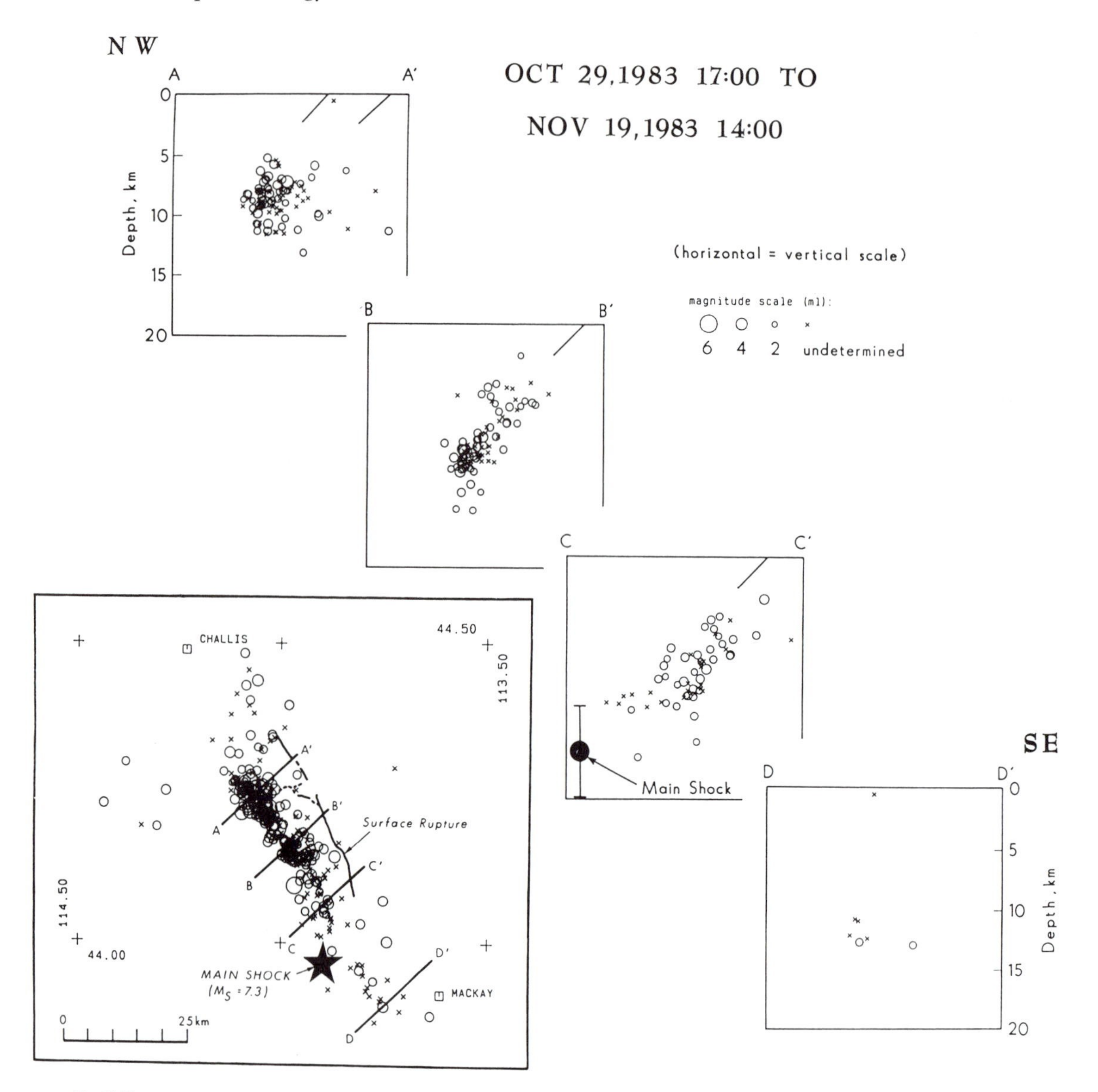

Figure 9–22. Map and cross sections of 374 aftershocks of the Borah Peak, Idaho, earthquake from 29 October to 19 November, 1983. Mainshock is deeper than nearly all aftershocks, and very few aftershocks are close to the mainshock. Most aftershocks are at depths of 4–12 km in a zone parallel to the surface rupture along the Lost River fault (solid and dashed lines on map, tick marks on cross sections). Surface rupture and aftershocks define a zone dipping about 45° southwest. After Richins et al. (1987).

Benchmarks in the epicentral area that were surveyed in 1933 and 1948 were releveled in 1983 and 1984, following the earthquake (Stein and Barrientos, 1985; Barrientos et al., 1987). The changes shown in Figure 9-18 are relative, but the profile is extended far enough northeast and southwest of the fault that there is no significant change in the relative elevation changes between benchmarks for a distance of 5 km at each end of the profile. The fault plane responsible for the deformation can be modeled from these geodetic data, because the pattern of displacements is sensitive to the slip and orientation of the fault. The geodetically constrained fault plane is consistent with the seismically determined plane and with the surface trace of coseismic rupture (Fig. 9-16). Models of curved faults fit the geodetic data more poorly, suggesting that the fault is planar.

The Thousand Springs segment ruptured previously in middle to early Holocene time (Scott et al., 1985), and the style and amount of surface displacement accompanying the prehistoric earthquake was similar to that in 1983. Slip rate on the Lost River fault at the Thou-

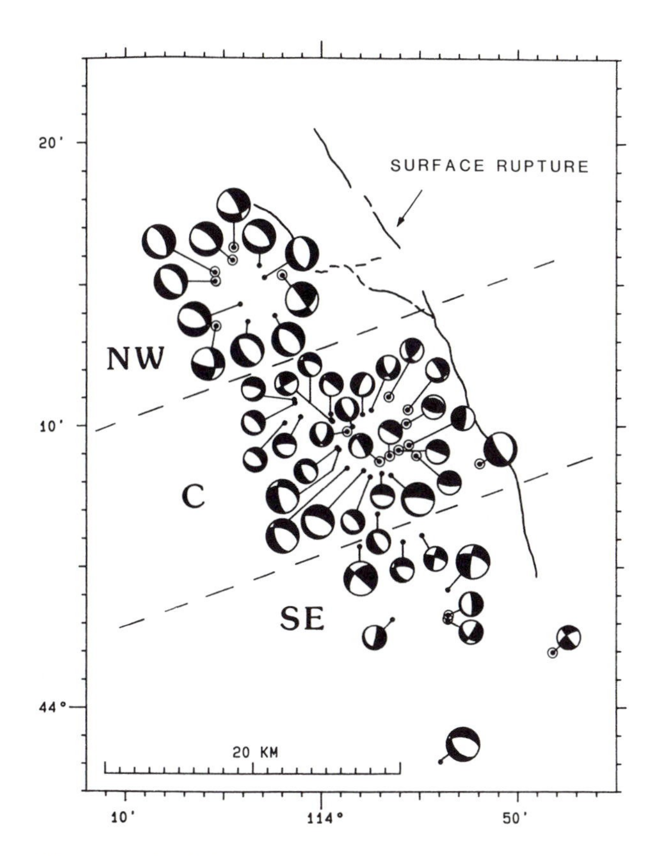

Figure 9–23. Fault-plane solutions of Borah Peak mainshock and well-located aftershocks. Larger focal spheres are for aftershocks of M ≥ 3.5. Southernmost focal sphere is that of mainshock. Dark quadrants are compressional. Dashed lines separate northwest (NW), central (C), and southeast (SE) regions of aftershock series. Surface rupture shown as solid and dashed lines. After Richins et al. (1987).

sand Springs segment is about 0.3 mm/yr (Scott et al., 1983). The latest prehistoric rupture on the Warm Spring segment occurred shortly before 5500-6200 years ago; this rupture was much larger than the sympathetic rupture that occurred in 1983. Both segments could have ruptured simultaneously in the prehistoric earthquake, or they could have ruptured in separate earthquakes. In contrast to the central segment of the Lost River fault, the Arco segment, close to the Snake River plain with half the structural relief at Thousand Springs, last ruptured about 30,000 years ago, giving a long-term slip rate there of 0.1 mm/yr.

Gulf of Corinth, Greece, Earthquakes of 1981

The Gulf of Corinth (Fig. 9-8) is one of the most seismically active regions in the eastern Mediterranean, and destructive earthquakes have been recorded there for the last 2500 years. The Gulf and the Peloponnesos Peninsula to the south are analogous to the Thousand Springs Valley and Lost River Range, respectively, except that seismicity is higher, and the Gulf is below sea level. The Gulf is an asymmetrical graben dominated by normal faulting on its south side. Surface rupture on one of these normal faults at the western end of the Peloponnesos during an earthquake in 1861 was described by Schmidt (1875), the first description of surface faulting by a trained contemporary observer.

At the eastern end of the Gulf, the graben may be divided into southern and northern sections (Fig. 9-25). The section south of the Perachora Peninsula, including Corinth and the Corinth Canal, accumulated marine sediments as faulting took place on the southern margin of the Gulf of Corinth (Vita-Finzi and King, 1985). Subsequently, the region north of the Perachora Peninsula was downdropped to form the Alkyonidhes Gulf. The marine sediments near Corinth were uplifted during this younger episode of faulting (Fig. 9-25). The 1981 earthquakes were an expression of this later period of faulting.

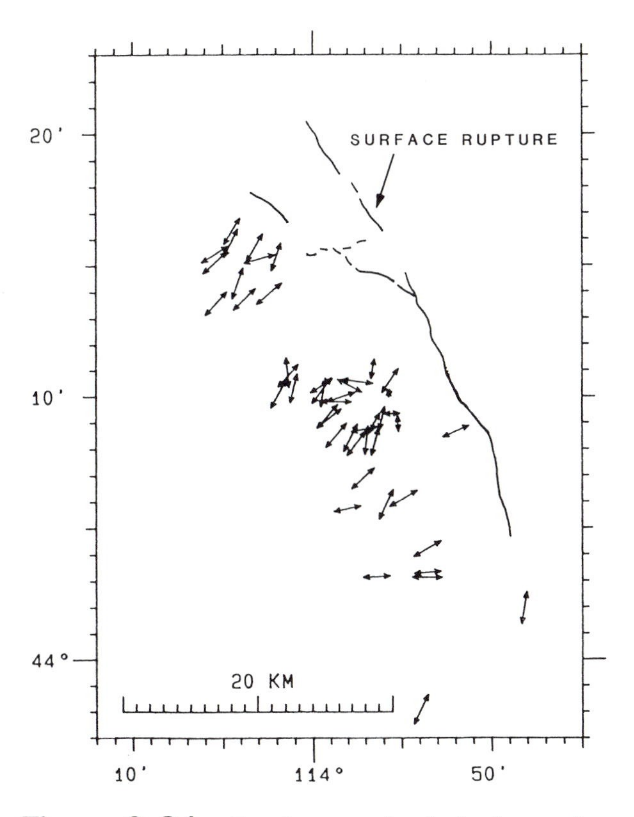

Figure 9–24. Tension axes for fault-plane solutions of Figure 9–23. Lengths of arrows are proportional to horizontal projections of tension axes. After Richins et al. (1987).

Three earthquakes were generated by slip on faults of this younger system in early 1981. Main parameters are summarized below, from Jackson et al. (1982):

Date	m_b	M_s	$M_0 \times 10^{25}$ (dyne-cm)	Depth (km)
24 February 1981	5.9	6.7	7.28	10
25 February 1981	5.5	6.4	1.68	8
4 March 1981	5.9	6.4	0.97	8

The three mainshocks, two major aftershocks, and surface ruptures accompanying the earthquakes appear in Figure 9-26. Surficial rupture on north-dipping faults was observed on the Perachora Peninsula after the 25 February earthquake, and new faulting on south-dipping faults north of Alkyonidhes Gulf appeared during the 4 March earthquake. Because the first two earthquakes occurred at night and were separated by less than six hours, it is unclear which earthquake produced the surface ruptures on the Perachora Peninsula. Surface rupture may have also occurred on the north-facing submarine escarpment on the southern side of the Gulf of Corinth (Fig. 9-26), but this cannot be substantiated (Taymaz et al., 1991).

Faulting on the Perachora Peninsula (Fig. 9-27) generally follows preexisting faults that had produced prominent north-facing limestone escarpments (Jackson et al., 1982). Displacements are commonly 50-70 cm, but are locally as large as 150 cm, with slip azimuths trending north. The 1981 faulting commonly follows the contact between limestone and alluvium or colluvium at the base of the escarpment. Faults of 1981 on the southern, higher escarpment are more continuous than those on other escarpments. At the western and eastern ends of 1981 faulting, surface rupture is more discontinuous with more variable trends. Only small parts of the northern, lower escarpment were reactivated, and new faulting there is discontinuous. Near the coast at Milokopi, a small south-facing fault was reactivated.

Faulting on the north side of the Gulf of Alkyonidhes consists of two continuous segments separated by a zone of discontinuous fractures (Fig. 9-27). Along the eastern segment, reactivated faults produced

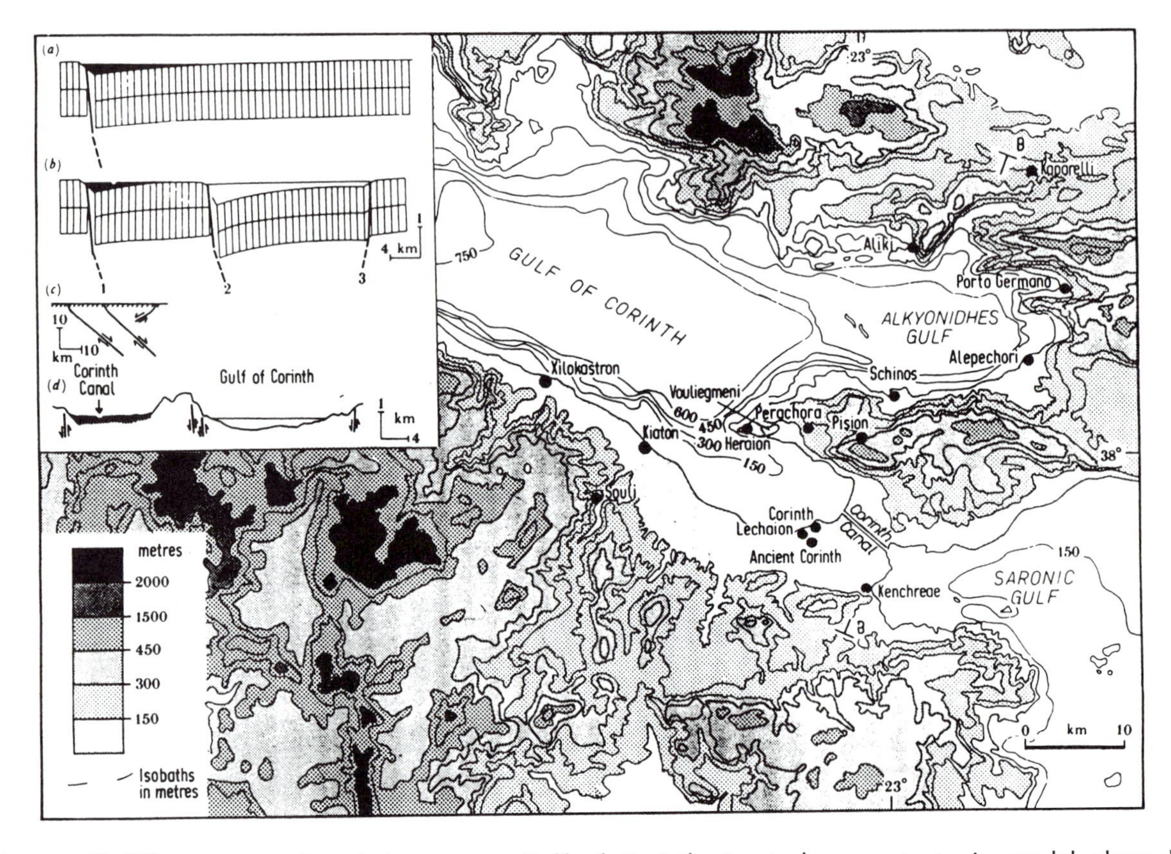

Figure 9-25. Topography of the eastern Gulf of Corinth. Inset shows a tectonic model along line B-B. Motion on the southern fault system is shown in (a); shading denotes marine sediment. In (b), displacement of faults forms the Alkyonidhes Gulf, uplifting marine sediments related to the earlier faulting. After Vita-Finzi and King (1985); inset interpretation proposed by Jackson et al. (1982) and King et al. (1985).

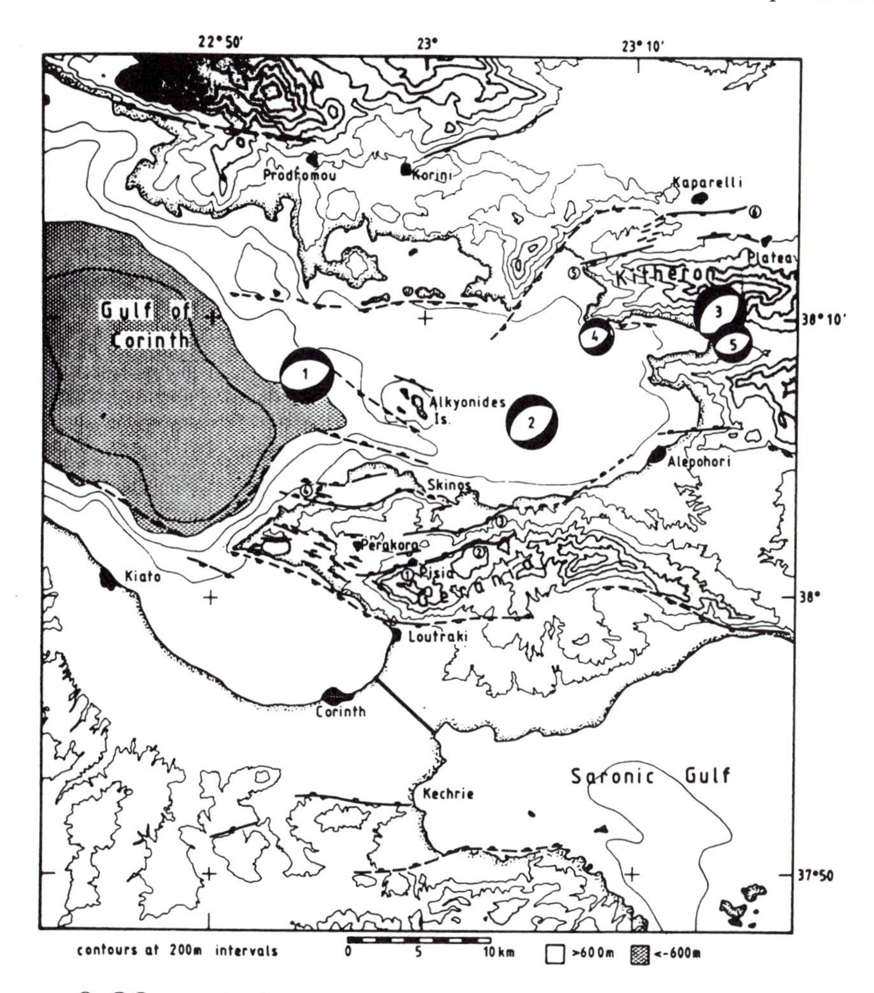

Figure 9–26. Fault-plane solutions (lower hemisphere) for three mainshocks (1, 2, 3) and two aftershocks (4, 5) of the 1981 Gulf of Corinth sequence. Normal faults with 1981 surface rupture have filled blocks in hangingwall and are numbered 1–6. Other normal faults, open blocks in hangingwalls. After Taymaz et al. (1991).

scarps as high as 3 m, although displacements averaged 50–70 cm. At its eastern end, the 1981 fault zone turns southeast and crosses valley-floor alluvium, departing from the older east-west scarp. Displacement diminishes in this section, and slip azimuths trend more easterly. At the west end of this segment, there is also a change in strike, and the 1981 fault zone breaks up into a series of discontinuous cracks and small scarps. Farther west, a near-continuous segment with displacements of 50–70 cm in a northerly direction continues to the coast and probably offshore.

The 1981 earthquakes were accompanied by significant changes in elevation, as noted by uplifted solution notches in limestone, raised beaches, and reports by local people of permanent changes in sea level (Vita-Finzi and King, 1985). Shorelines around the Alkyonidhes Gulf, located within the graben between the two 1981 faults, generally sank during the earthquake (Vita-Finzi and King, 1985), but raised Holocene shorelines around the Perachora Peninsula, largely in the upthrown block of the southern faults, show evidence of uplift (P. A. Pirazzoli, pers. comm., 1993). Uplift of dated archeological sites in the Perachora Peninsula shows 22 m of uplift in 30,000 years of the footwall block of the north-dipping fault system, an uplift rate of about 0.7 mm/yr. Change of elevation of Holocene shorelines and of archeological sites (Pirazzoli, 1993) allows the determination of long-term late Holocene slip rates on faults in the Gulf of Corinth for the last several thousand years.

Aftershocks, located to within 1 km horizontally and 2 km vertically (Fig. 9-28), do not define fault planes. Most lie in the hangingwall block common to the north-dipping and south-dipping faults and are between 3.6 and 10 km depth. However, most earthquakes in cross section C-C′ of Figure 9-28 are deeper and lie in the footwall block of the northern, south-dipping fault.

Focal-mechanism solutions (Figs. 9-26, 9-28) show mainly N-S extension with a minor component of strike slip. Nodal-plane dips are moderate and show no tendency to flatten with depth (Fig. 9-28b). However, by using fault-plane orientations based on focal mechanism solutions of mainshocks together with displacements based on surface data, King et al. (1985) constructed a velocity-space diagram that suggests a

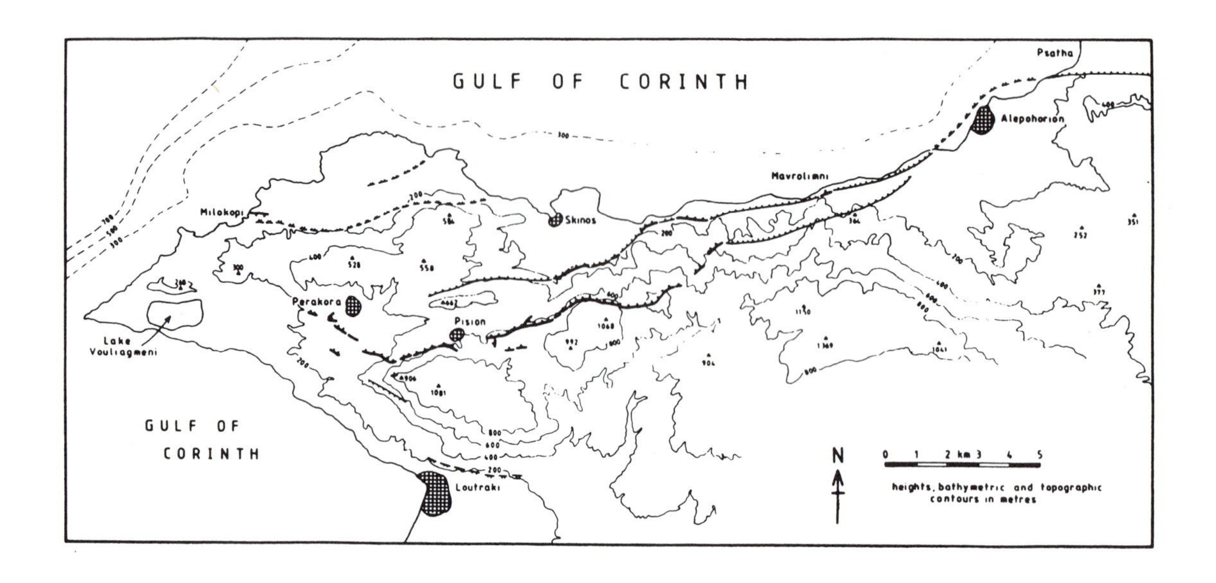

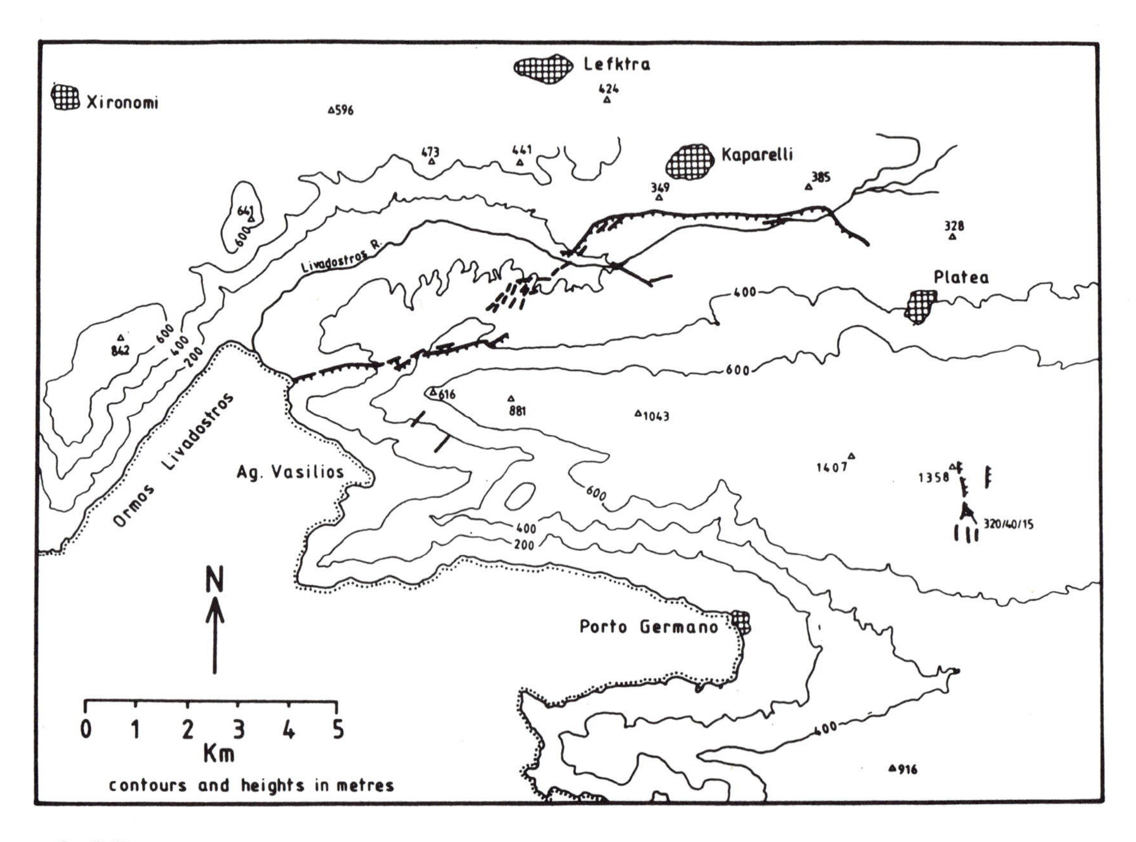

Figure 9–27. Maps of surface rupture accompanying 1981 earthquakes (top) in Perachora Peninsula and (bottom) northeast of Alkyonidhes Gulf. After Jackson et al. (1982).

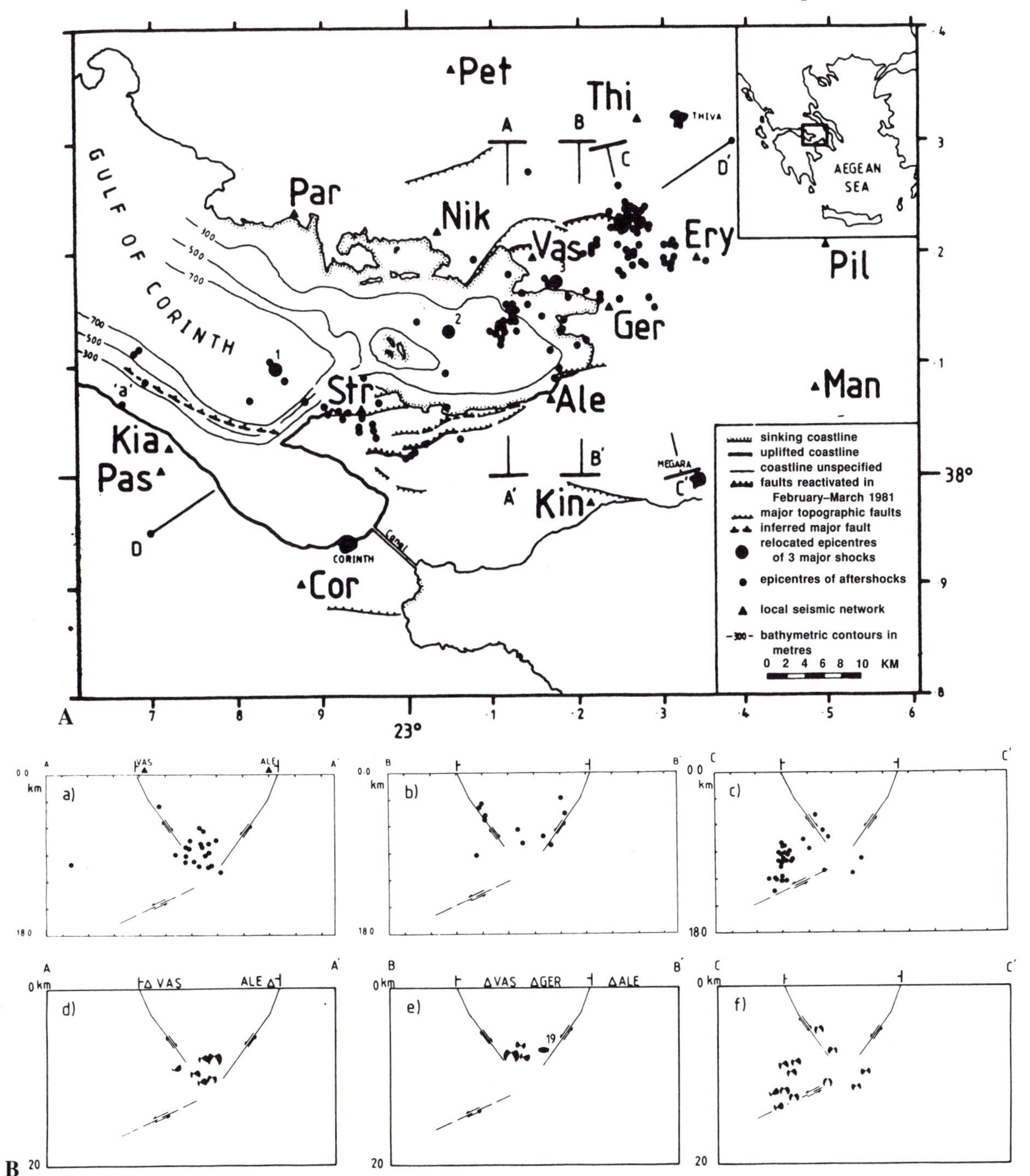

Figure 9–28. (a) Mainshock and aftershocks of the 1981 Gulf of Corinth sequence. Large filled circles locate the earthquakes of 24 February (1), 25 February (2), and 4 March (3); smaller circles locate aftershocks. A-A′, B-B′, C-C′ locate seismicity cross sections; north to left. Surface faulting of 1981 in heavy lines. Solid triangles locate seismic stations. After King et al. (1985) and Taymaz et al. (1991). (b) Cross sections of earthquakes associated with 1981 Gulf of Corinth sequence. Beach balls show fault-plane solutions as viewed in cross section. After King et al. (1985).

deeper, aseismic fault with lower dip than faults closer to the surface (Fig. 9-29). This interpretation suggests that the south-dipping faults north of Alkyonidhes Gulf are secondary and antithetic to the north-dipping faults on Perachora Peninsula.

Edgecumbe, New Zealand, Earthquake of 2 March, 1987

In contrast to the Borah Peak and Gulf of Corinth earthquakes, the Edgecumbe earthquake ($M_s = 6.6$)

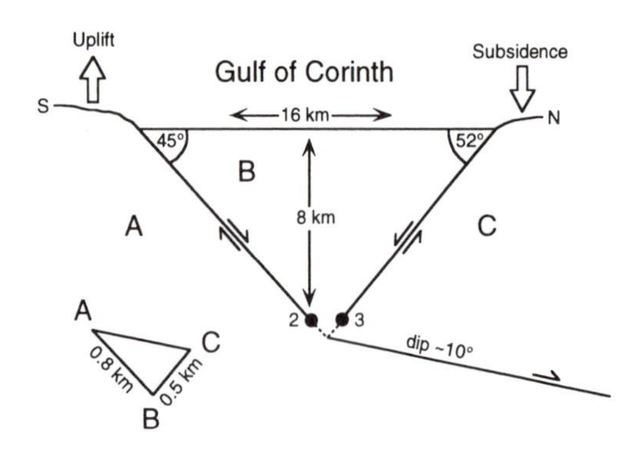

Figure 9–29. Block model of faulting close to section A-A′ of Figure 9–28. Solid dots show mainshocks of second and third earthquakes in sequence. Dip angles are from mainshock fault-plane solutions, and fault displacement is from surface data. Note the uplift of the south coast and subsidence of the north coast. Inset is a velocity-space diagram calculating the displacement and orientation of deep aseismic fault separating blocks A and C, based on dip direction and displacement of faults between blocks A and B, and between B and C. After Jackson (1987), modified from King et al. (1985).

ruptured faults in a lowland rather than in a basin-and-range setting. Faulting took place in the Whakatane graben at the eastern edge of the Taupo Volcanic Zone, in a back-arc setting (Fig. 9-7). The Taupo Volcanic Zone has been active for the last 2 million years, and the Whakatane graben has been subsiding at a rate of 1-2 mm/yr for the last 300 ka and possibly for the last million years. The late Quaternary extension rate across the graben is at least 4-6 mm/yr (Beanland et al., 1990).

The description that follows is summarized from Anderson et al. (1990), and Beanland et al. (1990), who refer to more detailed descriptions in a special edition of the *New Zealand Journal of Geology and Geophysics* (vol. 32, 1989).

Unlike the two earthquakes described in the previous text, the Edgecumbe earthquake had precursors. An earthquake swarm began on 21 February 35 km northwest of the mainshock, and another swarm began a few days later, close to the subsequent mainshock hypocenter (Fig. 9-30). The largest event in the eastern swarm was M_L4.9 on 28 February. On 2 March, the main Edgecumbe earthquake sequence began with a foreshock of M_L5.2, 7.5 minutes before the mainshock.

Eleven faults in a zone 16 km long and 10 km across underwent surface rupture during the earthquake (Fig. 9-7). Most dip northwest, and most or all are associated with preexisting faults. The largest vertical and extensional offsets occurred on the Edgecumbe fault, with surface rupture for a distance of 7 km (Fig. 9-31), resulting in a maximum dip-slip displacement of 3.1 m. The Edgecumbe fault rupture is in part expressed as a surficial warp with a wide fissure at its crest (Figs. 9-32; 9-33, 9-34); elsewhere it has a free face more typical of a Basin and Range normal fault. The fault is arcuate in map view, but in detail, the fault trace is highly irregular as it crosses the Rangitaiki alluvial plain (Fig. 9-34). Fissuring, surface warping, and the irregular map pattern may be due to the fact that the faults cut soft sediments of the Rangitaiki plain rather than rupturing a range front, as was the case at Borah Peak and the Gulf of Corinth.

The aftershock zone trended northeast and was nearly 90 km long, about half of it offshore (Fig. 9-35), raising the possibility that additional surface faulting could have occurred offshore. Aftershocks followed a normal decay pattern except for the occurrence of additional earthquake swarms in the aftershock zone. The mainshock and most of the aftershocks yielded focal-mechanism solutions characteristic of normal faults, with a subset of strike-slip focal-mechanism solutions (Fig. 9-35). The aftershocks occurred at relatively shallow depths of 0.2-9.6 km, with most between 4 and 6 km, possibly a reflection of the high heat flow of the region, which results in a relatively shallow brittle-plastic transition (Fig. 9-36). The mainshock depth of 8 km is near the base of the aftershock zone, presumably

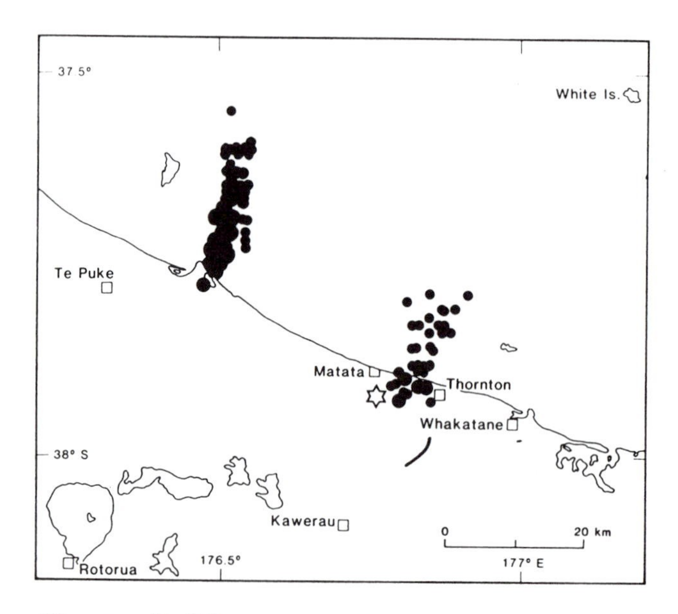

Figure 9–30. Earthquake swarms between 21 February and 1 March prior to the Edgecumbe earthquake (epicenter marked by star). Smaller circles have magnitude $3 < M_L < 4$, and larger circles $M_L > 4$. Edgecumbe fault of 1987 shown by arcuate line. For general location, see Figure 9–7. After Anderson et al. (1990).

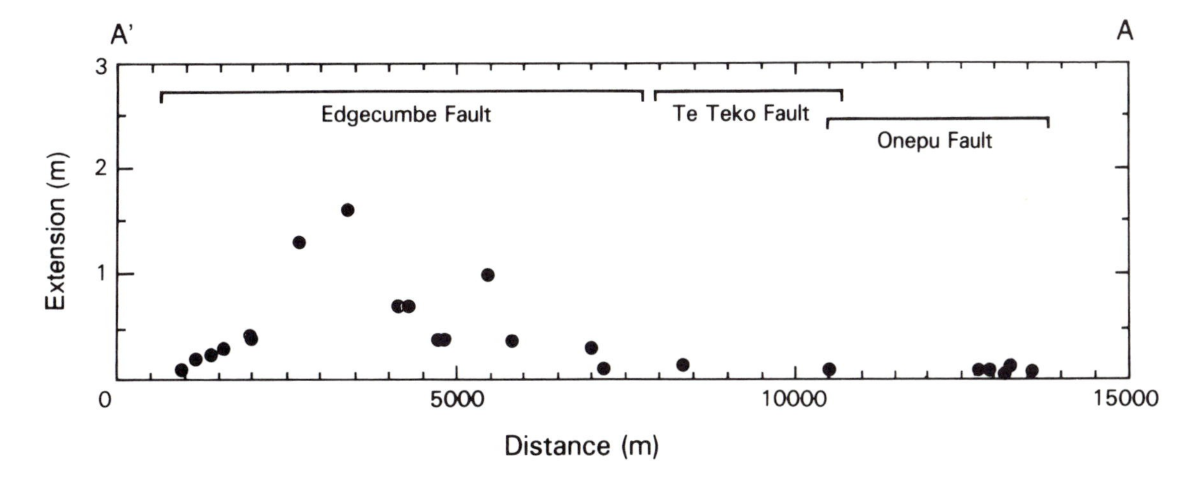

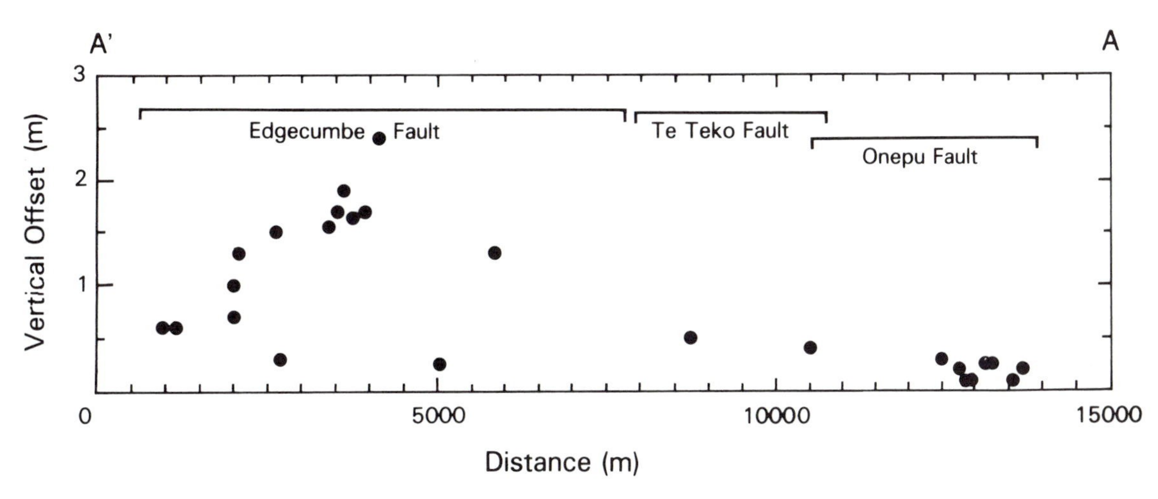

Figure 9–31. Coseismic extension (top) and vertical displacement (bottom) on the Edgecumbe, Te Teko, and Onepu faults during the Edgecumbe earthquake. Section is parallel to strike; for location, see Figure 9–7. After Beanland et al. (1990).

near the base of brittle crust. If the principal fault is defined by a line connecting the mainshock with the surface trace, the aftershocks are diffusely distributed in the footwall of this fault, suggesting that the footwall underwent internal deformation (Fig. 9-36). The Kawerau geothermal field has almost no aftershocks, although there are aftershocks southwest of this field. This suggests that temperatures are too high within this field for brittle fracture to occur.

After the earthquake, subsidence continued at a decreasing rate over a broad region (Figs. 9-37, 9-38). Maximum subsidence recorded was about 200 mm over 12 months near Edgecumbe, about 10% of the maximum coseismic subsidence. The post-seismic deformation is in the same sense as the coseismic deformation, but the greatest post-seismic subsidence is slightly northwest of the coseismic subsidence. At the Edgecumbe fault, about 60 mm vertical and 17 mm extensional displacement occurred between March and November 1987 (Fig. 9-38). The extension is undoubtedly tectonic, but some of the vertical changes may be due to consolidation during and after the earthquake.

STRUCTURAL AND GEOMORPHIC EXPRESSION

Relation between Fault Spacing and Thickness of Brittle Crust

There appears to be complete gradation in the characteristics of active normal faults at range fronts and

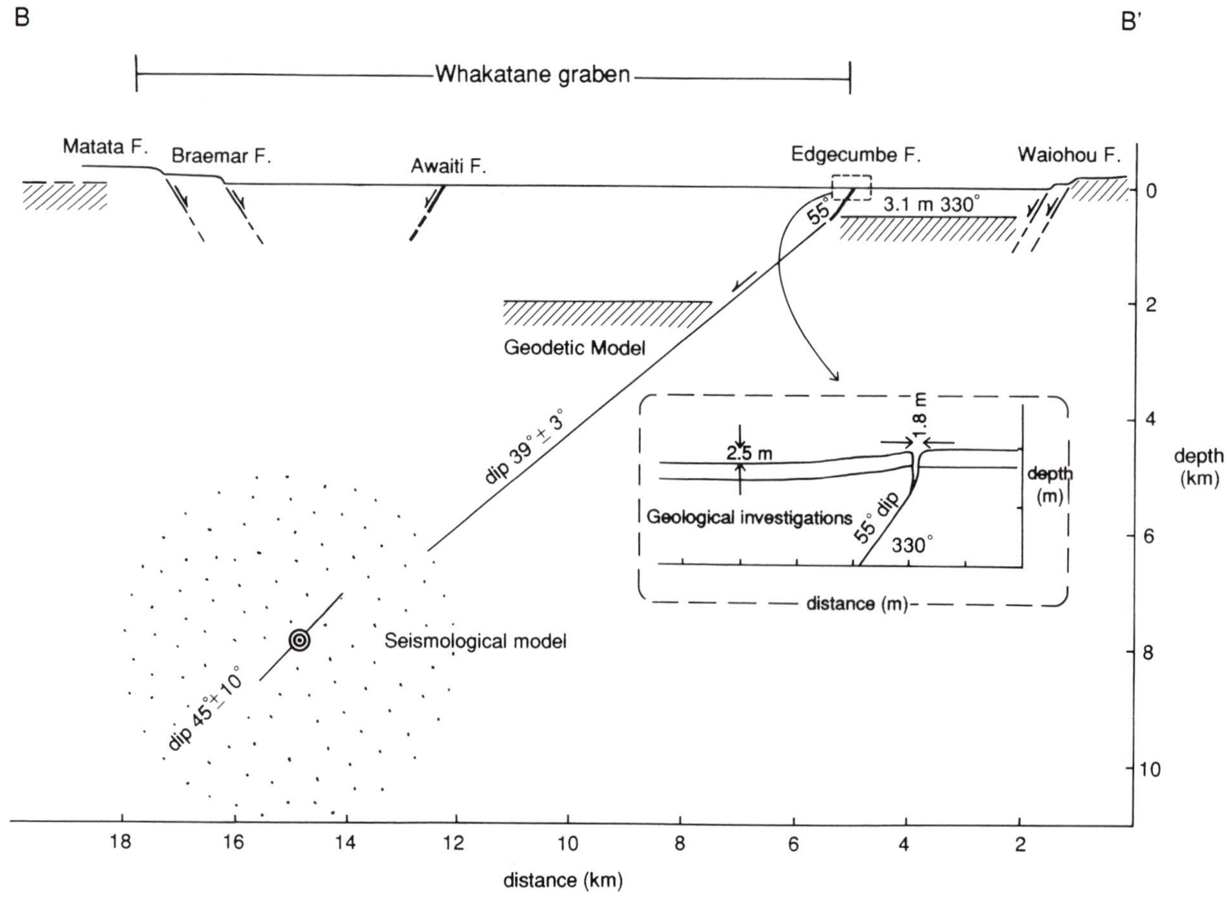

Figure 9–32. Cross section (located on Fig. 9–7) showing Edgecumbe fault based on location and focal-mechanism solution of mainshock, geodetic modeling, and surface geology. Heavier lines mark faults that ruptured in 1987. After Beanland et al. (1990).

normal faults occurring in areas of thin crust and active volcanism. Normal faults associated with sea-floor spreading and with volcanic eruptions tend to be closely spaced and to have relatively small displacements on individual faults. The gja, or gaping normal faults, of Iceland include tensile fissures, in which irregular fracture edges can be fit back together, as well as faults along which there has been shear displacement. Locally, the fissures are the sites of basaltic eruption, as illustrated by the Asal rift in Djibouti (Fig. 9-5) and by the East Rift Zone of Kilauea volcano on Hawaii (Fig. 9-6). Displacement rates may be more than an order of magnitude greater than rates on seismogenic faults in the Basin and Range, because the displacements are in many cases related to inflation of volcanic carapaces and the subterranean movement of magma, which are superimposed on long-term tectonic extension. Some extension in Iceland is recovered during times of deflation of volcanic carapaces due to collapse of subjacent magma chambers. Basin-and-range topography, with relief measured in hundreds of meters or kilometers, does not develop in tectonomagmatic environments, because the brittle crust is no more than a few kilometers thick, not enough to support high relief. Furthermore, fault scarps are commonly buried by lava flows. Tectonomagmatic faults are also not known to produce earthquakes larger than magnitude 6.5.

An intermediate tectonic environment is that characterized by active volcanism in relatively hot continental crust, such as the Taupo Volcanic Zone in New Zealand. Despite high heat flow, brittle crust is almost 10 km thick, enough to generate the Edgecumbe, New Zealand, earthquake of magnitude 6.6. However, brittle crust is probably too thin and hot, and individual faults are not long enough to generate earthquakes with magnitudes greater than 7. The dominant topography in these regions is related to volcanism, and basin-and-range topography is not strongly developed. Faults tend to be less than 10 km long, with cross-strike spacing of 5 km or less. Total displacement is measured in tens of meters.

In contrast, normal faulting in brittle continental crust at least 15 km thick develops basin-and-range topography with relief ranging from hundreds of meters to several kilometers. Basin-and-range topography may be accompanied by volcanic activity, as is the case in the East African rift valleys and the eastern range-front of the Sierra Nevada, California. Basins may be bounded by normal faults on one side (*half-grabens* or *fault-angle depressions*) or on both sides (*grabens*). The cross-strike spacing of normal faults is 10–15 km in grabens in central Tibet (Masek et al., 1994), up to 25 km in the Gulf of Corinth, 30–40 km in the Basin and Range province and in central Greece (Jackson and White, 1989), and around 50 km in the Baikal rift and in the Shanxi graben east of the Ordos Plateau. Where graben are 10–15 km across, as in central Tibet, only the brittle crust appears to be involved in extension and faulting, whereas in rifts 50 km across, the entire crust may be involved (Masek et al., 1994). Earthquakes of magnitude as large as 7.6 are to be expected where fault spacing is wide, as in the U.S. Basin and Range, which produced earthquakes at Pleasant Valley, Nevada, in 1915 (M7.6) and Borah Peak, Idaho, in 1983 (M7.3) and in the Altiplano, which produced an M7.3 earthquake at Ancash, Peru, in 1946. On the other hand, the 1959 Hebgen Lake, Montana, earthquake of M7.6 resulted from rupture of several short faults in an area adjacent to the active volcanism in Yellowstone National Park and not characterized by Basin and Range topography.

Even larger normal-fault earthquakes accompany bending of Mesozoic oceanic crust as it is flexed seaward of subduction zones in the western Pacific. The 1933 Sanriku earthquake off northeast Honshu, Japan,

Figure 9–33. The Edgecumbe fault, view to northeast. A preexisting scarp 1.3 m high was raised to 3.2 m high in 1987, accompanied by fissure. Much of vertical deformation is accompanied by warping to the left of fissure. Photo by Q. Christie, New Zealand DSIR.

Figure 9–34. Aerial view northeast of Edgecumbe fault, showing variability of strike. White patches are sandblows that erupted during the earthquake. Photo by D. L. Homer, New Zealand Institute of Geological and Nuclear Science.

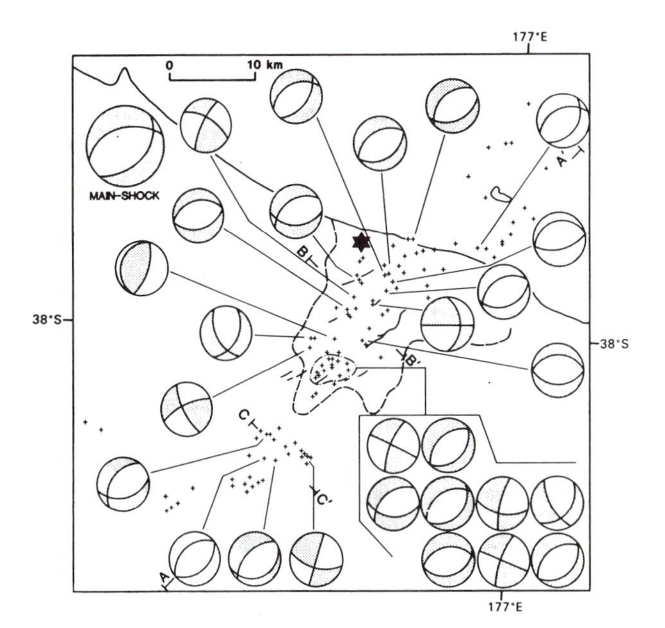

Figure 9–35. Fault-plane solutions of Edgecumbe aftershocks. Upper hemisphere projections, compressional quadrants shaded. Star shows mainshock. A-A′, B-B′, C-C′ locate cross sections of Figure 9–36. 1987 surface faults shown by solid lines. Compare with foreshocks shown in Figure 9–30, which are north and west of the mainshock. After Anderson et al. (1990).

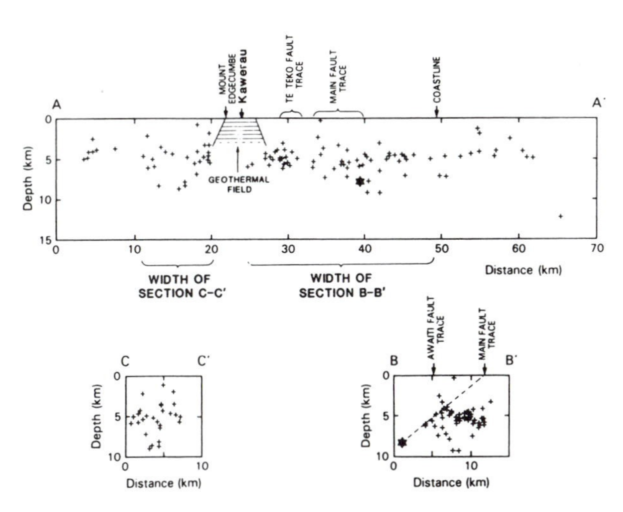

Figure 9–36. Cross sections through aftershock zone; sections located on Figure 9–35. After Anderson et al. (1990).

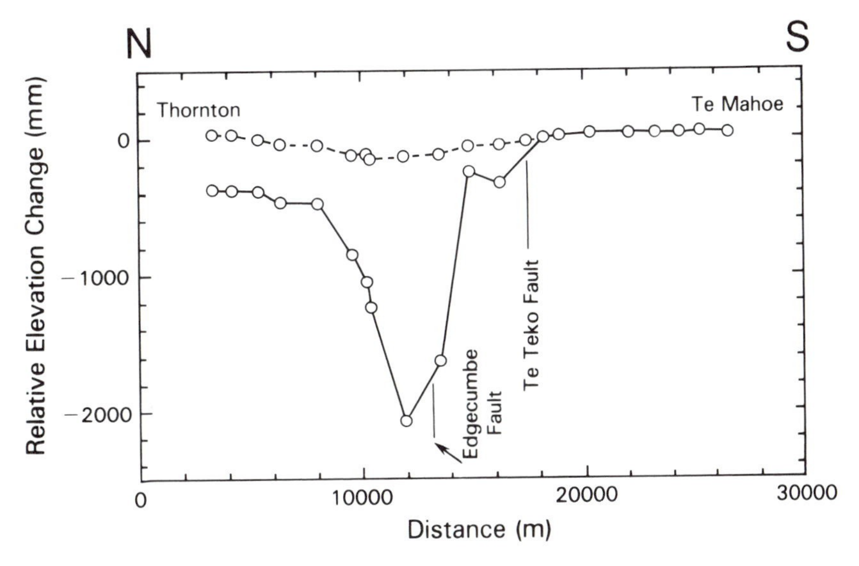

Figure 9–37. Coseismic (solid line) and post-seismic (dashed line) relative elevation changes on a leveling line between Thornton and Te Mahoe (located on Figure 9–38). Post-seismic changes shown were measured over 12 months. Diameter of circle indicates approximate error over an inter-benchmark interval. After Beanland et al. (1990).

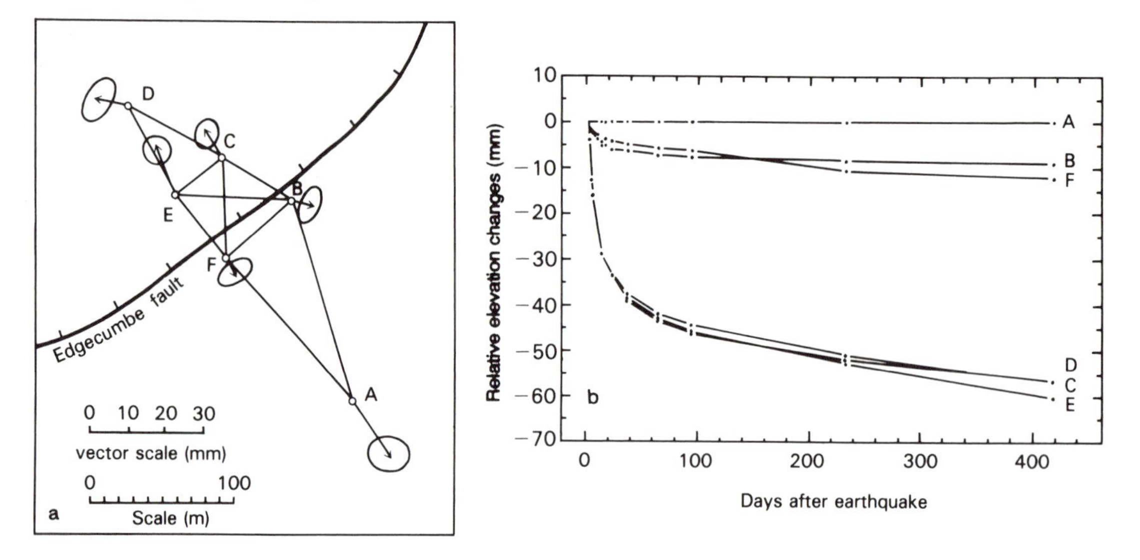

Figure 9–38. (a) Closely spaced geodetic array across Edgecumbe fault showing post-seismic displacement from 3 March to November, 1987. (b) Elevation changes of stations B to F relative to A in the period 3 March, 1987 to May, 1988. After Beanland et al. (1990).

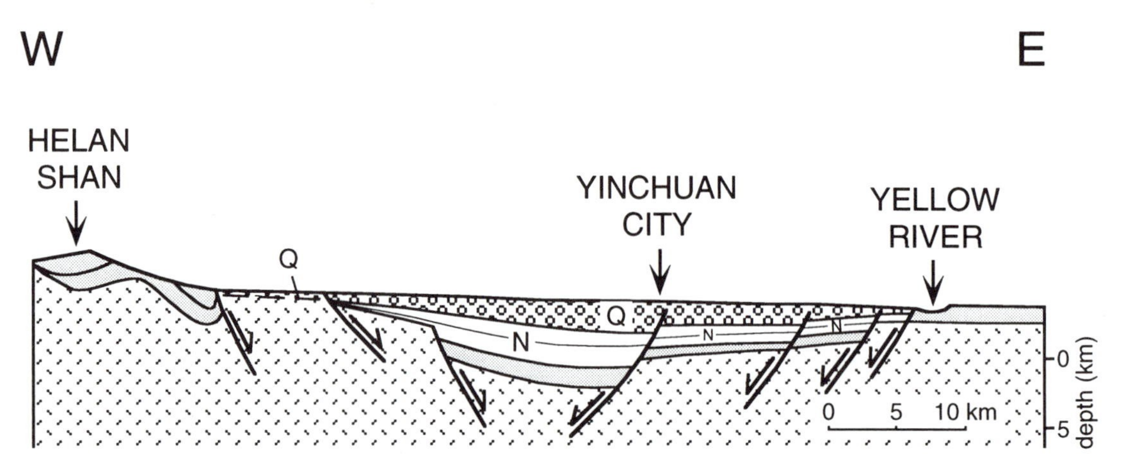

Figure 9–39. Cross section of Yinchuan basin on west side of Ordos Plateau between 38° and 39.5°N. In 1739, boundary fault on west side of basin offset the Great Wall of China 3 m vertically and 5.3 m horizontally. Nevertheless, the basin shows a typical normal-fault geometry, with faults bounding both sides of the graben. Shaded pattern: pre-Neogene strata. From Wang and Deng (1988) and Wang and Song (1989).

had a magnitude of 8.3, larger than any subduction-zone earthquake in this subduction zone in this century.

Oblique-Slip Normal Faults

There is also a complete gradation between pure dip-slip normal faults and pure strike-slip faults. Normal faults have an irregular map trace whereas the map traces of pure strike-slip faults tend to be more linear. Oblique-slip normal faults may have an irregular map trace, as illustrated by faults around the Ordos Plateau in China (Fig. 9-14). In cross section, they display a characteristic normal-fault geometry, with faults dipping 45-60° toward the center of the basin (Fig. 9-39), but streams crossing the fault may show strike-slip offset of their channels. *Pull-apart basins* develop along predominantly strike-slip faults containing en échelon stepovers tending to produce extension parallel to the strike-slip vector (discussed further in Chapter 8). Oblique-slip normal faults have produced earthquakes of magnitude 8 in basins surrounding the Ordos Plateau in China. These include the 1303 Hongdong earthquake in the Linfen graben on the southeast side of the Ordos, the 1556 Huaxian earthquake near Xian on the south side of the Ordos, and the 1739 Pinglu earthquake in the Yinchuan basin on the northwest side of the Ordos.

Map Patterns

Normal faults and reverse faults differ from strike-slip faults in that their surface traces are strongly irregular rather than straight, although there are exceptions to this rule (for example, the 1969 Alasehir, Turkey, earthquake fault described by Eyidogan and Jackson, 1985). This is probably because corrugations in the fault plane are preserved over many episodes of slip only if they are parallel to the slip vector. In addition to having irregular patterns in map view, normal faults may occur en échelon, with individual segments separated by unfaulted ground. However, when viewed parallel to their slip vector, both dip-slip and strike-slip faults seem to be fairly straight or broadly curved. Dip-slip faults appear straight in cross sections that include the slip vector (Fig. 9-39).

The surface ruptures of eight well-mapped Basin and Range normal-fault earthquakes are shown in Figure 9-40 to illustrate characteristic map patterns. All have traces with irregular strike. Most follow range fronts, indicating that the surface rupture is the continuation of a long-term process that uplifted the range and downdropped the adjacent basin. However, scarps of the Sonora earthquake (Fig. 9-40e), Dixie Valley and Fairview Peak earthquakes (Fig. 9-40b), and Rainbow Mountain earthquake (Fig. 9-40a) extend into valley fill. Some scarps of the Hebgen Lake earthquake occur entirely in Madison Valley alluvium (Fig. 9-40d). The Stillwater earthquake faults were largely within Carson Sink, far from any range front (Figure 9-40a), although this earthquake reactivated scarps formed the previous month at Rainbow Mountain. Historical fault scarps far from a range front may not have topographic expression, because they undergo displacement so rarely, or they may represent relatively new faults.

En échelon stepovers, with stepover widths from a few tens of meters to several kilometers, characterize the Pleasant Valley earthquake (Fig. 9-40c), the Sonora earthquake (Fig. 9-40e), and the Fort Sage earthquake surface ruptures (Fig. 9-40f). At seismogenic depths, these earthquakes may have been generated on single faults, and the stepovers may result from the tendency of faulting to follow older zones of weakness. The Pleasant Valley, Sonora, and Fort Sage scarps all face west, and the Dixie Valley scarps all face east, but there are west-facing and east-facing scarps associated with the Fairview Peak earthquake. Where faults face both directions, there may be a primary fault and a secondary, antithetic fault, as was the case for the 1981 Gulf of Corinth scarps.

Of the surface ruptures illustrated in Figure 9-40, only the Fairview Peak surface rupture is associated with a large component of strike slip. The Rainbow Mountain, Pleasant Valley, and Fort Sage earthquakes

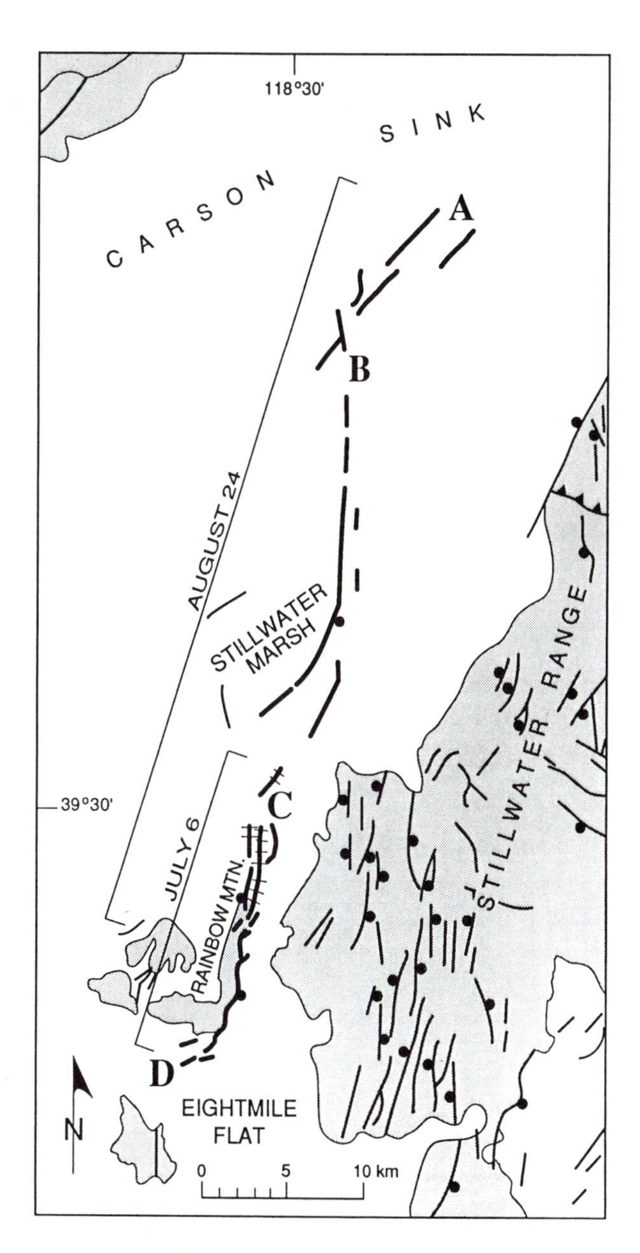

Figure 9–40. Rupture patterns of eight historic normal-fault earthquakes in the Basin and Range Province. Historic ruptures are shown in heavy lines, ball on downthrown side. Ranges are shaded, alluvium-filled valleys are unshaded. (a) Rainbow Mountain earthquake (6 July 1954, M_S6.3) and Stillwater earthquake (24 August 1954, M_S7).

produced mainly single ruptures. The Sonora earthquake produced a main rupture with shorter faults across strike within alluvium. The Stillwater, Dixie Valley, Fairview Peak, and Hebgen Lake earthquakes resulted from multiple ruptures. Multiple ruptures may represent secondary faults as well as the primary rupture, or they may express the diffusion of a primary rupture upward through relatively unconsolidated valley fill.

Cross-strike distance between faults that ruptured during the Hebgen Lake earthquake was as large as the length of coseismic rupture on any given fault. This was also true for the 1981 Gulf of Corinth earthquakes (Fig. 9-26) and the Edgecumbe earthquake (Fig. 9-7). However, as noted below, some of these faults were secondary or sympathetic surface ruptures. Cross-strike spacing between faults may reflect thickness of brittle crust; the Edgecumbe surface faults in relatively thin crust were much more closely spaced than faults in the Great Basin, the Aegean region, or northern China.

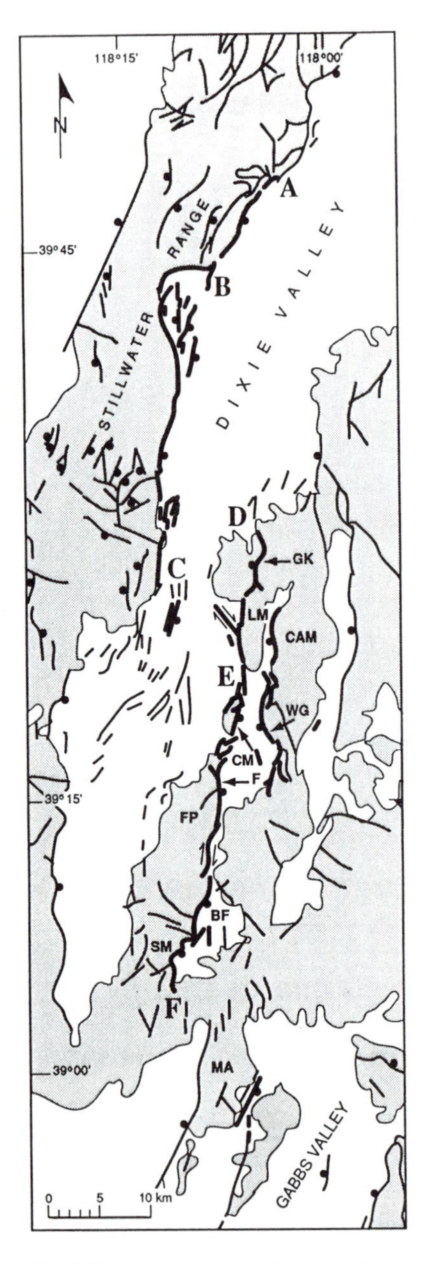

Figure 9–40. (b) Dixie Valley earthquake (ABC, M_S6.8) and Fairview Peak earthquake (DEF, M_S7.2), both 16 December 1954.

Segmentation

Map patterns lead to a subdivision of faults, especially long ones, into segments, with the objective of mapping geological structures along a fault that would constitute barriers to large earthquakes, or earthquake nucleation sites with rupture propagation in only one direction. An *earthquake segment* refers to "those parts of a fault zone or fault zones that have ruptured during individual earthquakes" (dePolo et al., 1991). A major objective of fault-zone mapping is the identification of those geological discontinuities along fault zones that may coincide with earthquake segment boundaries, that is, barriers to earthquake rupture.

There are three major types of discontinuity, with the first two geology-based (dePolo et al., 1991). *Geometric discontinuities* include abrupt changes in fault strike, en échelon stepovers, and gaps and salients in a fault zone. Stepovers and gaps in faulting that are only a few tens or hundreds of meters in dimension are not likely to be earthquake segment boundaries, because they are too small to be expected to continue downward to sites of nucleation of large earthquakes. Even if they do continue down to nucleation depths, modern observations suggest that they are too small to impede rupture propagation. *Structural discontinuities* include fault bifurcations or zones of increased structural complexity, intersections with other structures, and terminations at cross structures. *Behavioral discontinuities* include changes in slip rates, seismic behavior, senses of displacement, or creeping vs. locked behavior. As will be seen, some discontinuities do not serve as barriers to earthquake rupture.

The first segmentation model for a normal fault was devised for the Wasatch fault zone at the eastern edge of the Basin and Range Province, U.S.A., following suggestions by G. K. Gilbert in 1883 (Swan et al., 1980; Schwartz and Coppersmith, 1984). The most recent version of this model (Machette et al., 1991) is shown as Figure 9-41. Holocene movement is limited to the central part of the zone from the Brigham City segment to the Levan segment, where the fault zone has

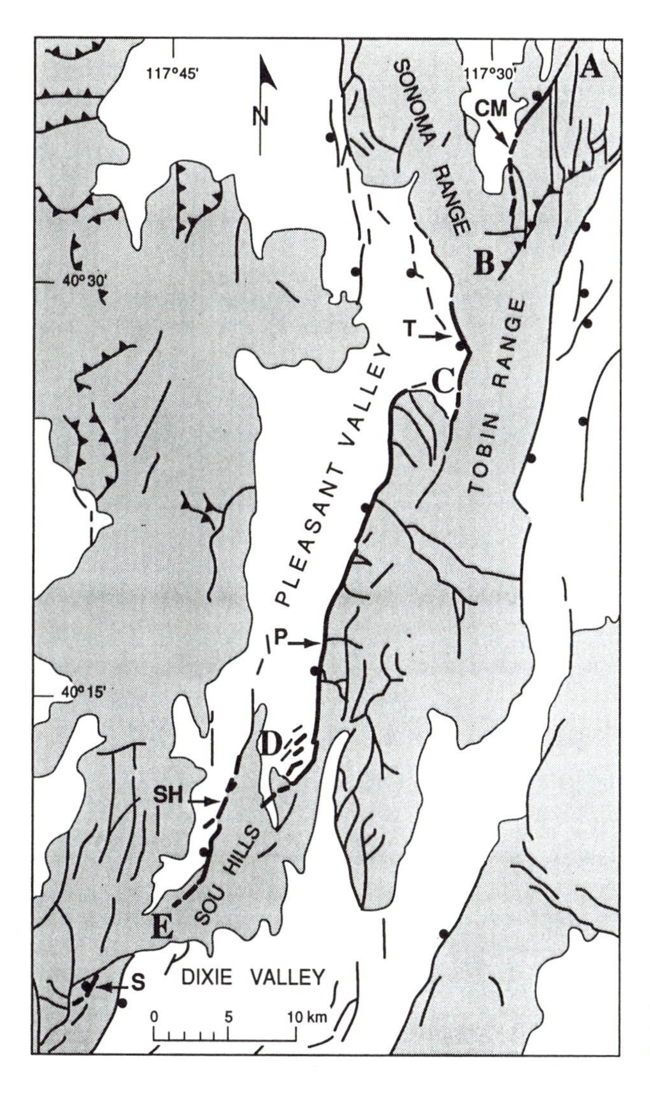

Figure 9–40. (c) Pleasant Valley earthquake (3 October 1915; M_S7.6).

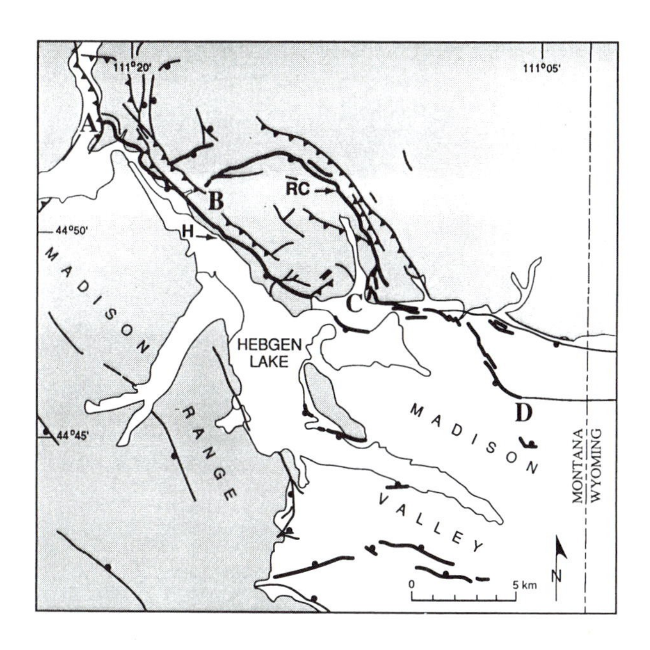

Figure 9–40. (d) Hebgen Lake earthquake (18 August 1959; M_S7.5).

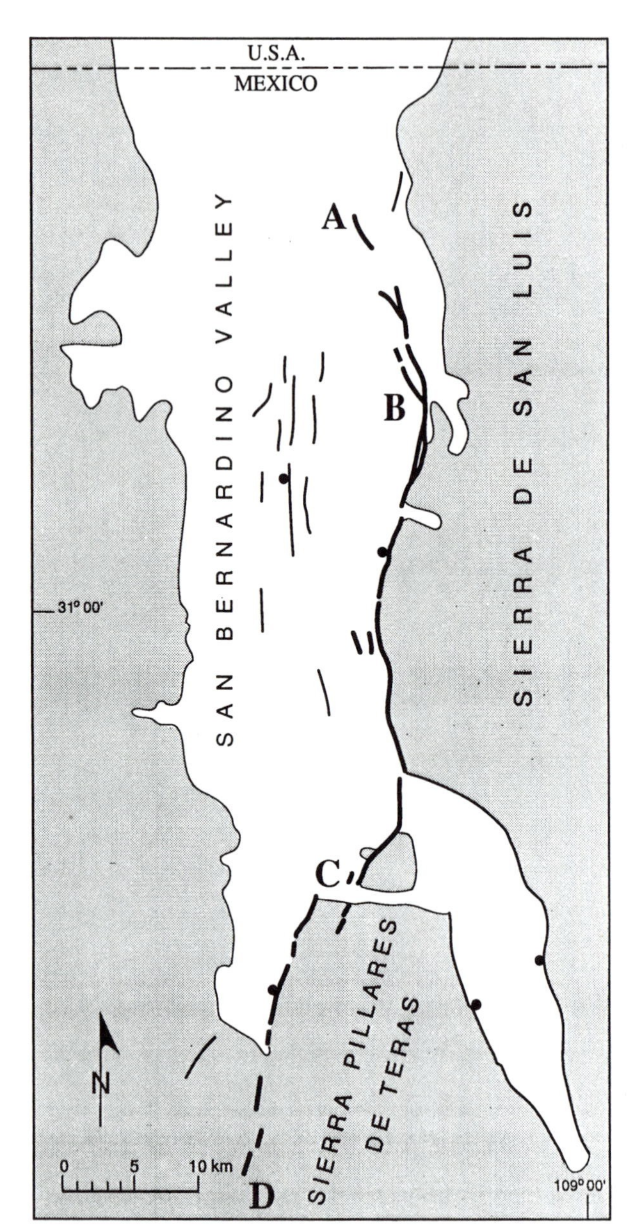

Figure 9–40. (e) Sonora earthquake (3 May 1887; M_W7.2–7.4).

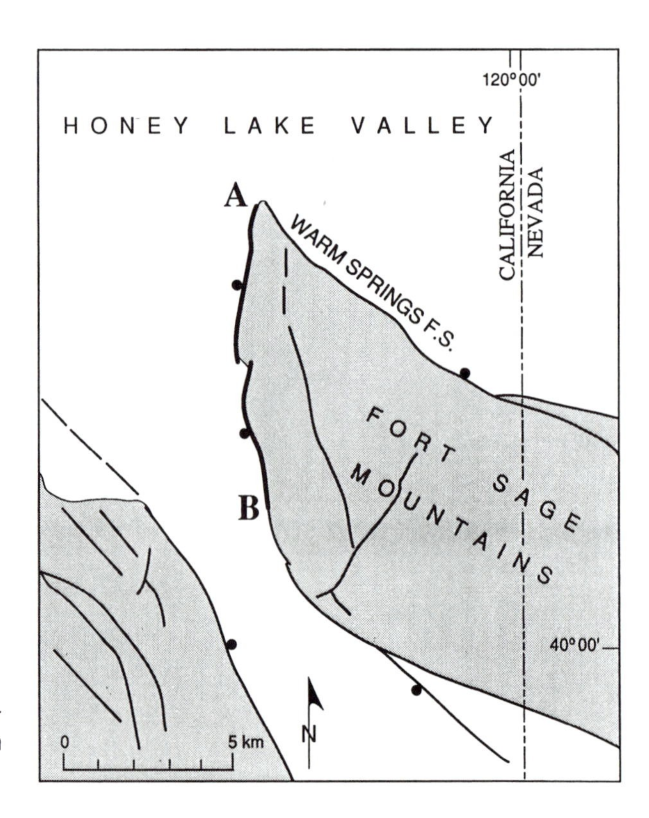

Figure 9–40. (f) Fort Sage earthquake (14 December 1950; M_L5.6. All parts of Fig. 9.40 are from DePolo (1991).

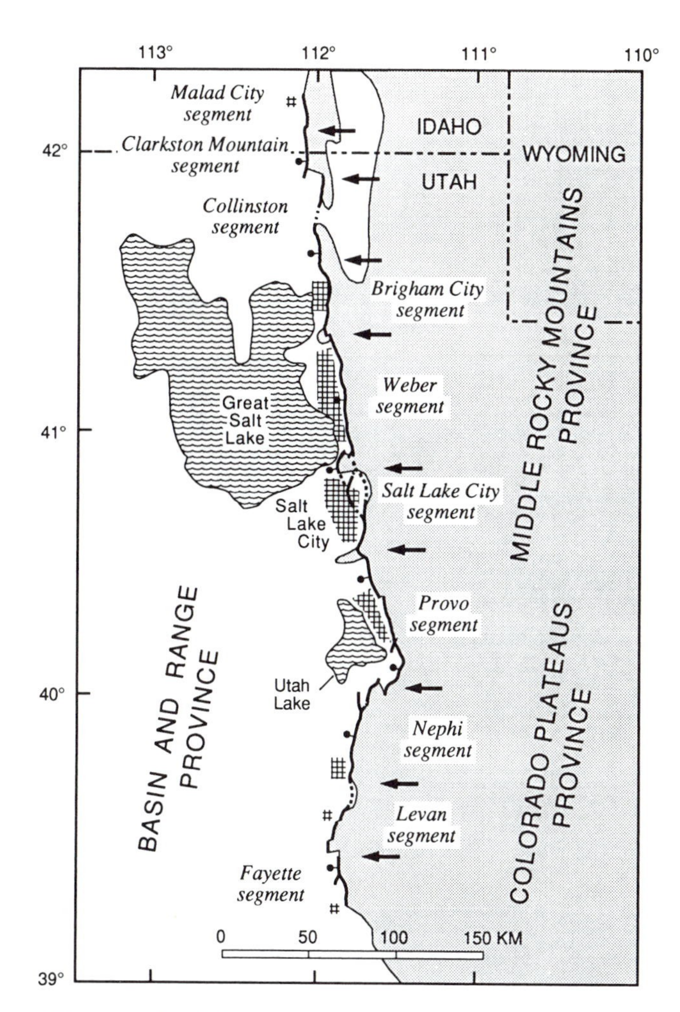

Figure 9–41. Segments of the Wasatch fault, U.S.A. Segment boundaries marked by solid arrows. After Machette et al. (1991).

its greatest topographic relief, and slip rates are highest. The high-slip-rate segments are significantly longer than segments at either end of the fault zone, with the Provo segment, 69.5 km measured along the fault trace, the longest. This is unusually long; rarely are faults continuous along strike for distances greater than 20–25 km (Jackson and White, 1989).

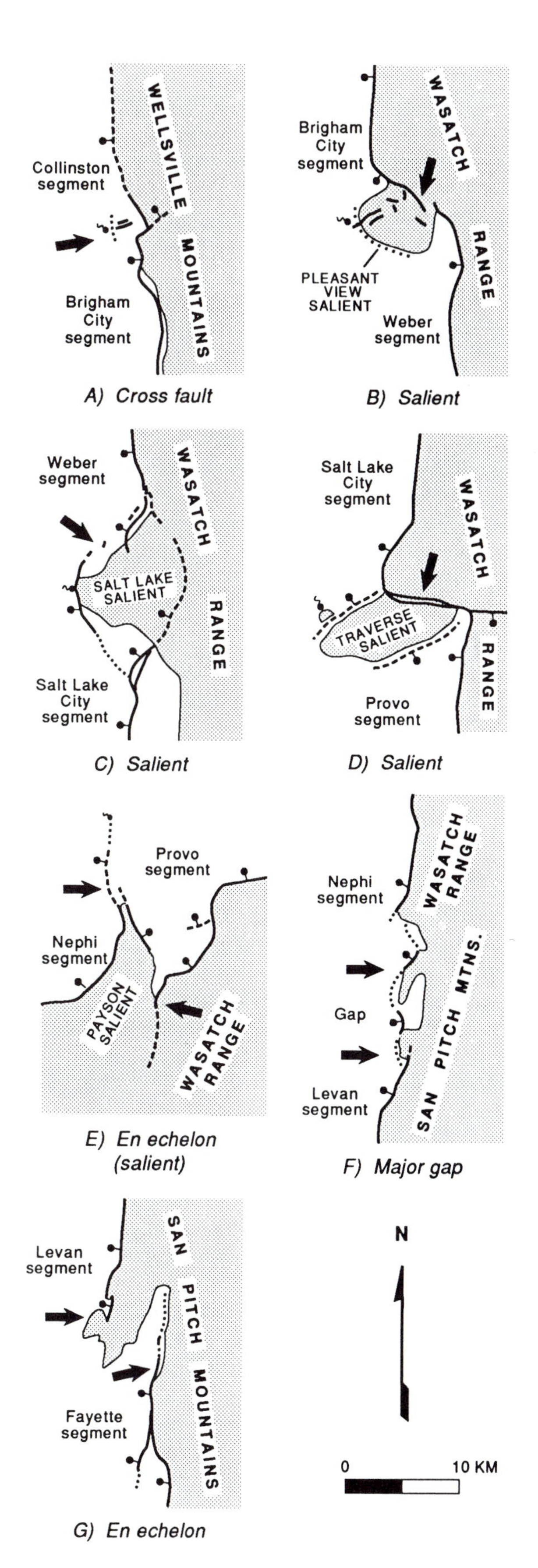

Figure 9–42. Map patterns at segment boundaries along the Wasatch fault zone in Utah, U.S.A. Bedrock is stippled; segment boundaries are shown by arrows. Bar and ball are on downthrown side. Figures are arranged from north to south. After Machette et al. (1991).

Map patterns at segment boundaries are illustrated in Figure 9-42. As described by Machette et al. (1991), salients are bedrock blocks that extend across the Wasatch fault and are generally bounded by Quaternary faults that are less active than those to the north and south. In the Chinese literature, these are called *push-up blocks*. Two en échelon stepovers are mapped as segment boundaries, a right-stepping boundary between the Nephi and Provo segments and a left-stepping boundary between the Levan and Fayette segments. The Traverse salient between the Salt Lake City and Provo segments marks an en échelon left step. The Nephi and Levan segments are bounded by a gap in late Quaternary faulting, and the Collinson and Brigham City segments, as well as the Salt Lake City and Provo segments, are bounded by cross faults.

The geologically-based segmentation model is in general agreement with the rupture zones of earthquakes over the last 6000 years based on extensive trenching (Fig. 4 of Machette et al., 1991); that is, the subdivisions of the fault appear to be earthquake segments. However, two earthquakes may have broken across the Provo-Nephi segment boundary (*leaky barrier* as defined by Crone and Haller, 1991, in which a barrier allows partial or sympathetic rupture of an adjacent segment); and one each across the Provo-Salt Lake City boundary and the Weber-Brigham City boundary.

How well do geologically-defined segment boundaries correspond to rupture boundaries of earthquakes on other normal faults? Before answering this question, we first recognize three types of surface rupture (dePolo et al., 1991). *Primary surface rupture* is considered to be directly connected to subsurface displacement on a seismic fault; such rupture defines an earthquake segment. *Secondary surface rupture,* on the other hand, is along a secondary fault, such as an antithetic normal fault intersecting the primary fault at relatively shallow depth (Figs. 9-28 and 9-29), or a fault branching off from the primary fault such as the faults in the Willow Creek Hills in the Borah Peak earthquake zone (Fig. 9-15). *Sympathetic surface rupture,* of concern to us here, is triggered slip along a fault isolated from the primary fault, as described previously for the range-front fault at the east edge of Warm Spring Valley, north of the Willow Creek Hills (Fig. 9-15). This is important because surface rupture may appear to cross a segment boundary, whereas only sympathetic surface rupture crosses this boundary. This is illustrated by an example from Dixie Valley, Nevada (Figs. 9-40b and 9-43; Zhang et al., 1991).

The 1954 Dixie Valley earthquake ruptured a relatively straight portion of the front of the Stillwater

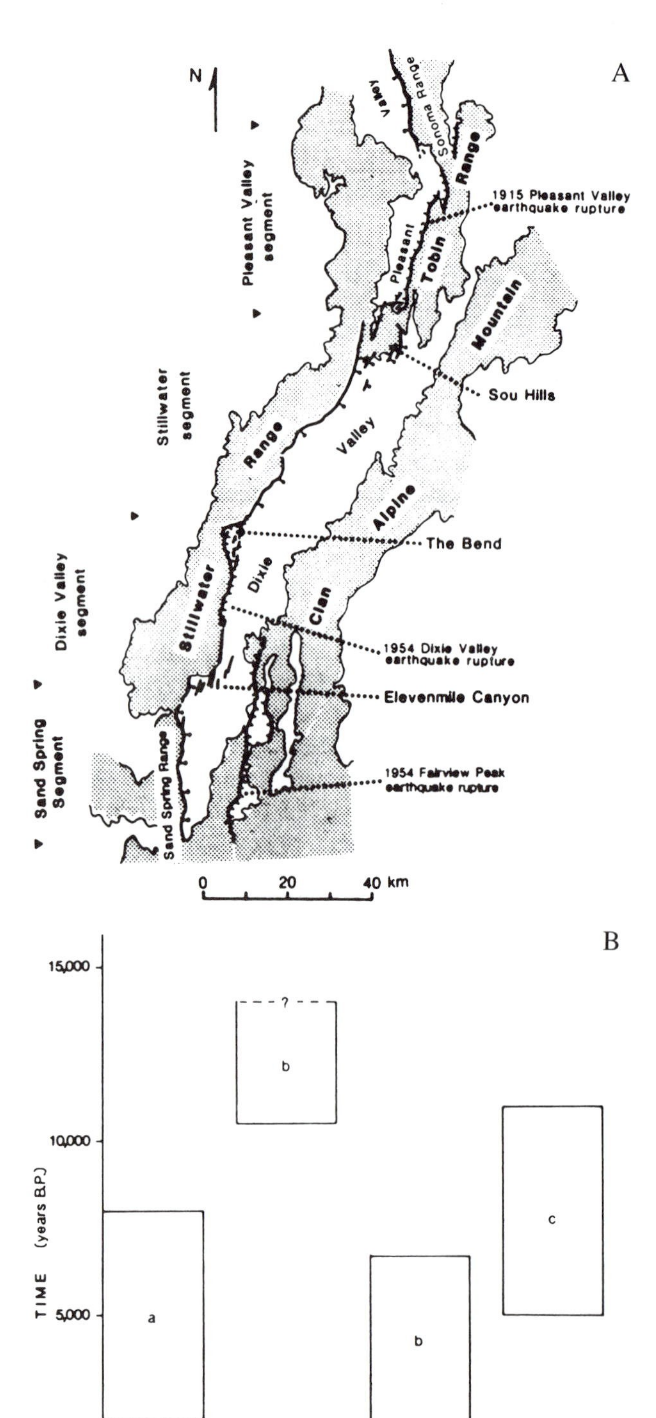

Figure 9–43. (a) Segments of the Pleasant Valley-Dixie Valley fault system. Compare with Figure 9–40B, C, which shows historic ruptures. (b) Earthquakes along the Pleasant Valley-Dixie Valley fault system for the last 15,000 years. Rectangles show the time range when the earthquake occurred, with a query marking uncertainty in the upper bound of the latest Pleistocene earthquake on the Dixie Valley segment. After Zhang et al. (1991).

Range that had not ruptured previously since the end of the Pleistocene (Fig. 9-43). However, small-displacement 1954 ruptures in the piedmont south of The Bend (B in Fig. 9-40b) reactivated faults that had undergone large displacement in the middle Holocene (Bell and Katzer, 1990), about the same time as the Stillwater segment to the north, which has not moved in historic time (Fig. 9-43). This suggests that the 1954 earthquake and the middle Holocene Stillwater-segment earthquake overlapped for a distance greater than 20 km in the vicinity of The Bend; indeed, 1954 rupture continued for another 14 km north of The Bend (AB in Fig. 9-40b). However, 1954 displacement north of The Bend is relatively small compared to that to the south. Zhang et al. (1991) argued that the 1954 rupture north of The Bend was secondary to the main rupture farther south, and the segment boundary is in the vicinity of The Bend, where there is a significant change in strike of the range front and a high degree of fault complexity.

The Sou Hills, a bedrock salient of great structural complexity between Pleasant Valley and Dixie Valley (Figs. 9-40 and 9-43a), marks the segment boundary between the Stillwater segment and the Pleasant Valley segment. Bruhn et al. (1987) pointed out for the Salt Lake City segment of the Wasatch fault that zones of structural complexity, involving subsidiary and bifurcating faults, tend to stop a propagating rupture because energy is diffused into a large volume of fractured rock rather than a small volume such as that of a geometrically simple primary fault. This diffusion of energy has the tendency to slow or stop the rupture, as illustrated for the northern end of the Thousand Springs Valley segment of the Borah Peak earthquake (Fig. 9-15).

The 1986 Kalamata earthquake (M_S 5.8) in the southern Peloponnesos of Greece did not rupture past a left-stepping en échelon segment boundary more than 14 km wide at the western range front of the Taygetos Mountains. But the 1915 Pleasant Valley, Nevada, earthquake (M_S 7.6) ruptured across three right-stepping segment boundaries (Fig. 9-41c) and terminated on the south at the Sou Hills salient and on the north at an abrupt change in strike of the western range front of the Sonoma Range (Fig. 9-41c). A large earthquake such as the 464 B.C. earthquake on the Sparta fault (M_S 7.2) might have ruptured across an en échelon stepover that would have been a barrier to the smaller Kalamata earthquake.

In summary, structural and geometric discontinuities may have value in predicting the boundaries of earthquake segments, but the process is not straightforward. A discontinuity may stop an earthquake of $M < 6$, but not an earthquake of $M > 7$. Determining whether rupture is primary or sympathetic is difficult and subject to interpretation. And, finally, to approach the problem, it is necessary to know details about late Cenozoic history and the Holocene paleoseismic history of the entire fault zone.

Folds Related to Normal Faults

Strong folding of competent strata is a response to contraction, and is, therefore, incompatible with the extensional environment. However, broad warping at the surface may accompany normal faulting at depth, and recognition of this warping is necessary to determine the total vertical displacement, including warping and faulting.

Rollover is the response by folding of strata in hangingwall blocks to displacement on nonplanar normal faults (Xiao and Suppe, 1992). If a fault is nonplanar, deformation must occur within the hangingwall as it moves around the bend in the fault (Fig. 9-44). If the fault is *listric,* that is, if dip decreases with increasing depth, displacement on the fault at depth will produce a tendency for the hangingwall to move away from the more steeply dipping part of the fault closer to the surface. This tendency results in a bending of strata in the hangingwall toward the fault, or in partial collapse of the hangingwall. Listric faults and accompanying rollovers are common in sedimentary basins, particularly major deltaic basins such as the Mississippi and Niger deltas, but they appear to be limited to sedimentary rocks within these basins, not penetrating into underlying basement. Because they are limited to weak sedimentary rocks in the uppermost crust, their movement is probably aseismic (Jackson and White, 1989), as is the case for active normal faults in and near Houston, Texas.

If nonplanar normal faulting is of crustal scale, then hangingwall response may be a set of antithetic secondary faults and a diffuse distribution of aftershocks within the hangingwall of the fault, as illustrated for aftershocks of the 1981 Gulf of Corinth earthquakes (King et al., 1985; Figs. 9-28a, 9-29). Such a crustal-scale fault may be blind, and its surface expression may be a syncline, as proposed for the Straits of Messina, Italy, and an earthquake of M7.5 there in 1908 (Valensise and Pantosti, 1992). This interpretation is based on dislocation modeling of releveling data and warped Pleistocene marine terraces.

Recognition of surficial warping is necessary to ascertain the true vertical displacement at the surface during an earthquake. Maximum displacement of surface faults accompanying the 1950 Fort Sage, California, earthquake (Fig. 9-40f) was 20 cm, but Gianella (1957) recognized that if warping of alluvium was taken into

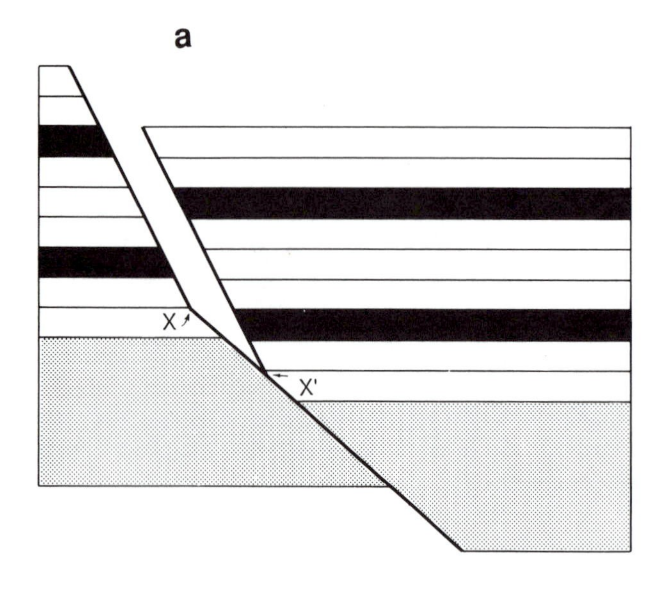

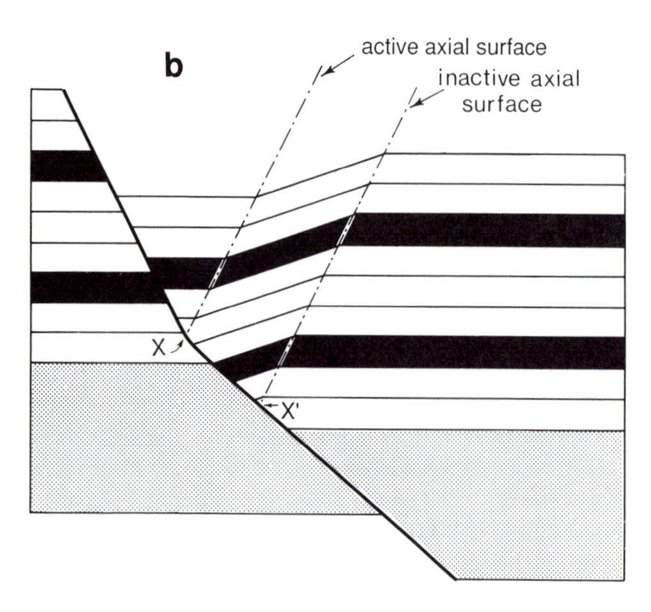

Figure 9–44. Balanced cross section showing development of rollover on listric normal fault. Decrease of dip with depth causes hangingwall to collapse onto footwall, producing a rollover fold. After Xiao and Suppe (1992).

account, the maximum offset would be 60 cm. Shorelines of Hebgen Lake were tilted toward the Red Canyon and Hebgen faults during the 1959 Hebgen Lake earthquake. The tilting extended beyond surface faulting, suggesting that surface rupture may have changed along strike to blind normal faults expressed at the surface by monoclinal warps facing in the direction of the downthrown block of the blind fault at depth.

The surface expression of buried normal faults is in part related to the depth of the top of the fault, the fault dip, and the nature of horizontal layering (Withjack et al., 1990). Vita-Finzi and King (1985) pointed out that the surface expression is also controlled by the length of straight-line segments of primary faults. The longer the straight segment of a fault is, the more likely it is to produce a surface scarp. Two nonparallel faults will display decreases of slip toward their tips, and the area between the faults may experience secondary faulting and folding.

Vita-Finzi and King also pointed out that the surface expression is influenced by the amount of slip per earthquake, as illustrated in Figure 9–45. Surficial deformation above the fault tip absorbs some of the slip at depth as a zone of breccia and crushed rock, and, as faulting continues, it must drive through this deformed zone. They noted that ruptures of displacement greater than 1 m are more likely to be expressed at the surface as a fault scarp, whereas for individual displacements smaller than 1 m, the surface expression will be folds and secondary faults. For even smaller displacements, surface expression will commonly be undulations or zones of monoclinal tilt.

Evidence that blind normal faults are relatively common includes the fact that the length of the aftershock zone may be considerably greater than the length of surface rupture, as was the case for the 1987 Edgecumbe, New Zealand, earthquake and the 1978 Thessaloniki, Greece, earthquake (for the latter, compare Mercier et al., 1979 and Soufleris et al., 1982). In the Basin and Range, most earthquakes of $M < 6$ are not accompanied by surface rupture at all.

Fault-Zone Features

Hancock and Barka (1987), Stewart and Hancock (1991), and Stewart (1993) have studied details of several active normal-fault zones in Greece and Turkey, all involving Mesozoic carbonate rocks faulted against Quaternary sediments. The features they observed are shown diagrammatically in Figure 9–46 and are in contrast to fault-zone features in unconsolidated sediments in the Basin and Range province (cf. Fig. 7–1). From the footwall to the hangingwall, Hancock and Barka recognize six components: (1) unbrecciated bedrock with bedding recognizable, (2) brecciated and shattered rock with calcite cement and void space, (3) thin sheets of compact zone-parallel mineralized and cemented fault breccia, called *subslip-plane breccia sheets* (illustrated in Fig. 9–47), (4) corrugated *slip planes* on the upper surfaces of breccia sheets, with the main plane the fault contact between footwall breccia and hangingwall colluvium (corrugations illustrated in Fig. 9–48), (5) linear trails or tabular sheets

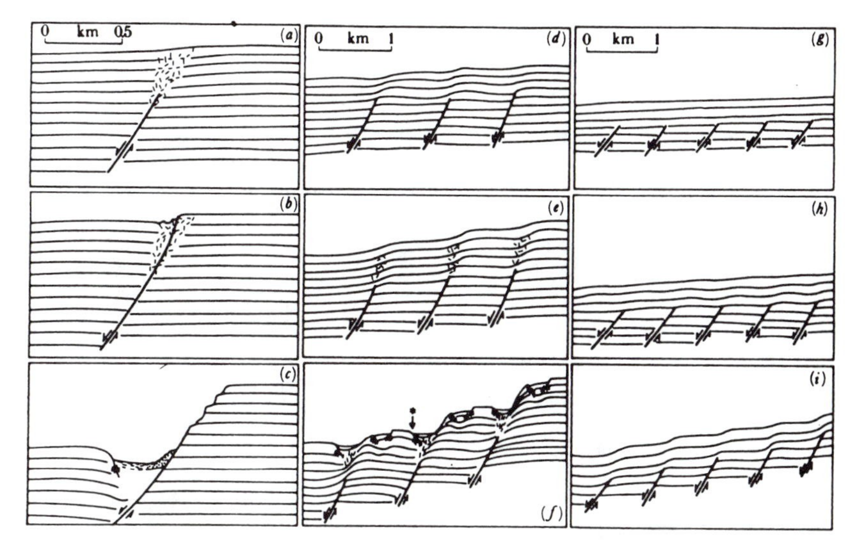

Figure 9–45. Surface expression of normal faults producing earthquakes of different amounts of slip. Sequence (a) to (c) is for a fault producing slip events at depth much greater than one meter. The fault cuts upward through its own near-surface shattered zone and forms a surface scarp. Sequence (d) to (f) shows faults with displacements somewhat less than one meter per event. The surface is marked by secondary structures, including folds, but the main faults do not reach the surface. Sequence (g) to (i) shows faults with displacements much less than one meter. The surface expression consists of undulations superimposed on a regional tilt. After Vita-Finzi and King (1985).

of brecciated colluvium, and (6) unbrecciated colluvium. These features are developed in carbonate rocks, and much of the fault-zone breccia is cemented and is resistant to erosion, in contrast to scarps in alluvium in the Basin and Range. Even where the fault zone is within colluvium or alluvium, the fault scarp does not degrade like a Basin and Range scarp, but rather like a scarp developed in bedrock (Stewart, 1993). Scarps several thousand years old may have a free face preserved in cemented colluvium, whereas uncemented colluvium farther away may be deeply eroded, producing a ragged appearance of the fault scarp.

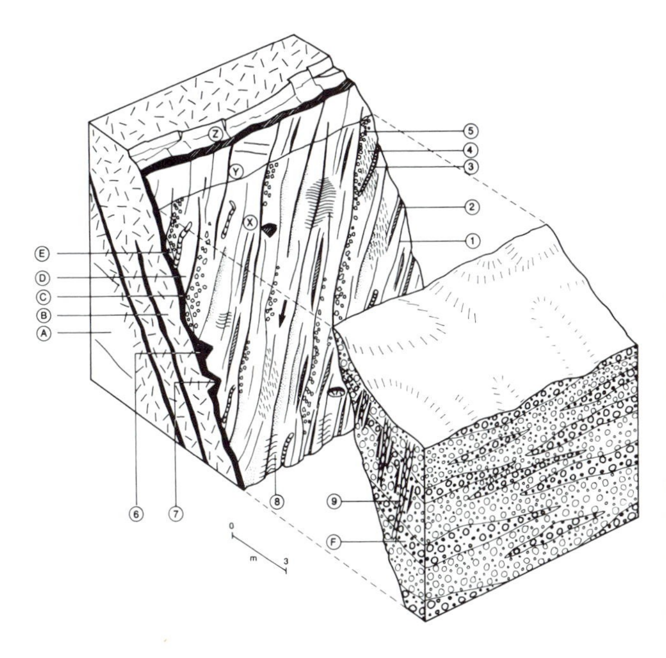

Figure 9–46. Diagrammatic exploded block diagram of a normal-fault zone based on the Yavansu fault in western Turkey. A, unbrecciated bedrock; B, coarse fault-precursor breccia; C, subslip-plane breccia sheet; D, corrugated slip plane; E, brecciated colluvium; F, unbrecciated colluvium. X, artificially exhumed fault plane; Y, fresh fault scarp above level of exhumation; Z, degraded fault scarp. 1, corrugation axis; 2, gutter; 3, tool track; 4, frictional-wear striations; 5, trail of brecciated colluvium (open circles) or mud smear (stipple); 6, spall mark; 7, pluck hole; 8, comb fracture traces; 9, reverse fissure-fault in colluvium. Scale bar gives approximate dimensions. After Hancock and Barka (1987).

Figure 9–47. Limestone normal-fault scarp near Corinth, Greece. The carapace of compact breccia is breached by solution pipes (holes) and stress-release fractures (lower left) to reveal an underlying zone of incohesive breccia. The soil-stained strip along the base of the scarp was exposed in the 1981 Gulf of Corinth earthquakes. Photo from Stewart (1993).

Fresh exposures of slip planes may be smooth (Fig. 9-49) to highly polished (examples also at Dixie Valley, Nevada). French workers call these planes *miroirs de failles,* and they would include slickensides. The corrugations on slip planes, with undulations up to several meters in wavelength, are indicators of relative slip direction; note that corrugations illustrated in Figure 9-46 have a component of right slip, whereas those in Figure 9-48 are predominantly dip slip. In time, weathering removes the polish and the details of the fault surface. The corrugation troughs may contain trails of brecciated colluvium or mud smears. The breccia contains fragments of footwall and hangingwall rock.

Other linear features that may be preserved on slip planes include (1) scratch marks, striations, and minor grooves, (2) tool marks, in some cases flanked by levee-like ridges, and (3) gutters, or rectilinear, steep-sided, flat-floored channels incised into subslip-plane breccia sheets. Subslip plane breccias may be cut by closely spaced fractures that are normal to the slip planes and close to normal to the corrugations. These are called *comb fractures* because of their resemblance to teeth on a comb in cross section. Comb fractures may be joints, or they may be small faults with displacement of a few centimeters. Pluck holes and spall marks cut into the underlying compact breccia sheet, with pluck holes penetrating the sheet into un-

Figure 9–48. Corrugated normal-fault plane at the base of the Furnofarango range front in south-central Crete. The lineation marked by the corrugations is in the direction of slip. The range front above the corrugations is eroded into a triangular facet. Photo from Stewart (1993).

Figure 9–49. Fault surface in the Wasatch fault zone at Salt Lake City, Utah. Quaternary gravel on the left is faulted against limestone. Striations pitch 70–80° in fault surface. Horizontal fractures partly caused by intersection of bedding with fault. Photo by Ronald Bruhn.

derlying rock (Hancock and Barka, 1987; Stewart and Hancock, 1991).

Not all of the features described above are of the same age. The footwall of the Dixie Valley fault, Nevada, includes plutons and hydrothermal mineral assemblages providing evidence that some of the fault structures formed millions of years ago at several kilometers depth and were uplifted during subsequent faulting and exhumed by erosion. Thus, in addition to juxtaposing rocks of different ages, the fault may bring together fault rocks formed at very different temperatures and pressures.

Stewart and Hancock (1991) describe map-pattern discontinuities at scales as large as hundreds of meters. These include (Fig. 9-50) *stepover zones* or *fault jogs* (Sibson, 1987), also called relay zones or en échelon stepovers. *Step-up zones* are local, high-angle faults that bring a local block of footwall rock to the surface. *Step-down zones* are concave depressions within the footwall block and may superficially resemble slumps; they may be bounded on both sides by faults. *Footwall cross faults* and *hangingwall cross faults* have already been discussed under *Segmentation;* these discussed here, however, are at scales of meters or tens of meters.

Tectonic Geomorphology

Much of the preceding discussion has dealt with normal faults in which the footwall is bedrock. Normal faults cutting unconsolidated sediments in the Basin and Range province form a free face, debris slope, and wash slope, as described in Chapter 7 and illustrated in Figure 7-1. With time, the free face and, later, the debris slope disappear, leaving a low, rounded wash-controlled slope.

The general features of normal-fault scarps in the Basin and Range Province were described by Gilbert (1890), and these features were later named by Slemmons (1957; Fig. 9-51). In a *simple fault scarp,* the dip of the fault steepens as it passes upwards from bedrock to surficial material, and a prism of surficial material is attached to the footwall. There may be a gaping fissure at the base of the fault scarp which quickly fills with colluvium. In a *subsidence zone,* the fissure is closed by sagging of the unconsolidated material. In a *gravity graben,* an antithetic fault cuts through the surface material, and a prism of sediment settles between this fault and the main fault, forming a near-surface graben. *Longitudinal step faults* form when the main fault splays upward as more than one surface trace; the surface materials between the faults may dip toward the main fault as a surficial rollover, or away from it. Slemmons (1957) noted that Gilbert's descriptions fit very well the surficial fault features produced by the 1954 Fairview Peak and Dixie Valley, Nevada, earthquakes.

Gilbert's observations pointed out the importance of taking into account surficial warping (Fig. 9-51b)

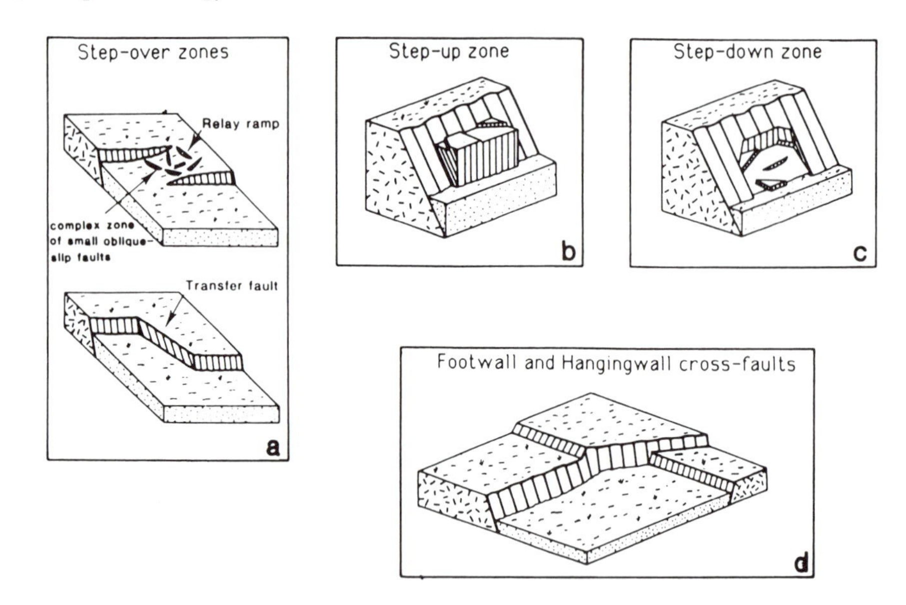

Figure 9–50. Schematic diagram showing small-scale structural complexities in normal-fault zones in the Aegean region. After Stewart and Hancock (1991).

and displacement on secondary faults (Fig. 9-51c, d) in determining the true vertical displacement on the range-front fault. In Figure 9-51b, fault displacement *bf* would be larger than the total slip across the fault zone due to sagging of the hangingwall. This must be taken into account in determining the true vertical displacement *bd*. In Figure 9-51c, fault displacement *bk* is too high, because the hangingwall block closest to the fault is a gravity graben. Displacement *hi* and the sagging of the gravity graben toward the main fault must be subtracted to determine the true displacement across the fault zone.

As the footwall of a normal fault undergoes continued uplift through many earthquake cycles, the original steep fault plane is dissected by streams and reduced in slope by erosion. As illustrated in Figures 7-8, 9-48, and 9-52, that part of the mountain front that has not been dissected by streams produces a *triangular facet*. The apex of the triangle is the crest of the divide between dissecting streams, and the base of the triangle is the faulted mountain front. The average slope of the facets is 30°, whereas the average dip of the range-front fault is 60°, indicating that the facet has been eroded back considerably from the original fault plane. The streams themselves carve canyons that are broad upstream and narrow at the range front, called *wine-glass canyons* from their resemblance to a wine glass when viewed from the valley in front of the range. Figure 9-52 shows triangular facets on the west side of the Tobin Range, Nevada, produced by long-term displacement on the range-front fault.

Triangular facets and wine-glass canyons are not themselves diagnostic evidence for active faulting. Late Pleistocene marine abrasion platforms and sea-cliffs cut into sediments in the Calabrian peninsula,

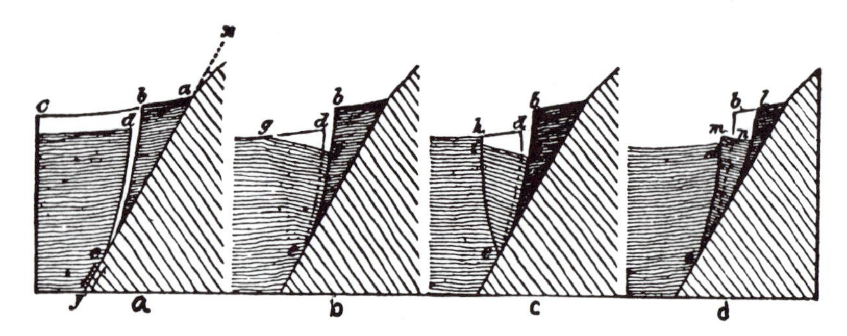

Figure 9–51. Gilbert's four diagrams. (a) simple fault scarp; (b) subsidence zone; (c) gravity graben; (d) longitudinal step faults. From Gilbert (1890) and Slemmons (1957).

Figure 9–52. Triangular facets sloping 28–35° on the west flank of the Tobin Range, central Nevada. Rupture accompanying the 1915 Pleasant Valley earthquake marked by bright band (fault ribbon, or *nastro di faglia*) near base of range. Photo by Robert E. Wallace.

Italy, have developed pronounced triangular facets as the former sea cliffs are dissected by high-gradient streams. These facets are unrelated to faulting.

Description of normal-fault scarps by Gilbert, Wallace, and others in the Basin and Range Province is applicable to normal faults in the arid regions of north China (see, for example, Zhang et al., 1986). However, Aegean and Apennine normal faults develop differently, as discussed in the preceding section, principally because colluvial materials in the fault zone tend to be cemented by carbonate and therefore behave as bedrock (Stewart, 1993).

Faulting produced by the 1915 Pleasant Valley, 1954 Fairview Peak, 1954 Dixie Valley, and 1983 Borah Peak earthquakes produced bright bands of less-weathered colluvium or bedrock near the range front (Fig. 9-52). These bands were used as evidence for historic rupture; Holocene non-historic scarps between the Dixie Valley and Pleasant Valley scarps lack such bright bands. Similar bands are common in Greece (Fig. 9-53) and in the Apennines of Italy, and in Italy they are called *nastro di faglia,* or fault ribbons. In these regions, *nastro di faglia* are not considered as evidence of Holocene rupture because in limestone terrain, compact breccia carapaces may be preserved for many thousands of years (Fig. 9-53). The 1915 Avezzano, Italy, earthquake resulted in pronounced *nastro di faglia* similar to those produced at Pleasant Valley that same year, but nearby *nastro di faglia* at Tre Monti were produced by a prehistoric earthquake that must have occurred more than 2500 years ago.

Expression in Trench Excavations

Trenches, whether dug by bulldozer, by backhoe, or by hand, offer little information about regional fault dip, because near-surface dip is strongly influenced by topography and by heterogeneity of the Quaternary deposits cut by the fault. But trenches offer the possibility of determining the late Quaternary fault history (that is, the paleoseismology) based on the stratigraphic relations between faults and late Quaternary sediments. In selecting a trench site, therefore, one attempts to maximize the possibilities for resolution, discrimination, and dating of slip events. Trenching a coarse gravel deposit, for example, may yield no more than one unit, and bedding and faulting may be obscured by the large clast size. It may be better to trench near the end of a fault where displacement is small, and both blocks are in alluvium, such as north of letter B in Figure 9-40e. Bell and Katzer (1990) obtained evidence critical to determining the relations of the Dixie Valley segment to the Stillwater segment by trenching the piedmont faults south of The Bend (B in Fig. 9-40b). Trenching close to a stream crossing a fault is more likely to reveal a more continuous

Figure 9–53. Limestone normal-fault scarp west of Lastros, Crete. The scarp is 4–5 m high and possesses a well-defined compact breccia carapace. Offset glacial deposits suggest that the scarp is late Pleistocene in age. Such scarps in the Apennines are called *nastri di faglia,* or fault ribbons. Photo by Paul Hancock.

section than trenching far from the stream if overbank rather than channel deposits can be exposed.

In the trench illustrated in Figure 9-54, a pebble bed in Unit Q_2 is displaced more than the overlying crosshatched unit, indicating displacement prior to deposition of the crosshatched unit (assuming no strike-slip component that would have juxtaposed different thicknesses of a wedge-shaped Q_2). The pebble bed in Unit Q_3 is offset the same amount as the cross-hatched unit, indicating displacement prior to development of the soil zone at the surface. Note that these relations by themselves do not identify individual earthquakes; each displacement could have been formed aseismically or by more than one earthquake.

Figure 9-55 directly documents at least two ages of faulting. Fault F_4 cuts unit 8, with a thermoluminescence date of around 57 ka, and is overlain unconformably by unit 9 dated as 22 ka. Fault F_3 cuts at least the bottom half of unit 9, and cracks extend to the surface, but the gravelly nature of this unit makes it unclear that the fault postdates unit 9. Fault F_1 is overlain unconformably by unit 11, which may predate unit 6, but correlation is uncertain. Fault F_5 cuts Holocene unit 10, dated as 1.3 ka, but the top of this unit, which is covered by a soil, appears to bury the fault scarp.

The presence of a *colluvial wedge,* the wedge-

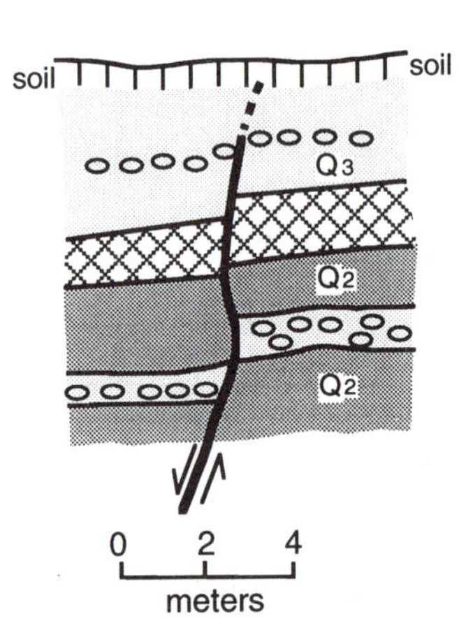

Figure 9–54. Trench across normal fault in Weihe basin, China. The crosshatched unit is offset the same amount as the pebble bed in overlying unit Q_3, indicating a displacement prior to the development of the surface soil. The pebble bed within unit Q_2 below the crosshatched unit is displaced a larger amount, indicating a displacement prior to the deposition of the crosshatched unit. This interpretation cannot be made if there is a large component of strike slip, because a thicker part of unit Q_2 could have been juxtaposed against a thinner unit by strike slip. From Figure 5–15 of Wang and Deng (1988).

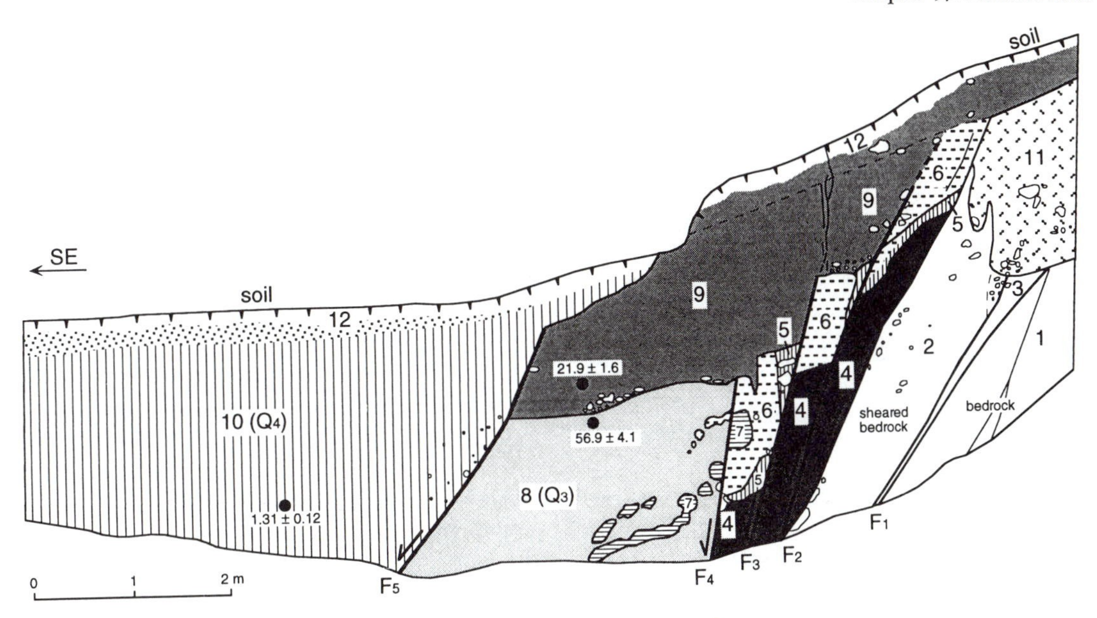

Figure 9–55. Trench log of fault north of Xihongzhan in the Huailai-Zhuolu basin in the northeast end of the Shanxi graben. Fault F_1 ruptured before deposition of the channel of unit 11, which cannot be correlated across the fault. Fault F_4 moved prior to deposition of unit 9. Fault F_3 cuts unit 9, but because of the gravelly nature of unit 9, it is not clear whether the faulting occurred during deposition of unit 9 or afterwards, even though cracks are mapped to the surface. The fault just to the right appears to cut through unit 9 if the correlation of gravelly unit 6 with its hangingwall equivalent is accepted. Fault F_5 cuts most of unit 10 (Q_4), but the upper part of unit 10 appears to bury its scarp. The numbers are thermoluminescence dates in years. From Figure 3 of Ran et al. (1991).

shaped deposit underlying the debris slope formed at the base of an eroding scarp (Fig. 7-1) is generally considered to be evidence for formation of a fault scarp. The wedge-shaped deposit, commonly gravelly, generally does not have a counterpart on the upthrown side. Figure 9-56 shows a trench providing evidence of three colluvial wedges, and, therefore, three earthquakes on the Helanshan fault adjacent to the Yinchuan basin (Fig. 9-39). A thermoluminescence date near the base of a colluvial wedge would date the time the material was covered during the formation of the debris slope, and this should be close to the date of the earthquake. However, a date near the top of the wedge would provide only a maximum age for the next younger event. The trench in Figure 9-56 is across a piedmont fault rather than the range-front fault, increasing the probability of preservation of deposits critical to dating faults.

Gilbert (1890) recognized that a fissure could form as faulting took place in unconsolidated sediments (Fig. 9-51a). This fissure would fill with detrital deposits during or immediately following an earthquake, producing what the Chinese call a *filled wedge*. As shown in Figure 9-57, the sediment in the trangular wedge between the main fault and antithetic fault F_3 is not found on either side of the wedge. The wedge was deposited as a fissure fill. The antithetic fault is overlain unconformably by unit 5, which is cut by another antithetic fault, F_4, which bounds another filled wedge. This fault is unconformably overlain by unit 6. At the base of the trench, unit 2, which is cut by the older filled wedge, unconformably overlies faults F_1 and F_2, documenting a still-earlier displacement.

Figure 9-58, on a boundary fault of the Weihe basin on the northern margin of the Qinling Mountains, documents four earthquakes using different techniques. Earthquakes I, II, and III are based on evidence provided by the presence of colluvial wedges. Earthquake IV is based on the presence of a filled wedge that cuts the youngest colluvial wedge. In addition, faults F_2 and F_3 are overlain unconformably by unit 1, which predates the oldest colluvial wedge.

Slip Rates and Recurrence Intervals

Determination of long-term slip rates is difficult for most dip-slip faults, because one block is being uplifted and eroded. This means that units that might be used to determine offset amounts are commonly either not deposited or not preserved on the upthrown side of the fault. In the Pleasant Valley area, Nevada,

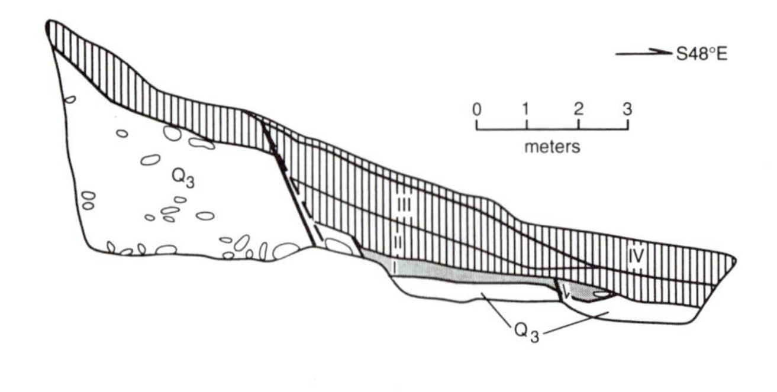

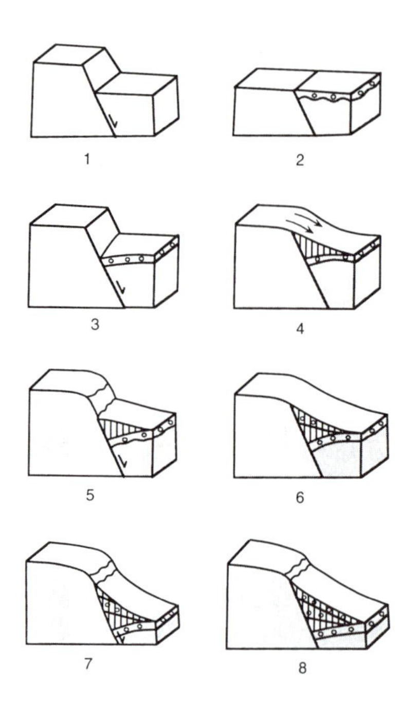

Figure 9–56. (a) Formation of three colluvial wedges (II, III, IV) in a trench across a piedmont fault in front of Helanshan Mountain flanking Yinchuan basin. Qaf, alluvial fan. (b) Block diagrams illustrating formation of the three colluvial wedges. After Wang and Deng (1988).

a basalt flow of Miocene age (10 Ma) is tilted toward the 1915 fault scarp and is not found on the uplifted block, an indication of a minimum slip rate of 0.1–0.2 mm/yr if displacement began soon after the flow was emplaced. A rough estimate of vertical separation on the Wasatch fault is made by the vertical distance between the altitude of the crest of the Wasatch Range and that of the valley floor beneath unconsolidated sediments on the downthrown side. The most active segments of the Wasatch fault, with the highest vertical separation rate, are adjacent to the highest part of the Wasatch Range crest (Schwartz and Coppersmith, 1984; Fig. 9–59; see following discussion).

Similarly, the central part of the Lost River fault, which ruptured in 1983, is adjacent to Borah Peak, the highest point in Idaho. This segment is more active than segments at the northern and southern ends of the fault, where slip rates are lower, and earthquake recurrence intervals are longer. If normal faulting occurs near sea level, as it did in the Gulf of Corinth, vertical displacement of the Stage 5e marine terrace, about 125,000 years old, or other dated terraces, can be used for a long-term vertical separation rate.

Holocene rates are generally obtained from trenches, but one must be able to correlate units across the fault, and commonly this is not possible in a trench. Also, it is necessary to ensure that the trenches span the entire fault zone.

Most information about slip rates and recurrence intervals for normal faults comes from northeast China, the Basin and Range Province of the western United States, the Apennines of Italy, and the Taupo Volcanic Zone of New Zealand, where extensive

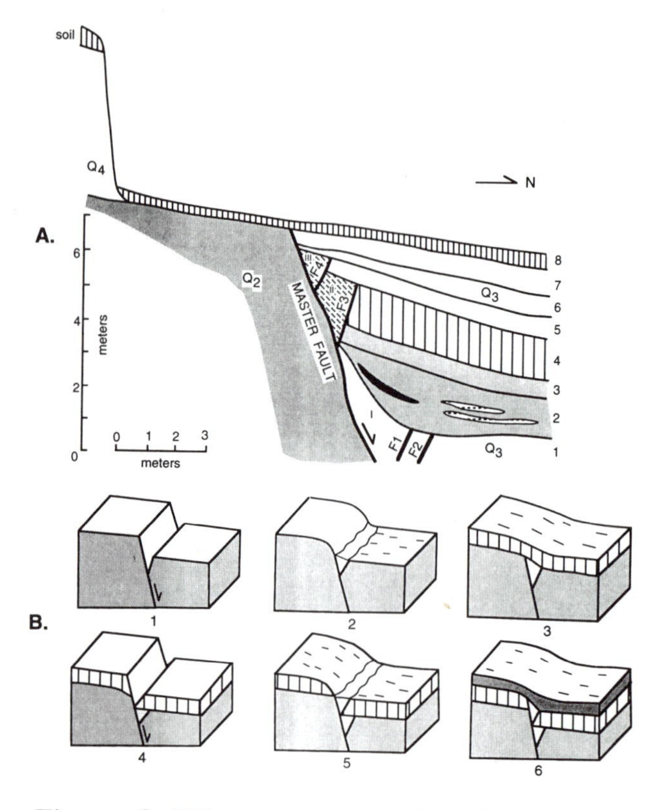

Figure 9–57. (a) Formation of two filled wedges (cf. Fig. 9–51a) in a trench across the piedmont fault zone of Huashan Mountain in the Weihe basin south of the Ordos block. (b) Block diagrams illustrating formation of filled wedges. After Wang and Deng (1988).

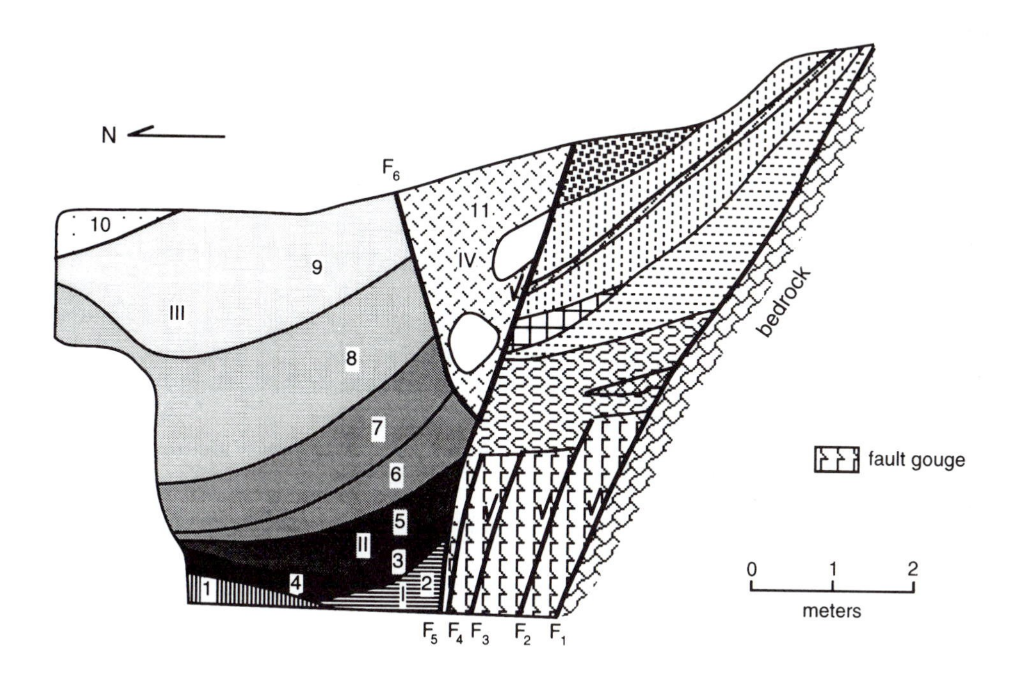

Figure 9–58. Trench log at Taipingkou, Huxian County, in Weihe basin adjacent to Qinling Mountains. Trench contains colluvial-wedge evidence for three earthquakes (I, II, III) and filled-wedge evidence for a fourth earthquake (IV). After Zhang et al. (1991).

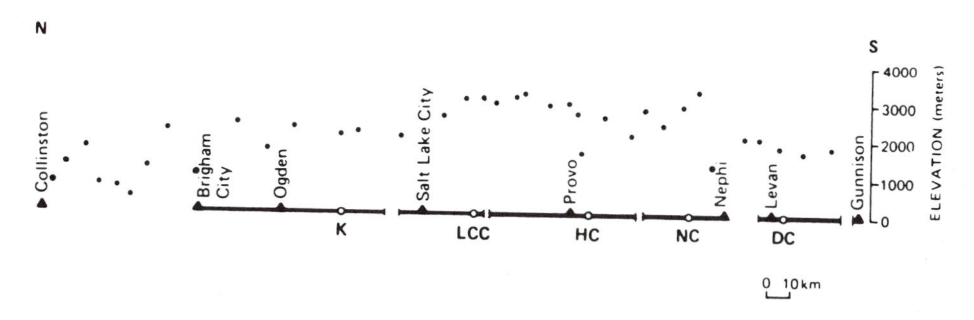

Figure 9–59. North-south topographic profile along the crest of the Wasatch Range. Stippled bands are segment boundaries of Schwartz and Coppersmith (1984), differing slightly from those of Machette et al. (1991) shown in Figure 9–41. Open circles are trench sites. Solid line along base of profile outlines the extent of surface faulting since 13,500 years ago. Highest slip rates on the Wasatch fault are where the Wasatch Range is highest. After Schwartz and Coppersmith (1984).

trenching has been done. The central (and most active) part of the Wasatch fault has a Holocene slip rate of 1-2 mm/yr with earthquake recurrence intervals of 2000-4000 years on a given segment in the central part of the fault. The less-active ends of the fault zone have slip rates less than 0.5 mm/yr, and they have not ruptured in the last 10,000 years (Machette et al., 1991). The Dixie Valley earthquake zone, despite its greater historical activity, has a slip rate of 0.5 mm/yr during the Holocene and 0.2 mm/yr for the last 200,000 years (Bell and Katzer, 1990). Each segment of the Dixie Valley-Pleasant Valley zone has sustained one or two earthquakes in the last 15,000 years (Fig. 9-43b), a recurrence interval averaging about 10,000 years. The 1887 Sonora, Mexico, earthquake produced up to 5.1 m displacement on a fault that had not moved for the preceding 200,000 years, an indicator that a low slip rate does not necessarily indicate low slip per event.

Slip rates reported for northeast China are 0.2 mm/yr for a fault at the northern edge of the Yanqing basin near Beijing (Cheng et al., 1991), 0.8-0.9 mm/yr for the northern Wutaishan piedmont fault in the northern Shanxi graben (Liu et al., 1991), and 1.2-2.5 mm/yr for a fault at the northern edge of the Huailai-Zhuolu basin (Ran et al., 1991). Slip rate on the Irpinia fault in the central Apennines of Italy is 0.3 mm/yr, and earthquakes recur on that fault about once in 2000 years, considerably longer than would be inferred from historical records (Pantosti and Valensise, 1990; D'Addezio et al., 1991). In the Taupo Volcanic Zone of New Zealand, slip rates are 1 mm/yr for the Edgecumbe fault and 3.3 mm/yr for the Paeroa fault (Berryman and Beanland, 1991). Archeological evidence indicates 22 m of motion on the north-dipping fault system in the Perachora Peninsula, Greece, in the last 30,000 years, a vertical separation rate of 0.7 mm/yr (Vita-Finzi and King, 1985).

In summary, slip rates on normal faults vary from <0.5 mm/yr to >3 mm/yr. Very high slip rates occur in regions of contemporaneous volcanism, where part of the displacement may be a response to movement of subterranean magma chambers. There is a relation between slip rate and recurrence interval, but not between slip rate and earthquake size; large earthquakes have ruptured faults with very low slip rates. Recurrence intervals are irregular, and there is evidence of earthquake clustering in northeast China, the Wasatch fault, and the Dixie Valley-Pleasant Valley fault system, the last clustering there being historical. With irregular recurrence intervals varying from 10^3 to 10^5 years, long-range forecasting of normal-fault earthquakes is a formidable task.

Fault Dip: Do Normal Faults Become Listric with Depth?

The dip on a seismogenic fault is determined in several ways: (1) direct measurement at the surface or in trenches or boreholes, (2) orientation of the zone of aftershocks (Figs. 9-13, 9-22), (3) orientation of the focal planes of the mainshock (Figs. 9-16, 9-28, 9-32) and major aftershocks, and (4) geodetic modeling (Figs. 9-16, 9-32). Perhaps the least reliable dip is that measured at or near the surface, because this dip is strongly affected by the Quaternary near-surface deposits that are cut by the fault. For example, surface thrust faulting with lengths of a few hundred meters was found accompanying the Borah Peak (Fig. 9-20) and Hebgen Lake normal-fault earthquakes. However, general orientation of fault planes based on different data sets: orientation of zones of aftershocks, position of mainshock with respect to surface rupture, and geodetic modeling, are surprisingly similar, as illustrated for Borah Peak by Figure 9-16 and for Edgecumbe by Figure 9-32. The evidence is clear not only that faults are more or less planar, but that their dips are mostly in the range 30-60° (Jackson, 1987; Jackson and White, 1989). There is no direct seismographic or geodetic evidence for low-angle normal faulting in the brittle continental crust, although multichannel seismic profiles show that young fault scarps in southwestern Utah merge into a low-angle detachment at about 3.5 km depth (Crone and Harding, 1984). The near-horizontal seismogenic normal faults beneath the island of Hawaii (Wyss and Koyanagi, 1992), a huge topographic edifice constructed on oceanic crust, are a special case.

This result is, at first glance, surprising for two reasons. One reason is the recognition of low-angle to subhorizontal normal faults, or *detachments,* associated with metamorphic core complexes in western North America and elsewhere in the world (Davis, 1987; Gans, 1987; Wernicke et al., 1987). Where are the present-day seismogenic counterparts of these Tertiary structures?

The second reason, as pointed out by Jackson (1987) and Jackson and White (1989), is that normal faults and the blocks they cut must rotate about a horizontal axis. If this were not so, a huge gravity anomaly of long wavelength should be produced, and the Moho discontinuity would be depressed, and neither of these effects is observed. Normal faulting on a crustal scale results in crustal extension. The analogy is the "thinning" of a set of books on a bookshelf as books slump when a bookend is removed. The slumping from a vertical to a low-angle position is analogous to this rotation. An outcome of tilting is that the faults will rotate to too low an angle for continued movement. In such cases, new high-angle normal faults will cut across the now-inactive low-angle faults. This has been observed in the Basin and Range province and in the Shanxi graben of China, the latter involving Pleistocene strata.

There is some evidence for coseismic low-angle normal faulting below the seismogenic zone. Figure 9-29 suggests that the direction and amount of displacement on two of the mainshocks of the 1981 Gulf of Corinth earthquakes are best explained by a decrease in dip with depth of the north-dipping master fault, which also accounts for internal deformation of the hangingwall block of this fault, as documented by aftershocks (Fig. 9-28). In addition, the 1969 Gediz and 1970 Alasehir earthquakes in western Turkey produced seismic signals originating after the faulting had ruptured the upper crust; these signals may have been caused by movement on normal faults dipping less than 10° (Eyidogan and Jackson, 1985). Jackson (1987) reasoned that the mainshock near the base of the seismogenic zone, in addition to rupturing toward the surface, may propagate downward into a transitional zone between brittle crust, where deformation is always frictional, and the lower crust, where deformation is always plastic or quasi-plastic (Fig. 3-9, taken from Scholz, 1988). In this transitional zone, deformation is plastic under low strain rates, but under high strain rates such as those accompanying an earthquake in the overlying brittle crust, deformation is brittle, analogous to toffee (Jackson, 1987) or to Silly Putty. This behavior is also implied in plots of depth distribution of slip by Tse and Rice (1986). This reasoning has been used to infer that high-angle normal faults in the eastern Basin and Range Province change down-dip to low-angle faults, in part following older structures (Smith and Bruhn, 1984).

SUMMARY

Normal faults develop in regions of extension; the maximum principal compressive stress is the load stress. The commonest setting for extension is at oceanic spreading centers, but here the crust is too thin and hot to generate large earthquakes. A continuum exists between fault swarms in active volcanic regions and range-bounding faults in continental regions; the faults in continental regions do generate large earthquakes and are the focus of this chapter. Normal faults are commonly associated with arc volcanism (Kyushu, Japan), but more commonly they are found in continental equivalents of back-arc basins (North Island of New Zealand, Aegean region). High plateaus behind collision zones undergo extension (Andean Altiplano, Tibet), and this extension is, in part, related to gravitational effects of high topography. However, the extension in Tibet normal to the Himalayan collision zone is geometrically similar to foreland extension in northwest Europe, northern Arabian Peninsula, and Patagonia. Intracontinental rift zones include the East African rift valleys, the Rhine Valley, the Rio Grande rift of southwestern United States, and the Baikal rift of Siberia. Broader, more diffuse intracontinental rift systems include the Basin and Range Province in the western United States and the graben systems around the Ordos Plateau of northeastern China.

Map patterns of normal faults are highly irregular, but in cross section, normal faults are straight, or at most, broadly curved. Normal faults with a large component of strike slip tend to be more straight in map view than pure dip-slip faults, but, in cross section, they look rather similar. Most large normal faults occur at range fronts. But in the Apennines, normal faults occur in the mountains, and basins in the downthrown blocks are small, probably because normal faulting has a shorter history there than in the Aegean or the Basin and Range. Fault traces are characterized by stepovers arranged en échelon, cross faults in both hangingwall and footwall, and piedmont faults downslope from the range front. In China, these piedmont faults have been found to be more recently active than the range-front fault. Short piedmont faults may bring up bedrock downslope from the range front (step-up zone), or faults may embay the range front (step-down zone). Bedrock blocks may lie across the faults (salients, or push-up blocks), bounded by faults less active than simpler range-front faults on either side.

These irregularities permit a long fault to be divided into segments. Segments based on geology may correspond to earthquake rupture zones, permitting a forecast of the moment magnitude of a future earthquake. Geological discontinuities that have been shown to be segment boundaries include large en échelon stepovers, gaps in faulting, salients, sharp changes in fault strike, and zones of great structural complexity. However, some segment boundaries may be "leaky"; earthquake rupture may cross these boundaries in some cases but not others. A segment boundary may stop a small earthquake, but not a very large one. In working out the history of earthquakes along some faults, it is seen that some surface ruptures are primary, that is, they connect directly to the seismogenic fault, and others are sympathetic (triggered) ruptures, on the far side of the segment boundary from the primary rupture.

Folds associated with normal faulting include

rollovers associated with listric faults in sedimentary basins. These listric faults are generally not seismogenic. In many cases, the surface expression of earthquake faulting is warping in addition to (or in place of) surface faulting. Relatively short faults with small displacement per earthquake are more likely to be expressed by folding, commonly associated with minor secondary faulting.

Fault-plane solutions, aftershock distribution, the relation of mainshock location with respect to surface rupture, and geodetic modeling indicate that normal-fault earthquakes occur on planar faults that dip mostly betwen 30° and 60°. However, some evidence from Greece and western Turkey suggests that earthquakes may propagate downward from the mainshock along a low-dipping fault in the transition zone between brittle and quasi-plastic failure. Large normal faults should rotate to lower dip with time, and the presence of tilted fault blocks is evidence for this. However, the transition from the observed steep seismogenic faulting and the postulated low-angle faulting in the quasi-plastic zone is poorly understood.

Normal-fault slip rates vary from less than 0.5 mm/yr to more than 3 mm/yr, with the higher rates commonly in regions with contemporaneous volcanic activity. The low slip rates of most normal faults are an indication that return times of earthquakes on normal faults are relatively long, measured in 10^3 to 10^5 years. Recurrence intervals are irregular, and there is evidence for earthquake clustering in northeast China and in the Basin and Range Province. The historical earthquake sequence in the Central Nevada Seismic Zone is a modern example of such clustering.

Suggestions for Further Reading

Coward, M. P., Dewey, J. F., and Hancock, P. L., eds. 1987. Continental Extensional Tectonics: Geological Society Spec. Pub. 28, 637 p.

Jackson, J. A., and White, N. J. 1989. Normal faulting in the upper continental crust: Observations from regions of active extension. Jour. Structural Geol. 11:15-36.

Roberts, A. M., Yielding, G., and Freeman, B., eds. 1991. The Geometry of Normal Faults: Geological Society Spec. Pub. 56, 264 p.

Ziegler, P. A. ed. 1992. Geodynamics of rifting, volume 1, case history studies on rifts: Europe and Asia. Tectonophysics, 208:1-363.

10

Reverse Faults and Folds

Structural geology began with the study of thrust faults, and much of the geological literature of the nineteenth century resounded with controversies about the presence of older rocks overlying younger rocks in the Alps. Later reports documented similar relations in the Scottish and Scandinavian Caledonides and in the Appalachians, and the geology of thrust belts became an important topic for research in structural geology. However, except for subduction zones, the literature on active tectonics prior to the 1970s focused mainly on strike-slip faults and normal faults, with only a few references to active reverse faults and folds. In this chapter, we discuss active reverse faults and folds that occur primarily in continental crust, leaving the discussion of reverse faults in subduction zones to Chapter 11.

The first description of coseismic reverse faulting was by Yamasaki (1896), following the 1896 Rikuu earthquake in northern Japan. Surface rupture was described from the 1929 Murchison earthquake in New Zealand and the 1952 Arvin-Tehachapi earthquake in southern California, but the rupture patterns from these earthquakes were not well understood. Not until the 1970s did it become clear that reverse faults constitute a major earthquake hazard worldwide. This became apparent because four large earthquakes of that decade were produced by reverse faults. The San Fernando, California, earthquake of 1971 was followed by the Caucete, Argentina, earthquake of 1977, the Tabas-e-Golshan, Iran, earthquake of 1978, and the El Asnam, Algeria, earthquake of 1980. In the previous decade, the Meckering, Australia, earthquake of 1968 provided evidence that reverse-fault earthquakes can occur in a most unlikely place—stable continental crust of Precambrian age. In the subsequent decade, the Coalinga, California, earthquake of 1983 demonstrated that earthquakes with fault-plane solutions indicative of reverse faults may not produce surface rupture, but may, instead, be expressed by active folds. Then in 1994, the Northridge earthquake provided evidence that a reverse fault that does not reach the surface could generate an earthquake capable of catastrophic damage amounting to billions of dollars.

The regions worldwide that are believed to have potential for reverse-fault earthquakes are located on Figure 10-1.

Reverse faults and flexural-slip folds form in an environment where the maximum principal compressive stress is horizontal, and the minimum stress is vertical, the load stress (Fig. 10-2a). In this stress field, a structurally isotropic material with a low coefficient of static friction has a plane of maximum shear stress that dips 45° and strikes parallel to the direction of intermediate compressive stress. The coefficient of static friction of most rock results in faults forming at an angle closer to 30° to the maximum compressive stress rather than 45° (Fig. 2-14). Other reasons for departure of the dip of a fault from 45° are a nonvertical minimum-stress direction due to topographic relief (Fig. 10-2b) and anisotropies due to bedding or preexisting faults or joints.

Reverse faults dip less than 90°, and, by definition, have a hangingwall block that has moved up with respect to the footwall block. Reverse faults dipping 45° or less are called *thrust faults*. Displacement is predominantly in the direction of dip. The term *detachment fault* is sometimes used to refer to large-scale thrust faults, but we restrict this term to low-angle normal faults in which the hangingwall has moved downward with respect to the footwall. A low-angle thrust fault which brings strongly-deformed upper-plate rocks over relatively-undeformed lower-plate rocks is called a *décollement*. A décollement may be the basal thrust of a set of imbricate thrust faults and folds called a *fold-and-thrust belt*. Such relations are an example of *thin-skinned tectonics*.

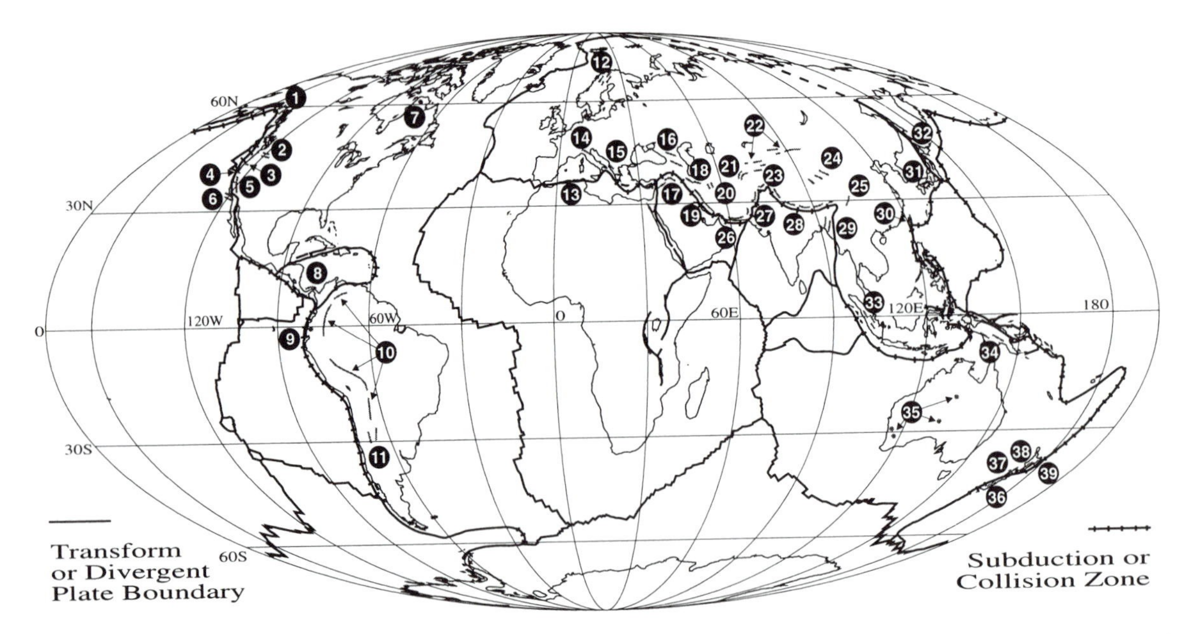

Figure 10–1. Locations of major active onshore reverse fault and fold systems. 1. Pamplona-Kayak Island zone, Alaska (largely offshore); 2. Yakima fold belt, Washington; 3. Cape Arago, Oregon (partly offshore); 4. northern California coast (partly offshore); 5. Coast Ranges, California; 6. western Transverse Ranges, California (partly offshore); 7. Ungava, Canada; 8. North Panama deformed belt (largely offshore); 9. Inter-Andean depression, Ecuador; 10. Subandean Cordillera; 11. Pampean Andes, Argentina; 12. northern Scandinavia; 13. Tellian Atlas, north Africa; 14. southern margin of Po Plain, Italy; 15. Epirus, Greece; 16. Greater and Lesser Caucasus; 17. Bitlis foldthrust belt and Palmyride fold belt; 18. Alborz Mountains, Iran; 19. Zagros Mountains; 20. Khorassan, Iran; 21. Tajikistan; 22. northern range front of Tianshan; 23. southern range front of Tianshan, China; 24. Qilian Shan-Qaidam basin, China; 25. Lungmen Shan-Min Shan, China; 26. Makran (partly offshore); 27. Kirthar and Sulaiman Ranges, Pakistan; 28. Himalayan frontal thrust; 29. Arakan fold belt, Burma; 30. Taiwan coastal plain; 31. Kinki Triangle, Japan; 32. Niigata fold belt, northern Japan (partly offshore); 33. central Sumatra; 34. Papua New Guinea; 35. Australia; 36. Central Otago, New Zealand; 37. northwest Nelson, New Zealand; 38. Wanganui folds, New Zealand; 39, East Coast Fold Belt, New Zealand (partly offshore).

TECTONIC SETTING

As we have in the previous chapters on strike-slip and normal faults, we divide our discussion of reverse faults by tectonic environment. The megathrusts of most convergent plate boundaries in which at least one plate is oceanic are considered separately in Chapter 11. These enormous structures juxtapose continental and oceanic crust, account for most of the seismic moment release on earth, and are known principally from remotely sensed data, that is, from seismic-reflection profiles, seismicity, and bathymetry.

In this chapter, we have separated the remaining active reverse faults into four categories: (1) active reverse faults that constitute the boundaries between plates of continental crust, in which the boundaries are nearly perpendicular to the direction of relative plate motion, such as the reverse faults of the Himalaya, (2) reverse faults that occur continentward of the volcanic arc in association with a subduction zone, such as those of northern Honshu, Japan, (3) reverse faults that are associated with strike-slip faults, such as those of the California Transverse Ranges, northwest China, and New Zealand, and (4) reverse faults in continental shield areas. A fifth category, reverse faults in deforming forearcs of subduction zones, is discussed in Chapter 11.

Orthogonal Convergence at Continental Plate Boundaries

The southern margin of Eurasia is colliding with four plates that are moving northwards: Australia, India, Arabia, and Africa (Figs. 1–4, 10–1). The collisions of the India and Arabia plates with Eurasia juxtapose continent against continent. The India-Eurasia boundary is marked by the Himalayan collision zone (no. 28, Fig. 10–1) flanked on the east by the Arakan fold-thrust belt of

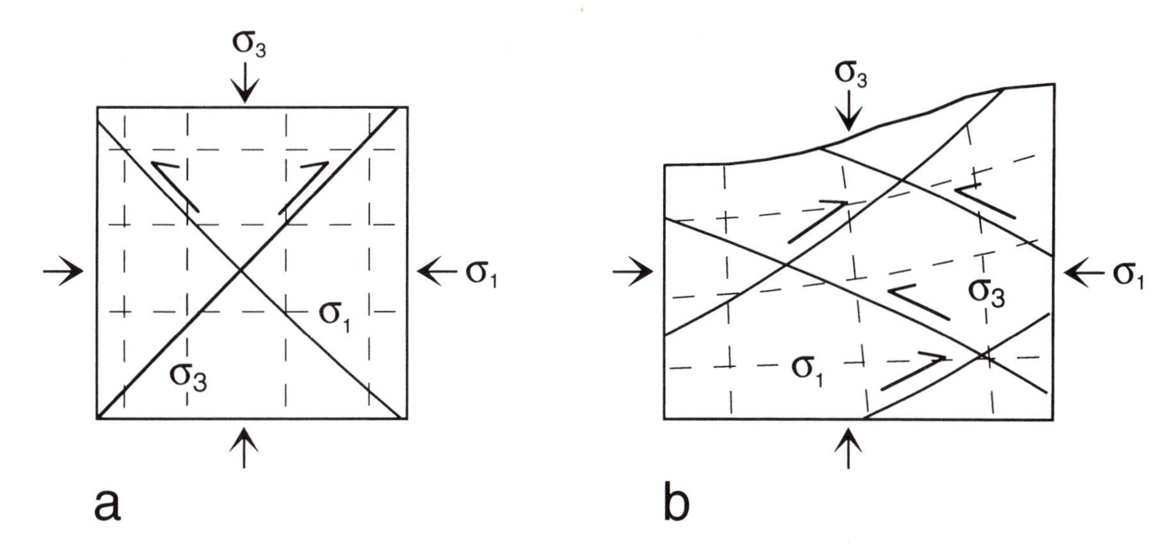

Figure 10–2. (a) Orientation of reverse faults parallel to two planes of maximum shear stress, assuming minimum principal compressive stress is vertical. Dashed lines mark stress trajectories. (b) Orientation of reverse fault parallel to a plane of maximum shear stress assuming one of principal stresses is perpendicular to a sloping surface adjacent to a range front. This would occur in the case where the range is being uplifted as the hangingwall of an active reverse fault. The stress trajectories reflect the perturbation by topography. In this case, the plane of maximum shear stress has a lower dip near the surface than at depth, resulting in a convex-upward fault trace in cross section.

Burma (no. 29, Fig. 10–1) and on the west by fold-and-thrust belts of the Sulaiman Range and Kirthar Range of Pakistan (no. 27, Fig. 10–1). The Arabia-Eurasia boundary is marked by the Zagros fold belt of Iran and Iraq (no. 19, Fig. 10–1), and the Bitlis thrust zone and Palmyride fold belt of Syria and Turkey (no. 17, Fig. 10–1).

These continental collision zones are separated by the Makran coastal region of Iran and Pakistan (no. 26, Fig. 10–1), where oceanic crust is subducting beneath Eurasia. South and east of the Indian collision zone, oceanic crust is subducting beneath continental crust in the Java-Sumatra trench. Still farther east, in eastern Indonesia and New Guinea (Hamilton, 1979), the Australian continent has collided with island arcs that contain small microcontinents. As a result, a fold-and-thrust belt has developed in New Guinea (no. 34, Fig. 10–1). West of the Arabian collision zone, the African continent is about to collide with the European continent, but for the most part, these continents are still separated by oceanic crust beneath the Mediterranean Sea (Le Pichon et al., 1988). Locally, however, this impending collision is manifested on land as compressional zones in northwestern Greece (no. 15, Fig. 10–1), along the southwestern margins of the Adriatic Sea and the Po plain of Italy (no. 14, Fig. 10–1), and in the western Mediterranean from the Atlas Mountains of Algeria eastward to Sicily (no. 13, Fig. 10–1).

Himalaya

The Himalaya marks the largest active continent-continent collision zone on earth. The collision between the Indian subcontinent and Eurasia began in early Tertiary time on the Indus suture zone (Fig. 10–3), which is delineated by serpentine and mafic igneous rocks (*ophiolite*) that are the relics of oceanic crust which formerly separated the two continents. Subsequently, the collision boundary shifted southward, and the northern edge of the Indian plate was thrust back onto itself, first along the Main Central thrust and later along the Main Boundary thrust (Fig. 10–3). The Himalaya was uplifted in the process. The Lesser Himalaya, between the Main Central thrust and the Main Boundary thrust, consists mainly of layered and in part metamorphosed rocks of Precambrian to Tertiary age. At the present time, the principal tectonic displacement zone is the Himalayan Frontal Fault System (Nakata, 1989; Yeats et al., 1992), which includes the Himalayan Front fault at the edge of the Indo-Gangetic Plains and a set of active anticlines and synclines to the north that are the surface expression of displacement on a buried décollement fault.

These folds between the Main Boundary thrust and the Himalayan Front fault comprise the Subhimalaya. The folded and faulted rocks of the Subhimalaya are Miocene to Pleistocene nonmarine strata that were deposited by alluvial river systems similar to the modern Ganga River and Indus River and their tributaries. South of the Himalayan Front fault, the modern rivers are filling a *foreland basin* or *foredeep* along the northern edge of the Indian shield, which is depressed by the weight of advancing thrust sheets of the Himalaya. The active folds overlie blind thrusts which merge northward and down-dip into a décollement. Beneath the décollement, the underlying Indian shield is not being deformed internally (Seeber and Armbruster, 1981; Yeats and Lillie, 1991).

The convergence rate between the stable Indian plate and the Himalaya is 10–15 mm/yr (Lyon-Caen and Molnar, 1985; Baker et al., 1988), only about one-fourth of the convergence rate between the Indian and Eurasian plates. This is consistent with the hypothesis that a major fraction of the plate convergence is accommodated by the eastward "escape" of central Asia along large strike-slip faults, tectonic shortening across the Tianshan, and by internal deformation accompanying uplift in Tibet and the Himalaya itself (Molnar, 1990; 1992; Avouac et al., 1993; Fig. 10–22). Active faults are present within the Himalaya of Nepal (Nakata, 1989) and Pakistan (Nakata et al., 1991), and tilted stream terraces are present along some major rivers crossing the Himalaya (Delcaillau, 1986).

Small earthquakes in India and Nepal are concentrated in a zone 50 km wide that extends south from the Main Central thrust across the Lesser Himalaya (Fig. 10–4). Most fault-plane solutions for earthquakes in this zone indicate fault planes dipping 5–30° north (Ni and Barazangi, 1984), for the most part steeper than the 6° dip of the main décollement. This suggests that the décollement may steepen abruptly in this region, forming a *thrust ramp* that enters the basement (cf. Molnar, 1990). The thrust ramp is characterized by high uplift rates measured geodetically (Jackson and Bilham, 1994) and by high stream gradients (Seeber and Gornitz, 1983).

This zone of high seismicity may mark the intersection of the top of basement with the décollement thrust fault. If the basement is considered to include the metasedimentary rocks of the Lesser Himalaya, the intersection of the top of the basement in the hangingwall with the thrust fault would be at (or in the air south of) the surface trace of the Main Boundary thrust, so that the actual displacement on the décollement would be at least 80–100 km (Molnar, 1990, illustrated in Fig. 10–4).

The foothills of the Himalaya were struck by great earthquakes in 1897, 1905, 1934, and 1950 (Fig. 10–3). These four earthquakes, all $M \geq 8$, are the largest known to have occurred on continental thrust faults, and they are comparable in scale to the great earthquakes of subduction zones. These earthquakes were not accompanied by surface rupture, but comparison of geodetic leveling surveys conducted before and after the 1905 earthquake suggests that this earthquake occurred on the décollement (Chander, 1988) and that the earthquake was accompanied by folding at the surface (Yeats and Lillie, 1991).

Although most of the seismicity of the southern flank of the Himalaya is attributed to the décollement, there is one exception. The Shillong Plateau, in eastern India, south of the Himalaya, is seismically active today and was underlain by the rupture zone of the great 1897 earthquake. Precambrian crystalline rocks are at the surface, and the crystalline block is cut on its south side by the high-angle Dauki fault, which is probably active. However, there is no evidence for surface rupture on the Dauki fault in 1897.

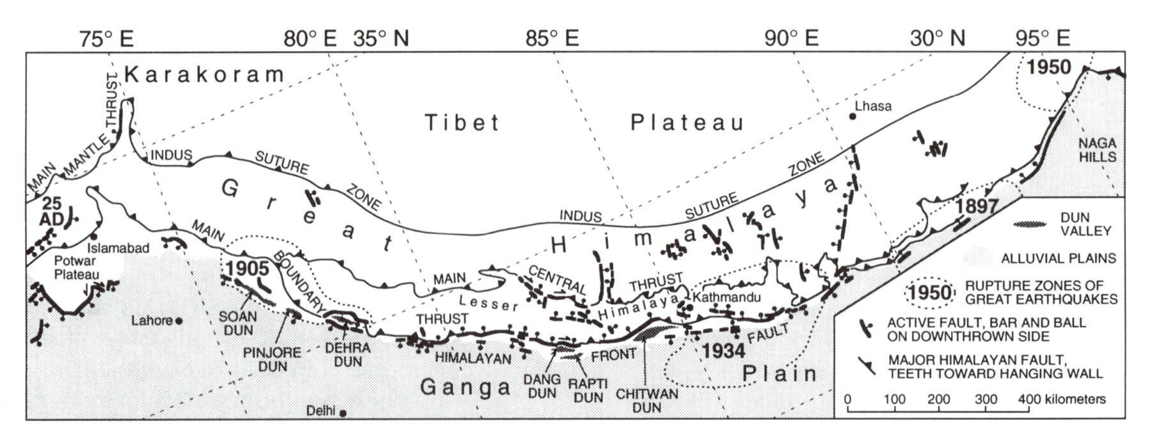

Figure 10–3. Active-fault map of Himalaya, compiled by Himalayan Active Fault Subcommittee of IGCP 206, from Yeats and Lillie (1991) and Yeats et al. (1992). Inset: rupture zones of great earthquakes.

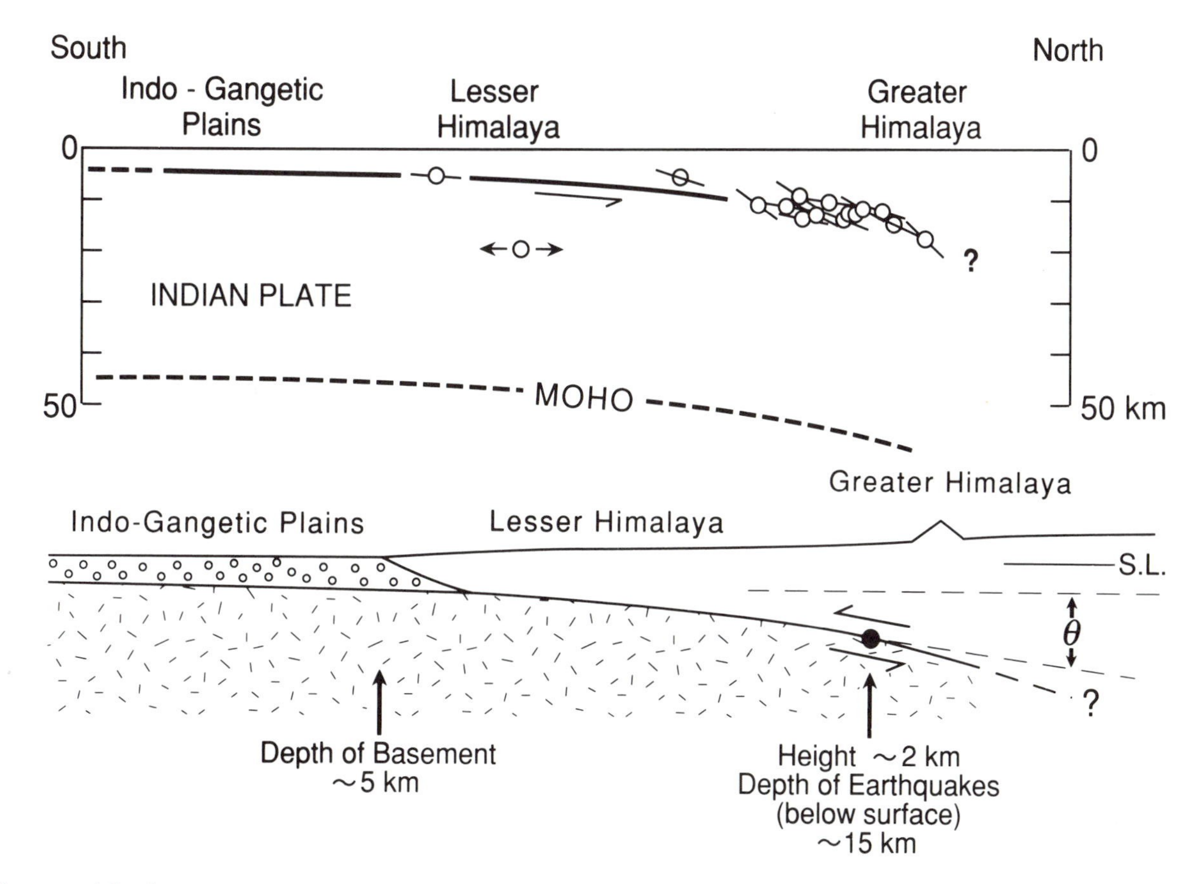

Figure 10–4. Cross section across Himalaya showing (a) that the nodal planes of intermediate-size earthquakes are generally steeper than the dip of basement as constrained by these earthquakes and by drill holes at the Himalayan front, and (b) illustrating idea that the décollement has displaced the leading edge of the Lesser Himalaya, which consists predominantly of Precambrian rocks, from the zone of earthquakes where the décollement enters basement, giving a minimum of 80–100 km displacement. The active décollement continues farther south, deforming the Subhimalaya. Great earthquakes on the décollement would be at or south of the zone of intermediate-size earthquakes, whereas surface deformation would occur at the Himalayan front. Modified from Molnar (1990).

Zagros

West of the Makran coast, where oceanic crust is subducting beneath Eurasia, the collision of the Arabian shield with Iran has uplifted the Zagros Mountains. The original collision zone is marked by the Main Zagros thrust, which, like its Himalayan counterpart, the Indus suture zone, is marked by ophiolite (Fig. 10-5). Subduction on the Main Zagros thrust has now ceased, and it is seismically inactive (Ni and Barazangi, 1986) except for the northern Zagros, where the surface trace of the thrust has been reactivated as the right-slip Main Recent fault (Berberian, 1981; 1995; Jackson and McKenzie, 1988), marked by arrows on Figure 10-6. Southwest of the Main Zagros thrust, the Zagros fold belt involves a 5- to 10-km-thick section of sedimentary rocks of the Arabian continental margin. The base of this section consists of as much as 1000 m of latest Precambrian salt and evaporite deposits. Evaporites are also present higher in the section in Permian, Triassic, Jurassic, Eocene, and Miocene strata (Berberian, 1981). The onset of deformation is established by the presence of late Miocene to Pleistocene synorogenic conglomerate (Berberian, 1981). The evaporite deposits, particularly the salt, act as decoupling layers above which sedimentary strata of the Arabian continent are sliding southwestward over Arabian continental basement rocks. The surface expression of this sliding consists almost entirely of folds rather than faults. Many of these folds are cored by salt. The major anticlines are southwest-vergent and are marked by southwest-facing topographic escarpments (Berberian, 1995).

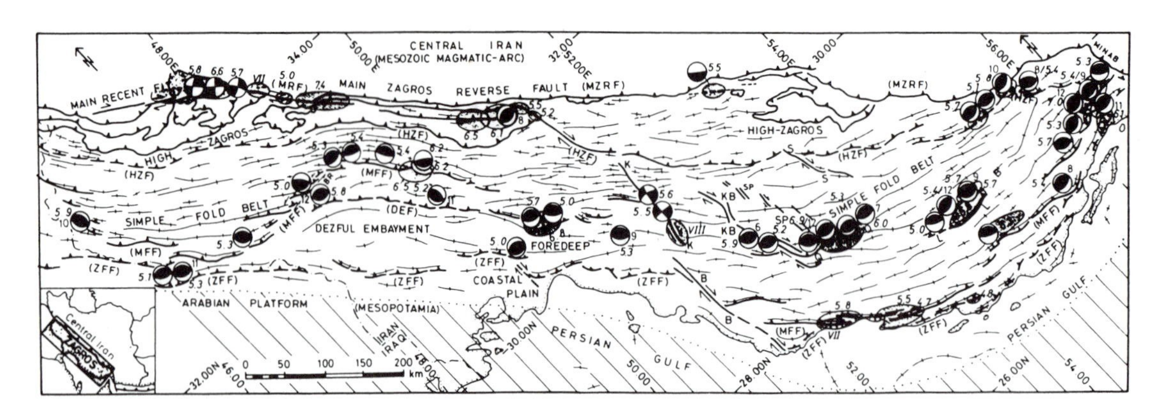

Figure 10–5. Tectonic map of the Zagros active fold belt of Iran and Iraq, showing basement faults that do not cut the overlying folds, fault-plane solutions and meizoseismal zones (gray ellipses) of selected earthquakes. Numbers in italics show local magnitude. Thin lines with tick marks are fold axes. After Berberian (1995).

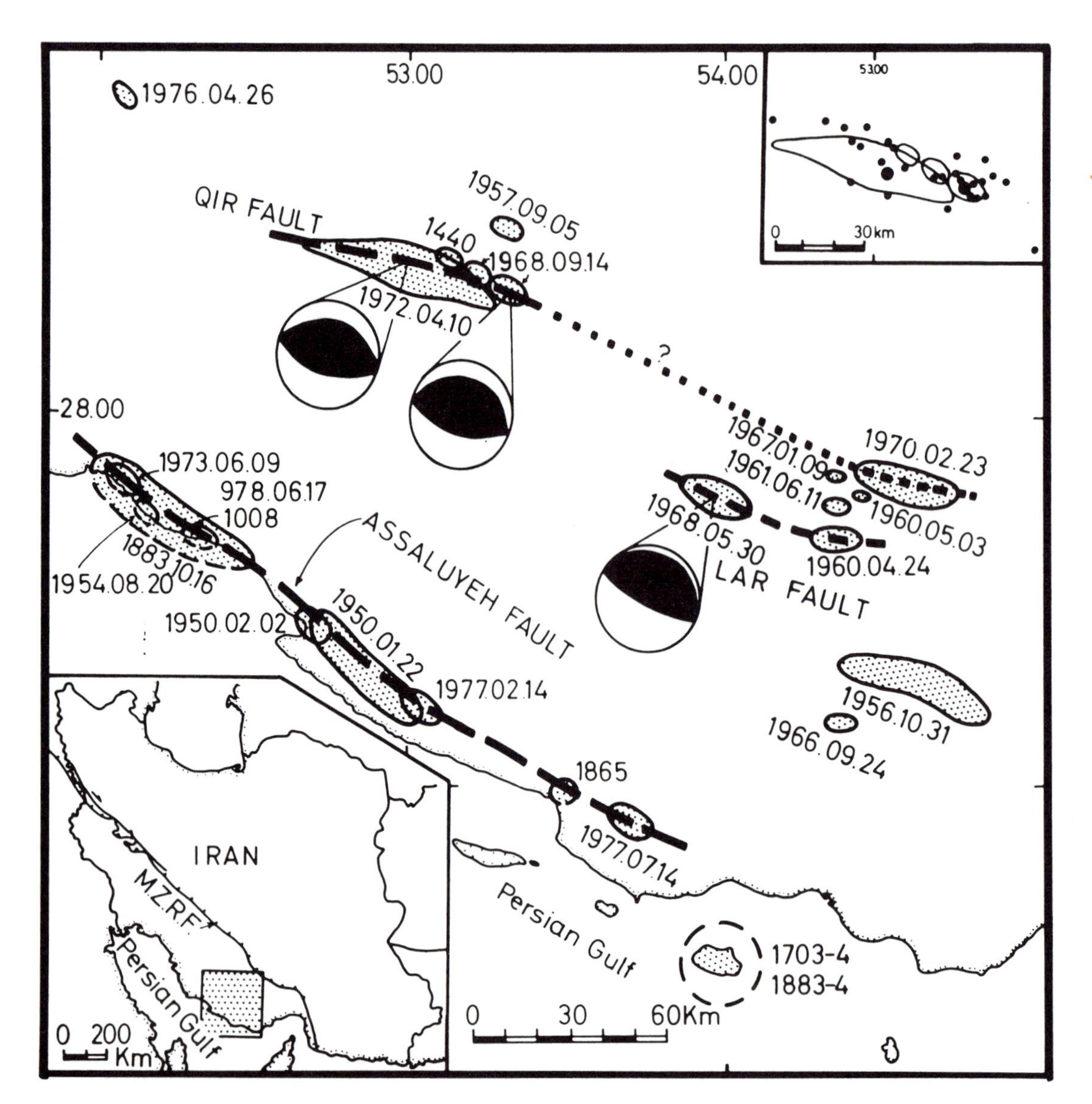

Figure 10–6. Linear pattern of aftershock zones of sub-Zagros reverse-fault earthquakes defines blind reverse faults beneath the Zagros fold belt. Only folds are found at the surface. MZRF = Main Zagros reverse fault, inactive except in northwest, where it is reactivated as a strike-slip fault. After Berberian (1981).

Holocene activity on some of these folds, documented by uplift of ancient watercourses and Holocene marine terraces, results in uplift rates of 2–7 mm/yr (summarized by Vita-Finzi, 1986 and Berberian, 1995).

The Zagros fold belt is largely aseismic, but earthquakes are common in the basement *beneath* the fold belt. Fault-plane solutions of these earthquakes indicate displacement mainly on low- to high-angle reverse faults (Ni and Barazangi, 1986). These basement faults do not reach the surface, but their geometry is expressed at the surface by the southwest-vergent anticlines (Berberian, 1995) and is further defined by the elongated isoseismals of large earthquakes (Berberian, 1981; 1995; Figs. 10–5, 10–6). These elongated isoseismal patterns contrast with the nonlinear isoseismals of great Himalayan earthquakes, which reflect very-low-angle source faults. The historical behavior of the Zagros décollement differs from that of the Himalayan décollement in that it has not produced large earthquakes. Instead, the largest earthquakes of the region are predominantly $M \leq 7$ and have originated on sources beneath the décollement. The Zagros décollement is not seismically active, because it is within low-strength salt, and, unlike the Himalayan décollement, it does not become deeper as it approaches the Main Zagros thrust. The Potwar Plateau and Salt Range of Pakistan are also underlain by salt of the same age as that in the Zagros Mountains, and the sedimentary rocks in this part of Pakistan are also largely aseismic. However, the décollement in this part of Pakistan deepens to the north, as it does in India and Nepal, and it may have produced a great earthquake in the northernmost Potwar Plateau in A.D. 25.

Syria-Turkey Border Region

To the northwest, the Zagros fold belt merges with the Bitlis fold-and-thrust belt of Syria and Turkey. This, in turn, curves southwestward toward the Dead Sea fault, which marks the western boundary of the Arabian plate (Ben-Avraham and Grasso, 1991; Lyberis et al., 1992, Fig. 10–7). As continental lithosphere of Arabia drives northward into Eurasia, the Anatolian plate is escaping westward toward oceanic lithosphere of the eastern Mediterranean Sea (Jackson and McKenzie, 1984; 1988; Fig. 10–7). However, the existence of active folds and thrust faults along the southwestern part of the East Anatolian fault, which marks the Anatolian-Arabian plate boundary (Lyberis et al., 1992), indicates that the rate of westward excape of Anatolia is insufficient to accommodate the continental collision fully. Farther south, within the Arabian plate, the Palmyride fold belt of Syria may be the locus of a new north-dipping zone of reverse faulting

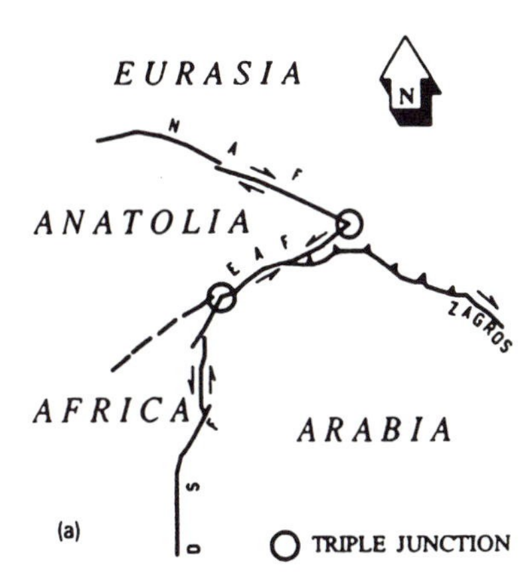

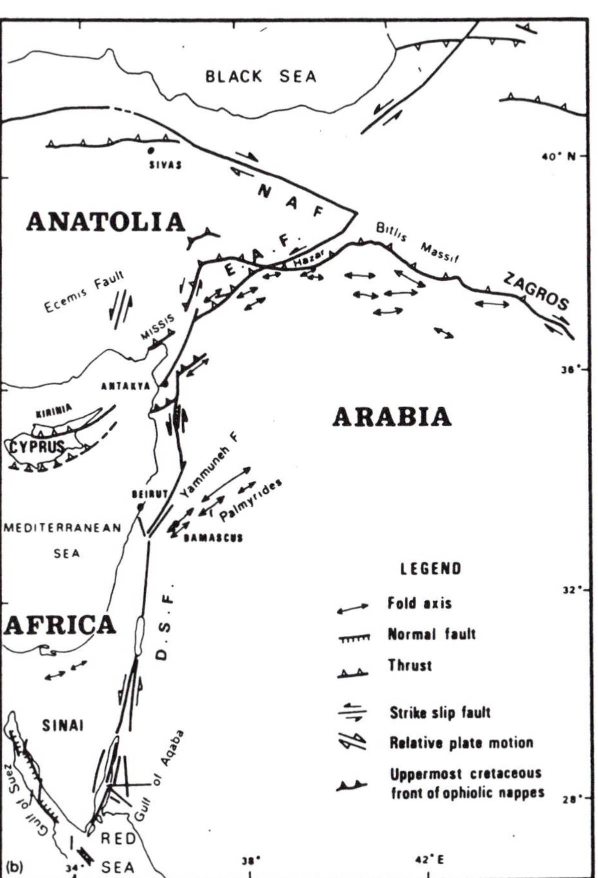

Figure 10–7. (a) Plate boundaries at the junctions of the African, Anatolian, Arabian, and Eurasian plates. (b) Folds and reverse faults between the Dead Sea fault (DSF) and the Zagros, including the East Anatolian fault system (EAF), Bitlis thrust zone, and Palmyride folds. (NAF) North Anatolian fault. From Lyberis et al. (1992).

(Rotstein and Kafka, 1982; Ben-Avraham and Grasso, 1991) possibly related to the Yammuneh (Yammouneh) restraining bend (cf. Chapter 8) in the Dead Sea fault in the vicinity of Damascus, Syria.

Tellian Atlas

The African plate is colliding with Europe in the western part of North Africa, but the existence of several small continental and oceanic plates in the western Mediterranean region results in complicated tectonics there (Le Pichon et al., 1988; Dewey et al., 1989; Rebaï et al., 1992). As summarized by these authors, Africa is rotating toward Europe around a pole of rotation west of the Straits of Gibraltar. According to global plate-tectonic models, convergence rates in North Africa increase eastward from Morocco to Tunisia. Despite low levels of instrumental seismicity in comparison with the eastern Mediterranean (Jackson and McKenzie, 1988), the Tellian Atlas in Algeria has been struck by large reverse-fault earthquakes, most recently in 1954 and 1980.

Behind-the-Arc Contraction Zones

Some convergent plate boundaries between continental and oceanic lithosphere include not only subduction zones but also zones of reverse faulting on the continent side of the volcanic arc. Inactive structures of the eastern foothills of the Canadian Cordillera, the Wyoming Rocky Mountains, the Valley-and-Ridge and Allegheny Plateau provinces of the Appalachians, and the Helvetic zone of the Alps originated in such a setting. These and their active analogues exhibit two principal structural styles:

(1) fold-and-thrust belts in which sedimentary rocks are decoupled from continental basement and thrust back onto the continent along a décollement (*thin-skinned tectonics*). and

(2) reverse faults that do involve crystalline basement (*thick-skinned tectonics*). In the North American Cordillera, the Wyoming Rocky Mountain province is an example of thick-skinned tectonics, whereas deformation in the adjacent Idaho-Wyoming thrust belt is an example of thin-skinned tectonics (Rodgers, 1987).

Zones of convergence on the continent side of volcanic arcs are present on the eastern side of the Andes from Argentina to Venezuela, in New Guinea, in eastern Indonesia (struck by a large earthquake north of the Island of Flores in 1992), in the North Panama deformed zone of Panama and Costa Rica, and in northern Honshu, Japan. Except in Japan, northwest Argentina, and, more recently, in Costa Rica, little attention has been focused on the active tectonics of these fold-and-thrust belts and thick-skinned reverse-fault regions. Most of our knowledge comes from petroleum exploration and regional geophysical surveys.

Eastern Foothills of the Andes

The Nazca oceanic plate is subducting N75°-80°E beneath the South American continent at a rate of 75 mm/yr in Ecuador and 87 mm/yr in central Chile

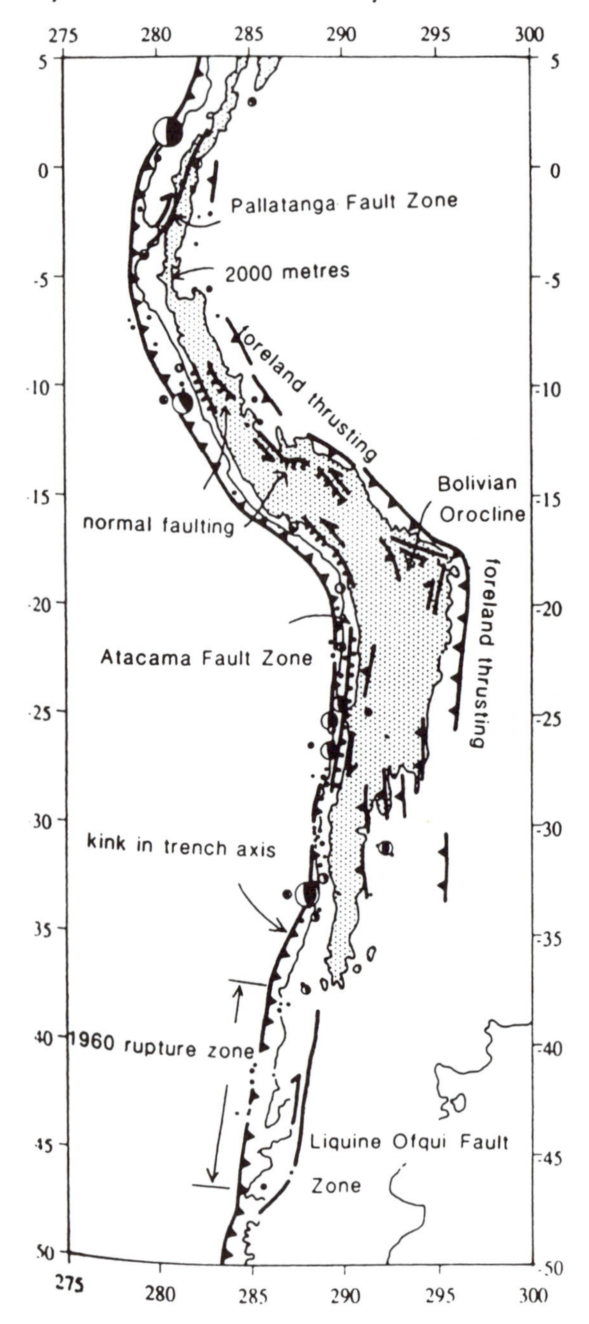

Figure 10–8. Distribution of active faulting in the Andes. Reverse faults shown with triangles toward hangingwall; normal faults with hachures toward hangingwall. Arrows show direction of strike slip. Foreland thrusting in the Subandean belt is thin-skinned, but between latitude 25° and 34°S, reverse faulting involves basement rocks. Shading shows terrain higher than 2000 m. Modified from Dewey and Lamb (1992).

(Stein et al., 1986). Variability in strike of the plate boundary results in variable senses and amounts of a strike-slip component to convergence in much of the Andes (Dewey and Lamb, 1992; Fig. 10-8). An active east-verging, thin-skinned foreland fold-and-thrust belt (Subandean belt) has developed in the eastern foothills of the Andes (Fig. 10-8). This is part of a deformation front that has been migrating eastward during much of the Cenozoic Era, as documented by synorogenic alluvial-fan deposits that are younger eastward (Ramos, 1988). The presently active fold-and-thrust belt is largely post-Miocene in age (Suárez et al., 1983; Jordan et al., 1983), and Holocene deposits are deformed. The fold-and-thrust belt is shortening at a rate of about 10 mm/yr (Dewey and Lamb, 1992), which is about one-eighth of the rate of plate convergence. The cross section in Fig-

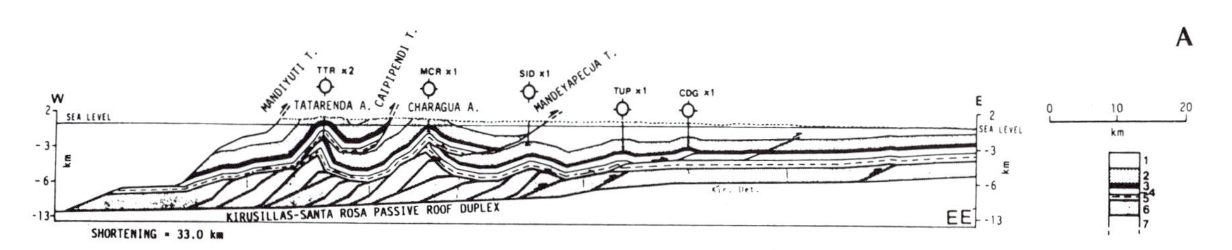

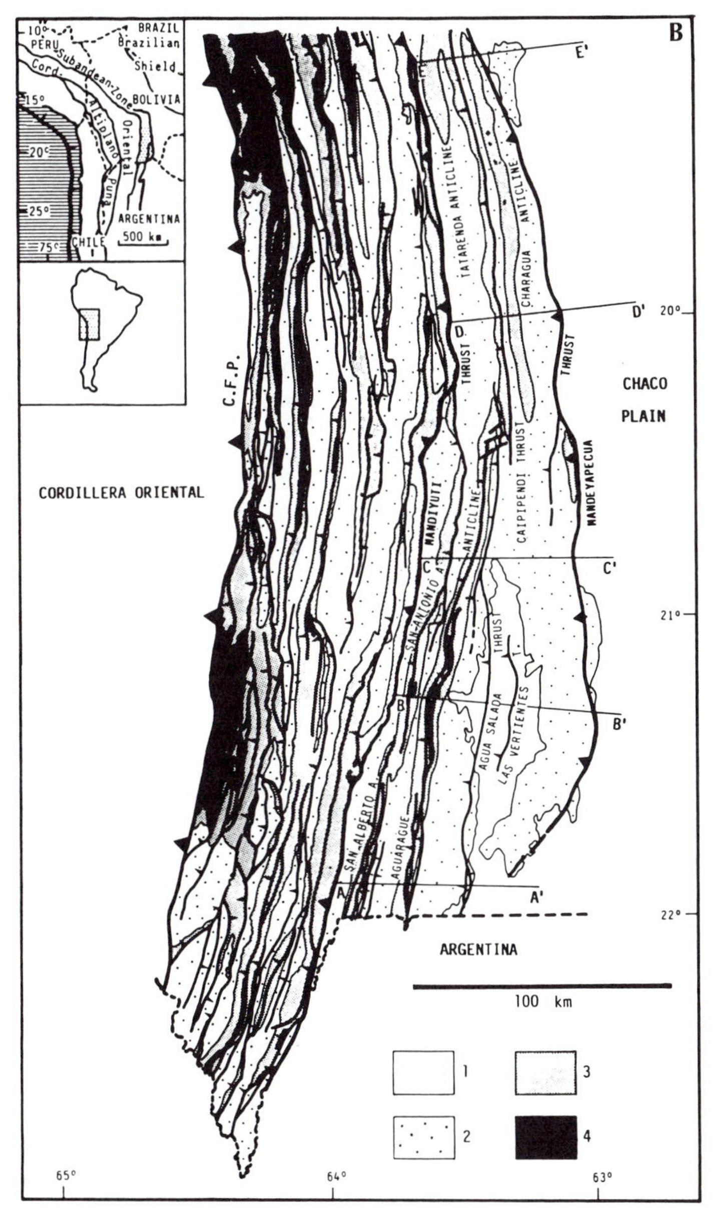

Figure 10–9. (a) Balanced cross section through the eastern foreland fold-and-thrust belt of the Subandean belt of southern Bolivia. Stratigraphic units: 1, Tertiary; 2, Mesozoic-Permian-Carboniferous; 3, 4, 5, Devonian; 6, Silurian-Devonian; 7, basement. Circles locate oil-well control. (b) Map of part of fold-and-thrust belt locating cross section E-E′ and showing large aspect ratio of structures. After Baby et al. (1992).

ure 10-9, constrained by well data and seismic-reflection profiles, illustrates the deformation in the Subandean fold-thrust belt in southern Bolivia (Baby et al., 1992). The map in Figure 10-9 shows that the ratio of length to width (aspect ratio) of individual thrust plates of the Subandean belt is very large, as it is in the Valley and Ridge Province of the Appalachians.

In northern Argentina, the Andean Precordillera is a thin-skinned fold-thrust belt overlying a décollement. However, reverse faulting is thick-skinned beneath the décollement and farther east in the Pampean Ranges (Smalley et al., 1993; Fig. 10-10). Mountain ranges 75-100 km long and 25-75 km wide are uplifted several kilometers relative to adjacent basins along range-front reverse faults, analogous to those in the Wyoming Rocky Mountain Province (Jordan et al., 1983). Thick-skinned reverse faulting is best developed where the dip of the subducting Nazca plate is nearly horizontal beneath the South American continent, and there is no active volcanic arc. In contrast, the thin-skinned Subandean fold-and-thrust belt has developed above a subducting slab that dips 30° east and is accompanied by active volcanoes.

The 1977 Caucete, Argentina, earthquake (M_s 7.4) originated on a thrust fault with a westward dip of 35°. Geodetic measurements suggest that the top of the rupture was 17 km beneath the surface (Kadinsky-Cade et al., 1985). West of the Pampean Ranges, the 1944 San Juan, Argentina, earthquake (M_s 7.4) may have occurred on a similar reverse fault cutting basement beneath the eastern edge of the Precordillera (Smalley et al., 1993).

Sedimentary rocks in the Subandean belt in Peru are 6-10 km thick above the décollement. Most of the seismicity is beneath the western part of the Subandean belt, although evidence for Quaternary deformation is largely in the eastern part of the belt. This suggests that Quaternary deformation is produced by low-angle faults that reach seismogenic depths far to the west. However, this interpretation is complicated by the fact that fault-plane solutions of earthquakes in this region indicate dips of 30-60°, much steeper than the dip of the décollement. Furthermore, the earthquakes are deeper than the décollement, although they are still within upper crust (Suárez et al., 1983). If historical seismicity is representative of the longer-term history, the shallow décollement beneath the fold-thrust belt may slip aseismically during earthquakes on reverse faults cutting basement farther west. The basement beneath the décollement, in contrast to that beneath the foreland fold-thrust belts of the Appalachians and Canadian Cordillera, is cut by seismogenic reverse faults that, like those in the Zagros, do not cut the overlying deformed sedimentary rocks. Thus the deformation of Quaternary sediments in the eastern part of the Subandean belt may reflect aseismic shortening at a shallower level than that producing earthquakes in crystalline crust beneath the décollement.

Papua New Guinea and Irian Jaya

Prior to early Miocene time, the island of New Guinea was a passive continental margin along the northern

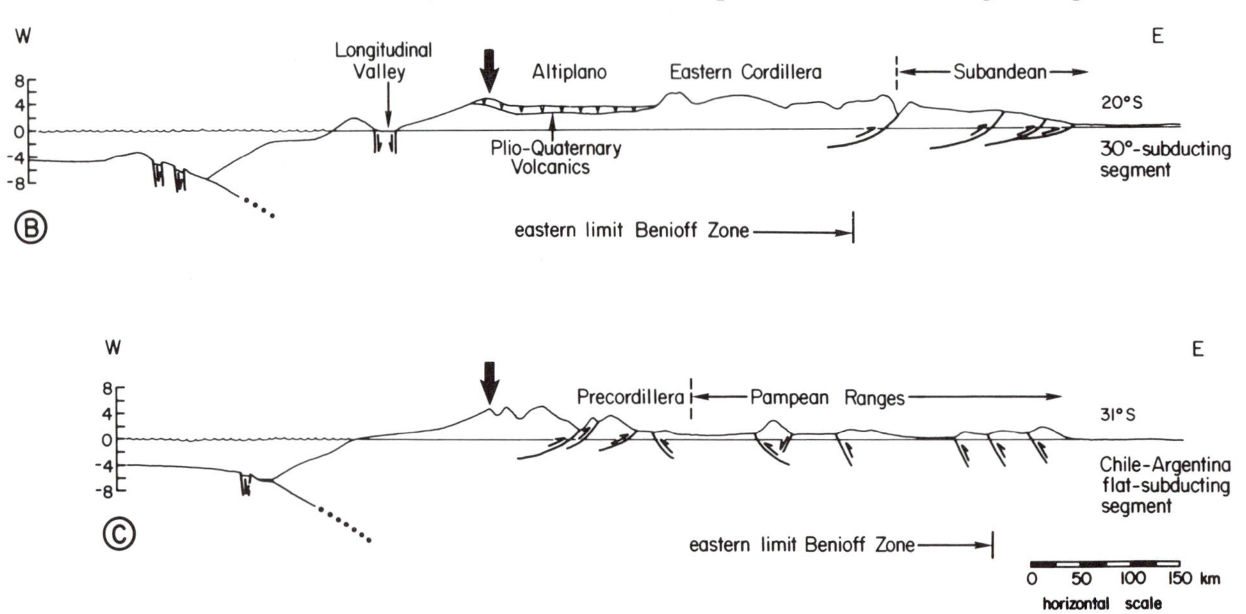

Figure 10-10. Two topographic profiles across the Andes comparing structure above an oceanic slab dipping 30° (thin-skinned tectonics in Subandean belt) and above an oceanic slab that is near flat beneath the continent (thick-skinned tectonics in Pampean Ranges). After Jordan et al. (1983).

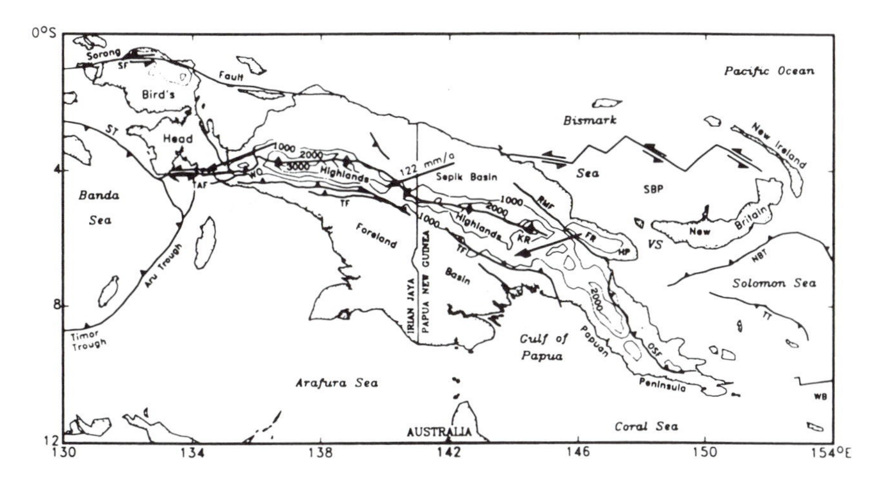

Figure 10–11. Tectonic map of New Guinea showing Pacific-Australian convergence vector, topography with contour interval 1000 m, major thrust faults with barbs on hangingwall side, and Medial Suture zone (barbs on both sides). DP, Darai Plateau; FR, Finisterre Range; HP, Huon Peninsula; KR, Kubor Range; MA, Muller anticline; OSF, Owen Stanley fault; NBT, New Britain trench; RMF, Ramu-Markham fault; SBP, South Bismarck plate; SF, Sorong fault; ST, Seram trough; TAF, Tarera-Aiduna fault; TF, thrust front; TT, Trobriand trough; VS, Vitiaz Strait; WB, Woodlark Basin; WO, Weyland overthrust. Modified from Abers and McCaffrey (1988; 1994).

edge of the Australian continent (Milsom, 1991). The boundary between the Pacific and Australian plates was farther north, and it was characterized by subduction of oceanic lithosphere of the Australian plate beneath island arcs south of the Pacific plate. In the early Miocene, the island arcs of this region began to collide with Australian continental lithosphere. The South Bismarck plate and Solomon Sea plate are remnants of this collision between continental crust and oceanic subduction zones (Cooper and Taylor, 1987; Abers and McCaffrey, 1988; Fig. 10–11).

Between the Aru trough and the Gulf of Papua, the Australian and Pacific plates are converging at a rate of 12 cm/yr in a N70°E direction (Minster and Jordan, 1978), oblique to the collision zone, especially in the Irian Jaya sector of New Guinea (Fig. 10–11). Although the volcanoes of the South Bismarck plate and the eastern Papuan Peninsula still have oceanic crust of the Solomon Sea plate descending beneath them and are active, the remainder of Papua New Guinea and Irian Jaya lacks active volcanoes. The late Cenozoic volcanoes of the Highlands of Irian Jaya and Papua New Guinea do not have a typical island-arc composition and have no evidence of recent eruptions (Milsom, 1991).

The Highlands of New Guinea are marked by a southward-verging, thin-skinned fold-and-thrust belt that is overriding the sediments of a foreland basin to the south (Hamilton, 1979). At the thrust front in Papua New Guinea, the décollement is at 3-5 km depth in Mesozoic or early Cenozoic strata. Farther north, the décollement steps down to 10–15 km (Hill, 1991). Reverse-fault earthquakes in the area of the fold-and-thrust belt are beneath the décollement, probably in crystalline basement, and their fault-plane solutions indicate movement on faults that dip 35-60°, which is too steep to be related to the décollement (Abers and McCaffrey, 1988; Fig. 10–12). These authors suggest that some surface structures

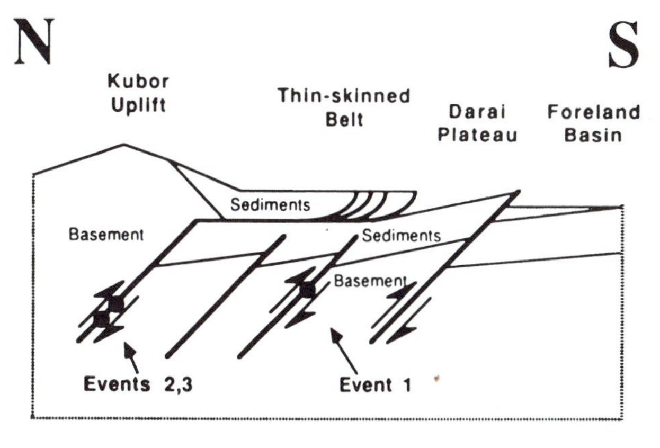

Figure 10–12. Relation of thin-skinned fold-and-thrust belt to seismic reverse faulting in basement in Papua New Guinea. After Abers and McCaffrey (1988).

south of the thrust front may be related to basement-involved reverse faulting, including a fault bounding the Darai Plateau on the south (Fig. 10–12). Northwest of the Darai Plateau, the Muller anticline is much larger than structures above the décollement, and this structure is also related to basement faulting. These reverse faults in basement are reactivated Mesozoic normal faults (Hill, 1991). Figure 10–12 shows an interpretation of the relation of thin-skinned thrusting to basement faulting that produces the earthquakes.

The high obliquity of the Australian-Pacific collision requires a large component of left-lateral slip in the collision zone, particularly in Irian Jaya, where the angle between the convergence vector and the strike of the collision zone is smallest. This is manifested primarily as large strike-slip earthquakes. Furthermore, many small earthquakes in the Highlands are produced by left-lateral strike slip. These tend to be shallower than the reverse-fault earthquakes (Abers and McCaffrey, 1988). However, the fault-plane solutions of earthquakes tend to be either pure reverse-slip or pure strike-slip-fault events rather than oblique-slip events. This suggests a partitioning of the strain release between strike-slip-fault and reverse-fault earthquakes, as illustrated in Figure 10–13. This is discussed later in the chapter.

North Panama Deformed Belt

The Cocos plate is subducting beneath Central America at the Middle America trench, and arc volcanism is active as far east as western Panama (Fig. 10–14). On the northeast side of the volcanic arc, Costa Rica and western Panama are being thrust northeast over the Caribbean plate along a fold-thrust belt, the North Panama Deformed Belt (Silver et al., 1990). This fold-thrust belt reaches the surface mainly offshore except at its western end in Costa Rica, the site of an earthquake of M_s7.5 in April, 1991. Figure 10–14 contains a cross section that intersects the aftershock zone and shows both the volcanic arc and the North Panama Deformed Belt. However, the active volcanic arc is not present farther east where the deformed belt is best developed offshore.

The North Panama Deformed Belt may not extend west of the landward extension of a strike-slip fault on the Hess Escarpment. West of this strike-slip fault, the tectonic environment in the volcanic region is extensional and strike-slip, not compressional (Fig. 10–14). Other complexities include the aseismic Cocos Ridge and the triple junction with the Nazca plate. East of the triple junction, motion on the plate boundary south of Panama is largely right-lateral strike slip.

The earthquake of 1991 produced no observed surface rupture, but measurements of uplifted shoreline features revealed coseismic growth of an anticline, which is best explained as the surface manifestation of slip at depth on a southwest-dipping reverse fault (Plafker and Ward, 1992; Fig. 10–14).

Northeastern Japan

Northeastern Japan provides another example of reverse faulting on the continent side of a subduction zone. The Pacific plate is subducting westward beneath northeastern Japan at a rate of 10.5 cm/yr. A new plate boundary has formed between northeastern Japan and oceanic crust of the Japan Sea, and northeastern Japan may now be part of the North American plate (Nakamura, 1983; Uyeda, 1991; Fig. 10–15). In northeastern Japan, the direction of maximum horizontal stress is east to east-southeast. Repeated measurements of the distance between a VLBI station on the east coast of Honshu and a station on the North American craton indicate that northeast Japan is undergoing tectonic shortening (Heki et al., 1990). This is expressed at the surface by two groups of reverse faults: (1) a thick-skinned fold-and-thrust belt, mainly offshore in the Japan Sea, but extending onshore at its southern end near Niigata (Fig. 10–16), and (2) a region within the active volcanic arc of northern Hon-

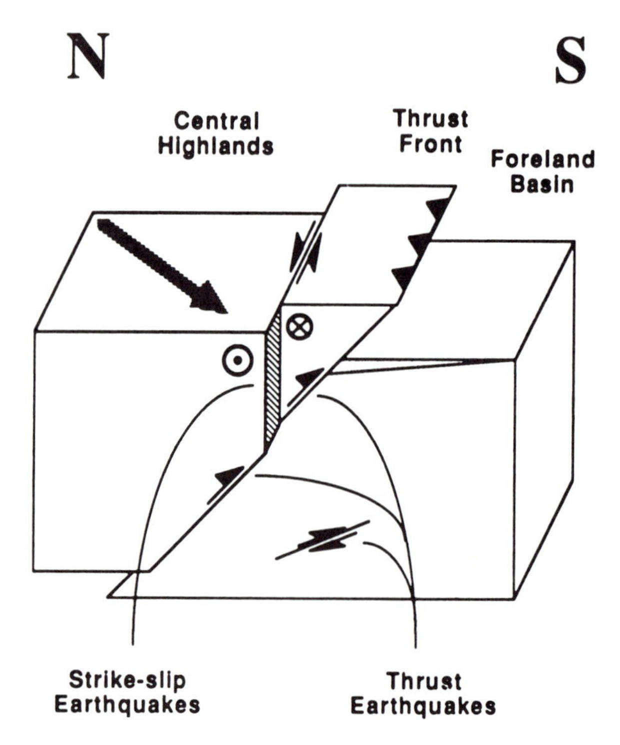

Figure 10–13. Partitioning of strain between strike-slip and reverse faulting in Irian Jaya due to highly oblique convergence between Australian and Pacific plates. After Abers and McCaffrey (1988).

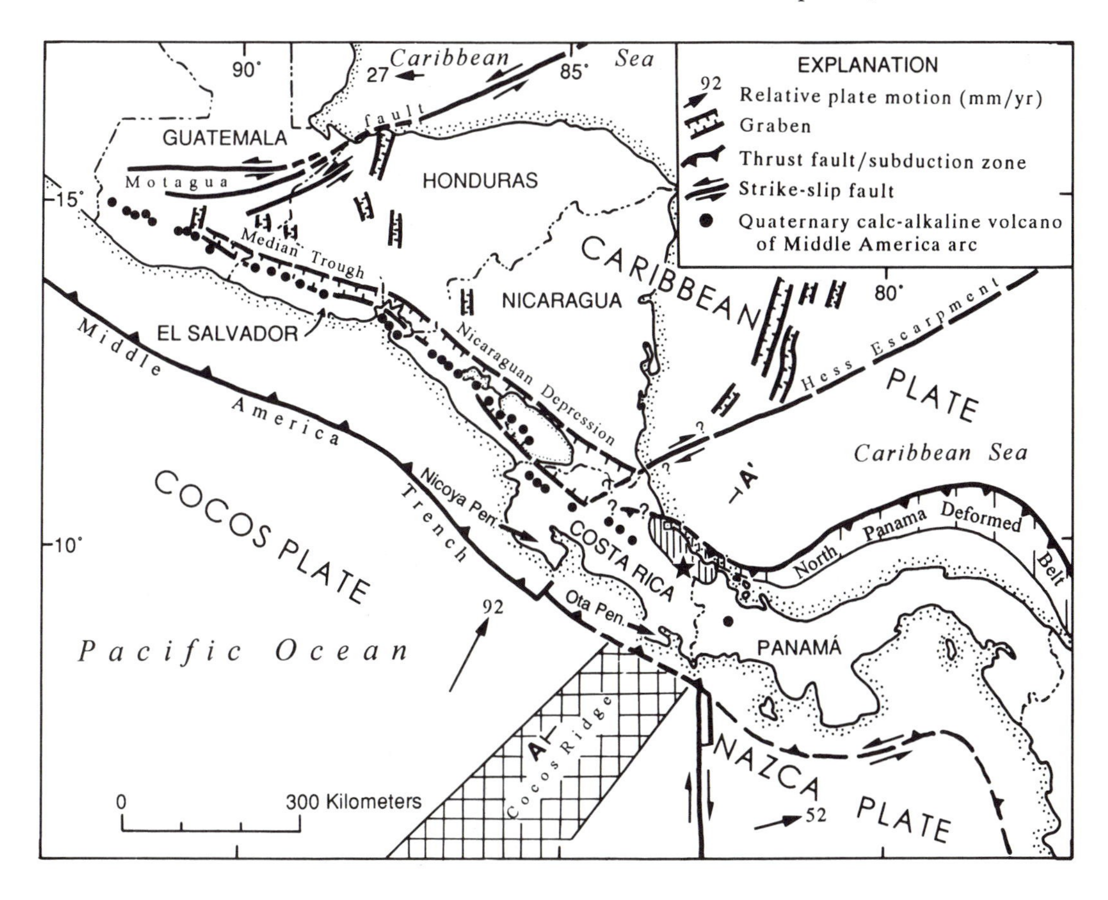

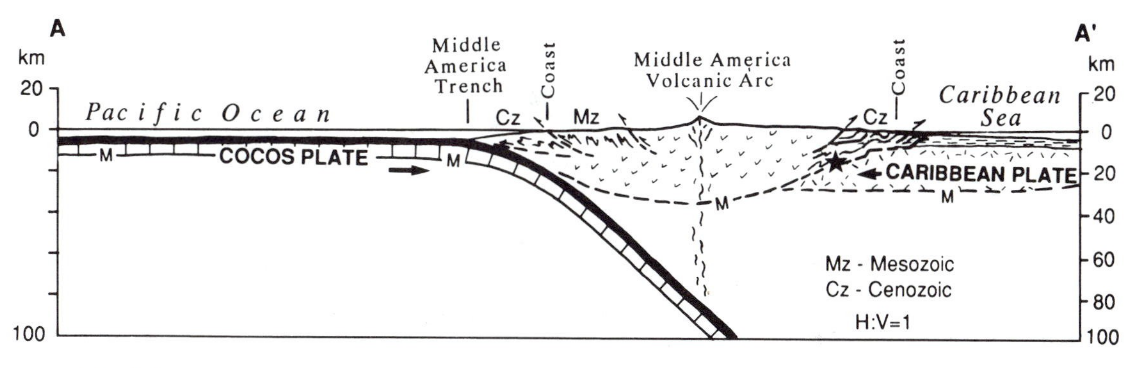

Figure 10–14. (Top) Tectonic setting of the April, 1991, earthquake (mainshock shown by star, aftershock zone shown as vertical line pattern) at the west end of the North Panama Deformed Belt. (Bottom) Cross section through A-A′ showing relations among subduction zone, volcanic arc, and North Panama Deformed Belt, with 1991 earthquake shown by star. From Plafker and Ward (1992).

shu with alternating ranges and basins, that has range fronts delineated by reverse faults (Figs. 10–15; 10–41).

Large earthquakes have struck both regions. The 1964 Niigata earthquake (M7.5) and the 1983 Japan Sea earthquake (M7.8; Ishikawa et al., 1984; Satake, 1985) occurred beneath the fold-thrust belt. Farther north, the 1993 Hokkaido-Nansei-oki earthquake (M_S7.8) off the west coast of Hokkaido may be related to convergence between Hokkaido and the Japan Sea. The 1896 Rikuu earthquake occurred on the Senya range-front fault separating the Mahiru Range from the Yokote basin (Yamasaki, 1896; Ikeda, 1983; Fig. 10–44).

The 1964 Niigata earthquake produced 3–5 m of upwarping of the sea floor, but no significant surface faulting (Satake and Abe, 1983; Kawasumi, 1973), and Awashima Island was tilted and uplifted 1.5 m (Stein and King, 1984; Fig. 10–16). Folds in Quaternary sediments on the sea floor are the result of slip on underlying blind reverse faults in crystalline basement, indicating that the Niigata earthquake was not related to deformation on a thin-skinned fold-thrust belt (Okamura et al., 1994). To the east, folded river terraces have been described by Ota (1969) and Ota and Suzuki (1979).

Another zone of active folding is in the Ishikari Lowland of southwestern Hokkaido where the convergence rates are considerably less than in the fold-thrust belt on the west side of northern Honshu. The Stage 5e marine terrace (125 ka) shows effects of deformation across this zone (K. Okumura, unpublished data).

Reverse Faults in Strike-Slip Environments

Most intraplate regions, including stable continental shields, are dominated by *in situ* stress fields in which

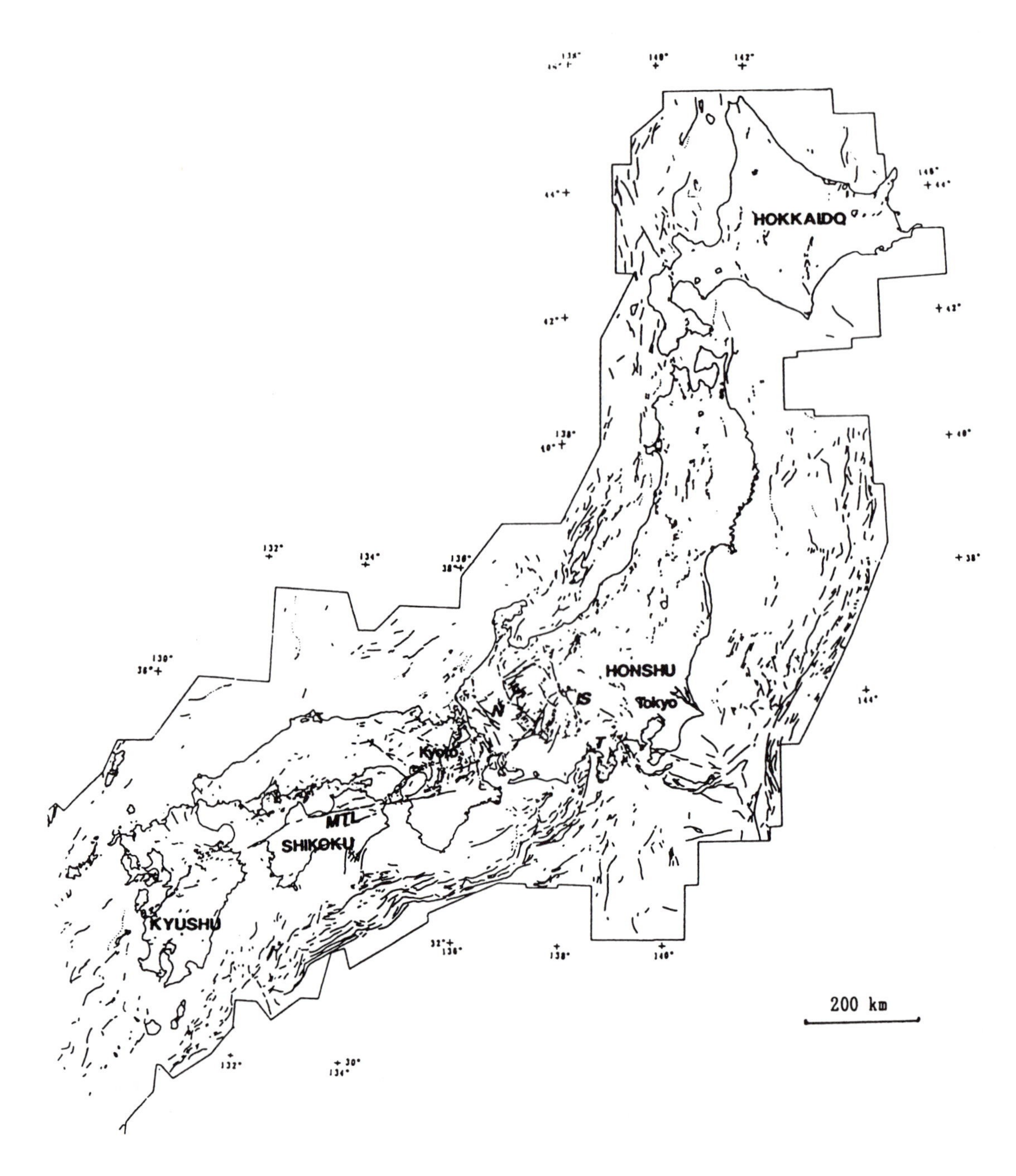

Figure 10–15. Active faults of Japan. IS, Itoigawa-Shizuoka Tectonic Line; MTL, Median Tectonic Line. From Matsuda and Kinugasa (1991) and Research Group for Active Faults of Japan (1992).

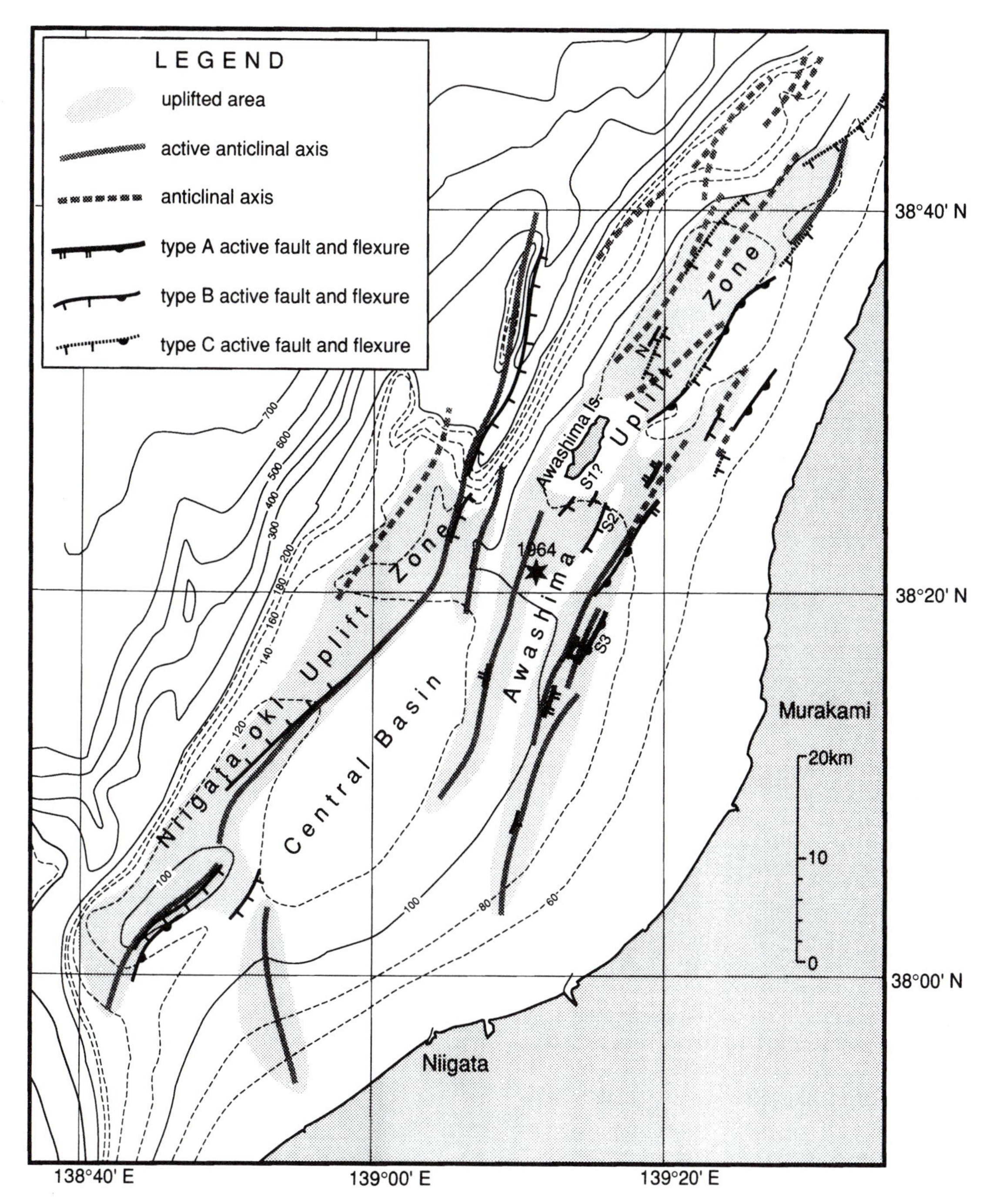

Figure 10–16. Awashima uplift, northern Japan, raised by the 1964 Niigata earthquake. Surface expression mainly broad folds and secondary faults; seismic source faults are largely blind. Note low ratio of length to breadth of individual structures. Niigata-oki uplift was unaffected by the 1964 earthquake. Type A, B, and C structures shown by decreasing geomorphic evidence of recent activity. N, S1, S2, and S3 locate faults based on seismicity. After Okamura et al. (1994).

the maximum compressive stress is horizontal (Zoback, 1992). Accordingly, it is not surprising that several zones of convergence are found in settings other than convergent plate boundaries. In New Zealand and California, convergence zones have formed near transform faults. In western China, Iran, the Caucasus, and the Columbia Plateau, convergence zones are associated with broader zones of faulting in continental interiors, including, but not limited to, large strike-slip faults. In a subsequent section, we discuss active reverse faults within continental shields.

California Coast Ranges

The San Andreas fault is the principal fault marking the transform boundary between the Pacific and North American plates in California. The fault strikes about N40°W in northern and central California, about 5° counterclockwise to the slip vector between the two plates. In southern California, the strike of the San Andreas fault is N65°W, even farther from the trend of the plate slip vector. This departure results in major contractional structures in the Transverse Ranges.

In the Coast Ranges of central California, fold-and-thrust belts have developed on both sides of the San Andreas fault. The vergence of structures in these belts is away from the San Andreas fault. The geometry of these fold-and-thrust belts produces a paradox. Right-lateral strike slip on a NW-striking fault implies a N-S orientation of σ_1, whereas the orientation of folds and thrusts implies a NE-SW orientation of σ_1. Wilcox et al. (1973) and Harding (1976) tried to resolve this problem by considering the folds as secondary structures to the strike-slip faults, a tectonic style they described as *wrench tectonics*. However, borehole breakouts show that σ_1 is oriented nearly perpendicular to the San Andreas fault (Fig. 5-30; Zoback et al., 1987; Mount and Suppe, 1992). A sequence of earthquakes between 1982 and 1985, including the 1983 Coalinga earthquake (M6.7; discussed in the following text under Case Histories), also indicates tectonic transport at nearly right angles to the San Andreas fault, based on earthquake fault-plane solutions and geodetic modeling of fault planes (Fig. 10-17; Stein and Ekström, 1992). Geodetic and seismicity data are compatible with geologic studies of Coast Range folds adjacent to the Great Valley of California using seismic-reflection and borehole data (Namson and Davis, 1988; Medwedeff, 1989; Wentworth and Zoback, 1989). The Coast Range folds in the vicinity of Coalinga, which are developed in strata of Mesozoic and Cenozoic age, appear to be underlain by a northeast-tapering wedge of metamorphosed Mesozoic accretionary-wedge deposits (Franciscan Formation) overlying crystalline basement (Wentworth and Zoback, 1990; Fig. 10-17).

Zoback et al. (1987) resolved the paradox by considering the San Andreas fault to be a very weak fault. Strain is apparently partitioned between strike-slip earthquakes on the San Andreas fault such as the 1857 earthquake (M_w 7.9) and dip-slip earthquakes beneath the Coast Ranges fold-thrust belt, such as the 1982–1985 earthquake sequence.

Transverse Ranges of California

In southern California, the anomalous N65°W strike of the San Andreas fault is referred to as the "Big Bend." The ranges flanking the "Big Bend" strike E-W, sharply truncating the NW-SE grain of the Coast Ranges to the north (Crowell, 1987; Fig. 10-18). The San Gabriel, Topatopa, and Santa Ynez mountains are bounded on the south by north-dipping range-front reverse faults, part of a discontinuous system of faults

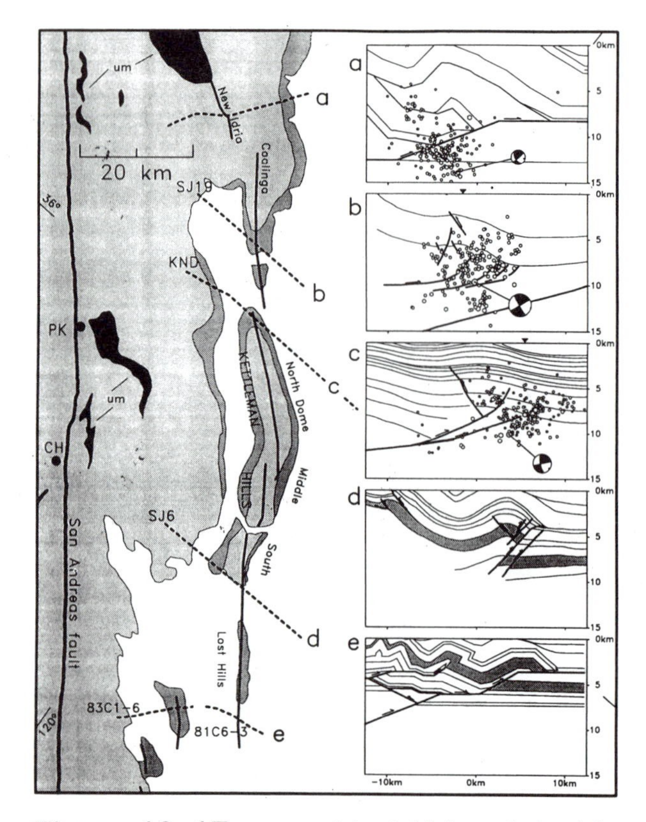

Figure 10–17. Part of the fold-thrust belt of the California Coast Ranges east of the San Andreas fault with balanced cross sections constrained by well and seismic-reflection data. Cross sections a, b, and c show back-hemisphere fault-plane solutions and aftershocks of earthquakes of the 1982–1985 sequence. Arrowheads on cross sections b and c locate peak coseismic uplift accompanying the 1983 Coalinga and 1985 Kettleman Hills earthquakes. Intermediate-weight solid lines mark anticlinal axes. Cross section a from Namson and Davis (1988), b from Wentworth and Zoback (1989), c from unpublished Ph.D. dissertation by A. Meltzer, d from unpublished work by R. Bloch, and e from Medwedeff (1989). Figure after Stein and Ekström (1992).

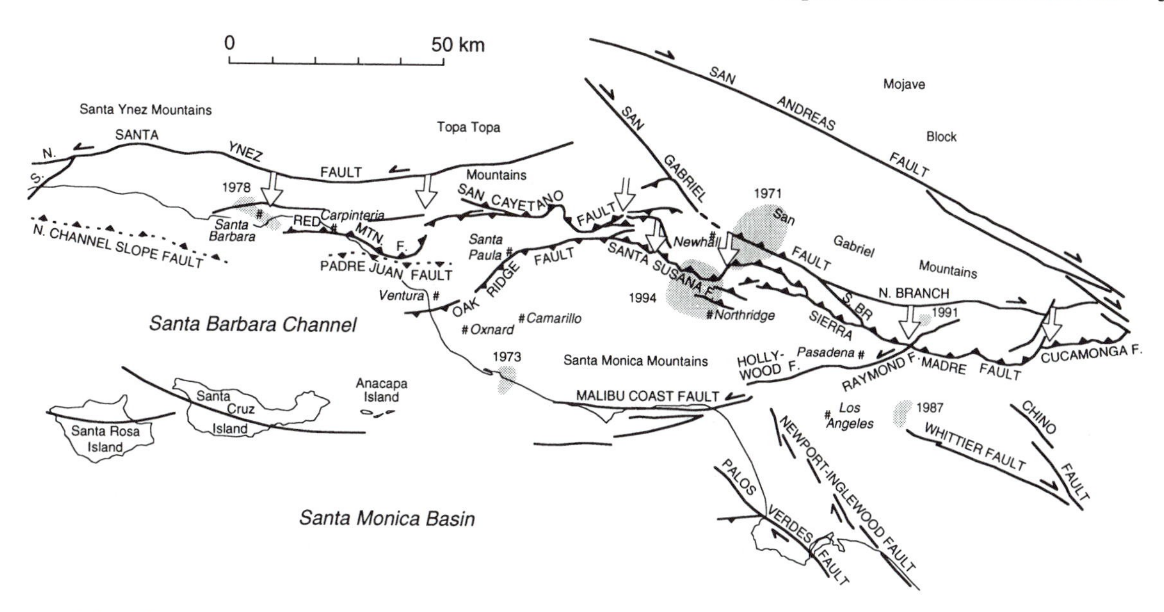

Figure 10–18. Reverse faults (teeth toward hangingwall) and earthquakes (dot pattern; numbers give year) of western Transverse Ranges, California. Half arrows mark strike-slip faults. Open arrows mark proposed segment boundaries.

that extends from the Santa Barbara Channel eastward to an intersection with the San Andreas fault at the eastern end of the San Gabriel Mountains. Other important reverse faults include the Pleito fault in the southern margin of the South San Joaquin basin, the south-dipping Oak Ridge fault in the Ventura basin which extends eastward to the San Fernando Valley as a blind thrust that produced the 1994 Northridge Earthquake, and a blind reverse-fault system beneath the Santa Monica Mountains north of the Los Angeles basin (Fig. 10-18). The basins within and flanking the Transverse Ranges have thick accumulations of Pliocene and Pleistocene strata. The Pleistocene section in the Ventura basin, caught between north-dipping and south-dipping reverse faults, is as much as 5 km thick, the thickest Pleistocene section in the world (Yeats, 1983).

Major earthquakes generated by these reverse-faults include the 1952 Kern County earthquake in the South San Joaquin Valley (M_s 7.7), the 1971 San Fernando earthquake at the eastern edge of the Ventura basin (M_w 6.7), the 1978 Santa Barbara earthquake in the western Ventura basin (M_L 5.9), the 1987 Whittier Narrows earthquake in the Los Angeles basin (M_L 5.9), the 1991 Sierra Madre earthquake at the southern edge of the San Gabriel Mountains northeast of Los Angeles (M_L 6.0), and the 1994 Northridge earthquake in the San Fernando Valley (M_w 6.7). Of these, only the 1952 and 1971 earthquakes produced surface rupture. The Northridge earthquake was the most costly earthquake in U.S. history.

Slip rates on the Oak Ridge fault in the Ventura basin are 5 mm/yr, and slip rates on the Red Mountain and San Cayetano faults are probably as high (Yeats, 1988; 1993). GPS satellite geodesy confirms the high convergence rate across the Ventura basin due to slip on these faults. The Ventura basin is closing at a rate of 7-10 mm/yr (Donnellan et al., 1993), with somewhat slower rates farther west in the Santa Barbara Channel (Larson and Webb, 1992). Convergence rates across the northern Los Angeles basin are about 5 mm/yr based on GPS data (Feigl et al., 1993).

Borehole breakouts and fault-plane solutions of earthquakes indicate that σ_1 is nearly normal to the strike of the "Big Bend" of the San Andreas fault rather than normal to the structural grain of the Transverse Ranges (Mount and Suppe, 1992). However, the Transverse Ranges seem to be more than a simple byproduct of the big bend of the San Andreas fault. The west-trending ranges extend considerably farther west than would be predicted by secondary effects of the "Big Bend."

New Zealand

Like California, New Zealand is crossed by a major transform fault, the Alpine fault (cf. Chapter 8; Figs. 8-3, 8-4). Unlike California, the slip vector between the plates (Pacific and Australian) is becoming more divergent from the trend of the Alpine transform fault. Currently, the relative plate motion vector is at an angle of more than 30° with the Alpine fault in the South Island of New Zealand. Accordingly, the Alpine fault is chang-

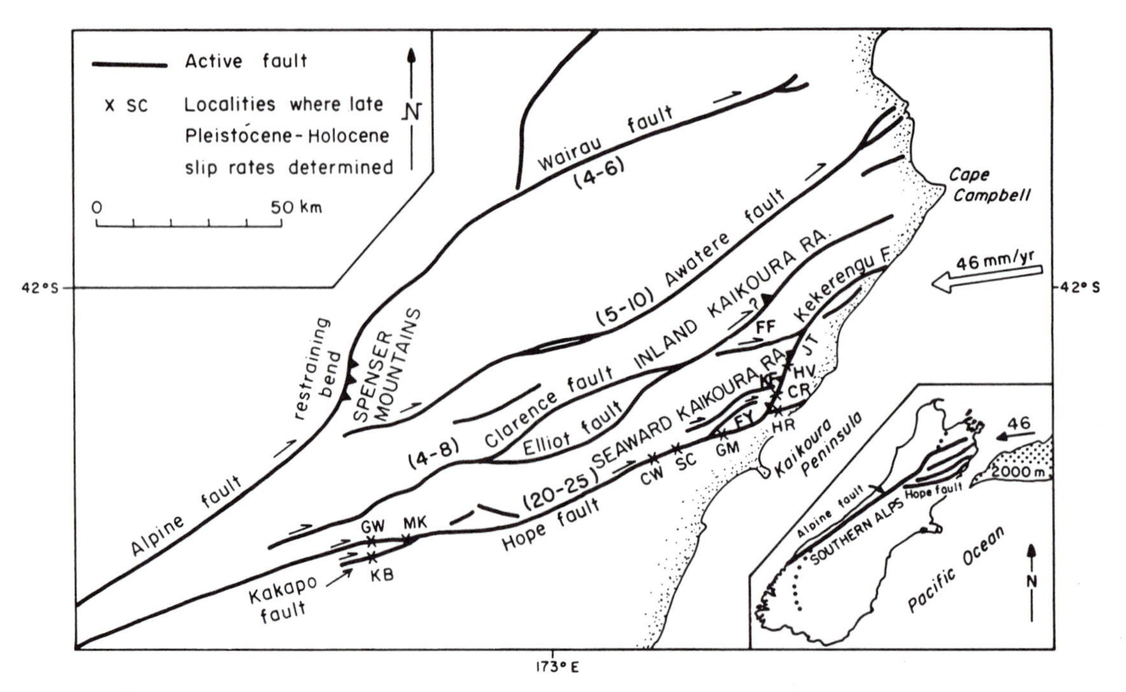

Figure 10–19. Map of northeastern South Island, New Zealand, showing Marlborough strike-slip faults with slip rates in mm/yr. Teeth marks point toward hangingwalls of reverse faults, which form in restraining bends of strike-slip faults. JT = Jordan thrust, which is uplifting the Seaward Kaikoura Range at a rate of 6–10 mm/yr. After Van Dissen and Yeats (1991).

ing to an oblique-slip reverse fault on which the Southern Alps are ramping upward over the Australian plate at rates of more than 4 mm/yr (inset to Fig. 10-19).

Reverse-fault provinces are present on both sides of the Alpine fault. In the Nelson area, on the northwest side, N- to NNE-striking reverse faults produced the 1929 Murchison earthquake (M7.8) and the 1968 Inangahua earthquake (M7.1), both of which were accompanied by surface rupture (Yeats and Berryman, 1987; Fig. 10-20). Deformation associated with the reverse faults has produced ranges and basins that are oblique to the Alpine fault; the system could be explained by wrench faulting. In Central Otago and adjacent areas, the reverse faults are nearly parallel to the Alpine fault, but range-and-basin topography has also developed there (Berryman and Beanland, 1991; Figs. 10-21). The seismicity of this region is low, but paleoseismic excavations have revealed evidence of Holocene movement on several of these faults.

Reverse faulting is known elsewhere in New Zealand. The strike-slip faults in the Marlborough region in the northeasternmost South Island are closer to parallel to the Pacific-Australian slip vector than is the Alpine fault to the southwest. Near the coast, the Hope fault bends northeastward into the Jordan

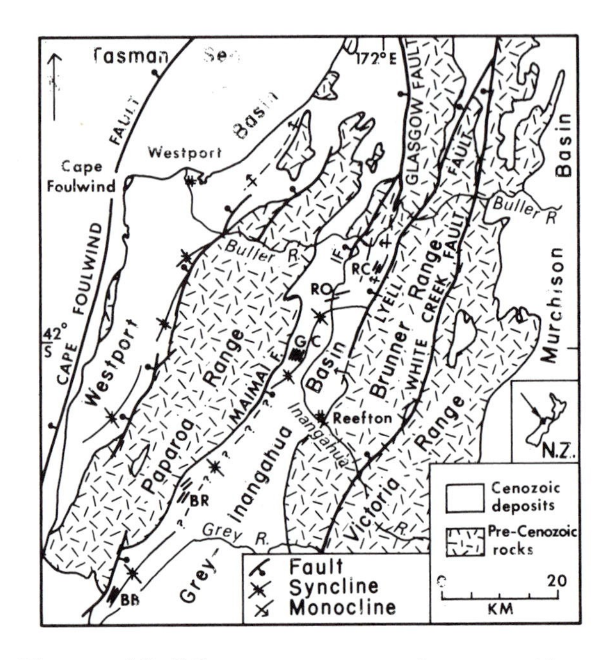

Figure 10–20. Tectonic map of range-and-basin province of northwest Nelson, South Island, New Zealand, location of 1929 and 1968 reverse-fault earthquakes. RO, RC, flexural-slip faults active in 1968; BB, BR, GC, prehistoric flexural-slip faults (see Fig. 10–50a). Arrow on monocline points toward downwarped side. After Yeats (1986).

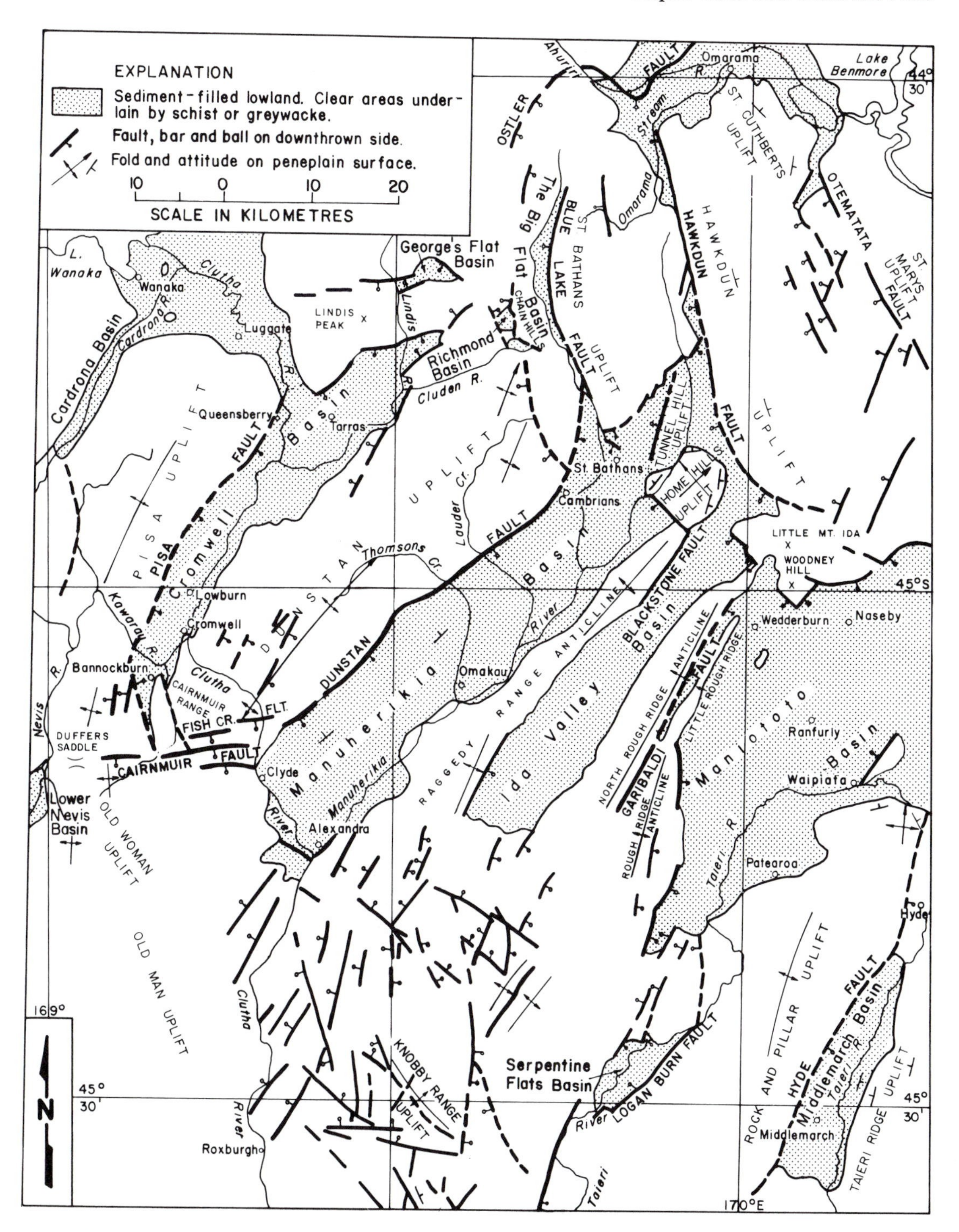

Figure 10–21. Tectonic map of range-and-basin province of Central Otago, South Island, New Zealand, based on Landsat. After Yeats (1987).

thrust, along which the Seaward Kaikoura Range is ramped upward at a rate as high as the rate of uplift in the Southern Alps (Van Dissen and Yeats, 1991; Fig. 10–19). In the Wanganui basin of the North Island, active folds have developed parallel and oblique to a set of strike-slip faults (Te Punga, 1957; Pillans, 1990).

Central Asia

Contractional structures formed as the principal result of continent-continent collision have been discussed in the previous text for the India-Eurasia and the Arabia-Eurasia collisions. North of these collision zones in western China and adjacent parts of Kazakhstan, Kyrgyzstan, and Tajikistan, the Eurasian plate has broken up into several large blocks that are moving east and west, out of the way of the rigid, northward-driving Indian and Arabian continental plates (Molnar and Tapponnier, 1975; Avouac and Tapponnier, 1993; Fig. 8-9). The most prominent block boundary faults are strike slip, but where the boundaries are irregular and trend east-west, fold-and-thrust belts have developed. Several of the more important fold-and-thrust belts north of the Indian plate are mentioned in this section and shown in Fig. 10-22. Fold-and-thrust belts north of the Arabian plate are discussed in the following section.

The Tianshan (Tien Shan) is the largest mountain range in central Asia north of the Himalaya. It extends from Kyrgyzstan and Tadjikistan eastward for 2000 km across Xinjiang Province, China, and separates two relatively stable blocks, the Tarim block on the south and the main mass of Eurasia on the north (Fig. 10-23). Evidence of active folding and thrusting is interpreted from Landsat and Spot imagery and high seismicity in the northern foothills adjacent to the Junggar (Dzungaria) basin (Avouac et al., 1993) and in the southern foothills adjacent to the Tarim basin (Tapponnier and Molnar, 1979; Fig. 10-24). The Tianshan is wedge shaped, tapering eastward, due to clockwise rotation of the Tarim basin to the south (Avouac et al., 1993). This rotation causes more shortening in the western Tianshan than in the east, and seismicity and active deformation are also greater toward the west.

The fold-thrust belt in the southern foothills is about 1000 km long in Xinjiang Province, and it continues westward into Kyrgyzstan and Tadjikistan, where late Quaternary folding and reverse faulting have been documented (Trifonov, 1978). Active, asymmetric anticlines are present in Xinjiang, with steeper south limbs accompanied by north-dipping reverse faults; the southernmost row of folds is the youngest (Feng et al., 1991; Avouac et al., 1993). Seismicity is high along this zone, including an earthquake of $M \approx 8$ near Kashgar in 1902 (Tapponnier and Molnar, 1979).

In the northern Tianshan foothills, three rows of northward-verging folds are developed, and the two northernmost rows of folds are the youngest (Feng et al., 1991; Avouac et al., 1993). The anticlines are re-

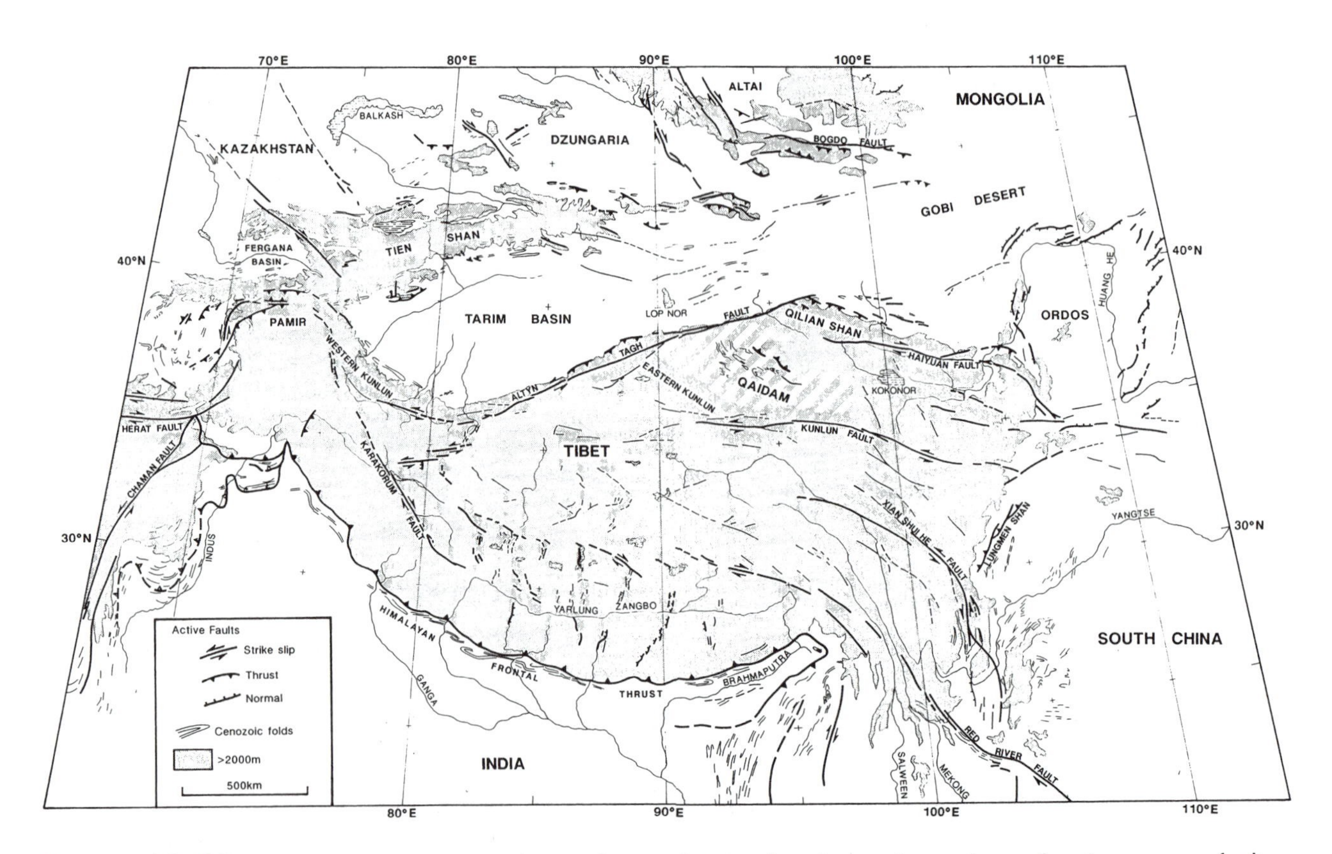

Figure 10–22. Active tectonic map of part of central Asia, showing major regions of active reverse faulting. These include the Himalayan frontal thrust system (cf. Fig. 10–3), the Pamirs, the southern and northern foothills of the Tien Shan (Tianshan), the Qilian Shan between the Altyn Tagh and Haiyuan strike-slip faults, and the Lungmen Shan at the eastern edge of the Tibetan Plateau. After Avouac and Tapponnier (1993).

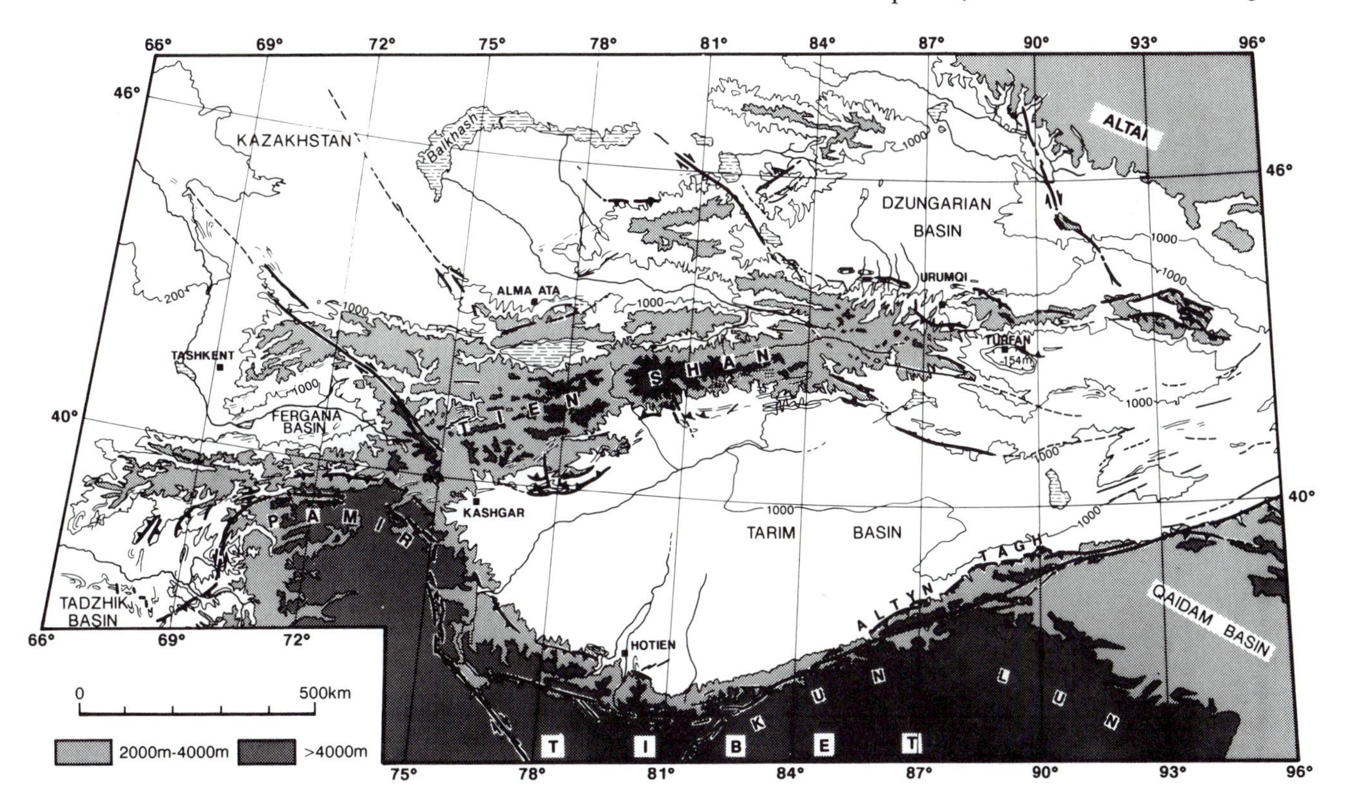

Figure 10–23. Tectonic map of the Tien Shan (Tianshan) of central Asia, which separates stable Eurasia on the north and the stable Tarim basin on the south. Both the Tarim basin and Dzungarian (Junggar) basin have maximum sediment thicknesses near the range front. Active reverse faults and folds characterize the north flank of the Tien Shan flanking the Dzungarian basin and the south flank northeast of Kashgar. After Avouac (1991).

lated to thrust faults at depth, some of which reach the surface; the anticlines and south-dipping thrusts flatten into a décollement at 7–8 km depth (Deng et al., 1991a, b; Fig. 10–24). The entire Tianshan is interpreted as a crustal-scale ramp anticline with an associated downwarp due to crustal thickening (Avouac et al., 1993). Large earthquakes have occurred in the eastern Tianshan (Tapponnier and Molnar, 1979), one of which produced surface rupture on a fault within one of the foreland anticlines in 1906 (Avouac et al., 1993).

East of the Tianshan, the ENE-striking Altyn Tagh strike-slip fault turns abruptly to become the ESE-striking Haiyuan strike-slip fault parallel to the northern margin of the Qilian Shan (Fig. 10–22). Where the strike changes, a set of ranges and basins bounded by reverse faults has developed (Tapponnier and Molnar, 1977). These include, from northeast to southwest, the Yumu Shan, Qilian Shan, and several ranges comprising the Nan Shan (Fig. 10–25). These ranges appear to be related to basement ramps rising from a décollement at the base of brittle crust (Tapponnier et al., 1990). Farther southwest, the Qaidam basin contains active folds and appears to be a cross-strike continuation of this reverse-fault zone. This fold-and-thrust belt was the site of the 1927 Gulang reverse-fault earthquake (M8) and the 1932 Changma earthquake (M_s 7.5), a complex event with both reverse-fault and strike-slip surface rupture (Peltzer et al., 1988; Meyer, 1991; Hou, 1992). An earthquake of M7.5 in A.D. 180 resulted from rupture of a thrust fault at the northern piedmont of the Yumu Shan (Tapponnier et al., 1990). The earthquakes in A.D. 180 and 1927 had relatively short surface rupture length compared to their magnitude.

At the eastern edge of the Tibetan plateau, the Lungmen Shan marks an isolated region of active thrust faulting and folding (Fig. 10–22; Chen et al., 1994). This convergence zone may be related to eastward motion of the Tibetan plateau along the Kunlun strike-slip fault.

Caucasus

North of the Bitlis and Zagros collision zones, the Caucasus region is converging northward against the Russian platform, closing an ocean basin that formerly extended from the Black Sea to the southern Caspian

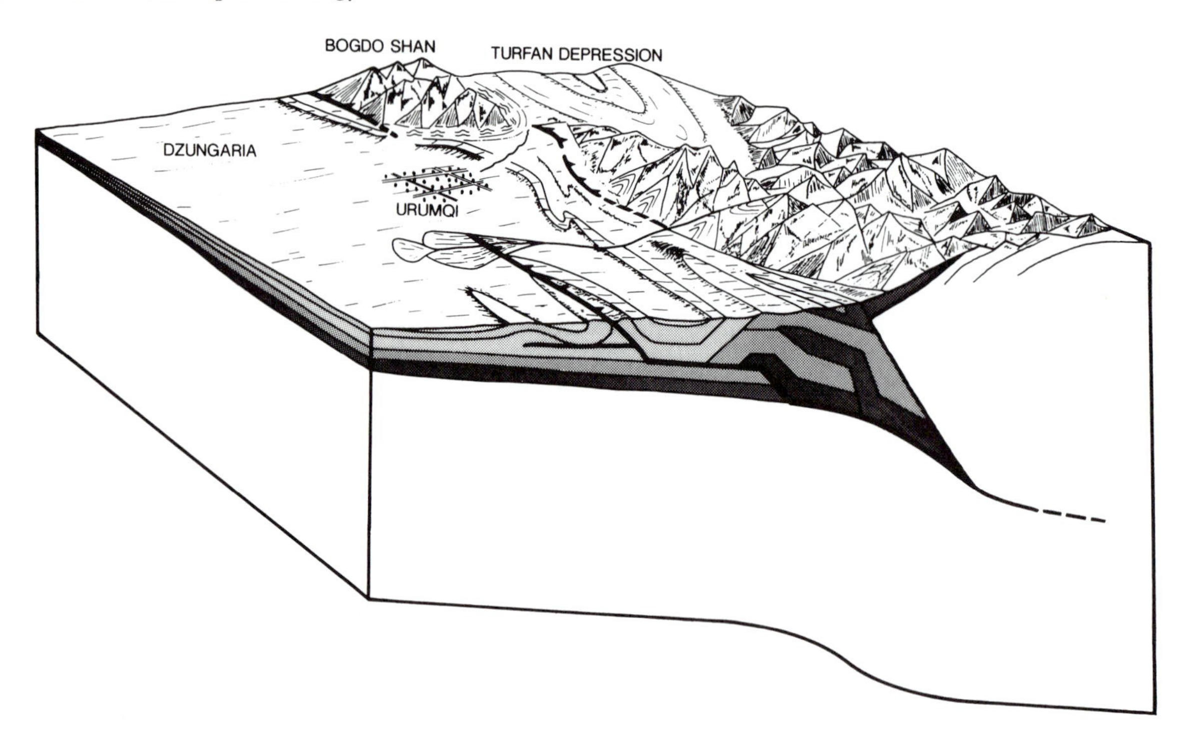

Figure 10–24. Block diagram showing balanced cross section of northern Tien Shan near Urumqi (located on Fig. 10–23). Active anticlines are interpreted as surface expression of thrust faults that ramp upward toward the surface and migrate toward the basin with time. The Tien Shan itself rides upward on a ramp that involves the entire brittle crust. Sedimentary rocks are Permian and younger, with the lightest pattern representing Upper Cretaceous and younger strata. After Avouac et al. (1993).

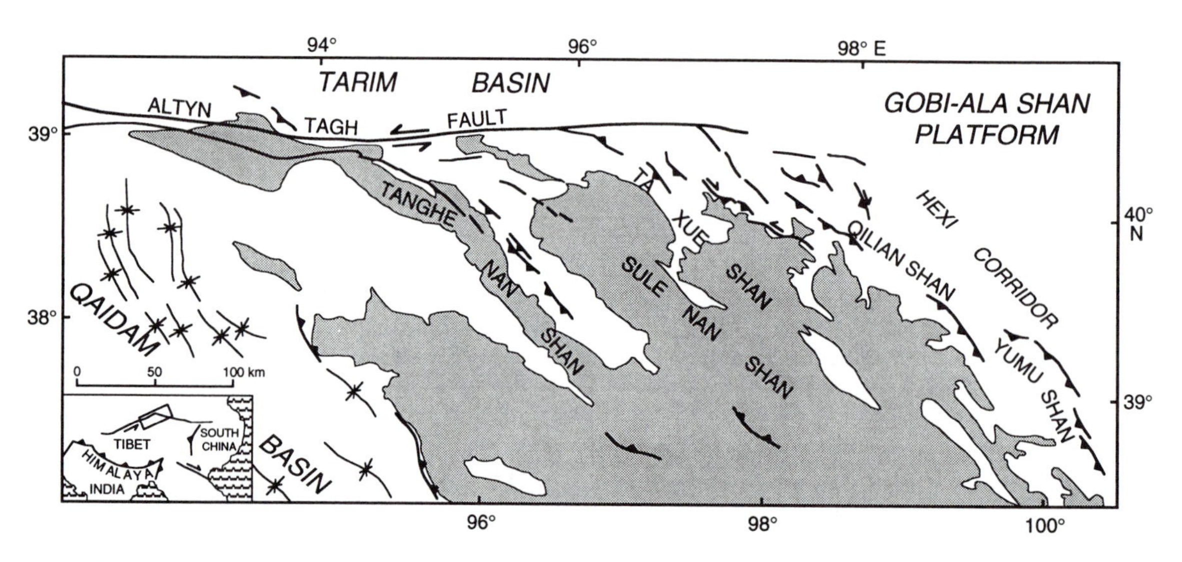

Figure 10–25. Ranges and basins bounded by reverse faults at the northeastern end of the Altyn Tagh left-slip fault, western China. Strike slip on the main part of the Altyn Tagh fault (cf. Fig. 10–22) diminishes northeastward due to shortening in the Nan Shan, Qilian Shan, and Yumu Shan. Areas higher than 3660 m (12,000 feet) are shaded. Synclinal axes show subsurface active folds in the Qaidam basin. After Tapponnier et al. (1990).

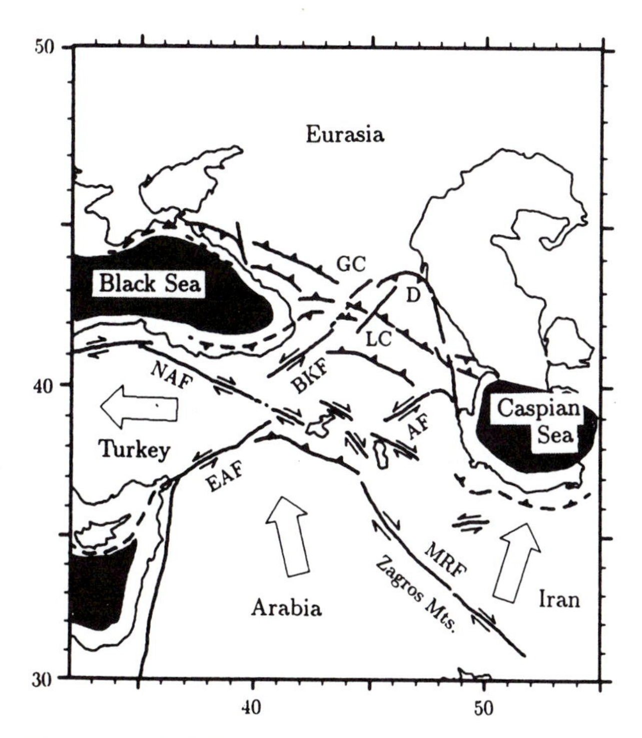

Figure 10–26. Tectonic setting of the Great Caucasus (GC) and Lesser Caucasus (LC) region, showing reverse faults (solid line, teeth toward upper plate) and strike-slip faults (solid line, arrows) that have developed as a result of Arabia-Eurasia collision. Large open arrows show motions of Arabia, Turkey, and Iran relative to Eurasia. AF, Araxes River fault; BKF, Borzhomi-Kazbeg fault; D, Dagestan; EAF, East Anatolian fault; NAF, North Anatolian fault. From Jackson (1992) as modified from Philip et al. (1989).

Sea. As described by Philip et al. (1989) and illustrated in Figure 10–26, the Caucasian collision has now split the ocean basin in two. The Anatolian (Turkey) block is being pushed westward, out of the way, and deformation in the Caucasus is being accommodated by north-facing and south-facing thrust faults and by left-slip and right-slip faults (Fig. 10–26). In the southeast Caucasus near the Caspian Sea, a fault cuts Holocene sediments and shows evidence of oblique reverse and right-lateral offsets (Trifonov, 1978).

The 1991 Racha earthquake in the Great Caucasus (M_S 7.0) occurred at relatively shallow depths on a thrust fault dipping 20–30°, producing no detectable surface rupture (Triep et al., 1995). The 1988 Spitak, Armenia, earthquake in the Lesser Caucasus produced reverse- and right-slip displacements (see discussion in the following text). The thrust belt is offset to the Alborz Mountains by a strike-slip fault south of the Caspian Sea, where the Rudbar-Tarom earthquake (M_S 7.7) occurred in 1990. However, even though this earthquake occurred in a region characterized by reverse faults and folds, it was a strike-slip event, which is evidence that strain is partitioned in this region (Berberian et al., 1992; Jackson, 1992).

Northeastern Iran

The Khorassan region is less seismically active than the Zagros, but it has been struck by many earthquakes, some with surface rupture. Faults striking E-W have left-slip, faults striking N-S have right-slip, and faults striking NW-SE are range-bounding reverse faults (Berberian, 1981; Jackson and McKenzie, 1984; Fig. 10–27). The 1978 Tabas-e-Golshan earthquake (M_S 7.7) was associated with a surface rupture 85 km long at the front of the Shotori Range, and the Ferdows area was struck by reverse-fault earthquakes in 1947 and 1968 (Berberian, 1979).

Kinki Triangle, Japan

The 350-km-long Median Tectonic Line (MTL, Fig. 10–15) is the longest strike-slip fault in Japan. The active fault follows an old zone of weakness, a tectonic boundary of Mesozoic and Tertiary age, also called

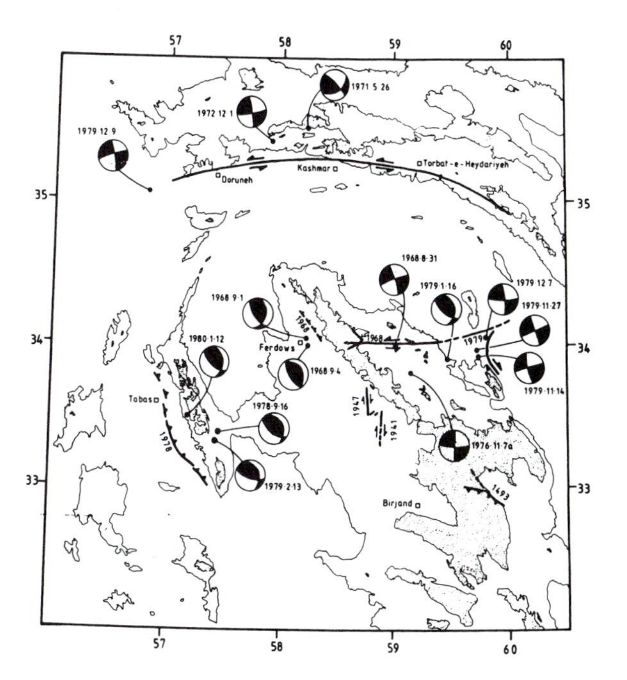

Figure 10–27. Active faults and earthquake focal mechanisms, with dates of earthquakes, in northeastern Iran. Contours at 4000 and 6000 feet; shaded above 6000 feet. Except for Doruneh fault, solid lines show faults with surface rupture in year given; reverse fault, teeth toward hangingwall; strike-slip fault, arrows. From Jackson and McKenzie (1984).

MTL, between two contrasting assemblages of metamorphic rocks. The older MTL continues east from Shikoku and turns northward to join the active Itoigawa-Shizuoka Tectonic Line (IS, Fig. 10-15). However, the degree of activity on the MTL decreases eastward, and it is last recognized as an active fault south of the Nara basin (Fig. 10-28). North of the MTL, shortening in the Kinki Triangle along reverse faults bounding Osaka plain, Nara basin, Lake Biwa, and Ise Bay has formed an alternation of basins and ranges (Sangawa, 1986). In contrast, very few active faults are mapped south of this part of the MTL (Matsuda and Kinugasa, 1991; cf. Fig. 10-15). Because one side of the MTL undergoes contraction, and the other side does not, strike slip on the MTL must increase from near zero south of the Nara basin westward to a maximum on Shikoku. In the Kinki Triangle, there are right-slip faults parallel to reverse faults, as shown on Figure 10-28 west of Lake Biwa.

Yakima Fold Belt, Columbia Plateau, Northwestern U.S.A.

The Miocene Columbia River Basalt Group contains a fold belt developed under north-south compression. As described by Reidel et al. (1989), the fold belt comprises asymmetrical, ridge-forming anticlines separated by broad synclinal valleys. Vergence is predominantly to the north except for the Columbia Hills structure near the Columbia River, which verges southward. The fold belt is transected by the NW-SE-trending Olympic-Wallowa lineament, which is locally called the Cle Elum-Wallula deformed zone and Rattlesnake-Wallula alignment (Fig. 10-29). This lineament may conceal a throughgoing strike-slip fault. Surface structure bears a resemblance to a thin-skinned fold-thrust belt, but subsurface data are inadequate to preclude the possibility of basement involvement. The folds began to develop during eruption of the Columbia River Basalt Group, but growth of anticlines has continued into the late Quaternary at Toppenish Ridge (Campbell and Bentley, 1981) and the eastern end of Saddle Mountains.

The Yakima fold belt is seismically more active than other parts of the Columbia Plateau, with some of the earthquakes occurring as swarms. Most earthquakes are less than 8 km deep (Ludwin et al., 1991), indicating that they are within the Columbia River Basalt Group. Fault-plane solutions indicate reverse faulting

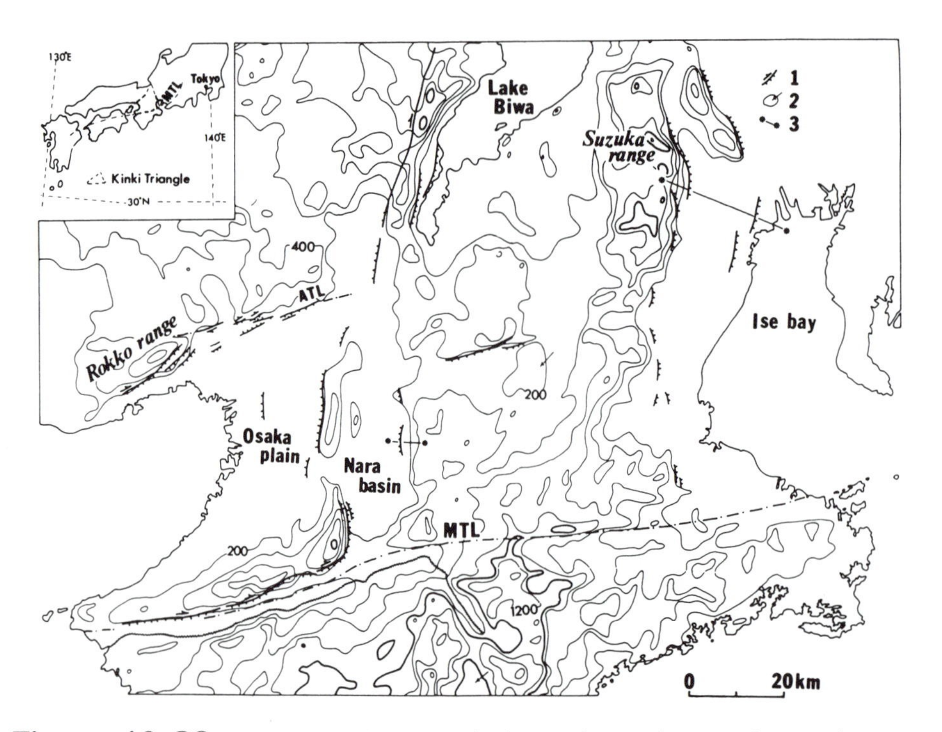

Figure 10–28. Relation of reverse faults in the Kinki triangle, southwestern Japan, to the Median Tectonic Line (MTL), a right-slip fault. Degree of activity decreases east of Nara basin. Active faults are shown with solid heavy line, tick marks on downthrown side. Strike-slip faults are shown with arrows. Fault at base of Rokko Range ruptured in the 1995 Hyogo-ken Nanbu (Kobe) earthquake. From Sangawa (1986).

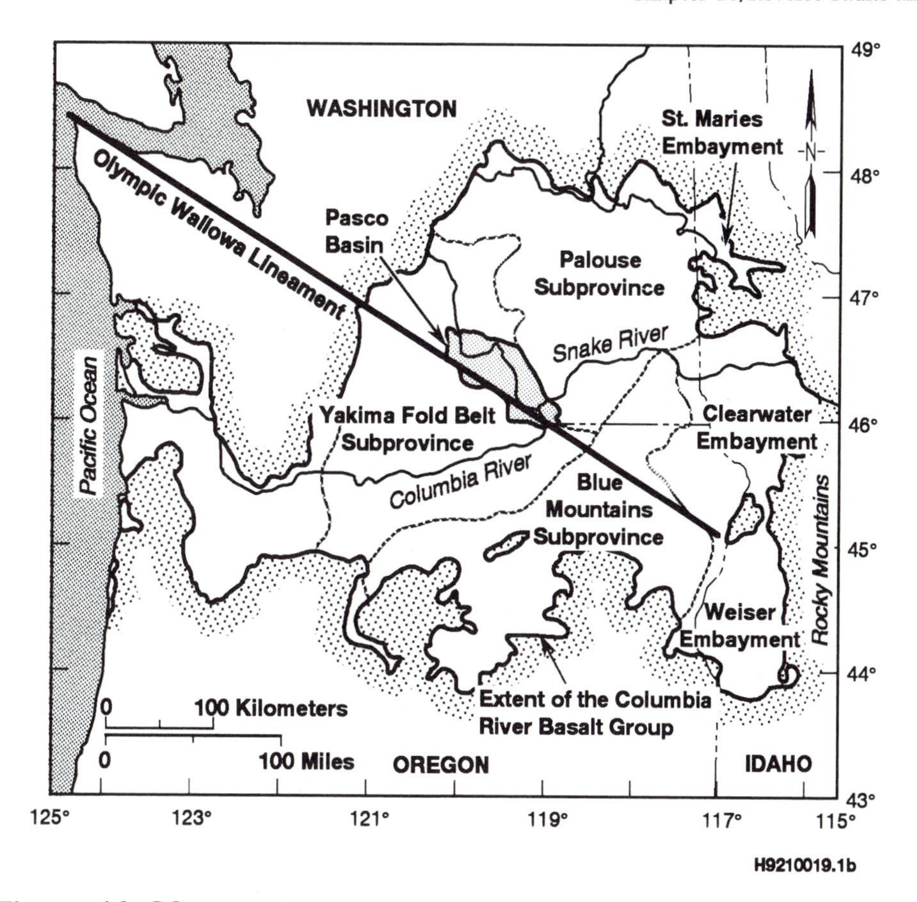

Figure 10–29. (a) Location map showing distribution of Columbia River Basalt Group. After Reidel et al. (1989).

with maximum compressive strain axes oriented N-S, which is consistent with the E-W trend of the folds. The greatest concentration of earthquakes is beneath the Saddle Mountains anticline east of the Columbia River.

Reverse Faults in Shield Areas

The continental shields of the earth have not undergone major deformation for hundreds of millions of years, yet they are characterized by a pattern of diffuse seismicity (Johnston and Kanter, 1990). The 1968 Meckering, Australia, earthquake (M_s 6.8) demonstrated that large earthquakes accompanied by surface rupture do occur in otherwise stable shields, although infrequently. Since 1968, reverse-slip surface faulting has accompanied large earthquakes elsewhere in Australia, in northern Canada, and in peninsular India. In addition, Holocene reverse faults have been described in northern Scandinavia and in Scotland. Only reverse faulting is discussed here; continental-shield earthquakes characterized by other styles of faulting, most notably strike slip, are discussed elsewhere (cf. Chapter 8).

The Australian continent was struck in 1968 by the Meckering earthquake (discussed below), in 1970 by the Calingiri earthquake (M_L5.7), and in 1979 by the Cadoux earthquake (M_L6.0), all in Western Australia (Gordon and Lewis, 1980; Lewis et al., 1981). Subsequently, the Marryat Creek earthquake (M_s5.8) struck South Australia in 1986 (Machette et al., in press), and three earthquakes struck Tennant Creek, Northern Territory, in 1988 (M_s6.3–6.7). All these earthquakes were accompanied by surface rupture on reverse faults (Choy and Bowman, 1990; Crone et al., 1992). The shallow focal depths of these earthquakes, generally less than 6–8 km, suggest that only the upper part of the brittle crust was ruptured. Fault traces associated with these earthquakes tend to be irregular in map view, and at Tennant Creek, ruptures occurred on both north- and south-dipping faults. The surface rupture was 38 km long at Meckering, 32 km long at Tennant Creek, and 15 km long at Cadoux, but only 3.5 km long at Calingiri. Surface morphology and trench excavations indicate that earthquake recurrence intervals on these faults are exceptionally long, 10^5 to 10^6 years, or even longer (Crone et al., 1992;

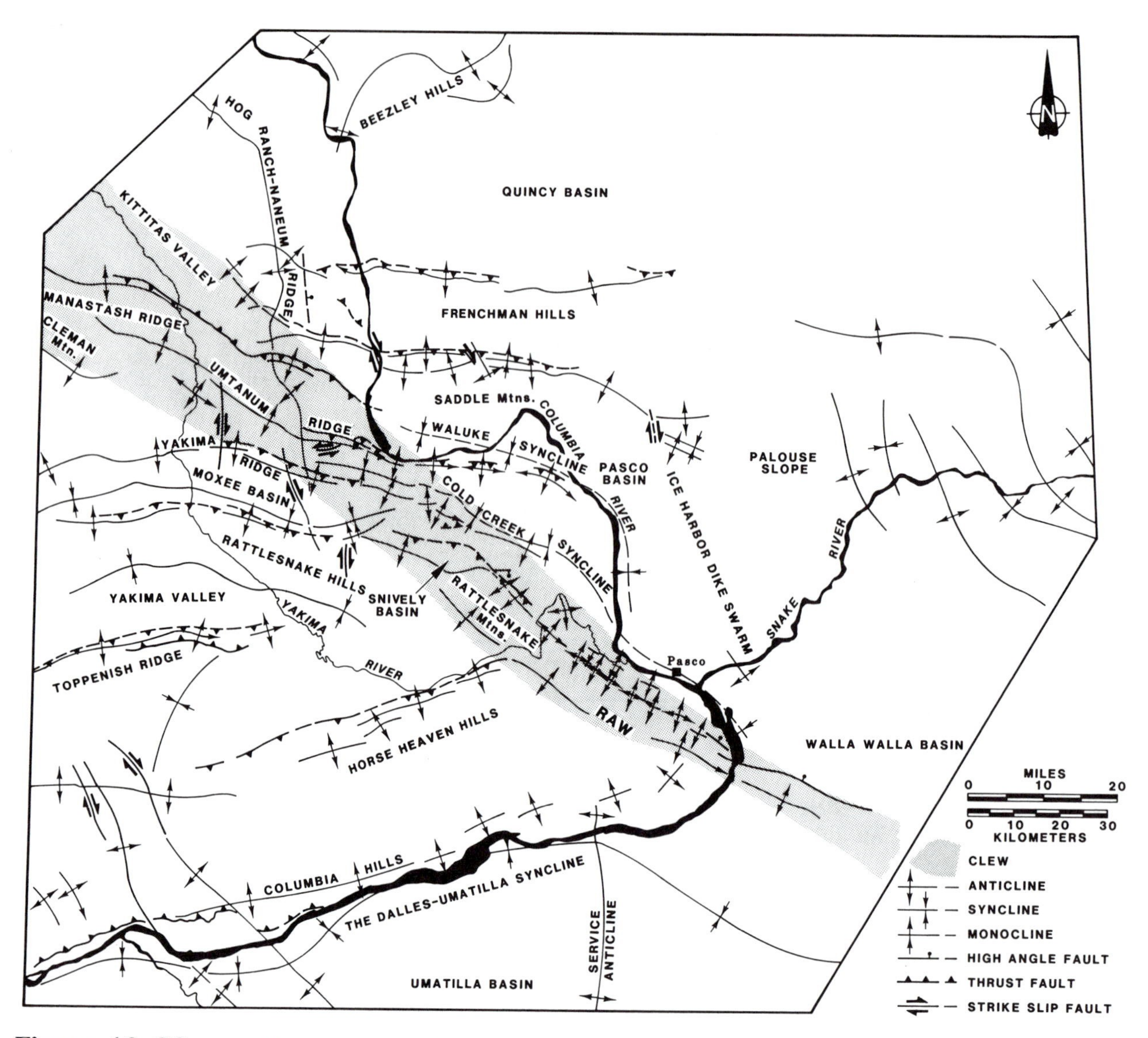

Figure 10–29. (b) The Yakima fold belt in the central Columbia Plateau, Washington and Oregon, U.S.A. Shading locates the Olympic-Wallowa lineament, which includes the Cle Elum-Wallula deformed zone (CLEW) and Rattlesnake-Wallula alignment (RAW).

Machette et al., 1993). One of the Tennant Creek ruptures followed a ridge formed by quartz veins, suggesting that it reactivated an old fault of unknown age. The Meckering scarp also followed a zone rich in quartz with abundant iron staining, also indicating a preexisting fault. A study of aerial photographs indicates several additional localities in Western Australia, well-preserved in the arid environment, where similar fault ruptures occurred in prehistoric time (Gordon and Lewis, 1980; Crone et al., 1992).

In September, 1993, a destructive earthquake of M6.4 struck peninsular India in a region in which active faulting was not previously suspected (Gupta, 1993). The fault-plane solution indicated reverse faulting along a fault striking normal to the previously determined maximum horizontal stress direction for this portion of the Indian shield. Length of surface rupture was at least 3 km (Seeber et al., 1993).

Several large earthquakes have struck the Canadian shield in the past few years, but only one, the 1989 Ungava, northern Quebec, earthquake (M_s 6.3) produced documented surface rupture (Adams et al., 1991). The rupture zone is at least 8.5 km long, with average throw of 0.8 m on a reverse fault with a left-slip component. Modeling of the deformed shoreline of a lake indicates that the rupture is unlikely to pen-

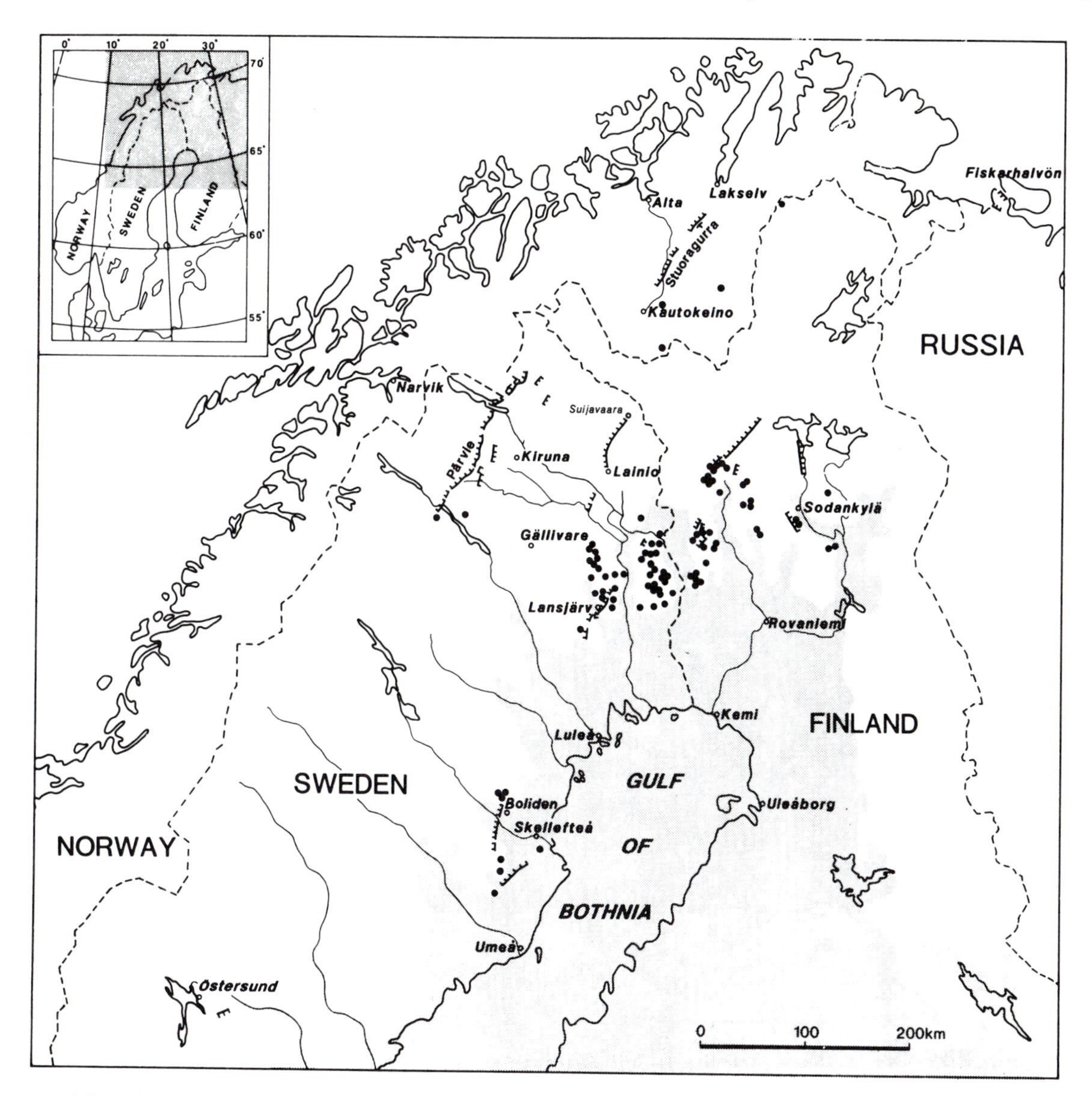

Figure 10–30. Location of late Quaternary fault scarps (solid lines with hachures on downthrown side) and landslides (solid dots) in northern Fennoscandia. The area beneath the highest uplifted shoreline, about 9000 years old, is shaded. After Lagerbäck (1992).

etrate deeper than about 5 km. There is a suggestion that the fault follows older structure.

Holocene faults have been mapped in northern Sweden, Norway, and Finland. As described by Lagerbäck (1992), the longest of these fault is the Pärvie fault of Sweden, with a length that is slightly more than 150 km (Fig. 10-30). Trenching studies of a fault at Lansjärv, Sweden, exposed evidence for reverse faulting that took place after local deglaciation but prior to deposition of Holocene marine sediments. This constrains the age of faulting to about 9 ka. Abundant landslides in the vicinity of the fault and liquefaction features in the excavations suggest that the faulting produced large earthquakes. Late Quaternary faults occur also in northwest Scotland along reactivated basement faults. Here, also, the faulting took place during and immediately following deglaciation (Davenport et al., 1989; Ringrose et al., 1991).

Late Quaternary reverse faults in shield areas appear to result from a variety of causes, not all of which are currently understood. Measured orientations of maximum horizontal stresses are consistent with the orientation of these faults (Zoback, 1992). Several of the faults are confined to the upper brittle crust. It is probable that all of these faults follow old zones of weakness in the crust: old faults, fractures, or anisotropy due to foliation. Recurrence intervals are long, at least hundreds of thousands of years in some cases. In northern

Europe, the timing of faulting appears to be related to isostatic rebound following deglaciation (Ringrose et al., 1991). However, activity on reverse faults in tropical shield areas must have a different cause.

CASE HISTORIES

In this section, we consider four reverse-fault earthquakes in greater detail. Each earthquake occurred in a very different tectonic environment, but each is representative of a class of reverse-fault earthquakes. The 1971 San Fernando earthquake resulted from rupture of a range-front fault; most of the surface deformation is due to faulting. The 1988 Spitak earthquake is more complex, because folding is involved, and some segments of the fault are represented at the surface entirely by folding. The surface expression of the fault that produced the Coalinga earthquake was essentially a fold overlying a deeply buried blind thrust fault. The Meckering earthquake was caused by a fault that ruptured a continental shield area and was restricted to the shallow part of the brittle crust.

San Fernando, California, 9 February 1971 (M_w6.7)

The San Fernando earthquake was accompanied by surface faulting on an irregular fault trace in the San Fernando Valley and at the range front of the San Gabriel Mountains (Figs. 5-11; 10-31). Aftershocks defined a reverse fault dipping 40°NE to a maximum depth of 15 km (Fig. 10-32a; Mori et al., 1995). Near-horizontal nodal planes of a few aftershocks led Hadley and Kanamori (1978) to speculate that the fault may flatten to near horizontal at the base of the brittle crust. However, the aftershocks do not define a horizontal fault at the base of brittle crust (Fig. 10-32a), and the fault may, instead, continue with a moderate dip into the lower crust as a ductile shear zone. The shallow portion of the thrust, including the surface ruptures, had relatively few aftershocks, presumably because the fault was in Neogene strata that were too weak to store significant amounts of seismic strain energy. The 1971 reverse fault terminated on the west at a NE-trending lateral ramp (called a "downstep" by Whitcomb et al., 1973) where the mainshock

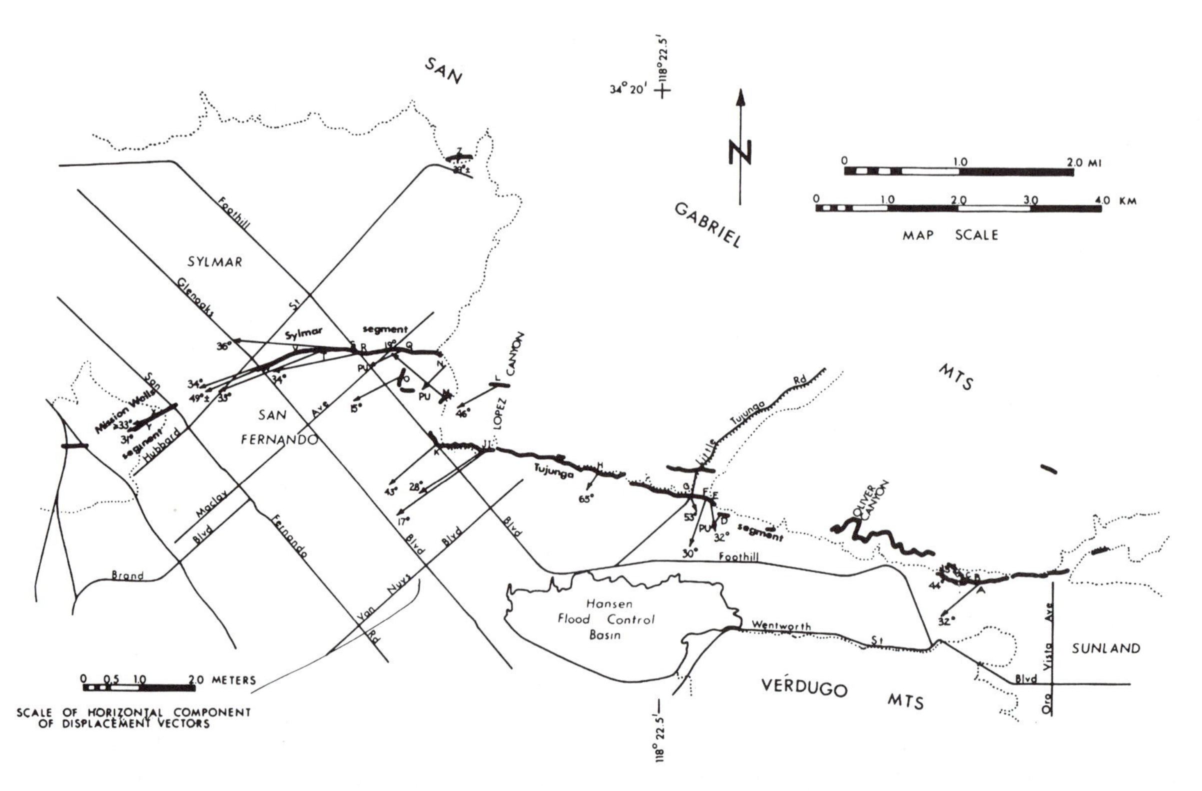

Figure 10–31. Horizontal movement of northern block with respect to southern block during 1971 San Fernando, California, earthquake. Lengths of arrows show horizontal component of calculated slip vectors; angle at head of arrow shows plunge. Numbers on north side of faults show fault dip. After Sharp (1975).

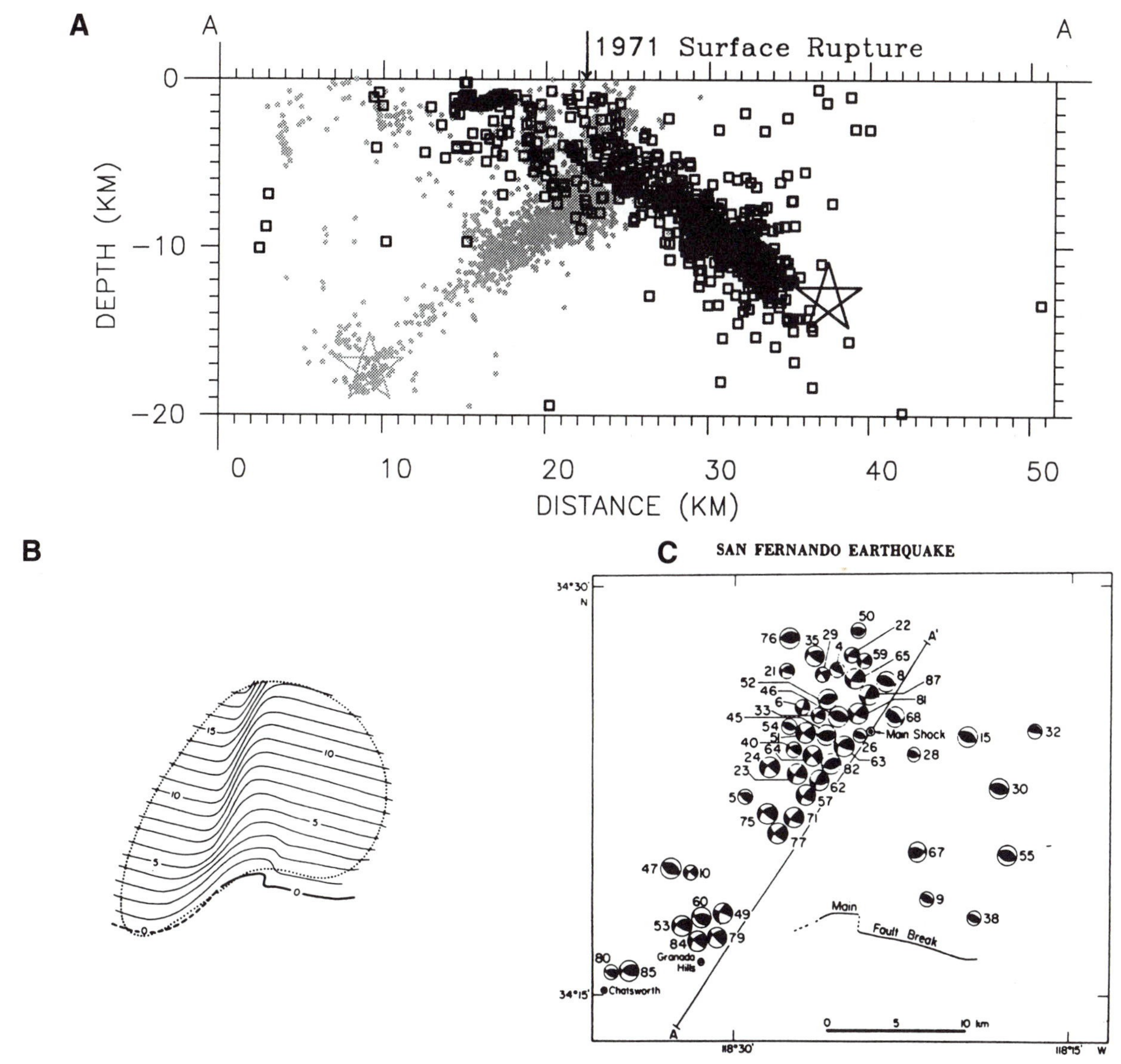

Figure 10–32. Tectonics of the 1971 San Fernando earthquake. (a) Cross section showing seismicity of 1971 San Fernando (open squares) and 1994 Northridge (shaded dots) earthquakes, illuminating reverse faults of opposite dip. The 1971 fault overrides the 1994 fault. (b) Structure contours in kilometers on 1971 fault surface based on seismicity showing a lateral ramp on which focal-mechanism solutions were primarily left slip. Heavy line with zero contour locates surface rupture. (c) High-quality focal-mechanism solutions for aftershocks of the San Fernando earthquake. Compressional quadrants are dark, dilatational are light. (a) After Mori et al. (1995); (b, c) after Whitcomb et al. (1973).

and numerous left-slip aftershocks occurred (Fig. 10–32b, c).

Surface rupture in 1971 was limited to the reverse fault east of the lateral ramp, even though young thrust faulting continues westward beyond the lateral ramp as the Santa Susana fault (Figs. 10–18; 10–47). Although fault-plane solutions for some aftershocks indicated reverse faulting west of the mainshock (Fig. 10–32), none of the reverse faults at the surface projection of these earthquake focal planes had surface rupture. Another range-front fault, north of the western (Sylmar) segment of 1971 faulting, also remained inactive except very locally (Fig. 10–31). Ikeda (1983) suggested that the *master boundary fault* (cf. Fig. 10–41) was active at the range front at the same time the Pliocene-Pleistocene sedimentary wedge was accumulating in the Sylmar basin immediately to the south (Figs. 10–31; 10–47). In 1971 and in prehistoric

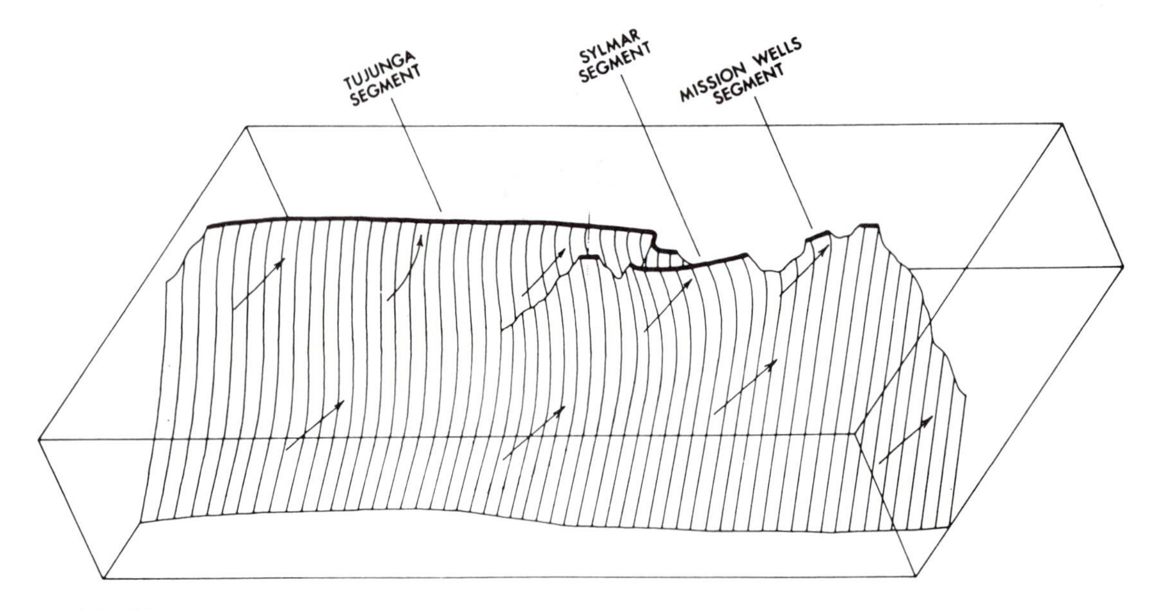

Figure 10–33. Schematic diagram of 1971 San Fernando fault surface, viewed from the northwest. Heavy lines show surface rupture. Lines on fault surface drawn in direction normal to average strike; arrows on surface show slip of northern relative to southern block. After Sharp (1975).

events, the fault propagated through the sedimentary wedge along bedding planes. Before the earthquake, late Quaternary displacement had been recognized on the San Fernando fault in the eastern (Tujunga) segment, but not in the Sylmar segment.

The lithologically distinctive Modelo Formation (thin-bedded silty to siliceous shale interbedded with sandstone) of Miocene age predates the compressional regime that now characterizes the Transverse Ranges, but it adds insight into the topography overlying the pre-1971 scarp. Along the Tujunga segment, the Modelo Formation is thrust over a 10-m-thick breccia of Modelo debris (Kahle, 1975). This breccia contains almost no clasts of crystalline rocks, even though the predominantly crystalline San Gabriel Mountains rise abruptly to the north. The breccia grades upsection into Quaternary alluvium and terrace gravel that contains predominantly crystalline clasts; these sediments have also been overridden by the hangingwall of the thrust. These relations are interpreted as evidence for a low, tectonic ridge of Modelo Formation immediately adjacent to the thrust on the hangingwall side which shielded the fan gravels downslope from it from receiving clasts of crystalline rocks. Farther downslope from the thrust, crystalline clasts predominate. Similar anticlinal ridges formed in the hangingwall of the 1968 Meckering, Australia, earthquake (discussed in the following text). Another explanation, favored by Kahle (1975), is that the Modelo was brecciated during faulting, and this tectonic breccia formed a colluvial apron that was overridden by later movement on the fault.

The displacement on tectonic ruptures was described by Sharp (1975), who noted that the surface trace and near-surface dip of the fault is more variable than the slip vectors (Figs. 10–31, 10–33). The fault dip was measured directly where a plane could be defined by surface expression or in trenches, or by the offsets of two nonparallel features such as city streets

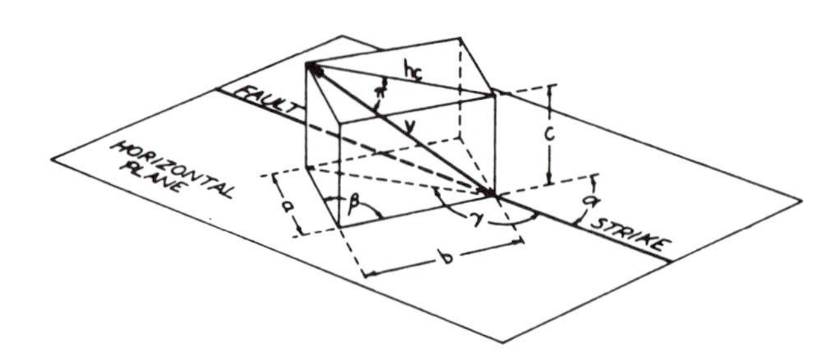

Figure 10–34. Three-dimensional view of a slip-vector cell. Net slip vector V is cell diagonal. Apparent lateral separation *a* of linear feature at angle α from fault strike; apparent lateral separation *b* of second linear feature at angle β from first; vertical component of slip is *c*; plunge angle of slip vector is π; azimuth angle of slip vector measured from fault strike is γ; length of vectors shown in Figure 10–31 corresponds to horizontal component of slip vector, *hc*. After Sharp (1975).

or fence lines that cross the fault (Fig. 10–34). The projected intersections of these nonparallel features in the hangingwall and footwall are connected by a line parallel to the slip vector of the fault. If the plunge of the line is greater than zero, this line and the strike line of the fault define the fault plane. Similarly, if a linear feature of known original length is cut by the fault, the piercing-point offset of this linear feature also comprises a slip vector. The problem is straightforward when the displaced features were at the same altitude prior to faulting, as is the case for city streets or drainage ditches. However, care had to be taken outside urban areas, where displaced surfaces in some cases included pre-1971 offset. The slip vectors, which were in part supported by measurement of slickensides, indicated a component of left-lateral slip in addition to dip slip (Sharp, 1975; Figs. 10–31, 10–33).

Slip vectors and fault dips are shown in Figures 10–31 and 10–33. Despite the simplicity of the fault pattern based on aftershocks, the near-surface fault configuration is complex, with steep to vertical dips along the Sylmar segment and relatively shallow dips in the Tujunga segment. The fault plane commonly follows bedding, indicating that the mechanical anisotropy related to bedding planes is at least as important as the orientation of planes of maximum shear stress. Because the

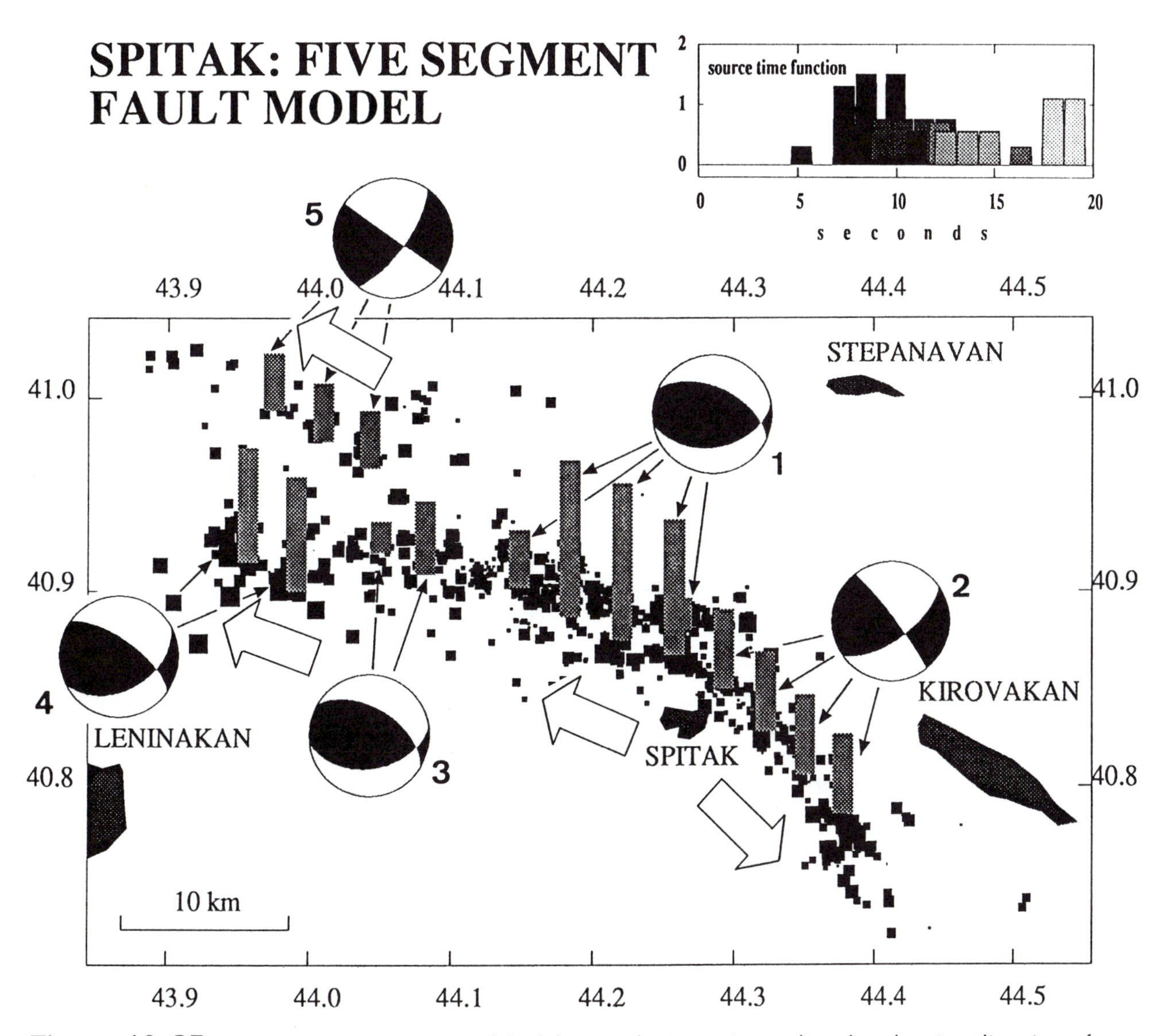

Figure 10–35. Five-segment source model of the Spitak, Armenia, earthquake, showing direction of rupture propagation (open arrows). Bars give relative seismic moment of each of 16 point sources. Time duration of the sources and their sequence is given in the box; earlier subevents are in darker gray shades. For fault-plane solution, dark quadrants are compressional. After Haessler et al. (1992).

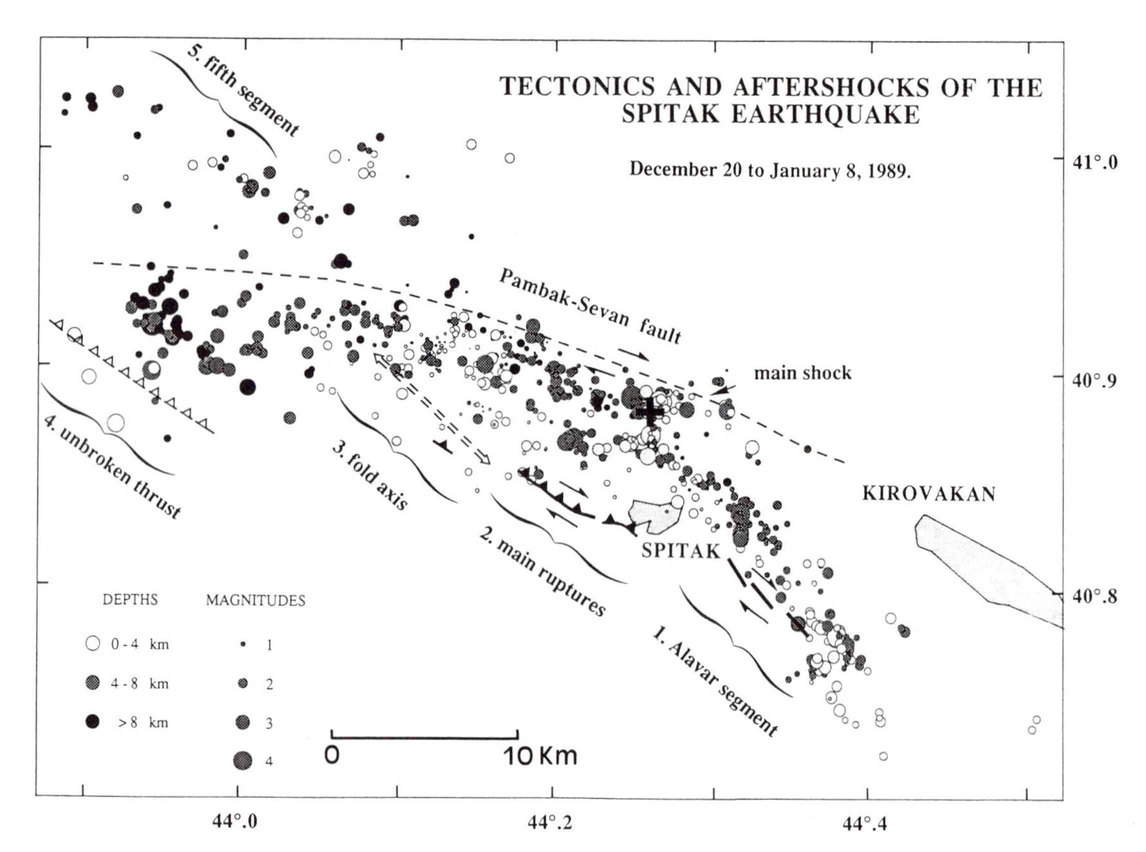

Figure 10–36. Distribution of aftershocks of the Spitak, Armenia, earthquake for the period 20 December 1988 to 8 January 1989. Cross locates mainshock. Heavy lines show surface faults; solid triangles on hangingwall of reverse fault; arrows indicate direction of strike slip. Broken double line is anticline that showed evidence of uplift in 1988. Dashed line with open triangles is reverse fault that did not rupture the ground surface in 1988. After Dorbath et al. (1992).

position of aftershocks was known, Sharp could infer that the steep dips of the Sylmar segment become more gentle with depth, resulting in a fault plane that is curved in the direction of slip (Fig. 10–33).

Other earthquakes similar to the San Fernando event in that they were accompanied by a long zone of surface rupture are the 1896 Rikuu, Japan, earthquake (M7.2), the 1911 Chon-Kemin, Kyrgyzstan, earthquake (M7.8), and the 1978 Tabas-e-Golshan, Iran earthquake (M7.5).

Spitak, Armenia, 7 December 1988 (M_s 6.9)

The Spitak earthquake struck the Lesser Caucasus (Fig. 10–26) in December, 1988. The earthquake had a focal depth of 5 km, which was one factor that contributed to the loss of more than 25,000 lives. The earthquake struck close to the Pambak-Sevan fault zone, a north-dipping reverse fault with right-slip component on the southern flank of the Lesser Caucasus (Fig. 10–26: reverse fault located southwest of LC). This fault zone includes a set of en échelon basins filled with Pliocene-Quaternary sediments; these en échelon basins are consistent with a right-lateral strike-slip sense of motion (Philip et al, 1992).

The mainshock source, which began to fail north of the town of Spitak, is modeled as a reverse fault dipping 65°N with a major right-lateral component (Haessler et al., 1992). The mainshock sequence is best modeled as five subevents that occurred during a period of 11 seconds (Fig. 10–35). The mainshock rupture initially propagated westward; two seconds later, a right-slip subevent began southeast of Spitak and propagated to the southeast. Meanwhile, the west-propagating mainshock broke into a southern branch on a north-dipping reverse fault (subevent 4), and a northern branch on a right-slip vertical fault (subevent 5; Haessler et al., 1992; note fault-plane solutions on Fig. 10–35).

The aftershocks, based on an array of seismographs that was installed 12 days after the mainshock, were subdivided into five segments, consistent with the mainshock sequence (Fig. 10-36; Dorbath et al, 1992). Three of these segments have surface expression (labeled 1, 2, and 3 in Fig. 10-36). The Alavar segment on the southeast is defined by a relatively narrow zone of aftershocks that have predominantly right-slip fault-plane solutions. Hypocentral depths along this segment increase to the northwest from 6-9 km and define a fault plane dipping 65°NE. Surface ruptures on this segment have an en échelon pattern that indicates right slip (Philip et al., 1992). The main segment near Spitak displays reverse-faulting focal mechanisms. The strike of the seismicity pattern changes from N40°W in the southeastern part of the segment to N60°W in the northwestern part. The dip of the fault is 55°N based on aftershock distribution, and the base of the aftershock zone is deeper than 10 km. Surface breaks reflect reverse faulting near Spitak, but farther west, the reverse fault, which is based on seismicity, is blind; it does not reach the surface (Philip et al., 1992). Instead, the surface expression in this area is a broad anticline that contains tensional features at its crest and a reverse fault on its south flank (Fig. 10-53); these features are secondary to the folding. Aftershocks extend still farther west, but there is no evidence for surface displacement on a reverse fault at the surface continuation of the aftershock pattern or on a fold northwest of the fold that was activated in 1988. Finally, a fifth segment, which is at the northwest end of the aftershock sequence, is characterized by right-slip focal-mechanisms; it showed no evidence of surface rupture.

This earthquake differed from the San Fernando event in two ways: (1) only a small part of the sub-

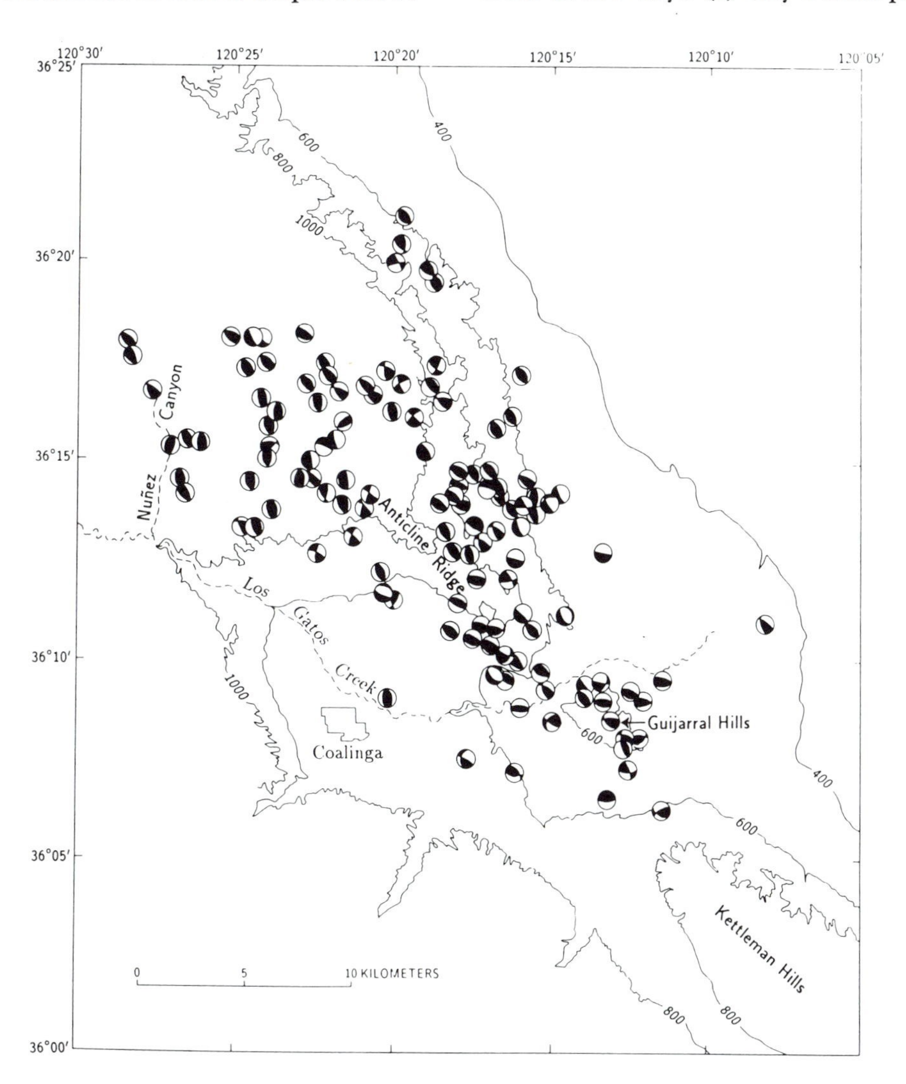

Figure 10–37. Fault-plane solutions for the Coalinga mainshock and larger aftershocks; compressional first arrivals dark, and dilatational arrivals white. Symbols are not scaled according to magnitude. After Eaton (1990).

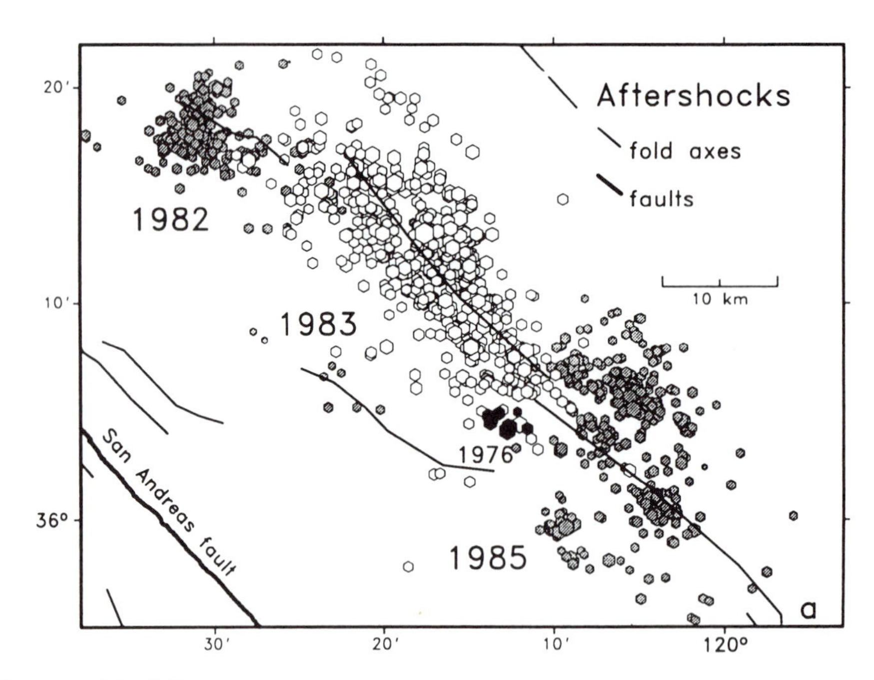

Figure 10–38. Location of aftershocks of the 1982 New Idria (shaded hexagons, northwest), 1983 Coalinga (open hexagons), 1985 Kettleman Hills (shaded hexagons, southeast), and the 1976 Polvadero (solid hexagons) earthquakes. After Stein and Ekström (1992).

surface rupture zone reached the surface, and (2) the earthquake consisted of both reverse-slip and strike-slip subevents, with most of the energy release occurring by reverse faulting. The 1932 Changma, China, earthquake (M7.6) also appears to have been accompanied by reverse-slip and strike-slip faulting. The 1980 El Asnam, Algeria, earthquake (M7.3) had a complex rupture pattern, and much of the surface rupture may have been secondary, related to folding. The 1992 M7.4 Suusamyr, Kyrgyzstan, earthquake consisted of several subevents, and the length of surface rupture was very short compared to the length of subsurface rupture (S. Ghose, unpublished data).

Coalinga, California, 2 May 1983 (M_s 6.7)

This earthquake, which is described in detail in U.S. Geological Survey Professional Paper 1487, was produced by faulting about 10 km beneath the Coalinga anticline. Three seconds after initiation of the mainshock, a subevent occurred on a fault about 5–6 km to the southwest (Choy, 1990). The fault-plane solution shows slip on either a reverse fault dipping steeply to the NE or on a thrust fault dipping gently to the SW (Fig. 10–17, cross section b). The aftershocks occur in a broad, 35-km-long zone, parallel to the Coalinga anticline, and 15–20 km wide (Eaton, 1990; Figs. 10–37, 10–38). The aftershock zone coincides in general with a broad band of abnormally high fluid pressures beneath the Coast Range fold belt (Yerkes et al., 1990). The aftershock zone is diffuse in cross section (Fig. 10–17), such that, in contrast to the San Fernando and Spitak earthquake faults, the distribution of aftershocks does not define either the NW- or the SE-dipping focal-mechanism plane as the earthquake-generating fault. The diffuse aftershock distribution may be related to the high fluid pressures which may cause a large volume of rock near the fault plane to be close to Coulomb failure and may contribute to folding of this rock volume in addition to slip on the main rupture. High stresses at the tip of the mainshock fault may be relieved by folding and by slip on low-displacement secondary faults, most of which do not reach the surface (Stein and Ekström, 1992). The rupture of one of these secondary faults did break the surface at the time of a M_L 5.2 aftershock on 11 June (Rymer et al., 1990). The Ventura Avenue anticline, which is characterized by small-displacement reverse faults (Fig. 2–24), may serve as a tectonic analog to the diffuse aftershocks beneath the Coalinga anticline (Stein and Yeats, 1989). In summary, one can consider the aftershocks as being triggered by the mainshock on a planar fault, but as occupying a volume of deforming rock rather than defining a fault plane.

Despite the lack of surface faulting associated with the mainshock, reoccupation of a leveling network after the earthquake produced geodetic evidence of uplift of the Coalinga anticline and subsidence in the syncline to the SW (Stein and King, 1984; Stein and Ekström, 1992; Fig. 10–39). The maximum uplift was 500 mm, and the area of maximum uplift coincides with the anticlinal axis and plunges in the direction of anticlinal plunge. The axis of subsidence also coincides with the synclinal axis. Uplift and subsidence continued at a sharply decreasing rate for several years after the earthquake (Stein and Ekström, 1992). The distribution and characteristics of the deformation are compatible with either a steeply NE-dipping reverse fault or a gently SW-dipping thrust fault. The SW-dipping fault is preferred, because (1) the geodetically modeled thrust fault is closer to the hypocenter than the high-angle reverse fault, (2) the postseismic uplift migrated NE, and the axis of uplift accompanying the 1985 Kettleman Hills earthquake (see the following text) is NE of the anticlinal axis, and (3) seismic-reflection profiles contain faint, SW-dipping reflections at 8–10 km depth beneath the Coalinga and Kettleman Hills anticlines (Stein and Ekström, 1992). The preferred structural interpretation, shown in cross section b of Figure 10–17, is a SW-dipping thrust fault changing updip to more steeply dipping splay faults and small reverse faults dipping NE and SW.

The Coalinga earthquake was part of a sequence of three earthquakes that began in New Idria in 1982 (M5.5) and concluded at Kettleman Hills North Dome in 1985 (M6.1). All three were reverse-fault earthquakes, and all three are compatible with a SW-dipping thrust fault beneath the Coast Ranges (Fig. 10–17). The aftershock patterns for the three earthquakes are contiguous, but do not overlap. The boundaries between aftershock zones coincide with en échelon offsets of anticlinal axes (Fig. 10–38). Focal mechanisms near the northwest end of the 1983 aftershock zone reveal a NE-striking band of right-lateral strike-slip aftershocks beneath the en échelon offset between the Coalinga and New Idria anticlines (Eaton, 1990; Stein and Ekström, 1992; Fig. 10–37). These observations suggest the presence of lateral ramps in blind seismogenic thrust faults that served to terminate individual earthquakes.

The Coalinga earthquake was the first well-documented example of a *blind-thrust earthquake,* in which the surface expression of the earthquake is a fold. The earthquake was followed in California by the 1987 Whittier Narrows earthquake (M5.9; Fig. 5–14) and the 1994 Northridge earthquake (M_w6.7), all on blind thrusts. The surface expression of the Northridge earthquake was upwarping of the hangingwall block (Yeats and Huftile, 1995), similar to the expression of the Coalinga earthquake. The 1991 Puerto Limón, Costa Rica, earthquake (M_S7.5) upwarped a section of coastline but did not produce surface rupture. The 1991 Racha, Georgia earthquake (M_S7.0) activated a thrust dipping no more than 31° at 10 km depth or less, yet produced no surface rupture (Triep et al., 1995). Reverse-fault crustal earthquakes that do not rupture the surface are relatively common, perhaps more common than the San Fernando type, which had surface rupture across most of the aftershock zone.

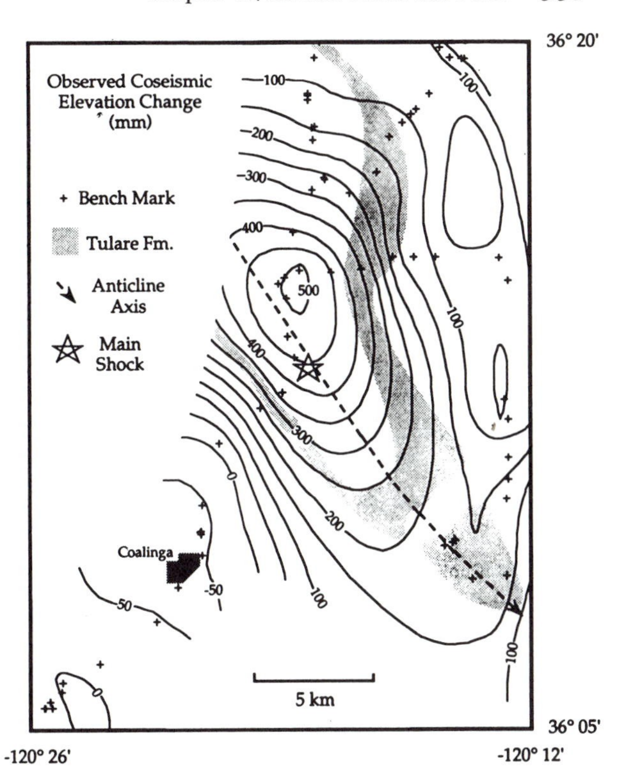

Figure 10–39. Contour map of coseismic elevation changes associated with the 1983 Coalinga earthquake. Contour interval is 50 mm. After Stein and Ekström (1992).

Meckering, Western Australia, 14 October 1968 (M_S6.8)

The Meckering earthquake occurred in a region with relatively high levels of upper-crustal seismicity called the South West Seismic Zone. This zone is characterized by relatively low topographic relief and is not associated with any major tectonic features. A study of body waves from the earthquake indicates a source depth of only 3 km on a reverse fault striking N19°W and dipping 37°E, with 3.3 m of east-west-directed dip slip. This slip direction resulted in a small component of strike slip (Vogfjörd and Langston, 1987). The fault-plane solution had a large non-double-couple compo-

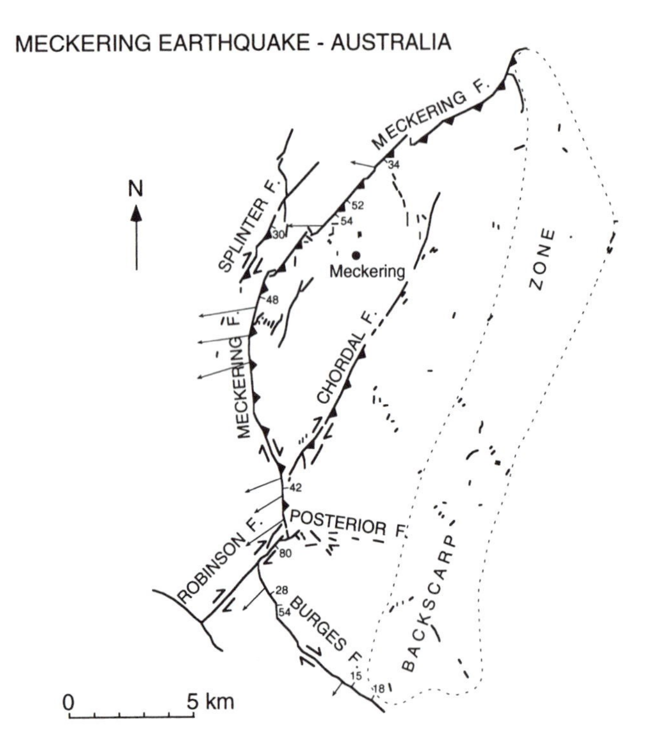

Figure 10–40. Surface faulting accompanying the 1968 Meckering, Australia, earthquake. Arrows show slip vector of eastern block with respect to western block. Numbers on hangingwall side show measured dip angle of fault; teeth are on hangingwall. Dots enclose Backscarp Zone, which subsided during the earthquake. After Gordon and Lewis (1980).

nent, suggesting that the fault is nonplanar. However, modeling of the fault based on earthquake wave forms gives best results for a planar fault, which indicates that most of the seismic energy was released on a fault with the parameters listed previously.

The largest amount of surface faulting produced by the earthquake was along the Meckering fault, an arcuate thrust fault 37 km long that had a significant right-slip component (Gordon and Lewis, 1980; Fig. 10–40). The displacement was greatest near the center of the fault: 2 m vertical, 1.5 m right slip. The fault dips 28–55°SE, averaging 43°. The Splinter fault is a frontal fault to the Meckering fault and is similar except that the Splinter fault dips generally < 30° and has displacements that are about one-fourth those on the Meckering fault. Slip vectors vary with strike, as if they emanated from a point east of the surface trace of the fault situated so that the fault would be annular to that point (Fig. 10–40). The Meckering fault contains four right steps, the largest of which is between the Meckering and Burges faults. These dextral steps are similar to the lateral ramp of the San Fernando fault except that at the right step to the Burges fault, ruptures extended east and west of the step as the Posterior and Robinson faults, respectively. Slickensides suggest two stages of movement on the Burges fault, right slip followed by reverse slip.

Ruptures along the Chordal fault formed during the six weeks following the earthquake. This fault underwent as much as 23 cm right slip and 15 cm normal dip-slip displacement. The Backscarp zone was a depressed linear zone about 4 km wide joining the two ends of the Meckering fault. It marked the eastern limit of faulting and contained numerous small tensional fractures.

Geodetic measurements indicated as much as 7.6 cm of depression west of the overthrust block. The maximum uplift of 1.6 m occurred about 2 km east of the fault trace. From this point eastward, uplift decreased to zero, and the Backscarp zone was depressed by the earthquake. In general, the earthquake was caused by uplift and eastward tilting of a disk-shaped body; the pattern of uplift, thrusting, and lateral slip indicates that the disk underwent counterclockwise rotation.

Vogfjörd and Langston (1987) questioned why the earthquake rupture was limited to the upper few kilometers of crust, especially because the low heat flow in this part of the Australian shield predicts a brittle-plastic transition depth of about 30 km. *In situ* stress measurements made after the earthquake indicated a high level of horizontal compressive stress, with maximum principal compressive stress oriented N77°E (Denham et al., 1980). The lowest compressive stress was measured near the epicenter. This suggests that the earthquake relieved the local horizontal stresses, which means that high horizontal stresses probably caused the earthquake. In this region, the load stress is the least principal stress (σ_3), which increases faster with increasing depth than does σ_1. Thus, with increasing depth, the difference between σ_1 and σ_3 is less, which would reduce the maximum shear stress (Vogfjörd and Langston, 1987).

STRUCTURAL AND GEOMORPHIC EXPRESSION

Map Patterns and Regional Topography

Range-and-Basin Province

The Basin and Range Province of western North America is characterized by fault-block mountains separated by basins (Figs. 9–11, 9–12). The mountains are uplifted, and the basins are downdropped along normal faults at range fronts, and the entire region is undergoing extension. However, several regions of fault-block mountains alternating with basins are un-

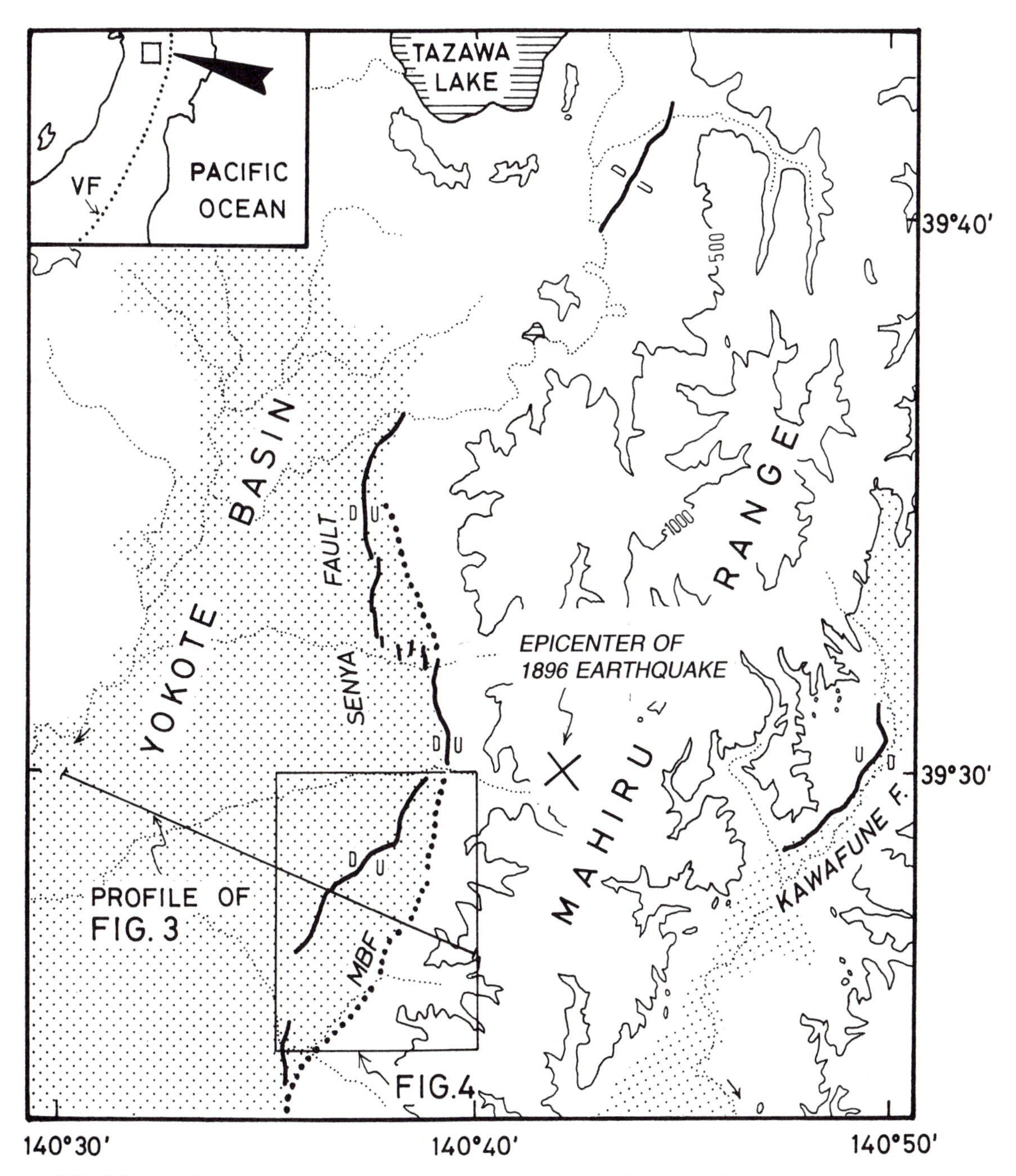

Figure 10–41. Surface rupture accompanying 1896 Rikuu, northern Honshu, earthquake (M7.2). Dotted pattern marks basins, including foothills. The 1896 surface faulting is shown in solid lines, inactive range-front reverse fault (master boundary fault, or MBF) in dotted lines. Kawafune reverse fault, on east flank of Mahiru Range, also moved in 1896. Contour interval 500 m. VF in inset indicates front of volcanic arc. After Ikeda (1983).

dergoing horizontal compression, and the range fronts are controlled by reverse faults, not normal faults. One such area is Central Otago, on the South Island of New Zealand, a region of ranges and basins in which the ratio of range length to range width (aspect ratio) is 5:1 or less, small in comparison to fold-thrust belts (Figs. 7-9, 10-21). This is referred to by New Zealand geomorphologists as a *range-and-basin province*. We use that term here to distinguish provinces dominated by reverse faults from those dominated by normal faults.

Range-and-basin topography also characterizes other reverse-fault regions in the northwestern part of the South Island of New Zealand (Fig. 10-20), in the Pampean Ranges of northwestern Argentina, in northern Honshu, Japan (Fig. 10-41), in the Kinki Triangle of southwestern Japan (Fig. 10-28), in northeastern Iran (Fig. 10-27), in the Nan Shan and Qilian Shan of western China (Fig. 10-25), and in the Transverse Ranges of California. In the Darai Plateau of Papua-New Guinea (Fig. 10-11) and the Shillong Plateau south of the Himalayan front in eastern India, a single range bounded by a reverse fault is adjacent to the front of a thin-skinned fold-thrust belt. In northeastern Iran, northwestern Argentina, the Kinki Triangle, and the Transverse Ranges, both reverse faults and intraplate strike-slip faults occur in the same region. In New Zealand and the Transverse Ranges, the range-and-basin terrain is adjacent to a major plate-bounding strike-slip fault. The Nan Shan-Qilian Shan ranges are terminated on the northwest by the intraplate Altyn Tagh strike-slip fault and on the southeast by the Haiyuan strike-slip fault.

The traces of range-bounding reverse faults are irregular, like those of normal faults and unlike those of strike-slip faults (Fig. 7-9). For this reason, active reverse faults are more difficult to recognize on satellite imagery than strike-slip faults. However, mountain-front sinuosity can be a useful guide to recency of faulting, because a broad pattern of curved or scalloped range fronts may reflect the irregularity of the fault trace and not simply the degree of erosion related to a mature, inactive range front (Fig. 7-9). Mountain-front sinuosity must be used cautiously, however, because through time, the active fault may migrate basinward and away from the range front (Ikeda, 1983; see discussion in the following text). In addition, active fault traces may be discontinuous, because a reverse fault at depth may be expressed at the surface by a fold or monoclinal flexure in addition to, or instead of, a surface rupture. Surface faulting accompanying historical reverse-fault earthquakes commonly is discontinuous, as exemplified by the Senya fault in northern Japan (Ikeda, 1983; Fig. 10-41), the Tabas fault in northeastern Iran (Berberian, 1979), and the San Fernando fault in southern California (Fig. 10-31).

In the semiarid climate of Central Otago, New Zealand, a widespread, well-preserved Cretaceous to early Tertiary erosion surface on the top of schist provides a datum that makes the tectonic configuration of uplifted ranges readily apparent (Figs. 7-9, 10-21; Yeats, 1987). The Dunstan and Pisa ranges are bounded by the Dunstan and Pisa faults, respectively, on their east sides only, so that the ranges have gentle northwest back slopes and steep southeast slopes. The flanking Cromwell and Manuherikia basins are similarly asymmetric, with basin axes close to the bounding faults. Each range and the basin flanking it on the northwest comprise a single block rotating about a horizontal axis. The Raggedy Range and Little Rough Ridge (Fig. 10-21) are elongate uplifts in which range-front surface faulting is absent or subordinate, although these uplifts may be controlled by faults that do not reach the surface. Faults may form on both sides of the range, as is the case for the Mahiru Range in northern Honshu (Fig. 10-41).

Other ranges and basins are also characterized by relatively low ratios of length to width. In central Iran, the ratio of length to width is about 4:1 (Fig. 10-27), and in northwestern China between the Altyn Tagh and Haiyuan faults, it varies from 4:1 to more than 7:1 (Fig. 10-25). In all these areas, the ranges and basins terminate along strike, commonly at a strike-slip fault as in Iran, northwestern China, and the Transverse Ranges.

Fold-and-Thrust Belts

In contrast to regions characterized by range-and-basin topography, fold-thrust belts are marked by ridges and basins that are relatively long with respect to their width, like the classical thin-skinned fold-and-thrust belts of the Appalachians and the Canadian Rocky Mountains. Long anticlinal ridges characterize the Sub-Himalaya between the Main Boundary thrust and the Himalayan Front fault. The Himalayan Front fault is discontinuous at the surface (Nakata, 1989) because it is in large part a blind thrust. The structure of the foothills is characterized by sets of thrust faults that dip beneath the mountains (imbricate thrusts) and are unaccompanied by folding. There, the zone of young deformation in the Sub-Himalaya is no more than a few kilometers wide, and the foothills rise abruptly from the Indo-Gangetic Plains.

Two segments of the Himalayan front, in northwestern India and in south-central Nepal, are characterized by south-verging folds and a broader zone of deformation, more than 100 km wide in northwestern India (Fig. 10-3). The range front itself may be marked by an anticline with discontinuous faulting on one or both sides, north of which lies a broad synclinal valley which is locally called a *dun* (Nossin, 1971; Fig. 10-42). Major Himalayan rivers enter the dun but flow parallel to the synclinal axes, because they are unable to maintain their courses across the rapidly rising anticlines at the range front (Fig. 10-42; Yeats et al., 1992).

Active anticlines rise up out of the alluvium in front of the Salt Range in Pakistan (Yeats and Lillie, 1991), the Sulaiman Range, Pakistan (Fig. 10-43), the Taiwan coastal plain (Suppe, 1987), the foothills of the Subandean Cordillera, the east flank of the California Coast Ranges (Fig. 10-16), and the northern and southern

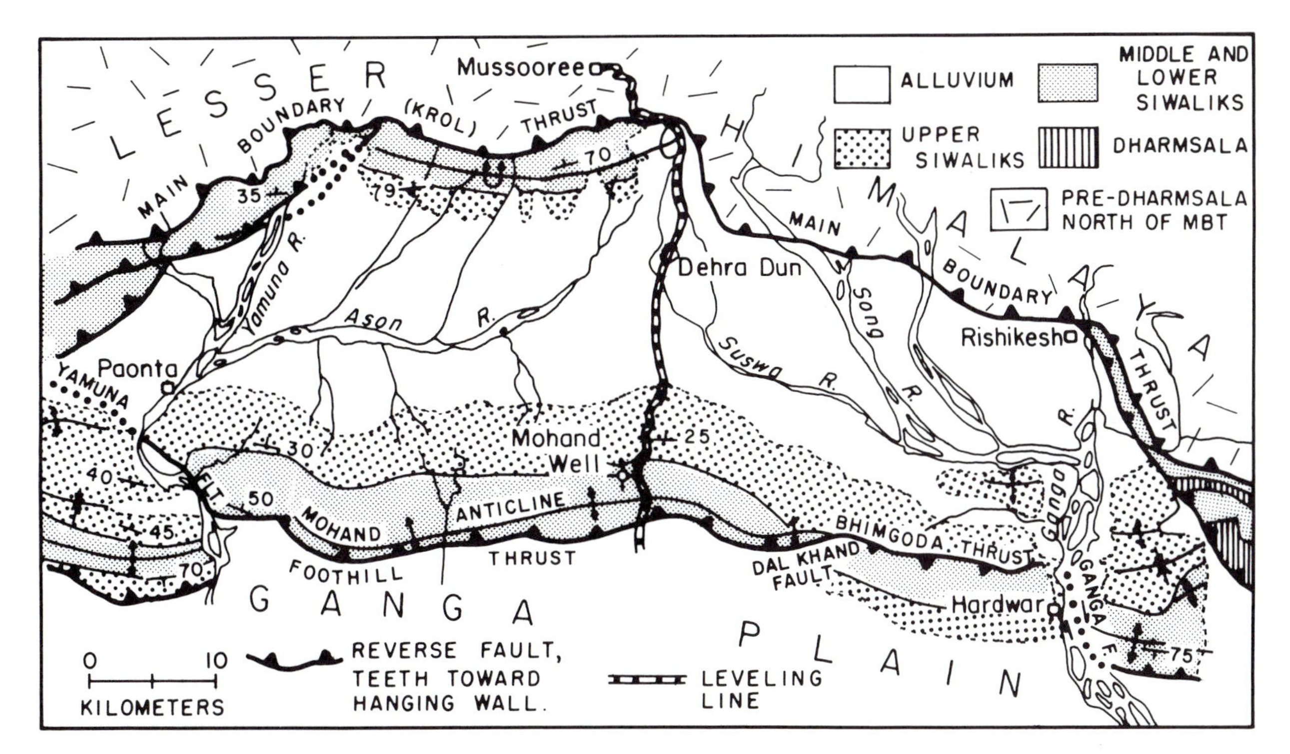

Figure 10–42. Geologic map of Doon (Dehra Dun) Valley. Siwaliks and Dharmsala are foredeep sediments related to Himalayan deformation and uplift. Note that two major rivers (Yamuna and Ganga) have a drainage divide within the dun and flow around the ends of the Mohand anticline rather than cross it. After Yeats and Lillie (1991).

Figure 10–43. Active anticlines at the front of the Sulaiman Range, Pakistan. Larger anticline is crossed by antecedent streams; smaller one at bottom diverts streams around the south end. Indus River floodplain at right; Indus is forced eastward by rising structures. Landsat image 2691-04555.

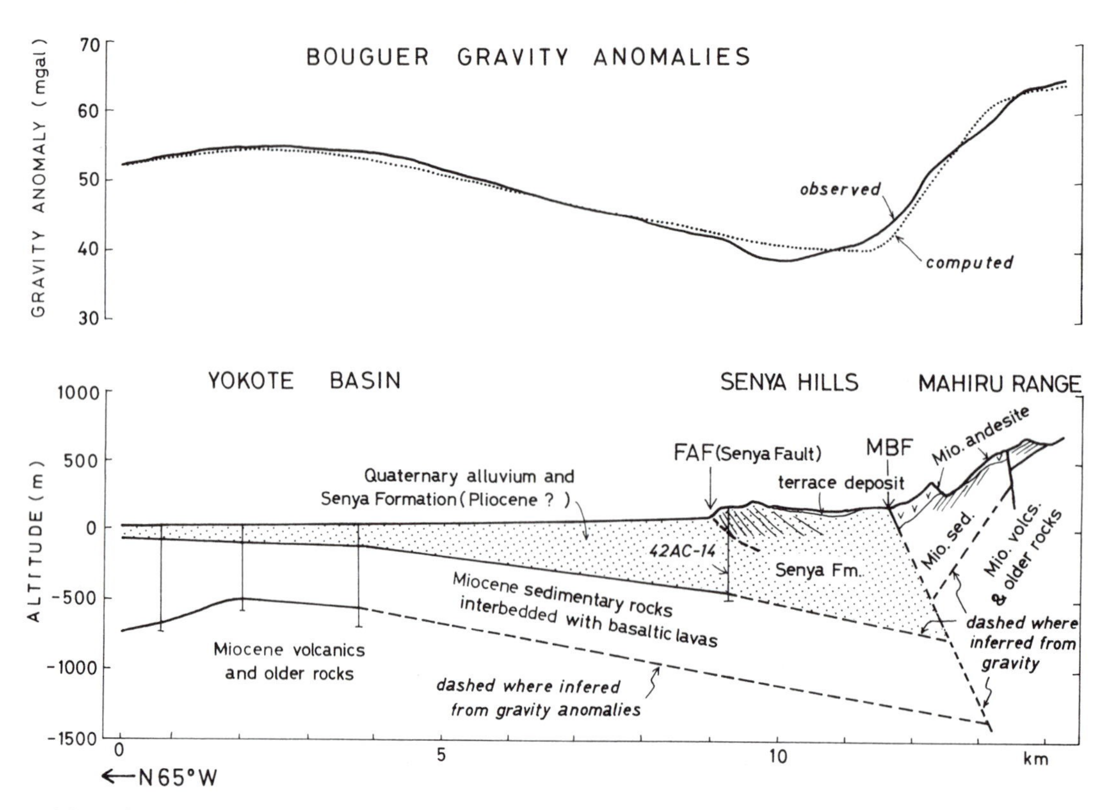

Figure 10–44. Geologic (bottom) and Bouguer-gravity (top) cross sections along the line of section shown in Figure 10–41. The Senya Formation thickens toward the range front, whereas the underlying Miocene sedimentary rocks do not, indicating that the range-front fault (MBF) formed during but not prior to deposition of the Senya Formation. FAF, Frontal active fault that ruptured in 1896. After Ikeda (1983).

foothills of the Tianshan in China (Feng et al., 1991; Avouac et al., 1993; Fig. 10–23). All of these regions have anticlines with a very large length-to-width ratio and are thus characterized as fold-and-thrust belts. The Yakima fold belt of southern Washington State may also be a fold-and-thrust belt, but here the relations are not as clear. Unlike the anticlines that block the Himalayan duns, antecedent streams cross frontal anticlines in the northern Tianshan, the Coalinga and Ventura Avenue anticlines in California, and the Sara El Maarouf anticline in the Tellian Atlas of Algeria, site of the 1980 El Asnam earthquake.

Cross-Strike Migration of Faulting

Reverse faults commonly migrate basinward with time. The best-studied example is the Senya fault in northern Honshu, Japan, which ruptured during the 1896 Rikuu earthquake (Ikeda, 1983). The Senya fault is near the eastern edge of the Yokote basin close to the foot of the Mahiru Range (Fig. 10–41). The 1896 earthquake was not accompanied by rupture of the range-front boundary fault (called by Ikeda the *master boundary fault,* or MBF) between an eastward-thickening wedge of Pliocene(?) strata (Senya Formation) in the Yokote basin and Miocene sedimentary and volcanic rocks of the Mahiru Range (Fig. 10–44). The range-front fault was active during deposition of the Senya Formation, but it is overlain unconformably by late Quaternary terrace deposits (Fig. 10–45). Movement has shifted basinward to the Senya fault (Ikeda, 1983). Faulting was accompanied by uplift of the Senya Hills, which were backtilted toward the Mahiru Range, and by range-facing reverse faults (Fig. 10–45). According to Ikeda (1983), basinward migration occurred because stress is concentrated at the east-dipping base of sedimentary rocks in the Yokote basin, which produces a bedding thrust. His model of basinward migration of reverse faulting is shown as Figure 10–46.

Basinward migration of faulting is also recognized in the Kinki Triangle (Ikeda, 1983; Sangawa, 1986), in central Otago (Beanland and Berryman, 1989), and in the Sylmar basin in the northern San Fernando Valley, where the western part of the 1971 San Fernando earthquake surface rupture occurred within the Sylmar basin, parallel to bedding, rather than at the range front (Ikeda, 1983; Figs. 10–31, 10–47). Faulting in

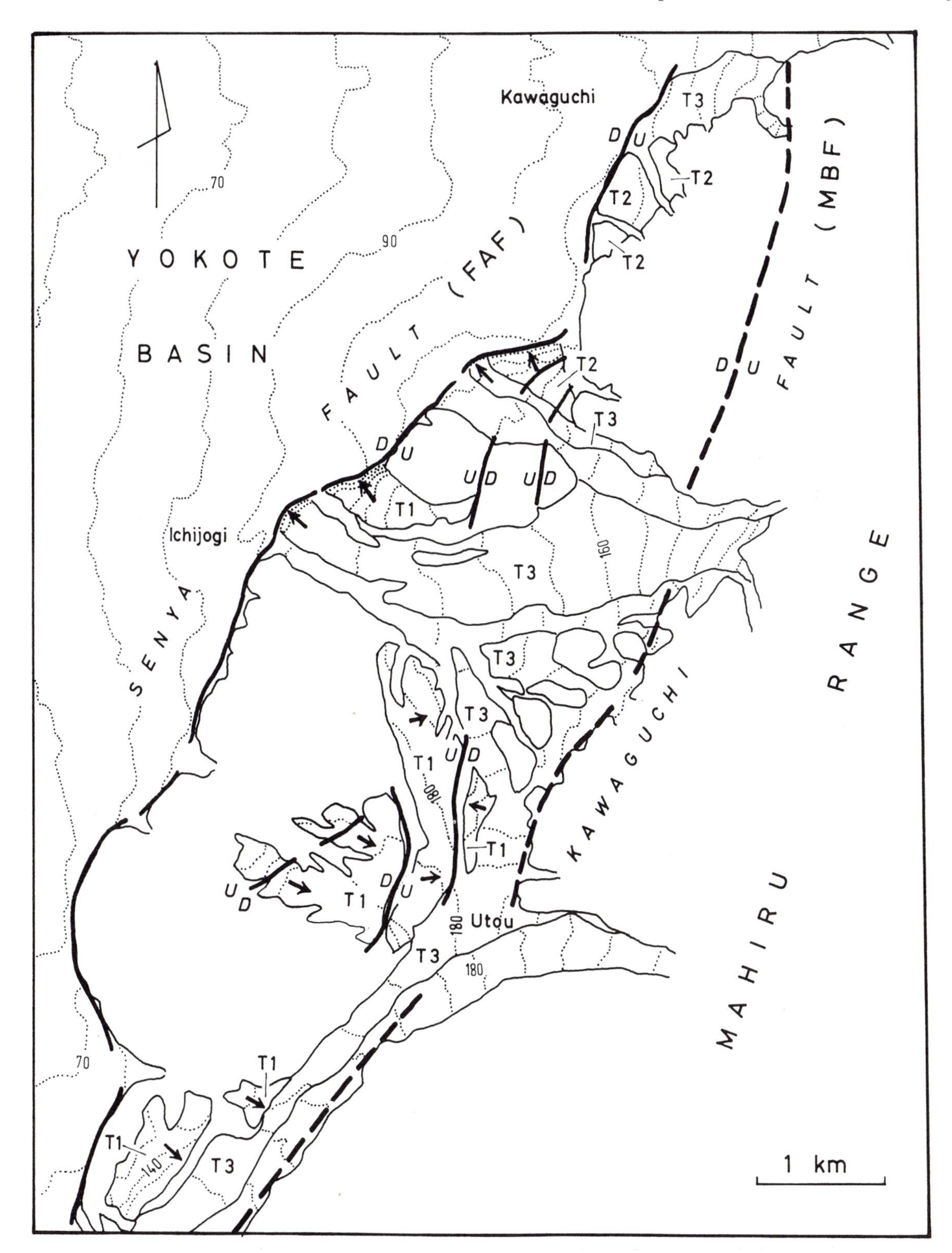

Figure 10–45. Late Quaternary deformation in western foothills of Mahiru Range; area located on Figure 10–41. Late Quaternary deposits, no pattern; bedrock, including Senya Formation. T1, T2, T3 show terraces in chronological order, older to younger; arrows show direction of tilt. Contour interval 10 m, shown only for terraces and alluvium. After Ikeda (1983).

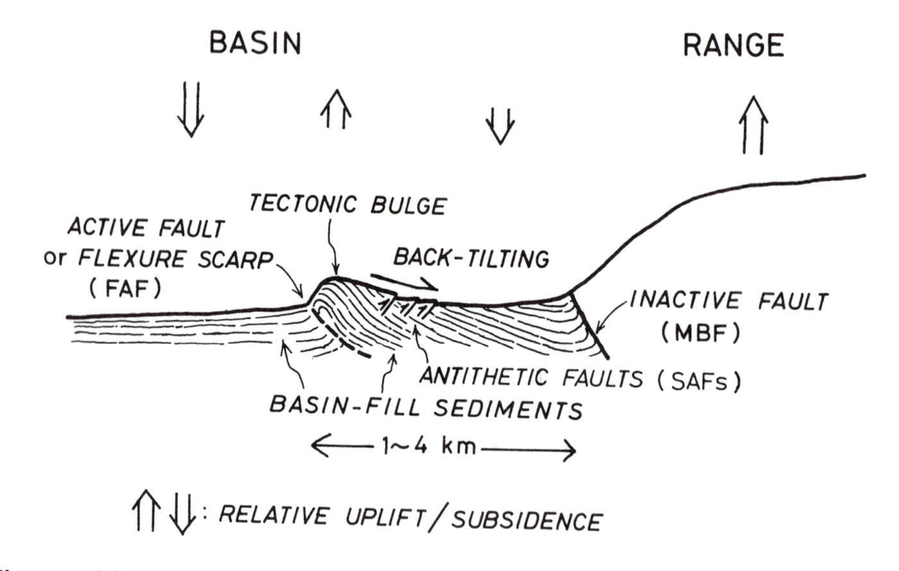

Figure 10–46. Schematic cross section showing thrust-front migration and associated deformation expected along reverse-fault range fronts. After Ikeda (1983).

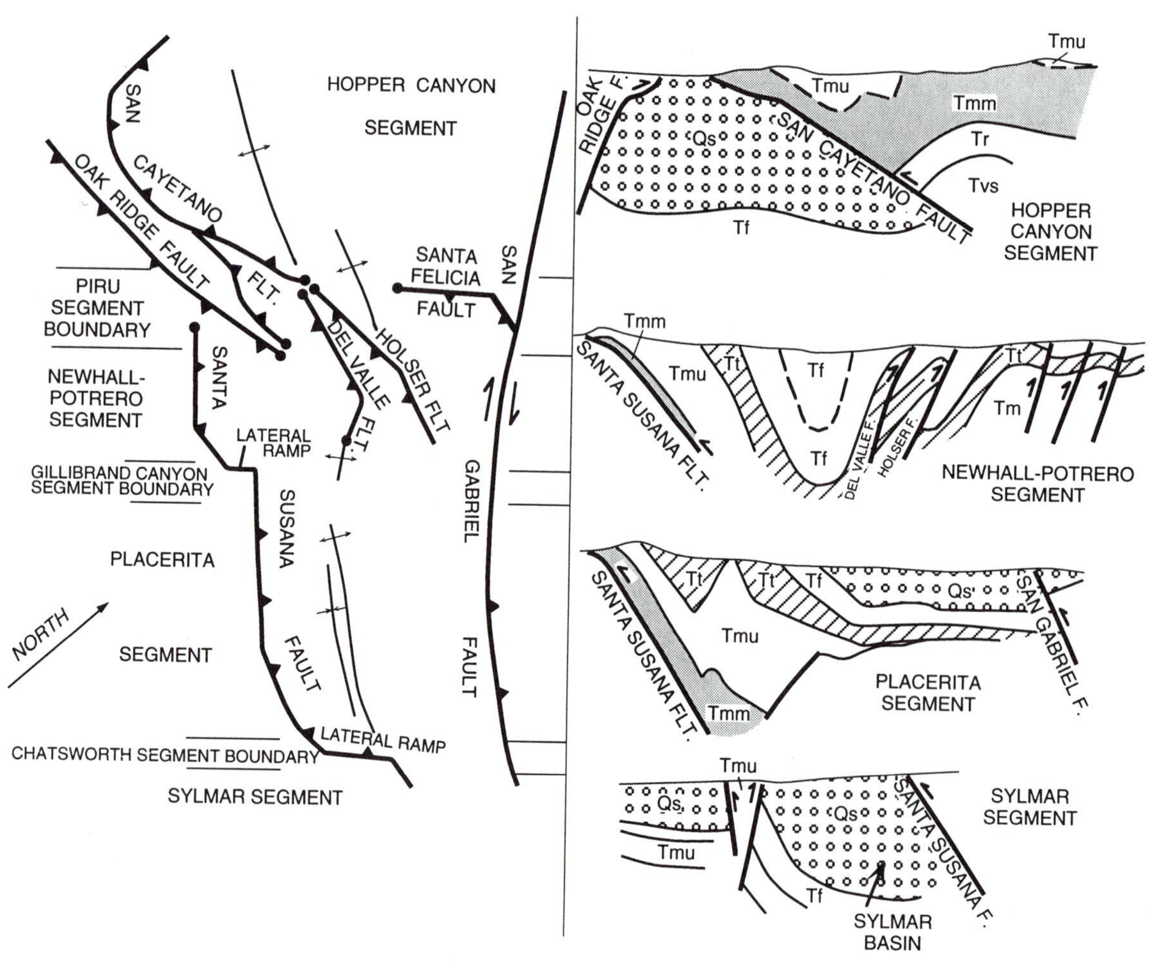

Figure 10–47. Segmentation of the Santa Susana-San Cayetano fault system in the east Ventura basin, California. Tvs, Paleogene strata; Tr, Tm, Tmm, Tmu, Tt, Miocene to earliest Pliocene strata; Tf, Pliocene strata; Qs, Pleistocene strata. Cross sections are schematic and are representative of entire segment. Note the contrast in hangingwall structure between segments. After Yeats et al. (1994).

1971 occurred near the base of the Pleistocene (Qs) in the Sylmar basin, and the Santa Susana fault was not reactivated. At the southern range front of the Santa Monica Mountains, the active fault steps southward into the Los Angeles basin, producing a fault scarp west of the Newport-Inglewood fault and a broad warp underlain by a blind thrust (Wilshire fault) east of the Newport-Inglewood fault (Hummon et al., 1994). In the northern and southern foothills of the Tianshan, China, and in most of the Himalayan front, the youngest folds coincide with the range front and deform late Quaternary alluvium.

However, west of the 1971 San Fernando fault, the most recently active strand of the Santa Susana fault is not along the Santa Susana Mountains range front, but is within the range. As the Santa Susana fault nears the ground surface, it curves to a near-horizontal attitude because it has been folded by the blind, south-dipping thrust that was activated by the 1994 Northridge earthquake (Yeats and Huftile, 1995; cf. Fig. 10-32a). A younger fault has cut through the curved fault and has emerged at the surface within the range (Fig. 10-48; Yeats et al., 1994). Out-of-sequence thrusting occurs in the Peshawar basin of Pakistan, far to the north of the Himalayan front (Fig. 10-3; Yeats and Hussain, 1989).

Segmentation

Is it possible on geological grounds to estimate how much of a fault zone will rupture during an earthquake? As in Chapter 9, we analyze the rupture zones of several earthquakes to look for a geological signature at the point of termination or initiation of earthquake rupture.

The 1971 San Fernando, California, earthquake nucleated near a lateral ramp on the Santa Susana fault, and faulting propagated eastward (Fig. 10-32). This lateral ramp is shown on Figure 10-47 as the Chatsworth segment boundary. To the west, the Santa Susana fault steps left in another lateral ramp, called the Gillibrand Canyon segment boundary, which generated a moderate-size earthquake in 1976 (Fig. 10-47). The Gillibrand Canyon lateral ramp is marked by a zone of pre-1994 seismicity that continues northeastward across the San Gabriel fault. The south-dipping aftershock zone of the 1994 Northridge earthquake terminated westward against this lateral ramp; aftershocks farther west delineated a north-dipping fault, probably the Santa Susana fault.

Still farther west, the Santa Susana fault dies out at the surface, and north-south shortening is accommodated on the San Cayetano and Oak Ridge faults. This displacement transfer zone may be a right-stepping lateral ramp at seismogenic depths, and it is called a segment boundary on Figure 10-47. This boundary marked the western edge of aftershocks of the 1994 Northridge earthquake.

Each of the segments defined in Figure 10-47 is characterized by a distinct structural style in the hangingwall; folds and faults within the segment terminate at the segment boundaries. For example, the Hopper Canyon and Sylmar segments are characterized by a thick Pleistocene nonmarine sequence in the footwall, whereas the intervening segments lack such a thick sequence.

The aftershock zones of the 1982 New Idria, 1983 Coalinga, and 1985 Kettleman Hills earthquakes are bounded by en échelon steps of the surface axes of the Coalinga and Kettleman Hills anticlines (Fig. 10-38), which suggests that the rupture zones of the 1982, 1983, and 1985 earthquakes were limited by lateral ramps in the blind thrusts controlling these folds (Stein and Ekström, 1992). However, the southeastern termination of the 1985 earthquake is not marked

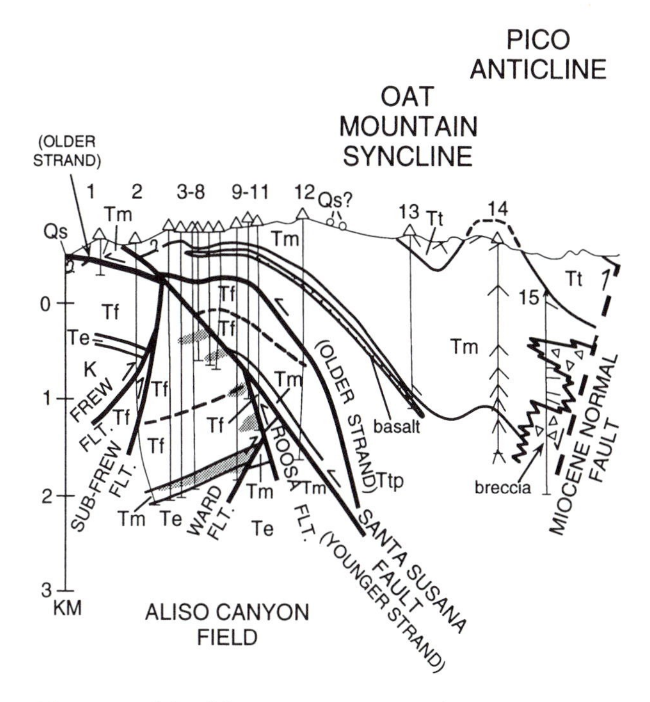

Figure 10–48. Geologic cross section of the Santa Susana Mountains (Placerita segment of Fig. 10–47) showing folded older strand of Santa Susana fault cut through by younger strand. Vertical lines with triangles show well control. K, Cretaceous strata; Te, Eocene strata; Tm, Miocene strata; Tf, Pliocene strata; Qs, Pleistocene strata. After Yeats et al. (1994).

by an obvious offset of the axis of the Kettleman Hills North Dome (Fig. 10-17). The fold-axis offset between the 1982 and 1983 earthquakes contains a northeast-trending zone of strike-slip aftershocks (Fig. 10-37), reminiscent of the strike-slip aftershocks of the San Fernando earthquake at the Chatsworth segment boundary (Fig. 10-32b). The reverse-faulting aftershocks northwest of both of the segment boundaries may be due to secondary or sympathetic rupture, such as that documented in the Willow Creek Hills and Warm Spring Valley at the northwestern end of the Borah Peak earthquake rupture zone.

The mainshock of the 1978 Tabas-e-Golshan earthquake nucleated down-dip from the southern end of surface rupture at the southern end of the Shotori Range (outlined by contours in Fig. 10-27; Berberian, 1979). The northern end of surface rupture, aftershocks, and damage equal to or greater than Intensity VII is located near the northern end of the Shotori Range. In this northern area, 1978 surface ruptures were discontinuous, and the northernmost ruptures dipped southwest, opposite to the dip of the main Tabas fault (Berberian, 1979). Damaging earthquakes occurred in this northern area in 1939 and 1974, whereas the main part of the Shotori Range had been seismically quiet for at least 1100 years. Similarly, the northern Honshu surface rupture of 1896 involved nearly all of the eastern edge of the Yokote basin adjacent to the Mahiru Range in northern Japan (Fig. 10-41). South of the 1896 rupture, the basin ends. An isolated northern fault southeast of Tazawa Lake ruptured in 1896, but total separation across this rupture is small compared to that on the main Senya fault, indicating that most previous ruptures were limited to the fault at the east edge of the Yokote basin. These earthquakes occurred on faults bounding ranges and basins that are a few tens of kilometers long. Earthquake rupture terminated near the point of topographic termination of the range and basin.

Segmentation of the Himalayan front is suggested by the rupture zones of the great 1905 earthquake, which had two centers of high intensity that coincided with reentrants in the Main Boundary thrust between the Ravi and Ganga rivers (Fig. 10-3). The 1905 rupture zone coincides with that part of the Himalayan foothills characterized by folding and dun valleys (Yeats et al., 1992). The southeastern edge of the rupture zone is located approximately at the site of a NE-trending basement ridge (Delhi-Hardwar Ridge in the underlying Indian plate. Between the rupture zones of the 1905 and 1934 earthquakes, the Himalayan front is marked mainly by an imbricate thrust zone.

Earthquakes on blind reverse faults beneath the Zagros fold belt are terminated at strike-slip faults that cut obliquely across the folds, such as the Kazerun, Borazjan, and Balarud faults (Berberian, 1995; Fig. 10-5K, B).

In summary, en échelon stepovers have served as earthquake rupture boundaries in the Transverse Ranges and Coast Ranges of California. In range-and-basin provinces in Iran and Japan, earthquakes ruptured the entire range front, limited by the terminations of the ranges and basins. In the Himalayan front, structural anisotropy in the subducting Indian continental plate may control the east end of the 1905 earthquake rupture. The 1905 and 1934 earthquake zones are characterized by a broad frontal fault zone with blind thrusts and dun valleys, whereas the area between the 1905 and 1934 ruptures is characterized by a narrow imbricate thrust zone.

Folds Related to Reverse Faults

Flexural-Slip Faults

Much of the near-surface deformation in contractile regions is expressed by flexural-slip (parallel) folding, in which a sequence of relatively stiff beds alternating with thin, more flexible, less-stiff layers is end-loaded, and buckling of the beds is accommodated by slip on mechanically weak bedding surfaces (Chapter 2; Fig. 10-49a; Yeats, 1986a, b). The layer thickness remains approximately constant across the fold hinge, and, except for the effects of compaction, original layer thickness is preserved. Such folds must form principally through layer-parallel slip and are referred to as *flexural-slip folds*. Slickensides on slip surfaces are typically perpendicular to the fold hinge. Upper layers slip over underlying layers toward anticlinal hinges and away from synclinal hinges. Slip is zero at fold hinges. On the limbs of the fold, the amount of slip on bedding faults depends on the maximum dip of the limbs and the thickness of each flexed layer. Because such bedding faults, or *flexural-slip faults,* do not produce stratigraphic separation, they are difficult to recognize unless there is a sequence overlying the flexural-slip fold wth angular unconformity, and continued folding causes deformation of these younger deposits as well (Fig. 10-49a). The younger deposits may be cut by faults that are parallel to bedding in the underlying folded sequence and upthrown toward the subjacent synclinal axis. The younger deposits are likely to be tilted toward the synclinal axis so that their dip is in the same direction as the underlying, more strongly folded beds.

Flexural-slip faulting was described by Suggate (1957) in the Grey-Inangahua depression in the northwestern South Island, New Zealand (Yeats, 1986a; Fig. 10-50a). A structural trough containing marine and

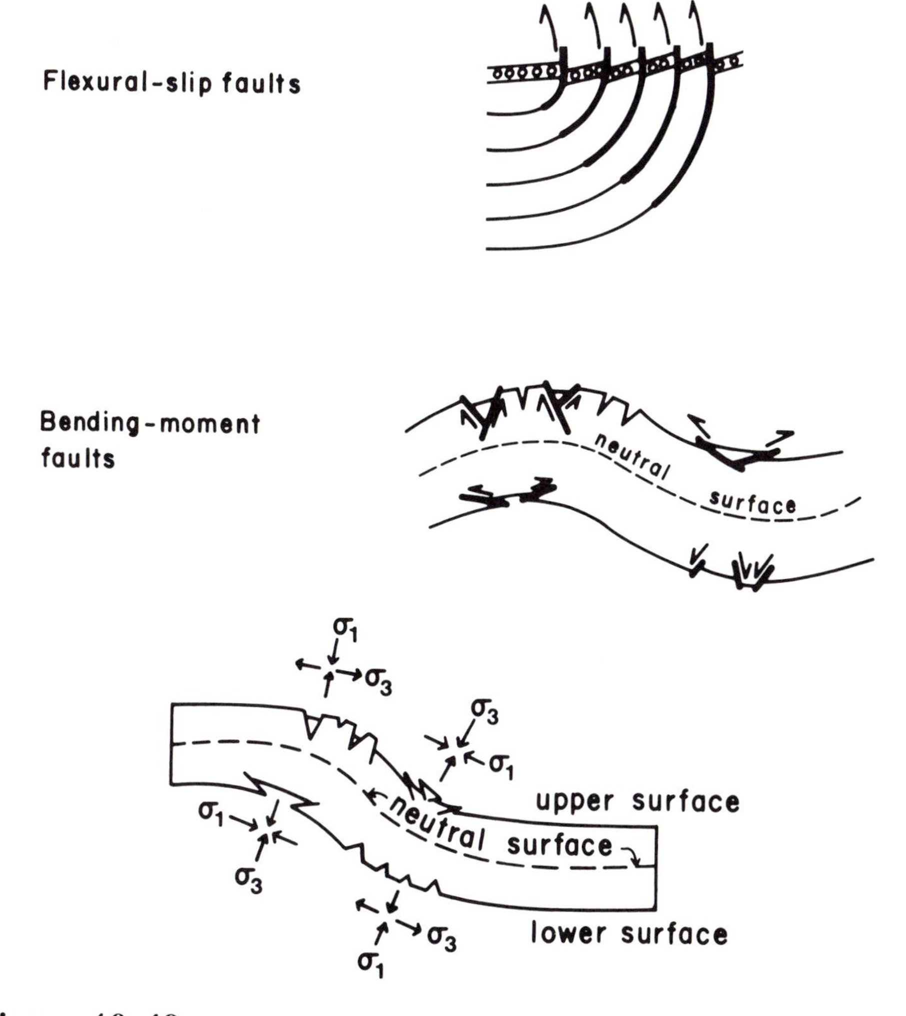

Figure 10–49. Secondary faulting accompanying flexural-slip folding. Top: Flexural-slip faults. Bottom: Bending-moment faults. Modified from Yeats (1986b).

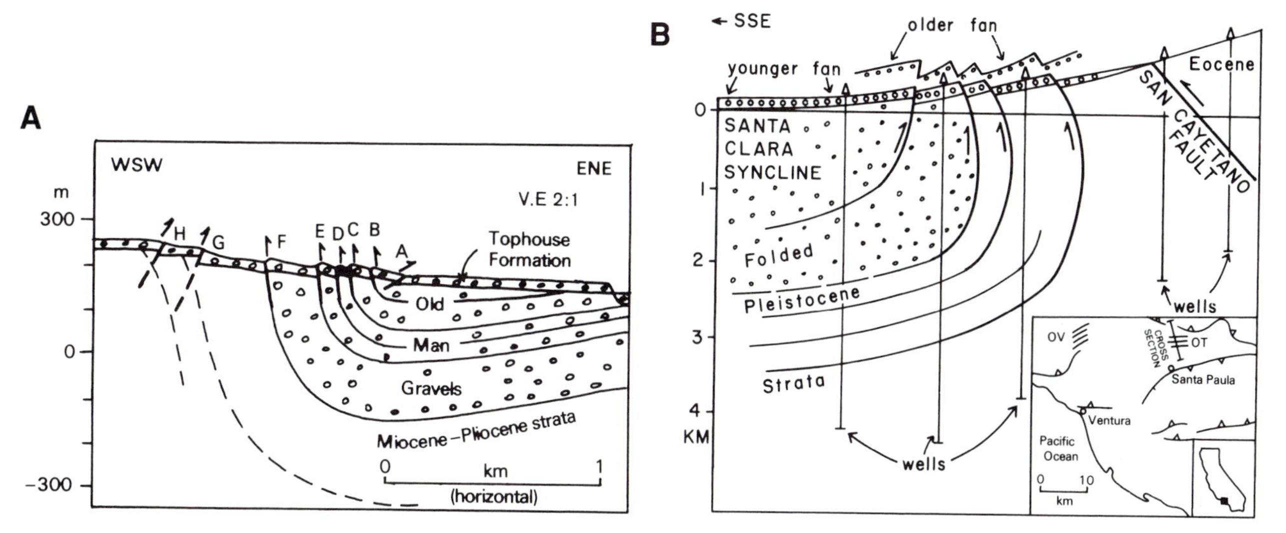

Figure 10–50. Flexural-slip faults. (a) Giles Creek, Grey-Inangahua depression, northwestern South Island, New Zealand. Vertical exaggeration 2:1. After Suggate (1957) and Yeats (1986a, b). (b) Santa Clara syncline, Ventura basin, California. No vertical exaggeration. After Rockwell (1983).

nonmarine clastic strata as young as early Pleistocene has been folded asymmetrically and faulted on its west side against the Paparoa Range (Fig. 10-20). The folded strata are overlain with angular unconformity by glacial outwash gravels (Tophouse Formation) that form prominent terraces above modern river level. The gravels are tilted toward the axis of the subjacent syncline in the clastic strata and are cut by reverse faults B through F that are parallel to bedding in the underlying folded strata and whose hangingwalls are upthrown toward the axis of the syncline.

A similar case occurs in the Ventura basin of southern California. There, Pliocene and Pleistocene strata in the north flank of the Santa Clara syncline are overlain with angular unconformity by late Quaternary alluvial-fan gravels of various ages (Rockwell, 1983). Here, also, the surfaces of the fans are tilted toward the synclinal axis and are cut by faults that are parallel to bedding in the underlying, strongly folded strata. The upthrown sides of the faults are on the side of the synclinal axis (Fig. 10-50b). The faults show normal separation where the subjacent bedding is overturned and reverse separation where the bedding is upright. Progressively older fan surfaces are tilted more and have increasingly higher scarps. Like the example from New Zealand, the faults cutting the alluvial-fan gravels are related to bedding slip during folding of underlying strata.

The structural relations between bedding-slip faults and faults cutting overlying, near-horizontal gravels are observed in a quarry near the Shinano River of northern Japan (Fig. 10-51; cf. Ota and Suzuki, 1979). Two of the four faults in the quarry extend upward into terrace material from steeply dipping contacts between conglomerate and fine-grained strata, suggesting that the location of flexural-slip faults is influenced by mechanical contrast between steeply dipping layers.

In 1981, a very shallow earthquake of $M_L 2.5$, possibly triggered by removal of overburden in a diatomite quarry, occurred near Lompoc, California, and was accompanied by at least 575 m of surface rupture on a bedding-plane reverse fault (Yerkes et al., 1983).

Bending-Moment Faults

Deformation also occurs within stiffer flexed layers of folds. This may be treated by considering a bed as an elastic plate that is bent by equal and opposite moments applied at its ends. The convex side is lengthened and placed in tension, and the concave side is shortened and placed in compression (Fig. 10-49b), producing a couple or bending moment. Between the part of the plate in compression and the part in tension, there is a neutral surface on which there is neither compression nor tension. The strain at any level within a bent plate is approximately proportional to the distance above or below the neutral surface and inversely proportional to the radius of curvature of the plate. On the convex surface, the minimum principal stress (σ_3) is tangent to the plate surface and perpendicular to the axis of bending, whereas on the concave surface, the maximum principal compressive stress (σ_1) has this orientation. If the neutral surface is located at the center of the plate, the differential stress (σ_1-σ_3) is zero at the neutral surface and is maximum at the outer convex and concave surfaces. If the plate is structurally isotropic and homogeneous, and folding is upright, initial rupture above the yield stress will be by extension fractures or normal faults at the convex surface and by reverse faults at the concave surface. These faults, called *bending-moment*

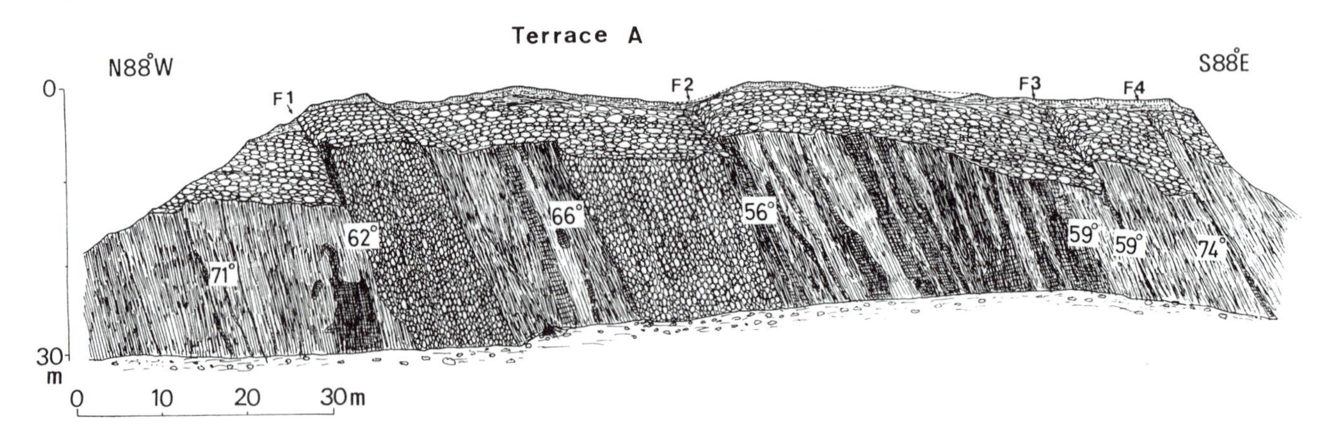

Figure 10–51. Quarry exposure of tilted and faulted Shinano River stream terrace unconformably overlying steeply-dipping Uonuma Group near Katakai, Niigata Prefecture, Japan. Bedding-plane faults have propagated upward into terrace deposits; faults F1 and F2 propagate from contacts between conglomerate and fine-grained strata of Uonuma Group. After Ota and Suzuki (1979).

faults (Yeats, 1986a, b), are commonly expressed geomorphically and are commonly observed in shallow excavations. The displacement on these faults is sometimes erroneously assumed to reflect displacement on the subjacent major fault under investigation.

Philip and Meghraoui (1983) presented evidence that normal faults and extension fractures on the hangingwall of the 1980 El Asnam thrust fault were produced by bending moment as the strata in the hangingwall were arched and folded (Fig. 10-52). Where the fault movement is purely dip-slip, the normal faults are parallel to the thrust-fault trace and to the anticlinal axis. Where faulting is oblique-slip, the normal-fault grabens develop in an en échelon pattern along the anticlinal axis.

Bending-moment faults are found on anticlinal crests in other regions. In the Yakima fold belt of central Washington State, the active Toppenish Ridge anticline has grabens along its anticlinal axis and reverse faults on the lower flank on the north side, close to the synclinal axis (Campbell and Bentley, 1981; Yeats, 1986b). Similarly, anticlines rising through the alluvium in the Qilian Shan of western China develop bending-moment faults parallel to anticlinal crests (Meyer, 1991; Fig. 10-59b).

The normal faults that bound the grabens are secondary faults that do not extend to significant depth. The folds that control them do not extend to seismogenic depth either, so these structures themselves are unlikely to produce large earthquakes.

Folds controlling flexural-slip and bending-moment faults may be related to major faults, as discussed in Chapter 2, and these faults may be seismogenic. Indeed, there are no known examples of flexural-slip faults or bending-moment faults that formed aseismically. On the other hand, there is paleoseismological evidence that the formation of the Ventura Avenue anticline in California and the Sara El Maarouf anticline in Algeria was accompanied by earthquakes. Flexural-slip folds and their accompanying flexural-slip faults and bending-moment faults provide useful information about timing of adjacent or subjacent faults that otherwise may not be evident.

There are several possible relations between folds and seismogenic faults. (1) The fold is in the footwall of an adjacent reverse fault, as in the Grey-Inangahua basin of New Zealand and the Santa Clara syncline in the Ventura basin of California (Fig. 10-50). (2) The fold is a fault-propagation fold that overlies a blind reverse fault at depth, as at Coalinga, California (Fig. 10-17b); (Chapter 2). (3) The fold is in the hangingwall of a flanking reverse fault, as in the 1988 Spitak, Armenia, earthquake (Philip et al., 1992) and the 1980 El Asnam, Algeria, earthquake (Philip and Meghraoui, 1983).

Tectonic Geomorphology of the Fault Trace

Reverse faults display a complex array of geomorphic expressions. Their morphology differs fundamentally from that of normal-fault scarps because the hangingwall block overrides the footwall block. The hangingwall may collapse onto the footwall, or it may roll over the footwall like the tread of a tank. In other cases, the fault flattens out below the surface instead of breaking the surface, and its surface expression is a broad warp. Because of these complexities, scarp-degradation criteria used to estimate the age of normal faulting in the Basin and Range Province of the western United States (Wallace, 1977) generally cannot be used for reverse-fault scarps.

Newly formed scarps on reverse faults can be categorized into three general initial configurations. In the vertical to overhanging model, the fault propagates to the surface at a moderate dip, and the overhanging part of the hangingwall immediately collapses onto the footwall (Fig. 10-53a, b). The collapse may produce a free face and a debris slope composed of the remains of the hangingwall. If the surface trace is in bedrock, the fault scarp may be preserved, although tensile cracks commonly form in the hangingwall (Fig. 10-53a). In some cases, however, the reverse fault steepens to near-vertical close to the surface (Fig. 10-53a), as observed on the Salt Range frontal thrust in Pakistan (Yeats et al., 1984), the reverse fault accompanying the Tabas-e-Golshan earthquake in Iran (Berberian, 1979), and reverse faulting accompanying the 1988 Tennant Creek, Australia, earthquake (Crone et al., 1992). Similarly, the dips on the Sylmar segment of the San Fernando earthquake fault are too steep to project in a plane downward into the aftershock zone (Fig. 10-33), which suggested to Sharp (1975) that internal deformation may have occurred in the hangingwall block, and that there was a larger component of uplift adjacent to the fault than there was farther away.

In the tank-tread model (Fig. 10-53c, d), near-surface materials such as turf and soil are entrained in the fault and override the ground surface on the hangingwall block. This style of faulting was observed in thrust faults that moved outward from the dome in the crater of Mt. St. Helens in 1980-1981 prior to dome-building events (Fig. 10-54). As the thrust-fault scarp advanced, individual fragments of partly consolidated volcanic tephra rode down the scarp face and were entrained and overridden by the fault at the base of the scarp, like the treads on a tank. On the South Island, New Zealand, the steepest part of the

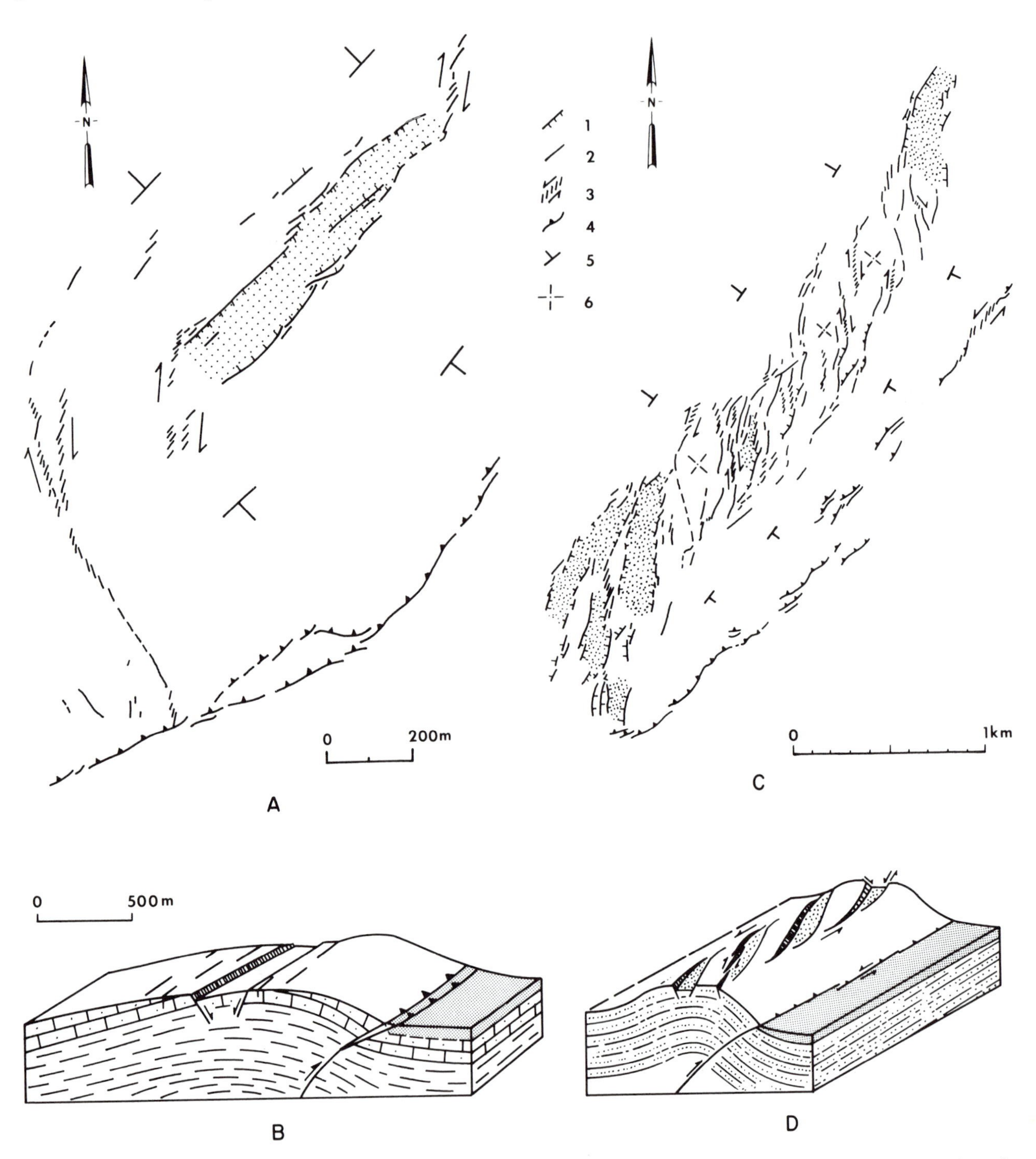

Figure 10–52. Coseismic bending-moment faults associated with 1980 El Asnam, Algeria, earthquake. Displacement on the main seismogenic thrust fault was accompanied by anticlinal folding in the hangingwall that resulted in the formation of grabens along the anticlinal crest. Grabens are shown by stipple pattern. These grabens are parallel to the anticlinal crest (a, b) or oblique and right-stepping along the anticlinal crest where the thrust faulting had a left-slip component (c, d). Map symbols: 1, normal fault; 2, tensile crack; 3, en échelon cracks related to strike-slip component of faulting; 4, thrust faults and pressure ridges; 5, attitude of dipping bed; 6, attitude of horizontal bed. After Philip and Meghraoui (1983).

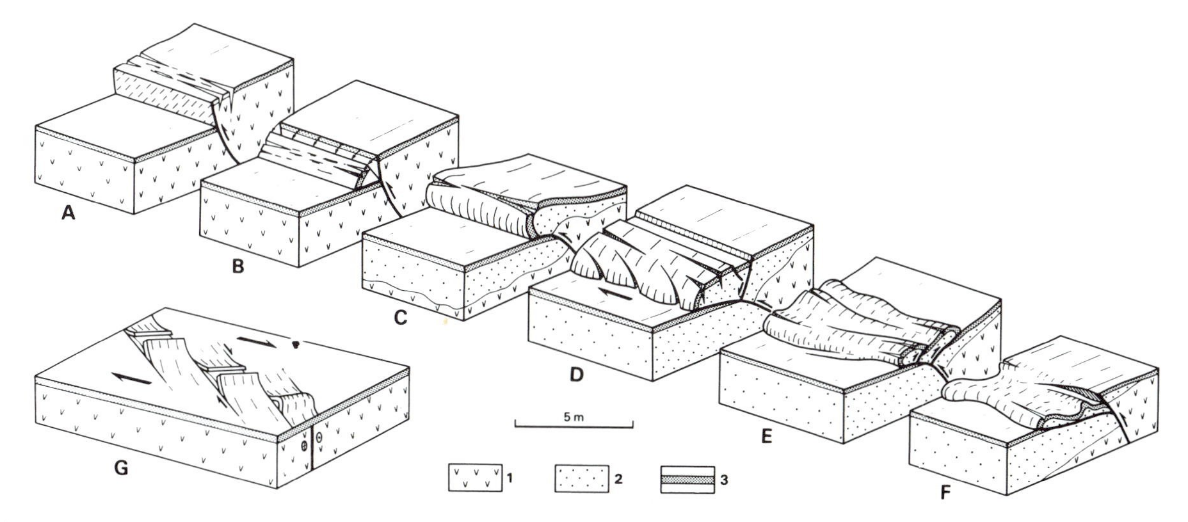

Figure 10–53. Fault-scarp features along the Spitak fault, Armenia. (a) simple thrust scarp; (b) hanging-wall collapse scarp; (c) simple pressure ridge; (d) dextral pressure ridge; (e) back-thrust pressure ridge; (f) low-angle pressure ridge: (g) en échelon pressure ridges. 1, bedrock; 2, soft Quaternary sediment; 3, turf. After Philip et al. (1992).

Ostler fault scarp is at the base of the scarp, suggesting entrainment of surface material during faulting (Fig. 10-55). Similar features were observed at Spitak, Armenia (Fig. 10-53c, d). The fault trace is commonly lobate, convex toward the footwall (Figs. 10-54, 10-55, 10-56).

In some cases, the fault may not reach the surface at all, and its surface expression is a broad, linear warp. Displacement in the uppermost layer may be distributed across a broad zone of unconsolidated materials, or the fault may become a bedding thrust at the base of the uppermost layer, with fault displacement being taken up near the surface by the steeper dip of the warp acting as an outcrop-scale fault-propagation fold. Broad warps were observed above several faults on the South Island, New Zealand, that later

Figure 10–54. Lobate reverse faults formed in 1981 around a growing dome in the crater of Mt. St. Helens (top of photo). Light-colored area is bounded on the dome side by a backthrust. From Chadwick and Swanson (1989).

Figure 10–55. Ostler fault, Mackenzie Country, South Island, New Zealand. The fault takes a lobate path across a glacial outwash surface 14,000 to 20,000 years old on the east side of the Southern Alps. Last movement was about 3500 years ago. Details of faulting include compressional wrinkles in foreground (left) and extensional or collapse features in background (above right). Power line in background gives scale. For location, see Figure 10–21. Photo by D. L. Homer, New Zealand Geological Survey.

trenching revealed to have their last displacement in latest Pleistocene. Age estimates of these scarps, using scarp-degradation criteria established for simple normal faults, would have predicted an erroneously great age for the faulting, because the initial scarp had a low slope angle produced by warping rather than by erosion of a fault scarp.

Gordon and Lewis (1980) made a detailed morphologic study of the 1968 Meckering, Australia, earthquake ruptures. The surface rupture occurred in a semiarid region of low relief where weathered Precambrian crystalline rocks are overlain by thin deposits of sand, alluvium, or soil covered by a thin grass mat. They recognized eight types of thrust scarps, shown in Figure 10–57. They did not observe tank-tread entrainment, but did note places where the grass mat and soil were rolled up and piled in front of the advancing thrust plate, with extensional cracks on the convex sides of these rolls (Fig. 10–57, examples 2, 4, and 6; see also Fig. 10–53f). A common feature at Meckering was a pressure ridge with thrust faults on both sides of the ridge; they called the main thrust a suprathrust, and the thrust facing the hangingwall block the subthrust (examples 3 and 4). (We prefer the term *backthrust* to subthrust.) The morphology of the scarp was controlled by the local orientation of the scarp with respect to the regional orientation of the fault, by the thickness and mechanical proper-

Figure 10–56. Oblique aerial photo of north-south limb of fault scarp formed during the March 30, 1986 Marryat Creek, Australia, earthquake. Throw of fault is about 0.5 m. Photo courtesy of Kevin McCue, Australian Seismological Centre, Bureau of Mineral Resources.

ties of the soil, and by the thickness of the vegetation mat. Extensional fractures were also observed at Spitak (Fig. 10–53d, f), on the Ostler fault in New Zealand (Fig. 10–55), on the Meers fault, Oklahoma, and on the Tennant Creek faults in Australia.

Map-pattern details of the ruptures at Meckering are shown in Figure 10–58. These patterns include en échelon thrust-fault scarps connected by extensional fractures that have a strike-slip component of displacement (Fig. 10–58a), en échelon fractures connected by pressure ridges (Fig. 10–58b; see also Fig. 10–53g), en échelon thrust-fault scarps connected by strike-slip faults (Fig. 10–58c), and pressure ridges bounded alternately by suprathrusts and backthrusts (Fig. 10–58d). In the first three patterns, the map trace of individual compressional features is oblique to the trend of the fault at depth, and the compressional features step to the right; the fault had a component of right-lateral strike slip as well as reverse slip. Double or multiple reverse-fault traces were also common, and the sum of displacement on multiple traces was the same as the displacement on nearby single traces. Commonly, the displacement on a single trace would diminish along strike, and displacement on another, parallel trace would progressively increase until it became the main trace, and the other scarp disappeared without visible connection to the parallel trace.

New structures rising out of the alluvium adjacent to the Qilian Shan of western China near the 1932 Changma earthquake surface rupture have a distinctive map pattern expressed by older alluvial-fan deposits (Meyer, 1991). The older fanglomerate tilted by warping and truncated by thrust faults forms a triangular pattern with the base of the triangle marked by the thrust fault in which the fanglomerate is faulted against recent fan material (Fig. 10–59a). For broader anticlinal uplifts not adjacent to a surface thrust, the outcrop pattern of the fanglomerate is diamond-shaped, reflecting the dip of the fanglomerate away from the anticlinal axis, which is marked by bending-moment normal faults (Fig. 10-59b). Meyer (1991) was able to recognize these features from SPOT satellite imagery.

In the Shinano River area near Niigata, Japan, the terrace is folded, and the anticlinal shape is preserved, as illustrated by the "Shinkansen fold" of Figure 10-60.

Expression of Reverse Faults and Folds in Excavations

Shallow excavations are useful for paleoseismological investigations (dating prehistoric faulting events and determining recurrence intervals for earthquakes) and for describing the internal structure of fault scarps and pressure ridges. Because of the variability of fault dip near the surface, there is considerable uncertainty about what one will find when a reverse-fault scarp is exposed in the shallow subsurface. The fault may intersect the surface partway down the scarp if the hangingwall has collapsed or sagged over the footwall, or it may reach the surface at the base of the scarp if the fault followed the tank-tread model (Fig. 10-53c). If the scarp reflects only warping at the surface, then an excavation may not expose a fault at all, or the fault may flatten below trench depth with displacement diminishing to zero. Where a fault cuts unconsolidated gravels, its only expression in an excavation may be a narrow zone in which shearing has

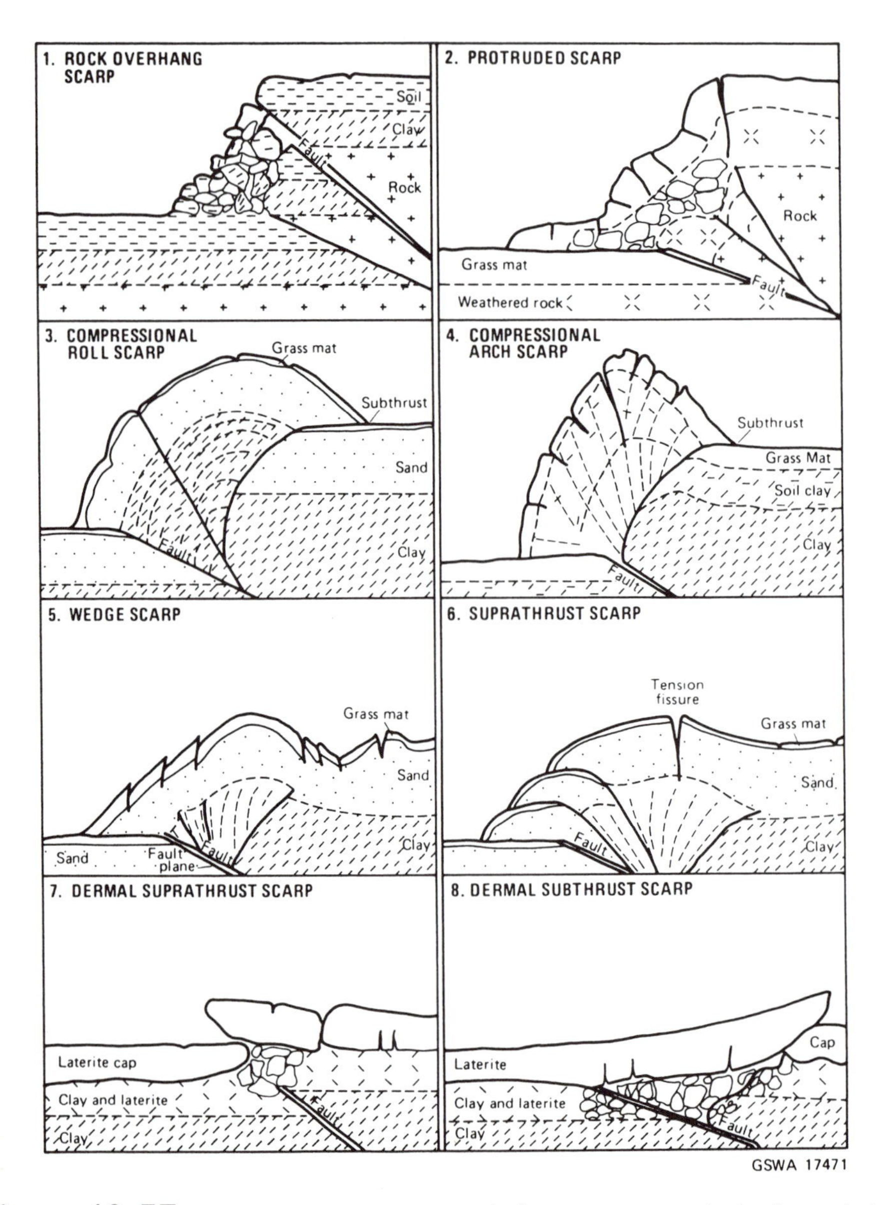

Figure 10-57. Variations in scarp morphology on reverse faults formed during the 1968 Meckering, Australia, earthquake. After Gordon and Lewis (1980).

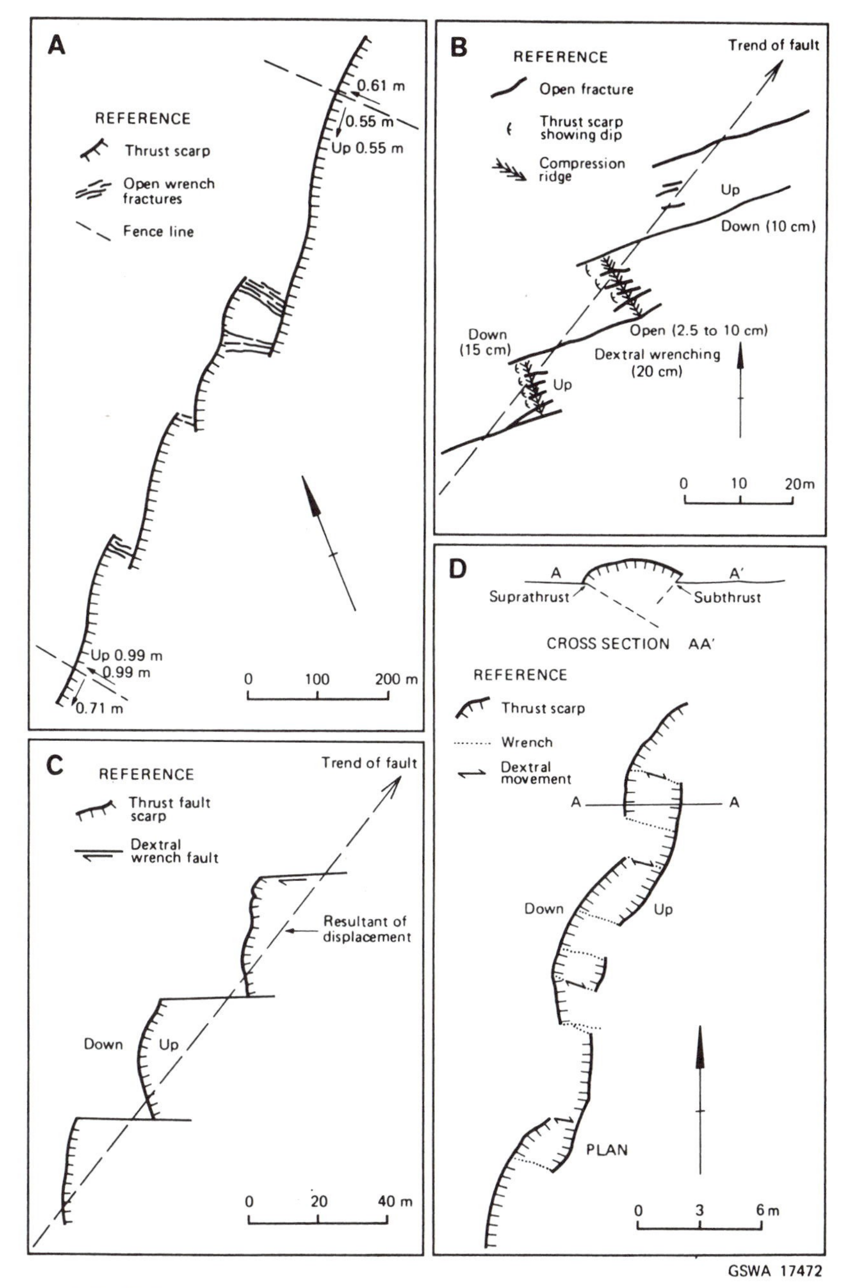

Figure 10–58. Typical fault patterns on the 1968 Meckering fault. After Gordon and Lewis (1980).

rotated flat, originally horizontal clasts into orientations parallel to the shear zone. In some cases, the zone of faulting becomes wider upsection within the gravels and has no direct expression other than the disruption of the top or base of the gravel unit.

Site selection of excavations for paleoseismologic investigations should seek to maximize the possibility of finding datable material and a useful stratigraphic sequence. An excavation in overbank stream sediment or in sediment ponded by a fault scarp is likely to yield more valuable information than a site in a stream channel or away from the stream drainage altogether. An excavation across a smaller scarp may be preferable to one across a larger scarp, because it may be easier to correlate the stratigraphy between hangingwall and footwall, and evidence for more prehistoric earthquakes may be found within trench depth. On the other hand, a small scarp may be the expression of fewer earthquakes than a larger one, and some events might be missed. In addition, the larger scarp may give

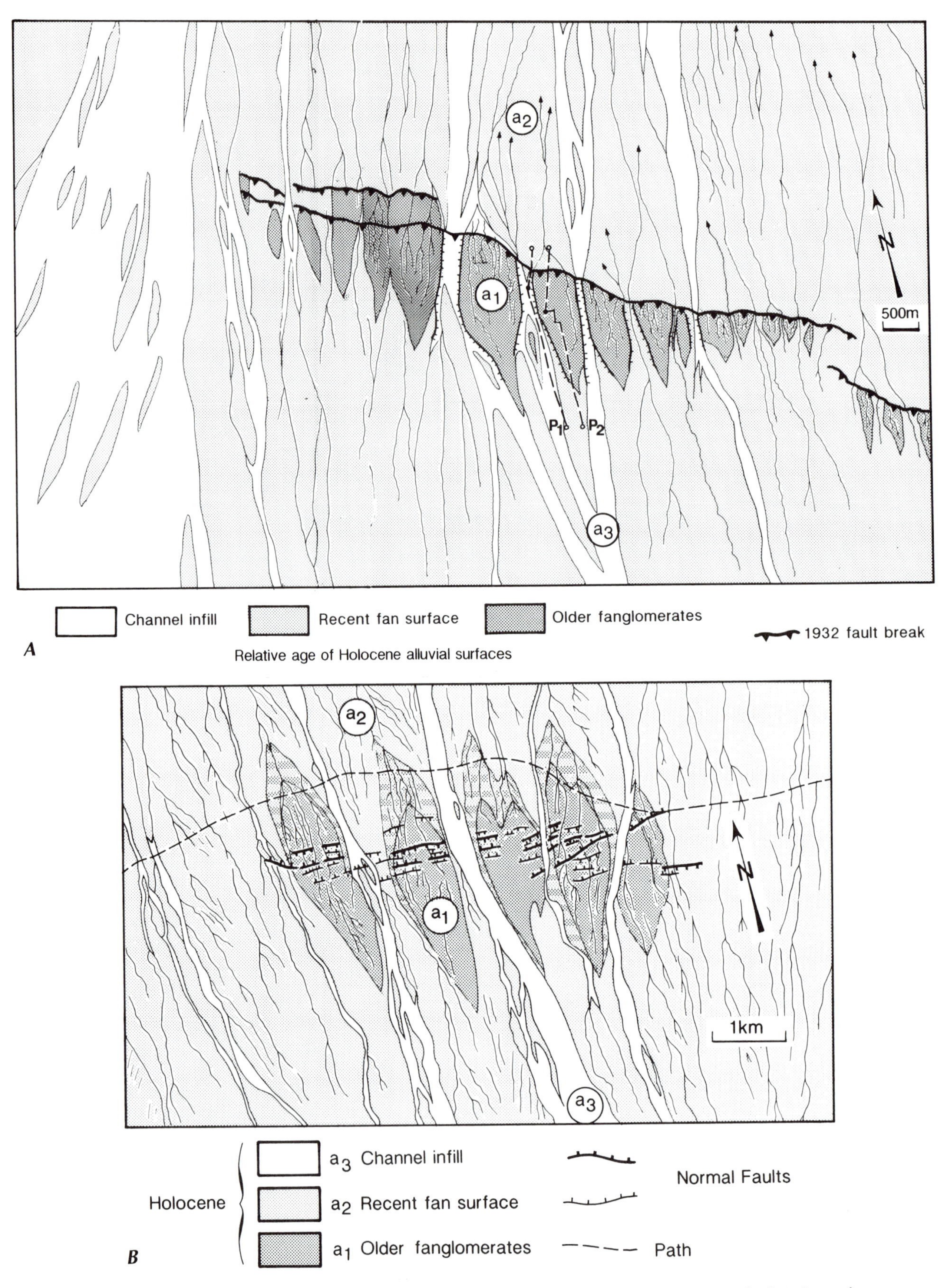

Figure 10–59. Map pattern of older fanglomerate uplifted along (a) active reverse fault (triangular pattern, with base of triangle along fault trace). (b) Active anticline (diamond pattern, with diamonds bisected by bending-moment faults along anticlinal crest. Examples are from piedmont of Qilian Shan in western China. After Meyer (1991).

Figure 10–60. Oblique aerial photo of river terrace folded into an anticline ("Shinkansen fold") at Nanokaichi, Shinano River area, near Niigata, Japan. After Ota and Suzuki (1979).

information about maximum slip in an earthquake, which is a measure of earthquake size.

Figure 10-61 shows a log of a bulldozer trench from the Dunstan fault zone in Central Otago, South Island, New Zealand (Beanland et al., 1986). Faults A and B follow bedding planes in Tertiary sedimentary rocks and cut an overlying sequence of fan gravels that are subdivided based on the abundance or the lack of graywacke clasts. Fault A cuts the basal graywacke-schist gravel, because the base of this unit is displaced,

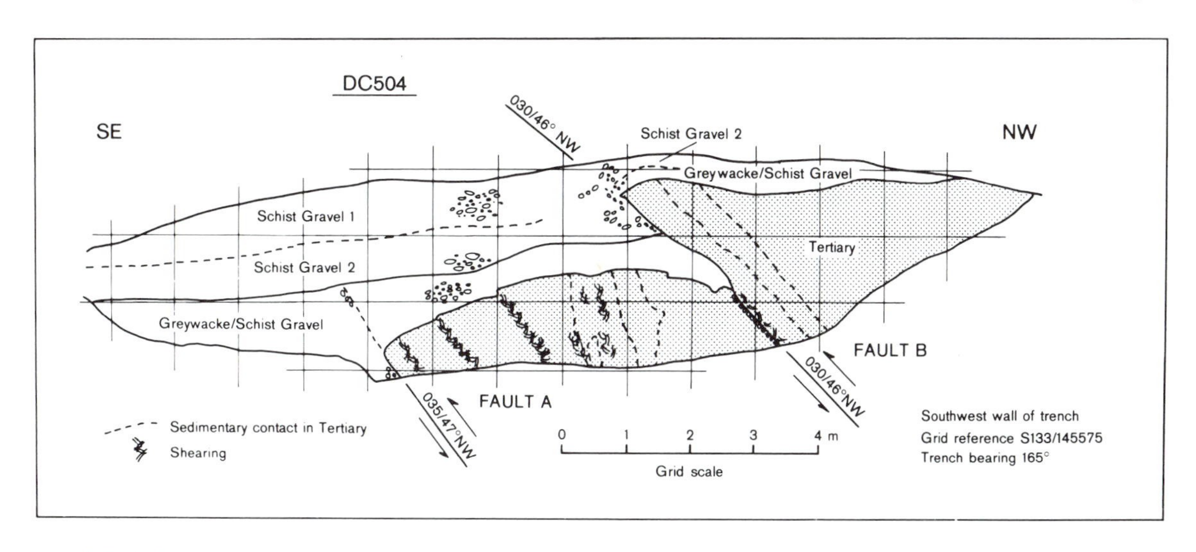

Figure 10–61. Trench log in Dunstan fault zone, Central Otago, New Zealand. For general location, see Figure 10–21. After Beanland et al. (1986).

but the base of the overlying schist gravel 2 is not displaced, indicating that the gravel postdates faulting. Fault B cuts all of the gravel units and is, therefore, younger than Fault A. However, the graywacke-schist gravel is thicker on the downthrown side of Fault B, suggesting that Fault B also moved at the same time as Fault A. Note that the Dunstan fault zone is within the Manuherikia basin rather than at the Dunstan range front. These fault scarps in the basin are smaller than those at the range front, and the Quaternary sediments in the basin are finer grained than those at the range front. Both of these factors made it preferable to excavate the scarps in the basin.

The Oued Fodda fault at the southeast flank of the Sara El Maarouf anticline ruptured during the 1980 El Asnam, Algeria, earthquake. Figure 10–62 is a photograph of a trench wall that exposed this fault, and Figure 10–63 is an interpretation by Meghraoui et al. (1988) identifying two pre-1980 rupture events on the fault. Units b and c are the same thickness on both sides of the fault, indicating that the fault rupture postdated deposition of unit c. Units d and e are each thicker on the downthrown side, which is evidence that faulting events E7 and E8 occurred during or after deposition of each of these units. Faulting in 1980 formed the scarp shown in the bottom drawing of Figure 10–63. With time, erosion may remove some or all of unit f from the hangingwall, as it did units d and e after the earlier events. Radiocarbon dates of these units constrain the age of the faulting event, but do not date the age of faulting directly.

Figure 10–64 is a map of the wall of an excavation (commonly called a *trench log*) across a segment of the Cardrona fault, New Zealand, presented because of the fault history revealed in the excavation. The trench cuts across a distinct 2.8–3.8-m-high east-facing fault scarp on a westward-sloping terrace surface. Westward bulging and tilting of the hangingwall block suggests that the deformation is distributed across a zone tens of meters wide. A bedrock ridge (not shown) protrudes through the terrace and shows no lateral offset across the fault. On the downthrown side, a fan constructed by a small stream has buried the terrace. In the trench, schist is overlain by a gravel

Figure 10–62. Photograph of trench wall exposing reverse fault at south edge of Sara El Maarouf anticline, Algeria; fault moved during 1980 El Asnam earthquake. Trench is located 1 km NE of canyon of Oued Cheliff River where it cuts through the anticline. Photo by M. Meghraoui; see Meghraoui et al. (1988).

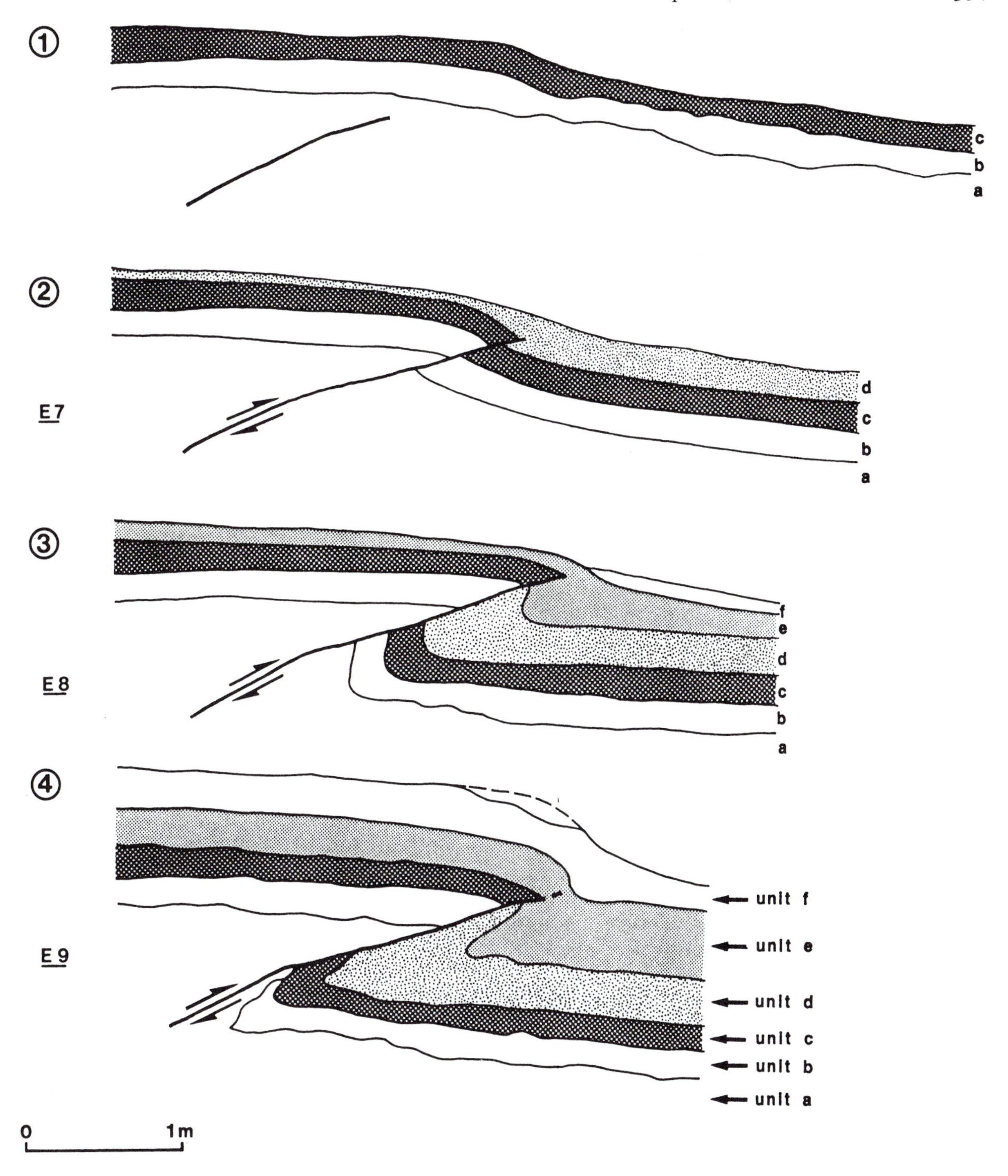

Figure 10–63. Interpretation of trench shown in Figure 10–62. Units are mostly gravels admixed with silty sands that are part of an alluvial fan. Movement episodes E7, E8, and E9 inferred from increased thickness of units on downthrown side. E9, which formed a fault scarp, formed during the 1980 earthquake. After Meghraoui et al. (1988).

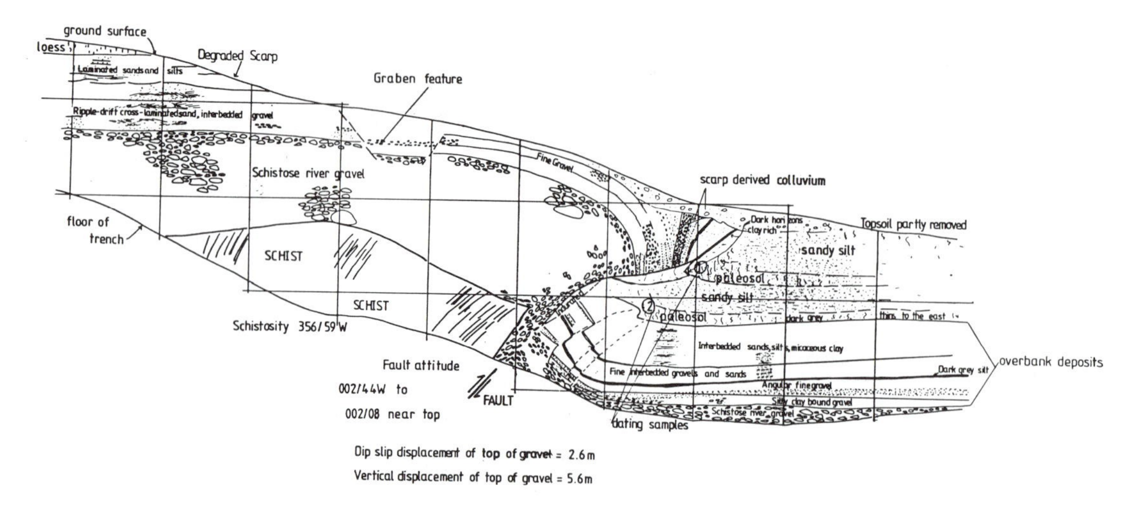

Figure 10–64. Trench log of Kawarau River segment of Cardrona fault in Central Otago, New Zealand. Grid is 2 m to the side. After Beanland and Barrow-Hurlbert (1988).

deposit of the Kawarau River. This gravel is overlain by 2 m of overbank sand, silt, and clay, which is overlain by a buried soil that is dated as 21,000 ±1600 years. This soil is buried by loess which is capped by another soil, which is dated as 10,500 ±600 years old, and by a younger loess. The youngest deposit is a poorly sorted, pebbly sand, which is interpreted as scarp-derived colluvium. The modern soil is developed on this unit. Figure 10–65 shows the interpretation by Beanland and Barrow-Hurlbert (1988). Before this latest Quaternary episode of faulting revealed in the excavation, the river terrace was abandoned. Following abandonment, a small tributary constructed an alluvial fan at the site of the excavation (Fig. 10–65a). Faulting event 1 (Fig. 10–65b) produced a scarp that blocked stream flow across this fan, leading to formation of a bog at the base of the new scarp. The radiocarbon age from bog sediments probably closely approximates the age of faulting event 1 or is slightly younger. Upon filling of the bog with sediment, drainage was restored, and loess accumulated on the lee side of the scarp. Faulting event 2 (Fig. 10–65c) initiated a similar cycle of sedimentation, including deposition of a younger loess. A third, undated event deformed the upper paleosol and produced the scarp-derived colluvium, upon which a young soil has formed. Note that the bogs and the scarp-derived colluvial wedge can be related directly to individual faulting events.

One of the most detailed trench logs of an active reverse fault was prepared by the Research Group for the Senya Fault (1986); for location of this fault, see Figure 10–41. The general stratigraphy (Fig. 10–66) consists of Tertiary mudstone unconformably overlain

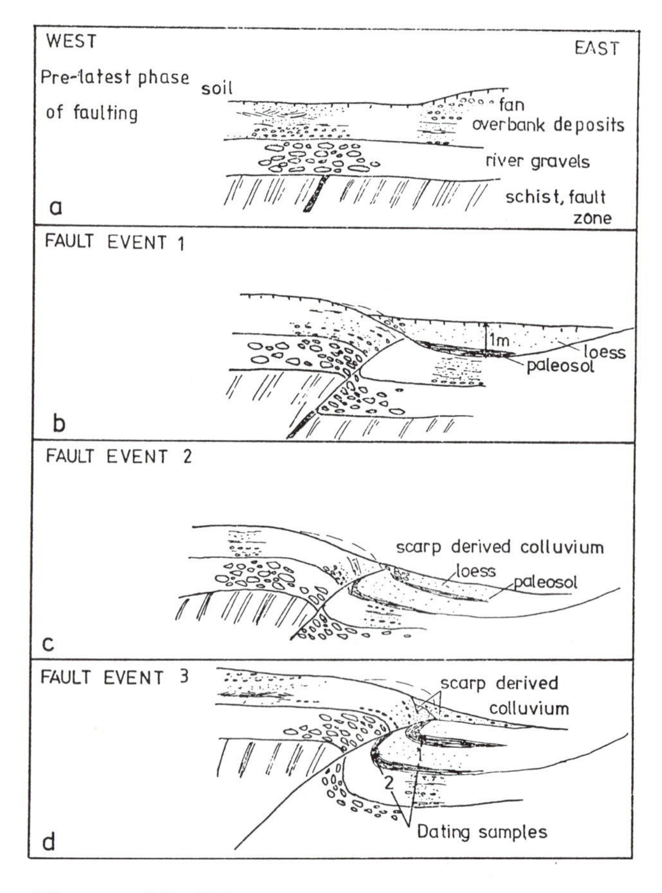

Figure 10–65. Sequence of fault events postulated by S. Beanland and D. Fellows (pers. comm., 1988; see Beanland and Barrow-Hurlbert, 1988) on data from Kawarau River trench.

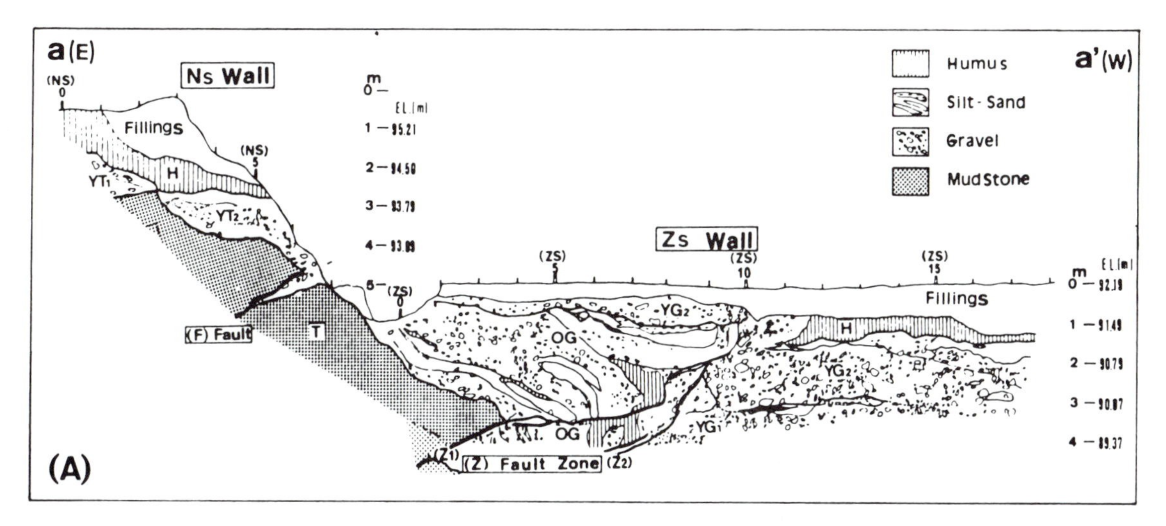

Figure 10–66. South wall of trench across the Senya fault, northern Honshu, Japan; the Senya fault ruptured the surface in 1896. T, Tertiary mudstone; OG, older gravel; YG_1, lower beds of younger gravel (YG); YT_1, higher terrace gravel equivalent to YG_1; YG_2, upper beds of the younger gravel (YG); YT_2, lower terrace gravel; H, soil, including pre-1896 rice-paddy material; Fillings, post-1896 fill. Scale divisions in meters. After Research Group for the Senya Fault (1986).

by old gravel, young gravel, and humus (soil), the last in part due to a rice paddy cultivated prior to the 1896 earthquake. The excavation is in the overlap zone between two en échelon reverse faults F and Z; fault Z consists of two strands, Z_1 and Z_2. Fault Z is nearly horizontal in the trench, and the old gravel (OG) has been dragged into the fault. Humus has also been entrained along the fault. A detail of the front of Fault Z (Fig. 10-67) shows that near the surface, the fault is lined with fine sand, which apparently followed a bed-

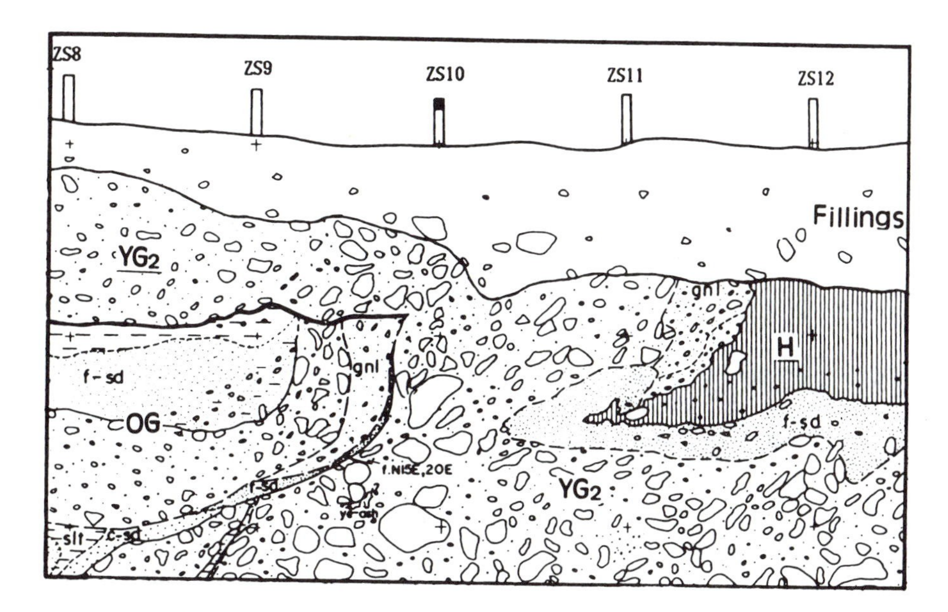

Figure 10–67. Detailed log of part of trench wall shown in Figure 10–66. Older gravel (OG) is folded prior to deposition of younger gravel (YG) and prior to last movement on fault. Humus and underlying sand are folded recumbently during 1896 earthquake. Note tilting of flat gravel clasts in YG adjacent to fault. Markers ZS8, ZS9, etc. are spaced one meter apart; no vertical exaggeration. After Research Group for the Senya Fault (1986).

ding plane in the older gravel. The humus layer in front of the fault was folded recumbently, probably in 1896. However, the angular unconformity in this trench shows that the folding of the old gravel (OG) must predate the deposition of the young gravel (YG).

Part of the north wall of a nearby trench cut across the same fault is shown as Figure 10-68. Tertiary mudstone (T) is thrust over gravel, and a crush zone has developed at the base of the mudstone. The gravel is tipped upward in the hangingwall, and bedding is overturned by drag folding in the footwall. Where gravel is faulted against gravel, the fault can be recognized only by the anomalous steep dips of platy clasts in the gravel, as shown at the far left.

Bending-moment and flexural-slip faults are commonly observed in excavations, and the orientation and displacement on these faults is sometimes erroneously assumed to be the same as the displacement on the subjacent major fault under investigation. Two examples are given from the Ventura basin, California.

The Ventura reverse fault forms a linear, south-facing scarp that trends east-west along the foot of the hills in the northern part of the city of Ventura. The fault extends eastward away from the hills and across a prominent alluvial fan (Sarna-Wojcicki et al., 1976; Fig. 10-69). Trench excavations across the scarp showed several normal faults and extensional fractures that cut all but latest Holocene sediments. These faults and fractures are concentrated in the flexed area updip from the subjacent Ventura fault and are interpreted as being produced by bending-moment deformation on the convex side of a monoclinal drape over the surface projection of the Ventura fault.

In the city of Camarillo, south of Ventura, excavations across a linear, south-facing scarp of the Camarillo fault (Fig. 10-70) revealed an increase in dip of sediments close to the scarp. Instead of a high-angle, north-dipping reverse fault, the trenches exposed a low-angle, north-dipping thrust and several high-angle, south-dipping reverse faults. These faults were interpreted as bending-moment faults on the concave side of a monoclinal bend in Quaternary sediments that are either draped over a buried fault or are part of a pressure ridge.

In these examples from the Ventura and Camarillo faults, the age of youngest sediments that are cut by secondary faults can be used to constrain the time of movement on the buried fault. However, the orientation and sense of displacement of bending-moment faults cannot be extrapolated to infer the sense of displacement on the buried fault. Thus, bending-moment faults accurately record the history of movement but do not indicate directly the amount or style of movement on the subjacent fault.

Meghraoui et al (1988) recognized prehistoric earthquakes on the Sara El Maarouf anticline, Algeria, in excavations across secondary faults. A bending-moment fault on the crest of the anticline produed a scarp 28 cm high in 1980 (Fig. 10-71a). Total separation of Unit a is 1.28 m, and both Unit a and Unit b are thicker on the downthrown side, which is interpreted as evidence of two or three earthquakes prior to 1980. In Figure 10-71b, exposure of a flexural-slip fault that produced a scarp 50 cm high in 1980 revealed progressively greater offsets of older horizons. During the 1980 earthquake, a flexural-slip fault ruptured the surface along a bedding contact between Pliocene calcareous deposits and Pleistocene gravel, very much like the flexural-slip

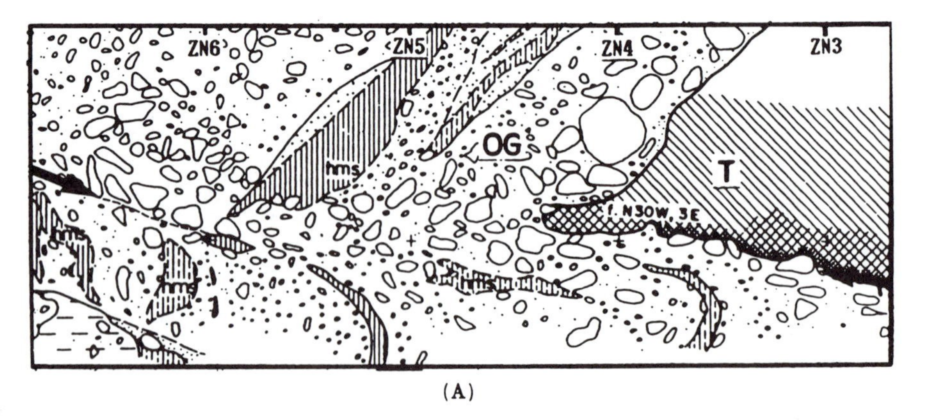

(A)

Figure 10–68. Part of another trench wall showing overturned soil layers in footwall of thrust. Fault is marked by crushed zone (cross-hatched) at base of Tertiary mudstone (T), but where gravel is faulted against gravel, the fault is recognized only by steeply tipped gravel clasts (left, at arrows). Markers ZN3, ZN4, etc. are spaced one meter apart, no vertical exaggeration. After Research Group for the Senya Fault (1986).

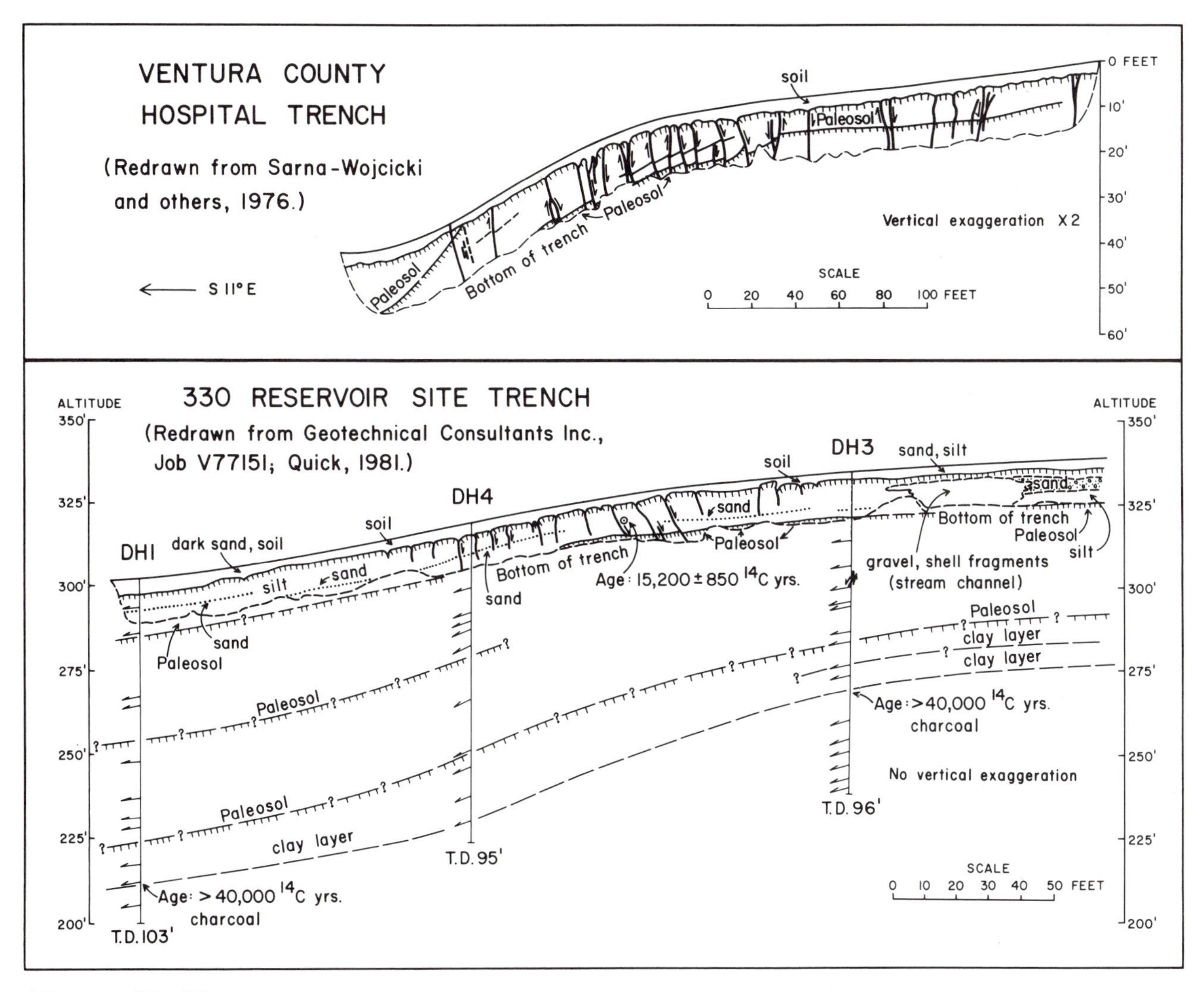

Figure 10–69. Two trench logs across Ventura fault, California. Ventura fault is expressed as flexure with slopes and dips steeper than areas to the north and south. In both trenches, normal faults or soil-filled extension fractures are concentrated in the area of maximum flexure, leading to an interpretation of bending-moment origin. After Yeats (1986b) and Sarna-Wojcicki et al. (1976).

faults whose locations are controlled by bedding anisotropy in northern Japan (Fig. 10–51). Logs of these trenches show that the secondary faults related to folding in 1980 had moved in earlier earthquakes and provided evidence that the major fold was formed by repeated earthquakes.

SYNTHESIS

Slip Rates and Recurrence Intervals

In this section, we summarize slip rates and earthquake repeat times on reverse faults based on paleoseismic investigations of both faulted strata and landforms.

The vertical displacement rate across the Senya fault, based on the offset of a 23,000-years-old surface, is 0.2–0.3 mm/yr, and the average recurrence interval of 1896-size earthquakes is 3000–4000 years (Matsuda et al., 1980). This recurrence interval is consistent with the 3500-year interval between the 1896 earthquake rupture and the age of the previous event, determined from stratigraphic evidence in excavations (Research Group for the Senya Fault, 1986). In the Kinki Triangle, Japan, the range-front fault at the eastern end of the Nara basin and the faults at the eastern edge of the Suzuka Range (Fig. 10–28) have vertical displacement rates of 0.3–0.6 mm/yr (Sangawa, 1986).

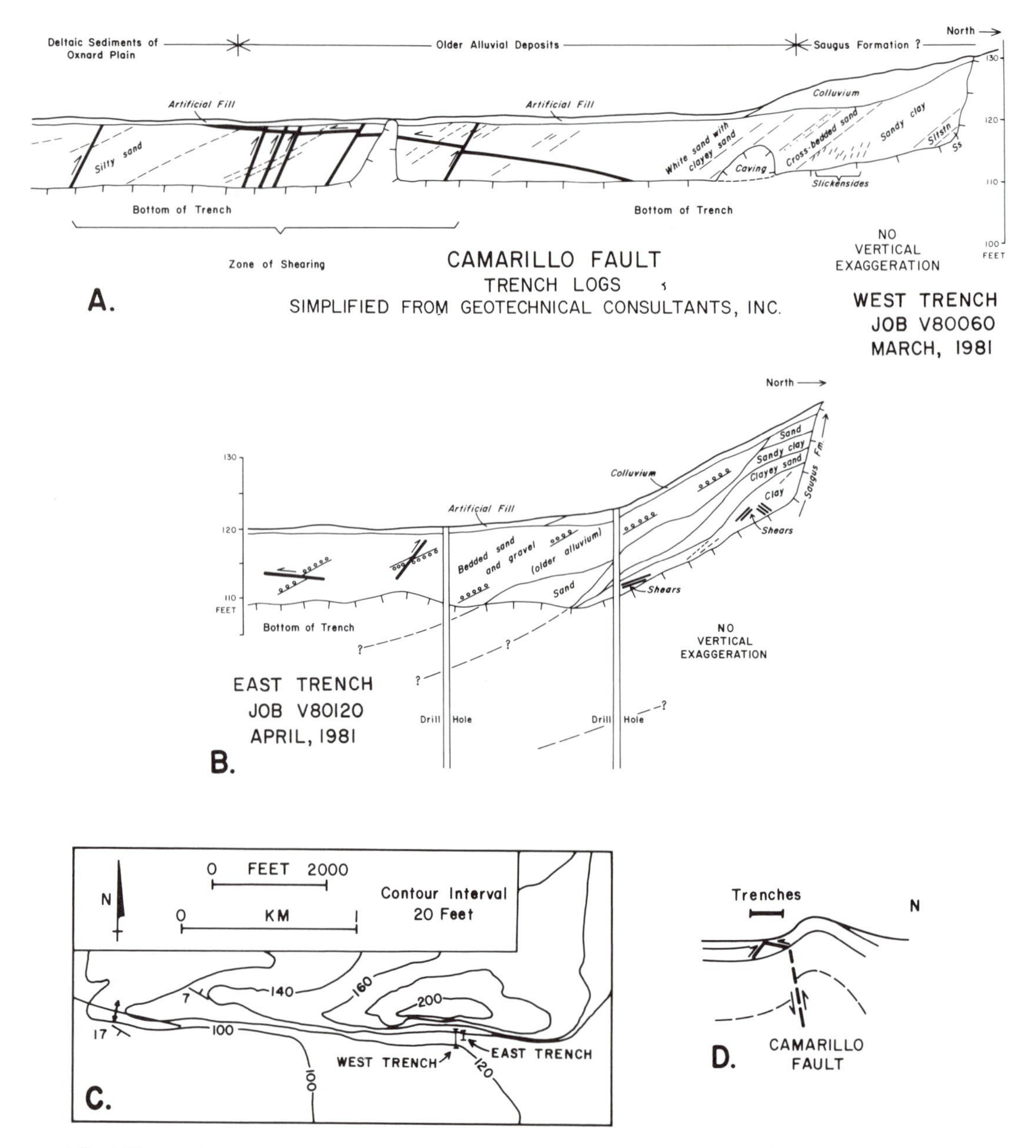

Figure 10–70. (a,b) Simplified sketches of trenches across Camarillo fault; no vertical exaggeration. (c) Topography of ridge in downtown Camarillo. Camarillo fault is presumed to control steep south flank of this ridge. (d) Sketch of interpreted relation between bending-moment faults in trenches and subjacent Camarillo fault. After Yeats (1986b).

Detailed mapping and paleoseismic excavations of deformed late Quaternary deposits along the Dushanzi-Anjihai fault in the northern Tianshan yield a vertical displacement rate of 0.5 mm/yr (Deng et al., 1991a, b) and an earthquake recurrence interval of 3000–4000 years based on two Holocene earthquakes dated in trenches.

Slip rate on the Oak Ridge fault in the Ventura basin, California, is 5 mm/yr, based on the offset of a 500-ka horizon (Yeats, 1988; 1993). Paleoseismic investigations are not yet available to calculate an average recurrence interval for earthquakes on the Oak Ridge fault. However, Sarna-Wojcicki et al. (1987) determined the earthquake recurrence interval of the Javon Canyon fault to the west based on natural exposures of colluvial wedges derived from the fault. These colluvial wedges provide paleoseismic evidence for a recurrence interval of 700 years. Rockwell (1988) determined a slip rate of about 9 mm/yr on the nearby San Cayetano fault, based on the

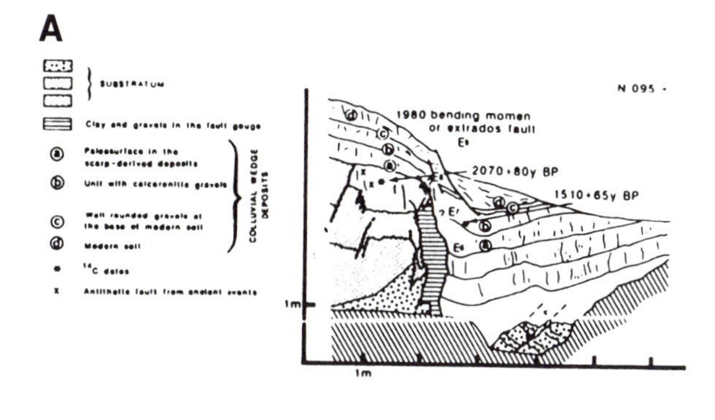

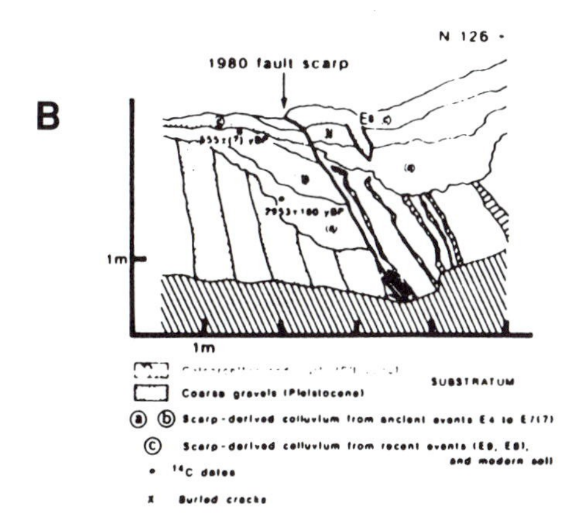

Figure 10–71. Trenches across (a) bending-moment fault and (b) flexural-slip fault, Sara El Maarouf anticline, Algeria. Holocene stratigraphic units a, b, and c are thicker on downthrown sides of faults, indicating earlier displacement. After Meghraoui et al. (1988).

amount of displacement on alluvial-fan surfaces of various ages. The slip rates for faults in the Ventura basin are high compared to other reverse faults in southern California, which is consistent with the unusually high rate of shortening across the Ventura basin, based on GPS satellite data (Donnellan et al., 1993).

Stratigraphic evidence from excavations shows that recurrence intervals based on long-term displacement rates must be viewed cautiously, because of *temporal clustering* of earthquakes. The El Asnam, Algeria, fault ruptured during several earthquakes in the last 1500 years, with an average recurrence interval of 300–500 years in the late Holocene. However, in the time interval between 1500 and 3500 years B.P., only one earthquake occurred (Meghraoui et al., 1988; Swan, 1988). Between 3500 and 4500 years B.P., a cluster of earthquakes occurred, whereas before 4500 years B.P., the fault was relatively quiet. Similarly, the Pisa fault in Central Otago, New Zealand, had two periods of activity, the last betweeen 70 and 35 ka (Beanland and Berryman, 1989). The period before 70 ka was quiet, and the fault has had no surface rupture since 23 ka.

The possibility of temporal clustering may explain a historical slip deficit in the Transverse Ranges. In the 200 years of recorded history in the Los Angeles-Ventura region, only two earthquakes as large as M_W 6.7 have occurred. However, convergence rates based on GPS measurements and geological data are so rapid that earthquakes of this size would be required every 11 to 16 years to relieve accumulating strain uniformly (Dolan et al., 1995). Two extreme scenarios illustrate the range of possible explanations. Perhaps large earthquakes of about M_W 6.7 occur in temporal clusters, and the period of historical record-keeping lies between a prehistoric and a future cluster. This scenario would be analogous to the Holocene history of normal faulting in the central Nevada seismic zone, which has been struck by several earthquakes during this century, but was relatively quiet during several prior millennia. Alternatively, the Los Angeles-Ventura region may be subject to occasional earthquakes as large as M_W 7.6, such as those that have struck northwestern Argentina, northeastern Iran, northern Honshu, Kyrgyzstan, and Costa Rica. Dolan et al. (1995) show that reverse-fault systems large enough to produce M_W 7.2 to 7.5 earthquakes do exist in the region, and they calculate that earthquakes within this magnitude range could be expected about every 140 years. Such a large-event scenario would not require temporal clustering.

The 1991 Racha, Georgia (M_s 7.0; Jibson et al., 1994; Triep et al., 1995), 1991 Costa Rica, and 1994 Northridge earthquakes occurred on blind thrusts, indicating that relatively large earthquakes may not rupture the surface and thus may go undetected based on subsurface excavations. The earthquakes that do rupture the surface may be only those that have greatest fault length and, accordingly, the greatest maximum slip. For this reason, it is necessary to look for evidence of warping in addition to evidence of fault displacement.

This is illustrated by the problem of determining the slip rate on the north-dipping Elysian Park blind thrust beneath the Santa Monica Mountains, California (Fig. 10–18; Davis and Namson, 1994), which is partially expressed at the surface as the Malibu Coast and Santa Monica faults. McGill (1989) found that the Stage 5e terrace (125 ka in age) is 48 m higher on the hangingwall side of this fault (locally named the Potrero Canyon fault) than on the footwall side. If the fault dips 45°, the dip-slip rate on this fault is 0.54 mm/yr. However, the present gradient of the marine abrasion platform is 1.5 to 6°, whereas the original gradient was probably no more than 1 or 2°. If the tilting of the platform is related to a blind thrust, the slip rate on the blind thrust must be added to the slip rate at the surface to determine the slip rate at seismogenic depths, which would account for the release of seismic moment accumulated by convergence. The combination of displacement on the Potrero Canyon fault and the blind thrust, as expressed by uplift and broad warping of the Santa Monica Mountains, results in a slip rate on the blind thrust of about 1 to 1½ mm/yr (Tsutsumi et al., submitted).

Partitioning between Reverse Faults and Strike-Slip Faults

The San Andreas fault presents two paradoxes to earth scientists trying to understand fault behavior (Zoback et al., 1987). (1) No heat-flow anomaly is associated with the fault, which is interpreted as evidence that no frictional heat is generated by the fault. This is surprising, considering the large number of earthquakes that have occurred on the fault. (2) The Coast Ranges adjacent to the fault in central California have undergone shortening at nearly right angles to the San Andreas fault (Fig. 10-17), which suggests that the fault is being squeezed from the side rather than being sheared obliquely. The 1982-1985 earthquake sequence in the California Coast Ranges also documented NE-SW shortening, as did borehole breakouts (Mount and Suppe, 1992; Fig. 5-30). Zoback et al. (1987) concluded that fault-normal compression and the absence of a heat-flow anomaly were evidence that shear stress at the San Andreas fault is very low, about 10 to 20 MPa. The orientation of the maximum horizontal stress far from the San Andreas fault in the Basin and Range province and in the Pacific plate is NNE (Zoback, 1992), but close to the fault, this orientation is rotated to the NE due to the low strength of the fault. Where the San Andreas fault changes strike in the Transverse Ranges, the maximum horizontal stress changes strike also, remaining nearly at right angles to the fault (Mount and Suppe, 1992).

The slip vector between the Pacific and North American plates trends about 5° clockwise from the strike of the San Andreas fault in central California, resulting in a component of compression normal to the fault. Reverse-fault focal mechanisms of the Coalinga earthquake sequence and earthquakes in the Transverse Ranges contrast with the strike-slip focal mechanisms of earthquakes on the San Andreas fault, indicating that *strain partitioning* must occur in the region (Lettis and Hanson, 1991). That is to say, faults tend to rupture in either dip-slip or strike-slip modes, but typically not with a combination of the two styles of deformation. In some cases, dip-slip and strike-slip faulting occur as subevents in the same earthquake, as in the 1932 Changma, China, 1957 Gobi-Altay, Mongolia, and 1988 Spitak, Armenia, earthquakes.

Mount and Suppe (1992) show that folds in oil districts of central and south Sumatra are characterized by borehole breakouts that show compression nearly at right angles to the Great Sumatran strike-slip fault. Evidence of fault-normal compression is also reported in structures adjacent to the Philippine and Alpine strike-slip faults (Mount and Suppe, 1992). Major reverse faults of the range-and-basin province of Central Otago, New Zealand, are parallel to the Alpine fault, which shows evidence of pure strike slip in southern Westland, southwest of the highest parts of the Southern Alps (Figs. 10-19 inset; 10-21). The Bitlis thrust belt and Palmyride fold belt in Syria and Turkey strike nearly parallel to the East Anatolian left-lateral fault (Fig. 10-7). Evidence of partitioning between strike slip and reverse faults is also reported in Irian Jaya (the Indonesian half of New Guinea), as shown in Figure 10-13 (Abers and McCaffrey, 1988). In central China, the active folds and reverse faults of the Nan Shan and Qilian Shan are nearly parallel to the Haiyuan left-lateral strike-slip fault, although they are oblique to the Altyn Tagh left-lateral strike-slip fault. Reverse faults on the west shore of Lake Biwa, Japan, are parallel to a major left-lateral strike-slip fault farther west (Sangawa, 1986; Fig. 10-28).

Finally, the 1990 Rudbar-Tarom earthquake (M_s7.7) occurred in the fold-thrust province of the Alborz Mountains of northwestern Iran, but focal-mechanism solutions from broad-band data indicate movement on a left-lateral strike-slip fault (Berberian et al., 1992). Surface ruptures showed evidence of folding and oblique reverse- and left-slip faults. The meizoseismal zone of the 1990 earthquake included the epicenter of an earthquake in 1983 (M_s5.0) that had a focal mechanism showing pure reverse slip. These two earthquakes were interpreted by Berberian et al. (1992) as evidence for strain partitioning.

The origin of strain partitioning is controversial, and many workers do not accept the idea of a weak San Andreas fault. However, there is geological and seismological evidence from many parts of the world that reverse slip can take place on faults that are parallel to strike-slip faults.

Earthquake Potential of Thin-Skinned and Thick-Skinned Fold-Thrust Belts

Thin-Skinned Fold-Thrust Belts

In a typical thin-skinned fold-thrust belt, such as the Appalachians or the Subhimalaya, a sequence of sedimentary rocks is decoupled and thrust over an unyielding, competent basement. (These rocks include accretionary prisms that are found at the base of continental slopes adjacent to subduction zones; these are discussed in Chapter 11.) Many of these fold-thrust belts show evidence of recent faulting and anticlinal growth, particularly near their leading edge (Yeats et al., 1984; Vita-Finzi, 1986; Suppe, 1987; Nakata, 1989). An example of this kind of deformation is the growth of an anticline at the front of the Himalayan décollement during the great 1905 Kangra earthquake (Yeats and Lillie, 1991). In contrast, the Zagros fold belt, which contains clear evidence of young surface deformation, is relatively aseismic; the reverse-faulting earthquakes that characterize this region are in the underlying basement (Berberian, 1981; Ni and Barazangi, 1986). This is true also for the fold-thrust belt in New Guinea (Abers and McCaffrey, 1988; Fig. 10-13). The probable reason for the absence of large earthquakes in rocks that are actively being deformed is that the décollement is in sedimentary rocks buried to only shallow depths, and the rock in and above the décollement is too weak to store enough strain energy to generate large earthquakes. However, the 1991 Racha earthquake (M_S7) in the Great Caucasus occurred on a low-angle thrust at depths probably not exceeding 10 km (Triep et al., 1995).

The leading edges of fold-thrust belts do not creep aseismically in the Himalaya, even though in Pakistan, the décollement is within salt. This is interpreted to mean that the thin-skinned décollements are locked in the rear at the point where they extend into high-strength rocks, as illustrated in Figure 10-4. Based on the location of maximum intensities in bedrock, the epicenter of the great 1905 Kangra earthquake was more than 50 km north of and downdip from the active Janauri anticline at the Himalayan front (Yeats and Lillie, 1991). Rupture then propagated southward along the décollement (Seeber and Armbruster, 1981).

The fault-plane solutions for earthquakes beneath the southern slopes of the Himalaya indicate a reverse fault steeper than the dip of the décollement (Ni and Barazangi, 1984; Fig. 10-4). Based on this evidence, Molnar (1990) proposed that this is the point where the décollement enters basement. Using nomenclature originating in the Valley and Ridge Province of the Appalachians, the zone of more steeply dipping faults is a *ramp,* and the low-dipping décollement extending to the south would be a *flat*. In the Appalachians, flats occur where the thrust is following bedding in relatively incompetent strata, and ramps occur where the thrust crosses stiffer rocks. The active ramp beneath the Great Himalaya is marked by uplift rates twice those of the flat underlying the Subhimalaya to the south (Jackson and Bilham, 1994) and by steeper gradients in longitudinal profiles of major rivers that cross the Himalaya as compared to gradients in the Lesser Himalaya (Seeber and Gornitz, 1983).

The Ventura Avenue anticline is a rootless fold that is related to slip on a décollement originating at the Oak Ridge fault, nearly 15 km away, which is blind where it is down-dip from the anticline (Yeats et al., 1988; Fig. 2-24). But Holocene-age erosional marine platforms on the flanks of the anticline (Sarna-Wojcicki et al., 1987) provide evidence that growth of the anticline occurs during earthquakes (Stein and Yeats, 1989). These earthquakes must occur on the blind Oak Ridge fault. An earthquake that produces surface displacement on a thin-skinned décollement would produce strong ground shaking at rock sites at the epicenter, which could be tens of kilometers away from surface displacement, as shown for the Kangra earthquake.

Thick-Skinned Fold-Thrust Belts

Aside from subduction zones, most reverse-faulting earthquakes occur on faults in the brittle upper crust, in many cases beneath shallower thin-skinned décollements. These faults commonly occur in sets, producing a range-and-basin style of topography such as Central Otago and northwest Nelson in New Zealand, northern Honshu and the Kinki Triangle in Japan, the Pampean Ranges in northwestern Argentina, and part of the Transverse Ranges in California. There are similar sets of faults beneath the décollements in the Zagros Mountains of Iran (Berberian, 1995; Fig. 10-6) and in New Guinea (Fig. 10-12). These sets of moderate- to high-angle reverse faults differ from low-angle fold-and-thrust belts principally in the aspect ratios of the ranges and basins associated with them. The Himalaya and the Appalachians contain structures that are very long compared to their width, whereas in Central Otago, the fault-bounded ranges and basins are relatively short compared to their width.

The Coalinga earthquake sequence showed that a fold belt consisting of structures that are long with respect to their width (large aspect ratio) can produce large earthquakes. Fault-plane solutions of the mainshocks and many aftershocks of the Coalinga earthquake sequence tend to be low angle (Fig. 10-17), suggesting that they merge with a low-angle thrust within or at the base of brittle crust. This is in contrast to the onshore western Transverse Ranges, in which aftershock zones of the 1971 and 1994 reverse-fault earthquakes show no tendency to flatten at the base of the seismogenic zone (Fig. 10-32a). However, the folds of the westernmost Transverse Ranges, including those of the Santa Ynez Mountains and the Santa Barbara Channel show a large aspect ratio like that of the Coast Ranges, and Shaw and Suppe (1994) have suggested that surface structures in the Santa Barbara Channel are related to low-angle blind thrusts at depth.

A thick-skinned fold-and-thrust belt can be modeled as a thin-skinned fold-and-thrust belt in which most or all of the brittle crust is above the décollement, and the plastic lower crust is below. In the Sulaiman Range of Pakistan, the décollement deepens westward to depths of 10–15 km (Jadoon et al., 1994), clearly deep enough to produce large earthquakes. A similar relation may be true for the Tianshan of western China (Avouac et al., 1993; Fig. 10–24). Crustal shortening based on this interpretation can be calculated using balanced cross sections (cf. Namson and Davis, 1988; Davis et al., 1989; discussion in Chapter 2). In this view, reverse faults flatten downward at the base of brittle crust, below which the crust is relatively weak compared to the relatively strong upper crust and upper mantle (Yeats, 1983). The alternation of periods of activity among several sets of faults in Central Otago, New Zealand, suggested to Beanland and Berryman (1989) that these faults are connected at a regional décollement beneath brittle crust (Fig. 10–72), even though ranges and basins in this region have a low aspect ratio. Massive thrust sheets composed predominantly of continental crystalline rocks are widespread in the Appalachian Piedmont, the Eastern Alps, the High Himalaya, and the San Gabriel Mountains of southern California. These may be uplifted and partially eroded examples of ancient thrust sheets at the base of brittle crust (cf. Hatcher and Hooper, 1992).

However, there does not seem to be a clear mechanical reason why reverse faults should flatten at the base of brittle crust unless the plastic layer between brittle upper crust and brittle upper mantle is very thin. The boundary between the brittle and plastic crust is probably a transitional zone that is brittle under high, coseismic strain rates and plastic under low, interseismic strain rates (cf. Chapter 3). Downward flattening of faults is a possibility if the base of brittle crust, as defined by earthquakes and the Moho discontinuity marking the top of the mantle, are both essentially horizontal. The presence of significant relief on the base of the seismogenic zone is difficult to explain in a model where the base of the brittle crust is a décollement.

Active reverse faults in the Ventura basin of southern California (Fig. 10–18) have been described as examples of faults that flatten at the base of brittle crust (Yeats, 1983). If so, a sum of slip rates on these faults would equal the rate at which the ranges south and north of the basin are converging. Summing the slip rates gives a convergence rate that is significantly higher (Huftile and Yeats, 1995) than the convergence rate based on GPS measurements during the past few years (Larson and Webb, 1992; Donnellan et al., 1993). The geologic convergence rates would be in better agreement with the geodetic convergence rates if the reverse faults continued downward into the lower crust as ductile shear zones, because the convergence rate would then be the sum of the products of the slip rates and the cosine of the fault dip (Yeats, 1993).

The Ventura basin contains the thickest section of Quaternary rocks in the world, and it also has the deepest crustal earthquakes in southern California (Bryant and Jones, 1992), evidence that the base of the seismogenic zone has considerable relief, forming a keel beneath the basin (Fig. 10–73). The Moho discontinuity is also depressed beneath the basin (Bryant and Jones, 1992). The dominant tectonic process in the Ventura basin during the past several million years

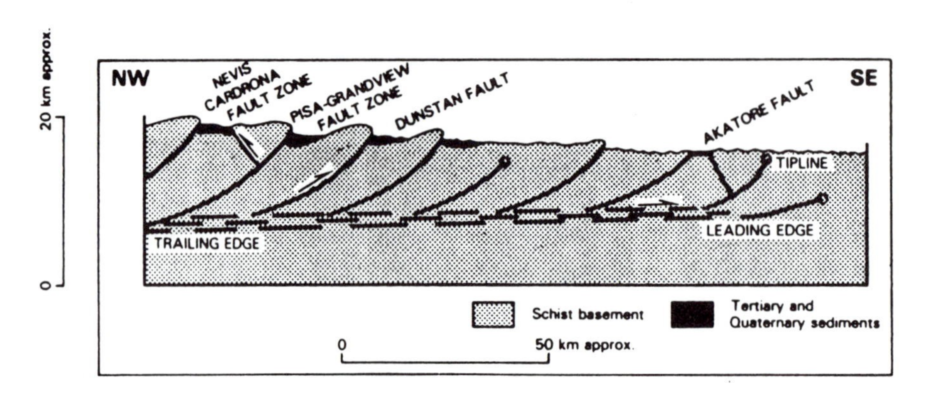

Figure 10–72. Central Otago, New Zealand, modeled as a thick-skinned thrust belt in which reverse faults in brittle crust flatten downward into a décollement. The décollement transmits stress such that at some times, one fault accommodates the shortening, and at other times, another fault does so. After Beanland and Berryman (1989).

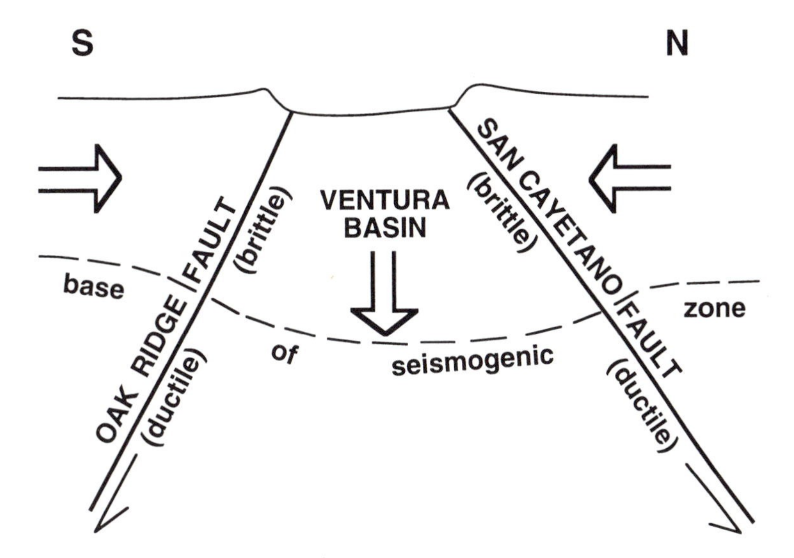

Figure 10–73. Tectonic model of Ventura basin, California. Faults in brittle crust continue as ductile shear zones in lower crust as the Ventura basin block is depressed. Open arrows show both the convergence of ranges toward the basin and the downward movement of crust within the basin. After Yeats (1993).

has been the downward displacement of the Ventura basin block between the Oak Ridge fault and San Cayetano-Red Mountain faults at the same time that the mountain ranges to the north and south have converged. The shortening of the brittle upper crust may be accompanied by plastic shortening and thickening of the lower crust as the Moho is depressed. The depression of the Moho may itself be related to a high-velocity ridge in the upper mantle beneath the Transverse Ranges (Hadley and Kanamori, 1977). The reverse faults would continue into the lower crust as ductile shear zones, accommodating the shortening (Yeats, 1993; Fig. 10-73). The deformation would be similar to that observed along high-angle ductile shear zones that are found in ancient mountain belts.

SUMMARY

Active reverse faults are found at continent-continent collision zones, forearc and back-arc contraction zones, intracontinental deformed zones, and in continental shield areas. They are more difficult to recognize in advance of an earthquake, because their surface traces may be irregular in plan, and in many places, their surface expression is by folding rather than faulting. The expression of reverse faults in surface exposure and in trenches is highly variable, ranging from distinct reverse faults to combinations of faulting and near-surface folding. In several regions, reverse faults occur in sets and produce a range-and-basin topography in which one or both basin margins are bounded by reverse faults. In some cases, such as the Zagros Mountains and New Guinea, the range-and-basin structural pattern is covered by a thin-skinned fold-thrust belt which is itself aseismic. Active reverse faults are commonly segmented, and the segment boundaries are typically marked by en échelon offsets and lateral ramps, although understanding the nature of reverse-fault segmentation is in its infancy.

Earthquakes on reverse faults have magnitudes as large as 7.8 in most regions, although in the Himalaya, great earthquakes of magnitude 8 and larger have occurred. The Himalayan case suggests that when low-angle fold-and-thrust belts are thick enough to include most of the brittle crust above the décollement, their failure may produce earthquakes much larger than those on steeply dipping reverse faults because of the greater fault area within brittle crust. Recurrence intervals on individual faults range from hundreds to thousands of years in active mountain belts, with considerable variability due to temporal clustering of earthquakes. In continental shield areas, recurrence intervals may be hundreds of thousands of years or longer. In several regions, reverse faults are parallel to strike-slip faults, which is evidence for the partitioning of strain between the two types of faults. In some cases, reverse faults may flatten to horizontal at or below the base of seismogenic crust, but in other cases, they may continue downward into plastic lower crust as ductile shear zones.

Suggestions for Further Reading

Davis, D., Suppe, J., and Dahlen, F. A. 1983. Mechanics of fold-and-thrust belts and accretionary wedges. Jour. Geophys. Res., 88:1153–72.

McClay, K. R., ed. 1992. Thrust Tectonics. London: Chapman & Hall, 447 p. Includes discussions on thrust mechanics and modeling together with examples from fold-thrust belts.

Narr, W., and Suppe, J. 1994. Kinematics of basement-involved compressive structures. American Journal of Science 294:802–60.

Research Group for the Senya Fault, 1986. Holocene activities and near-surface features of the Senya fault, Akita Prefecture, Japan—Excavation study at Komori, Senhata-cho. Bull. Earthquake Res. Inst., University of Tokyo 61:339–402 (Japanese, Eng. abs. and figure captions). This paper sets the standard for trench logging of a reverse fault.

Rodgers, J. 1987. Chains of basement uplifts within cratons marginal to orogenic belts. American Journal of Science, 287:661–92. Review of reverse faults, active and inactive, that involve the brittle crust.

Stein, R. S., and Yeats, R. S. 1989. Hidden earthquakes. Scientific American, 260(6):48–57.

Suppe, J. 1983. Geometry and kinematics of fault-bend folding. American Journal of Science, 283:684–721.

Suppe, J., and Medwedeff, D. A. 1990. Geometry and kinematics of fault-propagation folding. Eclogae Geologicae Helvetiae, 83:409–54.

Woodward, N. B., Boyer, S. E., and Suppe, J. 1989. Balanced geological cross sections. Am. Geophys. Union Short Course in Geology 6, 132 p.

Yeats, R. S. 1986. Active faults related to folding. *In* Active Tectonics, Studies in Geophysics. Washington, D.C.: National Academy Press, pp. 63–79.

Charles R. Darwin (1809-1882)

England

The *Beagle*, under the command of Captain FitzRoy, dropped anchor at Valdivia, a small Chilean town 800 kilometers south of Valparaiso. It was late February 1835, and the ship had been in Chile for more than seven months after an arduous winter passage through the Straits of Magellan. The shipboard scientist, Charles Darwin, had already taken several trips into the interior, near Valparaiso and in the dark, misty forests of the Isle of Chiloe farther south, collecting specimens of plants and animals and musing about the geology of the Andean Cordillera. He was about to celebrate his twenty-sixth birthday.

Compared to life on the ship, where Darwin frequently was seasick, the land expeditions were pure joy. A scramble to the top of a nearby peak for a view of the landscape was matched only by the thrill of finding a new species of plant or bird. But the mystery that absorbed him the most was the origin of the Andes, which he had first seen from the Patagonian plains as a shimmering wall of snowy peaks on the western horizon. How were the Andes raised to their present great height? And what was the meaning of sea shells found at an elevation of 400 meters near Valparaiso?

On the late summer morning of February 20, Darwin and his field assistant, Syms Covington, left the *Beagle* for field work ashore. They passed through apple orchards and had stopped in the forest to rest when at 11:40 A.M., the ground suddenly began to undulate and shake, and a breeze abruptly agitated the trees. On board the *Beagle*, it felt as if the ship had run aground. The earthquake lasted a full two minutes, but it seemed much longer. Darwin and Covington rushed back to the ship at Valdivia, where the damage was relatively minor. The *Beagle* raised anchor, and Darwin and Captain FitzRoy continued their survey northward, arriving at Talcahuano, the port of the town of Concepción, on March 4. There they found catastrophe.

From accounts gathered by Darwin and FitzRoy, great flights of sea birds were observed heading east toward the mountains about 10:00 A.M., just before the earthquake, and all the dogs left the town, heading for higher ground. At 11:40, there were a few foreshocks, then the mainshock arrived. People were unable to stand. The motion seemed to come from the southwest, as did a rumbling noise, like distant subterranean thunder. Concepción collapsed into rubble and ruin. At Talcahuano, the water left the harbor, grounding the ships at anchor, but 30 minutes later, a great wave swept across the harbor and onto the land to a height of 30 feet, carrying everything with it. This was followed by two additional waves, each greater than the last.

As Darwin surveyed the Bay of Concepción, he noted that "the most remarkable effect of this earthquake was the permanent elevation of the land; it would probably be far more correct to speak of it as the cause." The land around the bay had been uplifted several feet. And Captain FitzRoy, whose observations about the sudden uplift of the coastline were more complete than Darwin's, found that on a small island nearby, beds of dead mussels had been uplifted as much as ten feet above the present spring-tide high-water mark. However, Darwin had reservations about how permanent the elevation of the coastline really was. In July, he returned to Talcahuano and found that in places the land had subsided to its former elevation. The uplift by the earthquake was not everywhere permanent.

Still, Darwin had observed tectonic uplift almost as it happened, accompanied by a great earthquake, and the experience affected him profoundly. After leaving Concepción, he crossed the Andes via Portillo Pass east of Santiago. On Peuquenes Ridge, he found fossil shells at an elevation of more than 3700 meters, causing him to write "that nothing, not even the wind that blows, is so unstable as the level of the crust of the earth."

Upon returning to England the following year, Darwin collected his notes and those of others on the 1835 earthquake as well as one in 1837 and an earlier earthquake in 1751, all in Chile. He concluded that the area affected was elongated parallel to the Andes, and that at least some of the coseismic elevation was permanent, so that "the earthquake of Concepción on the 20th of February marked one step in the elevation of a mountain chain. . .. " The coincidence of the earthquake with eruptions of volcanoes in the Andes (he had observed Mt. Osorno in eruption on January 17) led Darwin to conclude that the volcanic activity and the earthquakes were related. The permanent elevation of the land caused him to question the general belief that "subterranean disturbances" were caused by the shrinking of the crust. "How [shrinking] can explain the slow elevation, not only of linear spaces, but of great continents, I cannot understand."

Suggestions for Further Reading

Barrett, P. H., ed. 1977. The collected papers of Charles Darwin. University of Chicago Press, v. 2, 277 p.

Darwin, C. 1840. On the connexion of certain volcanic phenomena in South America; and on the formation of mountain chains and volcanos, as the effect of the same power by which continents are elevated: Trans. Geol. Soc. London, 2nd ser., pt. 3, pp. 601–31.

Darwin, C. 1842. Journal of researches into the natural history and geology of the countries visited during the voyage of H.M.S. *Beagle* round the world, under the command of Captain FitzRoy, R. N. London: Nelson and Sons, 615 p.

Darwin, C. 1846. Third part of the geology of the voyage of the *Beagle* (2nd ed. 1876).

Keynes, R. D. ed., 1979. The *Beagle* record. Cambridge University Press, 409 p.

Moorehead, A. 1969. Darwin and the *Beagle*. New York and Evanston: Harper & Row, 280 p.

West, G. 1937. Charles Darwin the fragmentary man. London: George Routledge & Sons, 351 p.

11

Subduction-Zone Megathrusts

Large, shallow earthquakes on subduction-zone plate boundaries contributed about 90% of the total seismic moment released worldwide from 1900 through 1989 (Pacheco and Sykes, 1992). The 1960 earthquake off southern Chile alone accounted for 30–45% of the worldwide seismic moment released during this period. The great earthquakes of 1952 off Kamchatka and 1964 in the Gulf of Alaska were also characterized by large seismic moment release. Each of these three subduction zones (Kuril, Aleutian, Peru-Chile; located on Fig. 11-1) accounts for much larger moment release than any of the others (Pacheco and Sykes, 1992). Oddly, the four largest earthquakes, including the 1957 earthquake off the central Aleutian Islands, occurred within a period of 13 years (Uyeda and Kanamori, 1979).

Given the large contribution that subduction-zone earthquakes make to seismic-energy release worldwide, it is somewhat surprising that they have not caused more damage than has been observed to date. One reason is that these earthquakes are generally offshore, and therefore the greatest danger may be from tsunamis rather than seismic shaking. The subduction zones in Kamchatka, the Kuril Islands, southern Alaska, and southern Chile, which contributed so much to twentieth-century energy release, are far from densely-populated areas, as are subduction zones in the southwest Pacific.

However, several subduction zones, including the Java-Sumatra, Nankai, Northeast Japan, and Peru-Chile subduction zones are adjacent to heavily populated regions. The Cascadia subduction zone off the northwestern United States and southwestern Canada has had no great earthquakes since records began to be kept there, but Heaton and Kanamori (1984) and Atwater (1987) suggested that this region has the potential to produce a very large earthquake. This possibility is addressed in a later section of this chapter.

Earthquakes related to subduction zones are of four general types (Fig. 11-2): (1) shallow interplate thrust events caused by failure of the interface between the downgoing slab and overriding plate, (2) shallow earthquakes caused by deformation within the upper plate, (3) earthquakes at depths from 40–700 km within the downgoing oceanic slab, and (4) earthquakes seaward of the trench, caused mainly by flexing of the downgoing plate, but also by compression of the slab (Christensen and Ruff, 1988).

Although we discuss all four of these types, this chapter focuses predominantly on the first two: shallow interplate earthquakes and earthquakes within the upper and lower plates that are directly related to subduction. We have emphasized those features of subduction zones and their earthquakes that have geologic expression. In addition, we consider tectonic features of subduction zones that may influence earthquake nucleation and size as well as fault segmentation.

Several subduction zones that have been studied in detail and that have produced great historic earthquakes are highlighted. These include the southwest Japan, northeast Japan-Kuril Islands, Aleutian, and Peru-Chile subduction zones. These subduction zones show strong seismic coupling, referring to the force that acts across the thrust interface (Ruff and Kanamori, 1980; Lay et al., 1982; McCaffrey, 1993). Poorly coupled subduction zones that do not produce great earthquakes are discussed only to consider why their earthquake potential is less than that of strongly coupled zones. The Makran, Hikurangi, and Vanuatu subduction zones are important because they contain coseismically uplifted terraces that have bearing on their paleoseismology. Parts of the accretionary prisms of the southwest Japan, Cascadia, and Makran subduction zones are on land, more accessible to geological investigation, and these are discussed here. Continent-continent collision zones, in which the downgoing slab is at least in part continental, are treated in Chapter 10.

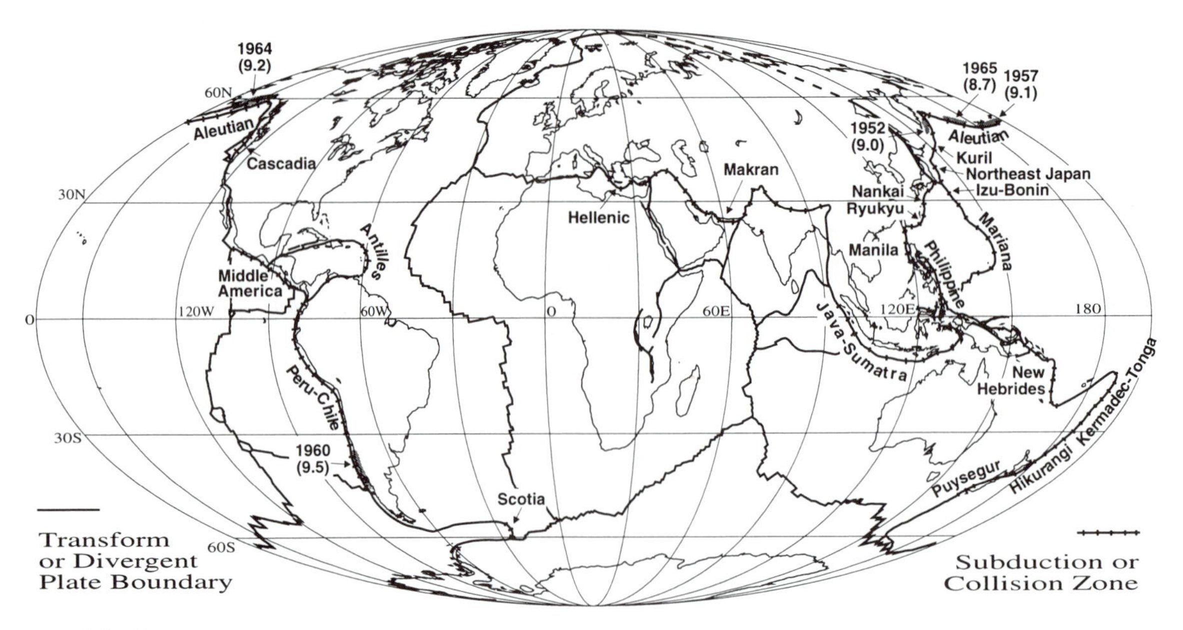

Figure 11–1. Subduction zones of the world in a plate-tectonic framework in which divergent and transform boundaries are shown together. The meizoseismal zones of the five greatest earthquakes of the twentieth century are shown in gray.

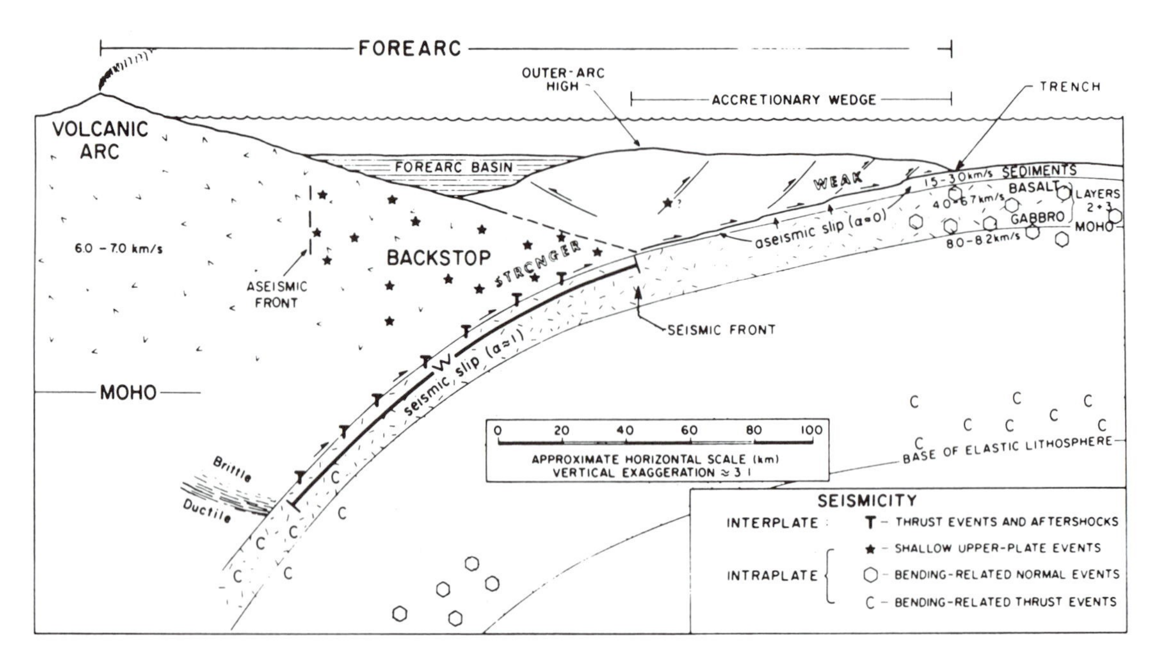

Figure 11–2. Idealized cross section of the upper 50 km of a subduction zone, identifying the trench, accretionary wedge, outer-arc high, forearc basin, and volcanic arc. Major features are characterized by their P wave velocities: sediment cover, oceanic crust (basalt underlain by gabbro), and the Moho discontinuity. α = the ratio of seismic slip to total slip; the value $\alpha = 0$ indicates all slip is released aseismically. The backstop consists of stronger materials that deform accompanied by earthquakes. W = the width of the seismogenic zone, which is the downdip extent of interplate earthquakes. From Byrne et al. (1988).

GENERAL CHARACTERISTICS OF SUBDUCTION ZONES

We discuss here the major features of subduction zones: the downgoing oceanic slab, the trench, the features of the upper plate between the trench and the volcanic arc, and the megathrust plate boundary (Fig. 11-2).

Not only are subduction zones characterized by the greatest earthquakes on earth, but they also contain the greatest extremes of topographic relief on earth, including depths greater than 10 km in some trenches in the western Pacific Ocean Trench depth is related to age, and thus temperature, of the downgoing slab; subducting cold Mesozoic crust leads to a deeper trench than more buoyant middle Cenozoic crust. In addition, the deepest trenches are virtually free of sediment. Trenches adjacent to major continental drainage systems are shallower due to sediment fill.

Proceeding toward the volcanic arc from the trench (Fig. 11-2), an *accretionary wedge* of deformed sediment may be encountered. (However, in several subduction zones, the accretionary wedge is poorly developed or absent.) The sediments of the accretionary wedge originate by scraping off the sedimentary cover of the incoming plate as well as sediments transported down the slope from the arc and continent. The accretionary wedge is analogous to a foreland fold-and-thrust belt on land in which reverse faults flatten into a décollement above oceanic crust. Thrusts may be landward-vergent as well as seaward-vergent, but are dominantly seaward-vergent. As in the case of active fold-and-thrust belts on land, the accretionary wedge may be close to critical failure throughout (Davis et al., 1983). As sediments are stuffed under the accretionary wedge, the thrust front may move outward, away from the arc, and older thrusts in the direction of the arc may be rotated upward to a steeper dip, producing a topographic bulge called the *outer-arc high*.

Byrne et al. (1988) concluded that the sediments of the modern accretionary wedge are too weak to produce large earthquakes as they deform. They suggested that the outer-arc high is the surface expression of that point on the subduction zone below which the materials of the upper plate are consolidated enough to deform by unstable stick slip, accompanied by earthquakes (Fig. 11-2). The weak accretionary wedge is analogous to the upper zone of velocity strengthening in a continental fault zone (Chapter 3), where rocks under low overburden pressure are relatively weak, and faulting takes place by stable sliding (Fig. 3-9; Scholz, 1988; 1990; Pacheco et al., 1993). However, Hyndman and Wang (1993) proposed that the subduction zone beneath the Cascadia accretionary wedge is strongly coupled (see discussion in the following text).

The zone in which rocks exhibit unstable frictional rate behavior is called the *backstop*. The top of the backstop may slope toward the trench, as shown in multichannel seismic profiles and by the top of the zone of earthquakes in the upper plate (Fig. 11-2; Byrne et al., 1988). The top of the backstop may be the contact between sediment and basement rock, or it may be simply an expression of the increase in rock strength with increasing confining pressure (cf. discussion in Chapter 3). Motion on the subduction zone beneath the backstop is accompanied by earthquakes, and the width of this zone (W in Fig. 11-2) is a critical parameter in estimating the area of rupture of a great subduction-zone earthquake. The down-dip limit of this seismic zone at the top of the downgoing plate is defined by the maximum depth of coupling of interplate earthquakes (Tichelaar and Ruff, 1993), which appears to correspond to the brittle-plastic transition in the upper plate. Below this transition, rocks in the upper plate reach temperatures high enough that their strength is reduced, and they exhibit stable frictional rate behavior and deform aseismically (Figs. 3-7; 3-8; 3-9). The critical rheology at the brittle-plastic transition is that of fault-zone rock rather than undeformed rock. As noted in Chapter 3, there may be a down-dip limit for *nucleation* of interplate earthquakes marked by the onset of quartz plasticity (300-350°C) and a deeper transitional zone through which interplate earthquakes may *propagate,* limited by the onset of feldspar plasticity at 450°C (Hyndman and Wang, 1993).

The interior of the downgoing plate beneath the subduction zone is still strong enough to generate earthquakes, because it includes crustal rock that may be cooler than the upper plate directly above it as well as mantle material that is still brittle at temperatures at which crustal material would already be plastic (Fig. 3-7). The fact that downgoing oceanic lithosphere is cooler than its surroundings generally results in low heat flow in the overlying accretionary wedge.

Beneath the zone of interplate earthquakes, faulting occurs in the downgoing plate due to the dehydration of serpentine to olivine to depths of 300 km and the transformation of olivine to a denser mineral, spinel, to depths of 700 km (Green, 1994). Earthquakes generated at depths of 300-700 km can exceed M8, as shown by an earthquake beneath Bolivia in June, 1994. Below 700 km, spinel decomposes to still denser phases, but this process is not accompanied by earthquakes.

Surface expression of the backstop may include a *forearc basin* formed by the arcward slope of the outer-arc high and the trenchward slope of the island arc itself. In several well-studied subduction zones, there is a decrease in instrumental seismicity in the direction of the volcanic arc due to the higher heat flow in the arc. The boundary between the seismogenic forearc basin and the relatively aseismic arc is identified as the *aseismic front* on Figure 11-2.

The state of stress in and behind the volcanic arc may be extensional or compressional. Examples of extension behind the arc include the back-arc basins of the Taupo Volcanic Zone of New Zealand (Fig. 9-7) and the Aegean Sea (Fig. 9-8). In addition, several island arcs in the western Pacific region, including the Izu-Bonin, Mariana, and Tonga arcs, are accompanied by back-arc basins, in which new oceanic crust is forming. In contrast, the northern Japan arc is characterized by contraction perpendicular to the arc. Back-arc contraction is also observed in Sumatra (Mount and Suppe, 1992), even though Sumatra contains a major strike-slip fault parallel to and within the arc.

These contrasting stress states can be described by considering plate motions with respect to a mantle reference frame rather than simply considering the relative velocities at plate boundaries. In Chapter 1, we pointed out that hot spots are used to define a mantle reference frame, because they are generated by thermal plumes rising up through the mantle. The base of a subduction zone, particularly a deep one that generates earthquakes at depths of hundreds of kilometers, can also be used to define a mantle reference frame. It is easy to move rigid plates over a plastic asthenosphere if the plates are horizontal. But a subduction zone protruding downward hundreds of kilometers into the upper mantle would act as a tectonic "sea anchor," making it difficult to move horizontally with respect to the lower mantle. On the other hand, the subduction zone may, over time, change dip as a result of mantle flow so that the position of the trench with respect to the lower mantle may change.

We consider this problem by identifying the forces that act on lithospheric plates at subduction zones (Forsyth and Uyeda, 1975; Fig. 11-3a). The negative buoyancy of the dense, cool slab of oceanic lithosphere that descends into the asthenosphere produces a vertical force F_{NB}. Part of this force is transferred to the plate as a *slab pull* force F_{SP}, which is opposed by a *slab resistance* force R_S, mainly at the base of the slab, and a *bending resistance* force R_B. Resistance is also produced by friction between the two plates (R_O), which, when overcome, produces interplate earthquakes. Depending on the velocity of the asthenosphere with respect to that of the lithosphere, there will be a drag resistance force R_{DO} beneath the subducting plate and R_{DC} beneath the overriding plate which may augment or impede subduction. This drag resistance may also contribute to the slab resistance force R_S. The overriding lithosphere in the upper plate may be placed in tension by the trench-suction force (F_{SU}). This force may be caused by a decrease in near-surface dip of the subduction zone due to *rollback* of the underthrusting plate (Fig. 11-3b) so that the overriding plate sags toward the trench. Alternatively, this force may be related to convection generated in the asthenosphere between the upper and lower plate.

Uyeda and Kanamori (1979) calculated the absolute-velocity vectors of the upper plates of Pacific

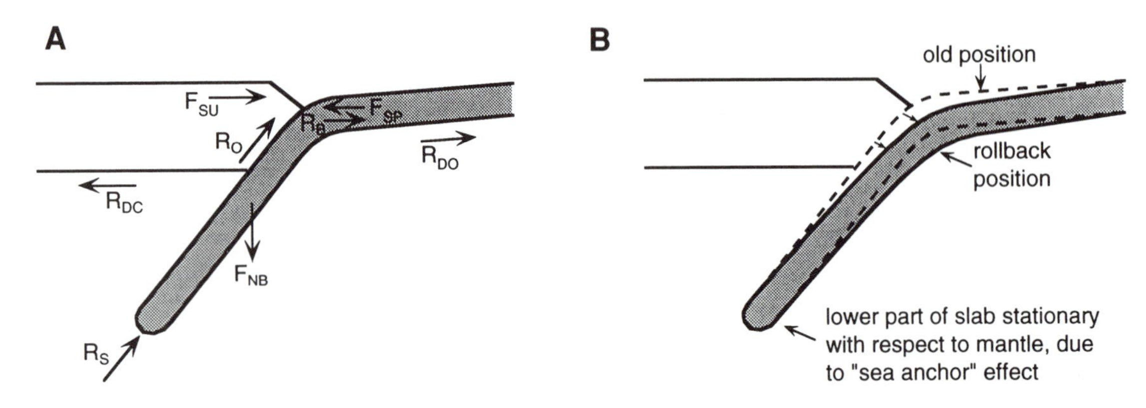

Figure 11-3. (a) Forces acting on a subducting slab. F_{NB} = force due to negative buoyancy of lithosphere, F_{SP} = slab-pull force, F_{SU} = trench-suction force, R_B = bending-resistance force, R_{DC} = drag resistance beneath overriding plate, R_{DO} = drag resistance beneath subducting plate, R_O = interplate frictional resistance force, R_S = slab-resistance force. (b) Illustration of rollback of trench due to change of dip of downgoing slab. Modified from Forsyth and Uyeda (1975).

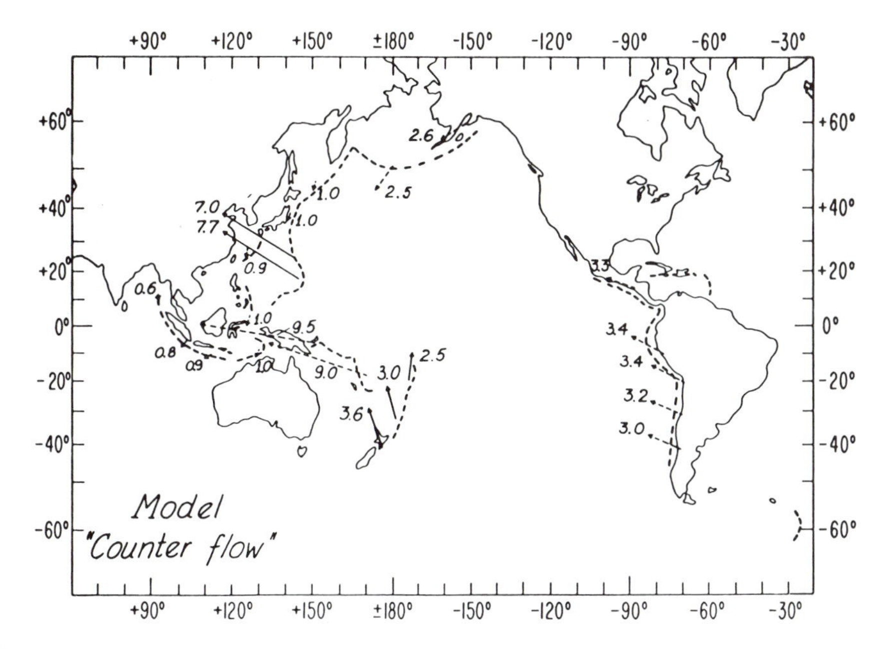

Figure 11–4. Velocity vectors of upper plates with respect to trenches, in centimeters per year. Because the subducting slab acts as a "sea anchor," this velocity is rather similar to the velocity of the upper plate with respect to the lower mantle. After Uyeda and Kanamori (1979).

subduction zones with respect to adjacent trenches (Fig. 11–4), which do not differ very much from the velocity vectors with respect to the lower mantle because of the "sea anchor" effect of the downgoing slab. The South American continent is moving slowly toward the Peru-Chile trench, but the oceanic upper plate west of the Izu-Bonin, Mariana, and Tonga subduction zones is moving *away* from the trench, and it is here that back-arc basins develop by sea-floor spreading. This means that in several subduction zones in the western Pacific region, subduction is accompanied by extension behind the arc. Royden (1993) referred to subduction zones in which the rate of subduction, controlled by slab pull by dense oceanic lithosphere, exceeds the rate of convergence as *retreating subduction boundaries*.

As shown in Figure 11–1, all of the greatest earthquakes of the twentieth century occurred on only a few subduction zones. A comparison between Figures 11–1 and 11–4 shows that the upper plates of subduction zones producing great earthquakes are all in compression. The Izu-Bonin and Mariana subduction zones south of Japan, with active back-arc basins, have produced smaller earthquakes. The Tonga-Kermadec trench, which is also characterized by back-arc basins, has produced only three earthquakes of $M > 8$, far fewer than the Kuril, Aleutian, or Peru-Chile subduction zones, and also none of the greatest earthquakes.

This led Uyeda and Kanamori (1979) to propose two end-member types of subduction zones, the *Chilean type* and the *Mariana type* (Fig. 11–5). The Chilean type is characterized by compression in the overlying plate whereas the Mariana type is characterized by tension, in which the upper plate moves away from the trench, producing a back-arc basin. The downgoing plate in Chilean-type subduction is characterized by a gentle dip so that it is strongly coupled to the overlying plate. This means that the down-dip width W (Fig. 11–2) is large, and great earthquakes may result. (In a later section, however, we will point out exceptions to the relation of great earthquakes to dip of the subduction zone which prevent a one-to-one correlation of dip to the presence of great earthquakes.)

Strong coupling causes end-loading of the subducting plate, producing a bulge (*outer rise*) seaward of the trench in which bending-moment normal-fault earthquakes occur, particularly following great subduction-zone earthquakes (Christensen and Ruff, 1988). As these authors point out, however, the accumulation of horizontal stress in strongly coupled regions also causes compressional earthquakes at the outer rise. The Mariana-type subduction zone, in contrast, lacks a bulge seaward of the trench, and only normal-fault earthquakes occur in the downgoing slab. The downgoing plate tends to have a much steeper dip due to slab pull by dense, oceanic lithosphere. Slab pull causes a weaker coupling between

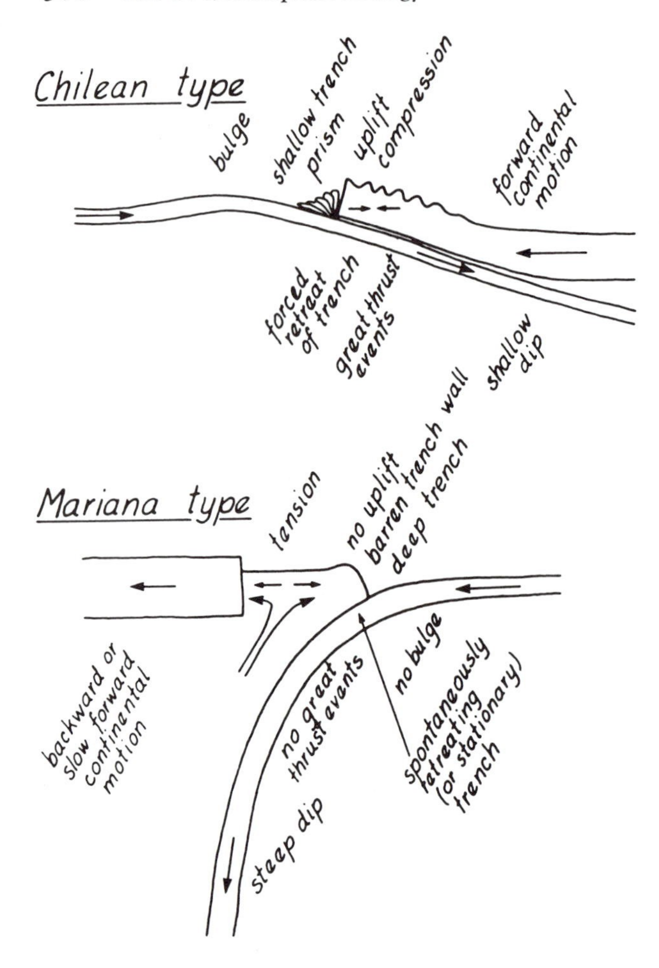

Figure 11–5. End-member types of subduction zones. The overriding plate is in compression in the Chilean type and in tension in the Mariana type subduction zone. Compression leads to a shallow dip of the subduction zone and a large area of contact between the plates in the seismogenic zone. It also leads to an accretionary prism, an outer-arc bulge, and great earthquakes. Tension leads to back-arc basins, a barren trench wall, and no great earthquakes. After Uyeda and Kanamori (1979).

plates so that much of the subduction takes place aseismically, and great earthquakes do not occur there.

The less-coupled Mariana-type subduction zone may have high interplate seismicity for small and intermediate earthquakes, whereas the Chilean-type zone may be seismically quiet at the plate interface because it is completely locked. This is analogous to the seismicity of the San Andreas fault (Chapter 8), which is high where the fault is creeping or where it ruptures in frequent, small earthquakes, but is low in the locked zones of the 1857 and 1906 earthquakes (Heaton and Kanamori, 1984; Fig. 4-10).

Finally, the Chilean-type subduction zone is most likely to contain an accretionary prism, whereas the Mariana-type subduction zone may contain a barren trench wall and relatively little sediment. This is a reflection of the lack of coupling so that the sedimentary cover of the downgoing plate can be subducted easily in comparison with the Chilean-type zone. Other reasons for the lack of an accretionary prism are that (1) sea-floor spreading in the active back-arc basin behind a Mariana-type subduction zone separates the arc and forearc from the continent which is the main source of sediment, and (2) the incoming oceanic plate generally contains only a thin sediment cover. However, we point out in a later section that some subduction zones under compression in the upper plate, such as northeast Japan, are characterized by bare trench walls and even by evidence of tectonic erosion of the edge of the upper plate.

Ruff and Kanamori (1980) compared subduction zones on the basis of the age of the oceanic plate being subducted and the convergence rate at the trench (Fig. 11-6). Then they plotted the largest magnitude (M_W) earthquake to occur on each subduction zone. This comparison showed that the greatest earthquakes have occurred along subduction zones characterized by a high convergence rate and relatively young subducting crust. Based on these criteria, Heaton and Kanamori (1984) predicted that the Cascadia subduction zone off the west coast of the United States and Canada is capable of sustaining earthquakes with M_W between 8.0 and 8.5, even though it has not done so during 175 years of record-keeping there.

The age of subducting crust is expressed in its buoyancy and its temperature. As pointed out in Chapter 1, oceanic spreading centers are relatively shallow, because newly formed lithosphere is hot and therefore of lower density than older lithosphere. Lower density causes the lithosphere to be more buoyant, thus relatively shallow. This buoyancy is the reason the lithosphere at a spreading center forms a ridge. As oceanic lithosphere moves away from the ridge crest, it cools, contracts, and sinks to a greater depth. Parsons and Sclater (1977) showed that the depth of oceanic crust younger than 80 Ma is directly related to the square root of its age. Crust older than 80 Ma continues to sink, but more slowly (Parsons and McKenzie, 1978). Oceanic crust subducting at the Izu-Bonin and Mariana trenches has an age of 150 Ma so that it forms a very deep trench. The low coupling is a reflection of great negative buoyancy. In contrast, oceanic crust subducting beneath southern Chile and southwest Japan is of middle Tertiary age (20 Ma). Crust this young is too buoyant to develop a deep trench, and its buoyancy contributes to its strong cou-

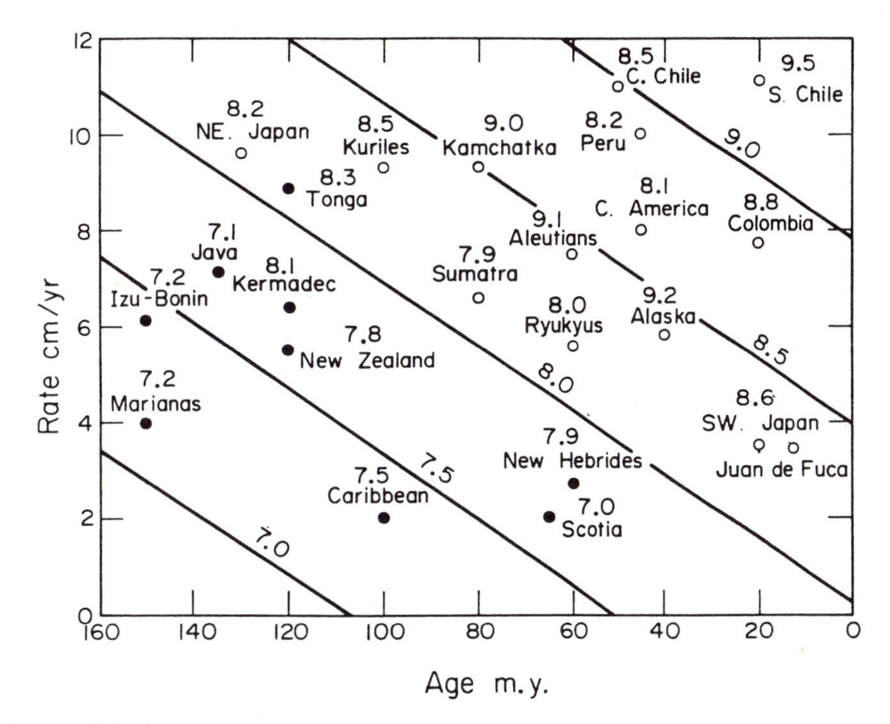

Figure 11–6. Relation of maximum earthquake size (M_W) to convergence rate at subduction zones, age of subducted lithosphere, and presence or absence of back-arc basins. Numbers show the largest earthquake (M_W) to occur on each subduction zone. Subduction zones with back-arc basins shown as solid circles; those without as open circles. This plot shows that the maximum earthquake size on a subduction zone increases as convergence rate increases and age of subducted oceanic lithosphere decreases. Note that subduction zones with back-arc basins are characterized by smaller maximum M_W earthquakes. Juan de Fuca (Cascadia) has not had a subduction-zone earthquake in 150 years of record-keeping, but the relations shown here suggest that earthquakes of M_W between 8.0 and 8.5 would be expected. After Heaton and Kanamori (1984).

pling. However, oceanic crust subducting beneath the Kuril and Kamchatka subduction zones is also of Mesozoic age (100 and 80 Ma, respectively), and here the effects of high convergence rate and compression of the upper plate appear to dominate.

As stated above, the maximum size of an earthquake generated at a subduction zone should, in theory, be related to the down-dip width W of the zone of coupling (Fig. 11-2). The base of W is controlled by the brittle-plastic transition, which is itself controlled by a critical temperature marking the depth of seismic coupling (Tichelaar and Ruff, 1993). If the convergence rate is high, then the subducting plate will penetrate to greater depths before it reaches the critical temperature, thereby increasing the length of W. Tichelaar and Ruff (1993) find that for most subduction zones, the depth of seismic coupling is 40 ±5 km. But at the junction between the Japan and Kuril trenches, coupling extends to 52-55 km, whereas off Mexico, this depth is only 20-30 km. The critical temperature itself may vary, although if shear stress is constant with depth, a critical temperature of 250°C would explain the variability in depths of coupling. Hyndman and Wang (1993), on the other hand, suggest that the down-dip limit of stick-slip behavior on the subduction zone is controlled by the onset of quartz plasticity at 300-350°C.

A relatively buoyant plate tends to be more strongly coupled to the upper plate. But it is also hotter to begin with, so that the down-dip edge of W would be expected at a shallower depth. Another factor influencing the length and position of W is the thickness of sediment cover on the incoming plate. Hyndman and Wang (1993) argue that for a plate with a thick sedimentary cover, the top of oceanic crust is at a higher temperature than the top of oceanic crust in a plate relatively free of sediment. The sediment acts as an insulator, inhibiting the escape of heat from the crust through hydrothermal circulation, and as a result of the geothermal gradient, the top of the basalt is hotter than that of an incoming plate that is sediment-free. A thickly sedimented plate has the same thermal effect as a young plate. The brittle-plastic transition is reached at a shallower level, thereby limiting the down-dip extent of W. The Cascadia and Nankai subduction zones are characterized by thickly sedimented incoming plates and by relatively high heat flow compared to subduction zones relatively free of sediment.

The topography of the subducting plate may be smooth and sediment-covered, or it may be rugged. Topographic relief may be due to seamounts, to normal faults formed as a result of flexural bending or initial sea-floor spreading, or to transform faulting. Lay et al. (1982) concluded that high relief in the subducting slab leads to a highly variable strength of the plate boundary, which itself tends to reduce the length of individual rupture zones. According to them, the largest earthquakes would develop in subducting slabs with low topographic relief.

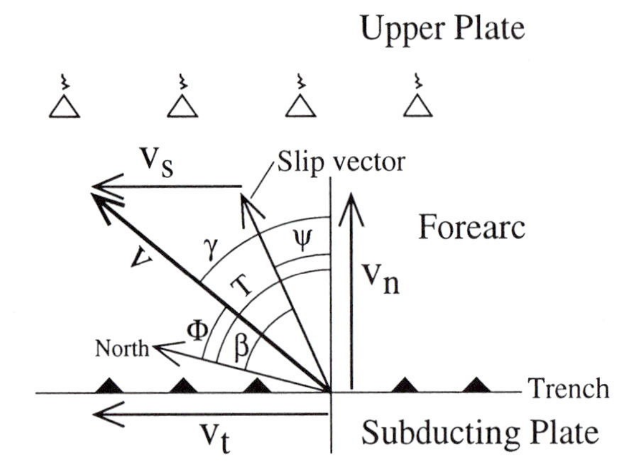

v = predicted plate convergence vector

T = trench-normal azimuth

Φ = plate motion azimuth

β = slip vector azimuth

ψ = trench normal - slip vector = T - β

Obliquity: $\gamma = T - \Phi$

Slip vector residual: $\Delta\beta = \beta - \Phi$

Forearc slip rate:

$v_S = v\,(\sin\gamma - \cos\gamma\ \tan\psi)$

Figure 11–7. Geometry of oblique convergence with respect to the upper plate behind the island arc (marked by triangles). The interplate slip vector V makes an angle γ with the dip direction of the trench, or trench-normal azimuth T. Internal deformation of the forearc causes it to move at a rate V_S with respect to the rigid upper plate. This causes the slip vector of interplate earthquakes to rotate with respect to the interplate slip vector by an angle called the obliquity. If the forearc deforms elastically, then $V_S = 0$, and the slip-vector residual is zero. V_n and V_t are the trench-normal and trench-parallel components of the plate vector V. After McCaffrey (1993).

High relief may produce *asperities* at the plate boundary, which are suggested by areas of high moment release within the rupture zones of great earthquakes (Thatcher, 1990). Areas of high moment release may also be influenced by the entrainment of sediment in the subduction zone or by the subduction of a spreading center. These variations in properties in the subduction zone suggest that a subduction zone may be complex, consisting of asperities in the unstable stick-slip field surrounded by areas of conditionally stable and fully stable sliding (Pacheco et al., 1993).

The foregoing discussion has inferred that the subduction vector points down-dip, normal to the trench. But the curvature of arcs is large with respect to variations in convergence directions of large plates controlled by spherical geometry (Fig. 1-5), which means that the expected interplate slip direction may differ from the dip direction by an angle referred to as *obliquity* (Fig. 11-7). McCaffrey (1993) observed that there may be a systematic deflection of slip vectors of interplate earthquakes from the slip vector predicted by plate tectonics (*slip-vector residual,* Fig. 11-7). Slip-vector residuals are best explained by internal deformation of the forearc region of the upper plate, expressed largely as strike-slip or oblique-slip faults.

Internal deformation of the forearc region means that only part of the convergence at the plate boundary is taken up at the plate interface; the rest is taken up by internal deformation. McCaffrey (1993) showed that the largest earthquakes occur on subduction zones characterized by relatively little internal deformation, as expressed geologically or as a small slip-vector residual. This means that a small slip-vector residual, in addition to high convergence rate and a young age of the subducting plate, is an indicator of large interplate earthquakes.

CASE HISTORIES

The subduction zones off the Pacific coasts of Japan and the United States are among the best-studied on earth, and three of them are adjacent to heavily populated areas. A large body of information about seismicity, crustal structure, onshore and offshore geology, and tectonics is available for each of these regions. In addition, the Nankai ("Southern Sea") subduction zone off southwest Japan has a recorded history of earthquakes covering more than thirteen centuries, including several repetitions of earthquakes on the same subduction zone segment. The Kuril and Aleutian subduction zones, including the Gulf of Alaska, accounted for three of the four greatest earthquakes of the twentieth century. The subduction zone off South America has a long historical record of earthquakes, is reasonably well studied, and the 1960 Chilean earthquake had the largest seismic moment release of the twentieth century. We describe these in a clockwise fashion around the Pacific Ocean, starting in the western Pacific.

In Japan, three subduction zones come together in a triple junction southeast of Tokyo (McKenzie and Morgan, 1969), the only triple junction of this kind on earth (labeled *och* in Fig. 11-8). Mesozoic oceanic crust of the Pacific plate is subducting westward beneath the Mariana, Izu-Bonin, Northeast Japan, and Kuril active volcanic arcs. South of the triple junction, the Mariana and Izu-Bonin arcs are bounded on the west by an actively-spreading back-arc basin (Mariana trough, marked by outward-facing arrows in Fig. 11-8). These subduction zones have produced none of the largest earthquakes (Uyeda and Kanamori, 1979), and they contain no accretionary prisms, in part because they are far from a sediment source, and there is little sediment on the subducting plate. Unlike the subduction zones south of the triple junction, the Northeast Japan subduction zone north of it has produced earthquakes with $M \geq 8$. These subduction zones are marked by a northward increase in convergence rate from the Marianas to northeast Japan and a decrease in the age of subducted crust from northeast Japan to Kamchatka (Fig. 11-6). The upper plate moves away from the trench south of the triple junction, but toward the trench farther north (Fig. 11-4).

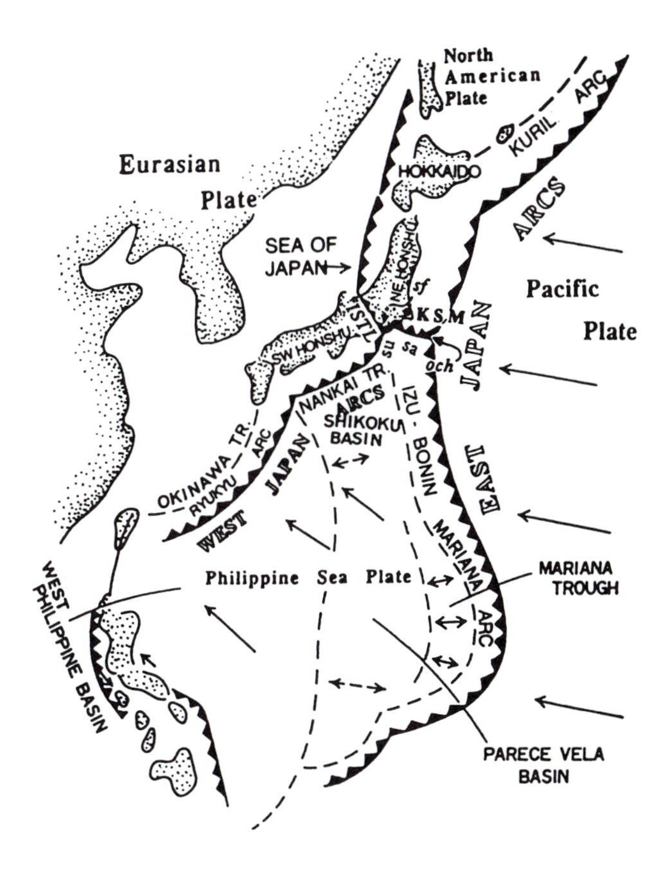

Figure 11–8. Plate-tectonic setting of Japan and surrounding regions. Barbs point toward upper plates of subduction zone; arrows show relative plate motion. Dashed lines mark volcanic arcs. *och,* Off Central Honshu triple junction; *sa,* Sagami trough; *sf,* South Fossa Magna triple junction; *su,* Suruga trough; *ISTL,* Itoigawa-Shizuoka Tectonic Line; *KSM,* Kashima VLBI station. From Uyeda (1991).

South of the triple junction, the Philippine Sea plate is subducting northwestward beneath southwest Japan and the Ryukyu arc. In contrast to northeast Japan, the Nankai trough is not accompanied by an active volcanic arc in southwest Honshu. However, volcanoes west of the cusp (or reentrant) at the triple junction (Fig. 11-9a) may be in part controlled by subduction of the Philippine Sea plate, and Pleistocene volcanoes in western Honshu may also be related to subduction. The Izu-Bonin arc is colliding end-on with the southwest Japan subduction zone beneath the island of Honshu. Oceanic crust west and east of the arc is being subducted, but the arc itself is a collision zone located at the Izu Peninsula and surrounding regions. The Northeast Japan and Izu-Bonin subduction zones contain earthquakes to depths of 600 km and are relatively simple in geometry (Fig. 11-9a). In contrast, the Southwest Japan subduction zone seems to be contorted, and the subducting slab is defined by earthquakes only to depths shallower than 90 km, if the Ryukyu subduction zone is excluded (Fig. 11-9b).

Southwest Japan Subduction Zone

The Philippine Sea plate is largely of Tertiary age, formed in part by back-arc spreading behind the Izu-Bonin and Mariana arcs (Fig. 11-8). The NNW-trending Kyushu-Palau Ridge (dashed line in Fig. 11-8) separates the older West Philippine basin on the west and the younger Shikoku Basin on the east. The Kyushu-Palau Ridge collides with the Eurasian plate

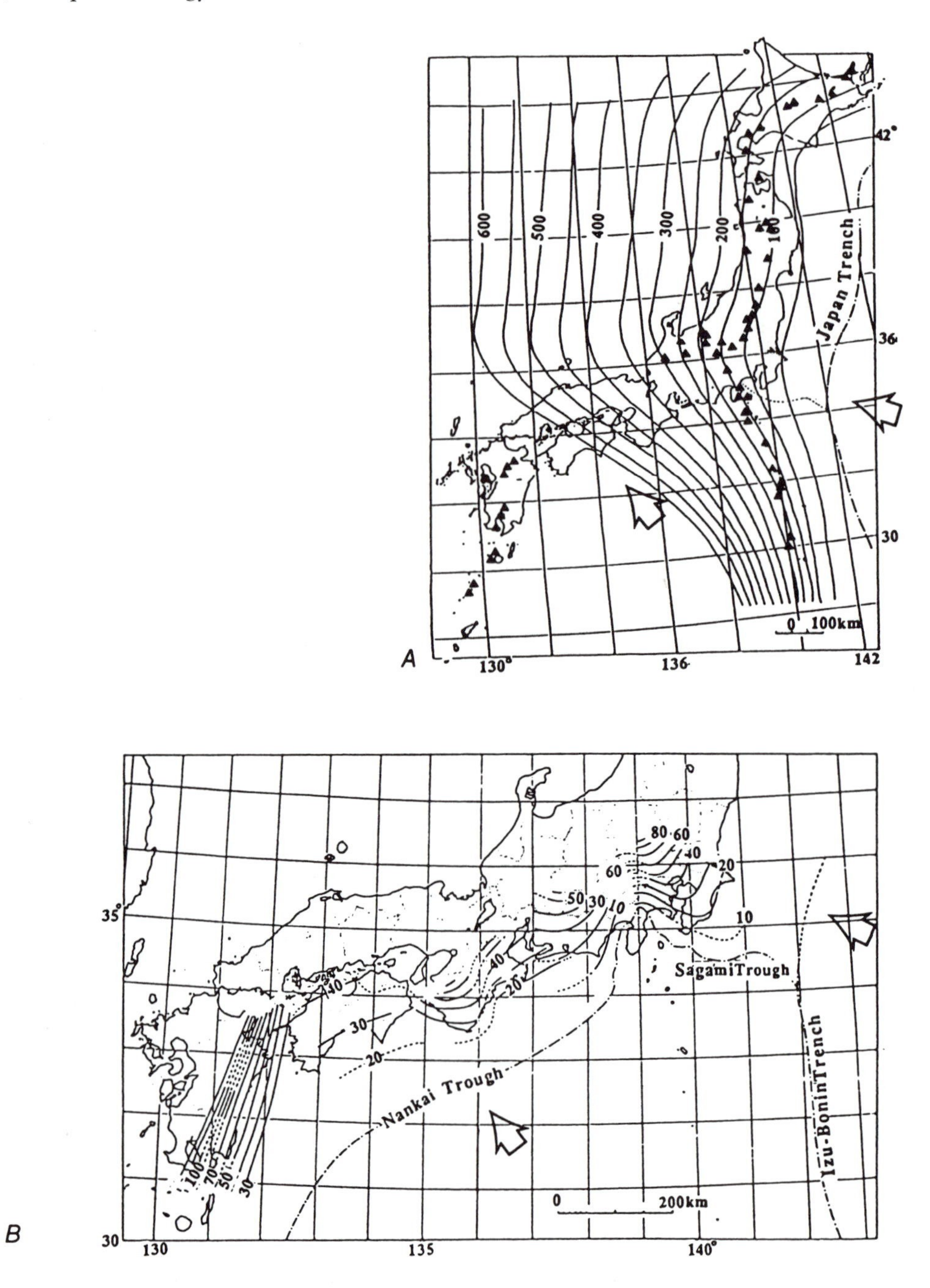

Figure 11–9. Structure contours in kilometers of the Wadati-Benioff zone above (a) the Pacific plate, (b) the Philippine Sea plate. Solid triangles in (a) locate active volcanoes. Note that the arc directly north of the Nankai trough is inactive, but that volcanoes west of the cusp between the Northeast Japan and Izu-Bonin arcs overlie an unusually deep part of the subducting Pacific plate, suggesting that they may be related to subduction of the Philippine Sea plate. Open arrows show the plate motions of the Pacific and Philippine Sea plates. After Uyeda (1991).

at a cusp between the Ryukyu trench and the Nankai trough (Fig. 11–8), two subduction zones that are very different in tectonic expression. The Ryukyu trench is deeper than 7 km, and oceanic crust of the West Philippine basin is about 60 Ma in age. The northern part of the Ryukyu arc, including the island of Kyushu, contains active volcanoes (Fig. 11–9a), and an incipient back-arc basin, the Okinawa trough (located on

Fig. 11-8), is forming to the northwest. The Wadati-Benioff zone beneath the Ryukyu arc is illuminated by earthquakes in the subducting slab to a depth of 120 km (Fig. 11-9b), but the subduction zone has produced no great earthquakes. The northern part of the Ryukyu arc-trench system resembles the Mariana-type subduction zone in that it is not strongly coupled seismically.

East of the Kyushu-Palau Ridge, the age of oceanic crust in the Shikoku basin is no older than late Oligocene or Miocene. The Shikoku basin has an average water depth of around 4000 meters, deepening to 4500-4900 m in the Nankai trough at the deformation front (Ashi and Taira, 1992). The sea floor is the top of a sedimentary sequence 1-2 km thick overlying oceanic crust. This sequence, as described at Site 808 of the Ocean Drilling Program (Taira et al., 1991; Fig. 11-10), consists of basal, seismically transparent, fine-grained hemipelagic sediment (shells of planktonic microfossils together with clays that have settled through the water column), overlain by seismically reflective sediment transported by bottom-flowing turbidity currents (Fig. 11-10) from a land source, principally the Fuji River. The young age of Shikoku basin crust and the relatively thick sedimentary section account for the absence of a trench and for the relatively high heat flow (more than 130 mW/m^2) observed at the toe of the accretionary prism of the Nankai trough (Taira and Pickering, 1991).

The Philippine Sea plate is subducting WNW beneath Japan at a rate of 3-5 cm/yr (Seno et al., 1993). The slip vector is oblique to the plate boundary, implying both a dip-slip component and a right-lateral strike-slip component. The accretionary wedge is cut by reverse faults and anticlinal folds spaced 0.5-1 km apart with vertical displacements of 5-20 m (Taira and Pickering, 1991). The most seaward of these faults, called *protothrusts,* provide evidence that the thrust front is migrating seaward into the Nankai trough along with the accretionary wedge (Fig. 11-10), which is building outward at a rate of about 3 mm/yr (Karig and Angevine, 1985). Mud volcanoes up to 2 km across and 200 m high occur in the Nankai trough adjacent to the deformation front and within the accretionary wedge, in part within a large landslide. The mud volcanoes on the floor of the trough follow NW-trending lineations, some of which extend into the accretionary prism. The reverse faults flatten downward into the décollement, which is imaged by seismic reflection as far as 30 km landward from the deformation front (Moore et al., 1990; Fig. 11-10).

Anticlinal ridges have trapped sediments to form *slope basins,* and the progressively greater landward tilt of deeper sedimentary layers in these basins may be due to uplift and landward tilting of the outer-arc ridges damming the sediment (Moore et al., 1990). Accretionary-prism strata are older and more strongly deformed landward, and the Shimanto Belt of Cretaceous to Miocene age exposed on land north of the slope basins is an ancient analog to the modern ac-

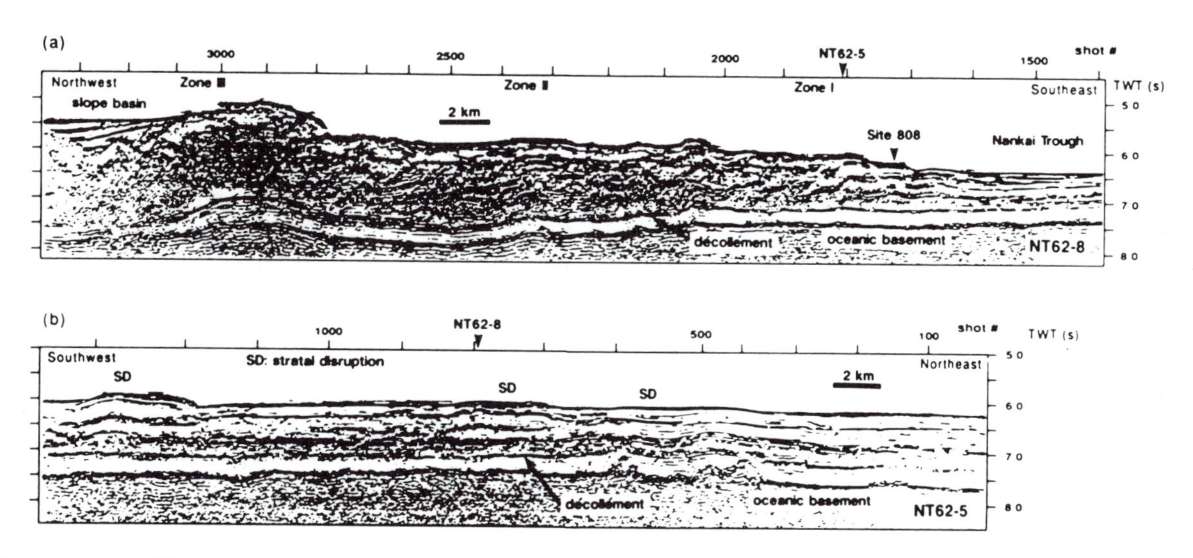

Figure 11-10. Seismic-reflection profile NT 62-8 from the Shikoku basin (Nankai trough; to the right of shot point 1700) across the Nankai accretionary prism. In the Shikoku basin, oceanic crust is overlain by a transparent hemipelagic unit and a reflective, land-derived turbidite unit. Stratigraphy based on a core hole by the Ocean Drilling Program at Site 808, located near the trench southeast of Shikoku. The transparent unit is traced to the left (northern) edge of the profile; it is overridden by deformed accretionary-wedge deposits along a décollement. From Ashi and Taira (1992) and Moore et al. (1990).

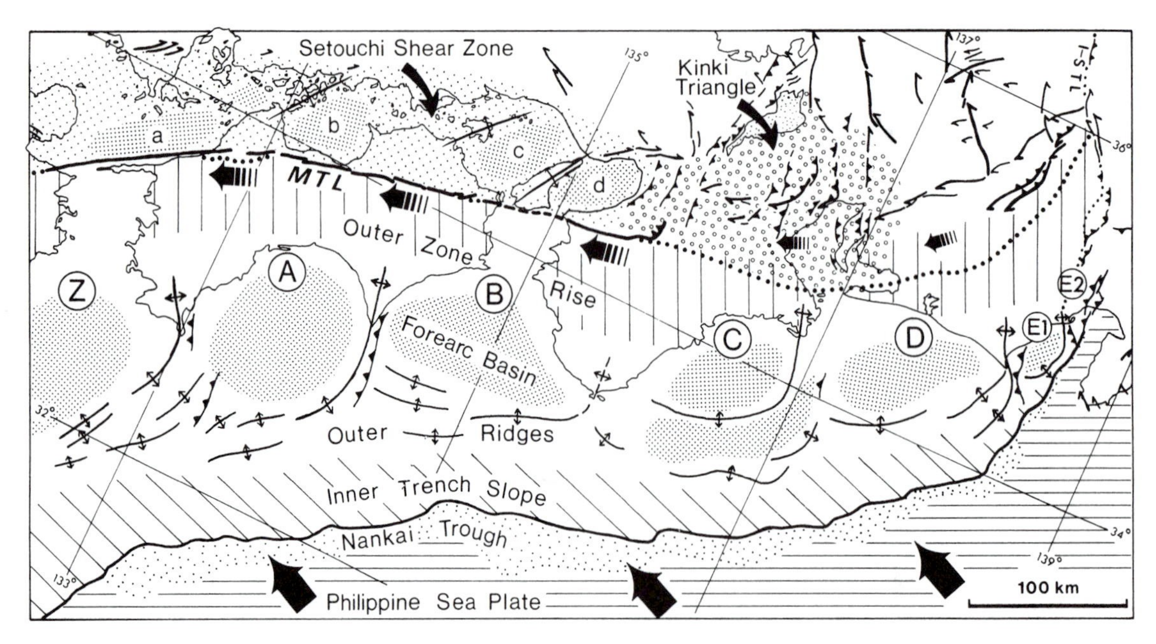

Figure 11–11. Active tectonic map of southwest Japan. Fine dot pattern delineates forearc basins (Z, A, B, C, D, E_1) and tectonic basins of the Setouchi shear zone (a, b, c, d). Large-dot pattern, Kinki Triangle, MTL, Median Tectonic Line, an active right-slip fault. Plate boundary at Nankai trough extends onshore at Suruga Bay (east end of map). After Sugiyama (1992).

cretionary prism (Taira and Ogawa, 1988). However, some of the thrusts above the Nankai deformation front appear to be active.

Five forearc-basin segments extend from the cusp with the Ryukyu trench eastward to the Suruga trough, where the plate boundary comes ashore (Sugiyama, 1992; 1994). These are labeled A through E in Figure 11-11. A sixth forearc basin, labeled Z on Figure 11-11, occurs at the cusp with the Ryukyu trench. These basins are separated by anticlines and faults striking at a high angle to the Nankai trough in which the west side of a fault is uplifted relative to the east side (Okamura, 1990; Sugiyama, 1992; 1994; Tsukuda, 1992). The largest fault, just offshore from Cape Muroto on the island of Shikoku (between basins A and B on Fig. 11-11), strikes almost due north. These oblique structures control the topography, including synclinal channels west and east of Shikoku Island and higher uplift rates in the mountains of Shikoku north of Cape Ashizuri (between basins Z and A on Fig. 11-11) and Cape Muroto (Tsukuda, 1992). The anticlines merge southward with the outer-arc ridges to produce structures shaped like an inverted "L" (Sugiyama, 1994; Fig. 11-11).

These forearc structures terminate south of the Median Tectonic Line (MTL), an active right-slip fault with a slip rate as high as 5-10 mm/yr on Shikoku (Okada, 1973; Tsutsumi and Okada, 1996) and 1-2 mm/yr farther east at the southern edge of the Kinki Triangle (Okada and Sangawa, 1978; Fig. 11-11). Very few active faults are found on land south of the MTL, but the Kinki Triangle north of the MTL contains one of the largest concentrations of active faults in Japan (Matsuda and Kinugasa, 1991; Research Group for Active Faults of Japan, 1992; Fig. 10-15).

North of the MTL, the Seto Inland Sea is an intra-arc depression underlain by the Setouchi Shear Zone, a set of tectonic basins alternating with anticlines (Sugiyama, 1992; 1994; Tsukuda, 1992). Each of the basins is elliptical in shape, elongated NE-SW at an angle of 30-45° to the MTL, producing an overall right-stepped arrangement consistent with a right-lateral component of slip on the Nankai subduction zone. This shear zone extends eastward on land as the Kinki Triangle, described in Chapter 10 (Fig. 10-28). The forearc structures, the MTL, and the Setouchi Shear Zone all respond to right-lateral shear accompanying oblique subduction of the Philippine Sea plate (Fig. 11-11), suggesting that the plate boundary is strongly coupled.

The eastern two basins, labeled c and d on Figure 11-11, are the best defined. They are separated by Awaji Island, an anticlinal uplift accompanied by right-lateral strike-slip faults, including the Nojima fault that ruptured during the 1995 Hyogo-ken Nanbu (Kobe) earthquake. Osaka Bay (labeled d) contains west-dipping sediments that are in fault contact with the uplifted rocks of Awaji Island. Farther east, in the

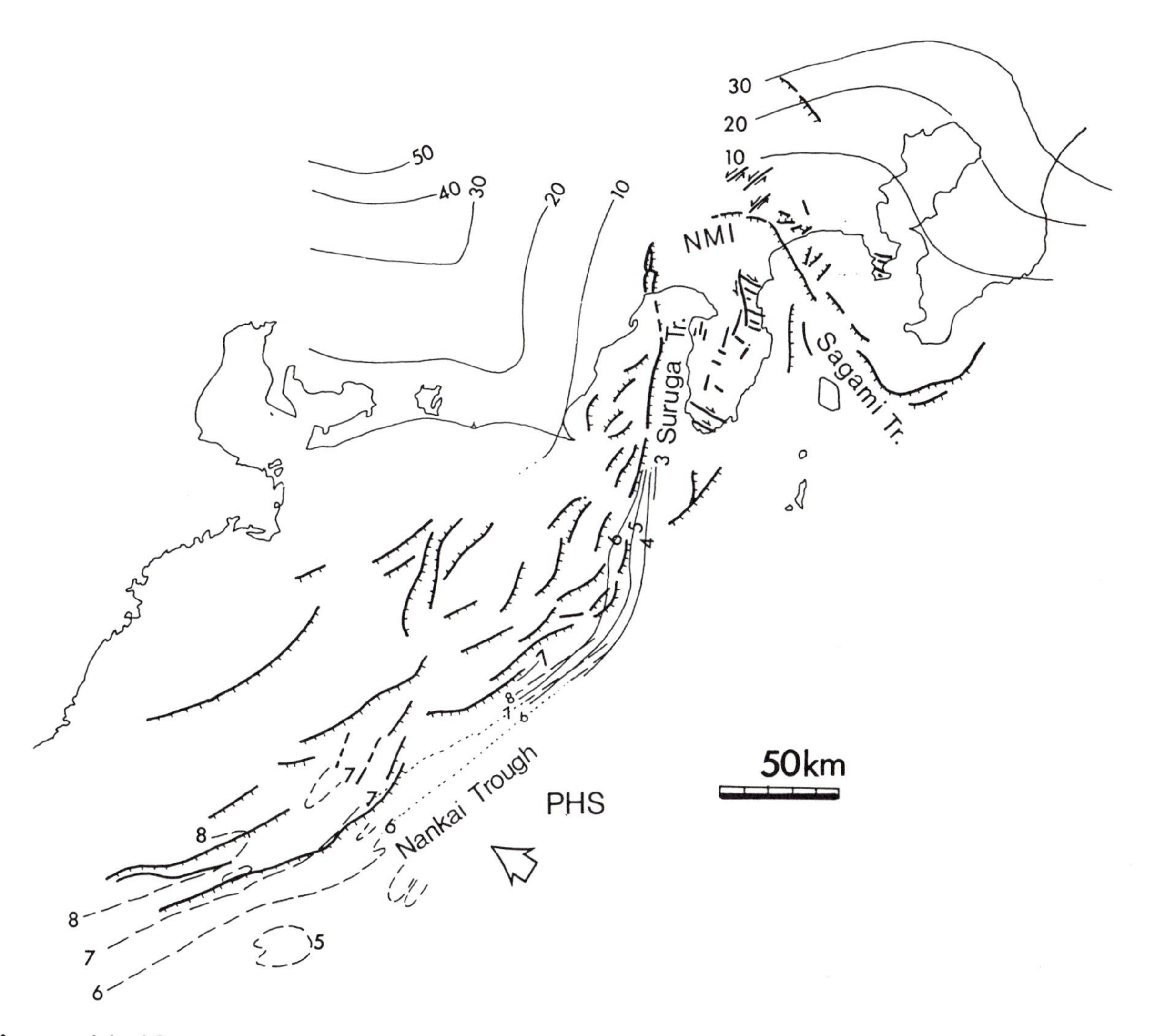

Figure 11–12. The Izu collision zone. The plate boundary extends from the Nankai trough through the Suruga trough to pass beneath Mt. Fuji, north of Izu Peninsula. NMI = northern margin of Izu at Mt. Fuji. To the east, the plate boundary follows the Sagami trough to a triple junction with the Pacific plate. Solid contours mark the top of the subducting Philippine Sea plate (PHS) in kilometers. Close to the plate boundary, dashed lines and solid lines mark the top of basement beneath the Nankai and Suruga troughs, respectively. Heavy lines mark active faults. After Yamazaki (1992).

Kinki Triangle, the Osaka, Nara, and Kyoto basins are bounded on the east by active reverse faults that absorb some of the right slip along the MTL to the south (Sangawa, 1986; Fig. 10-28).

At the east end of the Nankai trough, the plate boundary turns north into the Suruga trough and comes ashore, passing beneath Mt. Fuji; from there, it follows the Sagami trough down a submarine canyon to the triple junction with the Pacific plate (Figs. 11-12; 11-13; 11-14). The Izu-Bonin Ridge to the south is parallel to the Kyushu-Palau Ridge to the west (Fig. 11-8), but it is much larger and is in part subaerial, an active island arc. Between the Izu-Bonin Ridge and Mt. Fuji is the Izu Peninsula, part of the Philippine Sea plate; the plate boundary here is an arc-continent collision zone. The effects of the collision may be seen to the north by the rapid uplift and erosion of the Japan Alps, which produce the great volume of sediment carried by the Fuji River through the Suruga trough into the Shikoku basin.

The onshore faults north of Suruga Bay and northwest of Sagami Bay have high slip rates (Yamazaki, 1992; Fig. 11-14), but clearly these faults alone do not accommodate all of the convergence between the

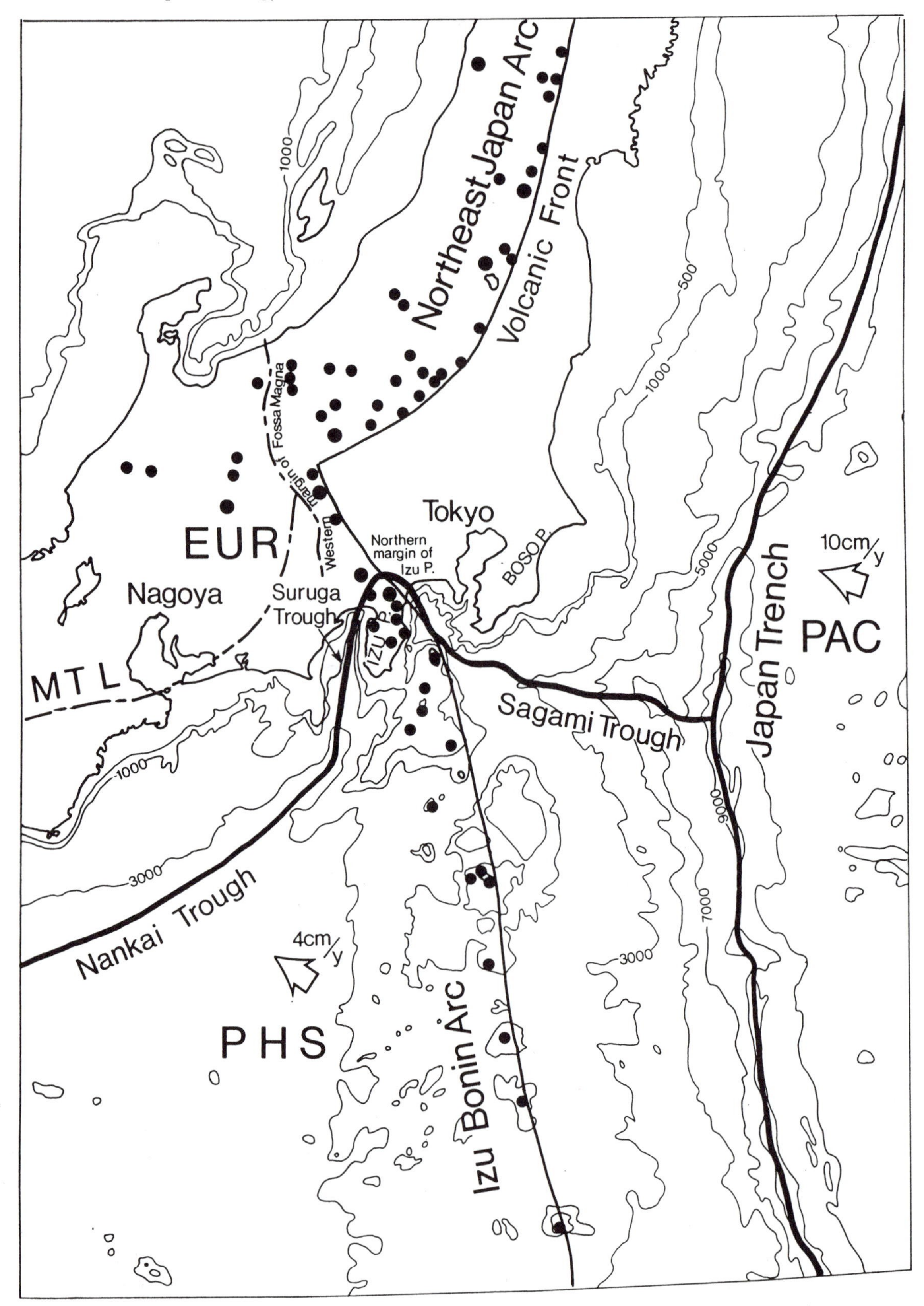

Figure 11–13. Plate tectonic setting around Izu Peninsula. Arrows show motion of Pacific (PAC) and Philippine Sea (PHS) plates relative to Eurasia (EUR). Solid circles show Quaternary volcanoes. MTL, Median Tectonic Line. After Yamazaki (1992).

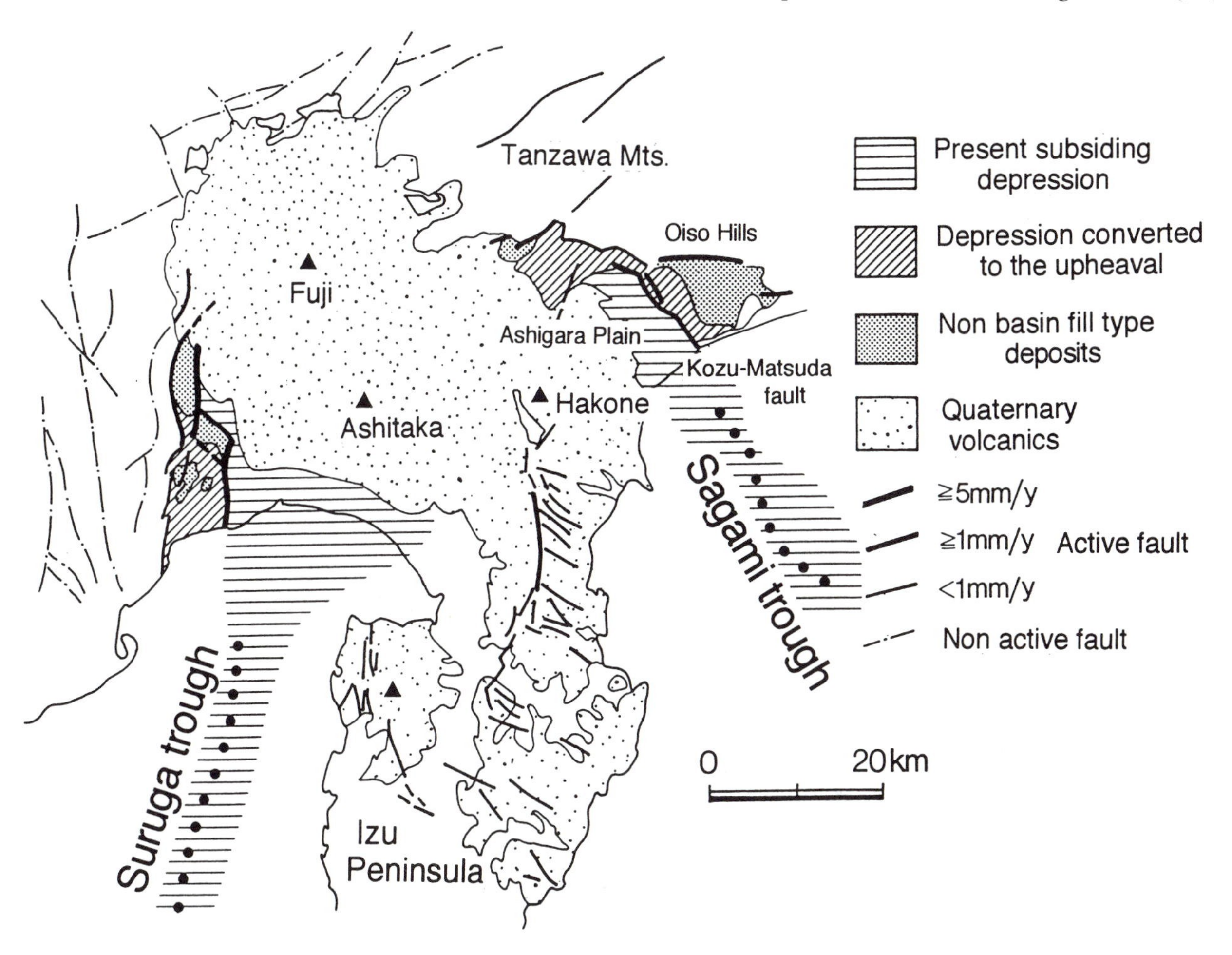

Figure 11–14. Onshore geology of the Izu collision zone, showing distribution of Quaternary sediments. After Yamazaki (1992).

Philippine Sea plate and the rest of Japan. The faults pass beneath Mt. Fuji, but that volcanic carapace shows no evidence of deformation. High uplift rates in the Japan Alps, north of the collision zone, suggest that some of the shortening is accommodated by crustal thickening. Some deformation is taken up within the Izu Peninsula itself (Fig. 11–14), and other faults lie offshore east and south of the Izu Peninsula (Fig. 11–14). The Zenisu Ridge, shown by bathymetric contours on Fig. 11–13, extends southwest from the Izu-Bonin ridge into the Shikoku basin south of the Nankai trough. This ridge is uplifted along an intraplate thrust, which is taking up part of the convergence and may soon take over as the main plate-boundary thrust (Le Pichon et al., 1987; Chamot-Rooke and Le Pichon, 1989). This would transfer the Izu Peninsula from the Philippine Sea plate to the rest of Japan.

Historical and archeological evidence documents interplate earthquakes for more than 1300 years on the Nankai portion of the southwest Japan subduction zone. This record is shown in Figure 11–15b, revised by Sangawa (1992), Ishibashi (1992), and Sugiyama (1994) from Ando (1975) on the basis of archeological evidence and historical records of tsunami runups and seismic shaking. The five regions A through E in Figure 11–15 correspond to the forearc basins A through E shown in Figure 11–11. Regions A and B generally rupture together, as do regions C and D.

The boundary between B and C limited the rupture zones of the 1944 Tonankai and 1946 Nankaido earthquakes; the mainshocks of these two events were close together on opposite sides of this boundary (Fig. 11–15a). Although the depth of the 1944 earthquake was 40 km according to Iwata and Hamada (1986), the epicenters of both the 1944 and 1946 earthquakes are located where the top of the Wadati-Benioff zone is at a depth of 20–30 km (compare Figs. 11–9b and 11–15a). Coseismic uplift of Cape Ashizuri and subsi-

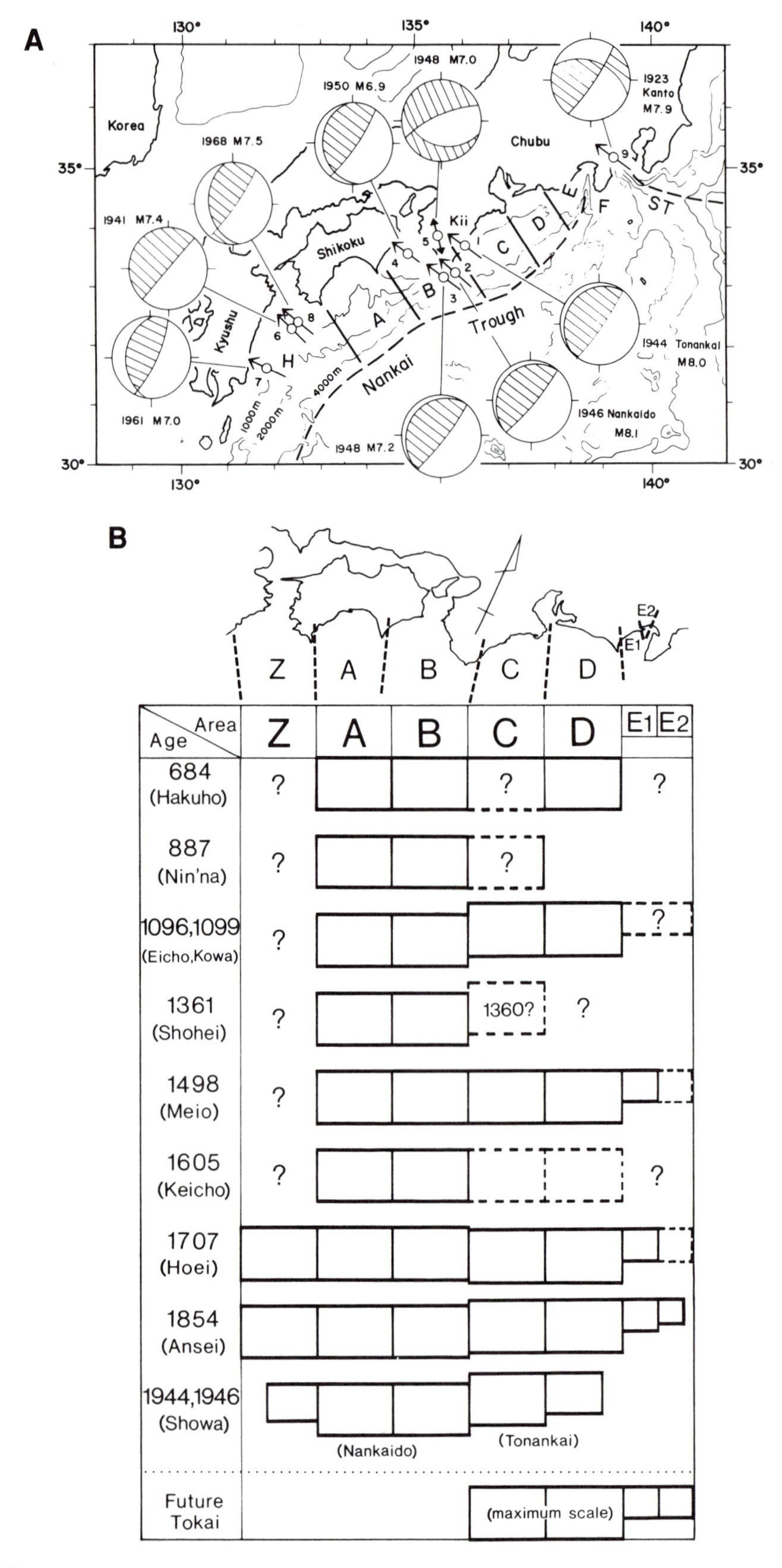

Figure 11–15. (a) Interplate earthquakes on the Southwest Japan subduction zone, with focal-mechanism solutions and slip vectors. Regions A through E comprise the Nankai segment of the subduction zone. Region H is the Ryukyu segment, and Region F is the Izu collision zone. From Shiono (1988). (b) Space-time distribution of subduction-zone earthquakes in Regions Z and A through E. From Sugiyama (1994).

dence of the Hyuga basin (Z, Fig. 11-11) show that the 1946 rupture continued west into region Z, with the remainder of region Z rupturing in earthquakes in 1941 (M7.4) and 1968 (M7.5; Fig. 11-15a). The 1946 earthquake did not rupture into region E, although this region ruptured in 1854, producing surface faulting in the Fuji River bed.

Two earthquakes (Ansei) ruptured the entire Nankai subduction zone 32 hours apart in 1854, separated by the B-C boundary. These were larger than the 1944 and 1946 events, rupturing east into region E and west across region Z, based on records of shaking in Hyuga Sea and northern Kyushu and uplift and subsidence records in region E (summarized by Sugiyama, 1994). The 1707 Hoei earthquake appears to have ruptured the entire Nankai subduction zone, which would make it one of the largest historical earthquakes in Japan. The failure of the 1944 earthquake to rupture as far northeast as the 1854 and 1707 events has caused great concern in Japan that region E (the Tokai gap) is about to rupture (Ishibashi, 1981). However, Uyeda (1991) doubts that such a large-magnitude earthquake will occur there because of the proximity of region E to the Izu collision zone (see discussion in the following text).

Rupture lengths of earthquakes prior to 1707 are more sketchy due to poorer historical records. The 1605 Keicho earthquake may have ruptured regions A through D. It produced relatively large-amplitude tsunamis relative to its shaking intensity, suggesting that it may have been a slow earthquake, or the epicenter was located closer to the Nankai trough than subsequent events (Ishibashi, 1989). The 1498 Meio earthquake had its source area in regions C and D, but liquefaction records from a site in region A suggests that the earthquake ruptured the entire subduction zone (although no data are available from region Z).

In summary, Nankai earthquakes recur at intervals of 1 to 1.5 centuries from 1361 onward, and intervals of 2 to 2.5 centuries from 684 to 1361. Regions A and B always rupture together, as do regions C and D; some earthquakes rupture across all four regions. Ruptures extend into regions Z and E in some cases, but not others, The boundary between B and C limited the 1096, 1099, 1854, 1854, 1944, and 1946 earthquakes but not the 1707, 1605, and 1498 earthquakes. (However, for these earlier earthquakes, timing data are not sufficient to preclude two closely spaced earthquakes such as the Ansei earthquakes of 1854.)

Figure 11-11 shows that regions A through E are bounded by transverse anticlinal ridges and faults. But the B-C boundary between the Muroto trough and Kumano trough, corresponding to the rupture limit of the 1946, 1944, and 1854 earthquakes, as well as earlier ones, is less pronounced than other transverse structures. The D-E boundary terminated the 1944 Tonankai earthquake on the northeast, but the 1854 Ansei-Tokai earthquake ruptured across this boundary. The 1707 Hoei earthquake was also thought to rupture across this boundary (Ishibashi, 1981), but recent work (Ishibashi, 1992) suggests that the 1707 earthquake did not rupture as far northeast as the 1854 event.

Subcrustal instrumental seismicity documents an oceanic slab (Fig. 11-16), but the subcrustal earthquakes marking the slab are cut off sharply at 40-90 km depth, as shown by the northern limit of contours in Figure 11-9b. The slab beneath Shikoku, as defined by seismicity, dips 8°N and is marked by earthquakes only to a depth of 40 km (Fig. 11-16). In contrast, the slab beneath Chubu (eastern Nankai subduction zone) dips 10°, steepening down-dip to 30°N, and it is marked by earthquakes to a depth of 50 km. The curvature of the top of the slab is strongly influenced by the Izu collision zone (Fig. 11-12). Slab dip is steepest beneath the Kii Peninsula, east of Shikoku (Fig. 11-9b), where the slab is marked by earthquakes to a depth of 60 km. The aseismic cutoff on the north side is closest to the Nankai trough beneath Kii Peninsula; the seismically marked slab is wider to the west and east (Shiono, 1988; Fig. 11-9b).

The contorted top of the downgoing slab (Fig. 11-9b) permits a comparison with the forearc segment boundaries between regions A through E. Although the boundary between regions A and Z overlies a sharp change in slab geometry between the Nankai and Ryukyu subduction zones, other boundaries show a poorer correspondence. The boundary between a gently dipping slab beneath Shikoku and a steeper slab beneath the Kii Peninsula occurs beneath the middle of region B, and a left step in the slab farther east occurs beneath the middle of region C.

Fault-plane solutions of microearthquakes recorded in recent years show that the slab is in tension either parallel to strike or down-dip (Fig. 11-17; Shiono, 1988). This is in contrast to fault-plane solutions of the great 1944 and 1946 interplate earthquakes, which were low-angle thrusts (Fig. 11-15a); the lack of recent interplate earthquakes prevents an accurate assessment of the depth of seismic coupling. The slab is aseismic farther north, where it has been mapped based on seismic tomography. Figure 11-16 also shows that the plate boundary close to the Nankai trough is also aseismic beneath the accretionary wedge. The down-dip width of the locked zone beneath the Nankai subduction zone has been estimated

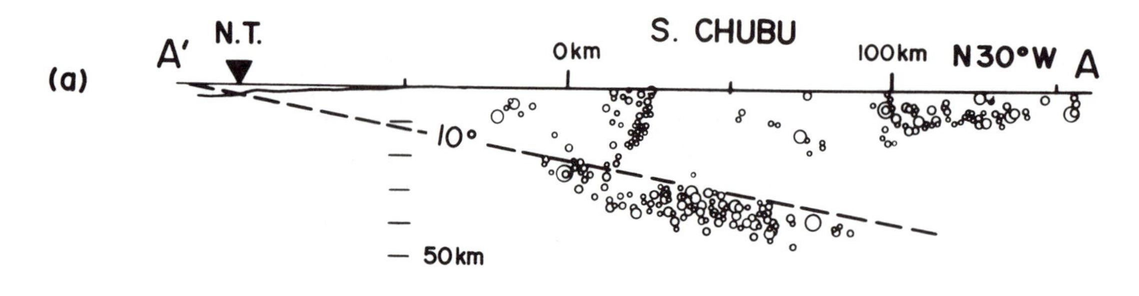

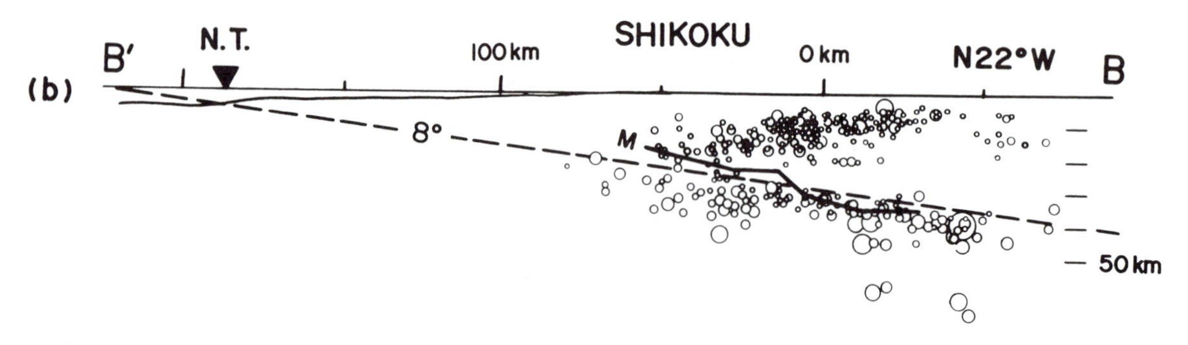

Figure 11–16. Cross sections of microseismicity beneath Chubu (eastern Nankai subduction zone) and Shikoku, with no vertical exaggeration. Inferred plate boundary is shown by dashed line. M shows estimated position of Moho discontinuity; N.T., Nankai trough. North is to right. After Shiono (1988).

by Hyndman et al. to be more than 100 km based on heat-flow, geodetic, tsunami, and seismicity data, but the locked zone is relatively shallow compared to other subduction zones in the western Pacific (Tichelaar and Ruff, 1993).

Geodetic data have been collected for southwest Japan since the 1890s, including leveling, triangulation, and tide-gauge data (Fitch and Scholz, 1971; Ando, 1975; Thatcher, 1984; Savage and Thatcher, 1992). These data now encompass nearly all of an interplate earthquake cycle, starting with the interseismic period between the 1854 and 1944 to 1946 earthquakes and now including 50 years of the present interseismic period. These records show that southern Shikoku Island and the Kii Peninsula subsided in the interseismic period, then were uplifted instantaneously during the 1944 and 1946 earthquakes, then subsided again rapidly in the first decade after the earthquake and then slowly and steadily afterwards. Although the post-seismic relaxation occurred fairly quickly at Cape Muroto, close to the trench, leveling data farther north show that the deformation migrated down-dip with time and became broader. The post-seismic relaxation phase took several decades to complete in comparison to only a decade at Cape Muroto (Thatcher, 1984).

Previous investigators showed a residual uplift after completion of an earthquake cycle, in large part based on uplifted Holocene terraces or benches at Capes Ashizuri and Muroto on Shikoku and also on the southern end of the Kii Peninsula (Yoshikawa et

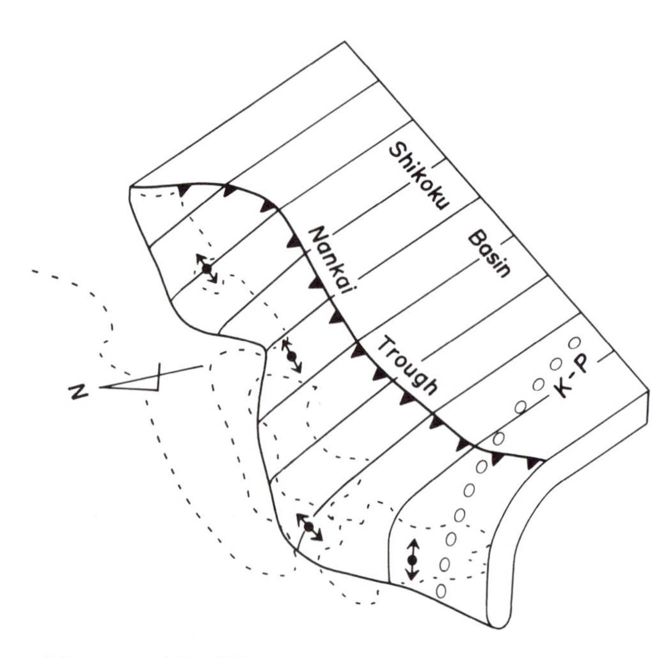

Figure 11–17. Schematic view of subducting Philippine Sea plate. Arrows show extensional axes of earthquakes. Coastline of southwest Japan shown in dashed lines. After Shiono (1988).

al., 1964). Maemoku (1988a) concluded, however, that there was no residual uplift after postseismic recovery following the 1707, 1854, and 1946 earthquakes. Nearly 6 m of cumulative uplift was recorded for these three earthquakes at Muroto port, but all this uplift was recovered by interseismic subsidence. Maemoku (1988a, b) and Maemoku and Tsubono (1990) also modeled the uplift pattern of Holocene terraces on Shikoku and the Kii Peninsula and concluded that terrace uplift and tilting were caused by faulting in the upper plate, either a splay from the megathrust or more local structures such as the north-trending Muroto fault directly offshore from Cape Muroto (Fig. 11-11). In addition, the 2.7-2.6 ka terrace has been offset by a small fault on land (Maemoku, 1988a). (The best geodetic model of Fitch and Scholz [1971] for the 1946 earthquake is also an upper-plate splay off the megathrust.)

The ages of the six Holocene terraces at Cape Muroto indicate a recurrence interval between terrace uplifts ranging from 300 to 1500 years, in contrast to 100 to 250 years for interplate earthquakes for the last 1300 years. Clearly, then, some of the interplate earthquakes did not produce coastal uplift preserved as Holocene terraces. This has led to a classification of Nankai earthquakes into Genroku type, where coseismic uplift of the Holocene terrace exceeds interseismic subsidence, resulting in permanent preservation of the terrace, and Taisho type, where coseismic uplift is recovered (Matsuda et al., 1978; Yonekura, 1979; Shimazaki, 1980; Sugiyama, 1994; Fig. 11-18). The Genroku-type earthquakes combine rupture of the plate boundary with internal deformation of the forearc, whereas Taisho-type earthquakes produce no internal deformation of the forearc. This indicates that Holocene coseismic benches (or subsided marshes, as at Cascadia) may represent maximum recurrence intervals, and all the subduction-zone earthquakes may not retain a permanent geological record.

East of the Izu collision zone, the continuous historical earthquake record covers less than 400 years. Closest to the collision zone, the Odawara area was struck by earthquakes in 1633 (M7.1), 1782 (M7), and 1853 (M6.5-6.7), in addition to the Genroku earthquake of 1703 and Kanto earthquake of 1923, both M8.2 (Ishibashi, 1985; 1992; Fig. 11-19). The focal mechanism of the 1923 earthquake was an oblique-slip reverse and right-lateral fault, consistent with the large strike-slip component predicted for the interplate slip vector with respect to the SE-trending plate boundary east of the collision zone (Fig. 11-8). The Odawara area has an earthquake recurrence interval of 70-79 years, and another earthquake is expected soon (Ishibashi, 1985), whereas farther away from the Izu collision zone, the recurrence interval is 220 years, and the earthquakes are an order of magnitude larger. Curiously, the collision zone itself seems to have its own independent earthquake history, with six earthquakes since the 1923 Kanto event. The 1703 earthquake was accompanied by permanent uplift and northward tilting on the Boso Peninsula, and Matsuda et al. (1978) mapped older Holocene benches, also tilted northward, with recurrence intervals longer than 1000 years between uplift events. Scholz and

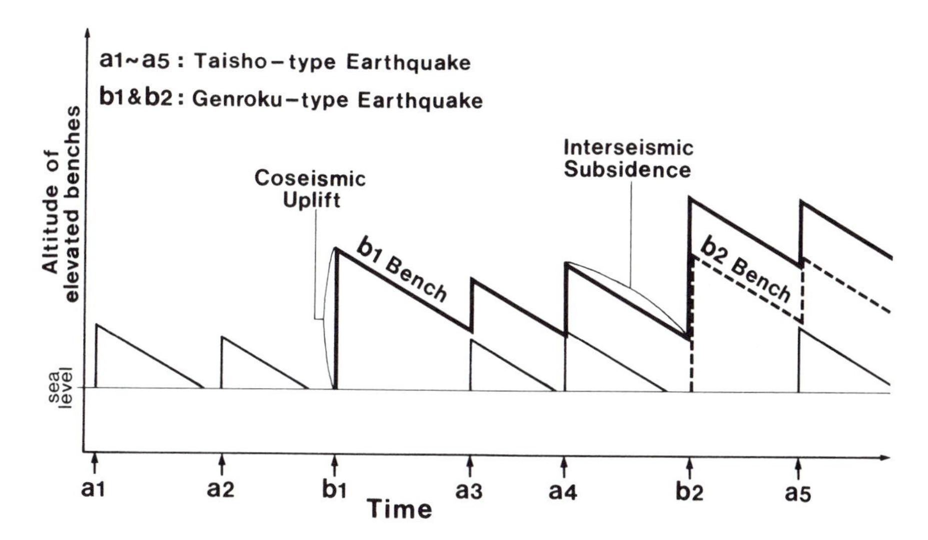

Figure 11–18. Formation of elevated benches by coseismic uplift associated with Genroku-type earthquake (permanent uplift) and Taisho-type earthquakes (coseismic uplift completely recovered in interseismic period. After Sugiyama (1994).

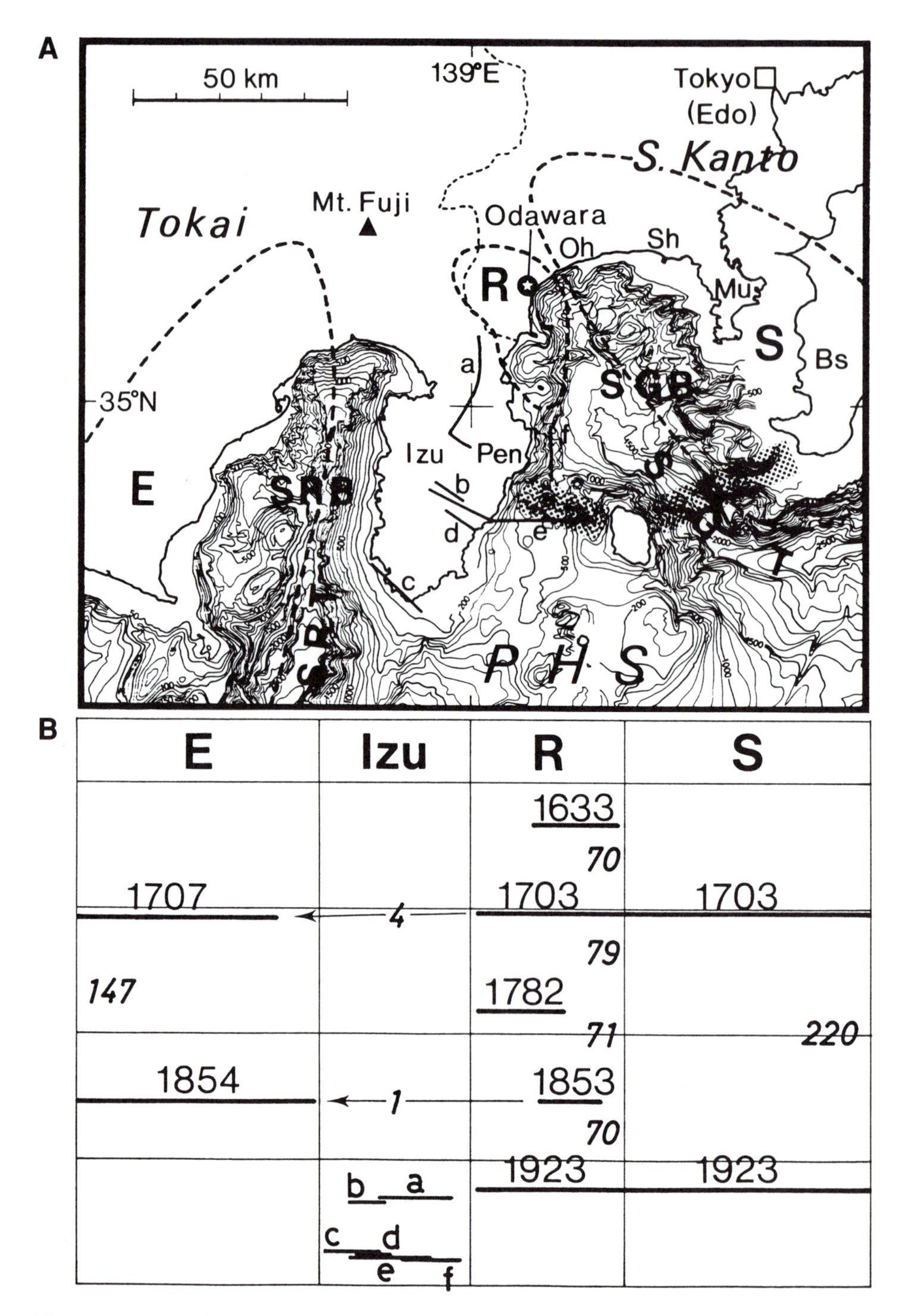

Figure 11–19. (a). Map of the area around Izu Peninsula. Part of the source regions are shown for great earthquakes in the Tokai district (E) and South Kanto district (S) as well as moderate earthquakes around Odawara (R). PHS, Philippine Sea plate; SGT, Sagami Trough, with Sagami Bay limited to region north of stippled band; SRB, Suruga Bay; SRT, Suruga Trough; a–f locates six earthquakes around Izu Peninsula since the 1923 Kanto earthquake. (b) Space-time diagram of the E, Izu, R, and S regions for the last four centuries. Numbers in italics show time in years between earthquakes. 1 and 4 in italics show that the last two great Tokai earthquakes were preceded by earthquakes in the Odawara area. After Ishibashi (1985).

Kato (1978) and Ishibashi (1985) reinterpreted these terraces as controlled by reverse faults in the upper plate. Vertical motions during and following the 1923 Kanto earthquake also showed offsets along faults in the upper plate (Scholz and Kato, 1978).

Northeast Japan and Kuril Subduction Zones

The Southwest Japan subduction zone is a Chilean-type zone based on its low dip, broad accretionary prism, and presence of great earthquakes. The Northeast Japan and Kuril subduction zones, however, do not fit either the Chilean or the Mariana end member. They lack broad accretionary prisms, and they have a moderate dip, like the Mariana type, but they also have an outer-arc bulge, and they are characterized by great earthquakes, like the Chilean type.

The Pacific plate is subducting WNW beneath the island arcs of the northwest Pacific region at nearly 10 cm/yr (Fig. 11-6). This rate is considerably higher than rates at the Izu-Bonin and Mariana subduction zones, where the convergence between the Pacific plate and Eurasia is partitioned between subduction zones at the west and east boundaries of the Philippine Sea plate, both with actively spreading back-arc basins (Fig. 11-8). The triple junction between the Izu-Bonin and Northeast Japan trenches is marked by a cusp, or abrupt change in strike, of the trench which is expressed at depth by a change in strike of the subducted slab to depths of 600 km (Fig. 11-9a). Another cusp lies between the Northeast Japan and Kuril trenches, and this, too, is expressed by changes in strike of the subducted slab at depth (Fig. 11-9a). The direction of convergence is WNW, so that the convergence is dip-slip at the Japan trench but oblique slip with minor right-lateral strike slip in the Kuril trench opposite Hokkaido (Fig. 11-20). Farther northeast, convergence is again dip-slip in the northern Kuril trench opposite Kamchatka (Minster and Jordan, 1978; Seno and Eguchi, 1983). The slip vector residual is relatively small throughout this zone (McCaffrey, 1993), suggesting little internal deformation of the forearc region of the upper plate.

Like the Izu-Bonin and Mariana trenches, the Northeast Japan and Kuril trenches are deep and relatively barren of sediment cover. This may be inherited from the Miocene. At that time, the Northeast Japan subduction zone resembled the Mariana and Izu-Bonin trenches in that the Japan Sea was an active back-arc basin. At the same time, the Kuril basin in the southern Sea of Okhotsk east of the island of Sakhalin was also opening behind the Kuril arc (Savostin et al., 1983). Orientation of metalliferous veins of Miocene age indicates that the Japan Sea side of northeast Japan was undergoing NNW extension at that time (Otsuki, 1990). The continental slope off the Sanriku coast of northeast Japan, now at 3 km water depth, was an outer-arc high called the Oyashio Paleoland (Kobayashi, 1983). The Miocene volcanic front was several tens of kilometers east of the Quaternary volcanic front (Sugi et al., 1983).

The Japan Sea and Kuril basins are not spreading now, even though they are both characterized by high heat flow. Instead, the region is now in E-W compression, including the Niigata fold belt off the west coast of Honshu and a range-and-basin reverse-fault province in northern Honshu (described in Chapter 10). Study of metalliferous veins indicates that horizontal compression began 5-7 Ma (Otsuki, 1990), although the Kitakami Mountains, the easternmost range in northern Honshu, may have been in compression since the early Miocene (Sugi et al., 1983). Crustal shortening on land in northern Honshu, including displacement on reverse faults and shortening on fold belts, amounts to 20-30 km, mostly in the Quaternary, resulting in a crustal shortening rate of 2-3 cm/yr (Otsuki, 1989, but see also Wesnousky et al., 1982). Terrestrial geodesy indicates compressional strain rates of about 1×10^{-7}/yr (cf. Otsuki, 1989). Displacement of the Kashima VLBI station (located on Fig. 11-8) with respect to VLBI stations on the North American continent was also explained by shortening within northern Honshu (Heki et al., 1990).

The development of a fold belt on the Sea of Japan side of northern Honshu, accompanied by large reverse-fault earthquakes in 1964, 1983, and 1993, suggests that a new convergent plate boundary has formed between northern Honshu and the Eurasian plate in the Sea of Japan (Fig. 11-8; Nakamura, 1983). Nakamura found no obvious plate boundary separating northeast Japan from the Bering Sea and Alaska, and he suggested that northeast Japan is part of the North American plate (Fig. 11-8). The southern end of this plate is a zone of active thrusting and strike-slip faulting (Uyeda, 1991) called the Itoigawa-Shizuoka Tectonic Line (ISTL; Fig. 11-8), which marks the western boundary of a late Cenozoic rift called the Fossa Magna. This rift has undergone strong folding and reverse faulting in the Quaternary.

The Pacific plate east of the Japan trench is cut by normal faults parallel to the trench that seem to be the result of bending moment as the Pacific plate is flexed downward beneath Japan (Fig. 11-21). Normal faulting was the apparent origin of the M8.3 Sanriku outer-rise earthquake of 1933 (Kanamori, 1972). Other normal faults striking N60°E are parallel to the Mesozoic magnetic lineations; they may be reactivated Mesozoic rise-crest normal faults (Cadet et al., 1987;

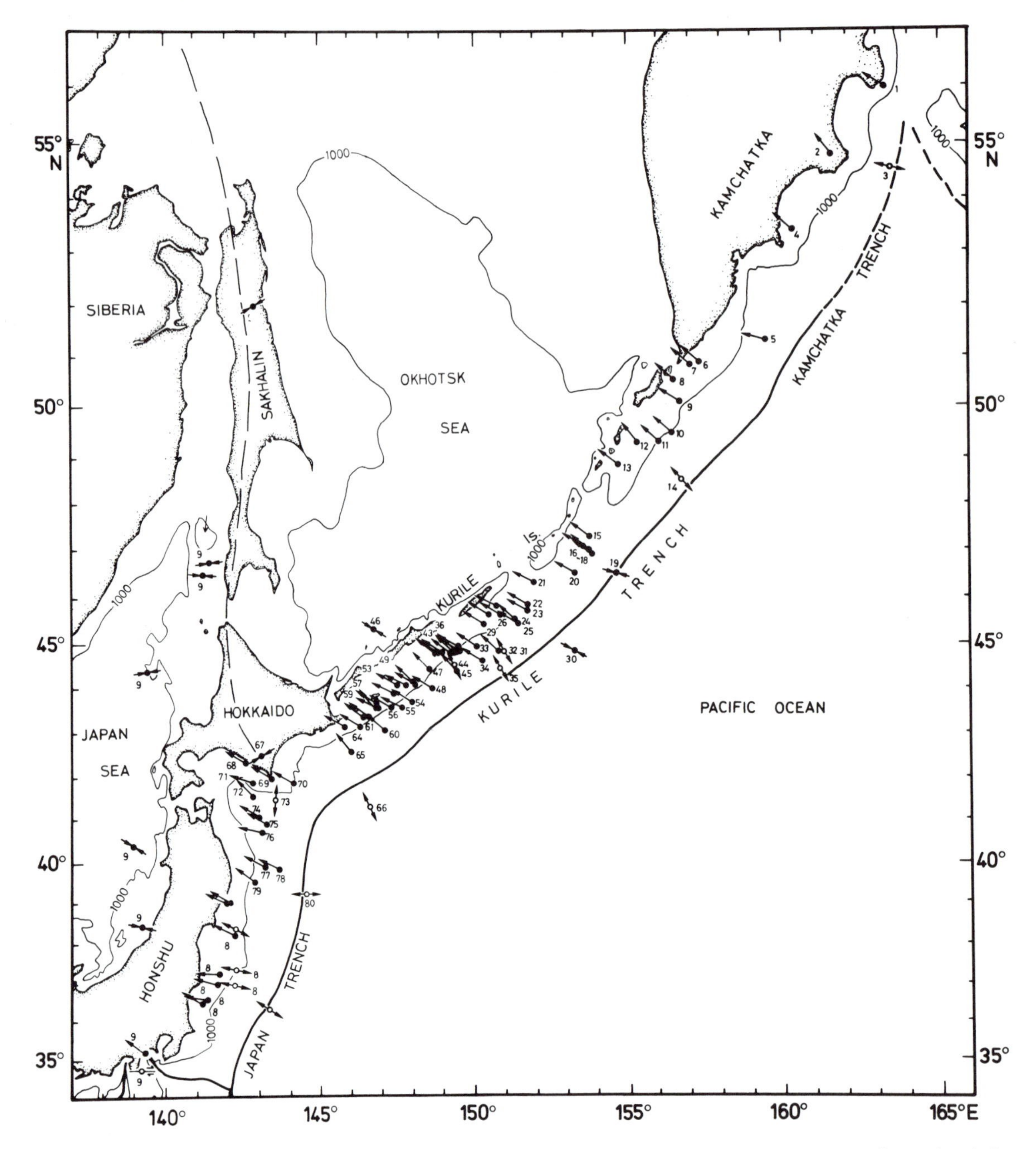

Figure 11–20. Focal mechanisms of shallow earthquakes at the Northeast Japan and Kuril subduction zones. Normal-fault events at or seaward of the trench are due to the bending moment of the incoming Pacific plate as it is loaded by the island-arc systems. After Seno and Eguchi (1983).

Fig. 11-21). The Kuril trench is parallel to these magnetic lineations, and so only the set of faults striking N60°E appears in the Pacific plate opposite this trench, possibly due to bending moment (Fig. 11-21). Unlike southwest Japan, the Japan trench is starved of sediment except near the triple junction where it receives sediment down a major submarine canyon in the Sagami trough (Taira and Pickering, 1991). The Kuril trench also has little sediment (Gnibidenko et al., 1983).

The continental slope off northeast Japan is not presently accumulating sediment; only about 1 km of sediment overlies the Cretaceous basement (Taira and Pickering, 1991). The trench wall is relatively barren, with a low heat flow of about 30 mW/m^2. The slope has a benched topography, marked by huge arcuate

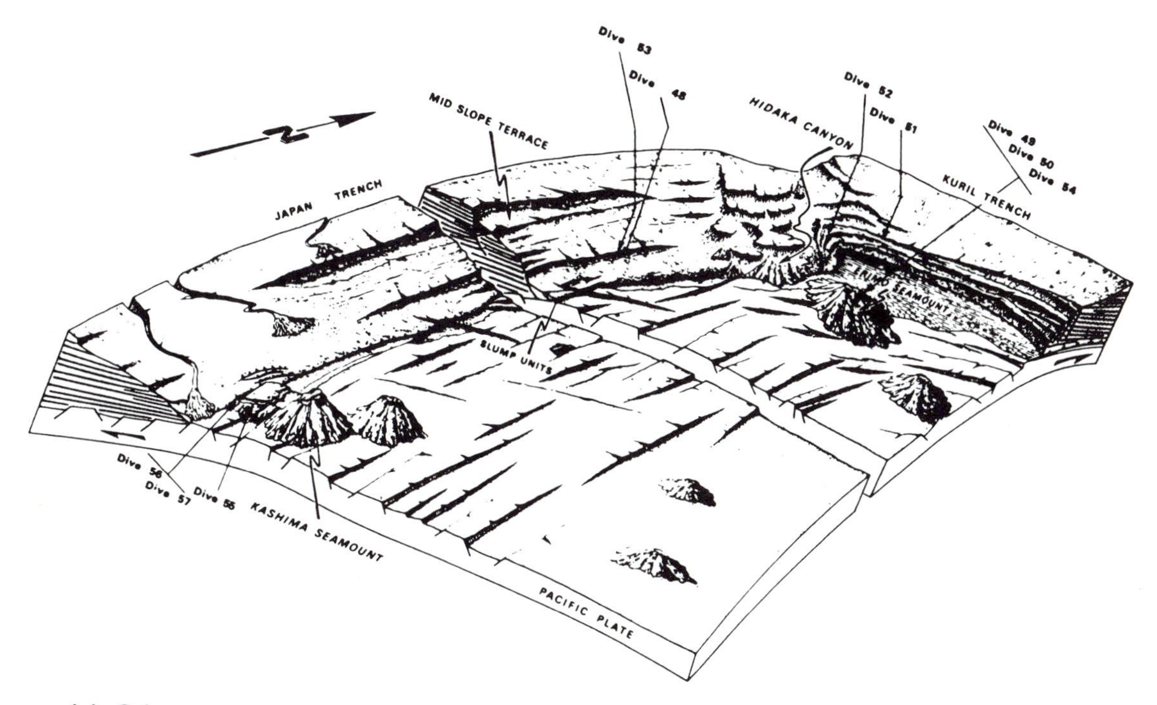

Figure 11–21. Block diagram of Northeast Japan trench viewed from the southeast. Pacific plate is cut by normal faults striking N10°E and N60°E and by seamounts, two of which are being subducted. The upper plate is marked by major slumps, some of which are collapsing into the Japan trench. The Kuril trench is stepped left at its change of strike with the Japan trench northwest of Erimo seamount. After Cadet et al. (1987).

slump scars (Fig. 11-21) exposing strata as old as Cretaceous. Some of the slumps are collapsing into the trench itself (von Huene and Culotta, 1989). Massive landslides are also common on the lower forearc slope of the Kuril trench (Gnibidenko et al., 1983). The cusp marking the change of strike between the Japan and Kuril trenches is close to the incoming Erimo seamount (Fig. 11-21), and another seamount has already subducted beneath the upper plate (Yamazaki and Okamura, 1989).

Earthquakes cease in the arcward direction several tens of kilometers east of the Quaternary volcanic front (Fig. 11-22), about coincident with the Miocene volcanic front. This was called the *aseismic front* by Yoshii (1979). Landward from this front, the base of the seismogenic zone is at a depth of about 20 km. Close to, but east of, the Quaternary volcanic front is a sharp cutoff of active-fault traces (Research Group for Active Faults of Japan, 1992; Fig. 10-15) referred to by Kinugasa (1991) as the *active-fault front*. Kinugasa suggested that the active-fault front may be related to a weaker crust on the landward side. (As noted earlier, the Median Tectonic Line marks another active-fault front in southwest Japan.)

Results from a local seismic network show that the slab descending beneath northeast Japan is marked by a double seismic zone (Hasegawa et al., 1978; Matsuzawa et al., 1990; Fig. 11-22). The upper zone is close to the plate boundary, and the upper part of this zone contains earthquakes characterized by reverse-fault focal mechanisms. The lower plane, 30–40 km deeper, is characterized by earthquakes with down-dip extension to depths of at least 200 km. The Wadati-Benioff zone dips about 30°WNW and is about 1200 km long, extending to depths of 600 km, where deep earthquakes show down-dip compressional fault-plane solutions (Seno and Eguchi, 1983).

The Northeast Japan subduction zone is characterized by great earthquakes, but seismic slip may account for only 20% of the relative plate motion (Seno and Eguchi, 1983; Pacheco et al., 1993). The down-dip depth of seismic coupling, based on the down-dip limit of aftershocks of interplate earthquakes, is 37–43 km, which corresponds reasonably well to the depth of the seismogenic zone in the upper plate, assuming a sharp decrease in geothermal gradient east of the volcanic front (Tichelaar and Ruff, 1993). This implies that deeper earthquakes in the upper seismic zone (Fig. 11-22) must occur within the downgoing plate rather than at the plate boundary, possibly responding to lower temperatures in the cool, downgoing slab. The double seismic zone may reflect the

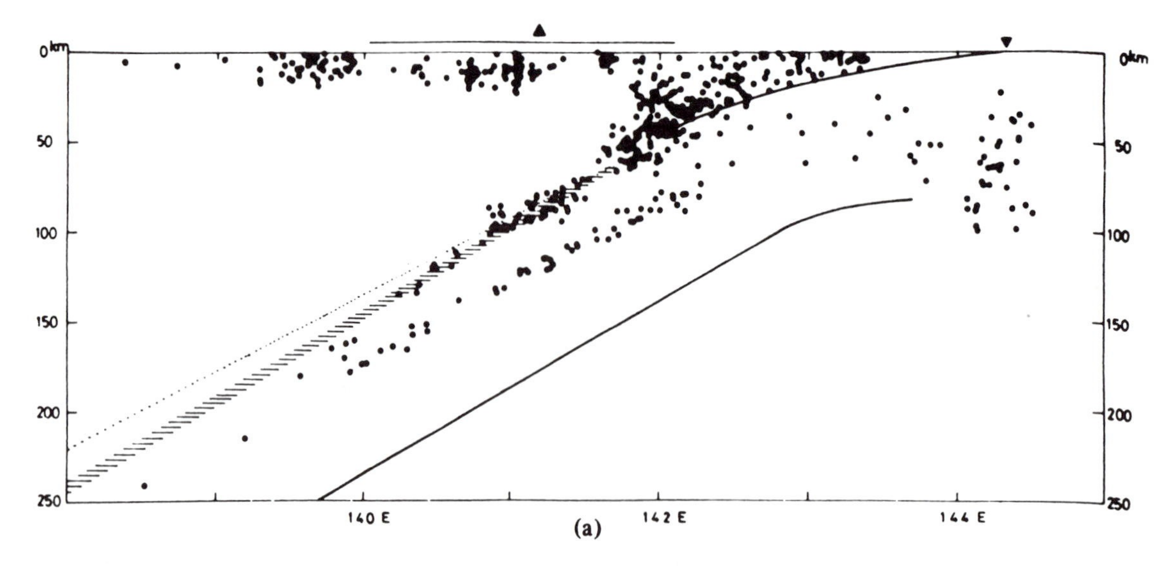

Figure 11–22. Earthquakes in the Tohoku District of northern Honshu projected on an E-W cross section, based on a velocity model with P and S waves in the downgoing slab 6% faster than those in the surrounding mantle. Upright triangle locates active volcanic arc; horizontal line locates the subaerial part of the upper plate on Honshu; inverted triangle locates trench. Note aseismic front of Yoshii (1979) east of volcanic arc; the outermost part of the upper plate near the trench is also aseismic. Top and bottom of subducting slab also shown. After Hasegawa et al. (1978).

inward-moving thermal front as serpentine dehydrates to olivine (Green, 1994).

The 1933 Sanriku normal-fault outer-rise earthquake is larger than any of the recorded subduction-zone earthquakes, including the 1968 Tokachi-oki event at the cusp with the Kuril Trench, with M = 8.2 (McCann et al., 1979; cf., Pacheco and Sykes, 1992; Fig. 11-23). All but the southern end of the subduction zone has been filled by earthquakes of M ≥ 7.8 in the past century (McCann et al., 1979). Seismic gaps may lie farther south; the 1953 Off-Boso earthquake of M7.5 might have been caused by a tear in the Pacific plate rather than rupture of the plate boundary (Ishibashi, 1992).

Farther north, the Kuril subduction zone produces larger earthquakes than northeast Japan, including one of the greatest earthquakes of the twentieth century in 1952 (M_W = 9) off Kamchatka (Fedotov, 1965; McCann et al., 1979; Rikitake, 1982; Figs. 11-1 and 11-24). The southern part of the subduction zone as far south as latitude 45°N has been filled in this century with earthquakes mostly of M > 8. This is separated from the rupture zone of the 1952 earthquake by a seismic gap (McCann et al., 1979). The mean recurrence time for earthquakes off Hokkaido for the past three centuries is about 85 years (Rikitake, 1982). This recurrence time is similar to that at the northern Kuril subduction zone, but historical records prior to the twentieth century are incomplete for much of this region.

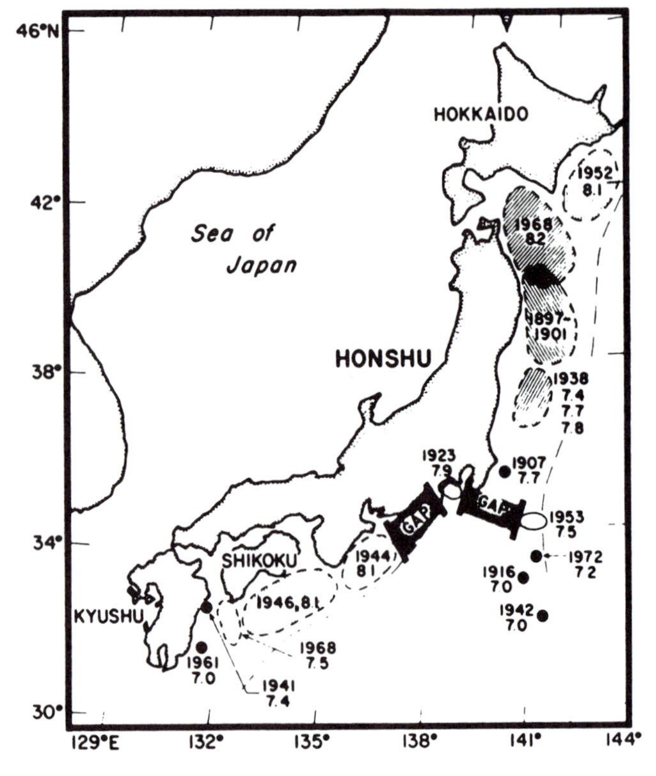

Figure 11–23. Recent large earthquakes near the east and south coasts of Japan. Hachured regions are rupture areas based on aftershocks. Year and magnitude shown next to earthquake location. 1933 Sanriku normal-fault earthquake not shown. After McCann et al. (1979).

Like northeast Japan, the Kuril subduction zone is

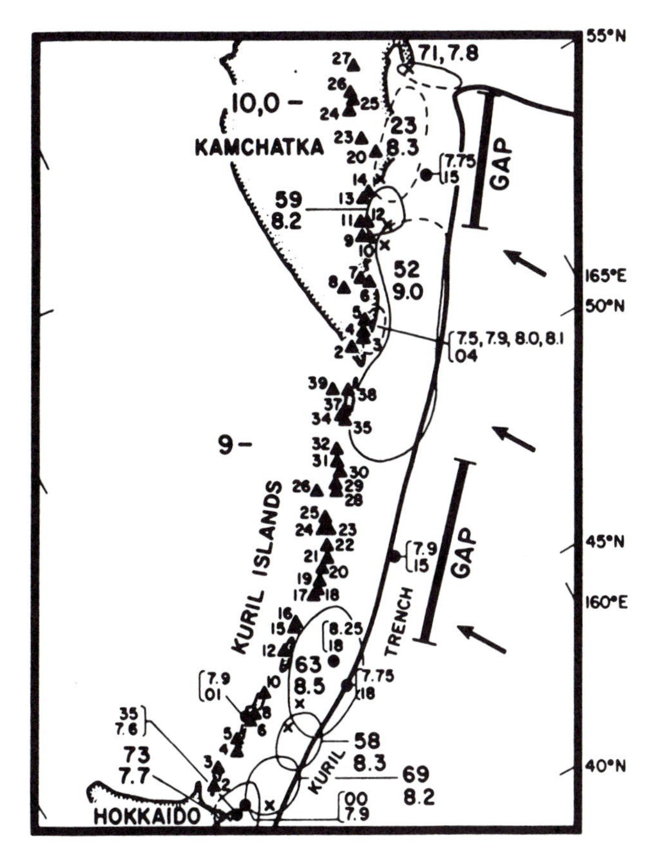

Figure 11–24. Great earthquakes on the Kuril subduction zone. Solid lines enclose rupture areas of recent great earthquakes based on aftershocks; dashed where based mainly on felt reports. Dots and X's show poor and good mainshock epicenter locations, respectively. Dates (twentieth century) and magnitudes of mainshocks given. Solid triangles locate Quaternary volcanoes. Arrows show relative plate motion. After McCann et al. (1979).

marked by a double seismic zone (Suzuki et al., 1983). The base of the zone of earthquakes in the upper plate slopes toward the Wadati-Benioff zone, reflecting the trenchward decrease in geothermal gradient. The depth of seismic coupling, based on the distribution of interplate aftershocks, is 37–43 km near the Kuril Islands and 38–40 km near Kamchatka; but the cusp between the Northeast Japan and Kuril subduction zones has seismic coupling to depths of 52–55 km (Tichelaar and Ruff, 1993).

Aleutian Subduction Zone

The Aleutian trench extends from a sharp cusp with the Kuril trench off the coast of Kamchatka east for 2900 km to the Gulf of Alaska. The trench marks the boundary between the subducting Pacific plate and the North American plate (Fig. 11-1). Convergence is NW and decreases eastward from 9.1 cm/yr south of Attu Island in the western Aleutian Islands to 6.3 cm/yr in the Gulf of Alaska. Subduction is dip-slip beneath the eastern Aleutians and the Alaska Peninsula. The curvature of the trench and arc causes subduction to be increasingly oblique westward (Figs. 11-1, 11-25) such that the plate slip vector is nearly parallel to the trench at its western end. At least part of the plate boundary is a right-lateral transform fault west of Attu Island (Fig. 11-25). Eastward, the trench terminates at the edge of the North American continent, marked by a collision zone and a transform boundary dominated by the Fairweather and Queen Charlotte fault zones.

The northern end of the Emperor Seamount Chain, with an age estimated at about 75 Ma, reaches the Pacific plate boundary at the cusp between the Kuril and Aleutian trenches. To the east, the Pacific plate is characterized by east-striking magnetic anomalies

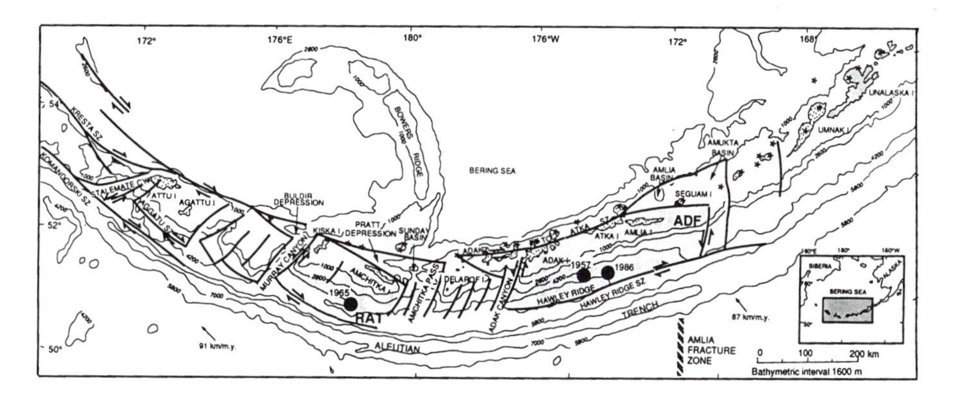

Figure 11–25. Map of the central Aleutian Islands showing left-lateral shear zones transverse to the arc, and right-lateral shear zones parallel to the arc. The high-strength Rat (RAT) and Andreanof (ADF) blocks are shaded. Epicenters of 1957, 1965, and 1986 earthquakes shown. After Ryan and Scholl (1993).

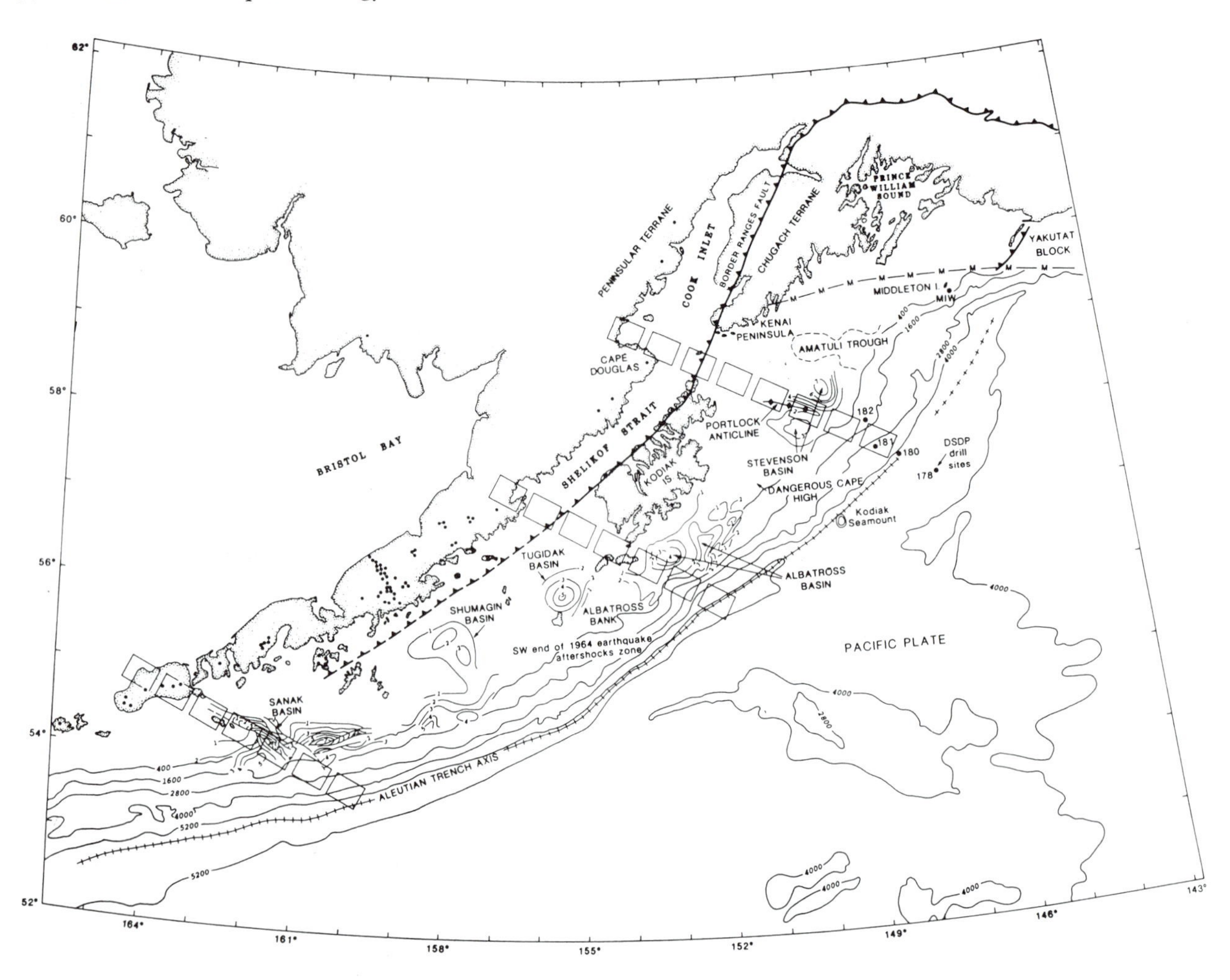

Figure 11–26. Tectonic map of the eastern part of the Aleutian subduction zone. Small stars locate Quaternary volcanoes; open dashed pattern marks geologically defined segment boundaries, M-M-M marks Yakutat-block slope magnetic anomaly, thick line marks Border Ranges fault separating forearc accretionary-wedge Chugach terrane from Peninsular terrane, with teeth in direction of presumed fault dip. After von Huene et al. (1987).

and north-striking fracture zones (such as the Amlia fracture zone in Fig. 11-25) related to sea-floor spreading between the Pacific plate and the mostly vanished Kula plate (Atwater, 1989). Southwest of Kodiak Island (Fig. 11-26), the magnetic anomalies change strike abruptly to nearly north, with east-striking fracture zones, a product of spreading between the Pacific plate and the already-subducted Vancouver plate. The age of crust of the Pacific plate opposite the trench is 45–55 Ma along most of the plate boundary, but is less than 40 Ma south of the Gulf of Alaska. Most of the Pacific plate opposite the Alaska Peninsula and Gulf of Alaska is covered by broad, deep-sea turbidite fan systems (Stevenson and Embley, 1989), interrupted by linear chains of seamounts.

All but the easternmost Aleutian Islands are built on oceanic crust, largely of Mesozoic age, transected by the strongly arcuate Bowers (Fig. 11-25) and Shirshov ridges, which include volcanic rocks of middle Tertiary age (Cooper et al., 1987). Oceanic crust of the Aleutian basin gives way eastward to the passive margin of the Alaska continental shelf, so that the eastern part of the arc is built on continental crust.

The Aleutian Islands are part of a broad platform called the Aleutian Ridge, containing volcanic rocks as old as 55 Ma (Scholl et al., 1987). Most of the ridge is capped by active volcanoes except for that part west of Murray Canyon (Fig. 11-25), where the strike-slip component of subduction is large. The ridge gives way southward to a slope basin (Aleutian Terrace), an outer-arc high marking the crest of a late Cenozoic accretionary wedge, and the flat-floored Aleutian trench.

The trench is 7.2 km below sea level in the central Aleutians, shallowing to 6.6 km south of the Komandorskiy Islands in the far west and to 4 km at its eastern termination in the Gulf of Alaska.

The summit platform of the ridge has subsided differentially since the late Miocene, a side effect of clockwise rotation due to oblique subduction of the Pacific plate (Geist et al., 1988). The ridge may be divided into clockwise-rotating blocks separated by submarine canyons that are tectonically controlled by NE-striking left-lateral faults (Fig. 11-25). The southern boundary of the Andreanof block, near the backstop of the subduction zone at the crest of the outer-arc high, is marked by right-lateral displacement related to oblique subduction (Ryan and Scholl, 1989; 1993).

The trench southeast of the Alaskan mainland is marked by an accretionary complex, in part incorporating submarine fans of the Pacific plate (Bruns et al., 1987). In addition to folding and thrusting above a gently dipping décollement, the accretionary prism deforms by right-slip and left-slip faults that strike at a low angle to the trench (Lewis et al., 1988). Forearc basins develop inboard of and parallel to a submarine outer-arc ridge and also at a 45° angle to the ridge, forming a segment boundary (Fig. 11-26). These basins are cut by normal faults, even though they are in the upper plate of the subduction zone. Farther from the trench, a second forearc ridge, including most of the Kenai Peninsula and Kodiak Island as well as small islands to the southwest (Fig. 11-26), consists predominantly of Mesozoic accretionary-wedge deposits intruded by early Tertiary plutons. The Alaska Peninsula with its active volcanoes lies still farther away from the trench.

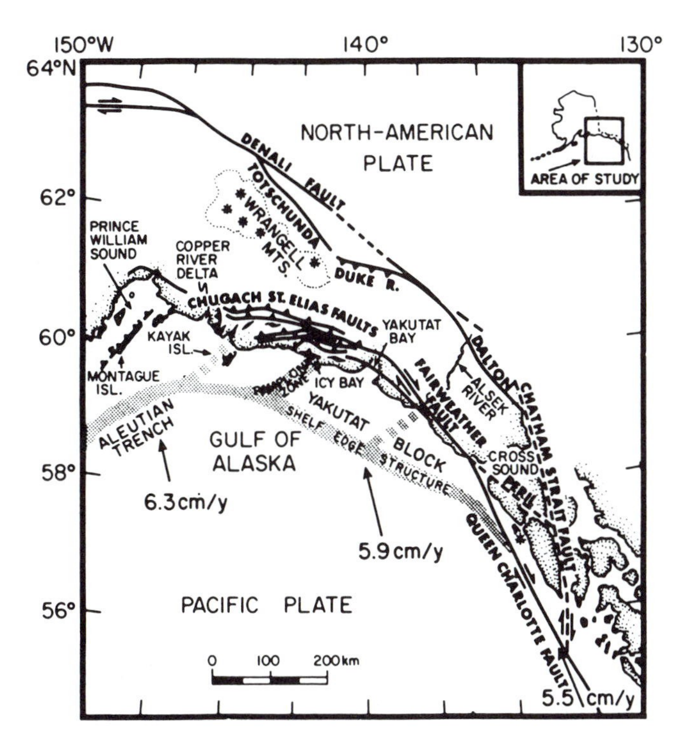

Figure 11–27. Tectonic map of transition between Aleutian subduction zone and transform faults marking the boundary between the Pacific and North American plates. Heavy solid and dashed lines show faults with late Cenozoic motion, with strike-slip faults marked by arrows and reverse faults marked by teeth in direction of fault dip. Shaded pattern marks Aleutian megathrust, Shelf Edge (Transition) fault marking southwest edge of Yakutat block, and Pamplona and Kayak Island convergence zones. Arrows show Pacific-North American slip vector. Asterisks locate Quaternary volcanoes. After Perez and Jacob (1980).

The Aleutian trench shallows at its eastern end (Fig. 11-26), and the plate boundary turns abruptly southeast along the Transition fault (Shelf Edge structure of Perez and Jacob, 1980; Fig. 11-27), in part strike slip (Plafker, 1987), but beneath which the Pacific plate is actively underthrusting with a slip vector based on earthquakes directed northeast (Perez and Jacob, 1980). Northeast of the Transition fault, the Yakutat block is separated from the North American plate by the strike-slip Fairweather fault, which turns west into the largely reverse-slip Chugach-St. Elias faults (Perez and Jacob, 1980; Fig. 11-27). The Yakutat block moves northwest against North America nearly as fast as does the Pacific plate, with convergence taken up along the Kayak Island and Pamplona structural zones, which seem to be fold-and-thrust belts (Bruns and Schwab, 1983; Plafker, 1987), and also along the Chugach-St. Elias faults (Fig. 11-27), north of which the St. Elias Range is uplifted to altitudes greater than 6 km. The boundary between the Yakutat block and the North American plate, therefore, is now a collision zone, analogous to that between the Izu Peninsula and the Japan Alps.

Earthquakes include bending-moment normal-faulting events near and seaward of the trench, interplate earthquakes with a maximum depth of seismic coupling of 35-41 km (Tichelaar and Ruff, 1993), earthquakes within the downgoing slab, and earthquakes within the upper plate between the aseismic accretionary wedge and the volcanic arc (Fig. 11-28). There is a double seismic zone, at least beneath the eastern Aleutians (House and Jacob, 1983). Earthquakes are recorded in the subducted slab to depths greater than 200 km in the central Aleutians but only to 150 km in the Gulf of Alaska (Davies and House, 1979; Fig. 11-28).

The Aleutian subduction zone sustained four great earthquakes in the twentieth century (Fig. 11-29).

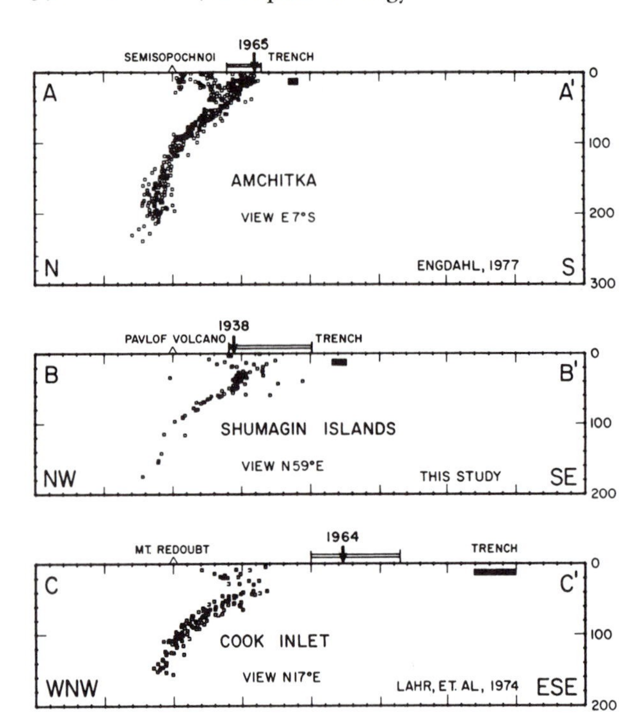

Figure 11–28. Three seismicity cross sections across Aleutian subduction zone, with earthquakes shown as open squares. Open triangles locate volcanic arc, and heavy solid line locates trench. Open double line with date lies above rupture zone of a great earthquake, with arrow marking epicenter. Note the increasing distance toward the east between the subduction zone marked by instrumental seismicity and the trench, an indication of the increasing width eastward of the accretionary prism. Note also that the rupture zones of the 1938 and 1964 earthquakes extend much farther trenchward than does the zone of instrumental seismicity. After Davies and House (1979).

From west to east, they are the 1965 Rat Island earthquake (M_W = 8.7), the 1957 Andreanof earthquake (M_W 9.1), the 1938 earthquake off the Alaska Peninsula (M_W = 8.2), and the 1964 Gulf of Alaska earthquake (M_W = 9.2).

The far western part of the plate boundary, near the Komandorskiy Islands, is in a seismic gap for great earthquakes, and it has lower instrumental seismicity than the rest of the zone. Earthquakes in the Komandorskiy region have reverse-fault mechanisms with slip nearly parallel to the trench (Cormier, 1975).

The 1965 earthquake nucleated near the eastern end of the Rat block south of Amchitka Island (Fig. 11-25), and the rupture zone extended from Amchitka Pass, east of this block, west to Attu and Agattu Islands and to Stalemate Canyon, west of which the Aleutian Arc is completely submerged for 350 km. The eastern and western ends of the 1965 rupture are controlled by left-lateral faults in Amchitka Pass and Stalemate Canyon, respectively. The Rat Island block, where the earthquake nucleated, is mechanically stronger than adjacent regions and may be related to asperities at depth on the main thrust zone (Ryan and Scholl, 1993). On the other hand, the earthquake ruptured across the Rat fracture zone in the Pacific plate and appeared to be uninfluenced by it.

No gap exists between the 1965 and 1957 rupture zones. The 1957 earthquake nucleated in the mechanically strong Andreanof block south of Atka Island (Ryan and Scholl, 1993) and ruptured across a zone of transverse faults around the Delarof Islands. The earthquake ruptured east of the Andreanof block and terminated near Umnak and Unalaska Islands in the eastern Aleutians, a region with no obvious transverse upper-plate structures or changes in the location of active volcanoes or instrumental seismicity (Fig. 11-25). The 1986 Andreanof earthquake (M_W = 7.9) nucleated near the 1957 mainshock south of Atka Island in the Andreanof block (Fig. 11-26) but did not propagate as far east or west as the 1957 earthquake did. The west end of the 1986 rupture is at the west end of the Andreanof block at Adak Canyon. The east end of the rupture is within the Andreanof block and is not marked by a prominent transverse structure. The eastern end of the 1986 rupture zone is near the Amlia fracture zone in the Pacific plate, however, which marks a sharp bend or offset in the subduction zone and an offset of the axis of active volcanoes (House and Jacob, 1983).

The northeastern end of the Aleutian subduction zone is dominated by the great 1964 earthquake, which terminated eastward at the eastern end of the Aleutian trench at the Kayak Islands structural zone, beneath the northwest end of the Yakutat block (Fig. 11-27). The earthquake had a modest left-lateral component, probably due to the fact that the plate slip vector locally is not orthogonal to the Aleutian trench. East of the Kayak Islands structural zone, the collision zone between the Yakutat block and the North American plate has not had any great earthquakes in historic time; this region is called the Yakataga seismic gap. Farther east, the collision zone was ruptured in the 1979 St. Elias earthquake (M_W = 7.2), with a slip vector trending south-southeast (Perez and Jacob, 1980). This rupture zone adjoins that of the 1958 Lituya Bay strike-slip earthquake along the Fairweather fault (M_s = 8.2).

The 1964 rupture zone terminated southwest of Kodiak Island where it adjoined the rupture zone of the great 1938 earthquake (Fig. 11-29b). The western ter-

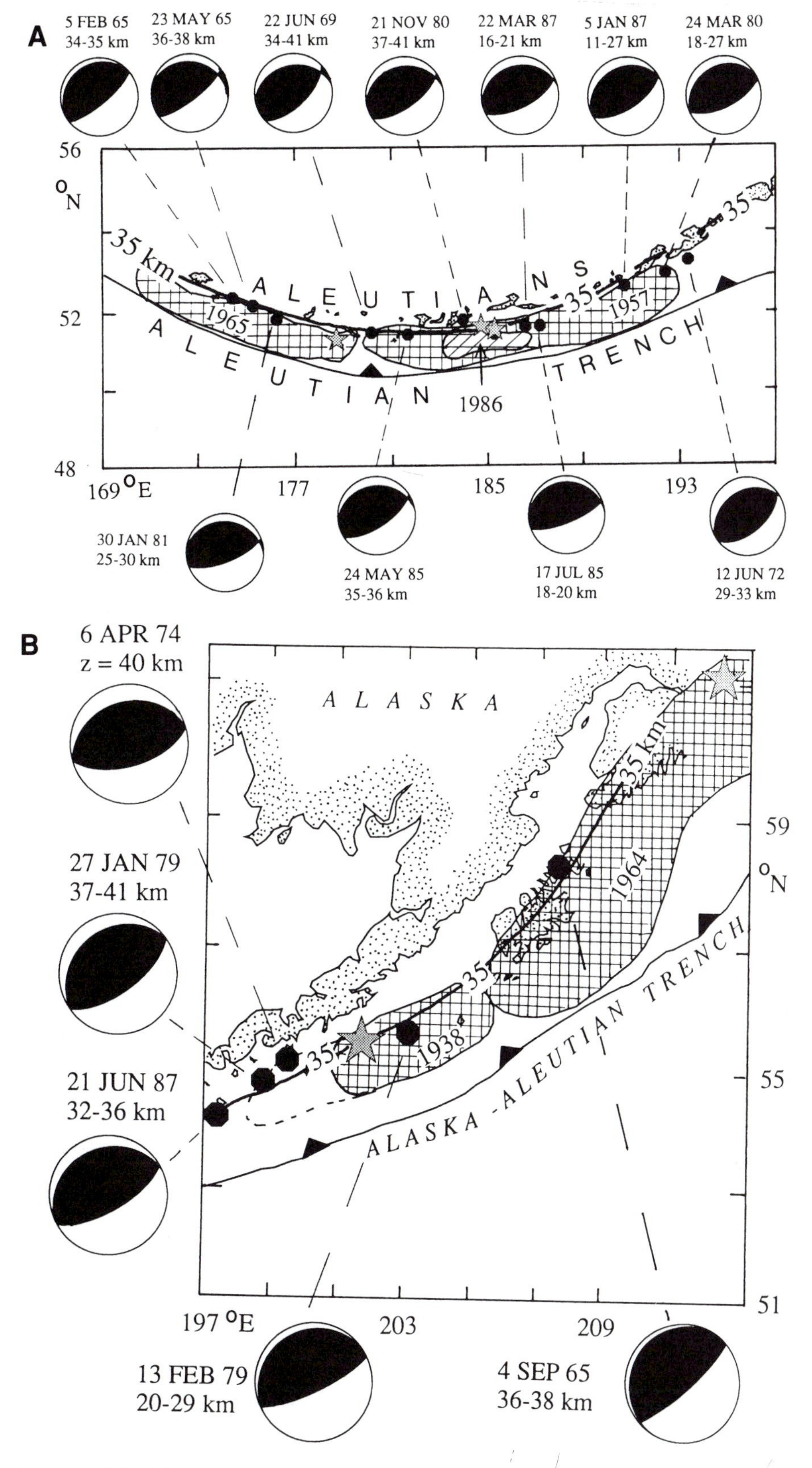

Figure 11–29. Rupture zones and focal mechanisms of great twentieth-century earthquakes at the Aleutian subduction zone with epicenters marked by stars. (a) Aleutian Islands, with seismic coupling to 35–41 km. (b) Alaska Peninsula and Gulf of Alaska, with seismic coupling to 37–41 km. After Tichelaar and Ruff (1993).

mination of the 1938 rupture zone, in turn, is east of a prominent transverse structure including the Sanak basin (Fig. 11-26). The Shumagin seismic gap separates the 1938 and 1957 rupture zones.

In summary, the Aleutian subduction zone is marked by three of the largest earthquakes of the twentieth century, with well-defined rupture zones. Some of the rupture zones end at prominent transverse geologic structures, but others do not. In addition, great earthquakes propagated across several conspicuous transverse structures in the upper plate as well as fracture zones in the Pacific plate. The 1957, 1965, and 1986 earthquakes in the western Aleutians nucleated in the strongest portions of the upper plate.

The 1964 earthquake was accompanied by up to 2.3 m of coastal subsidence of a region 800 km long parallel to the arc and 150 km across (Plafker, 1969; 1972; Fig. 11-30). This region included Kodiak Island, the Kenai Peninsula, and the Wrangell Mountains, which include several Quaternary volcanoes. The region between the subsiding zone and the trench was uplifted a similar amount, locally as much as 11.3 m, with the steepest gradient in change in altitude in the zone between subsiding and uplifting areas. Another uplifted region, less well defined, lies landward of the subsiding zone beneath the arc volcanoes northwest of Cook Inlet and south of the Alaska Range.

In most regions, altitude changes in 1964 are consistent with long-term changes throughout the Holocene, but there are exceptions, such as Kodiak Island. Vertical changes in 1964 are commonly opposite in sign to longer-term changes over the last 1300 years, suggesting that the short-term changes represent coseismic release of elastic strain. The long wavelength of the subsiding and uplifting zones is consistent with release of stored elastic-strain energy in an elastic upper plate that was loaded to failure along the subduction megathrust. In detail, however, the vertical changes require internal deformation of the upper plate, particularly the high uplift recorded on Montague Island (Fig. 11-30). This island and the sea floor to the southwest were cut by the high-angle-reverse Patton Bay fault, which may be an upper-plate thrust fault rising from the plate-boundary megathrust (Plafker, 1972).

Horizontal changes based on triangulation surveys show extension of the forearc region toward the trench by as much as 20 m, probably representing unloading of horizontal strain stored in the upper plate prior to the earthquake.

The presence of raised shorelines and drowned forests due to the 1964 earthquake led to a search for evidence for pre-1964 vertical changes. The shoreline of Middleton Island was uplifted 3.5 m during the 1964 earthquake. Five higher shorelines are found, with the highest at 46 m dated as 4300 years (Fig. 11-31). This yields an uplift rate of 9 mm/yr, apparently slowing toward the present. The time between shoreline uplifts ranges from 400 to 1300 years (Plafker, 1978; 1987; Fig. 11-31). If the plate convergence rate of 63 mm/yr were all accommodated by 1964-type subduction-zone earthquakes with horizontal slip of 20 m, the recurrence interval would be 330 years, suggesting that not all of the subduction-zone earthquakes produce uplift at Middleton Island.

Interbedded layers of peat and silt in the upper Cook Inlet suggest regional subsidence events similar to that recorded in 1964, with a recurrence interval of 200 to 800 years during the past 3200 radiocarbon years (Bartsch-Winkler and Schmoll, 1992). Not all of these subsidence events are demonstrated as having been caused by subduction-zone earthquakes. Uplift of the Middleton Island terraces may have resulted from local thrust faults within the upper plate (Plafker, 1978), as at Montague Island.

Cascadia Subduction Zone

The Cascadia subduction zone has the lowest instrumental seismicity of any subduction zone on the margin of the Pacific Ocean. This led to the general belief that subduction is aseismic there, posing no major seismic risk to the populated cities of the northwestern United States. This changed with the discovery of buried marshes and coastal forests that are best explained by subsidence accompanying a subduction-zone earthquake (Atwater, 1987). In addition, the low dip of the subduction zone, the absence of a trench, and the youthful age of subducting oceanic lithosphere suggested that the Cascadia subduction zone is of Chilean type, subject to great earthquakes (Heaton and Kanamori, 1984).

An oceanic remnant of the formerly large Farallon plate is subducting east-northeast beneath southwestern Canada and the northwestern United States (Fig. 11-32) at a rate of 40-50 mm/yr. The largest remnant is the Juan de Fuca plate, created at the active, but relatively low-seismic Juan de Fuca ridge. The Juan de Fuca ridge is offset by fracture zones to the north and south, forming smaller plates characterized by relatively high seismicity (Explorer, North Gorda, South Gorda) that seem to be rotating clockwise with respect to the North American continent. The Juan de Fuca plate is covered by up to 3 km of sediment, in large part deep-sea fans derived from sediments of the Columbia River (Fig. 11-33) and rivers flowing into the Straits of Juan de Fuca between Vancouver Island and the Olympic Peninsula. No topographic trench occupies the deformation front because of the thick sediment cover and the relatively young age of the in-

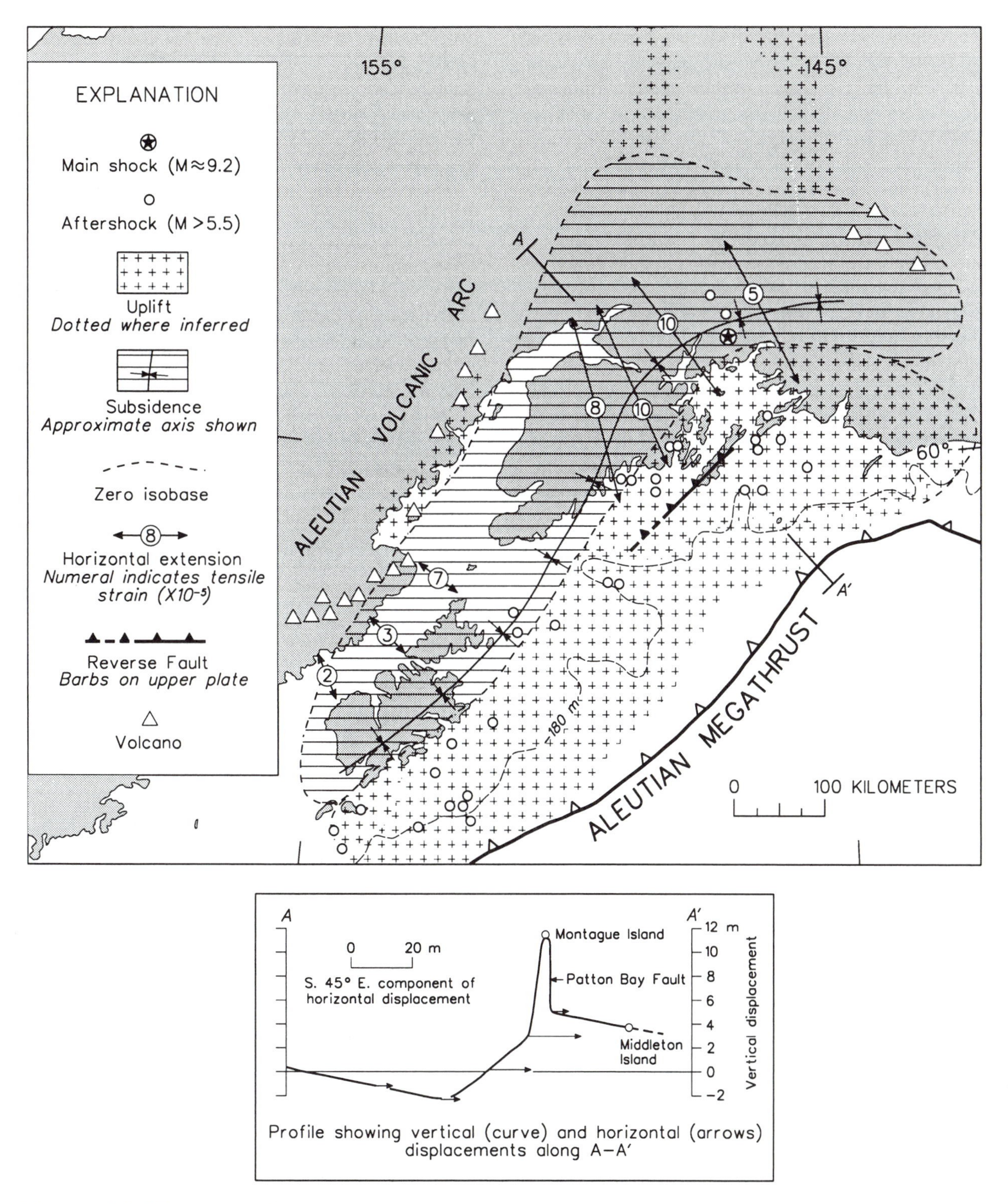

Figure 11–30. Tectonic displacements and seismicity accompanying the 1964 Alaskan earthquake relative to the Aleutian trench and the volcanic arc. From Plafker (1972).

coming oceanic plate, less than 20 Ma, which makes it relatively buoyant.

The continental margin includes a late Cenozoic fold-and-thrust belt in the accretionary wedge close to the deformation front. The accretionary wedge is broadest off the coast of Washington and the Straits of Juan de Fuca, narrowing to the north and south (Figs. 11-32; 11-33). Off the coast of Oregon, the wedge gives way eastward to Cenozoic marine strata overlying oceanic basalt of early Eocene age (Siletz terrane of Fig. 11-34) beneath the upper continental slope, continental shelf, and Coast Range (Snavely, 1987). The seaward edge of the Eocene oceanic crustal block off central Oregon is abrupt (Fig. 11-34; Tréhu et al., 1994). In the Olympic Peninsula of western Washington, the Eocene basaltic slab, considerably thinner than it is off central Oregon, is thrust over an accretionary-wedge sequence of Eocene to Miocene age (Tabor and Cady, 1978). This thrust probably extends offshore off the coast of Oregon (Snavely, 1987). The elongate crustal block including the Oregon Coast Range is relatively intact, probably because of the great thickness of Eocene oceanic crust. But in Washington, where the Eocene oceanic crust is thinner (Tréhu et al., 1994), the crustal blocks are smaller and are bounded by strike-slip faults accompanying clockwise tectonic rotation (Wells, 1989).

It was long assumed that the boundary between the Eocene oceanic crustal block and the accretionary prism was the seaward edge of the backstop, which would locate the locked zone for interplate earthquakes beneath the continental shelf and the Coast Range. However, in Washington, this boundary comes ashore and wraps around the Olympic Mountains, so that the locked zone would have to underlie the broad accretionary prism of the forearc. Hyndman and Wang (1993) pointed out that the combination of young age and thick sediment cover of the Juan de Fuca plate results in a relatively hot subduction zone. These high temperatures require that the locked zone be at relatively shallow depths on the plate interface, close to the deformation front, underlying the accretionary prism off Oregon as well as off Washington and Vancouver Island. This is in agreement with leveling changes on the southwest and northeast coasts of Vancouver Island, which are best explained by elastic-strain accumulation on a locked zone close to the deformation front.

In a general way, the Coast Range of Oregon, the Olympic Mountains of Washington, and Vancouver Island in Canada constitute an outer-arc high adjacent to a forearc basin on the east, the Willamette Valley, Puget Sound, and the Straits of Georgia, with the vol-

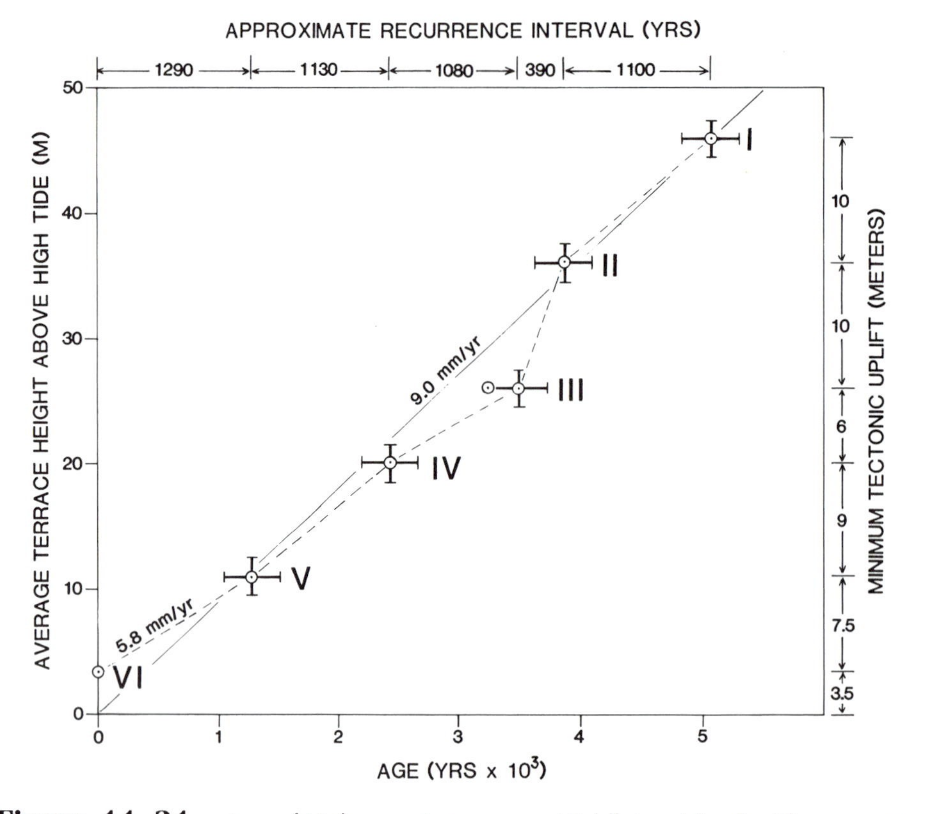

Figure 11–31. Age of Holocene terraces on Middleton Island with respect to terrace height, corrected for eustatic rise of sea level. Roman numerals identify earthquakes believed to have uplifted each terrace, with VI the 1964 earthquake; crosses indicate error bars in age and altitude. Uplift rate and recurrence interval also shown. After Plafker (1987b).

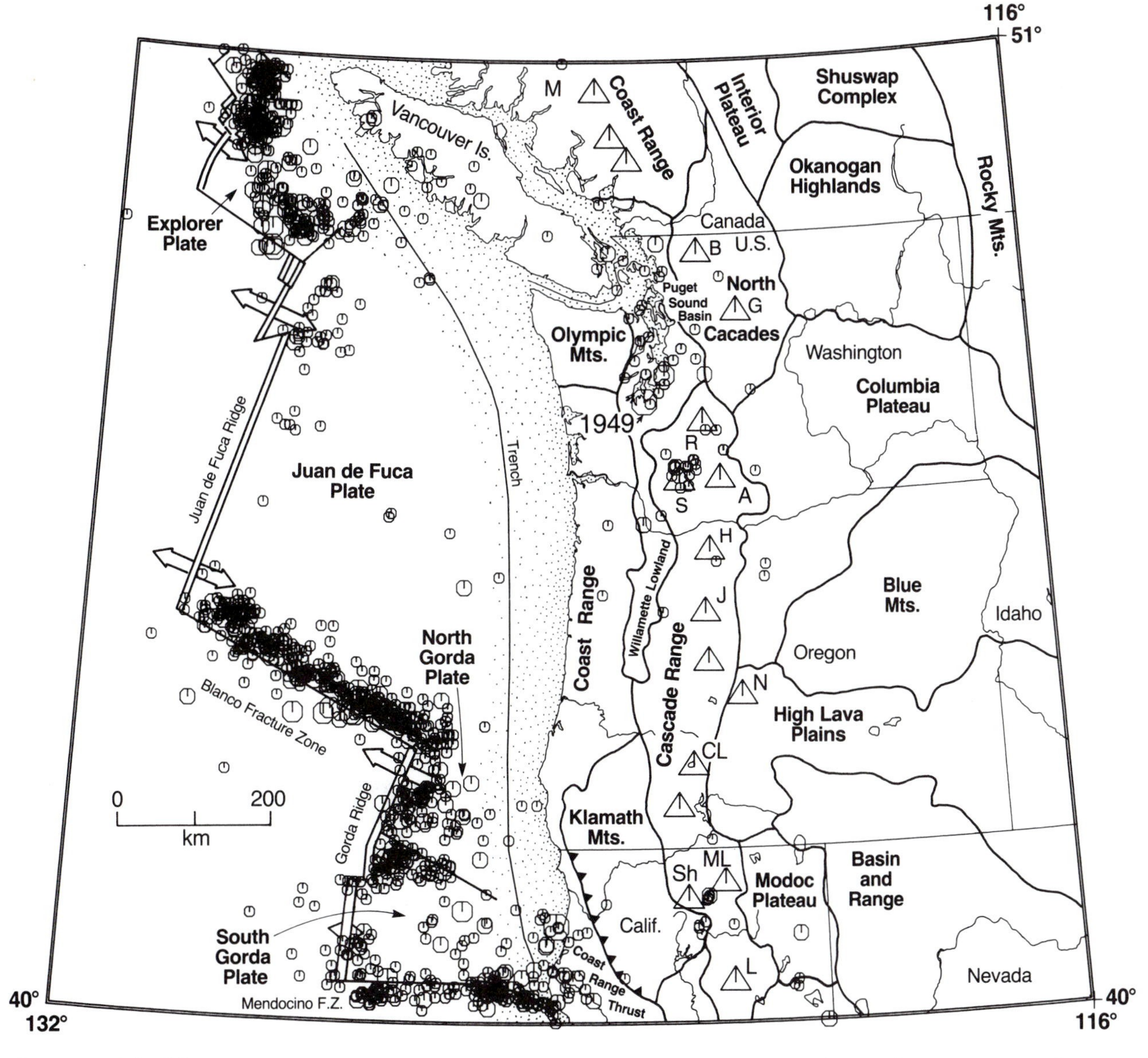

Figure 11–32. Plate boundaries, physiographic provinces, and seismicity for northwestern United States and adjacent parts of Canada. Octagons show earthquakes of M > 4 from National Oceanic and Atmospheric Administration catalog through 1985; largest earthquakes noted by year of occurrence. Open triangles show Quaternary volcanoes. From Shedlock and Weaver (1991).

canic Cascade arc still farther east (Fig. 11-32). However, Neogene basins also develop offshore, including the Tofino basin off Vancouver Island and the Astoria and Newport basins off Oregon (Fig. 11-33). The Pacific coast of the Olympic Peninsula is marked by Miocene mélange, with large blocks of sandstone and conglomerate in a mud matrix, suggesting fluid overpressures (Orange et al., 1993).

As noted above, the Cascadia subduction zone has the lowest seismicity of any convergent margin around the Pacific Ocean. There have been no great interplate earthquakes in 150-200 years of recordkeeping. Earthquakes occur in both the upper plate and lower plate (Fig. 11-35a), with lower-plate earthquakes as large as M7.1 in 1949 and M6.5 in 1965 in the Puget Sound region (Fig. 11-32). Contours on the subducting Juan de Fuca plate show a reverse curvature in the vicinity of the Olympic Mountains, concave toward the Pacific Ocean (Fig. 11-35b) rather than convex toward the ocean, as in most island arcs. This reverse curvature leads to internal stresses in the downgoing Juan de Fuca plate that generate earthquakes of M > 7. But there is little or no seismicity on the megathrust itself.

But on the basis of convergence rate and the age of subducting crust, the Cascadia subduction zone should generate earthquakes of M8–8.5 (Heaton and Kanamori, 1984; Fig. 11-6). Cascadia possesses the characteristics of a Chilean subduction zone: no trench, a large accretionary prism, and a gently dipping plate boundary. Moreover, bays and estuaries on the Pacific coast of Washington, Oregon, and northern California provide evidence that peat deposits, formerly high marshes and coastal forests, are overlain by marine sediments along a sharp contact, evidence of abrupt subsidence that is best explained as coseismic (Fig. 7-27; Atwater, 1987; Darienzo and Peterson, 1990). Some contacts between peat and overlying marine mud are marked by laminated fine-grained sand derived from the sea and considered to be the deposits of tsunamis. The subsidence was compared to that accompanying great earthquakes in Chile in 1960 and Alaska in 1964 (Plafker, 1972; Atwater, 1987). Radiocarbon dating of the youngest buried peat suggests that the coastline from Vancouver Island to northern California underwent abrupt subsidence, presumably caused by a subduction-zone earthquake, about 300 years ago (Atwater et al., 1991).

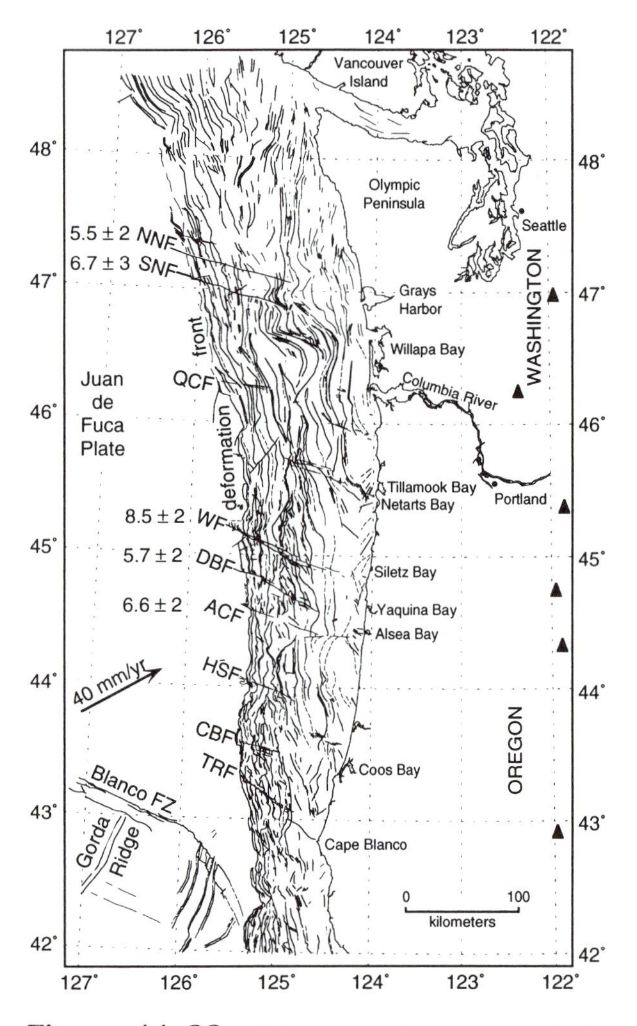

Figure 11-33. Pliocene and Quaternary faults and anticlines of the Oregon and Washington continental margin. 3-letter symbols (NNF, SNF, etc.) mark left-lateral faults with slip rates shown in mm/yr where known. Coseismically-buried marshes are found in Grays Harbor, Willapa Bay, Netarts Bay, and Coos Bay. Closed triangles locate Cascade volcanoes. Modified from Goldfinger (1994).

Further evidence that the Cascadia subduction zone is capable of generating great earthquakes comes from tectonic geodesy. The Olympic Peninsula is accumulating horizontal elastic strain (Savage and Lisowski, 1991; Fig. 11-36), and highway benchmarks in coastal Oregon show evidence of uplift of the southern Oregon coast and the mouth of the Columbia River relative to the northern Oregon coast and central Washington coast (Figs. 5-9a, 11-37; Adams, 1984; Mitchell et al., 1994). In addition, turbidites deposited on the abyssal plain west of the deformation front since the deposition of an ash layer 6850 years ago are best explained by earthquake triggering (Fig. 12-5 Adams, 1990; discussed further in Chapter 12).

The Cascadia offshore region is cut by numerous active faults and folds that deform Holocene deposits (Figs. 11-33, 11-38) and the latest Pleistocene lowstand surface. Structures include folds and thrusts in the accretionary wedge (MacKay et al., 1992), but also include a set of WNW-trending left-lateral strike-slip faults that cut the accretionary wedge, the plate boundary, and the Juan de Fuca plate (Goldfinger et al., 1992; Goldfinger, 1994; Fig. 11-33). These faults have slip rates of 5.5–8.5 mm/yr (Fig. 11-33; Goldfinger, 1994). Active faults and folds on the Oregon shelf may influence the position of Oregon bays that contain the evidence for tectonic subsidence, suggesting that these bays are downwarped by local structures rather than (or in addition to) flexural unloading of the upper plate (Goldfinger, 1994; L. McNeill, in prep.). The combination of folds and thrusts and strike-slip faults may be due to clockwise rotation of the continental margin caused by oblique convergence at the plate boundary (McCaffrey and Goldfinger, 1995).

If the marine sediments overlying peat deposits are due to subsidence accompanying local structures rather than flexural unloading of the plate boundary, does this negate the evidence for interplate earthquakes? The answer is no, for three reasons. First, the radiocarbon ages of buried peats in different bays are consistent with regional subsidence events 300 and 1700 years ago. If the local structures were independent of one another, then there should be no similarity of subsidence age between

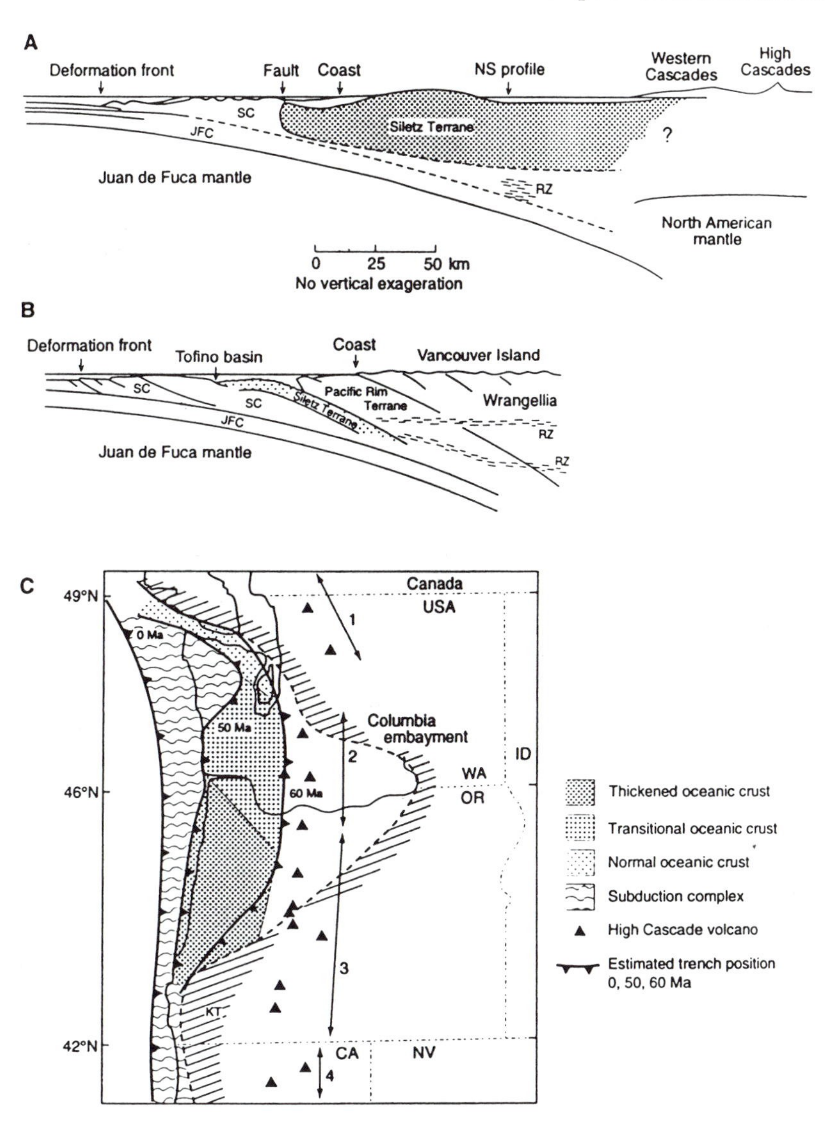

Figure 11–34. Cross sections, based on wide-angle reflection seismic data, across (a) central Oregon continental margin and (b) southwest Vancouver Island margin (center), showing difference in thickness of Eocene oceanic crustal block (Siletz terrane) and its abrupt termination off Oregon. (c) Map view of accretionary wedge and Eocene oceanic crust, showing how trench jumped westward between 60 Ma and today by incorporating to continent first the Eocene oceanic crust and then the accretionary wedge. After Tréhu et al. (1994).

bays. Second, the vertical change accompanying an individual subsidence event seems to be greater in Washington state, where forests are buried, than in northern Oregon, where only marshes are buried, a suggestion that vertical changes per event have an underlying cause at the plate boundary rather than a local structure. Third, if the local structures act independently, then there should be instrumental and historical seismicity like that in California, where in the past two centuries, local structures have generated earthquakes in the range M6 to 8. The absence of this seismicity signature suggests that the local structures along the coast are secondary to movement on the plate-boundary megathrust. This means that the age of a buried peat deposit provides evidence for the age of a plate-boundary earthquake, but the vertical changes cannot be used to model the rupture plane. In addition, the recurrence interval derived from marsh burials would be only a maximum recurrence interval for plate-boundary earthquakes, since, like the Nankai subduction zone, some of those earthquakes might not be expressed as marsh burials.

On the other hand, a large crustal earthquake apparently ruptured an east-trending fault near Seattle about 1100 years ago, producing a tsunami (Bucknam et al., 1992). Liquefaction features in the Chehalis Valley in southwestern Washington and in the mid-Willamette Valley are probably the result of large crustal earthquakes rather than the subduction-zone megathrust (Obermeier, 1995).

The general similarity of Cascadia to a Chilean-type subduction zone, which has generated an earthquake of M_w9.5, led to the suggestion that the Cascadia subduction zone is capable of generating earthquakes of M > 9, despite its low interplate seismicity. This seemed to be supported by the presence of 300-year-old buried peat at sites from central Washington to northern California.

Evidence against such large earthquakes, however, includes the following: (1) Uplift in the last few decades, recorded by highway releveling, is absent on the coast of central Oregon and southwest Washington (Figs. 5-9, 11-37; Mitchell et al., 1994), suggesting that elastic strain is not accumulating there. The rupture area of a plate-boundary earthquake beneath

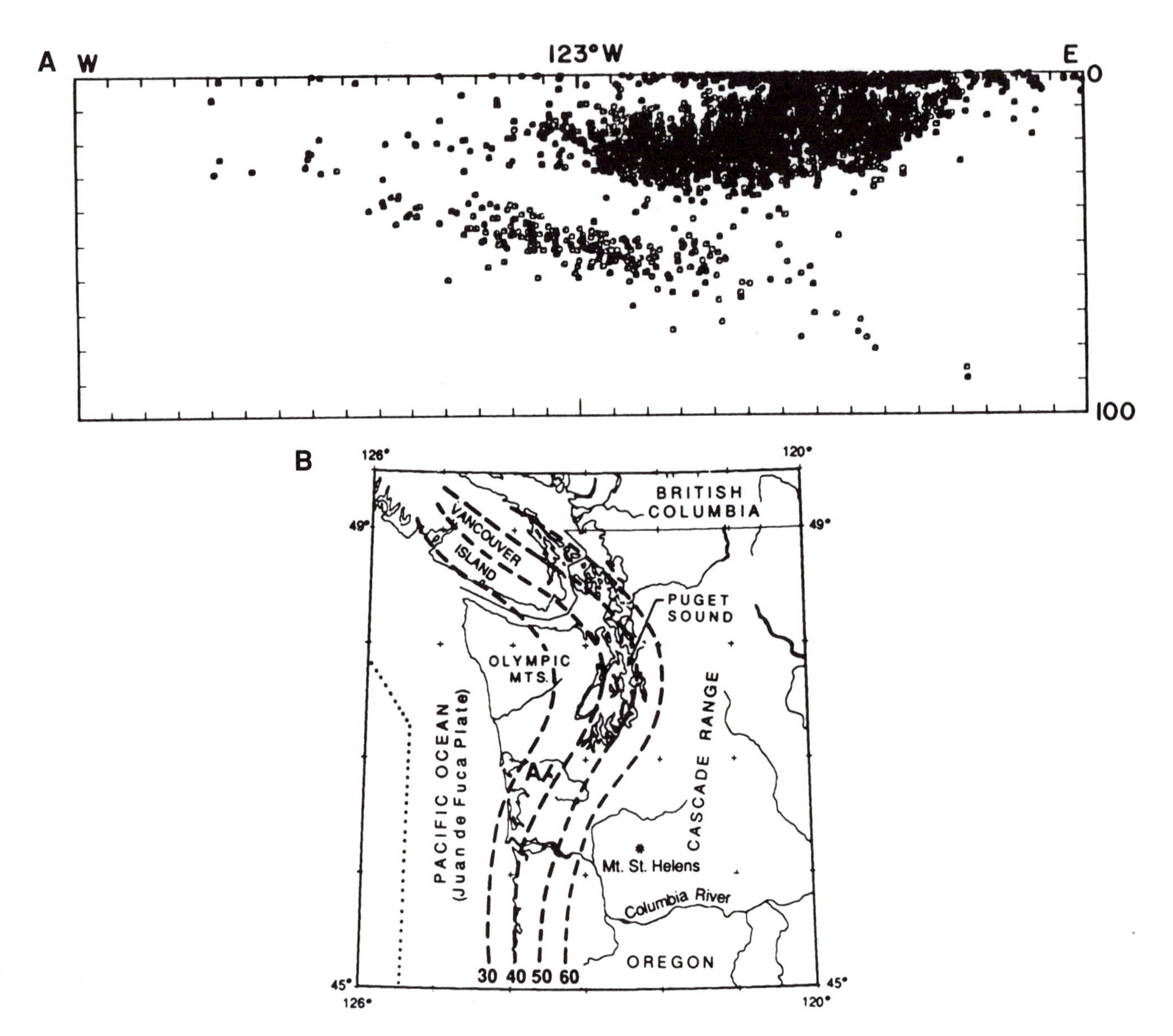

Figure 11–35. (a) Hypocenters of earthquakes in Washington projected onto a west-east section. Grid tick marks at 10-km intervals, no vertical exaggeration. Deeper earthquake belt is within Juan de Fuca plate and not at the plate boundary, which shows no earthquakes. Note lack of earthquakes in accretionary wedge. (b) Contour map, in kilometers, of Moho discontinuity in subducted Juan de Fuca plate. Dotted line locates approximately the base of the continental slope. From Crosson and Owens (1987).

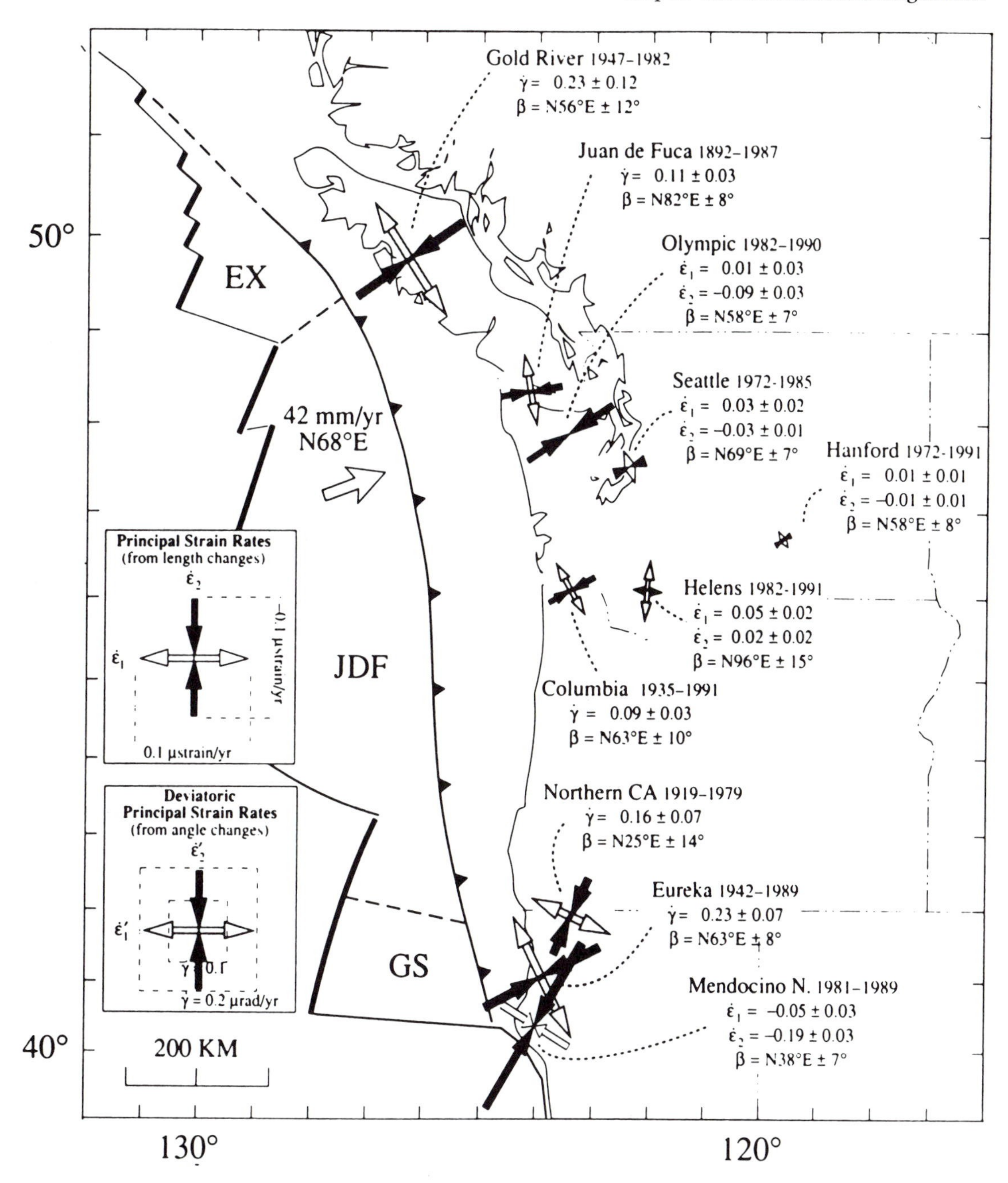

Figure 11–36. Average strain accumulation rates in ppm/yr above Cascadia subduction zone based on repeated geodetic surveys in several networks. EX = Explorer plate, JDF = Juan de Fuca plate, GS = South Gorda plate. Convergence vector of JDF relative to North America is from DeMets et al. (1990). Geodetic data from Drew and Snay (1989), Savage et al. (1991), Savage and Lisowski (1991), and unpublished U.S. Geological Survey and Geological Survey of Canada data.

regions of present-day coastal uplift would produce an earthquake closer to M8, unless the earthquake ruptured across an area where elastic strain is not accumulating. (2) The high slip rates of strike-slip faults at and near the subduction zone indicate considerable deformation of the forearc, which in other subduction zones is correlated statistically to a smaller maximum magnitude earthquake (McCaffrey, 1993; McCaffrey and Goldfinger, 1995). (3) A locked zone that is narrow, shallow, and close to the deformation front (Hyndman and Wang, 1993) would require a very large aspect ratio (ratio of length of the rupture zone to its width) if the entire Cascadia subduction zone, 1000 km long, ruptured at once. Such a large aspect ratio has not been observed elsewhere.

What, then, of the great extent of the 300-year-old buried peat horizon? Radiocarbon dating cannot separate earthquakes that are closely spaced in time, such as the 1944 and 1946 Nankai subduction-zone earthquakes (Bartsch-Winkler and Schmoll, 1992; Fig. 6-10). As Gary Carver put it, the instant burial of peats 300 years ago could have resulted from a "decade (or two) of terror" rather than an "instant of catastrophe."

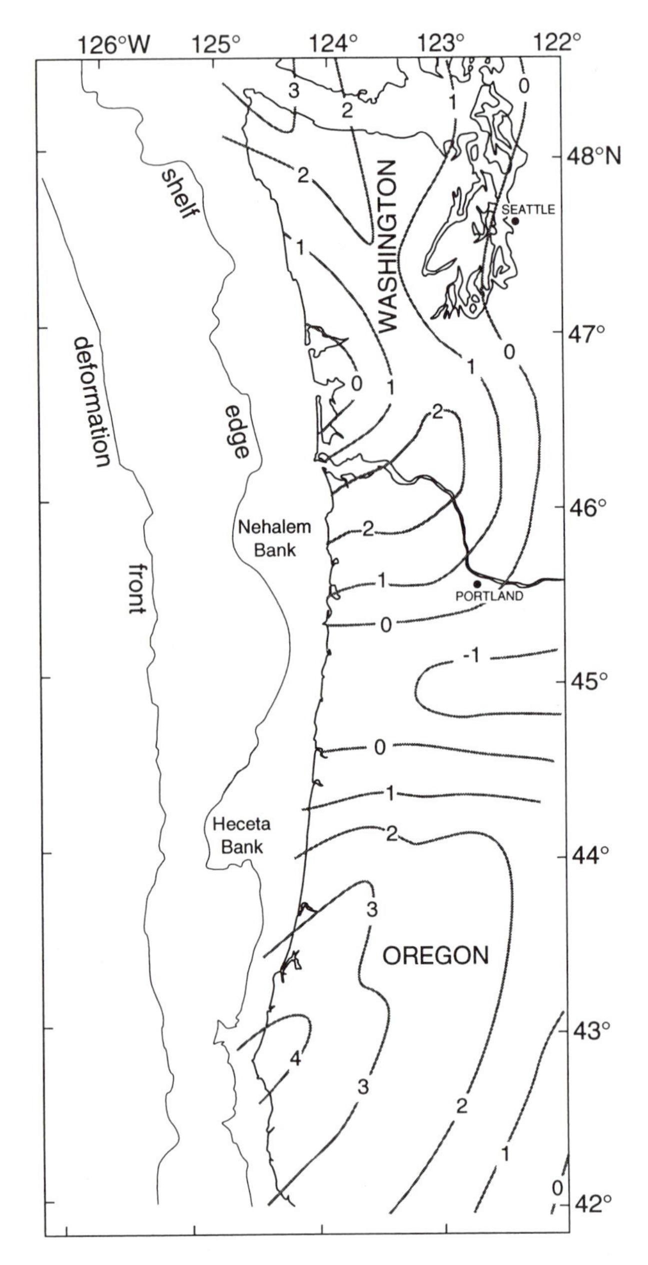

Figure 11–37. Tectonic uplift rates (in mm/yr) in western Oregon and Washington based on highway releveling data. See Fig. 5–9 from Mitchell et al. (1994).

Up to this point, we have considered only that part of the Cascadia subduction zone where oceanic crust is being subducted. However, like the eastern end of the Nankai subduction zone, the Cascadia subduction zone and accretionary wedge come ashore at their southern end in northern California, adjacent to the Mendocino fracture zone and transform fault (Clarke, 1992; Fig. 11-39). Southwest-verging folds and thrusts intersect the coastline at an oblique angle (Kelsey and Carver, 1988). A M7.1 earthquake in April, 1992, beneath Petrolia, California, may have occurred on the subduction zone near the triple junction between the Gorda, Pacific, and North America plates (Oppenheimer et al., 1993). The zone of deformation is 75-100 km wide, and convergence rates across this zone are 15-20 mm/yr (Carver and Burke, 1992), about half of the total convergence rate between the Gorda and North American plates. Trench excavations across thrust faults and core holes in growing synclines in coastal marshes show that much of the deformation is related to large earthquakes (Carver and Burke, 1992).

The Pacific-Gorda-North American triple junction and the Mendocino transform fault are moving northwestward relative to stable North America at about 55 mm/yr, and the northward migration of the Mendocino transform fault may account for the fact that most of the deformation of the northern California fold-thrust belt has taken place in the past million years. The Pacific plate is relatively rigid compared to the other two plates. The North American plate is slivered by NW-trending strike-slip faults of the San Andreas fault system (Kelsey and Carver, 1988). The Gorda plate is rotating clockwise and is deforming internally (Wilson, 1986), accompanied by large earthquakes with $M > 7$. Part of this internal deformation is a tectonic thickening of the Gorda plate, as defined by seismicity, against the buttress of the Pacific plate at the Mendocino transform fault (Carver and Burke, 1992; Smith et al., 1993; Fig. 11-39). This thickening elevates the accretionary prism above sea level, and the structures are rotated to become nearly parallel to the Mendocino transform fault where it turns southeast toward the San Andreas fault.

Peru-Chile Subduction Zone

The western continental margin of South America is a convergence zone, and from the Isthmus of Panama to latitude 46°S, the Nazca plate is subducting ENE beneath the continent at 78-84 mm/yr (DeMets et al., 1990; Fig. 11-40). The Nazca plate is composed of Tertiary oceanic crust, which is warmer and more

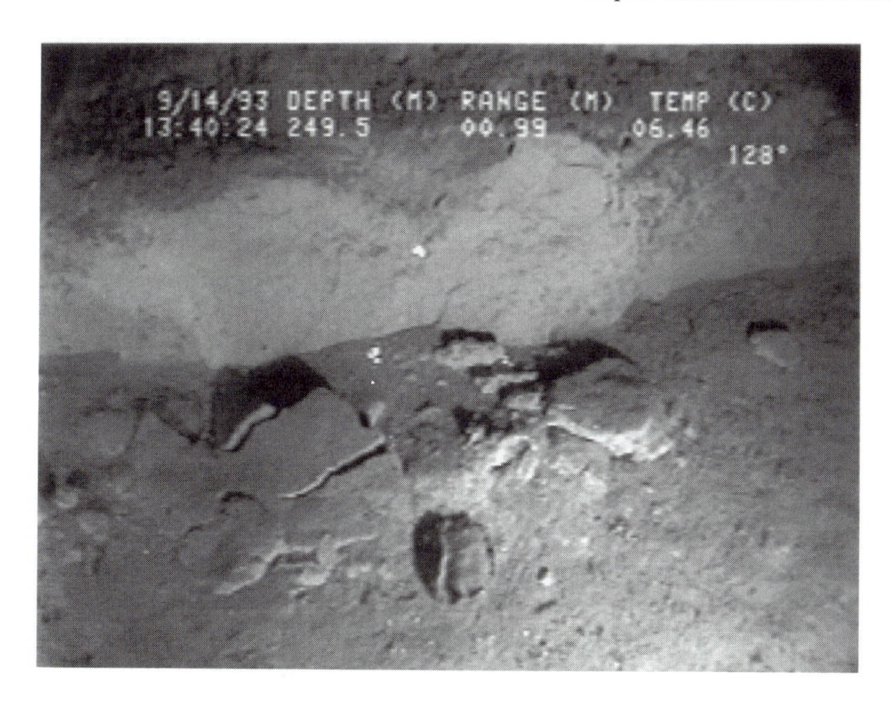

Figure 11–38. Active fault scarp at east end of Daisy Bank fault (DBF, Fig. 11–33), as viewed from DELTA submersible. Local vertical component of motion is up to the south; the fault dips steeply south. Holocene fine-grained sediment (dark gray) drapes horizontal surfaces underlain by latest Pleistocene clay, exposed in face of scarp (light gray). Scarp is about 0.5 meters high. Light laser dots are 20 cm apart. Photo by Gary Huftile.

buoyant than the older Mesozoic crust subducting beneath the island arcs of the western Pacific. For this reason, the trench is 2–3 km shallower than the trenches in the western Pacific. The deepest section, 7–8 km deep, lies off northern Chile, where the Nazca plate lithosphere is oldest (Schweller et al., 1981; Fig. 11–41). Unlike the Juan de Fuca plate and the northernmost Pacific plate, which are covered by deep-sea turbidite fans, the Nazca plate has only a thin hemipelagic sediment cover (Yeats and Hart, 1976).

The Carnegie, Nazca, and Juan Fernandez ridges intersect the trench and have a strong influence on segmentation of the subduction zone (Schweller et al., 1981; Nur and Ben-Avraham, 1981; Fig. 11–42). The northeast-striking Grijalva, Mendana, and Challenger fracture zones also intersect the trench south of latitude 3°S. (Fig. 11–40). Because the fracture zones separate oceanic crust of different ages and buoyancy, the fracture zones are marked by steep scarps.

The aseismic ridges allow subdivision of the Peru-Chile trench into a deep segment, virtually free of sediment, off northern Chile and southern Peru (Arica Bight) and shallower segments with trench-fill sediments north of the Nazca Ridge and south of the Challenger fracture zone at latitude 27.5°S, the latter boundary marked by a north-facing scarp about 1 km high (Fig. 11–41). The southern, sediment-filled segment is further subdivided into a segment with less sediment north of the Juan Fernandez ridge (D in Fig. 11–41) and a segment with a thick trench fill south of this ridge (E in Fig. 11–41). An outer-arc high (flexural bulge) is poorly developed except in the North Chile tectonic province at the Arica Bight, where the trench is deepest. The slope of the incoming Nazca plate west of the trench axis is marked by normal faults parallel to the trench axis, consistent with rupture accompanying bending of the oceanic plate downward into the trench. The presence of trench-fill sediment is largely controlled by climate: the northern and southern ends of the trench are adjacent to areas of high rainfall in the mountains, whereas the sediment-free trench adjacent to the North Chile tectonic province is adjacent to the Atacama Desert, one of the driest places on earth.

Surprisingly, the width of the continental slope does not change significantly with changes in depth of the trench. The reason for this is the poorly developed accretionary wedge. Off Peru, the Andes mountain chain, dominated by batholithic rocks, is succeeded seaward by inner shelf basins, an outer shelf high or a coastal range, and an outer set of basins close to the continental shelf edge (Thornburg and Kulm, 1981). The accretionary wedge occupies only the lower part of the continental slope and is only 15 km wide off central Peru (von Huene et al., 1988). Tectonic erosion must be taking place, as is the case off northeastern Japan, perhaps aided by massive submarine landslides such as those mapped off northernmost Peru (von Huene et al., 1989). Reverse faults are found at the base of the continental slope off southern Chile (Thornburg et al., 1990) and within the trench itself off Peru, where a 900-m basaltic ridge is uplifted along a reverse fault. But normal faults cut sediment of the shelf basins off Peru (Thornburg and Kulm, 1981) as well as Quaternary sediment on the coast of Peru (Mercier et al., 1992). In northern Chile, the Coastal Scarp, about 1 km high, extends for about 700 km, largely offshore, but it is onshore in the Mejillones Peninsula near Antofagasta (23°S), where active normal faulting is widespread (Armijo and Thiele,

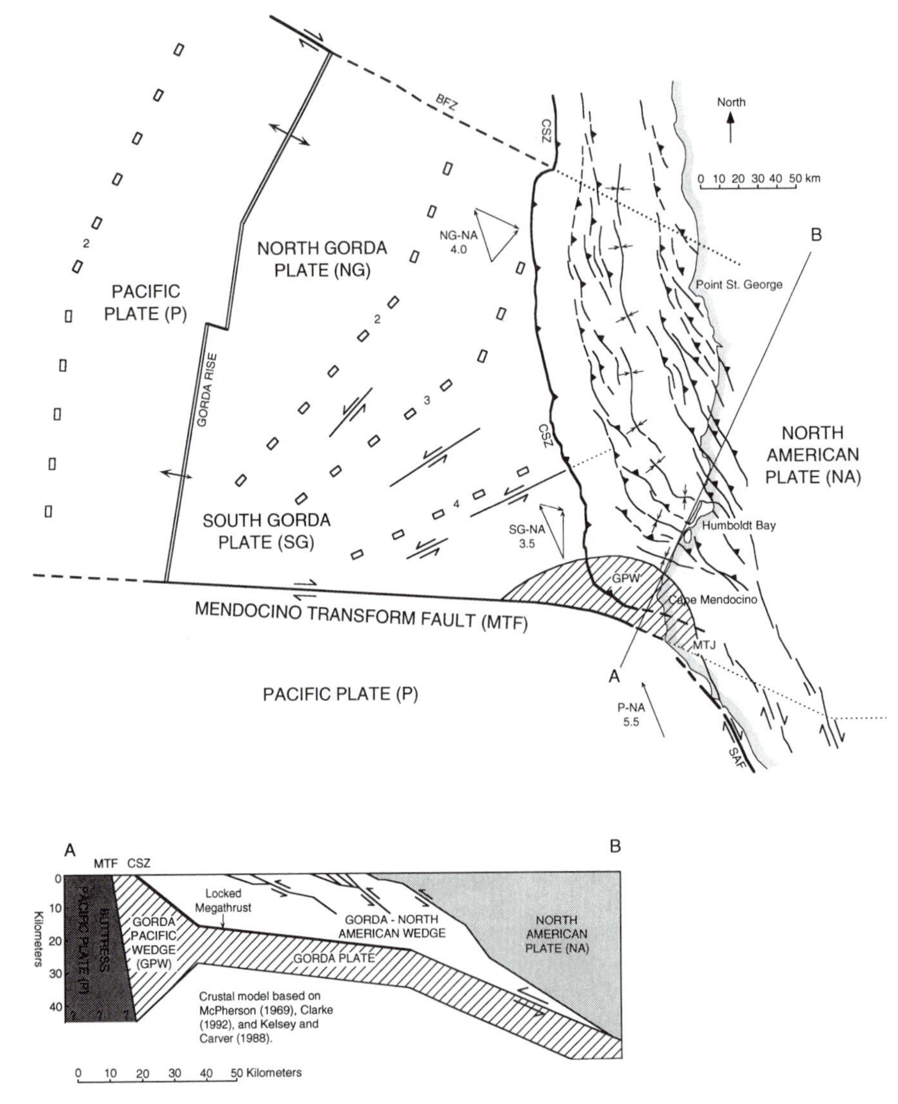

Figure 11–39. Tectonics of the southernmost Cascadia subduction zone (CSZ) and Mendocino transform fault (MTf) in northern California. Vector triangles show slip vectors among Pacific (P), North American (NA), Northern Gorda (NG), and Southern Gorda (SG) plates. Lines of open rectangles locate sea-floor magnetic anomalies which show rotation and decreased spreading rate of Southern Gorda plate. Diagonal lines locate wedge of thickening of Gorda plate (GPW), also shown along cross section A-B′. Other abbreviations: BFZ, Blanco fracture zone; SAF, San Andreas fault. After Carver and Burke (1992) and Smith et al. (1993).

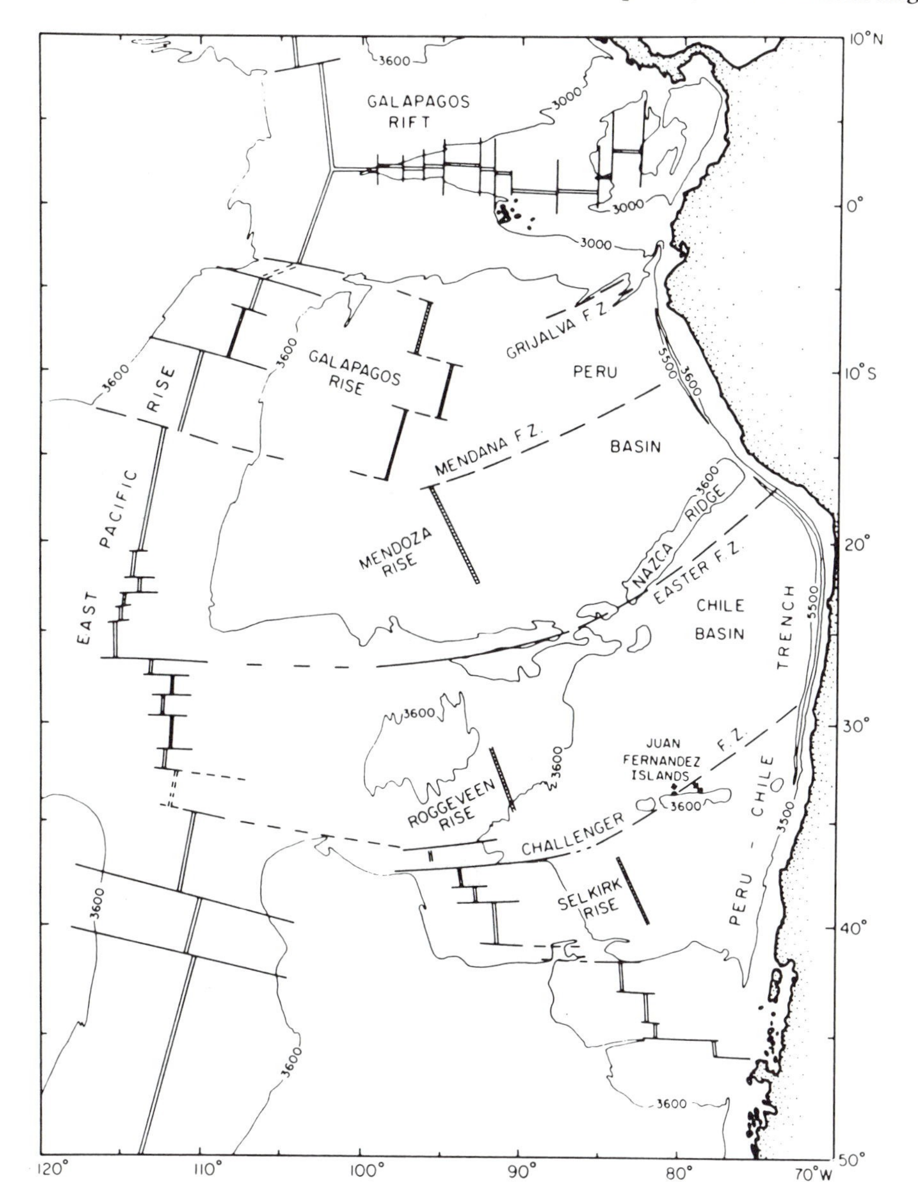

Figure 11–40. Map of Nazca plate, showing active spreading centers (open double lines), inactive spreading centers (double lines with hachures), major fracture zones, aseismic ridges, Peru-Chile trench, and South American coastline. Bathymetry in meters. After Schweller et al. (1981).

1990). Normal faulting farther east on the Altiplano is discussed in Chapter 9.

The South American coast contains several strike-slip faults that are nearly parallel to the trench (Fig. 10–8), and these have commonly been attributed to an oblique component of convergence between the Nazca Plate and South America (Dewey and Lamb, 1992; Fig. 10–8). These faults include the right-slip Pallatanga fault in Ecuador (Winter et al., 1993), and the Atacama and Liquine-Ofqui faults in Chile (Hervé and Thiele, 1987). The Atacama fault is generally considered a right-slip fault, but Armijo and Thiele (1990) have found evidence that its most recent movement was by left slip, a sense of slip opposite to that predicted by the oblique convergence.

The Andean volcanic arc is discontinuous, with gaps in active volcanism in the northern two-thirds of Peru and in north-central Chile. These gaps correspond to parts of the subducting slab that flatten beneath the overriding plate (Fig. 11–42), in contrast to other parts where the slab subducts at a steeper angle of 30° (Barazangi and Isacks, 1976). Boundaries of

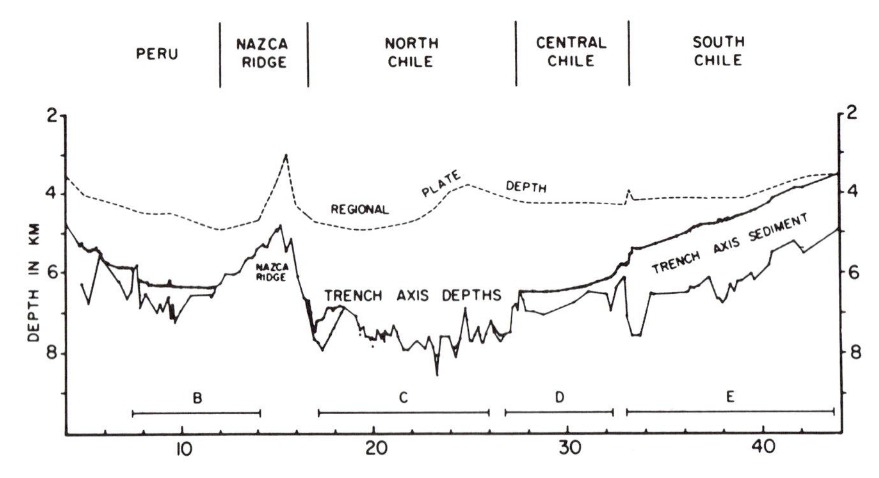

Figure 11–41. Longitudinal profile along Peru-Chile trench showing maximum trench-axis depths, thickness of trench-fill sediment, and regional depth of the Nazca plate 300 km west of the trench. Tectonic provinces of subduction zone identified at top. Juan Fernandez ridge separates Central Chile and South Chile provinces. Letters at bottom identify segments of Barazangi and Izacks (1976) in which B and D refer to flat subducting segments with no Quaternary volcanoes, and C and E refer to segments with a more steeply dipping subduction zone and active volcanoes. After Schweller et al. (1981).

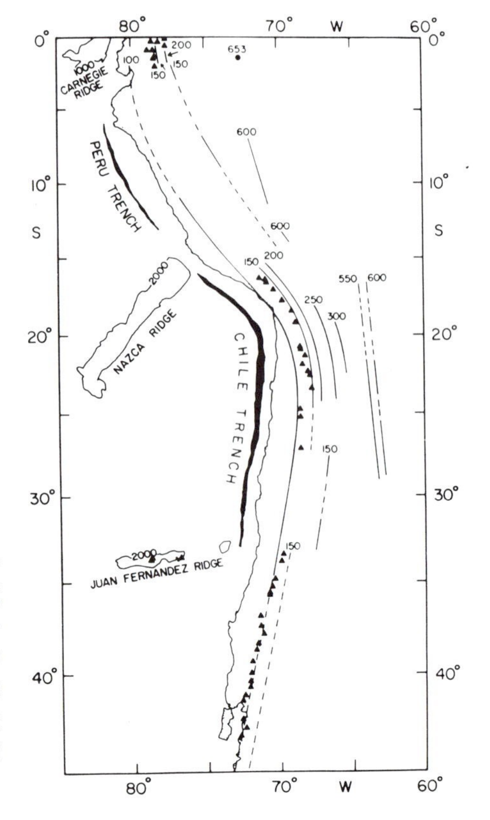

Figure 11–42. Map of Peru-Chile subduction zone showing segmentation into flat subduction segments with no active volcanoes and steeper dipping segments with active volcanoes (shown as black triangles). Solid lines with numbers are contours on top of subducting slab in kilometers. Also shown are deepest segments of trench (black), and aseismic ridges (enclosed by 2000-m contour). After Barazangi and Isacks (1976).

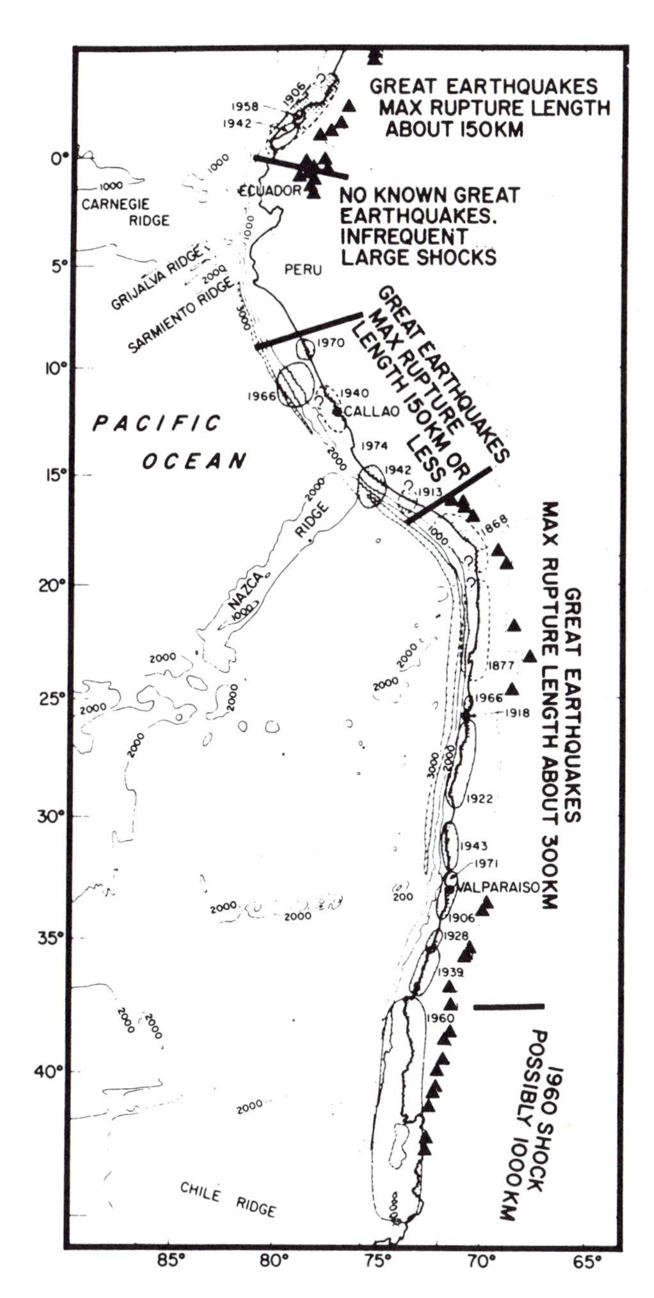

Figure 11–43. Rupture zones of large earthquakes identified by year of occurrence, and active volcanoes (solid triangles). After McCann et al. (1979).

the flattened segments may be tears in the subducting plate, but more likely they are flexures or lateral ramps, with the transition near latitude 30°S occurring over a distance of several hundred kilometers (Cahill and Isacks, 1992).

Except for northern Peru and southern Ecuador, all of the Peru-Chile subduction zone has been marked by great historical earthquakes, with maximum rupture lengths increasing from 150 km in Peru to 1000 km in southern Chile, the site of the great 1960 earthquake of M_W = 9.5 (McCann et al., 1979; Fig. 11-43). The absence of great earthquakes north of 7°S may be related to the subduction of the aseismic Carnegie ridge and the Grijalva fracture zone. Farther south, however, no broad gap exists opposite the Nazca ridge or Juan Fernandez ridge. A series of earthquakes in 1940, 1966, and 1974 terminated southward where the Nazca ridge intersects the trench, east of which flat subduction and no active volcanism change laterally southward to more steeply dipping subduction and active volcanism (compare Figs. 11-42 and 11-43). An earthquake in 1746 in the same area as the 1940-1974 earthquake sequence may have had a rupture length of 600 km, much longer than the more recent earthquakes there (Thatcher, 1990) or in northern Chile.

Farther south, the subduction zone is marked by earthquakes with long rupture zones, including earthquakes in 1868 and 1877 north and south, respectively, of the Arica Bight at 18°S, and an earthquake in 1922 at Copiapó in northern Chile (Fig. 11-43). The Arica Bight has not had a great earthquake since 1877 and is considered to be a seismic gap (McCann et al., 1979). Farther south, interplate earthquakes in 1966 at 25.5°S. (Taltal region) and in 1967 and 1985 at 30°S (Coquimbo region) suggest local warps or tears in the subducting slab, the latter possibly related to the Challenger fracture zone (Tichelaar and Ruff, 1991).

The 1960 earthquake of southern Chile caused vertical displacements for a distance of 1000 km, from 37°S to the triple junction with the Chile Rise south of 45°S (Plafker and Savage, 1970; Fig. 11-43). Up to 2 m of subsidence occurred in the Coastal Mountains and their southern continuation on Chiloé Island and in the Chonos Archipelago farther south (Fig. 11-44). To the west, offshore islands were uplifted as much as 6 m, and to the east, the coastal foothills of the Andes were uplifted up to a half meter. This pattern of vertical deformation was similar to that observed in the 1964 Alaskan earthquake (Plafker, 1972); the maximum subsidence occurred in a region characterized by long-term uplift. Tsunami records in Japan show that the 1960 Chilean earthquake was the largest in the Pacific region in at least 500 years, although an earthquake in 1837 broke all but the northern end of the 1960 rupture zone (Thatcher, 1990). The 1960 event seems to be the last of a southward-migrating sequence starting with an M_W = 7.6 shock in 1928 and an M_W = 8.2 shock in 1939. This same segment of the rupture zone was filled by a southward-migrating sequence in 1835 (witnessed by Charles Darwin) and 1837. Earlier sequences in the middle of

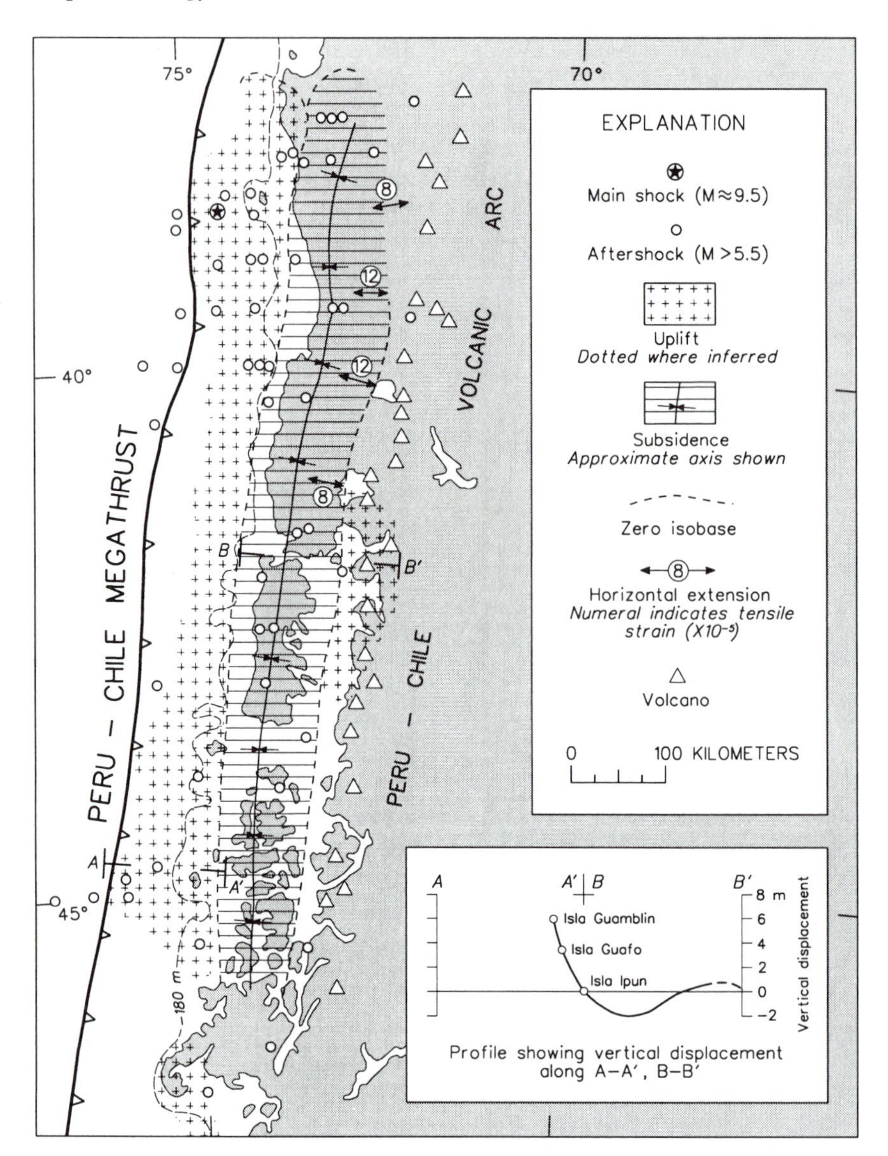

Figure 11–44. Tectonic displacements and seismicity associated with the 1960 Chilean earthquake. Compare with Figure 11–30. After Plafker (1972).

the eighteenth, seventeenth, and sixteenth centuries suggest a recurrence interval of 85–125 years (Thatcher, 1990).

Captain Fitzroy of the H.M.S. *Beagle* noted that uplift of Isla Mocha accompanying the 1835 earthquake was partly recovered by post-seismic subsidence a few months after the earthquake. Nelson and Manley (1992), on the other hand, described post-seismic uplift after the 1960 earthquake at a rate of 70 mm/yr. They observed the 1835 and 1960 strandline, but they concluded that other strandlines were uplifted aseismically, as is occurring today. The modeled fault responsible for the uplift is more likely a fault within the upper plate rather than the plate-boundary megathrust. Dislocation modeling of the coseismic 1960 deformation also suggests a reverse fault that reaches the surface within the upper plate (Plafker and Savage, 1970). Atwater et al. (1992) showed that three sites in the 1960 zone of submergence showed net emergence during the late Holocene. The work in Chile indicates that the geodetic changes are apparently related to upper-plate structures rather than to the megathrust, and that coseismic vertical changes commonly are not representative of long-term changes.

Other Subduction Zones with Active-Tectonic Expression

Makran

Between the continent-continent collision zones of the Zagros and the Himalaya (Fig. 10–1), oceanic lithosphere of the Arabian plate, 70–100 Ma in age, is being subducted beneath the Makran Ranges of Iran and Pakistan at a rate of several centimeters per year (Farhoudi and Karig, 1977; Jacob and Quittmeyer, 1979; Fig. 11-45). On the east, the Makran Ranges curve sharply northward into left-slip fault systems marking the western boundary of the Indian plate driving northward into Eurasia. A trench is not present, probably because the sedimentary sequence overlying the downgoing oceanic plate is locally > 6 km thick (White and Louden, 1982), comprising one of the world's largest forearcs. Arc volcanoes are present, but their trend is east-northeast, whereas the Makran structures strike east. The dip of the subduction zone is only a few degrees (White and Louden, 1982; Fig. 11-45).

Seismicity is low compared to the Zagros, especially the Iranian Makran. The distribution of earthquakes permits subdivision of the Makran subduction zone

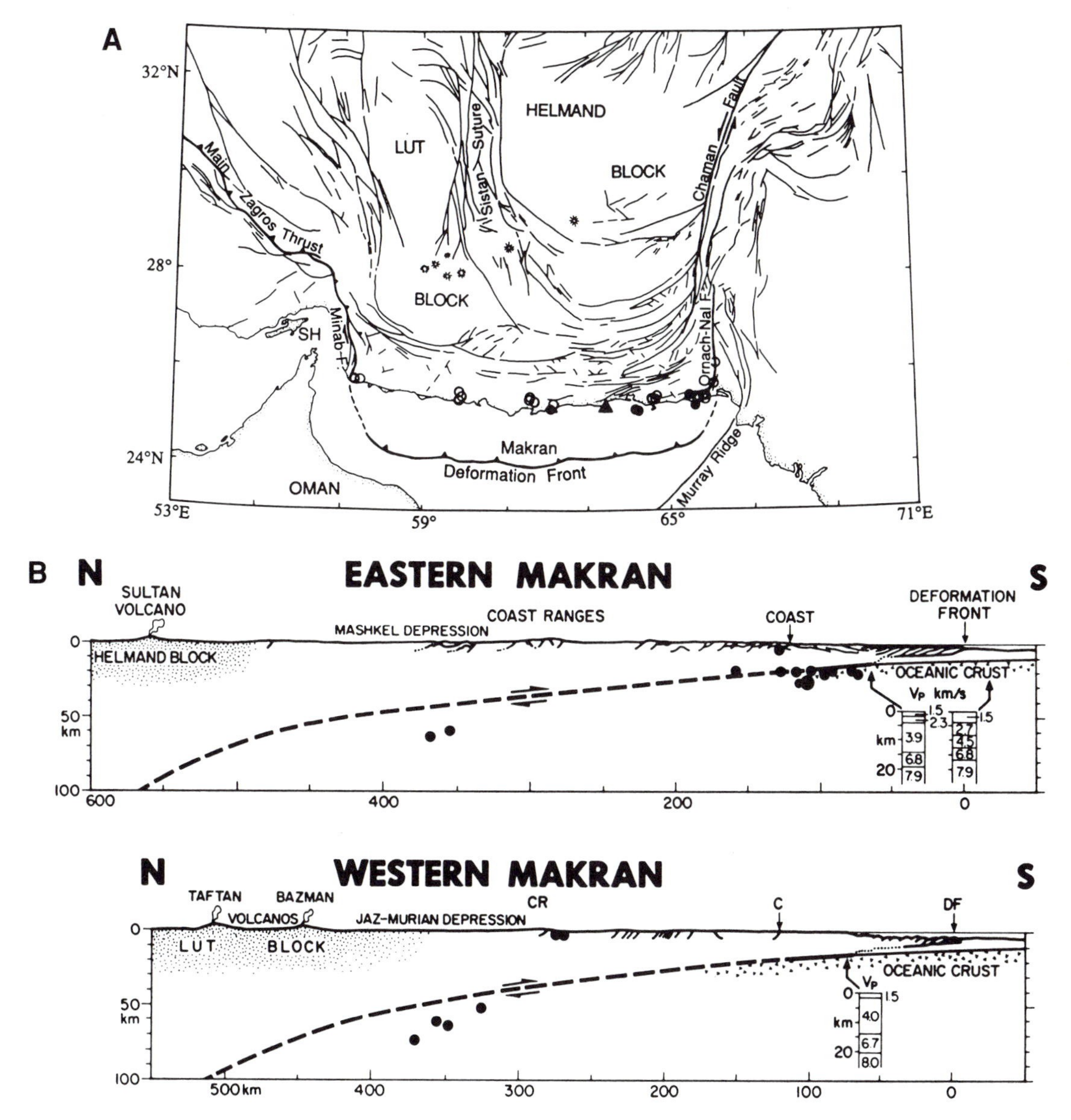

Figure 11–45. (a) Faults (solid lines) and Quaternary volcanoes (radiating hachures) of Makran subduction zone. Epicenter of 1945 earthquake shown as solid triangle. Mud volcanoes shown as open circles; those activated by 1945 earthquake shown in solid circles. Lut and Helmand blocks are older continental blocks separated by Sistan suture zone. SH, Straits of Hormuz. (b) North-south cross sections showing contrast in seismicity between eastern Makran, with interplate earthquakes near the coast, and western Makran, where interplate earthquakes are absent. In 1945, an earthquake ruptured the plate-boundary décollement from 90 to 180 km north of the deformation front. After Byrne et al. (1992).

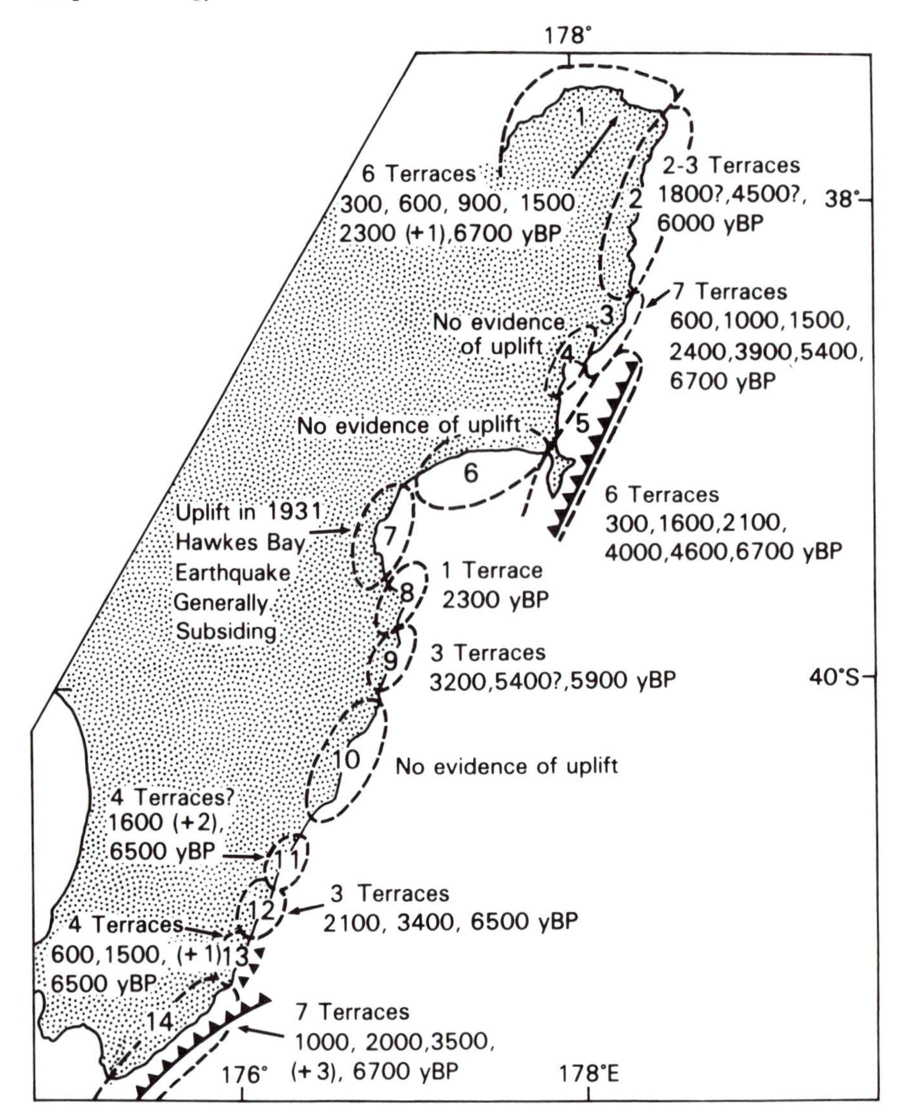

Figure 11–46. (a) Fourteen tectonic subregions on east coast of North Island, New Zealand, based on radiocarbon dates of Holocene uplifted terraces (illustrated in Fig. 7–25). Heavy lines with teeth show faults associated with coastal uplift.

into two segments divided by the Sistan suture between the Lut and Helmand blocks (Byrne et al., 1992; Fig. 11-45). In 1945, an interplate earthquake of $M_w = 8.1$ struck the Makran coast of Pakistan and produced 1-3 m of coastal uplift (Page et al., 1979). Other large earthquakes struck the Pakistan coast in 1765 and 1851. In contrast, the seismicity of the Makran coast in Iran is very low, and no major historical earthquake has been documented for this region, although an event in 1483 could have occurred there. But the Makran coast of Iran, like that of Pakistan, has undergone late Quaternary deformation (Vita-Finzi, 1981). Mud volcanoes on the Makran coast are evidence of overpressured sediment at depth; four mud volcanoes formed during the 1945 earthquake. Even though focal mechanisms of earthquakes near the coastline east of the Sistan suture indicate low-angle faulting, suggesting movement on the plate boundary (Byrne et al., 1992), focal mechanisms farther north show movement on more steeply dipping fault planes.

Hikurangi Subduction Zone

The southern end of the Tonga-Kermadec subduction zone turns southwest toward New Zealand as the Hikurangi subduction zone. The Pacific plate is subducting beneath New Zealand at a rate of about 50 mm/yr in a west-southwest direction, and the plate boundary curves closer to parallel with the slip vector southward. On the east coast of the North Island of New Zealand, part of a broad accretionary wedge is exposed on land.

The 1931 Hawke's Bay earthquake (M_s = 7.8) produced uplift of up to 2.7 m in an asymmetric dome > 90 km long and a maximum subsidence of 1 m southwest of the dome (Hull, 1990; Fig. 11-46). Oblique-slip secondary faulting was also observed. Fault modeling based on elevation changes in 1931 indicate that this causative fault cut upward from the megathrust at 15-20 km depth through the upper plate with reverse-slip and right-slip displacement (Haines and Darby, 1987). Vertical movements in 1931 were commonly opposite in sign to earlier Holocene movements (Hull, 1990). The 1855 Wairarapa earthquake (M = 8) also may have resulted from a listric fault that ruptured part of the plate-boundary megathrust and continued upward into the upper plate as a blind reverse fault and as a right-slip fault (Darby and Beanland, 1992).

Uplifted Holocene marine terraces on the North Island provide evidence for six clusters of prehistoric earthquakes in the past 2500 years, with irregular recurrence intervals ranging from 200 to 500 years (Berryman et al., 1989; Figs. 7-25, 11-46). Variations in terrace ages and distribution permit the subdivision of the east coast of the North Island into individual rupture zones with lengths ranging from 25 to 150 km, caused by earthquakes with M_w ranging from 7.3 to 8.0 (Berryman et al., 1989).

Western Taiwan

In the northern Philippines, oceanic crust of the South China Sea, which is part of the Eurasian plate, is being subducted beneath an island arc. Islands between

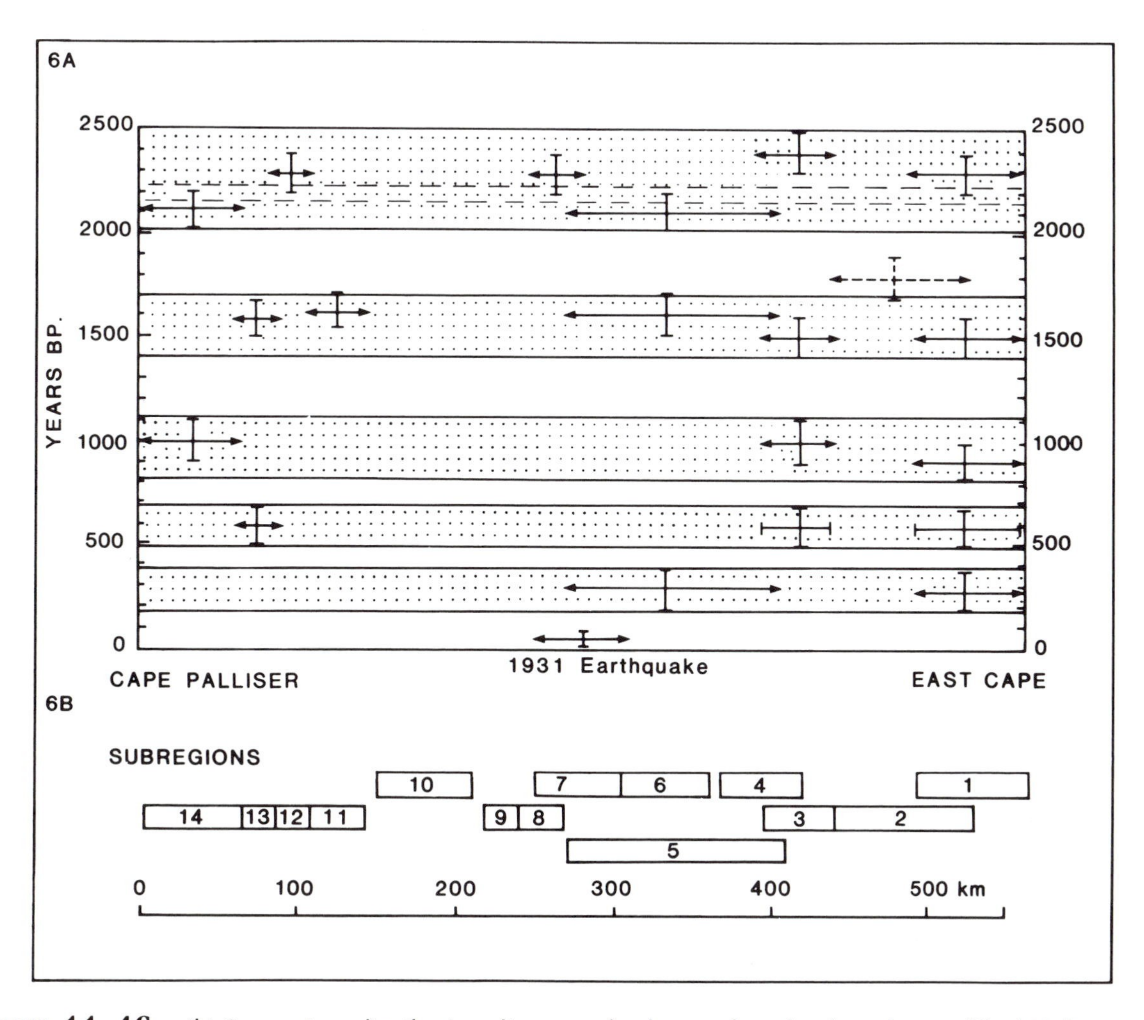

Figure 11–46. (b) Space-time distribution diagram of paleo-earthquakes based on uplifted Holocene marine terraces shown in (a). Six clusters of earthquakes (dot pattern) are separated by times of no earthquake occurrence. Rupture zone of 1931 Hawke's Bay earthquake also shown. Blocks at bottom represent segment boundaries, with blocks closest to the bottom those closest to the Hikurangi trench. After Berryman et al. (1990).

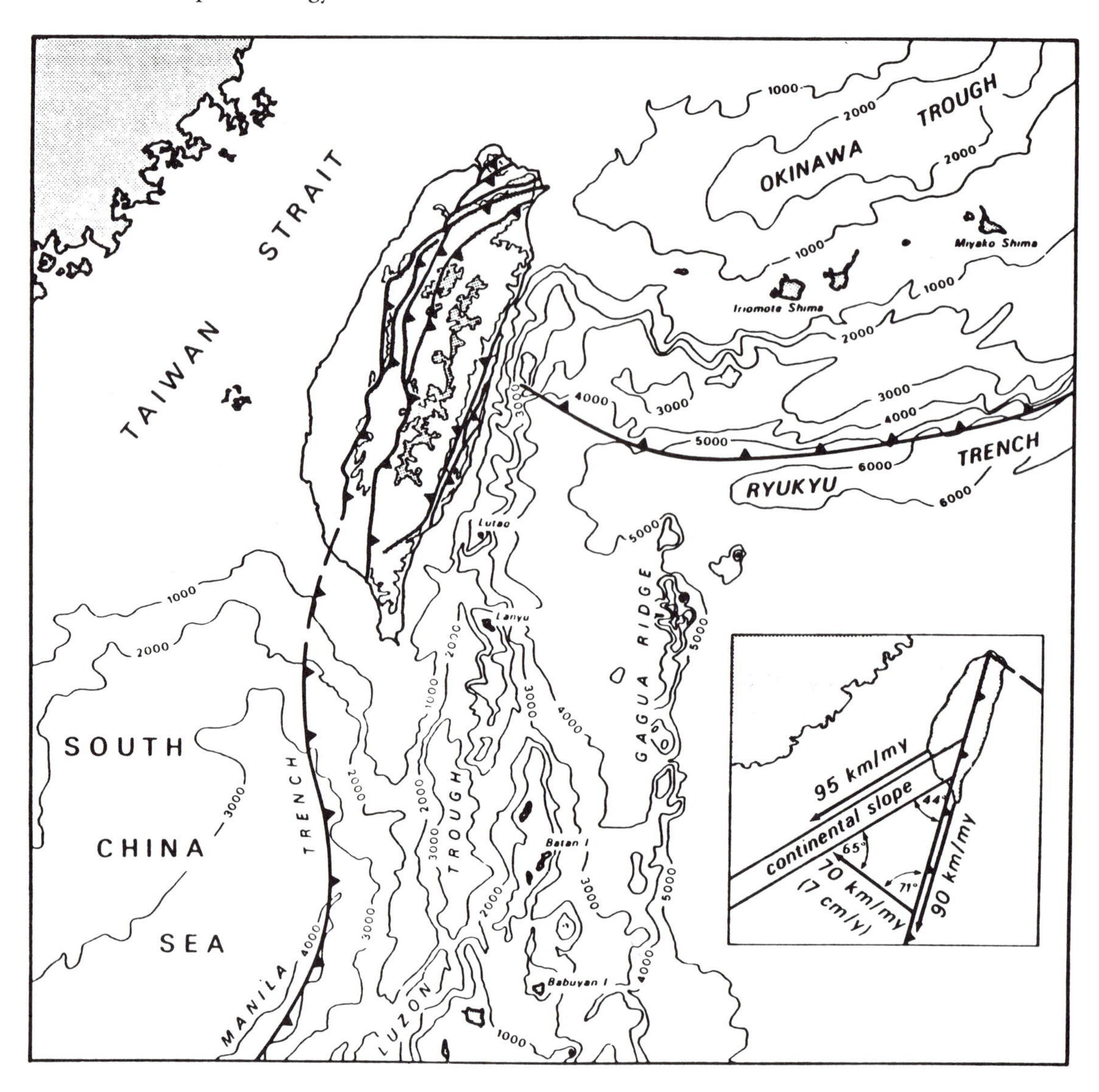

Figure 11–47. (a) Tectonic and bathymetric setting of Taiwan, with 1000-meter contours (submarine and Taiwan only). From Davis et al. (1983).

the Philippines and Taiwan are composed of Neogene volcanic rocks, and the Luzon trough to the west, which is a forearc basin, is separated from the Manila trench by an accretionary wedge (Fig. 11–47). To the north, in western Taiwan, the accretionary wedge is overriding the passive margin of the Asian continent. Because this continental lithosphere is more buoyant than the oceanic lithosphere of the South China Sea, the accretionary wedge is above sea level in western Taiwan (Suppe, 1987).

Synorogenic Pleistocene deposits increase in thickness eastward from less than 1 km in Taiwan Strait to more than 4 km at the edge of an onshore thin-skinned fold-thrust belt. The rate of sedimentation increased from 100 m/m.y. under stable continental-margin conditions in the Miocene to more than 1500 m/m.y. in the Pleistocene as Chinese continental basement was flexed downward. Thrust faults flatten into a décollement in Miocene or Pliocene strata, with considerable stratigraphic variation. The Pakuashan anticline is growing upward through late Quaternary alluvium of the Taiwan coastal plain (Suppe, 1987). This zone produced earthquakes with reverse-fault surface rupture in 1906 (M7.1), 1935 (M7.1), and 1946 (M6.7) (Bonilla, 1977), but the relations between these earthquakes and the thin-skinned fold-thrust belt are unclear.

On the east coast of Taiwan, up to eight elevated

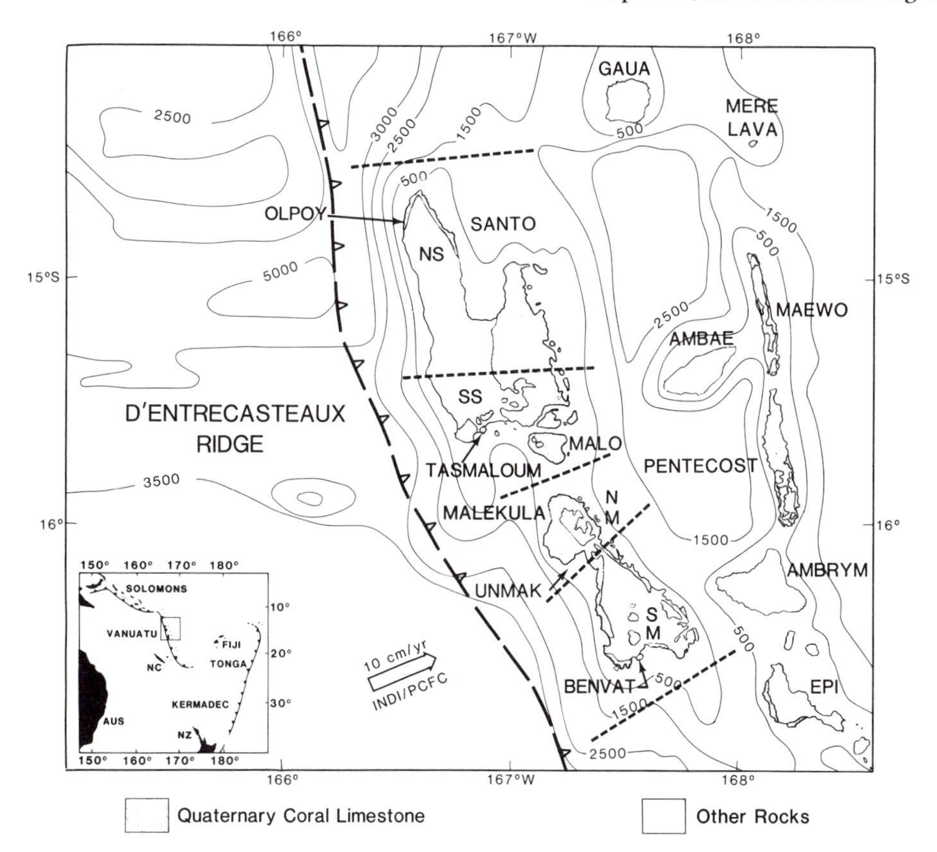

Figure 11–48. Central Vanuatu, southwestern Pacific, showing bathymetry and distribution of Quaternary coral limestone. Convergence rate does not include the contribution from back-arc spreading. Dashed lines show proposed segment boundaries based on dating of uplifted Holocene corals. After Taylor et al. (1990).

Holocene shorelines document uplift rates of 3–8 mm/yr (Liew et al., 1993). These shorelines are a response of coseismic uplift, with recurrence intervals between earthquakes not more than 1000 years.

Vanuatu

Oceanic crust of the Australian plate about 60 Ma in age is subducting beneath the Pacific plate at a rate of 100 mm/yr at the New Hebrides trench (Fig. 11-48). The Australian plate includes the aseismic D'Entrecasteaux ridge, which influences segmentation of the frontal island arc of Vanuatu. The historical record of Vanuatu is less than a century long, but precise ^{230}Th ages of coral undergoing coseismic uplift and the counting of annual growth layers in coral permit reconstruction of the earthquake history of Vanuatu for the past 6000 years (Taylor et al., 1990). Some of the uplift is coseismic, and some is aseismic. The uplift history, recurrence intervals, and amount of aseismic uplift are significantly different for different arc segments. In the northern three segments, aseismic slip is not continuous, but mostly occurs in the years of large earthquakes. Earthquake recurrence intervals are shortest in those segments where the percentage of seismic energy release is greatest. From one-tenth to one-third of slip associated with plate convergence is released by large earthquakes.

Other Regions

Tectonically uplifted terraces in other parts of the Pacific Rim are reviewed by Ota (1991), Ota and Kaizuka (1991), and Yonekura (1983). These include studies on the Ryukyu Islands (Pirazzoli and Kawana, 1986; Ota and Omura, 1991), northwestern Luzon in the Philippines (Maemoku and Paladio, 1992), and Timor (Vita-Finzi and Hidayat, 1991).

DISCUSSION

This chapter began with a general review of the characteristics of oceanic subduction zones, including structure, geomorphic expression, and seismicity. It continued with a consideration of the strength of the megathrust as a guide to the maximum size of individual earthquakes generated on it. This was followed by discussion of the major subduction zones on which the largest earthquakes are generated, and for which studies of late Quaternary geology and paleoseismology have been conducted. This was not a comprehensive review of all subduction zones, but only those for which the history of great earthquakes is relatively well known. It is not clearly established that subduction zones in the eastern Mediterranean region (Ionian, Hellenic) generate great interplate earthquakes. Marianas-type subduction zones in the western Pacific region are poorly coupled and, for the most part, do not generate earthquakes of $M > 7$. (The 1993 earthquake of $M_W = 7.8$ near Guam may be an exception; the plate slip vector there is at a relatively high angle to the trench.)

We now integrate the geologic and tectonic observations presented in the previous text, to address several questions. (1) Are uplifted terraces or downdropped marshes generated directly from rupture of a megathrust, or are they related to structures in the upper plate? (2) Do oceanic subduction zones and continental collision zones have different seismotectonic signatures? (3) Do geologically defined segments control the locations of mainshocks or terminations of rupture-zone boundaries of interplate earthquakes? (4) Do earthquakes tend to rupture the same segments of a subduction zone, and are recurrence intervals regular?

Holocene Vertical Tectonics, Megathrusts, and Upper-Plate Structure

Yoshikawa et al. (1964) described uplifted Holocene terraces on the southern coast of Shikoku in Japan and noted the similarity of uplift patterns to coseismic uplift accompanying great earthquakes in the Nankai subduction zone. Similar interpretations were made for Holocene terraces on the Boso Peninsula south of Tokyo (Matsuda et al., 1978) and on coral islands in Vanuatu in the southwest Pacific (Taylor et al., 1980). Atwater (1987) correlated marsh burials on the Washington coast with megathrust earthquakes in the Cascadia subduction zone, observing that great earthquakes produce a belt of subsiding terrain parallel to the plate boundary, accompanied by belts of uplift trenchward and landward from the subsiding terrain. Darienzo and Peterson (1990) extended Atwater's work to Oregon. If the vertical changes were related directly to slip on the megathrust, the boundaries of the rupture zones could be mapped to determine the amount of slip, leading to a determination of moment magnitude.

But modeling of coseismic elevation changes during the 1931 Hawke's Bay, New Zealand, earthquake on the Hikurangi subduction zone led to the conclusion that the uplift was directly related to a reverse fault near the coastline, not to the megathrust, although some movement down-dip on the megathrust surface may also have occurred (Hull, 1990). The 1855 Wairarapa earthquake was long considered to be a strike-slip event, but recent modeling of 1855 vertical changes by Darby and Beanland (1992) suggests that the earthquake was accompanied by rupture on a listric fault, including slip on both the Hikurangi low-angle megathrust and a steep reverse fault in the upper plate rising from the megathrust. The distribution of uplift on Holocene terraces on the Boso Peninsula requires faulting within the upper plate (Scholz and Kato, 1978). The uplifted terraces on Shikoku Island and on the Kii Peninsula are best explained by faults in the upper plate, by a single thrust cropping out on the continental slope or by a set of faults like the Muroto fault, which strikes at a high angle to the plate boundary. These Holocene terraces represent permanent upper-plate deformation, whereas uplift accompanying the 1707, 1854, and 1946 earthquakes was fully recovered after each earthquake, presumably because these earthquakes were limited to the subduction-zone megathrust and did not involve deformation within the forearc.

The 1964 Alaskan earthquake was accompanied by surface rupture on upper-plate reverse faults on Montague Island (Plafker, 1967). Vertical changes accompanying the earthquake were opposite in sign to prehistoric vertical changes. Similarly, the 1960 Chilean earthquake was accompanied by uplift on an offshore island that is best attributed to a fault within the upper plate (Nelson and Manley, 1992). In addition, areas of 1960 subsidence northeast of Chiloé Island were uplifted in prehistoric time (Atwater et al., 1992). At least some of the buried marshes in coastal bays on the Washington and Oregon coasts may be related to coseismic movement on faults and synclines on the continental shelf that extend to the coast, an interpretation that depicts these bays as tectonically downwarped features (L. McNeill, in prep.).

Even though local faults and folds may be responsible for coseismic uplift or subsidence, these structures are probably still secondary to movement on the megathrust. If local faults in the upper plate of the Cascadia subduction zone produced vertical changes independent of the megathrust, then the historical and instrumental seismicity would be more like that of California. In the Nankai subduction zone, terrace uplift accompanied the A.D. 887, 1361, and either the 1707 or the 1854 earthquakes (Maemoku, 1988a; Sugiyama, 1994). Other subduction-zone events, however, such as the highly tsunamigenic 1605 earthquake, did not produce permanent uplift of terraces; this earthquake may have ruptured only the megathrust. Terraces on Middleton Island, Alaska, show a recurrence interval of 400 to 1300 years, which is longer than the recurrence interval estimated for earthquakes on the Aleutian subduction zone. The recognition that some subduction-zone earthquakes do not leave a permanent signature in uplifted terraces or downdropped marshes implies that recurrence intervals based solely on uplifted or subsided coastal features, as at Cascadia, are maximum intervals, because earthquakes that involved only the megathrust would not leave a permanent record of uplift or subsidence.

The upper plates of the Nankai, Sumatra, Philippine, Hikurangi, and Peru-Chile subduction zones contain throughgoing strike-slip faults, more or less parallel to the trench, that take up some of the convergence (cf. Chapter 8). In the Aleutian and Cascadia subduction zones, the upper plate is broken up into rotating blocks; rotation may be characteristic of the Nankai subduction zone as well. The upper plate in northern Honshu, Japan, is shortening internally on reverse faults parallel to the Japan trench. An internally deforming upper plate means that not all of the shortening is taken up along the megathrust. Movement on the megathrust and on faults in the upper plate could be manifested as a multiple seismic event and a complex series of aftershocks, or as complex patterns of vertical and horizontal coseismic deformation, as was observed for the 1964 Alaskan earthquake.

Internal deformation of the forearc is also indicated by a difference between the slip vectors of interplate earthquakes and the slip vector predicted by plate tectonics, called the slip-vector residual (McCaffrey, 1993). The largest interplate earthquakes are found on those subduction zones with a low slip-vector residual. Absence of forearc deformation indicates a stronger forearc with a greater capacity for elastic-strain accumulation on the megathrust.

Comparison of Oceanic Subduction Zones and Continental Collision Zones

Several oceanic subduction zones grade laterally into collision zones, allowing a comparison of earthquake behavior where oceanic convergence and collision take place side by side. Three examples are discussed: (1) the northern edge of the Philippine Sea plate, marked by the Nankai and Sagami subduction zones and the intervening Izu collision zone, (2) the Cascadia subduction zone, which comes on land at its southern end in northern California, and (3) the Aleutian subduction zone, which grades to the east into a narrow collision zone south of the St. Elias Range. The earthquake signature of aseismic ridges that collide with the Peru-Chile trench is also discussed.

The northern edge of the subducting Philippine Sea plate is oceanic except for the Izu Peninsula, where the Izv-Bonin island arc collides end-on with continental rocks of southern Honshu, Japan. The plate boundary extends up Suruga submarine canyon, beneath Mt. Fuji north of the Izu Peninsula, and down Sagami submarine canyon to a triple junction involving the Philippine Sea, Pacific, and North American plates. The oceanic subduction zones west and east of the Izu Peninsula are marked by great earthquakes, but these earthquakes stop short of the Izu Peninsula itself. Izu has its own independent history; earthquakes tend to be smaller there and to follow a more complex rupture pattern. Several faults in Izu are active, indicating internal deformation of the peninsula (Fig. 8-9), but the sum of slip rates on individual faults in the collision zone is much smaller than the slip rate on the Nankai subduction zone. Several faults have been found offshore, with the most important being the Zenisu ridge. The plate boundary may be jumping from its present position north of Izu to the sea floor south of Izu at Zenisu ridge. In addition, part of the deformation is taken up by rapid uplift of the Japan Alps.

The source of the 1854 earthquake ruptured as far east as Suruga Bay, producing displacement on a small fault in the Fuji River bed. However, the 1944 earthquake did not rupture the Suruga Bay segment, raising fears that the Suruga Bay segment is in a seismic gap, known in Japan as the Tokai Preparation Zone. But the Suruga segment is transitional in crustal structure between oceanic subduction to the west and continental collision to the east. The seismotectonic behavior may also be transitional between oceanic and continental collision, and thus the Suruga segment may be no more a seismic gap than is the Izu collision zone farther east.

The Cascadia subduction zone comes ashore at its southern end, the triple junction between the Gorda, Pacific, and North American plates. Unlike the Juan de Fuca plate farther north, the Gorda plate is deforming internally. The seismicity of this part of the subduction zone is high, with frequent earthquakes of $M \geq 7$ at or near the plate boundary, whereas the seismicity of the rest of Cascadia farther north is low, with the last great earthquake 300 years ago. It is likely that this southernmost part of the Cascadia subduction zone has an earthquake history of relatively frequent, moderate-sized earthquakes, whereas the rest of Cascadia has relatively few, but much larger earthquakes. The 300-year event recorded in Washington and Oregon is also recorded at Humboldt Bay at the southern end of the subduction zone, however, leading to one hypothesis that the last great Cascadia earthquake ruptured all the way to the triple junction. An alternative hypothesis would be that the 300-year event ruptured only the oceanic part of Cascadia and triggered smaller events at the southern end of the zone, as was the case for Izu Peninsula. These events could be separated by several years, not long enough to distinguish by radiocarbon dating.

The 1964 Alaskan earthquake ruptured to the eastern end of that part of the Aleutian subduction zone in which only oceanic crust is being subducted. Farther east, the Pacific-North American plate boundary is a transition zone between oceanic subduction and transform faulting. Reverse-fault earthquakes occur in the transition zone, including the 1979 Mt. St. Elias earthquake of $M = 7.2$. The Yakutat block within the transition zone is deforming internally along zones of reverse faults. Part of the convergence between the Yakutat block and North America may be accommodated by rapid uplift of the St. Elias range, which contains some of the highest mountains in North America.

All these examples suggest a different seismotectonic behavior for continental collision zones than for oceanic subduction zones. In these examples, the upper plate is deforming internally, producing moderate-sized earthquakes of M7. At the southern end of Cascadia, the lower plate is also deforming internally. Earthquake size is smaller, and recurrence intervals are shorter in collision zones. A sum of slip rates across individual faults and folds in a collision zone may show a slip-rate deficit in comparison with slip rates on oceanic subduction zones; this slip-rate deficit is best illustrated in the Izu collision zone. An exception is the Himalayan continental collision zone, where great earthquakes of $M > 8$ are recorded. In that case, much of the slip is accommodated on faults farther north in central Asia and by uplift within the High Himalaya.

Finally, we compare the earthquake signatures of oceanic crust and of aseismic ridges of the Nazca plate subducting beneath South America. There have been no known great earthquakes in more than 400 years in northern Peru and southern Ecuador opposite the zone of collision with the Carnegie aseismic ridge and two major fracture zones to the south (Fig. 11-43). However, the fit between the zone lacking great earthquakes and the region of shallow oceanic crust is relatively poor. The southern end of the zone lacking great earthquakes is adjacent to oceanic crust more than 5 km deep (Figs. 11-40 and 11-41), and shallow oceanic crust extends north into a zone with great earthquakes (Fig. 11-43). The collision zone of the Nazca and Juan Fernandez aseismic ridges is characterized by great earthquakes like adjacent, more typical oceanic crust. The great 1960 Chilean earthquake ruptured south to the Chile rise at 46°S.

Segments Based on Geology and Segments Based on Earthquakes

The upper plate of the Nankai subduction zone may be divided into five domains (A, B, C, D, E), each of which contains a forearc basin and is bounded by faults and folds striking at a high angle to the subduction zone. Transverse structures between domains B and C controlled the rupture zones of the 1944 and 1946 earthquakes and two earthquakes in 1854, but the 1707 earthquake ruptured across this transverse structure. The epicenters of the 1944 and 1946 earthquakes were close to each other on opposite sides of the domain boundary. The boundaries between A and B and between C and D resemble the boundary between B and C, but no subduction-zone earthquake has been shown to terminate against either of these boundaries. The last three historical subduction-zone earthquakes in the western part of the Nankai subduction zone ruptured across the western boundary of domain A into domain Z in the northern part of the Ryukyu subduction zone. The 1946 earthquake terminated against the eastern boundary of domain D, but the 1854 and 1707 earthquakes ruptured across this boundary into domain E.

In the Aleutians, the 1986 earthquake terminated near an oceanic fracture zone and a transverse structure marking the boundary between two rotating blocks in the upper plate. However, the great 1957 and 1965 earthquakes were not constrained by any obvious transverse boundary, and rupture zones of other great earthquakes crossed rotating-block boundaries or fracture zones similar to those that constrained the 1986 earthquake.

In summary, geological segment boundaries are earthquake rupture boundaries in some cases, but there are many exceptions.

Are Subduction-Zone Earthquakes Characteristic?

An earthquake is characteristic if approximately the same segment of a fault ruptures as in previous earthquakes, and if slip distribution is also similar to that of previous events. To address the question of subduction-zone earthquakes being characteristic, we need to document several earthquakes on the same section of the megathrust. The Hikurangi, Cascadia, and eastern Aleutian subduction zones have evidence of prehistoric earthquakes based on uplifted terraces or downdropped marshes extending back over several thousand years, but, as discussed previously, terraces and marshes may respond to secondary structures, and some subduction-zone earthquakes may not be recorded by terrace uplift or marsh subsidence.

The Nankai subduction zone has a history extending over 13 centuries, and the Northeast Japan, Mexico-Central America, and Peru-Chile subduction zones have histories extending over 400 years. Hence, one must depend especially on these subduction zones to address the question of characteristic earthquakes based on historical data.

The Nankai subduction zone has earthquakes repeating over periods of 100–250 years, and the boundary between domains B and C has persisted over several earthquake cycles. But the 1946 earthquake did not rupture as far northeast as did the 1854 event, and the 1707, 1605, and 1498 earthquakes ruptured across the B-C domain boundary. Nankai earthquakes show some tendency to be characteristic, but at least as many instances of noncharacteristic behavior occur as do instances of characteristic behavior. Recurrence intervals show variability from less than 100 years to more than 250 years.

During the 1960 Chile earthquake, the subduction zone ruptured farther north than it did during the 1835 event, and the 1985 rupture zone was not the same zone as the 1837 rupture. But the 1960 and 1985 pair of earthquakes involved rupture of the same portion of the subduction zone as ruptured during the 1835 and 1837 earthquakes, so we can say that, taken together, the two earthquake pairs were characteristic through two episodes (Thatcher, 1990). Other earthquakes on the Aleutian, Mexico-Central American, and Peru-Chile subduction zones do not seem to be characteristic. Twentieth-century earthquake segments off Peru are about 150 km long, but an eight teenth-century earthquake involved a part of the subduction zone that may have been four times as long as the twentieth-century events.

Thatcher (1990) concluded that the tendency for earthquakes to rupture the same segment of a subduction zone is relatively weak, and recurrence intervals are highly variable.

SUMMARY

Most of the seismic energy released during the twentieth century was generated by subduction-zone earthquakes. In fact, more than one-third was generated by the 1960 Chile and 1964 Gulf of Alaska earthquakes. The maximum size of earthquakes generated on a subduction zone is influenced by age and temperature of the subducting oceanic crust and by rate of convergence. McCaffrey (1993) has shown an even stronger correlation between maximum earthquake size and internal deformation of the forearc and of subducting oceanic lithosphere; the more internal deformation, the smaller the maximum earthquake size. The Marianas subduction zone is characterized by old lithosphere and slow convergence, meaning that the megathrust is poorly coupled, and relatively little of the convergence is accommodated seismically. The southern part of the Peru-Chile subduction zone is characterized by young, buoyant crust and rapid convergence, resulting in strong coupling and great earthquakes. The Nankai, eastern Aleutian, and Makran subduction zones contain broad accretionary prisms and are characterized by great earthquakes, yet great earthquakes also characterize subduction zones with narrow prisms, such as northeast Japan, the Kurils, and the northern two-thirds of the Peru-Chile subduction zone.

Permanently uplifted Holocene terraces and downdropped marshes may be, for the most part, controlled by upper-plate structures that are probably secondary to displacement on the megathrust. Dating of these terraces and marshes allows the identification of subduction-zone earthquakes, but not all subduction-zone earthquakes trigger secondary forearc structures that uplift or downdrop coastlines. Rupture boundaries of great earthquakes have a geologic signature in some cases but not in others, and successive earthquakes commonly do not rupture the same segment. Boundaries between oceanic subduction zones and continental collision zones may be the most persistent earthquake segment boundaries. In addition, collision zones are characterized by smaller earthquakes and by partitioning of shortening among smaller faults distributed across a broad zone.

Knowledge of the geologic characteristics of subduction zones is still too rudimentary to allow successful forecasting of the size and frequency of future earthquakes. The seismic gap theory, which predicts that a segment of a subduction zone that has not moved in the longest time is the most likely to rupture in the future, has had some success in forecasting earthquakes (McCann et al., 1979), but the statistical significance of seismic-gap forecasting has been questioned by Kagan and Jackson (1991). Much needs to be learned about these great structural boundaries.

Suggestions for Further Reading

Byrne, D. E., Davis, D. M., and Sykes, L. R. 1988. Loci and maximum size of thrust earthquakes and the mechanics of the shallow region of subduction zones: Tectonics 7:833-857.

Dmowska, R., and Ekström, G. 1994. Shallow Subduction Zones: Seismicity, mechanics and seismic potential: New York, Springer-Verlag, 2:240 p.

Hilde, T. W. C., and Uyeda, eds. 1983. Geodynamics of the Western Pacific-Indonesian Region: Am. Geophys. Union Geodynamics Series 11, 457 p.

Monger, J. W. H., and Francheteau, J., eds. 1987. Circum-Pacific orogenic belts and evolution of the Pacific Ocean basin: Am. Geophys. Union Geodynamics Series 18, 165 p.

Talwani, M., and Pitman, W. C., III, eds. 1977. Island arcs, deep sea trenches and back arc basins: Am. Geophys. Union Maurice Ewing Series 1, 470 p.

Scholz, C. H., and Campos, J. 1995. On the mechanism of seismic decoupling and back arc spreading at subduction zones: Jour. Geophys. Res. 100:22, 103-22, 115.

Watkins, J. S., and Drake, C. L., eds. 1982. Studies in continential margin geology: convergent margins: American Assoc. Petroleum Geologists Memoir 34: 307-533.

Déodat Gratet de Dolomieu
(1750–1801)

France

The Aspromonte massif of Calabria forms the toe of the boot of Italy. It rises east of Sicily and the Straits of Messina, and it is flanked by a broad plain sloping northwest to the Tyrrhenian Sea. The region was settled by Greeks more than 2000 years ago, and in the last half of the eighteenth century, it was relatively prosperous, a countryside of small villages, olive groves, vineyards, and forests.

On February 5, 1783, this peaceful region was convulsed by a violent earthquake in which tens of thousands died, and entire villages at the foot of Aspromonte vanished. The following day, a large aftershock produced a tsunami that did great damage to the northeast tip of Sicily, and for the next seven weeks, additional aftershocks added to the destruction. The earthquake was the greatest catastrophe in the history of southern Calabria.

There happened to be in Italy at this time a young French nobleman, Déodat de Dolomieu, newly arrived from the Island of Malta. His parents had chosen a military career for him, arranging for his admission at the age of two into the Order of Malta. But he was poorly suited for soldiering; he was mainly interested in natural phenomena. He made the acquaintance of the Duc de Rochefoucault, a member of the Royal Academy of Sciences of Paris, which placed him in the company of the most brilliant earth scientists of his age. Among them were the geologist Jean Étienne Guettard, the first person to make a geologic map, and the naturalist G. Leclerc de Buffon, whose *Époques de la Nature* had appeared five years earlier. Dolomieu himself was admitted to the academy at the age of 25.

The real excitement in Paris in those days was the discovery by Guettard and by another geologist, Nicholas Desmarest, that certain rocks in central France had been formed by volcanoes similar to those active in Italy. The study of volcanoes excited Dolomieu, and prior to the Calabrian earthquake, he had visited volcanoes in Sicily, the Lipari Islands, and the areas Guettard and Desmarest had studied in central France. This was the age of scientific travel, of field work to learn how landscapes was formed. Nowhere were the effects of nature more dramatically displayed than the active volcanoes of Italy.

Dolomieu's arrival in Italy had terminated a career with the Order of Malta that had not been a happy one. He had come first to Malta in 1768, but he killed another member of the Order in a duel, which led to his imprisonment for several months. On his second stay, he rose in the ranks to second in command to the Grand Master. However, his abrasive personality led to serious tensions, and he resigned his position and moved to Italy in 1783.

Dolomieu was fascinated by the earthquake, just as he was fascinated by volcanoes. Was there some relationship between the earthquake and the volcanoes to the north and south? With a field assistant, de Godechart, he went to Calabria a year after the earthquake, in early 1784. He described the geology: a basement of granite in the Aspromonte massif, sedimentary layers with fossils in the plains to the northwest, but, surprisingly, no volcanic rocks.

The most remarkable features were massive landslides that took great tracts of ground downhill, damming streams and creating small lakes. These landslides caused much of the loss of life. In his report, he described the landslides in detail but he also wrote the following:

> "Along almost the entire length of the range, the rocks that were standing against the granite at the base of Mts. Caulone, Esope, Sagra, and Aspramonte [sic] slipped on that rigid core, whose slope is steep, and descended a little lower. Thus was created a fissure several feet wide, over a length of 9 to 10 miles between the rigid rock and the sandy rock, and this fissure prevails, almost without discontinuity, from San Giorgio, following the contour of the mountain base, as far as behind Santa Cristina[1]."

Had Dolomieu discovered surface faulting accompanying the earthquake? The fissure is located

[1]"Il s'ensuivit, que dans presque toute la longueur de la chaine, les terreins, qui étoient appuyés contre le granit de la base des monts Caulone, Esope, Sagra, & Aspramonte, glisserent sur ce noyeau solide, dont la pente est rapide, & descendirent un peu plus bas. Il s'établit alors une fente de plusieurs pieds de large, sur une longueur de 9. a 10. milles, entre le solide & le terrein sabloneux; & cette fente regne, presque sans discontinuité, depuis saint George en suivant le contours des bases, jusque derriere sainte Cristine."

by Dolomieu in the same place as normal faults at the foot of Aspromonte, in which Pleistocene marine sediments are in fault contact with granite. The meizoseismal zones of the February 5 earthquake and major aftershocks also follow this contact. Dolomieu could have been the first to map a surface rupture accompanying an earthquake, even though he and his colleagues at the Academy of Sciences had no concept of faulting; that would come later. He recognized the landslides as obviously resulting from the earthquake, and he proposed that the fissures he described were related to landsliding. Today, one can map a fault that could have produced the 1783 earthquake, but it has not been demonstrated that the fault had surface rupture in 1783.

Dolomieu's contribution, then, was that one could learn about an earthquake by field work. He described first the geologic setting of the earthquake, then the changes in the landscape that had been produced by the earthquake. Descriptions of the 1906 San Francisco earthquake and the 1983 Borah Peak, Idaho, earthquake would follow Dolomieu's example.

Dolomieu continued his field work and his publications, including trips to the Alps, visiting the range that now bears his name, the Dolomites. He would later be caught up in the fervor of the Revolution, becoming dismayed at the assassination or execution of many of his friends from the Academy, including La Rochefoucault, during the Reign of Terror. Finally, on returning from service with Napoleon in Egypt in 1799, his ship took refuge in Calabria in a storm. There he was jailed for 21 months in an incident related to his previous service on Malta. Although he was freed by intervention of his friends in Paris, his health was broken, and he died in 1801.

Suggestions for Further Reading

Cotecchia, V., Travaglini, G., and Melidoro, G. 1969. I movimenti franosi e gli sconvolgimenti della rete idrografica prodotti in Calabria del terremoto del 1783: Geologia Applicata e Idrogeologia 4:1–24 (English abs.)

Debelmas, J. 1988. A propos d'un bicentenaire: Hommage à Déodat de Dolomieu, "père" des Dolomites: La Montagne et Alpinisme 3:72.

Dolomieu, D. de, 1784. Memoire sur les tremblemens de terre de la Calabre pendant l'année 1783: Rome, Antoine Fulgoni, 70 p.

Morel, L. 1953. Un géologue dauphinois préromantique: Déodat Dolomieu (1750–1801): Soc. Dauph. de Ethnologie et Archéologie Bull., no. 224, 225, 226.

12

Secondary Effects

Secondary effects is a term used for nontectonic surface processes that are directly related to earthquake shaking or to tsunamis. Secondary effects are commonly the most spectacular expressions of an earthquake (Fig. 12-1); they cause much of the loss of life and property during earthquakes and are important for these reasons alone. However, it is not the intent of this book to assess the earthquake hazard due to secondary effects by themselves. Instead, we consider the geologic signatures of the secondary effects of earthquakes and their preservation in the geologic record as paleoseismic evidence for strong motion accompanying prehistoric earthquakes. The analysis of secondary effects is of particular value in considering the paleoseismology of continental shields and passive continental margins, where surface ruptures are commonly not found.

Certain geologic formations, such as sand dikes, can be said to serve as natural strong-motion accelerographs, recording the effects of high-intensity shaking. If a deposit can be clearly attributed to an earthquake and not to some other cause, this information, together with geologic evidence for tectonic displacements, can be used to work out earthquake paleointensities and recurrence intervals. However, as will be seen, many of the secondary effects observed during well-described historical earthquakes can also form by processes other than earthquakes. Prior to attributing a feature to an earthquake, the investigator must provide evidence that precludes other origins, one of the most challenging problems in paleoseismology.

Landslides are one of the most common geomorphic expressions of earthquakes. These are of several types, including rock falls, block slides, and slides of surficial materials. *Lateral spreads,* which are most commonly produced by liquefaction, occur on slopes as gentle as 0.1°. Shaking of offshore sediments dislodged by an earthquake may generate a *turbidite,* a bottom flow of dense, sediment-laden water that may spread for thousands of square kilometers. Other expressions of liquefaction are *sandblows* and *clastic dikes* and *sills,* which are the result of ejection of water and sediment during an earthquake.

Tsunamis (cf. Chapter 4) are high-amplitude, long-period waves that may rush much farther inland than normal storm waves or high tides and deposit sediment that is derived from the direction of the waves. Although not a direct effect of earthquake shaking, *tsunami deposits* are discussed briefly here. Just as in deducing an origin by earthquake-shaking, the investigator must search for evidence diagnostic of a tsunami origin rather than a nonseismic origin such as a very large storm.

LANDSLIDE DEPOSITS

Earthquakes are a major cause of landslides. Keefer (1984) posed the following questions that must be addressed in relating prehistoric landslides to earthquakes: (1) What types of landslides are caused by earthquakes? (2) What geological materials are most susceptible to earthquake-induced landslides? (3) Do earthquakes reactivate landslides that have nonseismic origins? (4) Can the number and distribution of landslides provide information about earthquake magnitude and intensities? To address these questions, Keefer (1984) studied 40 historical earthquakes in a variety of geologic, climatic, and seismic settings with magnitudes ranging from 5.2 to 9.5.

Table 12-1 shows Keefer's classification of landslides, based upon principles and terminology of Varnes (1978). This classification is based primarily on the earth materials comprising the landslide and on the character of movement, and secondarily on the degree of internal disruption and on water content. In this classification, *rock* means firm, intact bedrock, and *soil,* used in the engineering sense, means a loose, unconsolidated, or poorly cemented aggregate of particles that may or may not contain organic materials.

Figure 12–1. Landslide accompanying an earthquake of M_L7.1 on 11 May 1985 on the island of New Britain, Papua New Guinea. View is down the Bairaman Valley. Backscarp, 200 m high, is formed of weathered limestone. Dark streaks on flow are topsoil. Photo taken on 12 May by P. Lowenstein, Geological Survey of Papua New Guinea. From King et al. (1989).

Earthquakes with $M < 5.5$ generate tens of landslides, whereas earthquakes with $M > 8$ generate thousands (Keefer, 1984). The number of reactivated landslides is small compared to the total number of landslides. In part, this may be due to the lack of recognized criteria for a pre-earthquake origin, but in general, most earthquake-induced landslides occur in materials not previously involved in landslides.

Keefer concluded that the most abundant landslides accompanying the earthquakes he studied were rock falls, disrupted soil slides, and rock slides. These are especially susceptible to seismic shaking, and geologic environments in which these landslides are found are very common in seismogenic regions. Less abundant are soil lateral spreads, soil slumps, soil block slides, and soil avalanches. Still less abundant are soil falls, rapid soil flows, and rock slumps. Relatively uncommonly associated with earthquakes are slow earth flows, rock block slides, and rock avalanches.

Keefer's sample suggested that subaqueous landslides are also relatively uncommon, but he recognized that this may well be due to difficulty of observations on the sea floor. A submarine slump covering 15–20 km^2 in the eastern Gulf of Corinth, Greece, was triggered by the Corinth earthquakes of early 1981, described in Chapter 9 (Perissoratis et al., 1984). A landslide on a 0.25° slope off the northern California coast was a result of an M6.5–7.2 earthquake of November 8, 1980 (Field et al., 1982). The large number of subaqueous landslides found in the forearc of the Japan trench (Fig. 11-21) and in the Cascadia margin off the northwestern United States (Fig. 12-2) suggests that subaqueous landslides are relatively common, and, because of their proximity to subduction zones, they may be triggered by earthquakes.

The relative abundances of different types of landslides is affected by earthquake magnitude. Keefer found that earthquakes with $M < 6.5$ triggered pro-

Table 12–1

Characteristics of Earthquake Induced Landslides.

Name	Type of Movement	Internal Disruption	Velocity	Depth
		LANDSLIDES IN ROCK		
		Disrupted Slides and Falls		
Rock falls	Bounding, rolling, free fall	High or very high	Extremely rapid	Shallow
Rock slides	Translational sliding on basal shear surface	High	Rapid to extremely rapid	Shallow
Rock avalanches	Complex, sliding and/or flow	Very high	Extremely rapid	Deep

Table 12–1
Continued

Name	Type of Movement	Internal Disruption	Velocity	Depth
		Coherent Slides		
Rock slumps	Sliding on basal shear surface, component of headward rotation	Slight or moderate	Slow to rapid	Deep
Rock block slides	Transitional sliding on basal shear surface	Slight or moderate	Slow to rapid	Deep
		LANDSLIDES IN SOIL **Disrupted Slides and Falls**		
Soil falls	Bounding, rolling, free fall	High or very high	Extremely rapid	Shallow
Disrupted soil slides	Transitional sliding on basal shear surface or zone of weakened sensitive clay	High	Moderate to rapid	Shallow
Soil avalanches	Transitional sliding with subsidiary flow	Very high	Very rapid to extremely rapid	Shallow
		Coherent Slides		
Soil slumps	Sliding on basal shear surface, component of headward rotation	Slight or moderate	Slow to rapid	Deep
Soil block slides	Translational sliding on basal shear surface	Slight or moderate	Slow to very rapid	Deep
Slow earth flows	Translational sliding on basal shear surface with occasionally minor internal flow surges	Slight	Very slow to moderate with very rapid surges	Shallow, occasionally deep
		Lateral Spreads and Flows		
Soil lateral spreads	Translational on basal zone of liquefied gravel, sand, or silt or weakened, sensitive clay	Generally moderate, occasionally slight, occasionally high	Very rapid	Variable
Rapid soil flows	Flow	Very high	Very rapid to extremely rapid	Shallow
Subaqueous landslide	Complex, generally involving lateral spreading and/or flow; occasionally involving slumping and/or block sliding	Generally high or very high; occasionally moderate or slight	Generally rapid to extremely rapid; occasionally slight to moderate	Variable

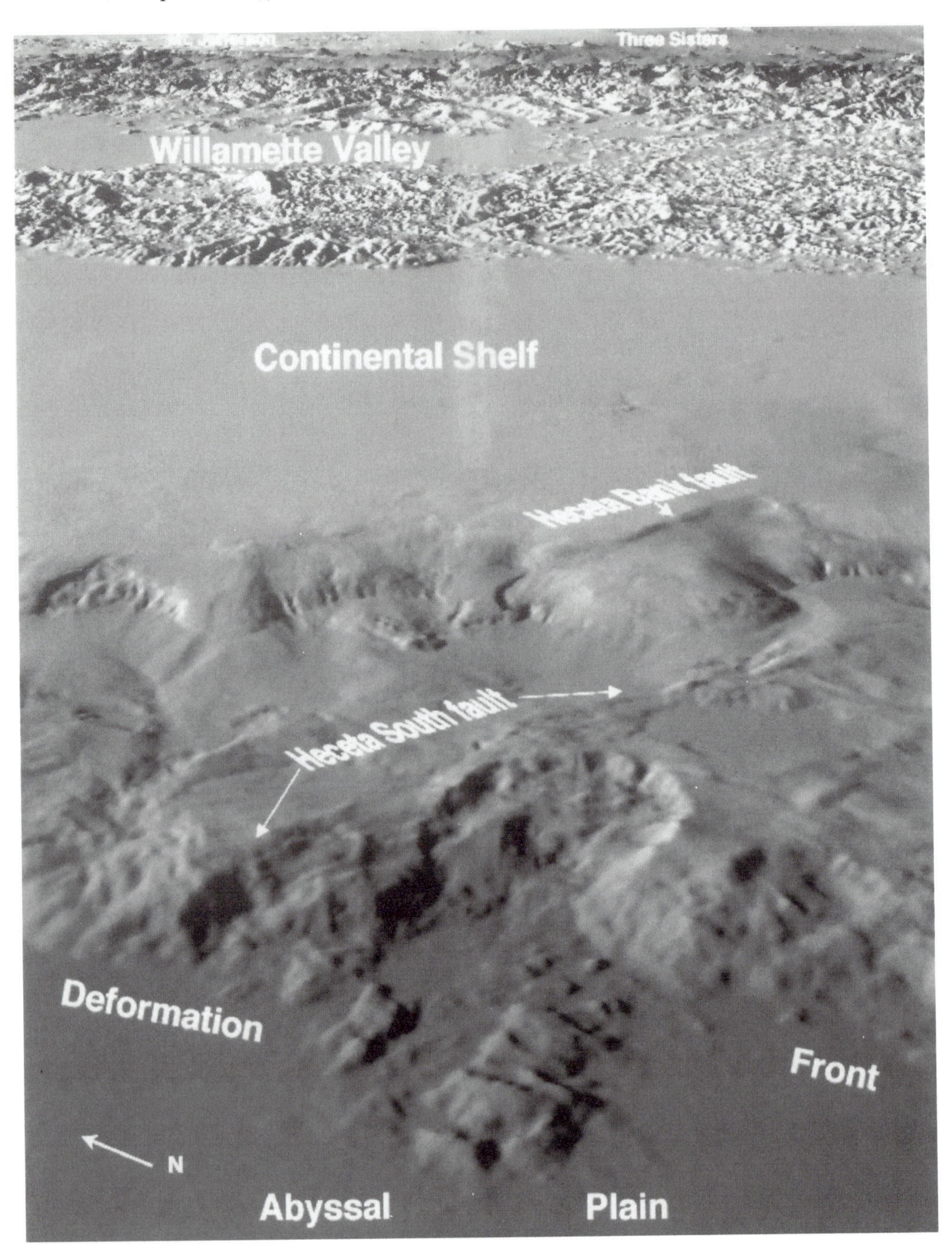

Figure 12–2. Large landslide at the base of the continental slope west of Florence, Oregon. Slide is 8 km across; debris has been transported across the deformation front onto the abyssal plain. The active Heceta South fault marks part of the northern side of the slide. Image created at Oregon State University from SeaBeam bathymetric maps of the National Oceanic and Atmospheric Agency and digitized land topography from the U.S. Geological Survey.

portionally more rock falls, rock slides, and soil falls. The smallest earthquake to produce landslides had M_L4; earthquakes of this size tended to produce rock falls, rock slides, soil falls, and disrupted soil slides. For soil slumps and soil block slides, the minimum M_L is 4.5; for rock slumps, rock block slides, slow earth flows, lateral spreads, soil liquefaction, and subaqueous landslides, the minimum M_L is 5.0; for rock avalanches, the minimum M_S is 6.0; for soil avalanches, the minimum M_S is 6.5.

However, there is controversy regarding the triggering of some very large landslides by small earthquakes of $M < 3.5$, including the Gros Ventre, Wyoming, rock avalanche of 1925 (Voight, 1978) and the Hope, British Columbia, landslide of 1965 (Mathews and McTaggart, 1978). At Gros Ventre, the controversy arises because of the 18- to 20-hour delay between the earthquake and the landslide, and at Hope, the seismic event may have been the result of initial rupture of the landslide shear surface, an effect rather than the cause.

Landslide-related phenomena are used in conjunction with other criteria to define levels of Modified Mercalli intensity (MMI, cf. Chapter 4; Wood and Neumann, 1931; Richter, 1958). To determine the lowest intensities for different landslide types, Keefer compared landslide type with its position within a MMI isoseismal contour based on all criteria. The lowest intensity for landslides is MMI IV. For disrupted slides and falls, the predominant minimum intensity is MMI VI. For coherent slides, lateral spreads, and flows, the predominant minimum intensity is MMI VII, with the absolute minimum MMI V. These minimum intensities may be one to five levels lower than indicated in descriptions accompanying the standard MMI scale. The formation of many landslide-dammed lakes during an earthquake requires shaking intensities of MMI IX or X (Adams, 1981).

A brief description of some of the more important types of earthquake-induced landslides from Table 12-1 follows. For a more complete discussion, see Keefer (1984).

Rock Falls

Individual boulders or disrupted rock masses bound, roll, or free-fall down slopes. These are the most abundant earthquake-induced landslides, and they produce great losses, particularly in heavily populated mountainous regions. Most rock falls studied by Keefer occurred in closely jointed or weakly cemented materials including pumice, tuff, shale, siltstone, sandstone, and conglomerate. Rock falls originate only on slopes steeper than 40°.

In addition to being abundant, rock falls are among the most spectacular manifestations of earthquakes. John Muir (1912), who was in Yosemite Valley, California, during the 1872 Owens Valley earthquake, wrote as follows:

> At half-past two o'clock of a moonlit morning in March, I was awakened by a tremendous earthquake, and though I had never before enjoyed a storm of this sort, the strange thrilling motion could not be mistaken, and I ran out of my cabin, both glad and frightened, shouting, "A noble earthquake! A noble earthquake!" feeling sure I was going to learn something. The shocks were so violent and varied, and succeeding one another so closely, that I had to balance myself carefully in walking as if on the deck of a ship among waves, and it seemed impossible that the high cliffs of the Valley could escape being shattered. In particular, I feared that the sheer-fronted Sentinel Rock, towering above my cabin, would be shaken down, and I took shelter back of a large yellow pine, hoping that it might protect me from at least the smaller outbounding boulders. For a minute or two the shocks became more and more violent—flashing horizontal thrusts mixed with a few twists and battering, explosive, upheaving jolts,—as if Nature were wrecking her Yosemite temple, and getting ready to build a still better one.
>
> I was now convinced before a single boulder had fallen that earthquakes were the talus-makers and positive proof soon came. It was a calm moonlight night, and no sound was heard for the first minute or so, save low, muffled, underground, bubbling rumblings, and the whispering and rustling of the agitated trees, as if Nature were holding her breath. Then, suddenly, out of the strange silence and strange motion there came a tremendous roar. The Eagle Rock on the south wall, about a half a mile up the Valley, gave way and I saw it falling in thousands of the great boulders I had so long been studying, pouring to the Valley floor in a free curve luminous from friction, making a terribly sublime spectacle—an arc of glowing, passionate fire, fifteen hundred feet span, as true in form and as serene in beauty as a rainbow in the midst of the stupendous, roaring rock-storm. The sound was so tremendously deep and broad and earnest, the whole earth like a living creature seemed to have at last found a voice and to be calling to her sis-

ter planets. In trying to tell something of the size of this awful sound it seems to me that if all the thunder of all the storms I had ever heard were condensed into one roar it would not equal this rock-roar at the birth of a mountain talus.

Rock Slides

Rock slides comprise material that is broken up into rock fragments and blocks that slide on surfaces where joints or bedding planes dip out of slopes. They are found in hillside channels on slopes steeper than 35°. The Bairaman landslide in Papua New Guinea (Fig. 12-1) began as a rock slide (King et al., 1989).

Rock Avalanches

These are landslides that break up into streams of rock fragments that may travel several kilometers on slopes of a few degrees at velocities of hundreds of kilometers per hour. All rock avalanches reported by Keefer had volumes of at least $0.5 \times 10^6 m^3$. One of the largest rock avalanches was triggered by an M_W 7.9 earthquake in Peru in 1970 (Plafker et al., 1971). A slab of rock and ice broke off a near-vertical cliff on Nevado Huascarán, fell 1000 m, disintegrated, and slid across a glacier, then overtopped morainal ridges downslope from the glacier and became airborne. After returning to earth, the mass separated into debris streams that entrained surface water and swept down the Río Shacsha valley at velocities of 280 km/hr. At a distance of 11 km from the source, some of this debris overtopped a low ridge and buried the city of Yungay, while at about the same time, the remaining debris struck the city of Ranrahirca, less than 4 minutes after initial slope failure (Fig. 12-3).

Figure 12–3. Rock avalanche triggered by a M_W7.9 earthquake in Peru in 1970. Avalanche began when a slab of rock and ice broke off the side of Nevado Huascarán, Peru's highest mountain, then continued as an avalanche, one lobe of which destroyed the city of Yungay and another which destroyed the city of Ranrahirca. From Plafker et al. (1971).

The 1959 Hebgen Lake, Montana, earthquake generated a rock avalanche containing $28 \times 10^6 m^3$ of broken rock that fell 400 m down the south wall of the Madison River canyon, accompanied by an air blast. Two-thirds of the slide crossed the Madison River, forming a lake, and some material was carried as much as 130 m up the far side (Hadley, 1964).

The Bairaman slide (Fig. 12-1) began as a rock slide, but rapidly was transformed into a rock avalanche with a volume of $180 \times 10^6 m^3$; the avalanche filled the Bairaman valley to a maximum depth of 200 m and created a lake behind the slide.

Kinetic energy for long-range transport of debris is provided by the initial fall from steep slopes. Nearly all of the avalanches reported by Keefer started on slopes undercut by active fluvial erosion or by Holocene or late Pleistocene glaciation. Most slopes producing rock avalanches were intensely fractured, with fractures spaced a few centimeters to tens of centimeters apart. Also common in rock-avalanche sources are planes of weakness dipping out of the slope, rock weathering, weak cementation of the rock, and evidence of previous landsliding.

Soil Block Slides

Soil block slides are deep-seated, and their basal shear surfaces are planar or gently curved. They may have grabens at their headwalls and pressure ridges along their toes, and some have internal fissures or grabens. The Government Hill slide in Anchorage, Alaska, which was formed during the 1964 earthquake, was

4 hectares in area, 27 m deep, and was caused by the failure of a river bluff 25 m high. The slide material was glacial outwash underlain by the Bootlegger Cove clay; the slide surface was in water-saturated clay (Hansen, 1966). Bluffs of the Mississippi River 16 to more than 60 m high failed during the New Madrid earthquakes of 1811–1812; the slides moved on layers of saturated sand and gravel underlain by impermeable clay (Jibson and Keefer, 1988).

Slow Earth Flows

These are tongue-shaped or teardrop-shaped masses of clay- to silt-size material bounded by lateral shear surfaces as well as basal shear surfaces. These move by displacement along these shear surfaces, with only minor internal deformation. The minimum slope reported for slow earth flows was 10°. A very large earth flow was initiated at least five days after the 1959 Hebgen Lake, Montana, earthquake and continued moving for at least a month. This earth flow reactivated an older landslide (Hadley, 1964).

Soil Lateral Spreads

Lateral spreads occur on slopes that may be as low as 0.1° and are large slabs of nonliquefied material that move downslope atop a liquefied layer of large areal extent. These may be accompanied by grabens at the headwall and reverse shear zones at the toe (Obermeier, 1987). These are discussed in the following text under Liquefaction Features.

DATING EARTHQUAKE-TRIGGERED LANDSLIDES

Landslide deposits can be used to date prehistoric earthquakes if it can first be determined that the landslide was, indeed, caused by an earthquake. One way to do this would be to determine that many landslides in an area are of the same age, although very large storms could also result in many landslides of the same age. If a landslide dammed a stream, producing a lake, the landslide would be slightly older than the basal lake deposits. A landslide would be younger than the youngest materials involved in the slide. If the landslide exposed a previously buried surface, the age of exposure of the surface would date the landslide.

Rock falls and rock slides may be dated by lichenometry (Nikonov and Shebalina, 1979; Smirnova and Nikonov, 1990; Bull et al., 1994). A fresh rock surface exposed after tumbling downhill during a rock fall caused by an earthquake is colonized by lichens with a known growth rate. Lichens pass through a colonization phase and a great-growth phase, followed by an exponentially declining growth-rate phase. Bull et al. (1994) measured the long axis of the largest *Rhizocarpon* subgenus *Rhizocarpon* on a large number of blocks in talus fields in the Southern Alps of New Zealand and the Sierra Nevada of California. They then compared their lichen measurements with dates for historical earthquakes affecting those regions. In New Zealand, lichen growth peaks appear to correlate with historical earthquakes in 1968, 1929, and 1881, although the 1888 Marlborough earthquake on the Hope fault did not correlate with a lichen growth peak. In the Sierra Nevada, regional rock fall events may correlate with major earthquakes in 1906, 1890, 1872, and 1857.

Adams (1981) noted that 11 small lakes were formed by landslides accompanying the 1929 earthquake in northern South Island, New Zealand. Most of these landslides were rock avalanches or rock falls. Lake Chalice, in the South Island, is blocked by a landslide dam 100 m high. Twigs from trees that were drowned when the lake rose behind the landslide dam were dated by radiocarbon at about 2170 and 2160 years B.P., indicating that the landslide was slightly older.

Three or four of six rock avalanches in the southeastern Olympic Mountains of Washington State were formed from 1000 to 1300 years ago, based on ^{14}C dates of outer rings of trees drowned by lakes formed behind the landslide dams (Schuster et al., 1992; Fig. 12–4). No rock avalanches have been formed in this region during the last century, even though there have been large storms and an earthquake of M7.1 in 1949, leading Schuster et al. (1992) to infer that the large avalanches were earthquake-induced. Three large tree-covered landslides were found in Lake Washington, east of Seattle. Trees drowned by the slides were dated by radiocarbon and dendrochronology, suggesting that all three landslides were formed at the same time, about 1000 years ago (Jacoby et al., 1992). The ages of the rock avalanches in the Olympic Mountains and the landslides in Lake Washington are the same as the age of a tsunami in Puget Sound (Atwater and Moore, 1992) and are consistent with the age of uplift accompanying abrupt displacement on the Seattle fault (Bucknam et al., 1992). The age correlation between a tectonic event and secondary effects in several regions strengthens the interpretation that all were produced by an earthquake.

In the Cascadia subduction zone off central Oregon, a landslide 32 km^3 in area has broken away from the leading edge of the overriding North American plate and blocked a distributary channel on the abyssal plain. The slump was dated as 10–24 ka based

Figure 12–4. Lake Jefferson, Olympic Peninsula, Washington, dammed by a rock avalanche about 1100 radiocarbon years ago. Trees in lake were drowned by the rock avalanche, which forms the far shore of the lake, now partially covered by forest. Note large boulders at right. Photo by Robert L. Schuster, U.S. Geological Survey.

on a ^{14}C date from a core taken from one of the slump blocks and from the thickness of pelagic sediments overlying the slump based on known sediment accumulation rates elsewhere on the abyssal plain (Goldfinger et al., 1992). This landslide cannot definitely be attributed to an earthquake.

Relative ages can be assigned to landslides based on degree of landscape degradation. Crozier (1992) subdivided New Zealand landslides into age groups based on degree of definition of landslide features, degree of stream dissection, soil development on top of the slides, and the age of overlying tephra deposits. This could be calibrated by dating organic material within each slide. McCalpin and Rice (1987) developed an age classification of 1200 landslides in the Rocky Mountains based on geomorphic expression. Jibson and Keefer (1989) showed that old, coherent landslides and earthflows in the New Madrid seismic zone were formed at the same time, consistent with triggering by the 1811–1812 earthquakes in that region.

GEOTECHNICAL EVIDENCE FOR A LANDSLIDE ORIGIN

Keefer (1984) noted that most landslides generated by earthquakes were not reactivated older slides. Accordingly, it should be possible to evaluate a possible earthquake origin on the basis of the engineering properties of the earth materials comprising the slide and the slide surface (Seed et al., 1983).

The state of looseness or denseness of cohesionless sediments is measured in place with the Standard Penetration Test (SPT) blow-count method (American Society for Testing and Materials, 1978). A sampling tube is driven into the ground by dropping a 63.5-kg (140-lb.) weight from a height of 76 cm (30 inches). The penetration resistance is the number of blows required to drive the sampler 30.5 cm (1 foot). Less-dense sand tends to liquefy more readily. Split-spoon samples taken during SPT testing are heavily disturbed by the sampling process, but are used to determine grain size, plasticity, water content, and color.

Undisturbed piston cores are necessary to determine soil shear strengths, as described by Jibson (1994). Two methods are used. In the direct shear method, the strain rate is slow enough to permit full drainage. In the consolidated-undrained triaxial shear method (CUTX), pore pressures are measured to allow modeling of drained conditions. The slope-stability analysis considers static (aseismic) and dynamic (coseismic) conditions. The geotechnical properties of the rock layers with potential for sliding are compared with the actual slope under the worst-case conditions in which the water table is as high as could reasonably be expected. A computer program is used to search for the slopes most likely to fail under aseismic conditions.

To consider failure under coseismic conditions, the dynamic displacement analysis method is used (Newmark, 1965; Seed, 1979; Jibson and Keefer, 1993). In the Newmark method, a landslide is modeled as a rigid block that begins to move only when the acceleration required to overcome frictional resistance and initiate sliding on an inclined plane is exceeded. The acceleration-time history of shaking is considered. Soils are presumed to behave in an undrained manner during earthquakes, because pore pressures generated by the transient ground deformation cannot dissipate during shaking.

For an example of a study of the seismic origin of landslides based on geotechnical data, refer to Jibson and Keefer (1993).

STRATIGRAPHIC EVIDENCE FOR EARTHQUAKES

In 1969, A. Seilacher referred to a deposit directly related to earthquakes as a *seismite*. Seilacher used this term to describe a graded bed in a depositional sequence that provided a record of an earthquake, a "paleoseismogram," as he put it. Seismites consist of sedimentary interbeds in a sequence of lake deposits or deep-marine sediments. Other stratigraphic evidence consists of sediments that were disturbed by earthquake shaking after deposition. Landslide deposits discussed in the previous text could also be included in this category if they are part of a stratigraphic record. Finally, the stratigraphic record may contain evidence of liquefaction and sand blows in the form of dikes and sills of sediment ejected during earthquake shaking.

Seismites

On 18 November 1929, a great landslide near the Grand Banks, off the eastern coast of Canada, snapped trans-Atlantic submarine cables and produced a large earthquake. (The seismogram written by the earthquake indicated that the earthquake was generated by the landslide, not the other way around; cf. Chapter 4.) Heezen and Ewing (1952) showed that the cables were broken by a sediment-laden current (*turbidity current*), heavier than water, that moved like a snow avalanche by turbulent flow down a submarine fan at speeds up to 65 km/hr. The turbidity current must have been triggered by the landslide.

Heezen and Ewing's work had little impact on the study of earthquakes, but it had a dramatic effect on sedimentology, because turbidity currents provided an explanation for *graded bedding*, or the upward decrease in grain size of some marine deposits. When a turbidity current comes to rest, the sediment particles settle to the sea floor according to Stokes' law, which states that the settling velocity is related to the radius of the particle. Sequences with repeated graded bedding are known throughout the world, predominantly in deep-marine sediments but also in lake deposits.

In southern California, for example, sandstones deposited in deep water in the Ventura basin were attributed to turbidity-current deposition (cf. Natland and Kuenen, 1951). Commonly these sandstones contain foraminifera characteristic of shallow water, but the intervening pelagic and hemipelagic fine-grained strata contain deep-water foraminifera, evidence that the sandstones had originally been deposited in shallow water, then were later reactivated as turbidity currents and redeposited in deep water.

It was noted in the literature that individual turbidites could be emplaced once every few hundred years, but the question of the triggering mechanism was not addressed. Because the turbidites could be related to submarine canyons in many cases, a common view developed that sediments could accumulate in canyon heads until there was enough sediment to trigger a turbidity current, either by a great storm providing a large influx of sediment, or by an earthquake.

In 1990, Adams (1990; Fig. 12-5) published a study of deep-sea cores collected in the Cascadia deep-sea channel off the coast of Oregon and Washington (Griggs and Kulm, 1970). Adams showed that the lower main Cascadia channel received 13 turbidites since the deposition of the Mazama volcanic ash dated at 6845 ± 5 radiocarbon years B.P. This is an average recurrence interval of 590 years, similar to recurrence intervals based on coastal marshes that had undergone repeated coseismic subsidence (Atwater, 1987). Turbidites were found in deep-sea channels originating from major rivers in the southern Oregon Coast Range in addition to channels originating in the Columbia River, evidence that the triggering event was regional and not related to a local event in a single drainage, such as a great storm.

Kastens (1984) correlated debris flows and turbidites from the Calabrian ridge off the coast of Italy between several submarine basins, evidence of a regional rather than a local origin, and presumably indicative of an earthquake source.

An earthquake in 1663 in Quebec was accompanied by high levels of silting in streams for several months after the event, leading Doig (1990) to look for silt layers in cores from Lake Tadoussac, 100 km northeast of Quebec City. Doig found more than 20

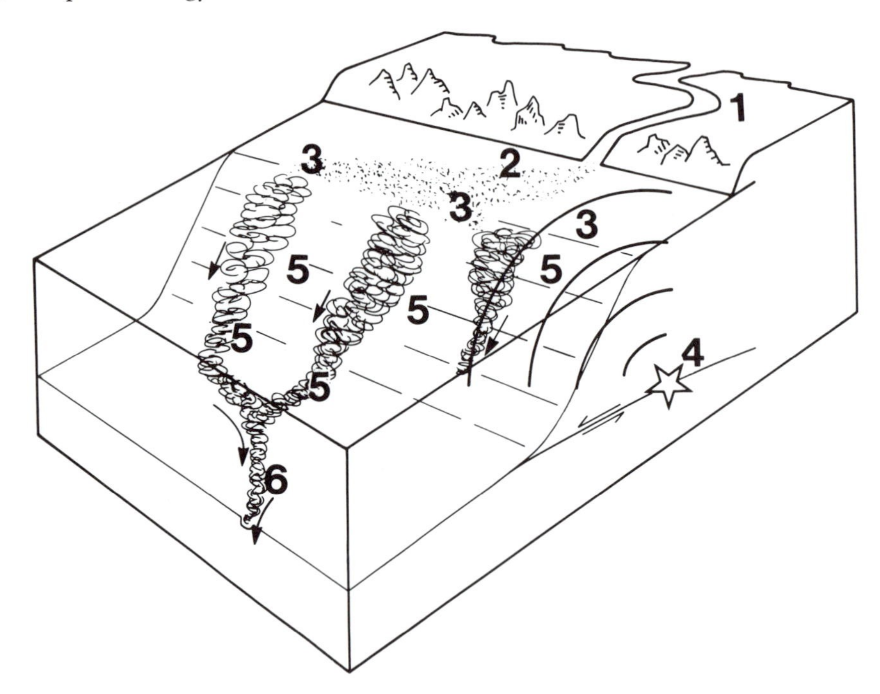

Figure 12–5. Generation of a turbidite off the mouth of the Columbia River (1). The river carries sediment to its mouth; the sediment drifts north along the continental shelf (2) and accumulates in the heads of submarine canyons (3). A subduction-zone earthquake (4) shakes the shelf and slope, causing sediments to liquefy and slump down the slope and form debris flows along submarine canyons (5). These debris flows become diluted and become turbidity currents that coalesce in the Cascadia Channel on the abyssal plain as one large turbidity current (6). From Adams (1990).

silt layers interbedded with highly organic sediments, and he correlated the 5 most recent silt layers with historical earthquakes, including the 1663 event.

Karlin and Abella (1992) found clay and silt turbidites with graded bedding in Lake Washington, near Seattle, with a high magnetic susceptibility in comparison with interbedded anoxic lake sediments. The ages of the turbidites were calibrated with a dated diatom horizon at A.D. 1800, a pollen horizon dating from the time of initiation of forest clear-cutting at A.D. 1880, and the age of opening of the Lake Washington Ship Canal at A.D. 1916. This enabled them to make a reservoir correction for radiocarbon dating of the turbidites (cf. Chapter 6). Using radiocarbon dating, they correlated a prominent silt turbidite with an earthquake about 1100 years ago dated by uplift of a Holocene terrace adjacent to the Seattle fault, which is the same age as radiocarbon dates from lakes dammed by rock avalanches in the Olympic Mountains and from tsunami deposits in Puget Sound. The size and distribution of the 1100-year silt suggested that there were at least three sediment sources that released into the lake almost simultaneously. Other event horizons were dated at 300–400 years, 1600–1700 years, 2200–2400 years, and 2800–3100 years B.P.; the last three dates correspond to dates of drowning of trees, presumably by multiple landslides. The 300–400-year event may correspond to a prominent event horizon recorded in marshes along the Washington coast that is thought to be the most recent Cascadia subduction-zone earthquake (cf. Chapter 11).

Liquefaction Features

Liquefaction is defined as "the act or process transforming any substance into a liquid." In cohesionless soils, this transformation from a solid to a liquefied state is a result of increased pore pressure which decreases effective stress during an earthquake (Fig. 2-9). Completely saturated soil, generally clean, cohesionless sand, which may include some gravel, may be liquefied during earthquake shaking by cyclic loading due to the upward propagation of shear waves (Obermeier, 1989; 1994). This can raise pore pressures to values as high as overburden pressure, effectively "floating" the sediment and its overburden.

Liquefaction accompanying earthquake shaking commonly originates at depths to about 10 m. The susceptibility of a sand layer to liquefaction (looseness or density) can be measured in place using the Standard Penetration Test (SPT) blow-count method (see the previous text under Landslides). In nearly all cases, liquefaction occurs in Holocene, generally late Holocene, deposits.

The shaking threshold to produce liquefaction features, even in highly susceptible sediments, is about 0.1g (National Research Council, 1985; Ishihara, 1985). Liquefaction features can develop at magnitudes as low as 5, but these features become relatively common at magnitudes of 5.5 to 6. The most important factors contributing to liquefaction are the amplitude of cyclic shear stresses (related to peak acceleration) and the number of applications of these shear stresses (related to duration of strong shaking).

The soil may be liquefied after earthquake shaking because of an increase in hydraulic gradient sufficient to produce an upward flow of water. The upward flow of a slurry of water and sand from an underlying layer of cohesionless soil with high pore pressure to the ground surface produces *sand blows* (Fig. 12-6). The conduit through which material is vented to the surface may be preserved as a *dike* or a *sill*. The flow commonly penetrates a less-permeable cap which tends to preserve high pore pressures.

In summary, criteria indicating an earthquake origin include (1) sedimentological evidence of an upward-directed hydraulic force, suddenly applied and of short duration, (2) more than one type of liquefaction feature present, including dikes, sills, vented sediment, or soft-sediment deformation, (3) groundwater settings where artesian conditions or nonseismic landsliding can be ruled out, (4) similar features at many locations within a few kilometers of one another, in similar geologic and groundwater settings, and (5) age evidence that the features formed in one or more discrete, short episodes over a large area, separated by long time periods during which such features did not form.

Lateral spreads are a result of liquefaction of a cohesionless soil which results in the downslope transport of its overburden cap. Because these are recognized on the basis of their geomorphic expression, they are mentioned briefly in the previous text under Landslide Deposits. In this section, we discuss lateral spreads in greater detail, followed by a discussion of liquefaction effects that are preserved in the stratigraphic record as dikes, sills, or disturbed sediments.

The Juvenile Hall landslide accompanying the 1971 San Fernando, California, earthquake is presented as an example of a lateral spread (Smith and Fallgren, 1975; Bennett, 1989; Fig. 12-7). The zone of ground displacement is 1220 m long with a maximum width of 275 m. The ground surface descends 12 meters along

Figure 12–6. Sand blows from the M6.6 Imperial Valley, California, earthquake of 15 October 1979, in a field north of Heber Dunes. The circular vents are aligned along a fracture in a lateral spread at Heber Road. Scale is 22 cm long. Photo taken 16 October by Kerry Sieh.

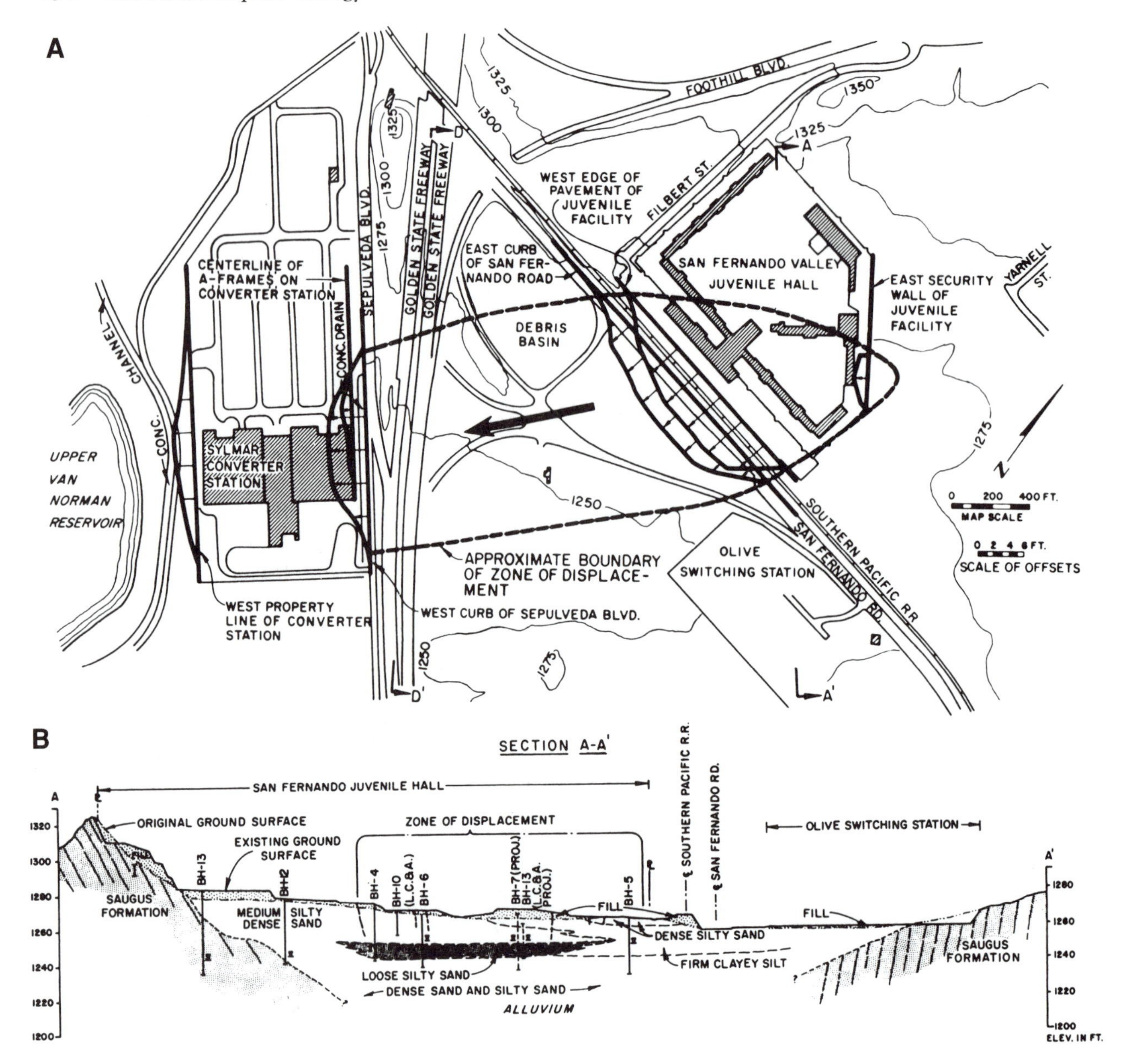

Figure 12–7. Lateral spread at Juvenile Hall and Sylmar Converter Station produced by 9 February 1971 San Fernando, California, earthquake. (a) Lateral displacements along four transects. Movement was from right to left. (b) Cross section along line A-A′, north to left. Zone of displacement coincided with a zone of soft, water-saturated silty sand that liquefied during the earthquake. From Smith and Fallgren (1975).

the length of the slide, a slope of about 1%. Ground ruptures are parallel to the direction of displacement and show evidence of right-lateral displacement along the northern edge and left-lateral displacement along the southern edge. Because the slopes are so low, these displacements could be mistaken for evidence for tectonic strike-slip displacement. However, trenches show that surface cracks tend to curve downward into the zone of displacement and to die out. The slide occurred in alluvial-fan material during a time when the water table was relatively high. Trench excavations showed that the slide was composed predominantly of dense, dry alluvial-fan sand and silt underlain by a soft, saturated zone of sandy silt and fine sand with a uniform texture which can be defined both on the basis of bucket-auger borings and soil-penetrometer surveys. The sediments making up the slide are estimated to be younger than 10 ka in age, whereas sediments beneath the failure zone are older than 35 ka (Bennett, 1989).

Hundreds of sand- and gravel-filled dikes in the

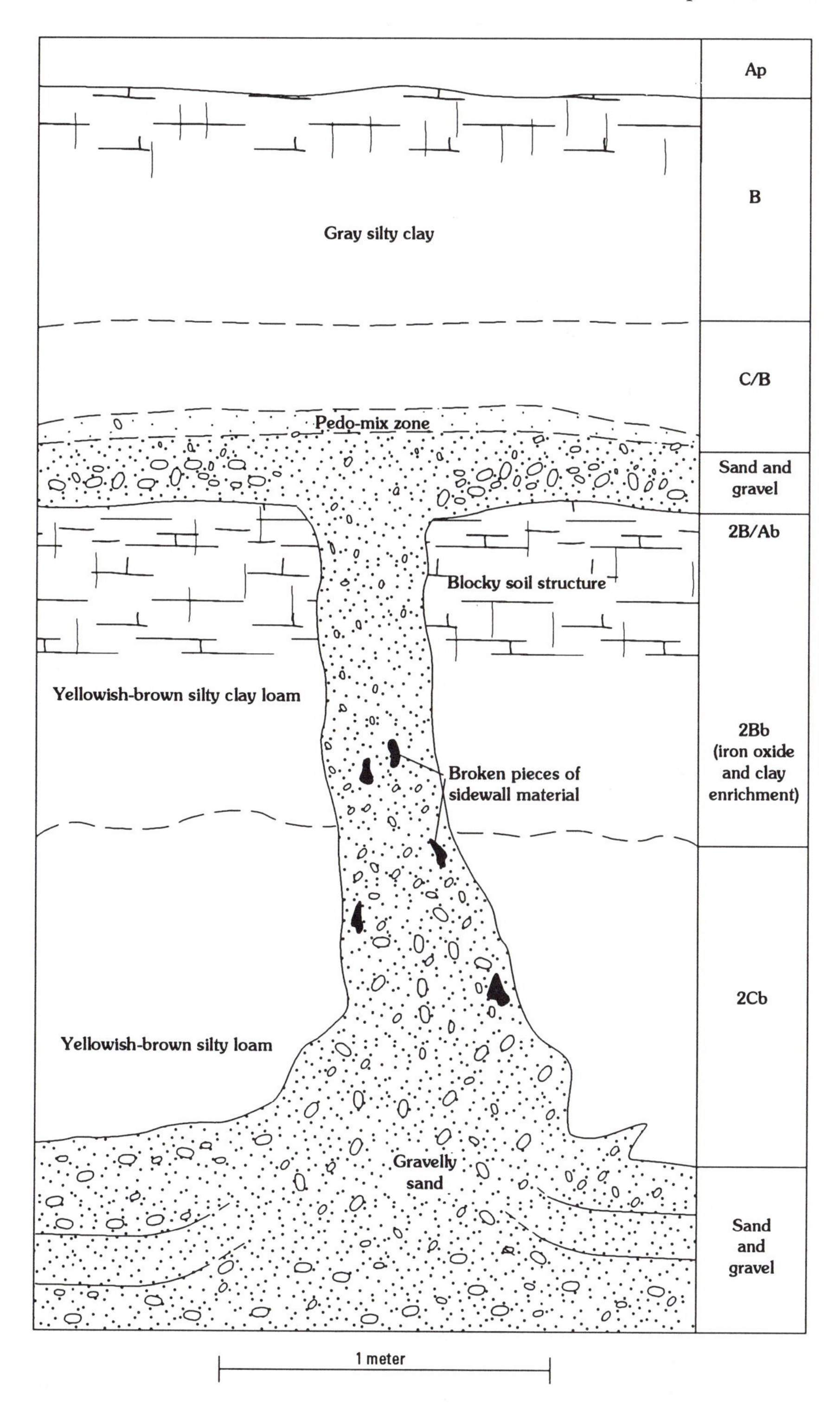

Figure 12–8. Diagrammatic vertical section showing buried sand-and-gravel-filled dikes and vented sediments along the Wabash River, southern Indiana and Illinois. Source beds are gravelly sands from Holocene or late Pleistocene alluvial point bars or braid bars. These are overlain by fine-grained overbank deposits that form a cap rock. Sediments show evidence of upward flowage and, in many cases, of upward decrease in grain size. Sand is mixed with soil where it flows out onto surface. Gray silty clay deposited after sand dike was emplaced. Letters on right are soil horizons. After Obermeier et al. (1993).

Wabash Valley of southern Indiana and Illinois, central United States, were interpreted as evidence for an earthquake 6100 ± 200 years ago with M_W 7.5, based on their areal extent (Obermeier et al., 1993).

These dikes are steeply dipping to vertical planar, sand- or sand-and-gravel-filled dikes that connect to a sediment source at depth (Fig. 12–8). Many dikes extend upward as much as 4 m above the source zone

Figure 12–9. Vertical aerial photo of part of the Mississippi floodplain in northeastern Arkansas showing long fissures through which sand vented (light linear features) and individual sand blows (light-colored spots), all formed by liquefaction during the 1811–1812 New Madrid earthquakes. Note how fissures formed parallel to point-bar alluvial deposits. From Obermeier et al. (1993).

of silty to gravelly sand. In many cases, sediment has vented to the surface to form sand blows, but in others, dikes pinch out below the surface. Erosionally truncated dikes are also common. Sidewalls of dikes tend to be parallel, particularly the larger dikes. Sediment within dikes is structureless, except that some elongate clasts derived from sidewalls tend to be vertical. Where gravel or coarse sand is present, the coarsest material shows a fining-upward tendency.

Sills are relatively uncommon, and they generally branch upward and irregularly from the main dike and crosscut sidewall stratigraphic horizons. Boundaries of sills are sharp and show little or no evidence of weathering after formation. This is in contrast to sand blows that spread out laterally from a vent; these commonly overlie a paleosol at their base (Obermeier et al., 1993).

Desiccation cracks filled by sand could be mistaken for sand dikes. Evidence for a sand dike origin would include flowage from below or of a source layer of cohesionless material, restriction of vented sand layers to a single stratigraphic horizon underlain by a paleosol, and a lack of internal structure within the dike except for a fining-upward tendency of the coarser clasts.

In map view, liquefaction features occur as long fissures or dikes or point-source sand blows (Fig. 12-9). Sand dikes formed during the 1811-1812 New Madrid earthquakes occupy fissures controlled by point-bar deposits of the Mississippi River floodplain.

An earthquake in Charleston, South Carolina, in 1886 produced a large number of sand-blow craters with evidence that fluidized sand vented explosively into the air, producing blankets of ejected sand and a central crater that was filled with sediment in the months following the earthquake. In cross section (Fig. 12-10), these sand blows are seen to be bowl-shaped or funnel-shaped craters containing both structureless sand and horizontal layers containing clasts of sand and soil (Obermeier et al., 1985). In contrast to the Wabash Valley, these sand blows contain evidence for both the ejection of sand and the later filling of the sand crater. Surprisingly, many of the sand blows excavated were formed by prehistoric earthquakes rather than the 1886 event, evidence that Charleston has been subjected to several large earthquakes in Holocene time (Obermeier et al., 1985).

When the Van Norman reservoir was drained after the 1971 San Fernando, California earthquake, Sims (1975) found that sediments formed behind the Van Norman Dam were deformed during the earthquake, forming small-scale folds, symmetrical load casts, heave-up structures, and contorted laminations. These structures were limited to a single horizon and could be correlated over great distances. In some cases, topographic relief formed by the deformational structures provided traps for sediment, evidence that the structures were formed near the sediment-water interface. Sims (1975) concluded that these deformed layers overlay sediments that became liquefied during the earthquake and were unable to support the layers. He also found older deformed horizons that he correlated with earthquakes in 1952 and 1930.

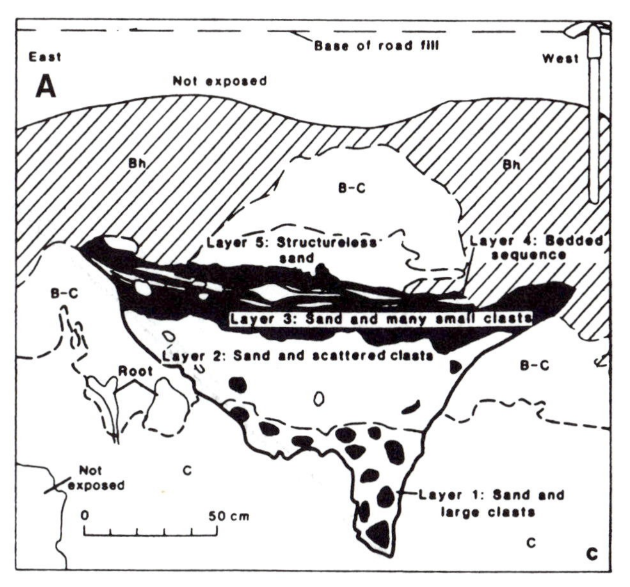

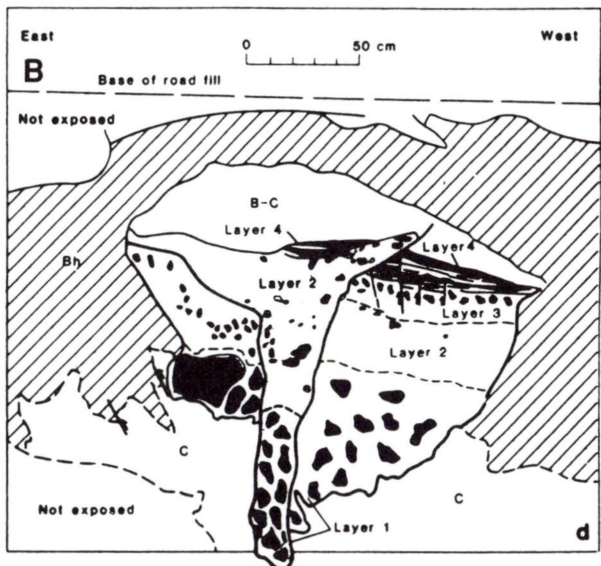

Figure 12-10. Sand-blow crater exposed in drainage ditch near Hollywood, South Carolina. Description of soil horizons: Bh: sand, brownish-black, massive; B-C: sand, pale reddish brown, friable, structureless; C: sand, yellowish-gray, friable, structureless. (a) Single sand-blow crater. (b) Cross-cutting sand-blow craters. All predate the 1886 Charleston earthquake. From Obermeier et al. (1985).

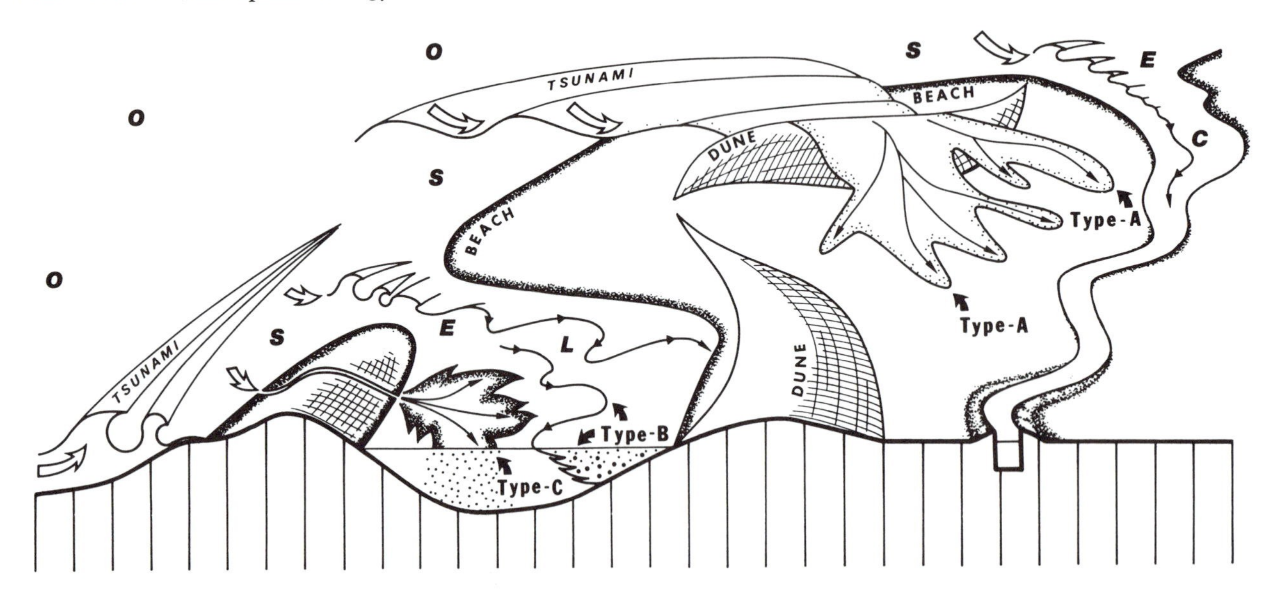

Figure 12–11. Depositional processes caused by a tsunami produced by the 26 May 1983 earthquake in the Japan Sea. Types A, B, and C explained in text. O = offshore. From Minoura and Nakaya (1991).

Marco et al. (1996) studied Pleistocene lacustrine sediments in the Dead Sea graben and found mixed layers that they interpreted as the results of fluidization of sediments on the Dead Sea floor during an earthquake.

Soft-sediment deformation structures and sand dikes (clastic dikes) are common in sedimentary rocks, but it is not clear that they are formed by earthquakes (Obermeier, 1994). For example, clastic dikes and soft-sediment deformation structures are common in fine-grained deposits of late Pleistocene age in the Pasco basin of south-central Washington. These deposits formed in ponded water and are related to catastrophic floods caused by rupture of a glacial ice dam and draining of a large glacial lake. Some of the dikes lens out downward, suggesting that they are fissures that have filled with sand, or they result from downward injection of water due to loading of flood waters. Others indicate a sediment source from below, as documented for the Wabash Valley sand dikes (Smith, 1993). Although some workers believe that the dikes are related to the catastrophic flooding, others have suggested that they are unrelated to the floods. The Pasco basin is characterized by moderate seismicity, and anticlines in the area show evidence of late Quaternary deformation (Fig. 10–29), so an earthquake origin would be reasonable. But at present, it is not demonstrated.

Sand boils and a sand dike 180 m long were found in 1983 adjacent to the active Garlock fault in the Mojave Desert of California, shortly after the region had been subjected to torrential rainfall. The sand boils came up through a preexisting ground fissure related to groundwater withdrawal. Holzer and Clark (1993) showed that sediment-laden surface runoff entered the upslope part of the fissure, flowed subhorizontally in the crack, and reemerged downslope as spectacular sand boils. Because they are parallel to an active fault, these sand boils could be interpreted as earthquake-related, but no recent earthquakes large enough to produce them have occurred in the area.

Another origin mechanism could be piping related to artesian water flow (Mansur et al., 1956). Flooding of the Mississippi River causes artesian flow beneath man-made levees that produce sand boils in nearby lowlands. These may have a cone shape, like earthquake-induced sand blows, but study of the internal shapes of the sand boils suggests that flow started slowly and steadily increased, in contrast to the sudden, violent flow accompanying earthquake-induced liquefaction.

Tree roots may decay and later fill with sand, mimicking sand dikes. These may show segregation of clay minerals due to weathering, but do not show the laminar bedding or the underlying graded zone characteristic of filled seismic craters (Obermeier, 1994). Other nonseismic features in subaqueous environments are load structures in muds, water-escape structures, and soft-sediment folding, the last which may result from deposition of turbidites on a low submarine slope.

TSUNAMI DEPOSITS

As described in Chapter 4, tsunamis cause great destruction and loss of life in coastal areas. These tsunamis may be caused by tectonic displacement of the sea floor accompanying a large earthquake, or they may be caused by more local sea-floor displacements accompanying submarine landslides (Plafker, 1969) or submarine volcanism (Watanabe, 1985). Tsunamis may leave characteristic deposits that could provide independent evidence for prehistoric earthquakes.

An earthquake of M7.7 struck beneath the Japan Sea on 26 May, 1983, producing a large tsunami that struck the Japan Sea coast of northern Honshu with waves up to 14 m high. Waves transported 10-ton blocks hundreds of meters inland, overtopped coastal sand dunes, and surged up rivers. Materials caught up in these waves were later carried seaward; 100 people lost their lives in this way.

Minoura and Nakaya (1991) noted three types of sedimentary processes associated with the tsunami accompanying the 1983 Japan Sea earthquake (Fig. 12-11). In type A, incoming waves transported material landward from beaches or dunes. Outgoing waves transported land materials seaward such that they accumulated in the shoreface region, forming submarine bars. In type B, waves invaded and eroded intertidal lagoons, transporting shells landward from their normal sites of accumulation. In type C, seismic shaking cracked sand dunes and beaches, and seawater rushed through and scoured out these cracks, linking interdune ponds to the open sea. Sand grains transported in suspension were deposited in ponds as a thin sand layer.

Cores in the interdune ponds revealed fine to coarse sand layers intercalated with black, organic sediment, the normal deposit found in the ponds. Minoura and Nakaya (1991) worked out the rate of deposition of the organic sediments and found that the sand layers occurred at the time of historical earthquakes off the coast of northern Japan in A.D. 1741, 1856, and 1983. The sand units have a sharp basal contact, with grain size coarsening upward.

Similar features were found by Minoura and Nakaya (1991) on the Sendai plain on the east coast of Honshu, where the sediments could be age-calibrated by a tephra layer dated as A.D. 870–934. Sand layers are medium- to fine-grained, well sorted, and appear to originate from coastal sand dunes. Two layers correlated with historical earthquakes off the coast in A.D. 869 and 1611.

Deposits on the coast of New South Wales, Australia, appear to be produced by tsunamis rather than large storms (Bryant et al., 1992). These include (1) large boulders with angular faces that have been tossed up onto raised beach platforms and locally jammed into crevices, (2) bimodal mixtures of isolated weathered, oblate boulders and sand deposited when sea level was too low to produce the deposit by large storm waves, and (3) isolated ridges, terraces, or mounds of chaotically sorted boulders, cobbles, and shell material. At one locality, isolated ridges of sand directly overlying estuarine mud are attributed to a tsunami (Bryant et al., 1992).

In the Cascadia subduction zone, an earthquake origin for the abrupt contact between peat deposits and overlying marine muds is strengthened by the occurrence of sheets of sand at the base of the muds (Darienzo and Peterson, 1990; Atwater, 1992; Atwater et al., 1995). Tsunami-derived sand sheets are finer grained as a function of distance from the sea; multiple sand layers, each characterized by fining-upward grain size, are attributed to a tsunami wave train. Heavy mineral assemblages and roundness of grains may distinguish a seaward-derived tsunami sand sheet from a land-derived fluvial deposit.

SUMMARY AND A CAUTIONARY WORD

Commonly the most spectacular expressions of an earthquake are the effects of shaking, which produce landslides, lateral spreads, sand blows, and other features not directly related to tectonic displacement. A description of the effects of an earthquake commonly includes a description of landslides, but it is more difficult to attribute a prehistoric landslide to an earthquake. One of the most successful techniques is to show that secondary effects in different environments formed at the same time, as was done for an earthquake on the Seattle fault about 1000 years ago. Another method is to date surfaces that are first exposed during an earthquake, such as those formed during a rock fall. If many surfaces are the same age, then the surface may have been formed by an earthquake rather than have a nonseismic origin such as a large storm. Contemporaneity of rock falls and rock slides is more indicative of an earthquake origin; however, contemporaneity of a large number of rapid soil flows may more likely be attributed to a large storm.

None of the types of landslide described by Keefer (1984) is formed only by earthquake shaking. An alternate origin is, up to now, always possible for a prehistoric landslide. An earthquake origin for some very large slides, the Gros Ventre and Hope landslides, has already been discussed. But there are many large landslides, such as a slide that dammed the Columbia River (Bridge of the Gods) with an unknown triggering mechanism. The origin of each of these large landslides is a subject of debate, with an earthquake origin only one of many under consideration.

The most definitive paleoseismological approach to establishing a seismic origin of a prehistoric landslide is geotechnical—a comprehensive slope-stability analysis of an individual landslide. If this analysis shows that landslide initiation would have been unlikely even under the worst set of non-seismic conditions, but that seismic shaking (cyclic loading) could generate a landslide, then it can be concluded that the landslide was generated by an earthquake.

It has been suggested that liquefaction features such as lateral spreads and sand dikes can only have been produced by earthquake shaking. But, as with landslides, all of the sedimentological features discussed previously can be explained in other ways. As Ricci Lucchi (1995) and Obermeier (1994) point out, apparent liquefaction features can be explained by rapid sediment loading, gas escape from buried sediments, cryoturbation, slumping, and other nonseismic processes. There is no "smoking gun" that unambiguously points to an earthquake origin.

Tsunami deposits have only been studied for a few years, but here, too, one must be careful to distinguish the deposits of tsunamis from deposits of 100-year or 500-year storms, those that occur rarely in a region but, when they do, leave a strong signature in the geologic record.

The challenge of using secondary effects to detect prehistoric earthquakes is to exclude all the nonseismic origins prior to assigning a particular feature to an earthquake.

Suggestions for Further Reading

Dawson, A. G. 1994. Geomorphological effects of tsunami run-up and backwash. Geomorphology, 10:83–94.

Ishihara K. 1985. Stability of natural soil deposits during earthquakes. Proceedings of the Eleventh International Conference on Soil Mechanics and Foundation Engineering, San Francisco, 1:321–76.

Jibson, R. W. 1994. Using landslides for paleoseismic analysis. U.S. Geological Survey Open-File Report 94-663, 57 p.

Keefer, D. K. 1984. Landslides caused by earthquakes. Geol. Soc. America Bulletin, 95:406–21.

National Research Council. 1985. Liquefaction of Soils during Earthquakes. Washington, D.C.: National Academy Press, 240 p.

Obermeier, S. F. 1994. Using liquefaction-induced features for paleoseismic analysis. U.S. Geological Survey Open-File Report 94-663, 72 p.

PART THREE

Living with Earthquakes

Historically, earthquakes have caused the world's most tragic natural disasters. No other natural phenomenon has produced loss of life as great as the 800,000 people killed in the Chinese earthquake of 1556. Even as recently as 1975, more than 250,000 people were killed in another Chinese earthquake, near Tangshan. Financial losses from even moderate earthquakes have become significant drains on national economies. The magnitude 6.7 Northridge, California, earthquake of 1994, for example, resulted in damages that exceeded 20 billion dollars, despite a relatively low loss of life. And financial losses from the Kobe, Japan, earthquake of 1995 are now estimated to have exceeded 200 billion dollars. Certainly one of the most important reasons for studying earthquakes is to temper the effects of such earthquakes in the future, through such practices as better land-use planning, and the design and construction of structures that will withstand severe ground shaking.

The earth-science contribution to living safely with earthquakes is primarily in the area of seismic hazard assessment, which, in turn, forms a fundamental basis for land-use planning, building-code formulation, design standards for critical structures, and disaster-mitigation strategies. The final chapter of this book is devoted to seismic hazard assessment and, particularly, to the role played by the earth scientist in this effort.

13

Seismic Hazard Assessment

Someday, short-term earthquake prediction, which is based on the recognition of physical precursors in the hours and days prior to fault rupture, may play an important role in seismic hazard mitigation, but nowhere in the world today are reliable short-term predictions being routinely issued. Furthermore, even if success in short-term prediction is eventually achieved, it will not solve the long-term problems of planning, and seismic design and construction, which ultimately determine whether structures will be sufficiently safe when the predicted earthquake does occur. In an analogous situation, even the most accurate prediction of a hurricane will not prevent damage and loss of life if proper preparatory efforts have not been made in the years prior to the event.

Similarly, the possibility of physically interfering with the earthquake mechanism, as, for example, by "triggering" a large event at a predetermined time or by causing the accumulated strain to be released in a series of smaller events instead of one large one, remains an intriguing long-term research goal, but it cannot at present be considered a practical mitigation technique.

Despite our current inability to predict earthquakes on a short-term basis, it must be recognized at the outset that earthquake hazard assessment is fundamentally a *predictive* effort—the attempt to forecast the likelihoods and effects of earthquakes in the years to come. As such, it is subject to all the uncertainties and frailties of any predictive science. In some ways, seismic hazard assessment is analogous to *long-term* earthquake prediction. Despite the wishes of the public and the advertisements of a few charlatans, there is no current "cookbook" method of seismic hazard evaluation in which we can have complete confidence; it remains a major field of ongoing research.

Earthquake hazard assessment is necessarily based almost solely on the old geological adage of *uniformitarianism,* often expressed as "the present is the key to the past." In our case, however, we explore the *past* as the key to the *future.* That is, earthquake processes of the recent past will in all likelihood be the same as those of the near future. In our case, however, past history is rather neatly divided into the historical and *pre*historical periods. The historical record of earthquake occurrences, together with the instrumental seismographic record, forms an invaluable body of knowledge that pertains directly to what might happen in the same region in the future. Obviously, the longer the historical and instrumental record is, the greater will be its predictive value. The prehistorical geologic record also has great relevance to seismic hazard evaluation, since its time span is everywhere greater than the historical record and is thus more meaningful statistically for extrapolating into the future. Probably this is the area where the most significant recent advances in seismic hazard evaluation have taken place, and many details of the paleoseismologic techniques are discussed in Chapters 8–11.

By all accounts, the greatest worldwide seismic *hazard* is the ground shaking that directly results from earthquakes, and the greatest *risk* is the structural damage that results from this ground motion. On the other hand, secondary effects, such as landslides triggered by the ground motion, have in some earthquakes caused far greater loss of life and property than seismic shaking (Chapter 12). More than 100,000 people were killed by massive landslides in loess triggered by the great 1920 earthquake in northern China ($M = 8.7$) (Close and McCormick, 1922), and the massive debris avalanche triggered by the 1970 Peruvian earthquake ($M_W = 7.9$) killed more than 15,000 people in one city alone (Plafker et al., 1971). Similarly, many subduction zone earthquakes have caused much greater loss of life from tsunami inundation than from the direct effects of shaking, as was true of the great Alaskan earthquake of 1964 ($M_W = 9.2$). Nevertheless, the emphasis in this chapter is on the assessment of the hazard of the earthquake occurrence itself, rather than that of the indirect secondary effects—

however important they may be in total risk assessment and mitigation. Even with this limitation, seismic hazard assessment is a very large field involving a number of different disciplines. This chapter necessarily restricts itself to those areas closely related to geology. For a broader treatment, particularly regarding the quantitative description of strong ground motion, the reader is referred to Reiter (1990). Many of the seismological and probabilistic subjects of this chapter are also discussed in quantitative but intriguing fashion by Lomnitz (1994).

HAZARD VS. RISK

A distinction is usually made between seismic *hazard* and seismic *risk*. The *hazard* is the physical phenomenon itself—the ground shaking, fault movement, liquefaction, etc.—that underlies the danger. The *risk*, on the other hand, is the likelihood of human and property loss that can *result* from the hazard. If no lives or property are at stake, the risk is nil no matter how great the hazard. For example, the 1989 Macquarie Ridge earthquake (south of New Zealand) produced no losses despite its magnitude of 8.3, whereas the 1960 Agadir, Morocco, earthquake, of magnitude 5.5, killed more than 12,000 people. It is in the area of evaluating the *hazard* where geologists and geophysicists make their primary contribution, whereas engineers, planners, and public officials are more concerned with evaluating and mitigating the *risk*. Clearly, however, evaluation of the risk cannot be made without prior understanding of the hazard, and a team effort is called for if the results are to be useful to society. Risk assessment is the intermediate step between risk perception and risk management or mitigation (Fig. 13-1).

Figure 13–1. Risk perception vs. risk assessment vs. risk management. In the analog to earthquakes, risk management might more appropriately involve (1) prohibiting people from standing beneath the boulder (*land-use planning*), (2) estimating the time of fall of the boulder (*earthquake prediction*), or (3) tying the boulder to the bedrock face (*earthquake engineering*). © 1993 by Sidney Harris—HEMISPHERES magazine.

PRODUCTS OF HAZARD ASSESSMENTS

Seismic hazard assessments are of many types and result in many different kinds of products. For example, a map of highly active faults is a simple type of hazard assessment product, as is a map of historical seismicity. Combining these maps with other parameters may result in seismic zoning maps in which different degrees of hazard are delineated. Specialized studies are often made of individual local hazards such as liquefaction, earthquake-induced landslides, or tsunami runup. Probabilistic approaches are becoming more commonplace, usually resulting in maps portraying the *likelihood* of earthquake occurrences or of specific parameters such as the probability of exceedance of given levels of ground shaking. The most extensive and thorough evaluations have been in connection with specific critical structures such as major dams and nuclear power plants, where public interest is high, and where licensing standards are very demanding. Tens of millions of dollars have been spent on seismic hazard and risk assessments for single facilities, and major controversies are not uncommon. Particularly for dams and nuclear facilities, public concern has often focused on seismic issues for the very reason mentioned above—that it is a predictive effort in which the uncertainties may still be very great. An important point, however, is that seismic hazard assessments are of many types, and it would be a mistake to assume that construction of realistic seismic zoning maps, for example, is the ultimate goal of every hazard assessment project.

ACTIVE VS. INACTIVE FAULTS

It has sometimes naively been assumed that all faults are either *active* or *inactive,* and that areas near active faults are therefore dangerous, and areas near inactive faults are safe. Unfortunately, nature isn't that simple! One of the major results of neotectonic field studies over the past two decades has been the documentation that faults have all degrees of activity: some, such as California's San Andreas fault and Turkey's North Anatolian fault produce large earthquakes once every few hundred years; others, such as the Dixie Valley fault in Nevada, produce large earthquakes on individual segments once every few thousands or even tens of thousands of years. Also, it is clear that some faults are highly active but produce only small earthquakes. Therefore, any distinction between active and inactive faults is simplistic and seldom very useful. Recent emphasis has, instead, been upon specifying the *degree of activity,* and then using this parameter to arrive at further quantitative assessments, as is discussed later in this chapter. In this sense, terms such as "Holocene active" are sometimes used to specify degree of activity through a stipulation of the time of most recent movement. Other less specific terms such as "potentially active" seem to have only locally recognized definitions.

For legal and regulatory purposes, different specific definitions relating to fault activity have arisen in recent years. For example, the U.S. Nuclear Regulatory Commission has, at least until recently, defined a *capable* fault as one that has undergone surface rupture once in 35,000 years or more than once in 500,000 years, in addition to a few other criteria. Similarly, the State of California, in legislation pertaining to the building of structures near fault lines, defines an active fault as one that has had Holocene surface displacement, meaning displacement within about the past 10,000 years (Hart, 1980). Such definitions may be necessary for legal purposes, but one should not confuse the necessity for legal distinctions with the geologic reality of a continuum of degrees of activity.

WHAT IS THE LIKELIHOOD OF "NEW" FAULTING?

Occasionally, the specter of "new" faulting is raised in hazard assessments. How can the geologist be confident that a completely new fault will not rupture through previously unbroken rock? The answer lies in the historical record of earthquake occurrences. With few exceptions, every major historic earthquake that has been associated with surface fault rupture has occurred on a fault with a demonstrable prior history of fault displacement, usually within late Quaternary time. Therefore, the probability of a major new fault breaking through a specific site is virtually nil, unless the site lies squarely within a very active and complex fault zone.

How then, one might ask, is a new fault ever "born"? Evidence suggests that faults originate as small fractures that then gradually lengthen and propagate during successive earthquakes or strain episodes. The Alpine fault of New Zealand, for example, did not come into being all at once during some gigantic prehistoric earthquake, but instead it presumably started as small fractures that gradually lengthened and coalesced over geologic time to the 1000-km-long fault we see today. Such incremental lengthening has, in fact, been observed: the 1968 Borrego Mountain earthquake in southern California was associated with fault rupture of the Holocene Coyote Creek fault (Appendix), but at the northern end of the rupture, new fractures penetrated previously unbroken Tertiary strata for several hundred meters still farther north, slightly extending the previous surface fault length.

PROBABILISTIC VS. DETERMINISTIC METHODS

An earthquake hazard assessment is said to be *deterministic* when it specifies a particular earthquake or level of ground shaking that is to be considered by the user, in terms of single-valued parameters such as magnitude, location, or peak ground acceleration. It generally does not specify how likely the event might be, except that it is considered "credible." Use of the word "credible" offers some leeway in not necessarily picking the absolute "worst case" situation, which some would interpret as being the maximum that is actually physically possible. What is "credible" and what is "worst case" are, like "beauty", subjective and in the eyes of the person making the assessment.

In a *probabilistic* assessment, on the other hand, numerical probabilities are assigned to earthquake occurrences and their effects during a specific time period such as the life of a given engineered structure. No attempt is made to define the "worst case" event, only to assign probabilities of different magnitudes of earthquakes and/or their effects. This assessment must recognize that, not only are there uncertainties in our knowledge of individual input parameters such as fault slip rates and earthquake magnitudes, but that nature itself displays some randomness. This intrinsic randomness is illustrated by the fact that, even if we knew perfectly the magnitude and distance of an earthquake, not all earthquakes of the same magnitude cause the same intensity of shaking at the same distance.

A hypothetical deterministic assessment for a "worst case" scenario might be to specify a magnitude 7.4 earthquake occurring on a north-trending vertical strike-slip fault whose closest approach to the site under consideration is 15 km, which will in turn give rise to a peak acceleration at the site of 0.4 *g*. A hypothetical probabilistic analysis, on the other hand, would produce a continuum of outputs, two levels of which might be represented by (1) a 50% probability of exceeding 0.3 *g* during the 30-year life of the structure under consideration, and (2) a 5% probability of exceeding 0.6 *g* during this time. These probabilities would not necessarily be determined by a single fault, but could include the contributions of a number of nearby faults with different parameters. Furthermore, the probabilities of exceeding certain threshold *g* values (particularly low ones) might not be controlled by the maximum earthquake on any one fault, but might instead be controlled by the much larger number of associated smaller events during the 30-year period.

Typically, a probabilistic assessment will portray the *total* hazard or risk from all of the relevant seismic sources, whereas the deterministic assessment is often limited to the single fault judged most likely to represent the "maximum credible" or "worst case" earthquake, with no quantitative statement of its likelihood. Thus it is not unusual, in a deterministic analysis, for the maximum ground motion at a specific site to be associated with a very local but relatively inactive fault, which, because of its very low slip rate, may make only a very small contribution to the total hazard or risk in a probabilistic assessment of the same site.

A probabilistic hazard assessment will usually leave the user with the choice of the appropriate level of hazard or risk to be considered. This choice will typically depend on the degree of conservatism desired. Normally one will choose a much more unlikely earthquake for a critical structure such as a nuclear power plant than for a less critical structure such as a parking garage. Thus, the concept of *acceptable risk* becomes important in the application of probabilistic hazard assessment methods. That is, just how unlikely does an earthquake have to be before it is no longer worthy of consideration for a particular project? For the nuclear plant, one might be willing, for example, to neglect the consequences of a damaging earthquake that would occur less frequently, on the average, than once in 10,000 years (annual probability of 10^{-4}), whereas earthquakes occurring more frequently than this would have to be accommodated in the design. For the garage, on the other hand, one might be willing to live with the 100-year earthquake (annual probability of 10^{-2}), simply because the consequences of failure are so much less than for the nuclear power plant. Although it might seem desirable to design critical structures for the maximum earthquake that is physically possible, this is simply not a realistic goal. For example, collision of an asteroid or comet with the earth, producing massive extinctions, has occurred worldwide once every hundred million years or so. Such collisions are clearly "possible," yet are deemed too infrequent to worry about in our planning—even if it were realistic to do so.

The concept of acceptable risk is closely related to the question of "How safe is 'safe enough'?" It is important to recognize, however, that this is basically a *social* question, not a technical one, and many scientists and engineers argue that they should not be put in the position of having to answer this question alone, without wider societal participation and acceptance of responsibility.

The deterministic hazard assessment method has the advantage that it does not require the presence of data bearing on time-dependent processes such as rate of earthquake occurrence or fault slip rate. An argument in its favor has been that such rate data are often simply unavailable, particularly in relatively unstudied areas, and therefore the judgment of the geologist or seismologist is paramount in making the arbitrary decision as to what is the appropriate earthquake magnitude or level of ground shaking to be considered for a particular project. Many users and clients would point out that the geotechnical team (geologists, seismologists, civil engineers) is in a much better position to judge appropriate comparative standards and licensing criteria than is the user or client. Another alleged advantage, sometimes cited—or quietly admitted—by government agencies and owners of critical facilities, is that they don't wish to face alone the very difficult and sometimes politically controversial decision as to the appropriate level of acceptable risk. At the same time, sometimes they would seemingly encourage the public naively to believe that the risk was thereby zero.[1]

In addition to placing the hazard assessment in a time framework, most probabilistic treatments have the advantage of making the details of the analysis more systematic, in the sense that every step in the decision process is traceable and recoverable, and uncertainties are specifically identified and usually quantified. Whereas individual judgments may be critical to both a deterministic and a probabilistic analysis, at least in the probabilistic approach, it is clear at what steps and in what ways judgment was offered. Sometimes there is a special use of *expert judgment,* using teams of experts whose opinions are elicited in a systematic and formalized fashion. At the same time, the mathematics of a sophisticated probabilistic analysis may be so complex that some nonmathematical users will feel that, rather than making the entire process more open and transparent, there is instead a real danger of hiding important assumptions and obfuscating the overall logic in a seeming mathematical fog. Certainly the intent of a probabilistic analysis should be exactly the opposite, but these dangers must be recognized by those making such analyses, and efforts must be made to overcome them and make the process understandable to the ultimate user.

[1]One of us (CRA) recalls a dam owner once telling him, in effect: "Don't give me all this gibberish about probabilities and acceptable risk; just tell me whether the dam is *safe* or not!"

Sometimes lost in the argument between determinists and probabilists is the fact that *both* techniques depend ultimately on the good judgments of those providing the input data. A simple deterministic statement by a single wise person may turn out to be far closer to the real truth than the collective opinion of an army of less-wise experts aided by the world's best mathematicians. As was pointed out by Reiter (1990), "Copernicus was right when he proposed that the earth revolves around the sun despite what others thought of him or his views." Indeed, Reiter succinctly summarizes the status of expert judgment as follows:

> The purpose of seismic hazard estimation is to provide practical answers to practical questions. Society does not have the luxury to wait for the answers until the "truth" is discovered. Given that restriction, there is no alternative to making careful use of expert judgments, and even opinions. What is also incumbent upon the experts and the analysts is to make sure that those who use the results of such analyses are aware that they do not rest upon the views of a Copernicus but rather upon a range of views which hopefully, but not necessarily, bound the truth when it eventually becomes evident.

DETERMINISTIC ASSESSMENTS

Definitions of Maximum Earthquakes

At the core of a deterministic earthquake hazard analysis, at least from the earth scientist's point of view, is the determination of the maximum earthquake or earthquakes that are to be used in subsequent analyses. Sometimes more than one earthquake is specified, such as a moderate nearby event and a more distant large event—each with its own engineering significance. The distant large event, for example, may have a smaller peak acceleration at the site, but a longer duration of shaking, which could be of overriding importance in the behavior of soils and embankments.

Exemplified by the deterministic method is the concept of the ***maximum credible earthquake,*** which has been defined as the maximum earthquake that is capable of occurring in a given area or on a given fault during the current tectonic regime. Readers who are geologists will immediately recognize that the "current tectonic regime" is not easily defined, and it is hardly surprising that major controversies have arisen over specifications of maximum credible earthquakes, particularly for critical projects where public feelings may be intense. The current tectonic regime in a given area might well be visualized as lasting for several millions of years, yet few people would favor designing a critical structure for an earthquake that would occur this infrequently—even if it were realistically possible to estimate its magnitude. The fact is that what is "credible" to one person may not be to another, and the inability to provide a workable definition for the term has led in recent years to a decreased emphasis on the "maximum credible earthquake" in hazard analyses.[2]

To avoid the problems of the maximum credible earthquake, other terms have sometimes been proposed. For example, the "maximum expectable earthquake" has sometimes been used, although such an event is not really "expectable" during the life of the project under consideration—only during some much longer and undefined future time period. In modern practice, it is increasingly common for the geologist or seismologist simply to prescribe the "maximum earthquake," with the assumption that his or her best judgment has been used in its application to the particular project under consideration. Ideally, of course, one could prescribe the maximum *probable* earthquake during a specified time period in the future, but this is the essence of the probabilistic approach that the deterministic method allegedly denies. In actual fact, however, any deterministic prescription of a maximum earthquake *must* involve some sort of a consideration of likelihood by the person making the estimate, and in this sense the deterministic approach entails, in practice, some elements of probability—whether admitted or not. That is, the idea of a maximum earthquake that is physically possible in a specific area or on a specific fault, within broad spans of geologic time, is simply not a concept that can be used practically.

Geologists and seismologists should generally avoid the term "design earthquake," because this term is often used by engineers in a specific engineering sense that may be distinct from that of the maximum earthquake under consideration.

[2]One of us (CRA) recently discovered that in a report written many years ago, he had, in a true Freudian typo, inadvertently labeled a summary column: "Maximum *dredible* earthquake" (italics added). What further definition of the term is necessary? To further underscore the problems in defining the maximum credible earthquake, it has sometimes been facetiously suggested that it is the earthquake that is only a shade smaller than the "minimum *in*credible earthquake."

In some cases, two levels of maximum earthquakes are considered. For example, the U.S. Nuclear Regulatory Commission defines the "Safe Shutdown Earthquake" (SSE) as that very rare ground-motion event for which the reactor is expected to shut down safely without significant risk to the public, whether or not there is damage to the plant. The "Operating Basis Earthquake" (OBE), on the other hand, is a smaller and more frequent level of ground shaking for which the plant is expected to continue operating during and after the event. The SSE is determined primarily by safety concerns, whereas the OBE is related more to economic factors of plant operation.

Techniques for Estimation of Maximum Earthquakes

Three principal lines of evidence have typically been used for estimating maximum earthquake magnitudes. First is the historical and instrumental record of earthquakes in the region under consideration or in other regions of similar seismotectonic settings. Second is the evidence from physical parameters of individual seismogenic faults in the area, particularly fault length and fault segmentation. Third is the paleoseismic evidence that may bear on maximum magnitudes of pre-historical events, particularly the displacement-per-event determinations that have been discussed in Chapter 8.

Especially in regions with long recorded histories of earthquakes, such as parts of China and the Middle East, the historical record is invaluable in estimating maximum magnitudes. For example, the fact that magnitude 8 earthquakes have occurred historically on parts of the North Anatolian fault of Turkey is obviously of great importance in estimating maximum earthquakes for this and other segments of the fault. In China, truly great earthquakes have taken place historically within the Shanxi graben and on the Tanlu fault. Although none of these events has occurred within the past 200 years, they clearly must be considered in assigning maximum earthquakes to the two regions for the future. They also serve as a cautionary flag to those evaluating seismicity in areas with shorter recorded histories, such as much of the United States. And in evaluating the seismicity of an area such as Nepal, it is obviously relevant to consider the seismicity of adjacent areas to the east and west that lie within the same Himalayan region. This represents the concept of the *seismotectonic province,* in which areas of similar seismic and tectonic characteristics are sometimes treated as units in hazard analyses. Typically, little is known about individual seismogenic structures within a province, so earthquakes are often assumed to occur randomly within the province. Seismotectonic provinces have sometimes necessarily been delineated on the basis of very subjective criteria such as basement rock terranes, potential field (gravity and magnetic) data, stress field data, and limited seismographic data, so it is not surprising that controversies have been particularly common when this technique has been used in estimating seismic hazard for critical facilities such as nuclear power plants. Certainly the emphasis today is instead on identifying and assessing individual seismogenic structures, but this is not yet realistically possible in all areas of the world that have experienced large historic earthquakes.

A traditional rule-of-thumb for assigning maximum earthquakes in regions of relatively short histories has been to add one-half magnitude unit to the largest known past event, but this technique is used simply to add conservatism and has no real logical basis. It may or may not be adequately conservative, and in some cases may be overly conservative.

Fault *length* is probably the most commonly used parameter in estimating maximum earthquake magnitude. Earthquake moment, which is the basis of the M_W scale (Chapter 4), is defined by the relationship,

$$M_o = \mu AD, \qquad (1)$$

where μ is the shear modulus (about 3×10^{11} dyne/cm^2), A is the rupture area, and D is the average slip. Since μ is usually treated as a constant, and the average slip, D, is limited to values of a few meters or tens of meters, the rupture area, A, is the dominant variable in influencing M_o and, thus, M_W, noting that if M_o is in dyne-CM,

$$M_W = (\text{Log } M_o - 16.05)/1.5 \qquad (2)$$

(Hanks and Kanamori, 1979). Furthermore, since fault depth is constrained by the thickness of the seismogenic crust and is not grossly different from one region to another, the area of rupture is primarily a function of *length* of rupture. A fault only one kilometer long simply cannot generate, in itself, a large earthquake, whereas great earthquakes are necessarily associated with ruptures of considerable length. For example, the largest earthquake of this century, the 1960 Chilean earthquake of $M_W = 9.5$, was associated with a rupture length of about 1000 km, and only by having a rupture of still greater length could even larger magnitudes be achieved, noting that the Chilean earthquake and other great subduction zone earthquakes occur on faults of relatively low dip. (As was pointed out in Chapter 4, a rupture of the entire earth's circumference would, in theory, correspond very roughly to an earthquake of $M_W = 10.6$, so that there really is, in effect, a practical upper limit to the magnitudes of earthquakes.)

Therefore, if a geologist can estimate the length of a given fault that is likely to rupture during a future earthquake, one can compare that projected rupture length with those that have actually been observed during historic earthquakes in order to estimate what associated level of magnitude is to be expected. Such regression relationships for worldwide earthquakes have been tabulated by many authors, most recently by Wells and Coppersmith (1994), who based their regressions on 216 historic earthquakes in which at least some source parameters were well documented. Their regression of surface rupture length on magnitude (M_W) for worldwide earthquakes of all slip types is shown in Figure 13-2.[3]

Just as regression relationships can be established between rupture length and magnitude, other fault parameters can also be used, such as maximum or average fault displacement. Subsurface rupture length, if it can be determined from aftershock distribution, may be more highly correlative than surface rupture length. Also, regressions may be made for different styles of faulting (strike-slip, thrust, or normal), although Wells and Coppersmith, (1994) concluded that the differences were not great. They also concluded that (1) average surface rupture lengths were about 75% of subsurface rupture lengths, (2) average surface displacements were about 50% of maximum surface displacements, and (3) average surface displacements were about equal to average subsurface displacements. Table 13-1 shows some of Wells and Coppersmith's regression relationships.

Although it is relatively simple to measure surface rupture length following an earthquake, it is much more difficult to estimate ahead of time the rupture length to be associated with the maximum earthquake on a given fault or fault zone. The rupture length generally cannot exceed the total length of the preexisting fault, but the 1992 Landers, California, earthquake (Chapter 8) dramatically illustrated that several unconnected, en échelon fault segments can break during the same earthquake. This greatly complicates the estimation of total fault length. How many segments

---

[3]Readers should note that, although each rupture length observation is associated with a corresponding observed magnitude, the regression of rupture length on magnitude (shown here) is not the same as the regression of magnitude on rupture length, even for the same data set. Thus, Figure 13-2 should be used to infer magnitude from rupture length but not vice versa. This seeming dichotomy results from the fact that in fitting the curve, the variances in the independent and dependent variables result in different best fits for the two cases.

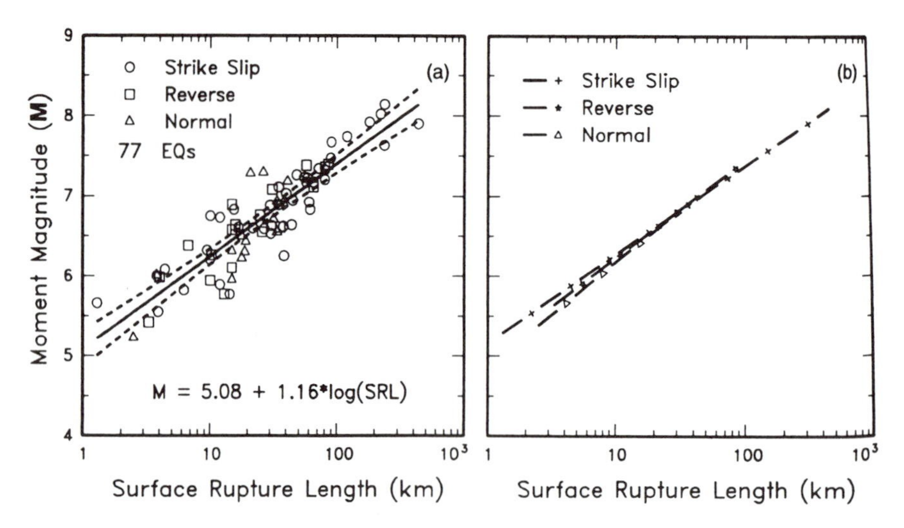

Figure 13–2. Regression of surface rupture length on magnitude for worldwide earthquakes of all slip types. Solid line is ordinary least-squares fit. Dashed lines indicate 95-percent confidence intervals. From Wells and Coppersmith (1994).

can break during a single earthquake? This question was discussed in Chapter 8, but it should be noted that in a deterministic hazard analysis, one must make a single-valued judgment, and to be conservative, ruptures on two or three individual segments are sometimes arbitrarily assumed to represent the maximum event. The probabilistic methodology has the advantage that one can assign judgmental probabilities to several scenarios. For example, one might estimate that, during the maximum earthquake, the probability is 10% that only one segment will break, 50% for two segments, 30% for three segments, and 10% for more than three segments.

Table 13–1
Selected regression relationships from Wells and Coppersmith (1994), for worldwide earthquakes and for all styles of faulting. M is moment magnitude (M_W), SRL is surface rupture length (km), AD is average displacement (m), and MD is maximum displacement (m). Data are valid only within specified magnitude, displacement, and fault-length ranges.

Relationship
Log(SRL) = −3.22 + 0.69 M
Log(MD) = −5.46 + 0.82 M
Log(AD) = −4.80 + 0.69 M
M = 5.08 + 1.16 Log(SRL)
M = 6.69 + 0.74 Log(MD)
M = 6.93 + 0.82 Log(AD)
Log(SRL) = 1.43 + 0.56 Log(MD)
Log(MD) = −1.38 + 1.02 Log(SRL)
Log(AD) = −1.43 + 0.88 Log(SRL)

PROBABILISTIC ASSESSMENTS

Probabilities are used in many ways in probabilistic hazard assessments. Traditionally, seismicity data have been the primary probabilistic tool for projecting future events, but in recent years, geological data—and fault slip-rate data in particular—have played an increasing and often dominant role. And with the increasing availability of Global Positioning Satellite (GPS) data, some of the most innovative assessments have been truly multidisciplinary, as represented by the recent study of seismic hazard in southern California by Ward (1994).

Use of Seismicity Data

The classic earthquake hazard analysis of Cornell (1968) placed major emphasis on seismic data, particularly that obtained from instrumental observations. Gutenberg and Richter (1954) had noted that, on a worldwide basis, earthquake magnitude and frequency had a systematic relationship to one another, namely that earthquakes of one magnitude interval were about ten times as frequent as those of one magnitude unit more. This was expressed in the Gutenberg-Richter *recurrence* relationship,

$$\text{Log } N = a - bM, \tag{3}$$

where N is the number of earthquakes of a given magnitude M or greater, and a and b are constants repre-

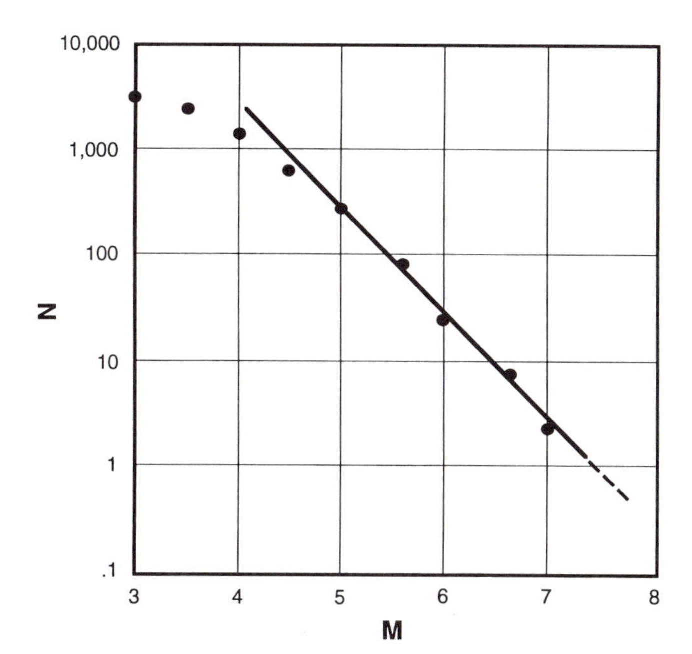

Figure 13–3. Recurrence curve for 30 years of seismic recording in the Imperial Valley of California. **M** is magnitude, and **N** is the number of earthquakes ≥ **M** per unit area, per unit time. Note fall off at lower magnitudes due to failure of the network to detect all small events. Dashed line represents extrapolation to higher magnitudes than those experienced in the recording period.

senting, respectively, the level of seismic activity, and the ratio of small to large events. N is usually normalized to some unit area and unit time (e.g., number of events per year per 1000 km^2), and b turns out to be nearly equal to 1.0 the world over (although small regional and temporal differences in b may be of tectonic significance in themselves). Figure 13-3 is an example of a recurrence curve. The straight-line fit is usually made either by eyeball, by least-squares, or by maximum likelihood methods (recognizing that different points represent different numbers of observations).

The recurrence curve has great initial appeal, because it seemingly allows extrapolations to magnitude levels higher than those included in the data set. If the curve is truly a straight line on this plot, then, for example, one might measure earthquakes of magnitude 3 to 6 over a period of only a few years to delineate the relationship, and then extrapolate the curve as a straight line out to magnitude 8 to determine how often such an event would occur.[4] What could be more "quantitative," and what could be simpler?

In reality, there are many pitfalls in the interpretation of recurrence curves, although they can be a valuable source of insight in some circumstances. Two specific problems are obvious: (1) The relationship would suggest that larger and larger earthquakes should occur with ever decreasing frequency, but there appears to be a practical high-end cutoff, inasmuch as earthquakes much larger than 9.5 simply don't occur. Unfortunately, the seismic data in themselves usually give no hint as to how or where this cutoff takes place, yet it is in this very part of the curve—at large magnitudes—where our interest primarily lies from the point of view of hazard assessment. Youngs and Coppersmith (1985) give an extensive discussion of this problem of the upper-bound earthquake, which is an important consideration. (2) The assumption in making the extrapolation to infrequent large events is that the data set is statistically representative of the longer time period, yet the very existence of temporal seismic gaps shows how grossly fallacious this assumption can be. Unless the sample period includes at least two of the largest events, how can one know whether the sample is indeed statistically meaningful? It is important to note that just because one has enough data to indicate a straight-line relationship, this by no means guarantees that the line accurately reflects the rate of occurrence of large events.

On the other hand, if the sample period is long, and if the area of coverage is large, a recurrence curve may have real predictive value. As was demonstrated for southern California by Allen et al. (1965), a detailed 30-year seismographic record for the entire region, including northern Baja California, Mexico, seemed to give meaningful results in estimating how often large events should occur somewhere within the broad region. But when smaller areas and specific faults were examined, the results were considerably less meaningful and in some cases exactly contrary to reason, as, for example, within the temporal seismic gap along the central San Andreas fault, which last broke during a great earthquake in 1857, but has had virtually no seismicity even at micro-earthquake levels during the present century.

Figure 13-4 shows a schematic example of a classical probabilistic hazard analysis based on the recurrence curve approach. In Step 1, individual areal and line (fault) earthquake sources are identified, and in Step 2, each is assigned a recurrence curve on the basis of recorded seismicity, with an estimated upper-bound earthquake. Step 3 shows the attenuation curves, and associated uncertainties, that are then used to estimate peak accelerations as a function of

[4]Since magnitude is itself a logarithmic function, a recurrence curve is essentially a log-log plot, and all scientists should have learned at an early age the dangers of extrapolations on log-log plots.

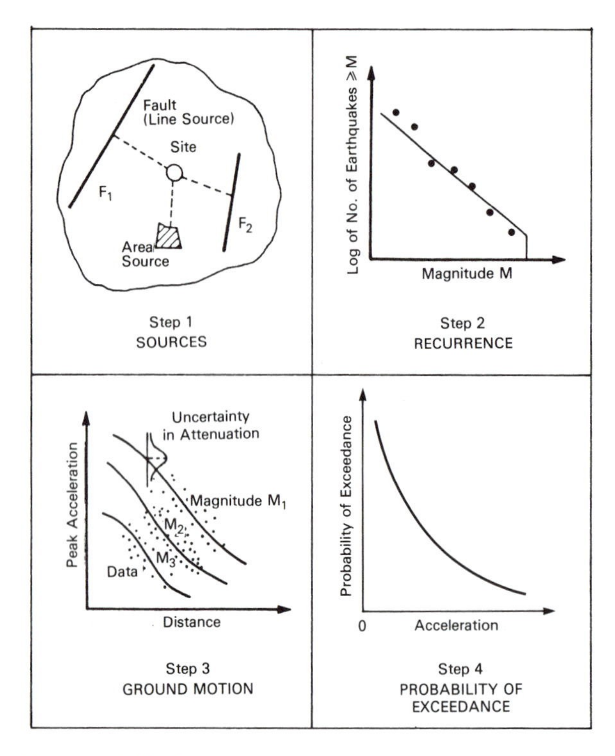

Figure 13–4. Steps in a classical probabilistic hazard analysis based on the recurrence curve approach. From *Earthquake Hazard Analysis* by Leon Reiter, after TERA Corporation (1978). Copyright © 1990 by Columbia University Press. Reprinted with permission of the publisher.

magnitude and distance. In Step 4, the results of Steps 2 and 3 are mathematically combined to give one curve showing the probability of exceedence of given levels of peak acceleration at a site during a specified time period. It is important to note that, unlike a deterministic analysis, events smaller than the maximum event may have a significant influence on the hazard assessment, particularly for smaller accelerations. It should also be noted that the only geological input is the specification of the areal and line sources; no information is used regarding degree of fault activity except that inferred from the seismological record. Furthermore, no account is taken of when the last major earthquake occurred on any of the specified faults or within any of the specified areas.

Figure 13-5 is an example of a regional probabilistic hazard map developed fundamentally by the method of Figure 13-4, in that it is based primarily on historical and recorded seismicity, although the presence of some major known faults was also considered. In this case, the end product is the peak horizontal ground velocities that have a 90% probability of not being exceeded in 50 years. Such maps are often the bases for regional seismic zoning maps, which are, in turn, frequently incorporated into local building codes and regulations. One such map is illustrated in Figure 13-6. Because maps of this type are heavily based on historical earthquakes, the zones of higher

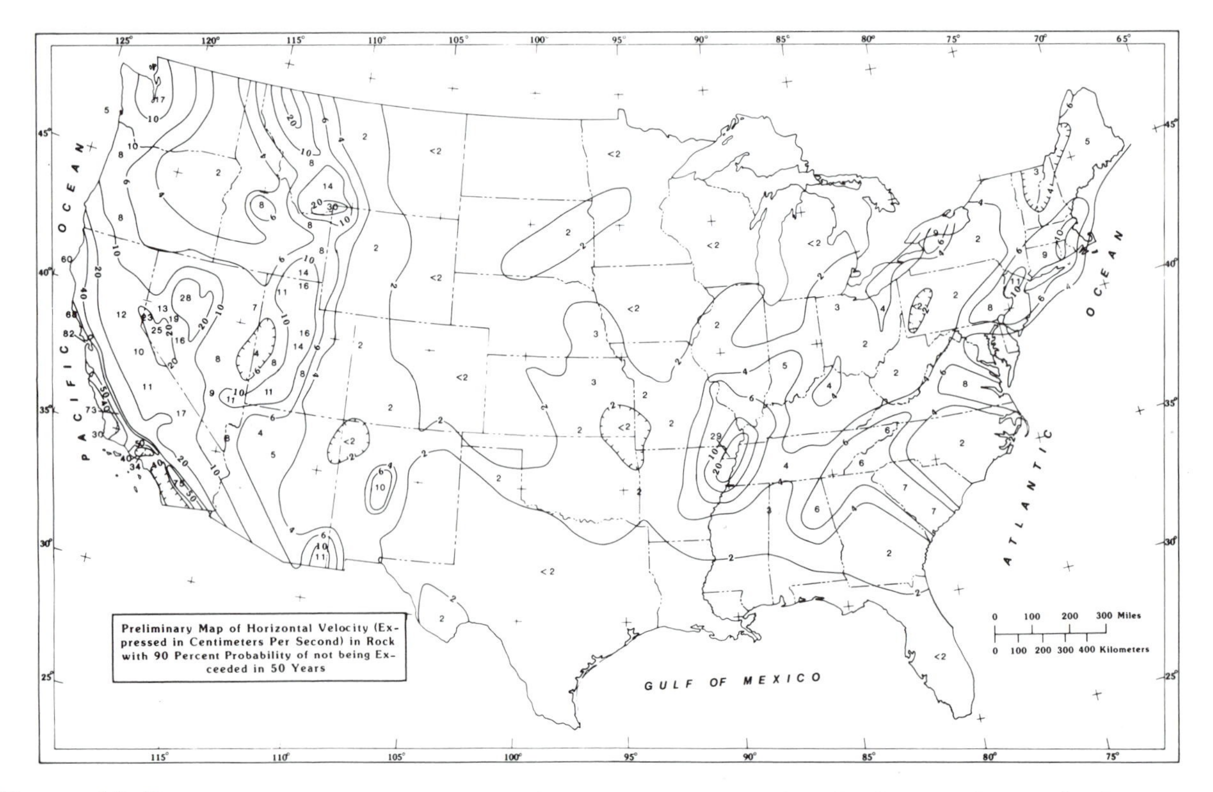

Figure 13–5. Map of estimated peak horizontal ground velocities (cm/sec) for the United States that have a 90% probability of not being exceeded in 50 years. From Hanks (1985), based on the work of Algermissen and others (1982).

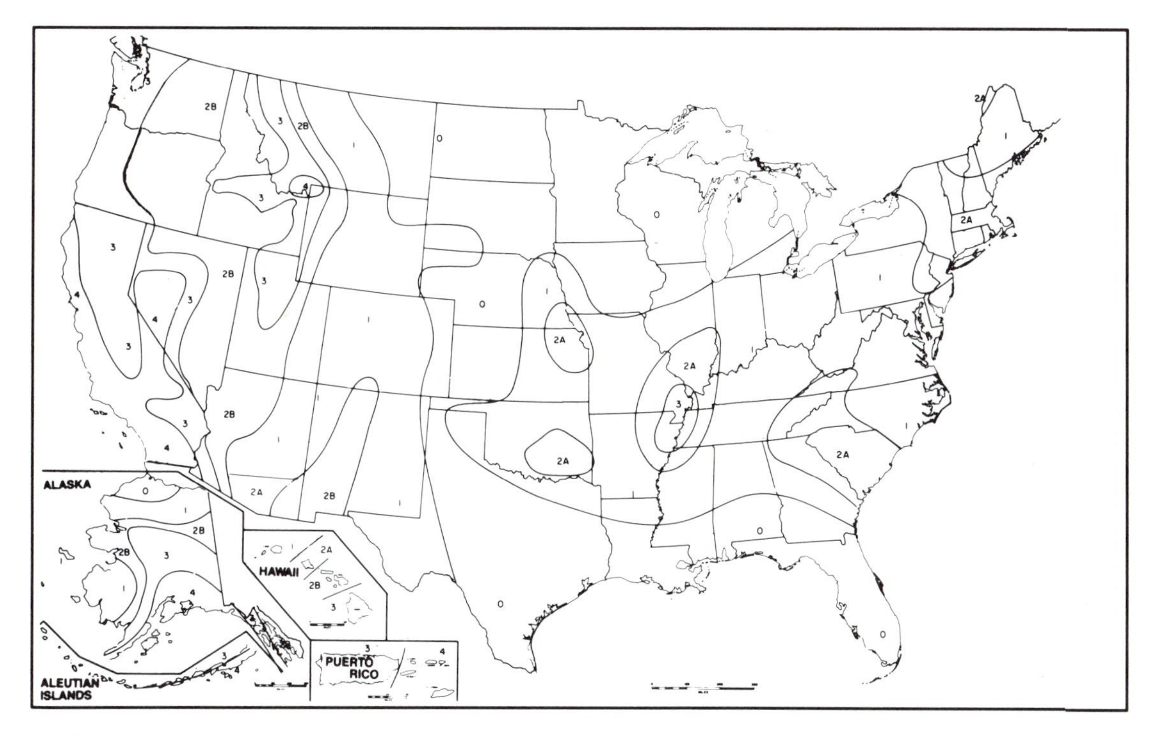

Figure 13–6. Uniform Building Code seismic zoning map of the United States. Higher numbers represent areas of higher hazard and, therefore, more stringent building requirements. Note similarity to Figure 13–5. Reproduced from the 1994 edition of *Uniform Building Code*™, copyright © 1994, with the permission of the publisher, the International Conference of Building Officials.

hazard tend to get broader with each new edition of the map—as earthquakes continue to occur in unexpected (previously quiescent) places. For example, earlier editions of Figure 13-6 did not have the zones of higher hazard in central Idaho and south-central Montana, which were a direct result of the 1959 Hebgen Lake, Montana, and 1983 Borah Peak, Idaho, earthquakes. Similarly, western Oregon has very recently been upgraded to zone 3 based on the perception of the earthquake hazard from the Cascadia subduction zone. This lesson of the broadening of zones with time should not be lost on those preparing such maps.

Another example of hazard maps derived from recurrence curve data, but with added geologic input, is shown for southern Italy in Figure 13-7, from Mayer-Rosa et al. (1993). In this case, all earthquakes within the numbered seismotectonic provinces were associated with one or two identified faults within each zone (although some Italian geologists would debate the single, long fault associations). That is, a and b values were calculated from all the earthquakes occurring from 1900 to 1980 within the individual zones, but these values were then assigned to the individual faults, maximum magnitudes were assigned,

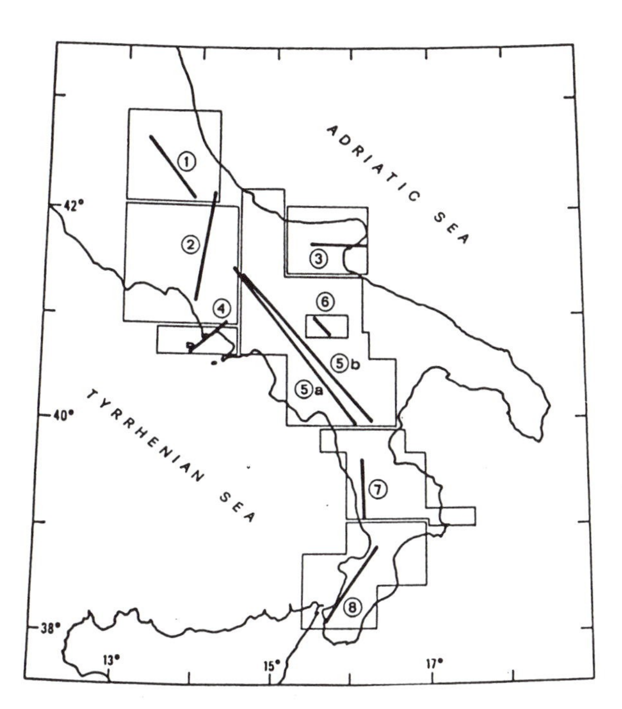

Figure 13–7. Seismotectonic provinces (numbered) and faults adopted for seismic zonation of southern Italy. From Mayer-Rosa et al. (1993).

and the resulting map of maximum accelerations (Figure 13-8) computed by the program of McGuire (1978), which distributed earthquakes along the identified faults. Although the technique has the potential advantage of recognizing individual seismogenic structures, it is subject to the same uncertainties as any other technique that assumes that a relatively short historical record is representative of long-term seismicity.

Occasionally one sees recurrence curves based on intensity instead of magnitude. Although intensity is sometimes the only measure of earthquake size available for older historic events, one must be careful in assuming that differences between intensity levels are integer units; this was not necessarily implied in the qualitative definitions of, for example, the Modified Mercalli intensity scale (Chapter 4).

The Characteristic Earthquake Concept

Although a recurrence curve of the type shown in Figure 13-3, may have long-term significance for a broad area, the application of recurrence curves to the seismicity of individual faults is more debatable, even for long time periods. In studying fault exposures in trenches along the Wasatch fault of Utah, Schwartz and Coppersmith (1984) noted that several segments of the fault showed recent displacements of 1-2 m, but no smaller displacements. From this and other examples, they concluded that individual faults have a tendency to produce repeated maximum earthquakes of a specific size that is characteristic of the particular fault or fault segment, but that smaller earthquakes do not occur systematically in the way that would be suggested by a simple recurrence curve such as that of Figure 13-3; otherwise smaller displacements would have been visible along the Wasatch fault in addition to the maximum ones.

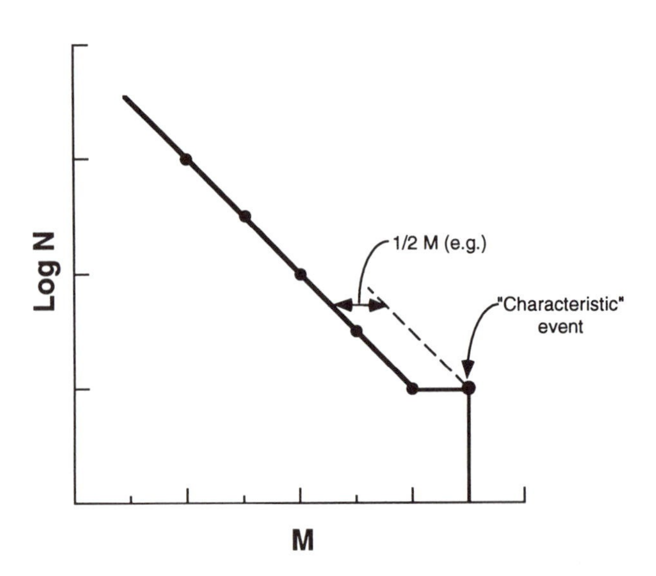

Figure 13–9. Schematic recurrence curve based on the characteristic earthquake model of Schwartz and Coppersmith (1984).

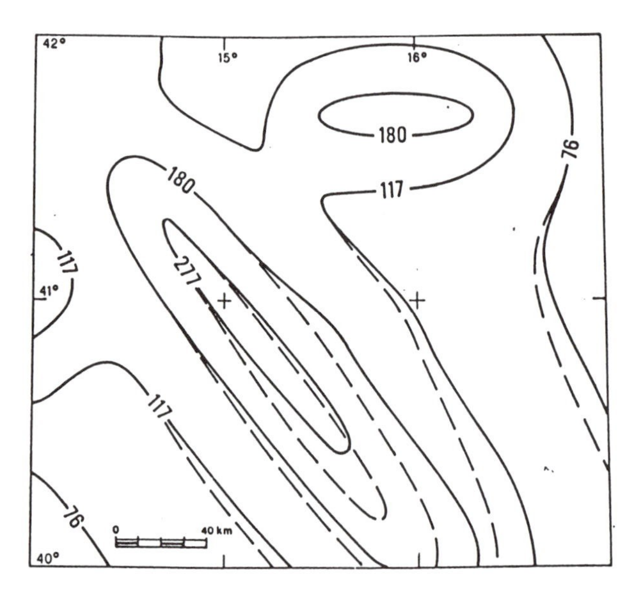

Figure 13–8. Maximum acceleration (in gal) expected not to be exceeded with 68% probability in 200 years, for area of Figure 13–7 from 40–42° N. and 14–17° E. From Mayer-Rosa et al. (1993). Dashed and solid lines represent two fault models.

This is the origin of the concept of the *characteristic earthquake,* and Schwartz and Coppersmith proposed that the recurrence curve for a characteristic earthquake might appear more like that of Figure 13-9, with an offset of perhaps one-half to one magnitude units between the line representing smaller earthquakes and the point representing the maximum "characteristic" event. Supporting evidence is given by the five or more seemingly characteristic earthquakes, each of about magnitude 6 and at nearly the same location, that have occurred semi-periodically during the past 150 years on the San Andreas fault at Parkfield, California. The model of Figure 13-3 would suggest that some 50 events of magnitude about 5 or greater should have accompanied these larger earthquakes, but this does not seem to have happened, at least in the same local area. Characteristic earthquakes have also been suggested by paleoseismic studies based on observations of repeated displacements of the same amount, such as along the Irpinia earthquake fault of southern Italy (Pantosti et al., 1993). Even some recent theoretical studies, such as that by Madariaga (1994) have supported the concept of the characteristic earthquake. It is important to note that the concept of the characteristic earthquake applies only to the earthquake's magnitude and associated displacement; it does not necessarily imply uniform or quasi-periodic recurrence intervals between characteristic events (Hecker and Schwartz, 1994).

Although the characteristic earthquake concept is one of continuing lively debate, it seems likely that Figure 13-9, representing the *characteristic earthquake* model, is much more representative of the seismicity of an individual fault than is that of Figure 13-3, representing the *exponential* model. This has important implications in the quantitative assessment of regional seismicity utilizing geologic data such as slip rates of individual faults.

Use of Geologic Data

A major objective of probabilistic seismic hazard assessments is the determination of the mean *recurrence interval* between large earthquakes on a specific fault or within a given region. Such results can, of course, be used directly to infer the probability of such an event occurring in the near future, particularly if one knows the time of the last such event. From geological field observations, however, it is seldom that numerous prehistoric earthquakes can be identified and dated with sufficient accuracy to provide a valid mean long-term recurrence interval. Nor, of course, is the historical record normally of sufficient length for such a determination, particularly for the large events with which one is primarily concerned in a hazard analysis. Instead, the primary geological input is usually that of the long-term *slip rate* on a fault, i.e., the slip rate measured over a sufficiently long time interval to encompass several characteristic earthquakes. Examples of slip-rate determinations are given in Chapters 8-11.

Ideally, fault slip rate should be readily converted to recurrence interval by the simple arithmetic relationship,

$$T_R = D/V, \quad (4)$$

where T_R is the mean recurrence interval, D is the mean displacement-per-event, and V is the long-term slip rate. However, several important assumptions are inherent in this seemingly straightforward relationship—all of them debatable: (1) It assumes the characteristic earthquake model, with similar-sized maximum events; (2) It assumes that, if D is to be obtained from the magnitude of the characteristic earthquake, an appropriate magnitude is assigned to that event, and it assumes a single-valued relationship between magnitude and fault displacement; (3) It assumes that all of the surface slip is represented by characteristic earthquakes and none by smaller ones; and (4) It assumes the basic tenets of the elastic rebound theory in that earthquakes are assumed to occur periodically when given strain levels are reached, i.e., the *time predictable* model (Reid, 1910; Shimazaki and Nakata, 1980).

An instructive example of the use of geologic data and of equation (4), as well as the treatment of uncertainties, is a recent study of 30-year probabilities[5] of large ($M \geq 7$) earthquakes on the various faults that traverse the San Francisco Bay area of California (Working Group on California Earthquake Probabilities, 1990). The specific geologic inputs were (1) the choice of fault segments, (2) the slip rate on each segment, (3) the expected magnitude of the next characteristic (segment-rupturing) earthquake on each segment, and (4) the time of the last characteristic earthquake on each segment. Except for the last item, major uncertainties accompanied each of these determinations, and it was important that these uncertainties be properly incorporated into the results.

The crux of the Bay Area probabilities study is indicated by Figure 13-10, which schematically portrays the probability density function as a function of time since the last segment-rupturing earthquake, which occurred at $t = 0$. If the mean annual recurrence time, T_R, were known exactly, and if nature faithfully subscribed to this, then the density function would be a simple spike at the year $t = T_R$. That is, the probability would be 100% that the earthquake would occur in that year. Similarly, if earthquakes occurred randomly and had no "memory" of previous events, the curve would be a horizontal line, and the probability of occurrence would be the same every year. However, in the time-predictable model that was used in this study, the probability density function of Figure 13-10 is a bell-shaped curve for two reasons: (1) Neither the displacements-per-event nor the long-term slip rates are known precisely, and this "parametric uncertainty" is reflected in the shape of the curve. (2) Even if the mean recurrence interval derived from these two parameters were known exactly,

[5]The choice of the 30-year time frame was arbitrary. Since the purpose of the study was to increase public awareness of earthquakes in the Bay area, the Working Group decided that a 1-year period would yield probabilities too low to attract attention, whereas a 100-year period was too long to be considered relevant, even with the higher probabilities. The period of 30 years was chosen, because it appeared to be within the realistic attention span of citizens and political leaders.

nature itself displays an "intrinsic uncertainty" in that there is a variability in the actual recurrence intervals. The broadness of the bell-shaped curve (i.e., its dispersion) reflects, in effect, the sum of these two uncertainties; if the uncertainties are low, the curve is highly peaked, whereas increasing uncertainties tend to broaden it. In the case of the Bay Area study, the geologists supplying the displacement-per-event and slip-rate data themselves estimated the parametric uncertainties, and the intrinsic uncertainty was estimated from the observed variability of characteristic recurrence times for circum-Pacific earthquakes (Nishenko and Buland, 1987). Furthermore, the curve was assumed to have a lognormal distribution. This and other details of the analysis, including the actual input data, are discussed in the Working Group report.

If T_e is the elapsed time since the last segment-rupturing earthquake (e.g., today), and ΔT is the time interval over which the probability is being calculated (in this case, 30 years), then probability of the earthquake occurring within the interval ΔT is given by the ratio of the dark-shaded area in Figure 13-10 to the sum of areas with light and dark shading. It is termed a "conditional" probability, in the sense that it is conditional upon no segment-rupturing earthquakes having occurred between $t = 0$ and $t = T_e$. As would be expected intuitively, and is obvious from Figure 13-10, if the test period, ΔT, occurs early in the recurrence cycle, the probability of recurrence during the period is low, whereas late in the cycle it becomes very high. Indeed, if the mean recurrence interval is, e.g., 100 years, and 200 years have passed since the last big earthquake, one must conclude either that the probability of its recurrence within the next year is very high, or (perhaps more likely) that the model is wrong.

The results of the Bay Area study are shown diagrammatically in Figure 13-11. Thirty-year probabilities from 1990, P, are shown for six specific fault segments, which had also been identified by geologists as the most likely locations of past and future characteristic events. These segment designations were relatively straightforward assignments, with the exception of the San Francisco Peninsula area, where uncertainties and differences of opinion on segment boundaries, as well as on appropriate slip rates and times of past events, necessitated a logic-tree analysis (see following section) to give a consensus opinion.

It is important to note that the numerical probabilities of Figure 13-11 reflect the consensus opinion of the Working Group, *contingent upon the model being correct,* namely the time-predictable model of Reid (1910) and Shimazaki and Nakata, (1980). While most members of the Working Group would probably agree that the model was the best that was avail-

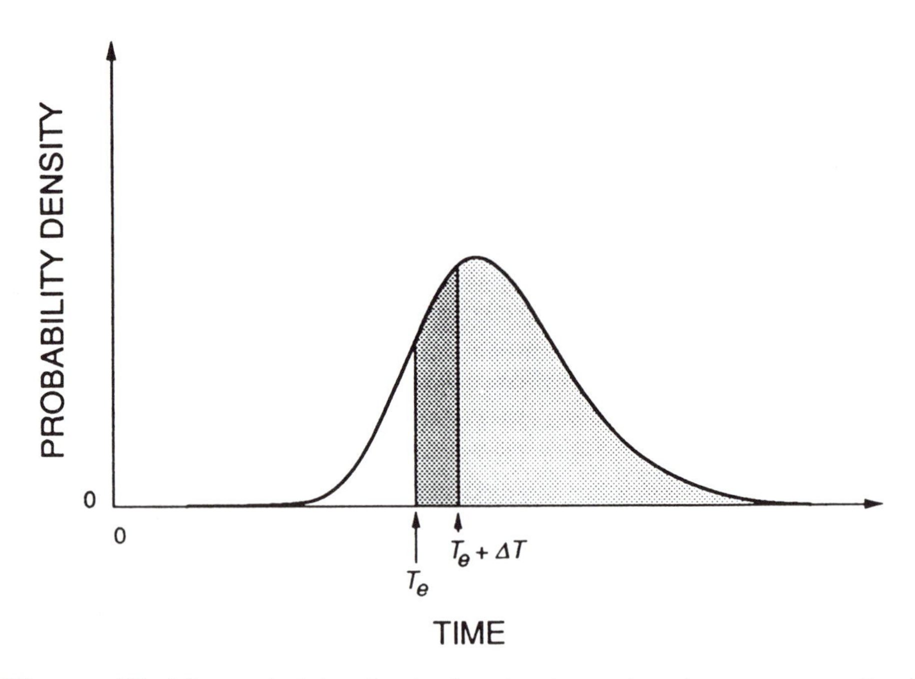

Figure 13–10. Probability density function for earthquake recurrence. Conditional probability for recurrence in time interval ΔT, given the elapsed time T_e since the last segment-rupturing earthquake, is the ratio of the area of dark shading to the sum of the areas with dark and light shading. From Working Group on California Earthquake Probabilities (1990).

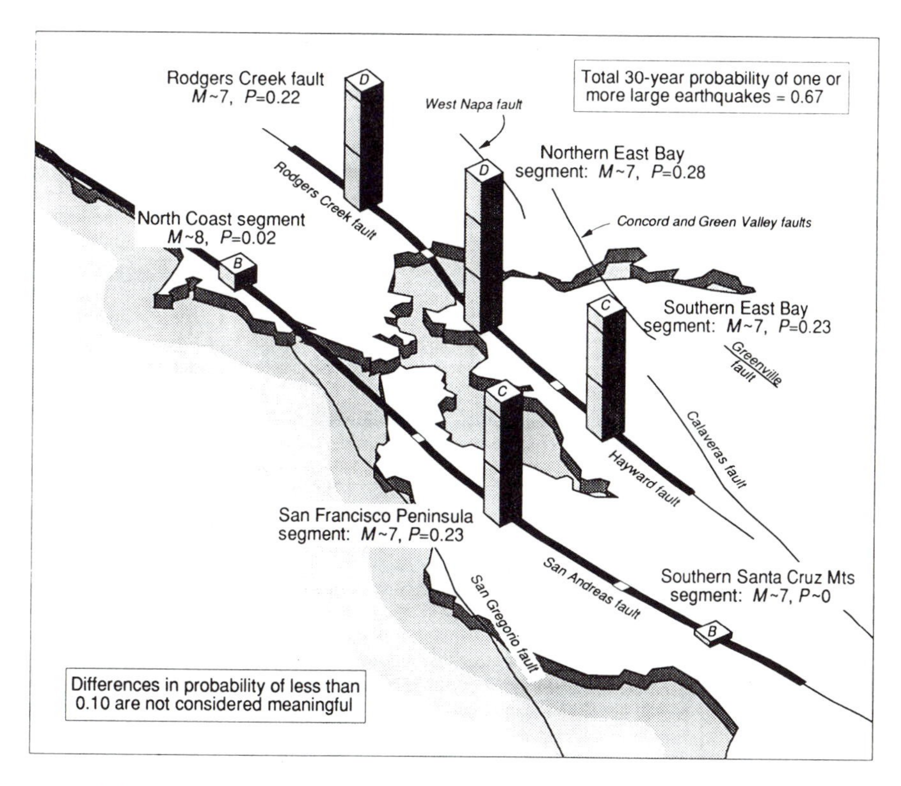

Figure 13–11. Thirty-year probabilities (P) of large earthquakes (M ≥ 7) in the San Francisco Bay region. From Working Group on California Earthquake Probabilities (1990).

able at that time, no one would claim it necessarily to be 100% correct, and all of the numbers shown are probably on the high side by some undetermined factor. Furthermore, as was specifically acknowledged in the Working Group report, the *total* 30-year probability for the Bay area (0.67 in Fig. 13-11) must represent a minimum probability since there may be other faults capable of producing M ≥ 7 earthquakes that have not been included. For example, the offshore San Gregorio fault was not included in the study simply because of the lack of relevant geologic data.[6]

But more importantly, recent paleoseismic studies of faults within the San Andreas fault system indicate that in many cases, ground-rupturing earthquakes have *not* occurred periodically, but instead have been clustered in times of increased seismicity (e.g., Grant and Sieh, 1994; Rockwell and Sieh, 1994). And some recent theoretical models based on considerations of different friction laws predict that, on a long fault system, large earthquakes will occur "irregularly in a completely non-periodic way" (Madariaga, 1994).

A somewhat similar study using fault slip rates was carried out by (Wesnousky, 1986) for all of onshore California, except that the results were carried through to indicate specific ground-motion parameters such as peak acceleration (Fig. 13-12). Furthermore, Wesnousky used *seismic moment rate* as a fundamental parameter. As was discussed in Chapter 4, seismic moment—as opposed to magnitude—is defined in terms of specific physical parameters of the rupture process and thus has many advantages, particularly where it can be calculated independently from the source spectra of body and surface waves (Hanks and Kanamori, 1979; Kanamori and Anderson,

[6]On November 24, 1987, two of the authors of this book were participating in the meeting of a committee charged with assigning probabilities to the occurrences of large earthquakes throughout California. Following a long discussion, it was decided to delete the Superstition Hills fault of southeastern California from the list, simply because we did not have adequate slip-rate data to assign meaningful probabilistic parameters. Less than three hours following this decision, the Superstition Hills fault ruptured in a magnitude 6.6 earthquake (Appendix), as if to "get even" with the committee! The lesson would seem to be that, at least in seismic hazard assessment, ignorance should not be equated with bliss.

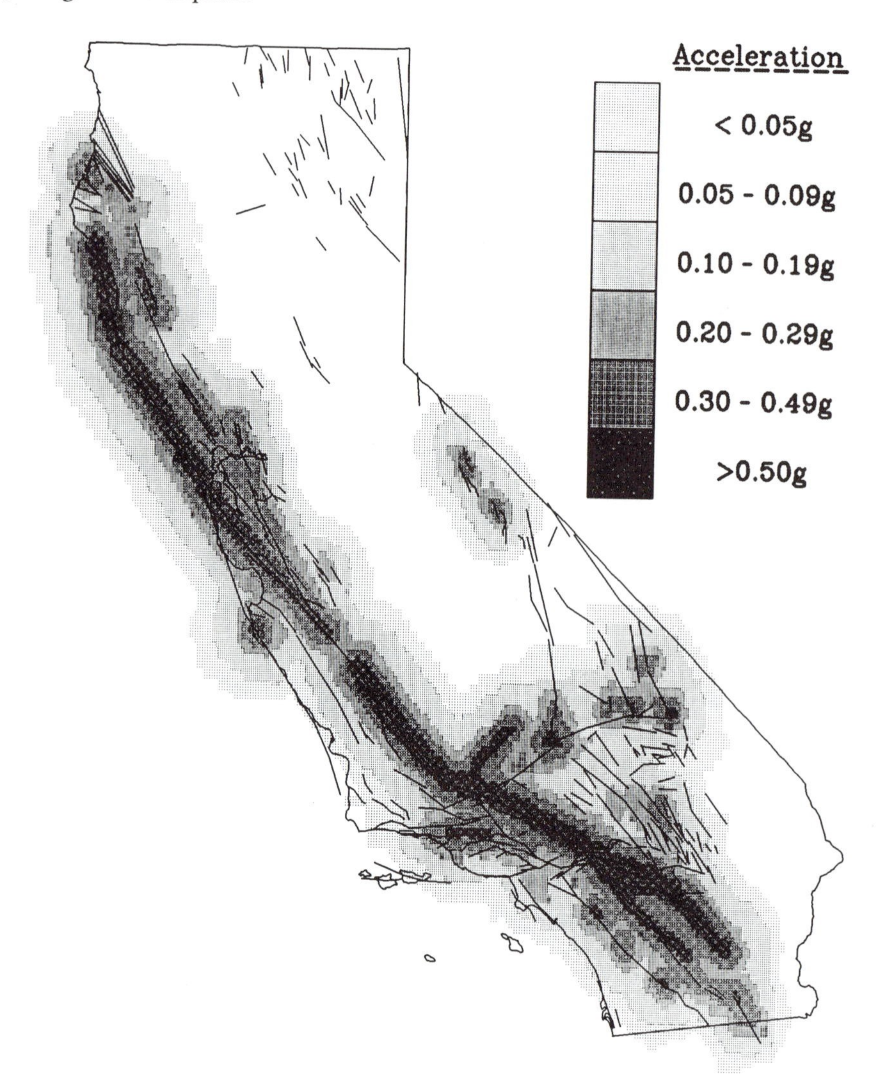

Figure 13–12. Peak acceleration expected to be exceeded on rock sites at the 10% probability level during a 50-year period due to earthquakes on Quaternary faults mapped onshore in California. From Wesnousky (1986).

1975). Analogous to equation (1), seismic moment rate is defined as

$$\dot{M}_o = \mu A V, \tag{5}$$

where V, as in equation (4), is the long-term slip rate. In analogy with equation (4), the mean recurrence interval is then

$$T_R = M_o^e / \dot{M}_o^g \tag{6}$$

where M_o^e is the seismic moment of the expected event, proportional to fault rupture length, and $\dot{M}_o^g$ is the moment rate, determined from the fault slip rate. Because of the absence of information on the time of the most recent characteristic earthquake on most of the 275 faults and fault segments that Wesnousky considered in his study, his maps (e.g., Fig. 13–12) necessarily assume that the occurrence of ground shaking is a Poisson process[7] and is independent of the recent history of large earthquakes.

Before leaving the subject of recurrence intervals, it should be pointed out that, particularly within "stable" continental regions, recurrence intervals between large earthquakes, or clusters of earthquakes, can be very long. Crone and Machette (1994) argue that such intervals as long as 100,000 years have typ-

[7]A Poisson process is one of uniform randomness, with no memory or time-dependence.

ified earthquakes on specific faults in the interior of Australia, and (Adams, 1974) suggest that the last event prior to the 1989 Ungava, Quebec, earthquake was more than 1 million years ago! Although the Ungava event occurred on a preexisting Archean fault, Reiter (1990) points out that there is such a multitude of similar faults throughout the Canadian shield, that the only realistic earthquake hazard assessment in the region must be based on the probability of a random or "floating" earthquake.

Logic Trees

An added refinement to many probabilistic assessments is the use of the *logic tree* technique, which is simply a means of formalizing and systematizing the thought process when several sequential decisions are to be made in a complex analysis, and each decision in the sequence is given an estimated probability of being correct. Logic trees are typically used in conjunction with expert judgment, where the probabilities are formally assigned by some sort of group consensus, but the technique can just as well be used by a single "wise" individual.

As a very simple example, consider the problem mentioned earlier in the chapter, that of the number of fault segments that are assumed to break during the maximum earthquake on a given fault. And then let us extend this to the magnitude of the resulting earthquake. In Figure 13-13, the four *branches* following the first *node* represent the four discretized possibilities, each with an estimated probability attached: only one segment rupturing, with an estimated probability of 10%; two segments, at 50%; three segments, at 30%; and four segments, at 10%. The probabilities assigned to the four branches must, of course, sum to 1.0. Now from each of the resulting four nodes, let us independently estimate the magnitude of the associated earthquake, assuming 25-km segment lengths and using magnitudes derived (approximately) from the data of the Appendix. And let us assume that we assign a 60% probability to the average magnitude value derived from Table 13-1, together with 20% probabilities for magnitude values $^1/_4$ units higher and lower. In the final column of Figure 13-13, each of the resulting twelve numbers is the estimated probability of that specific path being correct, obtained simply by multiplying the probabilities of each of the two sustaining branches. Since the probabilities for a magnitude 7 event, for example, result from three of the paths, its total probability is .38, which is the sum of the three individual probabilities. All of this is contingent, of course, on the individual probabilities having been estimated correctly, and the old adage of "garbage in, garbage out" applies just as well to logic trees as it does to other computer applications. Furthermore, the technique is dependent upon the ability to discretize the various choices, which in many cases may, in fact, represent a continuum of possibilities; and, of course, the choices must encompass the full range of realistic possibilities.

Figure 13-14 is the actual logic tree used in the geologic evaluation of the Hosgri fault for the seismic evaluation of the Diablo Canyon Nuclear Power Plant, California. Note that, as an example, only one path is carried through at each node following the first on Figure 13-14. If every path were to be completed, as

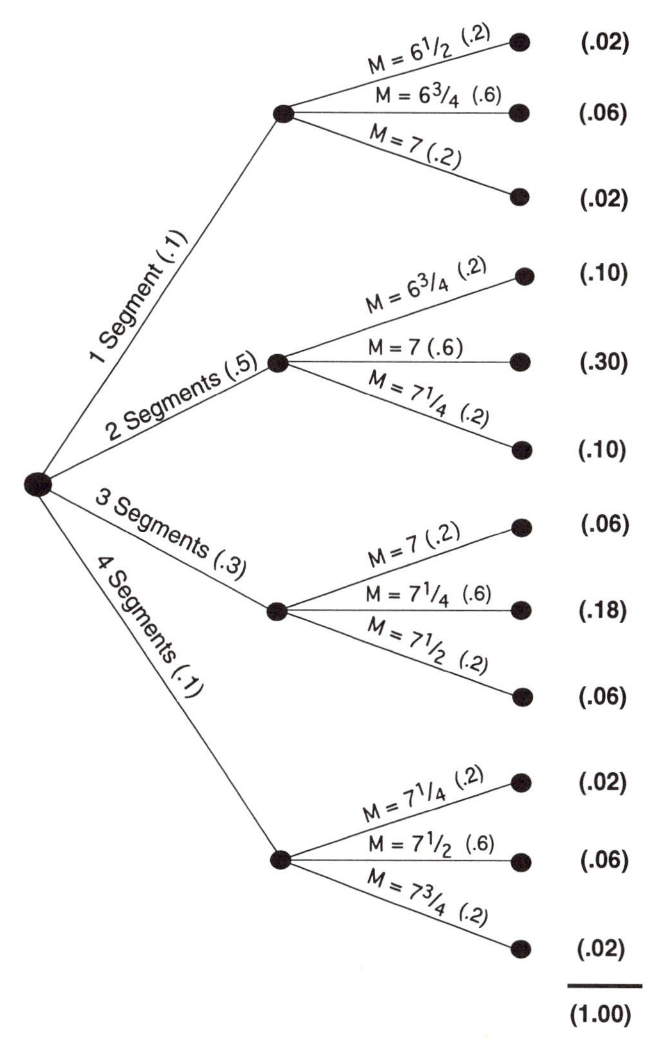

Figure 13–13. Hypothetical logic tree for estimating likelihoods of maximum earthquake magnitudes for a fault zone, depending on the numbers of 25-km-long segments rupturing during the characteristic earthquake. Estimated probabilities are in parentheses.

it was in the actual analysis, some 4000 individual branches would be present at the final column of the tree, and it is obvious why computer techniques are necessary in complex logic tree analyses.

The first step in the Hosgri fault analysis was the judgment of style of faulting, which was important because of the proximity of the fault to the reactor and of the dependence of ground-motion values on this parameter. The consensus of the expert team, based on extensive field work and seismological studies, was that the probability was high (65%) that the fault was strike slip, with lower probabilities assigned to oblique (30%) and thrust (5%) mechanisms, each category of which was specifically defined in terms of the rake of the slip vector. For each of these three sense-of-slip cases, probabilities were assigned, in turn, to the fault dip, and so forth through the eleven successive nodes. Since many of the probabilities were judged to be the same at each node in a given column, such as the choice between the "exponential" and "characteristic" magnitude distribution, the total number of individual decisions by the expert team was only a small fraction of the total number of branches in the last column. The resulting slip rates and magnitude distributions in the last columns were then used in the total hazard analysis, combined with contributions from other faults and earthquake sources.

It is important to note that the thrust-fault hypothesis, for example, contributed quantitatively to the final hazard estimate, albeit it in a minor way because of its low estimated probability. However, in a typical deterministic assessment, only a single path out of the ±4000 would have been declared "correct," with no weight given to *any* competing hypotheses. One could, on the other hand, use a combined approach in which, after construction of the logic tree, certain cases could be picked out or combined as the basis for a deterministic assignment.

From a mathematical point of view, the order of decisions in the logic tree makes no difference, as long as the decisions are independent. From the geological point of view, however, it is clear that some de-

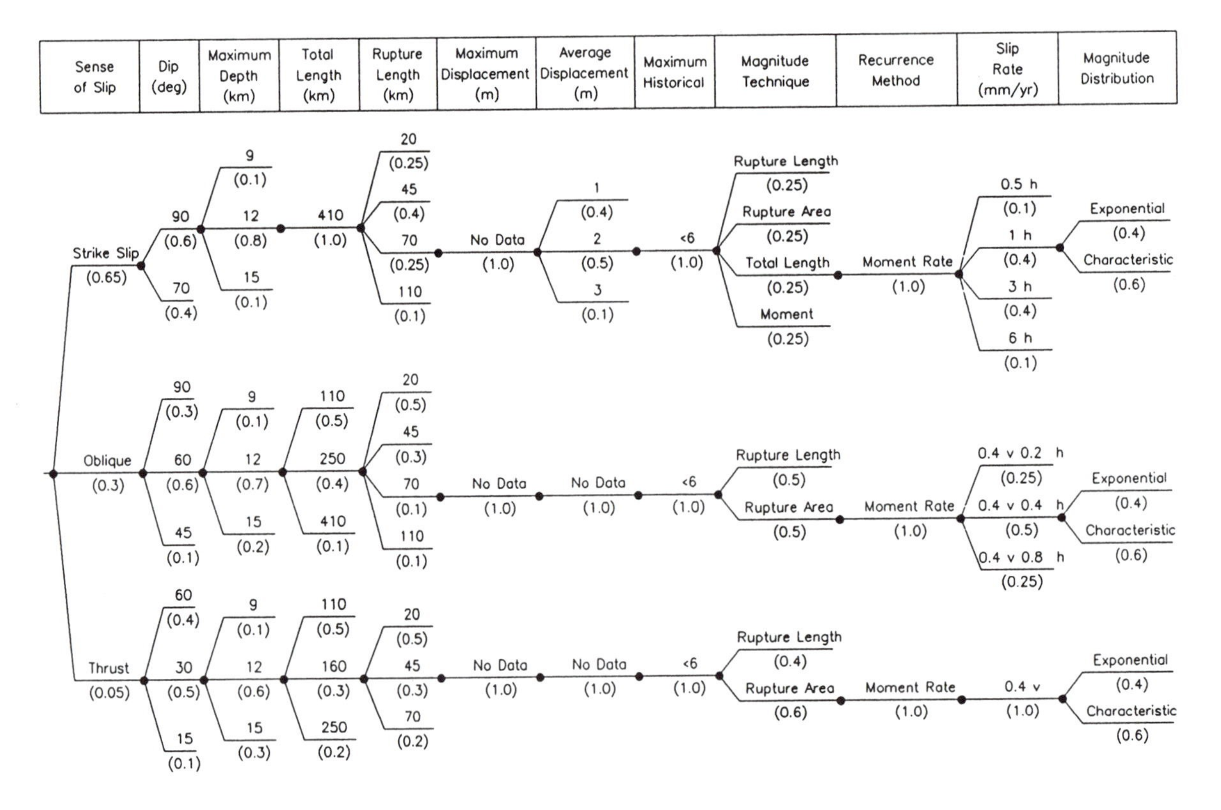

Figure 13–14. Logic tree used in assessment of the Hosgri fault for the Diablo Canyon Nuclear Power Plant, California. Note that in this example, only one branch is carried forward at most nodes. Otherwise, some 4000 branches would have been present in the right column! From Pacific Gas and Electric Company (1988).

cisions logically precede others. For example, fault-dip probabilities must be considered in the light of the fault type, i.e., its dominant sense of slip.

HAZARD ASSESSMENT OF BLIND THRUSTS

Blind thrusts represent a particularly difficult problem for seismic hazard assessment, because the geologist cannot directly examine the buried faults whose displacements cause such earthquakes. Furthermore, because of the low angle of dip of typical thrust faults, whether blind or not, earthquake hypocenters are much more difficult to correlate with individual structures than is the case for steeply dipping faults.

Blind thrusts are now recognized as significant sources of seismic hazard in many worldwide areas of active folding (Stein and Yeats, 1989), including parts of Taiwan, Iran, Argentina, India, Algeria, and California. Three recent damaging earthquakes in California were caused by displacements on blind thrusts, the $M = 6.3$ Coalinga earthquake of 1983, the $M = 5.9$ Whittier Narrows earthquake of 1987, and the $M = 6.7$ Northridge earthquake of 1994. All three have been extensively studied (e.g., Bennet and Sherburne, 1983; Hauksson and Jones, 1989; Scientists of the U.S. Geological Survey and the Southern California Earthquake Center, 1994). Fortunately, the three events are located in areas of oil production, so that both extensive drilling records and state-of-the-art seismic profiles are available to increase the understanding of geologic structures at shallow depth, although not, unfortunately, at the depth of primary seismic strain release.

The construction of balanced cross sections (Chapter 2) has been of paramount importance in gaining an understanding of blind thrusts and their seismic hazard, as exemplified by the work of Namson and Davis (1988), and Shaw and Suppe (1994). In particular, specific blind thrusts and thrust ramps have been identified as suspect, and long-term rates of tectonic convergence can be calculated from such reconstructions if stratigraphic control is adequate. Perhaps the most promising new data have come from measurements of the satellite-based Global Positioning System (GPS), which, within a period of only a few years, are now capable of accurately determining convergence rates across folded terranes, although the determination of surficial uplift rates of active anticlines (which are typically underlain by blind thrusts) is markedly less accurate than the horizontal measurements. Geomorphic studies are also promising, in that deformation of late Quaternary stream or coastal terraces may provide quantitative data on the uplift rates and locations of active folds. Indeed, the locale of the 1987 Whittier Narrows, California, earthquake was identified more than 60 years ago (Vickery, 1927) as an active anticline on the basis of warped geomorphic surfaces. On the other hand, it was the demonstrated *absence* of deformation of the 80,000-year-old marine terrace that virtually encircles the Diablo Canyon Nuclear Power Plant in California that led, in part, to the persuasive argument that active folds and underlying blind thrusts do not represent the dominant seismogenic process at that locale today.

Seismic hazard assessment of blind thrusts is still in its infancy, and some of the challenging specific problems are these:

1. In some areas, such as southern California, convergent folded terranes are also characterized by throughgoing strike-slip faults, and the relative role of each in hazard assessment is not clear. A balanced cross section necessarily assumes pure dip slip in its construction. Many dominantly thrust earthquakes in folded regions, such as the $M = 6.4$ San Fernando, California, earthquake of 1971, have had significant components of horizontal slip, as indicated by both focal mechanisms and surface observations.
2. Even if rapid convergence across an actively folding terrane can be documented, it is not clear how much of this convergence is being expressed seismically. Deeper detachments or fault structures, in particular, may deform primarily by creep processes.[8]
3. Balanced cross sections are generally not unique geometric interpretations, for a variety of reasons, and it is important to understand and quantify their uncertainties.

[8]Recent GPS measurements indicate that the borders of the folded Los Angeles basin are converging at the very fast rate of about 8 mm/yr, yet no large earthquakes have occurred within the area during the 150-year-long historical record. Either (a) much of the convergence is being manifested by nonseismic processes (e.g., creep at depth), or (b) the last 150 years represents a temporal seismic gap sandwiched between prehistoric and yet-to-come great earthquakes or much higher degrees of seismic activity. The solution of this quandary is obviously important to the residents of Los Angeles. Dolan et al. (1995) have argued in favor of alternative (b), noting the absence of ongoing surface deformation that would presumably have to accompany flow at depth, as well as the absence of high micro-earthquake activity in the basin, which might also be expected to accompany fault creep at depth.

4. Many regions of active folds are also characterized by active thrust faults that reach the ground surface, and, in a seismic hazard analysis, it is not always clear how these faults are related to underlying blind thrusts. In the M = 7.7 Kern County, California, earthquake of 1952, about one third of the total fault length (as indicated by aftershock distribution) was represented by surface rupture, and two thirds was blind (Appendix).
5. Segmentation of surface faults is a difficult enough problem (Chapters 8-11), but segmentation of blind thrusts is even more conjectural. Whereas many fold belts, such as those of Taiwan and Iran, extend for long distances, it is not at all clear how extensive are the individual seismogenic ramps or thrusts within the belt, and thus how great the potential is for truly large earthquakes.
6. Although balanced cross sections may provide accurate data on average deformation rates during the period of deposition and folding, modern earthquakes are primarily related to Holocene and late Quaternary deformation rates, which may or may not be similar to the longer-term rates.
7. Detailed seismological studies, utilizing dense seismographic networks, have thus far generally failed to recognize deeply buried detachments or individual blind thrusts, as, for example, through very shallowly dipping focal mechanisms. However, many regions have a relatively abrupt "floor" to micro-earthquakes that might represent a detachment surface, and the subject is one of intense study at the present time.

At the present time, therefore, considerable judgment is called for in assessing earthquake hazards related to blind thrusts, and uncertainties must be recognized as being very high. Regional comparisons and earthquake histories take on added importance, as, particularly, do geodetic observations (e.g., Ward, 1994). Recognizing that such earthquakes do, in fact, occur, but that individual hypocenters and magnitudes are difficult to specify, the *"floating earthquake"* hypothesis is sometimes used, wherein a duly conservative earthquake with appropriate fault parameters (e.g., depth, focal mechanism) is assumed to occur with equal likelihood *anywhere* within the suspect area, such as an area of folded rocks thought to be underlain by thrusts and ramps that have not been individually identified. The concept is somewhat akin to the seismotectonic province argument mentioned earlier.

TRIGGERED EARTHQUAKES

Only a few decades ago, it would have seemed ridiculous to suggest that significant earthquakes could be triggered by human activities. However, we now recognize that this has, in fact, happened in a number of ways. For example, earthquakes have been triggered by the filling and emptying of reservoirs, by the injection and withdrawals of fluids (particularly water), by underground nuclear explosions, and by mining operations. Furthermore, we also now realize that some natural earthquakes have triggered others at large distances. One can only conclude that many parts of the earth's crust, even in normally nonseismic "stable" areas, are in a state of quasi-equilibrium that is very close to the failure point, and that only a small perturbation can sometimes lead to failure. This was convincingly demonstrated in a controlled experiment at the Rangely oil field, Colorado, where small earthquakes were effectively turned on and off by the injection of water at various pressures (Raleigh and others, 1976). In all of these cases of triggered earthquakes, focal mechanisms indicate that failure has taken place by the usual faulting (i.e., shearing) process. That is, they do not represent a special "class" of earthquakes that is mechanically distinct from others.

Earthquake Triggered by Other Earthquakes

It is now known that earthquakes, particularly large ones, sometimes trigger other earthquakes at distances far greater than can be explained by the static stress changes associated with the fault dislocation of the causal event. Probably the best documented case is that of the 1992 Landers, California, earthquake (M = 7.3), which was followed, in the minutes and hours after the event, by an increase in seismicity throughout much of the western United States, particularly in areas of late Quaternary volcanism and geothermal activity, and at distances of up to 1250 km from the Landers epicenter (Hill et al., 1993). Noteworthy among these was a magnitude 5.6 earthquake at Little Skull Mountain, Nevada, at a distance of 240 km, that caused some local damage. Hill et al. (1993) conclude that the most likely triggering mechanism involves "nonlinear interactions between large dynamic strains accompanying seismic waves from the mainshock and crustal fluids (perhaps including crustal magma)." Why the triggering effects of the Landers earthquake were so much more noticeable than for other earthquakes of similar size is still a bit of a mystery, although the exceptionally strong directivity of the Landers seismic radiation toward the north—in the direction of most of the triggered activity—probably played a significant role.

The triggering of earthquakes by other earthquakes remains an intriguing seismological problem, but from the point of view of seismic hazard assessment, it makes little difference whether a local earthquake is triggered by a distant event or not. This is because our assessment of the earthquake potential for a given area or a given fault is based primarily on its history and, particularly, prehistory of earlier earthquakes at that site. It may well have been that these earlier earthquakes were triggered by distant large events, perhaps many times. But it is still the behavior of local faults that is the focus of hazard assessment, because they are the ones that cause the local shaking, *whether or not* their displacements are triggered by more distant events.

Reservoir-Triggered Earthquakes

Of earthquakes triggered by man, our focus herein is on those that are related to reservoirs, because this is the area of primary concern for triggered earthquakes in seismic hazard and risk assessment. At least three events triggered by reservoir filling have exceeded magnitude 6.0, and in two of these cases—Koyna, India (M = 6.5), and Xinfengjiang, China (M = 6.1)—hundreds of deaths resulted, and the associated dams were severely damaged and came uncomfortably close to disastrous failure.

Some reservoir-triggered earthquakes have commenced almost immediately upon the first filling of the reservoir, while others have been delayed for a number of years, following several cycles of reservoir filling and drawdown. The longer the response time, of course, the greater is the difficulty in demonstrating a true cause-and-effect relationship. Simpson et al. (1988) have argued that these two types of behavior represent two distinct causal mechanisms—one related to the elastic stress due to the load of the reservoir itself, and the other to the gradual diffusion of water from the reservoir to hypocentral depths. Direct loading can increase shear stresses at depth, and increased pore pressure can decrease the effective normal stress across potential planes of failure. Increased pore pressure, in turn, can be caused either by the volumetric strain component of the elastic field or by diffusion of pressure directly from the reservoir. It is important to recognize that in any case, the predominant energy released in the triggered earthquake is from preexisting tectonic strain, and that the added effects serve mainly to perturb this preexisting state. To emphasize this fact, the term "triggered earthquake" is preferred over "induced earthquake."

Figure 13-15, from Simpson (1986), shows several examples of the history of reservoir fillings and asso-

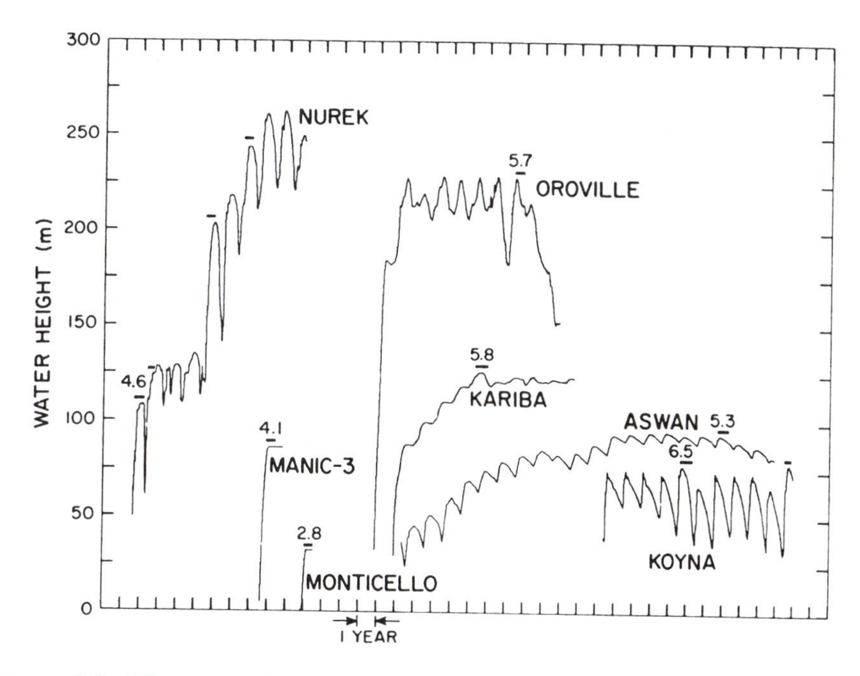

Figure 13-15. Water levels and times of main earthquake activity at selected reservoirs. Timescales for all plots are the same, but origins have been adjusted for clarity. Numbers indicate magnitudes of largest events; bars are times of significant increases in seismicity. From Simpson (1986).

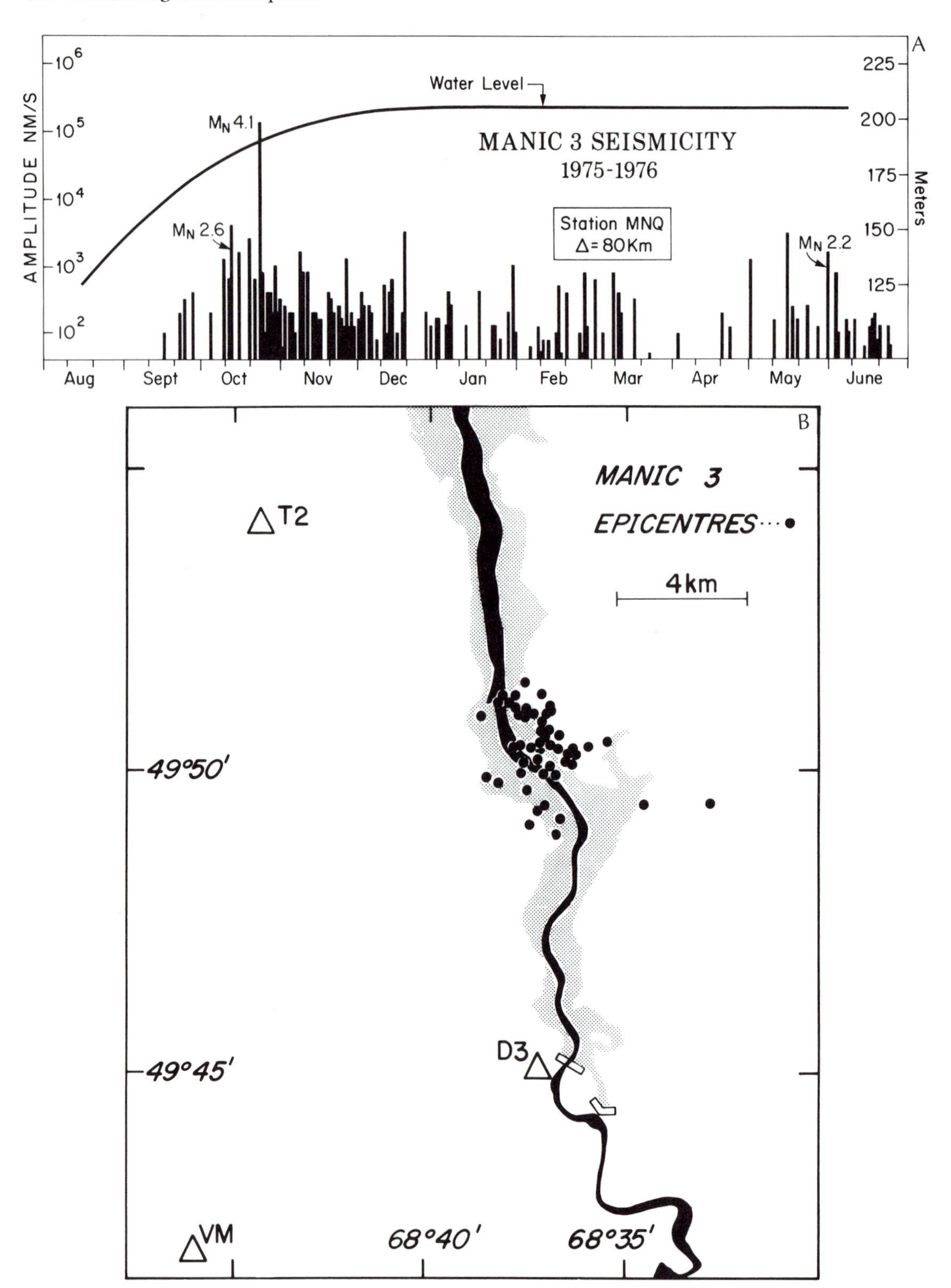

Figure 13–16. Seismic data from Manic-3 Reservoir, Quebec, Canada. (a) 1975–1976 seismic activity and water level. (b) Map of reservoir and principal epicenters. From Leblanc and Anglin (1978).

ciated earthquakes. Manic-3 Reservoir, in the Canadian shield of Canada, is a good example of almost immediate seismic response to reservoir filling, in an area virtually devoid of both previous seismic activity and of Quaternary faults (Leblanc and Anglin, 1978). Several thousand earthquakes followed the filling, with the largest (M = 4.1), occurring shortly before full water depth was attained (Fig. 13-16). Epicenters were concentrated beneath the reservoir about 10 km upstream of the dam, with average hypocentral depths of about 2 km.

On the other hand, the Oroville, California, earthquake of 1975 (M = 5.7) occurred 8 years following the initial reservoir filling, and the epicenter was 12 km south of the reservoir, on a previously identified Holocene fault. Similar-sized earthquakes have occurred historically in the same region, and therefore some scientists and engineers have argued that the earthquake had no direct cause-and-effect relationship to the presence of the reservoir. Others (e.g., Toppozada and Morrison, 1982) point out that (1) the Holocene fault on which the event occurred passes directly beneath the deepest part of the reservoir, which is very large and deep (230 m); (2) the earthquake followed the largest seasonal variation in water level since the first reservoir filling (Fig. 13-15); and (3) the 8-year delay may have been directly related to diffusion of pore pressure along the fault over the 12-km distance from the reservoir.

Aswan Dam (Lake Nasser), Egypt, is an intriguing example of delayed triggered seismicity, because it was not until 17 years following the commencement of filling that a M = 5.3 earthquake occurred, in 1981, beneath an arm of the reservoir (Kebeasy et al., 1987). However, the evidence for triggering is persuasive, because (1) in the 4000-year history of the upper Nile Valley, no earthquake approaching this size had been experienced (although they have been experienced farther downstream); (2) because of the very gradual reservoir filling (water has never yet gone over the spillway at Aswan Dam), only in 1975 was the epicentral area flooded, in a new embayment of the reservoir, and at that time water first had easy access to a major regional aquifer (the Nubian sandstone); (3) the 1981 event was near the time of maximum water level ever attained in the reservoir (Fig. 13-15), and it occurred just following the seasonal peak in water level, as did marked increases of seismicity in 1982, 1983, and 1984; and (4) subsequent to 1984, the water level again went below 80 m (and has not yet recovered), and earthquake activity effectively ceased. Most of the earthquakes occurred on a previously unrecognized late Quaternary strike-slip fault zone, without surface rupture, and, surprisingly, at depths as great as 30 km.

Out of thousands of dams built in recent years, only a small portion has been associated with significant earthquakes thought to have been triggered by filling of the associated reservoirs. Numerous studies have been carried out in the attempt to determine what unique physical or geological characteristics typify those sites where triggered events have and have not occurred (e.g., Baecher and Keeney, 1982; Castle et al., 1980; Packer et al., 1977; 1981; Stuart-Alexander, 1981), but only four significant correlations seem to emerge: (1) Triggered events are more likely beneath large and, particularly, deep reservoirs than beneath those of smaller size. (Note that in Fig. 13-17, only some 26 accepted or questionable triggered events occurred out of some 11,000 total reservoirs with water depths less than 92 m and with volumes less than 10^{10} m^3, whereas the ratio is vastly greater for deeper and larger reservoirs.); (2) Triggered events are more likely, at least during reservoir filling, in areas of normal and strike-slip faulting than in areas of reverse or thrust faulting[9]; (3) The largest triggered events have occurred predominantly in areas of late Quaternary faulting (Packer et al., 1981); and (4) Triggered events are more likely during periods of rapid changes in water level than at other times, as was documented at Nurek Dam, Tadjikistan, by Simpson and Negmatullaev (1981). There appears to be no significant correlation with geologic rock type. Well-documented small triggered events have sometimes occurred in association with reservoirs as shallow as 50 m, and in areas of stable, relatively unbroken rocks such as those of Precambrian shields, but triggered earthquakes large enough to be of real engineering concern have largely been limited to deep reservoirs and areas of late Quaternary faulting. There are perhaps some 250 reservoirs worldwide with water depths exceeding 80 m, at least 5 of which have triggered earthquakes in excess of M = 5.7. Thus, for a deep reservoir in an area of late Quaternary faulting, one might argue that the statistical chances of triggering an earthquake exceeding M = 5.7 are perhaps 5 in 250, or 1 in 50. Although this probability might appear to be low, surely it is not so low that engineers can dismiss the phenomenon when designing large dams, almost all of which would have truly catastrophic consequences of failure (Allen, 1982). Simpson (1986) has argued that the probability is particularly high in areas of long repeat times for naturally occurring earth-

[9]It is interesting that in at least one case (Tarbela Dam, Pakistan), minor seismicity in a thrust-faulted region has been observed to *decrease* during periods of reservoir filling, and a reasonable mechanical argument has been made for this phenomenon by Jacob et al. (1979).

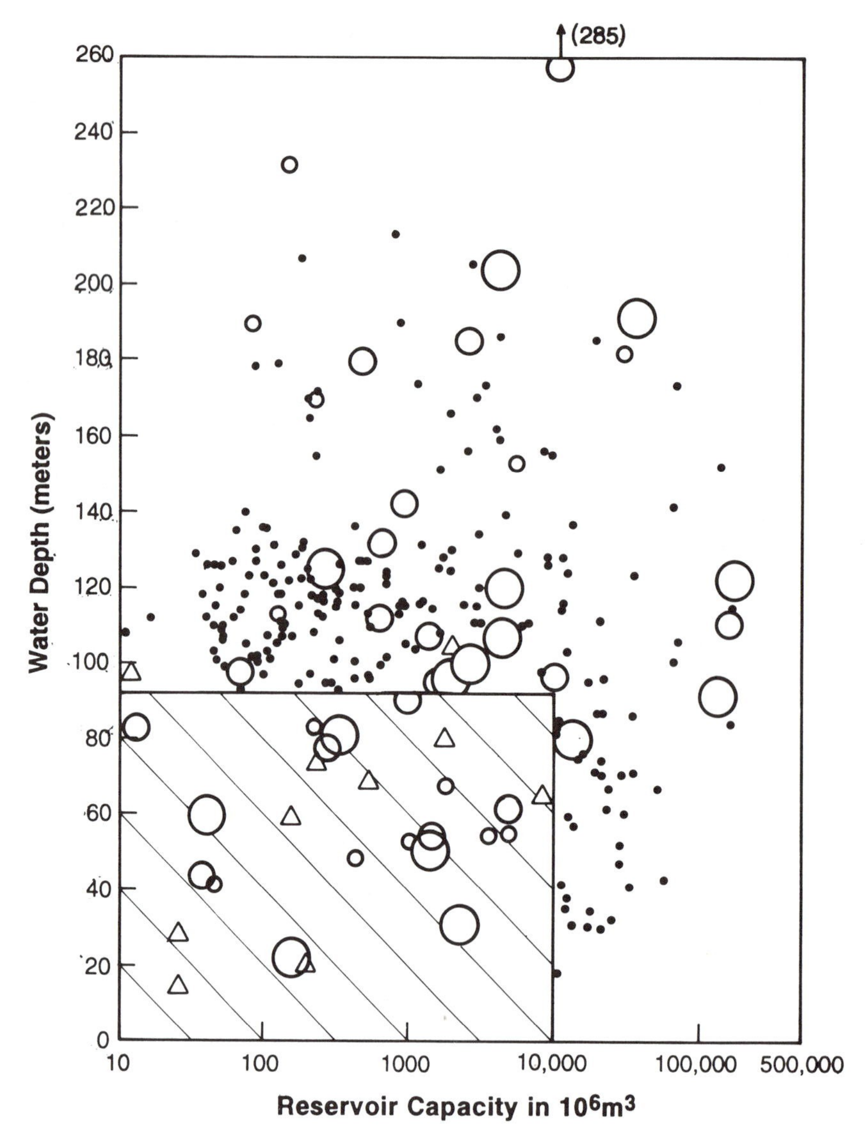

Figure 13-17. Relationship of depth and volume for reported cases of reservoir-induced (triggered) seismicity, RIS. About 11,000 reservoirs without reported RIS would plot within the diagonally-lined rectangle at lower left. Adapted from Baecher and Keeney (1982).

quakes. Some of the more significant earthquakes thought to be triggered by reservoir filling are listed in Table 13-2, although the correlation is sometimes debatable in areas where there is ongoing moderate natural seismicity.

At least two earthquakes generally viewed as reservoir triggered have been associated with surface displacements on the causative faults. The Koyna, India, earthquake of 1967 ($M = 6.5$) was associated with about 30 cm of surficial left-lateral displacement on a

Table 13–2

Earthquakes of magnitude 5.0 and greater reported to be reservoir-triggered, in order of decreasing magnitude. Some of these are of debatable correlation, as well as of debatable magnitude. Data from Simpson (1986), Packer et al. (1877), and Guha and Patil (1990).

Dam	Max. Mag.	Year	Max. Depth	Reference
Koyna, India	6.5	1967	100	Gupta and Rastogi (1976)
Kremasta, Greece	6.3	1966	120	Comminakis and others (1968)
Xinfengjiang, China	6.1	1962	100	Sheng and others (1973)
Kariba, Zambia-Zimbabwe	5.8	1963	123	Gough and Gough (1970a; 1970b)
Srinagarind, Thailand	5.8	1983	140	Ghose and Oike (1987)
Marathon, Greece	5.7	1938	60	Galanopoulos (1967)
Oroville, USA	5.7	1975	204	Toppozada and Morrison (1982)
Varragamba, Australia	5.4	1973	104	Guha and Patil (1990)
Aswan, Egypt	5.3	1981	110	Kebeasy and others (1987)
Coyote Valley, USA	5.3	1962	22	Toppozada and Cramer (1978)
Akasombo, Ghana	5.3	1964	109	Guha and Patil (1990)
Kinnersani, India	5.3	1969	62	Guha and Patil (1990)
Volte Grande, Brazil	5.1	1974	32	Guha and Patil (1990)
Benmore, New Zealand	5.0	1966	86	Adams (1974)
Eucumbene, New Zealand	5.0	1958	106	Cleary and others (1964)
Hoover, USA	5.0	1939	166	Rogers and Lee (1976)

fault with earlier Holocene displacement which probably passed through one arm of the reservoir (Cluff, 1977). And the Oroville, California, earthquake of 1975 (M5.7) was caused by displacement on the Cleveland Hill fault, (Clark et al., 1976) with 5-6 cm of normal displacement at the ground surface. The possibility of such displacement through a dam itself is obviously a matter of engineering concern, particularly for thin concrete dams (as opposed to embankment dams, which might better accommodate the displacement). Construction on at least one major project (Auburn Dam, California) was suspended when concern for this possibility, albeit remote, was expressed by geologists on the basis of neotectonic geologic studies at and near the damsite.

Because earthquakes associated with reservoirs—like other triggered events—primarily release preexisting tectonic strain, the argument has sometimes been made that the largest triggered earthquake at a given site can be no larger than the largest naturally occurring event; thus no special seismic considerations are necessary beyond those for which the dam would be designed even in the absence of the triggering phenomenon. While this argument may in part be true, the fact is that the reservoir represents a *changed* stress situation—one which may not have been present locally for tens of thousands of years or more. Thus the possibility exists of the reservoir triggering an event which otherwise might not have occurred for a very long time, if at all. In this sense, the *likelihood* of occurrence really has changed, even if the magnitude of the maximum event remains the same.

A CAUTIONARY CONCLUSION

As was emphasized at the beginning of this chapter, earthquake hazard assessment is fundamentally a *predictive* science, in that we are attempting to forecast the likelihoods and effects of earthquakes in years to come. Any such attempt is fraught with all the perils of using models that have large uncertainties—uncertainties that are probably larger than we have recognized and admitted in the past. Harking back to the time of the formulation of the elastic rebound theory, most such rupture models have been based on the simple concept of systematic strain buildup and systematic time-dependent strain release, which takes place abruptly during earthquakes. But detailed studies of large earthquakes, using a new generation of modern seismological instruments, have demonstrated in recent years that the physics of the rupture process is far more complicated than we ever envisaged, and that earthquakes are far more different from one to another than we ever imagined. In particular, the physics of the frictional resistance on fault surfaces is as yet ill-understood, and some elements of chaos theory may be more applicable than simplistic time-dependent models. And even if the physics of the rupture process were fully understood, the geological complexities of faults at depth, particularly in thrust-fault regimes, are far greater than we had previously thought.

All this does not mean to imply that current hazard assessments are meaningless, but it does suggest that our confidence in the results may sometimes have been overestimated. Just as the zones on seismic hazard maps seem to grow ever larger with each large earthquake, increasing realization of the complexities of earthquakes suggests that our confidence in existing models may also have been overestimated. The cautionary conclusion is that we must continue to be exceedingly conservative in earthquake hazard assessment, particularly for critical structures whose failure or loss of function would have serious—if not disastrous—public impact. The clear challenge is to improve the quantification of earthquake hazards in a realistic and usable way.

Suggestions for Further Reading

Allen, C. R. 1982. Reservoir-induced earthquakes and engineering policy. California Geology 35:248-50.

Cornell, C. A. 1968. Engineering seismic risk analysis. Seismol. Soc. America Bull. 58:1583-1606.

Dolan, J. F., Sieh, K., Rockwell, T. K., Yeats, R. S., Shaw, J., Suppe, J., Huftile, G. J., and Gath, E. M. 1995. Prospects for larger or more frequent earthquakes in the Los Angeles metropolitan region. Science 267:199-205.

Lomnitz, C. 1994. Fundamentals of Earthquake Prediction. New York: J. Wiley & Sons, 326 p.

Reiter, L. 1990. Earthquake Hazard Analysis—Issues and insights. New York: Columbia University Press, 254 p.

Scientists of the U.S. Geological Survey and the Southern California Earthquake Center. 1994. The magnitude 6.7 Northridge, California, earthquake of 17 January 1994. Science 266:389-97.

Shaw, J. H., and Suppe, J. 1994. Active faulting and growth folding in the eastern Santa Barbara Channel, California. Geol. Soc. America Bulletin. 106:607-26.

Simpson, D. W. 1986. Triggered earthquakes. Ann. Rev. Earth Planet. Sci. 14:21-42.

Stein, R. S., and Yeats, R. S. 1989. Hidden earthquakes. Scientific American 260(6):48-57.

Ward, S. N. 1994. A multidisciplinary approach to seismic hazard in southern California. Seismol. Soc. America Bull. 84:1293-1309.

Wells, D. L., and Coppersmith, K. J. 1994. New empirical relationships among magnitude, rupture length, rupture area, and surface displacement. Seismol. Soc. America Bull. 84:974-1002.

Youngs, R. R., and Coppersmith, K. J. 1985. Implications of fault slip rates and earthquake recurrence models to probabilistic seismic hazard estimates. Seismol. Soc. America Bull. 75:939-64.

Appendix:

Table of Historic Earthquakes with Surface Rupture

The surface effects of earthquakes were summarized by Lyell (1875), who thought that earthquakes, like volcanic eruptions, demonstrated important principles in the emerging field of geology. On the other hand, Montessus de Ballore (1925) and Richter (1958) discussed coseismic surface faulting as a clue to the earthquake process itself. Richter's descriptions of surface ruptures in the Balkans, east Africa, and the Indian subcontinent are still among the best available. Bonilla and Buchanan (1970) and Bonilla et al. (1984) tabulated world earthquakes known to have surface rupture to determine the relation of surface rupture length to magnitude; like Richter, they wanted to use surface rupture to learn more about the earthquake process. This table was expanded, with many additional case histories including the 1992 Landers, California, earthquake, by Wells and Coppersmith (1994), again with the purpose of learning more about the relation of surface rupture to magnitude and rupture area. In this table, magnitudes of Wells and Coppersmith (1994) have been used in preference to other sources, unless the other source revised a pre-instrumental magnitude based on geological data that resulted in a revision of earthquake moment (for example, Avouac et al. (1993), for the 1906 Manas, China, earthquake).

Information for the table in this book comes from two sources: (1) the inspection of surface faulting by qualified observers shortly after the earthquake, and (2) the correlation, long after the earthquake, of surface geological features with historical earthquakes, which we refer to as *historical paleoseismology*. Historical paleoseismology began with Lawson et al. (1908), who, in addition to describing surface rupture accompanying the 1906 San Francisco earthquake, reported on surface rupture accompanying an earlier earthquake (1868) on the Hayward fault.

Inspection of surface faulting soon after the event began with the 1855 West Wairarapa, New Zealand, earthquake (cf. Darby and Beanland, 1992), and the 1861 Egion earthquake south of the Gulf of Corinth in Greece (Schmidt, 1875). Systematic study of coseismic surface faulting began in New Zealand with the 1888 Marlborough earthquake, in Japan with the 1891 Nobi earthquake, and in the United States with the 1906 San Francisco, California, earthquake and the publication of descriptions of surface ruptures worldwide in the Bulletin of the Seismological Society of America. Yet even with inspection teams of trained observers, there may be controversy about whether disturbance of the ground surface is produced by tectonic rupture or by secondary effects such as seismic shaking, low-angle landsliding, or lateral spreading (see, for example, the two accounts of the 1992 Erzincan, Turkey, earthquake). The 1989 Loma Prieta, California, earthquake is also included even though there is disagreement (cf. Prentice and Schwartz, 1991) over the significance of presumed tectonic ruptures (cf. Aydin et al., 1992). Several listed earthquakes have surface ruptures recorded for a distance of a few kilometers or less; these ruptures may represent secondary effects.

A controversy has arisen over whether surface faulting is *primary*, part of the same rupture surface that includes the mainshock, or *secondary*, triggered on another fault by the earthquake simply because that fault was a zone of weakness and responds to differential shaking of competent blocks on both sides. The 1983 Borah Peak, Idaho, earthquake may have primary rupture only in the Thousand Springs segment. Tectonic ruptures also were found to the north in the Willow Creek Hills and in the Warm Springs Valley, but these may be secondary (cf. discussion in Chapter 9). Because the faulting north of the Thousand Springs segment has the same sense as faulting in that segment, both are included in the table. However, the 1979 Imperial Valley earthquake triggered up to 10 mm dextral slip on a 39-km-long section of the San

Andreas fault more than 90 km away from the seismogenic Imperial fault (Sieh, 1982); this is not included in the table.

It is now apparent that seismic faulting may warp the ground surface even though faulting is blind (does not reach the surface). These include the 1905 Kangra, India, earthquake of M8 and the 1977 Caucete, Argentina, earthquake of M7.4, both produced on blind reverse faults, and may include the Messina Straits, Italy, earthquake of M7.5, which Valensise and Pantosti (1992) interpreted as an earthquake on a blind normal fault. The 1983 Coalinga, California, earthquake on a blind reverse fault is not included in this table, but an aftershock on June 11 did rupture the ground surface, and it is included. Surface rupture reported in the table may be secondary, related to folding, rather than the surface expression of the main seismic fault, as illustrated for the 1980 El Asnam, Algeria, earthquake of M7.3 (Philip and Meghraoui, 1983), the 1988 Spitak, Armenia, earthquake of M6.8 (Philip et al., 1992), and the 1970 Uüreg Nuur, Mongolia, earthquake of M7 (Baljinnyam et al., 1993).

Historical paleoseismology is most closely identified with the work of N. N. Ambraseys in the eastern Mediterranean, the Middle East, and Africa. Programs are underway in China, Japan, and the United States to correlate surface ruptures based on paleoseismology with large historical earthquakes, and Task Group II-3 of the International Lithosphere Program has this as one of its main objectives throughout the world.

Recognition of pre-instrumental earthquakes on strike-slip faults relies only partly on construction of isoseismal maps; recognition of a linear zone of disturbed and disrupted ground is also necessary. Note, for example, that the M7.3 earthquake of 20 May 1202, probably in the Dead Sea fault zone, is not included (Ambraseys and Melville, 1988). This permits the recognition of rupture lengths of several pre-instrumental earthquakes on the North Anatolian fault (Ambraseys, 1970; 1975; Ambraseys and Finkel, 1988), but does not allow the identification of maximum strike-slip offset, because contemporary observers did not make note of this. The correlation of fresh topographic expression of fault offset was used by Nakata et al. (1990) to determine the rupture lengths of the 1645 and 1796 earthquakes on the Philippine fault. On the other hand, trenching on the Median Tectonic Line fault of Japan showed that this fault could have ruptured during an earthquake in 1956 (Okada et al, 1991), but there is not enough trenching along the fault trace to determine the length of this surface rupture (Tsutsumi and Okada, 1996). Similarly, tree-ring dating tied a California earthquake of 1812 to the San Andreas fault, but there is controversy about the length of 1812 surface rupture despite an extensive trenching campaign (Sieh et al., 1989; Salyards et al., 1992; Fumal et al., 1993). Trenching commonly does not yield evidence for strike-slip offset unless there are several closely spaced trenches that allow the mapping of, say, an offset channel from one side of the fault to the other.

Trenching is commonly able to provide evidence for maximum displacement on dip-slip faults, particularly normal faults which are less likely to be blind. Dip-slip faults produce a scarp, the height of which is more likely to be recorded by contemporary observers, as was the case for the 1899 Menderes Valley, Turkey, earthquake (Ambraseys and Finkel, 1987a). However, there is lively controversy over whether arcuate scarps with normal displacement are tectonic or are due to slope failure because the surface expression for both may be somewhat similar. This has been a particular problem for pre-instrumental (and even twentieth-century) ruptures in Greece (cf. Ambraseys and Jackson, 1990) and in Italy, particularly for the 1783 earthquake in Calabria. Another problem is the correlation of fresh geomorphic expression, such as the preservation of a free face, with isoseismal maps (cf. Armijo et al., 1991, for the earthquake that destroyed Sparta, Greece, in 464 B.C., and Liu, 1993, for normal faulting accompanying the 1895 Tashikuergan earthquake in Xinjiang, China). The Pleasant Valley, Nevada, and Avezzano, Italy, earthquakes of 1915 were accompanied by vegetation-free zones near the base of the range front, formed during the earthquake (Vittori et al., 1991), referred to by the Italians as *nastri di faglia*, or fault ribbons. In the central Nevada seismic zone, these fault ribbons mark the 1915 and 1954 surface ruptures, but not the middle Holocene rupture at the base of the Stillwater Range in northern Dixie Valley. *Nastri di faglia* are common in limestone terranes of Italy and Greece, and these are not easily correlated with historical earthquakes over more than 2000 years. Some are latest Pleistocene in age.

The Great Sumatran fault and Philippine fault are marked by isoseismals of large earthquakes that are distributed along most of their lengths, but these isoseismals are not considered sufficient evidence to include most of these earthquakes in this table. The danger of correlating isoseismal maps to a known Holocene fault is illustrated in the Apennines of Italy. Trenching on faults that ruptured the surface in the 1980 Irpinia earthquake showed that these faults did not rupture during an earthquake in 1964 with an even larger meizoseismal zone centered on the same region as in 1980 (Pantosti et al., 1993).

Table of Historic Earthquakes with Surface Rupture

NORTH AMERICA

Canada

Date	M	Name	Lat.	Long.	Strike	Type	L	H	V	Fault Name	Reference
19891225	6.3	Ungava (Quebec)	66.1N	73.6W	N30E	RE	8.5	0.1	1.8	Lac Turquoise	Adams et al 91

United States-Alaska

Date	M	Name	Lat.	Long.	Strike	Type	L	H	V	Fault Name	Reference
19580710	7.9	Lituya Bay	58.6N	137.5W	NW	RL	95–280		3.5–6.5	Fairweather	Plafker et al 78
19640327	9.2w	Prince William Sd.	59.8N	147.6W	N37E	RE	57		6–7	Patton Bay, Hanning Bay	Plafker 67, 69, Malloy 64

United States-California

Date	M	Name	Lat.	Long.	Strike	Type	L	H	V	Fault Name	Reference
18121208	7	"San Juan Capistrano" (Wrightwood)	34.3N	117.6W	WNW	RL	>25	6.25		San Andreas	Sieh et al 89, Salyards et al 92, Fumal et al 93, Jacoby et al 88
18360610	6.8	Hayward	37.9N	122.3W	NW	RL	61		1.5?	Hayward	Louderback 47, Toppozada et al 81, WGCEP 90, Jennings 94
183806??	7.2w	San Francisco	37.2N	122.0W	NW	RL	40			San Andreas	Louderback 47, Tuttle & Sykes 92
18570109	7.9	Fort Tejon	34.9N	119.1W	NW	RL	322	9.5		San Andreas	Sieh 78a,b, Grant & Sieh 93 Grant & Donnellan 94, Jennings 94
1861	6.4	San Ramon Valley	37.7N	122.0W	NNW	RL	13	?		Calaveras	Toppozada 81, Simpson et al 93
18681021	6.8	Hayward	37.5N	122.0W	NW	RL	48	0.9	0.3	Hayward	Lawson 0.8, Toppozada & Parke 82
18720326	7.6	Owens Valley	36.7N	118.1W	N20W	RN	108	10	4.4	Owens Valley	Beanland & Clark 94, dePolo et al 91
18900424	6.3	Chittenden	36.9N	121.3W	NW	RL	8	0.3?		San Andreas	Toppozada et al 81
19010302	6	Parkfield	35.8N	120.4W	NW	RL	?	?		San Andreas	Lawson 08, Brown et al 67
19060418	7.9	San Francisco	37.8N	122.6W	NW	RL	432	6.1		San Andreas	Lawson 08, Thatcher 75, Prentice & Schwartz 91
19100515	6	Temescal Valley	33.7N	117.5W	NW	RL	15	0.35		Elsinore	Toppozada & Parke 82, Rockwell 89
19220310	6.5	Cholame	35.8N	120.4W	NW	RL	0.4	?		San Andreas	Richter 58
19340607	6.0	Parkfield	35.8N	120.4W	NW	RL	20	?		San Andreas	Byerly & Wilson 35, Brown et al 67
19400519	7.2	Imperial Valley	32.8N	115.5W	NW	RL	64	5.8	1.2	Imperial	Ulrich 41, Jennings 94
19470410	6.2	Manix, Mojave	35.0N	116.5W	ENE	LL	1.6	0.08		Manix	Richter 58
19501214	5.6	Honey Lake Valley	40.1N	120.1W	N	NN	8.9		0.6	Fort Sage	Gianella 57, Jennings 94
19510123	5.6	Superstition Hills	33.0N	115.7W	NW	RL	3.2	?		Superstition Hills	Allen et al 65
19520721	7.7	Kern County	35.3N	118.7W	NE	LR	57	0.8	1.2	White Wolf	Buwalda & St. Amand 55, Stein & Thatcher 81, Jennings 94
19660628	6.4	Parkfield	35.8N	120.4N	NW	RL	37.8	0.2		San Andreas	Brown et al 67, Lienkaemper & Prescott 89
19680409	6.8	Borrego Mountain	33.2N	116.1W	NW	RL	31	0.38		San Jacinto (Coyote Cr.)	Clark 72
19710209	6.5	San Fernando	34.3N	118.4W	WNW	RE	16		2.5	San Fernando	Sharp 75
19750123	4.6	Brawley	32.9N	115.6W	N	NN	10.4		0.2	Brawley	Sharp 76
19750531	5.2	Galway Lake	34.5N	116.5W	N-N25W	RL	6.8	0.02		Galway Lake	Hill & Beeby 77
19750801	5.7	Oroville	39.4N	121.7W	N	NN	3.5		0.06	Cleveland Hill	Clark et al 76, Lahr et al 76
19780813	4.6	Stephens Pass	41.5N	121.9W	N	NN	2.0		0.3	Stephens Pass	Bennett et al 79
19790315	5.2	Homestead Valley	34.3N	116.4W	NNW	RL	3.3	0.1	0.06	Homestead Valley	Stein & Lisowski 83, Hill et al 80

Table of Historic Earthquakes with Surface Rupture Continued

Date	M	Name	Lat.	Long.	Strike	Type	L	H	V	Fault Name	Reference
19790806	5.9	Coyote Lake	37.0N	121.5W	NNW	RL	15	0.05		Calaveras	Reasenberg & Ellsworth 82 Herd et al 82
19790315	6.7	Imperial Valley	32.8N	115.4W	NW	RL	30.5	0.8		Imperial	Sharp 82
19800124	5.6	Livermore Valley	37.7N	121.7W	NW	RL	6.2	0.03		Greenville	Bonilla et al 80, Bolt et al 81
19800525	6.1	Mammoth Lakes	37.6N	118.8W	NNW	NN	20		0.25	Hilton Creek	Taylor & Bryant 80, Clark et al 82
19810426	5.6	Westmorland	33.0N	115.6W	NW	RL	17,16	0.08, 0.14		Imperial Superstition Hills	Sharp et al 86a
19810407	2.5	Lompoc Quarry	34.6N	120.4W	N84E	RR	0.6	0.1	0.2	unnamed	Yerkes et al. 83
19821001	5.2	Little Lake	35.7N	117.7W	N, NNE	RL	6	~0		Little Lake	Roquemore & Zellmer 83
19830611	5.4	Coalinga	36.3N	120.5W	N	RE	3.3	0.64		Nuñez	Rymer et al 90
19860708	6.0	N. Palm Springs	33.9N	116.6W	WNW	RR	9	<0.01		Banning	Sharp et al 86b
19860721	6.2	Chalfant Valley	37.4N	118.3W	NNW	RL	15.5	0.05		White Mts.	Kahle et al 86, dePolo and Ramelli 87
										Tableland	Lienkaemper et al 87
19871124	6.2	Elmore Ranch	33.1N	115.8W	NE	LL	9.0	0.28	0.03 (sum)	Elmore Ranch	Kahle et al 88, Sharp et al 89
19871124	6.6	Superstition Hills	33.0N	115.8W	NW	RL	27.0	0.92	0.25	Superstition Hills, Wienert	Kahle et al 88, Sharp et al 89
19891017	7.1w	Loma Prieta	37.2N	121.9W	NW	RR	2.7	0.1	0.2	San Andreas, Sargent	Aydin et al 92 Johnson & Fleming 93
19920628	7.3w	Landers	34.2N	116.4W	NNW	RL	85	6	>1	Johnson Valley, Homestead Valley, Emerson, Camp Rock	Sieh et al 93
19930517	6.2	Eureka Valley	37.2N	117.6W	N10E	NN	>4		0.02	unnamed	Peltzer & Rosen 93
19950817	5.4	Ridgecrest	35.6N	117.7W	N	RN	1	0.002		Airport Lake	Hauksson et al 95
19950920	5.8	Ridgecrest	35.6N	117.6W	N	RN	2.5	0.008	0.01	Airport Lake	Hauksson et al 95

General reference for California: Jennings 94

United States-Hawaii

Date	M	Name	Lat.	Long.	Strike	Type	L	H	V	Fault Name	Reference
19741130	5.5	Mauna Loa SE	19.4N	155.4W	NE	RL	2.2	0.5		Kaoiki	Jackson et al 92
19831116	6.6	Mauna Loa SE	19.5N	155.4W	N56E	RL	7	0.25		Kaoiki	Jackson et al 92

United States-Idaho

Date	M	Name	Lat.	Long.	Strike	Type	L	H	V	Fault Name	Reference
19831028	7.3	Borah Peak	44.2N	113.8W	NNW	NN	34	0.7	2.7	Lost River	Crone et al 87

United States-Montana

Date	M	Name	Lat.	Long.	Strike	Type	L	H	V	Fault Name	Reference
19590818	7.6	Hebgen Lake	44.8N	111.2W	N60W	NN	26.5		5.5	Red Cyn., Hebgen	Myers & Hamilton 64 Witkind et al 62

United States-Nevada

Date	M	Name	Lat.	Long.	Strike	Type	L	H	V	Fault Name	Reference
18691228	6.7	Olinghouse	39.5N	119.5W	NE	LL	23	3.65		Olinghouse	Sanders & Slemmons 79
19030903	6.0	Wonder	39.4N	118.1W	N5E	NN?	5.0		0.3?	Gold King	Slemmons et al 59
19151003	7.6	Pleasant Valley	40.3N	117.6W	NNE	NN	59		5.8	Tobin, Pearce, etc	Wallace 84
19321221	7.2	Cedar Mountain	38.8N	118.0W	NNW	RL	61	2.7		Stewart-Monte Cristo Valley	Gianella & Callaghan 34, dePolo et al 91
19340130	6.3	Excelsior Mts.	38.3N	118.4W	N65E	NN	1.4		0.13	unnamed	Callaghan & Gianella 35
19540706	6.3	Rainbow Mtn.	39.4N	118.5W	NNE	NN	18.0	0.9	0.30	Rainbow Mtn.	Tocher 56, dePolo et al 91 J. Caskey, unpub.
19540824	6.9	Stillwater	39.6N	118.5W	NNE	NN	34		0.76	Rainbow Mtn.	Tocher 56, dePolo et al 91
19541216	7.2	Fairview Peak	39.2N	118.2W	N5E	RN	67	3.7	3.8	Fairview Peak	Slemmons 57, J. Caskey, unpub.
19541216	6.8	Dixie Valley	39.6N	118.2W	N	RN	45		2.8	Dixie Valley	Slemmons 57, J. Caskey, unpub.

Table of Historic Earthquakes with Surface Rupture Continued

Date	M	Name	Lat.	Long.	Strike	Type	L	H	V	Fault Name	Reference
					United States-Utah						
19340312	6.6	Hansel Valley	41.7N	112.5W	N5E	NN	11.5	0.5		unnamed	Richter 58, Doser 89, McCalpin et al 87
General reference for Basin and Range: dePolo et al 91											
					Mexico						
18870503	7.4	Sonora	31.0N	109.2W	N5E	NN	75		5.1	Pitaycachi	Bull & Pearthree 88, Pearthree et al 90
18920224	7.1	Laguna Salada	32.5N	115.6W	NW	RN	>22	4	3.5	Laguna Salada	Mueller & Rockwell 95, Rockwell 89
19121119	6.9	Acambay	19.9N	99.8W	WNW	LN	35		40	Acambay	Moya et al 95
19341231	7.0	Cerro Prieto	32.2N	115.0W	NW	RL	100	3-4		Cerro Prieto	Allen et al 65, Anderson & Bodin 87, Doser 94
19560209	6.9	San Miguel	31.7N	15.9W	NW	RL	20	1.15		San Miguel	Shor & Roberts 58
					Guatemala						
18160722	7.6w	Sta.Maria Magdalena	15.4N	91.6W	E	LL	?	?		Chixoy-Polochic	White 85
19760204	7.5	Motagua	15.3N	89.1W	ENE-E	LL	230	3	0.2	Motagua	Bucknam et al 78, Plafker et al 76
19760204	5.8	Motagua	14.7N	90.6W	NNE	RN	>10	0.05	0.12	Mixco	Plafker et al 76
19820929	5.1	Chanmagua				NN	9		0.1		Wells & Coppersmith 94
					Nicaragua						
19310331	?	Managua	12.2N	86.3E	N36E	?	2	?	0.01	E	Brown et al 73, Sultan 31
19721223	6.2	Managua	12.1N	86.3E	N11E	LL	>5.9	0.38		A, B, C, D	Brown et al 73
					SOUTH AMERICA						
					Argentina						
19440115	7.4	San Juan	31.6S	68.5W		RE	6		0.6	unnamed	Castellanos 45, Groeber 44, Smalley et al 93
					Colombia						
19700926	6.1	Bahía Solano	6.2N	77.4W	NNW	NN	>1		0.07	Bahía Solano	Ramirez 71a, b, Restrepo 71
19830331	4.9	Popayán	2.5N	76.7W	NNW	NN	1.25		0.01	Pubenza	Lomnitz & Hashizume 85
					Peru						
19461110	7.3	Ancash	8.3S	77.4W	NW	NN	20		3.5	Quiches	Sébrier et al 88, Bellier et al 91
19690724	5.7	Huaytapallana	11.9S	75.1W	NW	LR	5.5	0.7	0.4	Huaytapallana	Philip & Mégard 77, Sébrier et al 88
19691001	6.2	Huaytapallana	11.9S	75.1W	NW	LR	9.5		2	Huaytapallana	Philip & Mégard 77, Sébrier et al 88
19860405	5.4	Cuzco	13.5S	71.9W	N50W	NN	6.5		0.1	Chincheros, Qoricocha	Mercier et al 92
					Venezuela						
19290117	6.9	Cumaná	10.4N	64.0W	W	RN	4				Paige 30
					AFRICA						
General reference for West Africa: Ambraseys & Adams 86											
					Algeria						
19801010	7.3	El Asnam	36.2N	1.5E	ENE	RE	31.2		6.5	El Asnam	Philip & Meghraoui 83, Yielding et al 81

Table of Historic Earthquakes with Surface Rupture Continued

Date	M	Name	Lat.	Long.	Strike	Type	L	H	V	Fault Name	Reference
19851027	5.9	Constantine	36.4N	6.8E	N55E	LL	3.8	0.11	0.04	Ain Smara	Bounif et al 87
19891029	6.1	Tipaza	36.6N	2.3E	W	RE	4.0		0.12	Mt. Chenoua	Meghraoui 91
					Djibouti						
19781107	5.3	Asal	11.6N	43.5E	NW	NN	4.8		0.5	unnamed	Abdallah et al 79, Lépine et al 80, Stein et al 91
					Ethiopia						
19610602	6.4	Kara Kore	10.5N	39.9E	N	NN	20		2.0	Borkenna graben	Gouin 79
19690405	6.1	Serdo	12.0	41.3E	N40W	LN	37	0.65	0.95	unnamed	Gouin 79
					Ghana						
19390622	6.4	Accra	5.5N	0.4W	NE	LN	20.0		0.4	Akwapim (zone)	Junner 41, Yarwood & Doser 90, Ambraseys & Adams 86
					Guinea						
19831222	6.2	Koumbia	11.9N	13.4W	WNW	RN	9.4	0.13	0.06	unnamed	Langer et al 87
					Kenya						
19280106	6.9	Subukia	0.2N	36.2E	NNW	NN	38		2.4	Laikipia escarpment	Ambraseys 91
					Sudan						
19661009	5.6	Jebel Dumbeir	12.6N	30.8E	N	LL	6.0	small	0	unnamed	Qureshi & Sadig 67
					ZAÏRE						
19660320	6.6	Toro	0.6N	30.2E	NNE	NN	40		2.5	Kitimba-Semliki	Loupekine 66, Maasha & Molnar 72
					EUROPE						
					Albania						
19050601	6.6	Shkodra	41.9N	19.4E	NE	?	7		1	Trush	Koçiaj & Sulstarova 80
19671130	6.6	Dibra	41.4N	20.4E	N40E	RN	>10		0.5	Vlora-Dibra	Sulstarova & Koçiaj 80
					Armenia						
19881207	6.8	Spitak	40.8N	44.2E	WNW	RR	24	2			Philip et al 92
					Bulgaria						
18580918		Sofiya	42.5N	23.3E	WNW	NN	20				Armijo et al unpub.
19040404	7.1	Krupnik	41.8N	23.2E	WNW	NN	60		2		Armijo et al unpub.
19280414	6.6	Chirpan-Plovdiv	42.1N	25.2E	W	NN	54		0.5		Bonchev & Bakalov 28, Jankov 45, Armijo et al unpub., Richter 58
19280418	6.9	Popovitsa-Plovdiv	42.2N	24.9E	WNW	NN	50		3.5		Bonchev & Bakalov 28, Jankov 45, Armijo et al unpub., Richter 58
					Greece						
−464	7.2	Sparta	37.0N	22.4E	NNW	NN	20	3–4?		Sparta	Armijo et al 91
18611226	7.0	Egion	38.2N	22.2E	WNW	NN	13	2.4		Helice	Schmidt *75*, Mouyaris et al 92
18940427	6.9	Atalanti	38.6N	23.2E	WNW	NN	47	1.7		Atalanti-Martinon	Skuphos *94*, Ambraseys & Jackson 90
19320926	7.0	Ierissos	40.5N	23.9E	E	NN	15	2.0		Stratoni	Pavlides & Tranos 91
19530905	5.7	Corinth	37.9N	23.1E	NNW	NN	3		0.8	unnamed	Stiros 95

Table of Historic Earthquakes with Surface Rupture Continued

Date	M	Name	Lat.	Long.	Strike	Type	L	H	V	Fault Name	Reference
19540430	6.7	Sophades	39.2N	22.2E	N60W	NN	30	0.9		unnamed	Ambraseys & Jackson 90, Papastamatiou & Mouyaris 86
19570308	6.5	Velestino	39.4N	22.7E	E	NN	1	?		unnamed	Ambraseys & Jackson 90, Papastamatiou 57
19660205	6.2	Kremasta	39.0N	21.6E	NE	NN	2	?		unnamed	Ambraseys & Jackson 90
19661029	5.9	Acarnania	38.8N	21.0E	WNW	NN	4	0.4			Ambraseys 75
19680219	7.1	Agios Efstratios	39.5N	24.9E	NE	RN	3	0.5		unnamed	Pavlides & Tranos 91
19700220	4.6	Yali	36.7N	27.2E	NW	NN	0.6		0.01	unnamed	Stiros & Vouyoukalakis, in prep.
19780620	6.4	Thessaloniki	40.6N	23.2E	W,WNW	NN11		0.2		unnamed	Mercier et al 79, Papazachos et al 79
19800709	6.4	Almyros	39.3N	22.8E	E	NN	9	0.2		unnamed	Ambraseys & Jackson 90, Papazachos et al 83
19810224	6.7	Gulf of Corinth	38.0N	23.0E	ENE	NN	15.3	1.5		unnamed	Jackson et al 82, Taymaz et al 91
19810304	6.4	Gulf of Corinth	38.2N	23.2E	ENE	NN	11	1.0		Kaparelli	Jackson et al 82
19830416	5.4	Akarnania	38.8N	20.9E	WNW	LL	2.4	0.15	0.05	unnamed	Koukis et al 90
19860913	5.8	Kalamata	37.0N	22.2E	N15E	NN	6.3	0.18		Kalamata	Lyon-Caen et al 88
19950513	6.6	Kozani-Grevena	40.0N	21.6E	N60E	NN	15		0.9	Aliakmonas River	Pavlides et al 95
19950616	6.1	Egion	38.2N	22.1E	E	NN	9		0.02	Egion	R. Collier unpub.

General reference for Greece: Ambraseys and Jackson 90

Iceland

Date	M	Name	Lat.	Long.	Strike	Type	L	H	V	Fault Name	Reference
1294		Rangarvellir	63.9N	20.0W	N	RL	?	?		S. Iceland Seis Z.	Einarsson et al. 81
1630		Land	64.0N	20.2W	N	RL	6	?		S. Iceland Seis Z.	Einarsson et al. 81, Einarsson 94
18960826	6.7	Skeid, Land	64.0N	20.2W	N	RL	9	0.8		S. Iceland Seis Z.	Einarsson et al. 81, Einarsson & Eiríksson 82, Stefánsson et al 93
19120506	7.0	Land	63.9N	20.0W	N	RL	20	3		S. Iceland Seis Z.	Bjarnason et al 93

Italy

Date	M	Name	Lat.	Long.	Strike	Type	L	H	V	Fault Name	Reference
19150113	7.0	Avezzano	42.0N	13.9E	NW	NN	24		0.9	Serrone	Serva et al 86, Ward & Valensise 89
19801123	6.9	Irpinia	40.8N	15.2E	N52W	NN	36		1.0	Irpinia	Pantosti & Valensise 90, Pantosti et al 93

WEST ASIA

Iran

Date	M	Name	Lat.	Long.	Strike	Type	L	H	V	Fault Name	Reference
14930110		Mu'minabad	32.9N	59.8E	NW	RE	30			Mu'minabad	Ambraseys & Melville 82
17210426		Southeast Tabriz	37.9N	46.7E	N50W		50			N. Tabriz	Berberian & Arshadi 77, Berberian 81, Ambraseys & Melville 82
17800108		Tabriz	38.2N	46.3E	N55W				6		Ambraseys & Melville 82
1838		Nosratabad	29.9N	60.0E	N		325				Ambraseys & Melville 82
19090123	7.0	Silakhur	33.3N	49.1E	NW	RE	65		2.5	Main Recent	Tchalenko et al 74, Ambraseys & Melville 82
19110418	6.2	Ravar	31.3N	57.1E	NNW	RR	18		0.5	Lakar Kuh	Berberian 81, Ambraseys & Melville 82
19290501	7.3	Kopet Dagh	37.6N	57.8E	N30W	RR	70		1		Tchalenko et al 74, Berberian 81, Ambraseys & Melville 82
19300506	7.4	Salmas	38.1N	44.7E	NW	RL	31	4	5	Salmas	Tchalenko & Berberian 74, Berberian & Tchalenko 76, Ambraseys & Melville 82

Table of Historic Earthquakes with Surface Rupture Continued

Date	M	Name	Lat.	Long.	Strike	Type	L	H	V	Fault Name	Reference
19331128	6.2	North Behabad	32.0N	55.9E	NW	RE?	5		1	Kuh Banan	Berberian et al 79, Berberian 81, Ambraseys & Melville 82
19410216	6.1	Muhammadabad	33.5N	58.9E	N	RR	10		0.5		Ambraseys & Melville 82
19470923	6.8	Dustabad	33.6N	58.7E	N10W	RR	20		1.3		Berberian 79, Ambraseys & Melville 82
19530212	6.5	Torud	31.0N	56.9E	ENE	RE	1		1.4	Torud	Abdalian 53, Berberian 76, Ambraseys & Moinfar 77
19571213	6.7	Farsinaj	34.6N	47.8E	N60W	RL	20		1	Main Recent	Ambraseys et al 73, Tchalenko & Braud 74
19580816	6.6	Firuzabad	34.3N	48.1E	N50W	RL	20		1.5	Main Recent	Ambraseys & Moinfar 74; Tchalenko & Braud 74
19620901	7.2	Buyin-Zara	35.6N	50.0E	N85W	LR	03	0.6	0.8	Ipak	Ambraseys 63
19680831	7.1	Dasht-e Bayaz	34.0N	59.0E	W	LL	80	4.5	2.5	Dasht-e Bayaz	Ambraseys & Tchalenko 69, Tchalenko & Ambraseys 70, Tchalenko & Berberian 75
19771219	5.8	Bob-Tangol	30.9N	56.6E	NW	RL	19.5		0.2	Kuh Banan	Berberian et al 79
19780916	7.5	Tabas-e-Golshan	33.3N	57.1E	NW	RE	85		3	Tabas	Berberian 79, 82
19791114	6.7	Kurizan	33.9N	59.8E	NNW	RR	17	0.9	0.6	Kurizan	Nowroozi & Mohajer-Ashjai 80, 85
19791127	7.1	Koli	34.1N	59.7E	W	LR	65	2.6	3.9	Dasht-e Bayaz	Nowroozi & Mohajer-Ashjai 80,
19810611	6.7	Golbaf	29.9N	57.7E	NNW, NNE	RR	15	0.05	0.06	Gowk	Berberian et al 84, Nowroozi & Mohajer-Ashjai 85
19810728	7.1	Sirch	30.3N	57.6E	NNW	RR	65	0.4	0.4	Gowk	Berberian et al 84, Nowroozi & Mohajer-Ashjai 85
19891120	5.7	South Golbaf	29.8N	57.7E	NNW	RR	1	0.01	0.004	Gowk	Berberian & Qorashi 94
19900620	7.7	Rudbar-Tarom	36.8N	49.5E	N80W	LR	84	0.6	1	Baklor, Kabateh, Zard Geli	Berberian et al 92

General reference for Iran: Ambraseys & Melville 82, Berberian 81

Lebanon

Date	M	Name	Lat.	Long.	Strike	Type	L	H	V	Fault Name	Reference
17591125	7.4s	Bekaa Valley	33.7N	35.6E	NNE	LL	100	?		Yammouneh	Ambraseys & Barazangi 89

Syria

Date	M	Name	Lat.	Long.	Strike	Type	L	H	V	Fault Name	Reference
14070429	7.0	Orontes	35.7N	36.3E	N	LL	?	?		Ghab	Ambraseys & Barazangi 89

Turkey

Date	M	Name	Lat.	Long.	Strike	Type	L	H	V	Fault Name	Reference
096709??	?	Honorias	40.8N	32.5E	ENE	RL	50	?		NAF	Ambraseys 70
0995????	?	Haykaberd-Claudias	38.7N	40.5E	ENE	LL	230	?		EAF	Ambraseys 70
103505??	?	Voukellarioi	40.7N	32.8E	ENE	RL	60	?		NAF	Ambraseys 70
1043????	?	Ekeghiatz	39.7N	39.0E	WNW	RL	150	?		NAF	Ambraseys 70
1050????	?	Gangra	40.4N	33.7E	ENE	RL	80	?		NAF	Ambraseys 70
16530223		Menderes Valley	37.9N	28.0E	W	NN	>70	2			Ambraseys & Finkel 87a
16680817	8.0	N. Anatolian	40.9N	35.0E	N70W	RL	450	?		NAF	Ambraseys & Finkel 88
17840723		N. Anatolian	39.7N	40.4E	WNW	RL	90			NAF	Ambraseys 75
18740503	7.1	Gölcük Gölü	38.5N	39.5E	ENE	LL	45	?	1-2	EAF	Ambraseys 89
18750327	6.7	Gölcük Gölü	38.5N	39.5E	ENE					EAF	Ambraseys 89
18990920	6.9	Menderes Valley	37.9N	28.0E	W	NN	40		1		Ambraseys & Finkel 87a
19090209	6.3	Zara-Imranli	40.1N	37.9E	W	?	15		0		Ambraseys & Finkel 87a
19120809	7.4	Saros-Marmara	40.8N	27.2E	ENE	RN	50		3	NAF N.Str.	Ambraseys & Finkel 87a, b, Ambraseys 88
19141003	7.1	Burdur	37.6N	30.1E	NE	RN	25		1.5		Ambraseys & Finkel 87a, Ambraseys 75, 88
19380419	6.8	Kirsehir	39.5N	34.0E	NW	RL	15	1	0.6		Ambraseys 70, Barka 92

Table of Historic Earthquakes with Surface Rupture Continued

Date	M	Name	Lat.	Long.	Strike	Type	L	H	V	Fault Name	Reference
19391226	7.8	Erzincan	38.0N	40.3E	WNW	RL	360	7.5	2.0	NAF	Barka 92, Barka & Kadinsky-Cade 88
19421220	7.2	Erbaa-Niksar	40.7N	36.5E	N60W	RL-N	70	1.7		NAF	Ambraseys 70, Barka 92
19431126	7.5	Ladik (Tosya)	41.0N	35.5E	W	RL	270	4.5		NAF	Ambraseys 70, Barka 92
19440201	7.5	Bolu-Gerede	40.9N	32.6E	ENE	RL	190	3.5		NAF	Ambraseys 70, Barka 92
19440625	6.0	Saphane (Ucak)	39.0N	29.4E	WNW	RN	19	0.3			Ambraseys 88
19460531	5.7	Ustükran	39.3N	41.2E	NW	RL	10	0.3		NAF	Ambraseys 88
19490817	6.9	Elmalidere	39.4N	40.7E	WNW	RL	60?	1.5		NAF	Ambraseys 75, Barka & Kadinsky-Cade 88
19510813	6.9	Kursunlu	40.7N	33.3E	ENE	RL	40	0.5			Pinar 53, Barka & Kadinsky-Cade 88
19530318	7.2	Yenice-Gönen	39.9N	27.4E	N50E	RL	50	4.35		NAF S.Str.	Ketin & Roesli 54
19570526	7.0	Abant	40.6N	31.0E	ENE	RL	40	1.6		NAF	Ambraseys 70, Barka 92
19641006	6.8	Manyas	40.0N	28.0E	N80W	NN	40		0.1		Ketin 66, 69, Öcal et al 68, Erentöz & Kurtman 64
19660819	6.8	Varto	39.2N	41.6E	WNW	RN	30	0.3	0.1	NAF	Ambraseys & Zátopek 68, Wallace 68
19670722	7.4	Mudurnu Valley	40.7N	31.2E	WNW	RL	80	1.9	1.2	NAF	Ambraseys & Zátopek 69
19671130	6.0	Pülümür	39.5N	39.6E	WNW	RL	4	0.20		NAF	Barka & Kadinsky-Cade 88
19680924	5.1	Kigi	39.2N	40.2E	NNW	?	6		0.25		Ambraseys 75
19690328	6.5	Alasehir	38.3N	28.5E	WNW	NN	36		0.82	Alasehir	Ketin & Abdüsselamoglu 69
19700328	7.1	Gediz	39.0N	29.4E	WNW	RN	45		2.8	Erdogmus-Hamamlar	Ambraseys & Tchalenko 72
19710512	6.2	Burdur	37.6N	30.1E	NE	RN	1		0.14		
19710522	6.7	Bingöl	39.0N	40.5E	NE	LL	35	0.25	0	EAF	Taymaz et al 91, Seymen & Aydin 72
19750906	6.7	Lice	38.4N	40.4E	W	RE	20		0.63		Arpat 77, Taymaz et al 91
19761124	7.3	Çaldiran	39.3N	44.1E	WNW	RL	55	3.5		Çaldiran	Arpat et al 77
19831030	6.9	Pasinler	40.4N	42.3E	NNE	LL	12	1.2		Horasan-Narman	Barka & Kadinsky-Cade 88
19920313	6.8	Erzincan	39.6N	39.5E	N60W	RN	62	0.1	0.2	NAF	Barka & Eyidogan 93, Trifonov et al 93

General reference for North Anatolian fault: Barka & Kadinsky-Cade 88, Barka 92; general reference for Turkey: Ambraseys 88
Abbreviations: NAF, North Anatolian fault; EAF, East Anatolian fault

North Yemen

Date	M	Name	Lat.	Long.	Strike	Type	L	H	V	Fault Name	Reference
19820929	6.0	Dhamer	14.7N	44.4E	NNW	NN	15		<0.05		Plafker et al 87

EAST ASIA

Afghanistan

Date	M	Name	Lat.	Long.	Strike	Type	L	H	V	Fault Name	Reference
15050706		Kabul	34.4N	69.0E	NNE	LV	60			Chaman	Heuckroth & Karim 70, Quittmeyer et al 79, Lawrence et al 92

India

Date	M	Name	Lat.	Long.	Strike	Type	L	H	V	Fault Name	Reference
18190616		Rann of Cutch	24.1N	69.1E	W	NN?	65		3	Allah Band	Oldham *98*, 28
19671210	6.5	Koyna	17.6N	73.8E	N	LL	>5		0.5	unnamed	Cluff 77
19930929	6.4	Latur	18.0N	76.4E	WNW	RE	3		1	unnamed	Seeber et al 93, Gupta 93

Pakistan

Date	M	Name	Lat.	Long.	Strike	Type	L	H	V	Fault Name	Reference
18921220		Chaman	30.9N	66.5E	NNE	LL	60	0.75	0.3	Chaman	Griesbach *93*, Davison *93*, Lawrence et al 92
19751003	6.4	Spinatizha	30.3N	66.3E	NNE	LL	5	0.04		Chaman	Farah 76

Table of Historic Earthquakes with Surface Rupture Continued

Date	M	Name	Lat.	Long.	Strike	Type	L	H	V	Fault Name	Reference
					Kyrgyzstan						
18870609	?	Verny	43.1N	77.0E	ENE	RE	?		?	Zaalai Alatau piedmont	Mushketov *90*, Tapponnier & Molnar 79
19110103	7.8	Chon-Kemin	42.8N	77.3E	W	RE	200	4		Chon-Kemin	Bogdanovich et al 14, Kuchai 69, Tapponnier & Molnar 79, Vilgelmzon 47
1920819	7.4	Suusamyr	42.2N	73.6N	N80E	RE	0.4		2.7	unnamed	Ghose et al 93
					Russia-Siberia						
18620112		Tsagan	52.3N	106.6E	WNW						Solonenko 68
19570627	7.9	Muya	56.2N	116.5E	NE	LN	35	1.5	5	Muya	Solonenko 65, Solonenko et al 66
19590829	6.8	Srednebaikalskoe				NN	10				Solonenko & Treskov 60
19950528	7.6	Neftegorsk (Sakhalin)	52.6N	142.8E	NE	RL	35	2	1.5	unnamed	Leith 95, Pavlenov & Semyonov unpub.
General reference for Siberia: Tapponnier and Molnar 79											
					Tajikistan						
19890122	5.6	Gissar					1.4	0.07	0.1		Zerkal & Vinnichenko 90
					Turkmenistan						
19481005	7.3						>12	3	0.5		Tchalenko 75
19830314	5.7	Kumdag					7.5	0.05	0.38		Trifonov et al 86
19840222	6.0						10.3	0.08	0.08		Trifonov et al 86
					Mongolia						
17611209		W. Mongolia(Ar Hötöl)	47.5N	91.8E	N30W	RR	>215	5-7	2	Hovd	Khil'ko et al 85, Trifonov 88
19030201	7.5	Üneyt	43.3N	104.5E	NE,ESE	VV	10,20		0.2	unnamed	Khil'ko et al 85
19050709	7.8	Tsetserleg	49.5N	97.0E	N60E	LR	130	2.5	2.5	Tsetserleg	Molnar & Deng 84
19050723	8.2	Bulnay	49.2N	96.0E	W,N, NW	LL,RL, RR	375	11	3	Bulnay, Düngen Teregtiyn	Florensov & Solonenko 65, Khil'ko et al 85, Trifonov 88
19571204	8.3	Gobi-Altay	45.1N	100.1E	N80W	LR	250	8	9	Bogd. Toromhon	Florensov & Solonenko 65
19580407	6.9	Bayan Tsagaan	45.1N	98.7E	W	LL	15	>1	1	unnamed	Florensov & Solonenko 65, Khil'ko et al 85
19601203	7.0	Buuryn Hyar	43.1N	104.5E	NW	RL	18	0.1	0.2	unnamed	Natsag-Yüm et al 71, Khil'ko et al 85
19670105	7.8	Mogod	48.2N	102.9E	N	RL	36	3.2	3.5	Mogod, Gurvan-Bulag	Natsag-Yüm et al 71, Khil'ko et al 85
19670120	6.7	Mogod	48.1N	103.0E	NW	RE	9	<0.5	3.5	Tüleet Uul	Natsag-Yüm et al 71, Khil'ko et al 85
19700515	7	Üüreg Nuur	50.1N	91.3E	W	RL	8		2	unnamed	Khil'ko et al 85
19740704	7	Tahiynshar	45.1N	94.0E	ENE	LL	17	0.4		unnamed	Khil'ko et al 85
General reference for Mongolia: Baljinnyam et al 93											
					China-Xinjiang						
1716	8	Tekes	43.2N	81.0E	N70E	LR	70	8	8	Tekes	Yang et al 88
18120308	8	Nileke	43.7N	83.5E	N80W	RL	100	4		Kaxhe	Feng 87
19061223	8	Manas	43.5N	85.0E	N85W	RE	146		0.58	Tugulu-Dushanzi	Avouac et al 93, Zhang et al 94
19310810	8	Fuyun	46.9N	90.1E	NNW	RL	180	14.6	3.6	Keketuohai-Ertai	Ding 85, Deng & Zhang 84, Baljinnyam et al 93

Table of Historic Earthquakes with Surface Rupture Continued

Date	M	Name	Lat.	Long.	Strike	Type	L	H	V	Fault Name	Reference
					China-Tibet-Qinghai						
0180	7.5	Gaotai (Luotuoshen)	39.4N	99.5E	WNW	RL	30	1-2	3.2	Yumushan	Tapponnier et al 90, Guo et al 93
14110929	8	Dangxiong	29.7N	90.2E	NE	LN	136	11-13	8-9	Nyanqing-tanggula	Wu et al 90a
15000104	7.5-8	Yiliang	25.0N	103.2E	N10E	LL	81	10-11	3	Xiaojiang	Cao et al 94
15110601	7.5	Yongsheng	26.7N	100.7E	N	LN	42		1.2	Chenghai	Guo et al 88
16090712	7.25	Hongyapu	39.2N	99.0E	WNW	LR	60	2-3	1-2	Fedongmiao-Hongyapu	Guo et al 93
17091014	7.25	Zhongwei	37.4N	105.3E	WNW	LR	30	7.4	1.1	Xiangshan-Tianjingshan	Zhang et al 88
17250801	7	Selaha	30.0N	101.9E	NW	LL	>30	0.45	0.3-0.4	Selaha	Li et al 92a
17330802	7.7	Dongchuan	26.3N	103.1E	N15W	LL	150	10.5	2	Xiaojiang	Zhu 88
17860601	7.7	Kangding	29.9N	102.0E	N25W	LL	>70	5-6	2-5	Xianshuihe	Long & Deng 90, Allen et al 91
18161208	7.5	Luhuo	31.3N	100.7E	NW	LL		3.6		Xianshuihe	Allen et al 91
18330906	8	Songming	25.0N	103.1E	N10E	LL	110-120	9-10	2	Xiaojiang	Chen & Li 88
18500912	7.5	Xichang	27.7N	102.4E	NW	LL	90	7	5.6	Zemuhe	Ren & Li 89
18881102	6.3	Jingtai	37.1N	104.2E	N80W	LL	38	1.7-2.3	0.2-0.6	Laohushan	Yuan et al 94, Deng et al 92
18930829	7	Qianning	30.6N	101.5E	NNW	LL	>40			Xianshuihe	Allen et al 91
18950705	7.5	Tashikuergan	37.7N	75.1E	NNW	RR	30	3.6	3.9	Taghman	Bai 86, Liu 93
19040830	7	Daofu	31.0N	101.2E	NW	LL				Xianshuihe	Allen et al 91
19201216	8.6	Haiyuan	36.7N	104.9E	WNW	LR	237	10-11	4.7	Haiyuan	Deng et al 86, 90
19230324	7.25	Daofu (Renda)	31.5N	101.0E	N45W	LL	>60	3		Xianshuihe	Qian et al 84, Allen et al 91
19270523	8	Gulang	37.6N	102.6E	N60W	NN	21		5.7	Taerzhuang-Shuixiakou	Liu et al 95
					N55W	RE	23		7.4	Xiafangzhai-Yanjiaxinzhuang	
19321225	7.6	Changma	39.7N	96.7E	N80W	LR	148.5	6.2	1	Changma	Meyer 91, Meyer et al 91, Hou 92, Peltzer et al 88
19341215	7.0	Gyaring Co	31.0N	89.0E	N55W	RN	>35	5	1.5	Gyaring Co	Armijo et al 89
19370107	7.5	Tuosuohu	35.5N	97.6E	WNW	LR	230	8	6	Kusaihu-Maqu	Jia et al 88
19470317	7.7	Dari	33.3N	99.5E	N50W	LR	90	5	5-6	Richa-Keshoutan	Dai 83
19511118	7.3	Danxiong	31.1N	91.1E	N55W	RL	81	7.3	1.5	Bengco	Tapponnier et al 81, Armijo et al 89, Wu and Deng 89
19520818	7.5	Naqu (Gulu)	31.3N	91.5E	NNE	LN	57.7	5	5-5.5	Gulu	Armijo et al 86, 89, Wu et al 90b
19540211	7.25	Shandan	39.0N	101.3E	NW	RN	18	2.9	0.95	Baodehe	Guo et al 93
19550414	7.5	Kangding	30.0N	101.8E	N40W	LL	30	3.1	1	Zheduotang	Allen et al 91, Li et al 92b
19700105	7.7	Tonghai	24.1N	102.6E	N60W	RL	60	2.7	0.5	Qujiang	Zhang 88
19730206	7.6	Luhuo	31.3N	100.7E	N55W	LL	90	3.6	0.5	Xianshuihe	Tang et al 76, Deng & Zhang 84, Allen et al 91
19770102	6.5	Mangya	38.2N	91.2E	WNW	RE	21		0.3	Mangya-Youkuangshan	Xu & Deng 96
19800222	6.8	Gyaring Co	30.8N	88.7E	N55W	RL	10	<1		Gyaring Co	Armijo et al. 89, RSBX 80
19810124	6.9	Daofu	31.0N	101.2E	N55W	LL	44	0.24		Xianshuihe	Qian et al 84, Deng & Zhang 84, Allen et al 91
19850823	7.4	Wuqia	39.4N	75.6E	W-NW	RE	15	1.55	1.7	Kazikeaerte	Feng et al. 88, Feng 94
19881105	6.8	Tanggula	34.4N	91.9E	W	LL	>9	4		Wulanwulahu-Gangqiqu	Wu et al 94
19881106	7.6	Langcang	22.9N	99.8E	NW	RL	50	2.2	0.3	Muga, Datangxi	Yu et al 91
19881106	7.2	Gengma	23.1N	99.4E	NNW	RN	14	0.98	0.9	Hangmuba	Zhou et al 90
					Eastern China						
13030917	8	Hongdong	36.3N	111.7E	NNE	RN	45	7.2-8.6	3.5-5	Huoshan	Xu & Deng 90

Table of Historic Earthquakes with Surface Rupture Continued

Date	M	Name	Lat.	Long.	Strike	Type	L	H	V	Fault Name	Reference
15560123	8	Huaxian	34.5N	109.7E	E	LN	30		4	Huashan N Piedmont	Zhang et al 89
16680728	8.5	Tangcheng	34.8N	118.5E	NNE	RR	120	7-9	2-3	Tanlu	Gao et al 88
16790902	8	Sanhe-Pinggu	40.0N	117.0E	NNE	RN	>10		3.16	Xiadian	Xiang et al 88
16950518	7.5	Linfen	36.0N	111.5E	NW	LN	>18		1.0	Linfen	Wang et al 91
17390103	8	Pinglu	38.8N	106.5E	NNE	RN	88	1.45	0.95	Helanshan	Deng et al 84, Zhang et al 86
19750204	7.3	Haicheng	40.7N	112.7E	NW	LL	28	0.55	0.2	Dayanghe	Deng et al 76
19760728	7.8	Tangshan	39.4N	118.0E	NNE	RN	10	1.53	0.7	Tangshan	Guo et al 77

General reference for China: Deng et al 92, Xu & Deng 96

Taiwan

Date	M	Name	Lat.	Long.	Strike	Type	L	H	V	Fault Name	Reference
19060317	7.1	Chiayi	23.6N	120.5E	ENE	RR	13	2.4	1.2	Meishan	Shieh 87, Omori 07
19350421	7.1	Taichung-Miaoli	24.3N	120.8E	NE	RR	53	1.5	3	Tuntzuchiao, Shintsoshan, Shihtan	ERI 36, Bonilla 77, Hsu & Chang 79, Shieh 87
19461205	6.7	Tainan	23.1N	120.3E	ENE	RR	12	2	0.76	Hsinhua	Bonilla 77, Chang et al 47
19511022	7.1	Hualien	23.9N	121.7E	NNE	LR	>7	2	1.2	Meilun	Bonilla 77, Hsu 62
19511125	7.3	Longitudinal Valley	23.4N	121.4E	NE	LL	40	1.63	1.3	Yuli	Bonilla 77, Hsu 62
19720424	6.9	Juishui (Hualien)	23.5N	121.5E	N25E		2.5	?	0.7		Hsu 76
19901213	6.2	Hualien	23.8N	121.6E	NNE	LR					Person 91

General reference for Taiwan: Hsu & Chang 79

Japan

Date	M	Name	Lat.	Long.	Strike	Type	L	H	V	Fault Name	Reference
0841????			36.2N	137.9E	NNW	LL		7.5		Gofukuji (ISTL)	Okumura et al 94
0841????	7	Jowa-Izu	35.1N	138.9E	NNE	LL	?	?	?	Tanna	TFTRG 83
08680803	7.1	Harima	34.9N	134.8E	WNW	LL	?	?	?	Yasutomi-Yamasaki	Okada et al 87
08781101	7.4	Gangyou	35.5N	139.3E	N	RE?	?	?	1.5	Isehara	Matsuda et al 88
15860118	7.8	Tensho	36.0N	136.9E	NNW	LL	65	?	?	Miboro	Sugiyama et al 91
15960905	7.5	Keicho	34.6N	135.6E	N80E	RL	?	?	?	Median Tectonic Line	Okada et al 91 Ishibashi 89 Okada 92
18470508	7.4	Zenkoji	36.7N	138.2E	N12E	RE	43		1.5-2.7	Zenkoji	Imamura 30, Omori 13
18540709	7.25	Iga-Ueno	34.8N	136.1E	ENE	RL	3		1.5	Iga-Ueno	Imamura 13
18541223	8.4	Ansei-Tokai	34.0N	137.8E	N	?	>3		1.5	Iriyamase	Yamazaki 92, Omori 13
18580409	7.1	Hietsu	36.4N	137.3E	N60E	RR	?	?	?	Atotsugawa	Okada et al. 89, Hirooka 91, Takeuchi & Sakai 85
18911028	8.0	Nobi	35.6N	136.6E	N45W	LL	80	1.2-8	1.8-6	Neodani	Matsuda 74
18941022	7.0	Shonai (Sakata)	38.9N	139.9E	N55E	?	10			Yadarezawa	Omori 00, Richter 58
18960831	7.2	Rikuu	39.5N	140.7E	NNE	RE	36		2-3.5	Senya	Yamasaki *96*, Matsuda et al 80
19181111	6.5	Omachi	36.5N	137.9E	N30E	?	1.1		0.15	SW of Omachi	Omori 22, Tsuboi 22
19230901	7.9	Kanto	35.1N	139.5E	N70W	RE	130	1.2-6	0.65-3	Shitaura, Sagami (submarine)	Imamura 24, 30, Yamasaki 25
19250523	6.8	Tajima	35.6N	134.8E	N30E	LV	1.6		1	Tai	Imamura 28, Yamasaki 27
19270307	7.3	Kita-Tango	35.5N	135.2E	NW	LL	18	3	1.2	Gomura	Yamasaki & Tada 28,
					N65E	RR	8	0.8	0.7	Yamada	Watanabe & Sato 28
19301126	7.3	Kita-Izu	35.1N	139.1E	NNE	LL	25.5	3.5	2.4	Tanna	Ihara & Ishii 32, TFTRG 83
					N70W	RL	6	1.2	0.8	Himenoyu	
19380529	6.1	Kussharo	43.6N	144.5E	N40W	LV	20	2.6	0.85	S. of L. Kussharo	Matumoto 59, Tsuya 38
19430910	7.2	Tottori	35.5N	134.1E	N80E	RL	45.5	2.5	1.1	Shikano	Tsuya 44, Kanamori 72, Okada et al. 81
19450113	6.8	Mikawa	34.7N	137.1E	N, E	RV,LV	21	1.5	2	Fukozu	Tsuya 46, Ando 74
19480628	7.1	Fukui	36.2N	136.2E	N10W	LR	25	2	0.7	NE of Fukui	Tsuya 50, Kanamori 73
19590131	6.2	Tesikaga	43.4N	144.4E	WNW	?	2		0.1	unnamed	Matumoto 59
19640616	7.5	Niigata	38.4N	139.2E	NNE	RE	20+		6	Awashima uplift	Abe 75, Kamata et al. 66; Okamura et al 94

Table of Historic Earthquakes with Surface Rupture Continued

Date	M	Name	Lat.	Long.	Strike	Type	L	H	V	Fault Name	Reference
1965(1966)	5.4	Matsushiro swarm	36.5N	138.2E	N55W	RL	4	0.57	0.15	Matsushiro	Matsuda 67, Tsuneishi & Nakamura 70
19740509	6.9	Izu-Hanto-oki	34.6N	138.8E	NW	RL	>6	0.45	0.25	Irozaki	Kinugasa 76, Kakimi et al 77
19780114	7.0	Izu-Oshima-kinkai	34.8N	139.3E	W	RL	21	1.83	0.35	Neginota	Tsuneishi et al 78, Yamazaki et al 79
19950117	6.9	Hyogo-ken Nanbu	34.6N	135.0E	NE	RL	9.5	1.9	1.2	Nojima	Yomogida & Nakata 95

General references for Japan: Yamazaki, H., 1985, Features of earthquake faults, in Earthquakes and Active Faults: ISU Company, Tokyo, 237-442 (J)
The Research Group for Active Faults of Japan, 1991, Active faults in Japan, sheet maps (1:100,000) and inventories: Univ. Tokyo Press, 437 p. (J).
The Research Group for Active Faults of Japan, 1992, Maps of active faults in Japan with an explanatory text: Univ. Tokyo Press

Indonesia

Date	M	Name	Lat.	Long.	Strike	Type	L	H	V	Fault Name	Reference
18920517		Tapanuli	0.7N	99.6E	NNW	RL		4		Great Sumatran	Reid 13, Richter 58
19430609	7.6	Padang Highlands	0.7S	100.7E	NNW	RL	60	3		Great Sumatran	Untung et al. 85

Philippines

Date	M	Name	Lat.	Long.	Strike	Type	L	H	V	Fault Name	Reference
16451130	8	Central Luzon	15.4N	121.4E	N40W	LL	>40	?	?	Philippine-Gabaldan	Hirano et al. 86, Nakata et al. 90
17961105	7.2	San Manuel	16.1N	120.7E	N60W	LL	33	?	?	Philippine-San Manuel	Hirano et al 86, Nakata et al 90
19730317	7.0	Ragay Gulf	13.5N	122.8E	N40W	LL	>30	3.2	0.21	Philippine	Allen 75, Morante & Allen 74
19900716	7.8	Central Luzon	15.7N	121.1E	N25W	LL	125	6.2	2.0	Philippine-Digdig	Nakata et al 90
19941115	7.0	Mindoro	13.5N	121.1E	N20W	RL	>15	3.0	1	unnamed	PHIVOLCS, unpub.

SOUTHWEST PACIFIC

Australia

Date	M	Name	Lat.	Long.	Strike	Type	L	H	V	Fault Name	Reference
19681014	6.8	Meckering	31.6S	117.0E	NNE	RR	37	1.5	2	Meckering	Gordon & Lewis 80
19700311	5.7	Calingiri	31.1S	116.5E	N08E	LR	3	0.14	<0.4	Calingiri	Gordon & Lewis 80
19790606	6.4	Cadoux	30.8S	117.2E	N20E	RR	28	0.7	< 1.4	Robb	Lewis et al 81
19860330	5.8	Marryat Creek	26.2S	132.8E	N, W	RR,LR	13	0.8	0.9	Marryat Creek	Machette et al 93
19880122	6.3	Tennant Creek	19.8S	133.9E	WNW	RE	10.2	0.25	0.1-0.2	Kunayungku	Crone et al 92
19880122	6.4	Tennant Creek	19.8S	133.9E	ENE	RE	6.7	0.1	1.1	W. Lake Surprise	Crone et al 92
19880122	6.7	Tennant Creek	19.8S	134.0E	WNW	RE	16	0.4	1.7	E. Lake Surprise	Crone et al 92

General reference for Australia: Crone et al 93

NEW ZEALAND

Date	M	Name	Lat.	Long.	Strike	Type	L	H	V	Fault Name	Reference
18550124	8.2	West Wairarapa	41.2S	175.3E	N45E	RL	85-105	13.5	1.5	Wairarapa	Ongley 43, Darby & Beanland 92
18880901	7.3	Marlborough	42.5S	172.6E	ENE	RL	50	1.5		Hope	McKay *90*, Cowan 91
19220610	6	Kaiapo	38.6S	175.9E	NNE	NN	12		3	Kaiapo, Whakapo, Whangamata	Grange 32, Sissons 79, Grindley & Hull 86
19290617	7.6	Murchison	41.8S	172.2E	NNE	LR	>8	2.5	4.9	White Creek	Henderson 37, Berryman 80
19310203	7.8	Hawke's Bay	39.2S	176.7E	NE	RR	15	1.8	4.6	Poukawa, Bridge Pa	Henderson 33, Marshall 33, Hull 90
19320916	7	Wairoa				RR	1		0.5	Wairoa	Ongley 37
19680524	7.1	Inangahua	41.9S	172.0E	NNE	LR	>2	0.2	0.4	Inangahua (Glasgow)	Lenson & Suggate 68, Lenson & Otway 71, Anderson et al 94
198307	3.4	Kaiapo	38.6S	175.9E	NNE	NN	1.2		0.05	Kaiapo	Hull & Grindley 83, Grindley & Hull 86
19870302	6.3	Edgecumbe	38.0S	176.5E	NE	NN	18		2.5	Edgecumbe	Beanland et al 90

Explanation **Date:** year, month, day; BC dates use negative symbol. **M:** magnitude; assume ML if not specified; other subscripts: b, body wave; s, surface wave; w, moment; t, tsunami. **Type:** RE, reverse; NN, normal; LL, left lateral; LR, left lateral and reverse; LN, left lateral and normal; RL, right lateral; RR, right lateral and reverse; RN, right lateral and normal; LV, left lateral; dip direction unclear; RV, right lateral; dip direction unclear; VV, dip slip, dip direction unclear. **L:** Length in km. Total length of rupture zone, including unbroken sections. **H:** Maximum horizontal offset in meters. **V:** Maximum vertical offset in meters. **Reference:** Only last two numbers of year given. Years prior to 1900 in italics.

Glossary

abrasion—The mechanical wearing, grinding, scraping, or rubbing away of rock surfaces by friction and impact.

acceptable risk—The maximum risk which one is prepared to tolerate. Any risk still higher would be intolerable and unacceptable.

accretionary wedge—A prism of deformed sediments in the upper plate of a subduction zone near the deformation front, formed in part by accretion of sediments of the downgoing plate, but also including sediments derived from the interior of the upper plate.

active continental margin—A continent-ocean basin transition marked by an active plate boundary, in most cases a subduction zone.

active fault—A fault that has had sufficiently recent displacements so that, in the opinion of the user of the term, further displacements in the foreseeable future are considered likely. *See also* **Holocene active fault** and **Pleistocene active fault.** There are all gradations of fault activity.

active tectonics—Tectonic movements that are expected to occur or have occurred within a time span of concern to society.

adhesion—The molecular attraction between contiguous surfaces.

aeolian—Said of deposits or landforms (such as sand dunes or loess) whose constituents were transported by atmospheric currents.

aftershock—A smaller earthquake that follows the mainshock of a series, originates within less than one fault length of the mainshock rupture, and is part of a flurry of earthquakes that occurs at a rate higher than the regional rate of seismicity before the mainshock.

alluvial fan—A fluvial deposit whose surface forms the segment of a cone that radiates downslope from the point where the stream leaves a source area.

alluvial river—A river that flows between banks and on a bed composed of sediment that is transported by the river, is sensitive to changes of sediment load, water discharge, and variations of valley-floor slope.

amplitude—Maximum height of a wave crest or depth of a wave trough.

angle of internal friction—The angle ϕ, taken to be positive, between the straight line defining the Coulomb fracture criterion and the abscissa, where the tangent of ϕ equals the coefficient of internal friction.

antecedent stream—A stream that was present before the formation of a fold, fault scarp, or other tectonic landform, and which maintains its course despite subsequent earth deformation.

anticline—A fold, generally convex upward, whose core contains the stratigraphically older rocks.

anticlinorium—A composite anticlinal structure of regional extent composed of lesser folds.

antithetic—Pertaining to minor normal faults that are oriented opposite to the major fault with which they are associated.

aragonite—A form of $CaCO_3$ that is a major component of shallow marine muds, unaltered coral reefs, and of mollusks.

aseismic front—In a subduction zone, the boundary between the seismogenic forearc basin and the aseismic volcanic arc.

asperity—Irregularity on the fault surface that retards slip. Region of relatively high shear strength on a fault surface.

asthenosphere—The layer or shell of the earth below the lithosphere, which is weak and in which isostatic adjustment takes place, magmas may be generated, and seismic waves are strongly attenuated.

asymmetrical fold—A fold whose axial surface is not perpendicular to the enveloping surface. A fold whose limbs have different angles of dip relative to the axial surface.

attenuation—The reduction in amplitude of a wave with time or distance traveled.

axial surface—A surface that connects the hinge lines of the strata in a fold.

B.C.E.—Before the Common Era, that is, before A.D. 1. A more generally acceptable designation than B.C., which refers dates in antiquity to the birth of Christ.

"b" value—A coefficient describing the ratio of small to large earthquakes within a given area and time period, often shown to be the same over a wide range of magnitudes. It is the slope of the curve in the Gutenberg-Richter recurrence relationship.

back-arc basin—Zone of extension developed behind the volcanic arc, on the other side of the trench. Synonym: **marginal basin**.

backlimb—The less steep of the two limbs of an asymmetrical, anticlinal fold.

backstop—That part of the upper plate of a subduction zone exhibiting unstable frictional rate behavior.

balanced cross section—A geologic cross section of rock in which all deformation has taken place within the plane of the section and bed lengths and cross-section area are conserved.

basalt—A dark-colored mafic igneous rock, commonly extrusive but locally intrusive (e.g., as dikes), composed chiefly of calcic plagioclase and clinopyroxene. The principal volcanic constituent of oceanic crust.

base level—The theoretical limit or lowest level toward which erosion of the earth's surface constantly progresses but seldom, if ever, reaches.

basin-and-range topography—An alternation of block-faulted ranges and valleys, with one or both sides bounded by normal faults. First used in the Basin-and-Range Province in the western United States.

beheaded channel—Portion of a channel downstream from a fault separated from its headwaters by strike-slip movement along the fault.

bending-moment fault—Fault formed due to bending moment applied to a flexed layer during folding. Normal faults characterize the convex side, placed in tension, and reverse faults characterize the concave side, placed in compression.

bilateral—Said of a fault rupture in which propagation of the rupture front away from the hypocenter occurs in two directions. Antonym: unilateral.

biostratigraphy—Stratigraphy based on the biological constituents of rocks, or stratigraphy with paleontological methods; the separation and differentiation of rock units on the basis of the description and study of the fossils they contain.

bioturbation—The disturbance of a sediment by animals or plants.

blind fault—A fault that does not, and never has, extended upward to the earth's surface. It usually terminates upward in the axial region of an anticline.

blind thrust—A blind reverse fault with a dip $< 45°$.

body wave—A seismic wave that travels through the interior of an elastic material.

bottomset bed—A horizontal or gently inclined layer of silt and/or clay deposited from suspension on a sea floor or lake bottom in front of the advancing margin of a delta or bar and progressively buried by the foreset beds of that delta or bar.

braided stream—A stream that divides into or follows an interlacing or tangled network of several small branching and reuniting shallow channels separated from each other by branch islands or channel bars, resembling in plan the strands of a complex braid.

breccia—A coarse-grained clastic rock composed of angular broken rock fragments in a fine-grained matrix.

brittle—Said of a rock that fractures at less than 3–5% deformation or strain.

brittle-ductile transition—A zone within the earth's crust that separates superjacent brittle rocks from subjacent ductile rocks. Commonly identified as the zone defining the deepest earthquakes in the crust. More precisely stated as **brittle-plastic transition**.

calcrete—A surficial sand and gravel cemented into a hard mass by calcium carbonate precipitated from solution and redeposited through the agency of infiltrating waters, or deposited by the escape of carbon dioxide from water above this water table.

caldera—A large basin-shaped volcanic depression.

capable fault—A fault along which it is mechanically feasible for sudden slip to occur.

cataclasis—Rock deformation accomplished by fracture and rotation of mineral grains or aggregates.

characteristic earthquake—The maximum earthquake that is thought to occur repeatedly on a given fault or fault zone, and occurring more frequently than would be deduced from simple extrapolation of smaller earthquakes using a Gutenberg-Richter recurrence relationship.

clastic—Said of a rock or sediment composed principally of broken fragments (clasts).

clay—A rock or mineral fragment or a detrital particle of any composition (often a crystalline fragment of a clay mineral), smaller than a very fine silt grain, having a diameter less than 4 microns.

cluse—A narrow deep gorge, trench, or water gap, cutting transversely through an otherwise continuous ridge, especially an antecedent valley crossing an anticlinal limestone ridge in the Jura Mountains of the European Alps.

coda—The concluding train of seismic waves recorded on a seismograph following the principal part of an earthquake.

coefficient of internal friction—The slope of the straight/ line, taken to be positive, defining the Coulomb fracture criterion.

cohesion—Resistance to shear fracture on a plane across which the normal stress is zero.

colluvial wedge—A prism-shaped deposit of colluvium at the base of (and formed by erosion of) a fault scarp or other slope, commonly taken as evidence in outcrop of a scarp-forming event.

colluvium—A mass of soil or sediment, commonly poorly sorted, loose, and heterogeneous, deposited by gravitational collapse, soil creep, or minor transport by water at the base of a slope.

comb fracture—A fracture subtending an angle of 60–90° with a slip plane which, in profile, resembles the teeth of a hand-held comb.

compaction—Densification by loss of porosity, commonly resulting from burial or dessication of earth material.

concentric fold—Parallel fold.

continent—Any of the earth's largest landmasses, characterized by lighter crustal densities than those of the ocean basins and, therefore, exposed primarily above the surface of the oceans.

continental crust—That type of the earth's crust which constitutes the continents and the continental shelves. Continental crust ranges in thickness from about 35 km to as much as 60 km under mountain ranges. The density of the upper layer of the continental crust is ~2.7 g/cm^3, and the speeds of compressional seismic waves through it are less then ~7.0 km/sec.

core (of earth)—The central part of the earth, liquid below a depth of 2855 km and solid below a depth of about 5144 km. Composed mainly of iron, with subordinate amounts of nickel and some lighter elements.

coseismic—During an earthquake.

Coulomb fracture criterion—A criterion of brittle shear failure based on the concept that shear failure will occur along a surface when the shear stress acting in that plane is large enough to overcome the cohesive strength of the material and the frictional resistance to movement.

craton—A large part of a continent that has been little deformed for hundreds of millions of years.

creep (along a fault)—*See* **fault creep**.

crust—The outermost layer or shell of the earth, defined as that part of the earth above the Mohorovicic discontinuity in speed of seismic waves.

crystalline rock—An inexact but convenient term designating an igneous or metamorphic rock as opposed to a sedimentary rock.

cylindrical fold—A fold that can be described geometrically by the rotation of a line through space parallel to itself.

damping—Loss of energy in wave motion due to transfer into heat by frictional forces.

debris flow—A high-density flow of mud or other debris.

debris slope—The surface at the base of and inclined away from the free face of a scarp or slope formed by the gravitational collapse of unconsolidated materials from the free face.

décollement—A plane between highly deformed overlying strata and less-deformed underlying rock or sediment. Synoynm: **detachment**, although this term is commonly restricted to low-angle normal faults.

deflation—The erosional lowering of the surface of a clastic surface by aeolian removal of finer-grained components.

delta—A fan- or triangular-shaped landform produced at or near the mouth of a river or stream by the deposition of debris in the stream as it enters a standing body of water.

deltaic—Pertaining to or characterized by a delta.

dendrochronology—The development of climatic, seismic, or other time-series by the analysis of the annual growth rings of trees.

density—Mass per unit volume.

design earthquake—A term used by engineers in the actual design of a structure. It may or may not be equivalent to the maximum earthquake prescribed by the geologist or seismologist.

deterministic hazard assessment—An assessment that specifies single-valued parameters such as maximum earthquake magnitude or peak ground acceleration, without consideration of likelihood.

detrital—Said of material (detritus) that has been broken and subsequently transported to its current location. For example, detrital charcoal in a paleoseismic excavation.

dextral—Synonymous with **right-lateral.**

diachronous—Said of a landscape or rock unit that varies in age from place to place. Antonym: **isochronous**.

diapir—A dome or anticlinal fold formed by the buoyant rise of subjacent material through denser superjacent material. Diapirs commonly contain cores of salt, shale, or serpentine.

dike—A tabular igneous or sedimentary intrusion that cuts across the bedding or foliation of the host rock.

dip—The maximum vertical angle between an inclined plane and a horizontal plane.

dip-slip fault—A fault which experiences slip only in the direction of its dip, that is, perpendicular to strike.

dislocation—Line defect in a crystal.

ductile—Said of a rock that is able to sustain, under a given set of conditions, 5–10% deformation before fracture or faulting.

dun—A broad synclinal valley in the Himalaya separated from the broad Gangetic Plain to the south by an active anticline.

earthquake cycle—For a particular fault, fault segment, or region, a period of time that encompasses an episode of strain accumulation and its subsequent seismic relief.

earthquake segment—That part of a fault zone or fault zones that has ruptured during individual earthquakes.

elastic limit—The greatest stress that can be sustained by a material without permanent deformation remaining when the stress is released.

electron-spin resonance—Resonance that occurs when electrons that are undergoing transitions between energy levels in a substance are irradiated with electromagnetic energy of a proper frequency to produce maximum absorption.

eluviation—The downward movement, by groundwater percolation, of soluble or suspended material from the A horizon to the B horizon of a soil.

en échelon—Said of geologic features such as faults that are in an overlapping or staggered arrangement. Each feature is relatively short, but collectively they form a linear zone, in which the strike of the individual features is oblique to that of the zone as a whole.

epicenter—The point on the earth's surface that is directly above the focus (hypocenter) of an earthquake.

epoch—A subdivision of geologic time, subordinate to a geologic period. For example, the Holocene and Pleistocene Epochs constitute the nearly 2-million-year-long Quaternary Period.

era—A subdivision of geologic time that encompasses more than one period. For example, the Quaternary and Tertiary Periods constitute the 65-million-year-long Cenozoic Era.

escape-block tectonics—The displacement, generally toward oceanic crust, of large fault-bounded blocks of continental crust laterally out of the way of collision of two continental plates. Applied principally to the collisions of the Indian and Arabian plates with Eurasia.

eustatic—Pertaining to worldwide changes of sea level that affect all the oceans. Eustatic changes may have various causes, but the changes dominant in the past few million years were caused by additions of water to, or removal of water from, the continental icecaps.

evaporite—A nonclastic sedimentary rock composed primarily of minerals produced during the extensive or total evaporation of a lake or sea.

event horizon—A bedding plane within a stratigraphic series that represents the ground surface at the time of a paleoseismic event.

expert judgment—The judgment of an individual or group of individuals considered knowledgeable in a field of special concern, often elicited in a formal and interactive manner.

extensometer—An instrument for measuring minute deformation of material under stress.

fault—A fracture or a zone of fractures along which displacement has occurred parallel to the fracture.

fault-angle depression—An arrow, downdropped crustal block bounced by a fault on only one side. Synonym: **half-graben**.

fault-bend fold—A fold within the hangingwall block of a fault, formed by bending of the hangingwall block as it slips over a nonplanar fault surface.

fault creep—Steady or episodic slip on a fault at a rate too slow to produce an earthquake.

fault length—(1) The total length of a fault or fault zone. (2) The rupture length along a fault or fault zone associated with a specific earthquake, representing either the observed surface rupture length or the rupture length at depth, usually determined from the aftershock distribution of the earthquake.

fault-propagation fold—A fold formed in front of a fault surface as the fault surface lengthens over a period of geologic time. Commonly associated with the upward termination of a thrust fault.

fault scarp—A slope formed by offset of the surface of the earth by a fault.

fault slip rate—The rate of slip on a fault averaged over a time period involving several large earthquakes. The term does not necessarily imply fault creep.

feldspar—A group of abundant rock-forming minerals of general formula: $MAl(Al,Si)_3O_8$, where M = K, Na, Ca, Ba, Rb, Sr, or Fe. Feldspars are the most widespread of any mineral group and constitute 60% of the earth's crust.

felsic—Said of an igneous rock having abundant light-colored minerals.

filled wedge—A wedge-shaped fissure filled with detrital deposits commonly recognized in paleoseismic studies as evidence for ground deformation.

first motion—On a seismogram, the direction of motion at the beginning of the arrival of a P wave. Conventionally, upward motion indicates an initial compression of the ground; downward motion, an initial dilatation.

fission track—The path of radiation damage made by product of nuclear fission in a mineral or glass by the spontaneous fission of radioactive atoms.

fissure swarm—A group of normal faults or fissures, all of small displacement or separation.

flexural-slip fault—A bedding fault formed by layer-parallel slip during flexural-slip folding.

flexural-slip folding—A process of folding in which deformation occurs by slip along bedding or foliation planes. In such folds, no change in thickness of individual strata occurs, so that the resultant folds are parallel.

floating earthquake—A conceptual earthquake that is considered equally likely to occur at any point within a specified region.

fluid inclusion—A tiny cavity in a mineral containing liquid and/or gas, formed by the entrapment of fluid during crystallization.

fluvial—Of or pertaining to a river or rivers. Produced by the action of a stream or river.

focal depth—The depth of the focus below the surface of the earth.

focus—The place within the earth where an earthquake commences and from which the first P waves arrive.

fold-and-thrust belt—A set of imbricate thrusts and folds overlying a décollement.

footwall—The underlying side of a nonvertical fault surface.

foreland—A stable area adjacent to an orogenic belt, toward which the rocks of the orogenic belt move, generally on the edge of a craton in continental crust.

foreland basin—A basin in a foreland formed by depression from the weight of advancing thrust sheets. Synonym: **foredeep**.

foreset bed—A stratum deposited on the steep, advancing front of a delta or bar.

foreshock—An earthquake that immediately precedes the mainshock of a series and originates within the region of the hypocenter of the mainshock. Currently recognized as such only after the occurrence of the mainshock.

fracture zone—An elongate zone of unusually irregular topography on the sea floor, which represents the inactive extension of a ridge-ridge transform fault.

free face—Surface of a scarp exposed directly by fault slip or by erosion.

frequency—Number of oscillations per unit time. One cycle per second = 1 Hertz.

friction—The resistance to motion of a body sliding past another body along a surface of contact.

geodesy—The science concerned with the determination of the size and shape of the earth and the precise location of points on its surface.

geodimeter—An electronic-optical instrument that measures distance with great precision by measuring the travel time of light between points.

geomorphology—The investigation of the earth's landforms; specifically the study of the nature, origin, and development of present landforms, their relationships to underlying structures, and of the history of geologic changes as recorded by these surface features.

geothermal gradient—The rate of change of temperature in the earth as a function of depth.

gja—Icelandic term for gaping fracture, including fissure or gaping normal fault.

Global Positioning System (GPS)—A geodetic surveying system in which multiple satellites are used to establish locations of ground stations.

gneiss—A rock formed by regional metamorphism in which bands or lenticles of granular minerals alternate with bands and lenticles characterized by minerals having flaky or elongate prismatic form.

gouge—A thin layer of fine-grained, highly cataclastic material within a fault zone. Synonym: **pug**.

graben—A crustal block, generally longer than it is narrow, that has dropped relative to adjacent rocks along bounding faults.

graded stream—A stream in which transporting capacity is equal to the amount of material supplied to it. Thus, a stream in which degradation and aggradation in the stream channel are in equilibrium.

granite—A medium- to coarse-grained rock which crystallized several kilometers below the earth's surface and in which quartz constitutes 10–50% of the light-colored mineral components, and in which the alkali feldspar to total feldspar ratio is between 65 and 90%. Broadly applied, any completely crystalline, quartz-bearing plutonic rock.

gravity graben—A prism of sediment settling between the main normal fault and an antithetic fault that intersects the main fault at a shallow depth, forming a near-surface graben.

Griffith crack—A small, penny-shaped or slitlike crack that in cross section is much longer than it is thick and that has a very small radius of curvature at its tips. Commonly an imperfection within the crystal lattice of crystal grains in a rock, or an intergranular or grain-boundary crack. Commonly modeled as an extremely flattened ellipsoid.

Gutenberg-Richter recurrence relationship—The observed relationship that, for large areas and long time periods, numbers of earthquakes of different magnitudes occur systematically, with the relationship $M = a - bN$, where M is magnitude, N is the number of events per unit area per unit time, and a and b are constants representing, respectively, the overall level of seismicity and the ratio of small to large events.

hangingwall—The overlying side of a nonvertical fault surface.

hazard—*See* **Seismic hazard**

heat flow—The amount of heat energy leaving the earth per cm^2/sec.

heave—The horizontal component of separation or displacement on a fault.

hinge—The locus of maximum curvature or bending in a folded surface, usually a line.

Holocene—The past 10,000 years; the latest epoch of the Quaternary period.

Holocene active fault—A fault which has had displacement on it within the Holocene epoch.

horst—An elongate crustal block, uplifted relative to adjacent blocks along faults bounding its longer sides.

hot spot—A volcanic center, 100 to 200 km across and persistent for at least a few tens of millions of years, thought to be the surface expression of a persistent rising plume of mantle material. Hot spots are not linked with volcanic arcs, but some are associated with oceanic ridges.

hydrolytic weakening—The reduction of strength in silicates due to reaction with pore fluids, principally water.

hydrostatic pressure—Stress that is uniform in all directions, e.g., beneath a homogeneous fluid, and causes dilation or compression rather than distortion in isotropic materials. The pressure exerted by the water at any given point in a body of water at rest.

hydrothermal—Of or pertaining to hot water, to the action of hot water, or to the products of this action.

illuviation—The accumulation, in a lower soil horizon, of soluble or suspended material that was transported from an upper horizon by the process of eluviation.

inactive fault—A fault which has had no displacement over a sufficiently long time period in the geologic past so that displacements in the foreseeable future are considered unlikely by the user of the term.

incompressibility—An index of the resistance of an elastic body, such as a rock, to volume change.

intensity (of an earthquake)—A measure of the severity of ground shaking, obtained by examination of the damage done to structures built by humans, changes in the earth's surface, and felt reports.

interferometer—An instrument for precise determination of wavelength, spectral fine structure, refraction, and very small linear displacements through the separation of light into two parts that travel unequal optical paths, and, when reunited, consequently interfere with each other.

interplate earthquake—An earthquake along a plate boundary.

intraplate earthquake—An earthquake within a plate.

invar—An iron-nickel alloy, characterized by an extremely small coefficient of linear expansion (approximately 10^{-6} inch per inch per degree centigrade) at ordinary temperatures.

island arc—An arcuate chain of volcanic islands, generally convex toward the open ocean and built upon the deep-sea floor.

isoseismal—Contour lines drawn to separate one level of seismic intensity from another.

isostasy—That condition of equilibrium, comparable to floating, of the units of the lithosphere above the asthenosphere.

isotropic—Said of a material whose properties are the same in all directions.

joint—A fracture or parting in a rock, characterized by no slip parallel to its surface. Often occurs with parallel joints to form part of a **joint set**.

lateral ramp—The steepening inclined segment of a thrust fault in which the strike of the inclined segment is parallel to the direction of tectonic transport.

lava—Magma or molten rock that has reached the surface.

leaky barrier—An earthquake segment boundary that is, from time to time, broken across by an earthquake.

left-lateral fault—A strike-slip fault across which a viewer would see the block on the other side move to the left.

leveling—Geodetic method of determining the differences in elevation between points on the earth's surface by determining a horizontal line between the points.

limb—That area of a fold between adjacent fold hinges.

limestone—A sedimentary rock consisting chiefly of calcium carbonate, primarily in the form of calcite.

liquefaction (of soils)—Transformation of a granular material from a solid state into a liquefied state as a consequence of increased pore-water pressures.

listric fault—A curvilinear fault, usually concave-upward, with steeper dips near the surface than at depth.

lithosphere—The outermost layer of the earth, distinguished from the subjacent asthenosphere by its greater rigidity and strength. Commonly includes the crust and part of the upper mantle, which are distinguished by differences in seismic wave speeds rather than rheological properties.

lithostatic pressure—The pressure at a point in the earth's crust equal to the pressure caused by the weight of a column of the overlying material.

loess—A widespread, homogeneous, commonly nonstratified, porous, friable, slightly coherent, usually highly calcareous, fine-grained blanket deposit consisting predominantly of silt, generally believed to be predominantly Pleistocene windblown dust.

logic tree—A formalized decision flow path in which decisions are made sequentially at a series of **nodes**, each of which generates **branches** flowing to subsequent nodes.

longitudinal step fault—A series of parallel faults formed when the main fault branches upward toward the surface into several subsidiary faults.

long-term earthquake prediction—A prediction with a time frame long enough so that only permanent responses such as those of land-use planning and building code revisions are realistic—usually representing tens, hundreds, or thousands of years. It is usually based on statistics of past earthquake occurrences or on physical models, not on the recognition of specific precursory phenomena.

Love waves—Seismic surface waves with only horizontal shear motion transverse to the direction of propagation.

mafic—Said of an igneous rock composed chiefly of one or more ferromagnesian, dark-colored minerals.

magma—Naturally occurring fluid rock, generated within the earth and capable of intrusion and extrusion. Igneous rocks are thought to have been derived through solidification and related processes.

magnetic declination—At any given point on the earth's surface, the horizontal angle between the magnetic line of flux and true north.

magnetic inclination—At any given point on the earth's surface, the vertical angle (plunge) between the magnetic line of flux and a horizontal plane. Downward and upward plunges are arbitrarily defined as positive and negative inclinations, respectively, in the Northern Hemisphere.

magnitude (of earthquakes)—Any of a number of measures of earthquake size at its source, based upon the amplitude of seismic waves recorded instrumentally.

mainshock—The largest earthquake within a closely-spaced and temporally-clustered series of earthquakes. Typically followed by aftershocks, smaller earthquakes which become less frequent with increased time since the mainshock.

mantle—A region of the earth distinguished from the overlying crust and subjacent core by differences in seismic wave speed.

maximum credible earthquake—The maximum earthquake that is capable of occurring in a given area or on a given fault during the current tectonic regime. "Credibility" is in the eyes of the user of the term.

maximum earthquake—The maximum earthquake that is thought. in the judgment of the user, to be appropriate for consideration in the location and design of a specific facility.

maximum probable earthquake—The maximum earthquake that, on a probabilistic basis, is likely to occur in a given area or on a given fault during a specific time period in the future.

mean recurrence interval—The mean time between earthquakes of a given magnitude, or within a given magnitude range, on a specific fault or within a specific area.

meander—One of a series of regular freely-developing sinuous curves, bends, loops, turns, or windings in the course of a stream.

meizoseismal region—The area of strong shaking and significant damage in an earthquake.

metamorphic facies—A set of metamorphic mineral assemblages, repeatedly associated in space and time, such that there is a constant and therefore predictable relation between mineral composition and chemical composition. It is generally assumed that the metamorphic facies represents the result of equilibrium crystallization of rocks under a restricted range of externally imposed physical conditions, e.g., temperature, confining pressure, H_2O pressure.

Miocene—An epoch of the Tertiary period between about 23 and 5 million years ago, after the Oligocene and before the Pliocene.

miroir de faille—A highly polished slip plane of a fault.

modulate—To vary a characteristic (such as amplitude, frequency, or phase) of a carrier wave or signal in a periodic or intermittent manner.

modulus of elasticity—The ratio of stress to strain under given conditions of load, for materials that deform elastically, according to Hooke's law.

Mohorovicic discontinuity—The boundary between the earth's crust and subjacent mantle, defined on the basis of a sharp discontinuity in seismic P-wave speed from 6.7–7.2 km/sec in the lower crust to 7.6–8.6 km/sec at the top of the upper mantle.

Mohr circle—A graphical representation of the state of stress at a particular position. The center of the circle is the mean normal stress. The circle is the path of a point that moves such that its radius is the maximum shear stress. Each point on the circle represents the shear stress and the normal stress on a plane with a given orientation and sense of shear with respect to the point.

moment—*See* **seismic moment**.

moment magnitude (M_w)—Magnitude of an earthquake estimated from the seismic moment.

monocline—A geological structure characterized by a unidirectional steepening of an otherwise uniformly dipping surface.

mullion—A columnar or linear structure along a fault plane, commonly elongate in the direction of dip.

mylonite—A very fine-grained rock with lineated and/or foliated fabric produced by extreme shear.

nastri di faglia—Bright bands of less-weathered colluvium or bedrock exposed along the scarp of an active normal fault. Synoynm: **fault ribbon**.

natural levee—A long, broad, low ridge or embankment of clastic debris, built by a stream on the floodplain adjacent to its channel in time of flood, when water overflows the channel.

neotectonics—(1) The study of the post-Miocene structures and structural history of the earth's crust. (2) The study of recent deformation of the crust, generally Neogene (post-Oligocene). (3) Tectonic processes now active, taken over the geologic time span during which they have been acting in the presently observed sense, and the resulting structures.

normal fault—A fault in which the hangingwall block has moved downward relative to the footwall.

normal magnetic polarity—Refers to a rock formed in a period when the earth's magnetic field was polarized as it is today, that is, magnetic lines of flux enter the north magnetic pole and exit the south magnetic pole.

normal stress—That component of stress which is perpendicular to a specified plane.

obliquity—The difference between the dip direction of a subducting plate and the expected interplate slip direction.

ocean basin—The area of the sea floor beyond the base of the continental margin, usually the foot of the continental rise.

olivine—An olive-green, grayish-green, or brown orthorhombic mineral: $(Mg,Fe)_2SiO_4$. A common rock-forming mineral of basic, ultrabasic, and low-silica igneous rocks such as basalt, peridotite, and dunite.

operating basis earthquake—A U.S. Nuclear Regulatory Commission term that specifies the maximum ground motion for which a reactor is expected to continue operating during and after the earthquake.

orogenic belt—A large, linear or arcuate region of the earth's crust that has been subjected to folding and other deformation during an **orogenic cycle**. Most later become mountain belts by post-orogenic uplift.

orthoclase—A mineral of the alkali-feldspar group: $KAlSi_3O_8$. A common rock-forming mineral occurring especially in granite, acid igneous rocks, and crystalline schists.

outer-arc high—A topographic bulge in the upper plate of a subduction zone trenchward of a **forearc basin**.

overturned fold—A fold in which one limb has tilted beyond the vertical plane.

P wave—The primary or fastest wave traveling away from an earthquake source, consisting of a train of compressions and dilatations parallel to the direction of travel of the wave.

paleoseismology—The investigation of individual earthquakes decades, centuries, or millenia after their occurrence.

paleosol—A buried soil formed during some period in the past.

parallel fold—A fold in which the thickness of individual beds is constant.

passive margin—A boundary between a continent and ocean basin not marked by an active plate boundary.

peat—An unconsolidated deposit of semicarbonized plant remains.

peneplain—A surface of regional extent, low local relief and low absolute altitude, produced by long-lived fluvial erosion.

peridotite—A coarse-grained plutonic rock composed chiefly of olivine with or without other mafic minerals such as pyroxene, amphibole, or mica, and containing little or no feldspar. Peridotite encompasses the more specific terms saxonite, harzburgite, lherzolite, wehrlite, and dunite. Accessor minerals of the spinel group are commonly present. Commonly altered to serpentine.

period—(1) The time interval between successive crests in a wave train; the period is the inverse of the frequency of a cyclic event. (2) The fundamental unit of the geological time scale, subdivided into epochs, and subdivisions of an era. Most earthquake geology deals with the latest period, the Quaternary, which began about 1.8 million years ago.

permeability—The property or capacity of a porous rock, sediment, or soil for transmitting a fluid. Measurement of permeability is in millidarcys.

piercing points—Two points, on opposite sides of a fault, that were, prior to slip on the fault, part of the same linear feature. Used to measure the offset across a fault.

pitch—*See* rake.

plagioclase—A group of triclinic feldspars of general formula $(Na,Ca)Al(Si,Al)Si_2O_8$. A common rock-forming mineral in continental and oceanic crust.

plate—A large, relatively rigid segment of the earth's lithosphere that moves in relation to other plates and the deeper interior of the earth.

plate tectonics—A global theory of tectonics in which an outermost sphere (the lithosphere) is divided into a number of relatively rigid plates that collide with, separate from, and translate past one another at their boundaries.

Pleistocene—A 1.8-million-year long epoch of the Quaternary period, following the Pliocene epoch and prior to the current 0.01-million-year long Holocene epoch. The Pleistocene epoch was characterized by multiple glaciations of continental landmasses.

Pleistocene active fault—A fault which has had displacement on it during the Pleistocene epoch.

Pliocene—An epoch of the Tertiary period, from about 5 million to about 1.8 million years ago, after the Miocene and before the Pleistocene.

plunge—The inclination of a fold axis or other linear structure, measured with respect to a horizontal line in the vertical plane.

pluvial lake—Any lake formed during an extended period of regionally wetter conditions, induced by either higher rainfall, lesser evaporation, or both. Common during the Pleistocene epoch in many parts of the world, but now mostly either partially or completely dried up.

Poisson process—A process of uniform randomness, with no memory or time dependence.

polygenetic—Resulting from more than one process of formation, such as a landscape formed by a combination of aeolian, fluvial, and glacial processes.

pore—A small opening or passageway in a rock.

porosity—The percentage of the bulk volume of a rock or soil that is occupied by pores (interstices), whether isolated or connected.

potentially active fault—A term used by different people in different ways, but sometimes referring to a fault that has had displacement on it within the late Quaternary period.

precursor—A geophysical phenomenon thought to be causally related to a later earthquake.

prediction (of earthquakes)—A forecast of the time, place, and magnitude of an earthquake.

preshock—An earthquake that occurs before the mainshock, but is not an unambiguous foreshock, either because it originated months or years before the mainshock or because it originated far from the epicenter of the mainshock.

pressure solution—(1) Solution occurring preferentially at the contact surfaces of mineral grains (crystals) where the external pressure exceeds the hydraulic pressure of the interstitial fluid, resulting in enlargement of the contact surfaces. (2) Tendency of a crystal subjected to a differential stress to dissolve more readily at surfaces on which the normal-stress component is greatest.

primary surface rupture—Surface rupture that is directly connected to subsurface displacement on a seismic fault.

principal stress—A stress that is perpendicular to one of three mutually perpendicular planes across which no shear stresses exist at a point within a stressed body.

probabilistic hazard assessment—An assessment which stipulates quantitative probabilities of the occurrences of specified hazards, usually within a specified time period.

probability—The ratio of the chances favoring a certain event to all the chances for and against it.

proper motion—The apparent change in position of a star that results from the projection on the celestial sphere of its motion with respect to the solar system.

protothrust—An incipient thrust fault formed in the downgoing plate seaward of the main deformation front.

pseudotachylite—A massive rock that frequently appears in microbreccias or surrounding rocks as dark veins of glassy or cryptocrystalline material, so named because of its macroscopic resemblance to **tachylite**, or basaltic glass. Characteristically contains a matrix of crystals less than 1 mm in diameter and/or small amounts of glass or devitrified glass cementing a mass of fractured material together.

pull-apart basin—A topographic depression produced by extensional bends or stepovers along a strike-slip fault.

pyroclastic flow—A turbulent flow of unsorted, mostly fine-grained volcanic material and high-temperature gas ejected explosively from a volcanic vent.

pyroxene—A group of dark rock-forming silicate minerals, closely related in crystal form and composition and having the general formula: $ABSi_2O_6$, where A = Ca, Na, Mg, or Fe^{+2} and B = Mg, Fe^{+2}, Fe^{+3}, Fe, Cr, Mn, or Al, with silicon sometimes replaced by aluminum.

quartz—Crystalline silica, an important rock-forming mineral with the formula SiO_2.

Quaternary—The 1.8-million-year-long second period of the Cenozoic era, following the Tertiary, consisting of the Pleistocene and Holocene epochs.

radiometric—Pertaining to the determination of geologic age by the measurement of radioactive isotopes and the products of their decay.

rake—The angle between the strike of a plane and any linear feature within that plane.

ramp—The steepening inclined segment of a thrust fault, especially where a bedding thrust changes from a stratigraphically lower to a higher layer.

range-and-basin topography—An alternation of ranges and valleys in which one or both of the valley margins is marked by reverse faults. First applied to Central Otago, South Island, New Zealand.

Rayleigh waves—Seismic surface waves with ground motion only in a vertical plane containing the direction of propagation of the waves.

recurrence interval—*See* **mean recurrence interval.**

recurrence relationship, recurrence curve—*See* **Gutenberg-Richter recurrence relationship.**

relief—An informal term for the physical shape, configuration, or general unevenness of a part of the earth's surface, considered with reference to variations of height and slope or to irregularities of the land surface; the elevations or differences in elevation, considered collectively, of a land surface.

remanent magnetization—That component of a rock's magnetization that has a fixed direction relative to the rock and is independent of moderate, applied magnetic fields such as the earth's magnetic field.

reservoir-triggered earthquake—An earthquake that is triggered by the filling of a reservoir or by water-level changes therein. It is thought mainly to represent the release of tectonic strain, with the reservoir being only a perturbing influence. Sometimes also referred to as a **reservoir-induced earthquake**.

reverse fault—A fault characterized by movement of the hangingwall block upward relative to the footwall block.

reversed magnetic polarity—Refers to a rock formed during a period when the earth's magnetic field was polarized opposite its present polarization, that is, magnetic lines of flux enter the south magnetic pole and exit the north magnetic pole.

rheology—The study of the deformation and flow of matter.

rift—Region of the crust where long, linear valleys have formed by tectonic depression accompanying extension, generally bounded by normal faults. Active rifts, such as the East African Rift and the Rhine Graben are usually marked by **rift valleys**.

right-lateral fault—A strike-slip fault across which a viewer would see the adjacent block move to the right.

rigidity—An index of the resistance of an elastic body to shear. The ratio of the shearing stress to the amount of angular rotation it produces in a rock sample.

risk—*See* **seismic risk.**

rollover—A feature of some normal faults in which the beds of the downthrown block dip toward the fault surface in an orientation opposite to that produced by drag.

S wave—The secondary seismic wave, traveling more slowly than the P wave and consisting of elastic vibrations transverse to the direction of travel.

safe shutdown earthquake—A U.S. Nuclear Regulatory Commission term that specifies the maximum ground motion for which a reactor is expected to shut down safely without significant risk to the public.

salient—An outward topographic or structural projection. Synonym: **push-up block** in Chinese literature.

sand—A rock fragment or detrital particle smaller than a granule and larger than a coarse silt grain, having a diameter in the range of 62 to 2000 microns (2 mm).

sand blow—A deposit of sand, a few centimeters to many meters in diameter or length, and its related fissures, dikes, pipes, or craters, which commonly resemble the form and plumbing of a volcano. Formed by the expulsion of liquefied sand from subsurface sources onto the ground surface during and immediately following moderate to large earthquakes.

sand boil—Sometimes used synonymously with **sand volcano** and **sand blow**, but originally a spring that bubbles through a river levee, with an ejection of sand and water, as a result of water in the river being forced through permeable sands and silts below the levee during flood stage.

sea-floor spreading—The splitting of oceanic crust and creation of new crust by convective upwelling of magma along the mid-oceanic ridges or world rift system.

secondary effects—Nontectonic surface processes that are directly related to earthquake shaking or to tsunamis.

secondary surface rupture—Rupture along a secondary fault, such as an antithetic normal fault intersecting the primary seismic fault at relatively shallow depths, or branching off from the primary seismic fault.

secular equilibrium—A condition reachable by a system of radiogenic isotopes and their radioactive products in which the products of each isotopes' concentration multiplied by its decay constant are equal.

secular variation—Changes in the earth's magnetic field over decades to centuries that are not related to the nondipole component of the magnetic field.

seismic gap—An area in an earthquake-prone region where there has been a below-average level of seismic energy that is thought to be temporary.

seismic hazard—The physical effects such as ground shaking, faulting, landsliding, and liquefaction that underlie the earthquake's potential danger.

seismic moment—The area of a fault rupture multiplied by the average slip over the rupture area and multiplied by the shear modulus of the affected rocks. It has the dimensions of moment, e.g., dyne-cm, N-m.

seismic moment rate—The long-term rate at which seismic moment is being generated.

seismic risk—The likelihood of human and property loss that can result from the hazards of an earthquake.

seismic wave—An elastic wave in the earth usually generated by an earthquake or explosion.

seismicity—The occurrence of earthquakes in space and time.

seismogenic—Produced by earthquakes.

seismogenic structure—One that is capable of producing an earthquake.

seismograph—An instrument for recording as a function of time motions of the earth's surface caused by seismic waves.

seismology—(1) The study of earthquakes. (2) The study of earthquakes, and of the structure of the earth, by both naturally and artificially generated seismic waves.

seismotectonic province—A region within which the active geologic and seismic processes are considered to be relatively uniform.

seismotectonics—That subfield of active tectonics concentrating on the seismicity, both instrumental and historical, and dealing with geological and other geophysical aspects of seismicity.

separation—The distance along a fault between two points on an offset plane dirsrupted by a fault. Not synonymous with **offset**.

serpentine—A group of common rock-forming minerals having the formula $(Mg,Fe)_3Si_2O_5(OH)_4$. Commonly derived by alteration of magnesium-rich silicate minerals (especially olivine) in mantle exposed along ridge-ridge transform fault zones.

shear stress—That component of stress which acts tangentially to a plane through any given point in a body.

shield volcano—A volcano in the shape of a flattened dome, broad and low, built by flows of very fluid basaltic lava or by silicic ash flows.

shoreline angle—The line of intersection between a sea cliff and a marine abrasion platform.

short-term earthquake prediction—A prediction within a few minutes, hours, days, or weeks of an earthquake, based upon the recognition of premonitory phenomena.

shutter ridge—A linear hill or scarp sloping in a direction opposite to the overall local topographic gradient, formed by strike-slip or oblique-slip offset of irregular topography. A shutter ridge tends to block the flow of water and debris across the fault, as shutters block the flow of light or air through a window.

silicic—Said of an igneous rock or magma relatively rich in silica.

silt—A rock fragment or detrital particle smaller than a very fine sand grain and larger than coarse clay having a diameter in the range of 4 to 62 microns.

similar fold—A fold in which the orthogonal thickness of the folded strata is greater in the hinge than in the limbs, but the distance between any two folded surfaces is constant when measured parallel to the axial surface.

sinistral—Synonymous with **left-lateral**.

slickensides—A polished and smoothly striated surface that results from slip along a fault plane.

slip rate—*See* **fault slip rate**.

slip vector—The magnitude and orientation of dislocation of formerly adjacent features on opposite sides of a fault.

slip-vector residual—The deflection of slip vectors of interplate earthquakes from the slip vector predicted by plate tectonics.

soil—(1) A roughly tabular stratum consisting of layers or horizons of clastic and/or precipitated mineral and/or organic constituents, formed by natural chemical and physical alteration (**pedogenesis**) of the earth's surface. (2) All unconsolidated materials above bedrock (engineering).

soil profile—A vertical section of a soil that displays all its horizons.

soil structure—The combination or aggregation of primary soil particles into aggregates or clusters (**peds**), which are separated from adjoining peds by surfaces of weakness.

soil chronosequence—A series of soils formed on similar geomorphic surfaces of different ages under similar topographic and climatic conditions, but during different periods of time.

stepover—Region where one fault ends and another en échelon fault of the same orientation begins. Described geometrically as being either right or left depending on whether the bend or step is to the right or to the left as one progresses along the fault.

stereographic projection—A perspective, conformal, azimuthal map projection in which meridians and parallels are projected onto a tangent plane, with the point of projection on the surface of the sphere diametrically opposite to the point of tangency of the projecting plane.

stick slip—A jerky, sliding motion associated with fault movement.

strain—Change in the shape or volume of a body as a result of stress.

strain partitioning—The tendency of faults in an environment where both strike-slip and dip-slip faults are subparallel to each other to rupture in either a dip-slip or a strike-slip earthquake rather than an oblique-slip combination of the two fault types.

stratigraphy—The description and interpretation of sedimentary layers, concerned not only with the original succession and age relations of strata but also with their form, distribution, lithologic composition, fossil content, geophysical and geochemical properties and their interpretation in terms of environment or mode of origin, and geologic history.

stream capture—The diversion of the headwaters of one stream into the channel of another stream. Synonym: **stream piracy**.

stress—Force per unit area.

stress drop—The sudden reduction of stress across the fault plane during rupture.

strike—The azimuth of a horizontal line drawn on

an inclined plane, which, along with the dip of the plane, provides a complete description of the orientation of the plane in space.

strike slip—That component of slip along a fault that is horizontal and in the plane of the fault, that is, parallel to the strike of the fault.

strike-slip fault—A fault on which the movement is parallel to the strike of the fault.

subaerial—Occurring at or near the interface between the atmosphere and the solid earth. Said of conditions and processes (such as erosion) that exist or operate on or immediately adjacent to the land surface, or of features and materials (such as aeolian deposits) that are formed and situated on the land surface.

subduction—The process of one lithospheric plate descending beneath another.

subduction zone—A long, narrow belt in which subduction takes place.

subslip-plane breccia sheet—Thin sheet of compact zone-parallel mineralized and cemented fault breccia.

surface-wave magnitude (M_s)—Magnitude of an earthquake estimated from measurements of the amplitude of surface waves with a period of 20 seconds.

surface waves—Seismic waves that follow the earth's surface only, with a speed less than that of S waves. There are two types of surface waves—Rayleigh waves and Love waves.

swarm—A spatially and temporally clustered group of earthquakes, in which no one earthquake is so much larger than the others as to be considered a mainshock.

symmetrical fold—A fold whose limbs have the same angular relationship to the axial surface.

sympathetic surface rupture—Triggered surface slip along a fault isolated from the primary fault.

syncline—A fold, generally concave upward, whose core contains the stratigraphically younger rocks.

tectonic geomorphology—The study of landforms that result from tectonic processes.

tectonics—A branch of geology dealing with the broad architecture of the outer part of the earth, that is, the regional assembling of structural or deformational features, a study of their mutual relations, origin, and historical evolution.

teleseism—An earthquake that occurs at a distant place, usually more than 1000 km away.

temporal clustering of earthquakes—The tendency for earthquakes to occur in temporal groups separated in time by a period of relatively low seismic strain release.

tephrochronology—The collection, preparation, petrographic description, and approximate dating of tephra (pyroclastic material from a volcano).

terrace riser—The vertical or steeply sloping erosional surface that separates a higher terrace surface from a lower one. Analogous to the vertical riser that separates two stair steps.

thermoluminescence—The property possessed by many substances of emitting light when heated. It results from release of energy stored as electron displacements in the crystal lattice.

thick-skinned tectonics—An interpretation that reverse faults and folds involve crystalline basement.

thin-skinned tectonics—An interpretation that folds and faults in an orogenic belt involve only the upper strata, lying on a décollement between which the structure differs or the rocks are undeformed.

throw—The vertical component of displacement on a fault.

thrust fault—A fault with a dip of 45° or less over much of its extent, on which the hangingwall block has moved upward relative to the footwall block.

time-predictable model—With reference to earthquakes, a recurrence model in which the time interval between two successive large earthquakes is proportional to the amount of seismic displacement of the preceding earthquake.

tomography—Construction of the image of an object or structure within the earth from measurements of seismic waves at the surface.

topography—The general configuration of a land surface or any part of the earth's surface, including its relief and the position of its natural and man-made features.

trace—That part of one geological surface where it intersects another surface, e.g., the trace of a bed on a fault surface, or the trace of a fault on the ground surface.

transcurrent fault—A nearly vertical strike-slip fault that cuts not only supracrustal sedimentary rocks, but also igneous and metamorphic rocks. Avoided in this text in favor of the more general, nongenetic term **strike-slip fault**.

transform fault—A strike-slip fault of plate-boundary dimensions that transforms into another plate-boundary structure at its terminus.

travertine—A dense, finely crystalline massive or concretionary limestone formed by rapid chemical precipitation of calcium carbonate from solution in surface and ground waters; also occurs in limestone caves, in veins, and along faults.

trench—(1) Bathymetric depression on the sea floor, commonly thousands of meters deep and thousands of kilometers long, which results from the bending of a plate of oceanic lithosphere as it begins its descent at a subduction zone. (2) Shallow excavation, up to several meters deep, excavated by equipment or by hand to reveal detailed stratigraphic and structural information about near-surface strata and faults.

trend—A general term for the direction or bearing of the outcrop of a geological feature of any dimension.

triangular facet—An inclined triangular erosional surface, commonly formed by erosion of a tectonically truncated ridge trending at a high angle to an active fault. May also form by wave erosion of a mountain front or by glacial truncation of a spur.

triangulation—A trigonometric operation for finding the location of a point by means of bearings from two fixed points a known distance apart.

triggered earthquake—An earthquake that occurs soon after or during a mainshock, but well outside the region of aftershocks.

triggered slip—Aseismic fault slip that occurs during or soon after seismogenic rupture of a nearby fault.

trilateration—A geodetic procedure for locating the relative positions of a set of points by the measurement of the distances between the points.

triple junction—Point where three plate boundaries meet.

tsunami—A long-period ocean wave usually caused by sea-floor movements during an earthquake, submarine volcanic eruption, or submarine landslide. In Japanese, means "harbor wave."

tufa—A chemical sedimentary rock composed of calcium carbonate, formed by evaporation as a thin, surficial, soft, spongy, cellular or porous incrustation around the mouth of a spring, along a stream, or along the shore of a lake; it may also be precipitated by bacteria or algae.

turbidite—The sedimentary deposit resulting from a turbidity current along the base of a body of water.

twin—A rational intergrowth of two or more single crystals of the same mineral in a mathematically describable manner, so that some lattices are parallel wheareas others are in reversed position. Twin gliding is crystal gliding that results in the formation of crystal twins.

ultimate strength—The maximum differential stress that a material can sustain under the conditions of deformation.

ultramafic—Said of an igneous rock composed chiefly of mafic minerals, e.g., pyroxene and/or olivine.

uniformitarianism—The geological concept that geologic processes are relatively constant in time, so that the geological processes observed today are a key to interpreting the earth's earlier history, as well as for visualizing likely processes in the future.

unilateral—Said of a fault rupture in which propagation of the rupture front away from the hypocenter occurs in one direction. Antonym: **bilateral**.

varve—A distinct sedimentary bed or sequence deposited in a body of still water within one year's time.

vergence—The direction toward which a fold is overturned or inclined.

viscosity—The property of a substance to offer internal resistance to flow; its internal friction. The ratio of the shear stress to the rate of shear strain.

volcanology—The study of volcanoes, including both their causes and phenomenology.

Wadati-Benioff zone—A narrow zone of earthquake foci that seismically illuminates a subduction zone. Commonly tens of kilometers thick.

wash slope—The aggradational part of a scarp upon which water-borne particles come to rest after transport from higher parts of the scarp.

wavefront—Imaginary surface or line that joins points at which the waves from a source are in phase.

wavelength—The distance between two successive crests or troughs of a wave.

weathering—The chemical and physical processes by which materials on or near the earth's surface are, by exposure to atmospheric agents, altered in color, texture, composition, firmness, or form, with little or no transport of the loosened or altered material.

wine-glass canyon—Canyon cut by a stream which is broad upstream but narrow at a range front cut by an active normal fault, where it is flanked by **triangular facets.**

wrench fault—A nearly vertical strike-slip fault that cuts not only supracrustal sedimentary rocks, but also igneous and metamorphic rocks. Avoided in this text in favor of the more general, nongenetic term **strike-slip fault.**

yield stress—The differential stress at which permanent deformation first occurs in a material.

Bibliography

Abdalian, S. 1953. Le tremblement de terre de Toroude, en Iran. La Nature 81(3222): 314–19.

Abdallah, A., Courtillot, V., Kasser, M., Le Dain, A.-Y., Lépine, J.-C., Robineau, B., Ruegg, J.-C., Tapponnier, P., and Tarantola, A. 1979. Relevance of Afar seismicity and volcanism to the mechanics of accreting plate boundaries. Nature 282:17–23.

Abe, K. 1975. Re-examination of the fault model for the Niigata earthquake of 1964. Jour. Phys. Earth 23:349–66.

Abers, G. A., and McCaffrey, R. 1988. Active deformation in the New Guinea fold-and-thrust belt: Seismological evidence for strike-slip faulting and basement-involved thrusting. Jour. Geophys. Res. 93:13,332–54.

Abers, G. A., and McCaffrey, R. 1994. Active arc-continent collision: earthquakes, gravity anomalies, and fault kinematics in the Huon-Finisterre collision zone, Papua New Guinea. Tectonics 13:227–45.

Acton, G. D., Stein, S., and Engeln, J. F. 1991. Block rotation and continental extension in Afar: a comparison to oceanic microplate systems. Tectonics 10:501–26.

Adams, J. 1981. Earthquake-dammed lakes in New Zealand. Geology 9:215–19.

Adams, J. 1984. Active deformation of the Pacific Northwest continental margin. Tectonics 3:449–72.

Adams, J. 1990. Paleoseismicity of the Cascadia subduction zone: evidence from turbidites off the Oregon-Washington margin. Tectonics 9:569–83.

Adams, J., Wetmiller, R. J., Hasegawa, H. S., and Drysdale, J. 1991. The first surface faulting from a historical intraplate earthquake in North America. Nature 352:617–19.

Adams, R. D. 1974. Statistical studies of earthquakes associated with Lake Benmore, New Zealand. Engineering Geology 8:155–69.

Agnew, D. C. 1986. Strainmeters and tiltmeters. Reviews of Geophysics 24:579–624.

Agnew, D. C., and Sieh, K. 1978. A documentary study of the felt effects of the great California earthquake of 1857. Seismological Society of America Bulletin 68:1717–29.

Aguirre, E., and Pasini, G. 1985. The Pliocene-Pleistocene boundary. Episodes 8:116–20.

Aki, K. 1966. Generation and propagation of G waves from the Niigata earthquake of June 16, 1964. Part 2. Estimation of earthquake movement, released energy, and stress-strain drop from the G wave spectrum. Bull. Earthquake Research Inst. 44:73–88.

Algermissen, S. T., Perkins, D. M., Thenhaus, P. C., Hanson, S. L., and Bender, B. L. 1982. Probabilistic estimates of maximum acceleration and velocity in rock in the contiguous United States: U.S. Geological Survey Open-File Report. 82–103.

Allen, C. R. 1975. Geologic criteria for evaluating seismicity. Geol. Soc. America Bulletin 86:1041–57.

Allen, C. R. 1982. Reservoir-induced earthquakes and engineering policy. California Geology 35:248–50.

Allen, C. R., Gillespie, A., Yuan, H., Sieh, K., Zhang, B., and Chengnan, Z. 1984. Red River and associated faults, Yunnan Province, China: Quaternary geology, slip rate and seismic hazard. Geol. Soc. America Bulletin 95:686–700.

Allen, C. R., Luo Z., Qian H., Wen X., Zhou H., and Huang W. 1991. Field study of a highly active fault zone: The Xianshuihe fault of southwestern China. Geol. Soc. America Bulletin 103:1178–99.

Allen, C. R., St. Amand, P., Richter, C. F., and Nordquist, J. M., 1965, Relationship between seismicity and geologic structure in the southern California region: Seismological Society of America Bulletin 55:753-797.

Allen, C. R., Wyss, M., Brune, J., Granz, A., and Wallace, R. 1972. Displacements on the Imperial, Superstition Hills, and San Andreas faults triggered by the Borrego Mountain earthquake, *in* The Borrego Mountain Earthquake, U. S. Geological Survey Prof. Paper 787:87-104.

Ambraseys, N. N. 1963. The Buyin-Zara (Iran) earthquake of September, 1962, a field report. Seismological Society of America Bull. 53:705-40.

Ambraseys, N. N. 1970. Some characteristic features of the Anatolian fault zone. Tectonophysics 9:143-65.

Ambraseys, N. N. 1975. Studies in historical seismicity and tectonics, *in* Geodynamics of Today. Royal Society of London 9-16.

Ambraseys, N. N. 1988. Engineering seismology. Earthquake Engineering and Structural Dynamics 17:1-105.

Ambraseys, N. N. 1989. Temporary seismic quiescence: SE Turkey. Geophys. Jour. 96:311-31.

Ambraseys, N. N. 1991. Earthquake hazard in the Kenya Rift: The Subukia earthquake 1928. Geophys. J. Int. 105:253-69.

Ambraseys, N. N., and Adams, R. D. 1986. Seismicity of West Africa. Annales Geophysicae 4B:679-702.

Ambraseys, N., and Barazangi, M. 1989. The 1759 earthquake in the Bekaa Valley: implications for earthquake hazard assessment in the eastern Mediterranean region. Jour. Geophys. Res. 94:4007-13.

Ambraseys, N. N., and Finkel, C. F. 1987a. Seismicity of Turkey and neighbouring regions, 1899-1915:. Annales Geophysicae 5B:701-26.

Ambraseys, N. N., and Finkel, C. F. 1987b. The Saros-Marmara earthquake of 9 August 1912. Earthquake Eng. Struct. Dyn. 15:189-211.

Ambraseys, N.N., and Finkel, C.F. 1988. The Anatolian earthquake of 17 August 1668. *In* Lee, W. H. K., Meyers, H., and Shimazaki, K., eds. Historical Seismograms and Earthquakes of the World. New York: Academic Press, pp. 173-80.

Ambraseys, N. N., and Jackson, J. A. 1990. Seismicity and associated strain of central Greece between 1890 and 1988. Geophys. J. Int. 101:663-708.

Ambraseys, N. N., and Melville, C. P. 1982. A History of Persian Earthquakes. Cambridge University Press, 219 p.

Ambraseys, N. N., and Melville, C. P. 1988. An analysis of the eastern Mediterranean earthquake of 20 May 1202, *in* Lee, W.H.K., Meyers, H., and Shimazaki, K., eds., Historical Seismograms and Earthquakes of the World. New York, Academic Press, Inc., pp. 181-200.

Ambraseys, N., and Moinfar, A. 1974. The seismicity of Iran: The Firuzabad (Nehavend) earthquake of 16 August, 1953. Annali di Geofisica 27:1-21.

Ambraseys, N. N., and Moinfar, A. 1977. The seismicity of Iran: The Torud earthquake of 12th February 1953. Annali di Geofisica 30:185-200.

Ambraseys, N. N., Moinfar, A., and Peronaci, F. 1973. The seismicity of Iran. The Farsinaj, Kermanshah, earthquake of 13th December, 1957. Annali di Geofisica 6:679-92.

Ambraseys, N. N., and Tchalenko, J. S. 1969. The Dasht-e Bayaz (Iran) earthquake of August 31st, 1968: A field report. Seismological Society of America Bulletin 59:1751-92.

Ambraseys, N. N., and Tchalenko, J. S. 1972. Seismotectonic aspects of the Gediz, Turkey, earthquake of March 1970. Geophys. Jour. Royal Astr. Soc. 30:229-52.

Ambraseys, N. N., and Zátopek, A. 1968. The Varto Ustukran (Anatolia) earthquake of 19 August 1966—summary of a field report. Seismological Society of America Bulletin 58:47-102.

Ambraseys, N. N., and Zátopek, A. 1969. The Mudurnu Valley, West Anatolia, Turkey, earthquake of 22 July 1967. Seismological Society of America Bulletin 59:521-89.

Anderson, E. M. 1942. The Dynamics of Faulting and Dyke Formation, with Applications in Britain. Edinburgh, Oliver and Boyd, 191 p.

Anderson, H., Beanland, S., Blick, G., Darby, D., Downes, G., Haines, J., Jackson, J., Robinson, R., and Webb, T. 1994. The 1968 May 23 Inangahua, New Zealand, earthquake: An integrated geological, geodetic and seismological source model. New Zealand Jour. Geol. Geophysics 37:59-86.

Anderson, H., Smith, E., and Robinson, R. 1990. Normal faulting in a back arc basin: Seismological characteristics of the March 2, 1987, Edgecumbe, New Zealand, earthquake. Jour. Geophys. Res. 95:4709-23.

Anderson, J. G., and Bodin, P. 1987. Earthquake recurrence models and historical seismicity in the Mexicali-Imperial Valley. Seismological Society of America Bulletin 77:562–78.

Anderson, R. 1990. Evolution of the Santa Cruz mountains by advection of crust past a San Andreas fault bend. Science 249:397–401.

Ando, M. 1974. Faulting in the Mikawa earthquake of 1945. Tectonophysics 22:173-86.

Ando, M. 1975. Source mechanisms and tectonic significance of historical earthquakes along the Nankai trough, Japan. Tectonophysics 27:119–40.

Andrews, D. J., and Bucknam, R. C. 1987. Fitting degradation of shoreline scarps by a nonlinear diffusion model. Jour. Geophys. Res. 92:12,857–67.

Angelier, J. 1979. Determination of the mean principal directions of stresses for a given fault population. Tectonophysics 56:717–26.

Anthony, E. Y., and Poths, J. 1992. ^{3}He surface exposure dating and its implication for magma evolution in the Potrillo volcanic field, Rio Grande rift, New Mexico, USA. Geochimica et Cosmochimica Acta 56:4105–08.

Appelgate, T. B., Jr., 1990, Volcanic and structural morphology of the south flank of Axial Volcano, Juan de Fuca Ridge: Results from a Sea MARC 1 side scan sonar survey. Jour. Geophys. Res. 95:12,765–83.

Arason, P., and Levi, S. 1990. Compaction and inclination shallowing in deep-sea sediments from the Pacific Ocean. Jour. Geophys. Res. 95:4501–10.

Archuleta, R. 1984. A finite faulting model for the 1979 Imperial Valley, California earthquake. Jour. Geophys. Res. 89:4559–85.

Armijo, R., Lyon-Caen, H., and Papanastassiou, D. 1991. A possible normal-fault rupture for the 464 B.C. Sparta earthquake. Nature 351:137–39.

Armijo, R., Lyon-Caen, H., and Papanastassiou, D. 1992. East-west extension and Holocene normal fault scarps in the Hellenic arc. Geology 20:491–94.

Armijo, R., Tapponnier, P., and Han T.-L. 1989. Late Cenozoic right-lateral strike-slip faulting in southern Tibet. Jour. Geophys. Res. 94:2787–2838.

Armijo, R., Tapponnier, P., Mercier, J.L., and Han T.-L. 1986. Quaternary extension in southern Tibet: field observations and tectonic implications. Jour. Geophys. Res. 91:13,803–72.

Armijo, R., and Thiele, R. 1990. Active faulting in northern Chile: Ramp stacking and lateral decoupling along a subduction plate boundary? Earth and Planet Science Letters 98:40-61.

Armstrong, C.F. 1979. Coyote Lake earthquake 6 August 1979. California Geology 32:248–50.

Arpat, E. 1977. 1975 Lice depremi. Yeryuvari Insan 2:15–28.

Arpat, E., Saroglu, F., and Iz, H. B. 1977. The 1976 Çaldiran earthquake. Yeryuvari Insan 2:29–41.

Ashi, J., and Taira, A. 1992. Structure of the Nankai accretionary prism as revealed from IZANAGI sidescan imagery and multichannel seismic reflection profiling: The Island Arc 1:104–15.

Atwater, B. F. 1987. Evidence for great Holocene earthquakes along the outer coast of Washington State. Science 236:942–44.

Atwater, B. F. 1992. Geologic evidence for earthquakes during the past 2000 years along the Copalis River, southern coastal Washington. Jour. Geophys. Res. 97:1901–19.

Atwater, B. F., Jiménez-Núñez, H., and Vita-Finzi, C. 1992. Net late Holocene emergence despite earthquake-induced submergence, south-central Chile. Quaternary International 15/16:77–85.

Atwater, B. F., and Moore, A. L. 1992. A tsunami about 1000 years ago in Puget Sound, Washington. Science 258:1614–17.

Atwater, B. F., Nelson, A. R., Clague, J. J., Carver, G. A., Yamaguchi, D. K., Bobrowsky, P. T., Bourgeois, J., Darienzo, M. E., Grant, W. C., Hemphill-Haley, E., Kelsey, H. M., Jacoby, G. C., Nishenko, S. P., Palmer, S. P., Peterson, C. D., and Reinhart, M. A. 1995. Summary of coastal geologic evidence for past great earthquakes at the Cascadia Subduction Zone. Earthquake Spectra 11:1–18.

Atwater, B. F., Stuiver, M., and Yamaguchi, D. K. 1991. Radiocarbon test of earthquake magnitude at the Cascadia Subduction Zone. Nature 353:156–58.

Atwater, T. 1970. Implications of plate tectonics for the Cenozoic tectonic evolution of western North America. Geol. Soc. America Bulletin 81:3513–36.

Atwater, T. 1989. Plate tectonic history of the northeast Pacific and western North America, *in* Winterer, E. Hussong, D., and Decker, R. The Eastern Pacific Ocean and Hawaii. Geol. Soc. America Decade of North American Geology N:21–72.

Avouac, J. P. 1991. Application des méthodes de morphologie quantitative à la néotectonique. Modele cinématique des déformations actives en Asie Centrale. Thèse, Univ. Paris VII, 200 p.

Avouac, J. P. Tapponnier, P., Bai M., You H., and Wang G., 1993. Active thrusting and folding along the northern Tien Shan and late Cenozoic rotation of the Tarim relative to Dzungaria and Kazakhstan. Jour. Geophys. Res. 98:6755-6804.

Avouac, J. P., and Tapponnier, P., 1993, Kinematic model of active deformation in central Asia. Geophys. Res. Lett. 20(10):895-98.

Aydin, A., Johnson, A. M., and Fleming, R. W., 1992, Right-lateral-reverse surface rupture along the San Andreas and Sargent faults associated with the October 17, 1989, Loma Prieta, California, earthquake. Geology 20:1063-67.

Baby, P., Hérail, G., Salinas, R., and Sempere, T. 1992. Geometry and kinematic evolution of passive roof duplexes deduced from cross section balancing: Example from the foreland thrust system of the southern Bolivian Subandean zone. Tectonics 11:523-36.

Baecher, G. B., and Keeney, R. L. 1982. Statistical examination of reservoir induced seismicity. Seismological Society of America Bulletin 72, p. 553-69.

Bai M. 1986. Active faults and strong motion earthquakes in Xinjiang. Acta Seismologica Sinica 8(1):79-91. (C, Eng. abs.).

Baker, D. M., Lillie, R. J., Yeats, R. S., Johnson, G. D., Yousuf, M., and Zamin, A. S. H. 1988. Development of the Himalayan frontal thrust zone: Salt Range, Pakistan. Geology 16:3-7.

Bakun, W. H., and Lindh, A. G. 1985. The Parkfield, California, earthquake prediction experiment. Science 229:619-24.

Baljinnyam, I., Bayasgalan, A., Borisov, B. A., Cisternas, A., Dem'yanovich, M. G., Ganbaatar, L., Kochetkov, V. M., Kurushin, R. A., Molnar, P., Philip, H., and Vashchilov, Y. Y. 1993. Ruptures of major earthquakes and active deformation in Mongolia and its surroundings. Geol. Soc. America Memoir 181, 62 p.

Barazangi, M.,, and Isacks, B. L. 1976. Spatial distribution of earthquakes and subduction of the Nazca Plate beneath South America. Geology 4:686-92.

Barberi, F., and Varet, J. 1977. Volcanism of Afar: small-scale plate tectonics implications. Geol. Soc. America Bulletin 88:1251-66.

Bard, E., Hamelin, B., Fairbanks, R. G., and Zindler, A. 1990. Calibration of the ^{14}C timescale over the past 30,000 years using mass spectrometric U-Th ages from Barbados corals. Nature 345:405-10.

Barka, A. 1992. The North Anatolian fault zone: Annales Tectonicae, Special Issue, Supp. 6:164-95.

Barka, A., and Eyidogan, H. 1993. The Erzincan earthquake of 13 March 1992 in eastern Turkey: Terra Nova 5:190-94.

Barka, A., and Kadinsky-Cade, K. 1988. Strike-slip fault geometry in Turkey and its influence on earthquake activity. Tectonics 7:663-84.

Barrientos, S. E., Stein, R. S., and Ward, S. N. 1987. Comparison of the 1959 Hebgen Lake, Montana and 1983 Borah Peak, Idaho, earthquakes from geodetic observations. Seismological Society of America Bulletin 77:784-808.

Bartsch-Winkler, S., and Schmoll, H. R. 1992. Utility of radiocarbon-dated stratigraphy in determining late Holocene earthquake recurrence intervals, upper Cook Inlet region, Alaska. Geol. Soc. America Bulletin 104:684-94.

Bassinot, F. C., Labeyrie, L. D., Vincent, E., Quidelleur, X., Shackleton, N. J., and Lancelot, Y. 1994. The astronomical theory of climate and the age of the Brunhes-Matuyama magnetic reversal. Earth Planet. Sci. Lett. 126:91-108.

Beanland, S., and Barrow-Hurlbert, S. A. 1988. The Nevis-Cardrona fault system, Central Otago, New Zealand: Late Quaternary tectonics and structural development. New Zealand Jour. Geol. Geophysics 31:337-52.

Beanland, S., and Berryman, K. R. 1989. Style and episodicity of late Quaternary activity on the Pisa-Grandview fault zone, Central Otago, New Zealand. New Zealand Jour. Geol. Geophysics 32:451-61.

Beanland, S., Berryman, K. R., Hull, A. G., and Wood, P. R., 1986, Late Quaternary deformation at the Dunstan fault, Central Otago, New Zealand. *In* Reilly, W. I., and Harford, B. E., eds., Recent Crustal Movements of the Pacific Region. Royal Soc. New Zealand Bull. 24:293-306.

Beanland, S., Blick, G. H., and Darby, D. J. 1990. Normal faulting in a back arc basin: Geological and geodetic characteristics of the 1987 Edgecumbe earthquake, New Zealand. Jour. Geophys. Res. 95:4693-4707.

Beanland, S., and Clark, M. 1994. The Owens Valley fault zone, eastern California, and surface rupture associated with the 1872 earthquake. U. S. Geol. Survey Bulletin 1982.

Bell, J. W., and Katzer, T. 1990. Timing of late Quaternary faulting in the 1954 Dixie Valley earthquake area, central Nevada. Geology 18:622-25.

Bellier, O., Dumont, J. F., Sébrier, M., and Mercier, J. L. 1991. Geological constraints on the kinematics and fault-plane solution of the Quiches fault zone reactivated during the 10 November 1946 Ancash earthquake. Seismological Society of America Bulletin 81:468-90.

Ben-Avraham, Z., and Grasso, M. 1991. Crustal structure variations and transcurrent faulting at the eastern and western margins of the eastern Mediterranean. Tectonophysics 196:269-77.

Ben Menahem, A. 1991. 4000 years of seismicity along the Dead Sea rift. Jour. Geophys. Res.96:195-216.

Bennett, J. H., and Sherburne, R. W., eds. 1983. The 1983 Coalinga earthquake, California. Calif. Div. Mines and Geology Special Publication, v. 66, 335 p.

Bennett, J. H., Sherburne, R. W., Cramer, C. H., Chesterman, C. W., and Chapman, R. H. 1979. Stephens Pass earthquakes, Mount Shasta—August 1978, Siskiyou County, California. California Geology 32(2):27-34.

Bennett, M. J. 1989. Liquefaction analysis of the 1971 ground failure at the San Fernando Valley Juvenile Hall, California. Assoc. Engineering Geologists Bull. 26:209-26.

Berberian, M. 1976. Documented earthquake faults in Iran. Geol. Surv. Iran Rep. no 39:175-83.

Berberian, M. 1979. Earthquake faulting and bedding thrusts associated with the Tabas-e-Golshan (Iran) earthquake of 16 September 1978: Seismological Society of America Bulletin 69:1861-87.

Berberian, M. 1981. Active faulting and tectonics of Iran. *In* Gupta, H. W., and Delany, F. M., eds. Zagros, Hindu Kush, Himalaya geodynamic evolution. American Geophys. Union Geodynamics Series 3:33-69.

Berberian, M. 1982. Aftershock tectonics of the 1978 Tabas-e-Golshan (Iran) earthquake sequence: A documented active 'thin- and thick-skinned tectonic' case. Geophys. Jour. Roy. Astr. Soc. 68:499-530.

Berberian, M. 1995. Master "blind" thrust faults hidden under the Zagros folds: Active basement tectonics and surface morphotectonics. Tectonophysics 241:193-224.

Berberian, M., and Arshadi, S. 1977. The Shibly rift system (Sahand region, NW Iran). Geol. Min. Surv. Iran 40:229-35.

Berberian, M., Asudeh, I., and Arshadi, S. 1979. Surface rupture and mechanism of the Bob-Tangol (southeastern Iran) earthquake of 19 December 1977. Earth and Planet Sci. Lett. 42:456-62.

Berberian, M., Jackson, J. A., Ghorashi, M., and Kadjar, M. H. 1984. Field and teleseismic observations of the 1981 Golbaf-Sirch earthquakes in SE Iran. Geophys. Jour. Roy. Astron. Soc. 77:809-38.

Berberian, M., and Qorashi, M. 1994. Coseismic fault-related folding during the South Golbaf earthquake of November 20, 1989, in southeast Iran. Geology 22:531-34.

Berberian, M., Qorashi, M., Jackson, J. A., Priestley, K., and Wallace, T. 1992. The Rudbar-Tarom earthquake of 20 June 1990 in NW Persia: Preliminary field and seismological observations and its tectonic significance. Seismological Society of America Bulletin 82:1726-55.

Berberian, M., and Tchalenko, J. S. 1976. Field study and documentation of the 1930 Salmas (Shahpur-Azarbaidjan) earthquake. Geol. Surv. Iran 39:271-342.

Berger, G. W. 1988. Dating Quaternary events by luminescence. Geol. Soc. America Special Paper 227:13-50.

Berggren, W. A., Burkle, L. H., Cita, M. B., Cooke, H. B. S., Funnell, B. M., Gartner, S., Hays, J. D., Kennett, J. P., Opdyke, N. D., Pastouret, L., Shackleton, N. J., and Takayanagi, Y. 1980. Towards a Quaternary time scale. Quaternary Research 13:277-302.

Berggren, W. A., Kent, D. V., Flynn, J. J., and Van Couvering, J. A. 1985. Cenozoic geochronology. Geol. Soc. America Bull. 96:1407-18.

Berggren, W. A., Hilgen, F. J., Langereis, C. G., Kent, D. V., Obradovich, J. D., Raffi, I., Raymo, M. E., and Shackleton, N. J. 1995. Late Neogene chronology: new perspectives in high-resolution stratigraphy. Geol. Soc. America Bull. 107:1272-87.

Bergman, E., and Solomon, S. 1988. Transform-fault earthquakes in the North Atlantic—Source mechanism and depth of faulting. Jour. of Geophys. Res. 93:9027-57.

Beroza, G. C., and Jordan, T. H. 1990. Searching for slow and silent earthquakes using free oscillations. Jour. Geophys. Res. 95:2485-510

Berryman, K. R., 1980, Late Quaternary movement on White Creek fault, South Island, New Zealand: New Zealand Jour. Geol. Geophysics 23:93-101.

Berryman, K. R., and Beanland, S. 1991. Variation in fault behaviour in different tectonic provinces of New Zealand. Journal of Structural Geology 13, p.177-89.

Berryman, K. R., Ota, Y., and Hull, A. G. 1989. Holocene paleoseismicity in the fold and thrust belt of the Hikurangi subduction zone, eastern North Island, New Zealand. Tectonophysics 163:185-95.

Berryman, K. R., Beanland, S., Cooper, A., Cutten, H., Norris, R., and Wood, P. 1992. The Alpine fault, New Zealand: Variation in Quaternary structural style and geomorphic expression. Annales Tectonicae Special Issue, Supp. 6:126-63.

Bevis, M., and Isacks, B. L. 1981. Leveling arrays as multicomponent tiltmeters: Slow deformation in the New Hebrides island arc. Jour. Geophys. Res. 86:7808-24.

Biasi, G., and Weldon, R., II. 1994. Quantitative refinement of calibrated ^{14}C distributions. Quaternary Research 41:1-18.

Bierman, P. R. 1994. Using in situ produced cosmogenic isotopes to estimate rates of landscape evolution: A review from the geomorphic perspective. Jour. Geophys. Res. 99:13,885-96.

Bilham, R. 1987. Space geodesy and earthquake prediction. Aerospace Century XXI, v. 64, Advances in the Astronautical Sciences 1637-59.

Bilham, R. 1991. Earthquakes and sea level. Space and terrestrial metrology on a changing planet. Reviews of Geophysics 29:1-29.

Bilham, R., Yeats, R., and Zerbini, S. 1989. Space geodesy and the global forecast of earthquakes. EOS (Trans. Amer. Geophys. Union) 70(5):65,73.

Birkeland, P.W. 1984. Soils and Geomorphology. New York: Oxford University Press, 372 p.

Bjarnason, I. T., Cowie, P., Anders, M. H., Seeber, L., and Scholz, C.H. 1993. The 1912 Iceland earthquake rupture: growth and development of a nascent transform system. Seismological Society of America Bulletin, v. 83, p. 416-35.

Blanpied, M. L., Lockner, D. A., and Byerlee, J. D. 1992. An earthquake mechanism based on rapid sealing of faults. Nature 358:574-76.

Bloom, A. L. 1991. Geomorphology: A Systematic Analysis of Late Cenozoic Landforms. 2nd edition. Englewood Cliffs, N.J.: Prentice Hall, 532 p.

Bloom A. L., and Yonekura, N. 1985. Coastal terraces generated by sea-level change and tectonic uplift. *In* Woldenberg, M. J., ed. Models in Geomorphology. London: Allen & Unwin, 139-54.

Bloom, A. L., and Yonekura, N. 1990. Graphic analysis of dislocated Quaternary shorelines, in Sea Level Change: Washington, D.C., National Academy Press, 104-13.

Bock, Y. 1994. A strategy for obtaining high spatial and temporal resolution of crustal deformation using GPS: U.S. Geol. Survey Open-File Report 94-176:227-35.

Bogdanovich, K. I., Kark, I. M., Korolkov, B. Ya., and Mushketov, D. I. 1914. Earthquake in the northern districts of the Tien Shan, 22 December 1910 (4 January 1911): Commission of the Geology Committee, Leningrad, USSR (R).

Bohannon, R., and Howell, D. 1982. Kinematic evolution of the junction of the San Andreas, Garlock and Big Pine faults, California. Geology 10:358-63.

Bolt, B. A. 1982. Inside the Earth: Evidence from Earthquakes. San Francisco: W. H. Freeman and Co., 191 p.

Bolt, B. A. 1993. Earthquakes. New York: W.H. Freeman and Co., 3rd ed., 331 p.

Bolt, B. A., McEvilly, T. V., and Uhrhammer, R. A. 1981. The Livermore Valley, California, sequence of January 1980. Seismological Society of America Bulletin 71:451-63.

Bomford, G. 1971. Geodesy. Oxford: Clarendon Press, 731 p.

Bonchev, S., and Bakalov, P. 1928. Les tremblements de terre dans la Bulgarie du Sud les 14 et 18 avril 1928. Rev. Soc. Géol. Bulgare.

Bonilla, M. G. 1977. Summary of Quaternary faulting and elevation changes in Taiwan. Geol. Soc. China Mem. 2:43-55.

Bonilla, M. G., and Buchanan, J. M. 1970. Interim report on worldwide historic surface faulting. U. S. Geol. Survey Open-File Report 70-34, 32 p.

Bonilla, M. G., Lienkaemper, J. J., and Tinsley, J. C. 1980. Surface faulting near Livermore, California associated with the January 1980 earthquakes. U. S. Geol. Survey Open-File Report 80-523, 31 p.

Bonilla, M. G., Mark, R. G., and Lienkaemper, J. J. 1984. Statistical relations among earthquake magnitude, surface rupture length, and surface fault displacement. Seismological Society of America Bulletin 74:2379-2411.

Boore, D. M. 1977. The motion of the ground in earthquakes. Scientific American 237:69-78.

Bounif, A., Haessler, H., and Meghraoui, M. 1987. The Constantine (northeast Algeria) earthquake of October 27, 1985: surface ruptures and aftershock study. Earth and Planet. Sci. Lett. 85:451-60.

Broecker, W. S. 1992. Defining the boundaries of the late glacial isotope episodes. Quaternary Research 38:135-38.

Brook, E. J., Brown, E. T., Kurz, M. D., Ackert, R. P., Jr., Raisbeck, B. M., and Yiou, F. 1995. Constraints on age, erosion and uplift of Neogene glacial deposits in the Transantarctic Mountains determined from in situ cosmogenic ^{10}Be and ^{26}Al. Geology. 23:1063–66.

Brown, N., Fuller, M., and Sibson, R. 1991. Paleomagnetism of the Ocotillo Badlands, Southern California, and implications for slip transfer through an antidilational fault jog. Earth and Planet. Sci. Lett. 102:277–88.

Brown, R. D. and Vedder, J. 1967. Surface tectonic fractures along the San Andreas fault. *In* The Parkfield-Cholame, California, earthquakes of June-August 1966. U.S. Geological Survey Prof. Paper 579:1–22.

Brown, R. D., Vedder, J. G., Wallace, R. E., Roth, E. F., Yerkes, R. F., Castle, R. O., Waanonen, A. O., Page, R. W., and Eaton, J. P., eds. 1967. The Parkfield-Cholame, California, earthquakes of June-August 1966—Surface geologic effects, water-resources aspects, and preliminary seismic data. U. S. Geol. Survey Prof. Paper 579, 66 p.

Brown, R. D., and Wallace, R. 1968. Current and historic fault movement along the San Andreas fault between Paicines and Camp Dix, California. *In* W. R. Dickinson and A. Grantz, ed. Conference on Geologic Problems of San Andreas Fault System, Proceedings 11: Stanford University Publications in the Geological Sciences, pp. 22–41.

Brown, R. D., Ward, P. L., and Plafker, G. 1973. Geologic and seismologic aspects of the Managua, Nicaragua, earthquakes of December 23, 1972. U.S. Geol. Survey Prof. Paper 838, 34 p.

Bruhn, R. L., Gibler, P. R., and Perry, W. T. 1987. Rupture characteristics of normal faults: an example from the Wasatch fault zone, Utah. *In* Coward, M. P., Dewey, J. F., and Hancock, P. L., eds. Continental Extensional Tectonics. Geological Society Special Publication 28:337–53.

Bruhn, R. L., Yonkee, W. A., and Parry, W. T. 1990. Structural and fluid-chemical properties of seismogenic normal faults. Tectonophysics 175:139–57.

Bruns, T. R., and Schwab, W. C. 1983. Structure maps and seismic stratigraphy of the Yakataga segment of the continental margin, northern Gulf of Alaska. U. S. Geol. Survey Miscellaneous Field Studies Map MF-1424, 2 sheets, scale 1:250,000, 25 p.

Bruns, T. R., von Huene, R., Culotta, R. C., Lewis, S. D., and Ladd, J. W. 1987. Geology and petroleum potential of the Shumagin margin, Alaska. *In* Scholl, D. W., Grantz, A., and Vedder, J. G., eds.. Geology and resource potential of the continental margin of western North America and adjacent ocean basins—Beaufort Sea to Baja California. Circum-Pacific Council for Energy and Mineral Resources Earth Science Series 6:157–89.

Bryant, A. S., and Jones, L. M. 1992. Anomalously deep crustal earthquakes in the Ventura basin, southern California. Jour. Geophys. Res. 97:437–47.

Bryant, E. A., Young, R. W., and Price, D. M. 1992. Evidence of tsunami sedimentation on the southeastern coast of Australia. Jour. Geology 100:753–65.

Buchanan-Banks, J. M., Castle, R. O., and Ziony, J. I. 1975. Elevation changes in the central Transverse Ranges near Ventura, California. Tectonophysics 24:113–25.

Bucknam, R. C., and Anderson, R. E. 1979. Estimation of fault-scarp ages from a scarp-height-slope-angle relationship. Geology 7:11–14.

Bucknam, R. C., Hemphill-Haley, E., and Leopold, E. B. 1992. Abrupt uplift within the past 1700 years at southern Puget Sound, Washington. Science 258:1611–14.

Bucknam, R. C., Plafker, G., and Sharp, R. V. 1978. Fault movement (afterslip) following the Guatemala earthquake of February 4, 1976. Geology 6:170–73.

Buhay, W., Schwarcz, H. P., and Grün, R. 1988, ESR dating of fault gouge: The effect of grain size. Quaternary Science Reviews 7:515–22.

Bull, W. B. 1977, The alluvial-fan environment. Progress in Physical Geography 1:222–70.

Bull, W. B. 1991, Geomorphic Responses to Climatic Change. New York: Oxford Press, 320 p.

Bull, W. B., and McFadden, L. D. 1977. Tectonic geomorphology north and south of the Garlock fault, California, in Doehring, D.O., ed., Geomorphology in Arid Regions. Proc. 8th Ann. Geomorphology Symposium, State Univ. New York, Binghamton: 115–38

Bull, W. B., and Pearthree, P. A. 1988. Frequency and size of Quaternary surface ruptures of the Pitaycachi fault, northeastern Sonora, Mexico: Seismological Society of America Bulletin 78:956–78.

Bull, W. B., King, J., Kong, F., Moutoux, T., and Phillips, W. M. 1994. Lichen dating of coseismic landslide hazards in alpine mountains. Geomorphology 10:253–64.

Burchfiel, B., Hodges, K., and Royden, L. 1987. Geology of Panamint Valley-Saline Valley pull-apart system, California: Palinspastic evidence for low-angle geometry of a Neogene range-bounding fault: Jour. N. Geophys. Res. 92:10,422-26.

Burr, N., and Solomon, S. 1978. The relationship of source parameters of oceanic transform earthquakes to plate velocity and transform length. Jour. Geophys. Res. 83:1193-1205.

Burke, W. H., Denison, R. E., Hetherington, E. A., Koepnick, R. B., Nelson, H. F., and Otto, J. B. 1982. Variation of seawater $^{87}Sr/^{86}Sr$ throughout Phanerozoic time. Geology 10:516-19.

Buwalda, J., and Richter, C. 1941. Imperial Valley earthquake of May 18, 1940 (abstract). Geol. Soc. America Bulletin 52:1944-45.

Buwalda, J., and St. Amand, P. 1955. Geological Effects of the Arvin-Tehachapi Earthquake. *In* G. Oakeshott. Earthquakes in Kern County California During 1952. San Francisco, Calif. Department of Natural Resources, Division of Mines Bulletin 171:41-56.

Byerlee, J. D. 1977. Friction of rocks. *In* Evèrnden, J. F., ed. Experimental Studies of Rock Friction with Applications to Earthquake Prediction. Menlo Park, CA, U.S. Geological Survey, p. 55-72.

Byerlee, J. D. 1978., Friction of rocks. Pure and Applied Geophysics 116, p. 615-26.

Byerly, P., and Wilson, J. T. 1935. The central California earthquakes of May 16, 1933 and June 7, 1934. Seismological Society of America Bulletin 25:223-46.

Byrne, D. E., Davis, D. M., and Sykes, L. R. 1988. Loci and maximum size of thrust earthquakes and the mechanics of the shallow region of subduction zones. Tectonics 7:833-57.

Byrne, D. E., Sykes, L. R., and Davis, D. M. 1992. Great thrust earthquakes and aseismic slip along the plate boundary of the Makran Subduction Zone. Jour. Geophys. Res. 97:449-78.

Cadet, J. P., Kobayashi, K., Lallemand, S., Jolivet, L., Aubouin, J., Boulègue, J., Dubois, J., Hotta, H., Ishii, T., Konishi, K., Niitsuma, N., and Shimamura, H. 1987. Deep scientific dives in the Japan and Kuril trenches. Earth and Planet. Sci. Lett. 83:313-28.

Cahill, T., and Isacks, B. L. 1992. Seismicity and shape of the subducted Nazca Plate. Jour. Geophys. Res. 97:17,503-29.

Callaghan, E., and Gianella, V. P. 1935. The earthquake of January 30, 1934, at Excelsior Mountains, Nevada. Seismological. Society of America Bulletin 25:161-68.

Campbell, N. P., and Bentley, R. D.1981. Late Quaternary deformation of the Toppenish Ridge uplift in south-central Washington. Geology 9:519-24.

Cao Z., Shen X., Song F., Wang Y., Yu W., and Li Z. 1994. Research on surface rupture zone of the 1500 Yiliang earthquake. Yunnan: Research on Active Fault 3:104-14 (C, Eng. abs.).

Carey, E. 1979. Recherche des directions principales de contraintes associées au jeu d'une population de failles. Review of Géol. Dyn. Géog. Phys. 21:57-66.

Carter, N.L., and Tsenn, M.C. 1987. Flow properties of continental lithosphere. Tectonophysics 136:27-63.

Carter, N., Friedman, M., Logan, J., and Stearns, D. 1981. Mechanical behavior of crustal rocks. American Geophysical Union Monograph 24.

Carter, W.E., and Robertson, D.S. 1986. Studying the Earth by very-long-baseline interferometry. Scientific American 255(5):46-54.

Carver, G. A., and Burke, R. M. 1992. Late Cenozoic deformation on the Cascadia Subduction Zone in the region of the Mendocino Triple Junction. *In* Burke, R. M., and Carver, G. A., coordinators. Pacific Cell, Friends of the Pleistocene Guidebook for the Field Trip to Northern Coastal California, pp. 31-63.

Castellanos, A. 1945. El terremoto de San Juan. *In* Cuatro Lecciones sobre Terremotos, pp. 79-242. Asociación Cultural de Conferencias de Rosario, Argentina.

Castle, R. O., Elliot, M. R., Church, J. P., and Wood, S. H. 1984. The evolution of the southern California uplift, 1955 through 1976. U. S. Geol. Survey Prof. Paper 1342, 136 p.

Castle, R. O., Clark, M. M., Grantz, A., and Savage, J. C. 1980. Tectonic state: Its significance and characterization in the assessment of seismic effects associated with reservoir impounding. Engineering Geology15:53-99.

Cerling, T. E., and Craig, H. 1993. Cosmogenic ^{3}He production rates from 39° N to 46°N latitude, western USA and France. Geochimica et Cosmochimica Acta 59:249-55.

Cerling, T. E., and Craig, H. 1994. Geomorphology and in-situ cosmogenic isotopes. Annual Review Earth Planet. Sci. 22:273-317.

Chadwick, W. W., Jr., and Swanson, D. A. 1989. Thrust faults and related structures in the crater floor of Mount St. Helens volcano, Washington. Geol. Soc. America Bull. 101: 1507–19.

Chamot-Rooke, N., and Le Pichon, X. 1989. Zenisu Ridge: Mechanical model of formation. Tectonophysics 160:175–93.

Chander, R. 1988. Interpretation of observed ground level changes due to the 1905 Kangra earthquake, Northwest Himalaya. Tectonophysics 194:289–98.

Chang L. S., Chow M., and Chen P. Y. 1947. The Tainan earthquake of December 5, 1946. Geol. Survey Taiwan Bull 1:17–20.

Chappell, J. 1974. Geology of coral terraces, Huon Peninsula, New Guinea: A study of Quaternary tectonic movements and sea-level changes. Geol. Soc. America Bull. 85:553–70.

Chase, C. 1978. Plate kinematics: The Americas, East Africa, and the rest of the world. Earth and Planet. Sci. Lett. 37:355–68.

Chen, J. H., Edwards, R. L., and Wasserburg, G. J. 1986. ^{238}U, ^{234}U and ^{232}Th in seawater. Earth and Planet. Sci. Lett. 80:241–51.

Chen, J. H., Curran, H. A., White, B., and Wasserburg, G. J. 1991. Precise chronology of the last interglacial period: $^{234}U^{230}Th$ data from fossil coral reefs in the Bahamas. Geol. Soc. America Bull. 103:82–97.

Chen R. and Li P., 1988. Slip rates and earthquake recurrence intervals of the western branch of Xiaojiang fault zone. Seismology and Geology 10, 2:1–13 (C).

Chen S. F., Wilson, C. J. L., Deng Q. D., Xiao L. Z., and Zhi L. L. 1994. Active faulting and block movement associated with large earthquakes in the Min Shan and Longmen Mountains, northeastern Tibetan Plateau. Jour. Geophys. Res. 99:24,025–38.

Chen, W.-P., and Molnar, P. 1983. Focal depths of intracontinental and intraplate earthquakes and their implications for the thermal and mechanical properties of the lithosphere. Jour. Geophys. Res. 88:4183–4214.

Cheng S., Fang Z., Yang Z., and Yang G. 1991. Characteristics of the active normal fault zone along the northern margin of the Yanqing basin at Yaojiaying in Beijing. Research on Active Fault 1:131–39 (C, Eng. abs.)

Chhibber, H. 1934. Geology of Burma. McMillan, London. 538 p.

Chinn, T. J. H. 1981. Use of rock weathering-rind thickness for Holocene absolute age-dating in New Zealand. Arctic and Alpine Research 13:33–45

Choy, G. L. 1990. Source parameters of the earthquake, as inferred from broadband body waves. *In* Rymer, M. J., and Ellsworth, W. L., eds., The Coalinga, California, earthquake of May 2, 1983: U. S. Geol. Survey Prof. Paper 1487:193–205.

Choy, G. L., and Bowman, J. R. 1990. Rupture process of a multiple main shock sequence: analysis of teleseismic, local, and field observations of the Tennant Creek, Australia, earthquakes of January 22, 1988. Jour. Geophys. Res. 95:6867–82.

Christensen, D. H., and Ruff, L. J. 1988. Seismic coupling and outer rise earthquakes. Jour. Geophys. Res. 93:13,421–44.

Clark, J. A., and Lingle, C. S. 1979. Predicted relative sea-level changes (18,000 years B.P. to present) caused by late-glacial retreat of the Antarctic ice sheet. Quaternary Research 11:279–98.

Clark, M. M. 1972a. Collapse fissures along the Coyote Creek fault. U. S. Geol. Survey Prof. Paper 787:190–207.

Clark, M. M. 1972b. Surface rupture along the Coyote Creek fault. U.S. Geol. Survey Prof. Paper 787:55–86.

Clark, M. M., Sharp, R. V., Castle, R. O., and Harsh, P. W. 1976. Surface faulting near Lake Oroville California, in August, 1975. Seismological Society of America Bulletin 66:1101–10.

Clark, M. M., Yount, J. C., Vaughan, P. R., and Zepeda, R. L. 1982. Map showing surface ruptures associated with the Mammoth Lakes, California, earthquakes of May 1980. U. S. Geol. Survey Misc. Field Studies Map MF-1396.

Clarke, S. H., Jr. 1992. Geology of the Eel River basin and adjacent region: Implications for late Cenozoic tectonics of the southern Cascadia Subduction Zone and Mendocino Triple Junction. American Assoc. Petroleum Geologists Bull. 76:199–224.

Clayton, R. W., and Hauksson, E. 1992. Partial support of joint USGS-CALTECH Southern California Seismographic Network. U. S. Geol. Survey Open File Report, OFR 92-258:213–18.

Cleary, J. R., Doyle, H. A., and Moye, D. G. 1964. Seismic activity in the Snowy Mountains region and its relationship to geologic structure. Geological Survey of Australia Journal 11:89–106.

Close, U., and McCormick, E. 1922. Where the mountains walked. National Geographic Magazine 41:445–64.

Cluff, L. S. 1977. Notes of a visit to Koyna Dam, India, January 14, 15, 16, 1977. Unpublished report to U.S. Bureau of Reclamation, 13 p.

Cohee, B., and Beroza, G. 1994. Slip distribution of the 1992 Landers earthquake and its implications for earthquake source mechanics. Seismological Society of America Bulletin 84:692-712.

Colman, S. M., and Pierce, K. L. 1981., Weathering rinds on andesitic and basaltic stones as a Quaternary age indicator, western United States. U.S. Geol. Survey Prof. Paper 1210, 56 p.

Colman, S. M., and Watson, K. 1983. Diffusion-equation model for scarp degradation. Science 221:263-65.

Comminakas, P., Drakopoulos, J., Moumoulidis, G., and Papazachos, B. 1968. Foreshock and aftershock sequences of the Cremasta earthquake and their relation to the waterloading of the Cremasta artificial lake. Annali di Geofisica 21:39-71.

Cook, E. R., and Kairiutstis, L. A., eds. 1990. Methods of Dendrochronology. London: Kluwer.

Cooper, P., and Taylor, B. 1987. Seismotectonics of New Guinea: A model for arc reversal following arc-continent collision. Tectonics 6:43-67.

Cooper, A. K., Marlow, M. S., and Scholl, D. W. 1987. Geologic framework of the Bering Sea crust. *In* Scholl, D. W., Grantz, A., and Vedder, J. G., eds., Geology and Resource Potential of the Continental Margin of Western North America and Adjacent Ocean Basins—Beaufort Sea to Baja California. Circum-Pacific Council for Energy and Mineral Resources Earth Science Series 6:73-102.

Coppersmith, K. J. 1982. Probabilistic evaluations of earthquake hazards. *In* Hart, E. W., Hirschfeld, S. E., and Schulz, S. S., eds. Proceedings of the Conference on Earthquake Hazards in the Eastern San Francisco Bay Area. Calif. Div. Mines Geol. Spec. Rept. 62:125-34.

Corbett, E. J., and Johnson, C. E. 1982. The Santa Barbara, California, earthquake of 13 August 1978. Seismological Society of America Bulletin 72:2201-26.

Cormier, V. F. 1975. Tectonics near the junction of the Aleutian and Kuril-Kamchatka arcs and a mechanism for middle Tertiary magmatism in the Kamchatka basin. Geol. Soc. America Bull. 86:443-53.

Cornell, C. A. 1968. Engineering seismic risk analysis. Seismological Society of America Bulletin 58:1583-1606.

Courtillot, V., Armijo, R., and Tapponnier, P. 1987. Kinematics of the Sinai triple junction and a two-phase model of Arabia-Africa rifting. *In* Coward, M. P., Dewey, J. F., and Hancock, P. L., eds. Continental Extensional Tectonics. Geological Society Special Publication 28:559-73.

Cowan, H. 1991. The North Canterbury earthquake of September 1, 1888. Journal of the Royal Society of New Zealand 21:1-12.

Cox, A., ed. 1973. Plate Tectonics and Geomagnetic Reversals. San Francisco: W.H. Freeman, 702 p.

Crampin, S., and Lovell, J. H. 1991. A decade of shear-wave splitting in the Earth's crust: what does it mean? what use can we make of it? and what should we do next? Geophys. J. Int. 107:387-407.

Crone, A. J. and Haller, K. M. 1991. Segmentation and the coseismic behavior of Basin and Range normal faults: Examples from east-central Idaho and southwestern Montana, U. S. A. Journal of Structural Geology 13:151-64.

Crone, A. J., and Harding, S. T. 1984. Relationship of late Quaternary fault scarps to subjacent faults, eastern Great Basin, Utah. Geology 12:292-95.

Crone, A. J., and Machette, M. N. 1994. Paleoseismology of Quaternary faults in the "stable" interior of Australia and North America—insight into the long-term behavior of seismogenic faults. U. S. Geol. Survey Open-File Report 94-568:40-42.

Crone, A. J., Machette, M. N., and Bowman, J. R. 1992. Geologic investigations of the 1988 Tennant Creek, Australia, earthquakes—implications for paleoseismicity in stable continental regions. U. S. Geol. Survey Bull. 2032-A, 51 p.

Crone, A. J., Machette, M. N., Bonilla, M. G., Lienkaemper, J. J., Pierce, K. L., Scott, W. E., and Bucknam, R. C. 1987. Surface faulting accompanying the Borah Peak earthquake and segmentation of the Lost River fault, central Idaho. Seismological Society of America Bulletin 77:739-70.

Crook, C. 1984. Geodetic Measurement of the Horizontal Crustal Deformation Associated with the Oct. 15, 1979 Imperial Valley (California) Earthquake. University of London, Ph.D. thesis.

Crook, R., Allen, C. R., Kamb, B., Payne, C. M., and Proctor, R. J. 1987. Quaternary geology and seismic hazard of the Sierra Madre and associated faults, western San Gabriel Mountains. U. S. Geol. Survey Prof. Paper 1339:27-63.

Crosson, R. S., and Owens, T. J. 1987. Slab geometry of the Cascadia Subduction Zone beneath Washington from earthquake hypocenters and teleseismic converted waves. Geophys. Research Lett. 14:824-27.

Crowell, J. C. 1987. Late Cenozoic basins of onshore southern California: complexity is the hallmark of their tectonic history. *In* Ingersoll, R. V., and Ernst, W. G., eds. Cenozoic Basin Development of Coastal California. Rubey Volume VI, Englewood Cliffs, N.J.: Prentice-Hall, pp. 207-41.

Crozier, M. J. 1992. Determination of paleoseismicity from landslides, in Bell, D. H., ed. Landslides (Glissements de terrain), International Symposium, 6th, Christchurch, New Zealand, 1992, Proceedings. Rotterdam, A. A. Balkema, v. 2, pp. 1173-80.

D'Addezio, G., Pantosti, D., and Valensise, G. 1991. Paleoearthquakes along the Irpinia fault at Pantano di San Gregorio Magna (northern Italy. Il Quaternario 4(1a):121-36.

Dahlstrom, C. D. A. 1990. Geometric constraints derived from the law of conservation of volume and applied to evolutionary models for detachment folding. American Association of Petroleum Geologists Bulletin 74:336-44.

Dai H.G. 1983. On the Dari earthquake of 1947 in Qinghai Province. Northwestern Seismological Journal 5(3):71-77. (C).

Dalrymple, G. B., and Lanphere, M. A. 1969. Potassium-Argon Dating. San Francisco: W.H. Freeman and Co., 258 p.

Damon, P. E. 1991. K-Ar dating of Quaternary volcanic rocks. *In* Morrison, R. B., ed. Quaternary Nonglacial Geology: Conterminous U. S. Geol. Soc. America Decade of North American Geology K-2:49-53.

Darby, D., and Beanland, S. 1992. Possible source models for the 1855 Wairarapa earthquake, New Zealand. Jour. Geophys. Res. 97:12,375-89.

Darienzo, M. E., and Peterson, C. D. 1990. Episodic tectonic subsidence of late Holocene salt marshes, northern Oregon, central Cascadia margin. Tectonics 9:1-22.

Davenport, C. A., Ringrose, P. S., Becker, A., Hancock, P., and Fenton, C. 1989. Geological investigations of late and post glacial earthquake activity in Scotland. *In* Gregersen, S., and Basham, P. W., eds. Earthquakes at North Atlantic Passive Margins: Neotectonics and Postglacial Rebound. Kluwer Academic Publ. 175-194.

Davies, J. N., and House, L. 1979. Aleutian subduction zone seismicity, volcano-trench separation, and their relation to great thrust earthquakes. Jour. Geophys. Res. 84:4583-91.

Davis, D., Dahlen, F. A., and Suppe, J. 1983. Mechanics of fold-and-thrust belts and accretionary wedges. Jour. Geophys. Res. 88:1153-72.

Davis, G. H. 1987. A shear-zone model for the structural evolution of metamorphic core complexes in southeastern Arizona. *In* Coward, M. P., Dewey, J. F., and Hancock, P. L., eds. Continental Extensional Tectonics. Geological Society Special Publication 28:247-66.

Davis, T. L., Namson, J., and Yerkes, R. F. 1989. A cross section of the Los Angeles area: seismically active fold and thrust belt, the 1987 Whittier Narrows earthquake, and earthquake hazard. Jour. Geophys. Res. 94:9644-64.

Davis, T. L., and Namson, J. S. 1994. A balanced cross section of the 1994 Northridge earthquake, southern California. Nature 372:167-69.

Davison, C. 1893. Note on the Quetta earthquake of Dec. 20, 1892. Geol. Mag. 10:356-60.

Dawson, A. G. 1994. Geomorphological effects of tsunami run-up and backwash. Geomorphology 10:83-94.

Delcaillau, B., 1986, Dynamique et évolution géomorphostructurale du piémont frontal de l'Himalaya: les Siwaliks du Népal oriental: Revue Geogr. Phys. et Geol. Dynamique. 27(5): 319-37.

Delmas, R. J. 1992. Environmental information from ice cores. Reviews of Geophysics 30:1-21.

DeMets, C., Gordon, R. G., Stein, S., and Argus, D. F. 1990. Current plate motions. Geophys. J. Int. 101:425-78.

Deng Q., Chen S., Song F. M., Zhu S., Wang Y., Zhang W., Burchfiel, B. C., Molnar, P., Royden, L., and Zhang P. 1986. Variations in the geometry and amount of slip on the Haiyuan fault zone, China and the surface rupture of the 1920 Haiyuan earthquake. Earthquake Source Mechanics, Geophysical Monograph 37:169-82.

Deng Q., Feng X., You H., Zhang P., Zhang Y., Li J., Wu Z., Xu X., Yang X., and Zhang H. 1991a. Characteristics and mechanism of deformation along the Dushanzi-Anjihai active reverse fault and fold zone, Xinjiang. Research on Active Fault 1:17-36 (Chinese, Eng. abs.).

Deng Q., Feng X., Yu H., Chen J., Li J., Zhang Y., Xu X., Wu Z., and Zhang H. 1991b. Paleoseismology and Late Quaternary activity of the

Dushanzi-Anjihai reverse fault zone, Xinjiang. Research on Active Fault 1:37-56 (C, Eng. abs.).

Deng Q., Wang T. M., Li J. G., Xiang H., and Cheng S. P. 1976. A discussion on source model of Haicheng earthquake. Scientia Geologica Sinica 3:195-204. (C).

Deng Q., Wang Y., Liao Y., Zhang W., and Li M. 1984. Fault scarps, colluvial wedges on the frontal fault of Mt. Helanshan and its active history during Holocene. Chinese Science Bull. 9:557-60. (C).

Deng Q., Yu G., and Ye W. 1992. Relationship between earthquake magnitude and parameters of surface ruptures associated with historical earthquakes. Research on Active Fault 2:247-65 (C, Eng. abs.).

Deng Q., and Zhang P. 1984. Research on the geometry of shear fracture zones. Jour. Geophys. Res. 89:5699-5710.

Deng Q., Zhang W., Wang Y., Zhang P., and Song F. 1990. Haiyuan active fault. Beijing: Seismological Press, 286 p. (C).

Denham, D., Alexander, L. G., and Worotnicki G. 1980. The stress field near the sites of the Meckering (1968) and Calingiri (1970) earthquakes, Western Australia. Tectonophysics 67:283-317.

DePaolo, D. J. 1986. Detailed record of the Neogene Sr isotopic evolution of seawater from DSDP Site 590B. Geology 14:103-06.

DePaolo, D. J., and Ingram, B. L. 1985. High-resolution stratigraphy with strontium isotopes. Science 227:938-41.

dePolo, C. M., and Ramelli, A. R. 1987. Preliminary report on surface fractures along the White Mountains fault zone associated with the July 1986 Chalfant Valley earthquake sequence. Seismological Society of America Bulletin 77:290-96.

dePolo, C. M., Clark, D. G., Slemmons, D. B., and Ramelli, A. R. 1991. Historical surface faulting in the Basin and Range province, western North America: Implications for fault segmentation. Journal of Structural Geology13:123-36.

Dewey, J., and Lamb, S. 1992. Active tectonics of the Andes. Tectonophysics 205:79-95.

Dewey, J. F., Helman, M. L., Turco, E., Hutton, D. H. W., and Knott, S. D. 1989. Kinematics of the western Mediterranean. *In* Coward, M. P., Dietrich, D., and Park, R. G., eds. Alpine Tectonics. Geol. Soc. Spec. Pub. 45:265-83.

Diament, M., Harjono, H., Karta, K., Deplus, C., Dahrin, D., Zen, M., Gerard, M., Lassal, O., Martin, A., and Malod, J. 1992. The Mentawai fault zone off Sumatra: A new key for the geodynamics of western Indonesia. Geology 20:259-62.

Ding G. Y., chief ed. 1985. The Fuyun earthquake fault zone in Xinjiang, China. Beijing: Seismological Press, 206 p. (C).

Dixon, T.H. 1991. An introduction to the Global Positioning System and some geological applications. Reviews of Geophysics 29:249-76.

Dmowska, R., and Ekström, G. 1994. Shallow subduction zones: seismology, mechanics and seismic potential. New York: Springer-Verlag, v. 2, 240 p.

Doig, R. 1990. 2300 yr history of seismicity from silting events in Lake Tadoussac, Charlevoix, Quebec. Geology 18:820-23.

Dokka, R. 1992. The Eastern California shear zone and its role in the creation of young extensional zones in the Mojave Desert region, *in* Walker Lane Symposium. Geological Society of Nevada, pp. 161-86.

Dokka, R., and Travis, C.1990. Late Cenozoic strike-slip faulting in the Mojave Desert, California. Tectonics 9:311-40.

Dolan, J. F., Sieh, K., Rockwell, T. K., Yeats, R. S., Shaw, J., Suppe, J., Huftile, G. J., and Gath, E. M. 1995. Prospects for larger or more frequent earthquakes in the Los Angeles metropolitan region, California. Science 267:199-205.

Donnellan, A., Hager, B. H., and King, R. W. 1993. Geodetic determination of rapid shortening across the Ventura basin, southern California. Jour. Geophys. Res. 98:21,727-39.

Dorbath, L., Dorbath, C., Rivera, L., Fuenzalida, A., Cisternas, A., Tatevossian, R., Aptekman, J., and Arefiev, S. 1992. Geometry, segmentation and stress regime of the Spitak (Armenia) earthquake from the analysis of the aftershock sequence. Geophys. J. Int. 108:309-28.

Doser, D. I. 1989. Extensional tectonics in northern Utah-southern Idaho, U.S.A., and the 1934 Hansel Valley sequence. Physics of the Earth and Planetary Interiors 54:120-34.

Doser, D. I. 1994. Contrasts between source parameters of $M > 5.5$ earthquakes in northern Baja California and southern California. Geophys. J. Int. 116:605-17.

.Drew, A. R., and Snay, R. A 1989. DYNAP: Software for estimating crustal deformation from geodetic data. Tectonophysics 162:331-43.

Duncan, R.A. 1991. Oceanic drilling and the volcanic record of hotspots. GSA Today 1:213-16, 219.

Dziewonski, A., and Woodhouse, J. H. 1983. An experiment in systematic study of global seismicity: Centroid-moment tensor solutions for 201 moderate and large earthquakes of 1981. Jour. Geophys. Res. 88:3247-71.

Earthquake Research Institute (ERI). 1936. Papers and reports on the Formosa earthquake of 1935. Tokyo Univ., Earthquake Res. Inst. Bull., Supp. 3, 238 p.

Eaton, J. P. 1990. The earthquake and its aftershocks from May 2 through September 30, 1983. *In* Rymer, M. J., and Ellsworth, W. L., eds. The Coalinga, California, Earthquake of May 2, 1983. U. S. Geol. Survey Prof. Paper 1487:113-70.

Eberhart-Phillips, D., and Michael, A. 1993. Three-dimensional velocity structure, seismicity, and fault structure in the Parkfield region, central California. Jour. Geophys. Res. 98:15,737-58.

Edwards, R. L., Chen, J. H., and Wasserburg, G. J. 1986. ^{238}U-^{234}U-^{230}Th-^{232}Th systematics and the precise measurement of time of the past 500,000 years. Earth and Planetary Sci. Lett. 81:175-92.

Einarsson, P. 1994. Holocene surface fractures in the South Iceland Seismic Zone (abs.). Earthquake Prediction Research in the South Iceland Test Area: State-of-the-art. Iceland Meteorological Office.

Einarsson, P., and Eiríksson, J. 1982. Earthquake fractures in the districts Land and Rangárvellir in the South Iceland Seismic Zone. Jökull 32:113-20.

Einarsson, P., Bjornsson, S., Foulger, G., Stefansson, R., and Skaftadottir, Th. 1981. Seismicity pattern in the South Iceland seismic zone. *In* Simpson, D., and Richards, P. G., eds. Earthquake Prediction: An International Review: Amer. Geophys. Union Maurice Ewing Series 4:141-51.

Eissler, H. K., and Kanamori, H. 1987. A single-force model for the 1975 Kalapana, Hawaii, earthquake. Jour. Geophys. Res. 92:4827-36.

Emiliani, C. 1955. Pleistocene temperatures. Jour. Geology 63:538-78.

Erentöz, C. and Kurtman, F. 1964. Rapport sur la tremblement de terre de Manyas survenir en 1964. Bull. Miner. Res. Expl. Inst. Turkey 63:1-8.

Eyidogan, H., and Jackson, J. A. 1985. A seismological study of normal faulting in the Demirci, Alasehir and Gediz earthquakes of 1969-70 in western Turkey: Implications for the nature and geometry of deformation in the continental crust. Geophysical Journal of Royal Astronomical Society 81:569-607.

Farah, A. 1976. Study of recent seismotectonics in Pakistan. Report CENTO Working Group on Recent Tectonics, Istanbul.

Farhoudi, G., and Karig, D. E. 1977. Makran of Iran and Pakistan as an active arc system. Geology 5:664-68.

Fedotov, S. A. 1965. Regularities of the distribution of strong earthquakes in Kamchatka, the Kuril Islands, and northeast Japan. Trudy Inst. Fiz. Zemli., Acad. Nauk, SSSR 36:66-94 (in Russian).

Feigl, K. L., Agnew, D. C., Bock, Y., Dong, D., Donnellan, A., Hager, B. H., Herring, T. A., Jackson, D. D., Jordan, T. H., King, R. W., Larsen, S., Larson, K. M., Murray, M. H., Shen, Z., and Webb, F. H. 1993. Space geodetic measurement of crustal deformation in central and southern California 1984-1992. Jour. Geophys. Res. 98:21.677-712.

Feng X. 1987. Paleoseismological study for Kaxhe fault zone, Xinjiang. Seismology and Geology 9, 2:74-77. (C, Eng. abs.).

Feng X. 1994. Surface rupture associated with the 1985 Wuqia earthquake, in Xinjiang. Research on Active Fault 3:45-55 (C, Eng. abs.).

Feng X., Deng Q., Shi J., Li J., You H., Zhang Y., Yu G., and Wu Z. 1991. Active tectonics of the southern and northern Tianshan and its tectonic evolution. Research on Active Fault 1:1-16.

Feng X., Luan C., Li J., and Zhang Y. 1988. The deformation zone of Wuqia earthquake of M = 7.4 in 1985. Seismology and Geology 10, 2:39-44. (C, Eng. abs.).

Field, M. E., Gardner, J. V., Jennings, A. E., and Edwards, B. D. 1982. Earthquake-induced sediment failures on a 0.25° slope, Klamath River delta, California. Geology 10:542-46.

Fitch, T. J. 1972. Plate convergence, transcurrent faults, and internal deformation adjacent to southeast Asia and the western Pacific. Jour. Geophys. Res. 77:4432-60.

Fitch, T. J., and Scholz, C. H. 1971. Mechanism of underthrusting in southeast Japan: A model of convergent plate interactions. Jour. Geophys. Res. 76:7260-79.

Fitzgerald, P. G., and Gleadow, A. J. W., 1990, New approaches in fission track chronology as a tectonic tool: examples from the Transantarctic Mountains. Nuclear Tracks and Radiation Measurements 17:351-57.

Florensov, N. A., and Solonenko, V. P. 1965. The Gobi-Altai earthquake. Moscow: Nauka, 1963 (R). English translation available from U.S. Dept. of Commerce, Springfield, VA, 1965.

Forman, S. L. 1989. Applications and limitations of thermoluminescence to date Quaternary sediments. Quaternary International 1:47-59.

Forslund, T. and Gudmundsson, A. 1991. Crustal spreading due to dikes and faults in southwest Iceland. Journal of Structural Geology 13:443-57.

Forsyth, D. W., and Uyeda, S. 1975. On the relative importance of the driving forces of plate motion. Geophys. Jour. Roy. Astr. Soc. 43:163-200.

Fowler, C. M. R. 1990. The solid earth, an introduction to global geophysics. Cambridge: Cambridge University Press, 472 p.

Fox, P., and Gallo, D. 1986. The geology of north Atlantic transform plate boundaries and their aseismic extensions. The Geology of North America M:157-72.

Frohlich, C.1994. Earthquakes with non-double couple mechanisms. Science 264:804-09.

Friedman, I., and Smith, R. L. 1960. A new dating method fusing obsidian—Part 1, The development of the method. Am. Antiquity 25:476-522.

Fuis, G., Mooney, W., Healy, J., McMechan, G., and Lutter, W. 1982. Crustal structure of the Imperial Valley Region, in The Imperial Valley, California, earthquake of October 15, 1979. U. S. Geol. Survey Prof. Paper 1254.

Fuller, M. 1912. New Madrid earthquake. U. S. Geol. Survey Bulletin 494:1-115.

Fumal, T. E., Pezzopane, S. K., Weldon, R. J., and Schwartz, D. P. 1993. A 100-year average recurrence interval for the San Andreas fault at Wrightwood, California. Science 259:199-203.

Galanopoulos, A. G. 1967. The influence of the fluctuation of Marathon Lake on local earthquake activity in the Attica basin area. Annales Geologiques des Pays Helleniques (Athens) 18:281-306.

Gallo, D., Fox, P., and Macdonald, K. 1986. A sea beam investigation of the Clipperton transform fault: the morphotectonic expression of a fast slipping transform boundary. Jour. Geophys. Res. 91:3455-67.

Gans, P. B. 1987. An open-system, two-layer crustal stretching model for the eastern Great Basin. Tectonics 6:1-12.

Gao W., Zheng L., and Lin Z. 1988. Earthquake-generating structures of the 1668 Tancheng earthquake with M = 8.5. Earthquake Research in China 4(3):9-15. (C).

Garfunkel, Z. 1974. Model for the late Cenozoic tectonic history of the Mojave Desert and its relation to adjacent areas. Geol. Soc. America Bull. 85:1931-44.

Gaudemer, Y., Tapponnier, P., and Turcotte, D. 1989. River offsets across active strike-slip faults. Annales Tectonicae 3:55-76.

Geist, E. L., Childs, J. R., and Scholl, D. W. 1988. The origin of summit basins of the Aleutian Ridge: Implications for block rotation of an arc massif. Tectonics 7:327-41.

Gere, J. M., and Shah, H. C. 1984. Terra Non Firma. New York: W. H. Freeman and Co., 2 03 p.

Ghose, R., and Oike, K. 1987. Tectonic implications of some reservoir-induced earthquakes in the aseismic region of western Thailand. Journal of Physics of the Earth 35; 327-45.

Ghose, S., Mellors, R. J., Korzhenkov, A. M., Omuraliev, M., Mamyrov, E., Pavlis, T. L., and Hamburger, M. W. 1993. Suusamyr earthquake of Kyrgyzstan, C. I. S. In prep.

Gianella, V. P. 1957, Earthquake and faulting, Fort Sage Mountains, California, December, 1950. Seismological Society of America Bulletin 47; 173-77.

Gianella, V. P., and Callaghan, E. 1934, The Cedar Mountain, Nevada, earthquake of December 20, 1932. Seismological Society of America Bulletin 24:345-77.

Gilbert, G. K. 1890. Lake Bonneville: U.S. Geol. Survey Monograph 1.

Gillard, G., Wyss, M., and Nakata, J. S. 1992. A seismotectonic model for western Hawaii based on stress tensor inversion from fault plane solutions. Jour. Geophys. Res. 97:6629-41.

Girdler, R. 1991. The Afro-Arabian rift system—an overview. Tectonophysics 197; 138-53.

Glen, W. 1982. The road to Jaramillo: Critical years of the revolution in Earth Science. Stanford University Press, 459 p.

Gnibidenko, H., Bykova, T. G., Veselov, O. V., Vorobiev, V. M., and Svarichevsky, A. S. 1983. The tectonics of the Kuril-Kamchatka deep-sea trench. *In* Hilde, T. W. C., and Uyeda, S., eds. Geodynamics of the Western Pacific-Indonesian Region. Am. Geophys. Union Geodynamics Series 11:249-85.

Goldfinger, C. 1994. Active deformation of the Cascadia forearc: Implications for great earthquake potential in Oregon and Washington. Corvallis, Oregon State University Ph.D. thesis, 202 p.

Goldfinger, C., Kulm, L. D., Yeats, R. S., Appelgate, B., MacKay, M. E., and Moore, G. F. 1992. Transverse structural trends along the Oregon convergent margin: implications for Cascadia earthquake potential and crustal rotations: Geology 20:141–44.

Goldfinger, C., Kulm, L. D., Yeats, R. S., Appelgate, T. B., Cochrane, G., and MacKay, M. E. 1995. Active strike-slip faulting and folding of the Cascadia plate boundary and forearc in northern and central Oregon. *In* Rogers, A. M., et al., eds., Assessing and reducing earthquake hazards in the Pacific Northwest. U. S. Geol. Survey Prof. Paper 1560. In press.

Gordon, F. R., and Lewis, J. D. 1980. The Meckering and Calingiri earthquakes October 1968 and March 1970. Geol. Survey of Western Australia Bull. 126:229 p.

Gordon, R. G. 1994. Present plate motions and plate boundaries. *In* Ahrens, T., ed. Handbook of Physical Constants. American Geophys. Union Monograph, in press.

Gordon, R. G. and Stein, S. 1992. Global tectonics and space geodesy. Science 256:333–42.

Gough, D. I., and Gough, W. I. 1977a. Stress and deflection in the lithosphere near Lake Kariba—I. Geophys. Jour. 21:65–78.

Gough, D. I., and Gough, W. I. 1977b. Load-induced earthquakes at Lake Kariba—II. Geophys. Jour. 21:79–101.

Gouin, P. 1979. Earthquake history of Ethiopia and the Horn of Africa. Ottawa, Canada, International Development Research Centre Monograph IDRC-118e, 258 p.

Goulty, N., and Gilman, R. 1978. Repeated creep events on the San Andreas fault near Parkfield, California, recorded by a strainmeter array. Jour. Geophys. Res. 83:5415–19.

Grange, L.I. 1932. Taupo earthquakes 1922. Rents and faults formed during earthquake of 1922 in Taupo district. New Zealand Jour. Sci. Technology 14:139–41.

Grant, L., and Donnellan, A. 1994. 1855 and 1991 surveys of the San Andreas fault: Implications for fault mechanics. Seismological Society of America Bulletin 84:241–46.

Grant, L., and Sieh, K. 1993. Stratigraphic evidence for seven meters of dextral slip on the San Andreas fault during the 1857 earthquake in the Carrizo Plain. Seismological Society of America Bulletin 83:619–35.

Grant, L. B., and Sieh, K. 1994, Paleoseismic evidence of clustered earthquakes on the San Andreas fault in the Carrizo Plain, California. Jour. Geophys. Res. 99:6819–41.

Green, H. W., II. 1994. Solving the paradox of deep earthquakes. Scientific American 271(3): 64–71.

Griesbach, C. L. 1893. Notes on the earthquake in Baluchistan on the 20th December 1892. Geol. Survey India Records 26, part 2:57–61.

Griggs, D., and Handin, J. 1960. Observations on fracture and a hypothesis of earthquakes. *In* Griggs, D., and Handin, J., eds. Rock Deformation. Geol. Soc. America Memoir 79:347–64.

Griggs, G. B., and Kulm, L. D. 1970. Sedimentation in Cascadia deep-sea channel. Geol. Soc. America Bull. 81:1361–84.

Grigsby, F. B. 1986. Quaternary tectonics of the Rincon and San Miguelito oil fields area, western Ventura basin, California. Corvallis: Oregon State University M. S. thesis, 110 p.

Grindley, G. W., and Hull, A. G. 1986. Historical Taupo earthquakes and earth deformation. Royal Society of New Zealand Bulletin 24: 173–86.

Groeber, P. 1944. Movimientos tectónicos contemporáneos. Univ. Nac. La Plata Notas Mus. Geol. 9:263–375.

Gudmundsson, A. 1987. Tectonics of the Thingvellir fissure swarm, SW Iceland. Journal of Structural Geology 9:61–69.

Guha, S. K., and Patil, D. N. 1990. Large water-reservoirárelated induced seismicity. Gerlands Beitr. Geophysik 99:S.265–88.

Guo S., Li Z., Cheng S., Chen X. C., Chen X., Yang Z., and Li R. C. 1977. Discussion on the regional structural background and the seismogenic model of the Tangshan earthquake. Scientia Geologica Sinica 4:305–21. (C, Eng. abs.).

Guo S., Xiang H., Zhang J., Hu R., and Zhang G. 1988. Discussion on the deformation band and magnitude of the 1511 Yongsheng earthquake in Yunnan Province. Jour. Seismological Research 11(2):153–62. (C. Eng. abs.).

Guo S., Xiang H., Wang Z., Ji F., and Li B. D. 1993. Active fault zones in Mt. Qilianshan and Hexizoulong regions. Beijing: Seismological Press, 285 p. (C).

Gupta, H. K. 1993. The deadly Latur earthquake. Science 262:1666–67.

Gupta, H. K. and Rastogi, B. K., 1976. Dams and earthquakes. New York, Elsevier Publishing Co., 229 p.

Gutenberg B., and Richter, C. F. 1954. Seismicity of the earth. Princeton: Princeton University Press, 440 p.

Hack, J. T. 1973. Stream-profile analysis and stream-gradient index. U. S. Geol. Survey Jour. Research 1:421-29.

Hadley, D. M., and Kanamori, H. 1977. Seismic structure of the Transverse Ranges, California. Geol. Soc. America Bull. 88:1469-78.

Hadley, D. M., and Kanamori, H. 1978. Recent seismicity in the San Fernando region and tectonics in the west-central Transverse Ranges, California. Seismological Society of America Bulletin 68:1449-57.

Hadley, J. B. 1964. Landslides and related phenomena accompanying the Hebgen Lake earthquake of August 17, 1959. *In* U. S. Geol. Survey, National Park Service, Coast and Geodetic Survey, and U. S. Forest Service, The Hebgen Lake, Montana, earthquake of August 17, 1959. U. S. Geol. Survey Prof. Paper 435:107-38.

Haessler, H., Deschamps, A., Dufumier, H., Fuenzalida, H., and Cisternas, A. 1992. The rupture process of the Armenian earthquake from broadband teleseismic body wave records. Geophys. J. Int. 109:151-61.

Hager, B. H., King, R. W., and Murray, M. H. 1991. Measurement of crustal deformation using the Global Positioning System. Annual Reviews of Earth and Planetary Sciences 19:351-82.

Haimson, B. C., and Fairhurst, C. 1970. In-situ stress determinations at great depth by means of hydraulic fracturing. *In* Somerton, W. H., ed. Rock Mechanics—Theory and Practice, Proceedings. 11th Symposium in Rock Mechanics, Berkeley, 1969. Chapter 28. Society of Mining Engineers of American Inst. of Mining Engineers, New York, pp. 559-84.

Haines, A. J., and Darby, D. J. 1987. Preliminary dislocation models for the 1931 Napier and 1932 Wairoa earthquakes. New Zealand Geol. Survey Rep. EDS 114.

Hamilton, W. 1979. Tectonics of the Indonesian region. U. S. Geol. Survey Prof. Paper 1078, 345 p., 1 pl.

Hancock, P. L., and Barka, A. A. 1987. Kinematic indicators on active normal faults in western Turkey. Journal of Structural Geology 9: 573-84.

Hancock. P. L. and Bevan, T. G. 1987. Brittle modes of foreland extension. *In* Coward, M. P., Dewey, J. F., and Hancock, P. L., eds. Continental Extensional Tectonics. Geological Society Special Publication 28:127-37.

Hanks, T. C., and Kanamori, H. 1979. A moment-magnitude scale. Jour. Geophys. Res. 84:2348-50.

Hanks, T. C., Bucknam, R. C., Lajoie, K. R., and Wallace, R. E. 1984. Modification of wave-cut and fault-controlled landforms. Jour. Geophys. Res. 89:5771-90.

Hansen, W. R. 1966. Effects of the earthquake of March 27, 1964, at Anchorage, Alaska. U. S. Geol. Survey Prof. Paper 542-A, 68 p.

Harden, J. W. 1982. A quantitative index of soil development from field descriptions: Examples from a chronosequence in central California. Geoderma 28:1-28.

Harding, T. P. 1976. Tectonic significance and hydrocarbon trapping consequence of sequential folding synchronous with San Andreas faulting, San Joaquin Valley, California. American Assoc. Petroleum Geologists Bull. 60:356-78.

Harris, R., and Day, S. 1993. Dynamics of fault interaction: parallel strike-slip faults. Jour. Geophys. Res. 98:4461-72.

Hart, E. W. 1980. Fault-rupture hazard zones in California. California Division of Mines and Geology, Special Publication 42, 25 p.

Hartzell, S. H., and Heaton, T. H. 1983. Inversion of strong ground motion and teleseismic waveform data for the fault rupture history of the 1979 Imperial Valley, California, earthquake. Seismological Society of America Bulletin 73:1553-83.

Hartzell, S. H., and Helmberger, D. 1982. Strong-motion modelling of the Imperial Valley earthquake of 1979. Seismological Society of America Bulletin 72:571-96.

Hasegawa, H. S., and Kanamori, H. 1987. Source mechanism of the magnitude 7.2 Grand Banks earthquake of November 1929; double couple or submarine landslide? Seismological Society of America Bulletin 77:1984-2004.

Hasegawa, A., Umino, N., and Takagi, A. 1978. Double-planed deep seismic zone and upper-mantle structure in the northeastern Japan arc. Geophys. Jour. Roy. Astron. Soc. 54:281-96.

Hatcher, R. D., and Hooper, R. J. 1992. Evolution of crystalline thrust sheets in the internal parts of mountain chains. *In* McClay, K. R., ed. Thrust Tectonics. London: Chapman & Hall, 217-33.

Hauksson, E., and Jones, L. M. 1989. The 1987 Whittier Narrows earthquake sequence in Los Angeles, southern California: Seismological and tectonic analysis. Jour. Geophys. Res. 94:9569-89.

Hauksson, E., Jones, L., Hutton, K., and Eberhart-Phillips, D. 1993. The 1992 Landers earthquake sequence: seismological observations. Jour. Geophys. Res. 98:19,835-58.

Hauksson, E., Hutton, K., Kanamori, H., Jones, L., Mori, J., Hough, S., and Roquemore, G. 1995. Preliminary report on the 1995 Ridgecrest earthquake sequence in eastern California. Seismological Research Letters 66(6):54-60.

Hays, J. D., Imbrie, J., and Shackleton, N. J. 1976. Variations in the Earth's orbit: Pacemaker of the Ice Ages. Science 194:1121-32.

Heaton, T. H., and Kanamori, H. 1984. Seismic potential associated with subduction in the northwestern United States. Seismological Society of America Bulletin 74:933-41.

Hecker, S., and Schwartz, D. P. 1994. The characteristic earthquake revisited: Geological evidence of the size and location of successive earthquakes on large faults. U. S. Geol. Survey Open-File Report 94-568:79-80.

Heezen, B. C., and Ewing, M. 1952. Turbidity currents and submarine slumps, and the 1929 Grand Banks earthquake. American Jour. Science 250:849-73.

Heim, A. 1949. Observaciones geologicas en la region del terremoto de Ancash de Noviembre de 1946. Soc. Geol. Peru V. Jub. 2(6), 28 pp.

Heki, K., Takahashi, Y., and Kondo, T. 1990. Contraction of northeastern Japan: Evidence from horizontal displacement of a Japanese station in global very long baseline interferometry networks. Tectonophysics 181:113-22.

Henderson, J. 1933. The geological aspects of the Hawkes Bay earthquakes. Department of Scientific and Industrial Research Bulletin 43.

Henderson, J. 1937. The west Nelson earthquake of 1929. New Zealand Jour. Sci. Technology 19(2):66-143.

Heubeck, C., and Mann, P. 1991. Geologic evaluation of plate kinematic models of the north American-Caribbean plate boundary zone. Tectonophysics 191:1-26.

Heuckroth, L E., and Karim, R. A. 1971. Earthquake history, seismicity and tectonics of the regions of Afghanistan. Seismological Center, Faculty of Engineering, Kabul University, 102 p.

Hervé, F., and Thiele, R. 1987. Estado de conocimiento de las megafallas en Chile y su significado tectonico. Comunicaciones 38:67-91.

Hickman, S. H. 1991. Stress in the lithosphere and the strength of active faults. Contributions in Tectonophysics. U.S. National Report 1987-1990. American Geophysical Union. 759-75.

Hilde, T. W. C., and Uyeda, S., eds. 1983. Geodynamics of the western Pacific-Indonesian region. American Geophys. Union Geodynamics Series 11, 457 p.

Hileman, J. A., Allen, C. R., and Nordquist, J. M. 1973. Seismicity of the southern California region 1 January 1932 to 31 December 1972. Seismological Laboratory, California Institute of Technology, Division of Geological and Planetary Sciences Contribution No. 2385, 404 p.

Hilgen, F. J. 1991. Astronomical calibration of Gauss to Matuyama sapropels in the Mediterranean and implications for the Geomagnetic Polarity Time Scale. Earth and Planet. Sci. Lett. 104:226-44.

Hill, K. C. 1991. Structure of the Papuan fold belt, Papua New Guinea. American Assoc. Petroleum Geologists Bulletin 75:857-72.

Hill, D. P., and 30 others. 1993. Seismicity remotely triggered by the magnitude 7.3 Landers, California, earthquake. Science 260:1617-23.

Hill, R. L., and Beeby, D. J. 1977. Surface faulting associated with the 5.2 magnitude Galway Lake earthquake of May 31, 1975, Mojave Desert, San Bernardino County, California. Geol. Soc. America Bull. 88:1378-84.

Hill, R. L., Pechmann, J. C., Treiman, J. A., McMillan, J. R., Given, J. W., and Ebel, J. E. 1980. Geologic study of the Homestead Valley earthquake swarm of March 15, 1979. California Geology 33:60-67.

Hirano, S., Nakata, T., and Sangawa, A. 1986. Fault topography and Quaternary faulting along the Philippine fault zone, central Luzon, the Philippines. J. Geography 95, 2:71-96. (J, Eng. abs.).

Hirooka, K. 1991. Quaternary paleomagnetic studies in Japan. The Quaternary Research 30:151-60.

Hobbs, W. H. 1910. The earthquake of 1872 in the Owens Valley, California. Beitr. zur Geophysik10:352-85.

Hodell, P. A., Mueller, P. A., and Garrido, J. R. 1991. Variations in the strontium isotope composition of seawater during the Neogene. Geology 19:24-27.

Holzer, T. L., and Clark, M. M. 1993. Sand boils without earthquakes. Geology 21:873-76.

Hou Z., ed. 1992. Changma active fault zone. Seismological Press, 219 p. (C).

Hough, S. 1994. Southern surface rupture associated with the M 7.3 1992 Landers, California, earthquake. Seismological Society of America Bulletin 84:817-25.

House, L. S., and Jacob, K. H. 1983. Earthquakes, plate subduction, and stress reversals in the eastern Aleutian arc. Jour. Geophys. Res. 88:9347-73.

Hsu, T. L.1962. Recent faulting in the Longitudinal Valley of eastern Taiwan. Memoir Geol. Soc. China 1:95-102.

Hsu, T. L. 1976. Neotectonics of the Longitudinal Valley, eastern Taiwan. Bull. Geol. Survey of Taiwan 25:53-62.

Hsu, T. L., and Chang, H. C. 1979. Quaternary faulting in Taiwan. Memoir Geol. Soc. China 3:155-65.

Hu, Q., Smith, P. E., Evensen, N. M., and York, D. 1994. Lasing in the Holocene: extending the $^{40}Ar^{39}Ar$ laser-probe method into the ^{14}C age range. Earth Planet. Sci. Lett. 123:331-36.

Hubbert, M. K., and Rubey, W. W. 1959. Role of fluid pressure in mechanics of overthrust faulting. I. Mechanics of fluid-filled porous solids and its application to overthrust faulting. Geol. Soc. America Bull. 70:115-66.

Hudnut, K., and Sieh, K. 1989. Behavior of the Superstition Hills fault during the past 330 years. Seismological Society of America Bulletin 79:304-29.

Hudnut, K., Seeber, L., Rockwell, T., Goodmacher, J., Klinger, R., Lindvall, S., and McElwain, R. 1989. Surface ruptures on cross-faults in the 24 November 1987 Superstition Hills, California, earthquake sequence. Seismological Society of America Bulletin 79:282-96.

Huftile, G. J., and Yeats, R. S. 1995. Convergence rates across a displacement transfer zone in the western Transverse Ranges, Ventura basin, California. Jour. Geophys. Res. 100:2043-67..

Hull, A. G. 1990. Tectonics of the 1931 Hawke's Bay earthquake. New Zealand Jour. Geology and Geophysics 33:309-30.

Hull, A. G., and Grindley, G. W. 1983, Active faulting near Taupo: EOS 65:51-52.

Hummon, C., Schneider, C. L., Yeats, R. S., Dolan, J. F., Sieh, K. E., and Huftile, G. J. 1994. Wilshire fault: Earthquakes in Hollywood? Geology 22:291-94.

Hyndman, R. D., and Wang, K. 1993. Thermal constraints on the zone of major thrust earthquake failure: The Cascadia subduction zone. Jour. Geophys. Res. 98:2039-60.

Hyndman, R. D., Wang, K., and Yamano, M. 1995. Thermal constraints on the seismogenic portion of the southwestern Japan subduction thrust. Jour. Geophys. Res. 100:15, 373-92.

Ihara, K., and Ishii, K. 1932. The earthquake of northern Izu. Imp. Geol. Survey of Japan Rept. 112, 111p (J) + 7 p. (Eng.).

Ikeda, Y. 1983. Thrust-front migration and its mechanisms—evolution of intraplate thrust fault systems. Dept. Geography, University of Tokyo Bulletin 15:125-59.

Ikeya, M., Miki, T., and Tanaka, K., 1982, Dating of a fault by electron spin resonance on intrafault material. Science 215:1392-93.

Imamura, A. 1924. Preliminary note on the great earthquake of S.E. Japan on Sept. 1, 1923. Seismological Notes 6:1-22.

Imamura, A. 1928. The Tazima earthquake of 1925. Imperial Earthquake Investigation Committee Bull. 10:71-107.

Imamura, A. 1930. Topographical changes accompanying earthquakes or volcanic eruptions. Publications of the Earthquake Investigation Committee in Foreign Languages 25, 143 p.

Imbrie, J., Hays, J. D., Martinson, D. G., McIntyre, A., Mix, A. C., Morley, J. J., Pisias, N. G., Prell, W. L., and Shackleton, N. J. 1984. The orbital theory of Pleistocene climate: Support from a revised chronology of the marine d^{18} O record. *In* Berger, A. L., et al., eds. Milankovitch and Climate. Hingham, Mass.: D. Reidel, 269-305.

Irwin, W. 1990. Geology and Plate-Tectonic Development. *In* R. Wallace, ed. The San Andreas Fault System, California. Denver, CO: U. S. Geol. Survey Prof. Paper 1515:61-80.

Isacks, B., Oliver, J., and Sykes, L. R. 1968. Seismology and the new global tectonics. Jour. Geophys. Res. 73:5855-99.

Ishibashi, K. 1981. Specification of a soon-to-occur seismic faulting in the Tokai district, central Japan, based upon seismotectonics. *In* Earthquake Prediction—An International Review. American Geophys. Union Maurice Ewing Series 4:297-332.

Ishibashi, K. 1985. Possibility of a large earthquake near Odawara, central Japan, preceding the Tokai earthquake. Earthquake Prediction Research 3:319-44.

Ishibashi, K. 1989. The possible activation of the Median Tectonic Line during the Keicho earthquake and its influence on the 1605 Nankai trough tsunami earthquake. Seismol. Soc. Japan 1989, No. 1, p. 62 (J).

Ishibashi, K. 1992. Recurrence history of great subduction zone earthquakes along the Sagami and Suruga-Nankai troughs, Japan, and its tectonic implication. preprint, 11 p.

Ishihara, K.1985. Stability of natural soil deposits during earthquakes. San Francisco: Proc., Eleventh Internat. Conf. Soil Mechanics and Foundation Engineering 1:321-76.

Ishikawa, Y., et al. 1984. Focal process of the 1983 Japan Sea earthquake. Earth Monthly 6:11-17 (Japanese).

Iwata, T., and Hamada, N. 1986. Seismicity before and after the Tonankai earthquake of 1944. Zisin, Ser. 2, 39:621-34 (J., Eng. abs.)

Jackson, D. D., Lee, W. B., and Liu, C. C. 1980. Aseismic uplift in southern California: An alternative interpretation. Science 210:534-36.

Jackson, J. A. 1987. Active normal faulting and crustal extension. *In* Coward, M. P., Dewey, J. F., and Hancock, P. L., eds. Continental Extensional Tectonics. Geological Society Special Publication 28:3-17.

Jackson, J. A. 1992. Partitioning of strike-slip and convergent motion between Eurasia and Arabia in eastern Turkey and the Caucasus. Jour. Geophys. Res. 97:12,471-79.

Jackson, J. A., Gagnepain, J., Houseman, G., King, G. C. P., Papadimitriou, P., Soufleris, C., and Virieux, J. 1982. Seismicity, normal faulting, and the geomorphological development of the Gulf of Corinth (Greece): The Corinth earthquakes of February and March 1981. Earth and Planet. Sci. Lett. 57:377-97.

Jackson, J. A., and McKenzie, D. 1984. Active tectonics of the Alpine-Himalayan belt between western Turkey and Pakistan. Geophys. Jour. Roy. Astron. Soc. 77:185-264.

Jackson, J. A., and McKenzie, D. 1988. The relationship between plate motions and seismic moment tensors, and the rates of active deformation in the Mediterranean and Middle East. Geophys. J. 93:45-73.

Jackson, J. A., and White, N. J. 1989. Normal faulting in the upper continental crust: Observations from regions of active extension. Journal of Structural Geology 11:15-36.

Jackson, M., and Bilham, R. 1994. Constraints on Himalayan deformation inferred from vertical velocity fields in Nepal and Tibet. Jour. Geophys. Res. 99:13,897-912.

Jackson, M. D., Endo, E. T., Delaney, P. T., Arnadottir, T., and Rubin, A. M. 1992. Ground ruptures of the 1974 and 1983 Kaoiki earthquakes, Mauna Loa volcano, Hawaii. Jour. Geophys. Res. 97:8775-96.

Jacob, K. H., Pennington, W. D., Armbruster, J., Seeber, L., and Farhatulla, S. 1979. Tarbela Reservoir, Pakistan: A region of compressional tectonics with reduced seismicity upon initial reservoir filling. Seismological Society of America Bulletin 69:1175-92.

Jacob, K., and Quittmeyer, R. C. 1979. The Makran region of Pakistan and Iran: trench-arc system with active plate subduction. *In* Farah, A., and DeJong, K. A., eds. Geodynamics of Pakistan. Quetta:, Geol. Survey of Pakistan, 303-17.

Jacoby, G. C., Sheppard, P. R., and Sieh, K. E. 1988. Irregular recurrence of large earthquakes along the San Andreas fault: Evidence from trees. Science 241:196-99.

Jacoby, G. C., Williams, P. L., and Buckley, B. M. 1992. Tree ring correlation between prehistoric landslides and abrupt tectonic events in Seattle, Washington. Science 258:1621-1623.

Jadoon, I. A. K., Lawrence, R. D., and Lillie, R. J. 1994. Seismic data, geometry, evolution, and shortening in the active Sulaiman fold-and-thrust belt of Pakistan, southwest of the Himalayas. American Assoc. Petroleum Geologists Bull. 78:758-774.

Jaeger, J. C., and Cook, N. G. W. 1979. Fundamentals of rock mechanics, 3rd ed. London: Methuen, 593 p.

Jahns, R. 1972. Tectonic evolution of the Transverse Ranges province as related to the San Andreas fault system. Proceedings of the Conference on Tectonic Problems of the San Andreas Fault Sustem, Stanford University Publication, Geological Sciences, v.XIII, p. 149-70.

Jankov, D. 1945. Changes in ground level produced by the earthquakes of April 14 and 18, 1928, in southern Bulgaria. Tremblements de terre en Bulgarie, nos. 29-31, Institut météorologique central de Bulgarie, Sofia, 131-36 (in Bulgarian).

Jenkins, D. G., Bowen, D. Q., Adams, C. G., Shackleton, N. J., and Brassell, S. C. 1985. The Neogene: Part 1. *In* Snelling, N. J., ed. The Chronology of the Geological Record. Blackwell Scientific Publications: 199-210.

Jennings, C. W. 1994. Fault Activity Map of California and adjacent areas with locations and ages of recent volcanic eruptions, scale 1:750,000. California Division of Mines and Geology Geologic Data Map No. 6, with explanatory text. 92 p.

Jia Y., Dai H., and Su X. 1988. Tuosuo Lake earthquake fault in Qinghai province. *In* Research on Earthquake Faults in China. Xinjiang Seismological Bureau, ed. Xinjiang Press: 66-71. (C).

Jia Y., Su X., Liu H., Chen Y., Dai H., and Hou K. 1994. Active characteristics of the eastern segment (Qilian-Shuangta segment) of the Huangcheng-Shuanta fault zone since late Pleistocene. Research on Active Fault 3:170–79 (C, Eng. abs.)

Jibson, R. W. 1994. Using landslides for paleoseismic analysis. U. S. Geol. Survey Open-File Report 94-663, Chapter B, 33 p.

Jibson, R. W., and Keefer, D. K. 1989. Statistical analysis of factors affecting landslide distribution in the New Madrid seismic zone, Tennessee and Kentucky. Engineering Geology 27:509–42.

Jibson, R. W., and Keefer, D. K. 1993. Analysis of the seismic origin of landslides: Examples from the New Madrid seismic zone. Geol. Soc. America Bull. 105:521-36.

Jibson, R. W., Prentice, C. S., Borissoff, B. A., Rogozhin, E. A., and Langer, C. J. 1994. Some observations of landslides triggered by the 29 April 1991 Racha earthquake, Republic of Georgia. Seismological Society of America Bulletin 84:963–73.

Joffe, J. S. 1949. Pedology. New Brunswick, N.J.: Pedology Publ., 662 p.

Johnson, A. M. 1970. Physical processes in geology. San Francisco: Freeman, Cooper and Co., 577 p.

Johnson, A. M., and Fleming, R. W. 1993. Formation of left-lateral fractures within the Summit Ridge shear zone, 1989 Loma Prieta, California, earthquake. Jour. Geophys. Res. 98:21,823–37.

Johnson, C., and Hadley, D. 1976. Tectonic implications of the Brawley earthquake swarm, Imperial Valley, California, January 1975. Seismological Society of America Bulletin 66:1133–44.

Johnson, C., and Hutton, L. 1982. Aftershocks and pre-earthquake seismicity. *In* The Imperial Valley Earthquake of October 15, 1979. U. S. Geol. Survey Prof. Paper 1254:59–76.

Johnson, S., and Jones, P. 1978. Microearthquakes located on the Blanco fracture zone with sonobuoy arrays. Jour. of Geophy. Res. 83:255–61.

Johnston, A. C., and Kanter, L. R. 1990. Earthquakes in stable continental crust. Scientific American 262(3):68–75.

Jordan, T. E., Isacks, B. L., Allmendinger, R. W., Brewer, J. A., Ramos, V. A., and Ando, C. J. 1983. Andean tectonics related to geometry of subducted Nazca plate. Geol. Soc. America Bull. 94:341-61.

Journal of Geophysical Research. 1980. Special Issue on Stress in the Lithosphere 85:6083–435.

Journal of Geophysical Research. 1992. Special Issue on the World Stress Map Project. 97:11,703-12,013.

Journal of Geophysical Research. 1995. Special Section: Mechanical Involvement of Fluids in Faulting. 100:12,831-13,132.

Julian, B. R. 1983. Evidence for dyke intrusion earthquake mechanisms near Long Valley caldera, California. Nature 303:323–25.

Junner, N. R. 1941. The Accra earthquake of 22nd June, 1939. Gold Coast Geol. Survey Bull. 13:1-67.

Kadinsky-Cade, K., Reilinger, R., and Isacks, B. 1985. Surface deformation associated with the November 23, 1977, Caucete, Argentina, earthquake sequence. Jour. Geophys. Res. 90:12,691-700.

Kagan, Y. Y., and Jackson, D. D. 1991. Seismic gap hypothesis: ten years after. Jour. Geophys. Res. 96:21,419-31.

Kahle, J. E. 1975. Surface effects and related geology of the Lakeview fault segment of the San Fernando fault zone. Calif. Div. Mines and Geology Bull. 196:121-35.

Kahle, J. E., Bryant, W. A., and Hart, E. W. 1986. Fault rupture associated with the July 21, 1986 Chalfant Valley earthquake, Mono and Inyo counties, California. California Geology 39:243–45.

Kahle, J. E., Wills, C. J., Hart, E. W., Treiman, J. A., Greenwood, R. B., and Kaumeyer, R. S. 1988. Preliminary report—surface rupture Superstition Hills earthquakes of November 23 and 24, 1987, Imperial County, California. California Geology 41:75–84.

Kakimi, T., Kinugasa, Y., Suzuki, Y., Kodama, K., and Mitsunashi, T. 1977. Geological researches on the Izu-Hanto-oki earthquake of 1974. Geol. Survey Japan Spec. Rep. 6:1-51(J, Eng. abs.).

Kamata, S., Hosono, T., Ito, K., and Hayakawa, M. 1966. A study on the geologic structures by sonic exploration around the epicenter of the Niigata earthquake. *In* Research Group of Niigata Earthquake. Report of the Geological Survey on the Niigata earthquake. Geol. Survey of Japan Spec. Rept. 3:32–42.

Kanamori, H. 1971a. Seismological evidence for a lithospheric normal faulting: the Sanriku earthquake of 1933. Phys. Earth. Planet. Int. 4:289-300.

Kanamori, H. 1971b. Faulting of the great Kanto earthquake of 1923 as revealed by seismological data. Bulletin of the Earthquake Research Institute, Tokyo University 48:115-25.

Kanamori, H. 1972. Determination of effective tectonic stress associated with earthquake faulting, the Tottori earthquake of 1943. Phys. Earth Planet. Interiors 5:426-34.

Kanamori, H. 1973. Mode of strain release associated with major earthquakes in Japan. Earth Planet. Sci. Ann. Rev. 1:213-39.

Kanamori, H., and Anderson, D. L. 1975. Theoretical basis of some empirical relations in seismology. Seismological Society of America Bulletin 65:1073-96.

Kanamori, H., Ekström, G., Dziewonski, A., Barker, J. S., and Sipkin, S. A. 1993. Seismic radiation by magma injection: an anomalous seismic event near Tori Shima, Japan. Jour. Geophys. Res. 98:6511-22.

Kanamori, H., and Given, J. W. 1982. Analysis of long-period seismic waves excited by the May 18, 1980, eruption of Mount St. Helens—a terrestrial monopole? Jour. Geophys. Res. 87:5422-32.

Kappel, E. S., and Normark, W. R. 1987. Morphometric variability within the axial zone of the southern Juan de Fuca Ridge: Interpretation from SeaMARC II, SeaMARC I, and deep-sea photography. Jour. Geophys. Res. 92:11,292-302.

Karig, D. E., and Angevine, C. L. 1985. Geologic constraints on subduction rates in the Nankai trough. Initial Reports, Deep Sea Drilling Project, 87:927-40.

Karlin, R. E., and Abella, S. E. B. 1992. Paleoearthquakes in the Puget Sound region recorded in sediments from Lake Washington, U.S.A. Science 258:1617-20.

Kastens, K. A. 1984. Earthquakes as a tariggering mechanism for debris flows and turbidites on the Calabrian ridge. Marine Geology 55:13-33.

Kastowo, and Leo, G. 1973. Geologic Map of the Padang Quadrangle, Sumatra. Bandung, Indonesia: Geological Survey of Indonesia, Directorate General of Mines, Ministry of Mines,

Kawai, N. 1984. Paleomagnetic study of the Lake Biwa sediments. *In* Horie, S., and Junk., W., eds. Lake Biwa. Dordrecht: Netherlands, pp. 399-416.

Kawasumi, H., ed. 1973. General report on the Niigata earthquake of 1964. Tokyo: Tokyo Electrical Engineering College Press.

Kearey, P., and Vine, F. J. 1990. Global Tectonics. Oxford: Blackwell Scientific Publications, 302 p.

Kebeasy, R. M., Maamoun, M., Ibrahim, E., Megahed, A., Simpson, D. W., and Leith, W. S. 1987. Earthquake studies at Aswan Reservoir. Journal of Geodynamics 7:173-93.

Keefer, D. K. 1984. Landslides caused by earthquakes. Geol. Soc. America Bull. 95:406-71.

Keller, E. A. 1986. Investigation of active tectonics: Use of surficial earth processes. *In* Wallace, R. E., ed. Active Tectonics. Washington, D.C.: National Academy Press, 136-47.

Kelsey, H. M., and Carver, G. A. 1988. Late Neogene and Quaternary tectonics associated with northward growth of the San Andreas Transform Fault, northern California. Jour. Geophys. Res. 93:4797-4819.

Ketin, I. 1966. 6 Ekim 1964 Manyas depremi enasinda ziminde meydana gelen tansiyon çatlaklari. Türkiye Jeoloji Kurumu Bülteni 10:1-2.

Ketin, I. 1969. Über die nordanatolische Horizontalverschiebung. Mineral. Res. Explor. Inst. Turk. 72:1-28.

Ketin, I., and Abdüsselamoglu, S. 1969. 23 Mart 1969 Demicri ve 28 Mart 1969 Alasehir Sarigöl depremleri Hakkinda Makro-Sismik Gözlemler. Maden Mecumuasi 4(5):21-26, Geol. Dept. Univ. Istanbul.

Ketin, I., and Roesli, F. 1954. Makroseismische Untersuchungen über das nordwestanatolische Beben wom 18 März 1953. Eclogae Geol. Helvetiae 46:187-208.

Khil'ko, S. D., Kurushin, R. A., Kochetkov, V. M., Balzhinnyam, I., and Monkoo, D. 1985. Strong earthquakes, paleoseismogeological and macroseismic data, in Earthquakes and the Bases for Seismic Zoning of Mongolia. The Joint Soviet-Mongolian Scientific Geological Research Expedition, Transactions 41, Moscow Nauka 19-83.

King, G. C. P., Stein, R. S. and Rundle, J. B. 1988. The growth of geological structures by repeated earthquakes: 1. Conceptual framework. Jour. Geophys. Res. 93:13,307-18.

King, G. C. P., Ouyang, Z. X., Papadimitriou, P., Deschamps, A., Gagnepain, J., Houseman, G., Jackson, J. A., Soufleris, C., and Virieux, J. 1985. The evolution of the Gulf of Corinth (Greece): An aftershock study of the 1981 earthquakes. Geophysical Journal of the Royal Astronomical Society 80:677-93.

King, J., Loveday, I., and Schuster, R. L. 1989. The 1985 Bairaman landslide dam and resulting debris flow, Papua New Guinea. Quarterly Jour. Engineering Geology 22:257-70.

Kinugasa, Y. 1976. The Izu-Hanto-oki earthquake of 1974 and Irozaki earthquake fault. Geol. Soc. Japan Mem. 12:139–49 (J, Eng. abs.).

Kinugasa, Y. 1991. Active faulting, seismicity and volcanic activities at a plate convergence margin as a base for natural hazard mitigation. Summaries of the International Seminar on Earthquake Prediction and Hazard Mitigation Technology, Tsukuba Science City, Japan March 5-March 8, 1991:543–48.

Kinugasa, Y., Tsukuda, E., and Yamasaki, H. 1992. Neotectonic Map of Japan. Geological Survey of Japan, Asakura Publishing Company, Ltd.

Kirby, S. H. 1983. Rheology of the lithosphere. Reviews of Geophysics and Space Physics 21: 1458–87.

Kneupfer, P. L. K. 1988. Estimating ages of late Quaternary stream terraces from analysis of weathering rinds and soils. Geol. Soc. America Bull.100:1224–136.

Kobayashi, K. 1983. Cycles of subduction and Cenozoic arc activity in the northwestern Pacific margin. *In* Hilde, T. W. C., and Uyeda, S., eds. Geodynamics of the western Pacific-Indonesian region. Am. Geophys. Union Geodynamics Series 11:287–301.

Koçiaj, S., and Sulstarova, E. 1980. The earthquake of June 1, 1905, Shkodra, Albania; intensity distribution and macroseismic epicentre. Tectonophysics 67:319–32.

Koto, B. 1893. On the cause of the great earthquake in central Japan, 1891. Journal of the College of Science, Imperial University of Japan 5(4):296–353.

Koukis, G., Rondoyanni, T., and Delibasis, N. 1990. Surface seismic strike-slip fault motions related to the 1983 Akarnania (western Greece) earthquakes of small magnitude. Annales Tectonicae 4:43–51.

Kuchai, V. K. 1969. Results of repeated examination of the remaining deformation in the pleistoseist of the Kebin earthquake. Geol. Geophys. 101–08 (R).

Kurz, M. D. 1986a. Cosmogenic helium in a terrestrial igneous rock. Nature 320:435–39.

Kurz, M. D. 1986b. *In-situ* production of terrestrial cosmogenic helium and some applications to geochronology. Geochimica et Cosmochimica Acta 50:2855–62.

Kurz, M. D., Colodner, D., Trull, T. W., Moore, R., and O'Brien, K. 1990. Cosmic ray exposure dating with *in-situ* produced cosmogenic ^{3}He: results from young Hawaiian lava flows. Earth and Planet. Sci. Lett. 97:177–89.

Kurz, M. D., and Brook, E. J. 1994. Surface exposure dating with cosmogenic nuclides. *In* Beck, C., ed. Dating in Exposed and Surface Context. Albuquerque: University of New Mexico Press, pp. 139–59.

Lacassin, R., Leloup, P., and Tapponnier, P. 1993. Bounds on strain in large Tertiary shear zones of south east Asia from boudinage restoration. Jour. of Struct. Geo. 15:677–92.

Lachenbruch, A., and Thompson, G. 1972. Oceanic ridges and transform faults; their intersection angles and resistance to plate motion. Earth and Planet. Sci. Lett. 15:116–22.

Lagerbäck, R. 1992. Dating of late Quaternary faulting in northern Sweden. Geol. Soc. London Jour. 149:285–91.

Lahr, K. M., Lahr, J. C., Lindh, A. G., Bufe, C. G., and Lester, F. W. 1976. The August 1975 Oroville earthquakes. Seismological Society of America Bulletin 66:1085–99.

Lajoie, K. R. 1986. Coastal tectonics. *In* Wallace, R. E., ed. Active Tectonics. Washington, D.C.: National Academy Press, pp. 95–124.

Lal, D. 1991. Cosmic ray labeling of erosion surfaces, *in situ* production rates and erosion models. Earth and Planet. Sci. Lett. 104:424–39.

Lambeck, K. 1988. Geophysical Geodesy: The Slow Deformation of the Earth. Oxford: Clarendon Press, 718 p.

Langbein, J. O., Burford, R. O., and Slater, L. E. 1990. Variations in fault slip and strain accumulation at Parkfield, California: Initial results using two-color laser geodimeter measurements, 1984–1988. Jour. Geophys. Res. 95:2533–52.

Langer, C. J., Bonilla, M. G., and Bollinger, G. A. 1987. Aftershocks and surface faulting associated with the intraplate Guinea, west Africa, earthquake of 22 December 1983. Seismological Society of America Bulletin 77:1579–1601.

Larson, K. M., and Webb, F. H. 1992. Deformation in the Santa Barbara Channel from GPS measurements 1987–1991. Geophys. Res. Lett. 19:1491–94.

Lawrence, R. D., Khan, S. H., and Nakata, T. 1992. Chaman fault, Pakistan-Afghanistan, in Bucknam, R. C., and Hancock, P. L., eds., Major active faults of the world. Results of IGCP Project 206. Annales Tectonicae Special Issue, Supplement to 6:196–223.

Lawson, A. C. 1906. The California earthquake. Science 23:961-67.

Lawson, A. C., chairman. 1908. The California earthquake of April 18, 1906—report of the State Earthquake Investigation Committee. Carnegie Institute, Washington, Pub. 87, v. 1.

Lay, T., Kanamori, H., and Ruff, L. 1982. The asperity model and the nature of large subduction zone earthquakes. *In* Earthquake Prediction Research 1:3-71. Tokyo, Japan: Terra Scientific Publishing Co.

Leblanc, G., and Anglin, F. 1978. Induced seismicity at Manic-3 reservoir, Quebec. Seismological Society of America Bulletin 68:1469-85.

Le Dain, A., Tapponnier, P., and Molnar, P. 1984. Active faulting and tectonics of Burma and surrounding regions. Jour. Geophys. Res. 89:453-72.

Le Dain, A., Robineau, B., and Tapponnier, P. 1979. Les effets tectoniques de l'événement sismique et volcanique de novembre 1978 dans le rift d'Asal-Ghoubbet. Soc. Géol. France Bulletin 22:817-822.

Lee, W. H. K., and Lahr, J. C. 1975. HYPO71 (Revised): A computer program for determining hypocenter, magnitude, and first motion pattern of local earthquakes. U. S. Geol. Survey Open-File Report 75-311, 114 p.

Lee, W. H. K., Johnson, C. E., Henyey, T. L., and Yerkes, R. F. 1978. A preliminary study of the Santa Barbara, California, earthquake of August 13, 1978 and its major aftershocks. U. S. Geol. Survey Circular 797, 11 p.

Leith, W. 1995. Sakhalin earthquake renews concerns about seismic safety in the former Soviet Union. EOS, Trans. Am. Geophys. Union 76:257-58.

Lenson, G. J., and Otway, P. M. 1971. Earthshift and post-earthshift deformation associated with the May 1968 Inangahua earthquake, New Zealand. Royal Soc. New Zealand Bull. 9:107-16.

Lenson, G. J., and Suggate, R. P. 1968. Preliminary reports on the Inangahua earthquake, New Zealand. Dept. of Scientific and Industrial Research Bull. 193:17-36.

Leopold, L. B., and Maddock, T. 1953. Hydraulic geometry of stream channels and some physiographic implications. U. S. Geol. Survey Prof. Paper 252, 57 p.

Le Pichon, X., Iiyama, T., et al. 1987. Nankai Trough and Zenisu Ridge: A deep-sea submersible survey.: Earth and Planet. Sci. Lett. 83:285-99.

Le Pichon, X., Bergerat, F., and Roulet, M.-J. 1988. Plate kinematics and tectonics leading to the Alpine belt formation; a new analysis. Geol. Soc. America Spec. Paper 218:111-31.

Lépine, J. C., Ruegg, J.-C., and Abdallah, A. M. 1980. Sismicité du rift d'Asal-Ghoubbet pendant la crise sismo-volcanique de novembre 1978. Bull. Soc. Géol. France 22:809-16.

Lewis, J. D., Daetwyler, N. A., Bunting, J. A., and Moncrieff, J. S. 1981. The Cadoux earthquake, 2 June 1979. Geol. Survey of Western Australia Report 11, 133 p.

Lewis, S. D., Ladd, J. W., and Bruns, T. R. 1988. Structural development of an accretionary prism by thrust and strike-slip faulting: Shumagin region, Aleutian trench. Geol. Soc. America Bull. 100:767-82.

Li T., Du Q., You Z., Zhang C., and Huang Q. 1992a. Activity of the Selaha-Kangding-Moxi (SKM) fault. Research on Active Fault 2:1-14 (C., Eng. abs.).

Li T., Du C., You Z., Zhang C., and Huang Q. 1992b. Recent activity of the Zheduotang fault and the earthquake of magnitude 7.5 in 1955. Research on Active Fault 2:15-23 (C., Eng. abs.).

Li, Y.-G., Teng. T.-L., and Henyey, T.L. 1994., Shear-wave splitting observations in the northern Los Angeles basin, southern California. Seismological Society of America Bulletin 84:307-23.

Liddicoat, J. C. 1991. Paleomagnetic dating. *In* Morrison, R. B., ed. Quaternary Nonglacial Geology: conterminous U. S.. Geol. Soc. America Decade of North American Geology K-2:60-61.

Liddicoat, J. C. 1992. Mono Lake excursion in Mono Basin, California, and at Carson Sink and Pyramid Lake, Nevada. Geophys. J. Int. 18:442-52.

Lienkaemper, J. J., and Prescott, W. H. 1989. Historic surface slip along the San Andreas fault near Parkfield, California. Jour. Geophys. Res. 94:17,647-170.

Lienkaemper, J. J., Pezzopane, S. K., Clark, M. M., and Rymer, M. J. 1987. Fault fractures formed in association with the 1986 Chalfant Valley, California, earthquake sequence: Preliminary report. Seismological Society of America Bulletin 77:297-305.

Liew, P. M., Pirazzoli, P. A., Hsieh, M. L., Arnold, M., Barusseau, J. P., Fontugne, M., and Giresse, P. 1993. Holocene tectonic uplift deduced from elevated shorelines, eastern Coastal Range of Taiwan. Tectonophysics 222:55-68.

Lin, J., and Stein, R. S. 1989. Coseismic folding earthquake recurrence, and the 1987 source mechanics at Whittier Narrows, Los Angeles basin, California. Jour. Geophys. Res. 94:9614-32.

Lindvall, S., Rockwell, T., and Hudnut, K. 1989. Evidence for prehistoric earthquakes on the Superstition Hills fault from offset geomorphic features. Seismological Society of America Bulletin. 79:342-61.

Liu G., Yu S., Zhang S., Dou S., Xu Y., and Fan J. 1991. The North Wutaishan piedmont active fault zone in Shanxi. Research on Active Fault 1:118-30 (C, Eng. abs.).

Liu H., Jia Y., Chen Y., Su X., Dai H., and Hou K. 1995. Surface rupture zone associated with the 1927 Gulang (M_s = 8) earthquake. Research on Active Fault 4:79-91.

Liu Q. 1993. Paléoclimat et contraintes chronologiques sur les mouvements récents dans l'ouest du Tibet: Failles du Karakorum et de Longmu Co-Gozha Co, lacs et pull-apart de Longmu Co et de Sumxi Co. Thése de doctorat, Univ. Paris VII, 358 p.

Lomnitz, C. 1994. Fundamentals of Earthquake Prediction. New York: John Wiley & Sons, 321 p.

Lomnitz, C., and Hashizume, M. 1985. The Popayán, Colombia, earthquake of 31 March 1983. Seismological. Society of America Bulletin 75:1315-26.

Long D., and Deng T. 1990. A preliminary study on the 1786 Kangding earthquake deformation characteristics. Jour. Seism. Res. 13(1):50-60. (C, Eng. abs.).

Louderback, G. D. 1947. Central California earthquakes of the 1830's. Seismological Society of America Bulletin 37:33-74.

Loupekine, I. S. 1966. The Toro earthquake of 20 March 1966. UNESCO Rep. RP/CON 0766, Paris.

Ludwin, R. S., Weaver, C. S., and Crosson, R. S. 1991. Seismicity of Washington and Oregon. *In* Slemmons, D. B., Engdahl, E. R., Zoback, M. D., and Blackwell, D. D., eds. Neotectonics of North America. Boulder, Colorado: Geol. Soc. America Decade Map 1:77-98.

Lyberis, N., Yurur, T., Chorowicz, J., Kasapoglu, E., and Gundoglu, N. 1992. The East Anatolian fault: An oblique collisional belt. Tectonophysics 204:1-15.

Lyell, C. 1875. Principles of Geology. 12th ed.

Lyon-Caen, H., and Molnar, P. 1985. Gravity anomalies, flexure of the Indian plate, and the structure, support, and evolution of the Himalaya and Ganga basins. Tectonics 4:513-38.

Lyon-Caen, H., Armijo, R., Drakopoulos, J., Baskoutass, J., Delibassis, N., Gaulon, R., Kouskouna, V., Latoussakis, J., Makropoulos, K., Papadimitriou, P., Papanastassiou, D., and Pedotti, G. 1988. The 1986 Kalamata (South Peloponnesus) earthquake: Detailed study of a normal fault, evidences for east-west extension in the Hellenic arc. Jour. Geophys. Res. 93:14,967-15,000.

Maasha, N., and Molnar, P. 1972. Earthquake fault parameters and tectonics in Africa. Jour. Geophys. Res. 77:5731-43.

Machette, M. N. 1982. Quaternary and Pliocene faults in the La Jencia and southern part of the Albuquerque-Belen basins, New Mexico: Evidence of fault history from fault-scarp morphology and Quaternary geology. Field Conference Guidebook, New Mexico Geol. Society 33:161-70.

Machette, M. N. 1985. Calcic soils ofthe southwestern United States. *In* Weide, D. L., and Faber, M. L., eds. Quaternary soils and geomorphology of the American Southwest. Geol. Soc. America Special Paper 203:1-21.

Machette, M. N. 1989. Slope-morphometric dating. *In* Forman, S. L., ed. Dating Methods Applied to Quaternary Geologic Studies in the Western United States. Utah Geol. Mineral Survey 89-7:30-42.

Machette, M. N., Crone, A. J., and Bowman, J. R. 1993. Geologic investigations of the 1986 Marryat Creek, Australia, earthquake—implications for paleoseismicity in stable continental regions. U.S. Geol. Survey Bull. 2032-B, 29 p.

Machette, M. N., Personius, S. F., Nelson, A. R., Schwartz, D. P., and Lund, W. R. 1991. The Wasatch fault zone, Utah—segmentation and history of Holocene earthquakes. Jour. Structural Geol. 13:137-49.

Machette, M. N., and Rosholt, J. N. 1991. Quaternary geology of the Grand Canyon, in Morrison, R. B., ed., Quaternary nonglacial geology: Conterminous U. S.. Geol. Soc. America Decade of North American Geology K-2:397-401.

Machida, H., 1991. Recent progress in tephra studies in Japan. The Quaternary Research 30:141-49.

MacKay, M. E., Moore, G. F., Cochrane, G. R., Moore, J. C., and Kulm, L. D. 1992. Landward vergence, oblique structural trends, and tectonic segmentation in the Oregon margin accretionary prism. Implications and effect on fluid flow. Earth and Planet. Sci. Lett. 109:477-91.

Mackin, J. 1948. Concept of the graded river. Geol. Soc. America Bull. 59:463-512.

Madariaga, R. 1994. The dynamic origin of earthquake complexity. EOS, Trans. Amer. Geophys. Union 75:426-27.

Madin, I. P., Lawrence, R. D., and ur-Rehman, S. 1989. The northwestern Nanga Parbat-Haramosh massif: Evidence for crustal uplift at the northwestern corner of the Indian craton. Geol. Soc. America Spec. Paper 232:169-82.

Maemoku, H. 1988a. Holocene crustal movement in Muroto Peninsula, southwest Japan. Geographical Review of Japan 61(Ser. A), 10:747-69 (J, Eng. abs.).

Maemoku, H. 1988b. Holocene crustal movement around Cape Ashizuri, southwest Japan. Geographical Sciences 43:231-40 (J, Eng. abs.).

Maemoku, H., and Paladio, J. 1992. Raised coral reefs at Bolinao, northwestern Luzon Island of the Philippines. Geographical Sciences 47:183-189.

Maemoku, H., and Tsubono, K. 1990. Holocene crustal movements in the southern part of Kii Peninsula, outer zone of southwest Japan. Jour. Geography 99:349-369 (J, Eng. abs.).

Malloy, R. J. 1964. Crustal uplift southwest of Montague Island, Alaska: Science 146:1048-49.

Mankinen, E. A., and Dalrymple, G. B. 1979. Revised geomagnetic polarity time scale for the interval 0-5 my BP. Jour. Geophys. Res. 84:615-26.

Mann, P., and Burke, K. 1984. Neotectonics of the Caribbean. Reviews of Geophysics and Space Physics 22:309-62.

Mansur, C., Kaufman, R., and Schultz, J. 1956. Investigation of underseepage and its control, Lower Mississippi River levees. Tech. Memo 3-242. Vicksburg, Mississippi, Army Corps of Engineers, Waterways Experiment Station, 421 p.

Marco, S., Stein, M., Agnon, A., and Ron, H. 1996. Long term earthquake clustering: A 50,000 year paleoseismic record in the Dead Sea graben: Jour. Geophys. Research 101:6179-91.

Marshak, S., and Mitra, G. 1988. Basic methods of structural geology. Englewood Cliffs, N.J.: Prentice Hall, 446 p.

Marshall, P. 1933. The geological aspects of the Hawkes Bay earthquakes. Department of Scientific and Industrial Research Bulletin 43.

Martinson, D. G., Pisias, N. G., Hays, J. D., Imbrie, J., Moore, T. C., and Shackleton, N. J. 1987. Age dating and the orbital theory of the ice ages: Development of a high-resolution 0 to 300,000-year chronostratigraphy. Quaternary Research 27:1-29.

Masek, J. G., Isacks, B. L., and Fielding, E. J. 1994. Rift flank uplift in Tibet: Evidence for a viscous lower crust. Tectonics, v. 13, p. 659-67.

Massonet, D., Rossi, M., Carmona, C., Adragna, F., Peltzer, G., Feigl, K., and Rabaute, T. 1993. The displacement field of the Landers earthquake mapped by radar interferometry. Nature 364:138-42.

Mathews, W. H., and McTaggart, K. C. 1978. Hope rockslide, British Columbia, Canada. *In* Voight, B., ed. Rockslides and Avalanches—1. Natural Phenomena. New York: Elsevier Publishing Co., 259-75.

Matsuda, T. 1967. Geological aspect of the Matsushiro earthquake fault. Earthquake Research Inst., Univ. Tokyo Bull. 45:537-50. (J, Eng. abs.).

Matsuda, T. 1974. Surface faults associated with Nobi (Mino-Owari) earthquake of 1891, Japan. Earthquake Research Inst., Univ. Tokyo, Spec. Bull. 13:85-126 (J, Eng. abs.).

Matsuda, T., and Kinugasa, Y. 1991. Active faults in Japan. Episodes 14:199-204.

Matsuda, T., Ota, Y., Ando, M., and Yonekura, N. 1978. Fault mechanism and recurrence time of major earthquakes in southern Kanto district, Japan, as deduced from coastal terrace data. Geol. Soc. America Bull. 89:1610-18.

Matsuda, T., Yamazaki, H., Nakata, T., and Imaizumi, T. 1980. The surface faults associated with the Rikuu earthquake of 1896. Earthquake Res. Inst., Univ. Tokyo Bull. 55:795-855.

Matsuda, T., Yui, M., Matsushima, Y., Imanaga, I., Hirata, D., Togo, M., Kashima, K., Matsubara, A., Nakai, N., Nakamura, T., and Matsuoka, K. 1988. Subsurface study of Isehara fault, Kanagawa prefecture, detected by drilling—depositional environments during the last 7000 years and fault displacement associated with the Gangyou earthquake in A.D. 878. Earthquake Res. Inst., Univ. Tokyo, Bull. 63:145-82.

Matsumoto, A., Uto, K., and Shibata, K. 1989. K-Ar dating by peak comparison method—new technique applicable to rocks younger than 0.5 Ma. Geol. Soc. Japan Bull. 40:565-79.

Matsu'ura, M., Iwasaki, T., Suzuki, Y., and Sato, R. 1980. Statical and dynamical study on faulting mechanism of the 1923 Kanto earthquake. Journal of Physics of the Earth 28:119-43.

Matsuzawa, T., Kono, T., Hasegawa, A., and Takagi, A. 1990. Subducting plate boundary beneath the northeastern Japan arc estimated from SP converted waves. Tectonophysics 181:123-33.

Matti, J. C., and Morton, D. 1993. Paleogeographic evolution of the San Andreas fault in southern California: A reconstruction based on a new cross-fault correlation. *In* The San Andreas Fault System: Displacement, Palinspastic Reconstruction and Geologic Evolution. Geol. Soc. America Memoir 178:107-60

Matumoto, T. 1959. Tesikaga earthquake of Jan. 31, 1959. Earthquake Res. Inst. Bull. 37:531-44 (J, Eng. abs.).

Mayer, L. 1986. Tectonic geomorphology of escarpments and mountain fronts. *In* Wallace, R. E., ed. Active Tectonics. Washington, D.C.: National Academy Press, 125-35.

Mayer-Rosa, D., Slejko, D., and Zonno, G. 1993. Assessment of seismic hazard for the Sannio-Matese area, southern Italy (Project "TERESA"). Annali di Geofisica 36:199-209.

McCaffrey, R. 1991. Slip vectors and stretching of the Sumatran forearc. Geology 19:881-84.

McCaffrey, R. 1993. On the role of the upper plate in great subduction zone earthquakes. Jour. Geophys. Res. 98:11,953-66.

McCaffrey, R., and Goldfinger, C. 1995. Forearc deformation and great subduction earthquakes: Implications for Cascadia offshore earthquake potential. Science 267:856-59.

McCalpin, J. P. 1982. Quaternary geology and neotectonics of the western flank of the northern Sangre de Cristo Mountains, south-central Colorado. Colorado School of Mines Quarterly 77, 97 p.

McCalpin, J. P., and Rice, J. B., Jr. 1987. Spatial and temporal analysis of 1200 landslides in a 900 km^2 area, middle Rocky Mountains. *In* International Conference and Field Workshop on Landslides, 5th, Christchurch, New Zealand, 1987. Proceedings, pp. 137-46.

McCalpin, J. P., Robinson, R. M., and Gan, J. D. 1987. Neotectonics of the Hansel Valley-Pocatello Valley corridor, northern Utah and southern Idaho. U. S. Geol. Survey Open-File Rept. 87-585, v.1, p. G1-G44.

McCann, W. R., Nishenko, S. P., Sykes, L. R., and Krause, J. 1979. Seismic gaps and plate tectonics: Seismic potential for major boundaries. Pure and Applied Geophysics 117:1090-1147.

McClay, K. R., ed. 1992. Thrust Tectonics. London: Chapman & Hall, 447 p.

McElhinny, M. W., and Senanayake, W. E. 1982. Variations in the geomagnetic dipole 1: The past 50.000 years. J. Geomagn. Geoelectr. 34:29-51.

McGill, J. T. 1989. Geologic maps of the Pacific Palisades area, Los Angeles, California. U.S. Geol. Survey Misc. Inv. Map I-1828, scale 1:4800, 2 sheets.

McGuire, R. 1978. FRISK: Computer program for seismic risk analysis, using faults as earthquake sources. U.S. Geol. Survey Open-File Report 78-1007.

McKay, A. 1890. On the earthquake of September 1888, in the Amuri and Marlborough districts of the South Island. New Zealand Geol. Survey Reports of Geological Exploration 1885-1889, 20:1-16.

McKenzie, D., and Morgan, J. 1969. Evolution of triple junctions. Nature 224:125-133.

Medwedeff, D. A. 1989. Growth fault-bend folding at southeast Lost Hills, San Joaquin Valley, California. American Assoc. Petroleum Geologists Bull. 73:54-67.

Meghraoui, M. 1991. Blind reverse faulting system associated with the Mont Chenoua-Tipaza earthquake of 29 October 1989 (north-central Algeria). Terra Nova 3:84-93.

Meghraoui, M., Jaegy, R., Lammali, K., and Albarède, F. 1988. Late Holocene earthquake sequences on the El Asnam (Algeria) thrust fault. Earth and Planet Sci. Lett. 90:187-203.

Meisling, K. 1980. Possible emplacement history of a sandblow structure at Pallett Creek, California. *In* Abbott, P. L., ed. Geological Excursions in the Southern California Area. San Diego, California: Department of Geological Sciences, San Diego State University:63-66.

Mercier, J. L., Mouyaris, N., Simeakis, C., Roundoyannis, T., and Angelidhis, C. 1979. Intraplate deformation: A quantitative study of the faults activated by the 1978 Thessaloniki earthquakes. Nature 278:45-48.

Mercier, J. L., Sebrier, M., Lavenu, A., Cabrera, J., Bellier, O., Dumont, J.-F., and Macharé, J. 1992. Changes in the tectonic regime above a subduction zone of Andean type: The Andes of Peru and Bolivia during the Pliocene-Pleistocene. Jour. Geophys. Res. 97:11,945-82.

Merritts, D., and Ellis, M. 1994. Introduction to special section on tectonics and topography. Jour. Geophys. Res. 99:12,135-.41.

Meyer, B., 1991. Mécanismes des grands tremblements de terre et du raccourcissement crustal oblique au bord nord-est du Tibet: Thèse, Université Paris VI, 129 p.

Meyer, B., Tapponnier, P., Gaudemer, Y., Mercier, N., Valladas, H., Suo S., and Chen Z. 1991. 1932 Chang Ma (M = 7.6) earthquake surface breaks and implications on regional seismic hazard. *In* Proceedings of the first I.N.S.U.-S.S.B. Workshop, Earthquakes from Source Mechanism to Seismic Hazard, October 22-25, 1991. Institut de Physique du Globe de Paris.

Middlemiss, C. S. 1910. The Kangra earthquake of 4th April, 1905. Geol. Survey India Memoir 37, 409 p.

Milsom, J. 1991. Oblique collision in New Guinea—implications for hydrocarbon exploration. *In* Cosgrove, J., and Jones, M., eds. Neotectonics and Resources. London: Belhaven Press, 257–67.

Minoura, K., and Nakaya, S. 1991. Traces of tsunami preserved in inter-tidal lacustrine and marsh deposits: Some examples from northeast Japan. Jour. Geology 99:265–87.

Minster, J. B., and Jordan, T. H. 1978. Present-day plate motions. Jour. Geophys. Res. 83-5331–54.

Mitchell, C. E., Vincent, P., Weldon, R. J., II, and Richards, M. A. 1994. Present-day vertical deformation of the Cascadia margin, Pacific Northwest, United States. Jour. Geophys. Res. 99:12, 257–77.

Mix, A. C. 1987. The oxygen-isotope record of glaciation. *In* Ruddiman, W. F., and Wright, H. E., Jr., eds. North America and adjacent oceans during the last deglaciation. Geol. Soc. America Decade of North American Geology K-3:111–35.

Molnar, P. 1990. A review of the seismicity and the rates of active underthrusting and deformation at the Himalaya. Journal of Himalayan Geology 1: 131–54.

Molnar, P., and Deng Q. 1984. Faulting associated with large earthquakes and the average rate of deformation in central and eastern Asia. Jour. Geophys. Res. 89:6203–27.

Molnar, P., and Tapponnier, P. 1975. Cenozoic tectonics of Asia: Effects of a continental collision. Science 189:419–26.

Molnar, P., and Tapponnier, P. 1978.Active tectonics of Tibet. Jour. Geophys. Res. 83:5361–75.

Monger, J. W. H., and Francheteau, J., eds. 1983. Circum-Pacific orogenic belts and evolution of the Pacific Ocean basin. Am. Geophys. Union Geodynamics Series 18:165 p.

Montessus de Ballore, F. de. 1924. La Géologie Sismologique. Paris: Librairie Armand Colin, 488 p., 118 fig., 16 pl.

Mooney, W. D., and Meissner, R. 1992. Multi-genetic origin of crustal reflectivity: A review of seismic reflection profiling of the continental lower crust and Moho. *In* Fountain, D. M., Arculus, R., and Kay, R. W., eds. Continental Lower Crust. Amsterdam: Elsevier, pp. 45–79.

Moore, G. F., Shipley, T. H., Stoffa, P. L., Karig, D. E., Taira, A., Kuramoto, S., Tokuyama, H., and Suyehiro, K. 1990. Structure of the Nankai trough accretionary zone from multichannel seismic reflection data. Jour. Geophys. Res. 95:8753–65.

Morante, E. M., and Allen, C. R. 1974. Displacement of the Philippine fault during the Ragay Gulf earthquake of 17 March 1973. Geol. Soc. America Abs. with Programs 7:744–45.

Mori, J., Wald, D. J., and Wesson, R. L. 1995. Overlapping fault planes of the 1971 San Fernando and 1994 Northridge, California earthquakes. Geophys. Research Letters 22:1033–36

Morrison, R. 1991. Introduction. *In* Morrison, R., ed. Quaternary Nonglacial Geology. Conterminous U.S. Geological Society of America Decade of North American Geology K-2:1–12.

Mount, V. S., and Suppe, J., 1992. Present-day stress orientations adjacent to active strike-slip faults: California and Sumatra. Jour. Geophys. Res. 97:11,995–12,013.

Mouyaris, N., Papastamatiou, D., and Vita-Finzi, C. 1992, The Helice fault? Terra Nova 4:124–29.

Moya, J. C., Langridge, R. M., Weldon, R. J., II, and Suarez, G. 1995. Preliminary paleoseismic investigation of the Acambay fault, and the effects of the 1912 Acambay earthquake, central trans-Mexican volcanic belt. Geol. Soc. America Abs. with Programs 27 (6):A-282.

Mueller, K. J., and Rockwell, T. K. 1995. Late Quaternary activity of the Laguna Salada fault in northern Baja California, Mexico. Geol. Soc. America Bull. 107:8–18.

Muhs, D. R. 1991. Amino acid geochronology of fossil mollusks. *In* Morrison, R. B., ed. Quaternary nonglacial geology: Conterminous U. S. Geol. Soc. America Decade of North American Geology K-2:65–68.

Muhs, D. R., Rosholt, J. N., and Bush, C. A. 1989. The uranium-trend dating method: principles and application for southern California marine terrace deposits. Quaternary International 1:19–34.

Muhs, D. R., Kelsey, H. M., Miller, G. H., Kennedy, G. L., Whelan, J. F., and McInelly, G. W. 1990.

Age estimates and uplift rates for late Pleistocene marine terraces: Southern Oregon portion of the Cascadia forearc. Jour. Geophys. Res. 95:6685-98.

Muir, J. 1912. The Yosemite. The Century Company, republished by Doubleday and Co., Inc., New York.

Müller, B., Zoback, M. L., Fuchs, K., Mastin, L., Gregersen, S., Pavoni, N., Stephanson, O., and Ljunggren, C. 1992. Regional patterns of tectonic stress in Europe. Jour. Geophys. Res. 97:11,783-803.

Muller, J, 1895. Nota betreffende de verplaasting van eenige traiangulatie pilaren in de residentie Tapanuli tgv de aardbeving van 17 Mei 1892. Natuurk. Tijdscht. v. Ned. Ind., 54:299-307.

Mushketov, I. V. 1890. Verny earthquake: 28 May (9 June) 1887. Commission of the Geology Committee, Leningrad, USSR, 154 p. (R).

Myers, W. B., and Hamilton, W. 1964. Deformation accompanying the Hebgen Lake earthquake of August 17, 1959. U. S. Geol. Survey Prof. Paper 435-I, 55-98.

Nábelék, J., Chen, W.-P., and Ye, H. 1987. The Tangshan earthquake sequence and its implications for the evolution of the North China basin. Jour. Geophys. Res. 92:12,615-128.

Naeser, C. W., Bryant, B., Crittenden, M. D., Jr., and Sorensen, M. L. 1983. Fission-track ages of apatite in the Wasatch Mountains, Utah. *In* Miller, D. M., Todd, V. R., and Howard, K. A., eds. Tectonic and Stratigraphic Studies in the Eastern Great Basin:. Geol. Soc. America Memoir 157:29-36.

Naeser, C. W., and Naeser, N. D. 1988. Fission-track dating of Quaternary events. *In* Easterbrook. D. J., ed. Dating Quaternary Sediments. Geol. Soc. America Special Paper 227:1-11.

Nakamura, K. 1969. Arrangement of parasitic cones as a possible key to a regional stress field. Bulletin of the Volcanological Society of Japan 14:8-20.

Nakamura, K. 1977. Volcanoes as possible indicators of tectonic stress orientation—principle and proposal. Journal of Volcanological and Geothermal Research 2:1-16.

Nakamura, K. 1983. Possible nascent trench along the eastern Japan Sea as the convergent plate boundary between Eurasian and North American plates. Bull. Earthquake Res. Inst. 58:711-22 (J).

Nakata, T. 1989. Active faults of the Himalaya of India and Nepal. Geol. Soc. America Spec. Paper 232:243-64.

Nakata, T., Tsutsumi, H., Khan, S.H., and Lawrence, R.D. 1991. Active faults of Pakistan: Map sheets and inventories. Research Center for Regional Geography, Hiroshima University, Spec. Pub. 21, 141 p.

Nakata, T., Tsutsumi, H., Punongbayan, R.S., Rimando, R.E., Daligdig, J., and Daag, A. 1990. Surface faulting associated with the Philippine earthquake of 1990. Jour. Geography 99(5):95-112. (J, Eng. abs.)

Namson, J., and Davis, T. 1988a. A structural transect across the western Transverse Ranges, California: Implications for lithospheric kinematics and seismic risk evaluation. Geology 16:675-79.

Namson, J.S., and Davis, T.L. 1988b. Seismically active fold and thrust belt in the San Joaquin Valley, central California. Geol. Soc. America Bull. 100:257-73.

Narr, W. 1992. Deformation of basement in basement-involved, compressive structures. Geol. Soc. America Special Paper 280:107-24.

Narr, W., and Suppe, J. 1994. Kinematics of basement-involved compressive structures. American Journal of Science 294:802-60.

Nash, D. B. 1980. Morphologic dating of degraded normal fault scarps. Jour. Geology 88:353-60.

National Research Council. 1985. Liquefaction of soils during earthquakes. Washington, D.C.: National Academy Press, 240 p.

Natsag-Yüm, L., Balzhinnyam, I., and Monkho, D. 1971. Earthquakes of Mongolia. *In* Seismic Regionalization of Ulan-Bator:54-82 (R).

Natland, M. L., and Kuenen, P. H. 1951. Sedimentary history of the Ventura basin, California, and the action of turbidity currents. Soc. Econ. Paleontologists and Mineralogists Special Pub. 2:76-107.

Nelson, A. R. 1992. Discordant ^{14}C ages from buried tidal-marsh soils in the Cascadia subduction zone, southern Oregon coast. Quaternary Research 38:74-90.

Nelson, A. R., and Manley, W. F. 1992. Holocene coseismic and aseismic uplift of Isla Mocha, south-central Chile. Quaternary International 15/16:61-76.

Nelson, A. R. and Personius, S. P. 1992. Earthquake recurrence and Quaternary deformation in the Cascadia subduction zone, coastal Oregon. U. S. Geological Survey Open-File Report 92-258:512-18.

Newmark, N. M. 1965. Effects of earthquakes on dams and embankments. Geotechnique 15 (2):139-60.

Ni, J. F., and Barazangi, M. 1984. Seismotectonics of the Himalayan collision zone: Geometry of the underthrusting Indian plate beneath the Himalaya. Jour. Geophys. Res. 89:1147-63.

Ni, J. F., and Barazangi, M. 1986. Seismotectonics of the Zagros continental collision zone and a comparison with the Himalayas. Jour. Geophys. Res. 91:8205-18.

Nikonov, A. A., and Shebalina, T. Y. 1979. Lichenometry and earthquake age determination in central Asia. Nature 280:675-77.

Nishenko, S. P., and Buland, R. 1985. A generic recurrence interval distribution for earthquake forecasting. Seismological Society of America Bulletin 77:1382-99.

Nishiizumi, K., Winterer, E. L., Kohl, C. P., Klein, J., Middleton, R., Lal, D., and Arnold, J. 1989. Cosmic ray production rates of ^{10}Be and ^{26}Al in quartz from glacially polished rocks. Jour. Geophys. Res. 94:17,907-16.

Nishiizumi, K., Kohl, C. P., Arnold, J., Klein, J., Fink, D., and Middleton, R. 1991. Cosmic ray produced ^{10}Be and ^{26}Al in Antarctic rocks: exposure and erosion history. Earth and Planet. Sci. Lett. 104:440-54.

Normark, W. R., Morton, J. L., and Ross, S. L. 1987. Submersible observations along the southern Juan de Fuca Ridge: 1984 *Alvin program*. Jour. Geophys. Res. 92:11,283-90.

Nossin, J. J. 1971. Outline of the geomorphology of the Doon Valley, northern U.P., India. Zeitschrift Geomorphologie, N.F. 12:18-50.

Nowroozi, A. A., and Mohajer-Ashjai, A. M. 1980. Faulting of Kurizan and Koli (Iran) earthquakes of November 1979, a field report. Bull. du Bureau de Recherches Geologiques et Minieres (2e series), Sec. IV, Geologie General 2:91-99.

Nowroozi, A. A., and Mohajer-Ashjai, A. M. 1985. Fault movements and tectonics of eastern Iran: Boundaries of the Lut plate. Geophys. Jour. Royal Astron. Soc. 83:215-37.

Nur, A., and Ben-Avraham, Z. 1981. Volcanic gaps and the consumption of aseismic ridges in South America. Geol. Soc. America Mem. 154:729-40.

Nur, A., Ron, H., and Scotti, O. 1986. Fault mechanics and the kinematics of block rotation. Geology 14:746-49.

Obermeier, S. 1987. Identification and geologic characteristics of earthquake-induced liquefaction features. *In* Crone, A. J., and Omdahl, E. M., eds. Directions in Paleoseismology. U.S. Geol. Survey Open-File Report 87-673:173-77.

Obermeier, S. F. 1989. The New Madrid earthquakes: an engineering-geologic interpretation of relict liquefaction features. U. S. Geol. Survey Prof. Paper 1336-B, 114 p.

Obermeier, S. F. 1994. Using liquefaction-induced features for paleoseismic analysis. U. S. Geol. Survey Open-File Report 94-663, chapter A, 58 p.

Obermeier, S. F. 1995. Preliminary limits for the strength of shaking in the Columbia River valley and the southern half of coastal Washington, with emphasis for a Cascadia subduction earthquake about 300 years ago. U. S. Geol. Survey Open-File Report 94-589, 40 p.

Obermeier, S. F., Gohn, G. S., Weems, R. E., Gelinas, R. L., and Rubin, M. 1985. Geologic evidence for recurrent moderate to large earthquakes near Charleston, South Carolina. Science 227:408-11.

Obermeier, S. F., Martin, J. R., Frankel, A. D., Youd, T. L., Munson, P. J., Munson, C. A., and Pond, E. C. 1993. Liquefaction evidence for one or more strong Holocene earthquakes in the Wabash Valley of southern Indiana and Illinois with a preliminary estimate of magnitude. U. S. Geol. Survey Prof. Paper 1536, 27 p.

Öcal, N., Uçar, S.B., and Taner, D. 1968. Manyas-Karacabey depremi 6 Ekim 1964. Istanbul Kandilli Rasathanesi. Sismoloji Yayinlari 11, Kandilli Observatory Internal Report.

Okada, A. 1973. Quaternary faulting along the Median Tectonic Line in the central part of Shikoku. Geogr. Rev. Japan 46:295-322 (J, Eng. abs.).

Okada, A. 1992. Proposal of the segmentation on the Median Tectonic Line active fault system. Geol. Soc. Japan Memoir 40:15-30. (J, Eng. abs.).

Okada, A., Ando, M., and Tsukuda, T. 1981. Trenches, late Holocene displacement and seismicity of the Shikano fault associated with the 1943 Tottori earthquake. Ann. Rept. Disaster Prevention Research Inst., U. Kyoto:105-26. (J, Eng. abs.).

Okada, A., Ando, M., and Tsukuda, T. 1987. Trenching study for Yasutomi fault of the Yamasaki fault system at Anji, Yasutomi Town, Hyogo Prefecture, Japan. J. Geography 96, 2:81-97. (J, Eng. abs.).

Okada, A., Matsuda, T., Tsutsumi, H., Morooka, T., and Mizota, K. 1991. Was the latest event of the Median Tectonic Line during the 1596 Keicho earthquake (M = 7.5)?—excavation study of the

Chichio fault belonging to the Median Tectonic Line in Shikoku. Seismol. Soc. Japan 1991, 2, p. 264. (Japanese).

Okada, A., and Sangawa, A. 1978. Fault morphology and Quaternary faulting along the Median Tectonic Line in the southern part of the Izumi Range. Geogr. Rev. Japan 51:385-405 (J, Eng. abs.).

Okada, A., Takeuchi, A., Tsukuda, T., Ikeda, Y., Watanabe, M., Hirano, S., Masumoto, S., Takehana, Y., Okumura, K., Kamishima, T., Kobayashi, T., and Ando, M. 1989. Trenching study of the Atotsugawa fault at Nokubi, Miyagawa village, Gifu Prefecture, central Japan. Jour. Geography 98:440-63. (J, Eng. abs.).

Okal, E., and Stein, S. 1987. The 1942 Southwest Indian Ocean Ridge earthquake: Largest ever recorded on an oceanic transform. Geophys. Res. Lett. 14:147-50.

Okamura, Y. 1990. Geologic structure of the upper continental slope off Shikoku and Quaternary tectonic movement of the outer zone of southwest Japan. Jour. Geol. Soc. Japan 96:223-37 (J, Eng. abs.).

Okamura, Y., Satoh, M., and Miyazaki, J. 1994. Active faults and folds on the shelf off Niigata and their relation to the 1964 Niigata earthquake. Zisin, Jour. Seismol. Soc. Japan 46:413-23. (J, Eng. abs.).

Okumura, K., Shimokawa, K., Yamazaki, H., and Tsukuda, E. 1994. Recent surface faulting events along the middle section of the Itoigawa-Shizuoka Tectonic Line—trenching survey of the Gofukuji fault near Matsumoto, central Japan. Jour Seismological Soc. Japan 46:425-38 (J, Eng. abs.).

Oldham, R. D. 1898. A note on the Allah Bund in the north-west of the Rann of Kucch. Geol. Survey India Mem. 28:27-30.

Oldham, R. D. 1899. Report on the great earthquake of 12th June 1897. Geol. Survey of India Memoir 29, 379 p.

Oldham, R. D. 1928. The Cutch (Kacch) earthquake of 16th June 1819, with a revision of the great earthquake of 12th June 1897. Geol. Survey India Mem. 46:71-147.

Omori, F. 1900. Notes on the Tokyo earthquake of June 20th, 1894. Imperial Earthquake Investigation Committee in Foreign Languages 4:25-33.

Omori, F. 1907. Preliminary note on the Formosa earthquake of March 17, 1906. Imp. Earthquake Inves. Comm. Bull. 2:53-69.

Omori, F. 1913. An account of the destructive earthquakes in Japan. Reports of the Imperial Earthquake Investigation Committee in Japanese Language (translation) 68B:1-180.

Omori, F. 1922. The Omachi (Shinano) earthquakes of 1918. Imperial Earthquake Investigation Committee Bull. 10:1-41.

Ongley, M. 1937. The Wairoa earthquake of 16th September 1932. 1. Field observations: New Zealand Jour. Sci. Technology 18:845-51.

Ongley, M. 1943. Surface trace of the 1855 earthquake. Royal Soc. New Zealand Trans. 73:84-99.

Oppenheimer, D., Beroza, G., Carver, G., Dengler, L., Eaton, J., Gee, L., Gonzales, F., Jayko, A., Li, W. H., Lisowski, M., Magee, M., Marshall, G., Murray, M., McPherson, R., Romanowicz, B., Satake, K., Simpson, R., Somerville, P., Stein, R., and Valentine, D. 1993. The Cape Mendocino, California, earthquakes of April 1992: Subduction at the triple junction. Science 261: 433-38.

Oppenheimer, D.H., Bakun, W.H., and Lindh, A.G. 1990, Slip partitioning of the Calaveras fault, California, and prospects for future earthquakes. Jour. Geophys. Res. 95:8483-98.

Orange, D. L., Geddes, D. S., and Moore, J. C. 1993. Structural and fluid evolution of a young accretionary complex: The Hoh rock assemblage of the western Olympic Peninsula, Washington. Geol. Soc. America Bull. 105:1053-75.

Ota, Y. 1969. Crustal movements in the late Quaternary considered from the deformed terrace plains in northeastern Japan. Japanese Jour. Geol. and Geophys. 40(204):41-61.

Ota, Y. 1991. Coseismic uplift in coastal zones of the western Pacific rim and its implications for coastal evolution. Z. Geomorph., N.F. 81:163-179.

Ota, Y., ed. 1994. Study on coral reef terraces of the Huon Peninsula, Papua New Guinea. A preliminary report on project 04041048, supported by Monbusho International Research Program, 188 p.

Ota, Y., and Kaizuka, S. 1991. Tectonic geomorphology at active plate boundaries—examples from the Pacific Rim. Z. Geomorph. N.F. Supp. 82:119-46.

Ota, Y., and Omura, A.1991. Late Quaternary shorelines in the Japanese islands. The Quaternary Research 30:175-86.

Ota, Y., and Suzuki, I. 1979. Notes on active folding in the lower reaches of the Shinano River, central Japan. Geogr. Rev. Japan 52:592-601 (Japanese).

Otsuki, K. 1990a. Neogene tectonic stress fields of northeast Honshu arc and implications for plate boundary conditions. Tectonophysics 181:151-64.

Otsuki, K. 1990b. Westward migration of the Izu-Bonin trench, northward motion of the Philippine Sea plate, and their relationships to the Cenozoic tectonics of Japanese island arcs. Tectonophysics 180:351-67.

Pacheco, J. F., and Sykes, L. R. 1992. Seismic moment catalog of large shallow earthquakes, 1900 to 1989. Seismological Society of America Bulletin 82:1306-49.

Pacheco, J. F., Sykes, L. R., and Scholz, C. 1993. Nature of seismic coupling along simple plate boundaries of the subduction type. Jour. Geophys. Res. 98:14,133-59.

Pacific Gas and Electric Company. 1988. Final report of the Diablo Canyon long term seismic program. San Francisco: Pacific Gas and Electric Company, 3 volumes.

Packer, D. R., Lovegreen, J. R., and Born, J. L. 1977. Reservoir induced seismicity, v. 6 *in* Woodward-Clyde Consultants, Earthquake evaluation studies of the Auburn Dam area, report to the U. S. Bureau of Reclamation.

Packer, D. R., Lovegreen, J. R., Harpster, R. E., Weaver, K. D., and Cluff, L. S., 1981, Reservoir induced seismicity—Active faulting at selected reservoirs: Woodward-Clyde Consultants, Geotechnical/ Environmental Bulletin, v. 14, no. 1, p.4-19.

Page, W. D., Alt, J. N., Cluff, L. S., and Plafker, G. 1979. Evidence for the recurrence of large magnitude earthquakes along the Makran coast of Iran and Pakistan. Tectonophysics 52:533-47.

Paige, S. 1930., The earthquake at Cumana, Venezuela, January 17, 1929: Seismological Society of America Bulletin 20:1-10.

Pantosti, D., Schwartz, D. P., and Valensise, G. 1993. Paleoseismology along the 1980 surface rupture of the Irpinia fault: Implications for earthquake recurrence in the southern Apennines, Italy. Jour. Geophys. Res. 98:6561-77.

Pantosti, D., and Valensise, G. 1990. Faulting mechanism and complexity of the November 23, 1980, Campania-Lucania earthquake, inferred from surface observations. Jour. Geophys. Res.. 95:15,319-41.

Pantosti, D., and Valensise, G. 1993. Irpinia earthquake fault based on field geologic observations. Annali di Geofisica 36:41-49.

Papastamatiou, I. 1957. The earthquakes of Velestino of 8 March 1957. Athens: Report Inst. Geologias and Erevnon Ypedaphos, 11 p.

Papastamatiou, D., and Mouyaris, N. 1986. The earthquake of April 30, 1954, in Sophades (Central Greece). Geophys. Jour. Roy. Astron. Soc. 87:885-95.

Papazachos, B. C., Mountrakis, A., Psilovikos, A., and Leventakis, G. 1979. Surface fault traces and fault plane solutions of the May-June 1978 shocks in the Thessaloniki area, north Greece. Tectonophysics 53:171-83.

Papazachos, B., Panagiotopoulos, D., Tsapanos, T., Mountrakis, D., and Dimoupoulos, G. 1983. A study of the 1980 summer seismic sequence in the Magnesia region of central Greece. Geophys. Jour. Roy. Astron. Soc. 75:155-68.

Parker, R., and Oldenburg, D. 1973. Thermal model of ocean ridges. Nature 242:137-39.

Parry, W. T., Hedderly-Smith, D., and Bruhn, R. L. 1991. Fluid inclusions and hydrothermal alteration on the Dixie Valley fault, Nevada. Jour. Geophys. Res. 96:19,733-48.

Parsons, B., and McKenzie, D. P. 1978. Mantle convection and the thermal structure of the plates. Jour. Geophys. Res. 83:4485-96.

Parsons, B., and Sclater, J. G. 1977. An analysis of the variation of the ocean floor bathymetry and heat flow with age. Jour. Geophys. Res. 82:803-27.

Paterson, M. S. 1978. Experimental Rock Deformation: The brittle field. Berlin: Springer-Verlag, 278 p.

Pavlides, S. B., and Tranos, M. D. 1991. Structural characteristics of two strong earthquakes in the North Aegean: Ierissos (1932) and Agios Efstratios (1968). Jour. Structural Geology 13:205-14.

Pavlides, S. B., Zouros, N. C., Chatzipetros, A. A., Kostopoulos, D. S., and Mountrakis, D. M. 1995. The 13 May 1995 western Macedonia, Greece (Kozani, Grevena) earthquake; preliminary results. Terra Nova 7:544-49.

Pearthree, P. A., Bull, W. B., and Wallace, T. C. 1990. Geomorphology and Quaternary geology of the Pitaycachi fault, northeastern Sonora, Mexico. Arizona Geol. Survey Spec. Paper 7:124-35.

Pelton, J. R., Meissner, C. W., and Smith, K. D. 1984. Eyewitness account of normal surface faulting. Seismological Society of America Bulletin 74:1083-89.

Peltzer, G., and Rosen, R. 1995. Surface displacement of the 17 May 1993 Eureka Valley, California, earthquake observed by SAR interferometry. Science 268:1333-36.

Peltzer, G., and Tapponnier, P. 1988. Formation and evolution of strike slip faults, rifts and basins during the India-Asia collision—an experimental approach. Jour. Geophys. Res. 93:15085-177.

Peltzer, G., Tapponnier, P., Gaudemer, Y., Meyer, B., Guo S., Yin K., Chan Z., and Dai H. 1988. Offsets of late Quaternary morphology, rate of slip, and recurrence of large earthquakes on the Chang Ma fault (Gansu, China). Jour. Geophys. Res. 93:7793-812.

Penick, J. 1976. The New Madrid Earthquakes of 1811-1812. Columbia, Mo., University of Missouri Press, 181 p.

Perez, O. J., and Jacob, K. H. 1980. Tectonic model and seismic potential of the eastern Gulf of Alaska and Yakataga seismic gap. Jour. Geophys. Res. 88:7132-50.

Perissoratis, C., Mitropoulos, D., and Angelopoulos, L. 1984. The role of earthquakes in inducing sediment mass movements in the eastern Korinthiakos Gulf. An example from the February 24-March 4, 1981 activity. Marine Geology 55:35-45.

Person, W. J. 1991. Seismological notes—November 1990-February 1991. Seismological Society of America Bulletin 81:2520-28.

Peterson, M. D., and Wesnousky, S. G. 1994. Fault slip rates and earthquake histories for active faults in southern California. Seismological Society of America Bulletin 84:1608-49

Philip, H., and Mégard, F. 1977. Structural analysis of the superficial deformation of the 1969 Pariahuanca earthquakes (central Peru). Tectonophysics 38:259-78.

Philip, H., and Meghraoui, M. 1983. Structural analysis and interpretation of the surface deformations of the El Asnam earthquake of October 10, 1980. Tectonics 2:17-49.

Philip, H., Cisternas, A., Gvishiani, A., and Gorshkov, A. 1989. The Caucasus: an actual example of the initial stages of continental collision. Tectonophysics 161:1-21.

Philip, H., Rogozhin, E., Cisternas, A., Bousquet, J.C., Borisov, B., and Karakhanian, A. 1992. The Armenian earthquake of 1988 December 7: Faulting and folding, neotectonics and palaeoseismicity. Geophys. J. Int. 110:141-58.

Pierce, K. L. 1986. Dating methods. *In* Wallace, R. E., ed. Active Tectonics. Washington, D.C.: National Academy Press, pp. 195-214.

Pierce, K. L. Obradovich, J. D., and Friedman, I., 1976. Obsidian hydration dating and correlation of Bull Lake and Pinedale glaciations near West Yellowstone, Montana. Geol. Soc. America Bull. 87:703-10.

Pillans, B. 1990. Vertical displacement rates on Quaternary faults, Wanganui Basin. New Zealand Jour. Geology and Geophysics 33:271-75.

Pinar, N. 1953. Etude geologique et macroseismique du tremblement de terre de Kursunlu du 13 aout 1951. Revue Facult. Sci. Univ. Istanbul, ser. A(18):131-141.

Pirazzoli, P. A. 1995. Tectonic shorelines. *In* Carter, R. W. G., and Woodroffe, C. D., eds. Coastal Evolution: Late Quaternary Shoreline Morphodynamics. Cambridge University Press (in press).

Pirazzoli, P. A., and Kawana, T. 1986. Détermination de mouvements crustaux quaternaires d'après la déformation des anciens rivages dans les îles Ryukyu, Japon. Revue de Geologie Dynamique et de Géographie Physique 27:269-78.

Pisias, N. S., Martinson, D. G., Moore, T. C., Shackleton, N. J., Prell, W., Hays, J., and Boden, G. 1984. High resolution stratigraphic correlation of benthic oxygen isotopic records spanning the last 300,000 years. Marine Geology 56:119-36.

Pisias, N. G., Mix, A. C., and Zahn, R. 1990. Nonlinear response in the global climate system: evidence from benthic oxygen isotopic record in core RC 13-110. Paleoceanography 5:147-60.

Plafker, G. 1967. Surface faults on Montague Island associated with the 1964 Alaska earthquake. U. S. Geol. Survey Prof. Paper 543-G, 42 p.

Plafker, G. 1969. Tectonics of the March 27, 1964, Alaska, earthquake. U. S. Geol. Survey Prof. Paper 543-I, 74 p.

Plafker, G. 1972. Alaskan earthquake of 1964 and Chilean earthquake of 1960: Implications for arc tectonics. Jour. Geophys. Res. 77:901-25.

Plafker, G. 1976. Tectonic aspects of the Guatemala earthquake of 4 February 1976. Science 193:1201-08.

Plafker, G. 1984. Model for the origin of the Yakutat block, an accreting terrane in the northern Gulf of Alaska—Comment. Geology 12:563.

Plafker, G. 1987a. Regional geology and petroleum potential of the northern Gulf of Alaska continental margin. *In* Scholl, D. W., Grantz, A., and Vedder, J. G., eds. Geology and resource potential of the continental margin of western North America and adjacent ocean basins—Beaufort Sea to Baja California. Circum-Pacific Council for Energy and Mineral Resources Earth Science Series 6:229-68.

Plafker, G. 1987b. Application of marine-terrace data to paleoseismic studies. *In* Crone, A. J., and Omdahl, E. M., eds. Directions in Paleoseismology. U. S. Geol. Survey Open-File Report 87-673:146-156.

Plafker, G., Agar, R., Asker, A. H., and Hanif, M. 1987. Surface effects and tectonic setting of the 13 December 1982 North Yemen earthquake. Seismological Society of America Bulletin 77:2018-37.

Plafker, G., Bonilla, M. G., and Bonis, S. B. 1976. Geologic effects. *In* Espinosa, A. F., ed. The Guatemalan Earthquake of February 4, 1976, A Preliminary Report. U. S. Geol. Survey Prof. Paper 1002:38-51.

Plafker, G., Ericksen, G. E., and Fernandez C., J. 1971. Geological aspects of the May 31, 1970, Peru earthquake. Seismological Society of America Bulletin 61:543-78.

Plafker, G., and Galloway, J. P., eds. 1989. Lessons learned from the Loma Prieta, California, earthquake of October 7, 1989. U. S. Geol. Survey Circular 1045, 48 p.

Plafker, G., Hudson, T., Bruns, T., and Rubin, M. 1978. Late Quaternary offsets along the Fairweather fault and crustal plate interactions in southern Alaska. Canadian Jour. Earth Sciences 15:805-16.

Plafker, G., and Savage, J. C. 1970. Mechanism of the Chilean earthquake of May 21, 1960. Geol. Soc. America Bull. 81:1001-30.

Plafker, G., and Ward, S. N. 1992. Backarc thrust faulting and tectonic uplift along the Caribbean Sea coast during the April 22, 1991 Costa Rica earthquake. Tectonics 11:709-18.

Plumb, R. A., and Hickman, S. H. 1985. Stress-induced borehole elongation: a comparison between the four-arm dipmeter and the borehole televiewer in the Auburn geothermal well. Jour. Geophys. Res. 90:5513-21

Pockalny, R., Detrick, R., and Fox, P. 1988. Morphology and tectonics of the Kane transform from Sea Beam bathymetry data. Jour. Geophys. Res. 93:3179-93.

Powell, R. E. 1993. Balanced palinspastic reconstruction of pre-late Cenozoic paleogeology, southern California: Geologic and kinematic constraints on evolution of the San Andreas fault system. Geol. Soc. America Mem. 178:1-106.

Prell, W. L., Imbrie, J., Martinson, D. G., Morley, J. J., Pisias, N. G., Shackleton, N. J., and Streeter, H. F. 1986. Graphic correlation of oxygen isotope stratigraphy application to the late Quaternary. Paleoceanography 1:137-62.

Prentice, C. S. 1989. Earthquake geology of the northern San Andreas fault near Point Arena, California. California Institute of Technology, Ph.D. thesis.

Prentice, C. S., and Schwartz, D. P. 1991. Re-evaluation of 1906 surface faulting, geomorphic expression, and seismic hazard along the San Andreas fauault in the southern Santa Cruz Mountains. Seismological Society of America Bulletin 81:1424-79.

Prescott, W. H., Lisowski, M., and Savage, J. C. 1981. Geodetic measurement of crustal deformation of the San Andreas, Hayward, and Calaveras faults near San Francisco, California. Jour. Geophys. Res. 86:10,853-69.

Prescott, W. H., Davis, J. L., and Svarc, J. L. 1989. Global positioning system measurements for crustal deformation: Precision and accuracy. Science 244:1337-40.

Press, F. and Siever, R. 1982, Earth, 3rd ed. New York: W. H. Freeman and Co., 613 p. .

Pringle, M. S., McWilliams, M., Houghton, B. F., Lanphere, M. A., and Wilson, C. J. N. 1992. $^{40}Ar^{39}Ar$ dating of Quaternary feldspar: Examples from the Taupo Volcanic Zone, New Zealand. Geology 20:531-34.

Putnam W. C. 1942. Geomorphology of the Ventura region, California. Geol. Soc. America Bull. 53:691-754.

Qian H., Tang Y., Zhang C., and Cao Y. 1984. Characteristics of ground fissures during the Daofu earthquake and movement of the faults in this earthquake area. Jour. Seism. Res. 7:53-60 (C).

Quennell, A. 1958. The structural and geomorphic evolution of the Dead Sea rift. Quarterly Journal of the Geological Society of London 114:1-24.

Quennell, A. 1959. Tectonics of the Dead Sea rift. *In* International Geological Congress, XX Session, Mexico pp. 385-405.

Quittmeyer, R. C., and Jacob, K. H. 1979. Historical and modern seismicity in Pakistan, Afganistan, northwest India, and southeast Iran. Seismological Society of America Bulletin 69:773-823.

Quittmeyer, R. C., Farah, A., and Jacob, K. H. 1979. The seismicity of Pakistan and its relation to surface faults. *In* Farah, A., and DeJong, K. A., eds. Geodynamics of Pakistan. Quetta: Geol. Survey of Pakistan 271-84.

Qureshi, I. R., and Sadig, A. A. 1967. Earthquakes and associated faulting in central Sudan. Nature 215:263-65.

Rabinowicz, E. 1965. Friction and wear of materials. New York: J. Wiley & Sons.

Raleigh, C. B., Healy, J. H., and Bredehoeft, H. D. 1976. An experiment in earthquake control at Rangely, Colorado. Science 191:1230-37.

Ramirez, J. E. 1971a. The destruction of Bahía Solano, Colombia, on September 26, 1970 and the rejuvenation of a fault. Seismological Society of America Bulletin 61:1041-49.

Ramirez, J. E. 1971b. La catastrofe de Bahía Solano del 26 de Spetiembre de 1970. *In* El Terremoto de Bahía Solano del 26 de Septiembre de 1970. Instituto Geofísico de los Andes Colombianos, Publ. Ser. A, Sismología 33:9-36.

Ramos, V. A. 1988. The tectonics of the Central Andes: 30° to 33° S latitude. Geol. Soc. America Spec. Paper 218:31-54.

Ran Y., Fang Z., Li Z., Wang J., and Li R. 1991. Characteristics of active faults around Huailai-Zhuolu basin. Research on Active Fault 1:140-54 (C, Eng. abs.).

Reasenberg, P., and Ellsworth, W. L. 1982. Aftershocks of the Coyote Lake, California, earthquake of August 6, 1979: A detailed study. Jour. Geophys. Res. 87:10,637-55.

Rebaï, S., Philip, H., and Taboada, A. 1992. Modern tectonic stress field in the Mediterranean region: Evidence for variation in stress directions at different scales. Geophys. J. Int. 110: 106-40.

Redfield, T., and Fitzgerald, P. 1993. Denali fault system of southern Alaska: An interior strike-slip structure responding to dextral and sinistral shear coupling. Tectonics 12:1195-1208.

Regional Seismological Bureau of Xizang, Tibet (RSBX) 1980. Map of macroseismic observations related to the magnitude 6.8 earthquake of February 22, 1980, in Xainza County, Tibet. Lhasa, People's Republic of China.

Reid, H. F. 1910. The mechanics of the earthquake. *in* Lawson, A. C., chmn., The California earthquake of April 18,1906. Carnegie Institute Washington Publication 87, v. 2, 192 p.

Reid, H. F. 1913. Sudden earth movements in Sumatra in 1892. Seismological Society of America Bulletin 3:72-79.

Reidel, S. P., Fecht, K. R., Hagood, M. C., and Tolan, T. L. 1989. The geologic evolution of the central Columbia Plateau. Geol. Soc. America Spec. Paper 239:247-64.

Reiter, L. 1990. Earthquake Hazard Analysis—Issues and Insights. New York: Columbia University Press, 254 p.

Ren J., and Li P. 1989. Earthquake-caused landforms and paleoseismic study on the northern segment of Zemuhe fault. Seismology and Geology 11(1):27-34. (C, Eng. abs.).

Research Group for Active Faults. 1991. Active faults in Japan: Sheet maps and inventories: Tokyo: University of Tokyo Press, 437 p. (J, Eng. abs.).

Research Group for Active Faults of Japan. 1992. Maps of active faults in Japan with an explanatory text: Tokyo: University of Tokyo Press (J, short English explanation).

Research Group for the Senya Fault. 1986. Holocene activities and near-surface features of the Senya fault, Akita Prefecture, Japan—excavation study at Komori, Senhata-cho. Earthquake Res. Inst., University of Tokyo Bull. 61:339-402 (J, Eng. abs.).

Restrepo, A. H. 1971. Zona de falla de Puerto Mutis en Bahía Solano. *In* El Terremoto de Bahía Solano del 26 de Septiembre de 1970. Instituto Geofísico de los Andes Colombianos Publ. Ser. A., Sismología 33;9-26.

Ricci Lucchi, F. 1995. Sedimentological indicators of paleoseismicity. *In* Serva, L., and Slemmons, D. B., eds. Perspectives in Paleoseismology. Association of Engineering Geologists Spec. Pub. 6:7-17

Richins, W. D., Pechmann, J. C., Smith, R. B., Langer, C. J., Goter, S. K., Zollweg, J. E., and King, J. J. 1987. The 1983 Borah Peak, Idaho, earthquake and its aftershocks. Seismological Society of America Bulletin 77: 694-723.

Richter, C. F. 1935. An instrumental earthquake magnitude scale. Seismological Society of America Bulletin 25:1-32.

Richter, C. F. 1958. Elementary Seismology. San Francisco: W. H. Freeman and Co., 768 p.

Richter, D., and Matson, N. 1971. Quaternary faulting in the eastern Alaska Range. Geol. Soc. America Bull. 82:1529-39.

Riedel, W. 1929. Zur mechanik geologischer bruch-erscheinungen: Centralbl. fur Mineral. Geol. uber Pal. 1929 B:354-68.

Rikitake, T. 1982. Earthquake Forecasting and Warning. Dordrecht, Netherlands: D. Reidel Publishing Co.

Ringrose, P. S., Hancock, P., Fenton, C., and Davenport, C. A. 1991. Quaternary tectonic activity in Scotland. *In* Forster, A., Culshaw, M. G., Cripps, J. C., Little, J. A., and Moon, C. F., eds. Quaternary Engineering Geology. Geol. Soc. Engineering Geol. Spec. Pub. 7:679-86.

Roberts, A. M., Yielding, G., and Freeman, B., eds. 1991. The Geometry of Normal Faults. Geological Society Special Publication 56, 264 p.

Rockwell, T. K. 1983. Soil chronology, geology and neotectonics of the north-central Ventura basin. PhD thesis, University of California, Santa Barbara, 424 p.

Rockwell, T. K. 1988. Neotectonics of the San Cayetano fault, Transverse Ranges, California. Geol. Soc. America Bull. 100:500-513.

Rockwell, T. K. 1989. Behavior of individual fault segments along the Elsinore-Laguna Salada fault zone, southern California and northern Baja California: Implications for the characteristic earthquake model. *In* Schwartz, D. P., and Sibson, R. H., eds. Fault Segmentation and Controls of Rupture Initiation and Termination. U. S. Geol. Survey Open-File Report 89-315:288-308.

Rockwell, T. K., Keller, E. A., and Dembroff, G. R. 1988. Quaternary rate of folding of the Ventura Avenue anticline, western Transverse Ranges, southern California. Geol. Soc. America Bull. 100:850-58.

Rockwell, T. K., and Sieh, K. 1994. Clustered ancient earthquakes in the Imperial Valley, southern California, preserved in coeval lacustrine strata. U. S. Geol. Survey, Open-file Report 94-568, p. 161.

Rodgers, J. 1987. Chains of basement uplifts within cratons marginal to orogenic belts. American Jour. Sci. 287:661-92.

Rogers, A. M., and Lee, W. H. K. 1976. Seismic study of earthquakes in the Lake Mead, Nevada-Arizona region. Seismological Society of America Bulletin 66:1657-81.

Roquemore, G. R., and Zellmer, J. T. 1983. Ground cracking associated with 1982 magnitude 5.2 Indian Wells Valley earthquake, Inyo County, California. California Geology 36(9):197-200.

Rosholt, J. N. 1985. Uranium-trend systematics for dating Quaternary sediments. U. S. Geol. Survey Open-File Report 85-298, 34 p.

Rotstein, Y., and Kafka, A. L. 1982. Seismotectonics of the southern boundary of Anatolia, eastern Mediterranean region: Subduction, collision and arc jumping. Jour. Geophys. Res. 87:7694-7706.

Royden, L. H. 1993. Evolution of retreating subduction boundaries formed during continental collision. Tectonics 12:629-38.

Ruff, L., and Kanamori, H. 1980. Seismicity and the subduction process. Physics Earth Planet. Interiors 23:240-52.

Ruff, L., Given, J., Sanders, C., and Sperber, C. 1989. Large earthquakes in the Macquarie Ridge Complex: Transitional tectonics and subduction initiation. Pure and Applied Geophysics129:71-130.

Russ, D. P. 1982. Style and significance of surface deformation in the vicinity of New Madrid, Missouri. U.S. Geol. Survey Prof. Paper 1236:45-114.

Ryan, H. F., and Scholl, D. W. 1989. The evolution of forearc structures along an oblique convergent margin, central Aleutian arc. Tectonics 8:497-516.

Ryan, H. F., and Scholl, D. W. 1993. Geologic implications of great interplate earthquakes along the Aleutian arc. Jour. Geophys. Research 98:22,135-46.

Rymer, M. J. 1992. The 1992 Joshua Tree, California, earthquake: tectonic setting and triggered slip. EOS, Transactions Am. Geophys. Union. 73:363,

Rymer, M. J., Kendrick, K. ., Lienkaemper, J. J., and Clark, M. M. 1990. Surface rupture on the Nuñez fault during the Coalinga earthquake sequence. *In* Rymer, M. J., and Ellsworth, W. L., eds. The Coalinga, California, earthquake of May 2, 1983. U. S. Geol. Survey Prof. Paper 1487:299-318.

Salyards, S. L., Sieh, K. E., and Kirschvink, J. L. 1992. Paleomagnetic measurement of nonbrittle coseismic deformation across the San Andreas fault at Pallett Creek. Jour. Geophys. Res. 97:12,457-70.

Sanders, C. O., and Slemmons, D. B. 1979. Recent crustal movements in the central Sierra Nevada-Walker Lane region of California-Nevada: Part III, the Olinghouse fault zone. Tectonophysics 52:585-97.

Sangawa, A. 1986. The history of fault movement sine late Pliocene in the central part of southwest Japan. *In* Reilly, W. I., and Harford, B. E., eds. Recent Crustal Movements of the Pacific Region. Royal Soc. New Zealand Bull. 24:75-85.

Sangawa, A. 1992. Earthquake Archeology. Tokyo: Chuko Shinsho, 251 p. (Japanese).

Sarna-Wojcicki, A. M., and Davis, J. O. 1991. Quaternary tephrochronology. *In* Morrison, R. B., ed. Quaternary nonglacial geology: Conterminous

U. S. Geol. Soc. America Decade of North American Geology K-2:93–116.

Sarna-Wojcicki, A. M., Lajoie, K. R., Meyer, C. E., Adam, D. P., and Rieck, H. J. 1991. Tephrochronologic correlation of upper Neogene sediments along the Pacific margin, conterminous U.S. *In* Morrison, R.B., ed. Quaternary nonglacial geology: Conterminous U.S. Geol. Soc. America Decade of North American Geology K-2:117–40.

Sarna-Wojcicki, A. M., Lajoie, K. R., and Yerkes, R. F. 1987. Recurrent Holocene displacement on the Javon Canyon fault—a comparison of fault-movement history with calculated average recurrence intervals. U. S. Geol. Survey Prof. Paper 1339:125–35.

Sarna-Wojcicki, A. M., Morrison, S. D., Meyer, C. E., and Hillhouse, J. W. 1987. Correlation of upper Cenozoic tephra layers between sediments of the western United States and eastern Pacific Ocean, and comparison with biostratigraphic and magnetostratigraphic age data. Geol. Soc. America Bull. 98:207–223.

Sarna-Wojcicki, A. M., Williams, K. M., and Yerkes, R. F. 1976. Geology of the Ventura fault, Ventura County, California. U. S. Geol. Survey Misc. Field Studies Map MF-781.

Saroglu, F., Emre, O., and Kuscu, I. 1992. The East Anatolian fault zone of Turkey. Annales Tectonicae Special Issue, Supp. 6:99–125.

Satake, K. 1992. Tsunamis. Encyclopedia of Earth System Science 4:389–92.

Satake, K. 1985. The mechanism of the 1983 Japan Sea earthquake as inferred from long-period surface waves and tsunamis. Phys. Earth and Planet. Inter. 37:249–60.

Satake, K., and Abe, K. 1983. A fault model for the Niigata, Japan, earthquake of June 16, 1964. Jour. Phys. Earth 31:217–23.

Satake, K., and Kanamori, H. 1990. Fault parameters and tsunami excitation of the May 23, 1989, Macquarie Ridge earthquake. Geophys. Res. Lett. 17:997–1000.

Sauber, J., Thatcher, W., and Solomon, S. 1986. Geodetic measurements of deformation in the central Mojave Desert, California. Jour. Geophys. Res. 91:12,683–93.

Sauber, J., Thatcher, W., Solomon, S., and Lisowski, M. 1994. Geodetic slip rate for the eastern California shear zone and the recurrence time of Mojave desert earthquakes. Nature 367:264–66.

Savage, J. C., Burford, R. O., and Kinoshita, W. T. 1975. Earth movements from geodetic measurements. California Division of Mines and Geology Bull. 196:175–86.

Savage, J. C., and Lisowski, M. 1991. Strain measurements and the potential for a great subduction earthquake off the coast of Washington. Science 252:101–03.

Savage, J. C., Lisowski, M., and Prescott, W. H. 1991. Strain accumulation in western Washington. Jour. Geophys. Res. 96:14,493–507.

Savage, J. C., and Thatcher, W. 1992. Interseismic deformation of the Nankai Trough, Japan, subduction zone. Jour. Geophys. Res. 97:11,117–135.

Savostin, L., Zonenshain, L., and Baranov, B. 1983. Geology and plate tectonics of the Sea of Okhotsk. *In* Hilde, T. W. C., and Uyeda, S. eds. Geodynamics of the western Pacific-Indonesian region. Am. Geophys. Union Geodynamics Series 11:189–221.

Schmid, S. M. 1982. Microfabric studies as indicators of deformation mechanisms and flow laws operative in mountain building. *In* Hsü, K.J., ed. Mountain Building Processes. London: Academic Press, pp. 95–110.

Schmidt, J. F. J. 1875. Studien über Erdbeben. Leipzig, Germany: Carl Scholtze.

Schneider, C.L., Hummon, C., Yeats, R.S., and Huftile, G.J. 1996. Structural timing and kinematics of the northern Los Angeles basin, California, based on growth strata. Tectonics, in press

Scholl, D. W., Vallier, T. L., and Stevenson, A. J. 1987. Geologic evolution and petroleum geology of the Aleutian Ridge. *In* Scholl, D. W., Grantz, A., and Vedder, J. G., eds. Geology and resource potential of the continental margin of western North America and adjacent ocean basins—Beaufort Sea to Baja California. Circum-Pacific Council for Energy and Mineral Resources Earth Science Series 6:123–55.

Scholz, C. H. 1988. The brittle-plastic transition and the depth of seismic faulting. Geol. Rundschau 77:319–28.

Scholz, C. H. 1990. The Mechanics of Earthquakes and Faulting. Cambridge: Cambridge University Press, 439 p.

Scholz, C. H., and Campos, J. 1995. On the mechanism of seismic decoupling and back arc spreading at subduction zones. Jour. Geophys. Res. 100:22,103–15.

Scholz, C. H., and Kato, T. 1978. The behavior of a convergent plate boundary: Crustal deformation in the south Kanto district, Japan. Jour. Geophys. Res. 83:783-97.

Schumm, S. A. 1986. Alluvial river response to active tectonics. *In* Wallace, R. E., ed. Active Tectonics. Washington, D.C.: National Academy Press, 80-94.

Schuster, R. L., Logan, R. L., and Pringle, P. T. 1992. Prehistoric rock avalanches in the Olympic Mountains, Washington. Science 258:1620-21.

Schwartz, D. P. 1988. Paleoseismicity and neotectonics of the Cordillera Blanca fault zone, northern Peruvian Andes. Jour. Geophys. Res. 93:4712-30.

Schwartz, D. P., and Coppersmith, K. J. 1984. Fault behavior and characteristic earthquakes: examples from the Wasatch and San Andreas faults. Jour. Geophys. Res. 89:5681-98.

Schweller, W. J., Kulm, L. D., and Prince, R. A. 1981. Tectonics, structure, and sedimentary framework of the Peru-Chile Trench. Geol. Soc. America Mem. 154:323-49.

Scientists of the U.S. Geological Survey and the Southern California Earthquake Center. 1994. The magnitude 6.7 Northridge, California, earthquake of 17 January 1994. Science 266: 389-97.

Scott, W. E., Pierce, K. C., and Hait, M. H., Jr. 1985. Quaternary tectonic setting of the 1983 Borah Peak earthquake, central Idaho. Seismological Society of America Bulletin 75:1053-66.

Sébrier, M., Mercier, L. L., Mégard, F., Laubacher, G., and Carey-Gailhardis, E. 1985. Quaternary normal and reverse faulting and the state of stress in the central Andes of south Peru. Tectonics 4:739-80.

Sébrier, M., Mercier, J. L., Macharé, J., Bonnot, D., Cabrera, J., and Blanc, J. L. 1988. The state of stress in an overriding plate situated above a flat slab: The Andes of central Peru. Tectonics 7:895-928.

Seeber, L., and Armbruster, J. G. 1981. Great detachment earthquakes along the Himalayan arc and long-term forecasting. *In* Simpson, D. W., and Richards, P. G., eds. Earthquake Prediction: An International Review. Amer. Geophys. Union Maurice Ewing Series 4:259-77.

Seeber, L., and Gornitz, V. 1983. River profiles along the Himalayan arc as indicators of active tectonics. Tectonophysics 92:335-67.

Seeber, L., Jain, S. K., Murty, C. V. R., and Chandak, N. 1993. Surface rupture and damage patterns in the Ms = 6.4, Sept. 29, 1993 Killari (Latur) earthquake in central India. Am. Geophys. Union 1993 Fall Meeting Program Addition, p. 222.

Seed, H. B. 1979. Considerations in the earthquake-resistant design of earth and rockfill dams. Geotechnique 29(3):215-63.

Seed, H. B., Idriss, L.M., and Arango, I. 1983. Evaluation of liquefaction potential using field performance data. Jour. Geotechnical Engineering, American Soc. Civil Engineering Div. 109:458-82.

Segall, P., and Pollard, D. 1980. Mechanics of discontinuous faults. Jour. Geophys. Res. 85:4337-50.

Seilacher, A. 1969, Fault-graded beds interpreted as seismites. Sedimentology 13:155-59.

Seno, T., and Eguchi, T. 1983. Seismotectonics of the western Pacific region. *In* Hilde, T. W. C., and Uyeda, S., eds. Geodynamics of the western Pacific-Indonesian region. Am. Geophys. Union Geodynamics Series 11:5-40.

Seno, T., Stein, S., and Griff., A. E. 1993. A model for the motion of the Philippine Sea plate consistent with NUVEL-1 and geological data. Jour. Geophys. Res. 98:17,941-48.

Serva, L., Blumetti, A. M., and Michetti, A.M. 1986. Gli effetti sul terreno del terremoto del Fucino (13 Gennaio 1915); tentativo di interpretazione della evoluzione tettonica recente di alcune strutture. Mem. Soc. Geol. Ital. 35:893-907 (Eng. abs).

Seymen, I., and Aydin, A. 1972. The Bingöl earthquake fault and its relation to the North Anatolian fault zone. Miner. Res. Explor. Inst. Turkey Bull. 79:1-8.

Shackleton, N. J., and Opdyke, N. D. 1973. Oxygen isotope and paleomagnetic stratigraphy of equatorial Pacific core V28-238: Oxygen isotope temperatures and ice volumes on a 10^5 year and 10^6 year scale. Quaternary Research 3:39-55.

Sharp, R. V. 1975. Displacement on tectonic ruptures. California Div. Mines and Geology Bull. 196:187-94.

Sharp, R. V. 1976. Surface faulting in Imperial Valley during the earthquake swarm of January-February, 1975. Seismological Society of America Bulletin 66:1145-54.

Sharp, R. V. 1982. Comparison of 1979 surface faulting with earlier displacements in the Imperial Valley. *In* The Imperial Valley, California, Earthquake of October 15, 1979. U.S. Geol. Survey Prof. Paper 1254:213-21.

Sharp, R., Budding, K., Boatwright, J., Ader, M., Bonilla, M., Clark, M., Fumal, T., Harms, K., Lienkaemper, J., Morton, D., O'Neill, B., Ostergren, C., Ponti, D., Rymer, M., Saxton, J., and Sims, J. 1989. Surface faulting along the Superstition Hills fault zone and nearby faults associated with the earthquakes of 24 November 1987. Seismological Society of America Bulletin 79:252-81.

Sharp, R. V., Lienkaemper, J. J., Bonilla, M. G., Burke, D. B., Fox, B. F., Herd, D. G., Miller, D. M., Morton, D. M., Ponti, D. J., Rymer, M. J., Tinsley, J. C., Yount, J. C., Kahle, J. E., Hart, E. W., and Sieh, K. E. 1982. Surface faulting in the central Imperial Valley. *In* The Imperial Valley, California, Earthquake of October 15, 1979. U. S. Geol. Survey Prof. Paper 1254, p. 119-43.

Sharp, R. V., Rymer, M. J., and Lienkaemper, J. J. 1986a. Surface displacement on the Imperial and Superstition Hills faults triggered by the Westmorland, California, earthquake of 26 April 1981. Seismological Society of America Bulletin 76:949-65.

Sharp R. V., Rymer, M. J., and Morton, D. M. 1986b. Trace-fractures on the Banning fault created in association with the 1986 North Palm Springs earthquake. Seismological Society of America Bulletin 76:1838-43.

Shaw, J. H., and Suppe, J. 1994. Active faulting and growth folding in the eastern Santa Barbara Channel, California. Geol. Soc. America Bull. 106:607-26.

Shaw, J. H., Bischke, R., and Suppe, J. 1994. Relations between folding and faulting in the Loma Prieta epicentral zone: Strike-slip fault-bend folding. *In* The Loma Prieta, CA, Earthquake of October 17, 1989. U. S. Geological Survey Prof. Paper 1550-F:F3-F21.

Shedlock, K. M., and Weaver, C. S. 1991. Program for earthquake hazards assessment in the Pacific Northwest. U. S. Geol. Survey Circular 1067, 29 p.

Sheng, C. K., Chen, H. C., Huang, L. S., Yang,, C. J., Chang, C. H., Li, T. C., Wang, T. C., and Lo, H. H. 1973. Earthquakes induced by reservoir impounding and their effect on the Hsinfengkiang Dam. Beijing, 44 p.

Shieh C. I. 1987. Engineering aspect of the Meishan fault in southern Taiwan. Geol. Soc. China Mem. 9:383-96.

Shimazaki, K. 1980. Uplift of the Holocene marine terraces, and intra- and inter-plate earthquakes. The Earth Monthly 2:17-24 (Japanese).

Shimazaki, K., and Nakata, T. 1980. Time-predictable recurrence model for large earthquakes. Geophys. Res. Lett. 7:279-82.

Shiono, K. 1988. Seismicity of the SW Japan arc—subduction of the young Shikoku basin. Modern Geology 12:449-64.

Shor, G., and Roberts, E. E. 1958. San Miguel, Baja California Norte, earthquakes of February, 1956—a field report. Seismological Society of America Bulletin 46:101-16.

Shroder, J. F., Jr. 1980. Dendrogeomorphology—review and new techniques of tree-ring dating. Progress in Physical Geography 4:161-88.

Sibson, R. H. 1977. Fault rocks and fault mechanics. Journal of Geological Society of London 133:191-213.

Sibson, R. H. 1983. Continental fault structure and the shallow earthquake source. Journal of Geological Society of London 140:741-67.

Sibson, R. H. 1984. Roughness at the base of the seismogenic zone: Contributing factors. Jour. Geophys. Res. 89:5791-99.

Sibson, R. H. 1985. Stopping of earthquake ruptures at dilational fault jogs: Nature 316:248-51.

Sibson, R. H. 1986. Earthquakes and rock deformation in crustal fault zones. Annual Review of Earth and Planetary Sciences 14:149-75

Sibson, R. H. 1987. Earthquake faulting as a structural process. Journal of Structural Geology 11:1-14.

Sibson, R. H. 1990. Conditions for fault-valve behaviour, in Knipe, R.J., and Rutter, E.H., eds., Deformation Mechanisms, Rheology and Tectonics: Geol. Soc. London Spec. Pub. 54:15-28.

Sibson, R.H. 1992. Implications of fault valve behavior for rupture nucleation and recurrence. Tectonophysics 211:283-93.

Sieh, K. 1978a. Pre-historic large earthquakes produced by slip on the San Andreas fault at Pallett Creek, California. Jour. Geophys. Res. 83:3907-39.

Sieh, K. 1978b. Slip along the San Andreas fault associated with the great 1857 earthquake. Seismological Society of America Bulletin 68:1421-28.

Sieh, K. 1981. A review of geological evidence for recurrence times of large earthquakes. *In* Earthquake Prediction—An International Review, pp. 181-207.

Sieh, K. 1982. Slip along the San Andreas fault associated with the earthquake. U. S. Geol. Survey Prof. Paper 1254:155-59.

Sieh, K. 1984. Lateral offsets and revised dates of large earthquakes at Pallett Creek, California. Jour. Geophys. Res. 89:7641-70.

Sieh, K., Bock, Y., and Rais, J. 1991. Neotectonic and paleoseismologic studies in west and north Sumatra. EOS, Trans. Amer. Geophys. Union 72:460.

Sieh, K., and Jahns, R. 1984. Holocene activity of the San Andreas fault at Wallace Creek, California. Geol. Soc. America Bull. 95:883-96.

Sieh, K., Jones, L., Hauksson, E., Hudnut, K., Eberhart-Phillips, D., Heaton, T., Hough, S., Hutton, K., Kanamori, H., Lilje, A., Lindvall, S., McGill, S.F., Mori, J., Rubin, C., Spotila, J.A., Stock, J., Thio, H., Treiman, J., Wernicke, B., and Zachiarasen, J. 1993. Near-field investigation of the Landers earthquake sequence, April to July, 1992. Science 260:171-76.

Sieh, K., Stuiver, M., and Brillinger, D. 1989. A more precise chronology of earthquakes produced by the San Andreas fault in southern California. Jour. Geophys. Res. 94:603-23.

Sieh, K., and Wallace, R. 1986. The San Andreas Fault at Wallace Creek, San Luis Obispo County, California. *In* Geol. Soc. America Centennial Field Guide CFG-1 (Cordilleran Section):233-38.

Silver, E. A., Reed, D. L., Tagudin, J. E., and Heil, D. J. 1990. Imiplications of the North and South Panama thrust belts for the origin of the Panama orocline. Tectonics 9:261-81.

Simpson, D. W. 1986. Triggered earthquakes: Annual Review of Earth and Planetary Science 14:21-42.

Simpson, D. W., Leith, W. S., and Scholz, C. H. 1988. Two types of reservoir-induced seismicity. Seismological Society of America Bulletin 78:2025-40.

Simpson, D. W., and Negmatullaev, S. Kh. 1981. Induced seismicity at Nurek reservoir, Tadjikistan, USSR. Seismological Society of America Bulletin 71:1561-86.

Simpson, G. D., Lettis, W. R., and Kelson, K. I. 1993. Segmentation model for the northern Calaveras fault, Calaveras Reservoir to Walnut Creek. U. S. Geol. Survey Open-File Report 93-195:672-81.

Sims, J. D. 1994. Stream channel offset and abandonment and a 200-year average recurrence interval of earthquakes on the San Andreas fault at Phelan Creek, Carrizo Plain, California. *In* Prentice, C., Schwartz, D., and Yeats, R., ed. Paleoseismology. : U. S. Geol. Survey Open-file Report 94-568:170-72.

Sims, J. D. 1975. Determining earthquake recurrence intervals from deformational structures in young lacustrine sediments: Tectonophysics 29:141-52.

Skuphos, T. 1894. Die zwei grossen Erdbeben im Lokris. Zeitschrift der Gesellschaft für Erdkunde 29:409-74.

Slemmons, D. B. 1957. Geological effects of the Dixie Valley-Fairview Peak, Nevada, earthquakes of December 16, 1954. Seismological Society of America Bulletin 47:353-75.

Slemmons, D. B., Steinbrugge, K. V., Tocher, D., Oakeshott, G. B., and Gianella, V. P. 1959. Wonder, Nevada, earthquake of 1903. Seismological Society of America Bulletin 49:251-65.

Smalley, R., Pujol, J., Regnier, M., Chiu, J.-M., Chatelain, J.-L., Isacks, B. L., Araujo, M., and Puebla, N. 1993. Basement seismicity beneath the Andean Precordillera thin-skinned thrust belt and implications for crustal and lithospheric behavior. Tectonics 12:63-76.

Smirnova, T. Y., and Nikonov, A. A. 1990. A revised lichenometric method and its application dating great past earthquakes. Arctic and Alpine Research 22:375-88.

Smith, G. A. 1993. Missoula flood dynamics and magnitudes inferred from sedimentology of slack-water deposits on the Columbia Plateau, Washington. Geol. Soc. America Bull. 105:77-100.

Smith, J. L., and Fallgren, R. B. 1975. Ground displacement at San Fernando Valley Juvenile Hall and the Sylmar Converter Station. California Div. Mines and Geology Bull. 196:157-63.

Smith, R. B., and Bruhn, R. L. 1984. Intraplate extensional tectonics of the eastern Basin-Range: inferences on structural style from seismic reflection data, regional tectonics, and thermal-mechanical models of brittle-ductile deformation. Jour. Geophys. Res. 89:5733-62.

Smith, S. W., Knapp, J. S., and McPherson, R. C. 1993. Seismicity of the Gorda Plate, structure of the continental margin, and an eastward jump of the Mendocino Triple Junction. Jour. Geophys. Res. 98:8153-71.

Snavely, P. D., Jr. 1987. Tertiary geologic framework, neotectonics, and petroleum potential of the Oregon-Washington continental margin. *In* Scholl, D. W., Grantz, A., and Vedder, J. G., eds. Geology and resource potential of the continental margin of western North America and adjacent ocean basins—Beaufort Sea to Baja California. Circum-Pacific Council for Energy and Mineral Resources Earth Science Series 6:305-35.

Solonenko, V. P. 1965. Tectonics of the Muya earthquake area. Izvestia Acad. Sci. USSR Geol. Ser. 4:58-70 (R).

Solonenko, V. P. 1968. Strong earthquakes according to seismostatistics. *In* Seismic Regionalization of the USSR. Nauka, Moscow, 60-66. (Russian).

Solonenko, V. P., Kurushin, R. A., and Khil'ko, S. D. 1966. Strong earthquakes, in Recent Tectonics, Volcanoes, and Seismicity of the Stanovoy Upland. Moscow, Nauka, 145-71 (Russian).

Solonenko, V. P., and Treskov, A. A. 1960. Central Baikal earthquake, 29 August 1959. Irkutsk, USSR: Irkutsk Book Publishers, 37 p (Russian).

Somerville, P. 1978. The accommodation of plate collision by deformation in the Izu block, Japan. Bulletin of the Earthquake Research Institute, University of Tokyo 53:629-48.

Soufleris, C., Jackson, J. A., King, G. C. P., Spencer, C. P., and Scholz, C. H. 1982. The 1978 earthquake sequence near Thessaloniki (northern Greece). Geophys. Journal Royal Astronomical Soc. 68:429-58.

Spell, T. L., and McDougall, I. 1992. Revisions to the age of the Brunhes-Matuyama boundary and the Pleistocene geomagnetic polarity timescale. Geophys. Res. Lett. 19:1181-84.

Spotila, J., and Sieh, K. 1995. Geologic investigations of the "slip gap" in the surficial ruptures of the 1992 Landers earthquake, southern California. Jour. Geophys. Res. 100:543-59.

Stefánsson, R., Bödvarsson, R., Slunga, R., Einarsson, P., Jakobsdóttir, S., Bungum, H., Gregersen, S., Havskov, J., Hjelme, J., and Korhonen, H. 1993. Earthquake prediction research in the South Iceland Seismic Zone and the SIL project. Seismological Society of America Bulletin 83:696-716.

Stein, M., Wasserburg, G. J., Aharon, P., Chen, J. H., Zhu, Z. R., Bloom, A., and Chappell, J. 1993. TIMS U-series dating and stable isotopes of the last interglacial event in Papua New Guinea: Geochim. et Cosmochim. Acta 57:2541-54.

Stein, R. S., and Barrientos, S. E. 1985. Planar high-angle faulting in the Basin and Range: geodetic analysis of the 1983 Borah Peak, Idaho, earthquake. Jour. Geophys. Res. 90:11,355-66.

Stein, R. S., Briole, P., Ruegg, J.-C., Tapponnier, P., and Gasse, F. 1991. Contemporary, Holocene, and Quaternary deformation of the Asal rift, Djibouti: implications for the mechanics of slow spreading ridges: Jour. Geophys. Res. 96:21,789-806.

Stein, R.S., and Ekström, G. 1992. Seismicity and geometry of a 110-km-long blind thrust fault. 2. Synthesis of the 1982-1985 California earthquake sequence: Jour. Geophys. Res. 97:4865-83.

Stein, R. S., and King, G. C. P. 1984. Seismic potential revealed by surface folding: 1983 Coalinga earthquake. Science 224:869-72.

Stein, R. S., King, G. C. P., and Rundle, J. B. 1988. The growth of geological structures by repeated earthquakes: 2. Field examples of continental dip-slip faults. Jour. Geophys. Res. 93:13, 319-31.

Stein, R. S., and Lisowski, M. 1983. The 1979 Homestead Valley earthquake sequence, California—control of aftershocks and postseismic deformation. Jour. Geophys. Res. 88:6477-90.

Stein, R. S., and Thatcher, W. 1981. Seismic and aseismic deformation associated with the 1952 Kern County, California, earthquake and relationship to the Quaternary history of the White Wolf fault. Jour. Geophys. Res. 86:4913-28.

Stein, R. S., and Yeats, R. S. 1989. Hidden earthquakes: Scientific American 260 6:48-57.

Stein, S., Engeln, J. E., De Mets, C., Gordon, R. G., Woods, D. R., Lundgren, P., Argus, D., Quibble, D., Stein, C., Weistein, S., and Wiens, D. A. 1986. The Nazca-South America convergence rate and the recurrence of the great 1960 Chilean earthquake. Geophys. Res. Lett. 13:713-16.

Stevenson, A. J., and Embley, R. 1987. Deep-sea fan bodies, terrigenous turbidite sedimentation, and petroleum geology, Gulf of Alaska. *In* Scholl, D. W., Grantz, A., and Vedder, J. G., eds. Geology and resource potential of the continental margin of western North America and adjacent ocean basins—Beaufort Sea to Baja California. Circum-Pacific Council for Energy and Mineral Resources Earth Science Series 6:503-22.

Stewart, I. S. 1993. Sensitivity of fault-generated scarps as indicators of active tectonism: some constraints from the Aegean region. *In* Thomas, D. S. G., and Allison, R. J., eds. Landscape Sensitivity. J. Wiley & Sons, pp. 129-147.

Stewart, I. S., and Hancock, P. L. 1991. Scales of structural heterogeneity within neotectonic normal fault zones in the Aegean region. Journal of Structural Geology 13:191-204.

Stiros, S. C. 1995. The 1953 seismic surface fault: implications for the modeling of the Sousaki (Corinth area, Greece) geothermal field. Jour. Geodynamics 20:167-80.

Stiros, S. C., and Vouyoukalakis, G. in prep. The 1970, Yali (SE edge of the Aegean volcanic arc) earthquake swarm: surface faulting associated with a small earthquake.

Stock, J., and Molnar, P. 1988. Uncertainties and implications of the late Cretaceous and Tertiary position of North America relative to the Farallon, Kula and Pacific plates. Tectonics 7:1339-84.

Strange, W. E. 1981. The impact of refraction correction on leveling interpretations in southern California. Jour. Geophys. Res. 86:2809-24.

Stuart-Alexander, D. E. 1981. Reservoir induced seismicity. U. S. Geol. Survey Open-File Report 81-167:461-64.

Stuiver, M. 1991. Radiocarbon dating. in Morrison, R.B., ed., Quaternary nonglacial geology: conterminous U.S.: Geol. Soc. America Decade of North American Geology K-2:46-49.

Stuiver, M. and Kra, R., eds. 1986. Proceedings of the 12th International Radiocarbon Conference, June 24-28, 1985. Radiocarbon 28:175-1030.

Stuiver, M., and Pearson, G. W. 1986. High-precision calibration of the radiocarbon time scale, A.D. 1950-500 B.C. Radiocarbon 28:805-838

Stuiver, M., Braziunas, T. F., Becker, B., and Kromer, B. 1991. Climatic, solar, oceanic, and geomagnetic influences on late-glacial and Holocene atmospheric $^{14}C^{12}/C$ change. Quaternary Research 35:1-24.

Suárez, G., Molnar, P., and Burchfiel, B. C. 1983. Seismicity, fault plane solutions, depth of faulting,and active tectonics of the Andes of Peru, Ecuador, and southern Colombia. Jour. Geophys. Res. 88:10,403-28.

Suggate, R. P. 1957. The geology of Reefton subdivision. New Zealand Geol. Survey Bull. n.s. 56, 146 p.

Sugi, N., Chinzei, K., and Uyeda, S. 1983. Vertical crustal movements of northeast Japan since middle Miocene. *In* Hilde, T. W. C., and Uyeda, S., eds. Geodynamics of the western Pacific-Indonesian region. Am. Geophys. Union Geodynamics Series 11:317-29.

Sugiyama, Y. 1994. Neotectonics of southwest Japan due to the right-oblique subduction of the Philippine Sea plate. Geofísica Internacional 33:53-76.

Sugiyama, Y., Awata, Y., and Tsukuda, T. 1991. Holocene activity of the Miboro fault system, central Japan, and its implications for the Tensho earthquake of 1586—verification by excavation survey. Seismol. Soc. Japan Programme and Abstracts, No. 2, p. 260. (J).

Sulstarova, E., and Kociaj, S. 1980. The Dibra (Albania) earthquake of November 30, 1967. Tectonophysics 67:333-43.

Sultan, D. I. 1931. The Managua earthquake. Military Engineer 23:354-61.

Suppe, J. 1983. Geometry and kinematics of fault-bend folding. American Journal of Science 283:684-721.

Suppe, J. 1985. Principles of Structural Geology. Englewood Cliffs, NJ: Prentice-Hall, 537 p.

Suppe, J. 1987. the active Taiwan mountain belt. *In* Schaer, J.-P., and Rodgers, J., eds. The Anatomy of Mountain Ranges. Princeton, New Jersey: Princeton University Press, pp. 277-93.

Suppe, J., and Medwedeff, D. A. 1990. Geometry and kinematics of fault-propagation folding. Eclogae Geologicae Helvetiae 83:409-54.

Suzuki, S., Sasatani, T., and Motoya, Y. 1983. Double seismic zone beneath the middle of Hokkaido, Japan, in the southwestern side of the Kurile arc. Tectonophysics 96:59-96.

Swan, F. H., III. 1988. Temporal clustering of paleoseismic events on the Oued Fodda fault, Algeria. Geology 16:1092-95.

Swan, F. H., III, Schwartz, D. P., and Cluff, L. S. 1980. Recurrence of moderate to large magnitude earthquakes produced by surface faulting on the Wasatch fault zone, Utah: Seismological Society of America Bulletin 70:1431-62.

Sykes, L. 1967. Mechanism of earthquakes and nature of faulting on the mid-ocean ridges. Jour. Geophys. Res. 72, p. 2131-53.

Sykes, L. 1971. Aftershock zones of great earthquakes, seismicity gaps and earthquake prediction for Alaska and the Aleutians. Jour. Geophys. Res. 76:8021-41.

Sylvester, A. G., 1988. Strike-slip faults. Geological Society of America Bulletin 100:1666-1703.

Sylvester, A. G., 1986. Near-field tectonic geodesy. *In* Wallace, R. E., ed. Active Tectonics. National Academy Press, pp. 164-80.

Szabo, B. J., and Rosholt, J. N., 1991. Conventional uranium-series and uranium-trend dating. *In* Morrison, R. B., ed. Quaternary nonglacial geology:

conterminous U.S.: Geol. Soc. America Decade of North American Geology K-2:55-60.

Tabor, R. W., and Cady, W. M. 1978. The structure of the Olympic Mountains, Washington—analysis of a subduction zone. U. S. Geol. Survey Prof. Paper 1033, 38 p.

Taira, A., and Ogawa, Y., eds. 1988. The Shimanto belt, southwest Japan—studies on the evolution of an accretionary prism. Modern Geology 12:1-4, 542 p.

Taira, A., and Pickering, K. T. 1991. Sediment deformation and fluid activity in the Nankai, Izu-Bonin and Japan forearc slopes and trenches. Phil. Trans. Roy. Soc. London 335:289-313.

Taira, A., Hill, I., Firth, J., et al. 1991. Proceedings of the Ocean Drilling Program, Initial Reports, 131. College Station, TX: Ocean Drilling Program.

Takeuchi, A., and Sakai, H. 1985. Recent event of activity along the Atotsugawa fault, central Japan—a paleomagnetic method for dating of fault activity. Active Fault Research 1:67-74 (Japanese).

Talwani, M., and Pitman, W. C., III, eds. 1977. Island arcs, deep sea trenches and back arc basins. Am. Geophys. Union Maurice Ewing Series 1, 470 p.

Tang C., Wen D., Deng T., and Huang S. 1976. A preliminary study on the characteristics of the ground features during the Luhuo M = 7.9 earthquake 1973 and the origin of the earthquake. Acta Geophysica Sinica 19(1):18-27 (C, Eng. abs.).

Tanna Fault Trenching Research Group (TFTRG). 1983. Trenching study for Tanna fault, Izu, at Myoga, Shizuoka Prefecture, Japan. Earthquake Research Inst., Univ. Tokyo, Bull. 58:797-830. (J, Eng. abs.).

Tapponnier, P., and Molnar, P. 1977. Active faulting and tectonics in China. Jour. Geophys. Res. 82:2905-30.

Tapponnier, P., and Molnar, P. 1979. Active faulting and Cenozoic tectonics of the Tien Shan, Mongolia, and Baykal regions. Jour. Geophys. Res. 84,:3425-59.

Tapponnier, P., and Peltzer, G. 1988. Formation and evolution of strike-slip faults, rifts and basins during the India-Asia collision—an experimental approach: Jour. Geophys. Res. 93:15085-117.

Tapponnier, P., Armijo, R., Manighetti, I., and Courtillot, V. 1990. Bookshelf faulting and horizontal block rotation between overlapping rifts in southern Afar: Geophys. Res. Lett. 17, p. 1-4.

Tapponnier, P., Mercier, J. L., Armijo, R., Han T., and Zhou J. 1981. Field evidence for active normal faulting in Tibet: Nature 294:410-14.

Tapponnier, P., Meyer, B., Avouac, J. P., Peltzer, G., Gaudemer, Y., Guo S., Xiang H., Yin K., Chen Z., Cai S., and Dai H. 1990. Active thrusting and folding in the Qilian Shan, and decoupling between upper crust and mantle in northeastern Tibet. Earth and Planet. Sci. Lett. 97:382-403.

Tapponnier, P., Lacassin, R., Leloup, P., Scharer, U., Zhong, D., Wu, H., Liu, X., Ji, S., Zhang, L., and Zhong, J. 1990. The Ailao Shan/Red River metamorphic belt: Tertiary left-lateral shear between Indochina and South China: Nature 343:431-37.

Tapponnier, P., Peltzer, G., Ledain, A., Armijo, R., and Cobbold, P. 1982. Propagating extrusion tectonics in Asia—New insights from simple experiments with plasticine: Geology 10:611-16.

Taylor, F. W., Edwards, R. L., Wasserburg, G. J., and Frohlich, C. 1990. Seismic recurrence intervals and timing of aseismic subduction inferred from emerged corals and reefs of the central Vanuatu (New Hebrides) frontal arc. Jour. Geophys. Res. 95:393-408.

Taylor, G. C., and Bryant, W. A. 1980. Surface rupture associated with the Mammoth Lakes earthquakes of 25 and 27 May, 1980. California Div. Mines and Geology Spec. Rept. 150:49-67.

Taymaz, T., Eyidogan, H., and Jackson, J. 1991. Source parameters of large earthquakes in the East Anatolian fault zone (Turkey). Geophys. J. Int. 106:537-50.

Taymaz, T., Jackson, J., and McKenzie, D. 1991. Active tectonics of the north and central Aegean Sea: Geophys. J. Int. 106: 433-90.

Tchalenko, J.1970. Similarities between shear zones of different magnitudes. Geol. Soc. America Bull. 81:1625-40.

Tchalenko, J. S. 1975. Seismicity and structure of the Kopet Dagh (Iran, USSR). Phil. Trans. Roy. Soc. London 278:1-25.

Tchalenko, J. S., and Ambraseys, N. N. 1970. Structural analysis of the Dasht-e Bayaz (Iran) earthquake fractures. Geol. Soc. America Bull. 81:41-60.

Tchalenko, J. S., and Berberian, M. 1974. The Salmas (Iran) earthquake of May 6th, 1930. Annales Geofisica (Rome) 27:151-212.

Tchalenko, J. S., and Berberian, M. 1975. Dasht-e Bayaz fault, Iran: earthquake and earlier related structures in bed rock. Geol. Soc. America Bull. 86:703-09.

Tchalenko, J. S., and Braud, J. 1974. Seismicity and structure of the Zagros (Iran): The Main Recent fault between 33 and 35 N. Phil. Trans. Roy. Soc. Lond.(1262)227:1-25.

Tchalenko, J. S., Braud, J., and Berberian, M. 1974. Discovery of three earthquake faults in Iran. Nature 248:661-63.

Te Punga, M. T. 1957. Live anticlines in western Wellington. New Zealand Jour. Sci. Technology B38(5):433-46.

Thelin, G. P., and Pike, R. J., 1991. Landforms of the conterminous United States: U.S. Geol. Survey Map 1-2220.

Thatcher, W. 1975. Strain accumulation and release mechanism of the 1906 San Francisco earthquake. Jour. Geophys. Res. 80:4862-72.

Thatcher, W. 1984. The earthquake deformation cycle at the Nankai trough, southwest Japan. Jour. Geophys. Res. 89:3087-101.

Thatcher, W. 1986. Geodetic measurement of active-tectonic processes. *In* Wallace, R. E., ed. Active Tectonics. National Academy Press, pp. 155-163.

Thatcher, W. 1990. Order and diversity in the modes of circum-Pacific earthquake recurrence. Jour. Geophys. Res. 95:2609-24.

Thatcher, W., and Lisowski, M. 1987. Long-term seismic potential of the San Andreas fault southeast of San Francisco, California. Jour. Geophys. Res. 92:4771-84.

Thelin, G. P., and Pike, R. J., 1991. Landforms of the conterminous United States: U.S. Geol. Survey Map 1-2200.

Thierstein, H. R., Geitzenauer, K. R., Molfino, B., and Shackleton, N. J. 1977. Global synchroneity of late Quaternary coccolith datum levels: Validation by oxygen isotopes. Geology 5:400-04.

Thomas, A. P., and Rockwell, T. K. 1996. A 300- to 550-year history of slip on the Imperial fault near the U. S.-Mexico border: Missing slip at the Imperial fault bottleneck. Jour. Geophys. Res. 101:5987-97.

Thomas, E., and Sieh, K. 1981. Quaternary development of the Elkhorn Hills along the San Andreas fault (abstract). Geological Society of America Abstracts with Programs 13: 534.

Thornburg, T. M., and Kulm, L. D.1981. Sedimentary basins of the Peru continental margin: structure, stratigraphy, and Cenozoic tectonics from 6°S to 16°S latitude. Geol. Soc. America Mem. 154:393-422.

Thornburg, T. M., Kulm, L. D., and Hussong, D. M. 1990. Submarine-fan development in the southern Chile Trench: A dynamic interplay of tectonics and sedimentation. Geol. Soc. America Bull. 102:1658-80.

Thouveny, N., and Creer, K. M. 1992. On the brevity of the Laschamp excursion: Bull. Soc. Géol. France 163:771-80.

Tichelaar, B. W., and Ruff, L. J. 1991.Seismic coupling along the Chilean subduction zone. Jour. Geophys. Res. 96:11,997-12,022.

Tichelaar, B. W., and Ruff, L. J. 1993. Depth of seismic coupling along subduction zones. Jour. Geophys. Res. 98:2017-37.

Tocher, D. 1956. Movement on the Rainbow Mountain fault: Seismological Society of America Bulletin 46:10-14.

Toppozada, T. R. 1981. Preparation of isoseismal maps and summaries of reported effects for pre-1900 California earthquakes. Calif. Div. Mines and Geol. Open-File Rep. 81-11.

Toppozada, T. R., and Cramer, C. H. 1978. Ukiah earthquake, 25 March 1978, seismicity possibly induced by Lake Mendocino. California Geology 31:275-81.

Toppozada, T. R., and Morrison, P. W. 1982. Earthquakes and lake levels at Oroville, California. California Geology 35:115-18.

Toppozada, T. R., and Parke, D. L. 1982. Area damaged by the 1868 Hayward earthquake and recurrence of damaging earthquakes near Hayward. *In* Hart, E. W., Hirschfeld, S. E., and Schulz, S. S., eds. Proceedings of the Conference on Earthquake Hazards in the Eastern San Francisco Bay Area. Calif. Div. Mines and Geol. Spec. Rep. 62:321-28.

Toppozada, T. R., Parke, D., and Higgins, C. T. Seismicity of California 1900-1931. Calif. Div. Mines and Geol. Spec. Rep. 135 39 p.

Toppozada, T. R., Real, C., and Parke, D. L. 1981. Preparation of isoseismal maps and summaries of reported effects for pre-1900 California earthquakes. Calif. Div. Mines and Geol. Open-File Report 81-11, 182 p.

Townsend, P. D., Rendell, H. M., Aitken, M. J., Bailiff, I. K., Durrani, S. A., Fain, J., Grun, R., Mangini, A., Me'dahl, V., and Smith, B. W. 1988. Thermoluminescence and electron-spin resonance dating: Part II—Quaternary Applications. Quaternary Science Reviews 7:243-536.

Tregoning, P., Brunner, F., Bock, Y., Puntodewo, S., McCaffrey, R., Genrich, J., Calais, E., Rais, J., and Subarya, C. 1994. First geodetic measurement of convergence across the Java Trench. Geophys. Res. Lett. 21:2135-38.

Tréhu, A. M., Asudeh, I., Brocher, T. M., Luetgert, J. H., Mooney, W. D., Nabelek, J. L., and Nakamura, Y. 1994. Crustal architecture of the Cascadia forearc. Science 266:237-43.

Triep, E. G., Abers, G. A., Lerner-Lam, A. L., Mishatkin, V., Zakharchenko, N., and Starovoit, O. 1995. Active thrust front of the Greater Caucasus: The April 29, 1991, Racha earthquake sequence and its tectonic implications. Jour. Geophys. Res. 100:4011-33.

Trifonov, V. G. 1978,. Late Quaternary tectonic movements of western and central Asia. Geol. Soc. America 89:1059-72.

Trifonov, V. G. 1988. Mongolia—an intracontinental region of predominantly recent strike-slip displacement: Active faults. *In* Neotectonics and Contemporary Geodynamics of Mobile Belt. Moscow, Nauka 239-72 (R).

Trifonov, V. G., Bayractutan, M. S., Karakhanian, A. S., and Ivanova, T. P. 1993. The Erzincan earthquake of 13 March 1992 in eastern Turkey: Tectonic aspects. Terra Nova 5: 184-89.

Trifonov, V. G., Vostriakov, G. A., Lykov, V., Orazsahatov, H., and Skobelev, S. F. 1986. Tectonic aspects of the 1983 Kumdag earthquake in western Turkmenia. Izvestia, USSR Academy of Science, Geol. Serial N5:3-16 (R).

Tse, S. T., and Rice, J. R. 1986. Crustal earthquake instability in relation to the depth variation of frictional slip properties. Jour. Geophys. Res. 91:9452-72.

Tsuboi, S. 1922. Note on the Oomati earthquakes of 1918. Reports of the Imperial Earthquake Investigation Committee in Japanese Language (translation) 98:13-21.

Tsukuda, E. 1992. Active tectonics of southwest Japan arc controlled by the westward translation of the forearc sliver. Geol. Soc. Japan Memoirs 40:235-50 (J, Eng. abs.).

Tsuneichi, Y., and Nakamura, K. 1970. Faulting associated with the Matsushiro Swarm earthquakes. Bull. Earthquake Research Inst., University of Tokyo 48:29-51.

Tsuneishi, Y., Ito, T., and Kano, K. 1978. Surface faulting associated with the 1978 Izu-Oshima-Kinkai earthquake. Bull. Earthquake Research Inst., University of Tokyo 53: 649-74.

Tsutsumi, H., Okada, A., Nakata, T., Ando, M., and Tsukuda, T. 1991. Timing and displacement of Holocene faulting on the Median Tectonic Line in central Shikoku, southwest Japan. Journal of Structural Geology 13:227-33.

Tsutsumi, H., and Okada, A. 1996. Segmentation and Holocene surface faulting on the Median Tectonic Line, southwest Japan. Jour. Geophys. Res. 101:5855-71.

Tsuya, H. 1938. Report on an investigation of the Kussharo earthquake of May 29, 1938. Zishin 10:285-313 (J).

Tsuya, H. 1944. Geological observations of earthquake faults of 1943 in Tottori Prefecture. Tokyo Imperial Univ. Earthquake Res. Inst. Bull. 22:1-32.

Tsuya, H. 1946. The Fukozu fault, a remarkable earthquake fault formed during the Mikawa earthquake of January 13, 1945. University of Tokyo Earthquake Res. Inst. Bull. 24:59-75.

Tsuya, H., ed. 1950. The Fukui earthquake of June 28, 1948. Tokyo, Special Committee for the Study of the Fukui earthquake, 197 p., 2 pl.

Tucholke, B., and Schouten, H. 1988. Kane Fracture Zone. Marine Geophysical Researches 10:1-39.

Turcotte, D. L., and Schubert, G. 1982. Geodynamics: J.Wiley & Sons, 450 p.

Tuttle, M. T., and Sykes, L. R. 1992. Re-evaluation of several large historic earthquakes in the vicinity of the Loma Prieta and Peninsular segments of the San Andreas fault, California. Seismological Society of America Bulletin 82:1802-20.

Twiss, R. J., and Moores, E. M. 1992, Structural Geology. New York: W. H. Freeman and Co., 532 p.

U. S. Geological Survey Staff. 1990. The Loma Prieta, California, earthquake: An anticipated event: Science 247:286-93.

Uemura, T., and Mizutani, S. 1984. Geological Structures: J. Wiley, 309 p.

Ulrich, F. 1941. The Imperial Valley earthquake of 1940. Seismological Society of America Bulletin 31:13-31.

Untung, M., Buyung, N., Kertapati, E., Undang, and Allen, C. R. 1985. Rupture along the Great Sumatran fault, Indonesia, during the earthquakes of 1926 and 1943. Seismological Society of America Bulletin 75:313-17.

Uyeda, S. 1978. The new view of the earth. San Francisco, W. H. Freeman, 217 p.

Uyeda, S. 1991. The Japanese island arc and the subduction process. Episodes 14:190-98.

Uyeda, S., and Kanamori, H. 1979. Back-arc opening and the mode of subduction. Jour. Geophys. Res. 84:1049-61.

Valensise, G., and Ward, S.N. 1991. Long-term uplift of the Santa Cruz coastline in response to repeated earthquakes along the San Andreas fault. Seismological Society of America Bulletin 81:1694-1704.

Valensise, G., and Pantosti, D. 1992. A 125-Kyr-long geological record of seismic source repeatability:

The Messina Straits (southern Italy) and the 1908 earthquake (M_s $7^1/_2$). Terra Nova 4:472-83.

Van Couvering, J. A., ed. The Pleistocene Boundary and the Beginning of the Quaternary. Cambridge University Press, in press.

Van Dissen, R., and Yeats, R. S. 1991. Hope fault, Jordan thrust, and uplift of the Seaward Kaikoura Range, New Zealand. Geology 19:393-96.

Varnes, D. J. 1978. Slope movement types and processes. *In* Schuster, R. L., and Krizek, R. J., eds. Landslides—analysis and control. National Academy of Sciences Transportation Research Board Special Report 176:12-33.

Vedder, J., and Wallace, R. E. 1970. Map showing recently active breaks along the San Andreas and related faults between Cholame Valley and Tejon Pass, California. U. S. Geol. Survey Misc. Geol. Investigations Map I-574, scale 1:24,000 (2 sheets).

Vickery, F. F. 1927. The interpretation of the physiography of the Los Angeles coastal belt. American Association of Petroleum Geologists Bulletin11:417-24.

Vilgelmzon, P. M. 1947. Kemino-Chuiskoe earthquake of June 21, 1938. Academy of Science of Kazak SSR, 39 p. (R).

Vincent, P. 1989. Geodetic deformation of the Oregon Cascadia margin. Eugene, University of Oregon M.S. thesis, 86 p.

Vita-Finzi, C. 1981. Late Quaternary deformation in the Makran coast of Iran. Zeitschrift für Geomorph. 40:213-26.

Vita-Finzi, C. 1986. Recent Earth Movements, an introduction to neotectonics. London: Academic Press, 226 p.

Vita-Finzi, C. 1991. First-order ^{14}C dating Mark II. Quaternary Proceedings No. 1. Quaternary Research Association, Cambridge: 11-17.

Vita-Finzi, C., and Hidayat, S. 1991. Holocene uplift in West Timor. Jour. Southeast Asian Earth Sciences 6:387-93.

Vita-Finzi, C. and King, G. C. P. 1985. The seismicity, geomorphology and structural evolution of the Corinth area of Greece. Philosophical Transaction of Royal Society London A314:379-407.

Vogfjörd, K. S., and Langston, C.A. 1987. The Meckering earthquake of 14 October 1968: A possible downward propagating rupture. Seismological Society of America Bulletin 77:1558-78.

Voight, B. 1978. Lower Gros Ventre slide, Wyoming, U.S.A. *In* Voight, B., ed. Rockslides and Avalanches—1. Natural phenomena. New York: Elsevier Publishing Co., 113-66.

von Huene, R., and Culotta, R. 1989. Tectonic erosion at the front of the Japan Trench convergent margin. Tectonophysics 160:75-90.

von Huene, R., Fisher, M. A., and Bruns, T. R. 1987. Geology and evolution of the Kodiak margin, Gulf of Alaska. *In* Scholl, D. W., Grantz, A., and Vedder, J. G., eds. Geology and resource potential of the continental margin of western North America and adjacent ocean basins—Beaufort Sea to Baja California. Circum-Pacific Council for Energy and Mineral Resources Earth Science Series 6:229-68.

von Huene, R., Suess, E., and Leg 112 Shipboard Scientists. 1988. Ocean Drilling Program Leg 112, Peru continental margin: Part 1, tectonic history. Geology 16:934-38.

Walcott, R. 1978. Geodetic strains and large earthquakes in the axial tectonic belt of North Island, New Zealand. Jour. Geophys. Res. 83:4419-29.

Wald, D., and Heaton, T. 1994. Spatial and temporal distribution of slip for the 1992 Landers, California, earthquake. Seismological Society of America Bulletin 84:668-91.

Wald, D., and Wallace, T. 1986. A seismically active section of the Southwest Indian Ridge. Geophys. Res. Lett. 13:1003-06.

Wallace, R. E. 1968. Notes on steam channels offset by the San Andreas fault, southern Coast Ranges, California. *in* Dickinson, W. R., and Grantz, A., ed. Conference on Geologic Problems of the San Andreas Fault System, Proceedings 11. Stanford University Publications in the Geological Sciences, pp. 6-21.

Wallace, R. E. 1968. Earthquake of August 19, 1966, Varto area, eastern Turkey. Seismological Society of America Bulletin 58:47-102.

Wallace, R. E. 1977. Profiles and ages of young fault scarps, north-central Nevada. Geol. Soc. America Bull. 88:1267-81.

Wallace, R. E. 1984a. Eyewitness account of surface faulting during the earthquake of 28 October 1983, Borah Peak, Idaho. Seismological Society of America Bulletin 74:1091-94.

Wallace, R. E. 1984b. Fault scarps formed during the earthquakes of October 2, 1915, Pleasant Valley, Nevada, and some tectonic implications. U. S. Geol. Survey Prof. Paper 1274-A.

Wallace, R. E., and Roth, E. 1967,. Rates and patterns of progressive deformation. *In* The Parkfield-Cholame California, Earthquakes of June-August 1966. U. S. Geol. Survey, pp. 23-39.

Wang Y. and Deng Q., eds. 1988. Active fault system around Ordos Massif. The Research Group on "Active Fault System around Ordos Massif", Beijing: State Seismology Bureau, 335 p.

Wang Y. and Song F. 1989. The active surrounding Ordos faults. *In* Ding G., ed. Atlas of Active Faults in China. Xi'an: Seismological Press, pp. 23-52.

Wang T., Li X., Zheng B., and Wang Y. 1991. Newly discovered geological evidences for the seismogenic structure of 1695 strong earthquake in Linfen, Shanxi province, China. Seismology and Geology 13(1):76-77. (C).

Ward, S. N. 1994. A multidisciplinary approach to seismic hazard in southern California. Seismological Society of America Bulletin 84:1293-1309.

Ward, S.N., and Valensise, G.R. 1989. Fault parameters and slip distribution of the 1915 Avezzano, Italy, earthquake observed from geodetic observations. Seismological Society of America Bulletin 79:690-710.

Watanabe, H. 1985. Comprehensive lists of tsunamis in Japan. Tokyo University Press, 206 p. (Japanese).

Watanabe, K., and Sato, H. 1928. The Tango earthquake of 1927. Imp. Geol. Survey of Japan Report 100, 102 p. (J), 16 p. (Eng.).

Watkins, J. S., and Drake, C. C., eds. 1982. Studies in continental margin geology: Convergent margins. American Assoc. Petroleum Geologists Memoir 34:307-533.

Webb, T. H., and Kanamori, H. 1985. Earthquake focal mechanisms in the eastern Transverse Range and San Emigdio Mountains, southern California and evidence for a regional decollement. Seismological Society of America Bulletin 75:737-57.

Wehmiller, J. F. 1993. Applications of organic geochemistry for Quaternary research: aminostratigraphy and aminochronology. Chapter 36 in Engel, M. H., and Macko, S. A., eds. Organic Geochemistry. New York: Plenum Press, 755-83.

Weldon, R. J. 1986. The late Cenozoic geology of Cajon Pass: implications for tectonics and sedimentation along the San Andreas fault. California Institute of Technology, Pasadena, Ph.D. thesis, 400 p.

Wells, D. L., and Coppersmith, K. J. 1994. New empirical relationships among magnitude, rupture length, rupture area, and surface displacement. Seismological Society of America Bulletin 84:974-1002

Wells, R. E. 1989. Mechanisms of Cenozoic tectonic rotations, Pacific Northwest convergent margin, U.S.A. *In* Kissel, C., and Laj, C., eds. Paleomagnetic Rotations and Continental Deformation. NATO ASI Series C, 254:313-25. Dordrecht, Netherlands: Kluwer Academic Publishers.

Wentworth, C. M., and Zoback, M. D. 1989. The style of late Cenozoic deformation at the eastern front of the California Coast Ranges. Tectonics 8:237-46.

Wentworth, C. M., and Zoback, M. D. 1990. Structure of the Coalinga region and thrust origin of the earthquake. *In* Rymer, M. J., and Ellsworth, W. L., eds. The Coalinga, California, earthquake of May 2, 1983. U. S. Geol. Survey Prof. Paper 1487:41-68.

Wernicke, B. P., Christiansen, R., England, P. C., and Sonder, L. J. 1987. Tectonomagmatic evolution of Cenozoic extension in the North American Cordillera. *In* Coward, M. P., Dewey, J. F., and Hancock, P. L., eds. Continental Extensional Tectonics. Geological Society Special Publication 28:203-21.

Wesnousky, S. G. 1986. Earthquakes, Quaternary faults, and seismic hazards in California: Jour. Geophys. Res. 91:12587-631.

Wesnousky, S. G., Scholz, C. H., and Shimazaki, K. 1982. Deformation of an island arc: Rates of moment release and crustal shortening in intraplate Japan determined from seismicity and Quaternary fault data. Jour. Geophys. Res. 87:6829-52.

Wesnousky, S. G., Scholz, C., Shimazaki, K., and Matsuda, T. 1984. Integration of geological and seismological data for the analysis of seismic hazard: A case study of Japan. Seismological Society of America Bulletin 74:687-708.

Westgate, J. A. 1989. Isothermal plateau fission-track ages of hydrated glass shards from silicic tephra beds. Earth and Planetary Sci. Lett. 95:226-34.

Whitcomb, J. H., Allen, C. R., Garmany, J. D., and Hileman, J. A. 1973. San Fernando earthquake series, 1971: Focal mechanisms and tectonics. Reviews of Geophysics and Space Physics 11:693-730.

White, R. S. 1985. The Guatemala earthquake of 1816 on the Chixoy-Polochic fault. Seismological Society of America Bulletin 75, p. 455-73.

White, R. S., and Louden, K. E. 1982. The Makran continental margin: structure of a thickly-sedimented convergent plate boundary. Amer. Assoc. Petroleum Geol. Memoir 34:499-518.

Whitehouse, I. E., McSaveney, M. J., Knuepfer, P. L. K., and Chinn, T. 1986. Growth of weathering rinds on Torlesse sandstone, Southern Alps, New Zealand. *In* Colman, S. M., and Dethier, D. P., eds. Rates of chemical weathering of rocks and minerals. Orlando, Florida: Academic Press, pp. 419-35.

Wilcox, R. E., Harding, T. P., and Seely, D. R. 1973. Basic wrench tectonics. Amer. Assoc. Petroleum Geol. Bull. 57:74-96.

Williams, P., and Magistrale, H. 1989. Surface ruptures on cross-faults in the 24 November 1987 Superstition Hills, California, earthquake sequence. Seismological Society of America Bulletin 79:390-410.

Williams, R., Odum, J., Pratt, T., Shedlock, K., and Stephenson, W.,1995. Seismic surveys assess earthquake hazard in the New Madrid area. The Leading Edge 14:30-34.

Wilson, D. S.1986. A kinematic model for the Gorda deformation zone as a diffuse southern boundary of the Juan de Fuca plate. Jour. Geophys. Res. 91:10,259-69.

Wilson, D. S. 1989. Deformation of the so-called Gorda plate. Jour. Geophys. Res. 94:3065-75.

Wilson, J.T. 1965. A new class of faults and their bearing on continental drift. Nature 207: 343-47.

Winter, T., Avouac, J.P., and Lavenu, A. 1993. Late Quaternary kinematics of the Pallatanga strike-slip fault (Central Ecuador) from topographic measurements of displaced morphological features. Geophys. J. Int. 115:905-20.

Wintle, A. G., and Prószynska, H. 1983. TL-dating of loess in Germany and Poland. PACT:9:547-54.

Withjack, M. O., Olson, J., and Peterson, E. 1990. Extensional models of forced folds. American Assoc. Petroleum Geol. Bull. 74:1038-54.

Witkind, I. J., Myers, W. B., Hadley, J. B., Hamilton, W., and Fraser, G. D. 1962. Geologic features of the earthquake at Hebgen Lake, Montana, August 17, 1959. Seismological Society of America Bulletin 52:163-80.

Wood, H. O., and Neumann, F. 1931. Modified Mercalli intensity scale of 1931. Seismological Society of America Bulletin 21:277-83.

Wood, P. R., and Blick, G. H. 1986. Some results of geodetic fault monitoring in South Island, New Zealand. Royal Society of New Zealand Bull. 24:39-45.

Woodcock, N. 1986. The role of strike-slip fault systems at plate boundaries. Philisophical Transactions of the Royal Society of London A 317:13-29.

Woodward, N. B., Boyer, S. E., and Suppe, J. 1989. Balanced geological cross-sections: an essential technique in geological research and exploration. Washington, D.C.: American Geophysical Union Short Course in Geology 6, 132 p.

Working Group on California Earthquake Probabilities (WGCEP) 1990. Probabilities of large earthquakes in the San Francisco Bay region, California. U. S. Geol. Survey Circular 1053, 51p.

Wu Z. and Deng Q.,1989. Deformation features and fracture mechanism of surface rupture of 1951 Bengco, Tibet M-8 earthquake. Seismology and Geology 11(1):15-26. (C, Eng. abs.).

Wu Z., Shentu B., Cao Z., and Deng Q. 1990a. The surface ruptures of Danxiong (Tibet) earthquake (M = 8) in 1411. Seismology and Geology 12(2):98-106. (C, Eng. abs.).

Wu Z., Cao Z., Shentu B., and Deng Q. 1990b. Preliminary research on the Nianqingtanggula mountain southeastern piedmont fault. J. Seism. Res. 13(1):40-50. (C).

Wu Z., Wang Y., Ren J., and Ye J. 1994. The active faults in the central Tibet plateau. Research on Active Fault 3:56-73. (C, Eng. abs.).

Wyss, M., and Koyanagi, R. Y. 1992. Seismic gaps in Hawaii. Seismological Society of America Bulletin 82:1373-87.

Xiang H., Fang Z., Xu J., Li R., Jia S., Hao S., Wang J., and Zhang W. 1988. Seismotectonic background and recurrence interval of great earthquakes in 1679 Sanhe-Pinggu M = 8 earthquake area. Seismology and Geology 10(1):29-37.

Xiao, H., and Suppe, J. 1992. Origin of rollover. American Association of Petroleum Geologists Bull. 76:509-29.

Xu X., and Deng Q. 1996. Nonlinear characteristics of paleoseismicity in China. Jour. Geophys. Res. 101:6209-31.

Xu X., and Deng Q. 1990. The features of late Quaternary activity of the piedmont fault of Mt. Huoshan, Shanxi province and 1303 Hongdong earthquake (Ms = 8). Seismology and Geology 12(1):21-30. (C., Eng. abs.).

Xu X., and Ma X. 1992. Geodynamics of the Shanxi Rift system, China. Tectonophysics 208:325-40.

Yamasaki, N. 1896. Preliminary report of the Rikuu earthquake. Report of the Imperial Earthquake Investigating Commission: 11:50-74 (J).

Yamasaki, N. 1925. Physiographical investigation of the Great Kwanto earthquake. Reports of the Imperial Earthquake Investigation Committee in Japanese Language (translation): 100B:11-54.

Yamasaki, N. 1927. On the cause of the Tajima earthquake of 1925. Reports of the Imperial Earthquake Investigation Committee in Japanese Language (translation): 101:31-34.

Yamasaki, N., and Tada, F.,1928. The Oku-Tango earthquake of 1927. Earthquake Research Institute 4:159-77.

Yamazaki, H. 1985. Features of earthquake faults. Chapter 4 of Earthquakes and Active Faults: Tokyo: ISU Company, pp. 237-442.

Yamazaki, H. 1992. Tectonics of a plate collision along the northern margin of Izu Peninsula, central Japan. Bull. Geol. Survey of Japan 42:603-57.

Yamazaki, H., Koide, H., and Tsukuda, E. 1979. Surface faults associated with the Izu-Oshima-Kinkai earthquake of 1978. Japan. Geol. Survey of Japan Spec. Rept. 7:7-35 (J, Eng. abs.).

Yamazaki, T., and Okamura, Y. 1989. Subducting seamounts and deformation of overriding forearc wedges around Japan. Tectonophysics 160:207-29.

Yang Z., Guo H., Ding G., and Xu D. 1988. Finding of Tekes-Zhaosu earthquake faults in Xinjiang and some discussion. Seismology and Geology 10(3):21-27 (C, Eng. abs.).

Yarwood, D. R., and Doser, D. I. 1990. Deflections of oceanic transform motion at a continental margin as deduced from waveform inversion of the 1939 Accra, Ghana, earthquake. Tectonophysics 172:341-49.

Yeats, R. S. 1981. Quaternary flake tectonics of the California Transverse Ranges. Geology 9:16-20.

Yeats, R. S. 1983. Large-scale Quaternary detachments in Ventura basin, southern California. Jour. Geophys. Res. 88:569-83.

Yeats, R. S. 1986a. Faults related to folding with examples from New Zealand. *In* Reilly, W. I. and Harford, B. E., eds. Recent Crustal Movements of the Pacific Region. Royal. Soc. New Zealand Bull. 24:273-92.

Yeats, R. S.,1986b. Active faults related to folding. *In* Wallace, R. E., ed., Active Tectonics. Washington, D.C.: National Academy Press, pp. 63-79.

Yeats, R. S. 1987. Tectonic map of Central Otago based on Landsat imagery. New Zealand Jour. Geology and Geophysics 30:261-271.

Yeats, R. S. 1988. Late Quaternary slip rate on the Oak Ridge fault, Transverse Ranges, California. Implications for seismic risk. Jour. Geophys. Res. 93:12,137-49.

Yeats, R. S. 1993. Tectonics: Converging more slowly: Nature 366:299-301.

Yeats, R. S. and Berryman, K. R., 1987. South Island, New Zealand, and Transverse Ranges, California: A seismotectonic comparison. Tectonics 6:363-76.

Yeats, R., Clark, M., Keller, E., and Rockwell, T. 1981. Active fault hazard in southern California: Ground rupture versus seismic shaking. Geol. Soc. America Bull. 92:189-96.

Yeats, R. S., and Hart, S. R., 1976. Introduction and Principal results, Leg 34, Deep Sea Drilling Project, in Yeats R. S., Hart, S. R., et al., Initial Reports of the Deep Sea Drilling Project, 34. Washington, D.C., U.S. Gov. Print. Off., 3-7.

Yeats, R. S., and Huftile, G. J. 1995. The Oak Ridge fault system and the 1994 Northridge earthquake. Nature 373:418-20.

Yeats, R. S., Huftile, G. J., and Grigsby, F. B. 1988. Oak Ridge fault, Ventura fold belt, and the Sisar decollement, Ventura basin, California. Geology 16:1112-16.

Yeats, R. S., Huftile, G. J., and Stitt, L. T. 1994. Late Cenozoic tectonics of the east Ventura basin, Transverse Ranges, California. Amer. Assoc. Petroleum Geologists Bull. 78:1040-74.

Yeats, R. S., and Hussain, A. 1989. Zone of late Quaternary deformation in the southern Peshawar basin, Pakistan. Geol. Soc. America Spec. Paper 232:265-74.

Yeats, R. S., Khan, S. H., and Akhtar, M. 1984. Late Quaternary deformation of the Salt Range of Pakistan. Geol. Soc. America Bull. 95:958-66.

Yeats, R. S., and Lillie, R. J. 1991. Contemporary tectonics of the Himalayan Frontal Fault System: Folds, blind thrusts, and the 1905 Kangra earthquake. Jour. Structural Geology 13:215-25.

Yeats, R. S., Nakata, T., Farah, A., Fort, M., Mirza, M. A., Pandey, M. R., and Stein, R. S.,1992. The Himalayan Frontal Fault System. Annales Tectonicae, Special Issue, supplement to v. 6:85-98.

Yeats, R. S., and Olson, D. J. 1984. Alternate fault model for the Santa Barbara, California, earthquake of 13 August 1978. Seismological. Society of America Bulletin 74:1545-53.

Yerkes, R. F., Ellsworth, W. L., and Tinsley, J. C. 1983. Triggered reverse fault and earthquake due to crustal unloading, northwest Transverse Ranges, California. Geology 11:287-91.

Yerkes, R. F., Levine, P., and Wentworth, C. M. 1990. Abnormally high fluid pressures in the region of the Coalinga earthquake sequence and their significance. *In* Rymer, M. J., and Ellsworth, W. L., eds. The Coalinga, California, earthquake of May 2, 1983. U. S. Geol. Survey Prof. Paper 1487:235-57.

Yielding, G., Jackson, J. A., King, G. C. P., Sinvhal, H., Vita-Finzi, C., and Wood, R. M. 1981. Relations between surface deformation, fault geometry, seismicity and rupture characteristics during the El Asnam (Algeria) earthquake of 10 October 1980. Earth Planet. Sci. Lett. 56:287-304.

Yonekura, N. 1979. Late Quaternary sea-level change and crustal movement in the intra- and circum-Pacific region. The Earth Monthly 1:822-29 (J).

Yonekura, N. 1983. Late Quaternary vertical crustal movements in and around the Pacific as deduced from former shoreline data. Am. Geophys. Union Geodynamic Series 11:41-50.

Yoshii, T. 1979. A detailed cross-section of the deep seismic zone beneath northeastern Honshu, Japan. Tectonophysics 55:349-60.

Yoshikawa, T., Kaizuka, S., and Ota, Y. 1964. Crustal movement in the late Quaternary revealed with coastal terraces on the southeast coast of Shikoku, southwestern Japan. Jour. Geodetic Soc. Japan 10:116-22.

Yoshikawa, T., Kaizuka, S., and Ota, Y. 1981. The Landforms of Japan. Tokyo: University of Tokyo Press, 222 p.

Youd, T., and Wieczorek, G. 1982. Liquefaction and Secondary Ground Failure. in The Imperial Valley, California Earthquake of October 15, 1979. U. S. Geol. Survey Prof. Paper 1254:223-46.

Youngs, R. R., and Coppersmith, K. J. 1985. Implications of fault slip rates and earthquake recurrence models to probabilistic seismic hazard estimates. Seismological Society of America Bulletin 75:939-64.

Yu W., Cai T., and Hou X. 1991. Deformation zone of M = 7.6 Lanchang earthquake. Seismology and Geology 13(4):343-52 (C).

Yuan D., Liu B., Lu T., He W., Liu J., and Liu X. 1994. Research on earthquake ruptures along the Laohushan active fault zone. Research on Active Fault 3:151-59 (C, Eng.abs.).

Zachariasen, J., and Sieh, K. 1995. The transfer of slip between two en echelon strike-slip faults: A case-study from the 1992 Landers earthquake, southern California. Jour. Geophys. Res. 100:15,281-302.

Zeitler, P. K. 1985. Cooling history of the NW Himalaya, Pakistan. Tectonics 4:127-51.

Zerkal, O. V., and Vinnichenko, S. M. 1990. Main regularities of the development of seismogenic features connected with the Gissar earthquake of 1989 Report submitted to meeting at Dushanbe May 29-June 2, 1990.

Zhang J.,1988. Characteristics of the 1970 Tonghai earthquake fault. *In* Research on Earthquake Faults in China. Wurumuqi, China: Xinjiang Seismological Bureau, eds. Xinjiang Press, pp. 72-70 (C).

Zhang A., Chong J., Mi F., Li S., and Wu J. 1991. Study on the activity and paleo-earthquakes in the fault zone along the northern margin of Qinling in Late Quaternary. Research on Active Fault 1:105-17 (C, Eng. abstract).

Zhang P., Deng Q., Xu X., Feng X., Yang X., Peng S., and Zhao R.,1994. Blind thrust and "fold earthquake"—the 1906 Manas, Xingjiang, northwestern China earthquake. Seismology and Geology 16:26-37 (C, Eng. abs.).

Zhang W., Jiao D., Chai Z., Song F., and Wang Y. 1988. Neotectonic features of the Xiangshan-Tianjingshan arc fracture zone and the seismic deformation zone of 1709 south of Zhongwei M = $7^1/_2$ earthquake. Seismology and Geology 10(3):1-20. (C, Eng. abs.).

Zhang B., Liao Y., Guo S., Wallace, R. E., Bucknam, R. C., and Hanks, T. C. 1986. Fault scarps related to the 1739 earthquake and seismicity of the Yinchuan graben, Ningxia Huizu Zhizhiqu, China. Seismological Society of America Bulletin 76:1253-87.

Zhang A., Mi F., and Chong J. 1989. Deformation relics of the 1556 Huaxian great earthquake and the study of paleoseismicity on the frontal fault zone of the Huashan Mts. Seismology and Geology 11(3):73-81.

Zhang P., Slemmons, D. B., and Mao F. 1991. Geometric pattern, rupture termination and fault segmentation of the Dixie Valley-Pleasant Valley active normal fault system, Nevada, U.S.A. Journal of Structural Geology 13:165-76.

Zhou R., Yu W., Gu Y., and Yao X. 1990. A study on rupture zone of 1988 Genma earthquake with magnitude 7.2 in Yunnan province. Seismology and Geology 12(4):291-302. (C, Eng. abs.).

Zhu C., Teng D., Duan J., and Wang Y. 1988. Ground rupture of 1733 Dongchuan earthquake. *In* Research on Earthquake Faults in China. Xinjiang Seismological Bureau, ed. Xinjiang Press:38-44. (C).

Zijderveld, J. D. A., Hilgen, F. J., Langereis, C. G., Verhallen, P. J. J. M., and Zachariasse, W.J. 1991.

Integrated magnetostratigraphy and biostratigraphy of the upper Pliocene-lower Pleistocene from the Monte Singa and Crotone areas in Calabria, Italy. Earth and Planet. Sci. Lett, 107:697–714.

Zoback, M. D., Moos, D., Mastin, L., and Anderson, R. N. 1985. Well bore breakouts and in situ stress. Jour. Geophys. Res. 90:5523–30.

Zoback, M. D., et al. 1987. New evidence on the state of stress of the San Andreas fault system. Science 238:1105–11.

Zoback, M. L. 1992. First- and second-order patterns of stress in the lithosphere: The World Stress Map Project. Jour. Geophys. Res. 97:11,703–28.

Zoback, M. L., and Zoback, M. D. 1980. State of stress in the conterminous United States. Jour. Geophys. Res. 85:6113–56.

INDEX

Page references to figures are in *italics*.
References to authors cited in the text and whose publications are listed in the Bibliography or Suggestions for Further Reading are in CAPITALS. If two authors are cited in the text, both are included herein; if the reference has more than two authors, only the first is listed. Individual historic earthquakes are mainly listed under the countries or U.S. states in which they occurred. The Appendix is not indexed.